GaN and Related
Alloys—1999

MATERIALS RESEARCH SOCIETY
SYMPOSIUM PROCEEDINGS VOLUME 595

GaN and Related Alloys—1999

Symposium held November 28–December 3, 1999, Boston, Massachusetts, U.S.A.

EDITORS:

Thomas H. Myers
West Virginia University
Morgantown, West Virginia, U.S.A.

Randall M. Feenstra
Carnegie Mellon University
Pittsburgh, Pennsylvania, U.S.A.

Michael S. Shur
Rensselaer Polytechnic Institute
Troy, New York, U.S.A.

Hiroshi Amano
Meijo University
Nagoya, Japan

Materials Research Society
Warrendale, Pennsylvania

Single article reprints from this publication are available through
University Microfilms Inc., 300 North Zeeb Road, Ann Arbor, Michigan 48106

CODEN: MRSPDH

Published by:

Materials Research Society
506 Keystone Drive
Warrendale, PA 15086
Telephone (724) 779-3003
Fax (724) 779-8313
Web site: http://www.mrs.org/

Library of Congress Cataloging-in-Publication Data

GaN and related alloys—1999 : symposium held November 28–December 3, 1999,
    Boston, Massachusetts, U.S.A. / editors, Thomas H. Myers, Randall M. Feenstra,
    Michael S. Shur, Hiroshi Amano
        p.cm.—(Materials Research Society symposium proceedings,
        ISSN 0272-9172 ; v. 595)
        Includes bibliographical references and indexes.
        ISBN 1-55899-503-X
            1. Gallium nitride—Congresses. 2. Gallium alloys—Congresses. 3. Semiconductors—
    Materials—Congresses. 4. Electroluminescent devices—Materials—Congresses.
    5. Lasers—Materials—Congresses. 6. Epitaxy—Congresses. I. Myers, Thomas H.
    II. Feenstra, Randall M. III. Shur, Michael S. IV. Amano, Hiroshi V. Series: Materials
    Research Society symposium proceedings ; v. 595
TK7871.15.G33 G37 2000
621.384'134—dc21   00-036166

Manufactured in the United States of America

CONTENTS

OPTICAL DEVICES

*Invited Paper

LATERAL EPITAXIAL OVERGROWTH

*Invited Paper

GROWTH, STRUCTURAL CHARACTERIZATION, SURFACE STUDIES, THEORY, DOPING

ELECTRONIC TRANSPORT AND DEVICES

ELECTRONIC AND STRUCTURAL
CHARACTERIZATION

GROWTH – MOCVD, HVPE, BULK

*Invited Paper

*Invited Paper

THEORY DOPING

CONTACTS, POINT DEFECTS, PROCESSING

*Invited Paper

DEVICES, OPTICAL CHARACTERIZATION, PROCESSING, CONTACTS, DEFECTS

QUANTUM DOTS, OPTICAL
CHARACTERIZATION, RARE EARTHS

*Invited Paper

PREFACE

This proceedings volume is a record of the submitted papers at Materials Research Society Symposium W, "GaN and Related Alloys," held November 28–December 3 at the 1999 MRS Fall Meeting in Boston, Massachusetts. The symposium consisted of nine half-day sessions (19 invited and 64 contributed talks), and two poster sessions (175 posters total). The attendance was very high for all sessions, with standing room only in a number of cases.

This continuation of the MRS symposium series on GaN and associated materials focused on advances in basic science, as well as the rapidly maturing technologies involving blue/green light-emitters, detectors and high power electronics. Nichia Chemical reported on the commercialization of a laser operating at 405 nm wavelength with a 4000 hour device lifetime. At 450 nm emission wavelength, significant reductions in lifetime are found, and are believed to arise from non-ideal properties of the InGaN alloy used in the active layer of the device. Transistors for microwave applications have achieved significant success in terms of device speed and high power capability. Improvements in the epitaxy of GaN were discussed, using both selective area growth techniques (lateral epitaxial overgrowth) and introduction of low-temperature intra-layers in the films. Advances in both molecular beam epitaxy and metal-organic vapor phase epitaxy were reported, including several studies of quantum dot formation in strained alloys. Hydride vapor phase epitaxy continues to show improvements, particularly for providing very thick films. As the material quality improves, advances in characterization (structural, optical, and electrical) have provided an increased understanding of the role of defects in the materials, and the effects of processing steps on material properties.

One of the highlights of the symposium was a report on Wide Band Gap Semiconductor Research in Europe presented by a panel including university and government scientists (V.A. Dmitriev, T.P. Chow, S. Porowski, M.G. Spencer, M.S. Shur, J.M. Zavada, J.C. Zolper).

It is with pleasure, and much optimism about the future of the III-V nitride field, that we present these proceedings for publication by the Materials Research Society.

Thomas H. Myers
Randall M. Feenstra
Michael S. Shur
Hiroshi Amano

January 2000

ACKNOWLEDGMENTS

The outstanding success of this symposium was due to both the speakers who presented their scientific work at the Meeting, and the symposium organizers who arranged the program and saw that it ran smoothly. We are very grateful for invaluable assistance in the program development from Philip Cohen, Steven Pearton, Christian Wetzel, and Alan Wright.

We would like to thank our invited speakers and session chairpersons, all whom contributed greatly to the success of the symposium:

Isamu Akasaki	John Northrup
Yasuhiko Arakawa	Steve Pearton
Fabio Bernardini	Klaus Ploog
Robert Davis	Linda Romano
Lester Eastman	Shiro Sakai
Thomas Frauenheim	Brian Skromme
Jung Han	Yuta Tezen
David Look	Wen Wang
Umesh Mishra	Christian Wetzel
Suzanne Mohoney	Colin Wood
Theodore Moustakas	Ed Yu
Shuji Nakamura	John Zolper

We also wish to thank the following organizations for their generous financial support of the symposium:

Carnegie Mellon University	SVT Associates
Office of Naval Research	West Virginia University
Oriel Instruments	

We are very grateful for the assistance of the editorial staff of the MRS Internet Journal of Nitride Semiconductor Research for their assistance in editing the proceedings.

Editor-in-Chief:	B. Monemar

Associate Editors:

C.R. Eddy, Jr	J. Neugebauer
K. Hiramatsu	B.J. Skromme
M. Kamp	J.S. Speck
D.C. Look	

The staff at Mira Digital Publishing (St. Louis, Missouri) provided efficient and professional support throughout the electronic submission and editing process.

We are also very thankful to the staff of MRS for their continuing assistance and support.

MATERIALS RESEARCH SOCIETY SYMPOSIUM PROCEEDINGS

MATERIALS RESEARCH SOCIETY SYMPOSIUM PROCEEDINGS

Prior Materials Research Society Symposium Proceedings available by contacting Materials Research Society

Optical Devices

Mat. Res. Soc. Symp. Vol. 595 © 2000 Materials Research Society

Room Temperature CW Operation of GaN-based Blue Laser Diodes by GaInN/GaN optical guiding layers

Masayoshi Koike, Shiro Yamasaki, Yuta Tezen, Seiji Nagai, Sho Iwayama and Akira Kojima
Optoelectronics, Toyoda Gosei Co.,Ltd., Haruhi-cho, Nishikasugai-gun, Aichi 452-8564, JAPAN

ABSTRACT

GaN-based short wavelength laser diodes are the most promising key device for a digital versatile disk. We have been improving the important points of the laser diodes in terms of optical guiding layers, mirror facets. The continuous wave laser irradiation at room temperature could be achieved successfully by reducing the threshold current to 60 mA (4 kA/cm^2). We have tried to apply the multi low temperature buffer layers to the laser diodes for the first time to reduce the crystal defects.

INTRODUCTION

Recently, III-V nitride semiconductor lasers have been improved to the grade of a digital versatile disk (DVD) application. The multi-quantum-well (MQW) separated confinement hetero-structure (SCH) laser diodes have been realized to more than 10000 hours lifetime under continuous wave (CW) irradiation at room temperature (RT) [1]. The GaInN/GaN MQW laser diodes are the key devices for the realization of an advanced DVD system, therefore many groups have been developing intensively.

Akasaki, et al. reported for the first time all over the world the stimulated emission from AlGaN/GaN/GaInN quantum well device by current injection at RT in 1995 [2]. After that, further progress has been achieved. The problem of high dislocation density was resolved by epitaxial lateral over-growth (ELO) GaN [3,4] or pendeo-epitaxial growth [5]. In this way, the several developments were necessary for the improvements of laser diodes such as reduction of the threshold current and extension of lifetime under CW irradiation.

In this paper, we report the improvements of characteristics of the laser diodes by applying the reduction of internal loss at cavity and the multi low temperature (LT) buffer layers, the high power operation and a RT CW irradiation of the MQW-SCH laser diodes.

EXPERIMENTS

One of the schematic structures of GaInN/GaN MQW-SCH lasers is shown in figure 1. A LT buffer layer is grown on a-plane sapphire substrate. An n-GaN layer, an n-$Al_{0.07}Ga_{0.93}N$ cladding layer, an n-GaInN/n-GaN optical guiding layer, a GaInN/GaN MQW active layer, a p-GaInN/p-GaN optical guiding layer, a p-$Al_{0.07}Ga_{0.93}N$ cladding layer, and a p-GaN contact layer are followed. The MQW consists of 4 GaInN wells (1 to 4 nm) and GaN barriers (5 to 10 nm). The conventional optical guiding layers consist of GaN. However, we applied GaInN/GaN optical guiding layers, because an optical confinement factor of GaInN optical guiding layer is superior to that of GaN.

Those GaN epitaxial layers are grown by metalorganic vapor phase epitaxy (MOVPE). The source gases of Al, Ga, In, and N are trimethylaluminum (TMA), trimethylgallium (TMG), trimethylindium (TMI), and ammonia (NH_3), respectively. The dopant source gases of Si and Mg are silane (SiH_4) and biscyclopentadienylmagnesium (bis-Cp_2Mg). The growth conditions are the same as the previous report [6,7]. Mg-doped p-AlGaN and p-GaN layers are activated by low energy electron beam irradiation after the growth. Dislocation density is evaluated by the method of KOH etch pit [8].

The laser diode is mesa geometry structure. The mesa widths are 2 to 5 μm, the cavity lengths are 300 to 700 μm, respectively. The mesa and the mirror facets are formed by reactive-ion-beam-etching (RIBE). TiO_2/SiO_2 dielectric multi layers are coated on both facets for the high-reflection (HR). From design, the reflectivity of a front and a back facets are 80% and 90%, respectively. The characteristics of the improved laser diode were measured under pulsed and CW current injection at RT. The pulse width was 100 ns and the duty ratio was 0.01%.

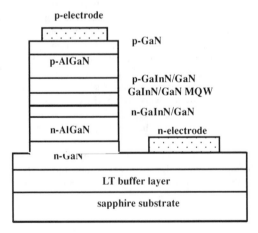

Figure 1. *Schematic structure of the GaInN/GaN MQW-SCH laser diode on sapphire substrate.*

RESULTS AND DISCUSSION

We introduced the GaInN/GaN guiding layers instead of GaN guiding layers to the conventional laser diodes and analyzed the internal loss [9]. Therefore, the optical confinement effect was improved and the internal loss should be reduced. The internal loss was 25 to 30 cm^{-1} as same level as the mirror loss. The I-L-V characteristics of the HR coated laser diode under pulsed operation are shown in figure 2. The mesa width and the cavity length were 5 and 500 μm, respectively. The threshold current reduced from 140 mA (5.6 kA/cm^2) to 60 mA (2.4 kA/cm^2) by the HR facet coating.

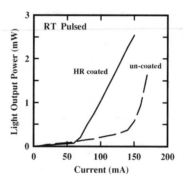

Figure 2. *I-V characteristics of laser diodes with HR coated and uncoated facets.*

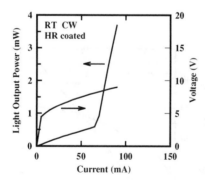

Figure 3. *I-L-V characteristics of GaInN/GaN MQW-SCH laser diode under CW irradiation.*

The CW irradiation of GaInN/GaN MQW-SCH laser diode was realized. The mesa width and the cavity length were 3 and 500 μm, respectively. The threshold current was 60 mA (4 kA/cm²) as shown in figure 3. The threshold current under CW irradiation is larger than that of pulsed operation due to thermal dissipation. RT-CW emission spectrum is shown in figure 4. The peak wavelength was 408 nm and its full width at half maximum (FWHM) was 0.03 nm.

On the other hand, the highest power of the laser diode was 47 mW as shown in figure 5. The mesa width and the cavity length were 3 and 700 μm, respectively. This LD did not use the a multi LT buffer layers. The output power increased without saturation as a function of the current. Therefore, the laser diodes can write DVD sufficiently.

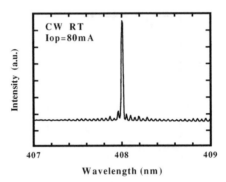

Figure 4. *Emission spectrum measured under CW irradiation.*

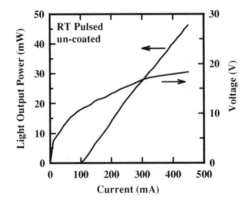

Figure 5. *I-L-V characteristics of laser diode under high light output power operation.*

The multi LT buffer layers [10] were effective for the reduction of dislocation. It is expected that a lifetime of blue laser diode can increase by reducing of defect. As shown in figure 6, the etch pit density (EPD) of 10 μm thickness GaN single layer surface using double LT buffer layers reduced from 1×10^8 cm^{-2} to 5×10^6 cm^{-2} . We applied the multi LT buffer layers to MQW-SCH laser diode epitaxial growth for the first time. The I-L-V characteristics of the laser diode using double LT buffer layers under CW irradiation is shown in figure 7. The mesa width and the cavity length were 3 and 500 μm, respectively. The threshold current was 64 mA (4.3 kA/cm^2) and almost same as the result of using single LT buffer layer (figure 3). We have been optimizing the laser diode structure and investigating the lifetime.

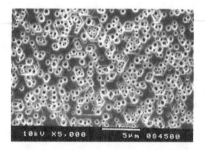

(a) *Conventional single low temperature buffer layer on sapphire substrate (EPD : 1×10^8 cm^{-2}).*

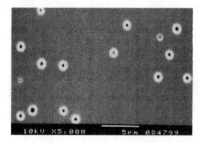

(b) *Double low temperature buffer layers on sapphire substrate (EPD : 5×10^6 cm^{-2}).*

Figure 6. *SEM images of etch pits on GaN surfaces.*

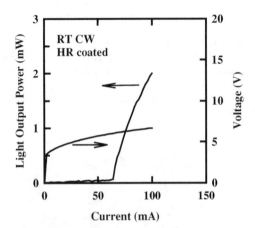

Figure 7. *I-L-V characteristics of the laser diode grown on double low temperature buffer layer under CW irradiation.*

CONCLUSION

We have succeeded in CW operation of GaInN/GaN MQW-SCH laser diode at RT. The high output power was more than 30 mW. The threshold current of 60 mA (4 kA/cm^2) was achieved by introducing the GaInN/GaN optical guide layer. We have succeeded in applying multi LT buffer layers method to low defect density laser diodes for the first time.

ACKNOWLEDGEMENTS

The authors would like to acknowledge Professors I. Akasaki and H. Amano of Meijo University for helpful suggestions and discussions.

REFERENCES

1. S. Nakamura, M. Senoh, S. Nagahama, N. Iwasa, T. Matsushita and T. Mukai, *Proc. 2nd Intern. Symp. On Blue Laser and Light Emitting Diodes, Chiba, Japan,* 1998, p.371 (Ohm-sha, Ltd. Tokyo, 1998).
2. I. Akasaki, H. Amano, S. Sota, H. Sakai, T. Tanaka and M. Koike, *Jpn. J. Appl. Phys.*, **34**, L1517 (1995).
3. A. Usui, H. Sunakawa, A. Sakai and A. A. Yamaguchi, *Jpn. J. Appl. Phys.*, **36**, L899 (1997).
4. S. Nakamura, M. Senoh, S. Nagahama, N. Iwasa, T. Yamada, T. Matsushita, H. Kiyoku, Y. Sugimoto, T. Kozaki, H. Umemoto, M. Sano, and K. Chocho, *Jpn. J.*

Appl. Phys., **36**, L1568 (1997).

5.   T. S. Zheleva, S. A. Smith, D. B. Thomson, T. Gehrke, K. J. Linthicum, P. Rajagopal, E. Carlson, W. M. Ashmawi and R. F. Davis, *MRS Internet J. Nitride Semicond. Res.*, **4S1**, G3.38 (1999).

6.   N. Koide, H. Kato, M. Sassa, S. Yamasaki, K. Manabe, M. Hashimoto, H. Amano, K. Hiramatsu and I. Akasaki, *J. Cryst. Growth*, **115**, 639 (1991).

7.   I. Akasaki, H. Amano, H. Murakami, M. Sassa, H. Kato and K. Manabe, *J. Cryst. Growth*, **128**, 379 (1993).

8.   T. Kozawa, T. Kachi, T. Ohwaki, Y. Taga, N. Koide and M. Koike, *J. Electrochem. Soc.*, **143**, 1, L17 (1996).

9.   S. Nakamura and G. Fasol, *The Blue Laser Diode*, P.250 (Springer-Verlag Berlin Heidelberg, 1997).

10.   M. Iwaya, T. Takeuchi, S. Yamaguchi, C. Wetzel, H. Amano and I. Akasaki, *Jpn. J. Appl. Phys.*, **37**, L316 (1998).

Mat. Res. Soc. Symp. Vol. 595 © 2000 Materials Research Society

Improved Characteristics of InGaN Multi-Quantum-Well Laser Diodes Grown on Laterally Epitaxially Overgrown GaN on Sapphire

Monica Hansen, Paul Fini, Lijie Zhao, Amber Abare, Larry A. Coldren, James S. Speck, and Steven P. DenBaars

Materials Department and Electrical and Computer Engineering Department, College of Engineering, University of California, Santa Barbara, California 93106

ABSTRACT

InGaN multi-quantum-well laser diodes have been fabricated on fully-coalesced laterally epitaxially overgrown (LEO) GaN on sapphire. The laterally overgrown 'wing' regions as well as the coalescence fronts contained few or no threading dislocations. Laser diodes fabricated on the low-dislocation-density wing regions showed a reduction in threshold current density from 8 kA/cm2 to 3.7 kA/cm2 compared the those on the high-dislocation 'window' regions. Laser diodes also showed a two-fold reduction in threshold current density when comparing those on the wing regions to those fabricated on conventional planar GaN on sapphire. The internal quantum efficiency also improved from 3% for laser diodes on conventional GaN on sapphire to 22% for laser diodes on LEO GaN on sapphire.

INTRODUCTION

InGaN-based laser diodes have potential in a number of applications such as optical storage, printing, full-color displays, chemical sensors and medical applications. Major developments in recent years have led to lifetimes in excess of 10,000 hours, demonstrating the viability of nitride laser diodes for commercial applications [1]. The implementations of lateral epitaxial overgrowth (LEO) and AlGaN/GaN modulation-doped strained-layer superlattices in the laser structure have led to increased lifetimes [2]. Strained-layer superlattices are used for strain relief to prevent cracking in the cladding layers [3], and the LEO technique has been shown to reduce the threading dislocation density in GaN grown by metalorganic chemical vapor deposition (MOCVD) [4-7] and hydride vapor phase epitaxy (HVPE) [8,9].

This reduction in dislocation density leads to benefits in device performance such as lower reverse-bias leakage current of p-n junction diodes [10] and LEDs [11,12] low gate leakage current in $Al_xGa_{1-x}N$ FETs [13] and low dark current, sharp cutoff $Al_xGa_{1-x}N$-based solar-blind photodetectors [14]. These results suggest the presence of mid-gap states associated with dislocations. For high quality coalesced LEO GaN, factors such as tilt misalignment between adjacent stripes must be carefully considered. The overgrown GaN wings have been shown to exhibit a tilt away from the seed region perpendicular to the stripe direction for both MOCVD [3,15] and HVPE [8,16] grown LEO GaN oriented in the $<1\bar{1}00>_{GaN}$ or $<11\bar{2}0>_{GaN}$ direction. When these tilts (typically greater than 1°) are present, a low-angle tilt boundary front with twice the magnitude of the average wing tilt is formed at the coalescence. Wing tilt has been correlated with the stripe aspect ratio in cross-section, specifically the ratio or overgrowth

(w) to height (h). This ratio is directly dependent on growth conditions (e.g. V/III ratio, temperature) and "fill factor" (the ratio of open width to pattern period); as w/h increases, wing tilt increases [17,18]. Minimizing wing tilt, minimizes tilt boundary formation and its associated dislocation density at the coalescence fronts. The coalescence of the stripes must be carefully controlled to prevent dislocations from forming at the coalescence fronts to allow a larger area of reduced dislocation density for device placement.

EXPERIMENTAL DETAILS

In this study, laser diodes were fabricated on high quality fully-coalesced LEO GaN leading to improved device performance. Lasers fabricated on the LEO wings, along with coalescence fronts containing few or no threading dislocations, exhibit a reduced threshold current density compared to those grown on conventional planar GaN on sapphire. The internal quantum efficiency increased from 3% to 22%.

InGaN multiple-quantum-well (MQW) laser diodes were grown by MOCVD in a two-flow horizontal reactor at both atmospheric and low pressure. In preparation for patterning a subsequent regrowth, a 2 μm thick GaN seed layer was grown on a c-plane sapphire substrate. A 2000 Å SiO_2 mask was patterned into stripes, oriented in the $<1\bar{1}00>_{GaN}$ direction, defining a 5 μm mask opening with a periodicity of 20 μm. After ~ 6 μm of LEO GaN growth on the SiO_2 mask, the GaN stripes grew laterally and coalesced, forming a flat surface. The conditions for growth and coalescence of the LEO GaN are described elsewhere [18]. Next, the InGaN MQW laser structure was grown on both LEO GaN and on 2 μm GaN on sapphire. The structure had an active region consisting of a 3 period $In_{0.13}Ga_{0.87}N$ (40 Å) /$In_{0.04}Ga_{0.96}N$:Si (85 Å) MQW followed by a 200 Å $Al_{0.2}Ga_{0.8}N$:Mg cap. The n and p-type cladding regions surrounding the active region consisted of 25 Å $Al_{0.2}Ga_{0.8}N$ / 25 Å GaN superlattices with a total thickness of 0.45 μm. The cladding regions were Si-doped for the n-cladding and Mg-doped for the p-cladding. A 0.1 μm GaN:Mg layer was used as a contact layer and a 0.1 μm $In_{0.05}Ga_{0.95}N$:Si layer was used beneath the lower n-type cladding as a compliance layer.

Laser diodes were fabricated above the SiO_2 mask in the nearly dislocation-free wing regions, above the coalescence fronts of the LEO GaN stripes, as well as above the dislocated window (seed) region. The laser cavity was oriented parallel to the direction of the SiO_2 stripes. Laser facets were formed by Cl_2 reactive ion etching (RIE) of 45 μm wide mesas of various lengths ranging from 400 μm to 1600 μm and p-contact stripes were patterned on these mesas with widths ranging from 5 μm to 15 μm. The structure was etched around the p-contact stripe through the p-cladding for index guiding. The n and p-contacts were formed by electron beam evaporation of Ti/Al and Pd/Au, respectively. Electrical testing was performed using 50 ns pulses with a 1 kHz pulse repetition rate.

RESULTS

Figure 1 shows cross-section TEM micrographs of the coalescence region. There are few or no threading dislocations (with a linear density $<4 \times 10^3$ cm^{-1}) generated

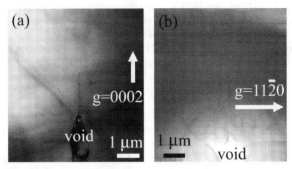

Figure 1. Bright-field cross-section TEM micrographs of a coalescence front viewed with (a) g = 0002 and (b) g = 11$\bar{2}$0 two-beam conditions.

at the coalescence fronts. This high quality coalescence results from low wing tilts of the laterally growing stripes. The LEO wings have a tilt of 0.1° relative to the underlying seed material, which was measured using x-ray diffraction as described earlier [18].

Figure 2 shows the typical light output per uncoated facet of 5×1200 µm² laser diodes grown on LEO GaN as a function of forward current under pulsed operation, The laser diodes are placed above the wing, window and coalescence front regions. The threshold current for the laser placed on the wing, window and coalescence front regions is 340 mA, 570 mA and 380 mA respectively, resulting in respective threshold current densities of 5.7 kA/cm², 9.5 kA/cm², and 6.3 kA/cm². The corresponding threshold voltages were 15.2 V, 17 V, and 16.1 V for lasers on the wing, window and coalescence fronts respectively. The minimum threshold current density was reduced by a factor of 2 from 8 kA/cm² for laser diodes placed on the wing region to 3.7 kA/cm² for laser diodes placed on the window region. Laser diodes grown on LEO GaN were also compared to those grown on sapphire. The minimum threshold current density was reduced by a factor of 2 from 10 kA/cm² for laser diodes grown on sapphire to 4.8 kA/cm² for laser

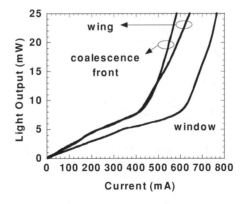

Figure 2. Typical light output per uncoated facet as a function of current for 5×1200 µm² laser diodes placed on the wing, window and coalescence front on LEO GaN.

diodes grown on LEO GaN on sapphire. The laser diodes on the LEO GaN showed this low threshold current density of 4.8 kA/cm^2 both above the SiO$_2$ mask regions and above the coalescence fronts of the LEO GaN. Unlike the threshold current density, the threshold voltage was not reduced as dramatically for the lasers grown on LEO GaN. The minimum threshold voltage was 21.8 V for lasers on LEO GaN compared to 17.4 V for lasers on GaN/sapphire. This reduction in threshold current density is attributed to a reduction in nonradiative recombination due to the lower dislocation density in the LEO GaN.

Figure 3 shows the reciprocal of external differential quantum efficiency as a function of length for laser diodes grown on sapphire and LEO GaN. The external differential quantum efficiency of the laser diode increases with increasing internal quantum efficiency or decreasing internal optical loss as seen in the following relationship [19],

$$\eta_d = \eta_i \frac{\alpha_m}{\langle \alpha_i \rangle + \alpha_m} \tag{1}$$

where η_d is the external differential quantum efficiency, η_i is the internal quantum efficiency, α_m is the mirror loss and α_i is the internal optical loss of the laser. The mirror loss can be defined as

$$\alpha_m = \frac{1}{L} \ln\left(\frac{1}{R}\right) \tag{2}$$

where L is the length and R is the facet reflectivity. R is estimated to be approximately 0.053 for RIE etched facets [20]. Substituting equation 2 into equation 1 and rearranging gives

$$\frac{1}{\eta_d} = \frac{1}{\eta_i} + \frac{\langle \alpha_i \rangle}{\eta_i \ln\left(\frac{1}{R}\right)} L \tag{3}$$

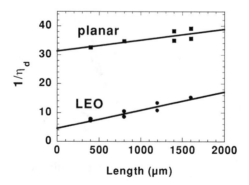

Figure 3. Inverse external differential quantum efficiency as a function of device length.

The internal quantum efficiency can be extracted from the y-intercept of figure 3 using equation 2. The increase in external differential quantum efficiency seen in the lasers on LEO GaN compared to those on sapphire is due to a increase in the internal quantum efficiency from ~3% to ~22%. As mentioned, the reduced reverse bias leakage current in p-n junction diodes suggests the presence of mid-gap states due to threading dislocations [21]. These mid-gap states provide nonradiative recombination centers thereby decreasing the internal quantum efficiency. Reducing the dislocation density, and hence the mid-gap states, will result in an increased internal quantum efficiency. The same effect is also seen in the spontaneous emission portion of L-I curve below threshold in figure 2 as well as in LEDs fabricated on LEO GaN [11,22], where the radiative efficiency increases with decreasing dislocation density.

CONCLUSIONS

In summary, InGaN multi-quantum well laser diodes have been fabricated on fully coalesced laterally overgrown GaN on sapphire. The wing regions as well as the coalescence regions of the LEO GaN contain few or no threading dislocations. The threshold current density was reduced by a factor of 2 from 10 kA/cm^2 for laser diodes grown on sapphire substrates to 4.8 kA/cm^2 for laser diodes grown on LEO GaN on sapphire. Laser diodes fabricated on the wing regions also showed a factor of 2 improvement as compared to those on the window regions from 8 kA/cm^2 to 3.7 kA/cm^2. The internal quantum efficiency also improved from 3% for laser diodes on conventional GaN on sapphire to 22% for laser diodes on LEO GaN on sapphire. This increase is attributed to a reduction of nonradiative recombination from a reduced dislocation density.

ACKNOWLEDGEMENTS

The authors will like to acknowledge funding support from DARPA (R. Leheny), NSF, ONR (Max Yoder, Colin Wood, Yoon-Soo Park) and Hewlett-Packard (S.Y. Wang) through the UC MICRO program.

REFERENCES

1. S. Nakamura, M. Senoh, S. Nagahama, N. Iwasa, T. Yamada, T. Matsushita, H. Kiyoku, Y. Sugimoto, T. Kozaki, H. Umemeto, M. Sano, and K. Chocho, Proc. Int. Conf. on Nitride Semicond., S-1, p. 444 (1997).
2. S. Nakamura, M. Senoh, S. Nagahama, N. Iwasa, T. Yamada, T. Matsushita, H. Kiyoku, Y. Sugimoto, T. Kozaki, H. Umemeto, M. Sano, and K. Chocho, Appl. Phys. Lett. **72**, 211 (1998).
3. K. Ito, K. Hiramatsu, H. Amano, I. Akasaki, J. Cryst. Growth **104**, 533 (1990).
4. D. Kapolnek, S. Keller, R. Vetury, R. D. Underwood, P. Kozodoy, S. P. DenBaars, and U.K. Mishra, Appl. Phys. Lett. **71**, 1204 (1997).
5. T. S. Zheleva, O.-H. Nam, M. D. Bremser, and R. F. Davis, Appl. Phys. Lett. **71**, 2472 (1997).

6. O.-H. Nam, M. D. Bremser, T. S. Zheleva, and R. F. Davis, Appl. Phys. Lett. **71**, 2638 (1997).
7. H. Marchand, J. P. Ibbetson, P. T. Fini, P. Kozodoy, S. Keller, J. S. Speck, S. P. DenBaars, and U. K. Mishra, MRS Internet J. Nitride Semicond. Res. **3**, 3 (1998).
8. A. Usui, H. Sunakawa, A. Sakai, and A. A. Yamaguchi, Jpn. J. Appl. Phys. **36**, L899 (1997).
9. A. Sakai, H. Sunakawa, and A. Usui, Appl. Phys. Lett. **71**, 2259 (1997).
10. P. Kozodoy, J. P. Ibbetson, H. Marchand, P. T. Fini, S. Keller, S. Keller, J. S. Speck, S. P. DenBaars, and U. K. Mishra, Appl. Phys. Lett. **73**, 957 (1998).
11. C. Sasaoka, H. Sunakawa, A. Kimura, M. Nido, A. Usui, and A. Sakai, J. Crystal Growth **189/190**, 61 (1998).
12. S. Nakamura, M. Senoh, S. Nagahama, N. Iwasa, T. Matushita, and T. Mukai, MRS Internet J. Nitride Semicond. Res. **4S1**, G1.1 (1999).
13. R. Vetury, H. Marchand, J. P. Ibbetson, P. Fini, S. Keller, J. S. Speck, S. P. DenBaars, and U.K. Mishra, Proceedings of the 25th Int. Symp. Comp. Semicond., Nara, Japan, 1998.
14. G. Parish, S. Keller, P. Kozodoy, J. P. Ibbetson, H. Marchand, P. T. Fini, S. B. Fleischer, S. P. DenBaars, U. K. Mishra, and E. J. Tarsa, Appl. Phys. Lett. **75**, 247 (1999).
15. H. Marchand, J. P. Ibbetson, P. Fini, S. Chichibu, S. J. Rosner, S. Keller, S. P. DenBaars, J. S. Speck, and U.K. Mishra, Proc. 25th Int. Symp. Comp. Semicond., Nara Japan, 1998.
16. K. Tsukamoto, W. Taki, N. Kuwano, K. Oki, T. Shibata, N. Sawaki, and K. Hiramatsu, Proc. 2nd Int. Symp. On Blue Laser and Light Emitting Diodes, Kisarazu, Chiba, Japan, 1998, p. 488-491.
17. P. Fini, J. P. Ibbetson, H. Marchand, L. Zhao, S. P. DenBaars, and J. S. Speck, to be published in J. Crystal Growth (1999).
18. P. Fini, L. Zhao, B. Moran, M. Hansen, H. Marchand, J. P. Ibbetson, S. P. DenBaars, U.K. Mishra, and J. S. Speck, Appl. Phys. Lett. **75**, 1706 (1999).
19. L. A. Coldren and S. W. Corzine, *Diode Lasers and Photonic Integrated Circuits (John Wiley & Sons*, Inc., New York, 1995), p. 53.
20. M. P. Mack, G. D. Via, A. C. Abare, M. Hansen, P. Kozodoy, S. Keller, J. S. Speck, U. K. Mishra, L. A. Coldren, and S. P. DenBaars, Electron. Lett. **34**, 1315 (1998).
21. For a review on physical properties of threading dislocations in GaN please see J.S. Speck and S. J. Rosner, to be published in Physica B (1999).
22. M. Hansen, P. Fini, A. C. Abare, L. A. Coldren, J. S. Speck, and S. P. DenBaars, presented at the 1999 Electronic Materials Conference, Santa Barbara, CA, 1999 (unpublished).

Mat. Res. Soc. Symp. Vol. 595 © 2000 Materials Research Society

Effect Of AlGaN/GaN Strained Layer Superlattice Period On InGaN MQW Laser Diodes

Monica Hansen, Amber C. Abare, Peter Kozodoy, Thomas M. Katona,
Michael D. Craven, Jim S. Speck, Umesh K. Mishra, Larry A. Coldren, and
Steven P. DenBaars
Materials and Electrical and Computer Engineering Departments
University of California, Santa Barbara, California 93106-5050

ABSTRACT

AlGaN/GaN strained layer superlattices have been employed in the cladding layers of InGaN multi-quantum well laser diodes grown by metalorganic chemical vapor deposition (MOCVD). Superlattices have been investigated for strain relief of the cladding layer, as well as an enhanced hole concentration, which is more than ten times the value obtained for bulk AlGaN films. Laser diodes with strained layer superlattices as cladding layers were shown to have superior structural and electrical properties compared to laser diodes with bulk AlGaN cladding layers. As the period of the strained layer superlattices is decreased, the threshold voltage, as well as the threshold current density, is decreased. The resistance to vertical conduction through p-type superlattices with increasing superlattice period is not offset by the increase in hole concentration for increasing superlattice spacing, resulting in higher voltages.

INTRODUCTION

GaN based laser diodes have potential in a number of applications such as optical data storage, printing, full-color displays and chemical sensors. Additionally, medical applications of nitride based lasers include laser ablation of tissue (e.g. angioplasty), optical detection and selective destruction of malignant tumors, and autoflourescence imaging for the early detection of cancer. Major developments in recent years have led to lifetimes in excess of 10,000 hours demonstrating the viability of nitride laser diodes for commercial applications [1].

Superlattices have been studied in the cladding for two reasons. First, it is difficult to grow thick AlGaN cladding layers needed for optical confinement due to cracking from lattice mismatch stresses during growth. AlGaN/GaN strained layer superlattices (SLSs) have been employed for strain relief of the cladding layer [2,3]. Secondly, Mg-doped AlGaN/GaN superlattices have also shown an enhanced hole concentration. Kozodoy et al. show that a periodic oscillation of the valence band edge, such as superlattices employing alloys with different valence band edge positions, can help overcome the poor doping efficiency of a deep acceptor by forming a two-dimensional hole gas (2DHG) [4]. This results from the ionization of acceptors when the band edge is far below the Fermi energy causing an accumulation of holes where the band edge is close to the Fermi level. Although the free carriers are separated into parallel sheets, their overall number may be much higher than in a simple bulk film. The average hole concentration of superlattice layers at room temperature increases by more than a factor of 10 over bulk films. This paper investigates the effect of vertical conduction of AlGaN/GaN superlattice cladding layers on InGaN laser diode structures.

EXPERIMENTAL DETAILS

InGaN multi-quantum well laser diodes were grown by metalorganic chemical vapor deposition (MOCVD) in a two-flow horizontal reactor on c-plane sapphire substrates using both atmospheric and low pressure. The device structure shown in Figure 1 has an active region consisting of a 3 period $In_{0.15}Ga_{0.85}N$ (40 Å)/$In_{0.05}Ga_{0.95}N$:Si (85Å) MQW with a 200 Å $Al_{0.2}Ga_{0.8}N$:Mg cap. The n- and p-type cladding regions surrounding the active region consisted of $Al_{0.2}Ga_{0.8}N$/GaN superlattices with a total thickness of 0.5 μm. The cladding regions were Si doped for the n-cladding and Mg doped for the p-cladding. The laser structures were grown with superlattice periods of 40 Å, 55 Å, 70 Å, 80 Å, 100 Å, 110 Å and 140 Å. The thickness of the AlGaN and GaN layers were the same yielding an average composition of 10% Al in the cladding. One structure was grown with a 0.5 μm thick bulk $Al_{0.1}Ga_{0.9}N$ cladding layer for comparison purposes. A GaN:Mg layer was used for a contact layer and an $In_{0.05}Ga_{0.95}N$:Si layer is used beneath the lower n-type cladding as a compliance layer. The compressively strained InGaN compliance layer works to counteract the tensile strain of the n-cladding layer, thereby reducing the overall strain within the structure.

Laser facets were formed by Cl_2 reactive ion etching of 45 μm wide mesas of various lengths ranging from 400 μm to 1600 μm. P-contact stripes were formed on the mesas with widths ranging from 5 μm to 15 μm. The structure was etched around p-contact stripe through the p-cladding for index guiding as shown in Figure 1. The n- and p-contacts were formed by electron beam evaporation of Ti/Al/Ni/Au and Pd/Au respectively. Electrical testing was performed using pulses of 50 ns with a 1 kHz pulse repetition rate. The surface topography of the epitaxial structure was measured in tapping

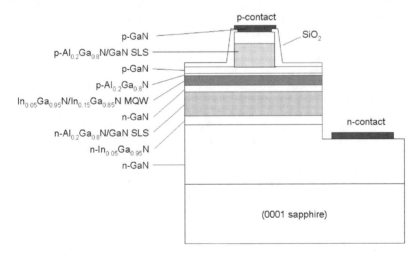

Figure 1 *Layer structure for laser diodes with superlattice cladding layers.*

mode using a Digital Instruments Dimension 3000 scanning probe microscope. X-ray rocking curves of the GaN (102) ω peak were recorded using Ge(220)-monochromatic Cu-Kα radiation in triple axis mode. Specimens for transmission electron microscopy (TEM) were prepared by wedge polishing followed by standard Ar$^+$ ion milling. Images were recorded on a JEOL 2000FX microscope operated at 200 kV.

DISCUSSION

Structural Characterization

The structural quality of laser structures with superlattice cladding were superior compared to a laser structure with the bulk $Al_{0.1}Ga_{0.9}N$ cladding as seen by atomic force microscopy (AFM) and off-axis x-ray diffraction (XRD). Figure 2 compares 5 x 5 μm AFM scans of a laser structure with bulk cladding layers and one with superlattices as the cladding. The surface morphology of the bulk cladding structure shows a poor step structure which appears to be a transition between two-dimensional step flow growth and layer by layer growth. The superlattice cladding sample has well defined steps formed by a step flow growth mode. The full width at half-maximum (FWHM) for the (102) GaN ω rocking curve is plotted versus strain layer superlattice period in Figure 3. The FWHM increases with increasing superlattice period and is greatest for the laser structure with bulk AlGaN cladding layers. Larger values of the (102) ω rocking curve FWHM indicate a higher presence of edge dislocations in the laser with bulk cladding [5].

Device Characterization

The effect of superlattice period is seen in the current-voltage (I-V) characteristics shown in Figure 4. The DC voltage of the devices increases with increased superlattice period and is dramatically higher for the bulk cladding. The threshold voltage of the laser diodes, which is the voltage measured at the threshold current, also increases with

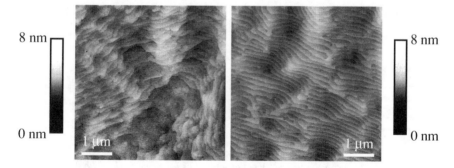

Figure 2 5 x 5 μm AFM micrographs of laser structures with (a) bulk AlGaN cladding layers and (b) AlGaN/GaN superlattice cladding layers.

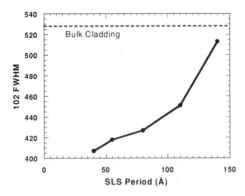

Figure 3 *Full width at half-maximum (FWHM) for the (102) GaN ωrocking curve versus strain layer superlattice period. The FWHM value for the laser structure with bulk AlGaN cladding layers is indicated by the dotted line.*

increasing superlattice period. Note that laser characteristics for the device with bulk claddings are not included since this sample did not lase. The voltage rise with increasing period can be explained by the lower probability for carriers to tunnel through the thicker AlGaN layers. The lateral resistivity of the superlattice layers decreases with increasing period as seen in the transmission line model (TLM) measurements of Figure 5a. Additionally, the threshold current density also increases with increasing superlattice period until it levels off for periods greater than 110 Å (Figure 5b). The increased threshold current density can be explained as an increase in surface recombination at the ridge as a result of the larger lateral conductivity of the larger period superlattice. Kozodoy *et al.* have shown that as the superlattice period increases, the spatially averaged hole concentration of AlGaN/GaN superlattices increases by more than

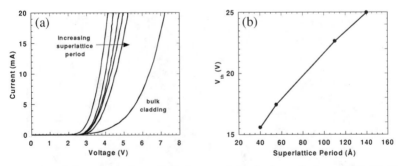

Figure 4 *(a) DC current-voltage (I-V) characteristics for laser diodes with 40, 55, 80, 110, and 140 Å period superlattice cladding layers and bulk cladding layers and (b) threshold voltage for laser diodes with varying superlattice period.*

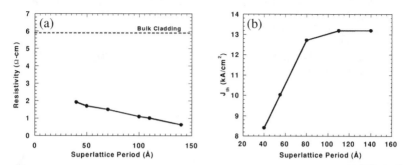

Figure 5 *(a) Lateral resistivity of cladding layers for various superlattice periods as well as bulk. The bulk cladding is indicated with a dotted line. (b) Threshold current density for laser diodes with varying superlattice periods*

ten times the value obtained for bulk AlGaN films [4]. However, the increase in hole concentration associated with increasing superlattice period does not offset the enhanced resistance to vertical conduction through the p-type superlattice cladding observed for larger superlattice periods, thus higher voltages are observed.

CONCLUSION

InGaN MQW laser diodes with strained layer superlattices as cladding layers were shown to have superior structural and electrical properties compared to laser diodes with bulk AlGaN cladding layers. As the period of the strained layer superlattices is decreased, the threshold voltage is decreased. The current density also decreases with decreasing superlattice spacing. Since the vertical conduction does not improve with a larger overall hole concentration from the periodic oscillation of the valence band edge in the superlattice layers, the advantage of the superlattices as laser diode cladding layers is only for strain relief.

ACKNOWLEDGEMENTS

The authors would like to acknowledge the support of DARPA (R. Leheny), NSF, ONR (Max Yoder, Colin Wood, Yoon-Soo Park) and Hewlett Packard through the UC MICRO program (S.Y. Wang).

REFERENCES

1. S. Nakamura, M. Senoh, S. Nagahama, N. Iwasa, T. Yamada, T. Matsushita, H. Kiyoku, Y. Sugimoto, T. Kozaki, H. Umemeto, M. Sano, K. Chocho. Proc. Int. Conf. on Nitride Semicond., S-1, p. 444 (1997).
2. K. Ito, K. Hiramatsu, H. Amano, I. Akasaki. J. Cryst. Growth 104, 533 (1990).

3.   S. Nakamura, M. Senoh, S. Nagahama, N. Iwasa, T. Yamada, T. Matsushita, H. Kiyoku, Y. Sugimoto, T. Kozaki, H. Umemeto, M. Sano, K. Chocho. Appl. Phys. Lett. **72**, 211 (1998).
4.   P. Kozodoy, M. Hansen, U.K. Mishra and S.P. DenBaars. Appl. Phys. Lett. **74**, 3681 (1999).
5.   B. Heying, X. H. Wu, S. Keller, Y. Li, D. Kapolnek, B. P. Keller, S. P. DenBaars, J.S. Speck. Appl. Phys. Lett. **68**, 643 (1996).

Mat. Res. Soc. Symp. Vol. 595 © 2000 Materials Research Society

Spatially Resolved Electroluminescence of InGaN-MQW-LEDs

**Veit Schwegler[1], Matthias Seyboth[1], Christoph Kirchner[1], Marcus Scherer[1]
Markus Kamp[1], Peter Fischer[2], Jürgen Christen[2], and Margit Zacharias[2]**
[1]Dept. of Optoelectronics, University of Ulm, 89069 Ulm, Germany
[2]„Otto-von-Guericke" University, Institute of Experimental Physics, Germany.

ABSTRACT

Electroluminescence (EL) is the most significant measure for light-emitting diodes since it probes the most relevant properties of the fully processed device during operation. In addition to the information gained by conventional spectrally resolved EL, scanning micro-EL provides spatially resolved information. The devices under investigation are InGaN/GaN-LEDs with single peak band-band emission at about 400 nm grown by MOVPE on sapphire substrates.

The μ-EL-characterization is performed as a function of injection current densities and the emission is investigated from the epitaxial layer as well as from substrate side. Spatially resolved wavelength images reveal emission peaks between 406 nm and 417 nm, corresponding either to In fluctuations of 1 %-1.5 % or local fluctuations of piezo electric fields. Beside the information on the emission wavelength fluctuations μ-EL is used to determine the temperature distribution in the LEDs and to investigate transparent contacts.

INTRODUCTION

EL is the major characterization technique for LEDs probing the active region of the whole device under operation. However, as an usually integral measurement it averages over the whole LED. An even deeper insight can be obtained employing spatially and spectrally resolved EL (μ-EL). This technique is used to gain valuable insights into GaN-based LEDs. It is a non-destructive and fast method which provides information about the full device including epitaxial growth and processing. The resulting EL data yield full spectra at each point of investigation at a 1 μm resolution. The recorded data set allows to extract the distribution of the peak wavelength, the intensity, the linewidth and the presence of other recombination paths (DAP, non-radiative, etc.). Those data can be correlated with a conventional microscope image to compare morphology and defects from growth and/or processing with the EL distribution. Furthermore, EL features such as peak wavelength and intensity are depending on local device temperature and current injection, respectively. The disclosed device properties might be used to improve spatial and spectral emission characteristics of LEDs. Within the present work EL-homogeneities, In-fluctuations, heat generation due to ohmic losses, current injection homogeneity, and transparent contacts are investigated.

Experimental

InGaN/GaN-LEDs emitting near the UV are grown by low pressure MOVPE in a horizontal reactor (AIXTRON AIX 200 RF) on c-plane sapphire substrates. The structure consists of 2000 nm GaN and 1000 nm Si-doped GaN followed by the active region, a

100 nm Mg-doped $Al_{0.08}Ga_{0.92}N$ layer, and finally 300 nm p-doped GaN. The active region containes either a 50 nm thick $In_{0.09}Ga_{0.91}N$-layer (double heterostructure) or 3 periods of $In_{0.09}Ga_{0.91}N/GaN$ (MQW structure). Conventional photolithography is used to define the mesa structure. Chemically-assisted ion-beam etching (CAIBE) transferred the pattern using a conventional photoresist mask. A second lithographic step defines the n- and p-contact area, using lift-off technique and subsequent Ni/Au metallization for ohmic contact formation. The resulting LEDs reveal narrow dominating peak emission at about 400-410 nm. The devices have series resistances of approximatly 30 Ω.

The spatially resolved μ-EL system is based on an optical microscope using UV-transparent lenses with a long working distance and computer controlled scanning stages. A DC-motor driven scanning stage enables the scanning with a resolution of 250 nm. The μ-EL is spatially resolved detected. For μ-EL an overall spatial resolution of 1 μm is obtained. The electroluminescence is dispersed in a 0.5 m spectrometer and detected by a liquid nitrogen cooled Si-CCD camera. The spectral resolution for μ-EL measurements reported here is 0.5 nm. All μ-EL measurements are performed at room temperature. LEDs are investigated from both the p-contact- and substrate-side (hereafter called front- and backside). A complete spectrum is recorded at each pixel (x,y) and stored during the scanning over 128 x 100 pixels. Typical examples of extracted information are local EL spot spectra, EL wavelength images in a certain window, i.e. mappings of the local emission peak wavelength, as well as sets of EL intensity images. For the μ-EL measurements also cf. [1].

RESULTS AND DISCUSSION

Figure 1 shows the comparison of a DH and MQW LED,revealing a major difference in the epitaxial quality of the active region and the subsequently grown p-AlGaN barrier layer. Figures 1a-f correspond to an imperfect DH-LED, whereas Figures 1g-l show results obtained from an improved MQW-LED. The micrographs of both circular LEDs are given in Figure 1a and 1g, showing the inner, non-transparent p-contact with the bondwire and the surrounding n-contact metallization. The integral EL spectra are given in Figures 1b and 1h. The DH-LED with the insufficient active region reveals a double peak EL where the emission at 400 nm is from the InGaN-layer and the EL at 440 nm originates from recombination within the p-GaN layer. Figure 1c shows the spatial distribution of the EL intensity from the InGaN (λ = 380-420 nm), whereas Figure 1d maps the emission intensity at about 440 nm (p-GaN) which is signifcantly more homogenously distributed. The peak emisson wavelength distribution is depicted in Figure 1e and 1f. Again the lower wavelength emission is more homogeneously distributed. The InGaN related emission (Figure 1e) reveals a certain inhomogeneity over the device. The observed wavelengths correspond to In-contents between 6-9%, using a bowing parameter of 3.2 eV. Since In-fluctuations are present but only of minor significance, the device is most severly limited by the inhomogenous intensity of the InGaN/GaN EL as can be seen from Figure 1c.

Once the quality of the InGaN/AlGaN active region is improved single peak emission is obtained (Fig. 1h). Figures 1i and 1j show the measured EL at current densities of 27 and 318 A/s, respectively. For the lower current densities the μ-EL reveals a uniform distribution of the EL intensity (1i) and the wavelength (1k). For higher

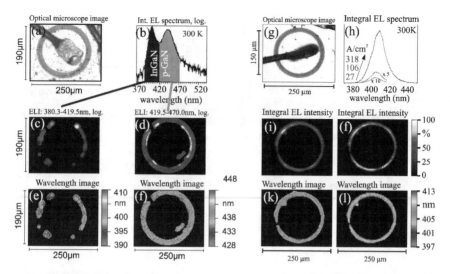

Figure 1 Micrographs (1a, 1g) and integral EL spectra (1b, 1h) of imperfect and improved LEDs, respectively. EL intensity distributions (1c, 1d, 1i and 1j) and wavelength distributions (1e, 1f, 1k and 1l) reveal strong differences between the LEDs related to the epitaxial quality.

current densities a localization of highest intensity at the mesa area takes place (Figure 1j). Low In fluctuations from 0.09 to 0.11 are found evidencing the good chemical uniformity of the InGaN layer (Figure 1l).

Spatial temperature distributions can be derived from μ-EL by fits to the spatially resolved spectra [2]. Figure 2 shows this examplarily at three points taken from the backside of the MQW-LED. The impact of such a temperature distribution can be seen at the wavelength image in Figure 2. A redshift beneath the p-contact due to increased temperature by elevated current density is observable. Heat drain effects of the bondwire can also be identified. For a more extended discussion of temperature distributions in InGaN MQW-LEDs cf. [3].

Transparent contacts are employed in LED technology to combine homogeneous current injection with good outcoupling efficiency. Therefore the contact geometry has to achieve homogeneous current injection with lowly absorbing materials. Figure 3 shows a transparent contact (Ni/Au 2 nm/6 nm) on the p-GaN of a LED. Although the spatial distribution of wavelength maxima seems quite homogeneous, the integral intensity distribution reveals a strongly increased EL-intensity at the mesa edge next to the n-contact which is also clearly visible in the intensity linescan. This can be explained by the different resistances of metal layer and n-GaN leading to a current flow pattern minimizing the overall resistance of the device. An additional high-resistivity undoped GaN layer between the active layer and the contact layer should spread the current more uniformly [4].

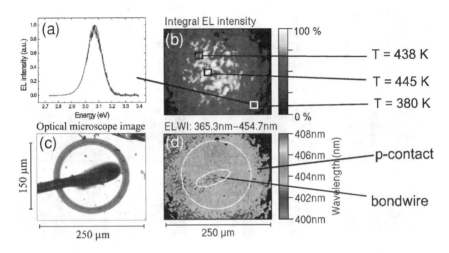

Figure 2 EL intensity distribution at j=320 A/cm⁻² (2b) and local temperatures derived using spectral fits to local spectra (2a). The EL wavelength distribution (2d) reveals the highest temperature (longest wavelength) beneath the p-contact.The bondwire (2c) acts as heat drain as can be seen from shorter wavelengths.

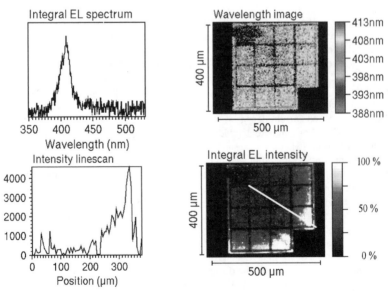

Figure 3 LED with transparent contact (including grid-shaped contact-enforcement) showing inhomogeneous brightness due to inhomogeneous current injection.

CONCLUSIONS

InGaN/GaN-LEDs have been comprehensively characterized by spatially resolved μ-EL. The non-destructive μ-EL probes the relevant properties of the optically active region of the devices under operation. It reveals all deficiencies from either epitaxial growth or processing and is found to be a very powerful technique to analyze LEDs. μ-EL provides direct access to emission from the InGaN active region as well as from the p-GaN. These data have an extremely high relevance since they are obtained from the optically active region of the device under operation. Device properties like current injection and heating can be disclosed enabling an improved device design.

ACKNOWLEDGEMENTS

Financial support by German Ministry of Science and Education (BMBF) and Osram Opto Semiconductors under contract No. 01 BS 802 is gratefully acknowledged by the group at Ulm University. The group at Magdeburg University acknowledges financial support by the Kultusministerium of Sachsen-Anhalt under contract No. 1432A/8386B.

REFERENCES

1. P. Fischer, J. Christen, M. Zacharias, V. Schwegler, C. Kirchner, and M. Kamp, Appl. Phys. Lett., **75**,. 3440 (1999).
2. K.G. Zolina, V. E. Kudryashov, A.N. Turkin, and A.E. Yunovich, Semiconductors, **31**, 901 (1997).
3. V. Schwegler, M. Seyboth, S. Schad, M. Scherer, C. Kirchner, M. Kamp, U. Stempfle, W. Limmer, and R. Sauer, submitted to Proc. of MRS, **595** (2000).
4. T. Mukai, K. Takekawa, S. Nakamura, Jpn. J. Appl. Phys., **37**,. L839 (1998).

Mat. Res. Soc. Symp. Vol. 595 © 2000 Materials Research Society

High reflectance III-Nitride Bragg reflectors grown by molecular beam epitaxy

H. M. Ng[*] and T. D. Moustakas
Electrical and Computer Engineering Department and Center for Photonics Research,
Boston University, 8 Saint Mary's St., Boston MA 02215, U.S.A.
*E-mail: hmng@lucent.com, present address: Bell Labs, Lucent Technologies, 600 Mountain Ave., Murray Hill, NJ 07974.

ABSTRACT

Distributed Bragg reflector (DBR) structures based on AlN/GaN have been grown on (0001) sapphire by electron-cyclotron-resonance plasma-assisted molecular-beam epitaxy (ECR-MBE). The design of the structures was predetermined by simulations using the transmission matrix method. A number of structures have been grown with 20.5 – 25.5 periods showing peak reflectance ranging from the near-UV to the green wavelength regions. For the best sample, peak reflectance up to 99% was observed centered at 467 nm with a bandwidth of 45 nm. The experimental reflectance data were compared with the simulations and show excellent agreement with respect to peak reflectance, bandwidth of high reflectance and the locations of the sidelobes.

INTRODUCTION

With the demonstration and commercialization of III-V nitride edge-emitting laser diodes [1], the emphasis in the research area is gradually shifting to the fabrication and demonstration of vertical cavity surface emitting laser (VCSEL) structures [2-4]. An important requirement for the operation of such a device is the use of high reflectance mirrors, usually in the form of distributed Bragg reflectors (DBRs). Honda and co-workers [2] have estimated the threshold current density in a GaN VCSEL structure and concluded that an increase in the peak reflectance of the DBR mirror from 90% to 99% results in more than an order of magnitude reduction in the threshold current density. Another important requirement for the fabrication of nitride VCSEL structures is the large bandwidth of the primary reflectance peak. This is important because the active region of the nitride lasers is based on InGaN heterostructures or multiple quantum wells, whose emission spectra are very sensitive to small variations in growth or process parameters. Specifically, the InGaN alloys were found to undergo both phase separation and ordering [5-6].

Two different approaches have recently been proposed for the fabrication of VCSEL structures based on III-V nitrides. In the one approach SiO_2/HfO_2 dielectric $\lambda/4$ stacks were used as top and bottom mirrors and flip-chip mounting techniques and laser-induced separation of the nitride structure from the sapphire substrate were employed to complete the device [4]. The alternative approach used DBRs based on 35-pairs of $Al_{0.34}Ga_{0.66}N/GaN$ as the bottom mirror and SiO_2/TiO_2 as the top mirror [3]. The second approach has the advantage that the nitride active region of the device can grow epitaxially on the top of the nitride DBRs.

The main difficulty in fabricating nitride DBRs with high reflectivity and large bandwidth is the small index of refraction contrast that can be obtained within the entire

AlGaN alloy composition. A number of groups have reported the fabrication of AlGaN/GaN DBRs with peak reflectance in the near-UV to blue-green region of the spectrum [7-11]. With the employment of 30-40 quarterwave periods, peak reflectivities at 390 nm of 96% were obtained with a bandwidth of about 14 nm. DBRs based on AlN/GaN quarterwave stacks have the potential for higher peak reflectance and larger bandwidth with approximately half the number of quarterwave periods.

We have previously reported that a 20.5 period DBR had peak reflectance of 95% at 392 nm [10]. The morphology and the reflectance of the structures were uniform across the 2-inch wafer. However, the simulation of the experimental data using the transmission matrix method required a variation of the thickness of the AlN (GaN) layers by as much as 30% of their nominal value. Thus in order to improve the performance of such DBRs, it is imperative to accurately control the thickness of the individual layers. Such control is more important in nitrides than in similar devices based on GaAs/AlAs because the thicknesses of the quarterwave layers in the nitrides are 400-600 Å while those for the arsenides are approximately 800-1000 Å.

In this paper, we report on the growth and characterization of high reflectivity AlN/GaN DBRs with peak reflectance up to 99% and bandwidth up to 45 nm.

SIMULATIONS

Simulations using the transmission matrix method [12] were performed in order to design the DBR structures. The transmission matrix formulation is given by

$$\begin{bmatrix} B \\ C \end{bmatrix} = \prod_{r=1}^{q} \begin{bmatrix} \cos(\delta_r) & \dfrac{i \cdot \sin(\delta_r)}{n_r} \\ i \cdot n_r \cdot \sin(\delta_r) & \cos(\delta_r) \end{bmatrix} \begin{bmatrix} 1 \\ n_{subs} \end{bmatrix} \tag{1}$$

$$\delta_r = 2\pi n_r \frac{d_r}{\lambda} \tag{2}$$

$$d_r = \frac{\lambda_d}{4 \cdot n_r} \tag{3}$$

where n_{subs} and n_r are the refractive indices of the substrate and the rth layer respectively, d_r is the geometrical thickness of the corresponding quarterwave layers and λ_d is the target wavelength for the peak of the high reflectance band. For the simulation, we have used the thicknesses of the AlN and GaN layers of 53.1 and 46.1 nm respectively while the refractive index values of AlN, GaN ($\lambda = 450$ nm) and sapphire used were 2.12, 2.44 and 1.78 respectively. We have assumed constant refractive index values for AlN and sapphire for the wavelength region of interest ($\lambda > 400$ nm). However, the dispersion of the GaN refractive index has to be taken into account in the simulation and we have modeled it using the Sellmeir equation [13].

The optical admittance is given by

$$Y = \frac{C}{B} \tag{4}$$

and the reflectance is defined by

$$R = \left(\frac{1-Y}{1+Y} \right) \overline{\left(\frac{1-Y}{1+Y} \right)} \tag{5}$$

Figure 1 shows the calculated peak reflectance of an AlN/GaN DBR (λ = 450 nm) with varying number of quarterwave periods. We see that a calculated peak reflectance of 99% or greater can be achieved using 16 or more periods in the DBR.

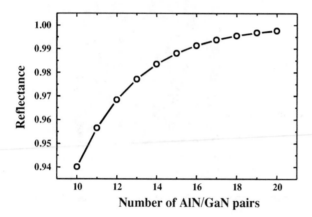

Figure 1: *The calculated peak reflectance of an AlN/GaN DBR (center wavelength = 450 nm) with varying number of periods from 10 to 20.*

EXPERIMENTAL METHODS

The DBR structures were deposited in a Varian Gen II MBE unit equipped with an ASTeX compact ECR microwave plasma source to activate the molecular nitrogen. The sapphire substrate was first subjected to a plasma nitridation at 150W microwave power at 800°C to convert the Al_2O_3 surface to AlN [14]. An AlN buffer layer of ~25 nm was deposited at 750°C. The quarterwave stack was then deposited at a growth rate of 0.35μm/h with the substrate temperature at 750°C. A total of 20.5 – 25.5 periods of AlN/GaN were deposited starting with AlN. Fluxes were calibrated from the growth rates of bulk layers.

The cross-section of thick AlN/GaN bilayers as well as of the DBRs were studied using a JEOL 6010 electron microscope. Thick AlN/GaN bilayers were grown in the growth run preceding the actual growth of the DBR structure to calibrate the growth rates of the two binaries. The reflectance of the DBRs was measured at normal incidence using a Xenon lamp as the light source. The reflected beam was dispersed through a 0.5m grating spectrometer and detected with a photomultiplier tube attached to the exit slit of the spectrometer. The reflectance spectrum of the DBR was normalized to the spectrum taken with a calibrated Al-mirror (Melles-Griot) in place of the sample.

RESULTS AND DISCUSSION

The cross-section SEM image of one of the DBR structures is shown in Fig. 2 where the entire 20 periods of the structure can be observed. The lighter layers represent

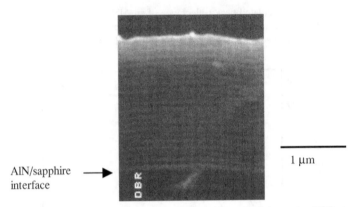

1 μm

AlN/sapphire
interface →

Figure 2: *Cross section scanning electron micrograph of the entire DBR stack.
The lighter layers represent AlN and the darker layers represent GaN.*

AlN while the darker layers represent GaN. The interface between AlN and sapphire is
also shown in the figure.

The normalized reflectance spectra of a number of DBR structures with peak
reflectance ranging from violet to green wavelength regions are shown in Fig. 3. We
have also previously reported a DBR with center wavelength at 392 nm (in the near-UV)
with peak reflectance of 95% [10]. For the samples shown in the figure, the highest peak
reflectance of 99% was measured for the sample with a center wavelength of 467 nm and
a bandwidth of 45 nm. The peak reflectance region can be tuned to different wavelengths
based on the thicknesses of the AlN and GaN quarterwave layers designed.

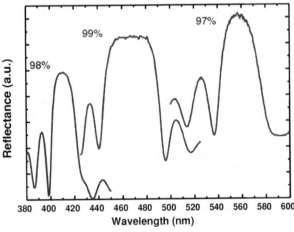

Figure 3: *Reflectance spectra of three AlN/GaN DBRs with peak reflectance of
98%, 99% and 97% at center wavelengths of 410, 467 and 560 nm
respectively*

In some samples, we have observed a network of cracks on the surface of the sample which we attribute to the tensile stress in the AlN layers. In order to prevent the cracking, we have grown some samples with an asymmetric structure using AlN layers thicker than the GaN layers to increase the mechanical yield strength of the AlN layers. We observed that the cracking has been reduced but we do not have empirical data to ascertain the strain state of the AlN layers at this time. The spectra on the left and right show smaller bandwidths compared to that of the one in the middle. Those two structures are asymmetric DBRs with the AlN layers three times thicker than the GaN layers. For such structures, the simulations show that the peak reflectance and the bandwidth will be reduced compared to the ideal case. To compensate for the reduced reflectance due to the asymmetry, the number of periods of the DBR can be increased. The samples on the left and right side of Fig. 3 have 25.5 and 20.5 periods respectively.

Figure 4 shows the results of the simulation in comparison to the measured reflectance spectrum of the DBR with peak reflectance in the blue wavelength region (shown in Fig. 3). The simulation shows excellent agreement with the experimental reflectance spectrum in terms of the peak reflectance, the bandwidth of the high reflectance region as well as the relative locations of the sidelobes.

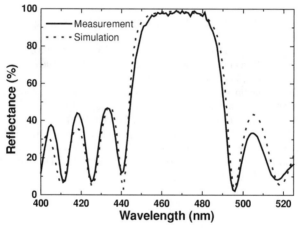

Figure 4: *Measured and simulated reflectance spectra of the AlN/GaN DBR with center wavelength at 467 nm and peak reflectance of 99%.*

SUMMARY

In conclusion, we have designed and fabricated high reflectivity DBRs using AlN/GaN quarterwave stacks. For the best sample, peak reflectance greater than 99% at 467 nm has been achieved using 20.5 periods. This paves the way for the fabrication of optoelectronic devices such as vertical cavity surface-emitting lasers, resonant-cavity light emitting diodes and photodetectors using the III-nitride system.

ACKNOWLEDGMENTS

The authors would like to thank Mr. E. Iliopoulos for helpful discussions and Mrs. Anlee Krupp for help with the SEM measurements. This work was partially funded by ONR under grant no. N00014-98-1-0213 (monitored by Dr. C.E.C. Wood).

REFERENCES

1. S. Nakamura and G. Fasol, *The Blue Laser Diode*, (Springer, Berlin, 1997).
2. T. Honda, A. Katsube, T. Sakaguchi, F. Koyama, and K. Iga, Jpn. J. Appl. Phys. **34**, 3527 (1995).
3. T. Someya, K. Tachibana, J. Lee, T. Kamiya, and Y. Arakawa, Jpn. J. Appl. Phys. **37**, L1424 (1998).
4. Y.-K. Song, H. Zhou, M. Diagne, I. Ozden, A Vertikov, A. V. Nurmikko, C. Carter-Coman, R. S. Kern, F. A. Kish, and M. R. Krames, Appl. Phys. Lett. **74**, 3441 (1999).
5. R. Singh, D. Doppalapudi, T.D. Moustakas, and L.T. Romano, Appl. Phys. Lett. **70**, 1089 (1997).
6. D. Doppalapudi, S.N. Basu, K.F. Ludwig Jr., and T.D. Moustakas, J. Appl. Phys. **84**, 1389 (1998); D. Doppalapudi, S.N. Basu, and T.D. Moustakas, J. Appl. Phys. **85**, 883 (1999).
7. M. Asif Khan, J.N. Kuznia, J.M. Van Hove and D.T. Olson, Appl. Phys. Lett. **59**, 1449 (1991).
8. T. Someya and Y. Arakawa, Appl. Phys. Lett. **73**, 3653 (1998).
9. R. Langer, A. Barski, J. Simon, N.T. Pelekanos, O. Konovalov, R. Andre and L.S. Dang, Appl. Phys. Lett. **74**, 3610 (1999).
10. H.M. Ng, D. Doppalapudi, E. Iliopoulos, and T.D. Moustakas, Appl. Phys. Lett. **74**, 1036 (1999) and erratum Appl. Phys. Lett. **74**, 4070 (1999).
11. I.J. Fritz and T.J. Drummond, Elec. Lett. **31**, 68 (1995).
12. H.A. Macleod, *Thin Film Optical Filters*, 2nd ed. (McGraw-Hill, New York, 1986).
13. M. Born and E. Wolf, *Principles of Optics*, 6th ed. (Pergamon, New York, 1980).
14. H.M. Ng, D. Doppalapudi, D. Korakakis, R. Singh and T.D. Moustakas, J. Crys. Growth **190**, 349 (1998).

Mat. Res. Soc. Symp. Vol. 595 © 2000 Materials Research Society

High-Sensitivity Visible-Blind AlGaN Photodiodes and Photodiode Arrays

J.D. Brown, J. Matthews, S. Harney, J. Boney, and J.F. Schetzina
Department of Physics - Box 8202
North Carolina State University
Raleigh, NC 27695-8202

J.D. Benson and K.V. Dang
Night Vision and Electronic Sensors Directorate
10221 Burbick Rd, St. 430
Fort Belvoir, VA 22060, USA.

Thomas Nohava, Wei Yang, and Subash Krishnankutty
Honeywell Technology Center
12001 State Highway
Plymouth, Minnesota 55441.

ABSTRACT

Visible-blind UV cameras based on a 32 x 32 array of backside-illuminated GaN/AlGaN p-i-n photodiodes have been successfully demonstrated. The photodiode arrays were hybridized to silicon readout integrated circuits (ROICs) using In bump bonds. Output from the UV cameras were recorded at room temperature at frame rates of 30-240 Hz. These new visible-blind digital cameras are sensitive to radiation from 285-365 nm in the UV spectral region.

INTRODUCTION

The III-V nitrides have developed rapidly over the past five years. This has lead to the commercialization of blue and green light emitting diodes, along with demonstrations of violet laser diodes and a variety of electronic devices [1-2]. In addition, a number of photodetectors based on photoconductive elements and arrays [3-6] along with junction devices have also recently been reported [7-9].

This paper reports new images from the first successful demonstration of ultraviolet (UV) digital cameras [8] based on an array of GaN/AlGaN heterostructure p-i-n photodiodes. These new (32 x 32 pixel) digital imagers are designed to sense radiation in the 285-365 nm wavelength band in the UV spectral region. Thus, the digital camera is visible-blind but is not solar-blind (250-280 nm). A discussion of photodiode properties is followed by a description of the experimental procedures employed to synthesize, process and study discrete photodiodes and photodiode arrays, and then by the experimental results obtained.

EXPERIMENTAL DETAILS

The photodiode structure employed in the present work is shown schematically in Figure 1. It consists of a base layer of n-AlGaN (~20% Al) followed by an undoped GaN layer and a p-GaN layer. The photodiode structure is deposited by MOVPE onto a

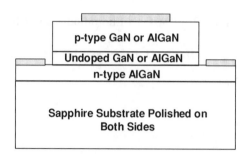

Figure 1 Schematic of p-i-n photodiode structure.

polished sapphire wafer to permit illumination of the device through the substrate. The photodiode structure employed in our initial experiments responds to UV light in the wavelength band from about 320 nm to 365 nm. At wavelengths shorter than 320 nm, the incoming light is absorbed in the thick AlGaN base layer (~20% Al) and the junction is not illuminated. Likewise, the diode does not respond to wavelengths greater than 365 nm, since this corresponds to the optical absorption edge of GaN at 300K. By increasing the Al content of the base layer it is possible to increase the optical bandwidth of the diode's UV responsivity. Likewise, by adding Al to the top layers, it is possible to change the diode UV responsivity band to other wavelength regions in the UV. Thus, UV detectors that sense different UV "colors" are possible.

Diode structures of the type shown were prepared by MOVPE both at North Carolina State University (NCSU) and at the Honeywell Technology Center using low-pressure, vertical-flow MOVPE reactors that employ high speed substrate rotation during film growth. The photodiode structures were deposited onto 2 in diameter c-plane sapphire substrates. The growth was initiated by depositing a thin AlN buffer layer at 500-650 °C; all subsequent layers were grown at 1050-1080 °C.

All device processing was completed at NCSU using standard semiconductor processing techniques which included photolithography using appropriately-designed masks, reactive ion etching to define mesa structures, and metallizations to provide ohmic contacts to the n-type and p-type layers of the device.

Spectral responsivity measurements [8] were completed at NCSU on selected discrete photodiodes. The 300K dynamic resistance of the photodiode at zero-bias, R_0, was measured for selected devices using a shielded low-noise enclosure and shielded probe tips. These measurements were combined with the device area A to obtain the R_0A product which was then used to obtain an estimate of the detector detectivity D* [8]. In order to use the GaN/AlGaN photodiodes as the basis for a new visible-blind UV digital camera it is necessary to employ flip-chip bonding techniques to hybridize an appropriately sized photodiode array to a silicon readout integrated circuit (ROIC). In the present case, NCSU designed and purchased a mask set for a photodiode array that matched a 32 x 32 ROIC chip provided by the Night Vision Laboratory (NVL) at Ft. Belvoir. Prior to hybridization at NVL, In bumps were deposited onto each of the mesas and n-contact layers of the photodiode array and onto the corresponding areas of the ROIC using NCSU facilities. Each 32 x 32 GaN/AlGaN photodiode array was hybridized to the silicon ROIC using facilities at NVL and then cemented onto a leadless chip carrier. Gold wire bonds were then used to link the various input/output channels from the ROIC to the leadless chip carrier. A closeup of a hybridized UV photodiode array is shown in Figure 2 mounted onto a printed circuit board in preparation for testing.

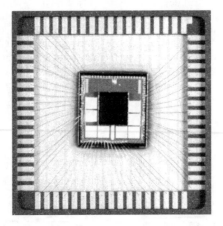

Figure 2 *Photograph of a hybridized UV focal plane array(FPA). The dark square in the middle is the GaN/AlGaN photodiode array which is In bump bonded to the larger ROIC chip. Wire bonds connect the ROIC outputs to the leadless chip carrier.*

The experimental setup used for testing the GaN/AlGaN UV FPAs is shown in Figure 3 below. It consists of an alphanumeric UV source, a quartz focusing lens, the UV digital

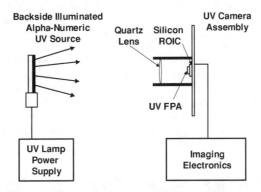

Figure 3 *Experimental setup for testing the UV digital camera.*

camera, and a computerized imaging system [8]. A UV Hg lamp fitted with a fluorescent filter was used as a UV light source. This produced a UV output centered at about 350 nm with a FWHM of ~40 nm. A computer-generated template was then attached to the front of the UV source generating a UV back-lighted scene. A fused quartz lens of focal length 25 mm was used to focus the UV scene onto the AlGaN FPA as shown in Figure 3.

RESULTS AND DISCUSSION

The room temperature spectral responsivity of two discrete AlGaN p-i-n photodiodes is shown in Figure 4. The first device consists of a base n-type layer of $Al_{0.2}Ga_{0.8}N$, an undoped GaN layer, and a p-type GaN:Mg layer. The device has a sharp cut-on beginning at about 365 nm, which corresponds to the optical absorption edge of GaN at room temperature. The responsivity reaches its maximum value of 0.2 A/W at a wavelength of 358 nm, corresponding to an internal quantum efficiency of ~82% [8]. $D^* = 6.3 \times 10^{13}$ cm $Hz^{1/2}W^{-1}$ for this type of UV photodiode. This is one of the largest D^* values ever obtained for any semiconductor photodiode at any temperature.

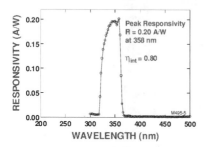 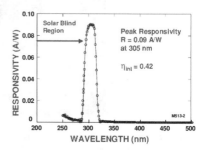

Figure 4. Spectral responsivities for two different p-i-n photodiodes.

The second device consists of a base n-type layer of $Al_{0.33}Ga_{0.67}N$, an undoped $Al_{0.16}Ga_{0.84}N$ layer, and a p-type $Al_{0.16}Ga_{0.84}N$ layer. This photodiode has a sharp cut on at ~320 nm and displays a peak responsivity of 0.09 A/W at 305 nm, corresponding an internal quantum efficiency of ~42%. Its responsivity band covers the region 285-320 nm. $D^* = 2.1 \times 10^{13}$ cm $Hz^{1/2}W^{-1}$ for this type of AlGaN heterostructure p-i-n photodiode.

Photodiode arrays have been successfully fabricated and tested for both of these types of devices. In addition, we have subjected selected devices to light pulses from a xenon lamp and have measured their transient response using an oscilloscope. From these measurements we estimate that the risetime of the nitride photodiodes is about 1 ns and the falltime is less that 1 μs. Thus, they are quite suitable for UV imaging applications.

Initial images from the 32 x 32 UV digital camera have already been reported [8] and will not be reproduced here. Rather, we will report new images obtained from several new UV scenes. The first set of UV images obtained from one of the 32x32 UV camera chips is shown below in Figure 5.

Figure 5. Alpha/Numeric images from visible-blind UV digital camera.

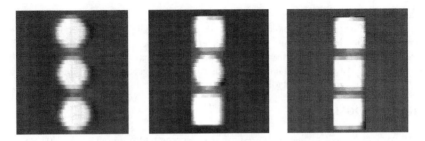

Figure 6 UV images of selected geometric shapes.

In Figure 6, UV images of several sets of geometric objects are shown. It is seen that the squares image reasonably well, but the circles suffer from the coarseness of the 32x32 pixel format.

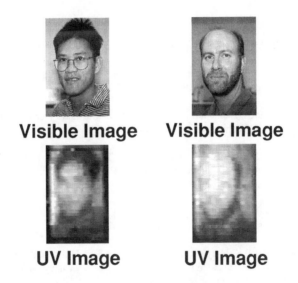

Figure 7 UV images of Dr. Dang (left) and Dr. Benson (right) of NVL

Figure 7 shows images of Drs. Dang and Benson, the UV images are produced with the back-lighting method described earlier. The quality of the UV images are similar to those obtainable from 32x32 pixel arrays of silicon photodiodes. Although the image resolution would greatly improve with the use of UV imagers based on larger 128x128 arrays of photodiodes, the two subjects are readily discernable using the 32x32 arrays.

SUMMARY AND CONCLUSIONS

These UV camera demonstrations represent another historical advancement in opto-electronic devices based on III-V nitride materials. When fully developed into large-format photodiode arrays, this new type of UV digital camera may be used in many wide-ranging applications including biological agent detection, missile and shellfire detection, atmospheric ozone-level detection, welding imagery, and flame sensing. In addition, large-format UV staring FPAs based on nitride photodiodes may play an important role in obtaining UV images of the stars and other astronomical objects of importance in understanding the creation and evolution of the universe.

ACKNOWLEDGMENTS

The work at NCSU is being supported by Army Research Office grants DAAH04-95-1-0627 and DAAD19-99-0113 administered by Dr. M. Dutta and by a Defense Advanced Research Projects Agency grant DAAD19-99-0010 under the Solar Blind Detector Program directed by Dr. R. Leheney. Honeywell would like to acknowledge the support and encouragement of Dr. Paul Schreiber at the Air Force Research Laboratories, Wright-Patterson Air ForceBase, Ohio.

REFERENCES

1. S. Nakamura, T. Mukai, M. Senoh, App. Phys. Lett. **64**, 1687 (1994).

2. Nakamura, G. Fasol, <u>The Blue Laser Diode</u>, published by Springer-Verlag (Heidelberg), 0 (1997). ISBN 3-540-61590-3, 343 pages.

3. Dennis K. Wickenden, Zhenchun Huang, D. Brent Mott, and Peter K. Shu, Johns Hopkins APL Technical Digest, 18(2) 217 (1997).

4. Z.C. Huang, D.B. Mott, P.K. Shu, CP420, Space Technology and Applications International Forum-1998, American Institute of Physics Conference Proceedings 1-56396-747 (1998).

5. T.Z.C. Huang, D.B. Mott, A. La, paper presented at SPIE Conference, Denver (July, 1999), to be published in SPIE conference proceedings.

6. Hadis Morkoc, Naval Research Reviews **51**, 24 (1999). This recent review article can be accessed on the World Wide Web at http://www.onr.navy.mil/onr/pubs.htm. It contains many photodetector references.

7. Wei Yang, Thomas Novova, Subash Krishnankutty, Robert Torreano, Scott McPherson, Holly Marsh, Appl. Phys. Lett. 73(8), 1086 (1998).

8. J.D. Brown, Zhonghai Yu, J. Matthews, S. Harney, J. Boney, J.F. Schetzina, J.D. Benson, K.W. Dang, C. Terrill, Thomas Nohava, Wei Yang, Subash Krishnankutty, MRS Internet J. Nitride Semicond. Res. **4**, 9(1999).

9. K. Linthicum, D. Hanser, T. Zheleva, O.K. Nam, R.F. Davis, Naval Research Reviews **51**, 44 (1999). This review paper on ELO and pendio-epitaxy can be accessed on the World Wide Web at http://www.onr.navy.mil/onr/pubs.htm.

Mat. Res. Soc. Symp. Vol. 595 © 2000 Materials Research Society

HIGH-QUALITY Al$_x$Ga$_{1-x}$N USING LOW TEMPERATURE-INTERLAYER AND ITS APPLICATION TO UV DETECTOR

M. Iwaya[1], S. Terao[1], N. Hayashi[1], T. Kashima[1], T. Detchprohm[2], H. Amano[1,2], I. Akasaki[1,2], A. Hirano[3] and C. Pernot[3,4]

[1]Department of Electrical and Electronic Engineering, Meijo University, 1-501 Shiogamaguchi, Tempaku-ku, Nagoya 468-8502, Japan
[2]High Tech Research Center, Meijo University, 1-501 Shiogamaguchi, Tempaku-ku, Nagoya, 468-8502, Japan
[3]Research and Development Department, Osaka Gas Co. Ltd., 17 Chudoji-Awata-machi, Shimogyo-ku, Kyoto 600-8815, Japan
[4]Invited researcher from Groupe d'Etude des Semiconducteurs, Universite Montpellier II, Place Eugene Bataillon, 34095 Montpellier Cedex 05, France.

ABSTRACT

Low-temperature (LT-) AlN interlayer reduces tensile stress during growth of Al$_x$Ga$_{1-x}$N, while simultaneously acts as the dislocation filter, especially for dislocations of which Burger's vector contains [0001] components. UV photodetectors using thus-grown high quality Al$_x$Ga$_{1-x}$N layers were fabricated. The dark current bellow 50 fA at 10 V bias for 10 μm strip allowing a photocurrent to dark current ratio greater than one even at 40 nW/cm^2 have been achieved.

INTRODUCTION

Although there is a large lattice mismatch of about 16% between GaN and sapphire substrate, device-quality GaN has been achieved with use of LT-buffer layer [1]. It was soon followed by the success of control of n-type conductivity by Si-doping [2], and achievement of p-type nitride films using Mg as a dopant [3] and the following special treatment [4]. These successes have led to the commercialization of nitride-based near-UV, blue and green light-emitting diodes and violet laser diodes.

There are another new field waiting for us. Optical devices in near-UV to vacuum-UV region are one of the most attractive target. The applications of which are chemical sensing [5], flame detection [6,7], ozone-hole sensing, remote sensing, high-density optical storage, excitation source for phosphors, fine lithography, etc. In order to realize such device applications, thick Al$_x$Ga$_{1-x}$N films with high-AlN molar fraction x and high-crystalline quality are essential.

There are several reports concerning the crystalline quality of Al$_x$Ga$_{1-x}$N films. Koide et al. reported that the crystalline quality of Al$_x$Ga$_{1-x}$N on a sapphire substrate covered with an LT-AlN buffer layer [8] is much improved in comparison with that of Al$_x$Ga$_{1-x}$N directly grown on sapphire. It is also reported that its crystalline quality progressively worsens with increasing x reported by Itoh [9]. He also reported that the crystalline quality of Al$_x$Ga$_{1-x}$N could be significantly improved by growing it on a GaN. At the same time, however, crack network generated with high-density if Al$_x$Ga$_{1-x}$N exceeded its critical thickness. Therefore, it had been quite difficult to achieve crack-free, high-quality and thick Al$_x$Ga$_{1-x}$N layers with a high AlN molar fraction.

Very recently, we reported that insertion of LT-deposited layer between HT-GaN reduces threading dislocation density [10,11]. LT-deposited layer inserted between HT-GaN is called the "LT-interlayer" are GaN film with the dislocation density as low as 5×10^6 cm^{-2} has been achieved. This LT-interlayer technique was applied to grow thick $Al_xGa_{1-x}N$. By using this technique, high quality in terms of narrow X-ray diffraction profile and crack-free $Al_xGa_{1-x}N$ with a whole compositional range was achieved [12-14]. The purpose of this work is to understand the relationship between the microscopic structure and the performance of the photodetectors based on these $Al_xGa_{1-x}N$ layers. For the application of flame sensing, the device should be blind to wavelength longer than 280 nm and have sensitivity on the order of 1 nW/cm^2 for shorter wavelength in order to detect only the flame luminescence. Furthermore, the photocurrent to dark current (PC/DC) ratio should be as large as possible in order to prevent misdetection when the dark current level varies due, for instance, to temperature fluctuation. A response speed on the order of a few ms is sufficient for safety application. Although several groups [15,16] have reported on the evolution of photocurrent with optical power, to our knowledge, there had been no discussion on $Al_xGa_{1-x}N$ based UV detector operated under low illumination intensity (< 1 μW/cm^2) with high PC/DC-ratio. The metal-semiconductor-metal (MSM) structure, with fabrication simplicity and need for only a single active layer, is useful tool for characterizing the quality of the detection layer. The results obtained by using LT-interlayered-$Al_xGa_{1-x}N$ are very promising for the development of UV high sensitivity detectors.

EXPERIMENT

Unintentionally doped $Al_xGa_{1-x}N$ films with were grown by organometallic vapor-phase epitaxy (OMVPE) at pressures around 200 hPa. Trimethylaluminum, trimethylgallium and ammonia were used as source gases. (0001) c-plane sapphire was used as the substrate. Figures 1(a) - 1(e) schematically show the structure of each sample. In all the growth, we used LT-buffer layer between nitrides and sapphire. Thickness of $Al_xGa_{1-x}N$ was fixed around 1 μm. Samples A was single $Al_{0.43}Ga_{0.57}N$ layer, which is the same as that reported by Koide et al. Sample B was $Al_{0.43}Ga_{0.57}N$ grown on HT-GaN, which is the same as that reported by Itoh et al. Sample C and D were $Al_{0.43}Ga_{0.57}N$ and GaN grown on HT-GaN, but the "LT-AlN interlayer" was inserted between them. Sample E was single HT-GaN. This reference sample is the underlying layer in samples B, C and D. The alloy composition x were precisely determined high-resolution X-ray diffraction. 2θ/ω scans of the (0002) and (20-24) diffractions were used to determine the lattice constants c and a at room temperature. For details, see ref. 17.

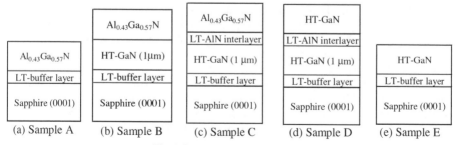

Fig. 1 Structures of the samples.

Plan-view and cross-sectional transmission electron microscopy (TEM) observations were carried out using a HITACHI H-9000NAR TEM system at an acceleration voltage of 300 kV. Ar^+-ion milling and focused Ga^+-ion beam milling were used for plan-view and cross-sectional TEM samples preparation.

The photodetector structure consists of interdigitated electrodes occupying an area of 1 mm^2. The fingers are 10 μm wide with 10 μm spacing. Using a conventional lift-off process, Ti/Au contacts were deposited by electron-beam and thermal evaporation, respectively.

RESULTS AND DISCUSSION

Figures 2(a) through 2(e) show surface SEM micrographs of five samples, respectively. In these five samples, cracks were formed only in sample B. It is identical to the results of Itoh [9]. Although cracks were not generated in sample A, atomic force microscopy showed that the surface was quite rough with a root mean square (RMS) roughness of 12.4 nm. In the image of sample C, which was $Al_{0.43}Ga_{0.57}N$ grown on "LT-AlN interlayer", no cracks are observed. The surface is much smoother than sample A, with the RMS roughness of about 0.4 nm. Both samples D and E are crack-free and have smooth surface.

(a) Sample A   (b) Sample B   (c) Sample C   (d) Sample D   (e) Sample E

Fig. 2 SEM micrographs of various samples

The mechanism of crack formation was studied by stress observation during growth using multi-beam stress sensor system technique [18]. In case of high-quality crystal, it is found that steep relief of the tensile stress, or in other words crack generation, occurred if thickness*tensile stress product exceeds 0.8 GPa*μm. The tensile stress during growth of $Al_{0.43}Ga_{0.57}N$ on "LT-AlN interlayer" shown as sample C is 0.05 GPa. Therefore, critical thickness is much thicker than 1 μm. In case of $Al_{0.43}Ga_{0.57}N$ grown on GaN shown as sample B, we could not observe the clear strain relief during growth because the critical thickness is too thin. Details was discussed elsewhere [13].

The crystalline quality of two types of crack-free $Al_xGa_{1-x}N$, that is, the same structure of sample A and sample C, is compared. In case $Al_xGa_{1-x}N$ was grown on sapphire using LT-buffer, when the same structure of sample A, the FWHM becomes wider with increasing x, which means the tilting component of the mosaicity increases rapidly with increasing AlN molar fraction. The same tendency has already been reported by Itoh [9]. In constant, when $Al_xGa_{1-x}N$ was grown on "LT-AlN interlyaer", the FWHM of XRC remained unchanged over the entire compositional range. This clearly indicates that the crystalline quality of $Al_xGa_{1-x}N$ film grown on GaN covered with "LT-AlN interlayer" is superior to that grown on sapphire covered with the LT-buffer layer.

Further characterization of the crystalline quality was carried out using TEM. Figures 3(a) and 3(b) show the plan-view TEM images of sample A and C, respectively. Grain boundaries of about 50-250 nm in size were observed in Fig. 3(a). In comparison, such grain boundaries could not be observed in Fig. 3(b). The structure of $Al_{0.43}Ga_{0.57}N$ shown in Fig. 3(b) is quite similar to that of the underlying GaN except for the density of threading dislocations. When the plan-view image shown in Fig. 3(b) was taken, the sample was slightly tilted from the [0001] axis in order to identify the type of threading dislocations. Using this observation technique, pure-edge and mixed dislocations exhibit strong contrast and can be easily distinguished. For details, see ref. 15. The magnified image in Fig. 3(b) shows that almost all the threading dislocations are pure-edge type ones. The densities of screw and mixed type dislocations are quite low. These results are further confirmed by cross-sectional dark-field image observations. The majority of threading dislocations is in contrast with g=[1-100] and out of contrast with g=[000-2]. From this result, it is concluded that the majority of threading dislocations in the top $Al_{0.43}Ga_{0.57}N$ layer has Burger's vector b=1/3[11-20]. Three types of threading dislocations propagate to the c-direction in hexagonal GaN were reported, that is, screw-type with Burger's vector b=[0001], edge-type with b=1/3[11-20] and mixed-type with b=1/3[11-23]. The results confirmed that the majority of threading dislocations in the uppermost $Al_{0.43}Ga_{0.57}N$ layer is pure-edge-type dislocations. Fig. 4 shows the compositional dependence of the density of threading dislocations in the same structure of sample A and sample C [14].

(a) Sample A

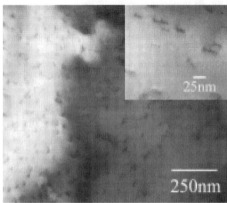
(b) Sample B

Fig. 3 Bright-field plan-view TEM image of the top surface

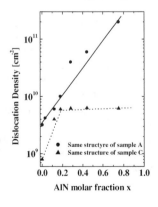

Fig. 4 Compositional dependence of the density of threading dislocations. Solid and dotted lines provide guides for the eye.

Table I. Characteristics of the investigated layers.

sample name	Threading dislocation density [cm^{-2}]				Dark current
	edge	screw	mixed	total	level at 10V
sample A	-	-	-	> 2×10^{11}	~ 40 mA
sample C	8-10×10^9	< 10^6	< 10^7	8-10×10^9	< 50fA
sample D	5×10^8	< 10^6	< 10^7	5×10^8	< 50fA
Sample E	1-2×10^9	7×10^7	5×10^8	2×10^9	~ 100 fA

The effects of each type of threading dislocations on the electrical and optical properties have not been clarified yet. Table I summarizes the density of each threading dislocations in these samples. Compared to the underlying GaN layer, Al$_{0.43}$Ga$_{0.57}$N grown on "LT-AlN interlyaer" contains higher density of edge type dislocations, although pure screw type and mixed type dislocation is much reduced. In this study, we fabricated and characterized photodetectors based on these four crack-free samples (samples A, C, D and E). Measurements were carried out at room temperature under a bias voltage of 10 V with and without illumination from mercury lamp (λ = 254 nm). The dark current for each sample is listed in Table I. Due to high dark current level, PC/DC of sample A was less than one even for 100 µW/cm^2. The other samples showed good uniformity from one detector another, and act as high sensitivity UV detectors with PC/DC greater than one even at very low weak illumination of 40 nW/cm^2. Using the mercury lamp at power density of 10 µW/cm^2 and mechanical shutter, we measured the photoconductive build-up and decay time under 10 V bias at room temperature. Before each experiment, the samples were kept in the dark for at least 12 hours. After the dark current reached a quasi-steady-state value, the excitation light was turned on for 2 hours. When starting measurement, we observed a transient response time as the time needed for the photoresponse signal to drop to 1 % of its maximum value (Fig. 5), compared to the 83 s needed for sample E of HT-GaN grown on LT-AlN buffer layer, samples grown on "LT-AlN interlayer" present much faster decay with response time of 1.6±0.05 s and 2.0±0.1 s, for sample D and C respectively. For sample A, the response time exceeds the measurement time. [7].

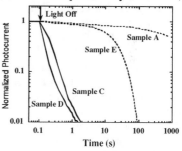

Fig. 5. The normalized photocurrent as a function of time shows a great improvement of the response speed for LT-AlN interlayer.

From these results, it is expected that pure-screw-type and mixed-type threading dislocations are electrically active and relates to deep traps, while pure-edge type dislocation is not so much.

CONCLUSIONS

The relationship between microscopic structure of Al$_x$Ga$_{1-x}$N and the UV photoconductive properties have been clarified. The "LT-AlN interlayer" is found to act as the dislocation filter, especially for dislocations of which Burger's vector contains [0001] components. The fabricated

MSM detectors show a very low dark-current level, below 50 fA at 10 V. These results show that "LT-interlayered-$Al_xGa_{1-x}N$" are very promising for the development of optical devices in the near-UV and vacuum UV region.

ACKNOWLEDGEMENTS

This work was supported in part by the Japan Society for the Promotion of Science "Research for the Future Program in the Area of Atomic Scale Surface and Interface Dynamics" under the project of "Dynamic Process and Control of the Buffer Layer at the Interface in a Highly-Mismatched System (JSPS96P00204)". and the Ministry of Education, Science, Sports and Culture of Japan, (contract number 11450131).

REFERENCE

1. H. Amano, N. Sawaki, I. Akasaki and Y. Toyoda: Appl. Phys. Lett., **48**, 353 (1986).
2. H. Amano and I. Akasaki: Mat. Res. Soc. Ext Abst., **EA-21**, 165 (1991).
3. H. Amano, M. Kitoh, K. Hiramatsu and I. Akasaki: J. Electrochem. Soc., **137**, 1639 (1990).
4. H. Amano, M. Kito, K. Hiramatsu and I. Akasaki: Jpn. J. Appl. Phys. **28**, L2112 (1989).
5. J. Han, M.H. Crawf, R.J. Shul, S.J. Hearne, E. Chason, J.J Figiel and M. Banas: MRS Internet J. Nitride Semicond. Res. 4S1, G7.7 (1999).
6. C. Pernot, A. Hirano, H. Amano and I. Akasaki: Jpn. J. Appl. Phys.**37**, L1202 (1998).
7. C. Pernot, A. Hirano, M. Iwaya, T. Detchprohm, H. Amano and I. Akasaki: Jpn. J. Appl. Phys. **38**, L487 (1999).
8. Y. Koide, N. Itoh, K. Itoh, N. Sawaki and I. Akasaki, Jpn. J. Appl. Phys. **27**, 1156 (1988).
9. K. Itoh, Doctor Thesis, School of Eng., Nagoya University, Nagoya, 1991.
10. M. Iwaya, T. Takeuchi, S. Yamaguchi, C. Wetzel, H. Amano and I. Akasaki: Jpn. J. Appl. Phys. **37**, L316 (1998).
11. H. Amano, M. Iwaya, T. Kashima, M. Katsuragawa, I. Akasaki, J. Han, S. Hearne, J. A. Floro, E. Chason and J. Figiel, Jpn. J. Appl. Phys. **37**, L1540 (1998).
12. H. Amano, M. Iwaya, N. Hayashi, T. Kashima, M. Katsuragawa, T. Takeuchi, C. Wetzel, I. Akasaki: MRS Internet J. Nitride Semicond. Res. **4S1**, G10.1 (1999).
13. M. Iwaya, S. Terao, N. Hayashi, T. Kashima, H. Amano and I. Akasaki: submitted to Appl. Surf. Sci.
14. T. Kashima, R. Nakamura, M. Iwaya, H. Kato, S. Yamaguchi, H. Amano and I. Akasaki: to be publised Jpn. J. Appl. Phys.
15. E. Muñoz, J. A. Garrido, I. Izpura, F. J. Sánchez, M. A. Sánchez-Garcia, E. Calleja, B. Beaumont and P. Gibart: Appl. Phys. Lett. **71**, 870 (1997).
16. F. Binet, J. Y. Duboz, E. Rosencher, F. Scholtz and V. Härlle: Appl. Phys. Lett. **71**, 1202 (1996).
17. H. Amano, S. Sota, T. Takeuchi, M. Kobayashi, I. Akasaki, J. Burm, W. J. Schaff and L. F. Eastman: Technical Report of IEICE, ED97-123, CPM97-110, 31 (1997-10).
18. J. Floro, E. Chason, S. Lee, R. Twesten, R. Hwang and L. Freund: J. Elec. Mat. **26**, 969 (1997).

Lateral Epitaxial Overgrowth

Mat. Res. Soc. Symp. Vol. 595 © 2000 Materials Research Society

Pendeo-epitaxial Growth and Characterization of GaN and related Materials on 6H-SiC(0001) and Si(111) Substrates

Robert F. Davis, T. Gehrke, K.J. Linthicum[+], T. S. Zheleva, P. Rajagopal[+], C. A. Zorman* and M. Mehregany*
Department of Materials Science and Engineering, North Carolina State University, Raleigh, NC 27695; *Department of Electrical and Computer Engineering, Case Western Reserve University, Cleveland, OH 44106

ABSTRACT

Discrete and coalesced monocrystalline GaN and $Al_xGa_{1-x}N$ layers grown via Pendeo-epitaxy (PE) [1] originated from side walls of GaN seed structures containing SiN_x top masks have been grown via organometallic vapor phase deposition on GaN/AlN/6H-SiC(0001) and GaN(0001)/AlN(0001)/3C-SiC(111)/Si(111) substrates. Scanning and transmission electron microscopies were used to evaluate the external microstructures and the distribution of dislocations, respectively. The dislocation densities in the PE grown films was reduced by at least five orders of magnitude relative to the initial GaN seed layers. Tilting in the coalesced GaN epilayers was observed via X-ray diffraction. A tilt of 0.2° was confined to areas of mask overgrowth; however, no tilting was observed in the material suspended above the SiC substrate. The strong, low-temperature PL band-edge peak at 3.45 eV with a FWHM of 17 meV was comparable to that observed in PE GaN films grown on 6H-SiC(0001). The band-edge in the GaN grown on AlN(0001)/SiC(111)Si(111) substrates was shifted to a lower energy by 10 meV, indicative of a greater tensile stress.

INTRODUCTION

It has been a necessity for investigators in the III-nitride community to grow films of GaN and related nitride materials using heteroepitaxial growth routes because of the dearth of bulk substrates of these materials. This results in films containing dislocation densities of 10^8-10^{10} cm^{-2} because of the mismatches in the lattice parameters and the coefficients of thermal expansion between the buffer layer and the film and/or the buffer layer and the substrate. These high concentrations of dislocations may also limit the performance of devices.

Several groups [2-10], including the present authors, have conducted research regarding selective area growth (SAG) and lateral epitaxial overgrowth (LEO) techniques for GaN deposition, specifically to reduce significantly the dislocation density. Increased emphasis in this research topic was fueled in part by the announcement by Nakamura, et al. [11-13] of the dramatic increase in projected lifetime of their GaN- based blue light-emitting laser diodes fabricated on LEO material. Using these approaches, researchers have been able to grow GaN films containing dislocation densities of $\approx 10^5$ cm^{-2} in the areas of overgrowth.

[+]Now with: Nitronex Corporation, Raleigh, NC 27606

However, to benefit from this reduction in defects, the placement of devices incorporating LEO technology is limited and confined to regions on the final GaN device layer that are located on the overgrown regions.

Recently we have pioneered a new approach to selective epitaxy of GaN and $Al_xGa_{1-x}N$ layers, namely, pendeo- (from the Latin: to *hang* or be *suspended*) epitaxy (PE) [1,14-19] as a promising new process route leading to a single, continuous, large area layer; multilayer heterostructures or discrete platforms of these materials. It incorporates mechanisms of growth exploited by the conventional LEO process by using an amorphous mask to prevent vertical propagation of threading dislocations; however, it extends beyond the conventional LEO approach to employ the substrate itself as a *pseudo-mask*. This unconventional approach differs from LEO in that growth does not initiate through open windows on the (0001) surface of the GaN seed layer; instead, it is forced to selectively begin on the sidewalls of a tailored microstructure comprised of forms previously etched into this seed layer. Continuation of the pendeo-epitaxial growth of GaN or the growth of the $Al_xGa_{1-x}N$ layer until coalescence over and between these forms results in a complete layer of low defect-density GaN or $Al_xGa_{1-x}N$. This is accomplished in one (GaN), two ($Al_xGa_{1-x}N$) or multiple (multilayer heterostructure) re-growth steps. And the need to align devices or masks for the growth of the subsequent layers over particular areas of overgrowth is eliminated, unless deposition only in certain areas is desired.

The following sections describe the experimental parameters necessary to achieve GaN and $Al_xGa_{1-x}N$ films via PE on SiC(0001) and Si(111) substrates. The microstructural and optical evidence obtained for the resulting films is also described, discussed and summarized.

EXPERIMENTAL PROCEDURES

Each pendeo-epitaxial GaN and $Al_xGa_{1-x}N$ film and the underlying GaN seed layer and the AlN buffer layer were grown in a cold-wall, vertical pancake style RF inductively heated metallorganic vapor phase epitaxy (MOVPE) system. Two distinct process routes were explored for growth on (i) on-axis 6H-SiC(0001) substrates and (ii) on-axis Si(111) substrates. In the former, each seed layer consisted of a 1 µm thick GaN film grown on a 100 nm thick AlN buffer layer previously deposited on a 6H-SiC(0001) substrate. Details of the experimental parameters used for the growth of these two layers are given in Ref. [20]. In the growth on the Si substrates, a 1 µm 3C-SiC(111) film was initially grown on a very thin 3C-SiC(111) layer produced by conversion of the Si(111) surface at 1360°C for 90 s via reaction with C_3H_8 entrained in H_2. The film was subsequently achieved by simultaneously decreasing the flow rate of the C_3H_8/H_2 mixture and introducing a SiH_4/H_2 mixture. Both the conversion step and the SiC film deposition were achieved using a cold-wall, vertical geometry, RF inductively heated atmospheric pressure chemical vapor deposition (APCVD) reactor. Details of the experimental parameters used for the conversion step and the growth of the 3C-SiC layer are given in Ref. [21]. A 100 nm thick AlN buffer layer and a 1 µm GaN seed layer were subsequently deposited in the manner described above for the 6H-SiC substrates.

A 100 nm silicon nitride growth mask was deposited on the seed layers via plasma enhanced CVD. A nickel etch mask was subsequently deposited using e-beam evaporation. Patterning of the nickel mask layer was achieved using standard photolithography techniques. The final, tailored, microstructure consisting of seed forms was fabricated via removal of portions of the nickel etch mask via sputtering and by inductively coupled plasma (ICP) etching of portions of the silicon nitride growth mask, the GaN seed layer and the AlN buffer layer. Critical to the success of the pendeo-epitaxial growth, the etching of the seed-forms was continued completely through the exposed GaN and AlN layers and into either the 6H-SiC substrate or the 3C-SiC layer, thereby removing all III- nitride material from the areas between the side walls of the forms. The seed forms used in this study were raised rectangular stripes oriented along the [1$\overline{1}$00] direction, thereby providing a sequence of parallel GaN sidewalls (nominally (11$\overline{2}$0) faces). Seed form widths of 2 and 3 µms coupled with separation distances of 3 and 7 µms were employed. The remaining nickel mask that protected the seed structures during the ICP etching process was removed using a wet etch. Immediately prior to pendeo-epitaxial growth, the patterned samples were dipped in a acid solution to remove surface contaminants from the walls of the underlying GaN seed structures.

A schematic of the pendeo-epitaxial growth of GaN is illustrated in Figure 1. There are three primary stages associated with the pendeo-epitaxial formation of this material: (i) initiation of lateral homoepitaxy from the sidewalls of the GaN seed, (ii) vertical growth and (iii) lateral growth over the silicon nitride mask covering the seed structure. Pendeo-epitaxial growth of GaN was achieved within the temperature range of 1050-1100°C using the same pressure and V/III ratio used for the deposition of the GaN seed layer, as described above. Additional experimental details regarding the pendeo-epitaxial growth of GaN and $Al_xGa_{1-x}N$ layers employing 6H-SiC substrates are given in Refs. [14–19, 22]. The morphology and defect microstructures were investigated using scanning electron microscopy (SEM) (JEOL 6400 FE) and transmission electron microscopy (TEM) (TOPCON 0002B, 200 KV). Determination of the degree of tilting in the PE GaN was achieved using X-ray diffraction (XRD) (Philips X'Pert MRD X-ray diffractometer) analysis. Optical characterization was performed via photoluminescence (PL) using a He-Cd laser (λ=325 nm).

RESULTS AND DISCUSSION

Growth on SiC substrates

The pendeo-epitaxial phenomenon is made possible by taking advantage of growth mechanisms identified by Zheleva et al. [8] in the conventional LEO technique and by using two additional key steps, namely, the initiation of growth from a GaN face other than the (0001) and the use of the substrate (in this case SiC) as a pseudo-mask. By capping the seed-forms with a growth mask, the GaN was forced to grow initially and selectively only on the GaN sidewalls. Common to conventional LEO, no growth occurred on the silicon nitride mask covering the seed forms. Deposition also did not occur on the exposed SiC surface areas at the higher growth temperatures employed to enhance lateral growth (the pseudo-mask effect). The Ga- and N-containing species more likely either diffused along the surface or evaporated (rather than having sufficient time to form GaN nuclei) from both the silicon nitride mask and the silicon carbide substrate. The pronounced effect of this is shown in Figure 3 wherein the newly deposited GaN has grown truly suspended (pendeo) from the sidewalls of the GaN seed structure. During the second PE event (ii), vertical growth of GaN occurred from the advancing (0001) face of the laterally growing GaN. Once the vertical growth became extended to a height greater than the silicon nitride mask, the third PE event (iii) occurred, namely, conventional LEO-type growth and eventual coalescence over the seed structure, as shown in Figure 3.

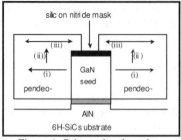

Figure 1. Schematic of pendeo-epitaxial growth from GaN sidewalls and over a silicon nitride mask.

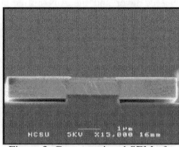

Figure 2. Cross-sectional SEM of a GaN pendeo-epitaxial growth structure with limited vertical growth from the seed sidewalls and no growth on the seed mask.

A cross-sectional TEM micrograph showing a typical pendeo-epitaxial growth structure is shown in Figure 4. Threading dislocations extending into the GaN seed structure, originating from the GaN/AlN and AlN/SiC interfaces are clearly visible. The silicon nitride mask acted as a barrier to the further vertical propagation of these defects into the laterally overgrown pendeo-epitaxial film. Since the newly deposited GaN is suspended above the SiC substrate, there are no vertically oriented defects associated with the mismatches in lattice parameters between GaN and AlN and between AlN and SiC. Preliminary analyses of the GaN seed/GaN PE and the AlN/GaN PE interfaces revealed evidence of the lateral propagation of the defects; however, there is yet no evidence that the defects reach the (0001) surface where device layers will be grown. As in the case of LEO, there is a significant reduction in the defect density in the regrown areas.

Figure 3. Cross-sectional SEM of a GaN/Al$_{10}$Ga$_{90}$N pendeo epitaxial growth structure showing coalescence over the seed mask.

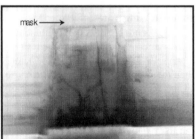

Figure 4. Cross-sectional TEM of a GaN pendeo-epitaxial structure showing confinement of threading dislocations under the seed mask, and a reduction of defects in the regrowth.

The continuation of the pendeo-epitaxial growth results in coalescence with adjacent growth fronts and the formation of a continuous layer of GaN, as observed in Figure 5. This also results in the practical elimination of all dislocations stemming from the heteroepitaxial growth of GaN/AlN on SiC. Clearly visible in Fig. 5(a) are the voids that form when adjacent growth fronts coalesce. Optimization of the pendeo-epitaxial growth technique should eliminate these undesirable defects.

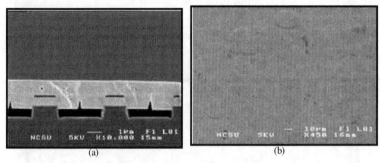

(a)   (b)

Figure 5. Micrographs taken via (a) cross-sectional SEM and (b) plan-view SEM of examples of pendeo-epitaxial growth with coalescence over and between the seed forms resulting in a single GaN layer.

Pendeo-epitaxial growth of individual forms and coalesced single layers of PE $Al_{10}Ga_{90}N$ alloys and PE GaN as well as forms containing heterostructures of alternating layers of GaN and $Al_{10}Ga_{90}N$ have been realized in this research as shown in Figure 3. Ref. [22] is discussing in more detail the process necessary to fabricate coalesced single layers of PE $Al_{10}Ga_{90}N$ alloys and their characterization.

Growth on Silicon Substrates

Research regarding the development of process routes leading to the growth of III-Nitride films on Si has been less relative to the number of studies concerned with growth on sapphire and SiC. This has been due in part to the three-dimensional nucleation and growth of GaN islands caused by the combination of significant mismatches in lattice parameters, the higher surface energy of GaN and the chemical reactivity of Si with the reactants in the growth environment. To address the above concerns we have developed a process route similar to those used for growth of GaN on 6H-SiC(0001), but replaced the 6H-SiC substrate with a 3C-SiC(111) transition layer grown on a Si(111) substrate. The atomic arrangement of the (111) plane of 3C-SiC is equivalent to the (0001) plane of 6H-SiC; this facilitates the sequential deposition of a high temperature 2H-AlN(0001) buffer layer of sufficient quality for the GaN seed layer, as discussed in Reference [20]. The 3C-SiC layer is also exposed between the GaN seed forms and the seed forms are capped with a silicon nitride growth mask to force the GaN to grow initially and selectively only on the GaN sidewalls in the manner described above for growth using the 6H-SiC(0001) substrates.

For the initial demonstrations of PE growth of GaN films on silicon, 0.5 and 2 μm thick 3C-SiC(111) layers were deposited on 50 mm diameter, 250 μm thick converted Si(111) substrates. All subsequent research described below used the ~2.0 μm barrier layer.

Figure 6 shows a cross-sectional SEM micrograph of a PE GaN layer grown laterally and vertically from raised GaN stripes etched in a GaN/AlN/3C-SiC/Si(111) substrate and over the silicon nitride mask atop each stripe.

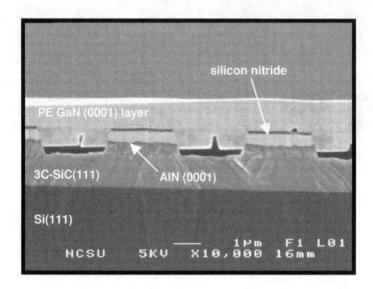

Figure 6. Cross-sectional SEM micrograph of a coalesced PE GaN epilayer deposited on a 3C-SiC/Si(111) substrate.

Tilting between adjacent growth fronts over the mask regions are commonly observed in coalesced GaN epilayers grown using the LEO technique. This phenomenon was also determined to be present in the PE GaN films.

The FWHMs of the XRD rocking curves for the (0002) reflection are indicative of the crystallographic tilt and are dependent on the crystal orientation, as shown in Figure 7. The XRD spectrum taken along the [$1\bar{1}00$], parallel to the stripes, consisted of one peak as shown in Figure 7(a), the spectrum along the [$11\bar{2}0$], perpendicular to the stripes, exhibited two superimposed peaks, separated by a tilt of 0.2°, as shown in Figure 7(b).

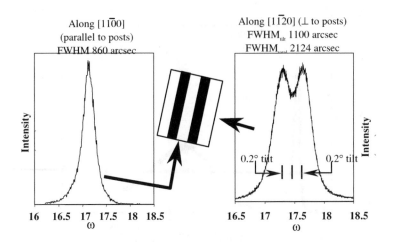

Figure 7. DCXRD analysis of the tilting in the coalesced PE films.

Unlike LEO, coalesced PE GaN epilayers contain two sets of coalesced growth fronts, namely, over the trenches and over the masks. A more sensitive method of evaluating the tilt in PE grown GaN is required to determine if it is present at both areas of coalescence. Selected area diffraction patterns were obtained using TEM and taken from a small area of coalescence in a trench region and near the region of coalescence over the silicon nitride mask. Analyses of these two patterns revealed no evidence of tilt in the laterally grown material over the trenches; however, significant tilt has occurred over the mask regions of the seed structures. Thus the mask has markedly influenced the crystallographic tilt in the overgrown film.

A typical room temperature PL spectrum for a PE GaN film grown on 2 μm wide raised stripes, oriented in the [11 00] direction, spaced 3 μm apart and capped with a silicon nitride mask is shown in Figure 8. The intensity of the spectrum is plotted on a linear scale. The band-edge emission peak and a deep level (yellow luminescence) emission peak were located at 363.6 nm (3.41 eV) and ~570 nm, respectively. The FWHM of the former was 109 meV.

The low temperature (14K) PL spectrum on a logarithmic intensity scale showed a strong near band-edge emission at 359 nm (3.45 eV), as shown in Figure 9. The FWHM was 17 meV. Also included in Figure 9 is a low temperature PL spectrum of PE GaN grown on a GaN/AlN/6H-SiC substrate. The band-edge peak in the latter spectrum is at 358 nm (3.46 eV) with a FWHM of 16 meV.

A comparison of these two spectra indicates that the optical character of the PE GaN grown on the 3C-SiC/Si substrate is nearly identical to the quality of the PE GaN grown on the 6H-SiC substrates. The band-edge of the PE GaN grown on Si is shifted by 10 meV to a slightly lower energy than that of the PE GaN grown on SiC, which is indicative of a slightly higher stress in the film of the former. As a further comparison, GaN(0001) films conventionally grown on AlN/6H-SiC substrates typically have a band-edge peak at 357.7 nm (3.466eV) and a FWHM of 4 meV.

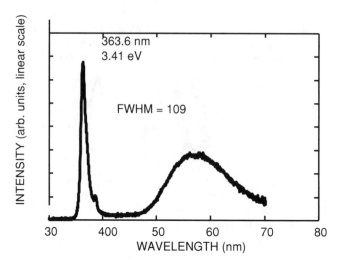

Figure 8. Room temperature photoluminescence of a coalesced layer of PE GaN grown on an GaN/AlN/3C-SiC/Si(111) substrate.

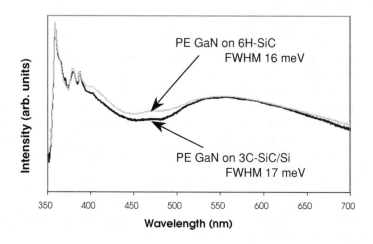

Figure 9. Comparison of low-temperature (14K) PL spectra of PE GaN grown on GaN/AlN/6H-SiC and GaN/AlN/3C-SiC/Si(111) substrates.

SUMMARY

Pendeo-epitaxy has been developed as an alternative and more simple approach of growing uniformly thin films with low densities of threading dislocations over the entire surface of a substrate. In particular, the growth of both discrete structures and coalesced GaN and $Al_xGa_{1-x}N$ films and multilayer heterostructures using pendeo-epitaxy on etched GaN seed layers previously grown on AlN(0001)/6H-SiC(0001) and AlN(0001)/SiC(111)/Si(111) substrates has been demonstrated. Tilting in the coalesced GaN epilayers of 0.2° was confined to areas of mask overgrowth; no tilting was observed in the material suspended above the trenches. The strong, low-temperature PL band-edge peak at 3.45 eV with a FWHM of 17 meV was comparable to that observed in PE GaN films grown on 6H-SiC(0001). The band-edge in the material on Si(111) was shifted by 10 meV to a lower energy, indicative of a greater tensile stress.

ACKNOWLEDGEMENTS

The authors acknowledge Cree Research, Inc. and Motorola for the SiC and the silicon wafers respectively. This work was supported by the Office of Naval Research under contracts N00014-98-1-0384 (Colin Wood, monitor) and N00014-98-1-0654 (John Zolper, monitor). The authors also acknowledge CWRU and the Georgia Technology Research Institute for use of their clean room facilities.

REFERENCES

1. Trademark of Nitronex Corporation, Raleigh, NC 27606
2. D. Kapolnek S. Keller, R. Vetury, R. Underwood, P. Kozodoy, S. Denbaars, and U. Mishra, Appl. Phys. Lett. **71**, 1204 (1997).
3. Y. Kato, S. Kitamura, K. Hiramatsu, and N. Sawaki, J. Cryst. Growth, **144**, 133 (1994).
4. O. Nam, M. Bremser, B. Ward, R. Nemanich, and R. Davis, Mat.Res. Soc. Symp. Proc., **449**, 107 (1997).
5. O. Nam, M. Bremser, B. Ward, R. Nemanich, and R. Davis, Jpn.J. Appl. Phys., Part 1 **36**, L532 (1997).
6. Sakai, H. Sunakawa, and A. Usui, Appl. Phys. Lett., **73**, 481 (1998).
7. H. Marchand, X. Wu, J. Ibbetson, P. Fini, P. Kozodoy, S. Keller, J. Speck, S. Denbaars, and U. Mishra, Appl. Phys. Lett., **73**, 747 (1998).
8. T. Zheleva, O. Nam, M. Bremser, and R. Davis, Appl. Phys. Lett., **71**, 2472 (1997).
9. O. Nam, T. Zheleva, M. Bremser, and R. Davis, Appl. Phys. Lett., **71**, 2638 (1997).
10. H. Zhong, M. Johnson, T. McNulty, J. Brown, J. Cook Jr., J. Schetzina, Materials Internet Journal, Nitride Semiconductor Research, **3**, 6, (1998).
11. S. Nakamura, M.Senoh, S. Nagahama, N. Iwasa, T. Yamanda, T. Matsushita, H. Kiyoku, Y. Sugimoto, T. Kozaki, H. Umemoto, M. Sano, and K. Chocho, Proc. of the 2nd Int. Conf. On Nitride Semicond., Tokushima, Japan, October, 1997.

12. S. Nakamura, M.Senoh, S. Nagahama, N. Iwasa, T. Yamanda, T. Matsushita, H. Kiyoku, Y. Sugimoto, T. Kozaki, H. Umemoto, M. Sano, and K. Chocho, Appl. Phys. Lett., **72**, 211 (1998).

13. S. Nakamura, M.Senoh, S. Nagahama, N. Iwasa, T. Yamanda, T. Matsushita, H. Kiyoku, Y. Sugimoto, T. Kozaki, H. Umemoto, M. Sano, and K. Chocho, Jpn. J. Appl. Phys., Part 2, **37** (1998), p. L309.

14. T. Zheleva, S. Smith, D. Thomson, K. Linthicum, T. Gerhke, P. Rajagopal, R.Davis, J. Electron. Matls. **28**, L5 (1999)

15. K. J. Linthicum, T. Gehrke, D.Thomson, E. Carlson, P. Rajagopal, T.Smith, R. Davis, Appl. Phys Lett. **75**, 196 (1999).

16. T. Gehrke, K. J. Linthicum, D.B.Thomson, P. Rajagopal, A. D. Batchelor and R. F. Davis, MRS Internet J. Nitride Semicond. Res. **4S1**, G3.2 (1999).

17. K. J. Linthicum, T. Gehrke, D.B.Thomson, K. M. Tracy, E. P. Carlson, T. P. Smith, T. S. Zheleva, C. A. Zorman, M. Mechregany and R. F. Davis, MRS Internet J. Nitride Semicond. Res. **4S1**, G4.9 (1999).

18. D.B.Thomson, T. Gehrke, K. J. Linthicum, P. Rajagopal, P. Hartlieb, T. S. Zheleva and R. F. Davis, MRS Internet J. Nitride Semicond. Res. **4S1**, G3.37 (1999).

19. T. S. Zheleva, D.B.Thomson, S. Smith, P. Rajagopal, K. J. Linthicum, T. Gehrke, and R. F. Davis, MRS Internet J. Nitride Semicond. Res. **4S1**, G3.36 (1999).

20. T. Weeks, M. Bremser, K. Ailey, E. Carlson, W. Perry, and R. Davis, Appl. Phys. Lett., **67**, 401 (1995).

21. C. A. Zorman, A. J. Fleischman, A. S. Dawa, M. Mehregany, C. Jacob, S. Nishino and P. Pirouz, J. Appl. Phys. **78**, 5136 (1995).

22. T. Gehrke, K.J. Linthicum, P. Rajagopal, E.A. Preble, E.P. Carlson, B.M. Robin, R.F. Davis, Int. Conf. on SiC and Related Materials (ICSCRM), Raleigh, NC, paper #269 (1999).

Mat. Res. Soc. Symp. Vol. 595 © 2000 Materials Research Society

Fabrication of GaN with Buried Tungsten (W) Structures Using Epitaxial Lateral Overgrowth (ELO) via LP-MOVPE

Hideto Miyake[1], Motoo Yamaguchi[1], Masahiro Haino[1], Atsushi Motogaito[1], Kazumasa Hiramatsu[1], Shingo Nambu[2], Yasutoshi Kawaguchi[2], Nobuhiko Sawaki[2], Yasushi Iyechika[3], Takayoshi Maeda[3] and Isamu Akasaki[4]

[1]Dept. of Electrical and Electronic Engineering, Mie University, 1515 Kamihama, Tsu, Mie, 514-8507, Japan
[2]Dept. of Electronic Engineering, Nagoya University, Furo, Chikusa, Nagoya, 464-8603
[3]Tsukuba Research Laboratory, Sumitomo Chemical Co., Ltd, 6 Kitahara, Tsukuba, 300-3294, Japan
[4]Dept. of Electrical and Electronic Engineering, Meijo University, 1-501 Shiogamaguchi, Tempaku, Nogoya 468-8502, Japan

ABSTRACT

A buried tungsten (W) mask structure with GaN is successfully obtained by epitaxial lateral overgrowth (ELO) technique via low-pressure metalorganic vapor phase epitaxy (LP-MOVPE). The selectivity of GaN growth on the window region vs. the mask region is good. An underlying GaN with a striped W metal mask is easily decomposed above 500 °C by the W catalytic effect, by which radical hydrogen is reacted with GaN. It is difficult to bury the W mask because severe damage occurs in the GaN epilayer under the mask. It is found that an underlying AlGaN/GaN layer with a narrow W stripe mask width (mask/window = 2/2 μm) leads the ELO GaN layer to be free from damage, resulting in an excellent W-buried structure.

INTRODUCTION

The group III-nitride semiconductors are promising materials for applications in opto-electronic devices such as light emitting diodes and laser diodes [1], and in field effect transistors[2]. Although these devices are fabricated mainly on sapphire substrate, it is not easy to achieve a high performance because of a high dislocation density in the crystal originating from the large differences in lattice constants between GaN and sapphire. An attempt to reduce the dislocation density of crystal using epitaxial lateral overgrowth (ELO) technique was employed in GaAs[3]. Recently, Usui *et al.* achieved a thick GaN epitaxial layer with a dislocation density as low as $10^7 cm^{-2}$ using ELO by hydride vapor phase epitaxy (HVPE) [4].

SiO_2 or SiN_x insulator has been normally used as the mask material for ELO-GaN. Besides those materials, ELO-GaN with tungsten (W) metal mask was investigated to fabricate optical and electronic devices with higher performance. It was reported that the ELO-GaN layers are successfully obtained with the W mask as well as SiO_2 or SiN_x masks via HVPE and have a better crystalline quality than that with SiO_2 mask[5].

In the ELO of GaAs, the W metal mask was employed and the W-buried structure was obtained to fabricate a permeable base transistor (PBT) or static induction transistor (SIT) [6]. The fabrication of the electronic devices such as PBTs or SITs via the ELO technique in GaN is very attractive for the use as high power and/or high frequency devices under crucial environments such as at high temperature or at highly radioactive environment. Recently, Kawaguchi et al. performed the selective area growth (SAG) of GaN using the W mask by metalorganic vapor phase epitaxy (MOVPE) and achieved high selectivity[7].

In this work, we demonstrate the ELO of GaN using the W mask by MOVPE and obtain the W-buried structure with GaN. The characteristics of ELO-GaN were also investigated by means of atomic force microscopy (AFM) and x-ray rocking curve (XRC).

EXPERIMENTAL PROCEDURE

MOVPE of GaN using a W mask was performed on two types of samples. The sample is a 4.0 μm-thick (0001) GaN epilayer, which had been grown on sapphire (0001) substrate using an GaN low temperature buffer layer by MOVPE. Another sample is a 40-100 nm thick AlGaN epilayer on the 4.0 μm-thick (0001) GaN epilayer. The 30 - 50 nm thick W mask was deposited by electron beam evaporation method. Stripe windows of 2 or 5 μm width with a periodicity of 4 or 10 μm, respectively, were developed in the <1100> direction of the GaN by conventional photolithographic method and then wet etching with a hydrogen peroxide (H_2O_2) and an ammonia solution. The source gases were trimethylgallium (TMG) and ammonia (NH_3), respectively. Hydrogen gas was used as a carrier gas during the growth process. ELO of GaN was performed by LP (low pressure) - MOVPE system with a horizontal reactor. Two steps ELO technique was carried out to bury W mask completely. In the first step, a low temperature (950°C) growth was performed for 30 min to prevent the underlying GaN from being decomposed. In the second step, a high temperature (1050°C) growth was done for 90 min to bury W mask. Here, it is expected in both steps that the GaN with {1122} facets are grown selectively in the first step and that the W mask is easily covered with GaN because of the fast ELO rate at a higher growth temperature in the second step, based on the results of ELO-GaN using SiO_2 mask[8].

RESULTS AND DISCUSSION

In order to investigate effects of the W mask which affects the underlying GaN and AlGaN layers in hydrogen ambient at a high temperature, we attempted thermal annealing at 500 to 700 °C of GaN and AlGaN with a striped W mask. Consequently, the surface of GaN window regions was roughened, including Ga droplets and GaN whiskers. This phenomenon occurred above 500°C. It is reported that an underlying GaN layer with a striped SiO_2 mask is decomposed higher than 900 °C in hydrogen ambient and there is no damage less than 900 °C[9]. Therefore, the W mask works as a catalyst to enhance decomposition of GaN. It is thought that W may produce radical hydrogen that can decompose GaN easily. On the other hand, the surface of AlGaN is not roughened even though the annealing temperature is 700°C. Any Ga droplets and GaN whiskers are not observed on the AlGaN layer. Hence, the catalytic effect of the W mask does not work in the AlGaN layer. Thus, it is found that the top AlGaN layer plays an important role to protect the bottom GaN from decomposition.

The ELO was performed using the W mask with stripe windows of 5 μm width. Figure 1 shows the cross sectional SEM images of ELO GaN with the underlying GaN (a) and AlGaN/GaN (b). About 5 μm-thick ELO GaN layers are obtained in both cases.

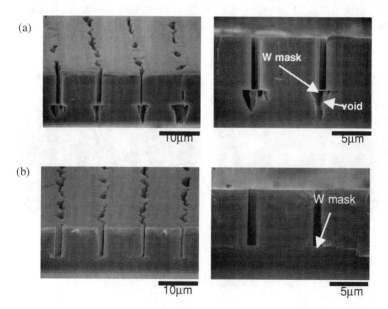

Figure 1 *The cross sectional SEM photographs of ELO GaN with underlying GaN (a) and AlGaN (b). The width of stripe window is 5 μm.*

Large voids occurs under the W mask as seen in Fig.1(a), indicating that the underlying GaN is decomposed during the ELO process owing the catalytic effect of the W mask. On the contrary, there are no voids in Fig.1(b). Therefore, the underlying AlGaN layer can prevent the bottom GaN from being decomposed.

The W masks are not buried completely after the growth process in the case of the 5 μm wide mask. The width of stripe was narrowed to bury the W mask completely. Figure 2 shows the cross sectional SEM images of ELO GaN with the underlying AlGaN/GaN layer. The widths of mask and window are 2 μm and 2 μm, respectively. This result clearly shows that the W mask is buried completely and the surface is smooth.

To characterize the surface morphology and the tilting of c-axis of the ELO GaN, we carried out x-ray diffraction and atomic force microscope (AFM). Figure 3 shows the results of AFM for the sample of Fig. 2. In this figure, atomic step structures with 0.1 μm-width and 0.3 nm-height are observed. This means that the step flow growth occurs in the ELO of the GaN layer. Figure 4 shows the XRCs of the (0004) reflection for $\phi = 0$ ° and 90 ° of the ELO GaN layer with the stripe directions of the masks along the direction of <1100>, where ϕ is the azimuth angle between the stripe direction of the mask and the scattering plane. The full width of half maximum (FWHM) is about 230 arcsec for each ϕ. There are no satellite peaks near the main peak, which suggests no tilting of c-axis. Thus, it is found that the ELO GaN with the W mask has high crystalline quality as well as a good surface morphology.

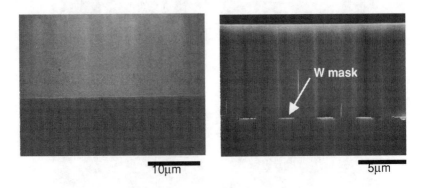

Figure 2 *The cross sectional SEM photographs of ELO GaN with underlying AlGaN. The width of stripe window is 2 μm.*

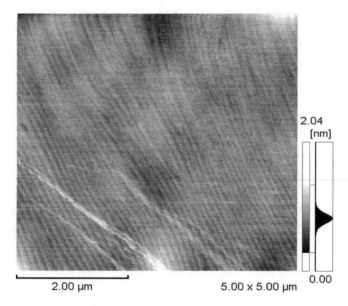

2.04
[nm]

2.00 µm

5.00 x 5.00 µm

0.00

Figure 3 *The results of AFM for the sample of Fig.2.*

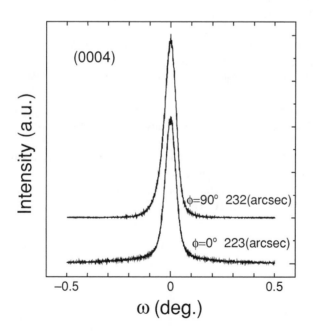

Figure 4 *The XRCs of the (0004) reflection of the sample of Fig.2.*

CONCLUSION

The two step ELO of GaN using W mask was performed by LP- MOVPE. GaN with a striped W metal pattern is easily decomposed above 500 °C by the W catalytic effect that causes radical hydrogen to attack the GaN layer. It is difficult to bury the W mask because severe damage occurs in the GaN epilayer under the mask. By employing the AlGaN/GaN underlying layer with the narrow W stripe mask of 2 μm, the excellent W-buried structure can be obtained free from damage.

ACKNOWLEDGEMENTS

This work was partly supported by "Research for the Future" program of Atomic Scale and Interface Dynamics of JSPS, "Light of the 21st Century" program of NEDO, the Proposed-Based R&D Program of NEDO (97S02-015) and Grant-in-Aid for Scientific Research from the Ministry of Education, Science, Sports and Culture (No.11450012).

REFERENCES

1 S. Nakamura, M. Senoh, S. Nagahama, N. Iwasa, T. Yamada, T. Matsushita, H. Kiyoku, Y. Sugimoto, T. Kozaki, H. Umemoto, M. Sano and K. Chocho, *Jpn. J. Appl. Lett.* **36** (1997) L1568.

2 R. Vetury, H. Marchand, G. Parish, P. T. Fini, J. P. Ibbetson, S. Keller, J. S. Speck, S. P. DenBaars and U. K. Mishra, *Inst. Phys. Conf. Ser.* No. 162: Chapter5 (1999) 177.

3 Y. Ujiie and T. Nishinaga, *Jpn. J. Appl. Phys.* **28** (1989) L327.

4 A. Usui, H. Sunakawa, A. Sakai and A. Yamaguchi, *Jpn. J. Appl. Phys.* **36** (1997) L899.

5 H. Sone, S. Nambu, Y. Kawaguchi, M. Yamaguchi, H. Miyake, K. Hiramatsu, Y. Iyechika, T. Maeda and N. Sawaki, *Jpn. J. Appl. Phys.* **38** (1999) L356.

6 H. Asai, S. Adachi, S. Ando and K. Oe, *J. Appl. Phys.* **55** (1984) 3868

7 Y. Kawaguchi, S. Nambu, H. Sone, M. Yamaguchi, H. Miyake, K. Hiramatsu, and N. Sawaki, *Jpn. J. Appl. Phys.* **37** (1998) L845.

8 H. Miyake, A. Motogaito and K. Hiramatsu, *Jpn. J. Appl. Phys.* **38** (1999) L1000.

9 K. Hiramatsu, H. Matsushima, H. Hanai and N. Sawaki : *Mat. Res. Soc. Symp. Proc.* **482** (1998) 991.

Mat. Res. Soc. Symp. Vol. 595 © 2000 Materials Research Society

Advanced PENDEOEPITAXY™ of GaN and $Al_xGa_{1-x}N$ Thin Films on SiC(0001) and Si(111) Substrates via Metalorganic Chemical Vapor Deposition

T. Gehrke, K. J. Linthicum[#], P. Rajagopal[#], E. A. Preble, R. F. Davis
Materials Research Center
North Carolina State University
Box 7919
Raleigh, NC 27695-7919

Abstract

Growth of GaN and $Al_xGa_{1-x}N$ thin films on 6H-SiC(0001) and Si(111) substrates with low densities of defects using the PENDEO™ process and the characterization of the resulting materials are reported. The application of a mask on the GaN seed structures hinders the vertical propagation of threading dislocations of the seed material during regrowth, but introduces a misregistry in the overgrowing material resulting in low quality crystal growth. This misregistry has been eliminated due to advanced processing and the exclusion of the masking layer. The new generation of samples do not show any misregistry, as shown by transmission electron microscopy.

Introduction

The PENDEO™ process is a new form of selective epitaxial growth that is dominated by the growth from sidewalls of rectangular stripes [1]-[5]. This process allows the growth of uniformly low defect density material over the entire surface of the semiconductor. Similar to LEO growth, a mask is employed to prevent vertical propagation of threading dislocations from the GaN seed structures into the regrown areas. The use of a mask can cause the formation of boundaries at the interface of coalescence of two growth fronts and a crystallographic tilt in the adjacent regions that have overgrown the mask. Electron-beam photolithography for the reduction in size of the GaN seed structures combined with the elimination of the masking material silicon nitride have been applied to achieve PENDEO™ growth of GaN and $Al_xGa_{1-x}N$ on the aforementioned substrates showing no formation of coalescence boundaries and no tilt in the overgrown regions. Microstructural results via scanning electron microscopy and transmission electron microscopy, as well as X-ray diffraction spectra, have been obtained and will be discussed in the following sections.

[#] Nitronex Corporation, Suite 104, 616 Hutton Street, Raleigh, NC, 27606

™ PENDEOEPITAXY and PENDEO are trademarks of Nitronex Corporation, Raleigh, NC 27606

For correspondence: e-mail: tgehrke@eos.ncsu.edu

Experimental Procedure

In the case of Si(111) as the substrate, a 3C-SiC buffer layer was deposited prior to any nitride growth, as described in Ref. [6]. On both substrates, 3C-SiC/Si(111) and 6H-SiC, films of GaN and $Al_xGa_{1-x}N$ have been grown via the PENDEO™-process on a 0.5 µm thick GaN(0001) seed layer grown on a high temperature, 100 nm thick AlN buffer layer. All nitride-based layers were grown using a cold-wall, vertical pancake style, RF-inductively heated metalorganic vapor phase epitaxy (MOVPE) system. The AlN buffer layers and the hexagonal GaN seed layers were each grown within the susceptor temperature ranges of 1080°C-1120°C and 980°C-1020°C, respectively, at a total pressure of 45 Torr. Triethylaluminum, triethylgallium, and NH_3 precursors were used in combination with a H_2 diluent. If employed, a 100 nm thick silicon nitride layer was used as a growth mask for blocking the continued threading dislocations during the PENDEO™ growth stage. The preparation of the GaN seed layer for this growth is explained in detail in Ref. [1]. Following these processing steps, the samples were degreased and cleaned prior to loading them into the MOVPE chamber. The PENDEO™ growth of the GaN and the $Al_xGa_{1-x}N$ layers were achieved within the susceptor temperature ranges of 1050-1100°C and 1080-1120°C, respectively, and at a total pressure of 45 Torr. Triethylgallium and NH_3 precursors were again used in combination with a H_2 diluent. The introduction of triethylaluminum into the growth chamber during PENDEO TM-growth produced $Al_xGa_{1-x}N$ layers having an atomic Al content of approximately 10%.

A JEOL 6400 FE scanning electron microscope (SEM) and a TOPCON 0002B, 200 KV, transmission electron microscope (TEM) were employed for microstructural analysis. A JEOL JAMP-30 high resolution scanning auger microprobe was utilized for measurements of the Al content of the as grown $Al_xGa_{1-x}N$ layers.

Results and Discussion

The first generation of GaN thin films grown by the PENDEO™-process are shown in Figure 1 for the substrate 6H-SiC and in Figure 2 for the substrate Si(111). A continuous film of fully coalesced GaN has been achieved by growing PENDEO™-GaN out of GaN seed posts covered with a silicon nitride mask to prevent vertical propagation of the threading dislocations of the seed post into the PENDEO™-GaN during regrowth.

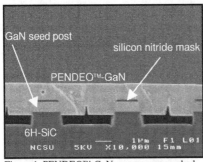

Figure 1. PENDEO™-GaN grown over masked GaN posts grown on a high temperature AlN buffer layer, on a 6H-SiC substrate.

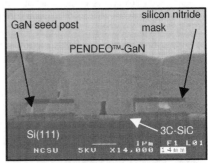

Figure 2. PENDEO™-GaN grown over masked GaN posts grown on a 2H-AlN/3C-SiC buffer layer, on a Si(111) substrate. A misregistry in the GaN above the mask is clear visible.

The SEM micrographs of Figure 1 and 2 indicate the influence of the silicon nitride masking layer on the quality of the PENDEO™-GaN overgrowing the mask. Figure 1 shows void formation above the silicon nitride mask, as known from the LEO process[7], which can be a result of a coalescence boundary formation, as presented in Ref. [8]. The formation of a coalescence boundary due to misregistry between the two growth fronts meeting above the silicon nitride mask is visible in Figure 2. Voids were being formed above the mask leading into a wavy surface morphology. The TEM micrograph shown in Figure 3 was prepared from a similar sample as presented in Figure 1. It shows the overgrowth region of PENDEO™-GaN above the silicon nitride mask. The growth fronts coming from both sides of the masking layer are highly defective and form a coalescence boundary, which may act as an additional source for generating defects. Similar results are well know for LEO of GaN[8]. Additional to the negative impact of the masking layer in the quality of the overgrowing GaN, in the case of $Al_xGa_{1-x}N$ the material tends to nucleate on the masking material, as presented in Figure 4 for $Al_{10}Ga_{90}N$ using a silicon nitride mask.

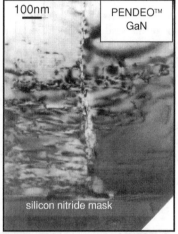

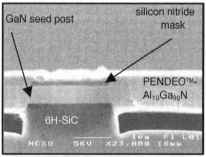

Figure 3. TEM micrograph of PENDEO™-GaN overgrown the silicon nitride mask. The coalescence boundary as a nucleation source for horizontal oriented dislocations is clear visible.

Figure 4. PENDEO™-$Al_{10}Ga_{90}N$ grown over a masked GaN post on a 6H-SiC substrate. The $Al_{10}Ga_{90}N$ nucleated on the silicon nitride mask.

Subject of ongoing studies with lateral overgrown GaN films is the tilting behaviour. We conducted an XRD characterization of PENDEO™-GaN grown from silicon nitride masked GaN seed posts on a 3C-SiC/Si(111) substrate. The rocking curves reveal one peak with a FWHM of 860 arcsec for a scan parallel to the posts and two peaks with a total FWHM of 2124 arcsec for a scan perpendicular to the posts[6]. This indicates that tilt is present in PENDEO™-GaN grown over masked seed posts. But what regions of the PENDEO™-GaN are responsible for the tilt? Small angle diffraction (SAD) pattern of coalescence regions of PENDEO™-GaN above the trench between two seed posts, and above the masked seed posts of the later sample were prepared using TEM, as shown in Figure 5. The SAD pattern of the coalescence region above the trench between two masked seed posts does not show any evidence of two crystallographic orientations tilted towards each other (see Figure 5(a)). In comparison, the SAD pattern of the coalescence region above the masking layer atop a GaN seed post shows clear evidence of the presence of two crystallographic orientations tilted towards each other (see Figure 5(b)). This behavior is also know from tilting studies of GaN grown via LEO[9],[10].

a)
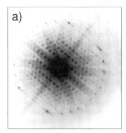

b)
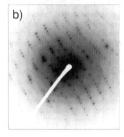

Figure 5. Small angle diffraction pattern of PENDEO™-GaN in the coalescence region (a) above the trench between two GaN seed posts and (b) above the silicon nitride mask atop a GaN seed post.

The application of a masking layer atop the GaN seed posts has the advantage of stopping the vertical propagation of threading dislocations of the seed post into the PENDEO™ regrown areas. But it also causes the formation of coalescence boundaries due to misregistry between the two coalescing growth fronts and tilt in the PENDEO™-GaN overgrowing the masking layer. Based on this facts a new generation of the PENDEO™-process has been developed. Electron beam (e-beam) lithography has been applied to reduce the width of the GaN seed posts into the submicron region and the masking layer has been eliminated. Figure 6 shows PENDEO™-GaN grown from submicron wide unmasked GaN seed posts on 6H-SiC substrate. No voids or misregistry above the seed posts are visible. The TEM micrograph in 6(b) shows a strong reduction in density of vertically oriented threading dislocations in the lateral and vertical grown PENDEO™-GaN compare to the seed post. Some dislocations were propagating vertically, but with reduced density. Noticeable is also that no dislocations were generated at the coalescence point of two growth fronts over the trench, above the void. SAD pattern from above the seed posts and above the voids did not show any evidence of tilt in this film.

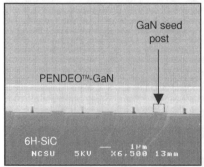
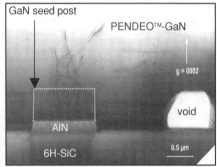

Figure 6(a). PENDEO™-GaN grown on unmasked submicron wide posts on a 6H-SiC substrate. No misregistry above the GaN seed post is visible.

Figure 6(b). TEM micrograph of the sample shown in (a). A strong reduction in density of threading dislocations in the PENDEO™-GaN compare to the GaN seed post is seeable.

The same has been observed for PENDEO™-GaN grown from unmasked GaN seed posts on 3C-SiC/Si(111) substrates, as shown in Figure 7. The TEM image in 7(b) reveals a strong reduction in density of vertically oriented threading dislocations in the PENDEO™-GaN grown region. Some horizontally oriented dislocations were being generated at the GaN/AlN interface of the GaN seed post, which do not propagate to the surface of the film.

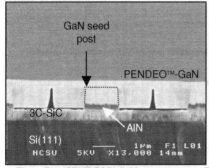 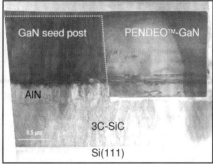

Figure 7(a). PENDEO™-GaN grown on un-masked GaN seed posts on a 3C-SiC/Si(111) substrate. No misregistry above the GaN seed post is visible.

Figure 7(b). TEM micrograph of the sample shown in (a). A strong reduction in density of threading dislocations in the PENDEO™-GaN compare to the GaN seed post is seeable.

Finally PENDEO™-$Al_xGa_{1-x}N$ with Al contents of about 10% were grown on submicron wide, unmasked GaN seed posts on 6H-SiC substrates, as shown in Figure 8. The Al content has been determined by high resolution Auger electron spectroscopy and a large area scan revealed a variation in the Al content of +/- 0.5%[11]. The elimination of the masking layer was necessary in this case after finding, that $Al_xGa_{1-x}N$ nucleates on the silicon nitride masking layer, as observed in Figure 4. The TEM micrograph in Figure 8(b) shows no lateral propagation of vertically oriented threading dislocations of the GaN seed post into the PENDEO™-$Al_{10}Ga_{90}N$ regions. Only a reduced number of threading dislocations is propagating vertically out of the GaN seed post into the regrown areas. No dislocations are being generated at the meeting point of the coalesced growth fronts

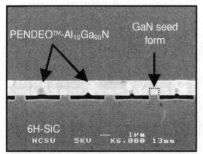 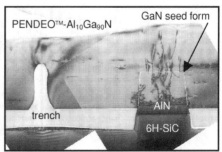

Figure 8(a). PENDEO™-$Al_{10}Ga_{90}N$ grown on unmasked GaN seed posts on a 6H-SiC substrate. No misregistry above the GaN seed posts is observable.

Figure 8(b). TEM micrograph of the sample shown in (a). A strong reduction in density of threading dislocations in the PENDEO™- $Al_{10}Ga_{90}N$ compare to the GaN seed post is obvious.

above the trench of the PENDEO™-Al$_{10}$Ga$_{90}$N film. SAD patterns taken via TEM of the PENDEO™-Al$_{10}$Ga$_{90}$N film above the GaN seed post and in the coalescence region above the trench do not show any evidence of tilt.

Conclusions

Growth of GaN and Al$_x$Ga$_{1-x}$N thin films on 6H-SiC(0001) and 3C-SiC/Si(111) substrates with low densities of defects using the PENDEO™ process have been reported. The disadvantages of applying a masked seed structure, namely formation of coalescence boundaries due to misregistry between the meeting growth fronts and tilt have been discussed and a new generation of the PENDEO™ process introduced. Overall, no evidence of coalescence boundaries and tilt were found in PENDEO™-GaN and PENDEO™-Al$_x$Ga$_{1-x}$N grown films on 6H-SiC and 3C-SiC/Si(111) substrates using the new process. This is a different finding compare to similar grown structures on sapphire substrates, as presented in Ref. [12].

Acknowledgements

The authors express their appreciation to C.A. Zorman and M. Mehregany from CWRU for the fruitful collaboration in the deposition of 3C-SiC on Si(111), as well as M. Johnson and W. Weeks from Nitronex Corporation for the helpful discussions The authors also thank K. Moore from Motorola for her help and assistance with processing samples via e-beam lithography.   This work was supported by the Office of Naval Research under the contracts N00014-96-1-0765 and N00014-98-1-0384 (C. Wood, technical monitor) and N00014-98-1-0654 (J. Zolper, technical monitor).

References

[1] T. Gehrke, K. J. Linthicum, D.B.Thomson, P. Rajagopal, A. D. Batchelor and R. F. Davis, MRS Internet J. Nitride Semicond. Res **4S1**, G3.2 (1999)

[2] K. J. Linthicum, T. Gehrke, D.B. Thomson, K.M. Tracy, E.P. Carlson, T,P.Smith, T.S. Smith, T.S. Zheleva, C.A. Zorman, M. Mehregany, and R.F.Davis, MRS Internet J. Nitride Semicond. Res **4S1**, G4.9 (1999)

[3] K. J. Linthicum, T. Gehrke, D.B. Thomson, E.P. Carlson, P. Rajagopal, T. Smith, and R.F. Davis, Appl. Phys. Lett., Vol. 75, No. 2, p. 196 (1999)

[4] D.B.Thomson, T. Gehrke, K. J. Linthicum, P. Rajagopal, P. Hartlieb, T. S. Zheleva and R. F. Davis, MRS Internet J. Nitride Semicond. Res **4S1**, G3.37 (1999)

[5] K.J. Linthicum, T. Gehrke, D. Thomson, C.Ronning, E.P. Carlson, C.A. Zorman, M. Mehregany, and R.F.Davis, MRS, Spring Meeting 1999, Symp. Y, 6.7

[6] T. Gehrke, K. J. Linthicum, E. Preble, P. Rajagopal, C. Ronning, C. Zorman, M. Mehregany, R. F. Davis, submitted for publication in J. of Electr. Mater.

[7] O.H.Nam, T.S. Zheleva, M.D. Bremser, D.B. Thomson, R.F. Davis, Mat. Res. Soc. Symp. Proc., Vol. 482, p.301 (1998)

[8] Z. Liliental-Weber, M. Benamara, W. Swider, J. Washburn, J. Park, P.A. Grudowski, C.J. Eiting, R.D. Dupuis, MRS Internet J. Nitride Semicond. Res. **4S1**, G4.6 (1999)

[9] A. Sakai, H. Suankawa, A. Usui, Appl. Phys. Lett., Vol. 73, No. 4, (1998) p. 481

[10] H. Marchand, N. Zhang, L. Zhao, Y. Golan, S.J. Rosner, G. Girolami, P.T. Fini, J.P. Ibbetson, S. Keller, S.P. DenBaars, J.S. Speck, U.K. Mishra, MRS Internet J. Nitride Semicond. Res. 4, 2 (1999)

[11] T. Gehrke, K.J. Linthicum, P. Rajagopal, E.A. Preble, E.P. Carlson, B.M. Robin, R.F. Davis, Int. Conf. on SiC and Related Materials (ICSCRM), Raleigh, NC, paper #269 (1999)

[12] P. Fini, H. Marchand, J.B. Ibbetson, B. Moran, L. Zhao, S.P. DenBaars, J.S. Speck, U.K. Mishra, MRS, Spring Meeting 1999, Symp. Y, 6.6

Mat. Res. Soc. Symp. Vol. 595 © 2000 Materials Research Society

A TEM study of GaN grown by ELO on (0001) 6H-SiC

P. Ruterana[1], B. Beaumont*, P. Gibart* and Y. Melnik**
Laboratoire d'Etudes et de Recherches sur les Matériaux, UPRESA 6004 CNRS, Institut de la Matière et du Rayonnement, 6 Boulevard du Maréchal Juin, 14050 Caen Cedex, France.
*Centre de Recherche sur l'Hétéroépitaxie et ses Applications, rue Bernard Grégory, Sophia Antipolis, 06560, Valbonne, France
**A.F. Ioffe Institute, St. Petersburg 194021, Russia,
present address: TDI Inc, Gaithersburg, MD 20877, USA.
1. Author for Correspondence Tel: 22 2 31 45 26 53, Fax: 33 2 31 45 26 60
email: ruterana@lermat8.ismra.fr

ABSTRACT

The misfit between GaN and 6H-SiC is 3.5 % instead of 16 % with sapphire, the epitaxial layers have similar densities of defects on both substrates. Moreover, the lattice mismatch between AlN and 6H-SiC is only 1%. Therefore, epitaxial layer overgrowth (ELO) of GaN on AlN/6H-SiC could be a route to further improve the quality of epitaxial layers. AlN has been grown by Halide Vapour Phase Epitaxy (HVPE) on (0001) 6H-SiC, thereafter a dielectric SiO_2 mask was deposited and circular openings were made by standard photolithography and reactive ion etching. We have examined GaN layers at an early stage of coalescence in order to identify which dislocations bend and try to understand why. The analysed islands have always the same hexagonal shape, limited by $\{10\overline{1}0\}$ facets. The **a** type dislocations are found to fold many times from basal to the prismatic plane, whereas when **a+c** dislocations bend to the basal plane, they were not seen to come back to a prismatic one.

INTRODUCTION

The epitaxial lateral overgrowth (ELO) technique has proven to be a relevant process to decrease the dislocation densities from 10^{10} to less than 10^7 cm^{-2} in GaN layers grown on sapphire. It has been implemented in both Metalorganic Vapor Phase Epitaxy (MOVPE) [1-3] and Hybrid Vapor Phase Epitaxy (HVPE) [4, 5], and considerable attention has been focussed on it, since a blue laser diode grown on an ELO GaN substrate, resulted in a lifetime of more than 10000 hours [6]. The conventional ELO method can be described as follows: first, a GaN layer of a few micrometers thick is grown, then a dielectric (SiO_2 or Si_xN_y) mask is deposited. Using standard photolithographic techniques, stripes are opened in the mask. During regrowth, either by MOVPE or HVPE, growth only occurs in the openings i.e., selective area epitaxy is achieved. A lateral growth over the mask leads to full coalescence. A smooth surface suitable for device fabrication is obtained if the ratio of the lateral to vertical growth rates is high enough. The standard way to realize ELO is to induce the lateral growth from the very beginning of the regrowth process[7]. Whereas the misfit between GaN and 6H-SiC is 3.5% instead of 16% on sapphire. The epitaxial layers have similar defect densities on both substrates. This effect has until now been attributed to the mosaïc growth of GaN on these substrates, with slightly misoriented islands bounded by threading dislocations. The lattice mismatch between AlN and 6H-SiC is about 1% which makes it most adequate as a buffer layer for the growth of GaN. In the following, we investigate the possibility of using HVPE AlN and 2S-ELO GaN in order to improve the quality of the active layers.

EXPERIMENTAL

Prior to epitaxy, the SiC was chemically cleaned and then H2 plasma treated, then a 100nm Al buffer layer was deposited by HVPE at 1100°C. The growth of GaN was next performed in a homemade MOVPE vertical reactor operating at atmospheric pressure. The carrier gas used in this study is a mixture of $H_2:N_2$. Trimethylgallium (TMGa) and NH_3 are used as precursors. A HeNe laser reflectometry set-up is used to monitor *in-situ* the growth process. First a 2µm-thick GaN layer is deposited directly on sapphire. In order to have large areas with electron transparency, it was necessary to prepare TEM samples by the tripod method. In this technique, two specimens are glued face to face using epoxy for cross section preparation. Then the first face is polished using a series of diamond coated plastic discs down to 0.5nm granulometry followed by silica colloïdal 0.05nm solution. Then 0.5 to 1nm slab is cut and glued on the tripod in order to grind the second face down to electron transparency. During this step, the thickness is monitored by use of silicon as one of the slabs. Therefore, when the sample thickness comes down to 10 µm, silicon becomes transparent and the polishing is stopped when the thickness fringes cross the interface and extend into the area of interest. In this process, the silicon is used to protect the specimen under study and is completely eliminated in the final stage of polishing. Usually a short cleaning in the ion mill is necessary before examination in the microscope. The specimens were analysed using a Jeol 2010 electron microscope operating at 200 kV.

RESULTS

At the initial stage of ELO, the use of circular windows has allowed us to determine the directions of fastest growth. Coalescence is only starting and the flat top islands are faceted (Figure 1a). In the Laue pattern, the mirror $\{11\bar{2}0\}$ planes of SiC show that the facets projections planes in the figure are $\{10\bar{1}0\}$(figure 1b). In this figure, there is an equal probability of finding one of the $6\{10\bar{1}0\}$ limiting facets for each island. This type of orientation was systematically used for TEM samples in order to make observations along a $<1\bar{2}10>$ directions.

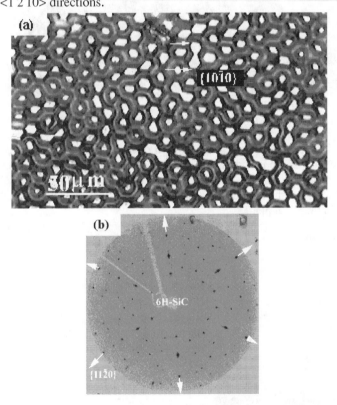

Figure 1: Surface morphology of the islands obtained by ELO during the first stage of growth (a), as well as the orientation of the surface (b).

As seen on figure 2, the use of a tripod allows to have a transparent area which extends to more than 100nm; this was used to investigate the various areas of the pyramidal islands. In particular, a simple diffraction experiment confirms the Laue orientation of figure 1 and shows that the interfaces between adjacent pyramids are along vertical {10$\bar{1}$0} facets.

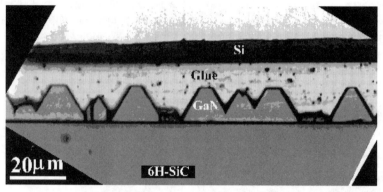

Figure 2. A TEM specimen of GaN obtained by the tripod polishing method

In particular a detailed analysis of the bending of dislocations has been carried out. From the center of the island, it can be seen in the 0002 dark field images that all the dislocations with **c** and **a** + **c** Burgers vectors bend to the basal plane (Fig. 3). Once this bending is accomplished, the dislocation will definitely stay in the basal plane.

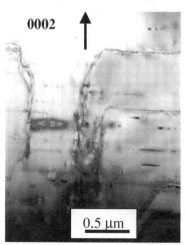

Figure 3. A 0002 weak beam image showing the bending of threading dislocations which have a **c** component in their Burgers vectors.

In the case of edge type <11$\bar{2}$0> threading dislocations, the behaviour appear to be slightly different. As shown of figure 4, which is a 11$\bar{2}$0 micrograph of the same area as in figure 3. This type of dislocations appear to fold many times from the basal to the prismatic plane in the center of the island. However, as one goes away of the central area, the dislocations have bent to the basal plane. This is in agreement with all the reported observations which show that ELO is able to lead to a drastic reduction of the threading dislocation densities.

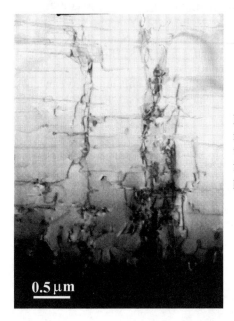

0.5 μm

Figure 4. A 1$\bar{1}$20 weak beam micrograph: threading dislocations with **a** component Burgers vectors fold more than once in basal and prismatic planes

Conclusion

In this work, we have shown that the threading dislocations bend to the basal plane on top of HVPE AlN on SiC. Based on these results and those available in the literature, it may be concluded that the lateral epitaxy is not a substrate driven mechanism. It is a consequence of the anisotropy of the properties of the crystals, such as dependence of the growth velocity on the crystallographic directions. Moreover, it is shown that the dislocation behavior is probably dependent on their Burgers vector. However more work is still needed in order to be able to make a more definite conclusion.

References

[1] D. Kapolnek, S. Keller, R. Vetury, R. D. Underwood, P. Kozodoy, S. P. Denbaars, and U. K. Mishra, Appl. Phys. Lett. **71**, 1204 (1997).

[2] O.-H. Nam, M. D. Bremser, T. S. Zheleva, and R. F. Davis, Appl. Phys. Lett. **71**, 2638 (1997).

[3] H. Marchand, X.H. Wu, J.P. Ibbetson, P.T. Fini, P. Kozodoy, S. Keller, J.S. Speck, S.P. DenBaars, U.K. Mishra, Appl. Phys. Lett. **73**, 747 (1998).

[4] A. Usui, H. Sunakawa, A. Sakai, and A. A. Yamaguchi, Jpn. J. Appl. Phys. **36**, L899 (1997).

[5] A. Sakai, H. Sunakawa, and A. Usui, Appl. Phys. Lett. **71**, 2259 (1997).

[6] S. Nakamura, M. Senoh, S. Nagahama, N. Iwasa, T. Yamada, T. Matsushita, H. Kiyoku, Y. Sugimoto, T. Kozaki, H. Umemoto, M. Sano, and K. Chocho, Proceedings of ICNS'97, Tokushima, Japan, edited by K. Hiramatsu (1997), p. 444 ; J. Cryst. Growth **189/190**, 820 (1998).

[7] B. Beaumont, M. Vaille, G. Nataf, A. Bouillé, J.-C. Guillaume, P. Vennéguès, S. Haffouz, P. Gibart, MRS Internet J. Nitride Semicond. Res, **Vol 3**, Art 20 (1998).

Mat. Res. Soc. Symp. Vol. 595 © 2000 Materials Research Society

Dislocation Mechanisms in the GaN Lateral Overgrowth by Hydride Vapor Phase Epitaxy

T. S. Kuan, C. K. Inoki, Y. Hsu, and D. L. Harris
Department of Physics, University at Albany, State University of New York, Albany, NY 12222
R. Zhang, S. Gu, and T. F. Kuech
Department of Chemical Engineering, University of Wisconsin, Madison, WI 53706

ABSTRACT

We have carried out a series of lateral epitaxial overgrowths (LEO) of GaN through thin oxide windows by the hydride vapor phase epitaxy (HVPE) technique at different growth temperatures. High lateral growth rate at 1100°C allows coalescing of neighboring islands into a continuous and flat film, while the lower lateral growth rate at 1050°C produces triangular-shaped ridges over the growth windows. In either case, threading dislocations bend into laterally grown regions to relax the shear stress developed in the film during growth. In regions close to the mask edge, where the shear stress is highest, dislocations interact and multiply into arrays of edge dislocations lying parallel to the growth window. This multiplication and pileup of dislocations cause a large-angle tilting of the laterally grown regions. The tilt angle is high (~8 degrees) when the growth is at 1050°C and becomes smaller (3-5 degrees) at 1100°C. At the coalescence of growth facets, a tilt-type grain boundary is formed. During the high-temperature lateral growth, the tensile stress in the GaN seed layer and the thermal stress from the mask layer both contribute to a high shear stress at the growth facets. Finite element stress simulations suggest that this shear stress may be sufficient to cause the observed excessive dislocation activities and tilting of LEO regions at high growth temperatures.

INTRODUCTION

The lateral epitaxial overgrowth (LEO) technique has recently been explored as a promising approach to reduce the dislocation density in GaN layers grown on a substrate. Various degrees of success using this method have been reported [1-4]. In GaN lateral overgrowth experiments using metalorganic vapor phase epitaxy (MOVPE) at 1000-1100°C, the dislocations originated from the seed layer are confined mostly over the growth windows [1, 2]. In a GaN lateral growth over a thin SiO_2 mask at 1000°C by hydride vapor-phase epitaxy (HVPE), however, most dislocations bend laterally into the overgrown regions [3, 4]. In either case, the defect density in the overgrown layer is reduced, since only the portion of defects in the seed layer not masked by the thin SiO_2 layer is able to propagate into the upper layer. During the LEO by either MOVPE or HVPE growth technique, arrays of edge dislocations are introduced at the mask edges, and they collectively tilt the c-axis of the LEO regions away from the [0001] epitaxial direction [2, 4]. The observed tilt ranges from less than ~0.2° [2] to larger than 10°. The correlation of c-axis tilt to growth technique and condition has not been established. In this study, we have carried out a series of LEO experiments and finite element stress simulations aimed at understanding the cause of LEO tilt and the associated dislocation mechanisms.

EXPERIMENTAL

The substrates used for our LEO experiments were 2.7-µm-thick GaN seed layers grown by MOVPE on a sapphire substrate. A 0.2-µm-thick SiO_2 mask layer was grown on top by tetraethoxysilane (TEOS)-based chemical vapor deposition at ~600°C, and growth windows were patterned by standard optical lithography and etching processes. The pattern consisted of 3- or 4-µm-wide stripe openings parallel to the [1$\bar{1}$00] direction separated by 8-µm-wide mask areas. In one sample (sample no. 3), the stripes (5.5 µm in width) were arranged in a radial pattern, allowing comparisons to be made between the LEO process along different lateral growth directions.

The samples were grown in a vertical HVPE reactor [5]. The samples were introduced into the growth region of the system under flowing NH_3 and N_2. The growth was initiated by the introduction of HCl to the Ga source region. The sample was kept at a constant temperature and gas phase ambient during growth. The five growth conditions investigated are listed in Table 1. Many factors influence the growth of HVPE GaN. We have altered the growth temperature in this study since it is the primary variable affecting the growth behavior.

Table 1 Growth conditions, growth rate, and defect structure

Sample	T_g (°C)	NH_3 flow rate (sccm)	HCl flow rate (sccm)	Vertical growth rate (µm/h)	Lateral growth rate (µm/h)	c-axis tilt
1	1050	600	18	8	3.6	7°
2	1050	1000	18	13	6	8°
3a	1075	1000	22.5	24	14	8°
3b	1075	1000	22.5	24	2.4	10°
4	1100	600	18	9.6	20	3°
5	1100	1000	18	5.8	20	5°

RESULTS AND DISCUSSION

A. Growth morphology

Triangular-shaped ridge growth was observed at 1050°C over 3-µm-wide stripe openings parallel to the [1$\bar{1}$00] direction in sample no. 1 [Fig. 1(a)]. The growth facets are curved before coalescing into a continuous film. In sample no. 2, higher NH_3 flux (1000 sccm) results in higher growth rate. The lateral/vertical growth rate ratio (= 0.45) and the ridge growth morphology remain the same as in sample no. 1, but coalescing has occurred, and the film has become continuous.

In sample no. 3a, higher (~2×) lateral and vertical growth rates with a lateral/vertical rate ratio of about 0.6 were observed on stripe windows parallel to the [1$\bar{1}$00] direction. The increase in growth rate is mostly due to the higher HCl flow rate. In sample no. 3b, the stripe windows are parallel to the [11$\bar{2}$0] direction, and the lateral growth rate is much smaller (2.4 µm/hr), resulting in a lateral/vertical rate ratio of ~ 0.1. From the scanning electron microscope images, we learned that on [11$\bar{2}$0]-oriented windows, the LEO is on ($\bar{1}$101) and (1$\bar{1}$01) facets, while on [1$\bar{1}$00]-oriented windows, the growth is on (11$\bar{2}$2) and ($\bar{1}$1$\bar{2}$2) facets. Our observations imply a much slower LEO rate on ($\bar{1}$101) and (1$\bar{1}$01) facets. Similarly strong dependencies of growth rate on facet

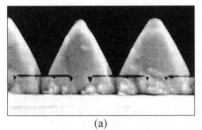
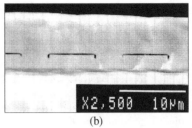

(a)   (b)

Fig. 1 Cross-sectional SEM images of GaN grown on windows parallel to the [1$\underline{1}$00] direction: (a) sample no. 1 grown for one hour. (b) sample no. 5 grown for one hour.

plane have also been reported for the lateral growth of GaN by MOVPE [1]. In samples no. 4 and no. 5, coalesced and smooth films were grown over parallel, [1$\underline{1}$00]-oriented windows at 1100°C [Fig. 1(b)]. The lateral growth rate at this temperature is estimated to be at least 20 μm/h from the film's surface curvature directly above the growth window.

B. Dislocation structure

Some morphological features such as the curved LEO facets observed in Fig. 1(a) can be attributed to the dislocation mechanisms involved during the growth. A cross-sectional TEM image of sample no. 1 in Fig. 2(a) reveals the dislocation structure after the sample is cooled to room temperature. A triangular ridge structure (region A) is grown first before the LEO occurs. This ridge structure contains vertical threading dislocations, mostly edge type with $b = 1/3<11\underline{2}0>$, propagating from the MOVPE seed layer [3, 6]. This initial structure has two facets extending at 66 degrees from the edges of the growth window, as marked by the dashed lines in Fig. 2(b). Since no low-index atomic planes parallel to [1$\underline{1}$00] are inclined at 66° from the (0001) plane, the observed angle may be determined by the lateral/vertical growth ratio [Table 1, tan^{-1} (8/3.6) = 66°]. As LEO proceeds from the two ridge surfaces, all dislocations bend toward the growth surfaces and propagate on horizontal (0001) planes and on inclined lattice planes such as (11$\underline{2}$2), (11$\underline{2}$3), and (11$\underline{2}$4), leaving the region B directly above the growth windows

(a)   (b)

Fig. 2 (a) A cross-sectional TEM image of a GaN ridge structure grown at 1050°C (sample no. 1). (b) Schematic of the dislocation glide behavior and final configuration in a GaN ridge structure observed in (a).

depleted of dislocations. The image force and/or the presence of shear stress close to the ridge surfaces may have caused the dislocation bending at high growth temperatures.

Fig. 2 indicates further that in LEO regions close to the mask edge (regions C), extensive dislocation multiplication has occurred, producing arrays of edge dislocations lying parallel to the [1$\underline{1}$00] ridge direction. These are perfect dislocations since no stacking fault contrast was observed on the slip planes. It is possible that the Frank-Reed type sources [7] are in operation for the generation of parallel dislocations during growth. These arrays of edge dislocations, confined mostly to the region C and constituting a series of tilt-type boundaries, tilt the c-axis of the crystal grown beyond region C away from the [0001] epitaxial orientation. A 7-degree c-axis tilt is measured from the image contrast of dislocations gliding on the tilted (0001) planes in region D in Fig. 2(a). As a result of the 7° tilt, the lower part of the (11$\underline{2}$2) facets on both sides of the ridge bends from the original 58° to a steeper 65° angle [Fig. 2(b)].

The crystal tilt remains in the rest of LEO films as growth continues, and, at the coalescence of growth facets, formation of a tilt-type grain boundary is unavoidable. The amount of crystal tilt can be measured accurately in a diffraction pattern taken from the bi-crystal at the coalescing boundary. The tilt angles measured for various growth conditions are listed in Table 1. In general, more parallel dislocations are observed in samples with higher tilt angles. Sample no. 3b has the largest c-axis tilt, and in it we find the highest density of edge dislocations in region C, as shown in Fig. 3(a). Fig. 3(b) shows the dislocation arrays observed near the SiO_2 mask edge in the sample no. 5. These parallel dislocations line up vertically into a series of about 15 low-angle tilt boundaries. With a measured vertical dislocation spacing D of about 50 nm and b = 0.3 nm, the tilt angle of each boundary is estimated to be $\theta \sim b/D \sim 0.35°$. The total tilt accumulated by the ~15 boundaries is about 5°, which is in good agreement with the diffraction measurement (Table 1).

C. Shear stress in LEO structures

The dislocation bending and c-axis tilting are most prominent in HVPE growth. The higher growth rates in HVPE result in LEO islands of high aspect ratio and large exposed surface area when compared to MOVPE growth. Since dislocations are

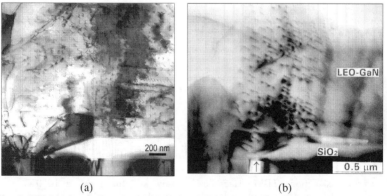

(a)   (b)

Fig. 3 TEM images of arrays of edge dislocation pileups parallel to the SiO_2 mask edge: (a) sample no. 3b and (b) sample no. 5.

observed to glide and multiply during the GaN lateral growth, substantial shear stress must be present in the growth island. A recent in-situ stress measurement [8] revealed that GaN grows under a constant tensile stress on a sapphire substrate by MOVPE at 1050°C. The tensile stress (0.2 to 0.6 GPa depending on growth conditions) is not relaxed by thermal cycling to room temperature or annealing at the growth temperature. If a growth island is bonded to a highly tensile substrate, a high shear stress is expected adjacent to the film edge [9]. If this shear stress resolved on the slip plane is higher than the critical stress needed for slip, then dislocation glide can be invoked to relax the stress. As the island grows vertically and laterally, the expanding materials boundary and geometry effect may necessitate the generation of arrays of dislocations and the onset of c-axis tilting. Since the generation and interaction of dislocations in a stressed crystal with moving boundaries are too complicated to simulate, we have set out to simulate just the static stress distribution in the triangular ridge structure at the growth temperature.

A periodic extension of an 8-μm-wide unit cell with a 3-μm-wide growth window at the center is used for the stress simulations. As the temperature rises from room temperature to 1050°C, a uniform horizontal expansion of the unit cell is imposed beyond thermal expansion so that the GaN seed layer is under a tensile stress at the growth temperature. Added on top of the uniform growth stress are local tensile stress fields due to the smaller thermal expansion of the oxide mask layer relative to GaN. Fig. 4 shows the simulated distribution of σ_{xx}, σ_{zz}, and σ_{xz} stresses in the 66° ridge structure at 1050°C. Here the x-axis is in the horizontal [11$\underline{2}$0] direction and the z-axis is in the vertical [0001] direction. The uniform growth stress σ_{xx} (set at 0.5 GPa) decreases rapidly to zero in the ridge island, and the thermal mismatch with the oxide mask layer contributes an extra 0.5 GPa in small regions at the window edges [Fig. 4(a)]. The vertical stress σ_{zz} and shear stress σ_{xz} close to the ridge surface [Fig. 4(b) and 4(c)] are linearly proportional to the GaN growth stress. Since the critical stress for dislocation glide, the Peierls stress, in a covalent crystal is typically 0.1% - 1% of the shear modulus and decreases with

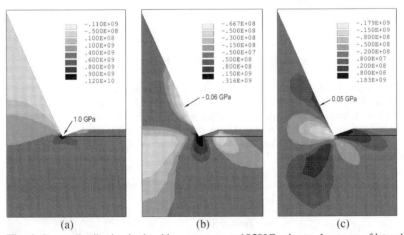

(a)   (b)   (c)

Fig. 4 Stress distribution in the ridge structure at 1050°C prior to the onset of lateral growth: (a) σ_{xx} along the horizontal x-axis, (b) σ_{zz} along the vertical z-axis, and (c) σ_{xz} shear stress on the (0001) plane.

increasing temperature, our simulations suggest that 0.05 Pa shear stress should be high enough for dislocation motions at 1050°C.

It is noted that a high density of edge dislocations ($> 2 \times 10^{10}/cm^2$) is generated locally in the high shear stress regions close to the mask edge, and dislocation pileups with Burgers vector $b = 1/3[11\underline{2}0]$ are observed experimentally. From elasticity considerations, a dislocation with $b = 1/3[11\underline{2}0]$ on the (0001) plane will be bent to the [1$\underline{1}$00] mask edge direction by the shear stress field. The observed pileups suggest that extensive stress relaxation is accomplished most likely via a Frank-Read type operation during the lateral growth.

CONCLUSION

We have investigated the dislocation mechanisms involved in the lateral growth of GaN by HVPE on a patterned substrate. The lateral growth rate is highly sensitive to the growth temperature and is dependent on the exposed growth facets. The growth morphology changes drastically from a narrow ridge to a smooth and continuous film as the growth temperature increases from 1050 to 1100°C. Threading dislocations from the seed layer bend into laterally grown regions and multiply into arrays of edge dislocations lying parallel to the growth window. This multiplication and pileup of dislocations lead to a large-angle c-axis tilting in the laterally grown regions. Finite element stress simulations indicate that at a high growth temperature, the tensile growth stress in the GaN seed layer and the thermal stress from the mask layer both contribute to a high shear stress at the growth facets. The relaxation of this shear stress is believed to be the driving force behind the excessive dislocation activities and the c-axis tilting occurring during the lateral growth.

ACKNOWLEDGMENT

This work was funded under the ONR MURI Grant #N00014-96-1-1179 through Dr. Colin Wood.

REFERENCES

1. O. -H. Nam, M. D. Bremser, T. S. Zheleva, and R. F. Davis, Appl. Phys. Lett. **71**, 2638 (1997).
2. H. Marchand, X. H. Wu, J. P. Ibbetson, P. T. Fini, P. Kozodoy, S. Keller, J. S. Speck, S. P. DenBaars, and U. K. Mishra, Appl. Phys. Lett. **73**, 747 (1998).
3. A. Sakai, H. Sunakawa, and A. Usui, Appl. Phys. Lett. **71**, 2259 (1997).
4. A. Sakai, H. Sunakawa, and A. Usui, Appl. Phys. Lett. **73**, 481 (1998).
5. R. Zhang and T. F. Kuech, Mat. Res. Soc. Symp. Proc., Vol. 512, 327 (1998).
6. W. Qian, M. Skowronski, M. De Graef, K. Doverspike, L. B. Rowland, and D. K. Gaskill, Appl. Phys. Lett. **66**, 1252 (1995).
7. F. C. Frank and W. T. Read, in *Symposium on Plastic Deformation of Crystalline Solids*, Carnegie Institute of Technology, 1950, p. 44.
8. S. Hearne, E. Chason, J. Han, J. A. Floro, J. Figiel, J. Hunter, H. Amano, and I. S. T. Tsong, Appl. Phys. Lett. **74**, 356 (1999).
9. M. Murakami, T. S. Kuan, and I. A. Blech, "Mechanical Properties of Thin Films on Substrates," in *Treatise on Materials Science and Technology*, Vol. 24, Academic Press, Inc., 1982, p.163-209.

Mat. Res. Soc. Symp. Vol. 595 © 2000 Materials Research Society

GaN and AlN Layers Grown by Nano Epitaxial Lateral Overgrowth Technique on Porous Substrates

M. Mynbaeva, A. Titkov, A. Kryzhanovski, A. Zubrilov, V. Ratnikov, V. Davydov, N. Kuznetsov, K. Mynbaev,
Ioffe Institute, St. Petersburg, Russia;
S. Stepanov, A. Cherenkov, I. Kotousova,
Crystal Growth Research Center, St. Petersburg, Russia;
D. Tsvetkov, V. Dmitriev,
TDI, Inc., Gaithersburg, MD, USA

Abstract

Defect density and stress reduction in heteroepitaxial GaN and AlN materials is one of the main issues in group III nitride technology. Recently, significant progress in defect density reduction in GaN layers has been achieved using lateral overgrowth technique. In this paper, we describe a novel technique based on nano-scale epitaxial lateral overgrowth.

GaN layers were overgrown by hydride vapour phase epitaxy (HVPE) on porous GaN. Porous GaN was formed by anodization of GaN layers grown previously on SiC substrates. Pore's size was in nano-scale range.

Thickness of overgrown layers ranged from 2 to 120 microns. It was shown that GaN layers overgrown on porous GaN have good surface morphology and high crystalline quality. The surface of overgrown GaN material was uniform and flat without any traces of porous structure. Raman spectroscopy measurements indicated that the stress in the layers grown on porous GaN was reduced down to 0.1 - 0.2 GPa, while the stress in the layers grown directly on 6H-SiC substrates remains at its usual level of about 1.3 GPa.

Preliminary experiments were done on HVPE growth of AlN layer on porous substrates. Improvement of surface morphology and crack density reduction has been observed.

Introduction

High defect density and residual stress in GaN heteroepitaxial layers grown on foreign substrates (sapphire, silicon carbide, and silicon) are the limiting factors for device applications [1]. High dislocation density negatively affects channel mobility in high power microwave transistors and laser diode life-time. Another sever problem related to stress in GaN-epitaxial structures is cracking. The cracking limits the thickness of GaN device layers grown on silicon and silicon carbide.

Defect density reduction in GaN heteroepitaxial materials has been demonstrated using epitaxial lateral overgrowth (ELOG) approach [2]. This technique allows one to grow GaN layers over a masked area. Dislocation density reduction in GaN grown over the mask was reported by a number of research teams. Recently, we have demonstrated defect density reduction in AlGaN alloys grown by HVPE (Fig. 1). In ELOG technique, mask fabrication requires rather precise allayment in certain crystallographic direction. Mask material may contaminate GaN overgrown layer. Additional stress and local material tilt may arise in GaN regions grown in the window area and over the mask.

In order to overcome this difficulties, we have proposed [3] to grow GaN layers on porous substrates. The porous substrate materials have been successfully applied to grow GeSi [4], SiC and AlN [5], $CoSi_2$ [6] on silicon. Porous SiC buffer has been used to reduce defect density in SiC epitaxial layers [7]. In this paper we describe preliminary results on GaN and AlN growth on porous substrates. Usually, pores have nanometer size. GaN and AlN layers were grown over these pores and we call this technique nano epitaxial overgrowth technique (NELOG). It is important that NELOG technique does not require any mask. This technique may be easily scaled for large area substrates.

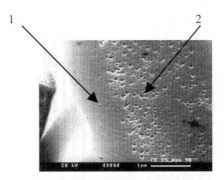

Figure 1. $Al_{0.3}Ga_{0.7}N$ alloy layer grown by HVPE technique using ELOG approach. After the growth, sample was treated by reactive ion etching to visualise defects. Area 1 corresponds to AlGaN layer grown over SiO_2 mask and area 2 corresponds to AlGaN grown directly on 6H-SiC substrate (window region). It is clearly seen that etch pit density in the alloy grown over the masked area (1) is much lower than that for AlGaN grown on non-masked area (2).

Experimental procedure and results

Porous GaN layers were formed by anodization of single crystal GaN epitaxial layers grown on SiC substrates by HVPE. 6H-SiC (0001)Si face oriented wafers were used as initial substrates. The HVPE technology employed for GaN layer deposition and properties of GaN layers grown by HVPE have been described elsewhere [8]. The layers were grown at 1000°C at atmospheric pressure. The anodization technique used to fabricate porous GaN layers has been described in ref. [9]. Thickness of porous GaN layers ranged form 1 to 10 microns. Pore size ranged from 50 to 200 nm.

GaN epitaxial layers up to 120 microns thick were grown by the same HVPE technique on porous GaN (Fig. 2). The layers had smooth surface; no traces of porous structure were detected. Reflection high-energy electron diffraction showed sharp Kikuchi lines indicating high crystal quality of the surface of grown layers. Results of Raman spectroscopy measurements revealed reduction of residual stress in GaN layer grown on porous GaN in comparison with GaN layer grown directly on 6H-SiC substrate (Fig. 3). The stress estimated by Raman spectroscopy measurements was about 1.3 GPa for layer grown on SiC and about 0.2 GPa for the layer grown on porous GaN.

Photoluminescence measurements detected a blue shift of the edge peak (Fig. 4) for GaN grown on porous GaN that also may be attributed to stress reduction.

X-ray diffraction measurements showed that GaN layers were of high crystal quality having the full width at a half maximum (FWHM) of ω-scan and ω–2Θ-scan x-ray rocking curves for (0002) reflections of about 400 arc sec and 40 arc sec, respectively.

Preliminary experiments have been done on AlN deposition on porous substrates. AlN layers were grown by HVPE technique [10] on porous SiC substrates. Porous SiC substrates were fabricated by surface anodization of commercial 6H-SiC (0001) wafers. Thickness of porous SiC layer ranged from 3 to 10 microns. AlN layers about 10 microns thick were deposited on untreated SiC wafer and porous SiC in the same epitaxial run. It was found that surface morphology of AlN was significantly improved by using of porous substrate (Fig. 5).

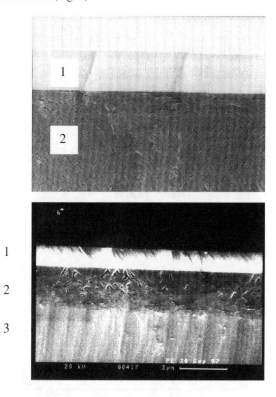

Figure 2. SEM images for two samples containing GaN layer (1) grown by HVPE on porous GaN (2) formed on 6H-SiC substrate (3). Thickness of GaN layer (1) is about 3 microns for top figure and about 2 microns for the bottom figure.

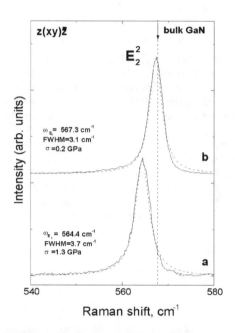

Figure 3. *Raman spectrum for GaN layer grown on porous GaN (b) and for GaN layer grown directly on 6H-SiC substrate (a).*

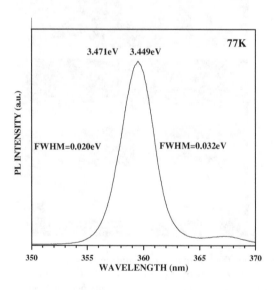

Figure 4. *Photoluminescence spectra (top) for GaN layer grown on porous GaN (dashed line) and for GaN layer grown directly on 6H-SiC substrate (77 K).*

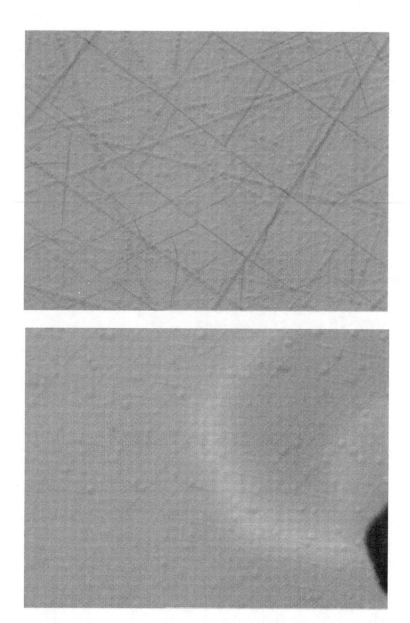

Figure 5. Surface morphology of AlN layer grown by HVPE on 6H-SiC substrate with (bottom) and without (top) porous buffer (x 1000). Black spot in the bottom picture is a micropipe penetrated from SiC substrate.

Surface of AlN grown on untreated SiC wafer contained a lot of defects caused by imperfections of surface treatment of initial SiC wafer. Such defects were not detected for AlN layers grown on porous SiC. AlN layers grown on porous SiC were much less cracked that AlN layers on standard SiC wafer.

Conclusions

GaN and AlN epitaxial layers were grown on porous substrates. Hydride vapor phase epitaxy was used for GaN and AlN deposition. GaN layers were grown on porous GaN and AlN layers were grown on porous SiC substrates. Reduction of residual stress in GaN layer and improvement in surface morphology of AlN layers were observed. The proposed NELOG technique is scaleable for large area substrates and does not require any masks.

Acknowledgements

Work at Ioffe Institute and CGRC was partly supported by INTAS programs # 96-1031 and 96-2131. Work at TDI was partially supported by ONR (contract monitor Colin Wood).

References

1   P. Fini, H. Marchand, J.P. Ibbetson, B. Moran, L. Zhao, S.P. DenBaars, J.s. Speck, U.K. Mishra, Mat. Res. Soc. Proc. **572**, 315-320 (1999).
2   T.S. Zheleva, S.A. Smith, D.B. Thomson, T. Gehrke, K.J. Linthicum, P. Rajagopal, E. Carlson, W.M. Ashmawi, and R.F. Davis, Mat. Res. Soc. Proc. **537**, G3.38 (1999).
3   M. Mynbaeva, A. Titkov, A. Kryzhanovski, I. Kotousova, A.S. Zubrilov, V.V. Ratnikov, V.Yu. Davydov, N.I. Kuznetsov, K. Mynbaev, D.V. Tsvetkov, S. Stepanov, A. Cherenkov, V.A. Dmitriev, Internet J. of Nitride Semicond. Research, **4**, 14 (1999).
4   S. Luryi, E. Suhir. *Appl.Phys. Lett.* **49**, 140-142 (1986).
5   D. Purser, M. Jenkins, D. Lieu, F. Vaccaro, A. Faik, M.A. Hasan, H.J. Leamy, C. Carlin, M.R. Sardela Jr., Q. Zhao, M. Willander, and M. Karlsteen, International Conference on Silicon Carbide and Related Materials 1999, Abstracts, 462 (1999).
6   Y.C. Kao, K.L. Wang, B.J. Wu, T.L. Lin, *Appl. Phys. Lett.* **51**, 1809-1811 (1987).
7   M. Mynbaeva, N. Savkina, A. Tregubova, M. Scheglov, A. Lebedev, A. Zubrilov, A. Titkov, A. Kryganovski, K. Mynbaev, N. Seredova, D. Tsvetkov, S. Stepanov, A. Cherenkov, I. Kotousova, V. Dmitriev, to be published in the Proceedings of the Int. Conf. on Silicon Carbide and Related Materials, October 10-15, 1999, NC.
8   N.I. Kuznetsov, A.E. Nikolaev, A.S. Zubrilov, Yu.V. Melnik, V.A. Dmitriev, Appl. Phys. Lett. **75**, 3138-3140 (1999).
9   M. Mynbaeva, D. Tsvetkov, *Inst. Phys. Conf. Ser.* **155**, 365-368 (1997).
10 Yu.V. Melnik, A.E. Nikolaev, S. Stepanov, I.P. Nikitina, K. Vassilevski, A. Ankudinov, Yu. Musikhin and V.A. Dmitriev, Materials Science Forum, **264-268**, 1121 (1998).

Mat. Res. Soc. Symp. Vol. 595 © 2000 Materials Research Society

Mass Transport, Faceting and Behavior of Dislocations in GaN

S. Nitta[1], T. Kashima[1], M. Kariya[1], Y. Yukawa[1], S. Yamaguchi[2], H. Amano[1, 2] and I. Akasaki[1, 2]
Department of Electrical and Electronic Engineering[1], High-Tech Research Center[2], Meijo University, Tempaku-ku, Nagoya 468-8502, Japan

ABSTRACT

The behavior of threading dislocations during mass transport of GaN was investigated in detail by transmission electron microscopy. Mass transport occurred at the surface. Therefore, growing species are supplied from the in-plane direction. The behavior of threading dislocations was found to be strongly affected by the mass transport process as well as the high crystallographic anisotropy of the surface energy of the facets particular to GaN.

INTRODUCTION

The Growth of GaN on sapphire using a low-temperature (LT-) deposited buffer layer [1] followed by the realization of p-type films [2] has lead to the fabrication of bright blue and green light-emitting diodes and violet laser diodes. Nevertheless, the GaN layer still contains high-density threading dislocations on the order of 10^8 to $10^{11} cm^{-2}$ [3]. Several efforts have been made to reduce the density of threading dislocations, such as epitaxial lateral overgrowth [4,5] or pendeo-epitaxy [6]. In order to achieve much higher performance light emitters and novel devices, reduction of the threading dislocation density together with understanding the growth mechanism on an atomic scale are still one of the most critical and important issues in nitride research.

Very recently, we found a new method of obtaining low-dislocation-density GaN. In the newly developed method, a LT-layer was deposited not only on the sapphire substrate, but also on high-temperature (HT)-grown GaN. Dislocation densities as low as mid-$10^6 cm^{-2}$ has been achieved by multiple insertion of a LT-layer. This layer deposited between HT-GaN is called the LT-interlayer [7]. High-temperature transmission electron microscopy (TEM) observation was performed to clarify the dynamical process of the LT-interlayer. The as-deposited LT-interlayer is composed of very fine crystallites with a size of about a few nm. First, atomic reconstruction and grain growth occurred in the initial stage of heating. After that, mass transport at the surface occurred. Then the surface became atomically flat. The mass transport leads to a bending of the threading dislocations, thereby achieving low-dislocation-density GaN films on top of the LT-interlayer.

The mass transport at the surface is not limited to the LT-layer, but should also be observed at the surface of HT-GaN. In the case of other III-V materials such as InP, GaAs and GaP, mass transport at the surface is well known and has been reported by several groups [8-12]. The essential procedure for the mass transport is the etching of the selected area and annealing in a group-V-source gas containing atmosphere without any supply of group-III source gas. Mass transport at the surface can be applied to fabricate such structures as p-n junctions, quantum dots, buried heterostructure lasers, and gratings for distributed feedback lasers and wafer fusion. Therefore, it is one of the most important key technologies for the fabrication of novel devices. However, there has been no report concerning the mass transport of HT-nitrides. We found that mass transport at the surface of HT-GaN occurs at around the growth temperature [13]. We also found that this phenomenon can be applied to grow dislocation-free areas in HT-GaN films on sapphire. In this study, a detailed TEM analysis was performed to clarify the behavior of threading dislocations in the mass-transported region, thereby allowing us to understand the mass transport process on an atomic scale.

EXPERIMENT

GaN 7 μm in thickness was grown by metal organic vapor phase epitaxy (MOVPE) on a sapphire (0001) substrate using a LT-AlN buffer layer. Trench stripes were patterned on the surface along the <11-20> direction by reactive ion etching. Several samples of trenches with different depths and widths were prepared. Then, in order to perform mass transport, these samples were annealed at 1100°C in a MOVPE reactor. The temperature was monitored using a thermocouple inserted into the graphite susceptor. During annealing, 0.20 mol/min of ammonia with nitrogen gas was supplied. No group-III alkyl source gas was supplied. After annealing for 12 min, the grooves were almost buried by mass transport processes. To analyze the mass transport process and behavior of dislocations, cross-sectional TEM study was carried out.

RESULTS

Fundamental mechanism of mass transport in shallow trench

The process of the mass transport of GaN is schematically shown in Fig. 1. Mass transport is generally thought to proceed toward the minimization of total surface energy [14]. Assuming a single mesa step, as shown in Fig. 1(a), at the convex part of the surface, the binding of crystal (Ga-N) is weakened by the higher surface energy. Therefore, GaN is decomposed and free Ga atoms diffuse to the surrounding. On the contrary, at concave part of the surface, Ga atoms are easily reincorporated into the solid and thus growth of GaN occurs. However, the high crystallographic anisotropy, which is characteristic of GaN, modulates the surface shape during mass transport. Therefore, it is hard to form round surface corners composed of higher index facets as observed in the case of InP [14]. {1-101} facets of GaN are energetically much more stable than other facets, except for {0001}, as has already been reported by other groups [15,16]. Thus, once the facets are formed at the upper trench edges, decomposition does not continue there. Therefore, the Ga species are supplied from the outer area of the surface, as shown in Fig. 1(b). Other stable facets such as {1-100} are formed before the {1-101} facet plane is completely formed, as shown in Fig. 1(c). Then, once {1-101} facets from both upper edges are connected to each other or reach the bottom of the groove, higher index facets are formed and the trench is gradually buried, as shown in Figs. 1(d) through 1(e).

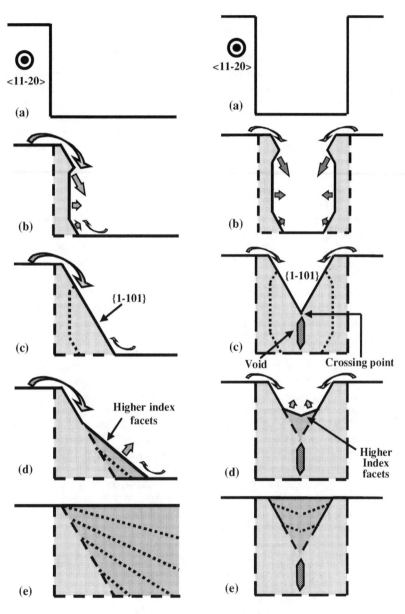

Fig. 1 Scheme of process of mass transport in GaN in shallow trench.

Fig. 2 Scheme of process of mass transport in GaN in deep trench. If the trench is deeper than the crossing point, a void is formed.

Mass transport in deep trench and mechanism of void formation

Fig. 2 schematically shows the process of mass transport into a deep trench. If the trench is deeper than the crossing point, a void is formed under the V-groove groove, as shown in Fig. 2(c). In the case of a deep trench, {1-101} facets from each upper edge are formed prior to other facet. Therefore, the connecting point is geometrically determined by the angle between (0001) and {1-101} facets. We should bear in mind that there is a small offset for {1-101} facets to form in both sides, because there is also growth toward [1-100] the direction. Once {1-101} facets are connected, forming a void under the V-groove, no more Ga species are supplied, resulting in void formation.

Behavior of threading dislocations

Fig. 3 shows the bright-field cross-sectional TEM image of the 4.5 μm wide trench after mass transport. This figure indicates several important aspects of mass transport. Dislocation-free regions were obtained in the upper areas of the left-half center and the right-half center of the trench. Threading dislocations reaching the bottom of these areas are bent. Therefore, they should not propagate to these regions. In the initial stage of mass transport, threading dislocations near the side wall continue to propagate vertically, and then bend horizontally when they reach the {1-101} facet surfaces which were formed during mass transport. The effect of this particular plane has already been reported by Sakai et al. [17]. However, during burying, some of these dislocations propagate toward the surface again. On the contrary, the direction of propagation of the threading dislocations that reach the bottom of the central part of the trench is changed by the higher index facet plane. Then, these dislocations merge to the center of the trench. In this case, crystallographic anisotropy of the higher index facet plane is not so high as in the case of {1-101}. Therefore, the dislocations bend toward the free surface in order that the energy should be minimum. From dark-field TEM images (not shown here) most of these dislocations are divided into two types with a Burger's vector of either $b_A=1/3<11-20>$ (type-A) or $b_B=1/3<11-23>$ (type-B). Each type-A and –B dislocation is indicated by a solid or open arrowhead, respectively. Fig. 4 schematically shows the behavior of each dislocation. A type-A dislocation is simply bent horizontally in the <11-20> or <1-100> direction, as shown in Fig. 4(a). On the contrary, a type-B dislocation propagates vertically until it reaches {1-101} facets. At the {1-101} facets the dislocation is bent horizontally in the <1-100> direction. However, when higher index facet growth appears, the dislocation propagates along the growth direction, as shown in Fig. 4(b). Both type-A and -B dislocations retain the Burger's vector direction even after bending.

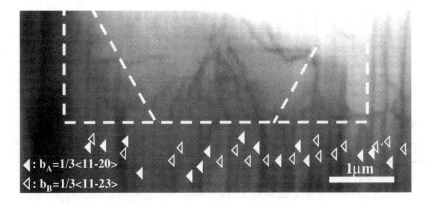

Fig. 3 Cross-sectional TEM image of shallow trench after mass transport.

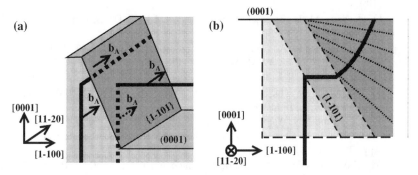

Fig. 4 Schematic images of behavior of dislocation
which has (a) $b_A=1/3<11-20>$ or (b) $b_B=1/3<11-23>$.

CONCLUSIONS

Processes of mass transport for the cases of shallow and deep trenches involving the mechanism of void formation were clarified. The cross-sectional TEM study indicated that threading dislocations which have a Burger's vector of either $b_A=1/3<11-20>$ or $b_B=1/3<11-23>$ were bent by the mass transport process. It was found that the behavior of threading dislocations is strongly affected by the mass transport process as well as high crystallographic anisotropy of the surface energy of facets particular to GaN. Mass transport is found to be very effective for the growth of low dislocation density GaN on sapphire. It is also useful for growing highly pure GaN, because this method does not require any masks, which sometimes act as an impurity source.

ACKNOWLEDGMENTS

This work was partly supported by the Japan Society for the Promotion of Science (JSPS) Research for the Future Program in the Area of Atomic Scale Surface and Interface Dynamics under the project "Dynamic Process and Control of the Buffer Layer at the Interface in a Highly-Mismatched System", and the Ministry of Education, Science, Sports and Culture of Japan (contract no. 11450131).

REFERENCES

1. H. Amano, N. Sawaki, I. Akasaki and Y. Toyoda, Appl. Phys. Lett. 48, 353 (1986).
2. H. Amano, M. Kito, K. Hiramatsu and I. Akasaki, Jpn. J. Appl. Phys. 28, L2112 (1989).
3. S. D. Lester, F. A. Ponce, M. G. Craford and D. A. Steigerwald, Appl. Phys. Lett. 69, 898 (1996).
4. A. Usui, H. Sunakawa, A. Sakai and A. A. Yamaguchi, Jpn. J. Appl. Phys. 36, L899 (1997).
5. O.-H. Nam, M. D. Bremser, T. S. Zheleva and R. F. Davis, Appl. Phys. Lett. 71, 2638 (1997).
6. T. S. Zheleva, S. A. Smith, D. B. Thomson, T. Gehrke, K. J. Linthicum, P. Rajagopal, E. Carlson, W. M. Ashmawi and R. F. Davis, Mater. Res. Soc. Internet J. Nitride Semicond. Res. 4S1, G3. 38 (1999).
7. M. Iwaya, T. Takeuchi, S. Yamaguchi, C. Wetzel, H. Amano and I. Akasaki, Jpn. J. Appl. Phys. 37, L316 (1998).
8. Z. L. Liau and J. N. Walpole, Appl. Phys. Lett. 40 (7), 568 (1982).
9. T. R. Chen, L. C. Chiu, A. Hasson, K. L. Yu, U. Koren, S. Margalit and A. Yativ, J. Appl. Phys. 54 (5), 2407 (1983).
10. M. Kito, N. Otsuka, S. Nakamura, M. Ishino and Y. Matsui, IEEE Photon. Technol. Lett., 8, 1299 (1996).
11. M. Imada, T. Ishibashi and S. Noda, Jpn. J. Appl. Phys. 37, L1400 (1998).
12. T. Ogawa, M. Akabori, J. Motohisa and T. Fukui, Jpn. J. Appl. Phys. 38, 1040 (1999).
13. S. Nitta, M. Kariya, T. Kashima, S. Yamaguchi, H. Amano and I. Akasaki, Proc. 3rd Intern. Symp. on Control of Semiconductor Interface, Karuizawa, Japan, (1999).
14. Z. L. Liau and H. J. Zeiger, J. Appl. Phys. 67 (5), 2434 (1990).
15. Y. Kato, S. Kitamura, K. Hiramatsu and N. Sawaki, J. Crystal Growth 144, 133 (1994).
16. S. Kitamura, K. Hiramatsu and N. Sawaki, Jpn. J. Appl. Phys. 34, L1184 (1995).
17. A. Sakai, H. Sunakawa, and A. Usui, Appl. Phys. Lett. 71, 2259 (1997); A. Sakai, H. Sunakawa, and A. Usui, Appl. Phys. Lett. 73, 481 (1998)

Mat. Res. Soc. Symp. Vol. 595 © 2000 Materials Research Society

Dislocation Arrangement in a Thick LEO GaN Film on Sapphire

Kathleen A. Dunn*, Susan E. Babcock*, Donald S. Stone*, Richard J. Matyi*, Ling Zhang† and Thomas F. Kuech†
*Materials Science and Engineering ; †Chemical Engineering
University of Wisconsin – Madison, Madison, WI 53706

ABSTRACT

Diffraction-contrast TEM, focused probe electron diffraction, and high-resolution X-ray diffraction were used to characterize the dislocation arrangements in a 16μm thick coalesced GaN film grown by MOVPE LEO. As is commonly observed, the threading dislocations that are duplicated from the template above the window bend toward (0001). At the coalescence plane they bend back to lie along [0001] and thread to the surface. In addition, three other sets of dislocations were observed. The first set consists of a wall of parallel dislocations lying in the coalescence plane and nearly parallel to the substrate, with Burgers vector (**b**) in the (0001) plane. The second set is comprised of rectangular loops with $\mathbf{b} = 1/3\,[11\bar{2}0]$ (perpendicular to the coalescence boundary) which originate in the coalescence boundary and extend laterally into the film on the $(1\bar{1}00)$. The third set of dislocations threads laterally through the film along the $[1\bar{1}00]$ bar axis with $1/3<11\bar{2}0>$-type Burgers vectors These sets result in a dislocation density of $\sim10^9$ cm^{-2}. High resolution X-ray reciprocal space maps indicate wing tilt of $\sim0.5°$.

INTRODUCTION

High dislocation densities plague GaN films because the commercially available bulk substrates all have substantial lattice mismatch (13.5% for Al$_2$O$_3$, 3.3% for SiC) and thermal expansion coefficient (TEC) mismatch with GaN (34% and 25% for Al$_2$O$_3$ and SiC, respectively). Presumably, eliminating one or both of these sources of mechanical stress is a step towards reducing the defect densities. Lateral epitaxial overgrowth (LEO) relieves components of the lattice mismatch stress by growing part of the GaN film over an amorphous material [1,2]. This report describes the dislocation microstructure in a 16μm thick coalesced film grown by the LEO method on a sapphire substrate. TEM images of the entire thickness of the film over five periods of the LEO mask pattern reveal a complex but characteristic arrangement of dislocations, many of which appear not to have their origin in the template.

EXPERIMENTAL PROCEDURES

The LEO substrate was fabricated using conventional CVD processes to grow a 1μm thick GaN template layer (MOCVD with TMGa and ammonia) and a ~0.1μm thick SiO$_2$ glass mask layer. Openings 2μm wide, oriented along $[1\bar{1}00]_{GaN}$ and separated by 10μm of glass, were patterned into the mask using conventional lithography. A 16μm thick LEO film was grown at 1100°C in two hours by MOVPE using diethyl-GaCl and ammonia as precursors with a V/III ratio of 1800 and hydrogen carrier, as described in [3]. Cross-sectional TEM specimens were prepared by tripod polishing with final thinning by Ar ion milling. Two-beam and weak-beam diffraction contrast TEM were used to image the dislocation arrangements in the film and determine Burgers vectors of the various sets of dislocations observed. Convergent probe microdiffraction and high-resolution x-ray diffraction reciprocal space maps of the symmetric 0002 reflection [4] were used to investigate the extent of lattice tilt in the LEO material.

RESULTS

Prior to coalescence, the cross-section of the growing LEO material was rectangular with smooth, vertical sidewalls on (11 20) that terminated in small, inclined bevels at the top surface. The bevels remained after coalescence, leaving ~2 μm deep V-shaped trenches where the coalescence boundaries intersect the surface.

High-resolution x-ray reciprocal space maps were recorded using the 0002 point with the stripes oriented both perpendicular (Fig 1a) and parallel (Fig 1b) to the diffractometer plane. The appearance of three distinct peaks in Fig 1a, but not 1b, indicates that the c-axis is tilted $\pm 0.5°$ about the stripe axis in a portion of the material, presumably over the glass [5]. The downward shift of the side peaks corresponds to a relative expansion of the lattice along [0001] in the regions where the c-axis is tilted, a result that is consistent with different levels of in-plane compressive stress existing in the tilted and untilted regions of the material. The significant difference in the diffuse scatter between the perpendicular and parallel maps suggests a highly anisotropic defect structure.

Figure 2a is a montage of TEM images showing four periods of the LEO structure. The schematic in Fig. 2b illustrates the components of the characteristic dislocation arrangement that are repeated in each period. Labeled with Roman numerals in Figs. 2a & b are: (I) a core of threading dislocations (TDs) that are copied from the template, (II) threading and longitudinally oriented dislocations in the coalescence plane (CP), (III) a V-shaped complex of dislocation loops, and (IV) longitudinal dislocations that run along the window axis in the upper part of the film. The arrows in the collection of unit cells in Fig 2c indicate the Burgers vectors (**b**) for each of these families of dislocations. The threading dislocations in the core have all of the "**a**" and "**a+c**" type Burgers vectors that typically populate epitaxial GaN films. The Burgers vectors of the dislocations in

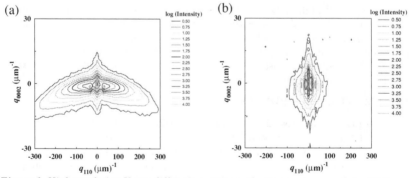

Figure 1. *High resolution X-ray diffraction reciprocal space maps recorded with the stripes perpendicular (Fig 1a) and parallel (Fig 1b) to the diffractometer plane. The three peaks in (a) indicate three distinct orientations for the c-axis, while the difference in the distribution of diffuse scatter between (a) and (b) indicates a highly anisotropic dislocation structure.*

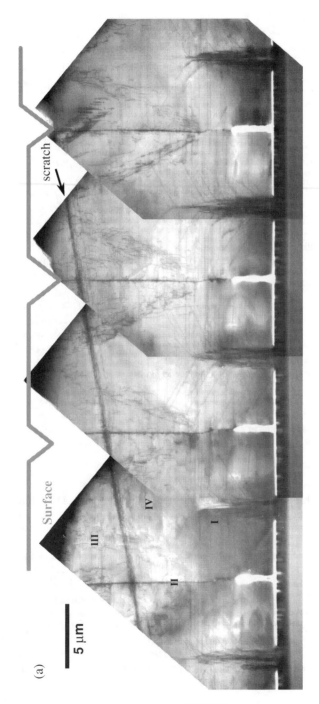

Figure 2. (a) Composite TEM image showing nearly four periods of the LEO structure. Four families of dislocations are labeled: (I) threading dislocations (TDs) copied from the template, (II) dislocations in the coalescence plane, (III) a V-shaped arrangement of rectangular loops, and (IV) longitudinal dislocations lying along the stripe axis.

W2.11.3

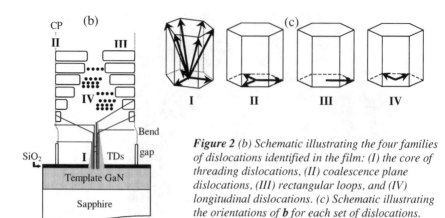

Figure 2 *(b) Schematic illustrating the four families of dislocations identified in the film: (I) the core of threading dislocations, (II) coalescence plane dislocations, (III) rectangular loops, and (IV) longitudinal dislocations. (c) Schematic illustrating the orientations of **b** for each set of dislocations.*

each of the other families are a specific subset of these **b**'s. Limitation to specific **b**'s indicates that these are introduced into the microstructure to fulfill a particular purpose.

(I) Threading Dislocations

Threading dislocations are copied from the underlying template in the region above the window. In this sample, within 8 μm of growth all the threading dislocations bend away from [0001]. All appear to bend in the plane normal to the stripe axis. Some, but not all, bend 90° to lie on (0001); others prefer one of several other habit planes [6]. As they approach the coalescence plane, at least a subset, if not all, of these dislocations bend again to thread to the surface in the coalescence plane (Fig. 3a). Even if there is no crystallographic tilt, net **b**, or disregistry at the coalescence plane, there is a dislocation structure in the boundary as a result of the bending threading dislocations. These dislocations do not disappear into the interface, nor do they continue into the next LEO stripe. They could act as sources or sinks for dislocations and point defects and/or provide a mechanism for mechanical deformation at the coalescence plane.

(II) Coalescence plane

In addition to the displaced threading dislocations, there is a wall of parallel dislocations (Fig 3b) that lie parallel or slightly inclined to the substrate and have Burgers vectors in the basal plane. These dislocations are present in somewhat dense sets with a local spacing that varies from 40-190 nm along the boundary. Their configurations are reminiscent of dislocation pile-ups in plastically deformed materials. The local dislocation spacings in this segment of the boundary correspond to a lower and upper estimate for the local misorientation across the coalescence plane of 0.16° and 0.43°, assuming $b=1/3[11\bar{2}0]$ for all of the dislocations. These angles will be 13% smaller if the **b**'s are $1/3[\bar{2}110]$ or $1/3[1\bar{2}10]$. Focused probe (micro-) diffraction, a technique with ~0.5° sensitivity, revealed no discernable tilt across this coalescence plane, consistent with the very small misorientation calculated from the dislocation spacing. These preliminary measurements were made near the top surface of the film. The inexact agreement with the x-ray diffraction data concerning the magnitude of the tilt (the x-ray results suggest ~1° of tilt across the coalescence plane) reaffirms the inhomogeneous nature of the local tilt throughout the sample.

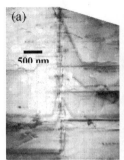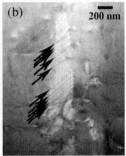

Figure 3. Structure of the coalescence plane. (a) Threading dislocations are displaced from the core to thread to the surface along the CP. (b) Walls of dislocations that lie in the CP with line direction approximately parallel to the substrate have configurations reminiscent of dislocation pile-ups in plastically deformed materials.

(III) Rectangular loops

The third set of dislocations consists of rectangular loops in the (1 $\overline{1}$00) plane, which give rise to the distinctive V-shaped arrangement of edge-type dislocations that is centered on each coalescence plane (Figs. 2 and 4). The configuration of the loops, particularly near the apex where the entirety of the loop is visible (Fig. 4a), indicates that they form at the coalescence boundary and propagate into the LEO material. Closer to the surface of the film where the loops have propagated as far as 5 μm, only the vertical segments of the loops are clearly visible in many instances. Considering that the TEM specimen is thin (0.1 - 0.2 μm thick), that the loop could lie anywhere through the specimen thickness, and that the sample may not be sectioned exactly on (1 $\overline{1}$00), the observed images are consistent with rectangular loops in the (1 $\overline{1}$00) plane. The dislocation images recorded under several different two-beam diffracting conditions were consistent with *all* of the loops having b = 1/3[11 $\overline{2}$0], the a-type Burgers vector that lies perpendicular to the coalescence plane. The reversal of the characteristic black-white-black g · b =2 contrast to white-black-white contrast at the coalescence plane indicates that the direction of the Burgers vector is opposite in the two legs of the "V." Due to the presence of the glass, the lattice parameter in this direction is not constrained by the substrate. Thus, these dislocations do not accommodate lattice mismatch, but rather must form in response to some other mechanical stress.

(IV) Longitudinal dislocations

The regions above the core of threading dislocations and between the legs of neighboring "V's" are populated by longitudinal dislocations of the type that have been observed in a number of films [7]. These dislocations have a-type Burgers vectors. The contrast produced where they intersect the surface of the TEM sample suggests that they have a screw component (i.e., Burgers vector other than that of the loops), but additional study of plan view samples is needed to characterize them more completely.

DISCUSSION

Four distinct dislocation families were observed in a thick LEO GaN film grown on sapphire. Each family has specific Burgers vectors, suggesting that they are introduced into the microstructure for a specific reason. The orientation of the Burgers vector in one of the dominant families, the rectangular dislocation loops (III), suggests that they form to in response to stresses other than lattice mismatch. An obvious candidate is TEC mismatch. An analysis of the deformation of the crystal that would be produced by the

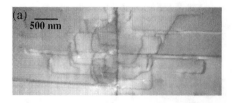

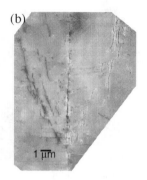

Figure 4. TEM images showing lateral extent of rectangular dislocation loops. Notice the contrast reversal of the dislocation image when the coalescence plane is crossed.

dislocation loops alone shows their net effect is to cause a contraction of the material in the bar by buckling it like an accordion. Other dislocations would be needed to satisfy compatibility requirements caused by the V-shaped configuration of the loops. Arguments can be made that the longitudinal dislocations in the coalescence plane and the LEO material satisfy these requirements, but further Burgers vector analysis and examination of the driving force would be needed to establish this mechanism unambiguously.

CONCLUSIONS

The dislocations observed in a thick, coalesced LEO GaN film on sapphire form a complex but characteristic arrangement, and do not appear to be solely copied from the template. While wing tilt and TD bending are commonly observed in LEO films, the observation of additional sets of dislocations with specific Burgers vectors suggests a large scale cooperative motion in response to the in-plane compressive stress or other driving forces.

ACKNOWLEDGEMENTS

This work was funded by the ONR-MURI on Compliant Substrates. Partial support for the electron microscopy facilities was provided by the NSF-MRSEC at UW-Madison.

REFERENCES

[1] T. S. Zheleva, O.-H. Nam, M. D. Bremser, and R. F. Davis, *Appl. Phys. Lett.* **71**, 2472 (1997).
[2] D. Kapolnek, S. Keller, R. Vetury, R.D. Underwood, P. Kozodoy, S. P. DenBaars, and U.K. Mishra, *Appl. Phys. Lett.* **71**, 1204 (1997).
[3] R. Zhang, L. Zhang, D. M. Hansen, M. P. Boleslawski, K. L. Chen, D. Q. Lu, B. Shen, Y.D. Zheng and T. F. Kuech, *MRS Internet J. Nitride Semicond. Res.* **4**: U473-U478, Suppl. 1 (1999).
[4] B.K. Tanner and D. K. Bowen, **High Resolution X-ray Diffractometry & Topography**, Taylor & Francis, Bristol, PA 1998.
[5] P. Fini, L. Zhao, B. Moran, M. Hansen, H. Marchand, J. P. Ibbetson, S. P. DenBaars, U. K. Mishra, and J. S. Speck, *Appl. Phys. Lett.* **75**(12), 1706 (1999).
[6] H. Marchand, X.H. Wu, J. P. Ibbetson, P.T. Fini, P. Kozodoy, S. Keller, J.S. Speck, S. P. DenBaars, U.K. Mishra, *Appl. Phys. Lett.* **73**(6), 747 (1998).
[7] Z. Liliental-Weber, M. Benamara, W. Swider, J. Washburn, J. Park, P.A. Grudowski, C.J. Eiting, R.D. DupuisMRS Internet J. Nitride Semicond. Res. 4S1, G4.6 (1999).

Growth, Structural Characterization,
Surface Studies, Theory, Doping

Mat. Res. Soc. Symp. Vol. 595 © 2000 Materials Research Society

POLARITY DETERMINATION FOR MOCVD GROWTH OF GaN ON Si(111) BY CONVERGENT BEAM ELECTRON DIFFRACTION

L. Zhao, H. Marchand, P. Fini, S. P. Denbaars, U. K. Mishra, J. S. Speck
Materials Department and Electrical and Computer Engineering Department
University of California, Santa Barbara, CA 93106

ABSTRACT

The polarity of laterally epitaxially overgrown (LEO) GaN on Si(111) with an AlN buffer layer grown by MOCVD has been studied by convergent beam electron diffraction (CBED). The LEO GaN was studied by cross-section and plan-view transmission electron microscopy (TEM). The threading dislocation density is less than $10^8 \, cm^{-2}$ and no inversion domains were observed. CBED patterns were obtained at 200 kV for the $<1\bar{1}00>$ zone. Simulation was done by many-beam solution with 33 zero-order beams. The comparison of experimental CBED patterns and simulated patterns indicates that the polarity of GaN on Si(111) is Ga face.

INTRODUCTION

In recent years it has been demonstrated that GaN and related alloys grown by metalorganic vapor phase epitaxy exhibit superb properties for the design and production of light-emitting diodes and lasers operating in the short wavelength of the visible spectrum. LEO GaN on sapphire and silicon carbide substrates has been shown to result in a significant reduction of the extended defect density (3-4 orders of magnitude) which, in turn, has led to improvements in device performance for blue lasers, light-emitting diodes, p-n junctions, and field-effect transistors. For a review of relevant topics, see ref. [1]. The success of LEO growth has revived interest in alternative substrates such as Si(111), which has potential advantages for device integration, thermal management, and cost issues. LEO GaN with low dislocation density on Si(111) substrates has recently been demonstrated [2].

The most common growth direction for GaN is normal to the {0001} basal plane. The basal plane has a polar configuration with two atomic sub-planes each consisting of either the cationic or the anionic element of the binary compound. Thus, in the case of GaN a basal plane surface should be either Ga or N terminated (See Fig.1). Note that the polarity is a bulk property. The identification of the polarity in GaN can be carried out by several techniques such as X-ray photoemission spectroscopy [3], convergent beam electron diffraction [4,5,6], and chemical etching [7]. For a review of GaN polarity determination, see ref. [8]. In this paper we report on CBED studies of LEO GaN on Si(111) substrate and show that the polarity is Ga face.

EXPERIMENTAL

Two inch-diameter Si(111) wafers were etched in buffered HF for one minute before growth. After heating to the growth temperature of 900°C under hydrogen, the TMAl and NH_3 precursors were introduced in the MOCVD growth chamber and the AlN buffer layer was deposited at a total pressure of 76 Torr. The thickness of the AlN layer

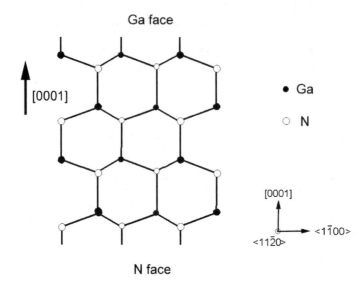

Figure 1. <11 $\bar{2}$ 0> projection of the GaN structure. [0001] is defined as the direction of Ga to N bond between Ga/N bilayers, as shown with the arrow in the figure.

was ~60 nm. The AlN layer was crack-free over the entire wafer, and the RMS roughness measured by AFM was on the order of 15 nm. The wafers were then coated with 200 nm-thick SiO_2 using plasma-enhanced chemical vapor deposition, and 5 µm-wide stripes oriented in the Si<11 $\bar{2}$ > direction were patterned using standard UV photolithography and wet chemical etching. The width of the SiO_2 mask regions was 35 µm. The LEO GaN stripes were obtained by performing a regrowth at ~1060°C using TMGa and NH₃.

TEM samples were prepared by wedge polishing followed by Ar⁺ ion milling. Diffraction contrast images and CBED patterns were obtained on a JEOL 2000FX microscope operated at 200 kV. The surface topography was imaged using a Digital Instruments Dimension 3000 AFM operating in tapping mode. Simulations were done by software package "Desktop Microscopist 2.0" [9]. The simulated CBED patterns were calculated by solution of the many-beam equation with 33 zero-order reflections. It is well known that the polarity of GaN/Al₂O₃ grown by MOCVD is usually Ga face. We also checked the polarity of our MOCVD grown GaN on Al₂O₃ by CBED and confirmed that it is Ga face.

RESULTS AND DISCUSSION

Cross-section TEM micrographs are shown in Figure 2. The seed region has a threading dislocation density on the order of ~10^{11} cm^{-2}, consisting predominantly of pure edge dislocations. In contrast, the LEO regions have a very low threading dislocation

Figure 2. Cross-section bright-field TEM micrographs, $g = 11\bar{2}0$. The irregular top and side surfaces are due to ion milling.

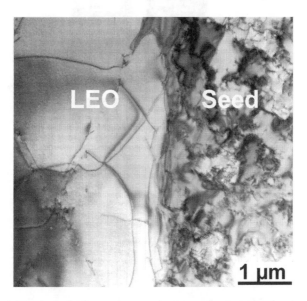

Figure 3. Plan-view bright field TEM image of LEO GaN/Si(111) clearly showing the seed (right) and overgrown region (left).

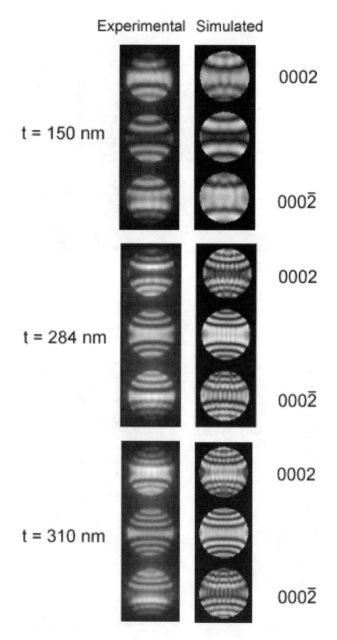

Figure 4. Experimental and simulated CBED patterns at different thickness. All patterns correspond to [1 $\bar{1}$ 00] zone axis.

density and show an essentially single-crystalline microstructure. The plan-view TEM image (Figure 3) further confirmed that the threading dislocation density was reduced several orders of magnitude in the LEO region. No inversion domain boundaries were observed under both cross-section and plan-view TEM. We reasonably assume that the GaN/Si(111) grown by MOCVD has one unique polarity.

Figure 4 shows the comparison of experimental and simulated CBED patterns. The experimental CBED patterns were taken under <1 $\bar{1}$ 00> zone-axis at 200 kV from different thickness region of the cross section sample. One should pay attention when indexing the diffraction pattern. For some models of transmission electron microscopes, there is a 180° inversion between image and diffraction pattern. This can be checked by underfocusing the diffraction lens to image a plane above the diffraction pattern. The image will not be inverted with respect to the diffraction pattern [10]. The simulations were calculated with 33 zero-order reflections and mean absorption coefficient 0.05. A series of CBED patterns were calculated with different thicknesses from 100 nm to 400

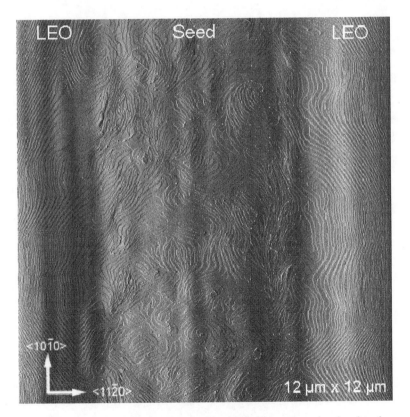

Figure 5. Surface topography measured by AFM. The contrast is related to the amplitude of the tip vibration during tapping-mode imaging.

nm in 10 nm steps. The pattern which best matched the experimental data was selected. From central disks comparison, the sample thickness was obtained. From diffraction disks comparison, the polarity was obtained. From the matching of experimental and calculated patterns, it is clear that the polarity of the LEO GaN on Si(111) is Ga face.

Figure 5 shows the surface topography of LEO GaN grown on Si(111) over 144 μm^2 area. Although the surface is undulated and shows spiral growth, the topography clearly consists of c/2-high atomic steps. In the seed region the pure screw and mixed-character threading dislocations ($\sim 2 \times 10^9$ cm^{-2}) are visible as surface depressions (~ 20 nm in diameter) that terminate atomic steps. The lack of such step terminations in the LEO GaN regions clearly shows that the threading dislocation density is reduced significantly. Overall the surface is quite smooth. Our results are consistent with the framework summarized by E. S. Hellman [8]: "smooth films grown by MOCVD are usually Ga face".

CONCLUSION

The polarity of MOCVD growth GaN/Si(111) with AlN buffer has been studied by convergent beam electron diffraction. Comparing the experimental patterns and simulated patterns, it is shown that the polarity of GaN film is Ga face.

ACKNOWLEDGEMENTS

This work was supported by the Office of Naval Research through a contract supervised by Dr. C. Wood and made use of the Materials Research Laboratory Central Facilities supported by the NSF under award No. DMR96-32716.

REFERENCES

[1]   J. S. Speck and S. J. Rosner, Physica B, in press.
[2]   H. Marchand, N. Zhang, L. Zhao, Y. Golan, S. J. Rosner, G. Girolami, P. T. Fini, J. P. Ibbetson, S. Keller, S. DenBaars, J. S. Speck, U. K. Mishra, MRS Internet J. Nitride Semicond. Res., **4**, Article 2, 1999.
[3]   T. Sasaki, T. Matsuoka , J. Appl. Phys. **64**, 4531 (1988).
[4]   Z. Liliental-Weber , J. Washburn, K. Pakula, and J. Baranowski, Microsc. Microanal., **3**, 436 (1997).
[5]   F.A. Ponce, D.P. Bour, W.T. Young, M. Saunders, J.W. Steeds, Appl. Phys. Lett. **69**, 337 (1996).
[6]   P. Vermaut, P. Ruterana, G. Nouet, Phil. Mag. A **76**, 1215 (1997).
[7]   J. L. Weyher, S. Müller, I. Grzegory, S. Porowski, J. Cryst. Growth **182**, 17 (1997).
[8]   E. S. Hellman, MRS Internet J. Nitride Semicond. Res. **3**, Article 11, 1998.
[9]   Available from Virtual Laboratories (Tel. 505 828 1640).
[10]   M. H. Loretto and R. E. Smallman, *Defect Analysis in Electron Microscopy*, (John Wiley & Sons Inc., New York, 1975), chapter 1.3.

Mat. Res. Soc. Symp. Vol. 595 © 2000 Materials Research Society

Structural Properties of Laterally Overgrown GaN

R. Zhang, Y, Shi, Y.G. Zhou, B. Shen, Y.D. Zheng
*Laboratory of Solid State Microstructures and Department of Physics, Nanjing
University, Nanjing 210093, CHINA*
T.S. Kuan
Department of Physics, State University of New York at Albany, Albany, NY 12222
S.L. Gu, L. Zhang, D.M. Hansen and T.F. Kuech
Department of Chemical Engineering, University of Wisconsin, Madison, WI 53706

ABSTRACT

Structural properties of epitaxially laterally overgrown (ELO) GaN on patterned
GaN 'substrates' by hydride vapor phase epitaxy (HVPE) have been investigated. The
epitaxially lateral overgrowth of GaN on SiO_2 areas is realized and a planar ELO GaN
film is obtained. Scanning electron microscope, transmission electron microscope (TEM)
and atomic force microscope (AFM) are used to study the structure and surface
morphology of the ELO GaN materials. AFM images indicate that no observable step
termination is detected over a 4 μm^2 area in the ELO region. TEM observations indicate
that the dislocation density is very low in the ELO region. No void at the coalescence
interface is observed. Lattice bending as high as 3.3° is observed and attributed to pileup
of threading dislocations coming from the underlying GaN "seeding layer" and tilting
horizontally and quenching at the coalescence interface.

I. INTRODUCTION

GaN and related compound materials have been intensively studied due to their
important practical applications and promising for potential applications [1-3]. The
existence of high density threading dislocations in GaN films is one of major factors to
impede further development of GaN optoelectronic and electronic devices [4]. So far the
most successful and practical technique for reducing the threading dislocation (TD)
density of GaN epilayers is the epitaxial lateral overgrowth (ELO). The TD density of the
GaN ELO region could be 4 or more orders lower than the conventionally grown GaN
films [5]. Both metalorganic vapor phase epitaxy (MOVPE)[6-9] and hydride vapor
phase epitaxy (HVPE)[10-11] have been used to produce GaN ELO materials with high
crystal quality. HVPE is a very useful technique to grow GaN films via both large-area
epitaxy and ELO. In HVPE ELO GaN materials, no observable void is found in the
coalescence interface and a variety of growth front geometries are observed [10].
Compared to a non-halide process, a halide process principally offers an advantage of a
high lateral-to-vertical growth rate ratio [12], which is a key factor influencing the growth
front and related dislocation motion in the GaN epilayer grown by ELO. HVPE offers a
high growth rate and high material quality for GaN growth [13,14]. The typical growth
rate can get 100-200 $\mu m/hr$. GaN produced by the HVPE technique does result in a
greatly reduced intensity of the defect-based luminescence referred to as the yellow band
(YL) when compared to the trimethyl gallium (TMG)-based MOVPE materials. The
ability to grow thick GaN film with high quality and obtaining low TD density in ELO
GaN films make it possible to get high quality self-standing GaN wafer in a single HVPE

growth chamber. In this paper the structural properties of HVPE ELO GaN films, especially coalesced flat-top ELO GaN films, are investigated systematically. The tilting of TDs from the vertical orientation to horizontal or inclined orientations in the GaN "seeding layer" is observed. The lattice tilt due to pileup of tilted TDs is found and the tilt angle is determined. This tilt angle is dependent on ELO growth conditions.

II. EXPERIMENTS AND RESULTS

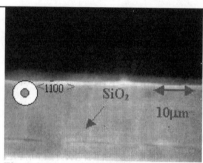

Fig. 1 Cross-sectional SEM picture of a coalesced flat-top ELO GaN sample grown by HVPE. There is no void observed at the coalescence interface.

The GaN 'substrate' for ELO growth is an MOVPE grown GaN film on a (0001) sapphire substrate coated with a ~100nm thick patterned SiO_2 layer. The pattern includes an array of 2-4 μm parallel stripe windows openings on a 12 μm period oriented along the $\langle 1\bar{1}00 \rangle$ orientation. The subsequent ELO growth of GaN is performed in a vertical HVPE system. In the HVPE process, the growth temperature is varied over the range of 1030°C to 1100°C. The input V/III ratio is controlled between 33 to 83 and the NH_3 mole fraction is over the range of $[NH_3]$=0.076-0.12. The nitrogen is used as the carrier gas. The operational pressure during the ELO growth is the atmospheric pressure.

The structural properties of the HVPE ELO GaN films are studied by using transmission electron microscope (TEM), scanning electron microscope (SEM) and high resolution X-ray diffraction (XRD). Atomic force microscope (AFM) is employed to characterize the surface morphology of the ELO GaN films. Fig. 1 shows a cross-sectional SEM picture of a coalesced flat-top HVPE ELO GaN sample. The growth temperature is 1100°C and $[NH_3]$=0.076, $[HCl]$=0.0023. It could be seen from the figure that the top is flat and the coalescence is perfect. There are no observable voids found at the coalescence interface.

Fig. 2 is an AFM image of the surface morphology in the ELO GaN area of the sample shown in Fig. 1, which clearly shows features characteristics of step-flow growth. There is no detectable step termination in the measured ELO GaN area, indicative of a low TD density

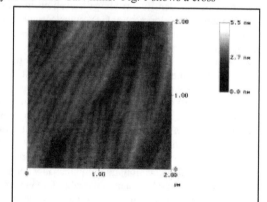

Fig. 2: A typical AFM image of the surface morphology of the ELO GaN area. No step termination, indicated by surface pits, is observable in the measure area.

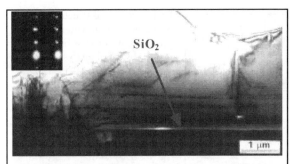

Fig. 3 A cross-sectional TEM picture of the HVPE ELO GaN film. The change in the direction of the TDs is observable in the picture. The insert is an electron diffraction pattern in the same area. Two sets of diffraction spots indicate that there is a lattice tilt in the ELO area due to pileup of TDs.

[15].

Fig. 3 presents a cross-sectional TEM picture of the same sample. The insert is an electron diffraction pattern in the same area. From Fig. 3, most TDs propagated from the MOVPE-grown GaN "seeding layer" along c-axis and bend from the vertical during ELO growth. This change in dislocation motion results in a reduced TD density zone over the masking regions of the ELO GaN film. A lattice tilt results from this change in dislocation direction leading to a new series of diffraction spots in the electron diffraction pattern. The angle between two spot lines is the angle of lattice tilt. In order to precisely determine this tilt angle, high- resolution ω-scan XRD measurements were performed as shown in Fig. 4. In Fig. 4, two separate, approximately symmetric, diffraction peaks appear on two sides of the major diffraction peak. The angle separation of the "tilt" diffraction peak to the major diffraction peak is ~3.3°. In this case, the full-width-at-half-maximum (FWHM) of the major diffraction peak is 643 arcsec. The

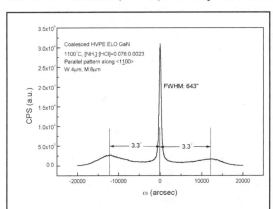

Fig. 4: High-resolution XRD rocking curve of the HVPE ELO GaN sample shown in Fig. 1. Two separated "tilt" diffraction peaks appear on both side of the major diffraction peak.

asymmetry of two "tilt" diffraction peaks may result from a slight misalignment in the XRD measurement.

Modifications to the ELO growth conditions can alter the TD distribution within the ELO films, as indicated by the intensity of the "tilt" diffraction peaks and their angle spacing from the major diffraction peak. Fig. 5 is a ω-scan XRD rocking curve of a HVPE ELO GaN sample grown using a $[NH_3]$=0.12, $[HCl]$=0.0022. From Fig. 5, two "tilt" diffraction peaks could be seen again as two wings of the major diffraction peak at an angle of ~2.4°, as compared to ~3.3° in the case of lower V/III ratio shown in Fig. 4. The FWHM of the major diffraction peak also decreased to 479 arcsec. The improved structural properties of the crystal coincide, in general, with conditions that

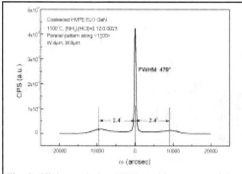

Fig. 5: High-resolution XRD rocking curve of the HVPE ELO GaN sample grown with mole fractions of $[NH_3]=0.12$, $[HCl]=0.0022$. Again, two separated "tilt" diffraction peaks appear on both side of the major diffraction peak.

lead to improved structural properties in non-patterned growth. This idea has been proved by TEM analysis. The TEM image of the HVPE ELO GaN sample shown in Fig. 5. reveals that many of the TDs initiating in the window area bend into the plane of growth or thread through the ELO region at an oblique angle. The ELO region, particularly in the regions far from the opening windows appear to have a low TD density and should give rise to a reduced XRD linewidth. No observable void is found at the coalescence interface in this case, too.

III. DISCUSSION

The resulting structure during the ELO process is known to be dependent on the gas phase transport of reactants to the growth front, which is dependent on the growth conditions and mask geometry. The growth chemistry also plays an important role in determining the local growth rates and local V/III ratio. These factors have been shown to have a primary influence on the facet formation in MOVPE growth. The addition of chloride-based species into the growth ambient will decrease the driving force or supersaturation and thus decrease the overall growth rate [12]. Reduced supersaturation has often been associated with higher surface transport lengths on the growth front. The specific supersaturation at the growing facet will have this additional factor, i.e. local supersaturation, which affects the local facet determination. The enhanced lateral growth rate of HVPE ELO technique assists in the elimination of the coalescence void at the merge interface [12].

The mechanism of the TD bending in the HVPE ELO region is unknown. The influence of a free-bounding surface on the motion of dislocations can play a role. In addition local strains arising from the initial strain state of the GaN "seeding layer" and the thermal expansion difference-based strains due to the presence of the masking materials will affect dislocation motion by developing local strain gradients in the film. The high lateral-to-vertical ratio and high rate of growth in HVPE, when compared to MBE or MOVPE, aid in the development of a merged front free of a coalescence void. Increased V/III ratio lead to improved materials properties, while maintaining a void-free coalescence, in ELO as in the non-masked growth.

IV. CONCLUSIONS

The dependence of the structural properties of HVPE ELO GaN films on gas phase V/III ratio has been investigated in this paper. A lower V/III ratio was found to lead to improved ELO materials, as determined by XRD measurement. No observable voids were present at the coalescence interface. A lattice tilt is seen in all of the HVPE samples characterizing the deflection of the <0001> planes between the GaN grown over the masking region and the initial window opening.

ACKNOWLEDGEMENT

R. Zhang, Y, Shi, Y.G. Zhou, B. Shen, Y.D. Zheng thank the financial support of the China "863" national high-tech development program and National Natural Science Foundation of China with contracts #69976014, #69636010, #69806006, #69987001. The authors would like to acknowledge the financial support of the ONR MURI on Compliant Substrates, and the facilities support of the NSF Materials Research Science and Engineering Center on Nanostructured Materials and Interfaces, USA.

REFERENCES

1. S.Nakamura, M.senoh, S.Nagahara, N.Iwasa, T.Yamada, T.Matsuahita, H.Kiyoku and Y.Sugimoto, *Jpn.J.Appl.Phys.* **135**, L74 (1996).
2. S.Nakamura, M.senoh, S.Nagahara, N.Iwasa, T.Yamada, T.Matsuahita, H.Kiyoku, Y.Sugimoto, T.Kozaki, H.Umemeoto, M.Sano, and K.Chocho, *Appl.Phys.Lett*, **72**, 2014 (1998).
3. S. J. Pearton, J. C. Zolper, R. J. Shul, F. Ren, J. Appl. Phys., **86**, 1(1999).
4. F.A. Ponce and D.P. Bour, Nature, 386, 351(1997).
5. H. Marchand, X.H. Wu, J.P. Ibbetson, P.T. Fini, P. Kozodoy, S. Keller, J.S. Speck, S.P. DenBaars, and U.K. Mishra, Appl. Phys. Lett. 73, 747 (1998).
6. Michael E. Coltrin, Christine C. Willan, Michael E. Bartram, Jung Han, Nancy Missert, Mary H. Crawford, Albert G. Baca, MRS Internet J. Nitride Semicond. Res. 4S1, G6.9(1999).
7. A. Sakai, H. Sunakawa and A. Usui, Appl. Phys. Lett., 71, 2259(1997).
8. O-H. Nam, M.D. Bremser, T.S. Zheleva and R.F. Davis, Appl. Phys. Lett., 71, 2638(1997).
9. H. Matsushima, M. Yamaguchi, K. Hiramatsu and Sawaki, Proc. 2 nd Int. Conf. Nitride Semiconductors, Tokushima, Japan, 1997, p492.
10. R. Zhang , L. Zhang, D.M. Hansen, Marek P. Boleslawski, K.L. Chen, D.Q. Lu, B. Shen, Y.D. Zheng and T.F. Kuech, MRS Internet J. Nitride Semicond. Res. 4S1, G4.7 (1999)
11. Yasutoshi Kawaguchi, Shingo Nambu, Hiroki Sone, Masahito Yamaguchi, Hideto Miyake, Kazumasa Hiramatsu, Nobuhiko Sawaki, Yasushi Iyechika and Takayoshi Maeda, MRS Internet J. Nitride Semicond. Res. 4S1, G4.1 (1999).
12. J-O Carlsson, Solid State & Mater. Sci., 16, 161(1990).
13. N.R. Perkins, M.N. Horton, Z.Z. Bandic, T.C. McGill and T.F. Kuech, Mat. Res. Sci. Symp. Proc. 395, 243 (1996).
14. R.J. Molnar, K.B. Nichols, P. Maki, E.R. Brown and I. Melngailis, Mater. Res. Soc. Symp. Proc., 378, 479 (1995).
15. H. Marchand, J.P. Ibbetson, Paul T. Fini , Peter Kozodoy , S. Keller, Steven DenBaars , J. S. Speck, U. K. Mishra, MRS Internet J. Nitride Semicond. Res. 3, 3(1998).

Mat. Res. Soc. Symp. Vol. 595 © 2000 Materials Research Society

Integration of PLZT and BST Family Oxides with GaN.

Andrei V. Osinsky, Vladimir N. Fuflyigin, Feiling Wang, Peter I.Vakhutinsky
and Peter E.Norris
NZ Applied Technologies, 14A Gill St.,Woburn, MA 01801, USA

ABSTRACT

Recent advances in the processing of complex-oxide materials has allowed us to monolithically grow ferroelectrics of lead lanthanum zirconate titanate (PLZT) and barium strontium titanate (BST) systems on a GaN/sapphire structure. High quality films of PLZT and BST were grown on GaN/c-Al$_2$O$_3$ in a thickness range of 0.3-5 μm by a sol-gel technique. Field-induced birefringence, as large as 0.02, was measured from a PLZT layer grown on a buffered GaN/sapphire structure. UV illumination was found to result in more symmetrical electrooptic hysteresis loop. BST films on GaN demonstrated a low frequency dielectric constant of up to 800 with leakage current density as low as $5.5 \cdot 10^{-8}$ A/cm^2.

INTRODUCTION

Combining ferroelectric oxide materials with III-nitride wide band gap semiconductors presents a special interest for a number of areas including high-speed image processing and high power microwave devices and circuits. For example, high-speed, high-density arrays of spatial light modulators based on ferroelectric (electro-optic) films can be controlled by GaN based field effect transistors (FETs). The advantages of AlGaN/GaN FETs compared to their Si-H based counterparts are high operation voltages and transparency in visible and near UV range. Stratified high-ε oxide/III-nitride structures can be also used as an advanced voltage variable capacitors (varactor) with wide tuning range for microwave applications. Fundamental properties of III-N semiconductors and their heterostructures, such as high thermal conductivity, break-down voltage and saturation velocity, chemical and mechanical stability, high density 2DEG in heterostructures make them superior to Si and GaAs for high power applications. Gallium nitride also possesses a distinctive advantage over arsenide or phosphide semiconductors with its higher thermal stability. For example, thermal annealing in nitrogen at 850°C indicated no deterioration of photoluminescent and electrical properties [1]. This allows fabrication compatibility and use of higher processing temperatures and more optimal growth conditions for oxide films on GaN.

In this work we report on growth and characterization of electric and electro-optic properties of high-quality oxide films of (Pb,La)(Ti, Zr)O$_3$ (PLZT) and (Ba,Sr)TiO$_3$ (BST) families on GaN/c-sapphire structures.

EXPERIMENTAL

Silicon doped n-type GaN layers, which have been used as a substrate for PLZT deposition, were grown by metalorganic chemical vapor deposition on c-oriented sapphire. The GaN layers, 5 μm thick, were fully relaxed, as evidenced by high resolution X-ray diffraction measurements with electron concentration and Hall mobility of $2 \cdot 10^{18}$ cm^{-3} and ~360 cm·V^{-1}·sec^{-1} respectively. The solution of lead, zirconium, barium, strontium and lanthanum carboxylates, and titanium isopropoxide in water-methanol mixture was used to grow PLZT and BST films. The temperature of the ferroelectric films' growth was kept between 650-700°C to avoid decomposition of GaN and to reduce the chemical interaction between GaN and PLZT in the interface layer. Thickness of the ferroelectric films was in 0.3-5 μm range. Deposition was performed using standard dip-coating technique. Details of the deposition procedure are described elsewhere [2].

XRD spectra were recorded on "Rigaku" diffractometer using $Cu_{K\alpha}$ radiation. Films thickness was measured using profilometry, i.e. "Dektak". Ferroelectric measurements were taken using standard Sawyer-Tower scheme. A HP4140A picoampermeter was used for I-V characterization of the grown films.

RESULTS AND DISCUSSION.

Ferroelectric properties of PLZT films on GaN.

Oxide films deposited on GaN with or without ITO layer showed good adhesion to the substrate, and passed the "scotch tape" test. X-ray diffraction (XRD) measurement indicated that both PLZT and BST films were single phase and polycrystalline without predominant texture. We deposited 20-40 nm thick layer of ITO on GaN in order to obtain a highly (110)-textured ferroelectric oxide films (Figure 1). Details of this approach are given in [3]. Orientation of the ferroelectric films appears to be one of the most important factor affecting their electro-optical properties. In PLZT phase with composition 8/65/35 ("rhombohedral" phase), polar axis is parallel to [111] direction. In (110)-oriented PLZT films the [111] direction lies in the substrate plane and is, therefore, perpendicular to the incident light beam. In this arrangement, maximum phase retardation between ordinary and extraordinary components of the light beam can be reached, if the material is used in transmission mode. Therefore, the control of the ferroelectric film orientation is very critical if electro-optic applications are implied.

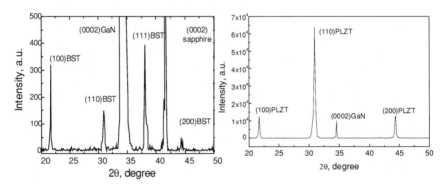

Figure 1. X-ray diffraction patterns of the 0.3 μm thick BST film on GaN/c-sapphire (left) and 3 μm thick PLZT film grown on ITO/GaN/c-sapphire. (right)

As it was shown by AFM study, PLZT films on GaN had a submicron grain size. This is smaller than the wavelengths used in optical communication systems, which are

1300 and 1550 nm. Hence, small optical loss related to the scattering on grain boundaries can be expected.

For characterization of electric properties of PLZT/n$^+$-GaN and PLZT/ITO/ n$^+$-GaN structures, vertical test capacitors were fabricated. Rectangular Au/Ti contact pads with 0.3x0.5 mm^2 area were deposited by e-beam evaporation on the top of the oxide films. Highly conductive n$^+$-GaN epilayer served as a bottom electrode to the ferroelectic film. Ohmic contact to the n$^+$-GaN bottom electrode was made using Ti/Al.

PZT and PLZT films exhibited a well-pronounced ferroelectric behavior with remnant polarization P$_r$ in the range of 20-25 μC/cm^2 (Figure 2). Ba$_{0.5}$Sr$_{0.5}$TiO$_3$ composition which was used in our experiments corresponds to high-ε paraelectric phase. Understandably the ferroelectric hysteresis loop was not observed for BST films grown on GaN/c-sapphire.

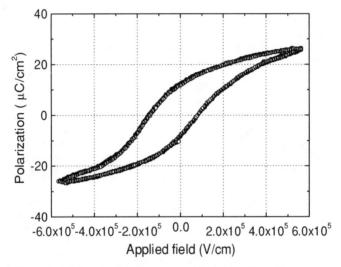

Figure 2. Ferroelectric hysteresis loop measured for PLZT film with composition 8/65/35 grown on ITO/GaN/c-sapphire.

Table I summarizes dielectric properties of the fabricated PLZT and BST films. All these materials possess relatively high dielectric constant. This speaks out for good crystallinity of high-ε layers. It is also worth mentioning that leakage current density measured on the grown films is under 6.0·10^{-8} A/cm^2 at 10V. This further proves high quality of the synthesized oxide layers.

Electro-optic properties of PLZT films on GaN.

To characterize the electrooptic properties of the PLZT films on the hexagonal GaN, the method of phase-modulated differential ellipsometry was employed in a reflection mode. Details of this method are described in [4]. The schematic in Figure 3

Table I. Dielectric properties of the PLZT and BST films deposited on GaN.

Material	ε	tan δ	J_c, A/cm^2 at 10V	Film thickness, nm
PLZT 8/65/35	700-900	0.03	$5.0 \cdot 10^{-8}$ A/cm^2	400-600
PLZT 0/52/48	1000-1200	0.03	$6.0 \cdot 10^{-8}$ A/cm^2	400-600
BST 50/50	600-800	0.01	$5.5 \cdot 10^{-8}$ A/cm^2	300-500

shows the arrangement of the optical path and electrodes. The phase modulated He-Ne laser beam entered the c-cut sapphire substrate from the side opposite to where GaN and PLZT were grown. The laser beam was transmitted through the substrate, the GaN layer, the thin ITO layer before entered the PLZT layer; upon reflection from the aluminum electrode, the beam again was transmitted through all the layers then collected by the phase sensitive detection system. A slow varying voltage signal was applied to the aluminum electrode and the conductive GaN or ITO/GaN layers. In this arrangement, the field-induced birefringence as a function of the applied field could be recorded. A theoretical model for calculating field-induced birefringence of the stratified structures was employed in processing the measured data.

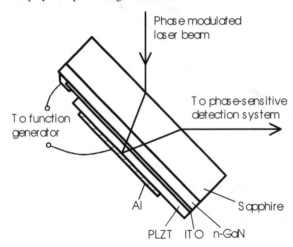

Figure 3. The structure of the device with ferroelectric/GaN integration and the light path during the characterization of the electrooptic properties of the PLZT layer. The thickness of the GaN layer and that of the PLZT layer are 5 and 3 microns, respectively.

The field-induced birefringence, Δn, of a PLZT layer on an n-GaN/c-sapphire wafer was measured as a function of the applied voltage. The thickness of the PLZT layer was 3 microns. As shown in Figure 4, the PLZT layer was highly electrorefractive. Under electric field strength of approximately 30 V/μm, a field-induced birefringence as high as 0.02 was measured, which was among the highest reported for all solid materials. The large field-induced birefringence was expected from the high dielectric polarization of the films measured under a similar external field. With the increase of the external electric field, a saturation behavior in the field-induced birefringence became obvious. Saturation of this kind is quite common in ferroelectric materials, resulting from the saturation in the material polarization vs. electric field hysteresis curve. Two other characteristics were also noticeable in the electrooptic response curve, i.e., the hysteretic behavior and the asymmetry of the loop. Although the selected PLZT composition normally exhibit 'slim-looped' electrooptic response, the rather 'fat' hysteresis loop was not very surprising.

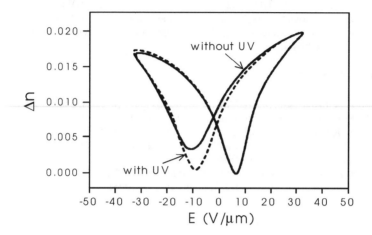

Figure 4. Field-induced birefringence of the PLZT layer in the ferroelectric/GaN integrated structure as a function of the external electric field.

It is well known that the polarization hysteresis loops of ferroelectric thin film materials can be altered severely by the properties of the substrates. In the absence of lattice matching, the difference between the films and their substrates in thermal expansion often create mechanical stress that in turn causes changes in their ferroelectric properties. Judging from the known thermal expansion properties of the materials, the PLZT films should be under a compressive stress. The asymmetry in the polarization hysteresis loop may be a result of several factors including; nonuniform stress distribution in the growth direction, different nature of the top and bottom electrode materials, and formation of the depletion region in semiconductor layers. A similar phenomenon was observed in an ITO/PLZT/Pt structure.

Figure 4. also shows an electro-optic hysteresis loop taken under UV illumination (dashed line). The measurements were taken from the described structure illuminated with UV light from xenon lamp. One can see that with UV illumination the hysterisis loop becomes more symmetric. The reason for it could be photoinduced space charges in the PLZT film or generation of non-equilibrium carriers in the GaN depletion layer. It appeared that photogenerated carriers reduce the width of the depletion region in semiconductor materials, which results in a more symmetric loop. The understanding of this phenomenon is in progress.

CONCLUSIONS.

High quality films of PLZT and BST were successfully grown by sol-gel technique on hexagonal conductive n+GaN epitaxial layers. We found that the introduction of a 20-

40 nm thick indium tin oxide (ITO) layer between PLZT and GaN results in highly (110)-textured PLZT layer with remnant polarization as large as 20-25 $\mu C/cm^2$ and strong electrooptic properties. BST films on GaN demonstrated low frequency dielectric constant as high as 800 with very low leakage currents. Field-induced birefringence of 0.02 was measured in Al/PLZT/ITO/GaN vertical capacitor structure. UV illumination from xenon lamp was found to result in a more symmetrical electrooptic hysteresis loop. We relate this to the photoinduced charge in the semiconductors depletion region at negative and positive biases.

REFERENCES

1. A.Davydov, T.J.Anderson, *in III-V Nitride Materials and Processes*, edited by T.D. Moustakas, Proc. volume 98-18 (Boston, MA, 1998), in press
2. V.Fuflyigin, K.K.Li, F.Wang, H.Jiang, S.Liu, J.Zhao, P.Norris, P.Yip, *in High-Temperature Superconductors and Novel Inorganic Materials*, edited by G.Van Tendeloo, (Kluwer Academic.Publ. 1999), p.279-284
3.V.Fuflyigin, A.Osinsky, F.Wang to be published in Applied Physics Letters.
4. F.Wang, E.Furman, G.H. Haertling, J.Appl.Phys. **78**, 9 (1995)

Mat. Res. Soc. Symp. Vol. 595 © 2000 Materials Research Society

HVPE and MOVPE GaN growth on slightly misoriented sapphire substrates

Olivier Parillaud, Volker Wagner, Hans-Jörg Bühlmann, François Lelarge, and Marc Ilegems
Institut de Micro- et Optoélectronique, Ecole Polytechnique Fédérale de Lausanne, CH-1015 Lausanne, Switzerland

ABSTRACT

We present preliminary results on gallium nitride growth by HVPE on C-plane sapphire with 2, 4 and 6 degrees misorientation towards M and A directions. A nucleation GaN buffer layer is deposited prior the growth by MOVPE. Surface morphology and growth rates are compared with those obtained on exact C-plane oriented sapphire, for various growth conditions. As expected, the steps already present on the substrate surface help to initiate a directed step-flow growth mode. The large hillocks, which are typical for HVPE GaN layers on (0001) sapphire planes, are replaced by more or less parallel macro-steps. The width and height of these steps, due to step bunching effect, depend directly on the angle of misorientation and on the growth conditions, and are clearly visible by optical or scanning electron microscopy. Atomic force microscopy and X-ray diffraction measurements have been carried out to quantify the surface roughness and crystal quality.

INTRODUCTION

(0001)-oriented sapphire is the main substrate used for the realization of GaN-based devices [1]. To our knowledge, very few reports deal with MOVPE of GaN on slightly misoriented sapphire [2-4], and only one concerns HVPE on such substrates [5]. However, for other III/V compounds GaAs or InP, it has been shown that the use of substrates misoriented by a few degrees can be very efficient to improve both crystal quality and surface morphology of the layers grown by HVPE. This is related to the fact that the supersaturation of the gas phase is quite small in the hydride technique as opposed to the conditions during MOVPE growth and growth occurs near the thermodynamic equilibrium as predicted by the Burton, Cabrera, and Frank theory [6].

EXPERIMENTAL DETAILS

Growth experiments have been carried out in two separate reactors. An horizontal home made reactor was used to grow low temperature GaN buffer layers prior to the growth by HVPE. Nitrogen and gallium precursors were ammonia (NH_3)

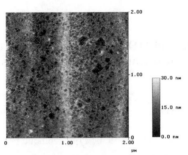

Figure 1 *AFM picture of the GaN buffer layer surface before annealing. Small GaN grain structure is clearly visible and a 10 nm high macro-step can be observed in the center of the image.*

and trimethylgallium (TMG) respectively and carrier gas was nitrogen.

Deposition temperature was 600°C after a nitridation step of the Al_2O_3 surface at 1100°C during 30 minutes. All the experiments were carried out at atmospheric pressure.

HVPE layers were subsequently grown in an horizontal Aixtron HVPE reactor. Details on the reactor geometry have been published in a previous paper [7]. GaCl resulting from the reaction of HCl with metallic gallium and NH3 are brought separately to the deposition zone. Nitrogen and mixed nitrogen / hydrogen were used as carrier gas. Source temperature and deposition temperature were kept constant at 900°C and 1050°C respectively for all the experiments presented here. All the samples were non intentionally doped.

Substrates used were exact C-axis oriented sapphire and C-axis with 2°, 4° and 6° misorientation towards M and A axis.

RESULTS

Low temperature MOVPE buffer layers.

Some authors have proposed a GaCl pre-treatment of the sapphire surface [8] or a sputtered ZnO intermediate layer [9,10] to facilitate the initial stage of nucleation of GaN. In our case, we use a thin (10nm) GaN buffer layer deposited by MOVPE at low temperature [11]. These layers have been characterized by X-Ray diffraction before and after annealing up to 1050°C in the HVPE reactor. This annealing was carried out under NH_3 atmosphere in order to prevent GaN decomposition. The Ω-rocking curves of the (002) GaN reflection after annealing develop into a broad peak which indicates that ordering of the small GaN grains is taking place. Atomic Force Microscopy was also performed on both exact and misoriented samples. The AFM picture presented in figure 1 clearly shows the grain structure of the buffer layer deposited on a 2° towards A oriented substrate before annealing. Pictures taken after annealing are similar and no clear evidence of a morphological change could be detected whatever the misorientation of the samples. A 30 nm high macro-step originating from the substrate surface is visible in the middle of the picture indicating that the final polishing of the sapphire surface does not lead to atomic steps.

Table I: Growth conditions, substrates used ,thickness and FWHM of XRD Ω-scan measured for the 3 different sets of experiments.

	Growth conditions	Substrates	Thickness	FWHM of the Ω-scan (002)
Set #1	V/III = 100 N_2 as carrier gas	C exact 2° towards M 2° towards A	9 µm 10 µm 12 µm	1071 1123 727
Set #2	V/III = 100 12 % H_2/N_2 as carrier gas	C exact 2° towards M 2° towards A	8 µm 10 µm 9 µm	746 654 496
Set #3	V/III = 10 N_2 as carrier gas	C exact 2° towards M 2° towards A	4 µm 4 µm 4 µm	849 905 852

HVPE layers

3 sets of experiments are presented here. The growth conditions, the samples used, the layer thickness and FWHM of XRD Ω-scan are listed in table I. The growth duration was 15 min. for all runs. Set #1 corresponds to the growth conditions usually used on exact sapphire. Sets #2 and #3 have been carried out under different vapor phase

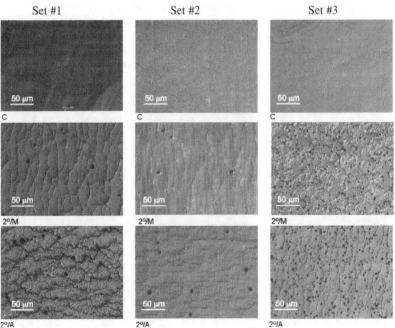

Figure 2 *Optical microscopy views of the surface of 9 samples listed in table I. For each set of experiment, exact, 2° towards M and 2°towards A are presented.*

composition. H_2 was introduced in carrier gas for set #2 and the V/III ratio was divided by a factor of 10 for set #3. Thickness was measured by cross-section optical microscopy after cleaving. Top view optical microscopy, Scanning Electron Microscopy (SEM) and AFM were used to check the surface morphology and evaluate the roughness.

Figure 2 shows optical micrographs of the samples listed in table I. Only layers grown on exact (0001) and 2° misoriented substrates are shown, as the surface morphology obtained on 4° and 6° misorientated substrates was increasingly degraded. Layers grown on exact sapphire exhibit typical hillocks and pyramids, depending on growth conditions. Cracks are also visible for all the exact samples. On misoriented substrates, a directed step-flow growth mode is dominant with steps perpendicular to the direction of misorientation. However, the expected smoothening of the surface does not occur due to large step bunching effects which lead to more or less parallel steps or a fish scale appearance. The introduction of hydrogen as part of the carrier gas (set #2) improves the surface morphology and can be a key issue for the use of misoriented substrates, since the surface diffusion is enhanced. All the misoriented samples exhibit hexagonal pits whose density increases dramatically for set #3.

The SEM pictures of the misoriented samples are presented in figure 3. In case of set #1, a high density of small hexagonal pits is observed along the step edges. It is not yet clear if the pits are the cause of or are caused by the macro-steps. Furthermore, we find that the crack density is greatly reduced in the direction parallel to the steps. AFM studies (figure 4) bring additional information on the surface roughness between the macro-steps and on the formation of these macro-steps. Typical atomic steps are detected between macro-steps. The macro-step height varies from 10 nm for set #1 and #2 to more than 100 nm for set #3. All layers exhibit a high density of dark spots, typically less than 1 μm in diameter, which are indicative of emerging dislocations. These dislocations tend to block the advance of the atomic steps and partly explain the formation of the macro-steps.

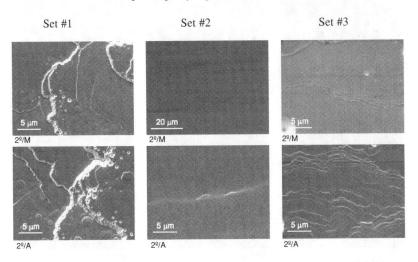

Set #1   Set #2   Set #3

5 μm   2°/M   20 μm   2°/M   5 μm   2°/M

5 μm   2°/A   5 μm   2°/A   5 μm   2°/A

Figure 3 SEM pictures of the 2° misoriented samples (top: towards M, bottom: towards A)

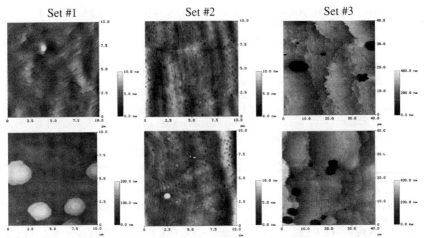

Figure 4 *AFM pictures of the 2° misoriented samples (top: towards M, bottom: towards A, set #1 and #2 10x10 µm², vertical scale 10 nm; set #3 40x40 µm², 400 nm)*

XRD measurements have been performed on all the samples. The FWHM of the Ω-rocking curve of the GaN 002 reflection is listed in table I. No clear improvement of the crystal quality can be deduced from these values. For the range of conditions explored, the samples misoriented towards A axis exhibit the lower values while values measured for exact and misoriented towards M samples are in the same range.

DISCUSSION AND CONCLUSION

As expected, the growth is affected by the misorientation of the substrates and a step-flow growth mode is evident. However, major step bunching effects occur, even at low angles of misorientation and whatever the direction of the misorientation (A or M). Furthermore, the surface morphology degrades dramatically for substrates with 4° and 6° misorientation. The causes suspected for this step bunching are strain relaxation and the large number of misfit dislocations generated at the sapphire / GaN interface that propagate up to the surface and block the steps, as evident on the AFM pictures. The presence of macro-steps on the substrate surface before the growth can also enhance this behavior. A slight improvement of the FWHM of the X-Ray rocking curve is observed for samples grown on 2° towards A misoriented sapphire and under mixed N2/H2 carrier gas. Overall, the results of this work do not suggest that significant improvement in crystalline quality or surface morphology can be expected by growth on misoriented (0001) substrates.

REFERENCES

1.   S. Nakamura, M. Senoh, S. Nagahama, N. Isawa, T. Yamada, T. Matsushita, H. Kiyoku, and Y. Sugimoto, Jpn. J. Appl. Phys., Part 2 35, L74 (1996).

2. K. Hiramatsu, H. Amano, I. Asaki, H. Kato, N. Koide, and K. Manabe, J. Crystal Growth 107, 509-512 (1991).
3. P. A. Grudowski, A. L. Holmes, C. J. Eiting, and R. D. Dupuis, Appl. Phys. Lett. 69 (24) (1996).
4. B. Pecz, M. A. di Forte-Poisson, F. Huet, G. Radnoczi, L. Toh, V. Papaioannou, and J. Stoemenos, J. Appl. Phys. 86, 11 (1999).
5. W. Seifert, G. Fitzl, and E. Butter, J. Crystal Growth 52, 257-262 (1981).
6. W. K. Burton, N. Cabrera, and F. C. Frank, Phil. Trans. Roy. Soc. 299 (1951).
7. O. Parillaud, V. Wagner, H. J. Bühlmann, and M. Ilegems, MRS Internet J. Nitride Semicond. Res. 3, 40 (1998).
8. K. Naniwae, S. Itoh, H. Amano, K. Itoh, K. Hiramatsu, and I. Agasaki, J. Crystal Growth 99, 381-384 (1990).
9. T. Detchprohm, K. Hiramatsu, K. Itoh, and I. Agasaki, Appl. Phys. Lett., 61 (22) (1992).
10. R. J. Molnar, K. B. Nichols, P. Malki, E. R. Brown and I Melngailis, Mat. Res. Soc. Proc. Vol. 378, 242-244 (1996).
11. V. Wagner, O. Parillaud, H. J. Bühlmann, and M. Ilegems, Phys. Stat. Sol. (a) 176, 429 (1999).

Mat. Res. Soc. Symp. Vol. 595 © 2000 Materials Research Society

Hydride Vapour Phase Homoepitaxial Growth
of GaN on MOCVD-Grown 'Templates'

T. Paskova[1], S. Tungasmita[1], E. Valcheva[1], E.B. Svedberg[1], B. Arnaudov[2],
S. Evtimova[2], P.Å. Persson[1], A. Henry[1] R. Beccard[3], M. Heuken[3] and B. Monemar[1]
[1] IFM, Linköping University, S-581 83 Linköping, Sweden
[2] Faculty of Physics, Sofia University, 5, J. Bourchier blvd., Sofia 1164, Bulgaria
[3] Aixtron AG, D-52072 Aachen, Germany

ABSTRACT

We report on an improved quality of thick HVPE-GaN grown on MOCVD-GaN
'template' layers compared to the material grown directly on sapphire. The film-substrate
interface revealed by cathodoluminescence measurements shows an absence of highly
doped columnar structures which are typically present in thick HVPE-GaN films grown
directly on sapphire. This improved structure results in a reduction of two orders of
magnitude of the free carrier concentration from Hall measurements. It was found that the
structure, morphology, electrical and optical properties of homoepitaxial thick GaN layers
grown by HVPE were strongly influenced by the properties of the MOCVD-GaN
'template'. Additionally the effect of Si doping of the GaN buffer layers on the HVPE-
GaN properties was analysed.

INTRODUCTION

Over the last years considerable improvements have been made in the crystalline
quality of thick GaN layers grown by HVPE [1,2] although many defect related issues of
thick films are still rather controversial. These materials still have a high-density of
threading dislocations and stacking mismatch boundaries due to heteroepitaxial mismatch
with the most commonly used sapphire substrates.

The development of pre-growth processes in MOCVD growth, involving either a
sapphire nitridation or GaCl pretreatment, or using different buffer layers, results in
optimisation of nucleation steps, which play a dominant role in determining the properties
of subsequently grown layers. Using low temperature AlN or GaN buffers, the MOCVD
growth successfully overcomes many drawbacks as cracks and high residual concentration
[3], however, the same approach was found to result in poor quality material in HVPE
growth. Promising results for the combination of MOCVD-grown GaN 'templates' with
subsequent homoepitaxially grown GaN by molecular beam epitaxy, have been recently
reported [4]. At the moment, only a ZnO buffer has been reported to be a suitable buffer
for growth of thick HVPE-GaN layers, even though it is largely desorbed at the early
stages of growth [1].

In this study, we concentrate on MOCVD-GaN 'template' layers as an alternative
buffer for GaN growth by HVPE. The material properties of the HVPE-GaN layers have
been investigated by different techniques, and their relationship with the buffer layer
properties has been exposed.

EXPERIMENTAL

The GaN layers were grown in a conventional HVPE system described previously
[5] on c-plane sapphire substrates. To ensure a better nucleation of the GaN, we utilised

MOCVD-GaN 'template' layers with a thickness of 2.5 μm grown in the Aixtron application laboratory. The MOCVD growth was carried out at 1170 °C. Two types of MOCVD 'templates', intentionally doped with Si with a free carrier concentration of 1.8×10^{18} cm^{-3} and intentionally undoped GaN layers with a free carrier concentration of 2×10^{17} cm^{-3} were used. The HVPE layers were grown at 1090 °C. NH$_3$, Ga and HCl were used as source materials under a stream of N$_2$ carrier gas. The substrates with MOCVD-GaN buffers were heated up to the growth temperature in a 40 % NH$_3$ atmosphere in order to protect the buffers. A layer without a buffer was grown using a nitridation pretreatment of the sapphire in 20% NH$_3$ for 15 minutes at the growth temperature, in the following referred to as a reference sample. High growth rates of 95-110 μm/h were used, and the thicknesses of the layers studied in this work were in the range of 22-25 μm.

The material properties of both MOCVD-GaN buffers and HVPE-GaN layers were investigated by a variety of techniques as x-ray diffraction (XRD) using a triple axis configuration providing a resolution of ~10 arcsec; atomic force microscopy (AFM) using a Nanoscope IIIa instrument operated in tapping mode; transmission electron microscopy (TEM); 7K cathotoluminescence (CL) using a Gemini 1550 LEO electron microscope with an accelerating voltage of 15 kV; 2K photoluminescence (PL) using the 244 nm wavelength of a frequency-doubled Ar ion laser; and variable temperature Hall effect measurements (T=80-300 K) using the Van-der-Pauw method.

RESULTS AND DISCUSSION

Doping of the MOCVD-GaN by Si to a concentration of $(3-8) \times 10^{18}$ cm^{-3} has been reported to reduce the biaxial stress in the layers as well as to reduce the density of vertical screw and edge dislocations along the c-axis [6,7]. Simultaneously, the density of misfit dislocations parallel to the interface has increased [7]. Such structural peculiarities have been found to promote an increase in mobility in the Si-doped epilayers [7]. That is why we investigate both type of MOCVD 'templates', slightly doped to a concentration of 1.8×10^{18} cm^{-3} and undoped layers as buffers for a consequent HVPE growth of thick layers. The material parameters of both MOCVD 'templates' are summarised in Table I, and a good agreement with previous results reported for similar samples can be seen [6,7].

Table I. Material parameters of both MOVPE-GaN buffer and HVPE-GaN layers.

	n_{Hall} (300K) (cm^{-3})	μ_{Hall} (300K) (cm^2/Vs)	ω FWHM (arcsec)	ω-2Θ FWHM (arcsec)	D (μm)	DBE FWHM (meV)	ΔE_{DBE} (meV)
buffer 1 (Si doped)	1.8×10^{18}	250	273	45	0.70	11.8	16.6
buffer 2 (undoped)	2×10^{17}	80	261	36	0.77	3.6	17.3
#1 (reference)	3×10^{18}	85	725	45	0.70	4.8	15.9
#2 (on buffer 1) measured	2.3×10^{17}	270	503	59	0.58	6.8	10.8
corrected	6.2×10^{16}	310					
#3 (on buffer 2) measured	3.6×10^{16}	515	227	17	0.98	2.2	9.3
corrected	2.0×10^{16}	630					

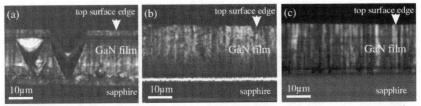

Fig.1 *Panchromatic CL images of cross-sections of the GaN layers grown at the same growth conditions on sapphire without a buffer (a); with Si-doped MOCVD-GaN (b); and with undoped MOCVD-GaN 'template' layers (c).*

Figure 1 presents CL images of cross-sections of the GaN layers with a comparable thickness of 22-25 μm. Fig.1(a) shows a cross-section of the layer grown by conventional nitridation pretreatment of sapphire and Fig.1(b) and (c) show the cross-sections of the layers grown on MOCVD 'templates' intentionally doped with Si and undoped, respectively. As shown in Fig.1(a) many bright columns can be seen, some of them even protruding to the top of the surface. On the other hand, the cross-sections of the films in Fig.1(b) and (c) are clean and columns do not occur. The spatially resolved CL spectra taken from the bright columnar region show a broad band typical of highly doped material [5,8]. The interface region in sample #2 appears as a bright stripe with a thickness of ~2 μm attributed to the Si-doped GaN 'template' with a carrier concentration of 1.8×10^{18} cm^{-3}. The bright columnar region and the columns protruding to the surface in the layer without a buffer were reported [9] to be responsible for the high residual Hall concentration in such layers typically higher than $(3-4) \times 10^{18}$ cm^{-3}. The removal of the columnar structure in the layers grown on MOCVD 'templates' results in a reduction of the free carrier concentration and an increase of the mobility determined by Hall effect measurements compared to the layers without a buffer. It should be mentioned that the measured carrier concentrations of HVPE layers grown on MOCVD 'templates' in both cases (samples #2 and #3) are about one order of magnitude lower than that of the underlying MOCVD layers. Thus there is a need of correction of these values using a two-layer model [10] and the corrected values are shown in Table I. We point out that both Hall-effect measured and corrected values of the mobility are lower than one can expect for the respective values of free carrier concentration, indicating compensation in the films. The latter is also evident from the temperature dependence of the electron concentrations and mobilities shown in Fig.2.

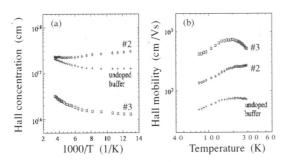

Fig.2 *(a) Hall-effect electron concentration vs reciprocal temperature and (b) Hall electron mobility vs temperature.*

The 2K PL spectra of both MOCVD buffers and HVPE layers shown in Fig.3(a) and (b) respectively, exhibit intense near-bandedge emission due to donor-bound excitons (DBE) at about 3.48 eV. The FWHM shown in Table I can not be directly correlated with the structural and electrical quality. However, the uniform Si doping in the MOCVD film (buffer 1) results in a clear broadening of the DBE line, and the non-uniform doping of the columnar structure in the reference sample (#1) results in a broad asymmetric band below the exciton lines. It is interesting to note that while MOCVD-GaN displays a well defined yellow luminescence band around 2.2 eV, the HVPE layers do not reproduce that feature and show no luminescence in this region, although a weak red emission around 1.9 eV identified in HVPE GaN layers without a buffer is still present in the PL spectra of layers grown on MOCVD 'templates'. Additionally, the exciton positions E_{DBE} are shifted to higher energies with respect to the energetic position of strain-free homoepitaxial grown GaN (3.471 eV [11]). The shift, ΔE_{DBE} (Table I), is slightly larger for the MOCVD-GaN undoped sample compared to the Si-doped one, due to a higher compressive stress in the layer and consistent with previously reported results [6,7]. The shift is much smaller in HVPE-GaN layers grown on MOCVD buffers, indicative of a presence of a relaxation mechanism in the structure.

The x-ray diffraction measurement was also applied to evaluate the GaN crystalline quality. The strong difference in FWHM values in both ω and $\omega/2\theta$ scans is reflected in the typical elliptic shape of (0002) reciprocal lattice points, which was reported for both MOCVD [12] and HVPE [8] grown layers. This shape is usually explained by mosaicity in the films. A more detailed description of HRXRD results for HVPE layers grown without a buffer can be found elsewhere [8]. The FWHM of the GaN films grown on MOCVD 'templates' were about 500 (#2) and 220 arcsec (#3) which is narrower than the 725 arcsec (#1) obtained for unbuffered growth (Table 1). These results suggest that a reduction in the mosaicity by the growth on MOCVD 'template' buffer is effective to improve the crystalline quality of the films. In addition, it was found that the introduction of Si reduces the size of the coherent scattering areas. The average size of the areas, D, using the Sherrer formula are shown in Table I. This effect was reproduced in the thick HVPE-GaN layers grown on the Si-doped buffer, while the thick GaN films grown on the undoped buffer exhibit a much narrower $\omega/2\theta$ line and an increased size of the coherent scattering areas. The difference in the effect of Si-doped and undoped MOCVD buffers on the quality of consequent thick layers may be attributed to the change of the dislocation structure and stress relaxation mechanism in the buffers by the Si doping. In

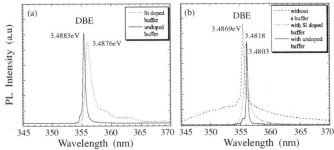

Fig.3 *PL spectra in the near-band-gap region of (a) the MOCVD-GaN 'templates and (b) the HVPE-GaN layers.*

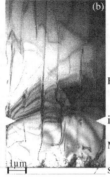

HVPE-GaN

interface

MOVPE-GaN

sapphire

Fig.4 Cross-sectional TEM
images of HVPE-GaN/
MOCVD-GaN/sapphire
interfaces for layers grown
on (a) Si-doped and (b)
undoped MOCVD-GaN
'templates'.

order to reveal their defect structure cross-sectional TEM micrographs of the films grown on doped and undoped buffer layers are shown in Fig.4(a) and (b), respectively. The dominant defects are dislocations due to the misfit strain introduced by the lattice mismatch. The doped buffer layer (Fig.4a) reveals dislocation lines oriented irregularly, and a high density of defects parallel to the interface being most probably basal plane dislocations and dislocation loops randomly distributed throughout the whole buffer layer. The overgrown HVPE layer inherits the same character of dislocation distribution, but with a much lower density of $\sim 7 \times 10^7$ cm^{-2} as a result of promoted interactions and thus annihilation of some of them. The undoped buffer and the overgrown HVPE layer show quite different defect structure (Fig.4b). A 500 nm region at the MOCVD-GaN/sapphire interface has a high density of stacking faults and half loops and at that thickness straight threading dislocations originate lying close to the [0001] growth direction. Some of them propagate in the overgrown HVPE layer. Others of the same character originate at the interface between the buffer and the overgrown layer. Between dislocations ($\sim 4 \times 10^7$ cm^{-2}), good quality regions of about 1-2 μm thickness are seen, keeping an almost constant value throughout the HVPE-GaN film.

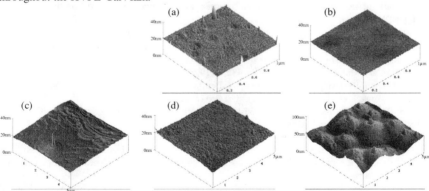

*Fig.5 Three-dimensional AFM images of the as-grown top surface of Si-doped MOCVD-GaN (a);
undoped MOCVD-GaN (b); HVPE-GaN without a buffer (c); HVPE-GaN on Si-doped MOCVD-GaN
'template' (d); HVPE-GaN on undoped MOCVD-GaN 'template' (e).*

The AFM images shown in Fig.5 reveal the microstructure of the top surfaces of the layers studied. The morphology of the undoped MOCVD 'template' (Fig.5b) appears very smooth with Rms of ~0.65 nm (from an area of $1 \times 1 \mu m^2$), while the surface of Si-doped MOCVD film is slightly rough (Rms = 1.55 nm) with well defined screw dislocations (Fig.5a). This surface appears to promote nucleation centres for subsequent HVPE growth and leads to a quite smooth microsurface of the consequent HVPE layer (Fig.5d). The reference HVPE sample exhibits a typical (for high growth rate, ~100 $\mu m/h$) growth surface of terraced hillocks mediated by screw dislocations and terrace bunching by edge dislocations (Fig.5c). No terrace bunching is evident in the layers grown on MOCVD buffers (Fig.5d, e) although hillocks with very smooth slopes are present in the HVPE layer grown on undoped buffer (Fig.5e). A change in the dominate growth mechanism is clearly observed in the HVPE-GaN layers using MOCVD buffers, most probably due to change of the structural defect arrangement.

The physical basis for the observation of an optimum buffer layer for physical properties of the main GaN films remains unclear. We speculate that achieving an optimum buffer is a balance between surface roughness and internal defect structure. Because the buffer with a rough surface evolves specific defects most of them will propagate into the subsequently grown layer. Therefore, the main film grown on a rough buffer layer cannot have high-quality. In contrast, the smooth buffer layer cannot supply a perfect crystallographic template for the main GaN epilayer, therefore the consequently grown GaN film cannot be morphologically perfect although owing better crystal quality.

CONCLUSION

Good-quality thick GaN layers have been deposited by HVPE on MOCVD-GaN 'template' layers. The defective columnar structure disappeared in the layers. The structural, optical and electrical properties are considerably improved compared to that obtained by growth directly on sapphire. The HVPE homoepitaxial overgrowth on MOCVD 'templates' appeared to be effective in reducing crystalline defects in thick HVPE-GaN films. Additionally, the GaN 'templates' were grown by the well developed and controllable MOCVD technique. Further investigations to clarify the HVPE-MOCVD nucleation mechanism will result in an optimisation of homoepitaxial regrowth.

REFERENCES

1. R.J. Molnar, W. Götz, L.T. Romano, N.M. Johnson, *J. Cryst. Growth*, **178**, 147 (1997).
2. T. Shibata, H. Sone, K. Yahashi, M. Yamaguchi, K. Hiramatsu, N. Sawaki, N. Itoh, *J. Cryst. Growth*, **189/190**, 67 (1990).
3. I. Akasaki, *Mat. Res. Soc. Symp. Proc.*, **482**, 3 (1998).
4. E.J. Tarsa, B. Heying, X.H. Wu, P. Fini, S.P.DenBaars, J.S. Speck, *J.Appl.Phys.* **82**, 5472 (1997).
5. P. Paskova, E.Goldys, B.Monemar, *J. Cryst. Growth*, **203**, 1 (1999).
6. S. Ruvimov, Z. Liliental-Weber, T: Suski, J.W. AgerIII, J. Washburn, J. Krueger, C. Kisielowski, E.R. Weber, H. Amano, I. Akasaki, *Appl. Phys. Lett.*, **69**, 990 (1996).
7. N.M. Shmidt, A.V. Lebedev, W.V. Lundin, B.V. Pushnyi, V.V. Ratnikov, T.V. Shubina, A.A. Tsatsul'nikov, A.S. Usikov, G. Pozina, B. Monemar, *Mater. Sci. & Enginer.* B, (1999) (in press).
8. T. Paskova, E.M. Goldys, R. Yakimova, E.B. Svedberg, A. Henry, B. Monemar, *J. Cryst. Growth,* (1999) (in press).
9. B. Arnaudov, T. Paskova, E.M. Goldys, R. Yakimova, S. Evtimova, I.G. Ivanov, A. Henry, B. Monemar, *J. Appl. Phys.*, **85,** 7888 (1999).
10. D. Look, R. Molnar, *Appl. Phys. Lett.* **70**, 3377 (1997).
11. K. Kornitzer, T. Ebner, K. Thonke, R. Sauer, C. Kirchner, V. Schwegler, M. Kamp, M. Leszczynski, I. Grzegory, S. Porowski, *Phys. Rev. B* **60**, 1471 (1999).

12. H.Sato, H. Takahashi, A. Watanabe, H. Ota, *Appl. Phys. Lett.*, **68,** 3617 (1996).

Mat. Res. Soc. Symp. Vol. 595 © 2000 Materials Research Society

THE NATURE AND IMPACT OF ZNO BUFFER LAYERS ON THE INITIAL STAGES OF THE HYDRIDE VAPOR PHASE EPITAXY OF GAN

Shulin Gu, Rong Zhang, Jingxi Sun, Ling Zhang, and T. F. Kuech

Department of Chemical Engineering, University of Wisconsin, Madison, WI 53706

Abstract

The nature and impact of ZnO buffer layers on the initial stages of the hydride vapor phase epitaxy (HVPE) of GaN have been studied by xray photoelectron spectroscopy (XPS), atomic force microscopy (AFM), xray diffraction (XRD) and photoluminescence (PL). During pre-growth heating, the surface ZnO layer was found to both desorb from ZnO-coated sapphire and react with the underlying sapphire surface forming a thin $ZnAl_2O_4$ alloy layer between ZnO and sapphire surface. This ZnO-derived surface promotes the initial nucleation of the GaN and markedly improves material surface morphology, quality and growth reproducibility.

Introduction

Hydride Vapor Phase Epitaxy (HVPE) is a promising technique to grow thick GaN materials. These thick layers can be used in device applications as well as a substrate for subsequent low-defect GaN growth by Metal-organic Vapor Phase Epitaxy (MOVPE) [1,2]. Due to the large lattice mismatch, the initial nucleation and growth of GaN on sapphire substrates determines the material properties of the subsequent epitaxial layer. In MOVPE, low temperature GaN or AlN buffer layers have been used to improve this initial nucleation for the subsequent high temperature GaN growth [3,4]. In the case of the HVPE process, low temperature buffer layers exhibit poor crystalline quality and lead to the deposition of polycrystalline material at high growth temperatures [5]. The initial nucleation behavior on sapphire, however, can be improved by the inclusion of a ZnO intermediate or buffer layer [1,2,5,6]. Based plane ZnO could provide a better lattice match between GaN and sapphire. In many cases, the ZnO buffer layer has been reported not to survive the initial heating and pre-growth treatment and is not reported to be present in the final epitaxial multi-layer structure [5]. Attempts to deposit highest quality ZnO buffer layer to avoid its dissociation from sapphire surface do not lead to further improvements in material quality but rather exhibit poorer properties in the subsequent GaN layer. In this case, some ZnO has been noted at the interface between the high temperature GaN and the sapphire [6]

The mechanism underlying the impact of the ZnO buffer layer on the initial stages of HVPE of GaN material is unknown, particularly in the absence of ZnO in the final structure [3,5,6]. ZnO can reactively diffuse into Al_2O_3 forming a spinel of $ZnAl_2O_4$ at high tempaeratures [7-9]. In previous studies, reactions between ZnO and Al_2O_3 powders, were limited in the absence of oxygen [9]. In this case, the gas phase transport of ZnO and its stability is affected by the oxygen activity. There is little data on the inter-diffusion between epitaxial ZnO and sapphire. $ZnAl_2O_4$ surface layer, if present at the growth front, could improve nucleation by lowering the surface energy and could provide an improved lattice match. In this paper, we have studied the formation, annealing and structure of ZnO epitaxial layers on sapphire used to promote nucleation and improve the subsequent GaN growth. In particular, the structure and composition of the ZnO-sapphire interface after high temperature annealing was investigated by atomic force microscopy (AFM), reflection high energy electron diffraction (RHEED), x-ray photoelectron spectroscopy (XPS), and x-ray diffraction (XRD) measurements.

Experiment Detail

ZnO buffer layers were initially deposited on sapphire surface by RF diode sputtering. Prior to ZnO deposition, the bare sapphire wafer was degreased with acetone and methanol followed by etching in a 3:1 sulfuric:phosphoric acid solution at 100°C and a final deionized water rinse. The wafer is then loaded into a RF diode sputtering system employing deposition conditions similar to those reported by Molnar et al [2]. ZnO thickness varied from 20 nm to 200 nm depending on the deposition time. RHEED measurements show this as-deposited ZnO film has a high degree of structural orientation. The FWHM of the XRD (002) rocking curves of these films was about 1.3 degrees. Prior to loading in the HVPE reactor, the ZnO coated wafer was again ultrasonically cleaned in acetone and methanol.

GaN was grown on these wafers in a vertical HVPE reactor. The interior structure of the reactor for gas distribution is similar to that of the horizontal system previously reported [10,11]. In the present system, an additional region called a backflow tube was added which allowed the sample to be heated to the growth temperature under a N₂ flow. After loading into the reactor, the ZnO-coated wafer was preheated to the growth temperature in the backflow tube for 10 mins under a flowing N2 environment and then introduced to the flowing growth stream containing the GaCl and NH₃ to initiate growth. The flow rate of HCl to Ga source controlled the growth rate over a range of 0.5 to 4 µm/min in this work.

Results and discussion

ZnO buffer layer will reactively dissociate under the HVPE pre-growth annealing conditions, leaving a specular surface since it is

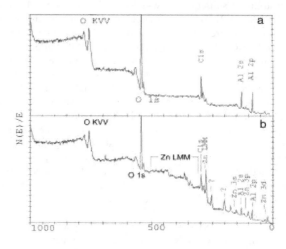

Fig.1 XPS spectra of Zn-derived sapphire surface after heat-treated under normal pre-growth conditions, (a) removing ZnO layer prior to annealing, and (b) removing ZnO layer after annealing.

thermodynamically unstable within an oxygen-free ambient. For buffer layer greater than 20 nm, PL and XRD studies reveal a residual film remains on sapphire surface despite an anneal time sufficient to dissociate this ZnO layer. The thickness of this residual layer depends on the initial ZnO thickness and annealing time. The chemical composition of the specular surface was initially examined by XPS. Prior to the XPS measurement, these samples receiving only the standard pre-growth conditions were dipped into an HCl solution to insure removal of any residual ZnO layer, if present. ZnO is readily etched in HCl. XPS measurements confirm the formation of a thin residual Zn-containing layer on the sapphire surface which is not ZnO. As a contrast, removal of the ZnO layer from sapphire surface prior to the pre-growth annealing results in no detectable Zn signal from the sapphire surface by XPS. These XPS results are shown in Fig.1, for both heat-treated under normal pre-growth conditions, spectrum (a) was obtained from the sample which has the ZnO layer removed prior to annealing and spectrum (b) from a sample etched in HCl after annealing. This figure indicates that some surface Zn remains as a result of a surface reaction that has occurred during pre-growth anneal, not during ZnO deposition process. Spectrum b is very similar to that obtained from $ZnAl_2O_4$ formed by annealing mixed ZnO and Al_2O_3 powder at high temperatures [12], suggesting a thin $ZnAl_2O_4$ layer formed during annealing. XRD measurements on this sample were not sufficiently sensitive to detect such a thin layer. In order to confirm the presence of a surface $ZnAl_2O_3$ layer by XRD, we annealed a thick ZnO layer on sapphire in air for 30 hrs. The XRD spectrum obtained from this sample shown in Fig.2, contains the (006) sapphire peak, as well as the (111) and (333) diffraction peaks of $ZnAl_2O_4$. $ZnAl_2O_4$ is a cubic spinel, and the presence of the (111) and (333) peaks indicates a degree of epitaxial alignment in this reactive diffusion couple. This well-oriented surface layer will have the (111) texture desired for the subsequent GaN growth.

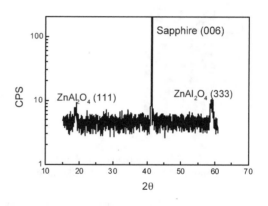

Fig.2 XRD spectrum of Zn-derived sapphire surface annealing at air ambience and for 30 hrs.

The surface reaction and compound formation modifies the sapphire surface morphology. Fig.3 is the AFM micrograph of the samples used to obtain the spectrum of Fig.1. The surface of sample a has no obvious morphological modification due to the pre-growth anneal and is similar to the bare sapphire surface after annealing in a N_2 ambient (not shown here). The surface of sample b is characterized by high density of small islands on the wafer surface. It is this ZnO-derived surface that promotes the enhanced initial nucleation and improved material quality in the subsequent GaN growth.

The surface reaction and compound formation modifies the sapphire surface morphology. Fig.3 is the AFM micrograph of the samples used to obtain the spectrum of

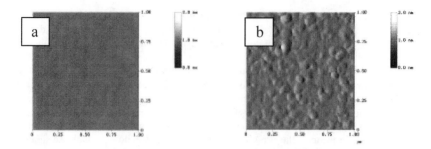

Fig.3 AFM micrograph of Zn-derived sapphire surface, (a) removing ZnO prior to annealing, (b) removing ZnO after annealing

Fig.1. The surface of sample a has no obvious morphological modification due to the pre-growth anneal and is similar to the bare sapphire surface after annealing in a N$_2$ ambient (not shown here). The surface of sample b is characterized by high density of small islands on the wafer surface. It is this ZnO-derived surface that promotes the enhanced initial nucleation and improved material quality in the subsequent GaN growth.

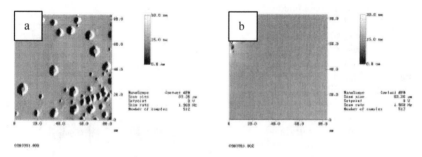

Fig.4 AFM micrograph of GaN material surface morphology (a) on bare sapphire, (b) on 20 nm ZnO buffer layer

GaN growth on the bare sapphire, often leads to a rough surface with inferior material properties. It is often difficult to reproducibly obtain high quality GaN possessing a smooth surface, narrow XRD linewidth, and low carrier concentration. In contrast, the use of ZnO buffer layers, routinely leads to smoother surfaces with few morphology features as well as improved materials properties. Fig.4 contains AFM micrographs of GaN surfaces grown on (a) bare sapphire and (b) an initial 20 nm ZnO buffer layer. The surface pit density is very high for GaN material grown on bare sapphire, which is often associated with the nucleation during initial stages of growth. A surface with the growth features related with step flow growth mode characterize the GaN layer grown on the ZnO buffer layer. XRD measurements on the GaN layers with an 8 μm in thickness possess a (002) rocking curve lindwidth as 400 to 500 arcsec, compared with the GaN layers of a similar thickness grown on bare sapphire normally a linwidth of 700 arcsec or more.

Room temperature PL spectra of GaN grown on bare sapphire, a 20 nm thick ZnO buffer layer, and a 80 nm thick buffer layer are shown in Fig.5. The PL spectrum of GaN layer on the 20 nm thick ZnO buffer layer has a slightly narrower bandage emission of 4 to 5 nm of FWHM compared to the FWHM of 7-8 nm for layers grown on bare sapphire. The GaN grown on the ZnO buffer layer also exhibits a weak blue band emission, which has been previously related to Zn doping [1,6]. This blue band emission centered at 420 nm increases in intensity in the PL spectrum of GaN material grown on a thick 80 nm buffer layer, also shown in Fig.5.

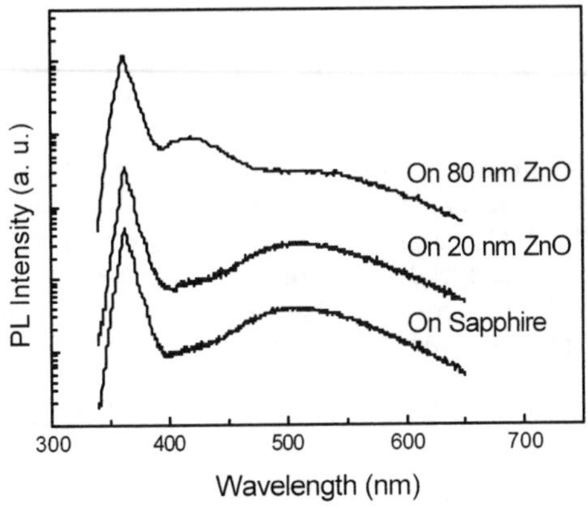

Growth on thick ZnO buffer layers can lead to delamination of the GaN layer from substrate, however, no such delamination occurs for materials grown on 20 nm buffer layer. Delamination is associated to the presence of a residual ZnO layer atop of the $ZnAl_2O_4$ layer which can persist after the pre-growth anneal for ZnO layers 80 nm or more in thickness. This residual ZnO layer is found to degrade the GaN quality, despite the potential to provide a better lattice matched layer. The line width of the GaN bandedge emission is broader than that found from GaN grown on the thin buffer layer (10 nm or more, depending on growth conditions), as found in a previous study [6]. For our specific system and pre-growth anneal the optimum thickness of the ZnO buffer layer is about 20 nm, which is thick enough to form the surface $ZnAl_2O_4$ layer but leaves no residual ZnO prior to starting the growth. Finally, the similar yellow band emissions, seen in Fig.5 for all three samples, is believed to be primarily related to the specific growth conditions used in the high temperature GaN layers. The intensity of

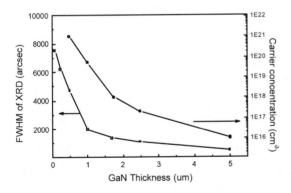

Fig.5 Room temperature PL spectra of GaN grown on different buffer layers

Fig. 6 FWHM of XRD (002) and carrier concentration of the GaN materials grown on 20 nm ZnO buffer layer versus material thickness.

YL can be decreased by a factor of one hundred by changing the growth conditions.

The change in material properties with GaN thickness has also been studied on the optimized ZnO buffer layer as shown in Fig.6. The FWHM of XRD and carrier concentration determining from CV measurements decrease quickly with material thickness. For a 5 μm thick GaN on 20 nm thick ZnO buffer layer, the FWHM of XRD is less than 500 arcsec with a surface carrier concentration as low as 1×10^{16}/cm^3. The GaN surface after 5 μm of growth is smooth with few morphology features. In contrast, GaN grown on bare sapphire exhibits a much slower drop in XRD linewidth and surface carrier concentration with thickness. To achieve a similar materials properties for GaN grown on bare sapphire a layer of 20 μm or more is required in our growth system under similar growth conditions.

Conclusion

ZnO buffer layers provide significant improvements in GaN material quality and system reproducibility by the HVPE method. The formation of a ZnAl$_2$O$_4$ surface layer by reactive diffusion between the ZnO and sapphire during pre-growth annealing process, can improve the initial nucleation and lead to a more rapid improvement in the GaN properties with thickness over layers grown on bare sapphire under similar conditions. The presence of this thin epitaxial ZnAl$_2$O$_4$ layer has been confirmed by XPS and XRD measurements.

Acknowledgement

This work is supported by DARPA-EPRI program on high power GaN devices.

References:

[1] T. Detchprohm, K. Hiramatsu, H. Amano, and I. Akasaki, Appl. Phys. Lett., 61, 2688(1992).
[2] R. J. Molnar, K. B. Nichols, P. Maki, E. R. Brown, and I. Melngailis, Mater. Res. Soc. Symp. Proc., 378, 479 (1995).
[3] S. Nakamura, Jpn. J. Appl. Phys., 30, L1705 (1991).
[4] I. Akasaki, and H. Amano, J. Crystal Growth, 163, 86 (1996).
[5] R. J. Monlar, W. Gotz, L. T. Romano, and J. M. Johnson, J. Crystal Growth, 178, 147 (1997).
[6] T. Ueda, T. F. Huang, S. Spruytte, H. Lee, M. Yuri, K. Itoh, T. Baba, and J. S. Harris Jr, J. Crystal Growth, 187, 340 (1998).
[7] J. T. Keller, D. K. Agrawal, and H. A. Mckinstry, Advanced Ceramic Materials, 3(4), 420 (1988).
[8] S. K. Sampath, and J. F. Cordaro, J. Am. Ceram. Soc., 81(3), 649 (1998).
[9] D. L. Branson, J. Am. Ceram. Soc., 48(11), 591 (1965).
[10] N. R. Perkins, M. N. Horton, Z. Z. Bandic, T. C. McGill, and T. F. Kuech, Mat. Res. Soc. Symp. Proc., 395, 243 (1996).
[11] R. Zhang, and T. F. Kuech, Mat. Res. Soc. Symp. Proc., 482, 709 (1998).
[12] B. R. Strohmeier, Surface Science Spectra, 3(2), 128 (1995).

Mat. Res. Soc. Symp. Vol. 595 © 2000 Materials Research Society

VISIBLE AND INFRARED EMISSION FROM GaN:Er THIN FILMS GROWN BY SPUTTERING

HONG CHEN*, K. GURUMURUGAN*, M.E. KORDESCH*, W.M.
JADWISIENCZAK**, AND H.J. LOZYKOWSKI**,
*Department of Physics and Astronomy
** School of Electrical Engineering and Computer Science, Ohio University, Athens,
OH 45701-2979, kordesch@ohio.edu

ABSTRACT

Erbium-doped films were grown on sapphire and silicon substrates by reactive sputtering, with different Er concentrations in the film. GaN films deposited at 800 K were determined to be polycrystalline by x-ray diffraction analysis, and retained their polycrystalline structure after annealing in nitrogen at 1250 K. The Er-doped films showed optical transmission beginning at about 360 nm, and the Er dose and film purity were determined with Rutherford backscattering spectroscopy. Photoluminescence and cathodoluminescence spectroscopy showed sharp emission lines corresponding to Er^{3+} intra 4f^n shell transitions over the range from 9 - 300 K. At above-bandgap optical and electron excitation, the $^4S_{3/2}$ and $^4F_{9/2}$ transition dominate, and are superposed on the "yellow band" emission. The infrared emission line at 1543 nm, corresponding to the Er $^4I_{13/2}$ to $^4I_{5/2}$ transition is also observed.

INTRODUCTION

Visible emission from Er-ion-implanted nitrides [1,2] has been reported, with application in electro-optical devices, and from sputtered amorphous AlN:Er thin films [3]. In the AlN:Er films, photoluminescence (PL) data could not be obtained; it is, however, possible to use above-bandgap excitation in sputtered GaN films. Sputtered GaN films have been deposited on sapphire and silicon substrates by reactive sputtering of Ga and Er metal targets in nitrogen. The Er is incorporated into the GaN film as an optically active inclusion (not strictly a dopant). The incorporation of a rare-earth luminescent center, such as Er, by sputtering may be an alternative to ion-implantation for large-scale device applications. The analysis of such films is detailed below.

EXPERIMENTAL

The GaN films were grown in a system described in reference [3]. Typical growth parameters for these films were: deposition temperature, 800 K, nitrogen-argon ratio, 4:1, total gas pressure, 5 mTorr, power, 120 Watts for Ga and 5Watts for the Er target. Film thickness was typically 0.2-0.3 µm. The Rutherford backscattering, x-ray diffraction and optical bandgap measurements were made with the facilities as described in [3].

Samples were given isochronal thermal annealing treatments (duration 30 min) at different temperatures 800 K, 1100 K, 1250 K and 1350 K in N$_2$ to optically activate the incorporated Er ions. The emission spectra presented are obtained from samples annealed 0.5 h at 1250 K in N$_2$, which seems to be the optimal annealing temperature.

For the cathodoluminescence (CL) and PL measurements, the samples were mounted on a cold finger cooled by a closed-cycle helium cryostat operating in temperature ranges from 10 K to 335 K. PL was performed using a He-Cd (325 nm, operating in cw mode 8 mW) and N_2 (337 nm, operating in pulse mode 1.3 mJ for PL kinetics measurement) lasers. The CL was excited by a 5 keV electron beam incident upon the sample at a 45^0 with an electron gun (Electroscan EG5 VSW) which was in a common vacuum ($1x10^{-7}$ Torr) with the cryostat. The PL and CL emitted light was detected by a monochrometer (ISA Hr-320) and detected with a CCD camera (Princeton Instruments TEA-CCD-512TK).

RESULTS

X-ray diffraction results for GaN:Er films grown on sapphire and silicon are shown in figures 1 and 2. In each case, the XRD spectrum typical for GaN c-axis normal to the substrate surface is observed, and no new or additional peaks are observed before or after annealing.

Figure 3 shows the optical bandgap measured for the GaN:Er films. The bandgap is measured by plotting the square of the absorption coefficient (α) versus the photon energy and extrapolating the linear portion of the curve to zero of the α^2 axis.

The films are transparent, and the optical bandgap determination shows the films to have a bandgap of about 3.6eV, so that the Er incorporation has little effect on the GaN bandgap. The results are shown in Figure 3.

The Erbium concentrations were determined by Rutherford backscattering spectroscopy, Figure 4, the ratio of Ga to N to Er in these films shows that the GaN films are stoichiometric, and that the Er concentration is about 1%.

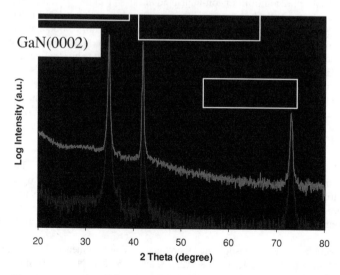

Figure 1: X-ray Diffraction spectra of Er-doped GaN films on Sapphire(0001). Top: annealed to 1250 K, Bottom: as grown, 800K.

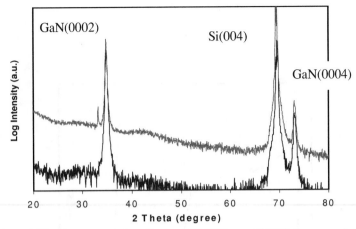

Figure 2: X-ray Diffraction spectra of an Er-doped GaN film on Si(001).Top: annealed to 1250 K, Bottom: as grown, 800K.

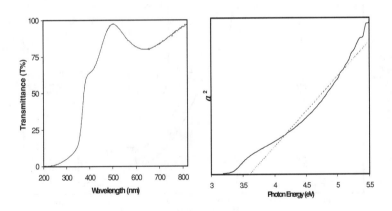

Figure 3.: GaN:Er film on sapphire. Right: solid curve, α^2, dotted curve, linearization of the α^2 plot . The bandgap is approximately 3.6 eV.

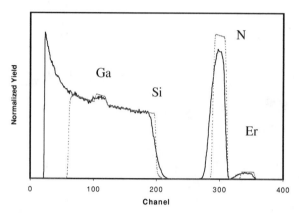

Figure 4. RBS spectra of GaN and Er-doped GaN films. In the doped film, the Ga:N:Er ratio is measured to be 1:1: 0.011; i.e. about 1% Er

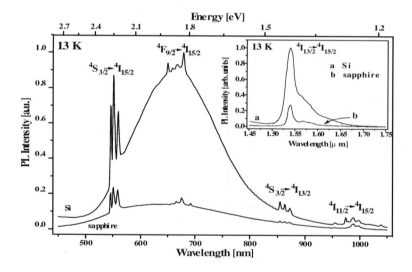

Figure 5: PL spectra of GaN:Er films on silicon and sapphire. Inset: Infrared emission in PL of GaN:Er films on silicon and sapphire. Spectra measured at 13 K

The PL spectrum in Figure 5 shows the "yellow band" on the silicon substrate to be more prominent. The green emission at about 550nm is split into several bands, indicating that the Er ions in these films are in several different locations in the crystal lattice. The IR band at 1543nm is more intense on the silicon substrate.

Figure 6 shows the dependence of the PL intensity on annealing temperature. The tendency for both substrates is similar, and the chioce of the optimal annealing temperature of 1250 K (950 °C) is clear from the data in the figure.

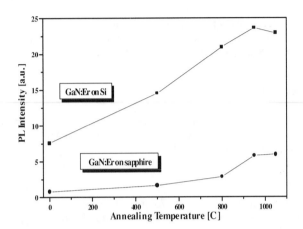

Figure 6. PL intensity for GaN:Er on silicon and sapphire as a function of temperature.

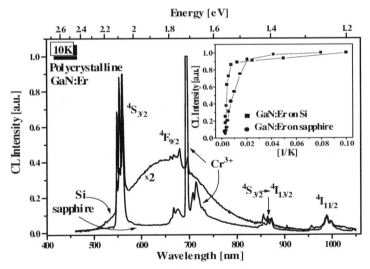

Figure 7. CL spectra for GaN:Er on silicon and sapphire. Inset: thermal quenching of the green emission line, the Er $^4S_{3/2}$ transition.

Figure 7 shows the CL spectra for the sputtered GaN:Er films on both silicon and sapphire substrates. The quenching of the green emission is very sharp, and is probably not due to quenching by other Er ions due to the low concentration of Er in these films.

CONCLUSIONS

The PL and CL data show that reactively sputtered GaN thin films on silicon and sapphire with Er incorporated into the films during growth by co-sputtering of Er can result in films that show visible and IR emission after annealing in nitrogen. The sputtering process is a viable alternative to ion implantation.

ACKNOWLEDGEMENTS

We thank D.C. Ingram and G.R. Harp for their technical contributions. This work was supported by the Ballistic Missile Defense Organization through grant N00014-96-1-0782 monitored by the Office of Naval Research.

REFERENCES

1. A.J. Steckl and R.Birkhahn, *Appl.Phys.Lett.*, **73**, p.1700 (1998).
2. H.J. Lozykowski, M.W. Jadwisienczak, and I. Brown, *Appl. Phys.Lett.*, **74**, p. 1129 (1999).
3. K. Gurumurugan, Hong Chen, G.R. Harp, M.W. Jadwisienczak, and H.J. Lozykowski, *Appl.*
   Phys. Lett., **74**, p. 3008 (1999).

Mat. Res. Soc. Symp. Vol. 595 © 2000 Materials Research Society

Growth kinetics of GaN thin films grown by OMVPE using single source precursors

R. A. Fischer*, A. Wohlfart, A. Devi, and W. Rogge
*Anorganische Chemie II, Ruhr Universtität Bochum, D-44801, Germany

ABSTRACT

We report the growth kinetics of GaN thin films using the single source precursor bisazido dimethylaminopropyl gallium (BAZIGA) in a cold wall reactor. Transparent, smooth, epitaxial (FWHM of the α-GaN 0002 rocking curve = 129.6 arcsec) and stoichiometric GaN films were grown on c-plane Al_2O_3 substrates in the temperature range of 870 – 1320K and high growth rates were obtained (up to 4000 nm/hr). Film growth was studied as a function of substrate temperature as well as reactor pressure. Although high quality films were obtained without using any additional source of nitrogen such as ammonia, we have investigated the effect of ammonia on the growth and properties of the resulting films. The films obtained were characterized by XRD, RBS, XPS, AES, AFM, SEM and the room temperature PL spectroscopy of GaN films grown exhibited the correct near band edge luminescence at 3.45 eV.

INTRODUCTION

The nitrides GaN, AlN, InN and their ternary alloy systems AlGaN and InGaN are very promising materials for applications in green/blue light-emitting diodes and semi-con-ductor lasers[1]. The commercial growth of high quality epitaxial group-13 nitrides by organometallic vapor phase epitaxy (OMVPE) is based on the co-pyrolysis of metal alkyls MR_3 (M = Al, Ga, In, R = CH_3, C_2H_5, ^tBu) and NH_3. But high temperatures are involved in this process for the effective activation of NH_3 (>773K for InN[2], >1173K for AlN and GaN[3]).

Many efforts have been undertaken to substitute the ineffective nitrogen source (NH_3) using alternative precursors based on the concept of preformed direct M-N bonds in the precursor molecule. Such precursors for the growth at a lower V/III ratio and milder growth conditions were achieved with many compounds for instance hydrazine (N_2H_4), 1-1-dimethyl hydrazine (M_2NNH_2), phenyl hydrazine ($PhNH-NH_2$), hydrazoic acid (HN_3)[6-8] and also alkyl amines (RNH_2, R = ^tBu, ^iPr)[9]. But the disadvantages of these precursors are their toxic and explosive nature as well as carbon incorporation into the resulting films. Besides these compounds, some group-III metal amido compounds such as $[R_2GaNH_2]_3$ (R = Et, Me) and $[Ga(NR_2)_3]_2$ (R = Et, Me) are interesting as single source precursors (SSP). However these precursors are only used with limited success in terms of the achieved film properties[10]. The azides such as $[R_2GaN_3]_x$ (R = Cl, H, Et, Me) are possible candidates as SSP and have been already used successfully to grow GaN[11]. The azide group combines a pre-formed strong Ga-N bond with a reduced number of undesired Ga-C and N-C bonds to minimize the carbon incorporation during film growth. But the disadvantage with these precursors are their air sensitivity and explosivity. To find an alternative to these problems we concentrated our research on the growth of epitaxial GaN films using the novel Lewis-base stabilized organometallic azide compound of the type $(N_3)_aM[(CH_2)_3NMe_2]_{3-a}$ (M = Al, Ga, In and a = 1, 2). These SSP's are non-pyrophoric, non-explosive and yield epitaxial GaN films under moderate conditions. The successful growth of GaN and InN films was already reported earlier using a simple horizontal hot wall CVD reactor[12-14]. In this paper we report the study on the growth kinetics of GaN from the precursor $(N_3)_2Ga[(CH_2)_3NMe_2]$ (bisazido dimethylaminopropyl gallium, BAZIGA) in a horizontal cold wall CVD reactor.

EXPERIMENT

For GaN film deposition, a horizontal low pressure cold wall CVD reactor (base pressure: 10^{-6} Torr) was fabricated (Fig. 1). It consists of a quartz tube about 50 cm in length and 2.5 cm in diameter. The substrates are placed on a SiC coated graphite susceptor at the center of the tube. For the substrate heating, an inductive heating arrangement was used monitored by a radiation pyrometer. During the heating process the outer wall of the central part of the quartz tube was cooled by water. Using this set up temperatures up to 1473K could be attained. Attached to the quartz tube are the precursor vaporizer heated by means of an air bath and a glas trap cooled with liquid nitrogen. To generate the necessary vacuum, a turbo-molecular pump backed up by a rotary pump was used. The reactor pressure during the CVD process was regulated using a motor driven throttle valve and mass flow controllers were used to control the flow of different gases. For GaN film deposition, epi-polished c-plane Al_2O_3 substrates as large as 1 x 1cm were used. Before loading it into the reactor, the substrates were degreased in trichloroethylene, etched in a mixture of H_2SO_4 / H_3PO_4 (3:1) at 353K and rinsed with water and acetone. After the etching and cleaning process the substrates were pre-treated under vacuum for several hours (12 h) and subsequently annealed at 1323K with hydrogen (10 sccm) for 1 hour. Following this annealing, the substrates were nitridated with ammonia (40 sccm) at 1223K for 30 min. The vaporizer temperature was maintained at 353K for all depositions. Finally the vaporizer valve was opened to start the deposition which lasted normally for 45 min. High purity nitrogen (40 sccm) was used as a carrier gas to transport the precursor to the reaction zone. The deposition temperature was varied in the range of

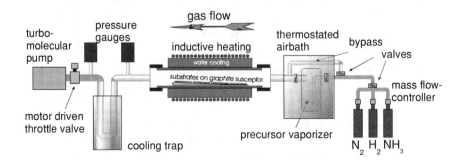

Fig. 1) Schematic of the cold wall CVD reactor

773 – 1323K, the reactor pressure between 0.080 – 100.0 mbar. To investigate the influence of ammonia as a reactive gas to the growth process, ammonia flow rates were varied at different temperatures and pressures. The crystalline properties of the obtained films were investigated in detail by x-ray diffraction studies using a D8-Advance Bruker axs diffractometer and a high resolution D8-Discover Bruker axs diffractometer. The 2Θ-Θ scans, rocking curves, pole figures, reciprocal space mappings and reflectometry measurements were carried out. The film composition was determined by Rutherford Back Scattering (RBS) using the instrument of DTL (Dynamitron Tandem Laboratory), X-ray Photoelectron Spectroscopy (XPS) as well as Auger Electron Spectroscopy (AES) using the equipment from FISONS. The optical properties were studied by room temperature Photoluminescence (PL). The surface morphology was investigated with Atomic Force Microscopy (AFM) using a Nanoscope Multimode III scanning probe microscope (Digital Instruments) and Scanning Electron Microscopy (SEM) using an equipment from LEO. The film thickness was analyzed by AFM, using a TOPOMETRIX mircoscope.

RESULTS AND DISCUSSION

The primary goal was to grow epitaxial GaN films using BAZIGA with N_2 as inert carrier gas but without any additional nitrogen source such as ammonia. In fact this was achieved successfully. The films were grown without any buffer layers except that the substrates were nitridated with ammonia. It was found that at temperatures below 973K the films were predominantly amorphous. X-ray diffractograms showed a broad 0002 α-GaN reflection with low intensity. At higher temperatures crystalline α-GaN was found with a sharp 0002 α-GaN reflection and high intensity. It was also found that the crystal quality and the signal intensity was clearly improved with increasing pressure due to the decreasing growth rate. At low reactor pressures (0.080 mbar) growth rates of the order of 4500 nm/hr were obtained whereas higher reactor pressures (8.000 mbar) resulted in growth rates of the order of 2000 nm/hr. Besides the reactor pressure, the growth rate strongly depends on the substrate temperature because of surface reactions and altered reaction pathways. From the Arrhenius-plot (Fig. 2) the temperature dependence of the growth rate can be clearly seen. The growth rate at lower temperatures indicates a thermally activated deposition kinetics. To determinate the activation energy (E_a) for the linear portion of the curve in the kinetic controlled part at lower temperatures, the Arrhenius equation $G = A \cdot \exp(-E_a/RT)$

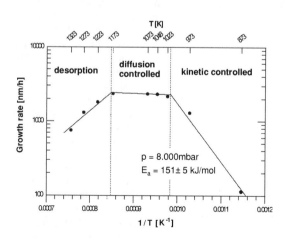

Fig. 2) Arrhenius-plot at 8.000 mbar without NH_3

was used, where G is the growth rate, A is the Arrhenius pre-exponential factor, R is the gas constant and E_a the activation energy that is the rate limitting process under the given conditions. For all depositions a reactor pressure of 8.000 mbar was adjusted. It was found that in the temperature range of 870 – 1020K the GaN film growth is associated with an activation energy of 151 ± 5 kJ/mol. The maximum growth rate of 2500 nm/h was obtained in the temperature range of 1020 – 1175K. In this range, the growth rate shows nearly no temperature dependence that means the growth is only controlled by the diffusion of the precursor molecule fragments to the substrate surface. At temperatures over 1175K the growth rate decreases because of desorption and depletion processes consistent with the typical behavior of a CVD process. To investigate the influence of ammonia as a reactive gas to the growth kinetics, an Arrhenius-plot under the same conditions (8.000 mbar reactor pressure and 363K vaporizer temperature) but with additional NH_3 (40 sccm) was carried out. Under these conditions some differences were observed and it was found that the slope of the kinetically controlled regime is lowered by more than half yielding an activation energy of $E_a = 65 \pm 5$ kJ/mol. A temperature range of 870–1060K for the kinetically controlled regime was obtained. Thereby the kinetically controlled regime is extended to higher temperatures than in the previous case where no NH_3 was used. The growth rate was also reduced and the highest growth rate of the order of 1000 nm/h was obtained in the range of 1060 – 1175K. The growth rate drops off again at temperatures greater than 1175K. In addition to the reduction in E_a and growth rates, the crystalline quality of the resulting films were poorer when compared to those obtained without NH_3 under similar CVD conditions. To study the effect of NH_3 on the crystalline quality, several α-GaN films were deposited at a reactor pressure of 8.000 mbar and a substrate temperature of 1073K with increasing ammonia flow. The film properties were determined with x-ray diffraction (XRD) and it was found that the obtained films were epitaxial due to the fact that the diffractograms showed only the 0002 reflection and also confirmed from their pole figure measurements (Fig. 4). To investigate the crystalline quality, rocking curves of the 0002 α-GaN reflection have been carried out. Fig. 3 shows the variation of FWHM (as determined by x-ray rocking curves) of the epitaxial α-GaN films with increasing ammonia flow rates. It is to be noted that the FWHM of the 0002 GaN rocking curves strongly depends on the flow rate of NH_3. With increasing NH_3 flow, the FWHM increases. Without ammonia a FWHM of 129.6 arcsec was achieved. This is to our knowledge the best ever reported value for the epitaxial films obtained from single source precursors without any additional source for nitrogen. As illustrated in Fig. 3, with increasing ammonia flow rates, the crystal quality becomes poorer. The pole figure measurement (Fig. 4) was

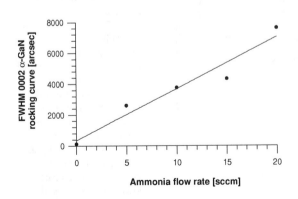

Fig. 3) Variation of FWHM with NH_3 flow rates

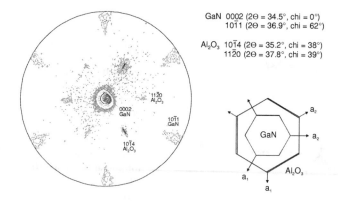

GaN 0002 (2Θ = 34.5°, chi = 0°)
10$\bar{1}$1 (2Θ = 36.9°, chi = 62°)

Al₂O₃ 10$\bar{1}$4 (2Θ = 35.2°, chi = 38°)
11$\bar{2}$0 (2Θ = 37.8°, chi = 39°)

Fig. 4) Pole figure measurement and orientation of a α-GaN film

setup with a fixed 2Θ value of 34.5° and ω = 17.25° and ψ = 0 – 65°, φ = 0 – 360°. The 2-dimensional plot shows both the reflections from α-GaN (0002) and (10$\bar{1}$1) and from the Al₂O₃ substrate (10$\bar{1}$4) and (11$\bar{2}$0) at different ψ and φ angles. From these reflections the orientation of the α-GaN layer relative to the sapphire substrate could be determined as 30°. By means of these pole figure measurements as well as 2Θ-Θ scans and reciprocal space mappings the hexagonal symmetry of the produced GaN films was confirmed.

To investigate the role of substrate nitridation, two amorphous low temperature films (grown at 773K and 8.000 mbar) were characterized by x-ray reflectometry. It was seen that during the nitridation of the bare Al₂O₃ substrates with NH₃ for 30 min, a thin layer of AlN (0.5 nm) was formed and this was verified by XPS measurements. The absence of the AlN layer on a non-nitridated Al₂O₃ substrate was also verified by XPS. The density of the grown α-GaN layer was found to be significantly lower (4.42 g/cm³) than crystalline bulk α-GaN (6.1 g/cm³). This could be caused by incorporation of residual precursor fragments as well as structural defects. The roughness of the GaN layer was determined as ~1 nm from reflectometry. This was also confirmed by AFM measurements. Crystalline high temperature films could not be characterized by x-ray reflectometry because of their higher surface roughness. The film composition was checked with RBS and XPS including nitrogen, gallium, carbon and oxygen to determine the Ga-N stoichiometry as well as the level of impurity. The RBS analysis revealed stoichiometric GaN in a "chemical sense" with a ratio of Ga and N of 1:1 without any additional oxygen. To determine possible impurities the films were analyzed by XPS. It was found that GaN films, grown at 973K, contained very small traces of carbon when grown without using ammonia and this can be accounted due to the fragmentation of the precursor leading to carbon incorporation. Whereas in GaN films grown at 1173K at 8.000 mbar reactor pressure and 363K vaporizer temperature with 20 sccm NH₃, carbon was absent (Fig. 5). It was found that very small amounts of ammonia were necessary to grow GaN films free of carbon.

CONCLUSIONS

It was demonstrated that it is possible to grow GaN films using BAZIGA as a single source precursor without any additional nitrogen source such as NH₃ with a high structural quality which is reasonable close to the materials grown with the standard process

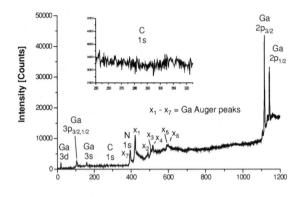

Fig. 5) XPS spektrum of a GaN film

and exhibiting the characteristic band edge luminescence of 3.45 eV. The use of NH_3 during the deposition has an important influence to the resulting film properties. It was found that the best crystalline α-GaN films were achieved at higher reactor pressures (above 10 mbar) without using any NH_3. In this case very small traces of carbon could be found by XPS. Moreover using very low flow rates of NH_3 (2-5 sccm) during deposition, the carbon incorporation was eliminated, but with increasing NH_3 flow rates (up to 20 sccm) the crystalline quality decreased drastically.

ACKNOWLEDGMENTS

The authors thank the Deutsche Forschungsgemeinschaft (DFG) for financial support. A. D. thanks the Alexander von Humboldt foundation, Germany for a fellowship.

REFERENCES

[1]   S. Nakamura, *MRS Bulletin* **23**, 1998, 37.
[2]   C. R. Eddy, T. D. Moustakas, *J. Appl. Phys.* **73**, 1993, 448.
[3]   S. Strite, H. Morkoç, *J. Vac. Sci. Technol.* **B10**, 1992, 1237.
[4]   D. Gaskill, N. Bottka, M. C. Lin, *Appl. Phys. Lett.* **48**, 1986, 1449.
[5]   S. Miyoshi, K. Onabe, N. Ohkouchi, H. Yaguchi, R. Ito, S. Fukatsu, S. Yoshida, *J. Cryst. Growth* **124**, 1992, 439.
[6]   H. Okumura, S. Misawa, T. Okahisa, S. Yoshida, *J. Cryst. Growth* **136**, 1994, 361.
[7]   D. G. Chtchkine, L. P. Fu, G. D. Gilliland, Y. Chen, S. E. Ralph, K. K. Bajaj, Y. Bu, M. C. Lin, *J. Chem. Soc.* **42**, 1995, 423.
[8]   C. J. Linnen, R. D. Coombe, *Appl. Phys. Lett.* **72**, 1998, 88.
[9]   B. Beaumont, P. Gilbart, J. P. Faurie, *J. Cryst. Growth* **156**, 1995, 140.
[10] A. C. Jones, S. A. Rushworth, D. J. Houlton, J. S. Roberts, V. Roberts, C. R. Whitehouse, G. W. Critchlow, *Chem. Vap. Dep.* **2**, 1996, 5.
[11] H. S. Park, S. D. Waezsada, A. H. Cowley, H. W. Roesky, *Chem. Mater.* **10**, 1998, 2251.
[12] R. A. Fischer, A. Miehr, O. Ambacher, T. Metzger, E. Born, *J. Cryst. Growth* **170**, 1997, 139.
[13] R. A, Fischer, H. Sussek, H. Pritzkow, E. Herdtweck, *J. Organomet. Chem.* **548**, 1997, 73
[14] A. Devi, W. Rogge, R. A. Fischer, F. Stowasser, H. Sussek, H. W. Becker, J. Schäfer, J. Wolfrum, *J. Phys. IV France* **9**, 1999, 589

Mat. Res. Soc. Symp. Vol. 595 © 2000 Materials Research Society

GROWTH AND CHARACTERIZATION OF GaN THIN FILMS ON Si(111) SUBSTRATES USING SiC INTERMEDIATE LAYER.

K. Y. Lim, K. J. Lee, C. I. Park, K.C. Kim, S. C. Choi, W.-H. Lee, E.-K. Suh, and G. M. Yang
Department of Semiconductor Science & Technology, Semiconductor Physics Research Center, Chonbuk National University, Chonju 561-756, Korea
K. S. Nahm
Department of Chemical Engineering, Semiconductor Physics Research Center, Chonbuk National University, Chonju 561-756, Korea

ABSTRACT

GaN films have been grown atop Si-terminated 3C-SiC intermediate layer on Si(111) substrates using low pressure metalorganic chemical vapor deposition (LP-MOCVD). The SiC intermediate layer was grown by chemical vapor deposition (CVD) using tetramethylsilane (TMS) as the single source precursor. The Si terminated SiC surface was obtained by immediately flow of SiH_4 gas after growth of SiC film. LP-MOCVD growth of GaN on 3C-SiC/Si(111) was carried out with trimethylgallium (TMG) and NH_3. Single crystalline hexagonal GaN layers can be grown on Si terminated SiC intermediate layer using an AlN or GaN buffer layer. Compared with GaN layers grown using a GaN buffer layer, the crystal qualities of GaN films with AlN buffer layers are extremely improved. The GaN films were characterized by x-ray diffraction (XRD), photoluminescence (PL) and scanning electron microscopy (SEM). Full width at half maximum (FWHM) of double crystal x-ray diffraction (DCXD) rocking curve for GaN (0002) on 3C-SiC/Si(111) was 890 arcsec. PL near band edge emission peak position and FWHM at room temperature are 3.38 eV and 79.35 meV, respectively.

INTRODUCTION

GaN related technology was developed, rapidly, because of their potential applications as short wavelength emitters and UV detectors. Recently, GaN based blue light emitting laser diodes have reached the level of long lifetime operating under the electrical continuous wave (CW) conditions.[1][2] Since bulk GaN substrates are not available, despite of the critical role which is substrate symmetry and dimensional lattice matching, the majority epitaxial technique of GaN films is grown by the technique of heteroepitaxy on sapphire substrate using the low temperature GaN or AlN buffer layers. However, the large lattice mismatch (~13%) and the differences of thermal expansion between the sapphire substrate and GaN epitaxial layer result in the generation of a lot of defects in the GaN layers. Hexagonal 6H-SiC, which has a much smaller lattice mismatch (~3%) with wurtzite GaN and a merit of high conductivity for the fabrication of top-bottom ohmic metal contact, is in principle a more suitable substrate. But SiC substrate is available only in small diameter and very expensive. So, the previous work has attempted to grow high quality GaN films on SOI substrates using SiC intermediate layer[3][4] and AlAs nucleation layer[5], stimulated by irreplaceable merits of Si wafer such as low cost, high surface quality, large area wafer availability, high conductivity and well-established processing techniques. In spite of enormous efforts, however, typically SiC films grown on Si substrate have a rough SiC/Si interface, a poor single crystal quality, a high concentration of lattice defects and high level of background impurities. Also, it is not easy to solve these problems because of the differences of thermal expansion coefficient and extremely large lattice mismatch (~20%) between SiC and Si. In spite of these problems, it was tried to grow the GaN films on Si substrate using SiC as intermediate layer, because of the SiC intermediate layers act on reduce the lattice mismatch and relax thermal expansion. In the previously reports, SiC films were successfully grown on directly Si(111) substrates.[6][7] SiC(111) planes is a better lattice match to subsequent growth of hexagonal GaN films than SiC(100) planes.

In this work, we grow the GaN films on Si(111) substrates using the SiC intermediate layer by low pressure metalorganic chemical vapor deposition (LP-MOCVD) and investigate the structural and optical characterization of GaN films. We compare properties of the GaN films growth on SiC/Si(111) with AlN or GaN buffer layer and without buffer layer. We also compare the results of the GaN films grown with and without high temperature thermal treatment of SiC intermediate layer in H_2 ambient.

EXPERIMENT

The SiC films on (111)-oriented Si substrates were grown by chemical vapor deposition (CVD) system using the tetramethylsilane (TMS) as the single source precursor. Before the loading into CVD system, the Si substrates were degreased in organic solvents and surface oxide was removed by etching in HF solution. The reactor

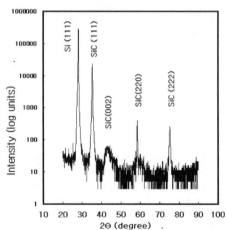

Fig. 1 *Optical microscope picture of the surface morphology of a 3-C SiC grown on Si(111) substrate.*

Fig. 2 *Wide angular range x-ray diffraction spectra of 3-C SiC grown on Si(111) substrate.*

was evacuated down to 5 torr pressure. The Si substrate was thermally cleaned at 1100 °C for 5 minute in H_2 ambient. Then, the reactor temperature was increased to 1250 °C for the SiC deposition. During the SiC growth, the TMS flow rate and H_2 flow rate were 1 sccm and 1000 sccm, respectively. The Si terminated SiC surface was obtained by immediately flow of SiH$_4$ gas after growth of SiC film at the same temperature. After 1 hour growth, the surface morphology and structural properties of the SiC films were investigated by optical microscope and x-ray diffraction (XRD), respectively. Figure 1 shows the optical microscope image. The surface morphology of SiC films grown on Si(111) is the shape of scales which comes from the surface of Si after thermal cleaning before the SiC growth. In figure 2, wide angle x-ray diffraction spectra of SiC film on Si(111) substrate is showed. The 2• peak of SiC(111) planes at 35.65 degree reveals that the SiC film has a 3C phase crystal structure.

The Si-terminated SiC/Si(111) substrate was cleaned again in organic solvents adding ultra sonic agitation. And the surface of SiC/Si(111) substrate was etched by dipping in 48% HF:H$_2$O for 3 sec and rinsed D. I. water. GaN layer was grown on SiC/Si(111) by low pressure MOCVD system, using H$_2$ carrier gas, trimethylgallium(TMG), trimethylaluminum (TMA) and Ammonia(NH$_3$) as Ga, Al and N precursors, respectively. Before the buffer layer growth, the substrate was thermally treated at 1150 °C for 10 min in H$_2$ ambient. Then two different buffer layers are grown. Thin GaN buffer layer was grown at 570 °C for 150 sec with TMG flow rate 80 • mol/min and NH$_3$ flow rate 4000 sccm at a reactor pressure 200 torr. Thin AlN buffer layer was grown at 1160 °C for 5 min with TMA flow rate 55 •mol/min and NH$_3$ flow rate 4000 sccm at the same reactor pressure. The thickness of GaN buffer and AlN buffer was calibrated 200 Å for the growth time of them. In order to grow GaN epilayer, the reactor temperature was increased up to 1160 °C under a flow of NH$_3$ + H$_2$ mixture. The duration of the temperature ramping and stabilization was about 4 min for the case of GaN buffer and 1 min for the case of AlN buffer. GaN epilayers were grown for 1 hour with TMG flow rate 120 •mol/min and NH$_3$ flow rate 4000 sccm at the temperature of 1160 °C.

After the growth, surface morphologies were observed with optical microscope and scanning electron microscopy (SEM). To evaluate the structural properties, we measured the wide angle 2• X-ray diffraction (XRD) and the full width at half maximum(FWHM) for the symmetric (0002) reflections of hexagonal GaN with double crystal X-ray diffraction (DCXD). Room temperature photoluminescence (PL) measurements were carried out using a 325 nm wavelength of He-Cd laser as an excitation source. The excitation power was about 30 mW. The PL-signals were detected by a photomultiplier tube.

RESULTS AND DISCUSSION

In order to find the optimal growth conditions of GaN epilayer on Si(111) substrate using the SiC intermediate layer, we investigated the dependence of various different buffer layers. There are significant difference in growth mechanism and morphology obtained from GaN films without GaN buffer layer and with GaN or AlN buffer layer on SiC/Si(111). Fig. 3 shows the photographs of the surface morphology of GaN films grown on SiC/Si(111) substrate. The sample in Fig. 3 (a), which was grown without buffer layer, reveals that GaN films were grown in 3-dimensional island mode. The GaN film grown with 200 Å GaN buffer layer shown in Fig. 3(b) reveals numerous individual grains and very rough surface. These surface morphologies indicate that GaN buffer layer is not good material to supplement 2-dimensional growth of GaN film. But, we have found that the surface morphology was improved significantly in GaN film grown with

AlN buffer layer. Fig. 3 (c) shows the GaN film grown with 200 Å AlN buffer layer with thermal cleaning processing in H_2 ambient before the growth AlN buffer layer. Fig. 3(d) shows that without thermal cleaning process. In spite of the enhancement of cracks due to thermal expansion, the growth mode of GaN epilayer with AlN buffer layer is fully 2-dimensional. And the thermal cleaning processing increases the density of hexagonal pits.

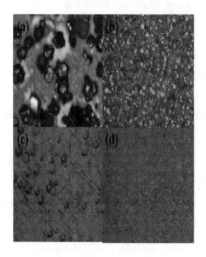

Fig. 3 *Optical microscope pictures of the surface morphology of GaN films grown on SiC/Si(111) substrate. (a) no buffer layer, (b) with 200 Å GaN buffer layer, (c) with 200 Å AlN buffer layer, (d) with 200 Å AlN buffer layer and no thermal cleaning before the growth*

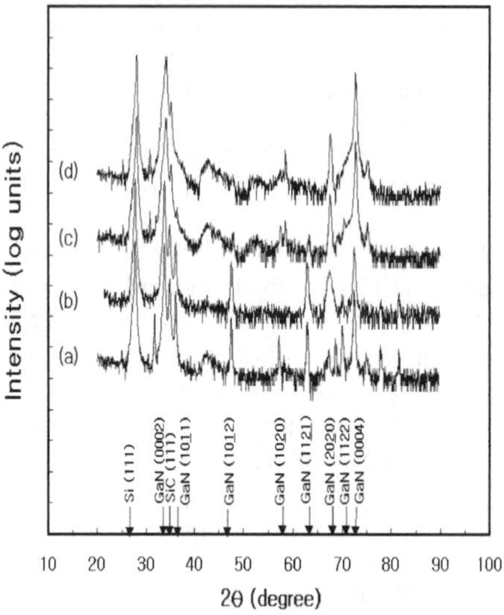

Fig. 4 *Wide angular range x-ray diffraction spectra of GaN films grown on SiC/Si(111) substrate. (a) no buffer, (b) 200 Å GaN buffer, (c) 200 Å AlN buffer, (d) 200 Å AlN buffer and no thermal cleaning before the growth.*

This means that Si atoms are desorbed from the SiC surface and the surface becomes more rough by the thermal cleaning. Further growth experiments to improve the surface quality and remove the cracks are under way.

XRD spectra of GaN epilayers grown on SiC/Si(111) substrates are shown in Fig. 4. We observed the strong peaks to be related the reflection of GaN(0002) plane and GaN(0004) plane at XRD values of 2•=34 degree and 2•=72 degree in sample (c) and (d). We also observed several weak peaks corresponding to the other GaN planes in sample (a) and (b). This indicates that a wurtzite GaN film can be grown with growth orientation along the c-axis in samples with AlN buffer layer. Also, the SiC (111) plane reflection peak at 2•=35.65 degree and the Si(111) plane reflection were shown in Fig. 4. The structural quality of GaN films was more clearly revealed in Fig. 5, which is the DCXD rocking curves of the reflection of GaN (0002) plane. The rocking curves of GaN films grown with GaN buffer and without buffer were not detected by DCXD. This indicates that these films have very poor structural quality. However, the GaN films grown with AlN buffer layer on SiC/Si(111) substrates exhibit a distinguished peak of DCXD corresponding to the reflection of GaN(0002) plane. Compared with GaN layers grown using a AlN buffer layer with thermal cleaning and without, the crystal qualities of GaN films without thermal cleaning processing are extremely improved. The intensity of DCXD peak of GaN grown without thermal cleaning is more intense and full width at half-maximum (FWHM) is smaller than one of the GaN grown with thermal cleaning. The FWHM of DCXD rocking curve of the GaN films grown at optimum growth conditions is 890 arcsec. The structural quality of GaN films grown on Si(111) substrate using the SiC intermediate layer is comparable to GaN grown on sapphire substrate.

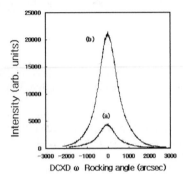

Fig. 5 *Double crystal x-ray diffraction spectra of GaN films grown on SiC/Si(111) substrate with 200Å AlN buffer layer. (a) with and (b) without the thermal cleaning at high temperature before the growth. FWHM of (a) and (b) are 960 and 890 arcsec, respectively.*

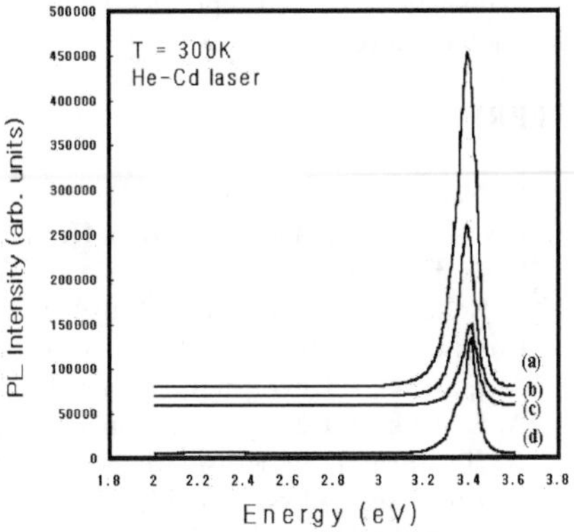

Fig. 6 *Room temperature PL spectra obtained under 30mW power excitation by the CW He-Cd laser from GaN films grown on SiC/Si(111) substrate. (a) with 200 Å AlN buffer layer and no thermal cleaning before the growth, (b) with 200 Å AlN buffer layer, (c) with 200 Å GaN buffer layer, (d) no buffer layer.*

The optical property of the GaN layers grown on SiC/Si(111) substrates was investigated by photoluminescence (PL) at room temperature. PL spectra are shown in Fig. 6. Strong near band edge emission at 3.38 eV with a FWHM in the range of 79 ~ 112 meV was observed from all films. In our opinion, line broadening of PL emission was caused by auto doping of Si and C impurities during the growth of GaN films. The yellow band emission at around 2.2 eV was not appeared. PL intensity of the GaN film grown with AlN buffer layer was more intense than one of GaN film grown with GaN buffer layer. These values of GaN films grown on SiC/Si(111) with AlN buffer layer, compared with one of GaN on sapphire, indicate the capability of good quality GaN films on Si substrate.

SUMMARY

In summary, the high quality hexagonal GaN layers have been successfully grown on Si(111) substrates using the SiC intermediate layer by LP-MOCVD. Structural quality and optical properties are improved by introducing an AlN buffer layer. The coalescence of initially formed grains during the growth of GaN films strongly increase by using AlN buffer layer. The cracks of GaN films, however, due to the residual thermal strain were enhanced with improving lateral homogeneous of GaN films. The high quality hexagonal

GaN film has a DCXD FWHM of 890 arcsec. The room temperature PL spectra of GaN showed an intense and sharp near-band-gap emission of hexagonal GaN at 3.38 eV with FWHM 79.35 meV. The yellow luminescence, which is corresponded to deep energy levels due to the imperfections of GaN, is very weak.

ACKNOWLEDGEMENTS

This work has been supported by the project of 1998-016-D0004 of the Ministry of Education of Korea.

REFERENCES

[1] S. Nakamura, M. Senoh, S. Nagahama, N. Iwasa, T. Yamada, T. Matsushita, H. Kiyoku, Y. Sugimoto, T. Kozaki, H. Umemoto, M. Sano and K. Chocho, Jpn. J. Appl. Phys. **37**, L1020 (1998)
[2] S. Nakamura, M. Senoh, S. Nagahama, T. Matsushita, H. Kiyoku, Y. Sugimoto, T. Kozaki, H. Umemoto, M. Sano and T. Mukai, Jpn. J. Appl. Phys. **38**, L226 (1999)
[3] A. J. Steckl, J. Devrajan, C. Tran and R. A. Stall, Appl. Phys. Lett. **69**, 2264 (1996)
[4] A. J. Steckl, J. Devrajan, C. Tran and R. A. Stall, J. Electron. Mater. **26**, 217 (1997)
[5] A. Strittmatter, A. Krost, M. Stra•burg, V. Turck, D. Bimberg, J. Blasing and J. Shristen, Appl. Phys. Lett. **74**, 1242 (1999)
[6] A. J. Steckl, C. Yuan, Q-Y. Tong, U. Gosele and M. J. Loboda, J. Electrochem. Soc. **141**, L66 (1994)
[7] Q. Wahab, R. C. Glass, I. P. Ivanov, J. Bircj, J.-E. Sundgren and M. Willander, J. Appl. Phys. **74**, 1663 (1993)

Mat. Res. Soc. Symp. Vol. 595 © 2000 Materials Research Society

OPTICAL PROPERTIES OF MANGANESE DOPED AMORPHOUS AND CRYSTALLINE ALUMINUM NITRIDE FILMS.

M.L. Caldwell*, H.H. Richardson*, M.E. Kordesch**
*Department of Chemistry and Biochemistry, Ohio University, Athens, OH 45701,
richards@helios.phy.ohiou.edu
**Department of Physics and Astronomy, Ohio University, Athens, OH 45701,
kordesch@helios.phy.ohiou.edu

ABSTRACT

An aluminum nitride (AlN) film deposited on silicon (100) was used as the substrate for growing manganese (Mn) doped AlN film by metal organic chemical vapor deposition (MOVCD). The (15.78 μm) under layer of AlN was grown at 615°C at a pressure of 10^{-4} Torr. The (2.1 μm) top layer of Mn-AlN was grown at the same temperature and pressure but doped with pulse valve introduction of the manganese decacarbonyl (100 ms on, 100 ms off). The film was then characterized *ex situ* with IR reflectance microscopy, X-ray diffraction, scanning electron microscopy imaging, cathodoluminescence, and X-ray fluorescence. The IR reflectance measurements showed a strong (A_1) LO mode for AlN at 920 cm^{-1} and 900 cm^{-1} with a shoulder at 849 cm^{-1}. X-ray Diffraction yielded three diffraction peaks at a 2θ position of 33, 36 and 38 degrees corresponding to 100, 002, and 101 lattice planes respectively. Cathodoluminescence results show strong visible emitted light from incorporated manganese. The relative percentage of manganese to aluminum was below the detection limit (0.01 %) of the X-ray fluorescence spectrometer. Amorphous Mn doped AlN films have also been grown using a low temperature atomically abrupt sputter epitaxial system. The amorphous Mn doped AlN showed no cathodoluminescence.

INTRODUCTION

Crystalline aluminum nitride (AlN) films have attracted considerable interest because they possess a very large band gap compared to the rest of the group III nitrides (6.2 eV) [1]. In addition, it has a high heat conductivity, excellent chemical and thermal stability; it will decompose rather than melt at 2400°C [2]. Amorphous AlN films have also attracted some interest because the band gaps of these semiconductors are close to those of the corresponding crystalline materials (5.6 eV vs 6.2 eV), with localized electronic states that do not extend into the bandgap. The amorphous AlN films have many of the desirable qualities of the crystalline materials, most importantly, they are not easily recrystallized. Recently, rare earth doped III-V semiconductors have been the subject of great interest because of their strong visible luminescence [3,4]. The films have been doped with rare earth elements, and light emission from doped crystalline and amorphous AlN in both the visible and infrared has been confirmed [5,6]. Manganese has been incorporated in powder samples of AlN [7-9], with the tetravalent manganese ion, Mn^{4+}, occupying tetrahedral sites. The incorporation of manganese into semiconductor hosts is of interest because the manganese activated AlN exhibits a maximum in the emission spectrum in the red region (~600 nm).

EXPERIMENTAL

Crystalline aluminum nitride (AlN) films were grown by metal organic chemical vapor deposition (MOCVD) in a high vacuum stainless steel reaction chamber. This system consists of a growth chamber, pumping unit, and a gas inlet. An Alcatel corrosive resistant turbo pump evacuates the chamber into the 10^{-5} Torr range. The source gases for the AlN were ammonia (NH_3) and trimethylaluminum (TMA). The gases were maintained in the chamber at a ratio of 2.5:1, respectively. Silicon (Si) (100) substrates, 5 mm. by 15 mm., were mounted onto a boron nitride ceramic heater and held in place by a molybdenum metal holder. A nichrome thermocouple was attached to the holder to determine the temperature during the deposition period. Pressures in the chamber were monitored with an ion gauge. The Si(100) substrates were flashed above 1050°C to remove the surface oxide coating. The substrates were then kept at a constant temperature around 850°C. A stoichiometric ration of 2.5:1 NH_3 to TMA was used for growth and manganese decacarbonyl, ($[Mn(CO)_5]_2$), the dopant source, was introduced into the flow reactor via a pulse valve.

Amorphous AlN-Mn thin films were sputtered onto quartz substrates using a RF sputter system. The target was an aluminum disk with an Al/Mn pellet pressed into the disk. Typical growth parameters were 8×10^{-4} torr N_2, 240 W rf power and approximately 0.6 Å/ second growth rates.

These films were characterized *ex situ* with IR reflectance microscopy, SEM imaging, X-ray fluorescence (XRF), X-ray diffraction (XRD), and cathodoluminescence (CL). An imagining fiber optic probe (boroscope) was placed ~5 mm away from the sample. Cathodoluminescence from the samples were collected by the fiber and focused onto a CCD camera. The cathodoluminescence spectrum was collected with a lock-in amplifier, dual grating monochromator and photomultiplier tube.

RESULTS

Sample 211 consists of a 2.1 μm Mn doped AlN film above of a 15.78 μm undoped AlN layer. A molybdenum mask was placed on half of the sample when the overlayer was grown. An image of sample 211, taken with the fiber optic probe, along with the schematic diagram of the sample is shown in Figure 1. IR reflectance microscopy was used to identify the thickness and chemical composition of the different layers. The entire sample (labeled A in figure 1) has an overall thickness of 17.88 μm while the pure AlN layer (labeled B) has a thickness of 15.78 μm. The IR reflectance measurements showed a strong (A_1) LO mode for AlN at 920 cm^{-1} and 900 cm^{-1} with a shoulder at 849 cm^{-1}. The relatively fast growth rates (3-5 μm per hour) produces multiple nucleation centers that lead to a very rough surface. X-ray Diffraction yielded three diffraction peaks at a 2θ position of 33, 36 and 38 degrees corresponding to 100, 002, and 101 lattice planes respectively. Films can be grown with the c-axis exclusively either parallel or perpendicular to the substrate but this particular film has texture both parallel and perpendicular to the substrate. The elemental composition of the film was examined with x-ray fluorescence (EDX). The amount of manganese in sample 211 was below the detection limit (0.01 %) of the X-ray fluorescence spectrometer.

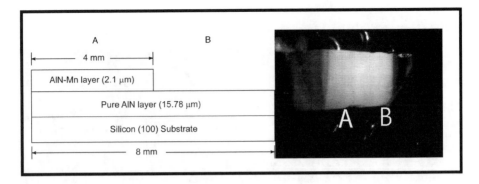

Figure 1. (left) Schematic diagram (side view) and (right) picture taken inside the CL chamber with the boroscope of sample 211.

This sample showed strong cathodoluminescence. A reverse intensity image of the cathodoluminescence taken with the boroscope (2.8 keV beam excitation) is shown in figure 2. The luminescence could be observed through a CaF_2 window and the region labeled A appeared yellow-orange to the eye while the region labeled B appear blue. The yellow-orange luminescent region is from the Mn doped overlayer while the blue luminescent region is from the "pure" AlN underlayer. The distance across the sample is ~8 mm and the dark stripe separating region A from B is ~250 μm. The topography of the sample is observable in figure 2. The oriented growth parallel to the substrate is observed as lines across the sample and epitaxial growth can be inferred from the image because the lines transverse across the entire sample including the region where the overlayer is grown on top of the underlayer.   A magnification of a region of figure 2 is shown in figure 3 (real image) where the topography of the sample is easily discernible.

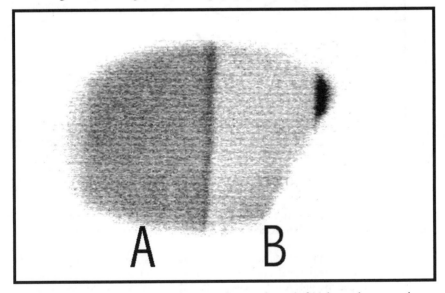

Figure 2. The cathode luminescence image of sample 211shows the two regions.

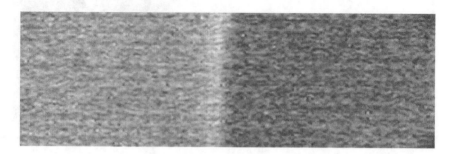

Figure 3. The real intensity CL image of sample 211 taken with the boroscope. The white vertical line in the center of the image is ~250 µm in width.

The CL spectrum taken from different regions of sample 211 is shown in figure 4. The spectrum of the entire image shown in figure 2 is presented in the top frame of figure 4. Three bands are observed. The bands are at 420 nm, 470 nm and 600 nm. The CL spectrum of region A (overlayer) is shown in the middle frame of figure 4. A band at 600 nm dominates the overlayer region (region A) with only minor intensity in the 420 nm band and 470 nm band. The underlayer region (region B) has two intense bands at 420 nm (blue) and 470 nm (blue-green) with a weaker band at 600 nm (red). The activation of Mn in AlN corresponds to a suppression of the 420 nm and 470 nm band with a corresponding increase in the 600 nm band.

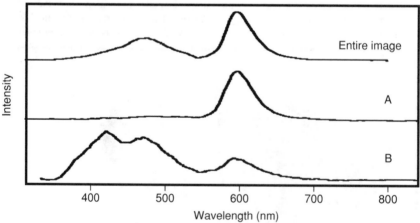

Figure 4. CL spectra of sample 211 from the entire image shown in figure 2, from the overlayer of AlN-Mn (A) and from the underlayer of pure AlN (B).

A previous study of Mn doped single crystals of AlN showed absorption bands in the 400 to 600 nm region and an emission band around 600 nm [8]. A theoretical analysis of the spectroscopic results suggested that Mn^{4+} ions are incorporated in tetrahedral lattice sites [8]. The energy level diagram derived from these spectroscopic results is shown in figure 5. The CL bands shown in Figure 4 agrees with the energy level diagram, presented in figure 5, for single crystal AlN doped with Mn. SIMS analysis on a different sample that has similar CL to sample 211 showed a higher level of Mn incorporation than any background impurity [5].

For cathodoluminescence, the direct gap excitation of AlN is accomplished with bombardment of the sample with 2.8 keV electrons. This energy is then transferred into the manganese ions incorporated into the lattice structure. The CL spectrum from the activated Mn sample (region A) is dominated by the lowest energy transition (600 nm). This implies that there is either a very effective energy transfer mechanism from higher to lower electronic states in activated Mn/AlN or that only energy transfer from the AlN populates the lowest electronic state of Mn. Region B must have small amount of Mn incorporated in the AlN because it is possible to observe cathodoluminescene. No cathodoluminescence is observed for thin films of pure AlN. Presumably, the concentration of Mn in region B is significantly less than that of region A. Observation of strong CL bands at 420 nm and 470 nm indicates that either the excited electronic states of Mn in this region must be comparable or that there is efficient energy transfer into all the upper electronic states of Mn.

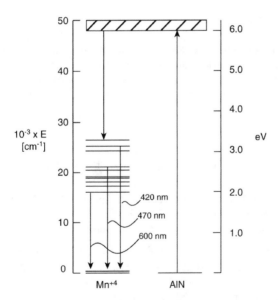

Figure 5. Energy levels for manganese centers in AlN, taken from F. Karel, J. Mareš, Czech. J. Phys. B 22, 847 (1972).

The CL properties of Mn incorporated into amorphous AlN was investigated. Amorphous AlN thin films (~5000 Å in thickness) incorporated with varying amounts of Mn (0.5% to 4%) were grown with RF sputtering at room temperature. Unfortunately, the amorphous AlN samples showed little or no cathodoluminescence. It is possible that the manganese is incorporated in the interstitial cavities of the lattice and energy transfer from the AlN lattice to the Mn is very inefficient. High temperature annealing of amorphous AlN does not significantly alter the properties of the film and a temperature region might exist where Mn can be activated (movement from interstitial to lattice sites) but the AlN film remains unchanged.

CONCLUSIONS

The incorporation of a small amount of manganese (<0.1%) in crystalline AlN gives rise to three emission bands at 420 nm, 470 nm, and 600 nm in the CL spectrum. These emission wavelengths correspond to previously published excitation and emission data for crystalline AlN films doped with manganese. The manganese activated crystalline AlN layer shows a strong emission band at 600 nm with minor intensity bands at 420 nm and 470 nm. The unactivated Mn crystalline AlN layer shows intnese bands at 420 nm, and 470 nm with a weaker band at 600 nm. The activation of Mn in crystalline AlN corresponds to a suppression of the 420 nm and 470 nm band with a corresponding increase in the 600 nm band. The amorphous AlN films incorporated with varying amounts of Mn showed little or no cathodoluminscence.

ACKNOWLEDGEMENTS

This work is supported by a BMDO URISP grant N00014-96-1-0782 entitled "Growth, Doping and Contacts for Wide Band Gap Semiconductors" and grant N00014-99-1-0975 entitled "Band-Gap Engineering of the Amorphous In-Ga-Al-Nitride Semiconductor Alloys for Luminescent Devices on Silicon from the Ultraviolet to the Infrared".

REFERENCES

1. F. Hasegawa, T. Takahashi, K. Kubo, Y. Nannichi, Jpn. J. Appl. Phys. 26, 1555 (1987).

2. D. V. Tsvetdov, A. S. Zubrilov, V. I. Nikolaev, V. A. Soloviev, V. A. Dmitriev, MRS Internet J. Nitride Semicond. Res. 1, 35, (1996).

3. A. J. Steckl, R. Birkhahn, Appl. Phys. Lett. 73, 1700 (1998).

4. H. J. Lozykowski, W. M. Jadwisienczak, Appl. Phys. Lett. 74, 1129 (1999).

5. R. C. Tucceri, M. L. Caldwell, H. H. Richardson, Mat. Res. Soc. Symp. Proc. 572, 413 (1999).

6. Gurumurugan et al., Appl. Phys. Lett. 74, 3008 (1999).

7. F. Karel, J. Pastrnák, Czech. J. Phys B. 19, 79 (1969).

8. F. Karel, J. Mares, Czech. J. Phys B. 19, 79 (1969).

9. A. D. Serra, N. P. Magtoto, D. C. Ingram, H. H. Richardson, Mat. Res. Soc. Symp. Proc. 482, 179 (1997).

Mat. Res. Soc. Symp. Vol. 595 © 2000 Materials Research Society

A STUDY OF THE EFFECT OF V/III FLUX RATIO AND SUBSTRATE TEMPERATURE ON THE In INCORPORATION EFFICIENCY IN In_xGa_{1-x}/GaN HETEROSTRUCTURES GROWN BY RF PLASMA-ASSISTED MOLECULAR BEAM EPITAXY

M. L. O'Steen, F. Fedler[a], R. J. Hauenstein
Dept. of Physics, Oklahoma State University,
Stillwater, OK 74078, U.S.A.

ABSTRACT

Laterally resolved high resolution X-ray diffraction (HRXRD) and photoluminescence spectroscopy (PL) have been used to assess In incorporation efficiency in In_xGa_{1-x}N/GaN heterostructures grown through rf-plasma-assisted molecular beam epitaxy. Average alloy composition over a set of In_xGa_{1-x}N/GaN superlattices has been found to depend systematically upon both substrate temperature (T_{sub}) and V/III flux ratio during growth. A pronounced thermally activated In loss (with more than an order-of-magnitude decrease in average alloy composition) is observed over a narrow temperature range (590–670°C), with V/III flux ratio fixed. Additionally, the V/III flux ratio is observed to further strongly affect In incorporation efficiency for samples grown at high T_{sub}, with up to an order-of-magnitude enhancement in In content despite only a minor increase in V/III flux ratio. PL spectra reveal redshifts as In content is increased and luminescence efficiency which degrades rapidly with decreasing T_{sub}. Results are consistent with In loss arising from thermally activated surface segregation + surface desorption processes during growth.

INTRODUCTION

Efficient In incorporation into epitaxial In_xGa_{1-x}N material is a widely-known but not well-understood problem for molecular beam epitaxial (MBE) and metalorganic chemical vapor deposition (MOCVD) growth methods[1–4], deriving fundamentally from the immiscibility in thermodynamic equilibrium of the InN-GaN system[5]. Recently, several papers[1,4,6–9] have begun to provide insights into various microscopic growth processes which might play a role in limiting In incorporation[1,7–9]. In this paper, we present a study which further elucidates the mechanisms of In loss by examining In incorporation efficiency in a set of In_xGa_{1-x}N/GaN superlattice samples grown by rf plasma-assisted molecular beam epitaxy (RF-MBE). The effect of both substrate temperature and incident V/III flux ratio on In incorporation efficiency is studied through the use of laterally resolved high resolution X-ray diffraction (HRXRD) and photoluminescence (PL) spectroscopy. More than an order-of-magnitude reduction of incorporated In is observed in the narrow temperature range 590°C–670°C as the result of strong thermally activated In-loss mechanisms. Additionally, at high growth temperatures, In incorporation efficiency is observed to increase by nearly an order of magnitude despite only a small increase in incident V/III flux ratio. Photoluminescence results indicate that optical quality of materials is improved through growth at temperatures above approximately 590°C and with higher V/III flux ratios.

EXPERIMENT

Two sets of $In_xGa_{1-x}N/GaN$ superlattices were grown in an SVT model 433R MBE system equipped with conventional solid sources for Ga and In, and an rf plasma source using ultrahigh purity N_2 gas to provide the flux of active nitrogen N*. Both metal sources were located diametrically opposite the N* source so as to provide a maximum gradient in N*/III flux ratio over a 2-in. wafer. The growth procedure is as follows: A thick GaN buffer layer ($\geq 1.2\mu m$) was deposited onto a nitrided Al_2O_3 (0001) substrate at a temperature of 750°C and a nominal growth rate of 0.8μm/h. Next, substrate temperature was lowered and Ga and N* fluxes were reduced for the reduced-temperature growth, as explained elsewhere[9]. Then, a 20-period superlattice was deposited with InGaN and GaN nominal layer thicknesses of 30Å and 200Å, respectively. N* flux was held constant throughout superlattice deposition, and substrates were *not* rotated during superlattice deposition so that dependencies on N*/III flux ratio could be examined. For the first set of superlattices, grown to identify the relevant growth phenomenology, incident fluxes were varied while T_{sub} was held nominally fixed (≈ 615°C). For the second set, T_{sub} was varied over the range, 540–670°C while all other growth parameters were held constant, with a nominal $In_xGa_{1-x}N$-layer composition of $x=30\%$. All samples were characterized with HRXRD through the use of a Philips Materials Research Diffractometer, and with low-temperature (10K) photoluminescence spectroscopy (PL) measurements using the 325-nm line of a HeCd laser as the excitation source.

RESULTS

A representative HRXRD θ-2θ scan from one of our $In_xGa_{1-x}N/GaN$ superlattices is presented in Fig. 1(a). A family of superlattice satellite peaks about the GaN (0002) symmetric Bragg reflection is observed. From a simple linear-regression analysis of superlattice peak positions [see inset of Fig. 1(a)] the superlattice period L and layer-averaged alloy compositions $\bar{x}$ are determined for each superlattice as a function of position on the wafer along the diagonal line (V/III flux-ratio gradient) depicted in Fig. 1(b). This latter figure also indicates the relevant source-substrate geometry during MBE growth. In addition to determining L and $\bar{x}$, it is straightforward to make an estimate of the separate (effective) GaN and InN growth rates as well by knowing in addition to the values of L and $\bar{x}$ the In-shutter duty cycle during MBE growth. Using our first sample set (fixed T_{sub}), lateral spatially resolved measurements of L, $\bar{x}$, and the GaN (InN) effective growth rates were used to examine qualitatively the salient phenomenological effects of N*/III ratio at fixed T_{sub}; an example is shown in Fig. 2(a)–(d). There we see overall trends as one moves from the near-metal to near-N* side of the wafer and for two different levels of N* flux (Φ_1 and $\Phi_2 > \Phi_1$). This corresponds to an increasing local N*/III ratio at fixed T_{sub}, moving from left to right in each figure. Of particular interest are the curves given in Fig. 2(c) and (d), three of which exhibit a maximum in growth rate (the fourth would have its maximum off the wafer). To the left (right) of each maximum we have growth rate limited by arrival of the N* (metal) species; the maximum itself corresponds evidently to the case of stoichiometric flux ratio. In contrast to the "1:1" flux ratio leading to optimal growth for GaN, $In_xGa_{1-x}N$ is often

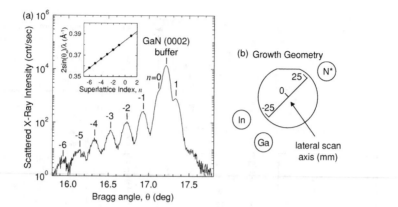

Figure 1. (a) Representative θ-2θ scan of a superlattice sample about the GaN (0002) diffraction peak. Inset: Linear regression analysis of the quantity $2\sin(\theta_n)/\lambda$ vs. superlattice peak index n, used to determine the superlattice period L and average Bragg-plane spacing $\overline{d}$. (b) Depiction of MBE growth geometry and in-plane axis used for laterally resolved X-ray measurements

claimed to need a significantly higher ratio (e.g., ~3:1)[6]. This is in evidence in Fig. 2(c) where excess N* conditions for GaN growth prevail over the entire wafer (Φ_2 case); thus, no GaN growth-rate maximum is observed. At the same time, the InN growth rate exhibits a clear maximum value near the center of the wafer. (Though not pursued here, in principle this method of laterally spatially resolved HRXRD analysis might perhaps be useful in facilitating a direct *quantitative* comparison of absolute N*/III flux ratios between the cases of GaN and InN growth.)

Figure 2(e) summarizes an important result: the strong dependence of In incorporation efficiency on MBE growth temperature at fixed N*/III flux ratio, and, a further strong dependence on flux ratio at high substrate temperatures. In the figure, the quantity $\overline{x}$ is plotted as a function of T_{sub} for each of three wafer positions (each corresponding to a particular N*/III flux ratio) for the superlattices in our second set (in which *only* T_{sub} was varied while all other growth parameters were carefully held constant from growth run to growth run). As seen from the figure, for fixed N*/III flux ratio, the dependence of In incorporation on T_{sub} as measured by $\overline{x}$ reveals two distinct regimes of behavior. At low substrate temperatures, $\overline{x}$ appears to saturate, and approaches the nominal value expected from the incident fluxes under conditions of unity sticking coefficient for both the Ga and In species. In contrast, at high temperature, the data provide clear evidence of a thermally activated In-loss mechanism. Over the range of data presented, the In incorporation efficiency drops by over an *order of magnitude* between the knee of the curve (~590°C for the growth rates used here) and the maximum T_{sub} value of this study (670°C). Additionally, Fig. 2(e) shows that even minor changes (± ~10%) in flux ratio may have a significant effect on the efficiency of In incorporation, especially at higher substrate temperatures. Evidently, slight alteration of the N*/III flux ratio (growth-surface stoichiometry) has a significant effect on the magnitude of any pertinent kinetic barrier(s) to In loss which may be present. Over the range of data

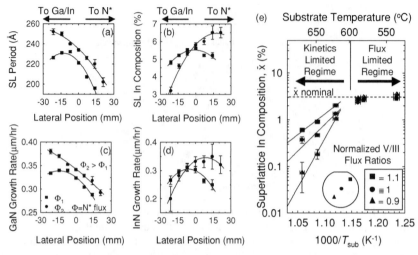

Figure 2. Spatial variations, due to local V/III flux ratio, in (a) superlattice period L, (b) mean alloy composition $\bar{x}$, (c) effective GaN growth rate, and (d) effective InN growth rate, for growth under two different N* fluxes. (e) Effect of T_{sub} on $\bar{x}$, for slightly differing V/III flux ratios. At low T_{sub}, $\bar{x}$ approaches the nominal value determined by incident fluxes while at high T_{sub}, $\bar{x}$ is limited via rate kinetics of thermally activated In-loss mechanisms (surface segregation + surface desorption), which freeze out here at $\approx$ 590°C. The kinetic barrier to In loss is extremely sensitive to N*/III ratio (growth-surface stoichiometry).

shown, the (effective) activation energies for the high-temperature portions range from 0.8–2.7eV.

A possible kinetic pathway for the loss of In during MBE growth may lie in thermally activated processes of In surface-segregation in combination with surface desorption. Previously, we have observed and have quantitatively modeled[10,11] a similar behavior for the case of N during plasma-assisted MBE growth of GaAsN/GaAs multilayers. That a strong alloy segregation should fundamentally occur is not surprising, given the predicted large miscibility gap of the (In)GaN ternary system[5], and reports by several groups of segregation and desorption of In during growth[1,4,6,7–9]. The observed decrease in In composition with increasing substrate temperature is clearly consistent with such a surface-segregation + surface-desorption mechanism. Moreover, the observed effect of the V/III flux ratio on $\bar{x}$ is consistent with this In-loss mechanism. For example, one possible effect of an increased V/III flux ratio is to decrease the steady-state surface density of Ga and In species in the first one (or two) surface adlayers[8], thereby reducing the surface mobility of Ga and In and hence slowing the rate In removal from the first subsurface layer via In-for-Ga exchange with Ga adatoms; this is consistent with recent reports by other groups[7,8]. Alternatively (or additionally), another possible consequence of the less metal-rich surface environment may be to increase the surface residence time of In, thus decreasing the likelihood of In loss to evaporation, and, provided that the surface In does not desorb on a monolayer-deposition time scale (nor re-segregate to the surface on the same time scale), increasing the likelihood of incorporation into the bulk.

Low temperature PL measurements from the second set of samples are also consistent with this segregation/desorption interpretation and provide insight into the effect of growth temperature and V/III flux ratio on the optical quality of the samples. The observed PL spectra from these samples are single, broad peaks, often with indications of Fabry-Perot resonances. Both the PL intensity and peak positions are strongly affected by the substrate temperature and incident V/III flux ratio. At temperatures below approximately 590°C, the luminescence intensity decreases rapidly; at the lowest temperature of this study (540°C), no PL spectra could be observed. The effect of the N*/III flux ratio on the PL spectra is illustrated in Fig. 3(a). The PL spectra are observed to shift to lower energy and the PL intensity is observed to increase for higher N*/III flux ratios (over the limited range of flux ratios used in this study). Fig. 3(b) shows the PL peak position as a function of the lateral position on the wafer (N*/III flux ratio) for temperatures above 590°C. Consistent with the segregation/desorption interpretation, the PL peak position is observed to move to lower energy as either the substrate temperature is decreased, or as the N*/III ratio is increased. In general, the measured PL peak positions appear at lower energies than expected based on estimates using reported[12] band-bowing parameters and the In compositions determined by HRXRD. This is most likely attributable to unintentional inhomogeneity within the InGaN layers. Lastly, a higher optical quality (luminescence efficiency) is observed for higher V/III flux ratios and higher growth temperatures.

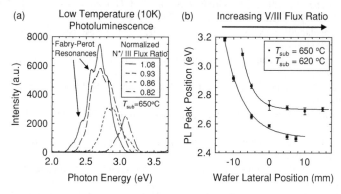

Figure 3. (a) Effect of incident N*/III flux ratio on photoluminescence spectra. Spectra are observed to shift to lower energies as the N*/III flux ratio increases. In general, a substantial improvement in optical quality (luminescence intensity) is observed upon going from flux-limited to kinetically-limited growth regime (see text).

SUMMARY

In$_x$Ga$_{1-x}$N/GaN superlattices structures have been grown by RF-MBE methods to study the effects of substrate temperature and incident V/III flux ratio on In incorporation efficiency. Below a critical substrate temperature ($\approx$ 590°C for growth rates on the order of 0.3μm/h), In incorporation is efficient but material quality as assessed by photoluminescence is very poor. Above this temperature, In incorporation efficiency is found to decrease sharply according to a Boltzmann factor, with an activation energy value which is affected by the N*/III flux ratio during growth. At fixed flux ratio, In incorporation efficiency decreases by over an order of magnitude between 590°C and 670°C, and at the latter temperature, a flux-ratio increase of only ~20% results in nearly an order of magnitude increase in incorporation efficiency. Optical quality markedly improves above the critical temperature, and to a lesser extent at high temperature improves with increasing flux ratio as well. Results are consistent with thermally activated processes of In surface segregation and surface desorption as the microscopic mechanism of In loss during MBE growth.

ACKNOWLEDGMENTS

The authors gratefully acknowledge the support of the Army Research Office under Grant No. DAAH04-94-G-0393. Additionally, one of us (M.L.O.) wishes to acknowledge financial support provided by the NASA Oklahoma Space Grant Consortium and the DoE EPSCoR Traineeship Program.

(a) Present Address: Laboratorium für Informationstechnologie, Universität Hannover, Schneiderberg 32, D-30167 Hannover, Germany.

REFERENCES

1. T. Böttcher, S. Einfeldt, V. Kirchner, S. Figge, H. Heinke, D. Hommel, H. Selke, and P. L. Ryder, Appl. Phys. Lett. **73**, 3232 (1998).
2. D. Doppalapudi, S. N. Basu, K. F. Ludwig Jr. and T. D. Moustakas, J. Appl. Phys. **84**, 1389 (1998).
3. R. Singh, D. Doppalapudi, T. D. Moustakas, and L. T. Romano, Appl. Phys. Lett. **70**, 1089 (1997).
4. B. Yang, O. Brandt, B. Jenichen, J. Müllhäuser, and K. H. Ploog, J. Appl. Phys. **82**, 1918 (1997).
5. I.-H. Ho and G. B. Stringfellow, Appl. Phys. Lett. **69**, 2701 (1996), and references therein.
6. J. R. Müllhäuser, O. Brandt, A. Trampert, B. Jenichen, and K. H. Ploog, Appl. Phys. Lett. **73**, 1230 (1998).
7. F. Widmann, B. Daudin, G. Feuillet, N. Pelekanos, and J. L. Rouvière, Appl. Phys. Lett. **73**, 2642 (1998).
8. H. Chen, A. R. Smith, R. M. Feenstra, D. W. Greve, and J. E. Northrup, MRS Internet J. Nitride Semicond. Res. **S1**, G9.5 (1999).
9. M. L. O'Steen, F. Fedler, and R. J. Hauenstein, Appl. Phys. Lett. **75**, 2820 (1999).
10. R. J. Hauenstein, D. A. Collins, X. P. Cai, M. L. O'Steen, and T. C. McGill, Appl. Phys. Lett. **66**, 2861 (1996).
11. Z. Z. Bandic, R. J. Hauenstein, M. L. O'Steen, and T. C. McGill, Appl. Phys. Lett. **68**, 1510 (1996).
12. M. D. McCluskey, C. G. Van de Walle, C. P. Master, L. T. Romano, and N. M. Johnson, Appl. Phys. Lett. **72**, 2725 (1998).

Mat. Res. Soc. Symp. Vol. 595 © 2000 Materials Research Society

Surface Morphology of GaN: Flat versus Vicinal Surfaces

M.H. Xie, S.M. Seutter, L.X. Zheng, S.H. Cheung, Y.F. Ng, Huasheng Wu, S.Y. Tong
Department of Physics, The University of Hong Kong, Pokfulam Road, Hong Kong

ABSTRACT

The surface morphology of GaN films grown by molecular beam epitaxy (MBE) is investigated by scanning tunneling microscopy (STM). A comparison is made between flat and vicinal surfaces. The wurtzite structure of GaN leads to special morphological features such as step pairing and triangularly shaped islands. Spiral mounds due to growth at screw threading dislocations are dominant on flat surfaces, whereas for vicinal GaN, the surfaces show no spiral mound but evenly spaced steps. This observation suggests an effective suppression of screw threading dislocations in the vicinal films. This finding is confirmed by transmission electron microscopy (TEM) studies. Continued growth of the vicinal surface leads to step bunching that is attributed to the effect of electromigration.

INTRODUCTION

Current intensive experimentation on III-V nitrides has led to rapid progress in blue light and high power device applications [1]. These achievements contrast greatly an apparent lack of good understanding of many fundamental issues related to growth and film properties of nitrides. To further improve the quality of epitaxial thin films and hereafter the performance of nitride-based optoelectronic and microelectronic devices, a better knowledge of its growth kinetics and surface dynamics is obviously needed.

In a non-equilibrium growth system such as molecular beam epitaxy (MBE), surface morphology contains important kinetic and dynamic information related to the growth process [2]. Therefore, by studying the morphological evolution of surfaces during MBE growth under various conditions, key kinetic parameters governing growth can be derived.

This paper presents recent observations of many novel morphologies during growth of GaN(0001) on SiC(0001) substrates without the use of buffer layers. Both nominally flat and vicinal substrates are used. In addition to the strong anisotropy in growth rates of two types of surface steps, step bunching during vicinal film growth is also observed. Furthermore, for a vicinal film, there is no spiral mound, which is the dominant feature for a flat surface.

EXPERIMENTS

MBE growth and scanning tunneling microscopy (STM) experiments of GaN films are conducted in a multi-purpose ultrahigh vacuum (UHV) system, with background pressures in the range of 10^{-11} mbar. The MBE system is equipped with a conventional effusion cell for gallium (Ga) and a radio-frequency plasma generator for N_2. For all the GaN growth reported in the paper, the Ga source temperature is set at 980°C, and the N_2 flow rate at 0.13 sccm (standard cubic centimeter per minute). The power of the plasma unit is set at 500W. These conditions correspond to an III/V flux ratio of approximately 2 and a GaN growth rate of 0.26 Å s^{-1} [3]. The substrates are nominally flat 6H-SiC(0001)

(Cree Research, Inc.) and vicinal 4H-SiC(0001) (Nippon Steel, Co.) misoriented towards [-1010] by 3.5°. Prior to GaN deposition, the substrates are deoxidized at 1100°C in a Si flux supplied from an *e*-beam evaporator. The chemical purity of the treated surface is monitored by *in situ* X-ray photoelectron spectroscopy (XPS), while the atomic order of the surface is analyzed by *in situ* low energy electron diffraction (LEED). Under a properly chosen Si flux condition, the SiC surfaces show the $(\sqrt{3} \times \sqrt{3})R30°$ surface reconstruction following the deoxidization process. STM images of the surface show atomically flat terraces of more than 1000 Å wide for the flat substrate and a step-terrace morphology for the vicinal substrate. GaN growth is initiated by simultaneously opening the source shutters of Ga and N and at substrate temperatures between 500°C and 650°C. Sample heating is achieved by passing through it a direct current (DC) along the long side of the rectangular substrate pieces (11 × 4 mm^2 in size).

To terminate growth, the source shutters are closed simultaneously and at the same time, the power of the plasma unit is switched off. Meanwhile, the sample is quenched rapidly by switching off the DC heating power. Due to the small sample size, the temperature drops quickly to below 400°C within seconds. The quenched surface is subsequently examined by *in situ* STM. Constant current mode STM experiments are conducted at room temperature at a tunneling current of 0.1nA and sample bias of -3V.

RESULTS AND DISCUSSIONS

GaN growth on both flat and vicinal substrates commences by three-dimensional (3D) island formation as indicated by reflection high-energy electron diffraction (RHEED) patterns. This is followed by a transition to two-dimensional (2D) growth after coalescence of 3D islands. It is noted that such transition from 3D to 2D growth is sooner for the vicinal film than flat ones. Continued growth on the two substrates leads to completely different morphologies, which is the subject of the following discussion.

Flat Surface

Fig. 1(a) shows a GaN surface following 1μm deposition at 650°C. The main observation is the dominance of spiral mounds on the surface. The density of such mounds is estimated from the image to be 2×10^9 cm^{-2}. As growth spirals originate from screw or mixed-type threading dislocations in the film [4], it is concluded that the density

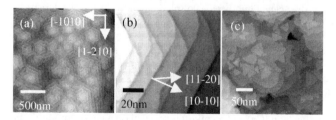

Figure 1. *STM images of GaN surfaces following deposition on flat SiC(0001) substrates at (a) and (b) 650°C, (c) 500°C. Note the scale difference between (a) and (b).*

of screw threading dislocations in the flat film is of the order of 2×10^9 cm^{-2}, which has been confirmed by transmission electron microscopy (TEM) investigations [5]. Similar morphologies have also been reported by others [6,7], therefore, it seems that it is characteristic to MBE-grown GaN films on flat substrates. Note that if a pair of growth spirals originates from growth at two vertical screw arms of a 'U'-shaped dislocation line, then the spirals in the pair ought to have opposite rotation direction. Indeed, both clockwise and anti-clockwise rotation spirals are seen.

Zooming-in one of the spiral mounds reveals a special step structure, as shown in fig. 1(b). Along one of the <10-10> crystallographic directions, steps are all double bilayer (BL) high, i.e., it equals $c = 5.2$ Å, the height of the unit cell of GaN in the [0001] direction. In between the two adjacent double BL steps, along the [11-20] direction, the steps are de-paired into single BL and a twilled structure results. Such a step structure is understood by considering the dangling bond characteristics of a wurtzite film [8]. For such a film's surface, there are two types of steps, namely type-A and type-B steps. For a type-A step, each edge atom has two dangling bonds whereas for type-B steps, the edge atom has only one dangling bond. Therefore, for a type-A step, there is a high density of kinks whose advance is limited by the arrival rate of adatoms by diffusion. On the other hand, for a type-B step, there are few kinks, as the creation of a kink site on such a step will expose the high-energy type-A step. As a result, the motion of a type-B step may involve nucleation of one-dimensional (1D) islands along the step edge, therefore, its speed of growth is reduced compared to the growth of a type-A step. For a wurtzite film, consecutive steps along a given <10-10> direction belong to type-A and type-B alternately, therefore, the fast-growing type-A step will ultimately catch up with the slower-growing type-B step underneath, and step-pairing occurs. On the same terrace, a step changes its character from type-A to type-B or vice versa upon turning 60°. This explains the formation of the twilled structure of single BL steps along the [11-20] direction.

Under the island nucleation growth mode at low substrate temperatures, e.g., 500°C, 2D single BL height islands form on terraces that show triangular shape, as depicted in fig. 1(c). The triangular shape is again due to the growth anisotropy of GaN. The fast-growing type-A step ultimately makes this step disappear and the island becomes bounded by the slower-growing, low energy, type-B steps. Unexpectedly, from the image of fig. 1(c), triangular islands oriented 180° with respect to each other on the same terrace are seen. To explain this, we assume atoms in some of the islands are wrongly stacked. For example, instead of the wurtzite ABAB... stacking, atoms take the cubic ABCA... stacking instead. The oppositely oriented triangular islands can also originate from twins in a pure cubic film, i.e., one with ABCA... and the other with ACBA... stacking. It has been shown by TEM studies that low growth temperature promotes cubic film formation and there is a high density of stacking faults and twin boundaries in such a film [9]. Therefore, the assertion of the 180°-oriented islands to stacking faults and twins is likely to be correct.

Vicinal Surface

Fig. 2 shows a GaN surface following a 0.5μm film deposition at 650°C on a vicinal SiC substrate. The most remarkable difference between this surface and the flat one (fig. 1(a)) is the absence of spiral mounds. As described earlier, spiral mounds reflect growth at screw threading dislocations. Therefore, the absence of spirals in fig. 2 implies a

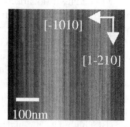

Figure 2. *STM image of a GaN film grown on a vicinal 4H-SiC(0001) substrate at 650°C. Note the absence of spiral mounds.*

[-1010]

[1-210]

100nm

suppression of threading screw dislocations in the vicinal film. This suggestion has been confirmed by TEM studies, which reveal a density of screw threading dislocation in the vicinal film of the order of 10^7 cm^{-2}. This is two-orders of magnitude less than that in the flat film. In addition to screws, TEM images also suggest a reduction of threading edge dislocations. Therefore, judging by these results, the quality of the vicinal epitaxial film is substantially improved. The cause of dislocation suppression is still under investigation, preliminary results indicate that it is likely related to the initiation and coalescence processes of islands when GaN is initially grown on SiC. Examining the individual steps reveals they are all double BL high, in agreement with the growth anisotropy described earlier.

Let's now consider the implication of the above step structure to step kinetics during MBE growth. Referring to the 1D model of a vicinal GaN surface as shown in fig. 3, X_A and X_B denote the positions of steps A and B belonging to type-A and type-B, respectively. Let W be the terrace width bounded by the two steps, while W_+ and W_- are the terrace widths in front of and behind the B and A steps, respectively (see fig. 3). If $k_{A,B}^{\pm}$ are adatom incorporation coefficients to steps from their respective upper (-) and lower (+) terraces, then the speeds of motion of the steps under the step-flow growth mode can be written as [10]:

$$
\frac{\partial X_A}{\partial t} = k_A^- W_- + k_A^+ W
$$

$$
\frac{\partial X_B}{\partial t} = k_B^- W + k_B^+ W_+
$$

(1)

In writing down equ. (1), we have ignored the lateral fluctuations of steps. The two terms on the right-hand side of the equation reflect contributions from the upper and lower terraces, respectively. Subtract the two equations and note that $W = X_B - X_A$, one obtains

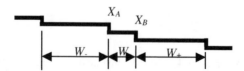

Figure 3. *A 1D model of a vicinal GaN surface. The consecutive steps belong to type-A and type-B, respectively, as required by the wurtzite crystal structure.*

W3.29.4

$$\frac{\partial W}{\partial t} = k_B^+ W_+ - k_A^- W_- + (k_B^- - k_A^+)W \tag{2}$$
$$\approx k_B^+ W_+ - k_A^- W_-$$

where we have used an approximation of $W \approx 0$ in deriving the second equality. It is valid when step-pairing has occurred (i.e., A step has caught up with the B step) and so there is no terrace between the two steps. Note further that statistically, $W_+ = W_-$, as the two terraces are equivalent. Therefore,

$$\frac{\partial W}{\partial t} \propto k_B^+ - k_A^- \tag{3}$$

In order to maintain the double BL step (DBS) structure during growth, one requires $\frac{\partial W}{\partial t} \le 0$, which leads to $k_B^+ \le k_A^-$. This mean that adatom capture by a type-A step from its upper terrace (k_A^-) is at least as efficient as adatom capture by a type-B step from its lower terrace (k_B^+).

Once a DBS has formed, which constitutes an A step on top and a B step below, the motion of such a DBS 'unit' is determined by adatom incorporation from the upper terrace behind step A and from the lower terrace in front of step B. If, as argued above, the rate of incorporation of adatoms from the upper terrace is higher than that from the lower terrace, continued growth will lead to further step bunching [10]. A complication comes, however, from the fact that adatom incorporation at a DBS may be different than that of an adatom diffusing down a step. At a DBS, adatoms need to cascade two BL height steps, A and B, in order to be successfully incorporated in the step. On the other hand, step bunching following continued growth of the vicinal film has indeed been observed [11]. An example of a bunched surface is shown in fig. 4, which is obtained after GaN deposition for another 2.5µm on top of the surface given in fig. 2. It is further shown that the bunching behavior depends on the DC direction that is applied across the sample in order to maintain the substrate temperature. Step bunching occurs only when the electric field points to the step down direction of the vicinal surface. When the field is reversed, the bunched steps tend to debunch. This observation, therefore, points to the electromigration effect [11,12]. If so, according to the field dependence, we infer that adatoms, which are likely N in this case when the growth is under Ga-rich condition, possess effective positive charges. At the moment, it is not clear whether step bunching is

Figure 4. STM image showing step bunching of the vicinal GaN film following prolonged growth at $650^\circ C$ and heated by a DC along the step-down direction (from left to right).

caused solely by the electromigration effect, or it is also caused by a higher rate of incorporation of adatoms from the upper terrace compared to that from the lower terrace. More studies are needed to separate out these two effects.

CONCLUSIONS

We have summarized our observations of surface morphology of GaN during its MBE growth. A comparison is made between the flat and vicinal surfaces, where the former shows spirals mounds due to screw threading dislocations, contrasting to vicinal films where there is no spiral mound. This suggests an effective suppression of threading dislocations. Furthermore, strong growth anisotropy and step bunching are observed, which can have important implications in the design of nitride-based device structures.

ACKNOLEDGEMENTS

This work is supported in part by HK RGC grants No. HKU7118/98P, 7117/98P, 7142/99P and 260/95P, US DOE grant No. DE-FG02-84ER 45076, US NSF grant No. DMR-9972958. The vicinal 4H-SiC substrate is kindly provided by Dr. N. Ohtani of Nippon Steel Corporation, Japan.

REFERENCES

1. S. Strite and H. Morkoc, *J. Vac. Sci. Technol.*, **B10**, 1237 (1992)
2. Z. Zhang and M. G. Lagally (eds.), *Morphological Organization in Epitaxial Growth and Removal*, World Scientific, Singapore, 1998
3. S. M. Seutter, M. H. Xie, W. K. Zhu, L. X. Zheng, Huasheng Wu, and S. Y. Tong, *Surf. Sci. Lett.*,(in press)
4. W. K. Burton, N. Cabrera, and F. C. Frank, *Phil. Trans. Roy. Soc.*, **243**, 299 (1951)
5. M. H. Xie, L. X. Zheng, Y. F. Ng, Huasheng Wu, N. Ohtani, and S. Y. Tong, *unpublished*
6. B. Heying, E. J. Tarsa, C. R. Elsass, P. Fini, S. P. DenBaars and J. S. Speck, *J. Appl. Phys.*, **85**, 6470 (1999)
7. A. R. Smith, R. M. feenstra, D. W. Greve, M. S. Shin, M. Skowronski, J. Neugebauer and J. E. Northrup, *J. Vac. Sci. Technol.*, **B16**, 2242 (1998)
8. M. H. Xie, S. M. Seutter, W. K. Zhu, L. X. Zheng, Huasheng Wu, and S. Y. Tong, *Phys. Rev. Lett.*, **82**, 2749 (1999)
9. M. H. Xie, L. X. Zheng, S. Y. Tong, *unpublished*
10. P. Bennema, and G. H. Gilmer, in *Crystal Growth: An Introduction* (ed. P. Hartman), North-Holland, 1973, p263
11. M.H. Xie, S.H. Cheung, L.X. Zheng, Y.F. Ng, Huasheng Wu, N. Ohtani, and S.Y. Tong, *unpublished*
12. A. V. Latyshev, A. L. Aseev, A. B. Krasilnikov, and S. I. Stenin, *Surf. Sci.*, **213**, 157(1989)

Mat. Res. Soc. Symp. Vol. 595 © 2000 Materials Research Society

GROWTH OF InN BY MBE

W.-L. Chen*, R. L. Gunshor*, Jung Han**, K. Higashimine***, and N. Otsuka***
*School of Electrical and Computer Engineering, Purdue University,
West Lafayette, IN 47907
**Sandia National Laboratories, Albuquerque, NM 87185
***Japan Advanced Institute of Science and Technology, Ishikawa, Japan

Abstract

A series of experiments were performed to explore the growth of InN by Molecular Beam Epitaxy (MBE). The growth conditions were optimized based on the study of RHEED during growth and InN dissociation experiments. Characterization of the InN thin films were performed by various techniques such as TEM and XRD.

Introduction

InN containing compounds such as the GaInN ternary are of importance for the fabrication of light emitting devices [1-3]. However, phase separation is known to prevent the homogeneous incorporation of high In-content GaInN for longer wavelength applications [4-5]. The behavior of pure InN growth might help to shed light on the mechanism of the formation of high concentration of InN in the ternary. InN has been prepared by various techniques [6-13]. However, InN normally shows poor optical properties, a high background carrier concentration and poor crystalline properties with indium clusters found from the x-ray diffraction spectrum.

For the MBE technique, the most important growth parameters are the flux levels and growth temperature. In the study reported here, RHEED behavior during growth and InN thermal dissociation experiments following growth have been performed to optimize the growth parameters. The results of the characterization of InN films by various means are presented in this report.

Experiment

A Perkin-Elmer 430 MBE system was used for the growth of InN. In flux was measured by a crystal monitor located at the growth sample position. An RF nitrogen plasma cell (Oxford CARS-25) was used to supply the active nitrogen. The substrates used were 2-2.5 μm thick GaN epilayers grown on c-plane sapphire by MOCVD at Sandia National Labs and, a 1μm thick molybdenum layer was evaporated on the backside of the substrates to absorb the radiation from the filament. Substrates were screw mounted on an unbonded molybdenum block. Immediately before being introduced into the growth chamber, the substrates were preheated in a preparation chamber to 350°C for 10 minutes to reduce outgassing from the block when it was heated in the growth chamber.

The growth temperature was measured by a thermocouple on the back of the substrate and was calibrated by the eutectic point of gold-germanium at 356°C. The plasma cell was in general operated at 500W with a nitrogen flow rate of 0.8-1.2 sccm. The chamber pressure was in the $2.5\text{-}5\times10^{-5}$ Torr range during growth. The nominal active nitrogen flux under this condition was measured by GaN RHEED intensity oscillations under N-limited growth conditions. The GaN growth rate was limited by the nitrogen flux at substrate temperatures lower than 750°C, above which dissociation became a factor. The available active nitrogen flux was determined to be 3.7×10^{14} atoms/cm²/sec which was equivalent to 0.36 ml/sec for InN if all the active nitrogen were incorporated. In the experiments, the In flux ranged from 0.02 to 0.3 ml/sec.

During the growth, RHEED provided real time observation of the growth front. Depending on the fluxes and substrate temperature, InN showed three distinct RHEED patterns during growth. Fig.1 illustrates three different RHEED patterns in the <11-20> direction. In general higher N/In flux ratio and lower temperature results in pure 3D bright spots. At a lower N/In flux ratio and higher temperature, the RHEED pattern is a streaky 2-fold pattern. For conditions in between, the RHEED pattern shows both streaky 2-fold streaks and bright 3D spots at the same time.

 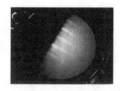 

Fig.1 Three RHEED patterns during the growth of InN: 1-fold spots, 2-fold + spots, and streaky 2-fold.

As soon as InN had been nucleated on the GaN surface, the RHEED changed from the GaN streaky pattern to 3D spots. After that, the RHEED continued with the 3D pattern or gradually changed to one of the two different RHEED patterns depending on the growth conditions. The "better" growth region judging from the optical absorption and X-ray Diffraction (XRD) characterization was within the pure 3D spots region.

All the better quality InN films, as judged by XRD and absorption characteristics, were grown with 3D RHEED patterns exhibited throughout the whole growth course. We find that the low N/In ratio and/or high growth temperature associated with streaky RHEED are ultimately detrimental to film quality. It is interesting to note that when In was deposited on InN at 350°C (our normal growth temperature) the RHEED observed was also streaky. We could not distinguish the streak spacing of InN from that of pure In. Therefore we speculated that the streaky 2 fold RHEED during the InN growth could also be associated with the presence of pure In at the growth surface. For temperatures above 430°C where InN dissociation becomes significant, an In rich surface results even for higher N/In ratios.

In the past the thermal dissociation of InN has been studied under high pressure and vacuum [14-16], with some disagreement between the reported results. The possible reason for the discrepancy could be the variations in crystal quality. Since the high nitrogen partial pressure over InN is one of the major reasons why InN is extremely difficult to grow, we performed a series of dissociation experiments in the MBE growth chamber.

The Quadrapole Mass Analyzer (QMA) has been previously used for the study of nitride growth in an MBE system previously [17]. In a dissociation experiment, an InN thin film (typically 0.3 um) was first grown on the GaN epilayer as described above. The sample was then gradually heated at a constant rate. The signals at atomic mass 14, 28, and 115 substrate temperature were recorded. At the same time, RHEED pictures were taken. Fig.2 illustrates the QMA signal intensities vs. temperature at different stages of the experiment. There are several interesting features in the plots:

Indium (115) and nitrogen signals (14 and 28) started to rise at about 430°C, as did chamber pressure. InN was previously reported to dissociate rapidly at 500°C [14]. In this experiment, as the pressure shows in Fig.2, there is considerable dissociation at 450°C. The nitrogen signals reached a local maximum at 570°C and then started to drop. However the In intensity kept increasing as the temperature rose. This drop in nitrogen likely represents the elimination of the InN crystal. The nitrogen escaping from the substrate started to drop after this point. However, In did not escape from the film surface immediately after the InN decomposed. The rapid increase of the nitrogen signals above 750°C can be attributed to the decomposition of GaN as well as the outgassing of the growth manipulator.

The RHEED evolution during the dissociation experiments provided further evidence that when InN dissociated, the In tended to stay on the surface. When the InN started to dissociate at 430°C, the RHEED remained as 3D spots with the intensity dropping as the temperature rose. The intensity kept dropping until the local maximum of nitrogen signals was reached at 570°C. At the same time, while the RHEED intensity dropped, the streaky pattern started to develop. The appearance of streaks is consistent with the idea that at this point most InN had decomposed and an accumulation of In on the surface lowered the RHEED intensity. It should be noted that when the RHEED of InN was studied during growth, we observed that the RHEED intensity of streaky patterns tended to decrease faster than did the 3D or mixed patterns. As a result, we attributed the decrease to the accumulation of In.

At a temperature 450°C, the loss of In due to dissociation was compatible to the In flux typically employed during growth. It should be recalled that the In flux during growth is limited by the available active nitrogen. This implied that the growth should be conducted at temperatures much lower than 450°C to avoid the decomposition of InN.

Results

For films grown with a spotty RHEED pattern, AFM studies showed grainy features which are consistent with the island growth mode. The diameter of the grains ranges from 0.2 to 0.3μm. AFM was also used to resolve the morphology of a sample exhibiting the mixed RHEED patterns (streaks and spots) described previously. The AFM picture shows that the surface is essentially a flat plane with a few islands of different height scattered over the surface. The RHEED streaks are interpreted as caused by the flat plane, while the spots were from the transmission type diffraction when electrons passed through the isolated islands.

Optical transmission experiments were conducted to evaluate the InN optical quality. Fig.3 shows the transmitted light intensity in the vicinity of bandgap of a reddish film.

(We found that the film color was useful in judging the relative quality of InN samples.) A relatively broad band edge was seen; the absorption characteristic is consistent with the red appearance of the film. A considerable absorption below the bandgap was also observed. Depending on the growth conditions, InN showed a red color with different degrees of transparence. Films grown at temperatures around 350°C with low In flux showed a brighter red color. Films grown under higher temperatures or lower N/In flux ratios tended to show a darker color tending to black when the below bandgap absorption is strong. The films with brighter red color also exhibited narrower XRD peak widths and improved microstructures as illustrated by the TEM images.

Fig.2 Dissociated In and N as a function of temperature.

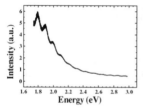

Fig.3 Optical absorption of a typical reddish InN.

Fig.4 Cross-sectional HRTEM of InN.

Table1 List of InN growth conditions and results.

Sample	In flux (ml/sec)	Temp (°C)	XRD FWHM	Growth interruption	Morphology/ Color
InN071398	0.05	355	1705	60secs per 12 ml	In balls, red
InN-8	0.05	330	1691	90secs per 12 ml	Wavy, red
InN-9	0.05	380	1613	90secs per 12 ml	dark
InN-10	0.05	405	1397	90secs per 12 ml	dark
InN-11	0.03	350	1314	No	dark
InN-12	0.03	350	1631	60secs per 12 ml	Red
InN-13	0.08	350	2539	60secs per 12 ml	dark
InN-14	0.02	350	1102	60secs per 12 ml	Red

The crystalline quality was analyzed by XRD and TEM. Table1 lists a group of films grown. The (0002) rocking curve diffraction was performed on a double crystal X-ray diffraction system. All films are 0.2-0.3 μm thick. Fig.4 is a High Resolution Transmission Electron Microscope (HRTEM) image taken from a cross-sectional sample of the InN layer grown under the optimized condition. The image shows that lattice fringes of basal planes of the InN crystal are seen continuously in the direction parallel to the growth plane from the bottom to the top of the layer, though these basal plane images are somewhat wavy. In the case of other InN layers whose growth conditions were not optimized, HRTEM images show distinct columnar-like structures, and lattice fringes of basal planes are discontinuous between neighboring columns with void-like images in boundary regions between columns. These HRTEM observations suggest that the optimization of the growth condition has improved the alignment of orientations of InN nuclei and lead to nearly single crystalline epitaxial layers as the growth has proceeded.

Discussion

Films grown at temperatures higher than 430°C are all black in appearance although the XRD width could be compatible to those with red color. For these films In droplets are usually found on the surface. The accumulation of In can be partly explained by the dissociation of InN at temperatures above 430°C as illustrated in Fig.2.

It was found that a high N/In flux ratio is required in order to obtain a reddish InN film. For a growth temperature of 355°C, In balls are seen on the surface although the N/In ratio is seven. To avoid the formation of In droplets, N/In needs to be increased above seven; the observation suggests that it is more energetically favorable for indium to form indium clusters rather than the InN compound.

A periodic growth interruption seems to effectively improve the film color. For every 12 monolayer deposition, the In shutter is closed for 60 or 90 seconds while the nitrogen flux is on. As seen in Table1, we find that the interruption appears more effective than reducing the In flux without periodic growth interruption.

Thermodynamic theory has been applied to build models for the growth of III-nitrides [18]. In the cited equilibrium theory, the formation of indium droplets is determined by comparing the equilibrium indium partial pressure to the partial pressure given by the vapor pressure of In over liquid In at the growth temperature. The approach implies the assumption that the formation of InN has priority over the formation of In droplets. The results in [18] showed that In condensation does not occur at regions where the N/In flux ratio is larger than one. However, our experimental data indicates that indium droplets can still be formed even when the N/In flux ratio is much greater than one. Therefore the higher nitrogen flux does not guarantee the lack of indium condensation. Kinetic processes such as the surface mobility of the adatoms or barriers for the formation of compound could be possible factors which affect the InN MBE growth. In general, growth temperature plays an important role in these complex kinetic processes. Higher growth temperature is preferred to overcome such energy barriers if present. However, higher decomposition rates also accompany higher temperatures, calling for more active nitrogen to compensate. Unfortunately it is very difficult to get abundant active nitrogen in a MBE chamber; the highest temperature is limited to values under 400°C in our system.

Conclusion

InN thin films have been grown on GaN/sapphire epilayers. The growth conditions are optimized based on a RHEED study and InN dissociation experiments. Films have been characterized by various techniques. The results indicate that seeking a stronger active nitrogen source is important to increase growth rate and growth temperature. Another direction to improve the film quality could possibly be to seek a possible surfacant for InN growth.

Acknowledgement

The authors acknowledge the technical support of Dr. A. Waag, Prof. A. Nurmikko, H. Zhou, Y.K. Song and Prof. A.K. Ramdas, M. Seong, Dr. J. Tsao and D. Lubelski. W.L. Chen and R.L. Gunshor are grateful for financial support from the Purdue Research Fundation.

References

[1] S. Strite, H. Morkoç, J. Vac. Sci. Technol. B 10, 1237 (1992).
[2] S. Nakamura, M. Senoh, N. Iwasa, S. Nagahama, T. Yamada, and T. Mukai, Jpn. J. Appl. Phys., 34, L1332 (1995).
[3] S. Nakamura, M. Senoh, S. Nagahata, N. Iwasa, T. Yamada, T. Matsushita, H. Kiyoku, Y. Sugimoto, Jpn. J. Appl. Phys. 35, L74 (1996).
[4] M. Shimizu, K. Hiramatsu, and N. Sawaki, J. Cryst. Growth 145, 209 (1994).
[5] R. Singh, D Doppalapudi, T.D. Moustakas, and L.T. Romano, Appl. Phys. Lett. 70, 089 (1997).
[6] K. Kubota, Y. Kobayashi, and K. Fujimoto, J. Appl. Phys. 66, 2984 (1989).
[7] C.R. Abernathy, S.J. Pearton, F. Ren, and P.W. Wisk, J. Vac. Sci. Technol. B11, 179 (1993).
[8] T.J. Kistenmacher, S.A. Ecelberger, and W.A. Bryden, J. Appl. Phys. 74, 1684 (1993).
[9] Y. Sato and S. Sato, J. Cryst. Growth 144, 15 (1994).
[10] Q. Guo, T. Yamamura, A. Yoshida, and N. Itoh, J. Appl. Phys. 75, 4927 (1994).
[11] W.E. Hoke, P.J. Lemonias, and D. G. Weir, J. Cryst. Growth 111, 1024 (1991).
[12] Y. Pan, W. Lee, C. Shu, H. Lin, C. Chiang, H. Chang, D. Lin, M. Lee, W. Chen, Jpn. J. Appl. Phys. 38, 645 (1999).
[13] S. Yamaguchi, M. Kariya, S. Nitta, T. Takeuchi , C. Wetzel, H. Amano, and I. Akasaki, J. Appl. Phys. 85, 7682 (1999).
[14] J.W. Trainor and K. Rose, J. Electron. Mater. 3, 821 (1974).
[15] A.M. Vorb'ev, G.V. Evseeva, and L.V. Zenkevich, Russ. J. Phys. Chem. 45, 1501 (1971).
[16] R.D. Jones and K. Rose, J. Phys. Chem. Solids 48, 587 (1987).
[17] R. Held, D.E. Crawford, A.M. Johnston, A.M. Dabiran, and P.I. Cohen, J. Electron. Mater. 26, 272 (1997).
[18] A. Koukitu and H. Seki, Jpn. J. Appl. Phys. 36, L750 (1997).

Mat. Res. Soc. Symp. Vol. 595 © 2000 Materials Research Society

Evidence from EELS of Oxygen in the Nucleation Layer of a MBE grown III-N HEMT

Tyler J. Eustis[1], John Silcox[2], Michael J. Murphy[3], and William J. Schaff[3]
[1]Department of Materials Science and Engineering, Cornell University, Ithaca, NY 14853
[2]School of Applied Engineering Physics, Cornell University, Ithaca, NY 14853
[3]School of Electrical Engineering, Cornell University, Ithaca, NY 14853

ABSTRACT

The presence of oxygen throughout the nominally AlN nucleation layer of a RF assisted MBE grown III-N HEMT was revealed upon examination by Electron Energy Loss Spectroscopy (EELS) in a Scanning Transmission Electron Microscope (STEM). The nucleation layer generates the correct polarity (gallium face) required for producing a piezoelectric induced high mobility two dimensional electron gas at the AlGaN/GaN heterojunction. Only AlN or AlGaN nucleation layers have provided gallium face polarity in RF assisted MBE grown III-N's on sapphire. The sample was grown at Cornell University in a Varian GenII MBE using an EPI Uni-Bulb nitrogen plasma source. The nucleation layer was examined in the Cornell University STEM using Annular Dark Field (ADF) imaging and Parallel Electron Energy Loss Spectroscopy (PEELS). Bright Field TEM reveals a relatively crystallographically sharp interface, while the PEELS reveal a chemically diffuse interface. PEELS of the nitrogen and oxygen K-edges at approximately 5-Angstrom steps across the GaN/AlN/sapphire interfaces reveals the presence of oxygen in the AlN nucleation layer. The gradient suggests that the oxygen has diffused into the nucleation region from the sapphire substrate forming this oxygen containing AlN layer. Based on energy loss near edge structure (ELNES), oxygen is in octahedral interstitial sites in the AlN and Al is both tetrahedrally and octahedrally coordinated in the oxygen rich region of the AlN.

INTRODUCTION

Within the last decade, the III-N semiconductors have experienced rapid development to the point that commercial optical devices are now available. Recently the importance of the piezoelectric properties of the III-N has come to light. With this realization, the knowledge and control of the polarity of III-N thin films has become paramount for device design. The piezoelectric and spontaneous polarizations in this system are large enough to induce a two-dimensional electron gas (2DEG) at $Al_{1-x}Ga_xN/GaN$ heterojunctions [1-7]. For Molecular Beam Epitaxy (MBE) material grown on sapphire substrates using a plasma nitrogen source, the nucleation layer determines the polarity of the resulting layers [8]. With this control of the polarity of the material, high quality MBE High Electron Mobility Transistor (HEMT) devices have been achieved [8]. In this paper, the composition and structure of the AlN nucleation layer in a HEMT device structure studied in the Cornell University Scanning Transmission Electron Microscope (STEM), will be presented.

The Cornell University STEM produces an electron probe approximately 2.1Å in diameter and thus is distinctly qualified for near atomic resolution interface investigations. As this probe is scanned across a TEM type sample, the scattered electrons are collected. Electrons scattered to high angles are collected by the Annular Dark Field (ADF) detector producing an image where the contrast depends mainly on atomic number (Z) and thickness. Thus ADF imaging is commonly referred to as Z-contrast imaging. Electrons scattered to small angles are either detected by the Bright Field Detector, producing an image analogous to normal bright field TEM, or are recorded as EELS by either the Serial or Parallel detector.

EELS provides information about the local bonding and electronic structure by probing unoccupied electronic states. The fast moving electrons from the STEM probe interact with the core electrons in the specimen, which are excited from their ground state to unoccupied states. Since the energy lost by the fast moving electrons in the probe is equal to the difference in energy between the core level and excited level, the measurement of the intensity of the probe electrons as a function of energy loss provides substantial details of the electronic structure and chemistry.

EXPERIMENTAL DETAILS

The III-N HEMT sample with an AlN nucleation layer was grown in a Varian GenII MBE at Cornell University. The specifics of the growth conditions, electrical characterization, and device results are given elsewhere [8]. The sample was prepared by standard tripod polishing techniques to form a wedge specimen [9]. The sample was ion milled with a BAL-TEC Res 010 for final thinning.

The Cornell University STEM has a maximum energy resolution of ~0.7eV over a large energy range (from 0 to 2keV). With high spatial and energy resolution, the STEM is particularly suited for sub-nanometer chemical and structural studies. With such a focused small probe, high spatial resolution EELS can be obtained. Thus compositional and structural information can be acquired with core-loss EELS. In addition, thickness information can be obtained on a nanometer scale from the ratio of the first plasmon intensity to the zero loss (electrons that have lost no energy) intensity. More detailed information concerning the STEM can be found elsewhere.

EELS spectra were acquired using the Parallel EELS detector. Specifically the nitrogen and oxygen K-edges were obtained simultaneously and individually by stepping across the GaN/AlN/Sapphire interfaces. The K-edges correspond to transitions from atomic 1s states to empty local and conduction band states with p-character. Spectra of the aluminum L_{23}-edge were also obtained. The L-edge corresponds to transitions from atomic 2p to empty local and conduction band states with d and s-character. The ADF signal is acquired in conjunction with the spectra. In addition, ADF images of the area of interest collected before and after spectrum acquisition indicated no significant spatial drift.

All data and images are acquired digitally. After acquisition, the spectra were smoothed and shot noise was removed. The background was subtracted using a standard power-law curve fit to the pre-edge. The intensities of the nitrogen and oxygen K-edges were integrated over 50 eV starting at threshold. Ratios of O/N+O were obtained using the standard equation involving integrated intensities and cross section [10]. Acquisition times for the K-edges were selected in an attempt to balance electron beam damage [11] and signal to noise. Thus, for the acquisition times used, some beam damage was observed in the sapphire substrate. The damage results in a very small artificial nitrogen signal in the sapphire.

DISCUSSION

The presence of oxygen is observed in the AlN nucleation layer upon examination of Figure 1. The oxygen signal is constant within the sapphire substrate. Moving across the AlN/sapphire interface (based on the ADF image in Figure 1(a)), the oxygen signal drops sharply. Once across the interface, the oxygen signal decays and reaches zero at approximately the AlN/GaN interface.

As stated above, the nitrogen signal within the sapphire results from beam damage. Therefore the ratio depicted in Figure 1(c) does not level off at one in the sapphire. On moving from the sapphire to the AlN, the nitrogen signal increases but does not peak until about half way through the AlN, as seen in Figure 1(b).

Upon calculation of the O/O+N ratio using the data in Figure 1(b), thickness effects are removed. The ratio is displayed in Figure 1(c). The ratio decreases at points moving across the AlN/sapphire as the oxygen signal decreases and the nitrogen signal increases.

Figure 2 is a bright field TEM image of the AlN/sapphire interface. The interface

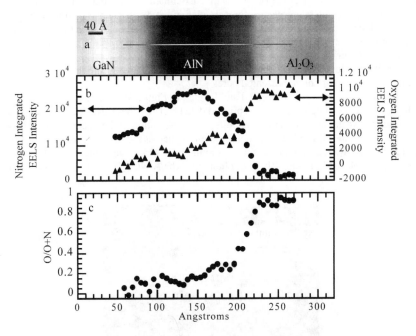

Figure 1. *(a) ADF image of the area of interest. (b) Integrated Intensity of N and O K-edge. (c) The ratio O/O+N vs. distance. The horizontal line represents where the data in (b) and (c) were taken.*

appears fairly abrupt giving no indication of any oxygen containing AlN. Thus un-calibrated bright field TEM imaging is inadequate to detect chemical profiles. Only through strict control of experimental conditions, has chemical mapping of semiconductor interfaces been achieved using bright field TEM images [12]. Others have stated based on TEM images that their observed interface is chemically sharp [13]. Without microanalysis, such claims are difficult to substantiate.

Now we turn to examination of the ELNES in an attempt to learn about the local environment of the individual constituents. Core-loss edges at significant points across the AlN/sapphire interface are displayed in Figure 3. The N K-edge in the oxygen poor AlN and in the oxygen rich AlN are displayed in Figure 3(a). The N K-edge for AlN compares well with previously published spectrum [11]. In Figure 3(a), notice the decrease in intensity of the first peak of the N K-edge from the oxygen rich region of the AlN relative to the other peaks as compared to the N K-edge from the relatively pure AlN. Based on the comparison of the two N K-edges with Multiple Scattering calculations, some of the oxygen in the AlN is in octahedral interstitial positions within the AlN lattice [11]. Abaidia et. al. determined the location of oxygen in AlN based on examination of extended energy loss fine structure. It was determined that oxygen would be in octahedral interstitial sides for $Al/(N+O) \simeq 0.8$ and in substitution positions for $Al/(N+O) \simeq 1$ [14]. Others determined the O in AlN was in substitution sites based on X-ray Absorption Spectroscopy [15]. The possibility of oxygen substituted for nitrogen in the AlN nucleation layer of the HEMT sample examined here cannot be ruled out. Unfortunately no core edge spectroscopy simulations of oxygen substituted for nitrogen are available with which to compare. In addition, it is important to note that the oxygen containing AlN layers examined by the above mentioned groups were grown in an oxygen-containing environment.

Small variations in the O K-edges were observed. However, little information is gained and thus will not be presented here.

Aluminum L-edges are presented and labeled in Figure 3(b). The edges obtained in AlN and in Al_2O_3 match well with published data [11,16]. The Al L-edge obtained at the interface between the AlN and the sapphire is a combination of the edges from AlN and Al_2O_3. In fact, the Al L-edges obtained by stepping across the interface using the parallel EELS detector were a smooth transition from the AlN Al L-edge to the Al_2O_3 Al L-edge. The edge displayed in Figure 3(b) was obtained at the AlN/Al_2O_3 interface. Calculations

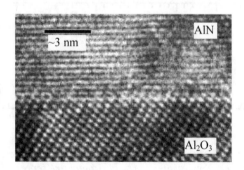

Figure 2. *Bright Field TEM image of the AlN/sapphire interface.*

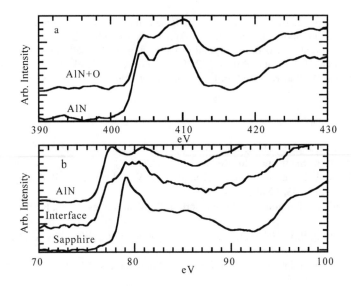

Figure 3. *EELS core-loss edges. (a) The N K-edge in the AlN and at the AlN/sapphire interface. (b) The Al L_{23}-edge in the AlN, at the AlN/sapphire interface, and in the sapphire.*

by Brydson et al. [17] confirm the validity of simply performing a linear sum of Al L-edges to determine the mount of different components. Thus, the Al at the interface is approximately 72% tetrahedrally bonded Al and 28% octahedrally bonded Al.

CONCLUSION

Oxygen has been observed by EELS in the AlN nucleation of a high quality MBE grown III-N HEMT device. The AlN nucleation layer is key to obtaining MBE Ga-face polarity material on sapphire. The oxygen appears to have diffused in from the sapphire substrate. The concentration profile suggests a sharp drop in oxygen crossing the AlN/sapphire interface. Once across the interface, the oxygen concentration decays approaching the GaN/AlN interface. Examination and comparison of the ELNES of the N K-edge and the Al L-edge reveals that oxygen in the AlN is in octahedral interstitial sites and that Al in the oxygen rich region of the AlN has both tetrahedral and octahedral coordination. Bright field TEM images fail to identify a chemical difference in the AlN. Although some information has been gained about the Ga-face polarity determining AlN nucleation layer, it is unclear at this point what role the oxygen plays.

ACKNOWLEDGEMENTS

This work was supported by the Office of Naval Research under MURI Contract No. N00014-96-1-1223 monitored by Dr. John C. Zolper. The Cornell STEM was acquired through the NSF (grant # DMR8314255) and is operated by the Cornell Center for Materials Research (NSF grant #DMR-9632275). The authors would like to acknowledge Mick Thomas and Dr. Earl Kirkland for technical support and helpful discussions.

REFERENCE

1. A. Hangleiter, I. Jin Seo, H. Kollmer, S. Heppel, J. Off, and F. Scholz, MRS Internet Journal of Nitride Semiconductor Research, **3** (15), **(1998)**.
2. P. M. Asbeck, E. T. Yu, S. S. Lau, G. J. Sullivan, J. Van Hove, and J. Redwing, Electronics Letters, **33**, 1230-1, (1997).
3. F. Bernardini, V. Fiorentini, and D. Vanderbilt, Physical Review B, **56**, R10024-7, (1997).
4. A. D. Bykhovski, B. L. Gelmont, and M. S. Shur, Journal of Applied Physics, **81**, 6332-8, (1997).
5. M. B. Nardelli, K. Rapcewicz, and J. Bernholc, Applied Physics Letters, **71**, 3135-7, (1997).
6. T. Takeuchi, H. Takeuchi, S. Sota, H. Sakai, H. Amano, and I. Akasaki, Japanese Journal of Applied Physics, Part 2, **36**, L177-9, (1997).
7. E. T. Yu, G. J. Sullivan, P. M. Asbeck, C. D. Wang, D. Qiao, and S. S. Lau, Applied Physics Letters, **71**, 2794-6, (1997).
8. M. J. Murphy, K. Chu, H. Wu, W. Yeo, W. J. Schaff, O. Ambacher, J. Smart, J. R. Shearly, L. F. Eastman, and T. J. Eustis, Journal of Vacuum Science & Technology B, **17**, 1252-4, (1999).
9. J. P. Benedict, R. M. Anderson, and S. J. Klepeis, in *Proceedings of the Twenty-Eighth Annual Technical Meeting of the International Metallographics Society*, Albuquerque, NM, USA, 1996 (ASM Int; Materials Park, OH, USA), p. 277-84.
10. R. F. Egerton, *Electron Energy-Loss Spectroscopy in the Electron Microscope*, 2nd ed. (Plenum Press, New York, 1996), p. 485.
11. V. Serin, C. Colliex, R. Brydson, S. Matar, and F. Boucher, Physical Review B, **58**, 5106-15, (1998).
12. A. Ourmazd, D. W. Taylor, J. Cunningham, and C. W. Tu, Physical Review Letters, **62**, 933-6, (1989).
13. M. Yeadon, M. T. Marshall, F. Hamdani, S. Pekin, H. Morkoc, and J. M. Gibson, Journal of Applied Physics, **83**, 2847-50, (1998).
14. S. Abaidia, V. Serin, G. Zanchi, Y. Kihn, and J. Sevely, Philosophical Magazine A, **72**, 1657-70, (1995).
15. M. Katsikini, E. C. Paloura, T. S. Cheng, and C. T. Foxon, Journal of Applied Physics, **82**, 1166-71, (1997).
16. S. D. Mo and W. Y. Ching, Physical Review B, **57**, 15219-28, (1998).
17. R. Brydson, Journal of Physics D, **29**, 1699-708, (1996).

Mat. Res. Soc. Symp. Vol. 595 © 2000 Materials Research Society

Formation of BN and AlBN During Nitridation of Sapphire Using RF Plasma Sources

A.J. Ptak[1], K.S. Ziemer[2], L.J. Holbert[1], C.D. Stinespring[2] and T.H. Myers[1,†]
[1] Department of Physics, [2] Department of Chemical Engineering
West Virginia University, Morgantown, WV 26506, [†] tmyers@wvu.edu

ABSTRACT

Evidence is presented that nitrogen plasma sources utilizing a pyrolytic boron nitride liner may be a significant source of B contamination during growth and processing. Auger electron spectroscopy analysis performed during nitridation of sapphire indicate the resulting layers contain a significant amount of BN. The formation of $Al_{1-x}B_xN$ would explain the observation of a lattice constant several percent smaller than AlN as measured by reflection high-energy electron diffraction. The presence of cubic inclusions in layers grown on such a surface may be related to the segregation of BN during the nitridation into its cubic phase.

INTRODUCTION

Nitridation of sapphire is an important step for many approaches to growing Group III-nitrides. As such, there have been many studies of the nitridation of sapphire by molecular beam epitaxy (MBE) using ammonia, [1,2] an electron cyclotron resonance (ECR) plasma source, [3] or an rf-plasma source. [4,5,6] The chemical evolution of the nitridation layer has been studied by x-ray photoelectron spectroscopy (XPS) [4], indicating an initially increasing nitrogen signal which ultimately saturates. A wide range of times have been reported necessary in order to obtain a completely nitrided surface, from ten to twenty minutes for an ECR source to several hours with an RF source. This range of nitridation times can be adequately explained by the different reactivity of the various active nitrogen species. [5] After nitridation, some authors find a completely relaxed surface with a final lattice constant equal to that of AlN, [1,3] while others find a lattice constant consistently smaller than that of AlN. [6] The formation of an $AlO_{1-x}N_x$ alloy has been invoked to explain the smaller lattice constant. However, $AlO_{1-x}N_x$ formation is unlikely based on thermodynamic considerations, and is contradicted by an in-situ transmission electron microscopy study of nitridation. [7] Another consideration is the common observation of the nucleation of cubic grains at the epilayer-sapphire interface. Widmann *et al.* [6] have related this nucleation of cubic grains to nitridation conditions. In this paper, we present results which indicate that B contamination from an rf-plasma source may explain the latter two phenomena.

EXPERIMENTAL DETAILS

Two rf-plasma sources are used in our lab used to produce various active species of nitrogen for the nitridation of sapphire and growth of GaN. These sources are an Oxford Applied Research (Oxfordshire, England) CARS-25 source and an EPI Vacuum Products

(St. Paul, MN) Unibulb source. The Oxford source features a removable pyrolytic boron nitride (PBN) liner and aperture plate. The EPI source contains the standard PBN Unibulb configuration with a 400-hole aperture

Nitridation experiments were performed in two separate systems. For chemical analysis of the evolution of the nitridation layer, nitridation was performed at 400 °C in a chamber connected to an Auger electron spectroscopy (AES) analysis system consisting of a Phi 545 Scanning AES Microprobe with Model 110A Cylindrical Electron Optics. Scans were performed with a 3 keV incident electron beam, a beam current of 2 µA and a nominal spot size of 3 µm. Reflection high energy electron diffraction (RHEED) measurements of the evolution of lattice constant with nitridation were performed in our MBE system, which has been described elsewhere. [9]

EVIDENCE FOR BORON DURING GROWTH

We have published several studies detailing the characterization of the active nitrogen flux produced by our two rf plasma sources for various operating conditions. [8,9,10] During the characterization involving mass spectroscopy, we saw a small but persistent signal indicating that the sources were also producing boron. Subsequent secondary ion mass spectrometry (SIMS) analysis of GaN grown using these sources verified B production. Figure 1 shows typical SIMS measurements from GaN layers grown under standard conditions using both our Oxford (a) and EPI (b) source. A significant background of boron was detected in each case. The flux of B atoms listed was determined from the growth rate of the GaN and assuming 100% incorporation of B. These results are also consistent with levels of B detected in nitrogen-doping studies of ZnSe and CdTe. [10]

The PBN liner is the likely candidate as the source of the boron from the plasma sources. Significant amounts of atomic nitrogen are produced inside the liner. Atomic nitrogen is highly reactive, and may be promoting decomposition of the PBN. We replaced the liner in the Oxford source after verifying the large B background and found similar levels of B in GaN grown with the new liner, indicating that it is an endemic problem. Source configurations can influence this effect as a significantly lower B flux was indicated for our EPI source. Operation almost continually over approximately three years has not changed the amount of B produced by the EPI source. Private discussions with other groups have indicated that B is potentially a universal issue for plasma sources.

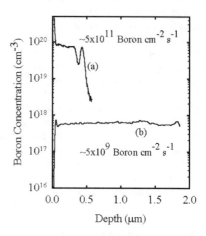

Figure 1. Comparison of Boron concentration from two rf-plasma sources.

RHEED STUDIES

RHEED was used to measure the lattice constant variation occurring during nitridation of sapphire using our EPI source, similar to studies reported previously. [1,6] The RHEED pattern was monitored continuously for sapphire exposure to the active nitrogen flux at temperatures of 200, 400 and 700 °C. In contrast to previous results, [6] we do not observe any evidence of nitridation at 200 °C. This result can possibly be explained to be due to differences in the fraction of active species found in different rf plasma sources. [5] Evidence of nitridation was readily observed at 400 and 700 °C. Immediately after active nitrogen exposure, streaks began to appear that are generally attributed to the replacement of oxygen by nitrogen in the sapphire lattice. An additional 10 to 15 minutes of exposure to active nitrogen resulted in the disappearance of the underlying sapphire pattern.

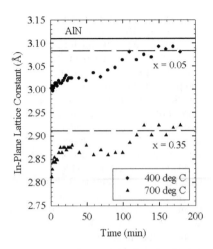

Figure 2. Variation of the in-plane lattice constant with nitridation time as observed using RHEED. The x-values represent the composition of $Al_{1-x}B_xN$ that would result in this lattice constant.

Figure 2 illustrates the evolution of the lattice parameter of the nitridation layer measured using RHEED. The measurements were calibrated using the bulk lattice parameters of sapphire. The variation in lattice constant appeared to reach an initial plateau, followed by a second increase to a final value that remained stable for longer times. In general, this final measured lattice constant was several percent lower than the 3.11Å lattice constant of AlN, in agreement with Widmann et al. [6] The lattice constant tended to be significantly smaller for the higher temperature nitridation. Faint and very diffuse rings appear superimposed over the streaks in the RHEED pattern at about the time of the second increase in lattice constant.

AUGER ANALYSIS DURING NITRIDATION

AES was used to perform a detailed examination of the chemical evolution of the nitridation layer as a function of active nitrogen exposure time at 400°C. The substrate heater in this system did not allow investigation of significantly higher temperatures, and nitridation was not observed for 200°C or below with the EPI source. As reported previously, [5] we observed significant differences in nitridation rates based on whether the active flux was predominantly atomic, ionic or neutral metastable nitrogen. Figures 3 and 4 show the evolution of the Al, N, and B signals measured as a function of nitridation time for two distinct cases. The data in Figure 3 were collected for nitridation under conditions which enhanced ion production with the Oxford source. After about 60 minutes, the Al, N and B signals had reached a plateau, with the B signal small. By assuming that this represented primarily an AlN layer, sensitivity factors relative to nitrogen for Al (0.72) and B (0.9) were obtained for our system to use in determining composition. Of particular significance to this paper is the large amount of B present in the nitridation layer shown in Figure 4, obtained using the EPI source. A previous study of the chemical evolution of a nitridation layer relied on XPS, [4] which is not as sensitive to B. The study used an Oxford source similar to ours, which produces primarily atomic and ionic active nitrogen and is not as efficient at capturing boron as illustrated by Figure 3. Focusing on Figure 4, it is readily apparent that both the B and N signal increase rapidly while the Al signal decreases. After about 100 minutes, the Al and B appear to reach equilibrium. We

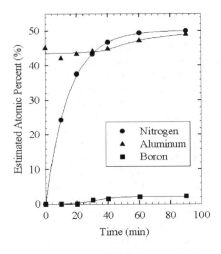

Figure 3. AES study of the change in surface composition upon nitridation with the Oxford source configured for a high ion flux.

believe that this trend is indicative of the initial formation of a BN layer, which then evolves into an alloy of AlBN during extended nitridation. The continuing increase of the N signal is likely related to the longer mean free path for electrons at this energy, ~11Å, vs. those of Al (6Å) or B (7Å). Thus, the Al and B signals would saturate for a relatively thin layer tickness while the increasing N signal implies a layer that is still growing with time. Note that the lines in both Figure 3 and Figure 4 are only meant as guides to the eye.

DISCUSSION

It is obvious that B can play a significant role during the long periods required for nitridation using an rf plasma source. AES analysis indicates a final layer that contains about 10 to 20% BN. This is consistent with RHEED lattice constant measurements which also imply significant amounts of BN. If this is due to alloy formation, the AES shown in Figures 3 and 4 would imply approximate x-values of 0.04 and 0.4 respectively. The RHEED studies indicated similar compositions are

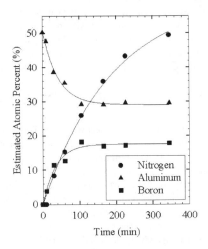

Figure 4. Nitridation with the EPI source.

possible assuming that $Al_{1-x}B_xN$ is being formed. RHEED analysis consistently indicates a lattice constant closer to that of pure AlN for lower temperature nitridation implying that the higher temperatures are more effective for B capture in the evolving nitridation layer.

Polyakov et al. [11] studied the growth of AlBN solid solutions by MOCVD. They found that for compostions larger than about x=0.01 phase segregation occurred into BN and low x-value AlBN. The implication is that high x-value AlBN is thermodynamically unstable. In Polyakov's study, the second B-rich phase appeared to be wurtzite BN. We cannot identify if this is occurring in our nitridation layers from the measurements made. It is possible that high x-value AlBN can be formed for the thin nitridation layers. However, it is also quite likely that phase separation is indeed occurring, particularly for the more B-rich layers found at higher temperatures. The onset of segregation may be indicated by the relatively abrupt second rise in lattice constant observed by RHEED, accompanied by the appearance of faint, diffuse rings. Indeed, the presence of both the faint ring pattern and the strong, streaky pattern is indicative of an inhomogeneous surface. Wideman et al. [6] found that lower nitridation temperatures led to reduced nucleation of cubic grains in subsequent AlN buffer layer growth. Based on our results, a possible mechanism explaining this result is the formation of larger x-value AlBN at higher temperatures, followed by segregation and nucleation of cubic BN inclusions once a critical thickness is reached. Polyakov et al.'s study suggests that the majority of the B-rich phase may maintain the wurtzite structure, thus higher temperatures and larger B concentrations may be necessary to nucleate the cubic phase.

CONCLUSIONS

Boron can be a significant contaminant originating in plasma sources utilizing PBN liners. While it is not clear what role a small background of B has on the properties of GaN, B may have a significant influence on the nitridation of sapphire using rf plasma sources. In particular, the formation of $Al_{1-x}B_xN$ alloys can explain observed lattice constants smaller than expected for AlN. Segregation of BN in large x-value alloys coupled with nucleation of the more thermodynamically stable cubic BN phase may explain one possible mode for the subsequent nucleation of cubic GaN or AlN on nitridated sapphire.

ACKNOWLEDGEMENTS

This work was supported by ONR Grant N00014-96-1-1008 and monitored by Colin E. C. Wood.

REFERENCES

1   N. Grandjean, J. Massies and L. Leroux, Appl. Phys. Lett. 69, 2071 (1996).
2   Z. Yang, L.K. Li, and W.I. Wang, J. Vac. Sci. Technol. B14, 2354 (1996).
3   T.D. Moustakas, R.J. Molnar, T. Lei, G. Menon and J.C.R. Eddy, Mater. Res. Symp. Proc. 242, 427 (1992).
4   C. Heinlein, J. Grepstad, T. Berge and H. Riechert, Appl. Phys. Lett. 71, 341 (1997).
5   A.J. Ptak, K.S. Ziemer, M.R. Millecchia, C.D. Stinespring, and T.H. Myers, MRS Internet J. Nitride Semicond. Res. 4S1, G3.10 (1999).
6   F. Widmann, G. Feuillet, B. Daudin and J.L. Rouviere, J. Appl. Phys. 85, 1550 (1999)
7   M. Yeadon, M.T. Marshall, F. Hamdani, S. Pekin, H. Morkoç, and J.M. Gibson, J. Appl. Phys. 83, 2847 (1998).
8   A.J. Ptak, M.R. Millecchia, T.H. Myers, K.S. Ziemer and C.D. Stinespring, Appl. Phys. Lett. 74, 3836 (1999).
9   T.H. Myers, M.R. Millecchia, A.J. Ptak, K.S. Ziemer and C.D. Stinespring, J. Vac. Sci. Technol. B17, 1654 (1999).
10 M. Moldovan, L.S. Hirsch, A.J. Ptak, C.D. Stinespring, T.H. Myers and N.C. Giles, J. Electron. Mater. 27, 756 (1998).
11 A.Y. Polyakov, M. Shin, W. Qian, M. Skowronski, D.W. Greve and R.G. Wilson, J. Appl. Phys 81, 1715 (1997).

Mat. Res. Soc. Symp. Vol. 595 © 2000 Materials Research Society

The Effect of the Buffer Layer on the Structure, Mobility and Photoluminescence of MBE grown GaN

Nikhil Sharma, David Tricker, Vicki Keast, Stewart Hooper[1], Jon Heffernan[1], Jenny Barnes[1], Alistair Kean[1], Colin Humphreys
Department of Materials Science and Metallurgy, University of Cambridge, Pembroke Street, Cambridge, CB2 3QZ
[1] Sharp Laboratories of Europe, Oxford Science Park, Oxford, OX4 4GA

ABSTRACT

Although GaN has been grown mainly by metal organic chemical vapour deposition (MOCVD), molecular beam epitaxy (MBE) offers the advantages of lower growth temperatures and a more flexible control over doping elements and their concentrations [1]. We are growing GaN by MBE on sapphire substrates, using a GaN buffer layer to reduce the misfit strain, thus improving the structural quality of the epilayer. The quality of the GaN epilayers (in terms of their photoluminescence, mobility and structure) has been investigated as a function of the buffer layer thickness and annealing time.

The investigation showed that increasing the buffer layer thickness improved the mobility of the material because the defect density in the GaN epilayer decreased. Optical characterisation showed that the ratio of the donor band exciton (DBE) peak (3.47eV) to the structural peak (3.27eV) in the photoluminescence spectrum, measured at 10K, increased with decreasing defect density. The unwanted structural peak can be considered to originate from a shallow donor to a shallow acceptor transition, which is clearly related to the structural defects in GaN. Thus by increasing the buffer layer thickness and annealing time the structural quality, mobility and photoluminescence improves in the GaN epilayers.

Structural characterisation by transmission electron microscopy (TEM) showed that the observed increase in the DBE to structural peak ratio in the photoluminescence spectra could be correlated with a decrease in the density of stacking faults in the GaN epilayers. The detailed structure of these stacking faults was investigated by dark field and high resolution TEM. Their effect on the electrical and optical behaviour of GaN may be assessed by determining the local change in the dielectric function in the vicinity of individual stacking faults.

INTRODUCTION

The effect of defects on the optical performance of heteroepitaxial grown layers of GaN has attracted much interest in recent years. However there has been very little conclusive experimental evidence supporting the claim that either dislocations or stacking faults in GaN introduce states in the band gap, hence decreasing the efficiency of direct band gap transitions. This work shows that optimisation of the buffer layer for the growth of high quality MBE GaN (as compared to previous work, [1],[2]) can be achieved by comparing the PL and electron mobility with the defect density in the epilayer. Optimisation of the buffer layer in MBE GaN has previously been demonstrated by Grandjean [1]. As Grandjean demonstrated increasing the buffer layer thickness improves the quality of the structure, mobility and photoluminescence of the epilayer. This improvement corresponds to a smooth and flat buffer layer surface. Thus the epilayer has a decrease in the dislocation and stacking fault density, hence increasing the recombinant DBE efficiency of GaN.

GROWTH AND ELECTRICAL PROPERTIES

GaN was grown on a (0001) sapphire substrate using a modified VG-semicom V80 research grade MBE machine; using an ammonia injection source with a high flux rate for the nitrogen and solid source for the Gallium to give a high III/V ratio (100:1) [3]. A low temperature (500 °C) GaN buffer layer was first deposited on the substrate to allow some of the large lattice mismatch (13.4% between Al_2O_3 and GaN) to be taken up by dislocations. The buffer layer was subsequently annealed (1050 °C) for 5-80 minutes in order to improve the growth surface morphology, which promotes 2-D growth of the homoepitaxial GaN epilayer. The epilayers were subsequently grown at $1000\ ^0\,C$ (showing no decomposition of GaN) until they were 4μm thick.

Seven samples were grown, with varying buffer layer thicknesses and annealing times. Table I shows the effect of these variables on the electron mobility, as measured by the Hall experiment. Increased buffer layer thickness improves the mobility of the epilayer by up to 50%. However, increasing the annealing time of the buffer layer has little effect on the mobility. When the buffer layer was annealed for an extended amount of time (>46mins) then the electrical properties of the bulk GaN was found to deteriorate.

Table I The effect of buffer layer thickness and anneal time on the mobility of MBE grown GaN

Sample	Buffer layer thickness (nm)	Annealing time of buffer (mins)	Mobility (cm^2/Vs)
N119	40	5	170
N125	95	15	254
N126	210	5	229
N129	20	15	155
N135	20	5	154
N139	20	46	223
N141	20	80	186

PHOTOLUMINESCENCE

PL was measured at 10K using a He-Cd (325 nm) laser to excite the sample and a mercury arc lamp to calibrate the spectrometer. All samples show a blue donor band exciton (DBE) transition at around 3.48eV [3] (value varies with strain in epilayer). At 3.27eV a "structural peak" appears [3], this unwanted transition is understood to be related to defects in GaN. A broad 2.2eV peak occurs at the yellow end of the spectrum, this may arise from Ga-O complexes, which are attracted by the strain field of edge dislocations in GaN [4]. High quality PL was categorised by a high DBE to structural peak intensity (figure 1).

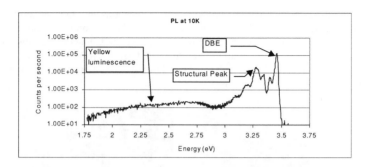

Figure 1 *Example of PL spectrum measured at 10K of MBE grown GaN, highlighting the important electron- hole transitions*

Table II highlights the effect of the buffer layer thickness and anneal time on the DBE/structural peak ratio. Increased buffer layer thickness and annealing time has a marked improvement on this ratio, and a small decrease of the yellow luminescence.

Table II The effect of buffer layer thickness and anneal time on the PL of GaN

Sample	Buffer layer thickness (nm)	Annealing time of buffer (mins)	DBE/Structural peak ratio
N119	40	5	2
N125	95	15	21
N126	210	5	168
N129	20	15	8.3
N135	20	5	0.7
N139	20	46	41
N141	20	80	43

TEM OF SAMPLES

TEM Bright field images (BF, straight through beam) and the more useful dark field images (DF, diffracted beam) were utilised to define the defect density of the epilayer. In the work presented three microscopes were used, the JEOL 200CX, JOEL 2000FX and the JOEL 4000EX. TEM samples were produced from MBE wafers using a precision ion polishing system (PIPS), which gives samples with a smooth surface and a large electron transparent area (thickness < 200nm). A detailed comparison was made between the thickness and anneal time of the buffer layer with the density and type of defects that were present.

Increased buffer layer thickness and annealing time gave a decrease in dislocation and stacking fault densities. Samples with high buffer layer thicknesses and annealing times showed little evidence of stacking faults and contain 2/3 edge (b =1/3<1-210>) and 1/3 mixed (b =1/3<11-23>) dislocations (figure 2 (a)). Samples with lower buffer layer thicknesses and annealing times showed evidence of both stacking faults, inversion domain boundaries and predominantly mixed dislocations (figure 2 (c)). N135 had an especially high defective structure, with horizontal dislocations and stacking faults appearing near the surface of the sample (figure 2 (d)). The horizontal dislocations were mixed in character and nucleated from the buffer layer, via threading dislocations. A reduction in dislocation density with growth thickness (figure 2 (b)) of the epilayer is related to a series of annihilation reactions caused by the dislocations bending round and interacting with adjacent dislocations.

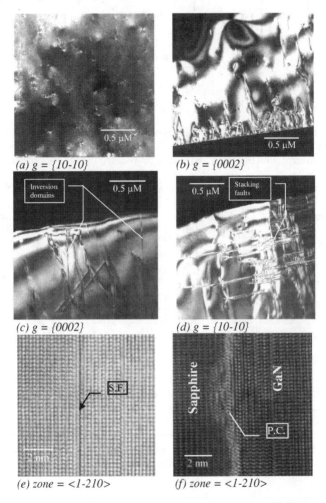

(a) g = {10-10}

(b) g = {0002}

(c) g = {0002}

(d) g = {10-10}

(e) zone = <1-210>

(f) zone = <1-210>

Figure 2 *DFTEM image of N125 in plan view (a) and in cross section (b), showing a higher density of dislocations near the interface due to horizontal dislocations; DFTEM images of inversion domains (c) and a stacking fault band (d) in N135 due to thinner buffer layer; HRTEM images of stacking faults (e) and of polycrystalline regions at the GaN/sapphire interface (f) due to a poor nucleation layer*

HRTEM was performed on the sapphire substrate/GaN interface with the JOEL 4000EX TEM. Samples with smaller buffer layer images show (figure 2(e)) evidence of a higher density of stacking faults near the substrate interface. This density is reduced for both an increase in buffer layer thickness and increase in annealing time. There is no evidence of an abrupt surface between the LT GaN buffer and the bulk GaN epilayer, however the growth on the sapphire substrate is not atomically perfect and small regions of polycrystallinity are located along the interface (figure 2(f)). TEM showed no sign of cubic material in any of the samples.

POLARITY DETERMINATION

Convergent beam electron diffraction (CBED) was employed to determine the polarity of the samples. A CBED pattern is produced by converging the electron beam to a small spot (around 100nm in diameter in the 2000FX) in the area of interest on the sample. An asymmetric CBED pattern is formed when the specimen is tilted such that the central beam is parallel to the <10-10> zone axis [5] (figure 3). Notice that the (0002) and (000-2) discs show very different contrast. This strong diffraction effect can be modelled convincingly using electron microscopy software (EMS, CIME, France).

A Bloch wave calculation was used to simulate the CBED pattern. Figure 3 shows how a small variation in the estimated thickness for the simulation gives a large variation of the intensities in the (0002) and (000-2) discs. It was found for all samples that the bulk material was N-faced ([000-1] direction towards the surface).

The absolute directions (either [0001] or [000-1]) outputted by EMS are ambiguous as sign errors can easily occur in the code. In order to confirm the correct directions a GaAs CBED pattern was modelled. The pattern has been studied in detail by Tafto *et al.* [6]. They found that by tilting the specimen to the [011] zone axis and then titling about 10^0 on the (200) plane a set of crosses appear in the (200) and (-200) discs. These crosses are either black or white due to destructive and constructive interference of reflecting beams. By modelling the GaAs reflections it was confirmed that the EMS code gave correct directions.

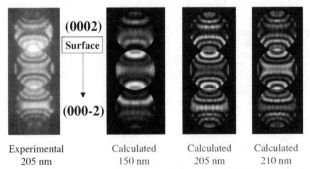

Experimental	Calculated	Calculated	Calculated
205 nm	150 nm	205 nm	210 nm

Figure 3 *Comparison between experimental and theoretical CBED at <10-10> zone axis of GaN*

CONCLUSIONS

Increasing the thickness of the buffer layer improves the PL, mobility and defect density of MBE grown GaN. An improvement of the defect density is also found for annealing of the buffer layer for up to 45 minutes at 1050 ^{0}C. The significance of these results is that the existence of both dislocations and stacking faults may introduce a shallow acceptor, which allows a 3.27eV electron-hole transition. This is supported by a large database of analysed MBE grown GaN (table 3), which show that optimisation of the buffer layer and hence structure of the bulk GaN has a marked improvement on the performance of the material. Further AFM measurements will link the buffer layer to the surface morphology. By employing electron energy loss spectroscopy (EELS) using the HB501 STEM the electronic structure in the vicinity of the stacking faults can be determined. Further work will be performed to improve the energy resolution and the zero-loss profile in order to directly extract the band gap energy of specific regions in a GaN sample. This will give direct evidence that states in the band gap are related to individual defects.

Table III Summary of results for measurements on MBE grown GaN grown on a LT GaN buffer layer

Sample	N119	N125	N126	N129	N135	N139	N141
Buffer layer thicknesses (nm)	40	95	210	20	20	20	20
Anneal time of buffer (mins)	5	15	5	15	5	46	80
Mobility (cm^2/Vs)	170	254	229	155	154	223	186
DBE/structural peak ratio	2	21	168	8.3	0.7	41	43
Dislocation density $(cm^{-2} \times 10^9)$	4.2	2.1	1.9	4.0	5.6	3.2	4.3
Polarity	N-face	N-face	N-face	N-face	N-face	N-face	N-face

ACKNOWLEDGEMENTS

N S would like to thank members of HREM group in Cambridge for their generous supervision and support. Also thanks to the GaN group at SHARP for their help and financial contribution to the research.

REFERENCES

1. Grandjean et al., J. Appl. Phys. volume 83(3), pp. 1379-1383 (1998)
2. Z. Yang, L. K. Li, W. I. Wang, Applied Physics Letters 67(12), 1686 (1995)
3. Gil, Group III nitride semiconductor compounds, Oxford Science Publications (1998)
4. J. Elsner, R. Jones, M. I. Heggie, P. K. Sitch, Phys. Rev. B. **58** (19), 12571 (1998)
5. Ponce *et al.*, Appl. Phys. Lett. **69** (3), 337 (1996)
6. J. Tafto *et al.*, J. Appl. Cryst. **25**, 60-64 (1982)

Mat. Res. Soc. Symp. Vol. 595 © 2000 Materials Research Society

MBE Growth Of GaN Films In Presence Of Surfactants: The Effect Of Mg And Si

Guido Mula[1,2], Bruno Daudin[2], Christoph Adelmann[2], Philippe Peyla[3]
[1] INFM and Dipartimento di Fisica, Università di Cagliari, Cittadella Universitaria, Strada Prov.le Monserrato - Sestu Km 0.700, 09042 Monserrato (CA), Italy.
[2] Département de Recherche Fondamentale sur la Matière Condensée - SP2M - Laboratoire de Physique des Semiconducteurs, CEA-Grenoble, 17, Rue des Martyrs, 38054 Grenoble cedex 9, France
[3] Laboratoire de Physique et Modélisation des Milieux Condensés, Maison des Magistères, Université Joseph Fourier-CNRS, BP 166, 38042 Grenoble cedex 9, France.

ABSTRACT

We present here a description and an analysis of the modifications in the growth behaviour of GaN induced by the presence of foreign species. The particular cases of Mg and Si are analysed. Profound changes, both in microscopic and macroscopic scales, occur in presence of Mg, even for fluxes of about $1/1000^{th}$ of the Ga flux. The growth rate can be increased by almost 50%, depending of the III/V ratio and on the amount of Mg. A theoretical model is proposed to describe the observed effect. It is found that Mg induces changes in the Ga and N diffusion barriers and acts as a surfactant. The effect is stronger on the α-GaN than on the β-GaN, where N is more tightly bonded. The effect of Si is by far less pronounced, probably because it is more easily incorporated than Mg, and its effect on the surface kinetics is then strongly reduced.

INTRODUCTION

The understanding of the kinetics of the growing species on the GaN surface is of paramount interest for optimal epitaxial growth. In particular, the growth by Molecular Beam Epitaxy (MBE) is still far from perfection, partially due to the strong dependence of the surface morphology on the III/V ratio. This situation is especially true for plasma-source MBE, where the growth in Ga-rich regime leads to a metal accumulation that eventually blocks the growth process, while the N-rich regime leads to rough surfaces. Therefore, the use of surfactants to modify the surface mobility of the chemical species can be a useful step in the search for optimisation. In fact, the determination of less critical growth conditions, where the surface can be flat and free of Ga droplets, should lead to better optical and structural properties.

A few possible candidates for surfactants have already been examined. Theoretical [1] and experimental [2] reports show that As presence on the β-GaN surface leads to preferential growth of the cubic phase and to flatter surfaces. The use of Mg and Si during metal-organic chemical vapour deposition (MOCVD) growth [3] is shown to improve structural quality by reducing the defect density. The addition of In also shows significant surfactant effect [4], increasing the surface mobility. Si seems to have an "anti-surfactant" effect on MOCVD-grown α-GaN [5]. What can be extracted from these results is that the GaN growth seems to be extremely dependent on the presence of foreign species. Mg and Si are the most commonly used materials for doping, and the knowledge of their influence on the growth parameter is clearly fundamental. We will present here a study on Mg- and Si-induced modifications of the growth kinetics at α- and β-GaN surfaces. A phenomenological model is proposed for the α-GaN phase.

EXPERIMENTAL DETAILS

All samples were grown by plasma-assisted molecular beam epitaxy (MBE) in a MECA2000 growth chamber using an EPI Unibulb source for nitrogen. The typical N_2 flow was about 0.5 sccm for a forward power of 300 W. The growth temperature, read by a thermocouple on the sample backside, ranged from 650 to 680°C. The substrates were both Metal-Organic Chemical Vapour Deposition (MOCVD) epiready GaN/sapphire or sapphire (0001)-oriented for α-GaN. For the β-GaN the substrates were (001)-oriented SiC deposited by CVD on (001)-oriented Si substrates. The samples were analysed in situ by Reflection High-Energy Electron Diffraction (RHEED) and ex-situ by optical microscopy (OM). The OM images were used to analyse the Ga-droplet distribution.

RESULTS AND DISCUSSION

The effect of Mg - α-GaN growth

Here the observed effects are by far more striking and interesting. Figure 1 shows the growth rate of α-GaN as a function of the Ga cell temperature T_{Ga} for a given N-plasma condition and for different impinging Mg fluxes. The reference curve obtained without Mg is also shown for comparison. The growth rates have been determined by RHEED oscillations. Eye-guides are added for an easier analysis. The "residual surface Mg" label refers to Mg atoms desorbed from the growth chamber walls in the few days following the use of the Mg cell, due to the high Mg vapour pressure. It is difficult to precisely evaluate this flux, but a reasonable estimate sets it to a few orders of magnitude lower than a standard doping flux. A first remarkable feature in figure 1 is the overall increase in growth rate observed using a very low Mg flux. This increase can be as high as about 40% in the case of the curve obtained with a Mg cell temperature of 230°C. It is important to stress that the maximum Mg flux used here is of the order of 1/1000th of the Ga flux. This implies that is impossible to explain the growth rate increase by the simple incorporation of Mg into the GaN matrix. The effect must then be related to significant Mg-induced changes in the surface kinetics.

Another even more striking feature is the very surprising "hollow" in the growth rate observed for low Mg fluxes: above a critical Ga flux there is a very sharp decrease in the growth rate, followed by a new increase at still higher Ga

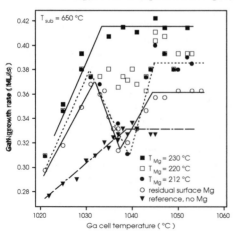

Figure 1. α-GaN growth rate as a function of the Ga cell temperature for various Mg cell temperatures. Eye-guides are added for an easier reading.

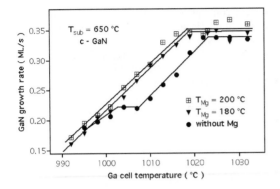

Figure 2. β-GaN growth rate as a function of T_{Ga} and for different temperatures of the Mg cell. The substrate temperature is 650°C.

fluxes. A more detailed analysis reveals that the growth rate slope in the Ga-limited regime is steeper in presence of Mg than without it, and that the new slope is in better agreement with the one expected from the Ga-cell flux behavior. For higher Mg cell temperatures (T_{Mg}), the hollow smoothens and eventually disappears. Above $T_{Mg} = 230$°C there was no further modification of the growth rate. Moreover, the RHEED pattern starts to degrade, indicating a rougher surface, and the RHEED oscillations were very difficult to observe. This behavior seems to be quite independent on the growth parameters. We do not observe qualitative changes increasing the N flow by a factor of 2 and/or varying the substrate temperature (650-680°C).

As far as the RHEED pattern is concerned, the most surprising effect is that the pattern remains streaky for Ga/N ratios quite far in the N-rich regime (the growth rate is decreased by almost 30%). The RHEED oscillations were intense and persistent below the critical T_{Ga} (from four-five clear oscillations for the lowest Ga flux to few tens near the critical point), slightly poorer within the hollow, where just 4-5 oscillations where visible, and improved again when further increasing the Ga flux. This seems to indicate two main phenomena: the first is that the surface seems to stay flat for growth in the Ga-limited regime, leading to a Ga droplets-free growth. The second is that something dramatic is happening in the vicinity of the critical Ga flux value, affecting the mobility of the species on the surface. These results will be discussed in more detail later in this article, when all the experimental details will be presented.

The effect of Mg – β-GaN growth

The growth of the cubic GaN phase is less affected by the presence of Mg than the hexagonal phase. In figure 2 we show the dependence of the growth rate on the Ga flux in the case of β-GaN, for two different impinging Mg fluxes. The reference curve without Mg is also shown for comparison. The observed behaviour is quite different from the α-GaN case. There is no longer a hollow, and the growth rate curve is quite regular. Adding Mg, the plateau around $T_{Ga} = 1005$°C disappears, the growth rate at higher T_{Ga} is increased, and no appreciable difference in the slope or in the final growth rate can be observed. Within the T_{Mg} range explored, the effect of Mg seems to be independent on the Mg flux value.

The effect of Si

There is a main difference in the behavior of Si and Mg atoms on the surface of GaN. Within the flux range we explored (doping level), the Si atoms are easily incorporated in the GaN matrix, limiting as a consequence the surface effect. As a matter of fact, to observe any effect on the kinetics of the Ga and N species we needed to create an accumulation of Si atoms on the surface, exposing the GaN surface to Si alone for a few seconds before starting the growth. The only observable effect was an increase of the growth rate in the plateau region as if we were not in the N-saturated regime. This is another clear indication, consistent with the results obtained with Mg, that the incorporation coefficient of N is definitely not unity. We tentatively attribute this behavior to a longer permanence time of N atoms on the surface, easing then their incorporation by the formation of GaN.

Optical microscopy and discussion

The observation of the hollow in the growth rate of α-GaN is quite a surprising phenomenon. How can a growth rate *decrease* when the amount of the available species *increases*? Before answering the question, a premise must be done about the reliability of RHEED oscillations for the measurement of growth rates. Several papers in the literature [6] point out that RHEED oscillations may be unreliable for determining growth rates, as in the case of step flow or bilayer growth. In our case, apart from the hollow, we observe a huge *increase* of the growth rate, ruling out the step flow argument. The regular variation of the growth rate excludes a possible frequency doubling effect. As far as the hollow is concerned, its appearance is so sharp that is difficult to believe it an artifact. In the following paragraphs we will discuss the possible meanings of the hollow, explaining at the same time why we strongly believe that this hollow corresponds to reality.

To understand the hollow, it is important to observe the full behavior of the growth rate, and in particular the fact that the slope of the curve in the Ga-limited regime is steeper in presence of Mg. It is well known that the excess Ga at the surface leads to the formation of Ga droplets, and it is a reasonable assumption that near the stoichiometry there are surface portions where the formation of Ga droplets becomes possible. This implies that the formation of Ga droplets should be also possible in the slightly N-rich regime. If we suppose that the presence of Mg on the surface stops the formation of Ga droplets, the slope of the growth rate should increase, because more Ga atoms would be available for the growth. The hollow in the growth rate can then be tentatively attributed to the sudden formation of Ga droplets above a critical Ga flux. To check our hypothesis, we performed OM measurements on the surface of α-GaN samples grown with and without Mg. The growth geometry was adapted to have a strong Ga flux gradient along the surface, so to be able to directly observe the onset of the Ga droplets formation as a function of the Ga flux. The Ga flux gradient was as high as 10%/cm. In figure 3 are shown two OM micrographs of the surface of samples without (left) and with (right) Mg. Below each micrograph there is the plot of the corresponding droplet counts over subsequent slices 0.1 mm wide and 1mm high. It appears clearly from the comparison of these two pictures that the threshold for Ga droplets formation is much sharper in presence of Mg than without it. On β-GaN (not shown), the droplet distribution with and without Mg is almost identical to the one observed for α-GaN without Mg.

We developed a phenomenological model for α-GaN, described in more detail in the reference [7], that takes into account in a simple way the Ga atoms capture cross sections for the formation of Ga droplets and for the incorporation into the GaN matrix. No specific assumptions are made on N, which is supposed to be available all the time. This means that our model is valid in the N-rich regime. Should the N sticking coefficient

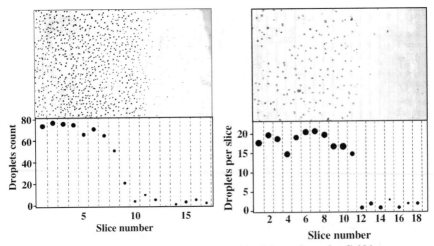

Figure 3. Optical micrographs (transmission mode) of the surface of α-GaN layers grown without (left) and with (right) Mg. The Ga flux gradient is about 10%/cm (right to left). Below each micrograph there is a plot of the corresponding droplet counting over slices 1mm wide and 0.1 mm high. The slice number is reported on the x-axis. The micrographs represent the measured area. The size of the markers is representative of the droplet size.

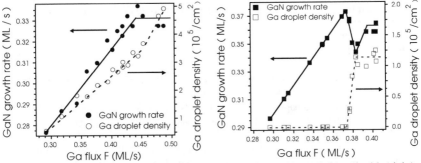

Figure 4. Growth rates (full symbols, left axis) for α-GaN without (left) and with (right) Mg. The corresponding Ga droplet densities (open symbols, left axis), obtained using our model, are also shown.

be lower than one, the number of available N atoms will be larger than the amount actually used, and our model will hold also for the Ga rich regime. We will not further discuss these issues here. In figure 4 we show the experimental growth rates (full symbols, left axis) obtained for α-GaN without (left) and with (right) Mg. The open symbols represent the corresponding Ga droplet density (right axis) obtained with our model through a fit to the experimental data. If we compare these results with those represented in the figure 3 we cannot but notice their remarkable agreement. First of all, the model explains the hollow with a sharp onset of the formation of Ga droplets, exactly as we observed experimentally. Second, the ratio of the droplet densities, with and without Mg, is about four in both the measurements and the theoretical fit.

These results clearly suggest that the Mg-induced kinetics modifications are a reduction of the Ga atoms mobility in the Ga-limited region (no droplets can be formed) and an increase of the mobility in the Ga-rich regime (the droplets are larger and less dense). Moreover, the amount of N available for the growth seems to be larger than what is usually assumed: the presence of Mg allows for a growth rate increase as high as about 30%. This strongly suggests that the sticking coefficient of N is not one and that it is highly dependent on the surface kinetics.

The effects on the β-GaN surface are considerably less pronounced, although the presence of Mg seems again to stop the Ga-droplet formation in the Ga-limited regime. As a matter of fact, we saw in figure 2 that the presence of Mg cancels the step in the growth rate present in the Ga-limited regime without Mg. The step is very reasonably related to the onset of Ga-droplet formation in the Ga-limited regime. This onset being very smooth, compared to the hollow of the α-GaN, it is very difficult to see the difference by means of optical microscopy of the growth with and without Mg.

CONCLUSIONS

The present study shows the deep changes induced by the presence of Mg and Si on the GaN surfaces during growth. Our experiments clearly indicate that the surface mobility on α-GaN can be dramatically affected by the presence of Mg. Moreover, several details strongly indicate that the sticking coefficient of N is probably far from unity and that more detailed studies on the MBE growth of GaN are necessary for improving samples quality.

REFERENCES

1.   T. Zywietz, Jörg Neugebauer, M. Scheffler, J. Northrup, C.G.Van de Walle, MRS Internet J. of Nitride Semicond. Res., **3**, Article 26, (1998).
2.   H. Okumura , H. Hamaguchi, G. Feuillet, Y. Ishida, S. Yoshida, Appl. Phys. Lett., **72**, 3056 (1998); G. Feuillet, H. Hamaguchi, K. Ohta, P. Hacke, H. Okumura, S. Yoshida, Appl. Phys. Lett., **70**, 1025 (1997).
3.   B. Beaumont, S. Haffouz, P. Gibart, Appl. Phys. Lett., **72**, 921 (1998); S. Haffouz, B. Beaumont, P. Gibart, MRS Internet J. of Nitride Semicond. Res., **2**, Article 8 (1998).
4.   F. Widmann, B. Daudin, G. Feuillet, N. Pelekanos, J.L. Rouvière, Appl. Phys. Lett., **73**, 2642 (1998).
5.   H. Hirayama, Y. Aoyagi, S.Tanaka, MRS Internet J. Nitride Semicond. Res., **4S1**, G9.4 (1999).
6.   E.S. Tok, J.H. Neave, F.E. Allegretti, J. Zhang, T.S. Jones, B.A. Joyce, Surf. Sci. **371**, 277 (1997); K. Sato, M.R. Fahy, B.A. Joyce, Surf. Sci. **315**, 105 (1994).
7.   Guido Mula, B. Daudin, P. Peyla, Proceedings of the 3rd Int. Conference on Nitride Semiconductors, Montpellier (France), July 1999, in press.

Mat. Res. Soc. Symp. Vol. 595 © 2000 Materials Research Society

The Effect of Al in Plasma-assisted MBE-grown GaN

Otto Zsebök, Jan V. Thordson, Qingxiang Zhao, Ulf Södervall,
Lars Ilver, and Thorvald G. Andersson
Applied Semiconductor Physics
Department of Microelectronics and Nanoscience
Physics and Engineering Physics
Chalmers University of Technology and Göteborg University
SE-412 96 Göteborg, Sweden

ABSTRACT

We have grown GaN, with addition of a 0.10 to 0.33 % Al, on sapphire(0001) substrates by solid-source RF-plasma assisted MBE. The Al-concentration was determined by secondary ion-mass spectrometry and Auger-electron spectroscopy, while the layer quality was assessed by photoluminescence and high-resolution scanning electron microscopy. Microscopy revealed a meandering pattern and a surface roughness varying with Al-content. The smallest surface roughness was obtained at 0.10 % Al. Photoluminescence revealed two main peaks attributed to the neutral donor-bound exciton. Its energy increased slightly with Al-concentration, which established a correlation between the Al-concentration and the band gap.

INTRODUCTION

Optoelectronic devices, based upon heterostructures of GaN and its ternary AlGaN, have attracted great interest during the recent years. By varying the Al-content the whole ultraviolet range around 3.4 - 6.2 eV can be covered [1,2]. It has earlier been shown that GaN surface morphology can be improved during GaN molecular beam epitaxy (MBE) growth by adding In as a surfactant, enhancing the migration of the impinging Ga-atoms on the surface [3,4]. Addition of alloy material in form of isoelectronic doping is easily made in the epitaxy process. Such small concentrations have definite effects by improving the crystalline quality in compound semiconductors in terms of deep levels, unintentional doping concentrations and dislocation density [4]. Earlier photoluminescence investigations on $Al_xGa_{1-x}N$ at low temperatures have shown that the neutral donor bound exciton emission $(D°,X)$ dominates in the PL spectrum [5,6].

To the best of our knowledge, no systematic investigation of (Al)GaN with Al at doping concentrations has been undertaken before. In this work we have grown a series of GaN samples containing small amounts of Al. The incorporated concentration ranged from 0.10 up to 0.33%. The aluminium concentration was controlled by the Al furnace temperature and measured by secondary ion-mass spectrometry (SIMS) and Auger-electron spectroscopy (AES). We assessed the layer quality by photoluminescence (PL), Hall effect measurements and high-resolution scanning electron microscopy (SEM). The PL-peaks were deconvoluted to determine the main transitions contributing to the PL signal. Our investigations revealed that adding a small amount of Al, improved the electrical, optical and even morphological properties of the GaN.

EXPERIMENTAL DETAILS

The structures were grown in a Varian GEN II MBE-system with a solid Ga-source and plasma-assisted activation of N_2, using a liquid nitrogen cooled Oxford CARS25 RF-source. The process gas was of 5N5-quality and additionally purified in an Aeronex catalytic gas purifier. The Ga-flux was calibrated by RHEED-oscillations during GaAs growth and converted to the growth rate of hexagonal GaN. The substrates were sapphire

(Al_2O_3), cut along the (0001)-plane, with a 0.5 µm thick Ti-layer on the backside and the growth temperature was measured by a pyrometer. Before insertion into the MBE-system the substrates were ultrasonically degreased in trichloroethylene, acetone and methanol and etched for 15 min at 48 °C in a mixture of $HCl:H_3PO_4$ (3:1). In the growth chamber the samples were first outgassed at 700 °C for 30 minutes, and thereafter nitridised by exposure to active nitrogen for another 30 minutes. At the same temperature a 110 nm thick GaN buffer-layer was subsequently grown at low growth rate 0.15 µm/h and with minimum nitrogen flow (0.6 sccm) and power (200 W) in order to maintain the RF-plasma of the nitrogen source. A (2x2) reconstruction pattern was obtained after a few minutes growth of the buffer layer, and this surface periodicity remained throughout the whole deposition. We denote the dilute $Al_xGa_{1-x}N$ alloy layers by (Al)GaN which are intermediate between normal alloys with x > 1% and isoelectronically doped layers, GaN:Al with x < 10^{-4}. The (Al)GaN bulk layer was grown (with the Al shutter open) using a nominal Ga-flux corresponding to a growth rate of 1 µm/h. The growth temperature was 750 °C, and RF-power and N_2-flux were 600 W and 3.0 sccm, respectively. The (2x2) reconstruction has been taken as a proof of a Ga-rich growth [7]. For a growth time of 1 h, cross-sectional SEM investigations indicated a thickness of only 600 - 700 nm which is a result of Ga desorption.

A measure of the Al-concentration in the samples was made by SIMS, using the peak intensity ratio between the Al and Ga-signals. The relation between the SIMS-signal and the Al-composition provided by the MBE-growth was fairly linear. The absolute calibration of about 3% Al in GaN was made by AES through the intensity ratio of the Al_{KLL} and Ga_{LMM} peaks. This value was used as the calibration value for the SIMS results. The AES was made with a Perkin-Elmer Physical Electronics PHI 590 equipment. An electron accelerating voltage of 3 keV at 1.5 µA beam current provided an analysis spot size of 3 µm.

For the PL investigation the samples were mounted on the cold finger of a closed cycle Cryogenic Lab System, CSW-204SLB cryostat, allowing the temperature to be varied between 6.5 and 350 K. The excitation was made by a Liconix He-Cd UV laser, giving a typical power of 25 mW at a wavelength of 325 nm. The monochromator was a Spex 270M single grating monochromator with a Photometrix AT200 CCD camera as detector. Hall effect measurements were obtained using 3x3 mm samples in the van der Pauw geometry. The contacting material was In and the contacts were annealed at 425 °C for 2 minutes. To characterise the sample surfaces a JEOL JSM-6301F scanning electron microscope was used.

RESULTS

Low-temperature PL-spectra (at 6.5 K) from two representative samples are shown in Figure 1. The upper spectrum is from pure GaN while the other is from (Al)GaN. The peak is non-symmetric with a low-energy tail, which is reduced with Al-concentration. The spectra have a typical peak width of ~20 meV. In order to resolve the intensity contributions, the peaks were fitted with Gaussian functions. Four Gaussian curves were extracted, except for the GaN-spectrum where 5 individual peaks were used. In this case a very good fit, better than 99%, was obtained to the measured results. This gives an estimated maximum error of ±2 meV in the extracted peak positions. In GaN, the two extracted peaks with the highest energies at 3.476 and 3.470 eV were clearly more intense and had a smaller line width than the others, 10 and 17 meV, respectively. These peaks are neutral donor-bound excitons ($D_1°,X$) and ($D_2°,X$), respectively [8]. The third deconvoluted peak is at 3.45 - 3.46, representing recombination from excitons bound to neutral acceptors ($A°,X$) [6,9]. The two further

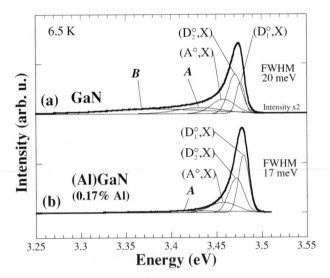

Figure 1. *Low-temperature photoluminescence spectra for a) pure GaN and b) (Al)GaN with 0.17% Al. Since the spectra clearly contain a low energy tail, especially for GaN, the peaks have been deconvoluted into 4 (Al)GaN or 5 (GaN) Gaussian peaks. The high-energy peaks were assigned to the (D_1°,X), (D_2°,X) and (A°,X), respectively. Peak A is from the recombination of a free hole with an electron bound on an impurity centre, identified with oxygen, while peak B is related to excitons deeply bound to defect centres at dislocations.*

peaks, A, at 3.41-3.43 eV and B around 3.36 eV, have clearly lower intensity and much larger line width compared to the previous ones. Peak A in the shoulder region is typical for free-to-bound emission involving shallow oxygen donors [10], while the very broad peaks around 3.36 eV can be attributed to localised excitons, deeply bound to defects, such as dislocations close to the interface [10,11]. At 0.17% Al-content the FWHM of the PL intensity was slightly reduced to 17 meV, due to the decreasing shoulder region and vanishing low-energy tail. This can be assigned to the decreased (A°,X) intensity and partly to the strongly reduced concentration of shallow oxygen donors related to the free-to-bound emisson. The peak B vanished by adding small amounts of Al to the GaN, which may be due to improved surface morphology as a result of the reduction of defects.

In Figure 2 the energy position of the resolved peaks (solid lines) are shown as a function of the Al-concentration. The dashed line represents the position of the PL-peak. Its energy increases linearly with Al-concentration given by

$$E_{peak}^{alloy} = E_{peak}^{GaN} + bx. \tag{1}$$

In our case we get b = 2.7±0.3 eV. This represents the linear change of the $Al_xGa_{1-x}N$ band gap energy. The GaN-matrix is distorted by the Al-atoms and therefore the band gap is modified already at the low compositions of ~0.10% Al. The rate of change for the (D_2°,X) and (A°,X) is similar, but somewhat smaller, to the one for the (D_1°,X), which strengthens the suggestion that the peak positions are due to the band gap change. The variation of the peak position for A has a well defined minimum at

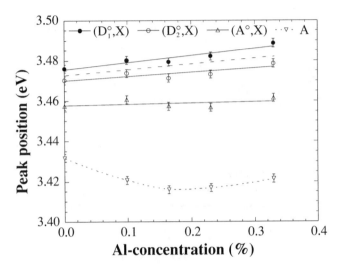

Figure 2. *The extracted peak positions as a function of Al-concentration. The lines represent the change of the (D_1°,X), (D_2°,X) and (A°,X) peaks which is parallel with the increase of the AlGaN-alloy band gap. The dashed line is the PL-peak position, while the dotted line represents the position of peak A.*

~0.17%. However, this may be a result of peak mixing, which can not be resolved at the fitting procedure we used. The relative area of the extracted peaks increased linearly with the Al-concentration, except the (D_1°,X)-peak, which had a definite maximum at ~0.1 %Al. The FWHM of the donor-bound excitons increased more pronounced above 0.1% Al.

In Figure 3, high-resolution SEM images demonstrate the familiar meandering pattern of the GaN surface when grown on sapphire. On the pure GaN-sample, Figure 3a, the presence of crystallites is significant. Adding small Al-concentrations smoothens the surface. One effect in PL was to reduce the peak B, which was attributed to excitons deeply bound to defect centres correlated to dislocations. Our sample surfaces are different from the results in an earlier investigation by SEM on surfaces of AlGaN, where hexagonally shaped depressions with some micrometre diameter were observed on otherwise smooth surfaces [12].

The Al-concentration in AlGaN films grown by MBE at high growth temperatures depends on Ga-desorption and the supply of nitrogen [13]. In our films the accuracy of the Al-concentration is determined within 10% based on SIMS measurements and the AES calibration. It is clear that adding a small, well controlled concentration of Al to the GaN has a definite effect on the PL-spectrum. We emphasise the difference in energy between the optical band gap and the excitonic peaks (D°,X) and (A°,X). These extracted peak positions are close to the measured peak maximum in Figure 1, which therefore follows the plotted lines in Figure 2 within ~ 5 meV. The (D°,X) transitions are believed to be only ~30 meV from the band edge. Therefore, this small difference between the band gap and the experimental peak with its excitonic transitions, support that the shift of the peak positions is due to the increase of the band gap despite the dilute alloy concentration.

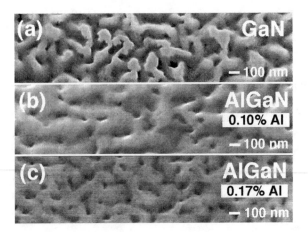

Figure 3. *High-resolution scanning electron microscopy images of the sample surfaces with increasing Al-content from GaN to (Al)GaN with 0.17% Al. In general there is an improvement of the surface morphology with Al-content in the layers.*

The variation of the band gap, or the PL peak position, with alloy concentration is usually described by

$$E(x) = E_{GaN} + b \cdot x + c \cdot x^2,$$ (2)

where $E(x)$ is the measured peak (band gap), E_{GaN} is the reference peak (the GaN band gap), and b and c are constants (c is the bowing parameter). When the position of the PL-emission follows the change in band gap, almost identical parameters, b and c, are used for band gap and exciton emissions in PL spectra. In some cases the band gap variation has been considered linear, i.e. c=0. For small concentrations, x = 0.05, the third term in the quadratic approximation is negligible. The values of b range from 0.64 [14] to 2.8 [15]. Most measurements have been made for x-values above 5% and typically much larger values. The slope in Figure 2 depends very much on the accuracy in determination of the Al-concentration and contributions to the band gap from strain due to differences in lattice constant and thermal expansion coefficients. The strain effect is believed to be limited at thin film thicknesses, below 1 μm, since the columnar GaN-structure is formed to relax strain energies, but peak shifts of 11 meV due to strain have been reported [6]. From these considerations we estimate b=2.7±0.3. The ~4 meV increase in band gap at only 0.10% Al is due to the relatively short distances between the isoelectronic atoms, ~18 Å which is enough to modify the band gap due to the shorter bonding length around the Al-atoms.

During growth of GaN by MBE, impurities from oxygen contamination of the nitrogen gas will be introduced together with the activated nitrogen and incorporates in the layer besides the column III material. The unintentional doping of the GaN by oxygen results in free electrons in the conduction band. Since both aluminium and oxygen are reactive, it is expected that the Al/O-ratio affects the incorporation in the alloy as well as the free electron concentration. The electrical characteristics change with Al-concentration, and details of this will be reported elsewhere.

As shown in the high-resolution SEM images in Figure 3, the concentration of the small dark areas, which are local hollows, have only a low or no dependence on the aluminium concentration. These are a result of the lattice mismatch which is almost constant. It is also clearly notable that the average surface smoothness improves with Al-concentration. This smoothness, and the presence of good RHEED patterns during growth, indicate very flat upper surfaces on the extended 10 - 100 nm wide terraces. A detailed analysis of the morphology shows a somewhat smoother surface for alloy concentrations giving improved electrical characteristics.

CONCLUSIONS

We have grown GaN and (Al)GaN by MBE, and characterised GaN with small Al-concentrations, in the range of 0.10 - 0.33%. The Al-concentration was determined by SIMS and AES. Low-temperature PL-emission indicated a linear increase of the band gap, $E(x) = E_{GaN} + 2.7 \cdot x$. The PL-peak intensity showed a minimum FWHM at 0.17% Al with 17 meV. The PL-peaks were deconvoluted into the $(D_1°,X)$, $(D_2°,X)$, $(A°,X)$ and shallow oxygen donor related peaks. High-resolution SEM analysis of the surfaces showed slightly improved morphology for alloy concentrations around 0.10%. This was confirmed by decreased PL intensities attributed to excitons bound to dislocation sites.

REFERENCES

1. S. Nakamura, *J. Cryst. Growth*, **170**, 11 (1997).
2. P. Shah and V. Mitin, *J. App. Phys.*, **81**, 5930 (1997).
3. F. Widmann, B. Daudin, G. Feuillet, N. Pelekanos, and J. L. Rouvière, *Appl. Phys. Lett.*, **73**, 2642 (1998).
4. C. K. Shu, J. Ou, H. C. Lin, W. K. Chen, and M. C. Lee, *Appl Phys. Lett.*, **73**, 641 (1998).
5. G. Steude, T. Christmann, B. K. Meyer, A. Goeldner, A. Hoffmann, F. Bertram, J. Christen, H. Amano, and I. Akasaki, *MRS Internet J. Nitride Semicond. Res.*, **4S1**, G3.26 (1999).
6. A. V. Andrianov, D. E. Lacklison, J. W. Orton, D. J. Sewsnip, S. E. Hooper, and C. T. Foxon, *Semicond. Sci. Technol.*, **11**, 366 (1996).
7. E. J. Tarsa, B. Heying, X. H. Wu, P. Fini, S. P. DenBaars, and J. S. Speck, *J. Appl. Phys.*, **82**, 5472 (1997).
8. D. Volm, K. Oettinger, T. Streibl, D. Kovalev, M. Ben-Chorin, J. Diener, and B. K. Meyer, *Phys. Rev. B*, **53**, 16543 (1996).
9. O. Lagerstedt and B. Monemar, *J. Appl. Phys.*, **45**, 2266 (1974).
10. S. Strauf, P. Michler, J. Gutowski, H. Selke, U. Birkle, S. Einfeldt, D. Hommel, *J. Cryst. Growth*, **189/190**, 682 (1998).
11. M. Smith, G. D. Chen, J. Y. Lin, H. X. Jiang, A. Salvador, B. N. Sverdlov, A. Botchkarev, and H. Morkoç, *Appl. Phys. Lett.*, **66**, 3474 (1995).
12. A. Y. Polyakov, A. V. Govorkov, N. B. Smirnov, M. G. Mil'vidskii, J. M. Redwing, M. Shin, M. Skowronski and D. W. Greve, *Solid State Electr.*, **42**, 637 (1998).
13. J. R. Jenny, J. E. Van Nostrand, and R. Kaspi, *Appl. Phys. Lett.*, **72**, 85 (1998).
14. M. D. Bremser, W. G. Perry, T. Zheleva, N. V. Edwards, O. H. Nam, N. Parikh, D. E. Aspnes, and R. F. Davis, *MRS Internet J. Nitride Semicond. Res.*, **1**, 8 (1996).
15. D. Korakakis, H. M. Ng, M. Misra, W. Grieshaber, T. D. Moustakas, *MRS Internet J. Nitride Semicond. Res.*, **1**, 10 (1996).

Mat. Res. Soc. Symp. Vol. 595 © 2000 Materials Research Society

Structure and Morphology Characters of GaN Grown by ECR-MBE Using Hydrogen-Nitrogen Mixed Gas Plasma

Tsutomu Araki, Yasuo Chiba and Yasushi Nanishi
Department of Photonics, Ritsumeikan University,
1-1-1 Noji-Higashi, Kusatsu 525-8577, Japan.

ABSTRACT

GaN growth by electron-cyclotron-resonance plasma-excited molecular beam epitaxy using hydrogen-nitrogen mixed gas plasma were carried out on GaN templates with a different polar-surface. Structure and surface morphology of the GaN layers were characterized using transmission electron microscopy. The GaN layer grown with hydrogen on N-polar template showed a relatively flat morphology including hillocks. Columnar domain existed in the center of the hillock, which might be attributed to the existence of tiny inversion domain with Ga-polarity. On the other hand, columnar structure was formed in the GaN layer grown with hydrogen on Ga-polar template.

INTROCUCTION

GaN and related III-V nitrides are promising semiconductors which can have applications in both optical devices including light emitting diodes in the blue-green and ultraviolet wavelength region, and electronic devices operating at high temperature, high frequency and high power. The recent commercial realization of light emitting diodes and the achievement of semiconductor lasers in the III-V nitrides have promoted much more attention in the field of research. These advances have been primarily achieved using metalorganic chemical vapor deposition (MOCVD) for the crystal growth. Molecular beam epitaxy (MBE) has been considered a suitable technique for the fabrication of high-quality device structure using III-V nitrides due to the high controllability of atomic order thickness, interface abruptness, composition and so on. One of the key issues plaguing the MBE growth of III-V nitrides has been a relatively low growth rate compared to other common growth techniques. Recently, high growth rates of GaN have been achieved by the development of nitrogen plasma sources. Fujita et al reported a growth rate of 1.4 μm/h in GaN growth using RF nitrogen plasma source [1]. Another approaches to achieve GaN growth with high growth rates is using an ammonia as the nitrogen source. Yang et al. [2] and Grandjean et al. [3] achieved growth rates of 1 μm/h and 1.2 μm/h respectively by MBE using the ammonia source. On the other hand, Zhonghai Yu et al. reported an increase in GaN growth rate by as much as a factor of 2 by the addition of atomic hydrogen during MBE growth [4]. We have also demonstrated the increase in the growth rate using hydrogen and nitrogen mixed gas plasma for electron-cyclotron-resonance (ECR) plasma-excited MBE [5]. In previous works [6,7], structural changes in GaN grown with hydrogen were investigated using electron microscopy, in which the dispersed columnar structure with a hexagonal pyramid surface was observed. The formation mechanism of the columnar structure was discussed by assuming nonuniformity of polarity of GaN surface. In this work, to develop a deeper understanding of the growth process of MBE-grown GaN with hydrogen, GaN growth by ECR-MBE using hydrogen and nitrogen mixed gas plasma were carried out on GaN templates with different polarity.

EXPERIMENTAL

Two types of GaN templates were grown by ECR-MBE for the GaN growth with hydrogen. The one (type A) was the GaN layer grown on (0001)sapphire after nitridation at 450°C for 5 min and growth of a low-temperature buffer layer at 450°C. The other one (type B) was the GaN layer grown directly on a thick GaN layer grown by hydride vapor phase epitaxy (HVPE). The growth temperature for the both GaN templates was 700°C. Ga cell temperature was 910°C. The microwave power and nitrogen flow rate were kept constant at 120 W and 30 sccm, respectively. After growth of the GaN templates, reflection high energy electron diffraction (RHEED) was used to determine the polarity of the GaN templates.

GaN epitaxial layers were grown on the GaN templates. Hydrogen gas with a flow rate of 6 sccm was added to the nitrogen plasma only during the growth of the epitaxial layers. Other growth conditions were the same as the conditions for the GaN templates.

The structure and morphological characteristics of the GaN layers were investigated using transmission electron microscopy (TEM). TEM specimens for cross-sectional observation were prepared by mechanical thinning and Ar-ion milling. The samples were observed with a JEM-2010 electron microscope operated at 200 keV.

RESULTS AND DISCUSSION

It has been well established by Smith et al. that RHEED reconstruction pattern showed different changes depending on the polarity of the GaN layer. When the substrate

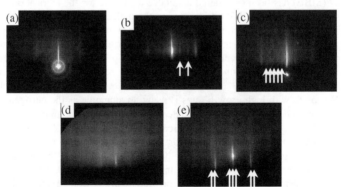

Figure 1. RHEED reconstruction patterns observed from the GaN templates. (a) 1×1 after growth, (b) 3×3 around 250°C and (c) 6×6 around 200°C for the template grown on (0001)sapphire after nitridation and growth of a low-temperature buffer layer (type A). (d) 1×1 after growth and (e) 1+1/6 around 250°C for the template grown directly on a thick GaN layer grown by HVPE (type B).

temperature is decreased, the surface reconstruction patterns with N-polarity are identified 1×1, 3×3 and 6×6 patterns [8] while a (1+1/6) reconstruction pattern corresponds to the Ga-polar surface [9]. Figure 1 shows observed RHEED reconstruction patterns at different substrate temperature for both the type A and B GaN templates. The type A templates showed the 3×3 pattern around 250°C and the 6×6 pattern around 200°C, indicating N-polarity. On the other hand, the type B templates showed the 1+1/6 pattern around 300°C, confirming Ga-polarity.

The growth of GaN epitaxial layer using nitrogen-hydrogen mixed gas plasma were carried out on these GaN templates with different polarity. Figure 2 shows a cross-sectional TEM image of the GaN layers grown on type A templates with N-polarity. The structure and surface morphology of the GaN layer are very different from that grown on sapphire substrate without template, which showed a dispersed columnar structure with hexagonal pyramids [6, 7]. Neither columnar structure and hexagonal pyramids were observed. However, a high density of stacking faults were observed in the GaN layer as well as the results in ref. [7]. Since the growth time for the template and the epilayer was same, it is evident that growth rate of the GaN grow with hydrogen was decreased compared to that of the template. The surface morphology was relatively flat although some hillocks existed as shown in Fig. 2 by an arrow. The density of the hillocks were around $10^9/cm^2$.

Figure 3 shows cross-sectional TEM image of the hillock observed using a [0001] diffraction vector. It is found that a columnar domain with a width of about 100 nm existed in the center of hillock. Rouviere et al. reported that when the GaN layer has a dominant N-polarity, the surface showed hexagonal pyramids and flat tops [10]. They confirmed the existence of inversion domains with Ga-polarity in the pyramids by convergent beam electron diffraction techniques, and attributed the pyramid shape to a different growth rate between Ga-polarity and N-polarity. Although we have not determined the absolute polarity of the inversion domains, it is reasonable also in our

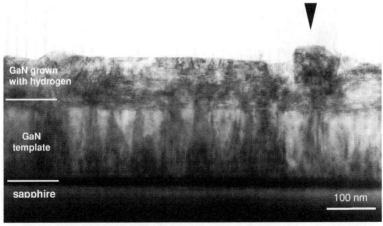

Figure 2. Cross-sectional TEM image along the [11$\bar{2}$0] of GaN layers grown with hydrogen on the type A GaN template with N-polarity.

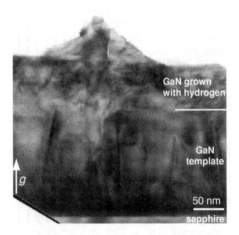

Figure 3. Cross-sectional TEM image observed using a [0001] diffraction vector, showing the hillock in the GaN layer grown with hydrogen on the type A GaN template. The columnar domain extended from the GaN template was clearly observed.

case that the columnar domain in the hillock was formed on the inversion domain with Ga-polarity, growing faster than the N-polar matrix.

Figure 4 shows a cross-sectional TEM image of the GaN layer grown with hydrogen on the type B GaN template with Ga-polarity. The growth rate of GaN grown

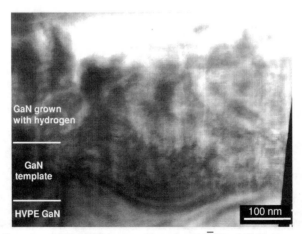

Figure 4. Cross-sectional TEM image along the [11$\bar{2}$0] of GaN layers grown with hydrogen on the type B GaN template with Ga-polarity.

with hydrogen was obviously higher than that of the GaN template grown for same time without hydrogen. It is evident that columnar structure was formed in the GaN layer, but the surface of the column did not have faceted planes. The results on the characters of the columnar domain is different from those of hexagonal columns with faceted surface [6], which is attributed to the difference in the polarity of the surface before growth.

From these results, it is evident that the structure and surface morphology of the GaN layers grown with hydrogen are strongly dependent of the polarity of the underlayer. Therefore, it should be very important to understand and control the polarity of the surface for the subsequent GaN epitaxial growth by ECR-MBE using hydrogen-nitrogen mixed gas plasma.

CONCLUSION

In conclusion, we have grown the GaN layers by ECR-MBE using hydrogen-nitrogen mixed gas plasma on GaN templates with both N-polar and Ga-polar-surface. Structure and surface morphology of the GaN layers were characterized using TEM. The relatively flat surface morphology including hillocks were observed in the GaN layer grown with hydrogen on N-polar template. It is found that the columnar domain existed in the center of the hillock. The columnar domain in the hillock was considered to be formed on the inversion domain with Ga-polarity, growing faster than the N-polar matrix. On the other hand, the GaN layer grown with hydrogen on Ga-polar template displayed columnar structure.

ACKNOWLEDGEMENT

This work was partly supported by Foundation for Promotion of Material Science and Technology of Japan (MST Foundation) and "Academic Frontier Promotion Project" from Ministry of Education, Science, Sports and Culture.

REFERENCES

1. N.Fujita, M.Yoshizawa, K.Kushi, H.Sasamoto, A.Kikuchi and K.Kishino, J.Cryst.Growth, **189/190,** 385 (1998).
2. Z.Yang, L.K.Li and W.I.Wang, Appl.Phys.Lett., **67,** 1686 (1995).
3. N.Grandjean, M.Leroux, M.Laugt and J.Massies, Appl.Phys.Lett., **71,** 240 (1997).
4. Zhonghai Yu, S.L.Buczkowsk, N.C.Giles, T.H.Myers and M.R.Richards-babb, Appl.Phys.Lett., **69, 2731** (1996).
5. Y.Chiba, T.Tominari, M.Nobata and Y.Nanishi, Inst.Phys.Conf.Ser., **162,** 651 (1998) 651.
6. T. Araki, Y. Chiba, M. Nobata, Y. Nishioka and Y. Nanishi, J.Cryst.Growth, (1999) (to be published).
7. T. Araki, Y. Chiba and Y. Nanishi, J.Cryst.Growth, (1999) (to be published).
8. A.R.Smith, R.M.Feenstra, D.W.Greve, M.-S.Shin, M.Skowronski, J.Neugebauer and J.E.Northrup, Appl.Phys.Lett. **72,** 2114 (1998).

9. A.R.Smith, R.M.Feenstra, D.W.Greve, M.-S.Shin, M.Skowronski, J.Neugebauer and J.E.Northrup, J.Vac.Sci.Technol. **B 16**, 2242 (1998).

10. J.L.Rouviere, M.Arlery, R.Niebuhr, K.H.Bachem and O.Briot, MRS Internet J. Nitride Semicond. Res. **1**, 33 (1996).

Mat. Res. Soc. Symp. Vol. 595 © 2000 Materials Research Society

ELECTRICAL PROPERTIES OF CUBIC InN AND GaN EPITAXIAL LAYERS AS A FUNCTION OF TEMPERATURE

J.R.L. Fernandez[*], V.A. Chitta[**], E. Abramof[***], A. Ferreira da Silva[***], J.R. Leite[*], A. Tabata[*], D.J. As[****], T. Frey, D. Schikora, and K. Lischka

[*]Instituto de Física da USP, C.P. 66318, 05315-970 São Paulo, SP, Brazil,
rafael@macbeth.if.usp.br
[**]Universidade São Francisco, Centro de Ciências Exatas e Tecnológicas,
13251-900 Itatiba, SP, Brazil
[***]Instituto Nacional de Pesquisas Espaciais (INPE-LAS), C.P. 515,
12201-970 São José do Campos, SP, Brazil
[****]Universität Paderborn, FB-6 Physik, Warburger Strasse 100, D-33095 Paderborn,
Germany

Cite this article as: MRS internet J. Nitride Semicond. Res. **595**, W3-40 (2000).

ABSTRACT

Carrier concentration and mobility were measured for intrinsic cubic InN and GaN, and for Si-doped cubic GaN as a function of temperature. Metallic n-type conductivity was found for the InN, while background p-type conductivity was observed for the intrinsic GaN layer. Doping the cubic GaN with Si two regimes were observed. For low Si-doping concentrations, the samples remain p-type. Increasing the Si-doping level, the background acceptors are compensated and the samples became highly degenerated n-type. From the carrier concentration dependence on temperature, the activation energy of the donor and acceptor levels was determined. Attempts were made to determine the scattering mechanisms responsible for the behavior of the mobility as a function of temperature.

INTRODUCTION

In the past few years nitride-based nanostructures have been successfully used in the fabrication of optoelectronic devices as well as in the development of high frequency and high-temperature electronic devices [1-4]. Most of the applications made so far are based on the hexagonal (h) phase of the nitride materials. However, cubic (c) GaN/GaAs (001) layers grown by metal organic chemical vapor deposition (MOCVD) and by plasma assisted molecular beam epitaxy (MBE) have recently been used to fabricate p-n junction light emitting diodes [5-6]. In order to improve on the performance of these devices, further studies of the optical and electrical properties of the nitride layers are required. Particularly, the improvement of the n- and p-type doping levels and carrier mobilities in these layers is a crucial task for the device technology, mainly for that involving the cubic phase of the materials [7,8].

In the present work measurements of carrier concentrations and mobilities of unintentionally doped c-InN and c-GaN MBE grown epitaxial layers are performed as a function of temperature. c-GaN samples doped with silicon are also investigated in the temperatures range 10 to 350K. The carrier concentrations and mobilities obtained for the cubic (zinc-blend) InN and GaN layers are compared with the results available for the corresponding hexagonal (wurtzite) nitride samples.

SAMPLES

The c-InN sample was grown on GaAs/InAs buffer layers firstly grown on GaAs (001) substrate by plasma-assisted MBE. We have used a Riber 32-system equipped with elemental sources of Ga, As, In, Si and an Oxford Research CARS25 radio frequency plasma source for the reactive nitrogen. After the growth of the GaAs buffer layer at a temperature of 610°C, we grew a 300nm layer of InAs at a temperature of 480°C under (2x4) reconstruction. The growth of the InN layer, with about 300nm width, was performed at a reduced temperature of 450°C. Further details about the growth of the c-InN layer are given in Refs. [9,10]. The same Riber 32-system was used to grow the intrinsic and the Si-doped c-GaN layers on GaAs (001) substrates. Elemental Si was evaporated at source temperatures between 750°C and 1100°C.

The widths of the c-GaN layers are of about 0.8 μm [8,11]. To perform the electrical characterization of the samples we have used Hall-effect measurements as a function of temperature at a 0.5T magnetic field. Ohmic contacts using In were accomplished according to Van der Pauw geometry. The characteristics of our samples are shown in Table I.

Table I – Characteristics of the cubic nitride samples: N_D is the nominal silicon concentration, n and p are the measured electron and hole concentrations at 300K, respectively.

Sample	N_D (cm^{-3})	n (cm^{-3})	p (cm^{-3})
c-InN	intrinsic	5×10^{20}	
c-GaN	intrinsic		1×10^{16}
c-GaN:Si			
#1	3.7×10^{17}		2.2×10^{16}
#2	1.2×10^{19}		7.0×10^{16}
#3	1.2×10^{19}		7.0×10^{16}
#4	2.6×10^{19}	1.1×10^{19}	
#5	5.5×10^{19}	2.6×10^{19}	
#6	1.2×10^{20}	7.2×10^{19}	
#7	2.6×10^{20}	8.0×10^{19}	
#8	2.6×10^{20}	1.0×10^{20}	
#9	2.6×10^{20}	1.2×10^{20}	

INTRINSIC c-InN

The measured carrier concentration in the unintentionally doped c-InN layer as a function of inverse temperature is shown in Fig.1. High-electron concentrations of the order of 10^{20} cm^{-3} were found in the range of temperature from 10 to 300K. This metallic behavior was also observed in MOCVD grown h-InN layers and attributed to the presence of native nitrogen vacancies in the sample [12]. The same behavior was also observed for metal organic MBE grown c-InN [13]. Fig. 1 shows that a sharp increase of the electron concentration occurs when the temperature raises above 200K. This behavior which was not observed in the h-InN layers may indicate that another deeper donor level, with activation energy of about 94,5 meV, exists in our c-InN layer. It is well known that ab initio calculation lead to zero or negative gap values for the InN binary compound. This fact has hindered severely our knowledge about native defects in this material from theory. On the other hand, very few attempts have been made so far to grow c-InN films [9]. Thus, systematic effort is required to identify the origin of the defects or impurities, which give rise to such high electron concentrations observed in our c-InN layer.

The electron mobilities measured by us in the c-InN layer as a function of temperature are shown in Fig.2. In the temperature range of 10 to 260K the measured mobilities are one order of magnitude smaller than those observed for the h-InN films [12]. This probably indicates the best quality of the hexagonal films. Above 200K a decrease of the mobilities follows the sharp increase in the concentrations. Attempts were made to adjust the observed behavior of the mobilities to different scattering and screening mechanisms. We expect that impurity scattering is involved in the changes of the mobility at lower temperatures and scattering by phonons is active at higher temperature, although the T dependencies obtained by us deviate from the expected ones for these mechanisms. Further investigations of both, h-InN and c-InN films are required to understand and control their electrical properties.

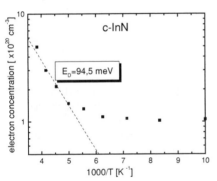 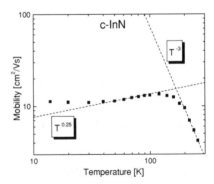

Figure 1: Electron concentration as a function of temperature for the intrinsic c-InN. The dashed line is used to calculate the activation energy E_D=94.5 meV.

Figure 2: Mobility as a function of temperature for the intrinsic c-InN.

INTRINSIC c-GaN

The Hall carrier concentration measured as a function of temperature for the nominally undoped c-GaN sample is shown in Fig. 3. Due to the presence of an intrinsic acceptor level a background hole concentration is obtained. This concentration increases from p=3.7 x 10^{13} cm^{-3} at 100 K (not shown in the Figure) to p=2.0 x 10^{16} cm^{-3} for T=350 K. The carrier concentration behavior near the room temperature region allows us to determine an activation energy of E_A=166 meV for the involved acceptor. Therefore, we can estimate an acceptor concentration of $N_A \approx 4 \times 10^{18}$ cm^{-3} for this nominally undoped sample. The measured hole mobility is presented in Fig. 4 as a function of temperature. The mobility initially increases with temperature reaching a maximum value of 1250 cm^2/Vs at T=120 K and then decreases to a value of 283 cm^2/Vs at room temperature. As can be seen in Fig. 4, the $T^{3/2}$ behavior, shown at low temperature, suggest that in this region the mobility is limited by ionized impurity scattering. Above 120 K the decrease of mobility, proportional to $T^{-3/2}$, is probably due to phonon scattering.

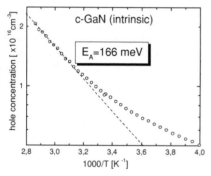

Figure 3: Hole concentration as a function of temperature for the intrinsic c-GaN. The dashed line is used to calculate the activation energy E_A=166 meV.

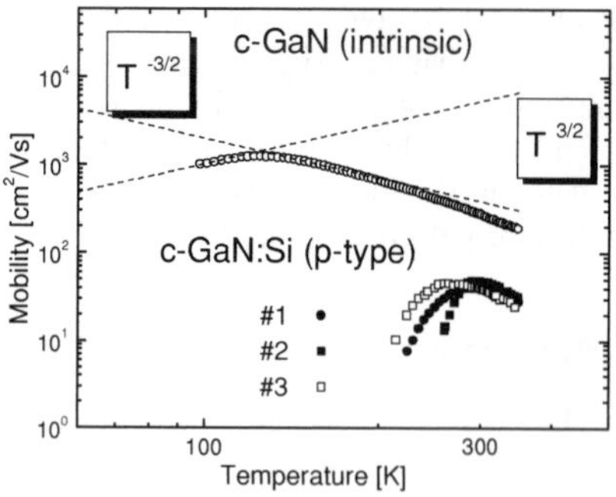

Figure 4: Mobility as a function of temperature for the intrinsic c-GaN (opened circle) and for the lightly doped c-GaN:Si (closed circles, closed squares, and opened squares).

LIGHTLY DOPED c-GaN:Si

We expect to obtain n-type conductivity for c-GaN when it is doped with Si. Contrary to that, p-type conductivity is measured for samples with a Si concentration below 1.2 x 10^{19} cm^{-3}. This is due to the fact that the acceptor background, present in intrinsic c-GaN, is not yet completely compensated for these concentrations. The hole concentrations measured for the samples #1, #2, and #3 are shown in Fig. 5. They all present the same behavior as a function of temperature, i.e., a decreasing of concentration as the temperature is lowered passing by a minimum and then increasing again when the temperature is further decreased. From the decrease of the concentration in the region near the room temperature an activation energy can be calculated yielding the values of 157, 154, and 171 meV for the samples #1, #2, and #3, respectively. All these values are very close to the activation energy measured for the acceptor level of the intrinsic c-GaN. This allows us to infer that the same acceptor level provides the measured high temperature hole concentrations in these samples. The increase of concentration at low temperature is probably due to a frozen of the carriers into a localized level. The shape of the mobility measured as a function of temperature (Fig. 4) is very similar to that presented by the intrinsic c-GaN. Nonetheless, the mobilities are much smaller and the $T^{3/2}$ (low temperature) and $T^{-3/2}$ (high temperature) dependence cannot describe their behavior anymore. This indicates that besides the scattering mechanisms described above, other mechanisms are involved.

HEAVILY DOPED c-GaN:Si

When the Si concentrations in c-GaN are higher than 2.5×10^{19} cm^{-3} a n-type conductivity is obtained. For these concentrations the acceptor level is completely compensated and we have a degenerated semiconductor. The electron concentration as a function of temperature measured for samples #4 to #9 is shown in Fig. 6. A metallic behavior can be observed, since the electron concentration does not change as the temperature is varied, which is typical of highly degenerated semiconductors.

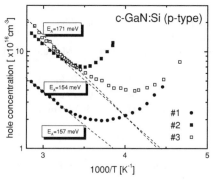

Figure 5: Hole concentration as a function of temperature for the lightly doped c-GaN:Si. The dashed lines are used to determine the activation energies.

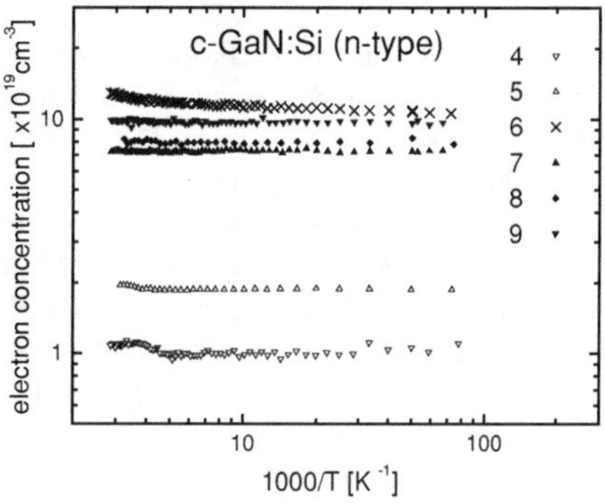

Figure 6: Electron concentration as a function of temperature for heavily doped c-GaN:Si.

CONCLUSIONS

Cubic InN and GaN binary semiconductor compounds were investigated by Hall effect measurements as a function of temperature. The InN sample shows a very high electron concentration ($n \cong 10^{20}$ cm^{-3}). At temperatures above 200K, a donor with an activation energy of $E_D = 94,5$ meV leads to a sharp increase of the electron concentration. The mobility at low temperature increases with $T^{1/4}$ while a decrease with T^{-3} is observed at a high temperature region. Nominally undoped c-GaN shows a p-type background concentration which varies from 10^{13} to 10^{16} cm^{-3} as the temperature is raised. The mobility dependence on temperature suggests that dislocations and ionized impurities are the scattering mechanisms at low temperature ($T^{3/2}$), while phonon scattering is the dominant one at a high temperature ($T^{-3/2}$). Si-doped c-GaN has p-type conductivity for Si-doping concentration below $N_D \cong 1.2 \times 10^{19} cm^{-3}$, and is highly n-type degenerated for $N_D \geq 1.2 \times 10^{19} cm^{-3}$. The behavior of the mobility as a function of temperature for the p-type c-GaN:Si is similar to that observed for the undoped sample, but its behavior is no longer described by $T^{3/2}$ and $T^{-3/2}$, indicating that other scattering mechanisms like dislocation scattering are taking place in our samples.

ACKOWLEDGEMENTS

This work was performed with partial support of a "CAPES/DAAD/PROBRAL" project within the Brazil/Germany scientific collaboration program. The authors also acknowledge financial support from FAPESP (Brazilian funding agency) and DFG (Deutsche Forschungsgemeinschaft).

REFERENCES

1. S. Nakamura and G. Fasol, *The Blue Laser Diode* (Springer, Berlin, 1997).
2. S. Nakamura, Semicond. Sci. Technol. **14**, R27 (1999).
3. M.A. Khan, J.N. Kuznia, D.T. Olson, W.J. Schaff, J.W. Burm, and M. Shur, Appl. Phys. Lett. **65**, 1121 (1995).
4. I.P. Smorchkova, C.R. Elsass, J.P. Ibbetson, R. Vetury, B. Heying, P. Fini, E. Haus, S.P. DenBaars, J.S. Speck, U.K. Mishra, J. Appl. Phys. **86**, 4520 (1999).
5. H. Yang, L.X. Zheng, J.B. Li, X.J. Wang, D.P. Xu, Y.T. Wang, X.W. Hu, and P.D. Han, Appl. Phys. Lett. **74**, 2498 (1999).
6. D.J. As, A. Richter, J. Busch, M. Lübbers, J. Mimkes, and K. Lischka, Appl. Phys. Lett. (in press).
7. L.K. Teles, L.M.R. Scolfaro, J.R. Leite, L.E. Ramos, A. Tabata, J.L.P. Castineira, and D.J. As, phys. stat. Sol. (b) **216**, 541 (1999).
8. D.J. As, D. Schikora, A. Greiner, M. Lübbers, J. Mimkes, and K. Lischka, Phys. Rev. **B54**, R11 118 (1996).
9. A.P. Lima, A. Tabata, J.R. Leite, S. Kaiser, D. Schikora, B. Schöttker, T. Frey, D.J. As, and K. Lischka, J. Cryst. Growth **201/202**, 396 (1999).
10. A. Tabata, A.P. Lima, L.K. Teles, L.M.R. Scolfaro, J.R. Leite, V. Lemos, B. Schöttker, T. Frey, D. Schikora, and K. Lischka, Appl. Phys. Lett. **74**, 362 (1999).
11. D.J. As, A. Richter, J. Busch, B. Schöttker, M. Lübbers, J. Mimkes, D. Schikora, K. Lischka, W. Kriegseis, W. Burkhardt, and B.K. Meyer, MRS Symp. Proc. Vol. **595** (2000) paper W3. 81.
12. N. Miura, H. Ishii, A. Yamada, M. Konagai, Y. Yamauchi, and A. Yamamoto, Jpn. Appl. Phys. **36**, L256 (1997).
13. W. Geerts, J.D. Mackenzie, C.R. Abernathy, S.J. Pearton, and T. Schmiedel, *Solid State Electronics*, **39**, 1289 (1996).

Mat. Res. Soc. Symp. Vol. 595 © 2000 Materials Research Society

ROLE OF ARSENIC HEXAGONAL GROWTH-SUPPRESSION ON A CUBIC GaNAs GROWTH USING METALORGANIC CHEMICAL VAPOR DEPOSITION

S. Yoshida*, T. Kimura**, J. Wu**, J. Kikawa*, K. Onabe**, and Y. Shiraki***
*Yokohama R&D Laboratories, The Furukawa Electric Co., Ltd.
2-4-3 Okano, Nishi-ku, Yokohama 220-0073, Japan
**Department of Applied Physics, The University of Tokyo,
7-3-1 Hongo, Bunkyo-ku, Tokyo 113-8654, Japan
***Research Center for Advanced Science and Technology, The University of Tokyo,
4-6-1 Komaba, Meguro-ku, Tokyo 153-8904, Japan

ABSTRACT

The hexagonal domain suppression-effects in cubic-GaNAs grown by metalorganic chemical-vapor deposition (MOCVD) is reported. A thin buffer layer (20 nm) was first grown on a substrate at 853 K using trimethylgallium and dimethylhydrazine (DMHy), and GaNAs samples were grown at different AsH_3 flow rates (0 ~ 450 µmol/min) at 1193 K. As a result, three types of surface morphologies were obtained: the first was a smooth surface ($AsH_3 = 0$ µmol/min); the second was a mirrorlike surface having small and isotropic grains (AsH_3 : 45 ~ 225 µmol/min); and the third involved three-dimensional surface morphologies (above 450 µmol/min of AsH_3 flow rate). Furthermore, it was confirmed using X-ray diffraction that the mixing ratio of hexagonal GaNAs in cubic GaNAs decreased with an increase of the AsH_3 flow rate. We could obtain GaNAs having a cubic component of above 85% at AsH_3 flow rates above 20 µmol/min. Therefore, the MOCVD growth method using AsH_3 and DMHy was mostly effective for suppressing hexagonal GaNAs. It was observed that the photoluminescence intensity of GaNAs was decreased with increase of arsine flow rate.

INTRODUCTION

III-N-V materials, such as GaNP and GaNAs, have a very large band-gap bowing and are very promising for light-emitting devices with a very wide wavelength region from ultraviolet to infrared [1,2]. They are expected to realize a wider range of bandgaps with a smaller lattice mismatch to GaN.

However, III-N-V type alloys can be hardly grown under an equilibrium condition, since they have a very large miscibility gap [3,4]. The gas-source molecular-beam epitaxy (GSMBE) growth of nitrogen (N) rich hexagonal (h) GaNAs [5] and GaNP [6,7] alloys on sapphire substrates and their optical properties have been recently reported. The maximum As and P concentrations of these alloys with a hexagonal structure are so far limited to 0.26% [6] and 8.3% [7], respectively. However, there has been no report on a metalorganic chemical vapor deposition (MOCVD) growth of cubic-GaNAs using the dimethylhydrazine (DMHy). In this paper, the As effect for hexagonal growth-suppression on cubic GaNAs growth is reported in MOCVD.

EXPERIMENTAL

GaNAs was grown on a GaAs (001) substrate using a horizontal MOCVD apparatus. The pressure in the reactor was 160 Torr. Trimethylgallium (TMG), DMHy [$(CH_3)_2NNH_2$] and arsine (AsH_3) were used as precursors of Ga, N, and As, respectively. A GaN buffer layer of about 20 nm in thickness was grown on the substrate using TMG and DMHy at 853 K. After that, a GaNAs layer was grown on the GaN buffer layer at 1193 K under the conditions of different AsH_3 flows. The growth temperature was constant at 1193 K, since when the growth temperatures were lower than 1193 K and also higher than 1223 K, the grown surface became rough. The surface morphologies of the samples were observed by scanning electron microscopy (SEM), and the surface roughness was measured by atomic force microscopy (AFM). The crystal structure was investigated by an X-ray diffraction (XRD) measurement. Secondary ion mass spectrometry (SIMS) was carried out to estimate As incorporation into the c-GaN films. To investigate the optical properties, a photoluminescence measurement of GaNAs was carried out using a He-Cd laser (325 nm) at 5K.

RESULTS AND DISCUSSION

In order to increase the As incorporation into cubic GaNAs, the AsH_3 flow rate was changed from 0 to 450 μmol/min and the flow rates of TMG and DMHy were fixed at 18 and 230 μmol/min, respectively. Without AsH_3 flow, the growth rate of c-GaN was 1.5 nm/sec. When the AsH_3 flow was increased to 450 μmol/min, the growth rate was reduced to 1.0 nm/sec. The thickness of the epitaxial layers was approximately 600 nm. Figure 1 shows the surface morphologies of the samples at different AsH_3 flow rates ranging from 0 to 450 μmol/min, observed by SEM. Figure 1 (a) shows GaN surface morphologies without AsH_3 flow. The shape of the grains is elongated in the [110] direction and the length of each grain is several microns. These grains have a tendency to form (111) B facets and (113) A facets [8]. Figure 1 (b) shows the surface morphologies in the case of AsH_3 flow under 225 μmol/min. The surface was smooth and the grains were smaller with more isotropic shapes. When the AsH_3 flow rate was, moreover, as large as 450 μmol/min, the GaNAs growth was more three dimensional and the grains were also isotropic and larger.

We also observed the AFM images in order to investigate the surface roughness of GaNAs layers grown at different AsH_3 flow rates (0 ~ 225 μmol/min). It was found that the surface became smoother with increasing the AsH_3 flow rate. That is, the root-mean-square (rms) showing a surface roughness became smaller than 22.7 nm with increasing the AsH_3 flow rate. The above-mentioned results of SEM and AFM indicate that with AsH_3 the grain size and shape was smaller and more isotropic, and that the surface was smoother, although the growth rate decreased wth increasing AsH_3 flow. This is considered to be due to As adsorption on the GaN surface during the growth process. It is considered that As adsorption on the GaN surface lowers the surface energy to stabilize the (001) surface, resulting in suppressing the generation of a hexagonal structure, as Foxon et al previously reported [9].

Figure 2 shows three-dimensional maps of XRD of the two grown layers with and without AsH_3 flow. The X-ray incident azimuth was the {1-10} direction. Two different diffraction peaks from h-GaN (1-101) were observed along with the main diffraction from c-GaN (002), as shown in Figure 2 (a). With this X-ray incident azimuth, the h-GaN (1-101) domains, holding on two different (111) B surfaces, were observed. The

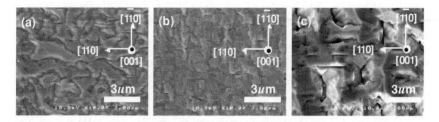

Figure 1. SEM images of samples with various AsH_3 flow rates ranging from 0 to 450 µmol/min. (a) AsH_3=0 µmol/min, ¶=AsH_3/(AsH_3+DMHy)=0, (b) AsH_3=225 µmol/min, ¶=0.49, (c) AsH_3=450 µmol/min, ¶=0.66.

Figure 2. Three-dimensional graphs of reciprocal space maps of (a) c-GaN and (b) c-GaNAs (AsH_3 flow rate was 90 µmol/min) with X-ray beams incident along the <1010> azimuth.

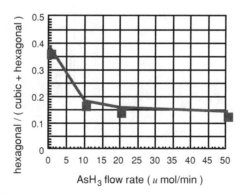

Figure 3. Concentration of the hexagonal phase in the c-GaN layers with various AsH_3 flow rates.

diffraction-peak intensity from h-GaN (1-101) on (111) B surfaces was much larger than from h-GaN on (111) A surfaces; this is consistent with that the grains at the c-GaN surface have (111) B facts and do not have (111) A facets. Interestingly, with AsH3 flow, the diffraction intensity from h-GaN (1-101) becomes significantly weaker. This tendency was common to these samples with AsH3 flow under 225 μmol/min. We estimated the concentration of the hexagonal phase in the c-GaN layers by comparing with the integrated diffraction intensities of c-GaN (002) and h-GaN (1-101) [10]. It is considered that with AsH3 flow during the c-GaN growth, As adsorbs on the GaN (001) surface and prevents the generation of (111) facets, resulting in maintaining the surface flatness. This smooth surface is favorable to c-GaN growth, because h-GaN is easy to grow with its c-axis parallel to the <111> directions. Figure 3 shows the concentration of the hexagonal phase in epilayers at different AsH3 flow rates. It was confirmed that with increasing the AsH3 flow the concentration of the hexagonal phase significantly decreased to 15%. That is, this growth condition is very effective for suppressing of hexagonal GaNAs.

Figure 4 shows SIMS depth profiles of the samples with AsH3 flow of 0 and 225 μmol/min. The As intensity of the samples with an AsH3 flow rate of 225 μmol/min became larger, as large as 20 in the same scale. The As counts increased according to the increase in the AsH3 flow rate up to 225 μmol/min. Therefore, As was actually incorporated in the c-GaN layers by the AsH3 flow. When the AsH3 flow rate was 450 μmol/min, GaNAs polycrystals were observed in the GaNAs layer. In order to further incorporate the As into GaNAs, the growth condition must to be further improved.

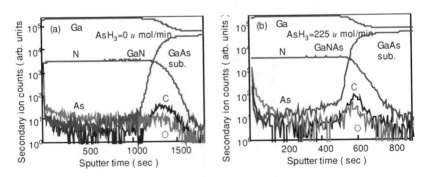

Figure 4. SIMS depth profiles of (a) c-GaN (without AsH3 flow), (b) c-GaNAs (AsH3 flow rate was 225 μmol/min).

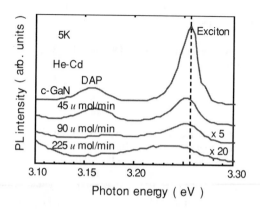

Figure 5. PL spectra of near-band edges with various AsH$_3$ flow rates at 5k.

We next carried out a photoluminescence (PL) measurement of GaNAs grown at different AsH$_3$ flow rates to investigate the optical properties. The two main peaks were observed. That is, the excitonic emission at 3.256 eV and the donor-acceptor pair recombination at 3.156 eV were observed in the PL spectrum of c-GaN. With increasing the AsH$_3$ flow rate, the excitonic emission became weak and the deep luminescence band around 2.8 eV became dominant. Figure 5 shows the near-band-edge emission of the GaNAs PL spectra. With increasing the AsH$_3$ flow rate, the exciton peak shifts toward lower energies and becomes broad. This is due to As incorporation in the c-GaN layer. The maximum value of the peak shift is 20 meV, which corresponds to approximately 0.1% As concentration, as estimated from the bowing parameter of h-GaNAs (19.6 eV) [5].

CONCLUSION

The hexagonal domain suppression-effects in c-GaNAs grown by MOCVD using DMHy were investigated. GaNAs samples were grown at different AsH$_3$ flow rates (0 ~ 450 μmol/min) at 1193 K. As a result, three types of surface morphologies were obtained: the first was obtained as a smooth surface (AsH$_3$ = 0 μmol/min); the second was a smooth surface having small and isotropic grains (AsH$_3$: 45 ~ 225 μmol/min); and the third involved three-dimensional surface morphologies (above 450 μmol/min of AsH$_3$ flow rate). Furthermore, it was confirmed using X-ray diffraction that the mixing ratio of hexagonal GaNAs in cubic GaNAs decreased along with an increase in the AsH$_3$ flow rate. We could obtained a GaNAs having a cubic component above 85% at AsH$_3$ flow rates above 20 μmol/min. Therefore, the MOCVD growth method using AsH$_3$ and DMHy was effective for suppressing hexagonal GaNAs.

ACKNOWLEDGMENTS

This work was supported by NEDO/JRCM's "The Light for the 21' Century" Japanese national project.

REFERENCES

1. L. Bellaiche, S. - H. Wei and A. Zunger, Appl. Phys. **70**, 3558 (1997).
2. S. - H. Wei and A. Zunger, Phys. Rev. **76**, 664 (1996).
3. G. B. Stringfellow, J. Electrochem. Soc. **119**, 1780 (1972).
4. G. B. Stringfellow, J. Cryst. Growth. **27**, 21 (1974).
5. K. Iwata, H. Asahi, K. Asami, R. Kuroiwa and S. Gonda, Jpn. Appl. **37**, 1436 (1998).
6. K. Iwata, H. Asahi, K. Asami and S. Gonda, Jpn. Appl. Phys. **35**, L1634 (1996).
7. H. Tampo, H. Asahi, K. Iwata, M. Hiroki, K. Asami and S. Gonda, Technical Report of Jpn. IEICE. **98**, ED98 (1998).
8. J. Wu, M. Kudo, A. Nagayama, H. Yaguchi, K. Onabe and Y. Shiraki, to be published in Phys. Stat. Sol. (b).
9. T. S. Cheng, L. C. Jenkins, S. E. Hooper, C. T. Foxon, J. W. Orton and D. E. Lacklison, Appl. Phys. Lett. 66, 1509 (1995).
10. H. Tsutchiya, K. Sunaba, S. Yonemura, T. Suemasu, and F. Hasegawa, Jpn. J. Appl. Phys. **36**, L1 (1997).

Mat. Res. Soc. Symp. Vol. 595 © 2000 Materials Research Society

Metal organic vapor phase epitaxy of GaAsN/GaAs Quantum wells using Tertiarybutylhydrazine

T. Schmidtling[*], M. Klein, U.W. Pohl, and W. Richter
Technische Universität Berlin, Institut für Festkörperphysik, Sekr. PN 6-1,
Hardenbergstr. 36, D - 10623 Berlin, Germany

ABSTRACT

GaAsN epilayers and quantum wells with a good structural quality and surface morphology were grown by low pressure metal organic vapor phase epitaxy using tertiarybutylhydrazine as a novel nitrogen source. The dependence of nitrogen incorporation on growth temperature was studied for epitaxy with arsine and tertiarybutylarsine precursors. A nitrogen content of 6.7 % was achieved using tertiarybutylhydrazine and tertiarybutylarsine at a low growth temperature of 530 °C. The observed room temperature luminescence shows an increasing redshift with increasing nitrogen contents of the wells.

INTRODUCTION

Devices for optical fibre communication require emission wavelengths between 1.3 and 1.55 µm. Bandgap tuning of GaAs based devices to reach this range can be achieved by alloying GaAs with nitrogen due to the large bandgap bowing of GaAsN [1]. A nitrogen content up to 4% leading to a near band edge PL at 1.25 µm was recently achieved in Metal Organic Vapor Phase Epitaxy (MOVPE) using dimethylhydrazine (DMHy, Me_2NNH_2) as nitrogen source [2,3]. To further lower the bandgap, the nitrogen content in the layers has to be increased. Due to a large miscibility gap [4], growth at low temperatures is required to attain a sufficient degree of nitrogen incorporation. We used tertiarybutylhydrazine (TBHy, $tBu(H)NNH_2$) as a novel nitrogen precursor for low pressure MOVPE of GaAsN. Due to a comparably weak tertiarybutyl-nitrogen bond strength, TBHy has a distinctly lower total decomposition temperature (around 350°C) compared to DMHy [5]. TBHy thus appears suitable to achieve a high nitrogen content in the layers. Furthermore, a decreased carbon incorporation was reported for GaN epilayers grown using TBHy [6].

---

[*] Corresponding author: phone: +49/30/31424820, fax: +49/30/31421769, email: kaethe@physik.tu-berlin.de.

EXPERIMENTAL PROCEDURE

Epitaxy was performed on GaAs(001) substrates at temperatures between 530 and 570 °C using TMGa, TBAs or AsH$_3$, and TBHy under hydrogen carrier gas. V/III ratios were adjusted to 15 and 200 for TBAs and AsH$_3$, respectively, using a constant TMGa flow of 20 µmol/min. The partial pressure ratio of the group V flows of the nitrogen and arsenic sources $x_v = P_{TBuHy}/(P_{TBuHy} + P_{As})$ was varied in the range 0.55 to 0.87. As a first step, we grew thick, relaxed layers with about 0.5 µm thickness to study the surface morphology. Based on these studies, we fabricated 20 nm thick GaAsN wells which were cladded by GaAs. Rocking curves were recorded using Cu K$_\alpha$ radiation in a high resolution double crystal diffractometer equipped with a double reflection Si(220) monochromator in (++-) arrangement. Thick layers were additionally characterised by Bragg Θ-2Θ scans using a single crystal diffractometer. Photoluminescence was excited using the 514 nm line of an argon laser, dispersed by a 50 cm monochromator with ~3.3 nm/mm linear dispersion and detected by a cooled Northcoast PIN diode.

RELAXED GaAsN EPILAYERS

The 0.5 µm thick GaAsN layers show smooth, mirrorlike surfaces. A hatch pattern is visible under a Nomarsky microscope, thereby indicating a relaxation by misfit dislocations even for layers with a low nitrogen content. The nitrogen concentration was estimated from X-ray spectra (figure 1). If we assume a complete strain relaxation and apply Vegard's law to lattice constants of 5.653 Å and 4.526 Å for GaAs and cubic GaN, respectively, the separation of layer and substrate reflections lead to nitrogen concentrations noted at the spectra in Fig. 1. These values [N]$_{max}$ represent upper bounds in case of a partial relaxation, as proved by a comparison with coherently strained, thin layers which were grown under comparable conditions. The layer with a high nitrogen content shows a broad reflection consisting of several superimposed peaks. This indicates that phase separation had occurred.

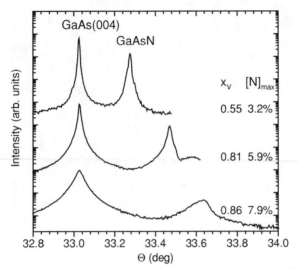

Figure 1 Rocking curves of the (004) reflection of 0.5 μm thick GaAsN/GaAs layers. x_V denotes the nitrogen part of the group V flux, $[N]_{max}$ is the upper bound of the nitrogen content in the GaAsN layers determined from the spectra.

The surface morphology was analysed by atomic force microscopy (AFM) in a 1 μm² scan region. The AFM image of a sample which was grown at 570 °C with 3.2 % nitrogen content shows smooth terraces with well defined steps of atomic height (figure 2a). The steps are believed to originate from a slightly misoriented substrate. The AFM image indicates a step flow growth mode which is also found for growth of pure GaAs at this temperature. For samples grown at a decreased temperature and a higher nitrogen concentration, a change of the surface morphology towards an island pattern which is superimposed on terraces is observed (figure 2b). Such a growth of two-dimensional islands is similar to that of GaAs grown under these conditions. The AFM study thus indicates that MOVPE growth of GaAsN is not significantly affected by the presence of the TBHy nitrogen precursor.

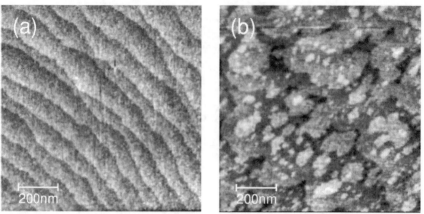

Figure 2 AFM images of 0.5 µm thick GaAsN layers, (a) grown at 570 °C with $x_v = 0.55$, $[N]_{max} = 3.2$ %, (b) grown at 530 °C with $x_v = 0.87$, $[N]_{max} = 7.9$ %.

GaAsN/GaAs QUANTUM WELLS

Using optimized growth conditions, we grew thin layers of about 20 nm thickness in GaAsN/GaAs double heterostructures. The good structural quality of the GaAsN quantum wells is proven by well pronounced fringes proven in X-ray diffraction (figure. 3). The nitrogen contents of the GaAsN layers were determined by dynamic simulation of the rocking curves. For simulation we assumed a poisson ratio of 0.366 for cubic GaN [7]. We find an increase of the nitrogen incorporation with decreasing growth temperature, similar to the findings reported for layers grown with DMHy [2].

At low partial pressure ratios x_V and consequently low nitrogen contents, the effect of using either TBAs or AsH_3 arsenic precursor on the nitrogen incorporation is found to be small. A high nitrogen content was particularly achieved using TBAs at a comparably high group V ratio x_V near 0.9 and a low growth temperature.

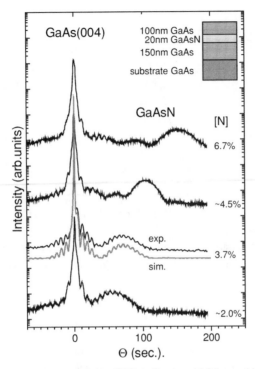

Figure 3 X-ray rocking curves of the (004) reflection of 20 nm thick GaAsN layers, cladded by GaAs. [N] represents the nitrogen contents determined by dynamic simulations.

Photoluminescence spectra recorded at room temperature (RT) show the expected redshift of the luminescence with increasing nitrogen content (figure 4). The redshift is accompanied by a gradual increase of the PL halfwidth. This indicates an increasing inhomogeneity of the nitrogen distribution. At high nitrogen contents, the luminescence is found to decrease and recovers after thermal activation, e.g. by a post growth annealing of the samples. Such a behaviour was also found for samples grown with dimethylhydrazine [2,8], and was assigned to a reduction of N-H complexes [8]. In our case annealing was achieved by laser irradiation.

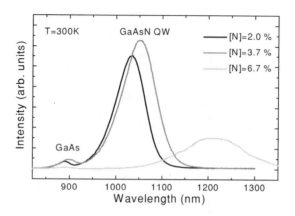

Figure 4 Room temperature photoluminescence spectra of 20 nm thick GaAsN/GaAs quantum wells, excited at 514 nm. [N] represents the nitrogen contents of the wells.

CONCLUSION

In summary, tertiarybutylhydrazine has proven to be a useful compound for the growth of GaAsN layers with good structural properties and high nitrogen concentrations. The nitrogen content was found to increase with decreasing growth temperature, and a concentration as high as 6.7 % was achieved using TBAs as arsenic precursor at a low growth temperature of 530 °C. Intensive luminescence of quantum well excitons is found for samples with moderate nitrogen composition (~ 4 %). Samples with high nitrogen content require a thermal activation of the PL. Emission wavelengths peaking slightly above 1.21µm were achieved by now.

REFERENCES

[1]   M. Weyers, M. Sato, H. Ando, *Jpn. J. Appl. Phys.* **31** (1992) L853. See also M. Sato, Mat. Res. Soc. Symp. Proc. **395** (1996) 285.

[2]   R. Bhat, C. Caneau, L. Salamanca-Riba, W. Bi, C. Tu , *J. Crystal Growth* **195** (1998) 427.

[3]   A. Ougazzaden, Y. Le Bellego, E.V.K. Rao, M. Juhel, L. Leprince, and G. Patriarch, *Appl. Phys. Lett.* **70** (1997) 2861.

[4] Y. Qui, S.A. Nikishin, H. Temkin, V.A. Elyukhin, Yu.A. Kudriavtsev, *Appl. Phys. Lett.* **70** (1997) 2831.

[5] U.W. Pohl, C. Möller, K. Knorr, W. Richter, J. Gottfriedsen, H. Schumann, K. Rademann, A. Fricke, *Mat. Sci. Eng. B*, **59** (1999) 20.

[6] U.W. Pohl, K. Knorr, C. Möller, U. Gernert, W. Richter, J. Bläsing, J. Christensen, J. Gottfriedsen, H. Schumann, *Jpn. J. Appl. Phys.* **38** (1999) L105.

[7] I. Akasaki and H. Amano, Crystal structure, mechanical properties and thermal properties of GaN, *Properties of group III-nitrides*, ed. J.H. Edgar (INSPEC, 1994) 30

[8] E.V.K. Rao, A. Ougazzaden, Y. Le Bellego, M. Juhel, *Appl. Phys. Lett.* **72** (1998) 1409.

Mat. Res. Soc. Symp. Vol. 595 © 2000 Materials Research Society

TEM Study of the Morphology Of GaN/SiC (0001) Grown at Various Temperatures by MBE

W.L. Sarney[1], L. Salamanca-Riba[1], V. Ramachandran[2], R.M Feenstra[2], D.W. Greve[3]
[1]Dept. of Materials & Nuclear Engineering, University of Maryland, College Park, MD
[2]Dept. of Physics, [3]Dept. of Computer & Electrical Engineering, Carnegie Mellon
University, Pittsburgh, PA

ABSTRACT

GaN films grown on SiC (0001) by MBE at various substrate temperatures (600° -
750° C) were characterized by RHEED, STM, x-ray diffraction, AFM and TEM. This
work focuses on the TEM analysis of the films' features, such as stacking faults and
dislocations, which are related to the substrate temperature. There are several basal plane
stacking faults in the form of cubic inclusions for samples grown at low temperatures
compared to those grown at high temperatures. The dislocation density is greatest for the
film grown at 600°C, and it steadily decreases with increasing growth temperatures.
Despite the presence of various defects, x-ray analysis shows that the GaN films are of
high quality. The double crystal rocking curve full width at half maximum (FWHM) for
the GaN (0002) peak is less than 2 arc-minutes for all of the films we measured and it
decreases with increasing growth temperature.

INTRODUCTION

GaN films grown on SiC usually have high defect densities. Typical defects in
GaN/SiC films include inversion domain boundaries, stacking faults, and unintended
polytype transformations. The presence of these defects emphasizes that even though
reducing the lattice mismatch improves film quality, other factors contribute to the nitride
film defect morphology [1]. Further defect density reduction may be achieved by
enhancing growth conditions. We examine the relationship between the SiC(0001)
substrate temperature and the GaN film quality.

EXPERIMENT

The damage produced by polishing was removed from Si-face 6H-SiC(0001)
substrates with ex-situ hydrogen etching [2]. The substrates were then placed into an ultra
high vacuum environment (pressure < 10^{-10} Torr) and outgassed at about 800°C for 30
minutes. In order to replenish any surface Si that may have been lost during oxide
removal, Si was deposited onto the substrate using an electron beam source. Oxide
desorption was done by annealing the substrate at about 1000° C until a 3x1 reflection
high energy electron diffraction (RHEED) pattern was obtained.

GaN films were grown by MBE onto the substrates using a Ga effusion cell and a
RF-plasma nitrogen source. The growth was a single step process with no nucleation
layer growth. We grew four samples with substrate temperatures of 600° C, 650° C, 700°

C, and 750° C. The temperature was monitored using a pyrometer and a thermocouple in contact with the back of the sample mounting stage. Growth was performed under highly Ga-rich conditions relative to the N concentration [3]. The films were characterized in-situ with RHEED and STM and ex-situ with AFM, HRXRD, and TEM. The TEM results were obtained on a JEOL 4000FX microscope operated at 300 kV. Cross-sectional TEM samples were prepared using tripod polishing and ion milling at room temperature. Low resolution TEM and a diffraction pattern with two-beam DF conditions were used to examine the defect morphology of the GaN films. HRTEM images and diffraction patterns allow detailed examination of the crystalline structure of the film.

RESULTS & DISCUSSION

Figure 1 shows TEM images of the sample grown at 600°C. The indexed diffraction pattern (Fig. 1a) indicates the presence of the SiC substrate, the 2H GaN film, and some 3C GaN regions with zone axes of $(2\bar{1}\bar{1}0)$, $(2\bar{1}\bar{1}0)$, and $(01\bar{1})$, respectively. Several of the low index spots from one of the three regions overlap with those from one or both of the other two regions, resulting in some very strong spots. The location of a diffraction spot is related to the lattice spacing. For instance, the small difference in lattice spacing for the $111_{3C\text{-}GaN}$ spot (d = 0.261 nm), the $0002_{2H\text{-}GaN}$ spot (d=0.259 nm), and the $0006_{6H\text{-}SiC}$ spot (0.251 nm) results in three spots that cannot be resolved by the TEM, and are labeled as spot 1 in Fig. 1a. Oblong spots (such as spots 2 and 3) are often two very closely neighboring spots. Cubic inclusions are easier to detect when we look at high index spots since the lattice spacing difference is greater. For instance, the $\bar{2}22_{3C\text{-}GaN}$ spot (labeled spot 8 in Fig. 1a) is distinguishable from the neighboring 6H SiC spot to the right. Spots due to twinning across the (111) plane in the cubic regions and spots due to multiple diffraction between twins or between cubic and hexagonal regions are labeled in Fig. 1a.

Figures 1b-1c show $(0002)_{2H\text{-}GaN}$ and $(01\bar{1}0)_{2H\text{-}GaN}$ dark field (DF) images of the sample grown at 600°C. Two beam conditions cause diffraction by specific hkl planes, and result in bright areas in the DF image where the hkl planes meet the Bragg condition [4]. The $(0002)_{2h\_GaN}$ two beam condition is nearly equivalent to the $(111)_{3C\text{-}GaN}$ condition, therefore the cubic inclusions do not cause contrast in the $(0002)_{2H\text{-}GaN}$ DF image shown in Fig. 1(b). The vertical defects visible in the (0002) image but invisible in the $(01\bar{1}0)$ image are screw dislocations with Burger's vector $\vec{b} = 0001$. The vertical defects visible in only the $(01\bar{1}0)$ image are edge-type and are probably double positioning boundaries or prismatic stacking faults, as we have observed in other GaN/SiC films [5]. Since the $(01\bar{1}0)_{2H\text{-}GaN}$ two-beam condition does not correspond to a two-beam condition for the cubic inclusions, we expect to see contrast due to the cubic inclusions. There are several horizontal bands in the $(01\bar{1}0)_{2H\text{-}GaN}$ DF image (Fig. 1c), which are basal plane stacking faults corresponding to 3C-GaN inclusions. High resolution TEM (Figure 1d) shows that the substrate/film interface marked by arrows is high quality and that the initial film nucleated is 2H GaN.

Diffraction patterns of the films grown at 650°C and 700°C (not shown here) have the same spots as the pattern for the film grown at 600°C (Fig. 1a). There are several regions of cubic GaN within the 2H GaN matrix, as shown in Fig. 2a. The cubic stacking in the inclusions and the abruptness of the transition from 2H to 3C GaN is shown in Fig 2b. The high-resolution image of the sample grown at 700°C (not shown here) is very similar to the image of the sample grown at 650° C, except that it has less 3C inclusions.

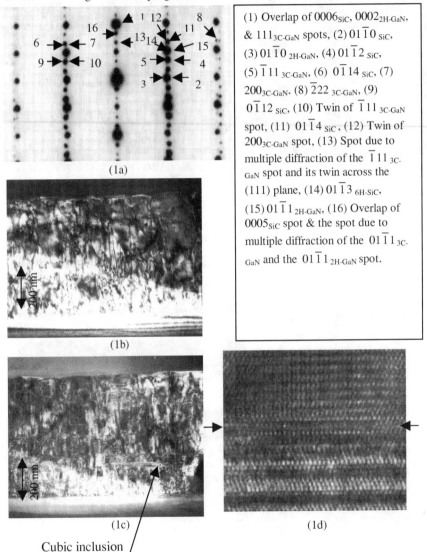

(1a)

(1b)

(1c)

Cubic inclusion

(1d)

(1) Overlap of 0006_{SiC}, $0002_{2H\text{-}GaN}$, & $111_{3C\text{-}GaN}$ spots, (2) $01\bar{1}0_{SiC}$, (3) $01\bar{1}0_{2H\text{-}GaN}$, (4) $01\bar{1}2_{SiC}$, (5) $\bar{1}11_{3C\text{-}GaN}$, (6) $0\bar{1}14_{SiC}$, (7) $200_{3C\text{-}GaN}$, (8) $\bar{2}22_{3C\text{-}GaN}$, (9) $0\bar{1}12_{SiC}$, (10) Twin of $\bar{1}11_{3C\text{-}GaN}$ spot, (11) $01\bar{1}4_{SiC}$, (12) Twin of $200_{3C\text{-}GaN}$ spot, (13) Spot due to multiple diffraction of the $\bar{1}11_{3C\text{-}GaN}$ spot and its twin across the (111) plane, (14) $01\bar{1}3_{6H\text{-}SiC}$, (15) $01\bar{1}1_{2H\text{-}GaN}$, (16) Overlap of 0005_{SiC} spot & the spot due to multiple diffraction of the $01\bar{1}1_{3C\text{-}GaN}$ and the $01\bar{1}1_{2H\text{-}GaN}$ spot.

Fig.1.(a) Diffraction pattern for sample grown at 600°C, (b) (0002) DF image of the sample grown at 600°C, (c) (01$\bar{1}$0) DF image of the sample grown at 600°C, (d) High resolution image of GaN/SiC interface.

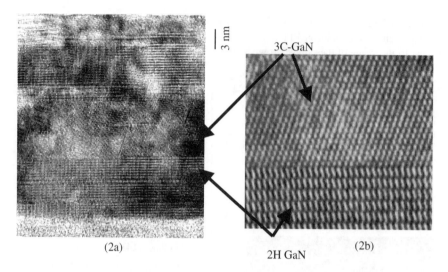

Fig. 2. (a) High resolution image of sample grown at 650° C. (b) High resolution image of the interface between a region of 2H GaN and 3C GaN.

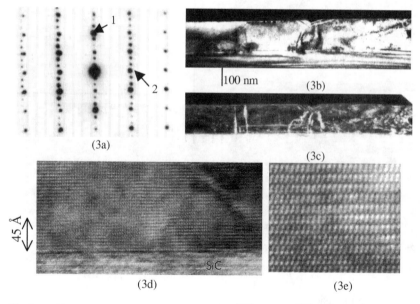

Fig. 3.   TEM of sample grown at 750° C. (a) $2\bar{1}\bar{1}0_{GaN} / 2\bar{1}\bar{1}0_{SiC}$ diffraction pattern. Label 1: 0006 SiC and 0002 GaN spots. Label 2: $(01\bar{1}0)$ SiC and $(01\bar{1}0)$ GaN spots. (b) (0002) DF image and (c) $(01\bar{1}0)$ DF image (d) High resolution image of GaN/SiC interface (e) High magnification image of the 2H GaN fringes.

The film grown at 750°C contains very few 3C inclusions. The 3C spots are not visible in the diffraction pattern (Fig. 3a) and only a few horizontal bands are seen in the (01$\bar{1}$0) DF image (Fig. 3c). The defect density of this sample is lower than the other three samples, as seen by the reduced number of visible dislocations in the DF images (Figs. 3b & 3c). High resolution images of the film grown at 750° C (Figs. 3d and 3e) show that this film is of higher quality the films grown at lower temperatures.

The total dislocation density, the density of threading dislocations which intersect

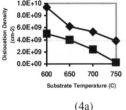 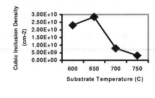

(4a)   (4b)

Fig. 4. (a) Density of dislocations plotted against substrate temperature. The upper line is the total dislocation density and the lower line is the density of threading dislocations which intersect the surface (b) Density of cubic regions plotted against substrate temperature.

the surface, and the density of cubic inclusions decreases with increasing substrate temperature, as shown in the TEM images and summarized in the Figs 4a and 4b. The film grown at 750°C has a total dislocation density of approximately 3.2×10^9 cm^{-2}. The density of threading dislocations which intersect the surface for the film grown at 750°C is approximately 2.1×10^8 cm^{-2}. The threading dislocations which intersect the surface for the film grown at 750°C are predominately edge dislocations.

Symmetric triple crystal radial (ω-2θ) scans (Figs. 5a and 5b) show that the FWHM decreases with increasing growth temperature, except for the film grown at 750°C. The increase in FWHM for the highest temperature sample is due to the thinness of this sample relative to the other three sample. This sample was grown for the same length of time as the other samples, but the growth temperature of 750°C is close to the temperature at which decomposition of GaN becomes significant (approximately 800°C). Decomposition of GaN during growth reduces the growth rate. The FWHM of the symmetric peak is as low as 30 arcseconds for the films grown at 700°C. The symmetric peak width is affected by defects which distort the interplanar spacing along the growth

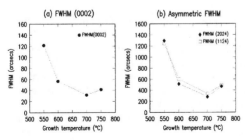

Fig. 5. HRXRD FWHM data for GaN films grown on 6H-SiC (0001) as a function of temperature: (a) Symmetric (0002) reflection (triple crystal ω-2θ scans), (b) asymmetric reflections (double crystal ω scans).

direction, notably the screw dislocations we see in the DF images. Therefore, the symmetric peak width is unaffected by certain edge dislocations and cubic inclusions ($d_{111\text{-}GaN} \approx d_{0002\text{-}GaN}$) since they do not distort the interplanar spacing along the growth direction. The asymmetric peak width is a better measure of total dislocation and cubic inclusion densities, which may explain its relatively large width compared to that of the symmetric peak. The x-ray data is discussed more thoroughly elsewhere [6].

Reciprocal space maps around the (0002) reflections show much greater elongation along the $k_{//}$ axis for the film grown at 600°C than for the film grown at 750°C. This implies a larger degree of tilt, or mosaicity, in the lower temperature films. Screw dislocations with $\vec{b} = 0001$ would cause tilt in a film. These results indicate a decrease in the number of screw dislocations with increasing growth temperature, in agreement with both the TEM and the x-ray results. Furthermore, AFM studies in our previous experiments show a decreasing number of spiral growth fronts with increasing growth temperature [7].

CONCLUSION

Increasing the growth temperature improves the crystalline quality of GaN grown on hydrogen-etched 6H SiC substrates. TEM shows that defect and cubic inclusion densities significantly decrease as the growth temperature is increased. Furthermore, x-ray k-space maps show that the mosaicity of the films decreases sharply with increasing growth temperature. Therefore, we conclude MBE at growth temperatures near the decomposition temperature of GaN improves the quality of wurtzite GaN grown on 6H SiC (0001).

ACKNOWLEDGMENTS

The work at Carnegie Mellon was supported by National Science Foundation, grant DMR-9615647, and the Office of Naval Research, grant N00014-96-1-0214.

REFERENCES
[1] R.F. Davis, T.W. Weeks, M.D. Bremser, S. Tanaka, R.S. Kern, Z. Sitar, K.S. Ailey, W.G. Perry, and C. Wang, *Mater.Res.Soc.Symp.Proc.* **395,** 3 (1996).
[2] V. Ramachandran, M.F. Brady, A.R. Smith, R.M. Feenstra, D.W. Greve, *J. Elec. Mat..* **27,** 308 (1998).
[3] A.R. Smith, V. Ramachandran, R.M. Feenstra, D.W. Greve, A. Ptak, T.H. Myers, W.L. Sarney, L. Salamanca-Riba, M.-S. Shin, M. Skowronski, *MRS Internet J. Nitride Semicond. Res.* **3**, 12 (1998).
[4] D. Williams & C. Carter, *Transmission Electron Microscopy*, Plenum Press, New York: 1996, p. 361.
[5] V. Ramachandran, R.M. Feenstra, W.L. Sarney, L.Salamanca-Riba, J.E. Northrup, L.T. Romano, D.W. Greve, *Appl. Phys. Lett.* **75**, (1999), 808.
[6] V. Ramachandran, R.M. Feenstra, W.L.Sarney, L. Salamanca-Riba, D.W. Greve, submitted to the proceedings of the 46th International Symposium of the American Vacuum Society, to appear in *J. Vac. Sci. Technol. A*.
[7] V. Ramachandran, A.R. Smith, R.M. Feenstra, and D.W. Greve, *J. Vac. Sci. Technol. A.* **17**, 1289 (1999).

Mat. Res. Soc. Symp. Vol. 595 © 2000 Materials Research Society

Microstructure and Physical Properties of GaN Films on Sapphire Substrates

Zhizhong Chen, Rong Zhang, Jianming Zhu, Bo Shen, Yugang Zhou, Peng Chen, Weiping Li, Yi Shi, Shulin Gu and Youdou Zheng
National Laboratory of Solid State Microstructures and Department of Physics, Nanjing University, Nanjing 210093, P.R.China

ABSTRACT

Transmission electron microscopy (TEM), x-ray diffraction (XRD), photoluminescence (PL) and Raman scattering measurements were applied to study the correlation between the microstructure and physical properties of the GaN films grown by light radiation heating metalorganic chemical vapor deposition (LRH-MOCVD), using GaN buffer layer on sapphire substrates. When the density of the threading dislocation (TD) increases about one order of magnitude, the yellow luminescence (YL) intensity is strengthened from negligible to two orders of magnitude higher than the band edge emission intensity. The full width of half maximum (FWHM) of the GaN (0002) peak of the XRD rocking curve was widened from 11 min to 15 min, and in Raman spectra, the width of E_2 mode is broadened from 5 cm^{-1} to 7 cm^{-1}. A "zippers" structure at the interface of GaN/sapphire was observed by high-resolution electron microscope (HREM). Furthermore the origins of TD and relationship between physical properties and microstructures combining the growth conditions are discussed.

I. INTRODUCTION

GaN is a promising material for optoelectronic and electronic devices such as blue-green light emitting diodes, laser diodes and high temperature, high frequency, high power transistors. Although blue current-injected laser diodes have been fabricated by S. Nakamura [1], there are still many problems under debate. The origins of the YL emission remain unclear while it is commonly observed in almost all undoped and n-doped GaN epilayer grown by conventional methods [2-5]. The generation of TD in the epilayer of which density is high as 10^8-10^{10} cm^{-2} [6-8], and the nature and the role of initial nitridation of sapphire surface and the low-temperature growth GaN (AlN, InN) buffer layer which affects the GaN-based materials epitaxy growth [9-13] are ambiguous too. Several groups reported the relationships between GaN microstructures and XRD, PL, and Raman spectra, respectively [11,14-17]. But there are few works to try to resolve above problems by studying the microstructures and physical properties combining the growth conditions, such as the initial nitridation time and buffer layer thickness, and so on. In this paper, different properties of two kinds of GaN samples are grown by metalorganic chemical vapor deposition (MOCVD) under different conditions. Two sets of GaN films in TEM, XRD, PL and Raman measurements are investigated. According to the results, the origins of TD and YL are discussed.

Table 1 Growth conditions of the sample A and B

	Sample A	Sample B
Nitridation time (min)	3	1.5
Buffer growth temperature (°C)	500	500
Buffer thickness (nm)	50	25
Epilayer Growth temperature (°C)	950	950
TMG flow (μmol/min)	20	28
NH_3 flow (l/min)	1.65	2.35
Epilayer thickness (μm)	0.9	1.7

II. EXPERIMENTS

The GaN films were grown by LRH-MOCVD on (0001) sapphire substrates. Details of the growth processes are given in Ref. 18. Table 1 shows the growth conditions of two typical samples which named as sample A and sample B.

Transmission electron microscope (high-resolution electron microscope) samples were prepared by the standard technique. Two pieces of GaN specimen were bonded by way of face-to-face by 504 glue. After mechanical grinding and polishing of the "sandwich" on the cross-sectional plane to a thickness of 20-30 μm, the specimens were further thinned to electron transparency with an Ar^+ ion mill. TEM (HREM) observations were carried out with a JEOL JEM-4000EX microscopy operated at 400 kV.

III. RESULTS AND DISCUSSION

Figure 1 shows typical cross-sectional TEM images of two samples near the

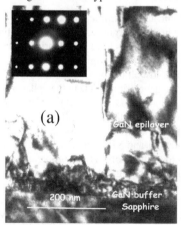 

Figure 1. Cross-sectional TEM images of GaN/sapphire. The inserts are the electron diffraction (ED) patterns of the GaN epilayers. The density of threading dislocation in (a) is an order lower than that in (b). (a) and (b) are corresponding to the sample A and B

GaN/sapphire interface. The inserts show the electron diffraction (ED) patterns of the GaN epilayers top regions. Corresponding to the growth processes, there are three zones in the GaN layer: buffer zone, "faulted" zone and "sound" zone. The buffer zone is above the interface between the GaN layer and the sapphire substrate, about 50 nm in figure 1(a) and 25 nm in figure 1(b), which were grown at low temperature about 500°C. The "faulted" zone is 0.4 μm thick from the top of buffer layer into GaN epilayer. There are some "haystack-like" domains in the very beginning. Among these domains, there are high density of dislocations and other defects, and just above these domains, the dislocation density reduced sharply. The area above the "faulted" zone, which is called "sound" zone, is with very low density of defects, uniform and high quality which can be concluded from ED patterns. In comparison, the sample A shows about an order lower density of threading dislocation (TD) than the sample B, its columnar diameters is larger and its ED pattern is sharper, too. All these results indicate that the crystal quality of the sample A is higher than the sample B.

In order to study the buffer layer difference between the sample A and B, we performed HREM observation. Figure 2 shows the HREM of the interfaces of GaN/α-Al$_2$O$_3$ of the sample B. We find "zippers" structure at the GaN/α-Al$_2$O$_3$ interface. The gears of the zipper are perpendicular to the interface, and the period is about 1.8 nm. There are more than one "zipper" standing side by side in some areas. The "zipper" zone fluctuates from 10 nm thick to 25 nm. In HREM images of the sample A, haze and inhomogeneous contrast in GaN layer is observed among nearly 100 nm above the interface of GaN/ α-Al$_2$O$_3$. It seems that crystal quality of buffer layer is poor. According to the study of the "zippers" structure in sample B [19], the amorphous-like growth pre-dominated at the beginning of the buffer layer growth. So did in the sample A. It can be seen that in table 1 that the sample A was grown with longer nitridation time and thick buffer layer. Decreasing the density of nucleation centers by elongating the initial nitridation advantages to the lateral growth when epitaxy growth begins. To increase the thickness of the buffer layer can increase the thickness of amorphous zone which is beneficial to relax the strain. The layers are believed to be nearly free of extended defects due to the solid-phase crystallization process that take place during the heating step prior to GaN deposition at higher temperature [15]. The truncated hexagonal pyramidal mesas in the amorphous zone may grow larger, which is also advantageous to the lateral growth [10-11].

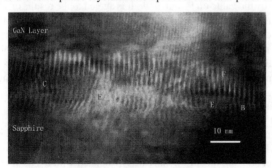

Figure 2. HREM images of the sample B. "Zipper" appears at the interface of GaN/ α-Al$_2$O$_3$. The gears of the "zipper" period is 1.8 nm

Figure 3 shows the x-ray diffraction spectra of the two epitaxial GaN films. The θ /2 θ scan of the GaN films on the basal-plane sapphire shows dominant peaks corresponding to (0002) peaks and their harmonic.

The FWHM $\delta\theta$ of a Bragg peak in a $\theta/2\theta$ scan can be expressed by the following

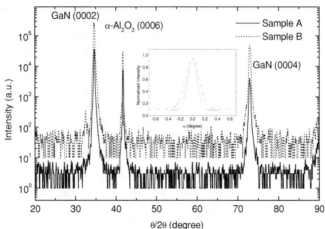

Figure 3. *The x-ray diffraction spectra of sample A and B. The average domain size and inhomogeneous strain of the sample A and B are calculated as 1753, 909 Å and 0.167%, 0.141% by the width and the diffraction angle of (0002) and (0004) peaks The insert is XRD rocking curve of GaN epilayers of (0002) reflection. The FWHM of the sample A and B are 11 min and 15 min, respectively.*

equation [20] ,

$$\delta\theta = \lambda/(2D\cos\theta + \varepsilon_{in}tg\theta) \qquad (1)$$

Here D is the average domain size, ε_{in} is the inhomogeneous strain. The values of $\delta\theta$ of GaN (0002), (0004) peaks in figure. 3 are 0.056°, 0.102° for sample A and 0.069°, 0.084° for sample B. From Equation (1), (D, ε_{in}) of sample A,B can be obtained (1750°, 0.167%), (900°, 0.141%), respectively by the value of θ and $\delta\theta$. It is agreed to the results of TEM, in which the sample A columnar diameters is higher and its ED pattern is shaper. V. Potin et al. and Ning et al. think the TD is originated from the low angle domain boundaries [6,8], which can be proved by our results. It seems that there are reactions of partial dislocations [7] since the ε_{in} in sample B is lower in sample A. The insert in figure 3 shows the x-ray rocking curve of the two samples. The FWHM from (0002) reflections of the two samples are 11 and 15 minutes (min), respectively. The FWHM of the sample A is 4 min lower than that of the sample B. This result may be due to the smaller average tilts of column crystallites in the epilayer of the sample A. If the c-axis is slightly tilted, screw dislocation with **b**=[0001] will be present [6]. What the c-axis screw TD and mixed TD broaden the FWHM of (0002) peak of XRD rocking curve is clarified by B. Heying et al. [16]. As reported by S. Christiansen et al. [21], the YL of GaN was associated with c-axis screw dislocations. We prefer the broadened FWHM due to screw TD after referring to the following photoluminescence (PL) measurement.

Room temperature PL spectra of GaN epilayers are shown in figure 4. Both spectra have three characteristic regions: band-edge and donor-acceptor pair emission (3.2-3.5 eV) which are not easy to be distinguished at room temperature, blue band emission (2.7-3.2

eV) and yellow band emission (1.9-2.5 eV). However, different samples have different types of spectra. The intensity ratio of the YL to the band-edge emission of the sample B is 3 orders higher than that of the sample A. Furthermore the YL of the sample B is stronger and has fine structure with two peaks at 2.10 eV and 2.29 eV, while that of the sample A is nearly invisible in the PL spectrum. As reported by F.A. Ponce [4], the intensity of the near band-edge PL emission of GaN is indicative of its optical quality. So the sample A has a higher optical quality than the sample B, which corresponds the high crystal quality concluded by the above TEM experiment. The YL intensity can vary over a wide range to GaN epilayers grown with different conditions. The GaN epilayers with high quality exhibits almost no YL emission. Although the origin of the YL is still being debated, it is well acknowledged that the YL corresponds to the

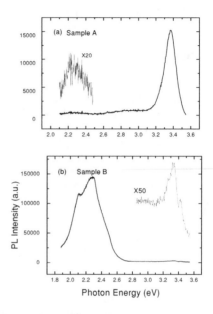

Figure 4. Photoluminescence of GaN epilayers, the intensity ratio of YL to band-edge emission in (b) is 3 orders of that in (a). There are fine peaks of YL at 2.10 and 2.29 eV in (b)

deformed crystal structure [5]. In our work, it seems that the YL is related to the screw TD. The growth condition may be concerned for the YL origin [5].

The Raman Spectra for the two GaN samples are measured. According to Ref.17, in the sample A, 417 and 568 cm^{-1} are assigned as A_{1g} mode of sapphire and E_2 mode of GaN. The sample B is assigned in the same way. In contrast to the E_2 mode of the sample A whose width is 5 cm^{-1}, that of the sample B is widened to 7 cm^{-1}, which means that in the sample B E_2 mode has shorter lifetime. It also shows that the sample B must be less perfect of crystal quality than the sample A.

IV. CONCLUSIONS

The crystalline structure of GaN epilayers has been studied using TEM and double crystal x-ray diffraction. When the density of TD increases about an order of magnitude, the FWHM of GaN (0002) peak of x-ray diffraction rocking curve is increased from 11 min to 15 min. The YL intensity is strengthened from negligible to two orders of band-edge emission intensity. The "zippers" structure is observed by HREM measurement of the GaN/α-Al$_2$O$_3$ interface. From the above results, we concluded the high quality GaN epilayer can be grown by optimum initial nitridation time and buffer layer thickness.

Acknowledgement

National High Technology Research and Development Project of China and

National Natural Science Foundation of China (contracts #69976014, #69636010, #69806006, #69987001) support this work.

REFERENCES

1. S. Nakamura et al., *Jpn. J. Appl. Phys.*, **35**, L74 (1996).
2. T. Ogino and M. Aoki, *Jpn. J. Appl. Phys.* **19**, 295 (1998).
3. J. Neugebauer and C.G. Van de Walle, *Appl. Phys Lett.*, **69**, 503 (1996).
4. F.A. Ponce, D.P. Bour, W. Götz and P.J. Wright, *Appl. Phys. Lett.*, **68**, 57 (1996).
5. R. Zhang and T.F. Kuech, *Appl. Phys. Lett.*, **72**, 1611 (1998).
6. X.J. Ning, F.R. Chien, P. Pirous, J.W. Yang and M. A. Khan, *J. Mater. Res.*, **11**, 580 (1996).
7. M.S. Hao, T. Sugahara, H. Sato, Y. Morishima, Y. Naoi, L.T. Romano, S. Sakai, *Jpn. J. Appl. Phys.* **37**, part2 (3A), L291 (1998).
8   V. Potin, P. Vermaut, P. Ruterana and G. Nouet, *J. Electron. Mater.*,**27**, 266 (1998).
9. S. Nakamura, Jpn. J. Appl. Phys. **30**, L1705 (1991).
10. I. Akasaki, H. Amano, Y. Koide, K. Hiramatsu and N. Sawaki, *J. Crystal Growth* **98**, 209 (1989).
11. G. Li, S.J. Chua, S.J. Xu, W. Wang, P. Li, G. Beaumont, and P. Gibart, *Appl. Phys. Lett.* **74**, 2821 (1999).
12. N. Grandjean, J. Massies, and M. Leroux, *Appl. Phys. Lett.* **69**, 2071 (1996).
13. S. Fuke, H. Teshigawara, K. Kuwahara Y. Takano, T. Ito, M. Yanagihara, K. Ohtsaka, *J. Appl .Phys.* **83**, 764 (1998).
14. S.J. Rosner, E.C. Carr, M.J. Ludowise, *Appl. Phys. Lett.*, **70**, 420 (1997).
15. F.A. Ponce, B.S. Krusor, J.S. Major, M.E. Plano, and D.F. Welch, *Appl .Phys. Lett*, **67(3)**, 410 (1995).
16. B. Heying, X.H. Wu, S. Keller, Y. Li, D. Kapolnek, B.P. Keller, S.P. DenBaars, and J.S. Speck, *Appl Phys. Lett.* **68(5)**, 643 (1996).
17. S.-C.Y. Tsen, D. J. Smith, K.T. Tsen, W. Kim and H. Morkoc, *J. Appl. Phys.* **82**, 6008 (1997).
18. B. Shen, Y.G. Zhou, Z.Z. Chen, P. Chen, W.P. Li, Y.D. Zheng, *Appl. Phys. A* **68**, 593 (1999).
18. Z.Z. Chen, J.M. Zhu, R. Zhang B. Shen,, Y.G. Zhou, P. Chen, W.P. Li (unpublished).
20. R.W. Vook, in *Epilayer Growth*, edited by J.W. Matthews, (Academic, New York, 1975), p.339.
21. S. Christiansen, M. Albrecht, W. Dorsch, *Mater. Sci. Eng.* **B43**, 296 (1997).

Mat. Res. Soc. Symp. Vol. 595 © 2000 Materials Research Society

Structural evolution of GaN during initial stage MOCVD growth

Chong Cook Kim, Jung Ho Je, Min-Su Yi[1], and Do Young Noh[1]
Department of Materials Science and Engineering, Pohang University of Science and Technology, Pohang, Korea
[1]Department of Materials Science and Engineering and Center for Electronic Materials Research, Kwangju Institute of Science and Technology, Kwangju, Korea

ABSTRACT

The structural evolution of GaN films during the initial growth process of metalorganic chemical vapor deposition (MOCVD) - low temperature nucleation layer growth, annealing, and high temperature epitaxial growth - was investigated in a synchrotron x-ray scattering experiment. The nucleation layer grown at 560°C that was predominantly cubic GaN consisted of tensile-strained aligned domains and relaxed misaligned domains. The hexagonal GaN, transformed from the cubic GaN during annealing to 1100 °C, showed disordered stacking. The atomic layer spacing decreased as the fraction of the hexagonal domains increased. Subsequent growth of epitaxial GaN at 1100 °C resulted in the formation of ordered hexagonal GaN domains with rather broad mosaicity.

INTRODUCTION

Gallium nitride (GaN) is a direct wide-band-gap semiconductor that has attracted considerable interests in optoelectronic devices operating in blue-green to ultraviolet regime [1,2]. Device quality GaN films are currently grown by a two-stage metalorganic chemical vapor deposition (MOCVD) process wherein growth of GaN nucleation layers on sapphire (0001) substrates at a low temperature is followed by growth of epitaxial GaN films at a high temperature [2,3]. Resulting electrical, optical, and structural properties depend sensitively on the conditions under which the nucleation layer is processed [2,4].

The structure of GaN films in the initial stage of the two-stage MOCVD growth is a focus of recent studies to elucidate the role of the nucleation layer in the final film quality. It has been revealed that the cubic GaN (c-GaN) with ABCABC stacking sequence and the hexagonal GaN (h-GaN) with ABAB stacking sequence coexist in the nucleation layers [5,6]. Upon annealing to a high temperature prior to the second stage growth, GaN nucleation layer mostly transforms into the h-GaN. The purpose of this study is to reveal the detailed structural changes accompanying the cubic to hexagonal transformation during the initial stage growth process.

EXPERIMENTAL DETAILS

The GaN nucleation layers were grown on sapphire (0001) substrates using MOCVD. The substrates were cleaned with solvents and subjected to *in-situ* pretreatment under H_2 flow at 1100 °C. The nucleation layers were grown at 560 °C for 4 minutes using trimethylgallium (TMGa) and ammonia (NH_3). The annealed nucleation layers were obtained by heating the nucleation layers at 1100 °C for 6 min in the growth chamber. The epitaxial GaN was grown on the annealed nucleation layer at 1100 °C for 5 min. The x-ray measurements were performed *ex-situ* on the samples taken out of the growth chamber at the three different stages of the growth.

The synchrotron x-ray scattering measurements were carried out at beamline 5C2 at Pohang Light Source (PLS) in Korea. The incident x-rays were focused vertically by a mirror. A double bounce Si(111) monochromator was used to monochromatize x-rays to the wavelength of 1.55 Å and to focus the beam in horizontal direction. Two pairs of slits in front of the detector provided an appropriate instrumental resolution of about 0.001Å^{-1} in reciprocal space. The experiment was carried out by measuring the x-ray scattering profiles along the $< 000l >$ direction and along the $< 10\bar{1}l >$ directions in reciprocal space using hexagonal coordinates. The detailed explanation of the scattering geometry is explained in Ref.6.

RESULTS AND DISCUSSIONS

We first examined the as-grown GaN nucleation layer deposited at 560 °C.

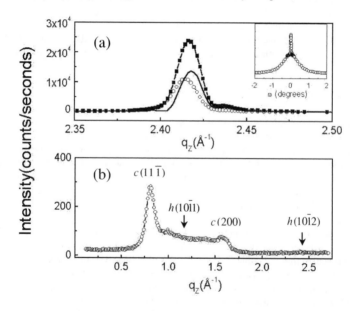

Figure 1. The x-ray diffraction profiles of the as-grown nucleation layer. (a) The longitudinal profiles of the GaN(0002)$_{hex}$ [or the GaN(111)$_{cub}$] reflection. Solid squares represent the total intensity along the substrate normal direction, while the open circles represent the intensity measured at 0.1 ° away from the normal direction. The solid line is the difference between the two that represents intensity from the aligned domains only. The inset of (a) represents the rocking curve of the peak. (b) The diffraction profile along the $< 10\bar{1}l >$ direction. The peaks indicated by c ($11\bar{1}$) and c(002) are originated from c-GaN.

Figures 1 (a) and its inset show the longitudinal profile and the rocking curve of the GaN(0002)$_{hex}$ [or the GaN(111)$_{cub}$] Bragg reflection, respectively. The rocking curve was composed of a sharp component and a broad component with 0.03° and 1° full width at half maximums (FWHM) respectively, indicating that there are domains with the crystalline axis aligned to the substrate normal direction (aligned domains) and domains with misaligned crystalline orientations (misaligned domains). The origin of the sharp component was attributed to the fact that the orientation of the domains grown in the very initial stage was constrained by the well-defined orientation of the sapphire substrate. Consequently the aligned domains are likely to be strained, while the misaligned domains are relaxed.

To examine the lattice strain of the as-grown nucleation layer, it is instructive to compare the c-axis lattice spacing of the aligned domains and that of the misaligned domains. The open circle in figure 1 (a) represents the diffraction profile of the misaligned domains measured at the rocking angle of 0.1° away from the peak in the rocking curve. The solid line represents the profile from the well-aligned domains calculated by subtracting the diffraction signal of the misaligned-domains from the total signal indicated by the filled squares. The c-axis layer spacing of the aligned and the misaligned domains calculated from the peak positions using $l_C = 2\pi / q_{PEAK}$ are 2.599Å and 2.603Å respectively. We note that the atomic layer spacing of the aligned domains, which are presumably more strained, is smaller than that of the misaligned domains by about 0.2%. The in-plane lattice spacing of the aligned domains is presumably larger than that of the misaligned domains due to the effect of the usual tetragonal distortion [7]. This indicates that the well-aligned domains are under tensile stress in the film plane, which is counter-intuitive considering that the in-plane lattice spacing of the sapphire substrate is smaller than that of GaN.

The tensile strain of the GaN nucleation layer can be understood in terms of the extended domain matching argument [8], wherein six Ga distances in GaN match to seven Al distances in sapphire. Although the in-plane lattice spacing of sapphire is smaller than that of GaN, seven times the Al distance (19.229 Å) in sapphire is larger than six times the Ga distance (19.134Å) in GaN by 0.5 %. The film would then be under tensile stress. The tensile strain of about 0.2% observed in the aligned nucleation layer indicates that the strain was partially relaxed. The extended domain matching condition, however, would be satisfied only when the mobility of growing GaN is large enough to form reasonably large in-plane domains, which is controlled by the growth temperature of the nucleation layer.

Interestingly the c-axis lattice constants of both the aligned and the misaligned domains estimated from figure 1(a) are larger than the bulk hexagonal GaN lattice spacing, although the film was under tensile stress in the film plane. We attribute the increase of the c-axis lattice constant to the cubic stacking sequence between the atomic GaN layers. It has been suggested that the layer spacing of c-GaN is larger than that of h-GaN [9]. In fact, the diffraction profile along the $< 10\overline{1}l >$ direction illustrated in figure 1(b) shows that the as-grown nucleation layer has the cubic stacking order. The reflections peaked at $q_Z = 0.81$ Å^{-1} and $q_Z = 1.56$ Å^{-1} correspond to the cubic $(11\overline{1})$ and the cubic (002) Bragg reflections respectively. The correlation length of the stacking order estimated from the width of the peak is 78 Å.

As the nucleation layer was annealed to 1100 °C, the stacking sequence changes from the cubic to hexagonal stacking as indicated by the appearance of the hexagonal

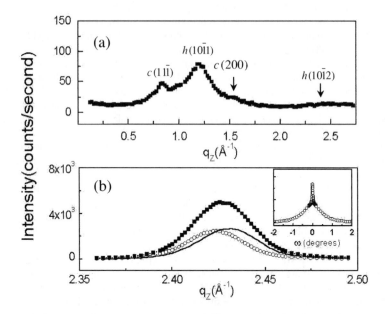

Figure 2. The x-ray diffraction profiles of the annealed GaN nucleation layer. (a) The diffraction profile along the $<10\bar{1}l>$ direction. $H(10\bar{1}1)$ represents the peak from h-GaN. (b) The longitudinal profiles of the GaN(0002). Solid squares represent the total intensity along the substrate normal direction, while the open circles represent the intensity measured at 0.1° away from the normal direction. The solid line is the difference between the two that represents the intensity from the aligned domains only. The inset of (b) represents the rocking curve at the peak.

(101) Bragg reflection at q_z=1.2 Å $^{-1}$ in the diffraction profile shown in figure 2(a). The intensity of the cubic $(11\bar{1})$ and the (002) reflections was greatly reduced, although they were still existent. The hexagonal $(10\bar{1}1)$ reflection was relatively broad, indicating that the correlation length of the stacking order was rather short, ~ 24Å. This suggests that the transformation was kinetically limited and many stacking faults were generated during the transformation.

The c-axis lattice constant decreased as the hexagonal stacking sequence developed during the annealing process as indicated by the shift of the GaN (0002) peak to a larger value in figure 2 (b). This is consistent with the fact that the layer spacing of the as grown c-GaN nucleation layer was larger than the bulk layer spacing of hexagonal GaN. The rocking curve of the peak shown in the inset of figure 2(b) demonstrates that the annealed GaN layer also consists of the aligned and the misaligned domains. By decomposing the longitudinal scan into two components similar to the case of the as-grown nucleation layer, we found that the aligned domains of the annealed GaN were still under tensile stress. The layer spacing of the aligned and misaligned domains was 2.586 Å and 2.593 Å respectively.

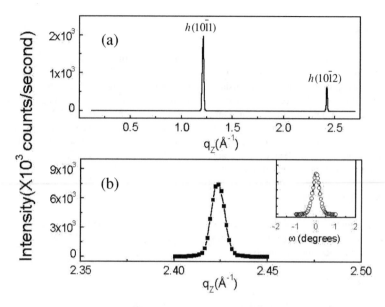

Figure 3. The x-ray diffraction profiles of the epitaxial GaN grown at 1100 °C on the annealed GaN nucleation layer. (a) The diffraction profile along the $<10\bar{1}l>$ direction. (b) The powder diffraction profiles in the <000l> direction. The inset of (b) represents the rocking curve of the GaN(0002).

From the similarities between the rocking curves of the as-grown and the annealed sample, we conclude that the aligned c-GaN domains transformed to the aligned h-GaN and the misaligned c-GaN domains transformed to the misaligned h-GaN. We note that the longitudinal profile of the annealed sample [figure 2(b)] is broader than that of the as-grown sample [figure 1(b)]. This is another indication that the transformation was kinetically limited, and the remaining cubic stacking sequences acted as the faults in hexagonally stacked GaN layers.

The epitaxial GaN grown at 1100 °C for 5 min on the annealed GaN nucleation layers was completely hexagonal. The diffraction profile in the $<10\bar{1}l>$ direction illustrated in figure 3(a) shows only the hexagonal $(10\bar{1}1)$ reflection without any trace of the cubic reflections. This suggests that the remaining cubic GaN in the annealed nucleation layer completely transformed to hexagonal GaN during the high temperature growth of the eptiaxial GaN. The peak was quite sharp and the correlation length of the stacking order was as large as 450 Å indicating that the hexagonal stacking in the epitaxial GaN was well ordered.

In the meanwhile, the rocking curve of the GaN (0002) reflection illustrated in the inset of figure 3(b) shows a single peak with 0.35°FWHM. The mosaic distribution was broader than that of the aligned domains, but sharper than that of the misaligned domains of the nucleation layer. We believe that the epitaxial GaN nucleated on the misaligned domains whose mosaicity improved during the lateral growth at high temperatures [10,11].

CONCLUSIONS

In conclusion, we have revealed the details of the structural evolution of GaN films during the initial process of MOCVD. The as-grown nucleation layer that was predominantly c-GaN consisted of tensile-strained aligned domains and relaxed misaligned domains. The tensile strain of the GaN nucleation layer was explained by the extended domain matching argument. During annealing to 1100 °C, the c-GaN transformed into disordered h-GaN. The atomic layer spacing decreased with increasing the fraction of the hexagonal domains, while the mosaic structure remained similar. The epitaxial GaN at grown 1100 °C on the annealed nucleation layer was ordered hexagonal GaN domains with rather broad mosaicity.

ACKNOWLEDGEMENTS

This study was supported by Korean Research Foundation made in the program year of 1998, and by KOSEF through ASSRC (1998). PLS is supported by Korean Ministry of Science and Technology.

REFRERENCES

1. H. Morkoc and S. N. Mohammad, *Science* **267**, 51 (1995)
2. S. Nakamura, *Jpn. J. Appl. Phys.* **30**, L1705 (1991).
3. H. Amano, N. Sawaki, Akasaki, and Y. Toyoda, *Appl. Phys. Lett.* **48**, 353 (1986)
4. S. Keller, B. P. Keller, Y. –F. Wu, B. Heying, D. Kapolnek, J. S. Speck, U. K. Mishra, and S. P. DenBaars, *Appl. Phys. Lett.* **68**, 1525 (1996).
5. X. H. Wu, D. Kapolnek, E. J. Tarsa, B. Heying, S. Keller, B. P. Keller, U. K. Mishra, S. P. DenBaars, and J. S. Speck, *Appl. Phys. Lett.* **68**, 1371 (1996).
6. A. Munkholm, C. Thompson, C. M. Foster, J. A. Eastman, O. Auciello, G. B. Stephenson, P. Fini, S. P. DenBaars, and J. S. Speck, *Appl. Phys. Lett.* **72**, 2972 (1998)
7. D. Y. Noh, Y. Hwu, J. H. Je, M. Hong, and J. P. Mannaerts, *Appl. Phys. Lett.* **68**, 1528 (1996)
8. N. Grandjean, J. Massies, P. Vennegues, M. Laugt, and M. Leroux, *Appl. Phys. Lett.* **70**, 643 (1997)
9. T. Kurobe, Y. Sekiguchi, J. Suda, M. Yoshimoto, and H. Matsunami, *Appl. Phys. Lett.* **73**, 2305 (1998)
10. X. H. Wu, P. Fini, S. Keller, E. J. Tarsa, B. Heying, U. K. Mishra, S. P. DenBaars, and J. S. Speck, *Jpn. J. Appl. Phys.* **35**, L1648 (1996).
11. B. Yang, A. Trampert, B. Jenichen, O.Brandt, and K. H. Ploog, *Appl. Phys. Lett.* **73**, 3869 (1998).

Mat. Res. Soc. Symp. Vol. 595 © 2000 Materials Research Society

Structural Properties of (GaIn)(AsN)/GaAs MQW Structures Grown by MOVPE

C. Giannini, E. Carlino, L.Tapfer, F. Höhnsdorf[*], J. Koch[*], W. Stolz[*],
PASTIS-CNRSM, I-72100 Brindisi, Italy;
[*]Materials Science Center, Philipps-University, D-35032 Marburg, Germany

ABSTRACT

In this work, we investigate the structural properties of (GaIn)(AsN)/GaAs multiple quantum wells (MQW) grown at low temperature by metalorganic vapour phase epitaxy. The structural properties, in particular the In- and N-incorporation, the lattice strain (strain modulation), the structural perfection of the metastable (GaIn)(AsN) material system and the structural quality of the (GaIn)(AsN)/GaAs interfaces are investigated by means of high-resolution x-ray diffraction, transmission electron microscopy (TEM), and secondary ion mass spectrometry. We demonstrate that (GaIn)(AsN) layers of high structural quality can be fabricated up to lattice mismatches of 4%. Our experiments reveal that N and In atoms are localized in the quaternary material and no evidences of In-segregation or N-interdiffusion could be found. TEM analyses reveal a low defect density in the highly strained layers, but no clustering or interface undulation could be detected. High-resolution TEM images show that (GaIn)(AsN)/GaAs interfaces are slightly rougher than GaAs/(GaIn)(AsN) ones.

INTRODUCTION

(GaIn)(AsN) alloys grown on GaAs substrates offer the unique possibility to realize optoelectronic devices for the wavelength emission in the range of 1300-1550 nm due to the large band gap bowing [1,2]. In particular, the realization of vertical-cavity surface-emitting lasers (VCSEL) emitting in the 1.3μm wavelength range at room temperature [3], solar cells in the 1.0-1.2 μm range [4] and resonant-cavity-enhanced (RCE) photodetectors operating near 1.3μm [5] was demonstrated recently. However, this material system exhibits a large miscibility gap under thermodynamic equilibrium conditions. Therefore, extreme non-equilibrium conditions at low growth temperatures are required in order to maintain homogeneous epitaxial deposition for the metastable (GaIn)(NAs) material system.

Due to the large miscibility gap of (GaIn)(AsN) the understanding of phase separation effects, as observed in Ga(AsN) and In(AsN) [6,7], is of great importance. Phase separation effects lead to large local strain fields and might be the origin of 3d-like quantum dot growth [8].

In this work, we investigate the structural properties of (GaIn)(AsN)/GaAs MQWs grown at low temperature (525°C) by metalorganic vapour phase epitaxy (MOVPE) using triethylgallium (TEGa), trimethylindium (TMIn), tetriarybutylarsine (TBAs) and 1,1-dimethylhydrazine (UDMHy). The structural properties were investigated by means of high-resolution x-ray diffraction (HRXRD), secondary ion mass spectrometry (SIMS) and transmission electron microscopy (TEM).

GROWTH AND CHARACTERIZATION

The (GaIn)(AsN)/GaAs multiple quantum well (MQW) heterostructures were grown by metal-organic vapor-phase epitaxy using an IR-heated horizontal reactor system (Aix 200). Triethylgallium (TEGa) and trimethylindium (TMIn) have been used as group-III sources and tetriarybutylarsine (TBAs) and 1,1-dimethylhydrazine (UDMHy) as group-V precursors. The growth was performed at a reactor pressure of 50 mbar under H_2 carrier gas at a total flow of 6800 sccm. All the samples here investigated were grown at the same temperature (525°C) on [100]-oriented GaAs substrates. The growth rates of the quaternary (GaIn)(AsN) wells and the GaAs barrier layers were 0.25μm/h and 1.0μm/h , respectively. In order to avoid strain relaxation processes and the generation of associated structural defects, the total thickness of the MQW structures was chosen to be smaller than the critical thickness, i.e. the MQWs consist of only 5 periods. The nominal thickness of the GaAs barriers and the quaternary (GaIn)(AsN) wells are the same for all the samples investigated, i.e. d_b=90nm and d_w=10nm, respectively. The growth parameters were optimized in order to have an In-content of x=0.33 in the quaternary layers. The N-content is constant in the individual well layers of one sample but it varies for the different samples of the investigated series in the range of y=0.01 and y=0.045.

The x-ray diffraction experiments were performed by using a high-resolution multi-crystal x-ray diffractometer. A 4-crystal channel-cut monochromator-collimator arrangement has been used to reduce the wavelength and angular dispersion of the beam incident on the sample, which are for the CuKα-radiation used 2.5×10^{-5} and 60μrad, respectively. The experimental x-ray diffraction patterns were simulated by using a dynamical scattering model in the recursive formalism [9]. The second-order approximation of the angular deviation was used in order to obtain precise values of the lattice strain [10].

High quality specimens for TEM experiments were prepared in [011] cross-section geometry by using mechanical pre-thinning and subsequent ion-milling (3.5-4.5 keV Ar ions) at liquid nitrogen temperature in order to minimize ion-induced damage and artefacts [11]. In particular , the use of a cooled stage specimen holder prevents the eventual element interdiffusion due to the local increases of the temperature during ion bombardment. The TEM specimens were analyzed by using a Philips CM30 TEM/STEM electron microscope operating at 300 keV with an interpretable resolution limit of 0.23nm. The micrographs were acquired digitally by using a Gatan 1024x1024 slow scan CCD camera.

SIMS analysis was carried out on a CAMECA ims 4f instrument using either a 2.5 keV O_2^+ or a 5.5 keV Cs^+ ions, with off-normal impacting angles of 55° (2 keV), and 42° (5.5 keV), respectively. The different conditions were chosen in order to improve the depth resolution, to better detect the indium and nitrogen signal and to reduce the matrix effects. The primary beam currents were chosen in the range of 20-60nA. The secondary ions were collected through an aperture which delimited the analyzed area to a section of 8μm diameter in the center of the eroded area (a square with a side of 250μm). Positive single ionized species were monitored under oxygen bombardment while molecular ion clusters MCs^+ (M=Ga, In, As, N) were detected when the cesium beam was employed [12]. The erosion rate was estimated by measuring the sputtered depth in the different matrices by using a stylus profilometer (TENCOR-alphastep).

EXPERIMENTAL RESULTS AND DISCUSSION

All the MQW samples (series of 6 samples) were analyzed in detail by HRXRD, SIMS and TEM. Here, for brevity sake we report on the results obtained on one selected sample which are representative for the whole series.

The HRXRD patterns close the (400) reciprocal lattice points of all the MQWs with different N-content exhibit a broad diffraction peak at lower angular position, about 20-30 mrad far from the GaAs substrate peak, which can be attributed to the quaternary (GaIn)(AsN) layers (see curve A of figure 1). This assumption is confirmed by kinematical simulation of the diffraction pattern and will hold if the well thickness is much smaller than the barrier thickness, i.e. $d_w<<d_b$. Consequently, the angular position between the quaternary layer peak and the GaAs substrate peak is related to the lattice strain in the (GaIn)(AsN) layers. The quaternary layers of all the samples here investigated are compressive-strained. In addition, the width of the quaternary layer peak is related to the thickness of the individual well layers in the sample. The high frequency interference fringes are due to finite thickness of the MQW and the fringes angular distance is related to the whole MQW thickness. The appearance of these interference fringes indicates a high structural quality and thickness homogeneity of the heterostructures. This evidence is also confirmed by additional measurements close to the (200) and asymmetric (422) and (511) diffraction peaks. In fact, the results of these measurements demonstrate that the heterostructures are in a pseudomorphic state, i.e. the in-plane strain is zero.

Figure 1 shows the experimental (curve A) and dynamically (curve B) simulated x-ray diffraction pattern of the selected MQW. The experimental curve is shifted with respect to the simulated one for clarity sake. The following parameters were used for the best simulation pattern: the well and barrier thickness are d_w=14.3nm and d_b=89nm, respectively, and the lattice strain along the [100] direction is $\varepsilon^{\perp}$=0.041.

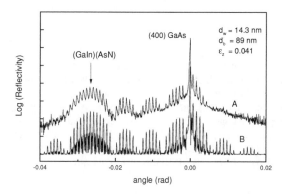

Figure 1 *Experimental (A) and simulated (B) (400) HRXRD pattern of the MQW considered.*

Unfortunately, the In and N concentration cannot be obtained from x-ray scattering experiments alone. The strain field depends on the In- as well as N-content and in addition, if the N-content is small (<30%) the influence on the structure factor will also be very small or experimentally not detectable. The incorporation behavior of In and N for the novel (GaIn)(AsN) metastable material system is very complex in MOVPE growth and not fully understood yet. However, if we assume that the In incorporation is independent from the N-incorporation [13], we will obtain for pseudomorphic layers of lattice strain ε^* the following relation between x_{In} and y_N:

$$y_N = \frac{\frac{1-v}{1+v}\varepsilon^* a_{GaAs} - x_{In}(a_{InAs} - a_{GaAs})}{x_{In}(a_{GaAs} - a_{GaN} - a_{InAs} + a_{InN}) + a_{GaN} - a_{GaAs}} \quad (1)$$

Here, a_j is the bulk lattice constants of material j and v is the Poisson ratio of GaAs. For In-concentration $x_{In}=0.33$, as estimated from the growth parameters used for the samples investigated here, we obtain the N mole fraction $y_N=0.015$. However, it is important to note that there are indications that the incorporation of In and N are interdependent [14]; in this case equation (1) should be used with care.

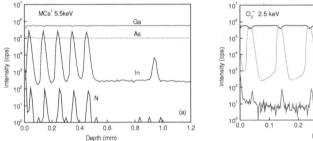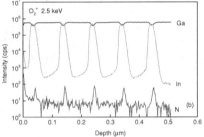

Figure 2 *SIMS profiles obtained by using 5.5keV Cs⁺ (a) and O_2^+ (b) primary beam.*

The In- and N-distribution as well as eventual segregation or interdiffusion of In and N in the MQWs were measured by SIMS. Figure 2 shows the SIMS profiles of sample A by using MCs⁺ 5.5keV ions (a) and O_2^+ 2.5 keV ions (b). The O_2^+ measurements allow a higher spatial resolution but are affected by a much stronger matrix effect. The results clearly show that N is localized in the quaternary layers and the N-signal is constant for all the 5 layers in the MQW. The In-concentration is the same for all the 5 well layers and no interdiffusion or segregation phenomena can be revealed [12]. These results are also confirmed by SIMS measurements carried out on the other samples.

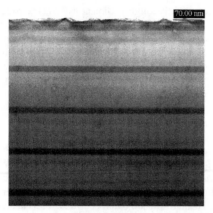

Figure 3 BF image of the considered sample showing all the layers of the MQW.

A deeper insight in the structural properties of the MQWs and the local structure of the (GaIn)(AsN)/GaAs interfaces has been obtained by TEM analyses. Figure 3 shows a bright field (BF) image of the sample A. The 5 (GaIn)(AsN)/GaAs periods are well observed and are uniform in thickness. The diffraction condition is chosen in order to minimize the crystallographic contrast and to show the quaternary layers darker with respect to the GaAs ones. The layer closer to the sample surface are brighter due to the decrease of the TEM specimen thickness and due to the related decrease of the high energy electron absorption in the relevant area. A relatively low density of extended defects has been observed. In particular, dislocations have been observed in the (GaIn)(AsN) layers with an average spacing of about (2 ± 1) μm along the [011] cross section geometry. These dislocations are probably caused by the high lattice strain in the quaternary layers as measured by x-ray diffraction and are less frequent in layers of lower mismatch and smaller thickness.

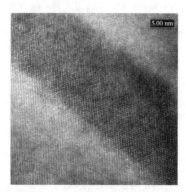

Figure 4 HRTEM image of one (GaIn)(AsN) layer of the MQW shown in figure 3.

Figure 4 is a high-resolution TEM micrograph in [011] zone axis of a (GaIn)(AsN) well shown in figure 3. The GaAs/(GaIn)(AsN) interfaces are sharp while the

(GaIn)(AsN)/GaAs interfaces are slightly rougher. Similar asymmetric roughness profiles were also observed in (GaIn)As/GaAs multilayers [15]. The roughness of the (GaIn)(AsN)/GaAs interfaces may also explain (i) the tail in the In-concentration at the (GaIn)(AsN)/GaAs interfaces as obtained by O_2^+ SIMS profiling (figure 2b), and (ii) the reduced amplitude of the interference fringes modulation in the high-resolution x-ray diffraction measurements (figure 1). It is also important to note that no clustering phenomena could be observed by TEM in the samples investigated here.

CONCLUSIONS

In summary, the structure of compressive strained (GaIn)(AsN)/GaAs MQWs grown by MOVPE was investigated by using HRXRD, SIMS and TEM. We show that (GaIn)(AsN) layers of high structural quality can be fabricated up to lattice mismatches of 4%. We found that N and In atoms are localized in the quaternary material and no evidences of In-segregation or N-interdiffusion could be found. TEM analyses reveal a low defect density in the highly strained layers, but no clustering or interface undulation could be detected. High-resolution TEM images show that (GaIn)(AsN)/GaAs interfaces are slightly rougher than GaAs/(GaIn)(AsN) ones.

ACKNOWLEDGEMENTS

The financial support of the Volkswagen-Stiftung (Hannover, Germany) is gratefully acknowledged.

REFERENCES

1. L. Bellaiche, S.H. Wie, A. Zunger, *Phys. Rev.*, **B54**, 17568 (1996).
2. S. Sakai, Y. Ueta, Y. Terauchi, Jpn. *J. Appl. Phys.*, **32**, 4413 (1993).
3. C. Ellmers, F. Höhnsdorf, J. Koch, C. Agert, S. Leu, D. Karaiskaj, M. Hofmann, W. Stolz, W.W. Rühle, *Appl. Phys. Lett.*, **74**. 2271 (1999).
4. S.R. Kurtz, A.A. Allerman, E.D. Jones, J.M. Gee, J.J. Banas, B.E. Hammons, *Appl. Phys. Lett.*, **74**, 729 (1999).
5. J.B. Heroux, X. Yang, W.I. Wang, *Appl. Phys. Lett.*, **75**, 2716 (1999).
6. M.K. Behbehani, E.L. Piner, S.X. Liu, N.A. El-Masry, S.M. Bedair, *Appl. Phys. Lett.*, **75**, 2202 (1999)
7. R.Beresford, K.S.Stevens, A.F.Schwartzman, *J.Vac Sci Technol.*,**B16**,1293 (1998).
8. H.P. Xin, K.L. Kavanagh, Z.Q. Zhu, C.W. Tu, *Appl. Phys. Lett.*, **74**, 2337 (1999).
9. L. Tapfer, M. Ospelt, H. von Känel, *J. Appl. Phys.*, **67**, 1298 (1990).
10. L. Tapfer, L. De Caro, C. Giannini, H.-P. Schönherr, K.H. Ploog, *Solid State Commun.*, **98**, 599 (1996).
11. D.G. Barber, *Ultramicroscopy*, **52**, 101 (1993).
12. C.Gerardi, C.Giannini, A.Passaseo, L.Tapfer, *J.Vac.Sci Technol.*,**B15**, 2037 (1997).
13. H.P. Xin, C.W. Tu, *Appl. Phys. Lett.*, **72**, 2442 (1998).
14. Z. Pan, T. Miyamoto, D. Schlemker, S. Sato, F. Koyama, K. Iga, *J. Appl. Phys.*, **84**, 6409 (1998).
15. Kuo-Jen Chao, Ning Liu, Chih-Kang Shih, D.W. Gotthold, B.G. Streetman, *Appl. Phys. Lett.*, **75**, 1703 (1999).

Mat. Res. Soc. Symp. Vol. 595 © 2000 Materials Research Society

Formation and stability of the prismatic stacking faultin wurtzite (Al,Ga,In) nitrides

P. Ruterana[1], A. Béré, and G. Nouet
Laboratoire d'Etudes et de Recherches sur les Matériaux –Institut des Sciences de la Matière et du Rayonnement, UPRESA-CNRS 6004, 6 Bd Maréchal Juin, 14050 Caen Cedex, France.
[1]Author for correspondance: email : ruterana@lermat8.ismra.fr, Tel: 33 2 31 45 26 53, Fax: 33 2 31 45 26 60

ABSTRACT

The formation of the $\{1\bar{2}10\}$ stacking fault, which has two atomic configurations in wurtzite (Ga,Al,In)N, has been investigated by high resolution electron microscopy and energetic calculations. It originates from steps at the SiC surface and it can form on a flat (0001) sapphire surface. A modified Stillinger-Weber potential was used in order to investigate the relative stability of the two atomic configurations. They have comparable energy in AlN, whereas the $1/2<10\bar{1}1>\{1\bar{2}10\}$ configuration is more stable in GaN and InN. In GaN layers, only the $1/2<10\bar{1}1>\{1\bar{2}10\}$ configuration was observed. The $1/6<20\bar{2}3>$ configuration was found in small areas inside the AlN buffer layer where it folded rapidly to the basal plane, and when back into the prismatic plane, it took the $1/2<10\bar{1}1>\{1\bar{2}10\}$ atomic configuration.

INTRODUCTION

Extensive research effort is being undertaken on wide band gap of GaN based semiconductors for their very promising device possibilities. Efficient laser diodes have been fabricated for emission in the blue range[1] in layers grown on sapphire by Metal Organic Vapor Epitaxy (MOVPE). The active GaN layers contain large densities of crystallographic defects[2,3], among which, one finds $\{1\bar{2}10\}$ planar defects which have recently been called double positioning boundaries (DPBs)[4,5], stacking mismatch boundaries (SMBs)[6] or inversion domain boundaries (IDBs)[7]. These faults have recently been investigated using High Resolution Electron Microscopy (HREM) and Convergent Beam Electron Diffraction (CBED), and it was shown that they are stacking faults on top of both sapphire and SiC[8-10]. In fact these planar defects have already been studied in the sixties and two displacement vectors have been measured by conventional microscopy[11,12]. In wurtzite ZnS, Blank et al[12] were the first to study the planar defects which folded from basal to prismatic $\{1\bar{2}10\}$ planes, and to interpret them as stacking faults whereas other authors considered them to be thin lamella of the sphalerite phase in CdS[13]. In this work, we have investigated these faults in (Al,Ga)N epitaxial layers by HREM and modified the Stillinger-

Weber[14] (SW) potential in order to analyze their stability in AlN, GaN, InN. The theoretical results were found to correlate with the HREM observations for AlN and GaN.

RESULTS

The stacking fault atomic structure

Two atomic models exist in the literature for the { 11$\bar{2}$0 } prismatic fault in wurtzite materials as originally characterized by conventional electron microscopy in the 60s by Drum[11] and Blank[12], respectively. Their projections are shown in figure 1 along [0001].

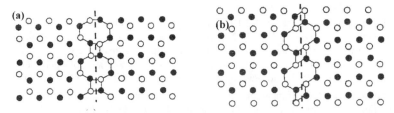

Figure 1. Atomic configurations of the { 11$\bar{2}$0 }stacking fault seen along [0001], a) The ½ [10$\bar{1}$1] displacement vector, b) The 1/6 [20$\bar{2}$3] stacking fault

Formation mechanisms

A. Formation on (0001) 6H-SiC surface

The [0001] SiC surface exhibits steps of various heights. If one ignores the mismatch along the c axis which is small, the GaN and SiC can be geometrically consid

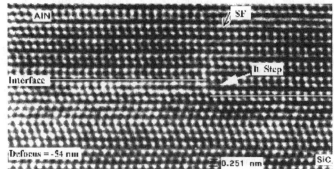

Figure 2. Generation of a prismatic stacking fault on a I1 type of step on the (0001) surface of 6H-SiC

A 6H-portion under a step can be decomposed into a faulted 2H stacking, and depending on the type of step, the decomposition may result in a displacement vector between the two crystallites which grow on the adjacent terraces (fig.2). In fact between the two polytypes, the only displacement vectors d which can result from the decomposition are those of the hcp lattice (the intrinsic I_1, R_{I1} = 1/6 < 20$\bar{2}$3 >, I_2 , R_{I2} = 1/3 <10$\bar{1}$0 > and the extrinsic E, R_E = 1/2 <0001>). Among them, I_1 and E have 1/2c component along the c axis and may lead to the formation of an extended defect in the epitaxial layer (fig.2). The displacement vector identical to I_2 is confined in the interface and may contribute to release the misfit strain. When all the step heights are taken account, it was found that the largest fraction of the faults that will be formed are I_1 as shown in table 1

Interface	Defect Character	Interface Configuration: Wurtzite Defect ratio: (%)
2H / 6H	I1 I2 E -	45 22 - 33

Table 1. Type of defects forming at an interface step between 2H-GaN and 6H-SiC

B. Growth on (0001) sapphire

The continuation of the two hcp anionic lattices may be described by three types of polyhedra at the interface. They are built with oxygen (O) and nitrogen (N) at their vertices, the metal (M) at the centre: one tetrahedron (3O, M, N) that leads to upward polarity, one tetrahedron (O, M, 3N) and one octahedron (3O, M, 3N) that lead to downward polarity. All these factors are taken into account in order to analyze how the resulting translation can be accommodated in the GaN layers by the formation of planar defects in the epitaxial layer: stacking faults (SF) or inversion domains boundaries (IDB). For the wurtzite structure, three models of IDBs have been proposed[15-18].

B.1. Growth on a single terrace

On an oxygen terminated surface, there are four ways to build a hcp stacking of nitrogen atoms, they are related to each other by one of the three stacking fault vectors, R_{SF}, of the wurtzite structure (R_{I1}, R_{I2} and R_E). Since two families of tetrahedral sites are available for the Ga^{3+} cation, this leads to eight GaN stackings linked by the operator W_{CE} = R_{SF} or W_{CE} = R_{SF} + m with opposite polarities(fig.3). In the case of Al-terminated surface, the number of stackings is reduced to three polyhedra due to the aluminium position in the Al_2O_3 structure. The systematic analysis of the polyhedra combinations allows to determine the relative proportions of the planar defects induced by the operator W_{CE}. For instance, if the interface polyhedra are exclusively downward tetrahedra, the R_{SF} will be only R_{II}, and combinations of downward tetrahedra and octahedra lead to the formation of R_{II} and R_{I2} faults in equal proportions which may explain the formation of stacking faults inside the epitaxial layers[19-21].

B.2. Growth on adjacent terraces

A step, between two terraces on which two GaN islands grow, introduces a translation T_S between them. Its component along c axis can be reduced in $R_E=1/2$ <0001> and a residual translation T_R. The number of R_E is calculated to minimize the absolute value of this residual translation. For instance, the step between two A terraces with a height of c/3 in Al_2O_3 (0.433 nm) is equivalent in GaN to 0.835 c_{GaN}, or : $T_S = 2R_E + T_R = T_R$ with $T_R = -0.165\ c_{GaN} \approx c_{GaN}/6$. By considering all the steps between A or B terraces (A or B layers of the hcp stacking), the T_S translation can be written as : $T_S = aR_E + T_R$ with a = 0 or 1 and T_R may take four values with the approximation $2c_{Al_2O_3} \sim 5c_{GaN}$(table.2) : So, the operator relating adjacent terraces must involve this additional parameter, T_R :

$$W_{CE} = R + m + T_R \text{ where } R = R_{SF} + aR_E. \qquad (1)$$

Steps	$h_{c_{Al_2O_3}}$ unit	$T_R\ c_{GaN}$ unit	T_R nm
A-A or B-B	nc (n: integer)	0	0
A-B	1/6, 5/6, 7/6 ,11/6	~ 1/12	~ 0.0432
A-A or B-B	1/3, 2/3	~ 1/6	~0.0863
A-B	1/2, 3/2	~ 1/4	~ 0.1295

Table 2. Residual translation in GaN as a function of the step height in Al_2O_3

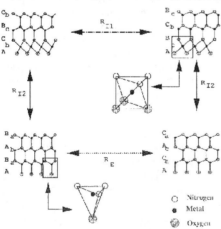

Figure 3. Four of the 8 possible stacking of wurtzite (Ga,Al)N on the (0001) sapphire surface, the polyhedra configurations at the interface are shown

As this residual translation is never equal to c/2, it cannot be simply accommodated by the formation of prismatic stacking faults. The operator describing IDBs can be written in order to point out the c/8 translation found inside some of the models in addition to the mirror operation (1). When compared to the residual translation T_R, the introduction of the IDB containing the c/8 translation may help in minimizing the shift along c at most of the steps; it is reduced to $1/24\ c_{GaN}$[20]. Therefore, in the case of steps, the residual translation cannot be minimized by introducing a stacking fault. During epitaxial growth of GaN layers on (0001) sapphire, the occurrence of basal and prismatic stacking faults is not a straightforward result of the steps at the substrate surface. They form on coalescence of islands whose growth has been initiated with different polyhedra[20].

Relative stability

The original SW potential is short and it cannot be used in order to predict the values of the non ideal c/a ratio and the vector u which characterizes the displacement along the c axis between the two hcp sublattices in the wurtzite structure. It is unable to reach the third nearest neighbour atom which is responsible of the deviation of c/a ratio from the ideal value. We have added a Gaussian term which was fitted in order to allow the interaction potential to reach the third nearest neighbour. It is given by:

$$v_g = \varepsilon C \exp\{-((d_{ij}-r_1)/\delta r_1)^2\}, \qquad (2)$$

where d_{ij} is the bond length between atoms i and j, r_1 the center of the Gaussian is situated between the second and third neighbours distances and Δr_1 is the width of the Gaussian function. According to the magnitude of r_1 and the sign of C, it is possible to adjust the c/a ratio with respect to the ideal value. The potential obtained in the previous section is applied to investigate the atomic structure of the $\{11\overline{2}0\}$ stacking fault. The above atomic models were taken into account with the starting configurations corresponding to the $1/6<2\overline{2}03>$ and $1/2<1\overline{1}01>$ stacking fault displacement vectors as discussed in the previous paragraphs. We applied the method of lattice relaxation to a set of 1984 atoms containing the defect for the two atomic models. As seen in table 3, the formation energy of each configuration depends on the compound for AlN, GaN and InN. In AlN, the two atomic configurations of the $\{11\overline{2}0\}$ stacking fault have a comparable value of formation energy. This is in complete agreement with our HREM observations in GaN layers and in the AlN buffers[22]

model	Amelinckx ($1/6<2\overline{2}03>$)	Drum ($1/2<1\overline{1}01>$)
AlN	100	100
GaN	78 (123)	22 (72)
InN	65	21

Table 3. Formation energy in meV/$\overset{0}{A}{}^2$ for the relaxed atomic configurations of the $\{11\overline{2}0\}$ prismatic stacking fault, in parenthesis are the values of reference 23

CONCLUSION

The two configurations of the stacking fault have already been investigated by conventional electron microscopy, one in AlN and the other in ZnS respectively. They both were found to fold easily from the basal to the prismatic plane of these wurtzite materials. In the basal plane both have the $1/6 < 20\bar{2}3 >$ displacement vector of the I1 stacking fault. In the $\{11\bar{2}0\}$ plane, they differ by $1/6 < 10\bar{1}0 >$, which is worth less than 4% in energetic balance between the two vectors (using the b^2 criterion) in the three nitrides compounds.

REFERENCES

[1] S. Nakamura, M. Senoh, S. Nagahama, N. Iwasa, T. Yamada, T. Mitsushita, H. Kiyoku, and Y. Sugimoto, Jpn. J. Appl. Phys. **35**, L74 (1996).

[2] F.A. Ponce, D. P. Bour, W. Götz, N.M. Johnson, H. I. Helava, I. Grzegory, J. Jun, and S. Porowski, Appl. Phys. Lett. **68**, 917 (1995).

[3] P. Vermaut, P. Ruterana, G. Nouet, A. Salvador, and H. Morkoç, Inst. Phys. Conf. Ser. **146**, 289 (1995).

[4] S. Tanaka, R. S. Kern, and R. F. Davis, Appl. Phys. Lett. **66**, 37 (1995).

[5] D. J. Smith, D. Chandrasekar, B. Sverdlov, A. Botchkarev, A. Salvador, and H. Morkoç, Appl. Phys. Lett. **67**, 1803 (1995).

[6] B.N. Sverdlov, G. A. Martin, H. Morkoç, and D. J. Smith, Appl. Phys. Lett. **67**, 3064 (1995).

[7] J. L. Rouvière, M. Arlery, R. Niebuhr, and K. Bachem, Inst. Phys. Conf. Ser. **146**, 285 (1995).

[8] P. Vermaut, P. Ruterana, G. Nouet, A. Salvador, and H. Morkoç, Mater. Res. Symp. Proc. 423, **551** (1996).

[9] P. Vermaut, P. Ruterana, G. Nouet, A. Salvador, and H. Morkoç, Phil. Mag. **A75**, 239, (1997).

[10] P. Vermaut, P. Ruterana, and G. Nouet, Phil. Mag. **A76**,1215, (1997).

[11] C. M. Drum, Phil. Mag., **11**, 313 (1964).

[12] H. Blank, P. Delavignette, R. Gevers, and S. Amelinckx, Phys. Stat. Sol. **7**, 747 (1964).

[13] L.T. Chadderton, A.F. Fitzgerald, and A.D. Yoffe, Phil. Mag. **8**, 167(1963)

[14] F.H. Stillinger, and T. A. Weber, Phys. Rev. **B31**, 5262 (1985).

[15] S.B. Austerman, and W.G. Gehman, J. Mater. Sci. **1**, 249(1966)

[16] D.B. Holt, J. Chem. Sol.**30**, 1297(1969)

[17] L.T. Romano, J.E. Northrup, and M.A. O'Keefe, Appl. Phys. Lett. **69**, 2394(1996)

[18] V. Potin, P. Ruterana, and G. Nouet, J. Appl. Phys. **82**, 2176(1997)

[19] B. Barbaray, V. Potin, P. Ruterana, and G. Nouet, Diamond and Related Materials, **8**, 314(1999)

[20] P. Ruterana, V. Potin, A. Béré, B. Barbaray, G. Nouet , A. Hairie, E. Paumier , A. Salvador, A. Botchkarev, and H. Morkoç, Phys. Rev. **B59**, 15917(1999)

[21] B. Barbaray, V. Potin, P. Ruterana, and G. Nouet Phil. Mag. A in press

[22] P. Vermaut, P. Ruterana, and G. Nouet, Appl. Phys. Lett. **74**, 694 (1999)

[23] J.E. Northrup, Appl. Phys. Lett. **72**, 2316(1998)

Mat. Res. Soc. Symp. Vol. 595 © 2000 Materials Research Society

Threading Dislocation Density Reduction in GaN/Sapphire Heterostructures.

A. Kvit, A. K. Sharma and J. Narayan
NSF Center for Advanced Materials and Smart Structures, Department of Materials Science and Engineering, North Carolina State University, Raleigh, NC 27695-7916.

ABSTRACT

Large lattice mismatch between GaN and α-Al_2O_3 (15%) leads to the possibility of high threading dislocation densities in the nitride layers grown on sapphire. This investigation focused on defect reduction in GaN epitaxial thin layer was investigated as a function of processing variables. The microstructure changes from threading dislocations normal to the basal plane to stacking faults in the basal plane. The plan-view TEM and the corresponding selected-area diffraction patterns show that the film is single crystal and is aligned with a fixed epitaxial orientation to the substrate. The epitaxial relationship was found to be $(0001)_{GaN}||(0001)_{Sap}$ and $[01\text{-}10]_{GaN}||[\text{-}12\text{-}10]_{Sap}$. This is equivalent to a 30° rotation in the basal (0001) plane. The film is found to contain a high density of stacking faults with average spacing 15 nm terminated by partial dislocations. The density of partial dislocations was estimated from plan-view TEM image to be 7×10^9 cm^{-2}. The cross-section image of GaN film shows the density of stacking faults is highest in the vicinity of the interface and decreases markedly near the top of the layer. Inverted domain boundaries, which are almost perpendicular to the film surface, are also visible. The concentration of threading dislocation is relatively low ($\sim 2 \times 10^8$ cm^{-2}), compared to misfit dislocations. The average distance between misfit dislocations was found to be 22 Å. Contrast modulations due to the strain near misfit dislocations are seen in high-resolution cross-sectional TEM micrograph of GaN/α-Al_2O_3 interface. This interface is sharp and does not contain any transitional layer. The interfacial region has a high density of Shockley and Frank partial dislocations. Mechanism of accommodation of tensile, sequence and tilt disorder through partial dislocation generation is discussed. In order to achieve low concentration of threading dislocations we need to establish favorable conditions for some stacking disorder in thin layers above the film-substrate interface region.

INTRODUCTION

GaN and other wide band group III-nitride semiconductors have been recently recognized as very important materials for the fabrication of optoelectronic devices emitting in the green - ultraviolet range. Moreover these materials have a large potential for electronic devices capable of operation under high-power, high- frequency and high-temperature conditions due to their superior physical properties. Important efforts have been made to improve the growth of these materials. The large differences in lattice parameters (16%) and thermal expansion coefficients between GaN and sapphire are the common reason for generation of very high concentration extended defects. Threading dislocations are the most common line defects observed for the films grown on hexagonal sapphire and SiC (0001) substrates. These dislocations are to found to originate at the substrate interface and caused by the inversion domain boundaries, stacking mismatch boundaries and double positioning boundaries [1,2]. Although GaN provides highly efficient devices the concentration of threading dislocations is usually very high (10^9-10^{10} cm^{-2}). The lowest densities of these defects lie in the mid 10^8 cm^{-2} range [3,4]. It is necessary to decrease the concentration of threading dislocations for further improvement of laser diode. I addition to relaxation of lattice structure, the generation of planar type of extended defects may terminate threading dislocations and prevent their propagation through active layers.

It is well known that the quality of final structure strongly depends on initial stages of growth. Several initial MOCVD growth conditions commonly used for GaN layer grown on (0001) α-Al$_2$O$_3$ substrate include: (1) high temperature nitridation of the sapphire surface; (2) deposition of low- temperature (600 °C) buffer layer followed by a high temperature (1050-1080 °C) epilayer growth; (3) low temperature AlN buffer growth followed by the high temperature GaN growth. The optimal growth temperature for the case of pulsed layer deposition of GaN is lower (720- 800 °C) owing to higher energy of grown species. The high average kinetic energy along with pulsed mode of laser ablation plasma is responsible for high quality films at lower substrate temperature. In the case of laser ablation, it is in principle possible to transfer the composition from the stoichiometric target to the substrate. In general, raising the growth temperature is effective in improving the crystallinity of epitaxial layers unless it causes thermal decomposition of the crystal [5].

In this investigation we have adopted two types of growth conditions: (1) pulsed layer deposition of "high- temperature" (850 °C) buffer layer and (2) "high- temperature" (850 °C) AlN buffer growth followed by the low temperature growth (750 °C) of GaN. The growth at high-temperature results in the formation of stacking faults that can buffer threading dislocations generated at optimized (stoichiometric) low temperature growth.

We have investigated defect reduction as a function of processing variables, where we show that the microstructure changes from threading dislocations normal to the basal plane to stacking faults in the basal plane. Finally the interaction between threading dislocation and partial dislocation and termination of threading dislocation have been demonstrated.

EXPERIMENTAL

The films were deposited in an ultra high vacuum chamber evacuated by a turbomolecular pump and cryopump to a base pressure of 5×10^{-10} Torr. The pulsed excimer laser (λ=248 nm, t_p=25 ns, 10 Hz) was used to ablate stoichiometric GaN (purity: 99.9%) target. The laser energy density used in these depositions was ~1.5-2 J/cm^2. The (0001) α-Al$_2$O$_3$ substrates were degreased in acetone and finally cleaned in methanol. The high temperature GaN (sample # *J2*) and AlN (sample #*J14*) buffer layers were deposited at 850°C and 800°C, respectively in vacuum. The final thicker layers of GaN were grown at 750°C in vacuum. The cooling was done slow (20°C/min) in N$_2$ gas flow.

Samples for cross-sectional transmission electron microscopy were prepared by gluing two samples film-to-film, mechanically grinding them to a thickness of ~50 μm and thinning up to 10 μm using model 656 Gatan dimpler. Thinning to electron transparency was accomplished in Gatan precision ion polishing system using Ar$^+$ ion milling with ion energies initially at 6 keV and then reduced to 2 keV. The beam was incident at near grazing incidence of 4°. The characterization of the films was carried out using transmission electron microscope Topcon 002B operating at 200 kV with point resolution 1.8 Å under imaging conditions close to the Scherzer focus.

RESULTS AND DISCUSSION

To grow epitaxial layers on widely mismatched systems, such GaN on α-Al$_2$O$_3$ (15%), we have to introduce domain matching epitaxy [6] where integral multiple of planes (7 for GaN and 6for α-Al₂O₃) match across the interface.

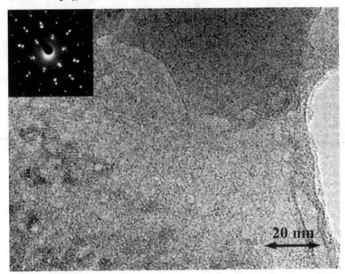

Figure 1. The plan-view TEM image of GaN 100 nm film grown on (0001) α-Al₂O₃ (sample #J2). The upper part without substrate is shown. High concentration of stacking faults is visible. On inset: the [0001] zone axis SADP of overlapping GaN and α-Al₂O₃. Double diffraction spots are visible.

The plan-view TEM image of GaN/α-Al₂O₃ (sample #*J2*) sample and the corresponding selected-area diffraction pattern show that the film is single crystal and is aligned with a fixed orientation to the substrate (Figure 1). We observed the double diffraction spots due to overlapping of films and substrate diffraction.

The epitaxial relationship was found to be $(0001)_{GaN}\|(0001)_{Sap}$ and $[01\text{-}10]_{GaN}\|[\text{-}12\text{-}10]_{Sap}$. This is equivalent to a 30° rotation in the basal (0001) plane, which is similar to relationship of epitaxial AlN grown on sapphire substrate [7]. Since the oxygen sublattice in sapphire is rotated 30° with respect to the unit cell of sapphire, the hexagonal plane of GaN is matched with the oxygen closed-packed plane of sapphire. This indicates that $Ga_{(GaN)}$ - $O_{(\alpha\text{-}Al2O3)}$ bonding at interface controls the epitaxial relationships. We found that this film contains a high density of stacking faults with average spacing 15 nm terminated by partial dislocations. The density of partial dislocations was estimated from plan-view TEM image to be 7×10^9 cm^{-2}.

Microstructure of the PLD films grown at temperatures 850 °C was found to be different from the above optimized growth conditions. The growth at this temperature really leads to loss of nitrogen in the grown film. However no trace of thermal decomposition and the formation of the cubic phase have been observed. The generation of high concentration stacking faults takes place under these growth conditions. The reason of this effect is the stoichiometry of PLD grown films is highly depends on stoichiometry of target. In our case the target was close to stoichiometry.

Figure 2. Cross-section image of GaN film grown on (0001) α-Al_2O_3 near [01-10] zone of sapphire (sample #*J2*).

Figure 2 shows a cross-section image of GaN film (sample #*J2*) grown at 850 °C and a high density of stacking faults running parallel to the film/substrate interface can be clearly observed. The density of these defects is highest in the vicinity of interface and decreases markedly near the top of the layer. Inverted domain boundaries [8], which are almost perpendicular to the film surface, are also visible. The concentration of threading dislocation is relatively low (~$2x10^8$ cm^{-2}).

Figure 3. Cross-sectional TEM microphotograph of GaN/AlN/α-Al_2O_3 structure in [01-10] zone of sapphire (sample #*J14*)

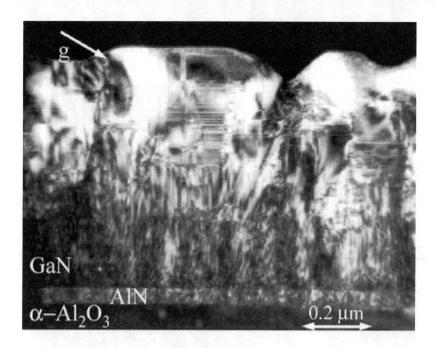

Figure 4. Cross-sectional TEM microphotograph of GaN/AlN/α-Al$_2$O$_3$ structure taken under g= [-21-12] two beam central dark field conditions. (sample #*J14*).

A somewhat different overall picture has been found for AlN buffer layer (sample #*J14*). Figure 3 shows that AlN layer is highly defective and concentration of threading dislocations as high as ~10^{10} cm^{-2}. As soon as they nucleated threading dislocations easily propagate through AlN/GaN interface. No stacking faults were observed in AlN layers grown at 850 °C. So Al - N bonds are stronger and the same nitrogen deficient effect should be expected at higher temperature. Note that lattice mismatch of α-Al$_2$O$_3$ with AlN (13%) is smaller than that with GaN (16.5 %). On the other hand the high concentration of stacking faults are observed in GaN layers region grown at 850 °C. Stacking faults could terminate threading dislocation. This effect demonstrated in Figure 4. A consider-able reduction in threading dislocation density is demonstrated in the layers above the stacking fault layers. This clearly shows buffering of threading dislocations by stacking faults.

Figure 5 shows high-resolution cross-sectional TEM image of GaN/α-Al$_2$O$_3$ interface. Contrast modulations due to the strain near misfit dislocations are seen. The average distance between these dislocations is 22 Å. This interface is sharp and does not contain any transitional layer. The interfacial region has a high density of Shockley and Frank partial dislocations. The Shockley partials accommodate changes in stacking sequence between adjacent regions [3,9], such as coalesced islands, and the Frank partials accommodate tilt disorder within the interfacial region. Since partial dislocations are associated with disorder in the stacking of closed-packed planes, their line directions are generally lie within the glide planes [9].

CONCLUSIONS

Reduction of threading dislocations is demonstrated by buffer layers containing a high density of stacking faults due to nonstoichiometric nitrogen deficient initial growth. The

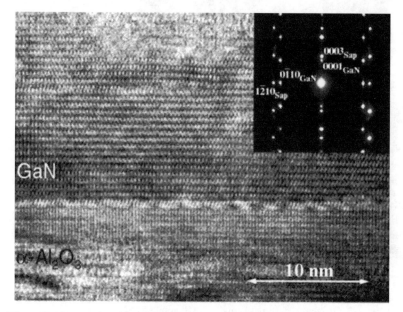

Figure 5. HRTEM image of GaN/α-Al₂O₃ in [01-10] zone of sapphire (sample #*J2*). On inset: superimposed [10-10] sapphire zone axis and [2-1-10] GaN zone axis SAD pattern.

stacking faults with tilted Burger vectors also help in the relaxation of lattice strain. In order to achieve low concentration of threading dislocations we need to establish favorable conditions for some stacking disorder in thin layers above the film-substrate interface region during the initial stages of film growth. The partials accommodate changes in stacking sequence between adjacent regions, such as coalesced islands, and tilt disorder within the near-interfacial region.

REFERENCES

1. B. N. Sverdlov, G. A. Martin, H. Morkoç, and David J. Smith, *Appl. Phys. Lett.* **67**, 2063 (1995).
2. David J. Smith, D. Chandrasekhar, B. Sverdlov, A. Botchkarev, A. Salvador, and H. Morkoç, *Appl. Phys. Lett.* **67**, 1830 (1995).
3. D. Kapolnek, X. H. Wu , B. Heying, S. Keller, B. P. Keller, U. K. Mishra, S. P. DenBaars, J. S. Speck, *Appl. Phys. Lett.* **67**, 1541 (1995).
4. P. Vennéguès, B Beaumont, S. Haffouz, M. Vaille, P.Gibart, *Journal of Crystal Growth* **187**, 167 (1998).
5. S. Oktyabrsky, K. Dovidenko, A. K. Sharma, J. Narayan, *Mat. Res. Soc. Symp. Proc.* **526**, MRS Spring Meeting, San Francisco USA (1998) 287.
6. J. Narayan, P. Tiwari, X. Chen, R. Chowdhary and T. Zheleva, *Appl. Phys. Lett.* 61, 1290 (1992); U.S. Patent **5**, 406, 123 (April 11, 1995).
7. K. Dovidenko, S. Oktyabrsky, J. Narayan, M. Razeghi, *J. Appl. Phys.* **79**, 2439 (1996).
8. J. Michel, I. Masson, S. Choux, A. George, *Phys. Status Solidi A* **146,** 97 (1994).
9. X.H. Wu, P. Fini, E.J. Tarsa, B. Heying, S. Keller, U.K. Mishra, S.P. DenBaars, and J.S. Speck, *J. Cryst. Growth*, **189/190**, 231 (1998).

Mat. Res. Soc. Symp. Vol. 595 © 2000 Materials Research Society

Microstructral Investigations on GaN Films Grown by Laser Induced Molecular Beam Epitaxy

H.Zhou, F.Phillipp, M.Gross* and H.Schröder*
Max-Planck-Institut für Metallforschung, Heisenbergstr.1, 70569 Stuttgart, Germany
*DLR, Institut für Technische Physik, Pfaffenwaldring 38-40, 70569 Stuttgart, Germany

ABSTRACT

Microstructural investigations on GaN films grown on SiC and sapphire substrates by laser induced molecular beam epitaxy have been performed. Threading dislocations with Burgers vectors of $1/3<11\underline{2}0>$, $1/3<11\underline{23}>$ and $[0001]$ are typical line defects, predominantly the first type of dislocations. Their densities are typically 1.5×10^{10} cm^{-2} and 4×10^{9} cm^{-2} on SiC and sapphire, respectively. Additionally, planar defects characterized as inversion domain boundaries lying on $\{1\underline{1}00\}$ planes have been observed in GaN/sapphire samples with an inversion domain density of 4×10^{9} cm^{-2}. The inversion domains are of Ga-polarity with respect to the N-polarity of the adjacent matrix. However, GaN layers grown on SiC show Ga-polarity. Possible reasons for the different morphologies and structures of the films grown on different substrates are discussed. Based on an analysis of displacement fringes of inversion domains, an atomic model of the IDB-II with Ga-N bonds across the boundary was deduced. High resolution transmission electron microscopy (HRTEM) observations and the corresponding simulations confirmed the IDB-II structure determined by the analysis of displacement fringes.

INTRODUCTION

Due to the large direct band gap and high thermal stability, GaN is a promising material for applications to light emitting diodes, laser diodes operating in visible and ultraviolet regions, as well as devices working at high temperatures. GaN thin films have been successfully grown by many epitaxial techniques, such as molecular beam epitaxy (MBE) [1], metalorganic chemical vapor deposition (MOCVD) [2] and pulsed laser deposition (PLD)[3]. In the present study, laser induced molecular beam epitaxy (LIMBE) [4, 5] was used as growth technique for epitaxial GaN films. This method is based on PLD. Alike the classic PLD configuration, a picosecond laser system was used.

Because of the potential applications of GaN in optoelectrics, the improvement of the quality of the epitaxial films and the structural characterization have been carried out recently [6, 7]. The majority of defects investigated are threading dislocations. Additionally, planar defects faceted along $\{1\underline{1}00\}$ and $\{1\underline{2}10\}$ planes have been observed. Only little effort has been devoted to the structural studies of the films grown by LIMBE. The aim of this paper is to study the microstructure of GaN films deposited by LIMBE method, in particular focusing on inversion domain boundaries (IDBs).

EXPERIMENTAL

GaN films were grown by laser induced molecular beam epitaxy on SiC (0001) and sapphire (0001) substrates. In addition to the chemical cleaning, the substrates were degreased and dried with N_2 gas, and heated in the growth chamber at 870 °C prior to growth. An average laser power of 3 W with a pulse rate of 1.6 kHz leads to a pulse energy of 1.3 mJ on the target. Experimental growth of GaN has been achieved at 730 °C and 5×10^{-2} mbar N_2 pressure with a growth rate of 100 nm/h. More details of the growth process are described elsewhere [4].

The surface morphology of the GaN films was studied by atomic force microscopy (AFM) and scanning electron microscopy (SEM). Cross-sectional and plan-view samples for transmission electron microscopy (TEM) were prepared by the conventional techniques. TEM and HRTEM observations were carried out on a Philips CM200 operating at 200 kV and a JEM 4000-FX microscope operating at 400 kV. Convergent beam electron diffraction (CBED) and HRTEM simulations were carried out by EMS software [8].

RESULTS AND DISCUSSION

The morphologies of the films grown on both SiC and sapphire exhibit similar features. Figure 1 shows a typical SEM micrograph of GaN/sapphire sample, in which some groves are in between rather flat areas. The root-mean-square (RMS) roughness value of the flat areas of GaN/SiC and GaN/sapphire are 9.6 nm and 14.6 nm, respectively. The depth ranges of groves are up to 10 nm and 50 nm, respectively.

Figure 2 shows cross-sectional TEM micrographs of GaN films imaged under g=0002 two-beam conditions. The orientation relationships shown by the insets are as following: $[1\underline{2}10]_{GaN}$ // $[1\underline{2}10]_{SiC,}$ $(0001)_{GaN}$ // $(0001)_{SiC}$, GaN/sapphire, $[1\underline{2}10]_{GaN}$ // $[1\underline{1}00]_{Sapphire}$ and $(0001)_{GaN}$ // $(0001)_{Sapphire}$. In addition to threading dislocations shown in figure 2 (a), planar defects IDBs are present in figure 2 (b). A groove generated by the coalescence of two islands is composed of two pyramidal planes with an angle of about 60°. The side walls can be indexed as $(\underline{1}011)$ and $(10\underline{1}1)$, respectively. The planar defects in the GaN/sapphire samples were observed in plan-view as shown in figure 3. The domains are bounded by $\{1\underline{1}00\}$ planes.

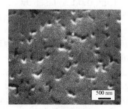

Figure 1. A typical SEM surface morphology of the GaN/sapphire film.

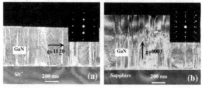

Figure 2. Cross-sectional TEM micrographs imaged with g=0002. (a) GaN/SiC along the $[1\underline{1}00]_{GaN}$ and (b) GaN/sapphire along the $[1\underline{2}10]_{GaN}$.

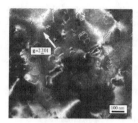

Figure 3. Plan-view TEM micrograph of domains in a GaN/sapphire sample imaged with **g**=$\underline{2}20\overline{1}$ close to the $[1\underline{1}04]_{GaN}$ zone axis.

The character of the domains was determined by the CBED technique. Figure 4 (a) and (b) show the CBED on-axis patterns taken along the $[1\underline{1}02]$ zone axis of GaN from a domain region and the adjacent region of matrix. The contrast fringes in the ($\underline{1}101$) and ($1\underline{1}0\overline{1}$) are reversed, which can be utilized to determine polarities. The same fringe at the zero spots means a nearly same thickness of the locations where CBED patterns were taken. For absolute determining the polarity, CBED-pattern simulations were performed. Figure 4 (c) shows a calculated pattern for a domain with a thickness of 50 nm. A good match of the simulated pattern with the experimental pattern has been obtained. The atomic configuration corresponding to the domain is shown in figure 4 (d). The polarity of the domain can be assigned to Ga-polarity, while the adjacent matrix is of N-polarity. In the following, $[000\underline{1}]$ (i.e., c-axis) is called the growth direction.

CBED experiments were also performed on a GaN/SiC sample. In contrast to the GaN/sapphire film, the GaN/SiC film has N-Polarity.

The GaN films grown on SiC and sapphire were deposited under identical

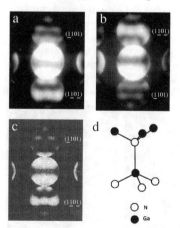

Figure 4. Convergent beam electron diffraction patterns (CBED) taken from an ID (a) and the adjacent matrix (b) in a GaN/sapphire sample. (c) simulated CBED pattern of the ID at a thickness of 50 nm. (d) atomic configuration of the crystal with Ga-polarity.

conditions without a buffer layer. The mismatch for the latter substrate is ~16% , which is much larger than that of the former substrate (3.4 %). Hence, the interface region of GaN/sapphire is easily subject to misfit strain, which would lead to steps along the interface. These steps could serve as sources for the formation of IDBs. The GaN/SiC film with Ga-polarity exhibits a relatively flat surface while the GaN/sapphire film with dominant N-polarity shows a rough surface, which is consistent with other observations [2].

A Si-terminted SiC substrate has been selected for the epitaxial growth, which readily leads to the Ga-polarity, because Si- and N-ions intend to incorporate. However, for the sapphire substrate, no nitridation treatment has been employed. This is believed to be the reason for non-unipolar nucleation. As evidenced in reference [10], the N/Ga ratio plays an important role in GaN epitaxy, because suitable ratio could preserve a pretended polarity and suppress the density of IDs. It appears that the N/Ga ratio for the growth of GaN/sapphire needs to be optimized to achieve unipolar crystal like in the case of GaN/SiC.

The GaN/sapphire films show truncated pyramids with some grooves located in between the rather flat top areas. IDs are distributed homogeneously in the film. In some reports, pyramidal steps and roughness of the GaN epilayer surface are attributed to the existence of IDs [2]. The assumption, however, of pronouncedly different growth rate between IDs and the adjacent matrix has not been supported by our observations.

In a titled plan-view sample, one ID can be viewed as being composed of two simple stacking faults, i.e., only the Ga-sublattice needs to be considered as the Ga atom has a larger scattering factor. However, an inversion operation should be taken into account. When $\mathbf{g}$=hki0 is excited, no strain or displacement fringe contrast has been observed, indicating that only displacement along the c-axis exists. Figure 5 shows the displacement fringes with $\mathbf{g}$=1$\bar{1}$01 excited. One can see that all IDs show asymmetric terminating fringes in bright-field, however, symmetric black fringes in dark-field. The same feature of the terminating fringes is observed for $\mathbf{g}$=$\bar{1}$102. However, no displacement fringe contrast was observed with $\mathbf{g}$=$\bar{3}$308. Our observations imply that the displacement along the c-axis might be ~c/8. Two possible IDB structures are shown in figure 6 for that case [9, 11-13].

From the displacement fringe contrasts using various diffraction reflections, it appears that all the IDBs in our samples behave the same as those studied in reference [14], in which dynamic calculations were performed and the atomic model shown in figure 6(b) was deduced. Hence, IDB-II structure is favorable for the samples.

Figure 5. Displacement fringes under two-beam conditions with $\mathbf{g}$=$\bar{1}$101 in (a) bright-field and (b) dark-field.

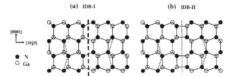

(a) IDB-I   **(b)** IDB-II

[0001]

[10$\bar{1}$0]

● N
○ Ga

Figure 6. Schematic presentation of atomic structures of the {1$\bar{1}$00} inversion domain boundary (IDB) in GaN along the [1$\bar{2}$10] projection. (a) a pure IDB structure IDB-I, (b) IDB-II structure with Ga-N bonds across the boundary.

    In order to differentiate the two models, we have conducted intensive HRTEM investigations on GaN samples along a <1$\bar{2}$10> zone axis. Three typical maxima of contrast were observed on the simulated images at about −10 nm, -50 nm and −90 nm with a thickness 10 nm as shown in figure 7(a)-(c), in which the atomic positions (Ga atom-smaller black spot; N atom-larger black spot) are superimposed on the simulated images. The white spots at defocus values of −10 nm and −90 nm represent the N atom, while at −50 nm they represent the tunnel images. Apparently, the stacking sequences highlighted by the dotted lines in figure 6 can be utilized to distinguish the two IDB models. It can be seen from figure 7(d)-(f) that the HRTEM experimental images exhibit a sequence identical to the simulated images. As mentioned above, IDBs in our samples are all of the same type. Hence, the result of our HRTEM investigations confirms the conclusion obtained from the displacement fringe contrast analyses, namely, only the IDB-II structure which is energetically favorable exists in the GaN/sapphire film grown by LIMBE.

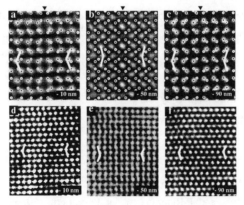

Figure 7. HRTEM calculated images of an IDB along a <1$\bar{2}$10> zone axis of GaN film with a thickness of 10 nm at defocuses (a) −10 nm, (b) -50 nm, (c) −90 nm; the correspondingly HRTEM experimental images at defocuses (d) −10 nm, (e) −50 nm, (f) −90 nm.

CONCLUSIONS

Microstructural investigations of GaN films grown on SiC and sapphire by laser induced molecular beam epitaxy have been performed. Our observations show that the majority of threading dislocations are of edge type. Additionally, inversion domains with Ga-polarity (with respect to the N-polarity of the matrix) are observed in the GaN/ sapphire samples. Misfit stress and initial substrate status are believed to be associated with the polarity selection. It appears that the N/Ga flux ratio needs to be optimized for the GaN/sapphire film to achieve Ga-polarity. Both, displacement fringe contrast and HRTEM studies indicate that the IDB-II model with Ga-N bonds across the boundary and with lower energy describes the most likely IDB-structure in the GaN samples.

ACKNOWLEDGMENTS

The authors would like to acknowledge Mrs. M. Kelsch and Mrs. A.Weisshardt for assistance in specimen preparations. Useful discussions with Dr. A.Rühm and Mrs. G.Henn are acknowledged.

REFERENCES

1. D.J.Smith, D.Chandrasekhar, B.Sverdlov, A.Botchkarev, A.Salvador and H.Morkoc, Appl. Phys. Lett. **67**, 1830 (1995) .

2. B.Daudin, J.L.Rouvière and M.Arlery, Mater. Sce. Eng. **B43** , 157 (1997).

3. D.Feiler, R.S.Williams, A.A.Talin, H.Yoon and M.S.Goorsky, J.Crystal Growth **171**, 12 (1997).

4. M.Gross, G.Henn and H.Schröder, Mater. Sci. Eng. **B50** , 16 (1997).

5. H.Zhou, F.Phillipp, M.Gross and H.Schröder, Mater. Sci. Eng. **B66**, (1999).

6. F.A.Ponce, D.Cherns, W.T.Young and J.W.Steeds, Appl. Phys. Lett. **69**, 770 (1996).

7. X.J.Ning, F.R.Chien, P.Pirouz, J.W.Yang and M.A. Khan, J. Mater. Res. **11** , 580 (1996).

8. P.A.Stadelman, Ultramicroscopy **21**, 131 (1969).

9. J.L.Rouvière, M.Arlery, B.Daudin, G.Feuillet and O.Briot, Mater. Sci. Eng. **B50**, 61 (1997).

10. L.T.Romano and T.H.Myers, Appl. Phys. Lett. **71**, 3468 (1997).

11. D.B.Holt, J. Phys. Chem. Solids **30**, 1297 (1969).

12. V. Potin, G. Nouet and P.Ruterana, Appl. Phys. Lett. **74**, 947 (1999).

13. J.E. Northrup, J. Neugebauer and L.T. Romano, Phys. Rev. Lett. **77**, 103 (1996).

14. D.Cherns, W.T.Young, M.Saunders, J.W.Steeds, F.A.Ponce and S.Nakamura, Phil. Mag. **A77**, 273 (1998).

Mat. Res. Soc. Symp. Vol. 595 © 2000 Materials Research Society

GaN Decomposition in Ammonia

D.D. KOLESKE, A.E. WICKENDEN, AND R.L. HENRY
Code 6861, Electronic Science and Technology Division,
Naval Research Laboratory, Washington, D.C. 20375

ABSTRACT

GaN decomposition is studied as a function of pressure and temperature in mixed NH_3 and H_2 flows more characteristic of the MOVPE growth environment. As NH_3 is substituted for the 6 SLM H_2 flow, the GaN decomposition rate at 1000 °C is reduced from $1x10^{16}$ $cm^{-2}s^{-1}$ (i.e. 9 monolayers/s) in pure H_2 to a minimum of $1x10^{14}$ $cm^{-2}s^{-1}$ at an NH_3 density of $1x10^{19}$ cm^{-3}. Further increases of the NH_3 density above $1x10^{19}$ cm^{-3} result in an increase in the GaN decomposition rate. The measured activation energy, E_A, for GaN decomposition in mixed H_2 and NH_3 flows is less than the E_A measured in vacuum and in N_2 environments. As the growth pressure is increased under the same H_2 and NH_3 flow conditions, the decomposition rate increases and the growth rate decreases with the addition of trimethylgallium to the flow. The decomposition in mixed NH_3 and H_2 and in pure H_2 flows behave similarly, suggesting that surface H plays a similar role in the decomposition and growth of GaN in NH_3.

INTRODUCTION

Metallorganic vapor phase epitaxy (MOVPE) is currently being used to grow GaN for the fabrication of blue light emitting diodes [1], lasers [2] and for high power electronic devices [3]. For MOVPE growth, NH_3 is typically used as the N source and high temperatures (> 1000 °C) are required to efficiently dissociate (i.e. 40-50 %) the NH_3 [4], because of the large N-H bond strength [5]. As a result, MOVPE growth temperatures are 100-500 degrees Celsius larger than the threshold temperature for GaN decomposition in vacuum [6,7] and in H_2 and N_2 [8,9]. The high rate of N_2 desorption is compensated by using large flows of NH_3 [10], however the extent of GaN decomposition that occurs during growth has not been measured.

The recent studies of Grandjean et al. [7] and Rebey et al. [11] have shown dramatic decreases in the GaN decomposition rate when small NH_3 flows are dosed onto GaN surface. For example, Grandjean et al. measure a GaN decomposition rate of 5 Å/s at 875 °C in vacuum, while under an NH_3 flux of $1.7x10^{17}$ $cm^{-2}s^{-1}$, the decomposition rate drops to 0.03 Å/s [7]. To explain the decrease in the GaN decomposition rate in NH_3, a site-blocking model has been proposed where the adsorbed NH_3 blocks sites necessary for N_2 formation and desorption [7]. A similar site blocking mechanism has also been proposed to explain reduced GaN growth when the NH_3 flux is increased [12]. In this paper we suggest that H also blocks sites on the GaN surface and H surface coverage effects must be considered in order to properly describe the GaN decomposition and growth kinetics.

EXPERIMENTAL DETAILS

Details of the GaN growth [13] and decomposition [8,9] are discussed elsewhere. The GaN films used in this study were grown and decomposed in a close-spaced showerhead MOVPE reactor. The growth and decomposition rates were determined from weight loss using an analytical balance [8]. The GaN films were grown at 1030 °C using 32 μmoles of trimethylgallium (TMGa), 2 SLM NH_3 and 4 SLM of

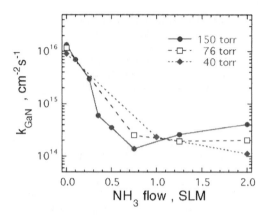

Fig 1. Plot of the GaN decomposition rate as a function of the NH₃ flow rate at three different pressures. For this plot the GaN was heated to a temperature of 992 °C using H_2 and NH_3 for a total flow of 6 SLM.

H_2 at pressures ranging from 40 to 300 torr. GaN decomposition was studied under similar flow conditions as the growth. The measured weights were converted to growth and decomposition rates per surface area (i.e. $cm^{-2}s^{-1}$) following Ref. 10. Expressed this way, a rate of 1.14×10^{15} $cm^{-2}s^{-1}$ corresponds to a thickness of 1 µm per hour. Temperature was calibrated by observing the melting point of 0.005" diameter Au wire on sapphire and correlating it to a thermocouple in direct contact with the susceptor underside [8]. After 2 years of use the set point temperature needed to melt the Au wire was reproducible to within 10 °C.

RESULTS

The change in the GaN decomposition rate, k_{GaN}, at a temperature of 992 °C as NH_3 is substituted for the H_2 flow is shown in Fig. 1. In Fig. 1, k_{GaN} is plotted as the NH_3 flow is increased from 0 to 2 SLM for pressures of 40, 76, and 150 torr. The total flow rate was kept constant at 6 SLM with the balance being H_2. Note that k_{GaN} decreases from $\approx$ 1×10^{16} $cm^{-2}s^{-1}$ to $\approx 1\times10^{14}$ $cm^{-2}s^{-1}$ as the NH_3 flow increases. At 150 torr, the minimum k_{GaN} occurs at a flow of 0.75 SLM of NH_3. At 76 torr, the k_{GaN} minimum occurs between 1.25 and 2.0 SLM of NH_3. At 40 torr no minimum in k_{GaN} is observed. At 150 torr and a flow of 2 SLM of NH_3, k_{GaN} is $\approx 4\times10^{14}$ $cm^{-2}s^{-1}$.

In Fig. 2, the data from Fig. 1 are replotted as a function of the NH_3 gas density, which depends on pressure. In Fig. 2, it appears that the k_{GaN} measured at different pressures have a common minimum at a NH_3 density of $\approx 1\times10^{19}$ cm^{-3}. This NH_3 density is the same order of magnitude as the calculated N desorption rate from GaN, which should be 9×10^{19} $cm^{-2}s^{-1}$ at 992 °C [10]. Two different dependencies of k_{GaN} on the NH_3 density are evident in Fig. 2. At lower NH_3 density, the k_{GaN} drops steeply as the NH_3 density increases from 3×10^{17} cm^{-3} to a value of near 1×10^{19} cm^{-3}. For NH_3 densities greater than 1×10^{19} cm^{-3}, k_{GaN} increases. This differs from the behavior observed by Grandjean *et al.*, where k_{GaN} only decreased for increasing NH_3 flow [7].

To determine the dependence of k_{GaN} on the NH_3 density, $[NH_3]$, separate fits were calculated for $[NH_3]$ both less than and greater than 1×10^{19} cm^{-3}. For $[NH_3] < 1\times10^{19}$ cm^{-3}, a functional form of $k_{GaN} = c[NH_3]^x$ was used and the data were fit by

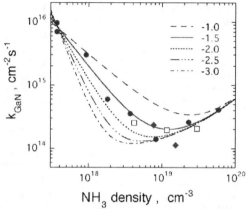

Fig. 2. Plot of the GaN decomposition rate measured as a function of the NH_3 density at 992 °C. The filled circles (red) were measured at 150 torr, the open squares (blue) were measured at 76 torr, and the filled diamonds (green) were measured at 40 torr. The lines are fits to the data using the expression $k_{GaN} = a + b[NH_3] + c[NH_3]^x$. For the fits the values of a and b are the same, while the value of x is fixed from −1.0 to −3.0 and c is varied for the best fit.

varying c, keeping x constant. For $[NH_3] > 1 \times 10^{19}$ cm^{-3}, a linear functional form, $k_{GaN} = a + b[NH_3]$, fit the data well. The series of lines shown in Fig. 2 are a combination of the two fits (i.e. $k_{GaN} = a + b[NH_3] + c[NH_3]^x$). For the combined fits, 5 curves were calculated for 5 values of x ranging from −1.0 to −3.0, keeping the linear fit constant. Clearly, the data are best fit using with $x = −1.5$ to −2.0.

Similar to NH_3, the GaN decomposition rate in N_2 is lower when compared to the rates measured in H_2 [9], however, in mixed N_2 and H_2 flows the rate is substantially larger than in mixed NH_3 and H_2 flows. This is shown in Fig. 3, where k_{GaN} is

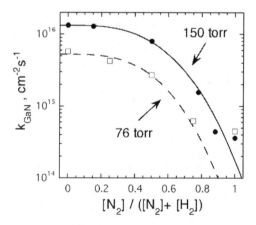

Fig. 3. Plot of the GaN decomposition rate vs. the ratio of the N_2 concentration to the total (i.e. $N_2 + H_2$) gas concentration. The GaN decomposition rate is shown at total pressures of 76 and 150 torr. The solid and dashed lines are cubic fits to the data.

plotted vs. the N_2 fraction of the total flow (i.e. $[N_2] + [H_2]$). For these measurements, the GaN films were annealed at 992 °C at pressures of 76 and 150 torr. In Fig. 3, k_{GaN} at 150 torr is reduced from 1.6×10^{16} cm^{-2}s^{-1} in pure H_2 to 3.5×10^{14} cm^{-2}s^{-1} in pure N_2 (factor of 45). For a 1:1 mixture of N_2 and H_2 at 150 torr, k_{GaN} decreases slightly to 8×10^{15} cm^{-2}s^{-1}, which is a factor of 2 compared to k_{GaN} in pure H_2. This is a significantly smaller decrease when compared to the decrease in a 1:2 mixture of NH_3 and H_2 (factor of 120). Note that k_{GaN} in pure N_2 and in mixed H_2 and NH_3 flows can be similar. For example in pure N_2, k_{GaN} is 3.5×10^{14} cm^{-2}s^{-1}, while in mixed H_2 and NH_3 at 150 torr, k_{GaN} is 4×10^{14} cm^{-2}s^{-1}, as shown in Fig. 1. The solid and dashed lines in Fig. 3 are cubic fits to the k_{GaN} vs. N_2 fraction. The cubic dependence is a result of the expected dependence of surface H coverage (i.e. $[H]^3$) for ammonia formation via the reaction $3H + N \rightarrow NH_3$. Previously, Thurmond and Logan also demonstrated NH_3 formation when GaN is heated in H_2 by titration of the basic exhaust gas [14].

In Fig. 4, the GaN decomposition and growth rates are plotted vs. pressure. In Fig. 4(a), k_{GaN} is plotted for GaN films annealed at 992 °C in H_2 [8]. Also in Fig. 4(b), the GaN growth rate at 1030 °C is plotted for conditions where 2 SLM NH_3, 4 SLM H_2, and 32 µmoles of TMGa were used. Finally, in Fig. 4(c) the GaN decomposition rate is plotted using the same conditions as (b) except no TMGa was used and hence decomposition was observed. Note that the decrease in the GaN growth rate as the pressure increases in Fig. 4(b) coincides with an increase in the GaN decomposition rate in Fig. 4(c). Also, the k_{GaN} shown in Figs. 4(a) and 4(c) have a similar shape as the pressure increases and these curves are nearly identical if the k_{GaN} in Fig 4(c) are multiplied by 30. This similarity in shape implies that surface H plays a similar role in the GaN decomposition for both pure H_2 and mixed NH_3 and H_2 gas environments.

DISCUSSION AND CONCLUSIONS

From the data presented in Fig. 1, the GaN decomposition rate is greater than 1×10^{14} cm^{-2}s^{-1} (i.e. $\approx$ 1/10 µm/hour) even in mixed NH_3 and H_2 flows. This is important for GaN growth because it suggests that some level of decomposition occurs during

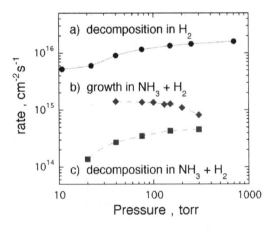

Fig. 4. Pressure dependence of the a) GaN decomposition rate in 6 SLM H_2 measured at T = 992 °C, b) GaN growth rate using, 32 µm TMGa, 2 SLM NH_3 and 4 SLM H_2 at T = 1030 °C, and c) GaN

decomposition rate using 2 SLM NH$_3$ and 4 SLM H$_2$ at T = 1030 °C. The only difference between b) and c) is the use of TMGa in b).

growth as previously speculated [10]. Currently, we are growing GaN at a pressure of 130 torr and a temperature of 1030 °C [13]. Under these growth conditions, the rates for growth and decomposition are 1.2x10^{15} cm^{-2}s^{-1} and 4x10^{14} cm^{-2}s^{-1} respectively as shown in Fig. 4. If the growth rate equals the incorporation rate minus the decomposition rate [10], this means that the incorporation rate is ≈ 4 times the decomposition rate under these growth conditions.

The decrease in k_{GaN} as the NH$_3$ density increases is due to NH$_3$ adsorption, which blocks sites needed for GaN decomposition. As shown in Fig. 2, the decrease in k_{GaN} depends on the –1.5 to –2.0 power of the NH$_3$ density. Since Ga desorption from GaN has been shown to be independent of H$_2$ pressure [8], reactions which remove N from the lattice probably influence the GaN decomposition rate more. This is clearly observed in the cubic dependence of k_{GaN} in Fig. 3 where NH$_3$ formation is favored at higher pressure. For N$_2$ formation and desorption one or both of the N atoms diffuse across the surface until they combine to form N$_2$. If open surface sites are necessary for N diffusion, blocking of these sites by NH$_3$ or H would decrease the hopping rate and as a consequence the N$_2$ formation rate would be decreased. If two (one) N must migrate for N$_2$ formation, the N$_2$ desorption kinetics would be second (first) order in the number of open surface sites. As the NH$_3$ density on the surface increases, the decrease in the GaN decomposition rate should be between first and second order, i.e. k_{GaN} α [NH$_3$]$^{-1}$ or k_{GaN} α [NH$_3$]$^{-2}$, depending on the details of N$_2$ formation and desorption. From Fig. 2, it is clear that the decrease in k_{GaN} vs. NH$_3$ density is closer to second order (power of –2) than first order.

At higher NH$_3$ densities (> 1x10^{19} cm^{-3}), the GaN decomposition rate increases linearly. This may be due to a decrease in the NH$_3$ site blocking suppression of N$_2$ desorption or a general increase in the H surface coverage. The increased H coverage could block sites necessary for NH$_3$ adsorption. Surface H has also been shown to aid in NH$_3$ adsorption and dissociation on GaN [15] and on Al [16]. In addition, large H coverage can favor NH$_3$ reformation and desorption by combining with adsorbed NH$_x$ species as suggested by Fig. 3. In contrast to NH$_3$, site blocking with H should lead to an increase in the decomposition rate.

Several groups have observed decreases in growth rate when H$_2$ is used in place of N$_2$ [17, 18], when the growth pressure is increased [19], and when higher NH$_3$ fluxes are used for growth [20]. In Fig. 4(b), the GaN growth rate decreases as the growth pressure increases. It is clear from Fig. 4(c) that the reason the growth rate decreases is because the increased GaN decomposition at higher pressures. However, to fully explain the reduction in the growth rate, the full effect of gas phase depletion of the TMGa also needs to be considered.

GaN grown in H$_2$, where GaN decomposition is enhanced compared to N$_2$, appears to have better crystalline order compared to GaN growth in N$_2$. Kistenmacher *et al* have shown that the FWHM of the GaN films grown in H$_2$ had narrower x-ray rocking curve linewidths and were better aligned compared (i.e. smaller mosaic dispersion) to GaN films grown in only N$_2$ [21]. Better alignment is also observed in laterally overgrown GaN when H$_2$ is used instead of N$_2$ [22]. Schön and coworkers find smoother morphologies and better electrical properties when growth is conducted in H$_2$ compared to N$_2$ [23]. Better electrical properties are observed for GaN grown at higher pressures where GaN decomposition is enhanced [13, 24, 25]. Recently, we have observed a near doubling of electron mobility in films grown at 150 torr compared to 76 torr, keeping all other growth parameters the same [13]. In this study, growth at higher pressure led to increased GaN grain size in the films [13], suggesting that the increased GaN decomposition at higher pressure plays a significant role in determining the grain size of the GaN film.

ACKNOWLEDGEMENTS
We thank V.A. Shamamian, V.M. Bermudez, R.J. Gorman, M.E. Twigg, J. Freitas, M. Fatemi, and J.C. Culbertson for discussions and characterization of the films. This work is supported by the Office of Naval Research.

REFERENCES
[1]   S. Nakamura, M. Senoh, N. Iwasa, S. Nagahama, T. Yamada, and T. Mukai, Jpn. J. Appl. Phys., **34**, L1332 (1995).

[2]   S. Nakamura, M. Senoh, S. Nagahama, N. Iwasa, T. Matushita, and T. Mukai, MRS Internet J. Nitride Semicond. Res. **4S1**, G1.1 (1999).

[3]   S.N. Mohammad, A.A. Salvador, and H. Morkoc, Proc. IEEE **83**, 1306 (1995).

[4]   S. S. Liu and D. A. Stevenson, J. Electrochem. Soc. **125**, 1161 (1978).

[5]   For H-NH$_2$ the bond strength is 4.8 eV. CRC Handbook of Chemistry and Physics, 66th edition, edited by R.C. Weast, (CRC Press, Cleveland, 1986).

[6]   R. Groh, G. Gerey, L. Bartha, and J. I. Pankove, Phys. Status Solidi A **26**, 353 (1974).

[7]   N. Grandjean, J. Massies, F. Semond, S. Yu. Karpov, and R.A. Talalaev, Appl. Phys. Lett. **74**, 1854 (1999).

[8]   D.D. Koleske, A.E. Wickenden, R.L. Henry, M.E. Twigg, J.C. Culbertson, and R.J. Gorman, Appl. Phys. Lett. **73**, 2018 (1998), erratum: ibid **75**, 2018 (1999).

[9]   D.D. Koleske, A.E. Wickenden, R.L. Henry, M.E. Twigg, J.C. Culbertson, and R.J. Gorman, MRS Internet J. Nitride Semicond. Res. **4S1**, G3.70 (1999); http://nsr.mij.mrs.org/4S1/G3.70/

[10]   For a discussion of the surface kinetics of GaN growth, see D.D. Koleske, A.E. Wickenden, R.L. Henry, W.J. DeSisto, and R.J. Gorman, J. Appl. Phys. **84**, 1998 (1998).

[11]   A. Rebey, T. Boufaden, B. El Jani , J. Cryst. Growth **203**, 12 (1999).

[12]   O. Briot, S. Clur, and R.L. Aulombard, Appl. Phys. Lett. **71**, 1990 (1997).

[13]   A.E. Wickenden, D.D. Koleske, R.L. Henry, R.J. Gorman, J.C. Culbertson, and M.E. Twigg, J. A. Freitas, Jr., Electron. Mat. **28**, 301 (1999).

[14]   C.D. Thurmond, R.A. Logan, J. Electrochem. Soc. **119**, 622 (1972); R. A. Logan and C.D. Thurmond, J. Electrochem. Soc. **119**, 1727 (1972).

[15]   M.E. Bartram and J. R. Creighton, MRS Internet J. Nitride Semicond. Res. **4S1**, G3.68 (1999).

[16]   C.S. Kim, V.M. Bermudez, and J.N. Russell, Jr., Surf. Sci. **389**, 162 (1997).

[17]   O. Ambacher, M.S. Brandt, R. Dimitrov, T. Metzger, M. Stutzmann, R.A. Fischer, A. Miehr, A. Bergmaier, and G. Dollinger, J. Vac. Sci. Technol B **14**, 3532 (1996).

[18]   M. Hashimoto, H. Amano, N. Sawaki, and I. Akasaki, J. Cryst. Growth **68**, 163 (1984).

[19]   M.A. Khan, R.A. Skogman, R.G. Schulze, and M. Gershenzon, Appl. Phys. Lett. **43**, 493 (1983).

[20]   O. Briot, S. Clur, and R.L. Aulombard, Appl. Phys. Lett. **71**, 1990 (1997).

[21]   T. J. Kistenmacher, D.K. Wickenden, M.E. Hawley, and R.P. Leavitt, Mat. Res. Soc. Symp. Proc. **395**, 261 (1996).

[22]   K. Tadatomo, Y. Ohuchi, H. Okagawa, H. Itoh, H. Miyake, and K. Hiramatsu, MRS Internet J. Nitride Semicond. Res. **4S1**, G3.1 (1999); http://nsr.mij.mrs.org/4S1/G3.1/

[23]   O. Schön B. Schineller, M. Heuken, and R. Beccard, J. Cryst. Growth **189/190**, 335 (1998).

[24]   L.D. Bell, R.P. Smith, B.T. McDermott, E.R. Gertner, R. Pittman, R.L. Pierson, and G.J. Sullivan, J. Vac. Sci. Technol. B **16**, 2286 (1998).

[25]   T.-B. Ng, J. Han, R.M. Biefeld, and M.V. Weckwerth, J. Electronic Mat. **27**, 190 (1998).

Mat. Res. Soc. Symp. Vol. 595 © 2000 Materials Research Society

Surface activity of magnesium during GaN molecular beam epitaxial growth

V. Ramachandran[1], R. M. Feenstra[1], J. E. Northrup[2] and D. W. Greve[3]
[1]Department of Physics, Carnegie Mellon University, Pittsburgh, Pennsylvania 15213
[2]Xerox Palo Alto Research Center, 3333 Coyote Hill Road, Palo Alto, California 94304
[3]Department of Electrical and Computer Engineering, Carnegie Mellon University, Pittsburgh, Pennsylvania 15213

ABSTRACT

Exposure of wurtzite GaN films grown on Si-polar 6H-SiC(0001) to magnesium during molecular beam epitaxy (MBE) has been studied. In the nitrogen rich regime of MBE growth, GaN films are known to grow with rough morphology. We observe on GaN(0001) that small doses of Mg act as a surfactant, smoothing out this roughness. An interpretation of this surfactant behavior is given in terms of electron counting arguments for the surface reconstructions. Previously, we have reported that larger doses of Mg lead to inversion of the Ga-polar GaN film to produce N-polar GaN. Several Mg-related reconstructions of the resulting GaN(000$\bar{1}$) surface are reported.

INTRODUCTION

Surfactant mediated film growth has been an important area of work for both film growers and surface scientists. For the epitaxy of thin films on lattice-mismatched substrates, surfactants provide a reduction in surface energy which balances the strain energy in the film, thereby enabling two-dimensional growth (without incorporation of the surfactant). Thus a potential application of surfactants is in the initial nucleation of GaN on mismatched substrates such as SiC or sapphire. Once nucleation has occurred and the film has grown thicker, surfactants may improve surface morphology of the film by affecting the growth mode or growth rates of different lattice planes, again favoring a more two-dimensional film. For high quality GaN devices, it is important to have both excellent bulk quality as well as a flat surface morphology. From this point of view, it is very important to explore the area of surfactants for GaN film growth. Various chemical species have been found to exhibit surfactant type effects during GaN growth, including magnesium (as seen in the growth rate of GaN on different crystallographic planes when it is doped with Mg) [1], arsenic [2], and indium [3].

In this work, we show evidence for a surfactant effect due to Mg during GaN growth. As has been noticed by several MBE growers, GaN growth in Ga-rich conditions produces smooth surfaces indicative of two-dimensional growth. But growth under N-rich conditions leads to three-dimensional, rough surfaces, for both Ga- and N-polar films [4,5]. Under N-rich conditions, we see that exposure of these Ga-polar rough films to sub-monolayer quantities of Mg leads to a smoothing of the film. An interpretation of this surfactant behavior of the Mg is given in terms of electron counting arguments for the surface reconstructions.

EXPERIMENT

The films are grown in a molecular beam epitaxy (MBE) chamber with base pressure less than 1×10^{-10} Torr, using Si-polar 6H-SiC(0001) substrates. The substrates are prepared *ex situ* by hydrogen etching [6]. The surface oxide is removed by Si pre-deposition and desorption [7] till a √3×√3R-30° reconstructed surface is obtained. Immediately after this, the substrate is brought to growth temperature and growth is initiated. Growth temperatures are in the range 600–700° C. Growth conditions are usually Ga-rich, for which we observe a streaky 1×1 RHEED pattern during growth. However, for our studies of the surfactant effect of Mg, we have grown films in the N-rich regime before Mg exposure. The hallmark of the N-rich regime is a spotty, bright RHEED pattern. Ga and Mg fluxes are produced by effusion cells, while an RF-plasma is used to excite N_2 molecules thereby facilitating reaction. Mg exposure is done *during* growth and is seen not to have any effect when performed during growth interrupts. This may have to do with the short residence times for Mg atoms on the GaN surface at the growth temperature. During growth, atoms on the surface are mobile enough for Mg to get incorporated in the crystal surface immediately upon their arrival, whereas even with prolonged exposures during interrupts little Mg sticks to the surface. Following growth, surfaces are characterized using *in situ* scanning tunneling microscopy (STM) and *ex situ* atomic force microscopy (AFM), as described elsewhere [8].

RESULTS
Surfactant Behavior of Mg

GaN films grown in the Ga-rich regime show a smooth two-dimensional (2-D) morphology. In Fig. 1(a), a RHEED pattern and AFM image obtained for a film grown under these conditions are displayed. The RHEED pattern is dim and streaky, since the growth is 2-D and at all times the surface is covered with excess Ga (at least 2 disordered ML) [8]. The AFM image displays deep pits and trenches (> 150 Å deep) which are characteristic of the columnar growth in the MBE process. In between these pits, however, the film shows a flat, 2-D morphology with atomic steps faintly visible. As the Ga flux becomes smaller, the RHEED pattern becomes brighter and eventually in the N-rich region of growth, the RHEED pattern begins to get spotty as shown in the inset in Fig. 1(b). The morphology of the film in the N-rich regime is shown in Fig. 1(b), and the roughening is seen as the granular appearance of the film between the deep pits.

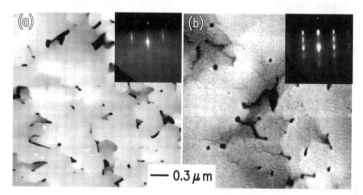

Figure 1 (a) AFM image of a film grown under Ga-rich conditions, (b) grown under N-rich conditions. In the insets we see corresponding RHEED patterns in the (11$\overline{2}$) azimuth.

During growth, films are briefly exposed to a Mg flux. A surfactant effect of Mg is seen on these Ga-polar films in the N-rich regime. Exposing the growth surface to as little as 0.2 ML of Mg under Ga-poor conditions leads to a reversal of the RHEED pattern from spotty to streaky. Also, when the growth is made very N-rich by reducing the Ga flux to about one half of that at the transition point, exposure to Mg often produces a streaky 2×2 pattern. At typical sample temperatures during growth (625° C for the sample shown below), the sticking coefficient of Mg on the GaN surface is expected to be rather low, therefore the 0.2 ML estimate we make from room temperature calibrations of the Mg source is an upper limit on the amount required to cause this surfactant effect. Figure 2 shows AFM images which present a sequence of surface morphologies during the smoothing of the surface. Figure 2(a) shows a rough film similar to the one in Fig. 1(b) grown in the N-rich regime. An exposure of about 0.2 ML of Mg leads to a film which shows areas such as those in Fig. 2(b). As indicated by the arrows, regions of 2-D growth start to nucleate on the rough film, where we can see atomic steps again. At about 0.7 ML Mg exposure, the surface is mostly smoothed out and the film shows morphology as in Fig. 2(c) and (d). The morphology of Fig. 2(d) is indistinguishable from that of Fig. 1(a), but small areas of roughness are still seen in Fig. 2(c). Closer inspection reveals that these patches of roughness follow the directions of steps, suggesting that the smoothing is a step-flow process.

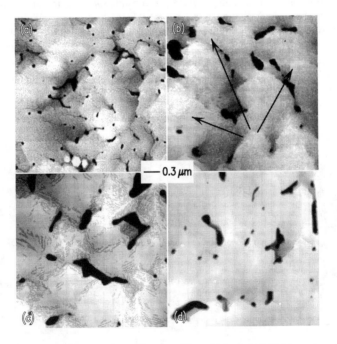

Figure 2 *AFM images showing the surfactant effect of Mg: (a) GaN film grown in N-rich regime without Mg, (b) and (c) intermediate stages of the smoothing process, (d) complete smoothing of surface as a result of Mg exposure. The arrows in (b) point to smooth areas beginning to appear interspersed amidst the rough regions.*

From our RHEED and AFM observations, it is apparent that Mg has a surfactant effect under Ga deficient growth conditions. The surfactant effect is seen to persist for some time after the exposure of Mg, confirming previous observations that Mg segregates to the surface during GaN growth [9]. However, extensive use of this surfactant effect is made difficult by the Mg-induced crystal polarity inversion reported previously [10], in which exposure of the surface to > 1 ML of Mg is found to change the film polarity from Ga-polar to N-polar.

Mg-derived reconstructions of GaN($000\bar{1}$)

We have also observed several Mg-induced low temperature reconstructions on the GaN surface after Mg has led to inversion, *i.e.* on the ($000\bar{1}$) face. These are obtained by exposing the GaN surface to a Mg flux while holding the substrate at a temperature around 250° C. Upon deposition of about 0.04–0.08 ML of Mg, we see a 5×5 reconstruction. With the addition of more Mg, we see the appearance of 4×4, 3×3 and 6×6 RHEED patterns. These patterns may be observed in the reverse order upon heating the 6×6 at around 400° C, and prolonged heating at this temperature leads to the disappearance of these patterns and the reappearance of the 1×1 usually seen on the N-polar face. Thus it is clear that these reconstructions arise from Mg atoms that are weakly bound to the surface. Figure 3 shows STM images corresponding to the 5×5 reconstruction. The structure is composed of a hexagonal array of corrugation maxima (probably adatoms), arranged in a 5×5 pattern. Some regions of the surface were covered with additional adsorbates (possibly excess Mg or Ga), forming the bright regions in Fig. 3(a). At this time we do not have detailed models for the structure of these various reconstructions, although it seems likely that they can be constructed by substituting Mg atoms for Ga atoms in the GaN($000\bar{1}$) adatom-on-adlayer structures [11]. For example, the 5×5 structure could correspond to having 2 Mg atoms/cell in adatom sites above the 1×1 adlayer.

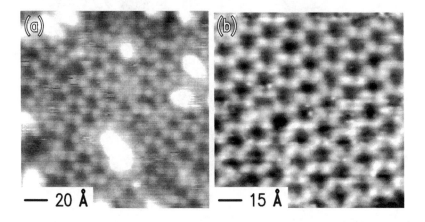

Figure 3 *STM images of 5×5 Mg-induced reconstruction on the N-polar GaN surface, acquired at sample voltages of (a) +2.0 V and (b) +2.5 V, and at a constant current of 0.1 nA. Grey-scale ranges are 1.4 and 0.5 Å for (a) and (b) respectively.*

DISCUSSION

As discussed above, the presence of small amounts of Mg on the GaN(0001) surface, growing under N-rich conditions, can lead to a conversion of the nominally rough surface (spotty RHEED pattern) to a smooth surface (streaky RHEED pattern). Concomitantly, a 2×2 RHEED pattern can be seen on the surface. Although we have not studied this 2×2 structure in detail (*e.g.* using STM), we can discuss its structure based on recent theoretical results [12]. Considering a variety of 2×2 arrangements, Bungaro *et al.* find that 1/4 ML of Mg substituting for Ga on the surface is energetically the most stable configuration (except for very Ga-rich conditions, which are not of concern here). In terms of electron counting, this surface would have 2 electrons available to fill dangling-bond derived surface states, so that a surface gap could develop and we would anticipate relatively low diffusion barriers compared to the Ga-terminated (0001) surface.

Another model for the Mg-induced 2×2 structure is possible, containing 3/4 ML of Mg substituting for surface Ga atoms. In this case, all dangling bonds on the surface would be completely empty, and again we would anticipate relatively low diffusion barriers. Based on total energy computations of the type described elsewhere [10], we find that this 3/4 ML model will be energetically favorable for Mg-rich and/or Ga-poor conditions. Explicitly, the energy difference between the 1/4 and 3/4 ML models may be expressed as

$$\Delta E \ (eV / 2 \times 2) = 0.05 \ eV - 2 \ (\Delta \mu_{Ga} - \Delta \mu_{Mg})$$

where $\Delta \mu_{Ga}$ and $\Delta \mu_{Mg}$ are the chemical potentials measured relative to their bulk values. A positive energy means that the 3/4 ML structure is preferred over the 1/4 ML structure. If $\Delta \mu_{Ga} = \Delta \mu_{Mg}$ (say, both are zero) then the 3/4 ML structure is preferred. In extreme N-rich conditions, we could have $\Delta \mu_{Ga} \approx -1.0$ eV. Under such conditions it is possible to stabilize the 3/4 ML structure for values of $\Delta \mu_{Mg} > -1$ eV. In our experiments we find that 0.2 ML of Mg can lead to the surfactant behavior, which would argue in favor of the 1/4 ML model as correctly describing the surface structure. On the other hand, the smooth areas of that surface extend over only a small portion of the total surface area (Fig. 2(b)), so that the 3/4 ML model may also be consistent with the data. Additional experimental study of the surface structure is required to more definitively determine the structure.

Thus, we believe that the presence of the Mg allows the surface to satisfy electron counting (at least for the 3/4 ML model), thereby leading to completely filled or completely empty surface state bands. Qualitatively, one expects such a surface to be non-reactive, and have relatively low diffusion barriers compared to the bare GaN surface [13]. Hence, a surfactant effect (smooth, 2-D growth) could be expected, in agreement with the experimental observations. A similar situation is believed to occur on arsenic exposed surfaces. In that case, near the critical Ga flux for the transition from smooth to rough morphology we find, in the presence of arsenic, an intense 2×2 streaky RHEED pattern, indicative of smooth morphology [14]. This pattern persists for a range of Ga fluxes both above and below the critical flux, thus demonstrating a surfactant behavior of the arsenic. Our interpretation of the 2×2 structure is that it arises from arsenic adatoms on the Ga-terminated surface (we believe that a number of groups have mistakenly identified this structure as arising from Ga adatoms, since the surface arsenic in their cases unintentionally arises from background arsenic contamination in their vacuum systems [14-16]). The 2×2 As-adatom surface satisfies electron counting, so again we expect relatively low diffusion barriers in that case.

SUMMARY

In summary, we have grown GaN films on hydrogen-etched 6H-SiC(0001) by MBE at growth temperatures of 600–700° C with brief exposures to Mg. The Mg exposures are seen to have a surfactant effect at low coverage, and at high coverage to invert the Ga-polar GaN film to a N-polar film. An interpretation of the surfactant effect is given in terms of electron counting arguments, and this effect is compared with that observed for arsenic overlayers. We have also observed low temperature reconstructions on the inverted surface, which may help in the understanding of Mg incorporation in N-polar GaN.

ACKNOWLEDGEMENTS

This work was supported by the Office of Naval Research, grant N00014-96-1-0214 monitored by Dr. Colin Wood, and by the National Science Foundation, grant DMR-9615647.

REFERENCES

[1]   B. Beaumont, S. Haffouz and P. Gibart, Appl. Phys. Lett. 72, 921 (1998).
[2]   Y. Zhao, F. Deng, S. S. Lau, and C. W. Tu, J. Vac. Sci. Technol. 16, 1297 (1998).
[3]   F. Widmann, B. Daudin, G. Feuillet, N. Pelekanos, and J. L. Rouvière, Appl. Phys. Lett. 73, 2642 (1998).
[4]   E. J. Tarsa, B. Heying, X. H. Wu, P. Fini, S. P. DenBaars, and J. S. Speck, J. Appl. Phys. 82, 5472 (1997).
[5]   A. R. Smith, V. Ramachandran, R. M. Feenstra, D. W. Greve, A. Ptak, T. H. Myers, W. L. Sarney, L. Salamanca-Riba, M.-S. Shin and M. Skowronski, MRS Internet J. Nitride Semicond. Res. 3, 12(1998).
[6]   V. Ramachandran, M. F. Brady, A. R. Smith, R. M. Feenstra and D. W. Greve, J. Electron. Mater. 27, 308 (1998).
[7]   V. Ramachandran, A. R. Smith, R. M. Feenstra and D. W. Greve, J. Vac. Sci. Technol. A 17, 1289 (1999).
[8]   A. R. Smith, R. M. Feenstra, D. W. Greve, M. S. Shin, M. Skowronski, J. Neugebauer, J. E. Northrup, J. Vac. Sci. Technol. B 16, 2242 (1998).
[9]   T. S. Cheng, C. T. Foxon, N. J. Jeffs, D. J. Dewsnip, L. Flannery, J. W. Orton, S. V. Novikov, B. Ya Ber and Yu A. Kudriavtsev, MRS Internet J. Nitride Semicon. Res. 2, 13 (1997).
[10] V. Ramachandran, R. M. Feenstra, W. L. Sarney, L. Salamanca-Riba, J. E. Northrup, L. T. Romano, and D. W. Greve, Appl. Phys. Lett. 75, 808 (1999).
[11] A. R. Smith, R. M. Feenstra, D. W. Greve, J. Neugebauer, and J. Northrup, Phys. Rev. Lett. 79, 3934 (1997).
[12] C. Bungaro, K. Rapcewicz, and J. Bernholc, Phys. Rev. B 59, 9771 (1999).
[13] T. Zywietz, J. Neugebauer, and M. Scheffler, Appl. Phys. Lett. 73, 487 (1998).
[14] V. Ramachandran, C. D. Lee, R. M. Feenstra, A. R. Smith, J. E. Northrup, and D. W. Greve, J. Cryst. Growth 126, to appear.
[15] K. Iwata, H. Asahi, S. J. Yu, K. Asami, H. Fujita, M. Fushida, and S. Gonda, Jpn. J. Appl. Phys. 35, L289 (1996).
[16] Q. K. Xue, Q. Z. Xue, R. Z. Bakhtizin, Y. Hasegawa, I. S. T. Tsong, T. Sakurai, and T. Ohno, Phys. Rev. Lett. 82, 3074 (1999).

Mat. Res. Soc. Symp. Vol. 595 © 2000 Materials Research Society

Simulations of Defect-Interface Interactions in GaN

J. A. Chisholm and P. D. Bristowe
Department of Materials Science and Metallurgy
University of Cambridge
Pembroke Street
Cambridge CB2 3QZ
United Kingdom

ABSTRACT

We report on the interaction of native point defects with commonly observed planar defects in GaN. Using a pair potential model we find a positive binding energy for all native defects to the three boundary structures investigated indicating a preference for native defects to form in these interfaces. The binding energy is highest for the Ga interstitial and lowest for vacancies. Interstitials, which are not thought to occur in significant concentrations in bulk GaN, should form in the $(11\bar{2}0)$ IDB and the $(10\bar{1}0)$ SMB and consequently alter the electronic structure of these boundaries.

INTRODUCTION

Vacancy and interstitial native defects are known to have a major influence on the electrical and optical properties of GaN [1]. For example, the N vacancy acts as single donor, the Ga vacancy as an acceptor and the interstitials act as amphoteric defects. The formation energies of native defects in bulk GaN have been calculated using first principles methods and their values are quite well established [2,3]. However, these formation energies will change in the neighbourhood of extended planar defects such as stacking mismatch boundaries and inversion domain boundaries. These boundaries are commonly observed in epitaxially grown GaN but so far the interaction between such interfaces and native point defects in GaN has not been investigated and it remains unclear whether the formation of interstitials and vacancies near the boundaries is encouraged or discouraged.

The calculation of defect-interface interactions requires computational cells containing several hundred atoms so as to avoid unwanted intercellular interactions when periodic boundary conditions are applied. This makes a first principles approach difficult and therefore we have used, in the first instance, a classical methodology which employs interatomic pair potentials that have been fitted to reproduce various bulk properties of GaN. This classical model is used to calculate the binding energy of Ga and N vacancies and interstitials to three commonly observed interfaces: the $(10\bar{1}0)$ stacking mismatch boundary (SMB), the $(10\bar{1}0)$ inversion domain boundary (IDB) and the $(11\bar{2}0)$ IDB. The atomic structures of these boundaries have been determined from transmission electron microscope observations [4-6] and are shown in figure 1. One particular point of interest has been the observation of two atomic structures for the $(10\bar{1}0)$ IDB [7]. The first boundary structure involves an inversion of the atomic species across the boundary. For the second boundary structure, referred to as IDB*, there is an additional translation of c/2 along the [0001]. First principles density functional calculations carried out by

Northrup at al [8] have shown that this second structure, which contains no dangling bonds, has the lower formation energy and we therefore focus on this structure in the present work. Analysis of the electronic structure also revealed that IDB* does not introduce any interface states in the forbidden gap but that the $(10\bar{1}0)$ SMB introduces an occupied state 1.1 eV above the valence band maximum.

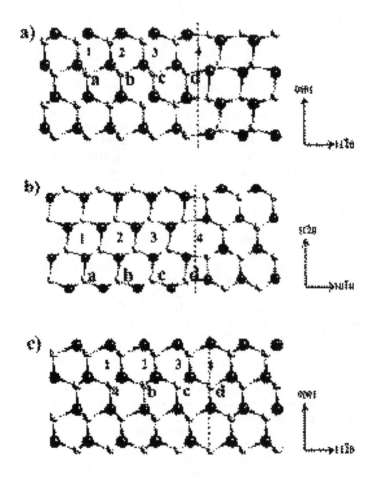

Figure 1. Observed boundary structures of (a) the $(10\bar{1}0)$ IDB*, (b) the $(11\bar{2}0)$ IDB and (c) the $(10\bar{1}0)$ SMB. Interstitial defects are placed in positions 1-4 and vacancy defects are placed in positions a-d. Dashed line indicates the location of the boundary.

METHODOLOGY

The computational methodology and pair potential scheme are well established (for a review see Harding [9]) and therefore only a summary is given here. The long range electrostatic energy is evaluated through the Ewald summation [10]. The interaction between ions at short range is described using Buckingham pair potentials which have the following form:

$$V_{ij} = A\exp(-r_{ij}/\rho) - Cr_{ij}^{-6} \tag{1}$$

In a previous study [11] the potential parameters A, ρ and C were fitted to experimental lattice parameters as well as elastic and dielectric constants. For the N ion, polarisation effects are taken into account using the Shell Model in which the electrons are modelled by a massless 'shell' with charge Y which is connected to the core by a spring with spring constant K [12]. The polarisability is then given by $\alpha = Y^2/K$.

The use of the supercell approach requires the inclusion of two boundaries per cell. We have carried out convergence tests and found that the following sizes of supercell are more than adequate to eliminate boundary-boundary interactions as well as interactions between native defects: $(10\bar{1}0)$ IDB* 576 atoms, $(11\bar{2}0)$ IDB 528 atoms and $(10\bar{1}0)$ SMB 416 atoms. For each calculation we allow full relaxation of the cell parameters and of the atomic coordinates by using a Broyden-Fletcher-Goldfarb-Shanno scheme. Our approach for calculating native defect binding energies is shown in figure 1. The binding energy is the difference between the formation energy of the defect in the bulk and the formation energy of the defect in the boundary. This change in formation energy is obtained simply from the change in the total energy of the supercell as the native defect is taken from the bulk environment and placed in the boundary.

$$E_{bind} = E_{tot}(\text{defect in bulk}) - E_{tot}(\text{defect in boundary}) \tag{2}$$

In the pair potential model the atoms are treated as ions in which the Ga atom is attributed a charge of +2 and the N atom a charge −2. The creation of a native defect therefore introduces an overall charge into the supercell. In order to maintain charge neutrality we create an additional native defect. For example, for a N vacancy we also create a Ga vacancy the position of which is kept fixed at the largest possible distance from the N vacancy. Again, the size of the supercells in each case is more than sufficient to eliminate any unwanted defect-defect interactions.

RESULTS

Calculated values for the boundary energies as well as native defect binding energies are shown in table I. The boundary energies compare well with those obtained by Northrup et al [8] who calculated values of 0.025 eV / Å^2 for the $(10\bar{1}0)$ IDB* and 0.105 eV / Å^2 for the $(10\bar{1}0)$ SMB.

Trends for the binding energies can be observed across the boundaries and across the native defects. We note first that for all boundary structures the binding energy is higher for interstitial defects than for vacancies. In bulk GaN, the interstitials have a particularly high formation energy of 5-10 eV (depending on the charge state [1]) and unlike the case for GaAs, interstitials are not thought to occur in significant concentrations. This result suggests that interstitial defects are more likely to form in the boundary. Indeed, a larger volume of space is available to accommodate interstitials for all three boundaries.

Table I. Calculated binding energies for Ga and N vacancies and interstitials to three different planar defects. The boundary energy of each planar defect is also given.

	$(10\bar{1}0)$ IDB*	$(11\bar{2}0)$ IDB	$(10\bar{1}0)$ SMB
Boundary energy $(eV / Å^2)$	0.017	0.215	0.095
Ga interstitial (eV)	1.46	5.31	5.28
N interstitial (eV)	1.04	3.95	4.40
Ga vacancy (eV)	0.44	1.61	2.39
N vacancy (eV)	0.43	1.90	2.79

In bulk GaN the introduction of a Ga interstitial causes an outward relaxation of the surrounding atoms. The same is generally true for an interstitial introduced into the boundaries. For the case of the $(10\bar{1}0)$ IDB*, one particular Ga atom in the boundary core is displaced outward by 1.0 Å because of the attraction of the Ga interstitial to three nearby N ions. This is shown in figure 2 (a). For the $(11\bar{2}0)$ IDB, two N atoms from the boundary are displaced inwards towards the Ga interstitial by 0.7 Å and 0.6 Å and a Ga atom in the boundary is pushed out by 0.8 Å. Also, for the $(10\bar{1}0)$ SMB, there is a small rearrangement whereby surrounding Ga atoms relax outwards. For this boundary there are no inward relaxations of surrounding N atoms due to the presence of N atoms located directly in the boundary core. A similar result holds for the N interstitial only that now surrounding Ga atoms relax inwards to meet the N interstitial and N ions are pushed out as shown in figure 2 (b) for the case of the $(10\bar{1}0)$ IDB*.

A careful comparison of lattice relaxations carried out within a radius of 7 Å for all native defects has shown that displacements are of the same magnitude both in the boundary and in the bulk. Also, it was seen that displacements associated with vacancies are less pronounced than those associated with interstitials. For the Ga vacancy in the $(10\bar{1}0)$ IDB*, two N ions relax outward by 0.4 Å and 0.2 Å and a Ga ion moves towards the vacancy by 0.3 Å . For the $(11\bar{2}0)$ IDB, N ions are observed to move out by 0.5 Å and 0.2 Å and there are significant displacements for two nearby Ga ions: 0.6 Å and 0.8 Å . For the SMB, where a Ga ion is only three-fold coordinated, surrounding N ions relax outward by 0.2 , 0.2 and 0.1 Å. Again the situation is similar for the N vacancy only now N ions are observed to move a fraction of an Ångstrom in towards the N vacancy and surrounding Ga ions relax outwards.

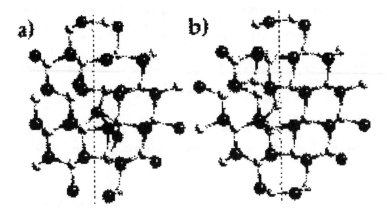

Figure 2. Relaxed structures of (a) the Ga interstitial and (b) the N interstitial in the (10 $\bar{1}$ 0) IDB*. Filled and open circles represent gallium and nitrogen atoms respectively. Dashed line indicates the location of the boundary.

The electronic structure of a boundary will be affected by the formation of native defects in its core. In the case of the (10 $\bar{1}$ 0) IDB*, the formation of vacancies and interstitials will cause the boundary to become electrically active. The concentration of such native defects in the boundary will depend on the point defect formation energy in the boundary environment. Defect-interface binding energies shown in table I indicate that the formation energies of interstitials are not significantly less than bulk values for this particular boundary. However, formation energies are significantly reduced for interstitials in the (11 $\bar{2}$ 0) IDB and the (10 $\bar{1}$ 0) SMB. The precise nature of such defects in the boundary requires a first principles approach and such a treatment is currently underway.

We note also that the binding energies follow a definite trend across the boundaries. For the case of the interstitials this can be explained in terms of the volume which is available to accommodate the defect. An interstitial can be thought of as being enclosed in a cage of atoms having a certain volume. Analysis shows that the volume of these interstitial cages increases in going from the (10 $\bar{1}$ 0) IDB* to the (11 $\bar{2}$ 0) IDB to the (10 $\bar{1}$ 0) SMB which correlates well with the corresponding increase in binding energies for the native defects across this series. The apparent smaller cage size of the SMB in figure 1 is due to the orientation of the figure.

CONCLUSIONS

We have employed a pair potential model to investigate the interaction between native defects and inversion domain boundaries and stacking mismatch boundaries in GaN. The binding energy is found to be positive for all the native defects and boundary structures considered. This indicates that native defects prefer to form in these types of boundary. The binding energy is highest for the Ga interstitial and lowest for the vacancies which can be explained by the larger volume available to accommodate interstitials in the boundary. We conclude that N and Ga interstitials, which do not form in significant concentrations in bulk GaN, should form in the $(11\bar{2}0)$ IDB and the $(10\bar{1}0)$ SMB. Investigations using first principles calculations are currently underway to confirm the trends in binding energies found in this study as well as gain information on the electronic structure of these boundaries in the presence of native defects.

ACKNOWLEDGEMENT

Support for this research was provided by the Engineering and Physical Sciences Research Council.

REFERENCES

1. J. Neugebauer and C. G. Van de Walle, Phys. Rev. B **50**, 8067 (1994).
2. P. Boguslawski, E. L. Briggs and J. Bernholc, Phys. Rev. B **51**, 17255 (1995).
3. J. Neugebauer and C. Van de Walle, Festkorperprobleme-Advances in Solid State Physics, **35**, 25 (1996)
4. L. T. Romano, J. E. Northrup and M. A. O'Keefe, Appl. Phys. Lett., **69**, (16), 2395 (1996).
5. Y. Xin, P. D. Brown, C. J. Humphreys, T. S. Cheng and C. T. Foxon, Appl. Phys. Lett., **70**, (10), 1308 (1997).
6. S. Ruvimov, Z. Liliental-Weber, J. Washburn, H. Amano, I. Akasaki, and M. Koike, MRS Symposium Proceedings, **482**, 387 (1998)
7. V. Pontin, G. Nouet and P. Ruterana, Appl. Phys. Lett., **74**, 947 (1999).
8. J. E. Northrup, J. Neugebauer and L. T. Romano, Phys. Rev. Lett., **77**, 103 (1996).
9. J. H. Harding, Rep. Prog. Phys., **53**, 1403 (1990).
10. P. P. Ewald, Ann. Phys., **64**, 253 (1921).
11. J. A. Chisholm, D. W. Lewis and P. D. Bristowe, J. Phys. Condensed Mat., **11**, L235 (1999).
12. B. G. Dick and A. W. Overhauser, Phys. Rev., **112**, 90 (1958).

Mat. Res. Soc. Symp. Vol. 595 © 2000 Materials Research Society

Effect of the Doping and the Al Content on the Microstructure and Morphology of Thin $Al_xGa_{1-x}N$ Layers Grown by MOCVD.

J.H. Mazur[1,2], M. Benamara[1], Z. Liliental-Weber[1], W. Swider[1], J. Washburn[1], C.J. Eiting[3,4], R. D. Dupuis[3].

[1]E.O. Lawrence Berkeley National Laboratory, MS 62-203, Berkeley, CA 94720

[2] New address: Philips Semiconductors, 9651Westoverhills Blvd., San Antonio, TX78251

[3]Microelectronics Research Center, The University of Texas at Austin, TX 78712

[4]New address: Air Force Research Laboratory, Materials and Manufacturing Directorate, WPAFB, Dayton, OH 45433-7707

ABSTRACT

$Al_xGa_{1-x}N$ {x=30% (doped and undoped), 45% (doped)} thin films were grown by MOCVD on ~2 µm thick GaN layer using Al_2O_3 substrate. These films were designed to be the active parts of HFETs with $n_s\mu$ product of about $10^{16}(Vs)^{-1}$. The layers were then studied by means of transmission electron microscopy (TEM) techniques. In this paper, it is shown that the $Al_xGa_{1-x}N$ layer thickness was non-uniform due to the presence of V-shaped defects within the $Al_xGa_{1-x}N$ films. The nucleation of these V-shaped defects has taken place about 20 nm above the $Al_xGa_{1-x}N$/GaN interface. Many of these V-shaped defects were associated with the presence of the threading dislocations propagating from the GaN/Al_2O_3 interface. We show that the density of these V-shaped defects increases with the doping level and also with the Al mole fraction in the films. The formation mechanism of the V-shaped defects seems to be related to the concentration of dopants or other impurities at the ledges of the growing film. This suggestion is supported by high resolution TEM analysis. The growth front between the V-shaped defects in the lower Al concentration thin films was planar as compared with the three-dimensional growth in the doped, higher Al concentration film. This interpretation of the origin of the V-shaped defects is consistent with the observed lowering of the Schottky barrier height in n-doped AlGaN/Ni Schottky diodes.

INTRODUCTION

In addition to applications in optoelectronics[1], GaN and related alloys are of interest for applications in high temperature, high frequency and high power electronic devices[2]. In many of these devices, the AlGaN layer is an active part of the device structure and the structural properties of this layer can affect the basic device characteristics. This has been demonstrated and inferred for example in the work of Shiojima, et al [3] who observed lowering of the Schottky barrier height and assigned it to the presence of the structural defects in the n-doped layer. In this work, several AlGaN/GaN HFET structures are analyzed using transmission electron microscopy (TEM).

EXPERIMENTAL

The GaN and the AlGaN layers were grown by low pressure MOCVD in a modified EMCORE D125 reactor system. Trimethylgalium (TMGa) and trimethylaluminum (TMAl) were used as column III precursors, while hydrogen-diluted silane (SiH_4) was used as the n-type dopant. High purity ammonia (NH_3) was used as the N source. The Al_2O_3 substrates were cleaned for 10 minutes in 10:1:1 H_2SO_4:H_2O_2:DI H_2O and then for 10 minutes in 2:9 HF:HNO_3. The cleaning was followed by a DI water rinse. Substrates were then loaded into the system and, after prepumping to a base pressure of 10^{-7} torr, were loaded into growth chamber. After a H_2 bake at 1050°C for 5 minutes, the substrates were cooled down and 25 nm-thick buffer layers were grown at 530°C. The wafers were then heated to 1050°C again and ~2 μm of GaN was grown in NH_3, H_2, and TMGa. A rapid switch to TMAl, TMGa, NH_3, and H_2 followed to grow undoped 30% AlGaN layers, as well as doped 30%

and 45% AlGaN layers. The samples were subsequently cooled to room temperature in H_2 and NH_3. TEM cross-sections were prepared along [11-20], and [1-1 0 0] directions normal to the growth direction [0001]. Tilting (bright field, weak beam and dark field) experiments were performed in the side entry stage JEOL 200 CX, while high resolution imaging was perform on JEOL ARM operating at 800 kV.

RESULTS

Figure 1 shows the low magnification image of the cross-section of an undoped 30% $Al_xGa_{1-x}N$ film. The $Al_xGa_{1-x}N$ layer was about 30 nm thick. Some undulation of the thickness of this layer was observed. These undulations were associated with the presence of the V-shaped defects, some of which extended almost halfway into the layer as shown in Fig. 2. Adding dopant to the $Al_xGa_{1-x}N$ layer while preserving the aluminum concentration at x=30% resulted in overall increase of the density of the V-shaped defects. Otherwise the morphology of the doped film was very similar to that of the undoped $Al_xGa_{1-x}N$ film. This is illustrated in Fig. 2 and Fig.3. By comparison, further increase in the concentration of Al to x=45% resulted in an increased density of the V-shaped defects, as well as in the changed morphology of the growth front from 2D to 3D. This is illustrated in Fig.5 and Fig.6.

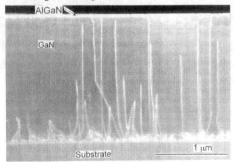

Fig. 1: Weak beam **g**=[0002] image of an undoped 30% AlGaN layer (sample **1170**) layer on about 1.7-2 μ GaN layer on a sapphire substrate. Note presence of V-shaped defects. In many cases these defects appear to be associated with the threading dislocations.

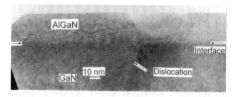

Fig. 2: High resolution image of an undoped AlGaN layer (sample 1170) layer on ~1.7-2.0 μm GaN on a sapphire substrate showing V-shaped defects. Note that the growth between V-shaped defects is planar (2D). The V- shaped defect to the left initiated about 200 nm away from the AlGaN /GaN interface.

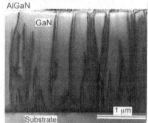

Fig. 3: TEM bright field image of the cross-section of a doped 30% AlGaN layer on ~2 μm GaN on a sapphire substrate (sample **1134B**).

Fig. 4: High resolution TEM image of the cross-section of a doped 30% AlGaN layer on ~2 μm GaN on a sapphire substrate (sample **1134B**).

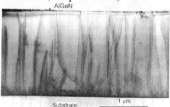

Fig.5: Bright field TEM image of the cross-section of a doped 45% AlGaN layer on ~2 μm GaN on a sapphire substrate (sample **1129A**).

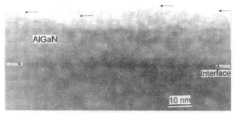

Fig. 6: High resolution TEM image of a doped 45% AlGaN layer on ~2 μm GaN on a sapphire substrate (sample **1129A**). Careful analysis of the growth front between V-shaped defects indicates 3D growth.

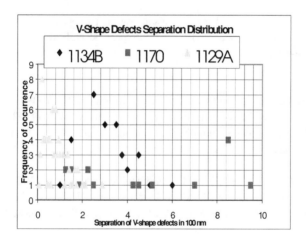

Fig.7: Frequency of the observation of the V-shaped defect as a function of the separation between two consecutive V defects is the highest for a doped sample with highest Al content and the lowest for the undoped lower Al content sample.

SUMMARY AND CONCLUSION

The density of the V-shaped defects increases with Al and dopant content. This is illustrated in Fig. 7 which summarizes the TEM results. This dependence on theAl concentration and on the dopant composition in the growing AlGaN film indicates that the poisoning of the ledges takes place and promotes V-shape defect formation. The local composition inhomogeniety due to poisoning of the ledges could be responsible for the observed lowering of the Schottky barrier height in n-doped AlGaN/Ni Schottky diode [3]. The role of threading dislocations in the formation of the V-shaped defects has to be further studied, because many, but not all, V-shaped defects were found in the proximity of these dislocations. The formation of the V-shaped defects presents a challenge for AlGaN/GaN electronic devices, as these defects will impact many electrical parameters and reduce device reliability. Therefore methods need to be developed to reduce the V-shaped defect density in AlGaN/GaN heterostructure.

ACKNOWLEDGEMENTS

Work at E.O. Lawrence Berkeley National Laboratory was supported by the Director, Office of Basic Science, Materials Science Division, U.S. Department of Energy, under contract No. DE AC03-76SF00098. The use of the facilities at the National Center for Electron Microscopy at E.O Lawrence Berkeley National Laboratory is greatly appreciated.
Work at the The University of Texas at Austin was in part supported by the ONR Contract N00014-95-1-1302 monitored by J. C. Zolper and NSF under Grants DMR-9312947 and CHE-89-20120.

REFERENCES
[1] Nakamura , S., Fasol, G. ,"The Blue Laser Diode," Springer, Berlin, (1997)
[2] Binari, S.C., "GaN electronic devices for future systems" in 1999 IEEE MTT-S International Microwave Symposium Digest, Matloubian, M., Ponti, E. Eds., (Cat. No.99CH36282) Anaheim, CA, USA, 13-19 June 1999.) Piscataway, NJ, USA: **3**, 1081-4 (1999)
[3] Shiojima, K., Woodall, J.M., Eiting, C.J., Grudowski, P.A., Dupuis, R.D. Journal of Vacuum Science and Technology B (Microelectronics and Nanometer Structures), J. Vac. Sci. Technol. B, Microelectron. Nanometer Struct. (USA), **17**, .2030-3 (1999)

Mat. Res. Soc. Symp. Vol. 595 © 2000 Materials Research Society

ELECTRICAL PROPERTIES OF OXYGEN DOPED GaN GROWN BY METALORGANIC VAPOR PHASE EPITAXY

R.Y. Korotkov and B.W. Wessels
Materials Research Center and Department of Materials Science and Engineering
Northwestern University, Evanston, IL 60208

ABSTRACT

Deliberate oxygen doping of GaN grown by MOVPE has been studied. The electron concentration increased as the square root of the oxygen partial pressure. Oxygen is a shallow donor with a thermal ionization energy of 27 ±2 meV. A compensation ratio of $\Theta = 0.3\text{-}0.4$ was determined from Hall effect measurements. The formation energy of O_N of $E^F = 1.3$ eV, determined from the experimental data, is lower than the theoretically predicted value.

INTRODUCTION

The origin of n-type conductivity in epitaxial GaN is of continuing interest. Residual electron concentration of undoped epitaxial GaN typically ranges from 10^{16} to 10^{19} cm^{-3}. Although the n-type conductivity was initially attributed to nitrogen vacancies, residual impurities such as oxygen and silicon are believed to be at least partially responsible for the high conductivity [1] . Oxygen substitutes for nitrogen behaving as a donor in GaN [1]. Several secondary ion mass spectroscopy (SIMS) studies of unintentionally doped GaN have observed large concentrations of O and Si [2-3,6]. The source of oxygen is believed to be in the ammonia used for metalorganic vapor phase epitaxial (MOVPE) growth. These observations are consistent with recent total energy calculations of van de Walle et al [1] that indicated that the solubility of O_N is relatively high in wurtzite GaN. Despite its importance, the nature of oxygen donors in GaN remains controversial [3,8]. Initial work on the deliberately oxygen doped GaN indicated that oxygen is a "shallow" deep donor with apparent activation energy of 78 meV as determined by optical measurements from the position of the donor bound exciton [4]. Since then, several groups have studied the electrical properties of both deliberate [3,5] and unintentionally oxygen doped materials [2,6-8,18]. Temperature dependent Hall effect measurements indicated that oxygen donors are shallow with a thermal ionization energy ranging from: 4-29 meV [2-3,6,7]. In contrast to these measurements, Chen *et al* subsequently indicated that substitutional oxygen is a deep donor with an activation energy close to 0.9 eV [8]. This was based on the observation of an infra-red photoluminescence (PL) emission band on an unintentionally doped GaN. In this paper we present results of the study of deliberate oxygen doping of high purity epitaxial GaN. Upon doping the electron concentration increased from 1×10^{17} to 3×10^{19} cm^{-3}. Oxygen is shown to behave as a simple donor in GaN. From defect equilibria studies, the formation energy of substitutional oxygen is calculated and compared to recent total energy calculations.

EXPERIMENTAL

Epitaxial GaN layers were grown by MOVPE onto the c-plane of sapphire substrates in an atmospheric pressure horizontal flow reactor using the reactants: trimethylgallium and ammonia. A Nanochem purifier was utilized to eliminate residual moisture and oxygen from the ammonia gas. Undoped as-grown GaN layers had free electron concentrations and mobilities of 0.9-1×10^{17} cm^{-3}

and 420-500 $cm^2V^{-1}s^{-1}$, respectively. Two oxygen-nitrogen gas mixtures were used as a dopant source (20 and 520 ppm of oxygen in nitrogen, respectively). The epitaxial layer consisted of a 20 nm GaN nucleation layer, a thin 50 nm undoped layer, and a two micron thick oxygen doped layer. Doped layers were grown at 1060°C. Hall measurements were performed using the van der Pauw geometry over the temperature range of 77-330 K. The ohmicity of the indium contacts were verified over all temperatures.

DEPENDENCE OF CARRIER CONCENTRATION ON OXYGEN PARTIAL PRESSURE

For substitution of oxygen on a nitrogen site the defect equilibrium equation is given by:

$$\frac{1}{2}O_2 \rightarrow O_N, \rightarrow K_O = \frac{[O_N]}{p_{O_2}^{1/2}} \tag{1}$$

where K_0 is the equilibrium constant and is given by $\exp[- G^F/kT]$. The free energy $G^F = E^F - TS^F$ for oxygen substitution can be obtained using first principles, total energy calculations. The value of E^F is given by [17]:

$$E^F\left(GaN:O_N^q\right) = E_{tot}\left(GaN:O_N^q\right) - \mu_O + \mu_N + qE_F \tag{2}$$

where E_{tot} is the energy of the neutral defect, μ_O and μ_N are the chemical potentials of oxygen and nitrogen , q is a charge state of defect and E_F is the Fermi energy. The chemical potential of oxygen is given by $\mu_O = kT\ln f_O = kT\ln(K_0P(O_2)^{1/2})$. The O_N concentration is thus given by:

$$[O_N] = N_{sites}\exp\left(\frac{S^F}{k}\right)\exp\left(-\frac{E^F}{kT}\right) = \exp(E_{tot} - \mu_N + qE_F)Kp_{O_2}^{1/2} = K*p_{O_2}^{1/2} \tag{3}$$

where N_{site} is the substitutional oxygen site density and T is the growth temperature. The entropy contribution is assumed to be small. The carrier concentration and its dependence on oxygen partial pressure can be obtained from the charge neutrality condition:

$$n = N_D - N_A = N_D(1-\Theta) = [O_N](1-\Theta)$$
$$and$$
$$n = K*(1-\Theta)p_{O_2}^{1/2} \tag{4}$$

-where Θ is the compensation ratio N_A/N_D. Since the oxygen donor is shallow, $n \approx N_D - N_A$. The carrier concentration can be calculated once the compensation ratio Θ is known. It has been shown for n-type GaN the compensation ratio is nearly independent of donor concentration and is of the order of 0.4 [10]. The theoretical expression of the free electron concentration versus oxygen partial pressure can be calculated at growth temperature [9-10] using Eqns. 1-4, where the site density $N_{site} = 4.4\times10^{22}$ cm^{-3} and the effective density of states is given by $N_c = 4.98\times10^{14}T^{3/2}$.

Fig. 1 shows the dependence of carrier concentration on oxygen dopant partial pressure. The carrier concentration increases as the square root of oxygen partial pressure up to 7×10^{18} cm^{-3}. The solid line is the calculated dependence of the free electron concentration on oxygen partial pressure using the formation energy as a fitting parameter. There is good agreement between theory and experimental data up to 7×10^{18} cm^{-3}, as seen in Fig. 1 for a formation energy of 1.3 eV. Since the measured electron concentration is directly proportional to the square root of the oxygen partial pressure and proportional to the oxygen donor concentration, it can be concluded that oxygen is a simple donor in GaN. The measured oxygen solubility, however, differs from theoretical prediction. According to calculations of Van de Walle *et al* the oxygen concentration should not exceed 10^{18} cm^{-3} for Ga rich growth conditions and 3×10^{16} cm^{-3} for N-rich conditions where E^F = 1.8 and 2.2 eV, respectively [1,17]. The calculated formation energy is the lowest under Ga-rich conditions. For these conditions the solubility of oxygen is limited by the formation of Ga$_2$O$_3$: $2\mu_{Ga}$ + $3\mu_O < \mu(Ga_2O_3)$. Since oxygen doped GaN epitaxial layers, however, are grown under N rich conditions, [21] the experimentally determined formation energy and the calculated value are not in good agreement. They differ by 0.9 eV. The theory underestimates the solubility of oxygen by two orders of magnitude.

The results of our oxygen doping experiments are, nevertheless, consistent with existing data on oxygen doping of GaN. Niebuhr et al [5] using N$_2$O as an oxygen dopant source showed that the electron concentration increased with N$_2$O partial pressure and saturated at a level of 4×10^{18} cm^{-3}. SIMS studies on oxygen contaminated epitaxial GaN by Forte-Poisson et al [2] indicated that the electron concentration is directly proportional to the amount of incorporated oxygen up to a concentration level of 6×10^{19} cm^{-3}, which is consistent with oxygen behaving as a simple shallow donor.

COMPENSATION OF OXYGEN DOPED THIN FILMS

To determine the compensation ratio the electron mobility of oxygen doped GaN was measured and plotted as a function of carrier concentration as shown in Fig. 2. For comparison, the Hall mobilities of undoped and oxygen and Si [11] doped samples are also presented. The theoretical calculation of mobility by Rode [10] is also shown for compensation ratios of $\Theta = 0.4, 0.8$. It can be seen from the plot that the O-doped samples have $\Theta = 0.4$ for concentration up to 8×10^{18} cm^{-3}. The compensation ratio increases however, from 0.4 to 0.6 for dopant concentrations in excess of 8×10^{18} cm^{-3}. The compensation ratio of 0.4 for oxygen doped

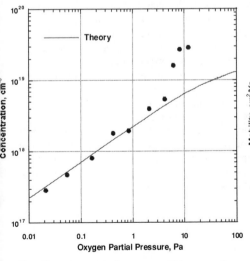

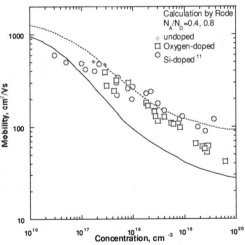

Fig. 1 Electron concentration versus partial pressure for oxygen-doped GaN

Fig. 2 Mobility versus carrier concentration for undoped, O and Si-doped films. Calculated mobilities of Rode are given by solid lines

samples is comparable to that of Si-doped and Se-doped material suggesting the same defect is compensating the films [11,16].

DETERMINATION OF IONIZATION ENERGY OF O_N

The temperature dependence of the Hall effect was measured to obtain the thermal ionization energy of the donors in both undoped and doped films. Analysis of the donor activation energy in GaN is complicated by the presence of a large contribution by impurity band conduction [6,18-19]. Typically, impurity band conduction in semiconductors is observed at low temperature (T < 10 K) when donor freezeout occurs. However the large unintentional doping and compensation in GaN, decrease electron mobility resulting in conduction via impurity bands even at 77 K. To determine the donor activation energy several techniques have been utilized to account for impurity band conduction at low temperature [2,12-13]. However, all these methods which are used to eliminate the effects of the low temperature tail of the impurity band can potentially lead to a large variation in the calculated activation energy. For the present study a single donor model was used for oxygen doped films, whereas a two donor, one compensating acceptor model was used for the undoped films.

The activation energy of the oxygen donors was determined from temperature dependent Hall effect measurements. The data is shown in Fig. 3. The carrier concentration for undoped samples was corrected for the Hall factor r = $<\tau(E)^2>/<\tau>^2$ found from the mobility data [9] . The activation energy and free electron concentration for the donor is calculated using the following equation [13]:

$$n + N_A = \frac{N_D}{1 + n/\phi} \tag{6}$$

- where $\phi = g_0/g_1 N_c T^{3/2} \exp(-E_D/kT)$, with the degeneracy of the unoccupied donor state $g_0 = 1$, the degeneracy of the occupied state $g_1 = 2$ and E_D is the activation energy of the donor. The acceptor concentration for undoped samples was found from mobility data. The measured electron concentration and activation energies are reported in Table 1. For undoped samples using a two-donor model, the activation energy of shallow and deep donors are 10 ± 2 and 52 ± 2 meV, respectively. Similar activation energies were obtained previously for the undoped samples [7,12]. A donor activation energy of less than 20 meV in GaN is attributed to a hydrogenic donor in the presence of screening [12]. The donor ionization energy in oxygen doped samples had an activation energy of 27 ± 2 meV. With increase in oxygen donor concentration impurity conductivity plays an ever more important role at low temperature, as seen in Fig. 3. At electron concentration $\sim 10^{19} \text{cm}^{-3}$ the slope of the carrier concentration versus reciprocal temperature decreases, suggesting that at this concentration GaN becomes degenerate.

To determine the concentration of compensating acceptors, the temperature dependent mobility was analyzed for undoped samples based on the approach taken by D. C. Look [10]. The mobility is calculated by solving the Boltzmann equation using the relaxation-time approximation. The mobility is given by $\mu_H = e<\tau(E)^2>/m^*<\tau(E)>$, where the brackets denote the average of relaxation time over electron energy E. The relaxation time $\tau(E)$ is given as:

$$\frac{1}{\tau(E)} = \frac{1}{\tau_{ac}(E)} + \frac{1}{\tau_{po}(E)} + \frac{1}{\tau_{pe}(E)} + \frac{1}{\tau_{ii}(E)} \tag{7}$$

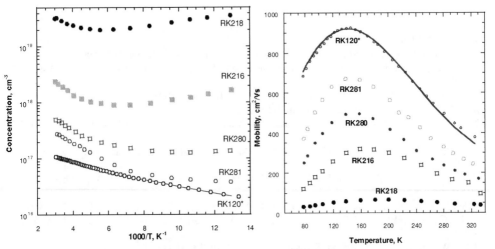

Fig. 3 Free electron concentration with respect to temperature data for doped and undoped * GaN. The caculated dependece is shown with solid lines

Fig.4 Mobility versus temperature for undoped * and doped GaN film. Theoretical curve is given by solid line

where the relaxation times τ_{ac} τ_{po} τ_{pe} τ_{ii} are due to an acoustic, polar-optical modes and piezoelectric, ionized scattering, respectively. The constants utilized in calculations are given in Ref. [12,20].

Table 1. Electronic characteristics of oxygen doped * and undoped samples calculated using Eqn.6-7.

Sample, #	E_{D1}, meV	E_{D2}, meV	N_1, N_2 $\times 10^{17} cm^{-3}$	N_A, $\times 10^{17} cm^{-3}$
RK120	52	10	1.1, 0.8	0.55
RK280*	27	-	40	-
Rk281*	27	-	10	-

From the fit, shown in Fig. 4 the acceptor concentration, and compensation ratio, $\theta \sim 0.3$ were determined and are reported in Table1. At low temperature the calculated mobility of oxygen doped samples differed significantly from the measured mobility due to large impurity band conduction.

CONCLUSIONS

The doping of GaN by oxygen prepared by MOVPE was studied. The free electron concentration increases as the square root of oxygen partial pressure up to a concentration of 7×10^{18} cm^{-3}. From the dependence of carrier concentration on oxygen partial pressure, it is concluded that oxygen behaves as a simple donor. The donor ionization energy is 27 ± 2 meV. Temperature dependent mobility and Hall measurements of O-doped samples indicate a constant compensation ratio of $\Theta = 0.3-0.4$. Based on the experimental data the calculated formation energy of O_N in GaN is 1.3 eV.

ACKNOWLEDGMENTS

This work is supported by the NSF GOALI program under grant ECS-9705134.

REFERENCES

1. C. G. van de Walle, C. Stampfl and J. Neugebauer, J. Cryst. Growth **189**, 505 (1998).
2. M. A. di Forte-Poisson, F. Huet, A. Romann, M. Tordjman, et al J. Cryst. Growth **195**, 314 (1998).
3. K. H. Ploog and O. Brandt, J. Vac. Sci. Technol. A **16**, 1609 (1998).
4. B-C. Chung and M. Gershenzon, J. Appl. Phys. **72**, 651 (1992).
5. R. Niebuhr, K. H. Bachem, U. Kaufmann, M. Maier, C. Merz, et al J. of Electr. Mat. 26, 1127 (1997).
6. W. Gotz, R. S. Kern, C. H. Chen, H. Liu, D. A. Steigerwald, et al, Mat. Sci. and Eng B **59**, 211 (1999).
7. V. A. Joshkin, C. A. Parker, S. M. Bedair, J. F. Muth, I. K. Shmagin, et al, J. Appl. Phys. **86**, 281 (1999).
8. W. M. Chen, I. A. Buyanova, Mt. Wagner, B. Monemar, et al, Phys. Rev. B **58**, R13351 (1998).
9. D. C. Look, Electrical Characterization of GaAs Materials and Devices (Wiley, New York, 1989).
10. J. W. Orton and C. T. Foxon, Semicond. Sci. Technol. **13**, 310 (1998).
11. S. Nakamura, T. Mukai and M. Senoh, J. J Appl. Phys. A **31**, 2883 (1992).
12. D. C. Look, J. R. Sizelove, S. Keller, Y. F. Wu, U. K. Mishraet al. Solid State Comm. **102**, 297 (1997).
13. D. C. Look, R. J. Molnar, Appl. Phys. Lett. **70**, 3377 (1997).
14. M. D. McCluskey, N. M. Johnson, et al, Phys. Rev. Lett. **80**, 4008 (1997)
15. R. A. Smith, Semiconductors (Cambridge, 1959)
16. G-C Yi and B. W. Wessels, Appl. Phys. Lett 69, 3028 (1996)
17. J. Neugebauer and C. G. Van de Walle, FESTKOR A S **35**: 25-43 (1996)
18. M. Ilegems and H. C. Montgomery, J. Phys. Chem. Solids **34**, 885 (1973)
19. N. F. Mott and W. D. Twose, Adv. Phys. **10**, 107 (1961)
20. To better fit the high temperature side of the mobility curve we used D. C. Look approach [12]. A near perfect fit to experimental data can be achieved when the acoustic deformation potential $E_1 = 17$ (9.2) eV of and $\varepsilon_0(\varepsilon_\infty^{-1} - \varepsilon^{-1}) = 0.104$ (0.0867) are used with literature values given in brackets.
21. R. Y. Korotkov and B. W. Wessels (unpublished).

Mat. Res. Soc. Symp. Vol. 595 © 2000 Materials Research Society

Optical and Electrical Properties of MBE Grown Cubic GaN/GaAs Epilayers Doped by Si

D.J. As[*], A. Richter[*], J. Busch[*], B. Schöttker[*], M. Lübbers[*], J. Mimkes[*], D. Schikora[*], K. Lischka[*], W. Kriegseis[**], W. Burkhardt[**], B.K. Meyer[**]
[*] Universität Paderborn, FB-6 Physik, Warburger Strasse 100
D-33095 Paderborn, Germany, d.as@uni-paderborn.de
[**] Universität Giessen, I. Physik. Inst., Heinrich-Buff-Ring 16,
D-35392 Giessen, Germany

ABSTRACT

Si-doping of cubic GaN epilayers grown by an rf plasma-assisted molecular beam epitaxy on semi-insulating GaAs (001) substrates is investigated by secondary ion mass spectroscopy (SIMS), photoluminescence (PL) and by Hall-effect measure-ments. SIMS measurements show a homogeneous incorporation of Si in cubic GaN epilayers up to concentrations of $5*10^{19}$ cm^{-3}. PL shows a clear shift of the donor-acceptor emission to higher energies with increasing Si-doping. Above a Si-flux of $1*10^{11}$cm^{-2}s^{-1} the near band edge lines merge to one broad band due to band gap renormalization and conduction band filling effects. The influence of the high dislocation density ($\approx 10^{11}$cm^{-2}) in c-GaN:Si on the electrical properties is reflected in the dependence of the electron mobility on the free carrier concentration. We find that dislocations in cubic GaN act as acceptors and are electrically active.

INTRODUCTION

Silicon is the prefered n-type dopant in the growth of GaN. In hexagonal GaN (h-GaN) controllable Si-doping has been demonstrated for concentrations form 10^{17} cm^{-3} to 10^{19} cm^{-3} [1]. At 300 K the luminescence intensity and the linewidth of the band-to-band transition increases monotonically with doping concentration [2,3]. Photoreflectance measurements further showed for heavily doped h-GaN a shrinkage of the energy gap due to band-gap renormalization (BGR) effects [4]. Detailed analysis of the electrical properties of Si-doped h-GaN further showed a significant influence of crystal defects [5] and of dislocations on the electron mobility [6, 7].

In this paper we summarize recent doping experiments of cubic GaN (c-GaN) epilayers by Si. Secondary ion mass spectroscopy (SIMS), photoluminescene (PL) at 300 K and 2 K and Hall-effect measurements are used to study the incorporation of Si, the optical and electrical properties of Si doped samples.

EXPERIMENTAL

The Si-doped cubic GaN epilayers are grown by an rf plasma assisted molecular beam epitaxy on semi-insulation GaAs (001) substrates at a substrate temperature of 720°C [8]. The growth rate is about 0.07 µm/h and the thicknesses of the layers are about 0.8 µm. Elemental Si is evaporated from a commercial effusions cell at source temperatures between 750°C and 1100°C, which correspondes to a variation of the Si-flux between $8.5*10^{5}$ cm^{-2}s^{-1} and $5.2*10^{11}$ cm^{-2}s^{-1}, respectively. The concentration and depth distribution of Si is measured by SIMS using Si implanted calibrated standards for quantification, and O^{2+} primary beam of 6 keV. Photo-luminescence measurements are performed at 300 K and in a He bath cryostat at 2 K. The luminescence was excited by a cw HeCd UV laser with a power of 3 mW and measured in a standard PL system. Hall-effect measurements were performed using square shaped samples at 300 K at a magnetic field of 0.3 T and with the samples in the dark. In was used for ohmic contacts.

RESULTS AND DISCUSSION

The incorporation of Si into c-GaN has been studied by SIMS. In Fig.1 the depth profiles of Si, GaN and As are depicted for a cubic GaN epilayer doped with elemental Si at a Si source temperature of T_{Si} = 1075°C. A homogeneous Si distribution is measured throughout the whole c-GaN epilayer and no accumulation neither at the interface nor at the surface is observed. The Si-concentration measured at a depth of about 0.4 µm (indicated by the arrow in Fig.1) is $2*10^{19}$ cm^{-3}. In Fig. 2 the Si-concentration measured by SIMS (full squares) and the free electron concentration measured by Hall-effect at room temperature (full triangles) are plotted versus the Si source temperature. As clearly can be seen, both the free electron concentration and the amount of incorporated Si exactly follow the Si-vapor pressure curve (full line in Fig.2) at $T_{Si} > 1000$°C [9]. This indicates that nearly all Si atoms are incorporated at Ga sites and act as shallow donors. At room temperature the maximum free electron concentration reached so far is about $5*10^{19}$ cm^{-3}. This clearly demonstrates the ability of controlled n-type doping of cubic GaN by Si up to concentrations which are necessary for the fabrications of laser diodes.

At room temperature the integrated intensity of the luminescence (full dots in Fig.2) also follows the Si-vapor pressure curve, indicating that at 300 K the optical properties are also determined by the Si-doping and that even at the highest free carrier concentration of $5*10^{19}$ cm^{-3} no quenching of the luminescence is observed. The optical properties of

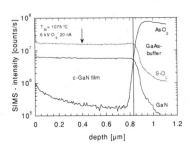

Figure 1. SIMS depth profiles of a Si doped cubic GaN epilayer.

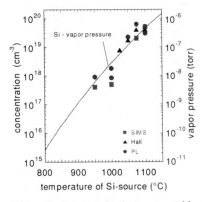

Figure 2. Si concentration measured by SIMS (full squares), free electron concentration measured by Hall-effect (full triangles) and integrated PL intensity (full circles) at 300 K versus Si source temperature. The full line represents the vapor pressure curve of Si [9].

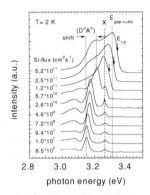

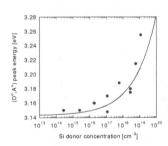

Figure 3. *Low temperature PL spectra of Si doped cubic GaN at different Si-fluxes.*

Si doped cubic GaN at low temperature are shown in Fig.3. At 2 K the spectrum of the sample grown with the lowest Si-flux $(8.5*10^5 \text{ cm}^{-2}\text{s}^{-1}$, $T_{Si} = 750°C)$ is dominated either by the excitonic transition X at 3.26 eV or by the donor-acceptor pair transition $(D°, A°)$ at 3.15 eV [10]. With increasing Si flux a clear shift to higher energies of the $(D°, A°)$ is observed. In contrast to that, the transition X stays at its position as expected for an excitonic line. The peak position of the $(D°, A°)$ as a function of the Si-concentration is plotted in Fig.4, where the Si concentration N_{Si} was calculated by dividing the measured Si-fluxes by the measured growth rates. Using a simple Coulomb-term model [11] the peak position of the $(D°, A°)$ transition can be estimated by

$$E_{(D^0,A^0)} = E_{gap} - (E_D + E_A) + \frac{e^2}{4\pi\varepsilon_0\varepsilon\,R} \qquad \text{with} \qquad R = \frac{1}{2}\cdot\frac{1}{\sqrt[3]{N_{Si}}}.$$

The full line in Fig.4 represents the model prediction and shows an excellent agreement with the experimental data (full dots).

Beyond a Si-flux $>1.2*10^{11}\text{cm}^{-2}\text{s}^{-1}$ $(T_{Si} > 1025°C)$ both lines merges to one broad band and the peak maximum shifts monotonically towards higher energies with increasing Si doping. Simultaneously the spectral shape of the main emission line becomes strongly asymmetric having a steep slope on the high-energy side and a smooth slope on the low

energy side of the spectra. Such a behavior is characteristic for momentum nonconserving (nonvertical) band-to-band transitions or to recombination of free electrons to local hole states [12] and has been observed in the spectra of GaAs heavily doped with Te [12] or Si [13]. The position at the steep high energy edge of the luminescence band is determined by the electron Fermi-level and it should shift to higher energies (Burstein-Moss shift [14]) as the conduction band fills with free electrons. In Fig.3 the position of half maximum $E_{1/2}$ is indicated by arrows. In the highest doped sample $E_{1/2}$ has a value of 3.342 eV. For a corresponding carrier concentration of about $5*10^{19}$ cm^{-3}, however, band filling of 240 meV is calculated using an electron effective mass of 0.2 in GaN

Figure 4. *Peak energy of the (D^0,A^0) band versus Si donor concentration at 2K. The full line was calculated by using a simple Coulomb-term model.*

[15]. This indicates that due to exchange interaction between free carriers the energy gap of cubic GaN has been shrinked from 3.305 eV without doping to 3.095 eV at an

electron concentration of $5*10^{19}$ cm^{-3}. As it is known for GaAs, the reduction in band-gap energy due to the so called band-gap renormalization (BGR) can be described by a $n^{1/3}$ power law [16]. From this we obtain a BGR coefficient of $-5.7*10^{-8}$ eVcm for cubic GaN, which is comparable to that observed in hexagonal GaN ($-4.7*10^{-8}$ eVcm) [17].

In Fig.5 the value $1/q.|R_H|$ is plotted as a function of the Si source temperature between 700°C and 1150°C. R_H is the Hall constant and q is the electronic charge. One clearly sees that for $T_{Si} \le 1025$°C all samples are p-type with a hole concentration of about $2*10^{16}$ cm^{-3}, which is nearly independent of the Si source temperature (open squares). Above $T_{Si} \ge 1025$°C the samples are n-type (full triangles) and, as already discussed in Fig.2, the measured free electron concentration exactly follows the Si-vapor pressure curve (dotted line). Temperature dependent Hall-effect measurements further show that these cubic GaN samples are totally degenerated. From this p to n-type transition at about 1025°C we conclude that a residual acceptor concentration of about $N_A = 4*10^{18}$ cm^{-3} exsists in our cubic GaN epilayers, and that the Si-donor has to compensate this residual acceptors. This value can be extrapolated from the dotted line (Si-vapor pressure curve) at 1025°C.

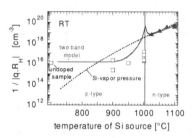

Figure 5. *$1/q.|R_H|$ versus Si source temperature measured at room temperature. The c-GaN layers were n-type above 1000°C and p-type below 1000°C.*

For a nominally undoped c-GaN sample a hole concentration of about $p = 1*10^{16}$ cm^{-3} and a hole mobility of $\mu_p = 283$ cm^2/Vs is measured at room temperature. This value is also included in Fig.5 (open square at the outmost left side). Temperature dependent Hall-effect measurements of this sample further showed that the involved acceptor has an activation energy of about $E_A \cong 0.166$ eV [18]. Due to the depth of this acceptor level only a few percent of the acceptors are thermaly activated at room temperature and contribute to the measured hole concentration. The estimated acceptor concentration of the nominally undoped sample is therefore $4*10^{18}$ cm^{-3}. This is exactly the same value, which we get from the Si-doping experiments above. Assuming a constant residual acceptor concentration and taking the room temperature values of the undoped samples for p and μ_p, a simple two-band conduction model [19] can be applied to calculate $1/q·|R_H|$ as a function T_{Si}. For the free electron concentrations n we use the Si-concentrations estimated from the Si-vapor curve (dotted curve in Fig.5) and for the electron mobilities μ_n we extrapolate the μ_n values measured at high T_{Si}, taking into account the increase of mobility with increasing free carrier concentration due to dislocation scattering. The two-band model prediction is shown in Fig.5 by the full curve and explains in a reasonable way the experimental observation.

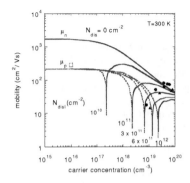

Figure 6. *300 K mobiliy versus free carrier concentrations of Si-doped c-GaN.*

Fig.6 shows the mobilities (μ_n, μ_p) of the Si-doped c-GaN epilayers versus the measured free carrier concentration (n, p) at room temperature. The full dots and the full triangles are samples of different series showing n-type conductivity (T_{Si} ≥ 1025°C). The open square is the nominally undoped p-type reference sample. The influence of the high dislocation density ($\approx 10^{11}$ cm^{-2}) on the electrical properties of c-GaN is reflected in the dependence of the electron mobility on the free carrier concentration. Similare to h-GaN [6] the mobility first increases with carrier concentration, reaches a maximum value of about 82 cm^2/Vs at an electron concentrations of $3*10^{19}$ cm^{-3} and decreases again. This behavior is characteristic for dislocation scattering, and shows that in cubic GaN threading edge dislocations are electrically active. Recently, a theory of charged-dislocation-line scattering has been developed and applied to h-GaN [7]. In Ref. 7 it has been shown that dislocations may well be charged and should have acceptor nature.

X-ray measurements and Rutherford Backscattering experiments on our cubic GaN layers showed that the dislocation density N_{disl} is in the order of $3*10^{11}$ cm^{-2} for 0.7 µm thick epilayers and that N_{disl} decreases with increasing epilayer thickness [20, 21]. Following Weimann et al. [6] and dividing N_{disl} by the lattice constant of c-GaN (a_{cub} = 0.452 nm) a residual acceptor concentration of about $6*10^{18}$ cm^{-3} is estimated. This value agrees within experimental error with the acceptor concentration which is necessary to explain the p- to n-type transition.

The curves in Fig.6 represent calculations of the room temperature mobility versus carrier concentration for different electrically active dislocation densities. In this calculation contributions of polar optical phonon scattering, acoustic phonon scattering, ionized impurity scattering and dislocation scattering have been taken into account. In addition, we have included that the compensation ratio $\Theta = N_D/N_A$ in n-type samples ($\Theta = N_A/N_D$ in p-type) changes if the incorporated donor concentration varies and that dislocation scattering is only active in n-type epilayers (in p-type GaN less than 1% of the acceptors are ionized at 300 K). The full lines are for n-type and the doted lines for p-type ranges. One clearly sees that the best agreement with experimental results is given for a dislocation density of about $3*10^{11}$ cm^{-2}. Thus, we believe that in cubic GaN threading edge dislocations are charged and act as compensating acceptors. This residual acceptor concentration has to be surpassed by the incorporated Si-donors to get n-type conductivity in c-GaN.

CONCLUSION

Si-doping of cubic GaN films grown by rf-plasma assisted MBE on semi-insulating GaAs (001) substrates is investigated, yielding n-type conductivity with a maximum electron concentration of $5*10^{19}$ cm^{-3}. Si is homogeneously incorporated into the epilayer and the amount follows exactly its vapor pressure curve. With increasing Si-concentration a continuous increase and broadening of the near-band luminescence is measured and a BGR coefficient of $-5.7*10^{-8}$ eVcm is obtained for c-GaN. Whereas the optical properties of Si-doped c-GaN samples are compareable to that of other III-V compounds the electrical properties of n-type c-GaN are strongly influenced by electrically active dislocations, which act as compensating acceptors. For advanced electrical and optical devices based on cubic group III-nitrides it will therefore be necessary to significantly reduce the dislocation density.

ACKNOWLEDGEMENTS

The authors acknowledge the support of DFG project number As (107/1-2).

REFERENCES

1. K. Doverspike and J.I. Pankove, in *Semiconductors and Semimetals* Vol. **50**, 259 (1998)
2. E.F. Schubert, I.D. Goepfert, W. Grieshaber, J.M. Redwing, Appl. Phys. Lett. **71** (7), 921 (1997)
3. E. Iliopoulos, D. Doppalapudi, H.M. Ng, T.D. Moustakas, MRS Symp. Proc. Vol. **482**, 655 (1998)
4. X. Zhang, S-J. Chua, W. Liu, K-B. Chong, J. of Cryst. Growth **189/190**, 687 (1998)
5. H. Tang, W. Kim, A. Botchkarev, G. Popovici, F. Hamdani and H. Morkoc, Solid-State Electronics **42** (5), 839 (1998)
6. N.G. Weimann, L.F. Eastman, D. Doppalapudi, H.M. Ng, T.D. Moustakas, J. Appl. Phys. **83** (7), 3656 (1998)
7. D.C. Look, J.R. Sizelove, Phys. Rev. Lett. **82** (6), 1237 (1999)
8. D. Schikora, M. Hankeln, D.J. As, K. Lischka, T. Litz, A. Waag, T. Buhrow and F. Henneberger, Phys. Rev. B **54** (12), R8381 (1996)
9. J.L. Souchiere, Vu Thien Binh, Surface Science **168**, 52 (1986)
10. D.J. As, F. Schmilgus, C. Wang, B. Schöttker, D. Schikora, and K. Lischka, Appl. Phys. Lett. **70** (10), 1311 (1997)
11. J.J. Hopfield, D.G. Thomas, M. Gershenzon, Phys. Rev. Lett. **10** (5), 162 (1963)
12. J. De-Sheng, Y. Makita, K. Ploog, H.J. Queisser, J. Appl. Phys. **53** (2), 999 (1982)
13. A.P. Abramov, I.N. Abramova, S. Yu. Verbin, I. Ya. Gerlovin, S.R. Grigorév, I.V. Ignatév, O.Z. Karimov, A.B. Novikov, and B.N. Novikov, Semiconductors **27** (7), 647 (1993)
14. E. Burstein, Phys. Rev. **83**, 632 (1954)
15. E.F. Schubert, *Doping in III-V Semiconductors*, Cambridge University Press (1993) p.38
16. H.C. Casey, Jr. and F. Stern, J. Appl. Phys. **47**, 631 (1976)
17. M. Yoshikawa, M. Kunzer, J. Wagner, H. Obloh, P. Schlotter, R. Schmitt, N. Herres, and U. Kaufmann, J. Appl. Phys. **86** (8), 4400 (1999)
18. J.R.L. Fernandez, A. Tabata, J.R. Leite, A.P. Lima, V.A. Chitta, E. Abramov, D.J. As, D. Schikora, K. Lischka, MRS Symp. Proc. Vol **595** (2000), W3.40
19. D.C. Look, in *Electrical Characterization of GaAs Materials and Devices*, Wiley, Chichester (1989), p.67
20. D.J. As, K. Lischka, phys. stat. sol. (a) **176**, 475 (1999)
21. J. Portmann, C. Haug, R. Brenn, T. Frey, B. Schöttker, D.J. As, Nucl. Instr. and Meth. in Phys. Res. B **155**, 489 (1999)

Mat. Res. Soc. Symp. Vol. 595 © 2000 Materials Research Society

Activation of Beryllium-Implanted GaN by Two-Step Annealing

Yuejun Sun, Leng Seow Tan, Soo Jin Chua and Savarimuthu Prakash
Centre for Optoelectronics, Department of Electrical Engineering,
National University of Singapore, 10 Kent Ridge Crescent, Singapore 119260,
Republic of Singapore

ABSTRACT

For the first time, p-type doping through beryllium implantation in gallium nitride was achieved by using a new annealing process, in which the sample was first annealed in forming gas (12% H_2 and 88% N_2), followed by annealing in pure nitrogen. Variable temperature Hall measurements showed that sheet hole concentrations of the annealed samples were about 1×10^{13} cm^{-2} with low hole mobilities. An ionization energy of 127 meV was estimated with a corresponding activation efficiency of ~ 100%. SIMS results revealed a relationship between the enhanced diffusion of Be and activation of the acceptors.

INTRODUCTION

A critical issue in the fabrication of gallium nitride (GaN) devices is the achievement of significant and controllable p-type doping. It still remains a challenge because of the high n-type autodoping background present in as-grown materials and the large ionization energy of acceptors, such as Mg, Zn and Cd.[1] The principal p-type dopant used for GaN is Mg with an ionization energy of 150-165meV.[2] Such a large acceptor ionization energy is problematic and two to three orders of magnitude higher atomic doping level of Mg must be incorporated into GaN in order to achieve the desired hole concentration at room temperature.[3] Although Zn and Cd are conventionally used as p-type dopants in the growth of other III-V compounds, Strite[4] suggested that the d-electron core relaxation in these elements is partially responsible for the enhanced depth, making efficient doping at room temperature impossible.

Beryllium (Be) was thought to be a shallower acceptor in GaN due to its large electronegativity and the absence of d-electrons. Ab initio calculations[5] predicted that Be behaves as a rather shallow acceptor in GaN, with a thermal ionization energy of 60 meV in wurtzite GaN. More evidence from photoluminescence (PL) spectra revealed that Be acts as an acceptor with an optical ionization energy ranging from 90-100 meV,[6,7] 150 meV,[8] to 250 meV.[9] However, the size of Be atoms is so small that it seems more probable for them to stay at interstitial sites (Be$_{int}$) rather than at substitutional sites (Be$_{Ga}$) in GaN. Theoretical calculations[10] also pointed out that the formation energy of Be$_{int}$ is much less than that of Be$_{Ga}$. Interstitial Be behaves like a double donor so that self-compensation is a significant drawback for the use of Be as an acceptor. That is why there is almost no achievement of electrical activiation of Be-doped GaN, except for Brandt et al. who obtained high mobility p-type materials from Be-O codoped cubic GaN by molecular beam epitaxy.[11]

In this paper, we introduce a new annealing process in which the Be-implanted GaN wafers were first annealed in forming gas, followed by annealing in a pure N_2 atmosphere.

Electrical activation of Be-implanted GaN was observed for the first time. The results confirmed the low activation energy of Be as an acceptor in GaN and also showed the possibility of p-type doping by Be implantation.

EXPERIMENT

The undoped GaN layers used in the experiments were 2μm thick and grown on c-plane sapphire substrates by MOCVD in a multiwafer rotating disk reactor at 1040 °C, with a ~20 nm GaN buffer layer grown at 530 °C in advance. The background n-type carrier concentrations were around 9×10^{16} cm^{-3}. The as-grown layers had featureless surfaces and were transparent with a strong near band-edge luminescence at 3.44 eV at room temperature. Be was implanted[+] into two pieces of the undoped GaN wafer at 40 keV. The doses were 3×10^{14} cm^{-2} and 1×10^{15} cm^{-2}, respectively. Samples of dimensions 6×6 mm^2 were cut from the wafers. Annealing was performed in a RTP system equipped with halogen-tungsten lamps. One set of the implanted samples was sequentially annealed at 900 -1100 °C for 45 s in flowing N$_2$ according to a face-to-face geometry. The other set of implanted samples was annealed in forming gas (12% H$_2$ and 88% N$_2$) first at temperatures ranging from 500 to 1100 °C and then in flowing N2. Hall effect measurements were conducted at room and variable temperatures with a magnetic field of 0.32 Tesla. Indium dots were alloyed at the corners of each sample according to the Van der Pauw geometry. Secondary ion mass spectrometry (SIMS) analyses were carried out in a CAMECA ims-6f microscope. An O_2^+ beam of 200 nA and 8 keV impact energy was used to sputter the samples. The quantification of the H concentration was not available due to the lack of reference. However, the H$^+$ intensities in all the samples have been normalized with respect to N$^+$, and therefore were comparable.

RESULTS AND DISCUSSION

From the variable temperature Hall measurements, a plot of the sheet carrier concentration/ temperature product ($p_s T^{-3/2}$) vs. reciprocal temperature can be constructed. Figure 1 shows such a plot for a Be-implanted GaN sample annealed at 600 °C and 10 s in forming gas first, followed at 1100 °C and 45 s in N$_2$. As shown by Gotz et al.,[12] the normal slope analysis in a linear region of the carrier concentration alone versus reciprocal temperature data could yield around 1/3 more than the real activation energy. We think the Arrhenius plot of the sheet carrier

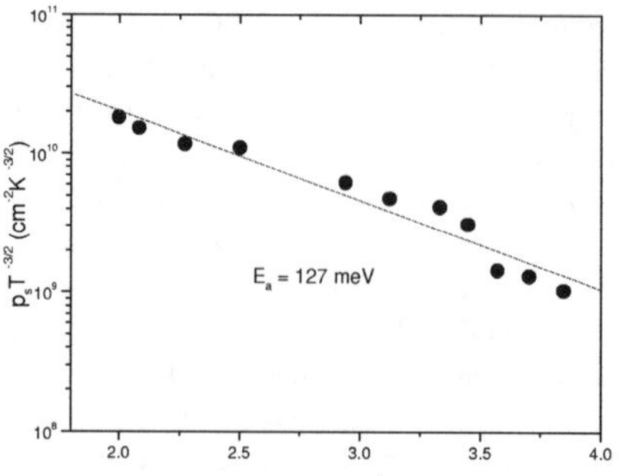

FIG.1 Arrhenius plot of the sheet carrier concentration/ temperature product of Be-implanted GaN annealed in forming gas first, followed in N$_2$, the dose is 3×10^{14} cm^{-2}

[+] The implantation was performed by Implantation Science Corporation, Wakefield, MA, USA

concentration/temperature product vs. the inverse temperature is more appropriate to estimate the ionization level of Be. Under nondegenerate conditions, the hole concentration can be expressed as the following relation: $p \propto p_o T^{3/2} exp(-E_a/kT)$, where p_o is the acceptor concentration and E_a is the activation energy of the acceptors. So, $ln(pT^{3/2})$ should be proportional to the inverse T. Due to non-uniformity of the implanted atoms in the substrate layer, here we use the sheet carrier concentration p_s to plot, instead of p. From Figure 1, the ionization energy of Be in GaN is calculated to be 127 meV, which is lower than that of Mg and Ca reported for implanted GaN.[2,13] Based on the ionization energy estimated above, only 0.73% of Be acceptors would be ionized at room temperature. The activation efficiency can thus be estimated to be around 100% for this sample (considering $p_s = 2.14 \times 10^{13}$ cm^{-2} at room temperature), if the activation efficiency is defined as sheet carrier concentration/(dose×ionization rate). The activation efficiency of Be is similar to that of Ca.[13]

The results of room temperature Hall measurement are summarized in Table I. Several remarkable things should be pointed out. (i) The implanted samples without post-annealing process still showed n-type conductive characteristics with a bit of reduction in electron concentrations, unlike most implanted GaN samples which had high resistivities due to defect states within the band gap that act as traps for the carriers. This may be the results of less damage produced by the smaller mass of Be, low implantation energy (40 KeV) and the resistance of GaN materials to damage by implantation, with considerable dynamic recovery of implantation-induced disorder. High-resolution x-ray diffractometry (HRXRD) was performed to detect the implantation damage. There was no difference in the rocking curves between the as-grown sample and the as-implanted sample. However, it should be noted that HRXRD results could not demonstrate conclusively that the implantation damage was not significant because HRXRD is not very sensitive to implant damage for the lower dose

TABLE I. Room temperature Hall effect data of GaN samples

Sample*	Annealing condition	Type & sheet carrier concentration (cm-2)	Mobility (cm^{2}V^{-1}s^{-1})
416G	as-grown	$n \sim 1.98 \times 10^{13}$	96
416I	as-implanted	$n \sim 8.4 \times 10^{12}$	63
416D5	1100 °C and 45 s in N$_2$	$n \sim 1.12 \times 10^{13}$	98
416D10	700 °C and 10 s in forming gas; then at 1100 °C and 45 s in N$_2$	high resistivity (p-type indicated from the thermal probe measurement)	
417G	as-grown	$n \sim 1.7 \times 10^{13}$	107
417I	as-implanted	$n \sim 8.6 \times 10^{12}$	122
417H5	1100 °C and 45 s in N$_2$	$n \sim 8.6 \times 10^{12}$	122
417H10	700 °C and 10 s in forming gas; then at 1100 °C and 45 s in N$_2$	$p \sim 2.0 \times 10^{13}$	3
417H14	600 °C and 10 s in forming gas; then at 1100 °C and 45 s in N$_2$	$p \sim 2.14 \times 10^{13}$	0.8

*The doses for Sample 416 series and 417 series are 1×10^{15} cm^{-2} and 3×10^{14} cm^{-2}, respectively

sample. (ii) The new annealing process for the high dose (1×10^{15} cm^{-2}) implanted samples produced high-resistivity material. This is probably because, besides the native donors and interstitial Be, more defects were introduced by the higher dose implantation, which compensated the effects of the activated acceptors. This is similar to the case of high Mg doping in GaN.[14] (iii) p-type doping was achieved for the lower dose (3×10^{14} cm^{-2}) implanted samples, with a sheet hole concentration around 10^{13} cm^{-2}. To confirm the realization of p-type doping, thermal probe measurements were performed. We used Mg-doped p-type GaN and undoped n-type GaN samples as references. The results were consistent with the Hall measurements, thus confirming that the Be-implanted and annealed samples were indeed p-type.

Figure 2 depicts the SIMS profiles of Be and H for different samples. Fig.2(a) shows the depth profiles of Be and H in the as-grown sample. The measured Be concentration was due to a background level noise governed by the intrinsic impurity and the detection limit of our SIMS measurement conditions. The depth profile of Be in the as-implanted sample exhibited a peak at a depth of 120 nm as shown in Fig.2(b). The H profile in this sample was almost the same as that in the as-grown sample. There was no significant channeling tail of Be, unlike that in the Be as-implanted samples (45 keV, 5 × 10^{14} cm^{-2}) in REF 15. We think the differences are from the implantation process, such as beam divergence or other random factors, but probably not from some factors like energy and dose, which are similar to ours. After annealing at 1100 °C for 45 s, the peak was nearer the surface than that in the as-implanted sample. The projected range was 73 nm (Fig.2(c)). The H profile remained similar. The Be redistribution in this case can be attributed to defect-assisted diffusion.[16] It is noted that some groups[15,17] did not see measurable redistribution of Be in GaN. In Fig. 2(d), the Be and H profiles of the sample annealed with the new process were substantially different from the others. In the top 230 nm layer, the Be profile showed a plateau-like shape, at a concentration of ~1.2×10^{19} cm^{-3}.

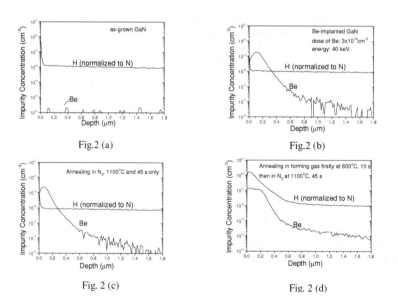

Fig.2 (a)

Fig.2 (b)

Fig. 2 (c)

Fig. 2 (d)

At depths beyond 0.5 μm, the Be level was much higher than that in the as-implanted sample and N_2-annealed sample, indicating enhanced in-diffusion of Be. In addition, accumulation of H was observed in the near-surface region, where the H concentration was one to two orders of magnitude higher than the background level. From the study of thermal stability of hydrogen in ^{2}H-implanted p-type GaN,[18] the diffusion of the deuterium started at a lower temperature of about 500-700 °C and Mg-H complexes were formed, but these were completely gone by 1000 °C. However, for our samples, even an 1100 °C annealing could not entirely break the bonds of potential H complexes and remove the H. From the analysis of Hall and thermal probe measurements, it seems that there exists some relationship between the enhanced diffusion of Be and the activation of the acceptors, similar to the case of Zn diffusion and the optical activation of acceptors in Zn-implanted GaN.[19]

The effect of the new annealing process on Be activation can be tentatively explained as follows. After implantation, there are many Be atoms at interstitial sites, which could easily form a large concentration of substitutional-interstitial (SI) Be-related complexes. Annealing in an atmosphere containing hydrogen might cause the formation of Be_{int}-H complexes first. The Be_{int}-H complexes can move into Ga vacancies (V_{Ga}) more easily and are then converted to Be_{Ga}-H-N complexes. The energy required to break the H-bonds in the Be_{Ga}-H-N complexes might be lower than that needed to move the Be_{int} directly to the substitutional sites during conventional annealing in N_2 only. The subsequent annealing in flowing N_2 will then depassivate the complexes and activate the implanted Be. An independent inference from a first-principles calculation[20] also pointed out that an annealing stage in an ambient including hydrogen before conventional annealing in N_2 would improve activation of p-type dopants in GaN.

CONCLUSION

In conclusion, we present a new annealing process and have observed significant improvement of activation in Be-implanted GaN for the first time, demonstrate the possibility of p-type doping in GaN through Be implantation and confirm that Be has a shallow acceptor level in GaN. Further optimization of annealing conditions will be investigated to improve the properties of Be-implanted GaN.

ACKNOWLEDGEMENTS

Yuejun Sun would like to thank the National University of Singapore for providing him with a Research Scholarship and Dr. Liu Wei for discussions. The author would also like to acknowledge Dr. David Look for useful discussions and manuscript preparation. This work is supported by the Singapore National Science and Technology Board through a grant NSTB/17/2/3 GR6471.

REFERENCE

1. S. Strite and H. Morkoc, *J. Vac. Sci. Tech. B,* **10**, 1237 (1992)
2   I. Akasaki, H. Amano, M. Kito and K. Hiramatsu, *J. Lumin.*, **48/49**, 666 (1991)
3. J. W. Orton, *Semicond. Sci. Tech.*, **10**, 101 (1995)
4. S. Strite, *Jpn. J. Appl. Phys.*, **33**, 699 (1994)
5. F. Bernardini, V. Fiorentini, and A. Bosin, *Appl. Phys. Lett.*, **70**, 2990 (1997)
6. D. J. Dewsnip, A. V. Andrianov, I. Harrison, J. W. Orton, D.E. Lacklison, G. B. Ren, S.E. Hopper, T. S. Cheng, and C. T. Foxon, *Semicond. Sci. Tech.*, **13**, 500 (1998)
7. F. J. Sanchez, F. Calle, M. A. Sanchez-Garcia, E. Calleja, E. Munoz, C. H. Molly, D. J. Somerford, F. K. Koschnick, K. Michael, and J. M. Spaeth, *MRS Internet J. of Nitride Semicon. Research*, V3, Article 19 (1998)
8. C. Ronning, E. P. Carlson, D. B. Thomson, and R. F. Davis, *Appl. Phys. Lett.*, **73**, 1622 (1998)
9. A. Salvador, W. Kim, O. Aktas, A.Botchkarev, Z. Fan, and H. Morkoc, *Appl. Phys. Lett.*, **69**, 2692 (1996)
10. J. Neugebauer and C. G. Van De Walle, *Appl. Phys. Lett.*, **85**, 3003 (1999)
11. O. Brandt, H. Yang, H. Kostial, and K. H. Ploog, *Appl. Phys. Lett.*, **69**, 2707 (1996)
12 W. Gotz, N. M. Johnson, C. Chen, H. Liu, C. Kuo and W. Imler, *Appl. Phys. Lett.,* **68**, 3144 (1996)
13 J.C. Zolper, R. G. Wilson, S. J. Pearton and R. A. Stall, *Appl. Phys. Lett.*, **68**, 1945(1996).
14 L. Eckey, U. von Gfug, J. Holst, A. Hoffmann, A. Kaschner, H. siegle, C. Thomsen, B. Schineller, K. Heime, M. Heuken, O. Schon and R. Beccard, *J. Appl. Phys.*, **84**, 5828 (1998)
15 J. C. Zolper, in : GaN and Related Materials, Ed. S. J. Pearton (Gordon and Breach, New York, 1997) ch.12

16   R. G. Wilson, J. M. Zavada, X.A. Cao, R.K. Singh, S. J. Pearton, H. J. Guo, S. J. Pennycook, M. Fu, J.A. Sekhar, V. Scarvepalli, R. J. Shul, J. Han, D. J. Rieger, J. C. Zolper and C. R. Abernathy, *J. Vac. Sci. Technol.*, A **17,** 1226 (1999).

17   C. Ronning, K. J. Linthicum, E. P. Carlson, P. J. Hartlieb, D. B. Thomson, T. Gehrke and R. F. Davis, *MRS Internet J. Nitride Semicond. Res.*, **4S1**, G3.17 (1999)

18   S. J. Pearton, R. G. Wilson, J. M. Zavada, J. Han and R. J. Shul, *Appl. Phys. Lett.*, **73**, 1877 (1998)

19   T. Suski, J. Jun, M. Leszczynski, H. Teisseyre, S. Strite, A. Rockett, A. Pelzmann, M. Kamp and K. J. Ebeling, *J. Appl. Phys.*, **84**, 1155 (1998)

20   F. A. Reboredo and S. T. Pantelides, *Phys. Rev. Lett.*, **82**, 1887 (1999)

Mat. Res. Soc. Symp. Vol. 595 © 2000 Materials Research Society

Co-doping Characteristics of Si and Zn with Mg in P-type GaN

K.S. Kim, C.S. Oh, M.S. Han, C.S. Kim, G.M. Yang*, J.W. Yang, C.-H. Hong, C.J. Youn, K.Y. Lim, and H.J. Lee
Department of Semiconductor Science & Technology and Semiconductor Physics Research Center, Chonbuk National University, Chonju 561-756, Korea
*Author to whom correspondence should be addressed gyemo@moak.chonbuk.ac.kr

ABSTRACT

We investigated the doping characteristics of Mg doped, Mg-Si co-doped, and Mg-Zn co-doped GaN films grown by metalorganic chemical vapor deposition. We have grown p-GaN film with a resistivity of 1.26 •cm and a hole density of 4.3×10^{17} cm^{-3} by means of Mg-Si co-doping technique. The Mg-Si co-doping characteristic was also explained effectively by taking advantage of the concept of competitive adsorption between Mg and Si during the growth. For Mg-Zn co-doping, p-GaN showing a low electrical resistivity (0.7 •cm) and a high hole concentration (8.5×10^{17} cm^{-3}) was successfully grown without the degradation of structural quality of the film. Besides, the measured specific contact resistance for Mg-Zn co-doped GaN film is 5.0×10^{-4} •cm^2, which is lower value by one order of magnitude than that for only Mg doped GaN film (1.9×10^{-3} •cm^2).

INTRODUCTION

The realization of high conducting p-type GaN film is one of the key factors to the success of GaN based light emitting devices such as light emitting diodes (LEDs) and laser diodes with low series resistance and high external quantum efficiency. Mg has been used as a typical acceptor dopant in GaN grown by metalorganic chemical vapor deposition (MOCVD), known to have the smallest ionization energy compared with other acceptor dopant sources. Nevertheless, it is difficult to get the p-type GaN with low resistivity and high hole concentration because it is still deep from valence band maximum. Recently, it was theoretically suggested that co-doping of n-type dopants (eg. Si, O, *etc.*) together with p-type dopants (eg. Mg, Be, *etc.*) in GaN is effective for the fabrication of high-conductivity p-type GaN[1, 2]. It drives us to the study of co-doping characteristics in p-type GaN.

In this work, we have studied the co-doping characteristics of Si with Mg as well as Zn with Mg.

EXPERIMENTS

All samples were prepared in a horizontal MOCVD reactor. The source materials of Ga, N, Mg, Si, and Zn are trimethylgallium (TMGa), ammonia (NH$_3$), bis-cyclopentadienylmagnesium (Cp$_2$Mg), 100 ppm monosilane (SiH$_4$), and diethylzinc (DEZn), respectively. Mg doped, Mg-Si, and Mg-Zn codoped GaN overlayers with thicknesses of 1 – 2 microns were grown in H$_2$ ambient at 1080 °C with various flow rates of Cp$_2$Mg, SiH$_4$, and DEZn. Other growth conditions are the same with those of the unintentionally doped GaN[3].

RESULTS AND DISCUSSION

Mg doped GaN

In order to investigate Mg doping characteristics in GaN, several kinds of GaN films with different hole concentrations obtained by varying the amount of Cp_2Mg flow rate were grown. Figure 1 shows typical Hall characteristics of the GaN films as a function of gas phase [Mg]/[Ga] ratio measured at room temperature after post-growth rapid thermal annealing (RTA) in N_2 ambient for 30 s at 900 °C. Electrical characteristics were evaluated using the conventional van der Pauw Hall measurement in which magnetic field of 0.5 Tesla and currents between 10 and 100 uA are applied. The GaN grown under a [Mg]/[Ga] ratio of 5.4 $\times$ 10^{-3} gives a high resistivity of 5.6 •cm and a low hole concentration of 7.3 $\times$ 10^{16} cm^{-3}. When the [Mg]/[Ga] ratio is reached to 7.6 $\times 10^{-3}$, the hole concentration and resistivity abruptly change and have the maximum value of 6.7 $\times$ 10^{17} cm^{-3} and the minimum value of 0.8 •cm, respectively. If the [Mg]/[Ga] ratio is larger than 7.6 $\times$ 10^{-3}, the hole concentration is gradually decreased and the resestivity is increased.

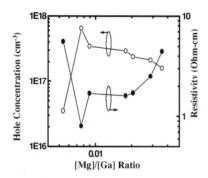

Figure 1, Dependence of resistivity (solid circle) and hole concentration (open circle) at room temperature on the gas phase [Mg]/[Ga] ratio during the growth of GaN.

Figure 2, The resistivities (solid circles) and hole concentrations (open circles) as a function of SiH_4 flow rate during the growth under the constant [Mg]/[Ga] ratio of 7.6 $\times$ 10^{-3}.

Mg-Si co-doped GaN

For the purpose of studying Si co-doping characteristics in Mg doped GaN films, some kinds of specimens with different SiH_4 flow rates were grown where the [Mg]/[Ga] ratio was kept constant to 7.6 $\times$ 10^{-3} thought to be the optimum condition. The Hall characteristics of the samples as a function of SiH_4 flow rate are exhibited in Fig. 2, which are carried out after the identical RTA treatment as mentioned earlier. The Si and Mg co-doped p-type GaN layers show anomalous electrical behaviors. If considering only absolute net acceptor and donor concentrations, one can easily predict that hole concentrations are gradually decreased with increasing SiH_4 flow rate, and finally type conversion arises. However, actual experimental data do not follow our expectation. In other words, the hole concentrations are increased as SiH_4 flow rates increase before occurring type conversion. It is also interesting fact that type conversion occurs at a SiH_4 flow rate of 0.13 nmol/min, because this value is the quantity corresponding to the doping level of 1×10^{17} cm^{-3} when applied in an undoped GaN. In addition, the maximum hole concentration (4.5×10^{17} cm^{-3}) of p-GaN co-doped with Si and Mg shows the lower value compared to that (6.7×10^{17} cm^{-3}) of only Mg doped GaN. This abnormal behavior could be figured out by introducing the concept of competitive adsorption between Mg and Si atoms because they get into the same Ga sublattice site, though the quantity of Si atom is much less than that of Mg atoms in gas phase. Actually, if we compare the bond length of Si - N and Mg – N, the bond length of Mg – N (0.1439 nm) is shorter than that of Si – N (0.1572 nm)[4]. Since the formation energy of Si – N bonding is smaller than that of Mg – N, the occupation by Si atom into Ga sublattice site is easier than that by Mg atom. Accordingly, for Mg and Si co-doped GaN films, we presume that somewhat larger amount of Si atoms and smaller amount of Mg atoms are introduced into the co-doped GaN film than those independently calibrated. Besides, for [SiH₄] < 0.13 nmol/min, it is reasonable that the hole concentration is increased with increasing SiH_4 flow rate because the increase of Si donors prevents from generating native defects leading to the self-compensation[1]. Therefore, this assumption can explain the experimental Hall data successfully and be proven by low temperature (100 K) measured PL spectra shown in Fig. 3.

Low temperature (100 K) photoluminescence (PL) spectra of GaN epilayers with different [Mg]/[Ga] ratios are shown in Fig. 3 (a). The peaks at 3.25 eV and 2.77 eV are attributed to donor-acceptor pair (DAP)[5] and deep donor-acceptor-pair[6] transitions, respectively. However, at present status, we can not explain the exact origins of transition lines peaking at 3.17 eV and 3.01 eV. It is apparent that blue emissions

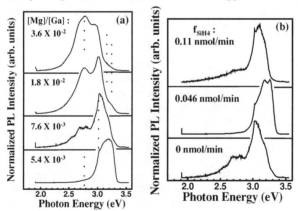

Figure 3, *Low temperature (100 K) measured PL spectra of (a) Mg doped GaN films with various [Mg]/[Ga] ratios, and (b) Mg-Si co-doped GaN films with different SiH₄ flow rates when keeping the [Mg]/[Ga] ratio constant to 7.6×10^{-3}.*

around 2.77 eV are increased and DAP transitions about 3.2 eV are quenched as [Mg]/[Ga] ratio increases. This means that the transition probability from the deep donor state to acceptor level is more probable and intense than that from the shallow donor state because the amount of deep donor is increased as Mg incorporation increases. This result shows good agreement with the electrical characteristics shown in Fig. 1.

The PL spectrum of Si and Mg co-doped GaN film with the SiH_4 flow rate of 0.046 nmol/min in Fig. 3 (b) is similar to the result of the PL spectrum of Mg doped GaN layer with the [Mg]/[Ga] ratio of 5.4 $\times$ 10^{-3} as shown in Fig. 3 (a). It means that the introducing SiH_4 source reduces the incorporation of Mg atoms, as we suppose.

Mg-Zn co-doped GaN

As we change Cp_2Mg (under constant DEZn flow rate of 0.616 nmol/min) and

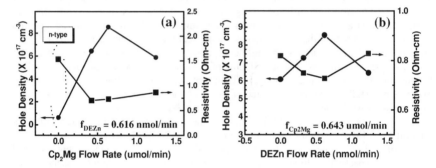

Figure 4, *Dependence of hole densities and resistivities for the Mg-Zn co-doped GaN films at the condition of (a) various Cp_2Mg flow rate and constant DEZn flow rate of 0.616 nmol/min , and (b) various DEZn flow rate and a constant Cp_2Mg flow rate of 0.643 umol/min.*

DEZn (under constant Cp_2Mg flow rate of 0.643 umol/min) flow rates, the variations of hole densities and resistivities of Mg-Zn co-doped GaN films at room temperature are shown in Fig. 4 (a) and (b), respectively. Theoretically, the co-doping of two different p-type dopants in GaN is not helpful to achieve a high conducting p-type GaN, because it forms a lot of native defect levels leading to hole compensation. However, in our prepared samples, we observed there exist optimum Cp_2Mg flow rate of 0.643 umol/min and DEZn flow rate of 0.616 nmol/min showing higher hole concentration (8.5 X 10^{17} cm^{-3}) and lower resistivity (0.72 •cm) than those for only Mg-doped GaN film.

Besides, from the high resolution x-ray diffraction (HRXRD) measurement as shown in Fig. 5, both (002) and (102) HRXRD full width at half maximum values are broadened by increasing Cp_2Mg flow rates not DEZn flow rates. So we can conclude that the incorporation of Zn atoms with Mg is the technique obtaining high conducting p-type GaN film without deterioration of the structural qualities of GaN films.

In order to compare Ohmic contact properties for the only Mg-doped (6.0 X 10^{17} cm^{-3}) and the Mg-Zn co-doped (8.5 X 10^{17} cm^{-3}) p-GaN films, we performed transmission line method (TLM) with ring contact geometry with outer ring radius of 200 um and gap spacings (5 – 45 um). The surface of p-type GaN specimens were treated using aqua regia[7], followed by the deposition of Pd(150 A)/Au(1400 A) Ohmic metals under a vacuum condition of 10^{-5} torr. After the metal deposition, the photoresist was

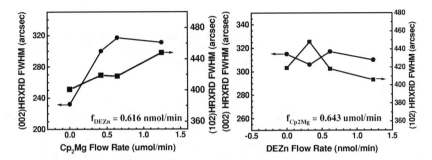

Figure 5, *Dependence of (002) and (102) HRXRD FWHMs for the Mg-Zn co-doped GaN films at the condition of (a) various Cp₂Mg flow rate and constant DEZn flow rate of 0.616 nmol/min , and (b) various DEZn flow rate and a constant Cp₂Mg flow rate of 0.643 umol/min.*

lifted off. Current-Voltage (I-V) characteristics of the prepared samples were measured by HP 4155 parameter analyzer.

Figure 6 (a) shows the I-V characteristics for both the only Mg-doped and the Mg-Zn co-doped GaN samples, measured between the Ohmic pads with a spacing of 5 um. The I-V curve obtained is nonlinear for the only Mg-doped sample, but linear for the Mg-Zn co-doped sample over the whole range of voltages. The specific contact resistivities

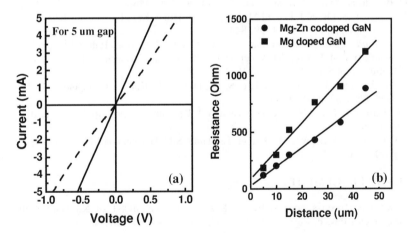

Figure 6, *(a) Comparison of the I-V characteristics measured between the contact pads with a gap spacing of 5 um. for the only Mg-doped (dashed line) and Mg-Zn co-doped p-type GaN films(solid line), (b) Variation of resistance as a function of gap spacing for the only Mg-doped (Solid square) and Mg-Zn co-doped (solid circle) p-type GaN films. solid lines are linear fits of experimental data.*

with correction factors [8] taken from the gradients and intercepts in the Fig. 6 (b) are determined [8] to be $1.9 \times 10^{-3} \cdot cm^2$ for the only Mg-doped sample and $5.0 \times 10^-$

$^{-4} \cdot cm^2$ for the Mg-Zn co-doped sample. Note that contact resistivity decreases by one order of magnitude by the co-doping technique. Therefore, the Mg-Zn co-doped GaN layer is expected to act as a good contact layer in device structure.

SUMMARY

We have found that Mg doped GaN layer shows high electrical conductivity, when the [Mg]/[Ga] ratio in gas phase is 7.6 × 10^{-3}. The Mg-Si co-doping characteristics were explained effectively taking advantage of the concept of competitive adsorption between Mg and Si during the growth. In other words, the occupation by Si atom into Ga sublattice site is easier than that by Mg atom, since the formation energy of Si – N bonding is smaller than that of Mg – N. We have also grown p-GaN film with a resistivity of 1.26 •cm and a hole density of 4.3 × $10^{17} cm^{-3}$ by means of Mg-Si co-doping technique. For Mg-Zn co-doping, p-GaN showing low electrical resistivity (0.7 •cm) and high hole concentration (8.5 × $10^{17} cm^{-3}$) was successfully grown without the degradation of structural quality of the film. Besides, the specific contact resistance for Mg-Zn co-doped GaN film measured by TLM is 5.0 × $10^{-4} \cdot cm^2$, which is lower value by one order of magnitude than that for only Mg-doped GaN film (1.9 × $10^{-3} \cdot cm^2$).

REFERENCES

1. T. Yamamoto and H.K. Yoshida, Jpn. J. Appl. Phys. **36**, L180 (1997).
2. O. Brandt, H. Yang, H. Kostial, and K.H. Ploog, Appl. Phys. Lett. **69**, 2707 (1996).
3. K.S. Kim, C.S. Oh, K.J. Lee, G.M. Yang, C.-H. Hong, K.Y. Lim, A. Yoshikawa, and H.J. Lee, J. Appl. Phys. **85** (1999) 8441.
4. *CRC Handbook of Chemistry and Physics*, 55th ed. (CRC, Boca Raton, FL, 1974).
5. U. Kaufmann, M. Kunzer, M. Maier, H. Obloh, A. Ramakrishnan, B. Santic, and P. Schlotter, Appl. Phys. Lett. **72**, 1326 (1998).
6. A.K. Viswanath, E.-J. Shin, J.I. Lee, S. Yu, D. Kim, B. Kim, Y. Choi, and C.-H. Hong, J. Appl. Phys. **83**, 2272 (1998).
7. J.K. Kim, J.-L. Lee, J.W. Lee, H.E. Shin, Y.J. Park, and T. Kim, Appl. Phys. Lett. **73**, 2953 (1998).
8. L.F. Lester, J.M. Brown, J.C. Ramer, L. Zhang, S.D. Hersee, and J.C. Zolper, Appl. Phys. Lett. **69**, 2737 (1996).

ACKNOWLEDGEMENTS

This work has been supported by the KOSEF and the MOST of Korea through the Semiconductor Physics Research Center at Chonbuk National University.

Mat. Res. Soc. Symp. Vol. 595 © 2000 Materials Research Society

Efficient Acceptor Activation in Al$_x$Ga$_{1-x}$N/GaN Doped Superlattices

I. D. Goepfert and *E. F. Schubert*
Department of Electrical and Computer Engineering
Boston University, Boston, MA 02215

A. Osinsky and *P. E. Norris*
NZ Applied Technologies, Woburn, MA 01801

Abstract

Mg-doped superlattices consisting of uniformly doped Al$_x$Ga$_{1-x}$N and GaN layers are analyzed by Hall-effect measurements. Acceptor activation energies of 70 meV and 58 meV are obtained for superlattice structures with an Al mole fraction of $x = 0.10$ and 0.20 in the barrier layers, respectively. These energies are significantly lower than the activation energy measured for Mg-doped GaN thin films. At room temperature, the doped superlattices have free hole concentrations of 2×10^{18} cm^{-3} and 4×10^{18} cm^{-3} for $x = 0.10$ and 0.20, respectively. The increase in hole concentration with Al content of the superlattice is consistent with theory. The room temperature conductivity measured for the superlattice structures are 0.27 S/cm and 0.64 S/cm for an Al mole fraction of $x = 0.10$ and 0.20, respectively.

Introduction

Magnesium is the most common acceptor used in GaN. The large activation energy of Mg of 150 meV to 250 meV[1,2,3] results in a low acceptor ionization probability. Other acceptors such as Be and Ca also have large activation energies of 150 meV[4] and 169 meV[5], respectively. The low acceptor activation in *p*-type GaN results in large series resistances and high operating voltages, thereby adversely affecting the performance of electronic and optoelectronic devices.

Recently, it has been theoretically demonstrated that doped superlattice (SL) structures[6] increase the free hole concentration as compared to homogeneous thin-film *p*-type GaN. Doped superlattices are doped ternary compound semiconductor structures with a spatially modulated chemical composition. The modulation of the chemical composition leads to a variation of the valence band energy and to a reduction of the acceptor activation energy. Acceptors in the GaN layers of the superlattice must be ionized by thermal excitation. However, acceptors in the Al$_x$Ga$_{1-x}$N barriers are ionized more easily because these acceptors are close to the GaN valence band edge.

Experimental

In this publication, the electronic properties of Al$_x$Ga$_{1-x}$N/GaN doped superlattice structures grown by MBE are analyzed. The superlattices consist of 20 periods of equally thick $Al_xGa_{1-x}N$ barriers (10 nm) and GaN wells (10 nm). The $Al_xGa_{1-x}N$ and GaN layers are uniformly doped with Mg at a level of $N_{Mg} \approx 10^{19}$ cm^{-3}. The Al mole fraction of the two different doped SLs are $x = 0.10$ and 0.20.

Hall-effect measurements were conducted on Mg-doped $Al_xGa_{1-x}N$/GaN superlattice samples. Photolithographically defined, 85 nm thick Ni contacts were deposited by electron-beam evaporation in the van der Pauw configuration. All samples display ohmic I-V characteristics. The samples were cooled in a Cryo Industries liquid nitrogen cryostat. Hall effect measurements, with a magnetic flux density of 0.5 T, were made between 150 - 400 K with 10 K intervals.

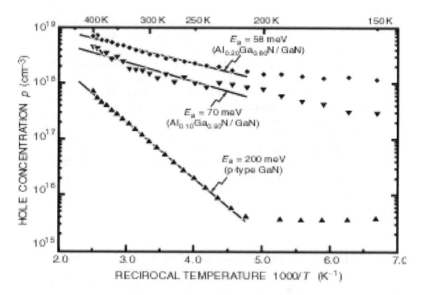

FIG. 1. Hole concentration verses reciprocal temperature for Mg-doped $Al_xGa_{1-x}N$/GaN superlattice samples with Al mole fraction in the barriers of $x = 0.10$ and 0.20. Also shown is a uniformly Mg-doped GaN thin-film sample with an acceptor activation energy of 200 meV.

Experimental results and discussion

Figure 1 shows the free hole concentration verses reciprocal temperature. The hole concentration displays an exponential dependence for temperatures in the range 220 - 400 K. The natural logarithm of the carrier concentration verses reciprocal temperature is fit with a least squares regression algorithm. The activation energy is determined using the relation

$$\ln p \quad \propto \quad -E_a/(k_B T) \tag{1}$$

where E_a is the effective acceptor activation energy, k_B is Boltzmann's constant, and T is the absolute temperature. The measured activation energies for the superlattices with $x = 0.10$ and $x = 0.20$ are 58 meV and 70 meV, respectively. This result clearly shows that doped $Al_xGa_{1-x}N$/GaN superlattice structures have a lower activation energy than homogeneous Mg-doped GaN thin films.

At room temperature, the carrier concentration of the doped SL structures are 2×10^{18} cm^{-3} and 4×10^{18} cm^{-3} for an Al mole fraction $x = 0.10$ and $x = 0.20$, respectively. The sample with 20 % Al shows a factor of two larger carrier concentration than the sample with 10 % Al. Clearly, the free hole concentration increases with an increase of Al mole fraction in the barriers of the doped SLs. This increase is consistent with theory. It should be noted that most free carriers reside in the GaN layers. Assuming that the carrier concentration in the barriers is much smaller[6] than in the GaN well layers, the actual free carrier concentration the GaN well layers of the superlattice is 8×10^{18} cm^{-3} which is the highest free hole concentration ever achieved in GaN.

Katsuragawa and coworkers[7] showed that as the Al content is increased in homogeneous $Al_xGa_{1-x}N$ from $x = 0.00$ to $x = 0.33$ the hole concentration decreases monotonically. Tanaka and coworkers[8] reported that for Mg-doped GaN to $Al_{0.08}GaN_{0.92}N$, the activation energy increases from 157 meV to 192 meV. In contrast, doped superlattice structures have smaller activation energies and larger carrier concentrations as the mole fraction of Al is increased in the superlattice barriers.

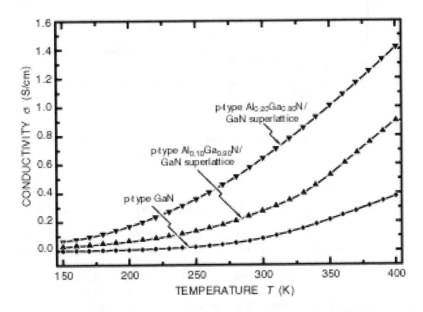

FIG. 2. Conductivity verses temperature for p-type $Al_xGa_{1-x}N$/GaN doped superlattices and homogeneous p-type GaN.

The conductivity verses temperature for the two SL structures are depicted in Figure 2. The room-temperature conductivity is 0.27 S/cm and 0.64 S/cm for the superlattice with an Al mole fraction of $x = 0.10$ and 0.20, respectively. At room temperature the conductivity for the SL with 20 % Al has a conductivity almost a factor of three greater than for the SL with 10 % Al. Note that the results shown in Figure 2 are the highest p-type conductivities achieved in a III-nitride semiconductor containing aluminum.

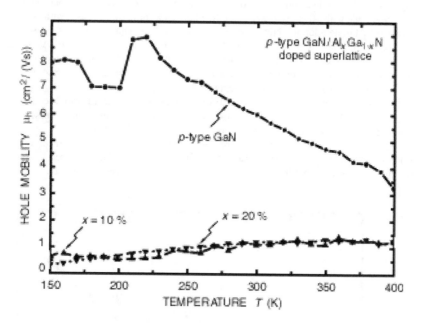

FIG. 3. Mobility verses temperature for p-type $Al_xGa_{1-x}N$/GaN doped superlattices and homogeneous p-type GaN.

Figure 3 depicts the mobility verses temperature for the doped superlattice samples. Note, that for an increase in Al content, the mobility does not change. The mobilities of the doped SL structures increase monotonically with temperature by more than a factor of three in the temperature range 150 - 400 K. This is in contrast to p-type GaN which exhibits a decrease in mobility as the temperature increases beyond 200 K. Due to the temperature dependence of the mobility, conductivity data without carrier concentrations[9] will result in an overestimation of the acceptor activation energy.

We have calculated the carrier concentration as a function of temperature for an $Al_xGa_{1-x}N$/GaN doped superlattice with x varied between $x = 0.00$ and $x = 0.30$. The thicknesses of the GaN and $Al_xGa_{1-x}N$ regions are both 10 nm. A uniform doping profile of $N_A = 1 \times 10^{18}$ cm^{-3} and an acceptor activation energy of $E_a = 200$ meV in the GaN and the $Al_xGa_{1-x}N$ is assumed. Using the parameters $\varepsilon_r = 9.0 - 0.5\,x$ for the relative permittivity[10] and $m_p^* = 0.8\ m_0$ for the effective hole mass[11], the hydrogenic model does not predict a significant deepening of the acceptor level in $Al_xGa_{1-x}N$ for $x \leq 0.20$. This is consistent with the experimental results of Katsuragawa et al.[7] who showed that the activation energy does not change markedly with the Al content. To determine the carrier concentration, the arithmetic average of the position dependent carrier concentration over one period of the superlattice is computed for a given Al content and temperature.

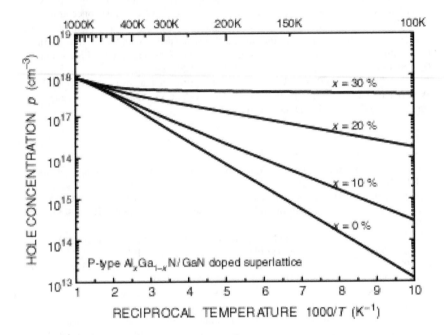

FIG. 4. Calculated hole concentration verses reciprocal temperature for p-type $Al_xGa_{1-x}N$/GaN doped superlattices with Al mole fractions of $x = 0.00, 0.10, 0.20$, and 0.30.

Figure 4 depicts the calculated carrier concentration verses $1/T$ for four doped SLs with different Al content in the $Al_xGa_{1-x}N$ barriers. Examination of the four curves shows that the carrier concentration increases with an increase in Al content in the barriers. Note, that in the freeze-out regime, the slope of each curve represents the effective activation energy of the acceptor, which decreases with increasing Al content in the barriers. Our experimental results shown in Figure 1 are consistent with the results of the theoretical model.

Conclusions

In conclusion, the electronic properties of Mg-doped superlattices consisting of uniformly doped $Al_xGa_{1-x}N$ and GaN layers are analyzed by Hall-effect measurements. Acceptor activation energies of 70 meV and 58 meV are reported for superlattice structures with an Al mole fraction of $x = 0.10$ and 0.20 in the barrier layers, respectively. These energies are substantially lower than the activation energy measured for Mg-doped GaN thin films. At room temperature, the doped superlattices have free hole concentrations of 2×10^{18} cm^{-3} and 4×10^{18} cm^{-3} for $x = 0.10$ and 0.20, respectively. The increase in hole concentration with Al content of the superlattice is consistent with theory. The room temperature conductivity measured for the superlattice structures are 0.27 S/cm and 0.64 S/cm for an Al mole fraction of $x = 0.10$ and 0.20, respectively.

The work at Boston University was supported in part by the ONR (Dr. C. E. C. Wood) and by the NSF (Dr. R. P. Khosla). The work at NZ Applied Technologies was supported by the NSF (Dr. D. Gorman).

References

[1] S. Strite and H. Morkoc, *GaN, AlN, and InN: A review*, Journal of Vacuum Technology **B10(4)**, 1237 (1992)

[2] H. Nakayama, P. Hacke, M. R.H. Khan, T. Detchprohm, K. Hiramatsu, and N. Sawaki, *Electrical transport properties of p-GaN*, Japanese Journal of Applied Physics **35**, L282 (1996)

[3] S. Fischer, C. Wetzel, E. E. Haller, and B. K. Meyer, *On p-type doping in GaN - acceptor binding energies*, Applied Physics Letters **67**, 1298 (1995)

[4] C. Ronning, E. P. Carlson, D. B. Thomson, and R. F. Davis, *Optical activation of Be implanted into GaN*, Applied Physics Letters **73**, 1622 (1998)

[5] J. W. Lee, S. J. Pearton, J. C. Zolper, R. A. Stall, *Hydrogen passivation of Ca acceptors in GaN*, Applied Physics Letters **68**, 2102 (1996)

[6] E. F. Schubert, W. Grieshaber, and I. D. Goepfert, *Enhancement of deep acceptor activation in semiconductors by superlattice doping*, Applied Physics Letters **69**, 3737 (1996)

[7] M. Katsuragawa, S. Sota, M. Komori, C. Anbe, T. Takeuchi, H. Sakai, H. Amano, I. Akasaki, *Thermal ionization energy of Si and Mg in AlGaN*, Journal of Crystal Growth **189/190**, 528 (1998)

[8] T. Tanaka, A. Watanabe, H. Amano, Y. Kobayasi, I. Akasaki, S. Yamazaki, and M. Koike, *p-type conduction in Mg-doped GaN and $Al_{0.08}GaN_{0.92}N$ grown by metalorganic vapor phase epitaxy*, Applied Physics Letters **65**, 593 (1994)

[9] A. Saxler, W. C. Mitchel, P. Kung, and M. Razeghi, *Aluminum gallium nitride short-period superlattces doped with magnesium*, Applied Physics Letters **74**, 2023 (1999)

[10] We use $\varepsilon_r = 9.0$ and $\varepsilon_r = 8.5$ for the relative dielectric constant for GaN and AlN, respectively. These values are compiled in Reference 1.

[11] The experimental effective hole mass in $Al_xGa_{1-x}N$ is unknown at this time. Although, Katsuragawa et al. in Reference 7 reported no significant change in the activation energy of Mg in $Al_xGa_{1-x}N$ indicating no significant change in the hole mass. See, for instance, Gil, *Group III Nitride Semiconductor Compounds* (Clarendon Press, Oxford, 1998)

Mat. Res. Soc. Symp. Vol. 595 © 2000 Materials Research Society

Structural and Optical Property Investigations on Mg-Alloying in Epitaxial Zinc Oxide Films on Sapphire

A.K. Sharma, C. Jin, A. Kvit, J. Narayan, J.F. Muth[1], C.W. Teng[1], R.M. Kolbas[1] and O.W. Holland[2]
Department of Materials Science and Engineering, North Carolina State University, Raleigh, NC 27695-7916.
[1]Department of Electrical and Computer Engineering, North Carolina State University, Raleigh, NC 27695-7911.
[2]Solid State Division, Oak Ridge National Laboratory, Oak Ridge TN, 37831-6048.

Abstract

We have synthesized single-crystal epitaxial MgZnO films by pulsed-laser deposition. High-resolution transmission electron microscopy, X-ray diffraction and Rutherford backscattering spectroscopy/ion channeling were used to characterize the microstructure, defect content, composition and epitaxial single-crystal quality of the films. In these films with up to ~ 34 atomic percent Mg incorporation, an intense ultraviolet band edge photoluminescence at room temperature and 77 K was observed. The highly efficient photoluminescence is indicative of the excitonic nature of the material. Transmission spectroscopy revealed that the excitonic structure of the alloys was clearly visible at room temperature. Post-deposition annealing in oxygen reduced the number of defects and improved the optical properties of the films. The potential applications of MgZnO alloys in a variety of optoelectronic devices are discussed.

Introduction

The intense interest in blue and ultraviolet light emitters and detectors has promoted enormous research efforts into the wide band gap semiconductors. In this context, the major efforts have been put into the development of high quality films of III-nitrides and their alloys. Recently, the successful syntheses of high quality GaN and its alloys have resulted in the commercialization of blue lasers, light-emitting-diodes (LEDs) and ultraviolet photodetectors [1]. With these successes, the LEDs in the visible range have been realized and the major concern now is to make improvements in these devices. Now, the immense interest has been generated in the development of ultraviolet light emitters, and truly solar blind photodetectors based on GaN material system [2-4], which are transparent to the visible and near UV portion of the spectrum. Compact ultraviolet sources and detectors could be useful to monitor or catalyze specific chemical reactions or to excite fluorescence in various proteins. As an alternative to the GaN material system, ZnO and its alloys [5,6] are of substantial interest. Alloying ZnO films with MgO or CdO may potentially permit the band gap to be controlled between 2.8 to 4 eV and higher. Zinc oxide or zincite (ZnO) is hexagonal, whereas magnesium oxide or periclase (MgO) is cubic. However, the similarity in ionic radii between Mg^{++} (1.36Å) and Zn^{++} (1.25 Å) allows some replacement in either structure. In the ZnO lattice, the solid solubility of Mg is limited to only 2% maximum [7]. In the present work, using pulsed laser deposition (PLD), we have achieved nonequilibrium phase space corresponding to 34 at. % Mg in ZnO, while maintaining the ZnO hexagonal structure with a lattice constant close to that of ZnO. These films also maintained favorable optical characteristics of wide band gap materials, including transparency in the visible and high excitonic binding energy (~60 meV) [8]. Direct observation of the excitonic nature of the films at room temperature was also observed, despite alloy broadening.

Experiment

The $Mg_xZn_{1-x}O$ films were synthesized by PLD using a KrF excimer laser (λ=248 nm, t_s=25 ns) in a high vacuum (~1x10^{-7} Torr) chamber. Several sintered targets containing different concentrations of MgO and ZnO were fabricated in the present work. The films were deposited in an oxygen partial pressure ~2x10^{-5} Torr, and a laser energy density of 2.5-3.5 J/cm^2 focused onto the target. The substrates (c-plane sapphire) were heated to 700-750 °C during the depositions. Some films were also annealed in oxygen for up to 10 hr. at 750 °C. In most cases, annealing in oxygen improved the optical and structural qualities of the films. The X-ray diffraction (XRD), high resolution transmission electron microscopy (HRTEM), Rutherford backscattering spectroscopy (RBS)/ion channeling, transmission and photoluminescence spectra were used to characterize these thin film heterostructures.

Results and Discussion

Figure 1 shows the XRD pattern of the $Mg_{0.34}Zn_{0.66}O$ film deposited at 750 °C. The XRD showed only (0001) reflections in all the films, with the position of the alloy films having a shift from the ZnO peak position in proportion to the MgO content. The lattice constant of ZnO was found to change by 1% (a increased by 1 % and c decreased by 0.9%), corresponding to the values of a=3.28Å and c=5.15 Å for the $Mg_{0.34}Zn_{0.66}O$ alloy. By XRD results, the optimum deposition temperature was found to be between 700-750 °C. The full-widths-at-half-maximum (FWHMs) of ω-rocking curves of the (0002) reflections were typically 0.25-0.40°, which is indicative of a high crystalline quality of the alloy films.

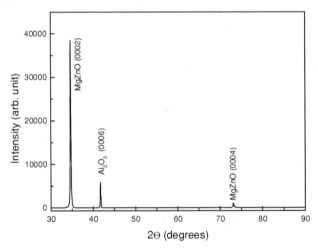

Figure 1. XRD pattern of a single crystal MgZnO film on (0001) α-Al$_2$O$_3$.

The RBS analysis of the samples was performed using a probe beam of 2.75 MeV He$^+$-ions and a standard, surface barrier detector positioned at 160°. Both random and aligned spectra were acquired to determine the composition and the crystalline quality of the deposited films. We could achieve Mg content in the films up to 34 at. %, which is more than an order of magnitude larger than the maximum value allowed by the phase diagram. The formation of metastable materials is possible by the nonequilibrium nature of pulsed laser deposition, which is clearly demonstrated in this experiment. The minimum ion-channeling yield (χ_{min}) of the film deposited at 750 °C was close to 5%, which is indicative of a single crystal film of high quality. Increasing Mg content to more than 34 at. % resulted in a phase separation of MgO. By further increasing the Mg content to more than 50 at. %, we were able to synthesize a cubic phase of MgZnO [9].

The cross section bright-field TEM micrograph from a Mg$_{0.34}$Zn$_{0.66}$O/α-Al$_2$O$_3$ sample is shown in Figure 2, with the corresponding diffraction pattern from the film-substrate (lower left inset) and the MgZnO film alone (lower right inset). The TEM results clearly show the formation of single-phase MgZnO, with no evidence of phase separation. The number density of threading dislocations was estimated to be ~10^{10} cm^{-2} close to the film-substrate interface, but decreased significantly with the film thickness. By subsequent thermal annealing at 750 °C in oxygen, the dislocation density decreased to 10^7-10^8 cm^{-2}. From the analysis of the diffraction patterns, the epitaxial relationship of the film was determined to be MgZnO[$\bar{1}2\bar{1}0$]‖(α-Al$_2$O$_3$[01$\bar{1}$0] and MgZnO[0001]‖(α-Al$_2$O$_3$[0001]. This in-plane epitaxial orientation relationship, similar to that for pure

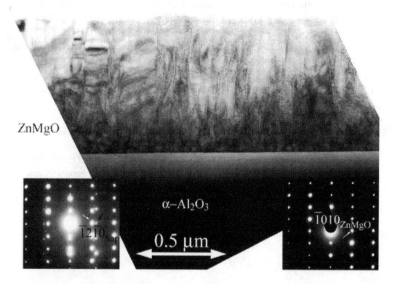

Figure 2. Cross-section TEM micrograph of a Mg$_{0.34}$Zn$_{0.66}$O film annealed in oxygen, depicting a high density of defects near the interface which decreases substantially towards the surface of the film. The left and right insets show electron diffraction from the film-substrate and the film alone, respectively.

ZnO/α-Al$_2$O$_3$ substrate, corresponds to 30° or 90° rotation of the film with respect to the substrate in the basal plane. The prevalent defects in MgZnO films were found to be threading dislocations, compared to stacking faults in pure ZnO epitaxial films on sapphire [10]. All the films with composition x varying between 0.19-0.34 were single crystal, having low defect content as compared to corresponding ZnO films, where a high density of stacking faults was observed near the interface [10]. It can be seen in Figure 2 that growing thicker films (more than 1 µm) will substantially reduce defect density towards the top of the film. Such defect-free films will be required for fabricating devices such as UV photodetectors. The specimens with four different concentrations of Mg in ZnO were studied by both plan-view and cross-section microscopy to characterize the quality of the crystals, defects and interfaces. The films with lower Mg content were found to have lower densities of threading dislocations.

We carried out transmission measurements using a Cary 5E UV-VIS-NIR spectrophotometer at room temperature and 77 K on the films deposited on double-side polished sapphire substrates. Figure 3 shows absorbance of the thin films scaled according to film thickness. Excitonic features are clearly apparent. In the ZnO film, even at room temperature, A and B excitons are clearly distinguished. In the MgZnO alloys, broadening due to alloying obscures the details of the individual excitons, but their net influence on the absorption spectra is still extremely significant. This is a result of the high binding energy, ~60 meV, of the exciton in MgZnO alloys [8]. With increasing Mg concentration in Figure 3, alloy broadening softens the slope of the absorption edge, but the exciton peak remains for all alloy compositions. Thus, for very thin films, transmission spectroscopy provides a sensitive indicator of material quality or the defect content of the films. The presence of excitons in the films is also apparent in the photoluminescence spectra excited with a large frame Ar$^+$ ion laser operating with deep UV optics (270-305 nm). The emission was collected and focused on a 0.64 m spectrometer equipped with an S-20 photocathode. The photoluminescence from the alloyed samples was very bright and of comparable or brighter intensity than unalloyed ZnO single crystal thin films [11] or single crystal ZnO grown hydrothermally. The photoluminescence spectra in Figure 4 have been normalized for clarity in discussion. The spectral positions of the photoluminescence peaks correspond to the excitonic band edge absorption shown in Figure 3. Furthermore, since the recombination process is excitonic, the FWHM values of the spectra do not broaden dramatically with increasing Mg concentration. This is in contrast with the previous report of the photoluminescence of MgZnO films taken at 4.2 K [5]. The defect photoluminescence is small compared to the band edge emission, and actually decreases with increasing Mg concentration.

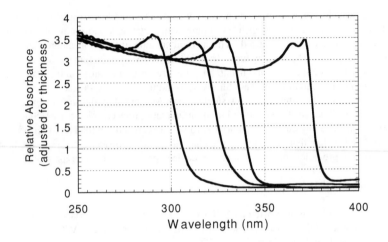

Figure 3. Absorbance of $Mg_xZn_{1-x}O$ films with x varying from 0-0.34 from right to left. The exciton peak is seen in all these films despite alloy broadening with increasing x.

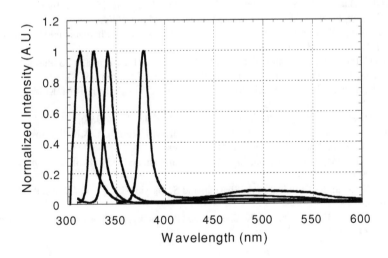

Figure 4. Room temperature PL from $Mg_xZn_{1-x}O$ films with x varying from 0-0.34 from right to left.

CONCLUSIONS

In conclusion, we report the growth of MgZnO alloy films grown by pulsed laser deposition. The epitaxial single crystal films with very bright photoluminescence at room temperature have been realized. The spectral position of the photoluminescence and optical band gap were successfully tuned from 3.3 to 4.0 eV by adjusting the Mg content, while preserving crystal quality and maintaining the presence of excitons in the material. The increasing Mg content in the ZnO films was found to reduce the defect photoluminescence. The ability to grow single crystal MgZnO alloy films opens up numerous possibilities for the construction of ultraviolet optoelectronic devices. Since the Mg-content achieved by our nonequilibrium growth method is more than an order of magnitude higher than that predicted by the phase diagram, further investigations are being conducted on segregation and its influence on device operation at high temperatures.

ACKNOWLEDGMENTS

This work was supported by NSF Center for Advanced Materials and Smart Structures. Work at ORNL was sponsored by DOE under contract DE-ACO5-960R22464 with Lockheed Martin Energy Research Corporation.

REFERENCES
1. S. Nakamura and G. Fasol, *The Blue Laser Diode* (Springer, Berlin, 1997).
2. G. Parish, S. Keller, P. Kozodoy, J.P. Ibbetson, H. Marchand, P.T. Fini, S.B. Fleischer, S.P. DenBaars, U.K. Mishra, and E.J. Tarsa, *Appl. Phys. Lett.* **75**, 247 (1999).
3. M. Razeghi, and A. Rogalski, *J. Appl. Phys.* **79**, 7433 (1996).
4. W. Wang, T. Nohova, S. Krishnakutty, R. Torreano, S. McPherson, and H. Marsh, *Appl. Phys. Lett.* **73**, 1086 (1998).
5. A. Ohtomo, M. Kawasaki, T. Koida, K. Masubuchi, H. Koinuma, Y. Sakurai, Y. Yoshida, T. Yasuda, and Y. Sewaga, *Appl. Phys. Lett.* **72**, 2466 (1998).
6. A.K. Sharma, J. Narayan, J.F. Muth, C.W. Teng, C. Jin, A. Kvit, R.M. Kolbas, and O.W. Holland, *Appl. Phys. Lett.*, **75**, 3327 (1999).
7. E.R. Segnit and A.E. Holland, *J. Am. Ceram. Soc.* **48**, 412 (1965).
8. C.W. Teng, J.F. Muth, R.M. Kolbas, Ü. Özgür, M.J. Bergman, A.K. Sharma, C. Jin, and J. Narayan, *Appl. Phys. Lett.*, (2000). (in press)
9. A.K. Sharma, A. Kvit, C. Jin, J.F. Muth, J. Narayan, and O.W. Holland, *Appl. Phys. Lett.* (2000). (submitted)
10. J. Narayan, K. Dovidenko, A. Sharma, and S. Oktyabrsky, *J. Appl. Phys.* **84**, 2597 (1998).
11. J.F. Muth, R.M. Kolbas, A.K. Sharma, S. Oktyabrsky, and J. Narayan, *J. Appl. Phys.* **85**, 7884 (1999).

Mat. Res. Soc. Symp. Vol. 595 © 2000 Materials Research Society

Doping Dependence Of The Thermal Conductivity Of Hydride Vapor Phase Epitaxy Grown n-GaN/Sapphire (0001) Using A Scanning Thermal Microscope

D.I. FLORESCU*, V.A. ASNIN*, L.G. MOUROKH*, FRED H. POLLAK*, and R. J. MOLNAR**

*   Physics Department and New York State Center for Advanced Technology in Ultrafast Photonic Materials and Applications, Brooklyn College of CUNY, Brooklyn, NY 11210

**   Massachusetts Institute of Technology, Lincoln Laboratory, Lexington, MA 02420-9108

ABSTRACT

We have measured the doping concentration dependence of the room temperature thermal conductivity (κ) of two series of n-GaN/sapphire (0001) fabricated by hydride vapor phase epitaxy (HVPE). In both sets κ decreased linearly with log n, the variation being about a factor two decrease in κ for every decade increase in n. $\kappa \approx 1.95$ W/cm-K was obtained for one of the most lightly doped samples ($n = 6.9 \times 10^{16}$ cm^{-3}), higher than the previously reported $\kappa \approx 1.7$-1.8 W/cm-K on lateral epitaxial overgrown material [V.A. Asnin $et\ al$, Appl. Phys. Lett. **75**, 1240 (1999)] and $\kappa \approx 1.3$ W/cm-K on a thick HVPE sample [E.K. Sichel and J.I. Pankove, J. Phys. Chem. Solids **38**, 330 (1977)]. The decrease in the lattice component of κ due to increased phonon scattering from both the impurities and free electrons outweighs the increase in the electronic contribution to κ.

INTRODUCTION

         Despite the considerable body of work, both experimental and theoretical, on the electronic, optical, and structural properties of group III nitrides [1] relatively little work has been reported on thermal conductivity κ. This quantity is of importance from both fundamental and applied perspectives. The lattice thermal conductivity is a function of the mean free path of the phonons and hence is determined by both intrinsic (phonon-phonon Umklapp scattering) and extrinsic (phonon-"defect", phonon-carrier scattering) factors [2]. Sichel and Pankove [3] determined κ of "bulk" hydride vapor phase epitaxy (HVPE) GaN as a function of temperature (25K<T<360K) with $\kappa \approx 1.3$ W/cm-K at 300K. More recently Asnin $et\ al$ [4] have performed high spatial resolution measurements on several lateral epitaxial overgrown (LEO) GaN/sapphire (0001) samples using a scanning thermal microscope (SThM) and found $\kappa \approx 1.7$-1.8 W/cm-K [4]. Slack has estimated an upper limit of 1.7 W/cm-K at 300K for GaN [5].

         We report high spatial resolution determination of κ at 300K on two sets of HVPE n-GaN/sapphire (0001) samples as a function of n. The measurements were made using a ThermoMicroscope's SThM Discoverer system [6], with a spatial resolution of $\approx$ 2-3 μm. Values of n were deduced from both 300K Hall effect and micro-Raman [longitudinal optical phonon-plasmon (LPP)] measurements. In both sets of samples κ decreased linearly with log n, the variation being about a factor two decrease in κ for every decade increase in n. $\kappa \approx 1.95$ W/cm-K was obtained for one of the most lightly doped samples ($n = 6.9 \times 10^{16}$

cm^{-3}), higher than previously reported κ [3,4].

Sample set A had unintentional n (6-800x10^{16} cm^{-3}) and thicknesses (t) in the range of 5-74 μm. For sample set B t was constant≈10 μm and 15x10^{16} cm^{-3}< n < 300x10^{16} cm^{-3}.

Our observation also helps to explain the results on the LEO material [4], which had $n \approx$ (10-20)x10^{16} cm^{-3} [7]. The decrease in the lattice component of κ due to increased phonon scattering from both the impurities and free electrons outweighs the increase in the electronic contribution to κ:

EXPERIMENTAL DETAILS

The GaN films were grown by the HVPE method in a vertical type reactor [8]. During this process, gallium monochloride is synthesized upstream by reacting HCl gas with liquid Ga metal at 800-900^{0}C.The GaCl is transported to the substrate where it is reacted with NH$_3$ at 1000-1100 ^{0}C forming GaN. All films were grown on (0001) sapphire. The carrier concentration n_H was determined by 300K Hall effect measurements. Several characteristics of the samples are listed in Table I.

The carrier concentration for $n \geq 40$x10^{16} cm^{-3} was also determined from the LPP modes observed in Raman scattering [9] and compared to the Hall effect results. Raman microprobe (≈ 2μm) measurements were made in the backscattering geometry using a triple grating spectrometer (Jobin-Yvon model T64000) and the 488 nm line of an Ar-ion laser as excitation. The Raman system was equipped with an Olympus BH2 microscope. Values of the carrier concentration (n_R) were deduced from Eqs. (2) and (3) in Ref. [8] (see Table I), using an electron effective mass (m_e^*) of 0.22 (in units of the free electron mass) [10] and a high frequency dielectric constant (ε_∞) of 5.5 [11].

The probe tip of the SThM system consists of a "V" shaped resistive thermal element incorporated at the end of a cantilever that enables atomic force microscopy-type feedback, as shown schematically in Fig. 1a. The arms of the cantilever are made of Wollaston process wire consisting of silver wire ≈ 75μm in diameter containing a platinum/10% rhodium core ≈ 3μm in diameter. The resistive element at its end comprises a 200μm length of platinum that has been exposed by removal of the silver and bent into a "V" shape (radius of curvature ≈ 1μm), which acts as the probe. The resistive element forms one leg of a Wheatstone bridge, as shown in Fig. 1b. A current is passed through the probe so that in air its temperature is about 40-50°C above ambient. There is a feedback loop to adjust the bridge voltage as necessary to keep the bridge balanced thus keeping the temperature of the probe constant. When the probe contacts the sample heat flows from the probe to the material, as shown in Fig. 1c. In the absence of feedback, this flow of heat would reduce the probe temperature, decreasing the resistance and causing the bridge to shift. The feedback senses this shift and increases the voltage applied to the bridge (U_{out}), returning the resistance to its set point. The thermal conductivity κ is proportional to the heat flow or (U_{out})^2, as shown in Fig. 1c.

Although initially designed to measure only relative spatial variations in U_{out}, Ruiz et al [12] developed a calibration procedure that makes is possible to evaluate absolute values of κ. Based on the results of Ref. [13] we estimate that the lateral/depth resolution is about 2 – 3 μm for materials with $\kappa \approx 1.5$-2 W/cm-K.

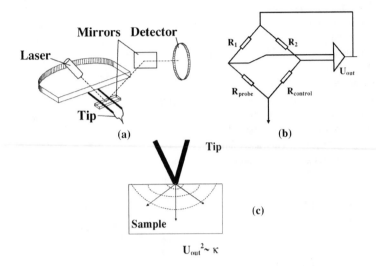

(a)   (b)

$U_{out}^2 \sim \kappa$

Fig. 1   (a) Schematic diagram of the SThM probe, (b) circuit diagram of the probe control, and (c) heat flow from the tip into the sample.

Table I   Summary of t, n_H/n_R, μ and κ of the two sets of samples.

Sample	t (μm)	n_H/n_R (10^{16} cm^{-3})	μ (cm^2/V-s)	κ (W/cm-K)
A1	74±5	6.3/-	921	1.82±0.1
A2	57±4	6.9/-	850	1.95±0.1
A3	23±2	24/-	633	1.26±0.07
A4	6±1	44/35	207	1.23±0.07
A5	13±2	190/150	300	0.76±0.04
A6	5±1	800/500	841	0.80±0.04
B1	10±1	14/-	470	1.62±0.08
B2	10±1	15/-	418	1.68±0.08
B3	10±1	16/-	485	1.65±0.08
B4	10±1	59/50	423	1.72±0.08
B5	10±1	100/85	369	1.38±0.07
B6	10±1	197/140	309	1.12±0.06
B7	10±1	300/250	276	1.10±0.05

EXPERIMENTAL RESULTS

Shown in Table I are the measured values of κ at 300K for samples A1-A6 and B1-B7. Note that samples A1 and A2 with $n = 6.3 \times 10^{16}$ cm^{-3} and $n = 6.9 \times 10^{16}$ cm^{-3} have $\kappa = 1.82 \pm 0.05$ W/cm-K and 1.95 ± 0.05 W/cm-K, respectively; the latter is the highest value of this parameter observed to date. Plotted in Figs. 2a and 2b are κ as a function of log n for the two sets of samples, respectively. Representative error bars are shown. The solid lines in the figure are least-square fits to a linear function. For both sets of samples $\kappa(n)$ is essentially the same.

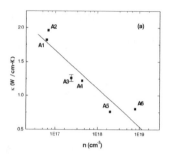

DISCUSSION OF RESULTS

From kinetic theory the lattice κ is given by [2]:

$$\kappa(T)=(1/3)v_s c(T)\ell(T)=(1/3)v_s^2 c(T)\tau(T) \qquad (1)$$

where v_s is the average velocity of sound (with only a weak temperature dependence), $c(T)$ is the lattice specific heat, $\ell(T)$ is the phonon mean free path, and $\tau(T)$ is the lifetime.

In almost all materials $\kappa(T)$ first increases with temperature, reaches a maximum (κ_{max}) at some characteristic temperature T_{ch}, and then decreases [4]. At low temperatures ℓ is relatively long and is dominated by extrinsic effects such as "defects"

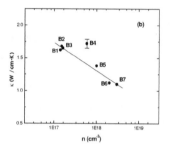

Fig. 2 Thermal conductivity as a function of carrier concentration, n, for samples (a) A1-A6 and (b) B1-B7.

and/or finite crystal size and $c(T) \propto (T/\Theta_D)^3$, where Θ_D is the Debye temperature. As the temperature increases $c(T)$ begins to saturate and intrinsic temperature dependent Umklapp processes become dominant, thus causing a decrease in ℓ. For GaN $T_{ch} \approx 200$K [3] and $\Theta_D \approx 600$K [14].

For $T < T_{ch}$ κ is very sensitive to "defect" density but still has some dependence in a range of T above T_{ch}. Since at 300K we are close to T_{ch}, the thermal conductivity will still be a function of n.

The observed dependence of κ on log n, as shown in Fig. 2, is difficult to account for in detail since we are in a regime where both extrinsic and intrinsic scattering processes are important. The scattering of the phonons from the impurity atoms can be described on the basis of mass-difference scattering, the relaxation time being given by [2]:

$$\tau_{imp}^{-1} \propto n(1 - M_{imp}/M_{av}) \qquad (2)$$

where M_{imp} is the mass of the impurity atoms and M_{av} is the average mass of the atoms in the material. Thus there will certainly be a dependence of κ on n. However, because of the Umklapp contribution, phonon scattering from the free carriers, and the contribution of the free carriers to the electronic component of κ; the particular function form is not immediately evident. Clearly more work in this area needs to be done.

Slack has estimated an upper bound of $\kappa \approx 1.7$ W/cm-K for GaN at 300K from the relation [5]:

$$\kappa = BM_{av}\delta\Theta_D^3/T\gamma^2 \tag{3}$$

where B is a constant, δ is the average volume occupied by one atom in the crystal, and γ is the Grüneisen parameter. By using the factor $M_{av}\delta\Theta_D^3$ as a scaling parameter he deduced the above value of κ for GaN at 300K. However, this analysis is limited since the above expression is applicable only for T $\gg \Theta_D$ ($\approx$ 600K in GaN).

Our observation also helps to explain the observed values of $\kappa \approx 1.7$-1.8 W/cm-K on the LEO material [6], which had $n \approx (10$-$20)$x10^{16} cm^{-3} [7] (see Fig. 2).

Certain GaN devices, such as high power field effect transistors, laser diodes, etc., would benefit greatly from GaN with higher thermal conductivity, as heat extraction from the device becomes more efficient with higher κ. Also, GaN has many potential applications in the area of high temperature electronics, where a large κ is very advantageous [15]. Our highest observed value of $\kappa \approx 1.95$-1.85 W/cm-K is somewhat smaller than single crystal AlN ($\approx$2.85 W/cm-K) [16] and is considerably higher than that of sintered AlN material [5]; the latter is often used as a heat sink material. Thus GaN based devices could be fabricated on HVPE GaN/sapphire material with the above thermal conductivities, thus avoiding costly processing steps.

SUMMARY

The doping dependence of the room temperature κ of two series of HVPE n-GaN/sapphire (0001) has been measured using a SThM. κ decreased linearly with log n, the variation being about a factor two decrease in κ for every decade increase in n in both sets of samples. The general behavior of $\kappa(n)$, i.e., decrease with increasing n, is similar to other semiconductors in a comparable temperature range. For one of the most lightly doped samples ($n = 6.9$x10^{16} cm^{-3}) $\kappa \approx 1.95$ W/cm-K, higher than the previously reported κ on several LEO samples and a thick HVPE material. The decrease in the lattice component of κ due to increased phonon scattering from both the impurities and free electrons outweighs the increase in the electronic contribution to κ. The implications of these findings for device applications and design are discussed.

ACKNOWLEDGEMENTS

The Brooklyn College work was supported by Office of Naval Research contract N00014-99-C-0663 (administered by Dr. Colin Wood) and the New York State Science and Technology Foundation through its Centers for Advanced Technology program. The Lincoln Laboratory work was sponsored by the US Air Force under Air Force contract #F19628-95-C-002. Opinions, interpretations, conclusions, and recommendations are those of the authors and are not necessarily endorsed by the US Air Force.

REFERENCES

[1]   See, for example, *Group III Nitride Semiconductor Compounds*, ed. by B. Gil (Clarendon, Oxford, 1998).

[2]   See, for example, *Thermal Conduction in Semiconductors* by C.M. Bhandari, and D.M. Rowe (Wiley, New York, 1988)

[3]   E.K. Sichel and J.I. Pankove, J. Phys. Chem. Solids **38**, 330 (1977).

[4]   V.A. Asnin, F.H. Pollak, J. Ramer, M. Schurman, and I. Ferguson, Appl. Phys. Lett. **75**, 1240 (1999).

[5]   G.A. Slack, J. Phys. Chem Solids **34**, 321 (1973).

[6]   ThermoMicroscopes, 1171 Borregas Avenue, Sunnyvale, CA 94089.

[7]   M. Pophristic, F.H. Long, M. Schurman, J. Ramer, and I.T.Ferguson, Appl. Phys. Lett. **74**, 3519 (1999).

[8]   R.J. Molnar, W. Gotz, L.T. Romano, and N.M. Johnson, J.Crystal Growth **178**, 147 (1997).

[9]   P.Perlin, J. Camassel, W. Knap, T. Taliercio, J.C. Chevin, T. Suski, I. Gregory, and S. Porowski, Appl. Phys. Lett. **67**, 2524 (1995).

[10]   M. Suzuki, and T. Uenoyama in *Group III Nitride Semiconductor Compounds*, ed. by B. Gil (Clarendon, Oxford, 1998), p. 307.

[11]   G. Popovici, H. Morkoç, and S.N. Mohammad in *Group III Nitride Semiconductor Compounds*, ed. by B. Gil (Clarendon, Oxford, 1998), p. 19.

[12]   F.Ruiz,W.D. Sun, F.H. Pollak, and C.Venkatraman, Appl. Phys. Lett. **73**, 1802 (1998).

[13]   A. Hammiche, H.M. Pollack, M. Song, and D.J. Hourston, Meas. Sci. Technol. **7**, 142 (1996).

[14]   I. Akasaki, and H. Amano in *Properties of Group III Nitrides*, ed. by J.H. Edgar (INSPEC, London, 1994) p. 30.

[15]   J.T. Torvik, M. Leksono, J.I. Pankove, and B. Van Zeghbroeck, MRS Internet J. Nitride Semicond. Res. **4**, 3 (1999).

[16]   G.A. Slack, R.A. Tanzilli, R.O. Pohl, and J.W. Vandersande, J. Phys. Chem. Solids **48**, 641 (1987)

Mat. Res. Soc. Symp. Vol. 595 © 2000 Materials Research Society

High Temperature Hardness of Bulk Single Crystal GaN

I. Yonenaga, T. Hoshi and A. Usui[1]
Institute for Materials Research, Tohoku University, Sendai 980-8577, Japan
[1]Opto-electronics and High Frequency Device Research Laboratories,
NEC Corporation, Tsukuba 305-8501, Japan

ABSTRACT

The hardness of single crystal GaN (gallium nitride) at elevated temperature is measured for the first time and compared with other materials. A Vickers indentation method was used to determine the hardness of crack-free GaN samples under an applied load of 0.5N in the temperature range 20 - 1200°C. The hardness is 10.8 GPa at room temperature, which is comparable to that of Si. At elevated temperatures GaN shows higher hardness than Si and GaAs. A high mechanical stability for GaN at high temperature is deduced.

INTRODUCTION

GaN (gallium nitride) and its alloys are promising as a wide band-gap, high temperature semiconducting material for application as blue- and ultraviolet-light-emitting devices, and high power/high frequency devices [1]. GaN materials are grown on various foreign substrates with large lattice mismatch and significant differences of thermal expansion coefficient. Such heteroepitaxial structures are inevitably accompanied by the introduction of various kinds of extended defects such as dislocations, prismatic stacking faults, etc. [2]. It is well recognized that the reduction in density of threading defects does lead to improvement of optical and electrical properties of GaN [3]. Thus, efforts are being made in controlling film stress to reduce the occurrence of detrimental defects during GaN hetero-epitaxial growth to obtain high performance devices. Understanding of dislocation behavior in this material is also indispensable.

The obstacle to conducting such an investigation has been the difficulty in growing bulk GaN crystal of a suitable size for deformation test. A few investigations are available on dislocation motion or hardness of GaN at room temperature. Drory et al. [4] measured the hardness and fracture toughness of GaN by a conventional hardness test. Nowak et al. [5] recently evaluated the yield strength of GaN by nano-indentation. Maeda et al. [6] observed the viscous motion of dislocations under electron irradiation using a transmission electron microscope operating at 200kV. Yonenaga [7] presumed empirically that the activation energy for dislocation motion in GaN is presumed to be 2 eV.

This paper reports the hardness of GaN at elevated temperature in comparison with those of other typical materials such as Si, GaAs, etc. As far as we know, this is the first information on the mechanical strength of GaN at elevated temperature and is important as a basis for the control of dislocation generation during crystal growth.

EXPERIMENT

GaN single crystals were prepared from a high quality GaN thick film grown on a 2-inch diameter (0001) sapphire substrate by using hydride vapour phase epitaxy (HVPE) together with the selective growth through SiO_2 windows by means of the facet initiated epitaxial lateral overgrowth (FIELO) technique. The details on the growth procedure are described elsewhere [8]. Finally, the thick grown layer was removed from the substrate. As a result, crack-free GaN single crystals of 0.5 mm thickness with mirror-like surfaces were successfully obtained and were sufficiently thick to be regarded as being bulk material. The density of grown-in dislocations was as low as 10^7 cm^{-2} [9]. Hardness measurements on the crystals were carried out by the Vickers indentation method using a pyramidal diamond indentor. The applied indentation load P was 0.5 - 5 N, and the dwell time 30 seconds, for every temperature tested in the range from room temperature to 1200°C in a high purity Ar gas atmosphere. Four indents were formed at every temperature for the Ga (0001) and N (000$\overline{1}$) basal plane surfaces.

RESULTS AND DISCUSSION

Figure 1(a) and 1(b) show an optical micrograph of the indent, formed for the N (000$\overline{1}$) surface of GaN at room temperature (RT) and at 200°C, respectively. Indents formed for the basal plane surfaces of the samples showed fracture characteristics for brittle materials with a pattern of radial cracks propagating from the indent corners at temperatures lower than 100°C, while more characteristic deformation without cracks occurred at elevated temperature as seen in figure 1(b).

Hardness H_v was estimated from the load P and diagonal lengths $2a$, measured by optical microscopy, of the indent using the following relation:

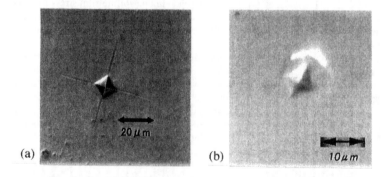

(a) (b)

Figure 1. Optical micrograph of Vickers indent of bulk single crystal GaN (a) at RT with an applied load of 1 N and dwell time of 30 s and (b) at 200°C with 0.5 N and 30 s.

$$H_v = P/(2a^2). \tag{1}$$

The fracture toughness K_c was also determined from the radial crack length c:

$$K_c = \xi (E/H_v)^{1/2}(P/c^{3/2}),$$ (2)

where E is the Young's modulus and ξ is a calibration constant ($= 0.016$) for brittle materials.

The hardness is comparable for the Ga (0001) and N (000$\bar{1}$) polar surfaces at all the temperatures investigated and is almost independent of or decreases slightly with an increase in the applied load.

At RT the hardness of GaN is estimated to be 10.8 GPa under the applied load 0.5 - 3 N, about twice the value of GaAs, which is similar to that 12 GPa reported by Drory et al. [4]. In their experiments the applied load was 2 N. Moreover, the fracture toughness is estimated to be 1.1 MPa√m under the applied load range 0.5 - 5 N using Eq. (2) with $E = 295$ GPa recently reported [5]. Table I summarises the hardness of various semiconductor (111) or (0001) surfaces, together with that of sapphire (0001) surfaces [10], at RT.

Figure 2 shows the hardness H_V of GaN, obtained with an applied load of 0.5N and dwell time of 30s, plotted against reciprocal temperature in comparison with other material (111) or (0001) surfaces. In the whole temperature range investigated, the hardness of GaN shows a gradual decrease from RT to 500°C, then something of a plateau in the range 500 - 1000°C and then a steep decrease. Such the temperature dependence is similar to that of 6H-SiC and sapphire, with the similar hcp-based structure, although the temperature range and hardness magnitudes of SiC or sapphire are higher than those of GaN. The plateau may appear in relation to the operation of different slip systems in the crystal structure. It is found that GaN is harder than GaAs in the whole temperature range investigated and that at temperatures lower than 600°C, the hardness of GaN is comparable to, or a little lower than, that of Si. Surprisingly, up to about 1100°C, GaN maintains its hardness and is harder than Si. Indeed, Si and GaAs show a steep decrease in hardness from 500°C and 200°C, respectively, with increasing temperature, which means the beginning of macroscopic dislocation motion

Table I. Vickers hardness of GaN and various semiconductors at room temperature or 300°C with an applied load of 0.5 N and dwell time of 30 s. The hardness of sapphire obtained with 2 N and 15 s is also included [10].

Material	Hardness (GPa)
GaN (0001)	10.2
GaAs (111)	6.8
Si (111)	12.0
6H-SiC (0001)	22.9 (300°C)
Al₂O₃ (0001)	28 (Ref. 10)

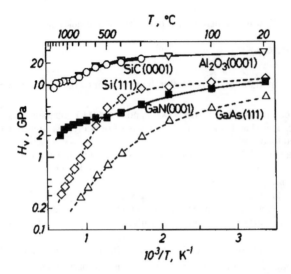

Figure 2. Vickers hardness of bulk single crystal GaN plotted against reciprocal temperature, with an applied load of 0.5 N and dwell time of 30 s, together with those of Si, GaAs, and 6H-SiC. The hardness of sapphire obtained with 2 N and 15 s is superimposed [10].

and plastic deformation. Thus, the result for GaN implies that such macroscopic dislocation motion and plastic deformation may start at around 1100°C.

The present results imply that GaN has a higher mechanical stability during device processing at high temperatures than Si, GaAs, and possibly other III-V compounds with the sphalerite structure. Unfortunately, a more complete physical understanding of hardness in order to derive the dynamic properties of dislocations, is still lacking in absence of sufficient theory and is a task in the future. However, we recognise that the present data provide a useful measure of material strength at elevated temperature. Further work is needed to determine dislocation mobilities within GaN bulk crystals under defined stress distribution.

CONCLUSION

A Vickers hardness for crack-free bulk single crystal GaN was determined in the temperature range 20 - 1200°C. The hardness of GaN is c10.8 GPa at room temperature and is omparable to that of Si at temperatures lower than 600°C. Up to about 1100°C, GaN maintains its harness, being mechanically stable in comparison with Si and GaAs.

ACKNOWLEDGEMENT

The authors are grateful to Prof. P. Pirouz of Casewestern Reserve University for the supply of SiC crystals.

REFERENCES

1. S. Nakamura, M. Senoh, S. Nagahama, N. Iwasa, T. Yamaha, T. Matsushita, H. Kiyohu, and Y. Sugimoto, *Jpn. J. Appl. Phys.*, **35**, L74 (1996).
2. P. D. Brown, in Proceedings of the 8th International Conference Defect-Recognition, Imaging and Physics in Semiconductors, Narita, Japan, September, 1999 (to be published).
3. S. Keller, B. P. Keller, Y. -E. Wu, B. Heying, D. Kapolnek, J. S. Speck, U. K. Mishra, and S. P. Den-Baars, *Appl. Phys. Lett.*, **68**, 1525 (1996).
4. M. D. Drory, J. W. Ager III, T. Suski, I. Grzegory, and S. Porowski, *Appl. Phys. Lett.*, **69**, 4044 (1996).
5. R. Nowak, M. Pessa, M. Suganuma, M. Leszczynski, I. Grzegory, S. Porowski, and F. Yoshida, *Appl. Phys. Lett.*, **75**, 2070 (1999).
6. K. Maeda, K. Suzuki, M. Ichihara, S. Nishiguchi, K. Ono, Y. Mera and S. Takeuchi, in Proceedings of the 20th International Conference on Defects in Semiconductors, Berkeley, USA, June, 1999 (to be published).
7. I. Yonenaga, *J. Appl. Phys.*, **84**, 4209 (1998).
8. A. Usui, H. Sunakawa, A. Sakai, and A. A. Yamaguchi, *Jpn. J. Appl. Phys.*, **36**, L899 (1997).
9. A. Sakai, H. Sunakawa, and A. Usui, *Appl. Phys. Lett.* **71**, 2259 (1997).
10. B. Ya. Farber, S. Y. Yoon, K. P. D. Lagerlöf, and A. H. Heuer, *Phys. Stat. Sol.* (a) **137**, 485 (1993).

Electronic Transport and Devices

Mat. Res. Soc. Symp. Vol. 595 © 2000 Materials Research Society

AlGaN/GaN High Electron Mobility Transistor Structure Design and Effects on Electrical Properties

E.L. Piner, D.M. Keogh[†], J.S. Flynn[†], and J.M Redwing[‡]
Epitronics / ATMI, 21002 North 19th Avenue, Suite 5, Phoenix, AZ 85027
[†]ATMI Ventures / ATMI, 7 Commerce Drive, Danbury, CT 06810
[‡]Current Address: Penn State University, Dept. of Materials Science and Engineering, University Park, PA 16802

ABSTRACT

We report on the effect of strain induced polarization fields in AlGaN/GaN heterostructures due to the incorporation of Si dopant ions in the lattice. By Si-doping (Al)GaN, a contraction of the wurtzite unit cell can occur leading to strain in doped AlGaN/GaN heterostructures such as high electron mobility transistors (HEMTs). In a typical modulation doped AlGaN/GaN HEMT structure, the Si-doped AlGaN supply layer is separated from the two-dimensional electron gas channel by an undoped AlGaN spacer layer. This dopant-induced strain, which is tensile, can create an additional source of charge at the AlGaN:Si/AlGaN interface. The magnitude of this strain increases as the Si doping concentration increases and the AlN mole fraction in the AlGaN decreases. Consideration of this strain should be given in AlGaN/GaN HEMT structure design.

INTRODUCTION

Strain induced piezoelectric polarization is a critical factor in the creation of the two-dimensional electron gas (2DEG) at the strained AlGaN / relaxed GaN interface in a high electron mobility transistor (HEMT) structure[1,2]. This strain is a result of the lattice mismatch between $Al_xGa_{1-x}N$ and GaN (x is typically between 0.15 and 0.30). To further enhance the sheet carrier concentration of the 2DEG, a Si-doped AlGaN layer may be grown that is typically separated from the channel by an undoped AlGaN spacer layer. An additional $1E13cm^{-2}$ sheet density of carriers can be obtained by doping the AlGaN at 0.5-$5E19cm^{-3}$ Si. This is about a factor of two increase over an undoped AlGaN/GaN HEMT structure.

By doping with Si, a perturbation of the lattice occurs due to the substitution of the smaller Si ion on the column III lattice site. This creates an additional contraction of the AlGaN wurtzite unit cell thus creating strain at the AlGaN:Si donor layer / AlGaN spacer layer interface. A similar effect in bulk GaN:Si epilayers grown on sapphire has been observed.[3] However, in this case, the effect of the strain from the Si-dopant was a source for stress relaxation due to the inherent compressive strain associated with GaN-on-sapphire caused by the differences in lattice constant, as well as, thermal coefficients of expansion between GaN and α-sapphire. A determination of the strain associated with incorporating Si as a substitutional dopant and the effects on the HEMT structure design are discussed.

EXPERIMENTAL

For an $Al_xGa_{1-x}N$:Si layer pseudomorphically strained on a relaxed $Al_xGa_{1-x}N$ layer, the strain, $\varepsilon_\perp$, is given by,

$$\varepsilon_\perp = 2x_{Al}\left(\frac{a_{(Al)GaN}}{a_{(Al)GaN:Si}} - 1\right), \tag{1}$$

where $a_{(Al)GaN}$ is the a-axis lattice constant for AlN content in the range of 0 – 100%. Vegard's Law of linear interpolation is assumed. $a_{(Al)GaN:Si}$ is the corresponding a-axis lattice constant at a given Si doping concentration. Table 1 summarizes the lattice constants as a function of Si-doping for GaN:Si and AlN:Si determined by substituting the atomic percent Si associated with the doping level for Ga and/or Al on the column III lattice site of the wurtzite unit cell. 100% activation of the Si dopant was assumed. The undoped GaN and AlN a-axis lattice constants were taken to be 3.1892 and 3.112A, respectively.[4]

Additional calculations were performed to determine the strain associated with AlGaN on a relaxed GaN layer. The equation for strain, in this case, would be,

$$\varepsilon_\perp = 2x_{Al}\left(\frac{a_{GaN}}{a_{AlN}} - 1\right) \cong 0.0495x_{Al}, \tag{2}$$

where a_{GaN} and a_{AlN} are the a-axis lattice constants that were defined earlier and x is the AlN mole fraction which ranges from 0 – 1. No relaxation of the AlGaN or AlGaN:Si was assumed. This would be the case for AlGaN layers below the critical thickness for relaxation.

RESULTS

Figure 1 illustrates the strain associated with doping $Al_xGa_{1-x}N$ with Si for doping concentrations of 5E18, 1E19, and 5E19cm^{-3} and as a function of the Al content ($0{\leq}x{\leq}1$). As would be expected, the strain is a strong function of the Si-doping concentration and increases with increasing doping. It should be noted that this strain is tensile since the Si-dopant creates a smaller unit cell compared to the equivalent Al mole fraction $Al_xGa_{1-x}N$

Table 1. Lattice constants (a-axis) determined for GaN and AlN at Si-doping concentrations of 5E18, 1E19, and 5E19cm^{-3}.

Si-doping conc. (cm^{-3})	GaN:Si	AlN:Si
5E18	3.1891	3.1119
1E19	3.1890	3.1117
5E19	3.1882	3.1111

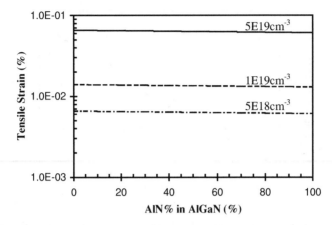

Figure 1. Si-doping induced strain as a function of mole fraction AlN in
AlGaN. Si doping levels are as noted on the graph.

undoped unit cell. (This strain is similar in nature to adding additional Al to an undoped
AlGaN in, for instance, an AlGaN/GaN HEMT structure.)

Figure 1 also indicates a slight decrease in the strain as the AlN mole fraction increases.
This is a result of the AlN unit cell being smaller than the GaN unit cell and, thus, the Si
substitutional impurity has less of an effect on the AlN lattice compared to the GaN
lattice. The Si-doping strain increases with: 1. Increasing Si concentration and 2.
Reducing AlN content of the AlGaN. The strain associated with doping GaN with
5E19cm^{-3} Si is 0.066% and is tensile.

DISCUSSION

Of greater interest to HEMT structure design is how this effect manifests itself in a
strained AlGaN / relaxed GaN system. Figure 2 is a graph of the strain associated with
pseudomorphically strained AlGaN on relaxed GaN. If the AlGaN is doped with Si, the
effect of the pseudomorphic and doping-induced strain is cumulative and increases the
overall strain as shown by the dashed line of figure 2. The strain from incorporating
5E19cm^{-3} Si into the lattice is equivalent to increasing the AlN content in AlGaN by
~1.4%. As the Si-doping level decreases, the dashed line would move closer to the solid
line and the doping-induced strain would decrease as shown in figure 1.

In a modulation doped AlGaN/GaN HEMT, it is common practice to incorporate a thin
undoped AlGaN spacer layer between the GaN and AlGaN:Si to spatially remove the
ionized donors from the channel. The resulting structure is schematically shown in figure
3. A Ga-faced crystal structure is assumed which is the common polarity in metalorganic
chemical vapor phase epitaxially grown GaN.[5] However, a similar argument could be
made for a N-faced crystal.

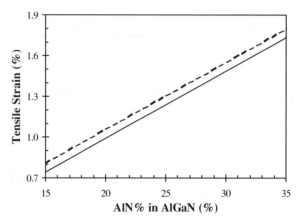

Figure 2. Calculated strain for undoped AlGaN (solid line) and AlGaN:Si at a doping concentration of $5E19cm^{-3}$ (dashed line).

Figure 3 indicates the formation of the 2DEG at the AlGaN spacer layer / relaxed GaN layer interface. The sheet carrier concentration of the 2DEG depends on both the spontaneous and piezoelectric polarization associated with the AlGaN/GaN heterostructure. The additional lattice mismatch associated with the Si-doping will result in a second source of strain localized at the AlGaN:Si donor layer / AlGaN spacer layer interface. This strain manifests itself by the creation of an additional piezoelectric polarization in the AlGaN:Si donor layer. Due to the tensile nature of this strain, the resulting sheet charge will be positive. This additional sheet charge may accumulate at the AlGaN:Si/AlGaN interface resulting in a second 2DEG channel. The piezoelectric polarization associated with $5E19cm^{-3}$ Si is $\sim0.35E12cm^{-2}$ which is an additional $1 - 4\%$ of the polarization induce sheet charge generated by the $Al_xGa_{1-x}N/GaN$ heterostructure (x is in the range of 0.15-0.35).

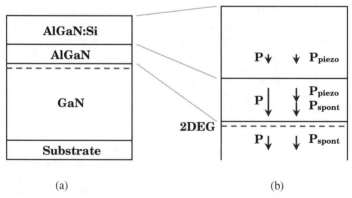

(a)   (b)

Figure 3. (a) modulation doped AlGaN/GaN HEMT structure and (b) the corresponding polarization fields. (Ga-faced crystal is assumed.)

CONCLUSIONS

Si-doping (Al)GaN causes a contraction of the wurtzite unit cell and can be a source of strain in doped AlGaN / GaN heterostructures. In a modulation doped AlGaN/GaN HEMT structure, in which the Si-doped AlGaN supply layer is separated from the 2DEG channel by an undoped AlGaN spacer layer, dopant-induced strain can create an additional source of charge at the AlGaN:Si/AlGaN interface. The magnitude of this strain increases as the Si doping concentration increases and the AlN mole fraction in the AlGaN decreases. The effect on the strain due to adding additional Si has a much greater effect than reducing the Al content in the AlGaN. Consideration of this strain should be given in AlGaN/GaN HEMT structure design.

ACKNOWLEDGMENTS

Support for this research was provided by the U.S. Army Ballistic Missile Defense Organization and the Air Force.

REFERENCES

1. P.M. Asbeck, E.T. Yu, S.S. Lau, G.J. Sullivan, J. Van Hove, and J.M. Redwing, Electron. Lett. **33**, 1230 (1997).
2. E.T. Yu, G.J. Sullivan, P.M. Asbeck, C.D. Wang, D. Qiao, and S.S. Lau, Appl. Phys. Lett. **71**, 2794 (1997).
3. C.S. Kim, D-K Lee, C-R Lee, S.K. Noh, I-H Lee, and I-H Bae, Mater. Res. Soc. Symp. Proc. **482**, 567 (1998).
4. *Properties of Group III Nitrides*, edited by J.H. Edgar (INSPEC, London, 1994).
5. O. Ambacher, J. Smart, J.R. Shealy, N.G. Weimann, K. Chu, M. Murhpy, W.J. Schaff, L.F. Eastman, R. Dimitrov, L. Wittmer, M. Stutzmann, W. Rieger, and J. Hilsenbeck, J. Appl. Phys. **85**, 3222 (1999).

Mat. Res. Soc. Symp. Vol. 595 © 2000 Materials Research Society

Negative Differential Conductivity in AlGaN/GaN HEMTs: Real Space Charge Transfer from 2D to 3D GaN States?

J. Deng, R. Gaska, M. S. Shur, M. A. Khan[1], J. W. Yang[1]
Department of ECSE and CIEEM, Rensselaer Polytechnic Institute
Troy, New York 12180, USA
[1]Department of ECE, University of South Carolina
Columbia, South Carolina 29208, USA

ABSTRACT

We report on non-thermal negative differential conductivity (NDC) in AlGaN/GaN HEMTs grown on sapphire substrates by low-pressure MOCVD. The sheet electron density was on the order of few times 10^{12} cm^{-2} and the Hall mobility was 1,000 cm^2/V.s. The HEMTs had threshold voltage close to zero and could operate at high positive gate bias up to 3 to 3.5 Volts, with a very low gate leakage current. NDC was observed at the gate bias larger than 1.5V and at the drain biases between approximately 0.5Vg and Vg. We excluded the possibility of self-heating as the cause, since the NDC occurs at relatively small power levels where self-heating effects are negligible.

An explanation we provided for the NDC effect is the new mechanism of real space charge transfer from 2D to 3D GaN states, which leads to a decrease in the channel mobility at large 2D electron gas densities. The observed low leakage can be explained by an enhanced molar fraction of aluminum at the heterointerface that results in a larger conduction band discontinuity. Our model that accounts for the piezoelectric and pyroelectric effects is consistent with the observed NDC effect. The Hall mobility dependence on the gate bias and sheet carrier concentration [1] is consistent with the real space transfer mechanism.

This NDC effect in GaN/AlGaN HEMTs may find applications in high-performance digital circuits at elevated temperatures.

INTRODUCTION

Recently, there has been much progress in the research and development of AlGaN/GaN High Electron Mobility Transistors (HEMTs). However, the physics of these novel devices has not yet been fully understood. In this paper, we report on a new mechanism of the negative differential mobility in AlGaN/GaN HEMTs.

Negative differential conductance (NDC) in HEMTs was previously observed [2-3]. However, in these publications, the NDC was attributed to either gate leakage current [2] or self-heating effect [3]. In [2], it was claimed that the two dimensional electron gas was heated in the channel. This heating caused the "real-space transfer" [4] between the GaAs conducting layer and the AlGaAs barrier layer. Self-heating might also lead to NDC [3] since with the maximum source-drain current of 1A/mm, the GaN devices grown on sapphire biased at 5V could have a temperature rise of 125K, which would significantly reduce the mobility and the maximum charge density of the channel.

In our samples, these mechanisms do not play an important role for the following reasons. Self-heating is not important since the power dissipation in our samples is only on the order of 0.5W/mm. For an estimated impedance of 25W/mm [3], this power dissipation should lead to the temperature rise of 12K, which is not enough to cause NDC.

As shown below, piezoelectric and pyroelectric effects in AlGaN/GaN heterostructures hinde the real space transfer into the AlGaN layer. However, as we discuss below, real space transfer int the 3D states in GaN is a possible mechanism explaining the observed NDC.

EXPERIMENTS

The $Al_{0.2}Ga_{0.8}N$/GaN HEMTs were grown on sapphire substrates by low-pressure metal organic chemical vapor deposition (MOCVD). The epilayer design of the samples was described ir [5]. The measured 2DEG sheet density was on the order of $10^{12}cm^{-2}$ and the Hall mobility was 1,000 cm^2/V.s.

Figure 1(a) shows the current-voltage characteristics of the 45 μm wide AlGaN-GaN HEMT. The source-drain spacing L_{ds} was 5 μm and the gate length L_g was 1.7μm. The applied gate bias was as high as 3.0 V. The negative differential conductivity (NDC) was observed at the gate biases $V_g > 1.5$ V. The drain current increased with V_{ds} to the peak value, I_{peak}, determined by the gate bias, and then decreased until saturation current (I_{sat}) has been reached. For HEMTs with the drain to-gate spacing of 1 μm and 2 μm the peak drain current was achieved at the source-drain voltages of approximately $V_{peak} = 1.5$ V and $V_{peak} = 2.5$ V, respectively. The NDC was observed at the source-drain voltages $V_{peak} < V_{ds} < V_{valley}$, where V_{valley} was the drain bias when the NDC effect stopped. The drain current reached its saturation beyond V_{valley}. In Figure 1(b), we show the gate bias dependence of V_{peak}, V_{valley} and the peak-valley current ration $R=I_{peak}/I_{valley}$. This figure demonstrates that V_{valley} increased with the gate bias with a slope of about 1, while the increase rat of V_{peak} was much smaller. On the other hand, R also increased almost linearly with V_g.

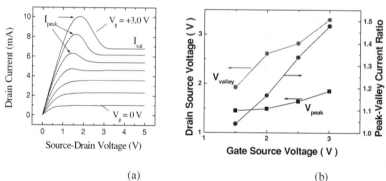

(a)   (b)

Figure 1. Drain current-voltage characteristics of the 45 μm wide AlGaN-GaN HEMT. (a) Dra current dependence on Vd, Vg varies from 0 to 3V with the step of 0.5V. (b) The V_{peak}, V_{valley} ar I_{peak}/I_{sat} as functions of the gate bias.

We measured NDC in the HEMTs with the source-drain spacing from 5μm to 15μm and the offset gate length from 2μm to 12μm. The source-to-gate and gate-to-drain distances were approximately the same and equal to 2μm and 1μm, respectively. Figure 2 demonstrates the normalized current drop as a function of the gate length for 95 μm wide HEMTs ($\Delta I_{neg} = I_{peak} - I_{sat}$) The obtained results clearly showed strong dependence of NDC on L_g. At the gate bias of +3.0 V, the increase in the gate length for about 6 times (source-drain spacing increases 3 times) reduces NDC by a factor of larger than 30. The saturation current for the same gate lengths decreases linearly only by one third, from 12 mA to 8 mA. The ΔI_{neg} dependence on the gate length can be described by expression $\Delta I_{neg} \sim L_g^{-\alpha}$, where α is from 1.5 to 2.

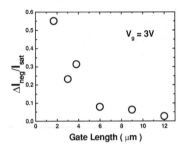

Figure 2. Gate length dependence of the current drop in NDC (normalized to the I_{sat})

DISCUSSION

The gate current versus drain voltage at different gate biases for the device in Figure 1 was measured. (See also [6]). Even for the zero source-drain voltage, the gate leakage current at high positive gate bias ($V_g = +3.0$ V) was less than 0.5 mA, whereas the drain current reduction in the NDC region for the same gate bias was more than 3mA (see Figure 1(a)). At the source-drain voltage of 1.5 V and higher (NDC region) the gate current dropped below 1μA. These results indicated that the gate current of the device were negligible compared to ΔI_{neg}.

As we mentioned before, the device had a maximum drain current on the order of 0.1~0.2A/mm, which would only raise the temperature by 10~20K at a thermal impedance of 25 K·mm/W [3]. Also, the self-heating NDC led to a monotonous decrease of the drain current in the saturation regime, whereas the NDC reported here took place in a limited drain voltage range (see Figure 1(a)). Therefore ,we concluded that both the gate current and self-heating were not the mechanism leading to the NDC in our samples.

The explanation we provided for the NDC was the mobility and sheet carrier density decrease due to the real space transfer from the 2D to 3D GaN states. First, we compared the mobility of the two-dimensional and the three-dimensional electron gas. The results are shown in Figure 3. Even at room temperature, the mobility of 2D-electron gas is around 50% larger than the mobility of 3D electrons. This difference is much larger at cryogenic temperatures.

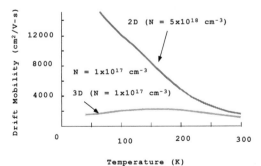

Figure 3. Calculated 2D and 3D electron mobility in GaN at different temperatures

In Figure 4, we compared the band diagram of HEMTs in different material systems – GaN-based and GaAs-based. Due to the piezoelectric and pyroelectric effects, the conduction band profile in the AlGaN layer does not have a minimum that is needed for the real space transfer from GaN into AlGaN. Also, in contrast to AlGaAs/GaAs system, the energy difference between the Fermi level in the 2D gas and the bottom of the conduction band in the GaN is relatively small due to the residual doping in GaN. This small difference made the electron transition into the GaN bulk more important.

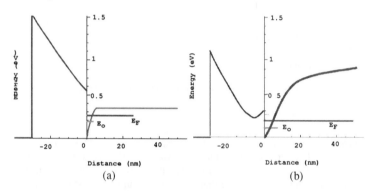

Figure 4. Comparison between the band diagram of GaN based HEMTs and GaAs based HEMTs (a) Calculated band diagram of GaN HEMTs at the doping level of 10^{17}cm^{-3} (b) Qualitative band diagram of GaAs HEMTs

Hence, we believe that this transfer of 2D electrons into 3D GaN states reduces the channel mobility and the 2D-electron sheet concentration and leads to the current drop in the NDC region. The Monte-Carlo simulation results for AlGaAs/GaAs HEMTs reported in [7] also pointed out to the possibility of such a mechanism (see Figure 5). The dash-dotted line in Figure 5 shows the electron concentration in the quantum well and the full lines show the bulk GaAs electron concentration. This figure demonstrates that the electrons inside the quantum well experience the transition into the GaAs buffer along the channel. However, due to the low barrier height of the GaAs transistors (Figure 4(b)), the gate current becomes dominant at higher gate biases, which makes it difficult to observe the NDC related to such a transfer in GaAs devices at room temperature.

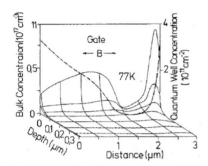

Figure 5. Monte Carlo simulation results on electron concentration in GaAs based HEMTs. Dashed dotted line, concentrations in a quantum well; full line, concentrations in bulk GaAs. After [7]

In order to illustrate this new mechanism of NDC, we calculated the current-voltage characteristics of our devices for different values of the electron mobility using the AIM-Spice [8], which helped to understand the device operation. In Figure 6(a), a mobility of 750 cm^2/V.s was used and reasonable agreement was reached in the low drain bias region. The mobility dropped to 400 cm^2/V.s in Figure 6(b), which give reasonable fit in the saturation region. The maximum sheet charge density for these two regions was also different. (See Table 1) These simulation results confirm that a mobility drop at the higher gate bias might be the cause of the NDC effect, as illustrated in Figure 3.

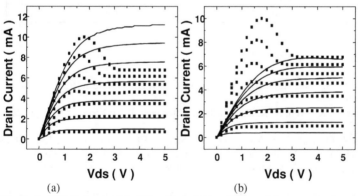

(a) (b)

Figure 6. AIM-Spice model of the above device. The model parameters are in Table 1. The symbols are the experimental results and the lines are from simulation.

CONCLUSIONS

In this paper, we report the non-thermal negative differential conductance in AlGaN/GaN HEMTs. With the help of device characterization and modeling, we excluded the possibility of self-heating and real space transfer into the AlGaN layer current as the causes of the NDC. We linked the NDC to the mobility reduction caused by the transfer from 2D to 3D states in the channel. This mechanism was supported by the device simulation and band diagram calculations.

ACKNOWLEDGEMENTS

The work has been supported by the Office of Naval Research (Project Monitor Dr. Colin Wood and Dr. John Zolper). We are also grateful to Dr. Knap and Dr. Roumiantsev for their useful discussions.

Table 1. AIM-Spice parameters for Figure 6.

Parameters	2D channel	3D channel	Unit
W_g (Gate width)	45	45	μm
L_g (Gate length)	1.7	1.7	μm
D_i (Distance to buffer layer charge)	30	30	nm
λ (Output conductance parameter)	0	0	V^{-1}
μ (Low-field mobility)	0.075	0.04	$m^{-2} V^{-1} s^{-1}$
n_{max} (Sheet charge density)	$1 \bullet 10^{17}$	$4 \bullet 10^{16}$	m^{-2}
Φ_b (Heterojunction barrier height)	1.1	1.1	V
R_s (Drain series resistance)	30	30	Ω
R_d (Source series resistance)	30	30	Ω
v_s (Saturation velocity)	40000	40000	m/s
M (Knee voltage parameter)	2	2	-
V_T (Zero-bias threshold voltage)	-0.6	-0.6	V

REFERENCES

1. R. Gaska, M. S. Shur, A. D. Bykhovski, A. O. Orlov, and G. L. Snider, Appl. Phys. Lett. 74, 287 (1999)
2. M. S. Shur, D. K. Arch, R. R. Daniels, J. K. Abrokwah, IEEE Electron Device Lett. 7, 78 (1986)
3. R. Gaska, A. Osinsky, J. W. Yang and M. S. Shur, IEEE Electron Device Lett. 19, 89 (1998).
4. K. Hess, H. Markoc, H. Shichijo and B. G. Streetman, Appl. Phys. Lett. 35(b), 469 (1979)
5. R. Gaska, Q. Chen, M. Asif Khan, A. Ping, I. Adesida and M. S. Shur, Electron Lett. 33, 125, (1997)
6. J. Deng, R. Gaska, M. S. Shur, M. A. Khan, J. W. Yang, submitted to Appl. Phys. Lett.
7. D. Widiger, K. Hess, J. J. Coleman, IEEE Electron Device Lett. 5, 266 (1984)
8. T. A Fjeldly, T. Ytterdal and M. Shur, Introduction to Device Modeling and Circuit Simulation (1997)

Mat. Res. Soc. Symp. Vol. 595 © 2000 Materials Research Society

Two-Dimensional Electron Gas Transport Properties in AlGaN/(In)GaN/AlGaN Double-Heterostructure Field Effect Transistors

Narihiko Maeda, Tadashi Saitoh, Kotaro Tsubaki, Toshio Nishida and Naoki Kobayashi
NTT Basic Research Laboratories, Physical Science Laboratory,
3-1 Morinosato Wakamiya, Atsugi-shi, Kanagawa, 243-0198, Japan

Two-dimensional electron gas transport properties have been investigated in nitride double-heterostructures. A striking effect has been observed that the two-dimensional electron gas mobility has been drastically enhanced in the AlGaN/GaN/AlGaN double-heterostructure, compared with that in the conventional AlGaN/GaN single-heterostructure. The observed mobility enhancement has been shown to be mainly due to the enhanced polarization-induced electron confinement in the double-heterostructure, and additionally due to the improvement of the interface roughness in the structure. Device operation of an AlGaN/GaN/AlGaN double-heterostructure field effect transistor has been demonstrated: a maximum transconductance of 180 mS/mm has been obtained for a 0.4 µm-gate-length device. In the double-heterostructure using InGaN channel, the increased capacity for the two-dimensional electron gas has been observed. The AlGaN/(In)GaN/AlGaN double-heterostructures are effective for improving the electron transport properties.

INTRODUCTION

AlGaN/GaN heterostructure field effect transistors (HFETs) have recently been attracting much attention because of their promising uses for high-voltage, high-power, and high-temperature microwave applications [1-7]. Improving device performances requires an understanding of how we can increase the mobility and the density of the two-dimensional electron gas (2DEG) in this material system. Desirable designs of heterostructures should be investigated for superior transport properties.

In nitride heterostructures with wurtzite crystal structures in (0001) orientation, there exist strong polarization effects, i.e., (i) the piezoelectric polarization effect which depends on the lattice strain [8], and (ii) the spontaneous polarization effect which is determined by the constituent materials of the heterostructure [11]. These polarization effects largely influence the electrical properties in the heterostructures such as the potential profile and the electron density [8-16].

The 2DEG transport properties in nitride double-heterostructure field effect transistors (DH-FETs) are very interesting, because influence of the polarization effects should be enhanced in DH-FETs, compared with that in conventional single-heterostructure field effect transistors (SH-FETs). In SH-FETs, only the positive polarization charges emerge at one heterointerface; on the other hand, positive and negative polarization charges emerge at two heterointerfaces in DH-FETs, so that the electric field in the channel should be increased in DH-

FETs. In this paper, we report on the 2DEG transport properties in DH-FETs and clarify their differences from those in SH-FETs. We propose AlGaN/(In)GaN/AlGaN DH-FETs are effective in enhancing the 2DEG confinement due to the polarization effects, and furthermore effective in increasing the 2DEG mobility.

EXPERIMENTAL DETAILS AND DISCUSSION

Comparison of GaN DH and GaN SH

To examine the influence of the enhanced polarization effects in DH-FETs on the transport properties, we have fabricated both conventional AlGaN/GaN SH-FETs and AlGaN/GaN/AlGaN DH-FETs, and have compared their electrical properties.

The sample structures of (a) conventional GaN SH-FETs and (b) GaN DH-FETs are (a) $Al_{0.15}Ga_{0.85}N$ (300 Å) /GaN (1 μm), and (b) $Al_{0.15}Ga_{0.85}N$ (300 Å) /GaN (200 Å) /$Al_{0.15}Ga_{0.85}N$ (1000 Å), respectively. The samples were grown on SiC(0001) substrates using AlN buffer layers by metalorganic vapor-phase epitaxy (MOVPE) at 300 Torr. In all samples, only the surface-side 300 Å $Al_{0.15}Ga_{0.85}N$ barrier layers were uniformly doped with Si.

We have performed Van der Pauw Hall effect measurements on the GaN SH- and GaN DH-FET samples, both of which are doped with the same Si doping concentration of 4×10^{18} cm^{-3}. Figure 1 shows the temperature dependencies of the 2DEG mobility and the 2DEG density from 20 to 380 K in the two samples. The striking feature shown in Fig. 1 is a drastic increase in the 2DEG mobility in the DH-FET sample. The 2DEG mobility in DH-FET reaches 8900 cm^2/Vs at 20 K, whereas that in SH-FET reaches 4600 cm^2/Vs. The 2DEG densities at 20 K in DH- and SH-FET samples are 6.5×10^{12} and 7.5×10^{12} cm^{-2}, respectively. This difference in the 2DEG density is not so large as to explain the observed large difference in the 2DEG mobility. Hence, the observed mobility enhancement in DH-FET is a unique phenomenon specific to nitride HFETs.

To clarify the dependencies of the 2DEG mobility on the 2DEG density in DH- and SH-FETs, we have performed Hall effect measurements under the gate-voltage application. In this experiment, we have used DH- and SH-FET samples with the Si doping concentration of 5×10^{18} cm^{-3}. The grown structures were mesa-etched to form a 400X400 μm^2 rectangular Van der Pauw geometry device, by using low-damage reactive ion etching (RIE) technique using chlorine-nitrogen mixed plasma [17]. The Al/Au ohmic metals were deposited on the corners of the rectangular mesa structure, and the Ni/Au Schottky gate metals was deposited on almost all the area of the device except for the ohmic terminals.

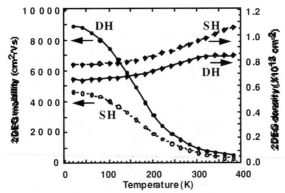

Figure 1. *Temperature dependencies of 2DEG mobility and density in GaN SH- and DH-FETs.*

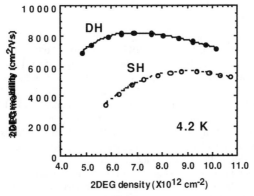

Figure 2. *Dependencies of 2DEG mobility on 2DEG density in GaN SH- and DH-FETs.*

For the above devices, we have performed Hall effect measurement at 4.2 K under the gate voltage (V_g) application. The V_g was stepped from 0 to -2.5 V in -0.25 V increments. In this V_g region, the 2DEG density has shown to be almost linearly controlled by V_g both in DH- and SH-FET samples.

Figure 2 shows the dependence of the 2DEG mobility on the 2DEG density in the DH- and SH-FET samples at 4.2 K. One feature observed in Fig. 2 is that the 2DEG mobility in DH-FET is larger than that in SH-FET. The enhancement in the mobility is very large especially at low 2DEG densities, which is consistent with the result shown in Fig. 1. Another feature observed in Fig. 2 is the shift of the 2DEG mobility peak position. In SH-FET, the maximum 2DEG mobility of 5700 cm^2/Vs is obtained at 9.0×10^{12} cm^{-2}; whereas in DH-FET, the maximum mobility of 8100 cm^2/Vs is obtained at a lower 2DEG density of 6.8×10^{12} cm^{-2}.

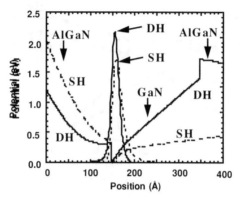

Figure 3. *Potential profiles and 2DEG distributions in GaN SH- and DH-FETs. The 2DEG dinsity is 6.0X10^{12} cm^{-2} in both structures. The Al composition is 0.15.*

To clarify the mechanism of the mobility enhancement in DH-FET, we have calculated the potential profiles and 2DEG distributions both in the SH- and DH-FETs by self-consistently solving the Schrodinger and Poisson's equations. Both piezoelectric and spontaneous polarization effects have been taken into account. Figure 3 shows the result calculated for the same 2DEG concentration of 6X10^{12} cm^{-2}. The 2DEG confinement is enhanced in DH-FET due to the enhanced electric field in the DH channel. Thus, the strongly confined 2DEG and the increased electron concentration in DH-FET lead to enhance the 2DEG mobility, as the result of (i) the enhanced screening effect against the ionized impurity scattering both in the channel and from the barrier layer, and (ii) the enhanced screening effect against the piezoelectric scattering in the channel. Figure 3 also explains why the 2DEG mobility in DH-FET assumes the maximum value at a lower 2DEG density. In DH-FET, the energy barrier height of the surface-side AlGaN layer is shown to be lower as the result of the enhanced electric field in the DH channel. Hence, electrons overflow into the AlGaN layer at the lower critical 2DEG density where the 2DEG mobility assumes the maximum value. This circumstance is ascribed to the feature mentioned above.

The observed features in Fig. 2 can be thus explained by the enhanced 2DEG confinement effect in DH-FET. However, we presume that some additional mechanisms should also have contributed to the mobility enhancement, since the observed enhancement is very large. We suppose that the improvement of the interface roughness in DH-FETs should have additionally contributed to the mobility enhancement, because surface morphology is observed to be improved in DH-FETs.

We demonstrate device operation of a GaN. The mesa isolation of the devices was performed by RIE. Ti/Al/Ti/Au and Ni/Au layers were used as the ohmic and Schottky metals, respectively. The gate length and width of the fabricated DH-FET were 0.4 and 20 μm, respectively. Figure 4 shows the I-V characteristics of the device. The maximum transconductance (g_m) was estimated to be 180 mS/mm. It should be noted that sufficient pinch-off characteristics have been obtained even in DH-FETs.

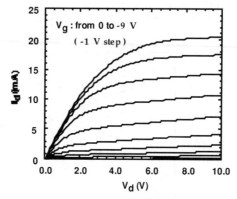

Figure 4. *I-V characteristics in GaN DH-FET. The gate length and gate width is 0.4 and 20 μm, respectively. The maximum transconductance is 180 mS/mm.*

Comparison of InGaN DH and GaN DH

It is of great interest in nitride DH-FETs to use InGaN channels instead of GaN channels, because the capacity for the 2DEG density is expected to be increased in InGaN DH-FETs as the result of the enhanced polarization effect (i.e., increased polarization charges) and the enlarged conduction-band discontinuity at the AlGaN/InGaN heterointerface. We compare the 2DEG mobilities and densities in GaN and InGaN DH-FETs. The $In_{0.06}Ga_{0.94}N$ layer is used as the channel in the InGaN DH-FET.

Figure 5 plots the 2DEG mobility data as a function of the 2DEG density, measured at 77 K in GaN and InGaN DH-FET samples. Here, two different Si doping concentrations (4×10^{18} and 6×10^{18} cm^{-3}) are used in both GaN and InGaN DH-FET samples. At lower 2DEG densities in Fig. 5, the 2DEG mobility in InGaN DH-FET is as high as that in GaN DH-FET: a high mobility of 7600 cm^2/Vs is obtained at 77 K for the 2DEG density of 7.6×10^{12} cm^{-2}. Thus, the 2DEG mobility is enhanced also in the DH-FET with the $In_{0.06}Ga_{0.94}N$ channel.

At higher 2DEG densities in Fig. 5, on the other hand, a distinct difference in the 2DEG mobility has been shown. In GaN DH-FET, a mobility of 4500 cm^2/Vs is obtained for a relatively high 2DEG density of 9.2×10^{12} cm^{-2}. In InGaN DH-FET, on the other hand, a mobility as high as 5600 cm^2/Vs is obtained for a high 2DEG density of 1.2×10^{13} cm^{-2}. Thus, the degradation of the 2DEG mobility along with the increase in the 2DEG density is smaller in InGaN DH-FET than in GaN DH-FET. This indicates that InGaN DH-FET has the larger capacity for 2DEG than GaN DH-FET. This is valid also at 300 K. Thus, high 2DEG mobilities and high 2DEG densities can be simultaneously obtained in InGaN DH-FETs.

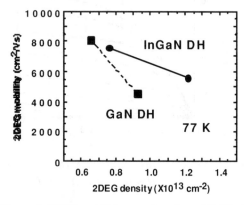

Figure 5. *2DEG mobilities as a function of 2DEG density in GaN and InGaN DH-FET samples. The In composition is 0.06.*

CONCLUSION

We have investigated two-dimensional electron gas transport properties in nitride double-heterostructures. A striking effect has been observed that the 2DEG mobility has been drastically enhanced in the AlGaN/GaN/AlGaN double-heterostructure, compared with that in the conventional AlGaN/GaN single-heterostructure. The observed mobility enhancement has been shown to be mainly due to the enhanced polarization-induced electron confinement in the double-heterostructure, and additionally due to the improvement of the interface roughness in the structure. Device operation of an AlGaN/GaN/AlGaN DH-FET has been demonstrated: a maximum transconductance of 180 mS/mm has been obtained for a 0.4 μm-gate-length device. In the double-heterostructure using InGaN channel, the increased capacity for the two-dimensional electron gas has been observed. The AlGaN/(In)GaN/AlGaN double-heterostructures are effective for improving the electron transport properties.

ACKNOWLEDGMENTS

The authors would like to thank Dr. Yoshiro Hirayama for his assistance in Hall effect measurements, Dr. Masaaki Tomizawa, Dr. Hirotaka Tanaka and Dr. Hideaki Takayanagi for their assistance in computer calculation, and Dr. Naoshi Uesugi and Dr. Sunao Ishihara for their encouragement throughout this work.

REFERENCES

1. M.A. Khan, Q. Chen, M.S. Shur, B.T. Mcdermott, and J.A. Higgins, IEEE Electron Device Lett. **17**, 325 (1996).
2. Y.F. Wu, B.P. Keller, S. Keller, D. Kapolnek, P. Kozodoy, S.P. Denbaars, and U. K. Mishra, Appl. Phys. Lett. **69**, 1438 (1996).
3. O. Aktas, Z.F. Fan, S. N. Mohammad, A. Botchkarev, and H. Morkoc, Appl. Phys. Lett. **69**, 3872 (1996).
4. S.C. Binari, J.M. Redwing, G. Kelner, and W. Kruppa, Electron. Lett. **33**, 242 (1997).
5. Q. Chen, J.W. Yang, R. Gaska, M.A. Khan, M.S. Shur, G.J. Sullivan, A.L. Sailor, J.A. Higgings, A.T. Ping, and I. Adesida, IEEE Electron Device Lett. **19**, 44 (1998).
6. Y.F. Wu, B.P. Keller, P. Fini, S. Keller, T.J. Jenkins, L.T. Kehias, S.P. Denbaars, and U.K. Mishra, IEEE Electron Device Lett. **19**, 50 (1998).
7. A.T. Ping, Q. Chen, J.W. Yang, M.A. Khan, and I. Adesida, IEEE Electron Device Lett. **19**, 54 (1998).
8. A. Bykhovski, B. Gelmont, and M. Shur, J. Appl. Phys. **74**, 6734 (1993).
9. A.D. Bykhovski, B.L. Gelmont, and M.S. Shur, J. Appl. Phys. **78**, 3691 (1995).
10. A.D. Bykhovski, B.L. Gelmont, and M.S. Shur, J. Appl. Phys. **81**, 6332 (1997).
11. F. Bernardini and V. Fiorentini, Phys. Rev. **B56**, 10024 (1997).
12. R. Gaska, J. W. Yang, A.D. Bykhovski, M.S. Shur, V.V. Kaminskii, and S. Soloviov, Appl. Pys. Lett. **71**, 3817 (1997).
13. R. Gaska, J.W. Yang, A.D. Bykhovski, M.S. Shur, V.V. Kaminskii, and S. Soloviov, Appl. Pys. Lett. **72**, 64 (1998).
14. P.M. Asbeck, E.T. Yu, S.S. Lau, G.J. Sullivan, J. Van Hove, and J. Redwing, Electron. Lett. **33**, 1230 (1997).
15. E.T. Yu, G.J. Sullivan, P.M. Asbeck, C.D. Wang, D. Qiao, and S.S. Lau, Appl. Phys. Lett. **71**, 2794 (1997).
16. N. Maeda, T. Nishida, N. Kobayashi, and M. Tomizawa, Appl. Phys. Lett. **73**, 1856 (1998).
17. T. Saitoh, H. Gotoh, T. Sogawa, and H. Kanbe, Mat. Res. Symp. Proc. **442**, 63 (1997).

Mat. Res. Soc. Symp. Vol. 595 © 2000 Materials Research Society

HIGH-TEMPERATURE RELIABILITY OF GaN ELECTRONIC DEVICES

Seikoh Yoshida and Joe Suzuki
Yokohama R&D Laboratories, The Furukawa Electric Co., Ltd.
2-4-3, Okano, Nishi-ku, Yokohama, 220-0073, Japan, seikoh@yokoken.furukawa.co.jp

ABSTRACT

High-quality GaN was grown using gas-source molecular-beam epitaxy (GSMBE). The mobility of undoped GaN was 350 cm^2/Vsec and the carrier concentration was 6×10^{16} cm^{-3} at room temperature. A GaN metal semiconductor field-effect transistor (MESFET) and an n-p-n GaN bipolar junction transistor (BJT) were fabricated for high-temperature operation. The high-temperature reliability of the GaN MESFET was also investigated. That is, the lifetime of the FET at 673 K was examined by continuous current injection at 673 K. We confirmed that the FET performance did not change at 673 K for over 1010 h. The aging performance of the BJT at 573 K was examined during continuous current injection at 573 K for over 850 h. The BJT performance did not change at 573 K. The current gain was about 10. No degradation of the metal-semiconductor interface was observed by secondary ion-mass spectrometry (SIMS) and transmission electron microscopy (TEM). It was also confirmed by using Si-ion implantation that the contact resistivity of the GaN surface and electrode materials could be lowered to 7×10^{-6} ohmcm^2.

INTRODUCTION

III-V nitrides, SiC and diamond, are very promising materials for electronic devices that can operate under high-temperature, high-power, and high-frequency conditions [1], since these materials have a high melting point, a wide bandgap, a high breakdown electric field, and a high saturation velocity [2]. Furthermore, they have very low on-resistance during operation compared with Si devices. Several groups have reported on GaN electronic devices [3-5]. Regarding high-temperature operation devices, there have only been a few reports concerning transistors that can operate at temperatures of 673 K [3-7], and the reliability of the GaN metal semiconductor field effect transistor (MESFET) at high-temperature. We have recently investigated the GaN MESFET using gas-source molecular beam epitaxy (GSMBE) [8-11]. However, until now, no life test of the FET by continuous current-injection at 673 K for over 1010 h has been reported.

Furthermore, there have been few reports concerning the bipolar junction transistor [12-15]. Pankov fabricated a hetero-bipolar junction transistor using n-type GaN and p-type SiC, since p-type GaN with a high carrier concentration was very difficult to grow. They obtained a high current gain and high-temperature operation at 533 K. We have also recently reported that a GaN n-p-n bipolar junction transistor can be operated at high temperature [15]. However, life test at 573 K for over 800 h has not been reported.

This paper reports on the high-temperature reliability concerning a life test of a GaN MESFET at 673 K for over 1010 h, a life test of over 850 h at 623 K of an n-p-n bipolar junction transistor and Si-ion implantation technique for obtaining a low-contact resistivity.

EXPERIMENT PROCEDURE

A GSMBE apparatus was described elsewhere [8]. Dimethylhydrazine (DMHy, $((CH_3)_2NNH_2)$) as a nitrogen source gas, and ammonia (NH_3) as a nitrogen source gas, as well as Knudsen effusion cells for a solid Ga source, Si as an n-type dopant, and Mg as a p-type dopant were also placed in this chamber. The buffer layers of GaN were grown using DMHy at substrate temperatures of 973 K, since DMHy was easily decomposed at lower temperatures (below 873 K) compared to that of ammonia. The Ga beam-equivalent pressure (BEP) was $5x10^{-7}$ Torr and the BEP of DMHy was $4x10^{-5}$ Torr. A thick film of undoped GaN was grown using Ga ($6x10^{-7}$ Torr) and ammonia gas ($5x10^{-6}$ Torr) on the GaN buffer layer at 1123 K. The BEP of Si was $5x10^{-9}$ Torr and that of Mg was $8x10^{-9}$ Torr. The growth rate was 500 nm/h. The carrier concentration of undoped GaN grown at 1123 K was $6x10^{16}$ cm^{-3} and the mobility was 350 cm^2/Vsec. The carrier concentration of Mg-doped GaN was $2{\sim}3x10^{17}$ cm^{-3} (p-type) and the mobility was about 20~30 cm^2/Vsec at room temperature. The activation ratio of Mg was smaller than 10^{-2} compared with the results of a secondary-ion mass-spectrometry (SIMS) analysis. In the case of Si doping, the n-type carrier concentration was controlled in the range from $1x10^{17}$ cm^{-3} to $5x10^{18}$ cm^{-3}. An etching technique using an electron cyclotron resonance (ECR) plasma was used for GaN device fabrications. Si ion implantation into GaN was carried out to obtain a lower contact resistivity. The structures of the MESFET and the bipolar junction transistor were made using GaNs grown by GSMBE.

RESULTS AND DISCUSSION

Si-ion implantation

The GaN surface layer with a Si concentration of over 10^{19} cm^{-3} is very important to obtain a very lower n-type contact resistivity. When we doped the Si with a carrier concentration of over 10^{19} cm^{-3} into the GaN layer using GSMBE, the surface morphology of GaN was very rough, and sometimes Si precipitates were observed on the grown GaN surface. Furthermore, when we formed an n-type ohmic electrode, the contact resistivity between the electrode materials and the GaN was very high. In our growth system, a high Si doping of over 10^{19} cm^{-3} into the GaN layer was very difficult. In order to reduce the contact resistivity, we investigated Si ion implantation instead of high Si doping based on growth. The Si-ion implantation acceleration voltage was 30 keV. The amount of Si dose was 1×10^{19} cm^{-3}. The depth of Si ion implantation was 10~30 nm. The substrate was heated at 873 K during Si-ion implantation. After ion implantation, GaN was annealed at 1123 K for 30 min an the ambient of nitrogen gas. Undoped GaN was used for implantation.

The sheet resistivity of undoped GaN without was 350 ohm/cm^2. After implantation, the sheet resistivity was reduced to 25 ohm/cm^2. Furthermore, the contact resistivity between the electrode materials and the GaN surface was measured. The n-

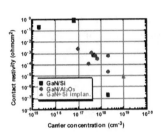

Figure 1. Contact resistivities versus GaN carrier concentrations.

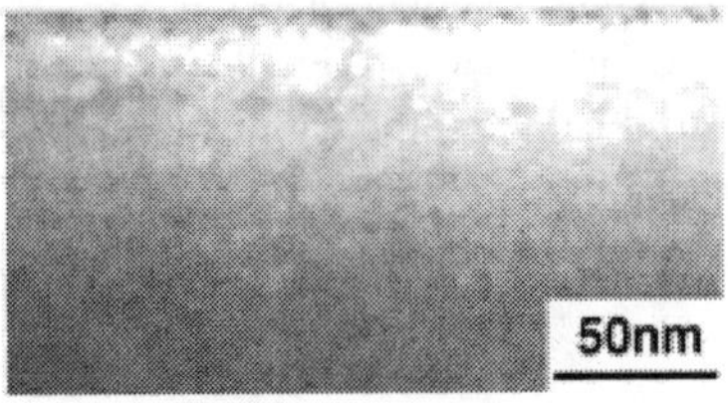

Figure 2. Cross-sectional TEM image of GaN after Si-ion implantation and annealing at 1123 K.

type ohmic electrode material was Al/Ti/Au (30nm/100nm/50nm). Before ion implantation, the contact resistivity of the undoped GaN was 3×10^{-3} ohmcm^2. After Si-ion implantation, the contact was 7×10^{-6} ohmcm^2 as shown in Figure 1. That is, it was confirmed that Si-ion implantation into the GaN layer was very effective for reducing the contact resistivity. Furthermore the cross-seat ion of Si-implanted GaN was observed by transmission electron microscopy (TEM). Figure 2 shows a TEM image of Si-implanted GaN after annealing at 1123 K. No damage based on Si-ion implantation was observed. Therefore, it was thus confirmed that Si-ion implantation is very effective for the contact resistivity.

GaN MESFET

The GaN MESFET structure is as follows [14]. A 50 nm-thick GaN buffer layer was grown on a sapphire substrate. An undoped 1000 nm-thick GaN layer was grown on an GaN buffer layer and an AlN layer was grown on the undoped GaN layer to isolate the active layer. Lastly, a 350 nm-thick Si-doped GaN active layer was grown on the AlN layer. The surface morphology of the grown GaN was smooth. The active layer had a carrier concentration of 3×10^{17} cm^{-3} and a mobility of 250 cm^2/Vsec at room temperature. Etching of the GaN was carried out by a dry-etching technique using an

ECR plasma to form the FET. The etching gas was a mixture of CH_4 (5 sccm), Ar (7 sccm), and H_2 (15 sccm) [10]. The etching rate of the Si-doped and undoped GaN layers was 14 nm/min. We formed a source and drain using Au/Ti/Al, and a Schottky-gate as Au/Pt on a patterned GaN sample using an ECR sputter-evaporation method, respectively. The gate length of the GaN MESFET was 2.5 μm and the gate width was 100 μm.

The FET property was obtained at room temperature. The breakdown voltage between gate-source was also about 80 V. The pinch-off voltage was about -8 V. The transconductanse (g_m) was about 25 mS/mm. Furthermore, to investigate the high-temperature reliability of a GaN MESFET, we carried out a life test of a FET by continuous current-injection at 673 K. The FET property at 673 K was measured by continuously injecting a current of I_{ds} at V_{ds}=20 V and V_{gs}=0 V. Figure 3 shows the aging time versus I_{ds} of GaN MESFET at 673 K. It should be noted that no change of I_{ds} was observed for over 1010 h. The g_m was also constant at 25 mS/mm. Figure 4 shows the characteristics of I_{ds} as a function of V_{ds} for increasing values of the gate-source voltage (V_{gs}) at 673 K. The pinch-off voltage was also about -8 V. The FET

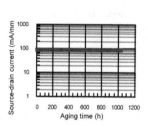

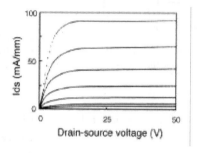

Figure 3. Aging property of I_{ds} of a GaN MESFET during continuous-current injection at 673 K.

Figure 4. Current-voltage characteristics (I_{ds}-V_{ds}) of a GaN MESFET at 673 K. The gate voltages (V_{gs}) was changed from 0 V to -5 V in steps of -1 V.

property was almost the same as that at room temperature, and good performance of the FET was maintained at 673 K. Using transmission electron microscopy (TEM), we also observed that the interfaces of the electrode materials and the GaN layer were not degraded.

Based on these results, the high-temperature reliability of a GaN MESFET was confirmed, since the electrode materials did not diffuse into the GaN, and the GaN layer was not deformed at 673 K.

GaN bipolar junction transistor

We investigated the reliability of a GaN bipolar junction transistor (BJT) at high temperature. The structure of the bipolar junction transistor using GaN is shown in Figure 5 The thicknesses of the emitter, base, and collector layers were 450 nm, 350 nm, and 500 nm, respectively. The carrier concentration of the emitter and collector was 5×10^{17} cm^{-3} and the base carrier concentration was 1.5×10^{17} cm^{-3}. The structure was also formed by ECR plasma etching using $CH_3/Ar/H_2$. The emitter size was 350x400 μm^2. The sizes of the base and collector were 200x150 μm^2 and 450x600 μm^2, respectively. The electrode materials of the emitter and collector were formed using Au/Ti/Al. The base contact was also formed using Au/Ti/Ni.

In order to investigate the operation of the bipolar junction transistor at 573 K, the current-voltage characteristics, as measured in the common emitter mode at 573 K, were measured. It was found that the bipolar junction transistor performance at 573 K remained unchanged, as shown in Figure 6, although I_c was slightly changed compared with that produced at room temperature. The current gain (dI_c/dI_b) was about 10. Figure 7 shows the p-n junction property between the base and the emitter electrodes during heating at 573 K. The breakdown voltage was over 10 V. Furthermore, a life test of a bipolar junction transistor was carried out. Figure 8 shows the result of an aging test at 573 K. Here, V_c was 4 V and I_b was 120 μA. Ic did not change for over 850 h. That is, it was confirmed that the bipolar junction transistor was operated for over 850 h. It was also found by SIMS that interfaces between the electrode materials and the GaN layer after heating at 573 K was abrupt and that the GaN was not affected by heating at 573 K. We thus confirmed the reliability of the GaN bipolar junction transistor at 573 K.

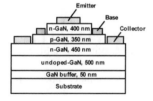

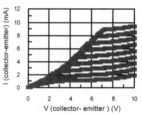

Figure 5. Schematic drawing of the structure of an n-p-n bipolar junction transistor using a GaN.

Figure 6. Current (I_c) - voltage (V_c) characteristics as measured in the common-emitter mode at 573 K. The base current (I_b) was changed from 20 μA in steps of 50 μA.

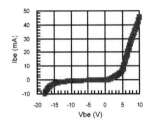
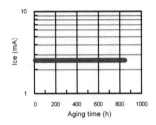

Figure 7. p-n junction property between of the base and emitter electrodes at 573 K.

Figure 8. Aging property of GaN BJT at 573 K.

CONCLUSION

A high-quality GaN was grown using GSMBE. A GaN MESFET and an n-p-n bipolar junction transistor were fabricated. The life test of a GaN MESFET at 673 K was examined by continuous current-injection at 673 K. We confirmed that the FET performance did not change at 673 K for over 1010 h. No degradation of the metal-semiconductor interface was observed by SIMS or TEM. Furthermore, the fabrication of a GaN bipolar junction transistor was carried out. The life performance of the bipolar transistor at 573 K was examined during continuous current-injection at 573 K. We confirmed that the performance of the bipolar transistor did not change at 573 K for over 850 h. No degradation of the metal-semiconductor interface was observed by SIMS or TEM. The reliability of a GaN BJT at high temperature was thus confirmed.

REFERENCES

1. T. P. Chow and R. Tyagi, IEEE Trans. Electron Devices **41**, 1481 (1994).
2. H. Morkoc, S. Strites, G. B. Gao, M. E. Lin, B. Sverdlov and M. Burns, J. Appl. Phys. **76**, 1363 (1994).
3. S. C. Binari, B. L Rowland, W. Kruppa, G. Kelner, K. Doverspike, and D. K. Gaskill, Electron. **30**, 1248 (1994).
4. M. A. Khan, M. S. Shur, J. N. Kuzunia, Q. Chin, J. Burm and W. Schaff, Appl. Phys. **66**, 1083 (1995).
5. A. Ozgur, W. Kim, Z. Fan, A. Botchkarev, A. Salvador, S. N. Mohammad, B. Sverdlov and H. Morko□, Electron. **31**, 1389 (1995).
6. O. Akutas, Z. F. Fan, S. N. Mohammad, A. E. Botchkarev and H. Morko□, Appl. Phys. **69**, 3872 (1996).
7. S. C. Binari, K. Doverspike, G. Kelner, H. B. Dietrich and A. E. Wickenden, Solid State Electron. **41**, 97 (1997).
8. S. Yoshida, J. Appl. Phys. **81**, 1673 (1997).
9. S. Yoshida, J. Cryst. Growth, **181**, 293 (1997).
10. S. Yoshida and J. Suzuki, Jpn. J. Appl. Phys. **37**, L482 (1998).
11. S. Yoshida and J. Suzuki, J. Appl. Phys. **84**, 2940 (1998).
12. J. Pankov, S. S. Chang, H. C. Lee, R. J. Molnar, T. D. Moustakas and B. Van

Zeghbroeck, IEDM Tech. Dig., 389 (1994).

13. L. S. Mc Carthy, P. Kozodoy, M. J. W. Rodwell, S. P. Denbaars and U. K. Mishra, IEEE Electron Device. Lett. **20**, 277 (1998).

14. S. Yoshida and J. Suzuki, J.Appl. Phys. **85**, 7931 (1999).

15. S. Yoshida and J. Suzuki, Jpn. J. Appl. Phys. **38**, L851 (1999).

Mat. Res. Soc. Symp. Vol. 595 © 2000 Materials Research Society

FABRICATION AND CHARACTERIZATION OF GaN JUNCTIONFIELD EFFECT TRANSISTORS

L. Zhang,* L. F. Lester,** A. G. Baca,* R. J. Shul,* P. C. Chang,* C. G. Willison,*
U. K. Mishra,*** S. P. Denbaars,*** and J. C. Zolper****
* Sandia National Laboratories, Albuquerque, NM 87185
** University of New Mexico, Albuquerque NM 87106
*** University of California, Santa Barbara, CA 93106
****Office of Naval Research, Arlington, VA 22217

ABSTRACT

Junction field effect transistors (JFET) were fabricated on a GaN epitaxial structure grown by metal organic chemical vapor deposition. The DC and microwave characteristics, as well as the high temperature performance of the devices were studied. These devices exhibited excellent pinch-off and a breakdown voltage that agreed with theoretical predictions. An extrinsic transconductance (g_m) of 48 mS/mm was obtained with a maximum drain current (I_D) of 270 mA/mm. The microwave measurement showed an f_T of 6 GHz and an f_{max} of 12 GHz. Both the I_D and the g_m were found to decrease with increasing temperature, possibly due to lower electron mobility at elevated temperatures. These JFETs exhibited a significant current reduction after a high drain bias was applied, which was attributed to a partially depleted channel caused by trapped electrons in the semi-insulating GaN buffer layer.

INTRODUCTION

Wide bandgap GaN and its related materials have great potential for high power and high temperature microwave electronic device applications owing to their low intrinsic carrier concentration, high breakdown field, high saturation velocity, and excellent chemical stability [1-3]. A significant amount of effort has been devoted to the development of various GaN-based field effect transistors (FETs) [4-8], which has led to the demonstration of high performance GaN/AlGaN MODFETs [9-12]. Compared to MESFETs and MODFETs, JFETs provide a higher gate voltage swing and a lower reverse gate leakage current due to a higher built-in potential of the p-n junction gate than the Schottky gate used in MESFETs and MODFETs. This is especially important for high temperature operation. In addition, the junction gate is metallurgically more stable and environmentally more robust than a Schottky gate since it is effectively buried beneath the surface and is subjected to high temperature during crystal growth. Therefore, JFETs are expected to withstand higher temperature operation than MESFETs and MODFETs. In this paper, we report the fabrication and characterization of epitaxially grown GaN JFETs.

MATERIAL AND FABRICATION

The epitaxial structure of GaN JFETs was grown by metal organic chemical vapor deposition (MOCVD) on a C-plane sapphire substrate. The layer structure consisted of a 4.2 µm semi-insulating GaN buffer layer, a 950 Å Si-doped n-GaN channel, a 60Å undoped GaN, and a 500 Å Mg-doped p-GaN. The sample was annealed at 850 °C for 15 sec in N_2 to activate the Mg dopant. Hall measurements showed a free carrier concentration of 1.3×10^{18}, 2.4×10^{18}, and 6.1×10^{14} cm^{-3} in the p-GaN, n-GaN, and SI-GaN, respectively. The electron mobility in the n-GaN active layer was 270 cm^2/V-sec. The fabrication process began with a mesa isolation etch in an inductively coupled BCl$_3$/Cl$_2$/Ar plasma. Next, a gate metal of Ni/Au/Ni was e-beam evaporated on top of the mesa and used as the mask for the self-aligned source-drain etch in a BCl$_3$/Cl$_2$/Ar ICP discharge. The device was completed with Ti/Al source and drain ohmic contact metallization. Post-metallization annealing was not performed. A transmission line method (TLM) measurement showed an as-deposited source and drain ohmic contact resistance of 4.2 Ω-mm, a specific contact resistance of 5×10^{-5} Ω-cm^2, and a sheet resistance of 4700 Ω/square, respectively. These values were relatively large, possibly due to plasma induced damage and an overetched source and drain region.

RESULTS AND DISCUSSION

Table I summarizes the DC and microwave result of a 0.8 μm X 50 μm GaN JFET with a source-drain spacing of 3 μm. A maximum I_D of 270 mA/mm and a maximum g_m of 48 mS/mm were measured at V_G=1 V. A R_S of 8.5 Ω-mm and a R_D of 13 Ω-mm were obtained using the end resistance measurement technique. From the above result, an intrinsic transconductance (g_{m0}) of 81 mS/mm was calculated. The channel was completely pinched off at a threshold voltage of V_G=-8 V, with an I_D=210 μA/mm at V_D=15 V. A gate-drain diode reverse breakdown voltage of 56 V was achieved, which corresponded to a breakdown field of 2.5×10^6 V/cm. The fact that the breakdown field approached the theoretical predicted breakdown field of GaN indicated that the relatively small breakdown voltage of the JFET as compared to the reported GaN MESFETs and MODFETs [4,12] was primarily due to the high doping concentration in the n-GaN. The gate leakage current of our JFETs was large as compared to GaN MODFETs [13,14], possibly due to the plasma induced damage to the junction [15, 16]. The forward turn-on voltage of the gate diode was ~ 1 V using a 1 mA/mm current criterion. This value is only 30% of the bandgap energy of GaN. Pernot, et. al. have also reported a low turn-on voltage of 1.2 V on GaN p-n diodes [17]. The cause of this low turn-on is not known at present but may result from defect levels in the GaN.

maximum drain current (I_D)	270 mA/mm
threshold gate voltage (V_T)	-8 V
knee voltage (V_{knee})	8 V
gate-drain diode breakdown voltage	56 V
gate turn-on voltage	~ 1 V
extrinsic transconductance (g_m)	48 ms/mm
parasitic source resistance (R_S)	8.5 Ω-mm
parasitic drain resistance (R_D)	13 Ω-mm
cut-off frequency (f_T)	6 GHz
maximum frequency (f_{max})	12 GHz

Table I: DC and microwave characteristics of a 0.8 μm X 50 μm GaN JFET.

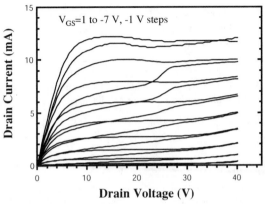

Figure 1: I_D-V_D characteristic showing the drain current collapse at $V_D < 25$ V due to a partially depleted channel by electrons trapped in the SI-GaN. The solid curves were measured individually with 5 min of illumination to release trapped electrons in the SI-GaN. The dashed ones were measured in rapid recession.

Similar to what Binari and co-workers have reported on GaN MESFETs [4], a significant reduction in I_D occurred in our JFETs after they were subjected to a high V_D. Figure 1 shows two sets of I_D-V_D curves measured up to V_D=40V under white-light illumination. The solid curves were measured individually with 5 minutes of illumination before each curve was taken while the dashed curves were measured in rapid succession. The upper dashed trace (V_G=1 V) is nearly identical to the upper solid trace since they were both subjected to a long illumination. The subsequent dashed curves exhibited a large decrease in I_D at V_D<25 V. This current collapse effect was not obvious unless V_D was increased above 20V. The reduction in I_D was caused by high-field injection and subsequent trapping of electrons in the SI-GaN, which depleted part of the active channel from the backside. We have demonstrated that these trapped electrons were located at the drain side of the channel where the electrical field was the highest [18]. At V_D<5 V, the transistor was below saturation and the current was limited by R_S, R_D, and the channel resistance, R_{CH}. Since the channel was partially depleted by the trapped electrons, a lower channel conductance was obtained for the dashed I_D-V_D curves as compared to the solid ones in Figure 1. Above the knee voltage, it was assumed that the channel consisted of a velocity-saturated section in parallel with a space-charge region caused by trapped electrons in the GaN buffer. The latter was evidenced by the nearly constant output conductance between the knee voltage and V_D=25 V. The rapid increase in I_D of the collapsed curves at V_D~25V may indicate a local breakdown of the space charge region. Therefore, regardless of illumination, I_D at V_D>25V was limited by an undepleted, velocity-saturated region on the drain side of the space charge layer.

The microwave performance of the JFETs was characterized using an HP 8510 network analyzer. A unity current-gain cutoff frequency (f_T) of 6 GHz and a maximum frequency (f_{max}) of 12 GHz were obtained at V_D=15 V and V_G=0 V from the small-signal S-parameters. These values were comparable to the reported f_T and f_{max} on GaN MESFETs and GaN/AlGaN MODFETs with a similar gate length [19, 20]. The intrinsic f_T of our JFETs would be higher by taking into account the large source and drain resistance.

Following the derivation made by Tasker and Hughes [21], we were able to calculate f_{T0} of the JFETs by de-embedding the effect of the parasitic R_S and R_D from measured f_T. From [21], the extrinsic cut-off frequency

$$f_T = \frac{g_{m0}/2\pi}{[C_{GS}+C_{GD}]\cdot[1+(R_s+R_D)/R_{DS}]+C_{GD}\cdot g_{m0}\cdot(R_s+R_D)}$$

where C_{GS}, C_{GD}, and R_{DS} are the gate-source, gate-drain capacitance, and output resistance, respectively. Rearranging this equation, the intrinsic cut-off frequency

$$f_{T0} \equiv \frac{g_{m0}}{2\pi(C_{GS}+C_{GD})} = f_T\cdot\left[1+\frac{R_s+R_D}{R_{DS}}+\frac{C_{GD}}{C_{GS}+C_{GD}}\cdot g_{m0}\cdot(R_s+R_D)\right]$$

where

$$C_{GS} = \frac{C_{gs0}}{\sqrt{1-V_{gs}/V_{bi}}}, \qquad C_{GD} = \frac{C_{gs0}}{\sqrt{1-V_{gd}/V_{bi}}}$$

respectively. C_{gs0} is the zero-bias gate-source capacitance, whereas $V_{gd}=V_G-I_DR_S$ and $V_{gd}=V_G-(V_D-I_DR_D)$ are the voltage-drop across the p-n junction on either side of the channel, respectively. From the above equations, we obtain

$$f_{T0} = f_T\cdot\left[1+\frac{R_s+R_D}{R_{Ds}}+\frac{g_{m0}\cdot(R_s+R_D)\cdot\sqrt{V_{bi}-(V_G-I_DR_S)}}{\sqrt{V_{bi}-(V_G-I_DR_s)}+\sqrt{V_{bi}-(V_G-V_D+I_DR_D)}}\right]$$

Since all the variables on the right hand side are directly measurable, this equation provides a simple method to extract the intrinsic f_{T0} from parasitic-limited FETs. An f_{T0} of 10 GHz was calculated for GaN JFETs using DC measurement results.

In order to estimate the ultimate speed that can be achieved in our JFETs, a calculation was made to simulate the change in g_m as L_g decreases. Figure 2 shows the calculated g_m vs. L_g curves for various saturation velocity (v_s). A 60% increase was obtained in g_m as the L_g reduced from 0.8 to 0.1 µm. By fitting the g_m of the GaN JFETs into this plot, a v_s of 6.6×10^6 cm/s was obtained. Notice that the gate capacitance decreases linearly as the gate length shrinks. In addition, the electron velocity can exceed the v_s in ultra-short gate length FETs due to velocity overshoot under high electrical field. Considering all these effects, an f_T in excess of 50 GHz could be expected in GaN JFETs upon further improving the device design and process techniques, and scaling down the gate length. Also shown in Figure 2 were the reported g_m values of GaN MODFETs and MESFETs. The data are somewhat scattered, possibly due to different material quality for different research groups. The g_m of GaN JFET is comparable to most of them, suggesting a reasonably good quality

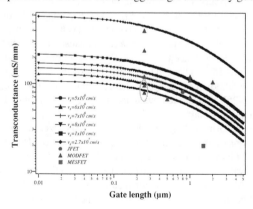

Figure 2*: Simulation result of g_m as a function of L_g for various v_s values. The transconductance of GaN JFETs and reported GaN MODFETs and MESFETs are also shown for comparison.*
of JFET material.

In the above analysis, the effect of non-ideal ohmic contact to p-GaN on the g_m and f_T has not been considered. For microwave operation, the large gate displacement current causes a significant voltage drop on the gate resistor, leading to a reduced g_m and f_T. In addition, the extra capacitance at metal-p-GaN interface caused by the Schottky-type Ni/Au to p-GaN contact results in a further reduction in the f_T. In order to optimize the performance of GaN JFETs, a low resistance metal contact to p-GaN with ideal ohmic I-V characteristic needs to be developed.

The DC performance of the GaN JFETs was also studied at elevated temperatures. The measurement was made on a hot plate in atmosphere. Figure 3 shows the plots of I_D and g_m as a function of V_G at different temperatures. A continuous deterioration in I_D and g_m was observed with increasing temperatures up to 200°C where the device failed after being heated in the air for ~2 hours. As the temperature was increased from 25°C to 200°C, I_D dropped from 270 mA/mm to 75 mA/mm and g_m was reduced from 48 mS/mm to 12 mS/mm, respectively. This was mainly caused by the reduction in electron mobility at elevated temperatures due to enhanced polar optical phonon scattering. However, the I_D-V_D curve exhibited an excellent pinchoff at a gate bias of –8 V, indicating a negligible gate leakage current. The failure of the JFETs after high temperature measurement was caused by the degradation of Ti/Al ohmic contact. An inspection of the devices after measurement showed a broken metal connection at the sidewall of the isolation mesa. The mesa profile was highly anisotropic, which resulted in a poor coverage of Ti/Al metal over the sidewall. Hence, the oxidation of Al upon prolonged heating in the air caused an open circuit.

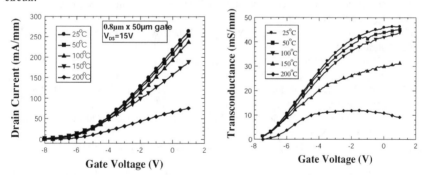

Figure 3*: Drain current and transconductance as a function of gate voltage and measurement temperature.*

CONCLUSION

In summary, JFETs were fabricated on an epitaxially grown GaN p-n junction. These devices exhibited excellent pinch-off and a breakdown voltage that agreed with theoretical predictions. An extrinsic transconductance of 48 mS/mm was achieved with a maximum I_D of 270 mA/mm. Drain current collapse was observed in these devices after a high drain bias was applied due to partially depleted channel by the trapped electrons in the SI-GaN. The microwave measurement showed an f_T of 6 GHz and an f_{max} of 12 GHz. A simple method was developed to extract f_{T0} from parasitic-limited FETs. An f_{T0} of 10 GHz was calculated for GaN JFETs using DC and rf results.

ACKNOWLEDGEMENT

This work was supported in part by the National Science Foundation CAREER Grant ECS-9501785. Sandia is a multiprogram laboratory operated by Sandia Corporation, a Lockheed Martin Company, for the United States Department of Energy under contract DE-ACO4-94AL85000.

References

1. V. A. Dmitriev, K. G. Irvine, C. H. Carter Jr., I. Kuznetsov, and E. V. Kalinina, Appl. Phys. Lett., **68**, 229 (1996).
2. B. Gelmont, K. Kim, and M. S. Shur, J. Appl. Phys., **74**, 1818 (1993).
3. S. N. Mohammad, A. Salvador, and H. Morkoç, Proc. IEEE **83**, 1306 (1995).
4. S. C. Binari, W. Kruppa, H. B. Dietrich, G. Kelner, A. E. Wickenden, and J. A. Freitas JR, Solid-St. Electron., **41**, 1549 (1997).
5. S. Yoshida and J. Suzuki, J. Appl. Phys., **84**, 2940 (1998).
6. U. K. Mishra, Y. F. Wu, B. P. Keller, S. Keller, and S. P. Denbaars, IEEE Trans. Microwave Theory Tech., **46**, 756 (1998).
7. M. A. Khan, Q. Chen, M. S. Shur, B. T. Dermott, J. A. Higgins, J. Burm, W. J. Schaff, and L. F. Eastman, Solid-St Electron., **41**, 1555 (1997).
8. J. C. Zolper, R. J. Shul, A. G. Baca, R. G. Wilson, S. J. Pearton, R. A. Stall, Appl. Phys. Lett., **68**, 2273 (1996).
9. C. H. Chen, S. Keller, G. Parish, R. Vetury, P. Kozodoy, E. L. Hu, S. P. Denbaars, U. K. Mishra, and Y. F. Wu, Appl. Phys. Lett., **73**, 3147 (1998).
10. Y. F. Wu, B. P. Keller, D. Kapolnek, P. Kozodoy, S. P. Denbaars, and U. K. Mishra, Appl. Phys. Lett., **69**, 1438 (1996).
11. Y. F. Wu, B. P. Keller, P. Fini, J. Pusl, M. Le, N. X. Nguyen, C. Nguyen, D. Widmen, S. Keller, S. P. Denbaars, and U. K. Mishra, Electron. Lett., **33**, 1742 (1997).
12. L. F. Eastman, K. Chu, W. Schaff, M. Murphy, N. G. Weimann, and T. Eustis, MRS Internet J. Nitride Semicond. Res., **2**, art. 17 (1997).
13. S. C. Binari, J. M. Redwing, G. Kelner, and W. Kruppa, Electron. Lett., **33**, 242 (1997).
14. Y. F. Wu, B. P. Keller, S. Keller, D. Kapolnek, P. Kozodoy, S. P. Denbarrs, and U. K. Mishra, Solid-St. Electron., **41**, 1569 (1997).
15. A. Osinski, S. Gangopadhyay, B. W. Lim, M. Z. Anwar, M. A. Khan, D. V. Kuksenkov, and H. Temkin, Appl. Phys. Lett., **72**, 742 (1998).
16. J. M. Van Hove, R. Hichman, J. J. Klaassen, P. P. Chow, and P. P. Ruden, Appl. Phys. Lett., **70**, 2282 (1997).
17. C. Pernot, A. Hirano, H. Amano, and I. Akasaki, Jpn. J. Appl. Phys., **37**, L1202 (1998).

18. L. Zhang, L. F. Lester, A. G. Baca, R. J. Shul, P. C. Chang, C. G. Willison, U. K. Mishra, S. P. Denbaars, and J. C. Zolper, IEEE Electron Device Lett., accepted.
19. O. Aktas, Z. F. Fan, A. Botchkarev, S. N. Mohammad, M. Roth, T. Jenkins, L. Kehias, and H. Morkoç, IEEE Electron Device Lett., **18**, 293 (1997).
20. Y. F. Wu, B. P. Keller, S. Keller, D. Kapolnek, S. P. Denbarrs, and U. K. Mishra, IEEE Electron Device Lett., **17**, 455 (1996).
21. P. J. Tasker and B. Hughes, IEEE Electron Device Lett., **10**, 291 (1989).

Electronic and Structural
Characterization

Mat. Res. Soc. Symp. Vol. 595 © 2000 Materials Research Society

The atomic structure of extended defects in GaN

P. Ruterana, and G. Nouet
Laboratoire d'Etudes et de Recherches sur les Matériaux , UPRESA 6004
CNRS, Institut des Sciences de la Matière et du Rayonnement, 6 Bd Maré-chal
Juin, 14050 Caen Cedex, France. email: ruterana@lermat8.ismra.fr

ABSTRACT

GaN layers contain large densities (10^{10} cm^{-2}) of threading dislocations, nanopipes, (0001) and $\{11\bar{2}0\}$ stacking faults, and $\{10\bar{1}0\}$ inversion domains. Three configurations have been found for pure edge dislocations, mainly inside high angle grain boundaries where the 4 atom ring cores can be stabilized. Two atomic configurations, related by a $1/6 < 10\bar{1}0 >$ stair rod dislocation, have been observed for the $\{11\bar{2}0\}$ stacking fault in (Ga-Al)N layers. For the $\{10\bar{1}0\}$ inversion domain boundaries, a configuration corresponding to the Holt model was observed, as well as another with no N-N or Ga-Ga bonds.

INTRODUCTION

Due to their direct bandgap, III-V nitrides are excellent candidates for optoelectronic application from red to ultra-violet (1.89 eV : InN, 6.2 eV : AlN). However the fabrication of devices is encountering intrinsic problems which are now motivating a world wide research effort. This has been exponentially increasing for the last ten years since the discovery of Mg for p doping[1]. Light emitting diodes were available from 1993 and laser emission was demonstrated at the end of 1995, all in layers containing up to 10^{10} cm^{-2} extended defects[2,3]. The large majority of the extended defects are threading dislocations which originate from the interface with the substrate and mostly cross the whole epitaxial layer[4-6]. Planar boundaries on the $\{11\bar{2}0\}$ prismatic planes have also been investigated and have led to some controversial reports. They have been characterized as double positioning boundaries[7] stacking mismatch boundaries[8] and inversion domain boundaries[9], as well as stacking faults[10-12]. In this report, we discuss our results on the atomic structure of the extended defects obtained in (Ga, Al) N layers grown by molecular been epitaxy on SiC and sapphire in the light of recent literature.

EXPERIMENTAL DETAILS

The investigated GaN layers were grown on the (0001) sapphire or 6H-SiC surfaces by electron cyclotron resonance (ECR) or NH$_3$ gas source MBE. HREM experiments were carried out along the $[11\bar{2}0]$ and [0001] GaN directions on a Topcon 002B microscope with a resolution of 0.18 nm (Cs = 0.4 mm).

CRYSTALLOGRAPHIC CONSIDERATIONS

Along the **c** axis, this misfit is quite small in the case of 6H-SiC, whereas it is close to 20 % on sapphire. Whereas wurtzite is a stacking of two interpenetrating hcp lattices, the 6H packing of SiC is more complex as it is deduced from fcc by twinning every three sequences. Recent work has shown that steps at the (0001) 6H-SiC connect terraces by a displacement vector which is either zero, or equal to one of the three stacking fault vectors of the hexagonal lattice[10]. When this vector has a component along [0001], an equivalent translation will exist between islands grown on the adjacent terraces.

For growth on (0001) sapphire, the situation is more complex. A simple geometrical analysis shows that even on a flat surface, there are eight possibilities for the growth of a GaN layer. They are either related by displacement vectors and/or an inversion operation[13]. Using 5c GaN ~ 2c sapphire, were identified four types of steps which lead to residual translations upon growth on adjacent terraces (table 1). The induced translation is better minimized by the formation of inversion domains on top of hexagonal terraces rather than prismatic stacking faults[14,15].

Steps	$h/c_{Al_2O_3}$ unit	T_{R}/c_{GaN} unit	T_R/nm
A-A or B-B	nc (n: integer)	0	0
A-B	1/6, 5/6, 7/6 ,11/6	~ 1/12	~ 0.0432
A-A or B-B	1/3, 2/3	~ 1/6	~0.0863
A-B	1/2, 3/2	~ 1/4	~ 0.1295

Table 1: The 4 values of the residual translation T_R due to steps between oxygen terraces of the (0001) surface of sapphire

THE DISLOCATIONS

The misfit dislocations

The interfacial area of GaN/sapphire can exhibit zones without extended defects, but even then, it is not easy to locate the misfit dislocations at the interface. A Fourier filtering shows that they are regularly spaced dislocations. And, if we take the core to be located at the interface, the measured average distance of 1.7 nm shows that they define a stepless relaxed area .

The threading dislocations

Conventional TEM has shown that these dislocations originate at the interface and the large majority (**a**) propagate to the sample surface. Their line is roughly parallel to the **c** growth axis and the large majority are edge type. It has been shown that **c** and **a+c** dislocations tend to bend and annihilate ; the density of such dislocations that reach the layer surface is only a few percent of that near the interface[9,16].

In MBE grown GaN layers, our HREM observations exhibit a typical contrast (Fig. 1), which is not explained by an eight atoms core. It agrees with a 5/7 atoms core, and from our observations, this configuration has more or less an equal frequency with the 8-atom ring core.

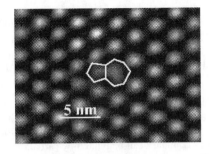

Inside some layers we have observed high angle grain boundaries rotated about the [0001] axis, in coincidence orientation relationships Although these may not be present in good quality epitaxial layers, they

Figure 1. HREM along [0001] of a $1/3 <11\bar{2}0>$ dislocation in GaN with 5/7 atom ring core

were found to exhibit 5/7, and 4 as well as the 8-atom ring dislocation cores[17].

The nanopipes

Such defects are empty or filled holes which exhibit a dislocation character, they mostly have a regular shape. They can extend up to a few tens of nanometres and are usually limited by $\{10\bar{1}0\}$ planes. They have been called nanopipes in analogy to SiC in which such features can measure more than 10 μm and are called macropipes. From our experience they can be classified into two classes, those which have an edge component and those which are pure screw dislocations. As shown in fig. 2, as larger as 2**a** edge component may be exhibited by such defects. It was noticed that the former class of nanopipes are confined inside the first 200 nm of the epitaxial layer[16]; they contain amorphous material in layers grown on 6H-SiC, whereas they are empty on top of sapphire[17].

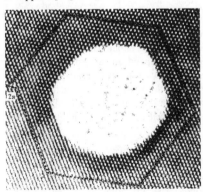

Figure 2. Nanopipe with a 2**a** Burgers vector component, limited by $\{10\bar{1}0\}$ facets

THE $\{11\bar{2}0\}$ STACKING FAULTS

During the last few years a number of reports have been published on the {11$\bar{2}$0} defects in GaN giving various names and atomic structures[7,8,9]. They have finally been shown to be identical[10,11] to those which have been studied in the early sixties by various investigators[13,19]. Along [0001], where the main results were obtained inside the GaN layers, the Drum model (½ [10$\bar{1}$0]) was found to agree with all our observations[10,11]. In cross section along the [11$\bar{2}$0] zone axis, on an image recorded inside the AlN buffer layers, it was possible to identify the two atomic configurations which were connected by a basal I1 stacking fault (fig. 4).Therefore, as this boundary can be limited by partial dislocations and that it continuously folds in the {0001} and {11$\bar{2}$0} planes as for example in 2 and 3 on fig.4, one can use the name stacking fault to designate it.

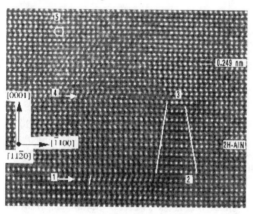

Figure 4. Simultaneous observation of the Amelinckx(2-3) and Drum (4-5) models of the {11$\bar{2}$0} stacking faults, they are a mere continuation of the basal I1, as seen from 1 to 2 and 3 to 4

THE INVERSION DOMAIN BOUNDARIES

Inversion domains are naturally possible in the wurtzite structure due to the two tetrahedral sites of the anion sublattice which cannot be simultaneously occupied by cation atoms. However their formation was found to depend on the growth parameters such as substrates used, buffer layer thickness and growth temperature[6]. In epitaxial GaN, it is now admitted that they tend to develop in N polar matrix in which they grow at a higher rate and usually reach the layer surface in the center of pyramidal features. In Ga polar layers, the inversion domains are confined to the interfacial area[6]. The domains that reach the epitaxial layer surface are limited by {10$\bar{1}$0} boundary planes and, although they originate from the sapphire surface. In the GaN films grown on SiC, inversion domains were not found which supports a strong substrate effect underlying their formation[20]. On top of sapphire, two types of samples with inversion domains were identified by their layer surface morphologies[21]. Some have flat surfaces, whereas others exhibit small pyramids with a surface roughness of about 100-200 nm. In the latter case, inversion domains up to 50 nm in width were located in the centers of the pyramids with V type atomic boundaries[22]. In the former, a high density of smaller Holt type inversion domains[23] could be found inside the layer, all less than

20 nm in size. In the model proposed by Holt[24], the Ga and N (fig.5a) atoms are only interchanged across the boundary, whereas in the "V" configuration, there is an additional $c/_2$ translation of one crystal in order to avoid the formation of Ga-Ga or N-N bonds (Fig. 5b). Our observations did not indicate any switching from one configuration to the other inside the same sample. In contrast, as shown in figure 6, it was possible to notice that in the two configurations the boundary plane could be located in the glide or shuffle positions, thus breaking one or two bonds per atom.

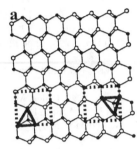
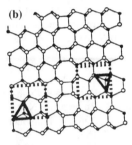

Figure 5. The two types of observed { 10$\overline{1}$0 } inversion domain boundary models, a,): Holt type, b): V or IDB*

Figure 6. An HREM image of an area where an IDB of "V" atomic configuration switches: A-B, two Ga-N bonds are in the boundary plane, B-C, only one bond is in the boundary plane.

SUMMARY

The extended defects in GaN layers are mostly threading dislocations which have been shown to exhibit three possible atomic configurations. The planar defects are either { 10$\overline{1}$0 }inversion domain boundaries or { 11$\overline{2}$0 }stacking faults; each of them has been shown to have two atomic configurations.

REFERENCES

1. Amano H., Kito M., Hiramatsu K., Aksaki I., 1989, Jpn. J. Appl. Phys. **28**, L 2112.

2. Ponce F.A., Major S.J., Plano W.E., and Welch D.F., 1994, Appl. Phys. Lett. **65**, 2302

3. Nakamura S., Senoh M., Nagahama S., Iwasa N., Yamada T., Mitsushita T., Kiyoku H., and Sugimoto Y., 1996, Jpn. J. Appl. Phys. **35**, L74.

4. Lester S.D., Ponce F.A., Crawford M.G., and Steigewald D.A., 1995, Appl. Phys. Lett., **66**, 1249.

5. Vermaut P., Ruterana P., Nouet G., Salvador A., Botchkarev A., and Morkoç H., 1995, Inst. Phys. Conf. Ser., **146**, 289.

6. Rouvière J.L., Arlery M., and Bourret A., 1997, Inst. Phys. Conf. Ser. **157**, 173.

7. Tanaka S., Kern R.S., and Davis R.F., 1995, Appl. Phys. Lett. **66**, 37.

8. Sverdlov B.N., Martin G.A., Morkoç H. and Smith D.J., 1995, Appl. Phys. Lett. **67**, 2063.

9. Rouvière J.L., Arlery M., Niebuhr R., and Bachem K., 1995, Inst. Phys. Conf. Ser. **146**, 285.

10. Vermaut P., Ruterana P., Nouet G., and Morkoç H., 1997, Philos. Mag. A**75**, 239.

11. Vermaut P., Ruterana P., Nouet G., 1997, Philos. Mag. A**76**, 1215.

12. Vermaut P., Ruterana P., Nouet G., Salvador A., and Morkoç H., 1997, Inst. Phys. Conf. Ser., **157**, 183.

13. P. Ruterana, V. Potin, A. Béré, B. Barbaray, G. Nouet , A. Hairie, E. Paumier , A. Salvador, A. Botchkarev, and H. Morkoç, Phys. Rev. 1999, **B59**, 15917-15925

14. Xin Y., Pennycook S.J., Browing N.D., Nellist P.D., Sivanathan S., Beaumont B., Faurie J.P. and Gibart P., 1998, Mat. Res. Symp. Proc. **482**, 781.

15. Potin V., Ruterana P., Nouet G.,. Pond R.C., and Morkoç H.,Phys. Rev. B, **61**, in press

16. Vermaut P., Ruterana P., Nouet G., Salvador A., and Morkoç H., 1996, MRS Int. J. Nitride Res. **1**, 42

17. Ruterana P., Vermaut P., Nouet G., Botchkarev A., Salvador A. and Morkoç H., 1997, Mater. Sci. Eng. **B50**, 72.

18. Drum C.M., 1965, Philos. Mag. A**11**, 313.

19. Blank H., Delavignette P., Gevers R., and Amelinckx S., 1964, Phys. Stat. Solid **7**, 747.

20. Ponce F.A., Bour D.P., Götz W., Johnson N.M., Helava H.I., Grezgory I., Jun J. and Porowski S., 1996, Appl. Phys. Lett. **68**, 917

21. Potin V., Nouet G., and Ruterana P., 1999, Appl. Phys. Lett. **74**, 947.

22. Northrup J.E., Neugebauer J., and Romano L.T., 1996, Phys. Rev. Lett. **77**, 103.

23. Potin V., Ruterana P., and Nouet G.,1997, J. Appl. Phys. **82**, 1276

24. Holt D.B., 1969, J. Phys. Chem. Solids **30**, 1297

Mat. Res. Soc. Symp. Vol. 595 © 2000 Materials Research Society

TEM Study of Bulk AlN Growth by Physical Vapor Transport

W.L. Sarney[1], L. Salamanca-Riba[1], T. Hossain[2], P. Zhou[2], H.N. Jayatirtha[2], H.H. Kang[1], R.D. Vispute[1], M. Spencer[2], K.A. Jones[3]

[1] Dept. of Materials & Nuclear Engineering, University of Maryland, College Park, MD,
[2] Materials Science Research Center of Excellence, Howard University, Washington, D.C.,
[3] U.S. Army Research Laboratory, Adelphi, MD

ABSTRACT

We are attempting to grow bulk AlN that would be suitable as a substrate for nitride film growth. Bulk AlN films were grown by physical vapor transport on 3.5° off-axis and on-axis 6H SiC seed crystals and characterized by TEM, x-ray-diffraction, Auger electron microscopy, and SEM. TEM images show that the bulk AlN does not have the columnar structure typically seen in AlN films grown by MOCVD. Although further optimization is required before the bulk AlN is suitable as a substrate, we find that the structural characteristics achieved thus far indicate that quality bulk AlN substrates may be obtained in the future.

INTRODUCTION

Despite the rapid progress made in nitride based semiconductor film growth, there are still no ideal substrates for high quality epitaxial growth. AlN and GaN are typically grown on SiC, which is expensive, or on Sapphire, which is cheaper but has a large lattice mismatch with the nitride based films. Minimizing the lattice mismatch and enhancing the growth conditions improves the nitride film morphology by lowering the density of defects such as misfit dislocations. Other defects, such as inversion domain boundaries, result from the film's non-isomorphism with the substrate rather than the lattice mismatch [1]. The ideal substrate needs to be as lattice matched and isomorphic to the film as possible. Nitride buffer layers are often grown on sapphire or SiC in an attempt to provide a lattice matched and isomorphic surface for film growth. While buffer layers have helped improve film quality, films grown on buffer layers are usually columnar and contain a higher density of dislocations than desired. AlN or GaN substrates are necessary to achieve further film quality improvement. Due to the high melting temperatures and high dissociation pressures of III-N compounds, bulk nitride crystal growth is difficult. Although the exact nitrogen dissociation pressure is not known, values cited by Landolt and Börnstein [2] indicate that the nitrogen dissociation pressure of AlN is orders of magnitude smaller than that of GaN or InN [3]. Bulk AlN, therefore, should be easier to grow than bulk GaN or InN. In this study we examine the structural quality of bulk AlN grown by physical vapor transport and compare it to AlN buffer layers grown by MOCVD.

EXPERIMENT

Bulk AlN was grown by physical vapor transport by the decomposition of AlN powder in the presence of ambient nitrogen. The growth temperature range was **2150°-**

2200° C with nitrogen pressures of 400~410 Torr. The separation between the seed and AlN powder was approximately 4 mm under a temperature gradient of 1 – 3 °C/mm. The growth rate was varied between 10 and 50 microns per hour. We investigated two different seed crystals, including singular 6H SiC and off-axis 6H SiC miscut 3.5° toward the [01$\bar{1}$0] direction.

This study focuses on the transmission electron microscopy (TEM) characterization of bulk AlN. We used a JEOL 4000FX TEM operated at 300 kV. Cross-sectional TEM samples were prepared using tripod polishing and room temperature ion milling. Further characterization was conducting with Auger electron microscopy, x-ray diffraction, and scanning electron microscopy.

RESULTS & DISCUSSION

TEM images of the bulk AlN grown on 3.5° off-axis and on on-axis 6H SiC are shown in Figs. 1 and 2, respectively. As seen in Fig. 1, the substrate appears to be off axis by a much larger angle than the expected 3.5°. The bulk AlN in this sample was grown at a temperature of 2150°C. Exposing the SiC to such a high temperature may have caused the steps to move and bunch together. Figures 1(a) and 1(b) are dark field images taken with (0002) and (01$\bar{1}$0) two beam conditions, respectively. The bulk AlN is non-columnar, as shown in Figs 1(a) and 1(b). The bands present near the substrate step suggest that there are also steps along the beam direction, normal to the page. There are several small defects imbedded in the AlN, one of which has been magnified and is shown in Fig. 1(d). The white contrast surrounding this defect may be a strain field. Although we do not currently know what these defects are, one possibility is that they are small oxygen or aluminum precipitates. Despite the step bunching of the substrate's surface, the high-resolution image (Fig. 1c) shows that the quality of the AlN/SiC interface is good and that the AlN is of good structural quality. Diffraction patterns, not shown here, indicate that the AlN is well-aligned with the substrate.

The bulk AlN grown on on-axis SiC is also non-columnar, as shown in the (0002) and (01$\bar{1}$0) DF images (Figs. 2a and 2b). Several small defects are imbedded at or just above the AlN/SiC interface, one of which is magnified and shown in Fig. 2(d). It is not clear whether these defects are of the same nature as those seen in the AlN grown on the off-axis SiC seed. These defects are approximately ten times smaller than similar looking defects in the AlN grown on off-axis SiC. Their location near the substrate suggest that they could be small precipitates of Si or C. Other possibilities include oxygen or aluminum precipitates, or dislocation loops. We did not see these defects in the high-resolution image shown in Fig. 2(c). This may be because they are too far apart and the small area examined in high resolution did not include any of these defects, or because these defects have no contrast at high magnifications. Further investigation is necessary to determine the nature and composition of these defects. Figure 2(c) shows the high quality of the AlN/SiC interface and the structure of the bulk AlN.

Auger electron microscopy was used to detect possible incorporation of impurities such as oxygen or carbon within the AlN. Ion beam sputtering caused the sample surfaces to become completely charged, making it impossible to obtain a spectrum. If the sample contained carbon or oxygen impurities, the charge would dissipate more easily and a spectrum would be obtainable. We obtained a spectrum with other lower-quality AlN samples, which are not presented in this paper. The spectrum for the low-quality AlN

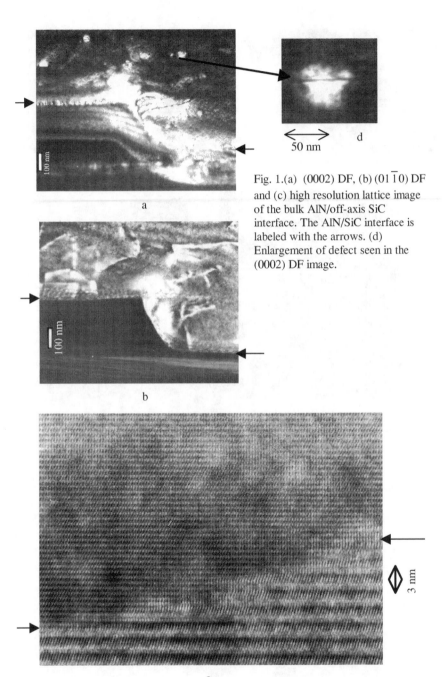

Fig. 1.(a) (0002) DF, (b) (01$\overline{1}$0) DF and (c) high resolution lattice image of the bulk AlN/off-axis SiC interface. The AlN/SiC interface is labeled with the arrows. (d) Enlargement of defect seen in the (0002) DF image.

a

b

c

d

$\xleftrightarrow{\hspace{1cm}}$ 50 nm

100 nm

100 nm

3 nm

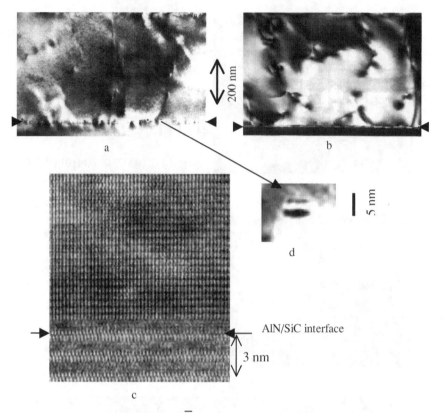

Fig. 2.   (a) (0002) DF, (b) ($01\overline{1}0$) DF and (c) high resolution lattice image of the bulk AlN/on-axis SiC interface. (d) Enlargement of defects seen just above the AlN/SiC interface in the (0002) DF image.

showed small amounts of carbon incorporation. The small number and size of the defects discussed in the preceding paragraphs and their location at or near the AlN/SiC interface may preclude them from dissipating the surface charge.

SEM images of the samples' surfaces are shown in Figs. 3(a)-3(d). The surface of the sample grown on the off-axis seed (Figs. 3a-3b) has a step morphology with several cracks running perpendicular to the steps. There are several highly faceted large hexagonal grains, as shown in Fig. 3(b). The surface of the sample grown on the on-axis seed (Figs 3c-3d) contains many hexagonal grains stacked on top of one another. The surface is flatter than the surface of the AlN grown on the off-axis seed, and the grains are not faceted. Additionally, the top grains are much smaller than those seen on the AlN grown on the off-axis seed.

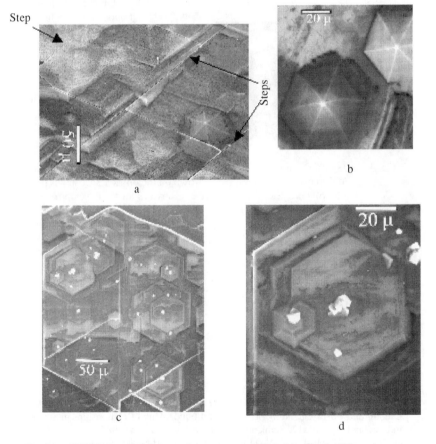

Fig. 3.   SEM images of the samples grown on 3.5° (a-b) off-axis SiC, and (c-d) on-axis SiC. The steps are labeled with arrows.

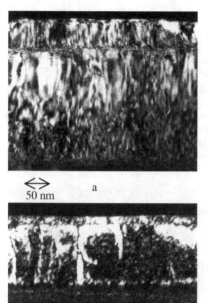

$\Longleftrightarrow$
50 nm

a

b

Fig. 4. (0002) DF images of the AlN films grown on (a) 3.5° off axis SiC and (b) on-axis SiC by MOCVD

X-ray results indicate that the AlN grown on on-axis SiC is of better quality than the AlN grown on off-axis AlN. Rocking curve measurements show that the full width half maximum (FWHM) of the (0002) peak for AlN grown on the on axis seed is 5.16 arc-minutes, while the FWHM of the AlN grown on the off-axis seed is 9.6 arc-minutes.

It is interesting to compare the morphology of the AlN grown by physical vapor transport to the AlN films we have grown by MOCVD. Figures 4(a) and 4(b) show (0002) DF images of AlN films grown on on-axis SiC and 3.5° off-axis SiC, respectively. Both films are highly columnar and contain an inversion domain boundary near the surface. These defects hinder the prospect of using these films as buffer layers for high quality nitride film growth.

CONCLUSION

The bulk AlN, particularly that grown on the on-axis SiC, shows promising characteristics for the development of this material as a substrate. Further optimization of the growth conditions and a polishing technique to improve the surface quality is needed to yield an AlN surface suitable for high-quality nitride film growth.

ACKNOWLEDGEMENTS

The authors gratefully acknowledge the support of the MRCP Army Grant No. DAAL 019523530.

REFERENCES

[1] B.N. Sverdlov, G.A. Martin, and H. Morkoç, Appl. Phys. Lett. **67**, 2063 (1995).
[2] Landolt and Börnstein, Numerical Data and Fundamental Relationships in Science and Technology, vol. 17, Semiconductors, Springer, Berlin (1984).
[3] G. Popovici, H. Morkoç, S.N. Mohammad, Group III Nitride Semiconductor Compounds, ed. B. Gil, Clarendon Press, Oxford, (1998).

Mat. Res. Soc. Symp. Vol. 595 © 2000 Materials Research Society

Thermal expansion of GaN at low temperatures - a comparison of bulk and homo- and heteroepitaxial layers

Verena Kirchner[1], Heidrun Heinke[1], Sven Einfeldt[1], Detlef Hommel[1], Jaroslaw Z. Domagala[2], Michal Leszczynski[3]
[1] University of Bremen, Institute of Solid State Physics, Bremen, Germany
[2] Polish Academy of Science, Institute of Physics, Dept. of X-ray Studies and Electron Microscopy, Warsaw, Poland
[3] High Pressure Research Center, Unipress, Warsaw, Poland

ABSTRACT

The thermal expansion of different GaN samples is studied by high-resolution X-ray diffraction within the temperature range of 10 to 600 K. GaN bulk crystals, a homoepitaxial layer and different heteroepitaxial layers grown by metalorganic chemical vapor deposition (MOCVD) and molecular beam epitaxy (MBE) were investigated. Below 100 K the thermal expansion coefficients (TEC) were found to be nearly zero which has to be taken into account when estimating the thermal strain of GaN layers in optical experiments commonly performed at low temperatures. The homoepitaxial layer and the underlying GaN substrate with a lattice mismatch of $-6 \cdot 10^{-4}$ showed identical thermal expansion. The comparison between the temperature behavior of lattice parameters of heteroepitaxial layers and bulk GaN points to a superposition of thermally induced biaxial strain and compressive hydrostatic strain.

INTRODUCTION

GaN epitaxy is commonly performed on substrates with high mismatch in both lattice parameters and thermal expansion. Whereas the first is relaxed by misfit dislocations within the first few nanometers, the latter results often in thermally induced strain of GaN epilayers as it is indicated by a variety of optical experiments [1,2]. Despite the known importance of thermally induced strain for GaN epitaxy, there are still a lot of open questions and uncertainties related with this subject. On the one hand, this concerns basic material parameters which are essential for a correct description of thermally induced strain in GaN layers, as the temperature behavior of the TEC of GaN in the full range from typical growth temperatures down to low temperatures of a few Kelvin as commonly used for optical experiments. Moreover, there were some indications that the TEC could depend on the characteristics of the material as the free electron concentration [3]. On the other hand, the formation of thermally induced strain in GaN epilayers in dependence on the growth procedures and parameters is far beyond to be fully understood [4].

In the present paper the thermal behavior of differently grown GaN is studied. This concerns the basic material parameters which could be investigated for bulk and homoepitaxal GaN samples of high crystalline perfection. The thermal expansion of these samples is compared to that of typical heteroepitaxial layers deposited by different growth techniques on c-plane sapphire.

EXPERIMENTAL

Different types of samples were investigated: bulk GaN, homoepitaxial GaN and heteroepitaxial GaN layers grown by MOCVD and MBE. The investigated GaN bulk crystals were grown at the High Pressure Research Center, Unipress, Warsaw by the high-temperature high-pressure method at 1800 K and 15 kbar [5]. They have hexagonal platelet shape with lateral sizes of 3 to 4 mm. One of the bulk crystals was Mg-doped, the other was nominally undoped and overgrown by MOCVD with a 3 μm thick undoped GaN layer [6]. For this, a substrate temperature of 1050°C was used similar to that applied for MOCVD of the two heteroepitaxial GaN layers deposited on c-plane sapphire. The growth conditions for these layers provided by S. DenBaars (University of California Santa Barbara) and S. Nakamura (Nichia Chemical Industries) can be found in ref. [7] and [8]. In addition, a 4.5 μm thick GaN layer grown by MBE with an electron cyclotron plasma source on c-plane sapphire at a substrate temperature of 820°C was investigated. In contrast to the MOCVD layers, this layer was deposited directly onto the nitridated substrate without growing a low-temperature buffer before.

A high-resolution X-ray diffractometer Philips X'Pert MRD equipped with a Cu sealed anode, a four crystal monochromator and a triple crystal analyzer was used for the measurements. This diffractometer was extended by a continous flow X-ray cryostat of Oxford Instruments enabling temperature dependent measurements from 10 to 630 K. The lattice parameters were calculated from the scattering angles directly measured by triple axis 2θ scans [9].

RESULTS

The progress in the growth of GaN bulk crystals over the last few years has resulted in crystal platelets with lateral extensions of up to 10 mm characterized by very narrow X-ray diffraction profiles [5] which are a prerequisite for the accurate determination of lattice parameters [9]. However, highly resolved reciprocal space maps (RSMs) for such crystals as shown in figure 1 for an overgrown GaN substrate indicate that even the small GaN crystals obviously can consist of several macroscopic grains. Within the area of about 1 mm² illuminated by the X-ray beam, we found typically two or three grains contributing to the scattered signal as indicated by corresponding sharp intensity maxima in the (00l) RSMs at different q_x values. From the difference in q_x the tilt of the crystallographic orientation between different grains can be calculated to be in the range of a few hundred arcseconds. Additionally, the grains are characterized sometimes by slightly different lattice parameters as it is indicated by the small shift of the intensity maxima A and B along q_z in figure 1. The changes in lattice parameter c by $8 \cdot 10^{-4}$ Å between different grains can be ascribed to different impurity concentrations pointing to a possible preferential accumulation of impurities in some grains.

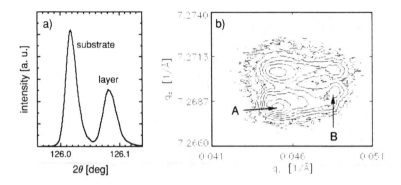

Figure 1. Triple axis 2θ/ω scan for the (006) reflection (a) and the corresponding RSM (b). The intensity maximums A and B are assigned to two grains of the substrate with different lattice parameter c.

The composition of the GaN bulk crystals of several grains with possibly slightly different lattice constants requires special efforts for an accurate determination of the temperature behavior of GaN lattice parameters. In our experiments, RSMs and triple axis ω scans were performed before measuring the scattering angle by 2θ/ω scans. In this way it was ensured that the lattice parameters of the same grain were determined at all temperatures.

The RSM in figure 1(b) clearly demonstrates that the grain structure of the GaN substrate is reproduced in the overgrown layer as expected. The shift of substrate and layer intensity maxima along the q_z axis corresponds to a difference in c lattice parameter of $2 \cdot 10^{-3}$ Å. In earlier reports this behavior was ascribed to native defects or to different free electron concentrations in substrate and layer material [3]. RSMs of asymmetric reflections showed the pseudomorphic state of the layer in the whole temperature range investigated.

The lattice parameters of homoepitaxial layers and GaN bulk crystals come close to each other if the bulk material is Mg doped during growth [5]. This is confirmed by figure 2 where the lattice parameters c of the intentionally undoped GaN bulk, the homoepitaxial layer deposited on it, and a Mg doped GaN crystal are plotted in dependence on the temperature.

Similarly to the GaN bulk and the homoepitaxial layer, several μm thick heteroepitaxial GaN layers on sapphire substrates were investigated. In figure 3 the results for the c lattice parameters of layers grown by MOCVD and MBE are compared with those for the GaN bulk material and the homoepitaxial layer. For better clarity the latter data are represented as lines. Clear differences are visible for the two MOCVD samples. Whereas the sample MOCVD II has a c lattice parameter which is clearly higher than that of the homoepitaxial layer in the whole temperature range, the c values from the MOCVD I sample are near to those of the homoepitaxial layer. For the MBE grown layer the lattice parameters are between those of the MOCVD samples. Obviously, the slopes of the curves differ clearly for the different samples as well.

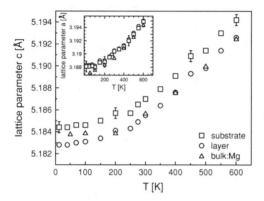

Figure 2. *Lattice parameters c in dependence on the temperature for an undoped bulk crystal (substrate) overgrown with a 3 μm thick MOCVD layer and a Mg-doped GaN bulk crystal. The error bars are representative for all lattice parameters given. The lattice parameters a are shown in the inset.*

DISCUSSION

From figure 2 the following informations can be obtained: (1) The temperature dependence of thermal expansion of GaN can be extracted. This has been done in ref. [10] assuming a linear expansion within each of the temperature ranges of 10 to 100 K, 100 to 250 K, and 250 to 600 K. (2) Within the experimental errors, there is no indication for negative TEC at low temperatures. However, the amount of negative TEC as reported for other materials is often too small to be reflected in a significant change in lattice constants [11]. (3) The difference in lattice parameters between the homoepitaxial layer and the underlying substrate is constant over the whole investigated temperature range with an accuracy of $2 \cdot 10^{-4}$ Å.

The differences in c lattice parameters for the heteroepitaxially grown layers can be partially attributed to different growth temperatures for the MOCVD and the MBE process. Since the growth temperature for the MOCVD samples is higher than that for the MBE sample, the thermal strain induced in the MOCVD layers should be higher. This is obviously the case for the MOCVD II sample for which the larger lattice parameters in growth direction indicate a stronger compressive biaxial strain. For the MOCVD I sample this model fails because the strain state seems to be the same as that of the MBE sample. A comparison of the a lattice parameters (not shown here) indicates that this is probably due to large hydrostatic strain components in the MBE and MOCVD I sample.

The increase of the lattice parameters c for all three heteroepitaxial layers is different from that of the bulk crystals and the homoepitaxial layer. Below 200 K the slope is close to zero as for the bulk and homoepitaxial GaN, but above 200 K the slope is clearly smaller than for the bulk material. This reflects directly the thermally induced

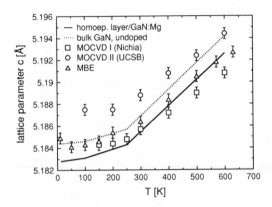

Figure 3. *Comparison of the lattice parameters c in dependence on the temperature for bulk GaN, homoepitaxial and different heteroepitaxial GaN layers.*

strain in the epilayers grown on sapphire. On the other hand, the crossing of the curves for the heteroepitaxial layers with those for the bulk material far below the growth temperatures points to a strong influence of compressive hydrostatic strains in these layers.

SUMMARY

The temperature dependence of the lattice parameters of bulk, homoepitaxial and heteroepitaxial GaN was investigated and compared. Neither an influence of the electron concentration on the thermal expansion nor a negative thermal expansion at low temperatures was found in our experiments for bulk and homoepitaxial GaN. The comparison of results for heteroepitaxial samples with bulk values points to a superposition of thermally induced biaxial strain and compressive hydrostatic strain in the heteroepitaxial layers.

ACKNOWLEDGEMENT

The authors would like to thank I. Grzegory and S. Porowski (High Pressure Center, Unipress, Warsaw) for the growth of the GaN single crystals and P. Prystawko for the MOCVD growth of the GaN homoepitaxial layer. The heteroepitaxial MOCVD layers were generously provided by S. DenBaars (University of California Santa Barbara) and S. Nakamura (Nichia Chem. Ind.) which is kindly acknowledged.

REFERENCES

1.	W. Rieger, T. Metzger, H. Angerer, R. Dimitrov, O. Ambacher and M. Stutzmann, Appl. Phys. Lett., **68**, 970 (1996).

2.   W. Shan, R. J. Hauenstein, A. J. Fischer, J. J. Song, W. G. Perry, M. D. Bremser, R. F. Davis and B. Goldenberg, Mat. Res. Soc. Symp. Proc., **449**, 841 (1997).

3.   M. Leszczynski, J. Bak-Misiuk, J. Domagala and T. Suski, Mat. Res. Soc. Symp. Proc., **468**, 311 (1997).

4.   C. Kisielowski, J. Krüger, S. Ruvimov, T. Suski, J. W. Ager III, E. Jones, Z. Liliental-Weber, M. Rubin, E. R. Weber, M. D. Bremser and R. F. Davis, Phys. Rev. B, **54**, 17745 (1996).

5.   S. Porowski, J. Cryst. Growth, **189/190**, 153 (1998).

6.   J. L. Weyher, P. D. Brown, A. R. A. Zauner, S. Muller, C. B. Boothroyd, D. T. Foord, P. R. Hagemann, C. J. Humphreys, P. K. Larsen, I. Grzegory and S. Porowski, J. Cryst. Growth, **204**, 419 (1999).

7.   B. Heying, X. H. Wu, S. Keller, Y. Li, D. Kapolnek, B. P. Keller, S. P. DenBaars and J. S. Speck, Appl. Phys. Lett., **68**, 643 (1996).

8.   S. Nakamura, M. Senoh, N. Iwasa, S. Nagahama, T. Yamada and T. Mukai, Jpn. J. Appl. Phys., **34**, L1332 (1995).

9.   P. F. Fewster and N. L. Andrew, J. Appl. Cryst., **28**, 451 (1995).

10.   V. Kirchner, H. Heinke, D. Hommel, J. Z. Domagala and M. Leszczynski, submitted to Appl. Phys. Lett.

11.   K. Wang and R. R. Reeber, Mat. Res. Soc. Symp. Proc., **482**, 863 (1998).

Mat. Res. Soc. Symp. Vol. 595 © 2000 Materials Research Society

THE ROLE OF THE MULTI BUFFER LAYER TECHNIQUE ON THE STRUCTURAL QUALITY OF GaN

M. Benamara, Z. Liliental-Weber, J.H. Mazur[+], W. Swider and J. Washburn.
Materials Science Division, Lawrence Berkeley National Laboratory, Berkeley CA 94720, 62/203
[+] *present address: Philips Semiconductors, 9651 Westoverhills Blvd., San Antonio, TX 78251*

M. Iwaya , I. Akasaki and H. Amano.
Meijo University, Dept. of Electrical and Electronic Engineering, 1-501 Shiogamaguchi, Tempaku ku, Nagoya 468-8502, Japan.

ABSTRACT

Successive growth of thick GaN layers separated by either LT-GaN or LT-AlN interlayers have been investigated by transmission electron microscopy techniques. One of the objectives of this growth method was to improve the quality of GaN layers by reducing the dislocation density at the intermediate buffer layers that act as barriers to dislocation propagation. While the use of LT-AlN results in the multiplication of dislocations in the subsequent GaN layers, the LT-GaN reduces dislocation density. Based upon Burgers vector analysis, the efficiency of the buffer layers for the propagation of the different type of dislocations is presented. LT-AlN layer favor the generation of edge dislocations, leading to a highly defective GaN layer. On the other hand, the use of LT-GaN as intermediate buffer layers appears as a promising method to obtain high quality GaN layer.

INTRODUCTION

GaN and related alloys are of particular interest since their ability to cover a wide spectral range [1] that is not possible with any combination of any other semiconductor materials. Since no cheap crystalline substrate with a lattice parameter close enough to that of GaN can be available, growth techniques have been improved in order to limit the defect density in the GaN layer. The most commonly used technique consists in the deposition of an AlN buffer layer to form the junction between the substrate and the GaN layer and leads to the formation of more than 10^9 dislocations/cm^2. Elaborated techniques have since emerged. The so-called "lateral epitaxial overgrowth" LEO technique appears today to give the best results. The lowest dislocation density is observed at the top of the GaN layers despite the presence of grain boundaries with mixed dislocations to compensate a misorientation [2]. Another technique is to first deposit a thick low-temperature (LT) GaN on the substrate. The first stage of the growth gives 3-D islands of GaN on the substrate. Afterwards, deposition of GaN at high-temperature (HT) makes the growth follow the same principles as those for LEO. These two latter techniques are based upon the faster growth along the $<\overline{2}110>$ directions. In this paper, we report on a Transmission Electron Microscopy study of GaN layers obtained by using the deposition of successive LT-GaN layers between HT-GaN layers.

EXPERIMENTAL

GaN layers were grown on (0001) sapphire substrates in an horizontal reactor as described in ref [3]. TMGa, TMAl and amonia were used as Ga, Al and N source respectively, at around 140 Torr growth pressure.

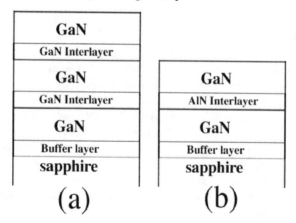

fig 1: Schematic drawing of the structure of the two studied samples

The structure of the samples used in this study is represented in fig. 1. A 1 μm thick GaN layer is deposited at high-temperature (HT) over a low-temperature LT-GaN or LT-AlN buffer layer. Afterwards, a second buffer layer is deposited at low-temperature (LT). This second buffer layer is called interlayer and its goal is to improve the quality of the subsequent 1 μm thick HT-GaN layer. It is a few nanometer thick AlN or GaN layer. For the investigations, the TEM samples were prepared by standard procedure using ion-milling at liquid nitrogen temperature and examined using a Jeol 200 CX operating at 200 kV. Before the observations, the samples were mounted on a gold covered grid and then dipped in a KOH (50%) solution for 5 min. This etch step was aimed at removing damage created during the ion milling.

RESULTS
A- GaN/GaN interlayer/GaN

Fig 2 is a dark field image of the sample the structure of which is represented in fig 1.a. The diffraction conditions were chosen so that all dislocations are visible in the image. It is noticed that the first HT-GaN layer is highly defective with a huge density of threading dislocations. Conventional diffraction contrast method was used to determine their Burgers vectors **b** using the invisibility criterion, **g.b**=0. Diffraction analysis showed that more than 80% of these dislocations have a Burgers vectors **b=a**. Their density is well above $10^{11}/cm^2$.

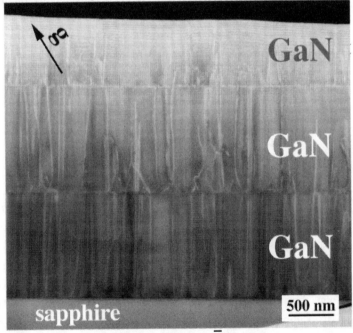

Fig 2: Dark field image of sample 1 with **g**=1$\bar{1}$02. The density of dislocations decreases towards the surface.

The insertion of the first thin LT-GaN nucleation layer as a buffer layer does not allow one to obtain a high-quality HT-GaN layer. The large lattice mismatch with the sapphire substrate can be held responsible for such a high defect density.

This figure also shows a significant improvement has taken place in the subsequent GaN layer after deposition of the second intermediate LT-GaN layer. This second HT-GaN layer has a defect density of about 4.10^9 cm^{-2}. This confirms the role of the nucleation layer that acts as a barrier to dislocation propagation. Since dislocations in nitrides are frozen after the growth, the efficiency of the LT-GaN layer is mainly related to the growth process and the dislocation annihilation has taken place during the first growth stages of this LT-layer.

The density of dislocations in the third HT-GaN layer is around 1.4×10^9 cm^{-2}. This density is three times smaller than in the second layer and it shows that the deposition of another LT-GaN layer on top of the second HT-GaN still benefits the technique. Nevertheless, this second intermediate LT-GaN layer is less efficient and this can be explained by the increasing quality of the GaN layer.

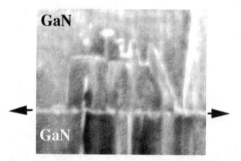

Fig.3: Dislocation half loop formed in the HT-GaN layer.

Dislocation half loops were observed with a high density in the vicinity of all LT-layer. They might be the result of lateral overgrowth of adjacent GaN islands, during which threading dislocations are forced to bend over into basal planes. Then their interaction with other dislocations often results in the formation dislocation half loops [4].

B- GaN/AlN interlayer/GaN

Fig 4 represents two weak-beam dark field images taken under different conditions. Fig 4.a. is taken with $1\bar{1}00$ reflecting planes and fig. 4.b with g=0002. Since the diffraction vector used on fig. 4.a is $\mathbf{g}=1\bar{1}00$, most of edge and mixed dislocations with Burgers vectors **b=a** and **b=a+c** appear in the image (edge dislocations with **b**=1/3[11 $\bar{2}$ 0] are out of contrast because g.b=0 and g.b∧u=0 simultaneously).

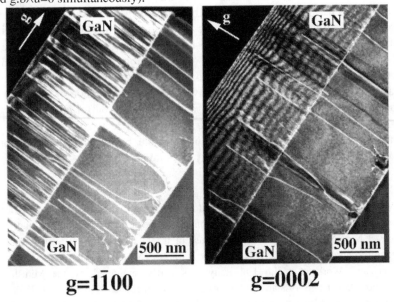

$$g=1\bar{1}00 \qquad\qquad g=0002$$

Fig. 4: Dark field images of the same area taken under different conditions (a) with **g**=1 $\bar{1}$ 00: note the high density of edge dislocations in the upper layer. (b) with **g**=0002, the upper layer appears defect free.

It is obvious that the use of the LT-AlN buffer layer that makes the junction with the sapphire substrate greatly improves the quality of the first GaN layer. The density of threading dislocations in this layer is about 2.10^9 cm^{-2}. Even though the quality of the first HT-layer depends on the growth conditions and especially its first stages, the first HT-GaN layer of this sample has a better crystalline quality than the first HT-GaN layer of sample 1 (where a LT-GaN buffer layer was used). This density of 2.10^9 cm^{-2} is typical for MOCVD grown GaN layers on sapphire using LT-AlN buffer layer. Nevertheless, the AlN buffer layer inserted between the two HT-GaN layers does not reduce the dislocation density. While the density of screw and mixed dislocations (10^8 cm^{-2}) remain the same in both GaN layers, the second AlN buffer introduces edge dislocations with a high density in the upper layer (Fig. 4.a).

SUMMARY

Successive growth of thick GaN layers separated by either LT-GaN or LT-AlN layer have been investigated by transmission electron microscopy techniques. One of the objectives of this growth method was to improve the quality of GaN layers by reducing the dislocation density at the intermediate buffer layers that act as barriers to dislocation propagation. While the use of LT-AlN results in the multiplication of dislocations in the subsequent GaN layers, the LT-GaN reduces dislocation density. Based upon Burgers vector analysis, the efficiency of the buffer layers for the propagation of the different type of dislocations is presented and discussed. LT-AlN buffer layers favors the generation of edge dislocations, leading to a highly defective GaN layer. On the other hand, the use of LT-GaN as intermediate buffer layers appears as a promising method to obtain high quality GaN layer.

ACKNOWLEDGMENT

Work at E.O. Lawrence Berkeley Laboratory was supported by the Director, Office of Basic Science, Materials Science Division, U.S. Department of Energy, under the Contract No. DE-AC03-76SF00098. The use of the facility at the National Center for Electron Microscopy at E.O. Lawrence Berkeley National Laboratory is greatly appreciated. Work at Meijo University was partly supported by the Ministry of Education, Science, Sports and Culture of Japan (contract Nos 09450133, 09875083 and High-Tech Research Center Project), Japan Society for the Promotion of Science (JSPS) Research for the Future Program in the area of Atomic Scale Surface and Interface Dynamics under the project of "Dynamical Process and Control of Buffer Layer at the Interface of Highly-Mismatched System".

REFERENCES

[1] S. Nakamura and G. Fasol, The blue Laser Diode, (Springer, Berlin, 1997).

[2] Z. Liliental-Weber, M. Benamara, W. Swider, J. Washburn, J. Park, P.A. Grudowski, C.J. Eiting and R. D. Dupuis, MRS Internet J. Nitride Semicond. Res. 4S1, G4.6 (1999)

[3] H. Amano, M. Iwaya, N. Hayashi, T. Kashima, M.Katsuragawa,T. Takeuchi, C.Wetzel and I.Akasaki,   MRS Internet J. Nitride Semicond. Res. 4S1, G10.1 (1999).

[4] S. Ruvimov, Z. Liliental-Weber, J. Washburn, Y. Kim, G. S. Sudhir, J. Krueger and E. R. Weber, Mat. Res. Soc. Proc. Vol. 572, 295 (1999).

Mat. Res. Soc. Symp. Vol. 595 © 2000 Materials Research Society

Physical Properties of Silicon Doped Hetero-Epitaxial MOCVD Grown GaN: Influence of Doping Level and Stress

P.R. Hageman, V. Kirilyuk, A.R.A. Zauner, G.J. Bauhuis and P.K. Larsen
Research Institute for Materials, University of Nijmegen, Toernooiveld, 6525 ED Nijmegen, The Netherlands

ABSTRACT

Silicon doped layers GaN were grown with MOCVD on sapphire substrates using silane as silicon precursor. The influence of the silicon doping concentration on the physical and optical properties is investigated. A linear relationship is found between the silane-input molfraction and the free carrier concentration in the GaN layers. The morphology of the samples is drastically changed at high silicon concentrations. Photoluminescence was used to probe bandgap variations as function of the silicon concentration. Increasing of the doping concentration led to a continuous shift of the exciton related PL to lower energies, while the intensity of the UV emission was found to increase up to a carrier concentration of $n=2.5 \times 10^{18}$ cm^{-3}.

INTRODUCTION

Especially for the use in high-temperature and high-power electronics, knowledge about the physical properties of n-type GaN layers is of utmost importance in order to minimise the electrical resistance of the layers, i.e. reducing thermal losses, and to account for the influence of build-in stress. In MOCVD growth of GaN, the n-type doping process with silicon as dopant is relatively simple and straightforward. However, it is well known that even relatively low concentrations of incorporated silicon can have a tremendous influence on the optical and mechanical properties of the material. In this paper the emphasis will be on the influence of the doping concentration on the physical and optical properties of the silicon doped, MOCVD grown GaN layers.

EXPERIMENTAL

The GaN layers were grown in a horizontal low-pressure MOCVD reactor at 1045°C on two-inch sapphire (0001) substrates using trimethylgallium (TMG) and ammonia (NH_3) as precursors. Further details on the growth can be found in [1]. Silane (SiH_4), diluted to 50 ppm with H_2, was used as silicon precursor. The grown layers had thicknesses between 1.4 and 1.8 μm. Hall measurements, employing the Van der Pauw configuration, were used to characterise the samples electrically. Photoluminescence (PL) measurements were carried out at 4K; the 325-nm line from a HeCd laser provided optical excitation with an excitation density of about 15 Wcm^{-2}. The luminescence was dispersed by a 0.6 m monochromator and detected by a cooled GaAs photomultiplier. The HR-XRD measurements were performed on a Bruker D8.

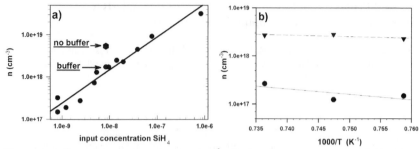

Figure 1. *a) Free electron concentration (cm^{-3}), Hall measurements, versus [SiH$_4$] for GaN layers grown at 1045°C with constant NH$_3$ and TMG partial pressures. The slope of the linear relationship is 0.95. b): free electron concentration (cm^{-3}), Hall measurements, versus the reciprocal growth temperature (K^{-1}) for two [SiH$_4$] ($8x10^{-10}$ and $1.92x10^{-8}$).*

RESULTS AND DISCUSSION

To evaluate SiH$_4$ as precursor for n-type doping of GaN, the dependency of the free electron concentration (n) on the input mol fraction silane ([SiH$_4$]) and growth temperature was studied. Figure 1 shows n (cm^{-3}) versus [SiH$_4$] as determined by Hall-Van der Pauw measurements at 300K in experiments where all parameters were kept constant except from [SiH$_4$]. In figure 1 n increases linearly with [SiH$_4$] with a slope of 0.95. Values up to $1x10^{19}cm^{-3}$ have been reached easily without observation of any saturation of the free carrier concentration. In contrast to III/V semiconductors like GaAs [2], saturation of n is not observed. A problem forms the drastic change in morphology (see figure 2) due to the large [SiH$_4$]. For example, no reliable values for the layer thickness could be obtained for [SiH$_4$] = $8.2x10^{-7}$ although the estimated value for n nicely fits in the graph. The hexagonal symbol in figure 1 represents a layer grown without a low temperature buffer leading to an N-face layer [3] as confirmed by free etching experiments using a 20% KOH solution [4]. As is well known, the donor incorporation in N-face layers is higher than in Ga-face layers [5]. This is clearly illustrated by the experiment in which two layers are compared grown with the same [SiH$_4$]. The N-face layer, grown without a low temperature buffer layer, has a 4 times higher silicon incorporation ($5.4x10^{18}$ cm^{-3} compared to $1.7x10^{18}$ cm^{-3}) compared to the epilayer grown on the buffer layer. This is illustrated in figure 1a.

To reveal the temperature behaviour of the silicon doping of GaN process using SiH$_4$, experiments were performed at different temperatures keeping [SiH$_4$] constant. Plotting n versus T^{-1} reveals the apparent activation energy of the process (see figure 1b). The obtained values, around 8 kcal/mol, point to a diffusion limited incorporation process, which is supported by the 1:1 relationship between n and [SiH$_4$].

The electron mobilities are a measure of the electrical quality of the material. Values ranging from 320 cm^{2}V^{-1}s^{-1} for n = $5.3x10^{17}$ cm^{-3} till 192 cm^{2}V^{-1}s^{-1} for n = $9.2x10^{18}$ cm^{-3} were obtained indicating a compensation grade of about 0.4 [6,7]. To distinguish between the contribution of a possible interfacial layer and the epilayer, temperature dependent Hall measurements have to be performed. The samples described in this paper consist of an epilayer deposited on a highly defective low temperature buffer layer with a thickness of 20 nm as is revealed by TEM measurements. The influence of this buffer layer on the

Figure 2. SEM photograph of the morphology of the GaN sample grown with [SiH₄] = 8.2x10⁻⁷, magnification 5200x, marker represents 1 μm.

electrical properties of the layers was evaluated using the two-layer model [8] which was developed for HVPE grown material which contains a highly conductive layer near the interface. It turned out that in the case of our samples, the buffer layer has hardly any influence on the overall electrical properties. C-V measurements were performed giving identical values ($5.2x10^{17}$ cm^{-3}); as the Hall measurements ($5x10^{17}$ cm^{-3}) confirming that the buffer layer does not influence the electrical properties of the layers.

Incorporation of silicon in the GaN lattice influences, to a certain extend, the crystal properties of GaN. In general, when grown on sapphire, GaN is under tensile strain at growth temperatures [9]. During cooling down it goes into compression due to thermal expansion mismatch between the GaN epilayer and the sapphire substrate. Incorporation of Si enhances this effect because the ionic radius of Si^+ (0.41 Å) is smaller than Ga^+ (0.62 Å). This Si substitution compresses the GaN layer in the c-axis direction thereby putting the basal GaN plane in tension. The incorporation of Si in GaN can lead to crack formation when a critical doping concentration (for a given thickness) is exceeded [7] due to a large tensile strain during growth. Additionally, the formation of dislocations due to the silicon incorporation may be another factor in the crack formation process.

To check whether an increased density of dislocations may be the cause for the crack formation, X-ray diffraction rocking curve measurements were performed on the silicon doped GaN layers. The FWHM of the (0002) and (10$\bar{1}$5) reflections increase with increasing [SiH₄], as given in figure 3, indicating that the crystal quality decreases as the amount of incorporated silicon increases which could indicate an increase in density of all type of threading dislocations. From figure 3 it can be concluded that there is a linear relationship between the incorporated Si concentration and the FWHM for both reflections for [SiH₄] up to 7.8x10⁻⁸. However, this does not apply for a sample with a [SiH₄] as large as 8.2x10⁻⁷ which shows a FWHM of 356 arcsec, a value comparable to the undoped samples. In contrast to the layers with less incorporated Si, this layer is very inhomogeneous and the morphology is dominated by separate islands deposited in a rough matrix, see figure 2. This can be caused by either the relaxation of the lattice due to island formation or because of deposition of $Si_{1-x}N_x$ during GaN growth, which is formed by the presence of large concentrations of SiH₄ and NH₃ [11].

Photoluminescence (PL) measurements at 4 K were used to investigate

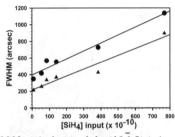

Figure 3. FWHM of the (0002) (circles) and the (10$\bar{1}$5) (triangles) reflections versus the input molfraction SiH₄ for layers grown at identical conditions.

spontaneous emission properties of Si-doped hetero-epitaxial GaN. As in the case of unintentionally doped layers the excitons bound to the neutral donors (D°BE) dominate the PL spectra of the moderately Si doped samples at low temperatures. Donor-acceptor pair recombination (DAP) with its LO-phonon replicas is present in all doped GaN layers. With increase of the doping concentration up n=2.5x10^{18} cm^{-3} to the band edge PL shifts to the lower energies (figure 4), probably caused by band gap narrowing. Another reason for the red shift of the exciton related PL could be a strain release caused by substitution of Ga by Si. The total UV emission first increases as the doping concentration increases, but starting from n=4.6x10^{18} cm^{-3} the band edge PL gets broadened and diminishes, which indicates the increasing role of the non-radiative Auger recombination. No considerable changes in "yellow" luminescence were found for different doping levels. The "yellow" luminescence in the samples with the lowest UV emission was however somewhat stronger.

The PL signal of the sample with the highest free electron concentration (n=3.2x10^{19} cm^{-3}) is dominated by two broad lines that can reflect the structural inhomogeneity of the sample, since this layer reveals very rough and even mosaic-like morphology structure (see figure 3). Splitting of the D°BE peak was also observed in the Si doped GaN layer grown directly on sapphire without a buffer layer, i.e. N-face orientation (figure 5). Both D°BE and DAP peaks in the N-face layer are blue shifted as compared with the Ga-face one, though both layers were grown using the same growth conditions.

The quantitative study of the band gap narrowing and correlation of the band gap with the lattice constants of the hetero-epitaxial layers is unfortunately very difficult. A reason for this is the non-uniform depth distribution of the charge in these layers. In contrast to Hall measurements, which give an integrated value of the free carrier concentration, the PL signal comes only from the upper part of the layer. Although the non-uniform dept distribution in carrier concentration is in an absolute sense not large, its impact on the PL measurements is significant. As the penetration depth of the excitation light is comparable with the thickness differences in the studied GaN layers (d ~ 1.4-1.8 μm), the PL emission is strongly dependent on the layer thickness and some discrepancies of the PL data with the Hall measurements appear.

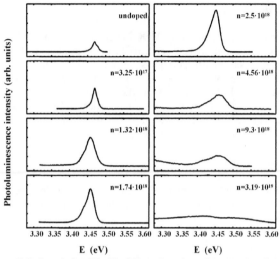

Figure 4. *Evolution of the bandedge PL (T=4K) as function of increasing free electron concentration n (cm⁻³).*

The qualitative comparison of the PL emission in the differently doped GaN samples shows that the silicon impurities of moderate concentrations (up to $n=2.5 \times 10^{18}$ cm⁻³) either eliminate non-radiative recombination centres or promote new ways for the radiative recombination in the hetero-epitaxial GaN making the UV emission more efficient. From both the PL and Hall data it follows that the free electron concentration is always higher in the N-face layers (see figures 1 and 5) in comparison with the Ga-face ones, although the same growth conditions are applied. The same phenomenon was found for the homo-epitaxial growth of GaN as well. Either the incorporation of the donor in the N-face layers is higher or the compensation ratio is smaller. The latter possibility is supported by recent suggestions about easier acceptor incorporation in the Ga-face GaN [12].

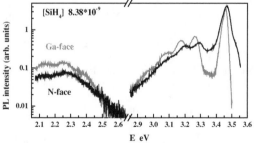

Figure 5. *Low temperature PL (4K) of hetero-epitaxial GaN of Ga- and N-face polarity. Although both layers were grown using the same [SiH₄], the free carrier concentration in these layers are different: $n_{N\text{-}face}=5.4 \times 10^{18}$ cm⁻³ and $n_{Ga\text{-}face}=1.7 \times 10^{18}$ cm⁻³.*

CONCLUSIONS

Doping and characterisation of MOCVD grown GaN on sapphire substrates is presented. An 1:1 relationship between [SiH$_4$] and the resulting carrier concentration is found, whereas the silicon incorporation showed to be hardly temperature dependent. For high [SiH$_4$] a strong influence on the morphology of the GaN layer is found. Increasing [SiH$_4$] broadened the FWHM of the X-ray diffraction rocking curves. The intensity of the UV PL signal is found to increase at increasing carrier concentrations up to n=2.5x10^{18} cm^{-3}. For the same carrier concentrations, photoluminescence studies revealed a shift of the conduction band towards lower energy positions.

ACKNOWLEDGEMENTS

This work is financially supported by the Dutch Technology Foundation (STW). Mrs. M. Moret and J.L. Weyher are greatly acknowledged for their contributions.

REFERENCES

1   F.K. de Theije, A.R.A. Zauner, P.R. Hageman, W.J.P. van Enckevort and P.K. Larsen, J. Crystal Growth **197**, 31 (1999)

2   X. Tang, H.G.M. Lochs, P.R. Hageman, M.H.J.M. de Croon and L.J. Giling, J. Crystal Growth **98**, 827 (1989)

3   P. Prystawko, M. Leszczynski, B. Beaumont, P. Gibart, E. Frayssinet, W. Knap, P. Wisniewski, M. Bockowski, T. Suski, S. Porowski, Phys. Status Solidi **(b) 210** 437 (1998)

4   J.L. Rouviere, J.L. Weyher, M. Seelmann-Eggebert and S. Porowski, Appl. Phys. Letters **73** (1998) 668

5   J.L. Rouviere, M. Arlery, B. Daudin, G. Feuillet, O. Briot, Mater. Sci. Eng. **B50** 61 (1997)

6   D.L. Rode, D.K. Gaskill, Appl. Phys. Letters **66,** 1972 (1995)

7   "Gallium Nitride (GaN II)", Semiconductors and semimetals, volume 57, Edited by J.I. Pankove and T.D. Moustakas, Academic Press (San Diego) 1999, p. 29.

8   W. Götz, L.T. Romano, J. Walker and N.M. Johnson, Appl. Phys. Letters **72,** 1214 (1998)

9   S. Hearne, E. Chason, J. Han, J.A. Floro, J. Figiel, J. Hunter, H. Amano, I.S.T. Tsong, Appl. Phys. Letters **74**, 356 (1999)

10   "Gallium Nitride (GaN II)", Semiconductors and semimetals, volume 57, Edited by J.I. Pankove and T.D. Moustakas, Academic Press (San Diego) 1999, p. 289.

11   P. Vennéguès, B. Beaumont, S. Haffouz, M. Vaille, P. Gibart, J. Crystal Growth **187** (1998) 167

12   M. Leszczynski, P. Prystawko, and S. Porowski, DataReview Series, **99**, B 2.3, (1999) 391

Mat. Res. Soc. Symp. Vol. 595 © 2000 Materials Research Society

Probing Nitride Thin Films in 3-Dimensions using a Variable Energy Electron Beam

Carol Trager-Cowan[1], D. McColl[1], F. Sweeney[1], S. T. F. Grimson[1], J-F. Treguer[1], A. Mohammed[1], P. G. Middleton[1], S. K. Manson-Smith[1], K. P. O'Donnell[1], W. Van der Stricht[2], I. Moerman[2] and P. Demeester[2], M. F. Wu[3], A. Vantomme[3], D. Zubia[4] and S. D. Hersee[4]
[1]Dept. Physics and Applied Physics, University of Strathclyde, Glasgow G4 0NG, UK.
[2]IMEC-INTEC, University of Gent, Gent 9000, Belgium.
[3]KULeuven, Leuven, Belgium.
[4]University of New Mexico, Center for High Technology Materials,1313 Goddard, SE Albuquerque, NM, USA.

ABSTRACT

In this paper we illustrate the application of electron beam techniques to the measurement of strain, defect and alloy concentrations in nitride thin films. We present brief comparative studies of CL spectra of AlGaN and InGaN epilayers and EBSD patterns obtained from two silicon-doped 3 µm thick GaN epilayers grown on an on-axis (0001) sapphire substrate and a sapphire substrate misoriented by $10°$ toward the m-plane $(10\bar{1}0)$.

INTRODUCTION

Thin films incorporating GaN, InGaN and AlGaN are presently arousing considerable excitement because of their suitability for UV and visible light emitting diodes and laser diodes. However, the films are of variable quality because of the lattice mismatch between them and presently used substrates (sapphire and α-SiC). We are presently using a number of electron beam analysis techniques namely cathodoluminescence (CL) imaging, CL spectroscopy and electron backscattered diffraction (EBSD), to investigate both the structural and optical properties of nitride films in 3-dimensions. Information in the 3rd dimension is extracted by acquiring data at different electron beam energies. Electron beams of energy between 1 and 20 keV are well matched to the length scales typical of nitride heterostructures. For example, a 2 keV beam deposits energy to a depth of ≈50 nm, a 10 keV electron beam deposits energy to a depth of ≈ 600 nm, while a 20 keV beam deposits energy to a depth of ≈ 2 µm in a GaN layer.

In this paper we illustrate the application of electron beam techniques to the analysis of nitride thin films by presenting brief comparative studies of
(i)   CL spectra of AlGaN and InGaN epilayers
and
(ii)   EBSD patterns obtained from two silicon-doped 3 µm thick GaN epilayers grown on an on-axis (0001) sapphire substrate and a sapphire substrate misoriented by $10°$ toward the m-plane $(10\bar{1}0)$.

EXPERIMENTAL DETAILS

All of the samples described in this paper were grown by metalorganic vapour phase epitaxy (MOVPE) on (0001) sapphire substrates (unless otherwise indicated). CL spectra were acquired using a home-built electron beam excitation system providing beams of energy up to 30 keV in a spot size of ≈ 200 µm at current densities up to 20 A cm^{-2} (but limited in the work to be described here to < 10 mA cm^{-2}). In the present work the front faces of the samples were positioned normal to the exciting electron beam and CL detected from the sample edge. Samples were cooled to a temperature of approximately 25 K using a closed cycle helium cryorefrigerator. Spectra were acquired using an Oriel InstaSpec^{TM} cooled 2-dimensional CCD array mounted at the output focal plane of a Chromex 0.5M monochromator.

A Cambridge 600S scanning electron microscope has been adapted by the addition of a home-built CL detection system (incorporating an Oxford Instruments parabolic mirror light collector) and a home-built EBSD imaging system. These allow the acquisition of comparative physical and luminescent images and EBSD patterns from the same area of a sample.

EXPERIMENTAL RESULTS AND DISCUSSION

CL profiling of AlGaN and InGaN epilayer

A variable energy electron beam has been used to investigate an AlGaN-on-GaN bilayer, grown on a sapphire substrate. The thickness of the AlGaN and GaN layers were determined by Rutherford backscattering (RBS) to be 0.48 µm and 0.44 µm respectively, RBS measurements also revealed the mean aluminium concentration to be 7%.

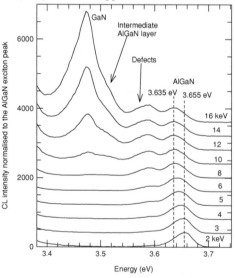

Figure 1 shows low temperature CL spectra acquired from the AlGaN/GaN/sapphire layers at electron beam energies from 2 keV to 16 keV. Spectra have been normalised in intensity and shifted vertically for clarity. It can be seen that the highest energy CL emission band, which we attribute to bound exciton emission from the AlGaN, red shifts from 3.655 eV to 3.635 eV as the electron beam energy increases from 2 keV to 16 keV. No beam current dependence of this peak shift is observed. At electron beam energies ≥ 3 keV we observe a band peaking at ≈ 3.59

Figure 1. CL spectra acquired at different electron beam energies from an AlGaN/GaN/Sapphire structure.

eV. This band increases in intensity with respect to the AlGaN bound exciton band as the electron beam energy increases. At electron beam energies ≥ 8 keV, an emission band at ≈3.475 eV becomes evident. We attribute this band to bound exciton emission from the underlying GaN epilayer. Also appearing at energies ≥ 8 keV is a shoulder on the high energy side of the GaN bound exciton band at ≈ 3.52 eV.

In order to interpret these results, it is necessary to know to what depth an electron beam of given energy penetrates the structure. Monte Carlo simulations of the electron trajectories [1] in an $Al_{0.07}Ga_{93}N$ layer, were therefore carried out. Figure 2 shows the results of these simulations.

Figure 2 illustrates that the electron beam penetrates deeper into the structure as its

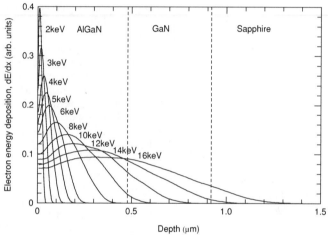

Figure 2. *Electron beam energy deposition calculated using Monte-Carlo simulations of the electron trajectories in an $Al_{0.07}Ga_{0.93}N$ layer.*

energy increases. It implies that electron beams with energy of 2 keV to 6 keV will address only the $Al_{0.07}Ga_{93}N$ layer, while electron beams with incident energy ≥ 8 keV break through into the GaN underlying layer while electron beams with energy ≥ 14 keV penetrate the sapphire substrate. This explains why electron beam energies ≥ 8 keV are required to excite significant emission from the GaN underlying layer. Some emission is just discernible from the GaN layer at electron beam energies ≤ 8 keV, we attribute this to secondary excitation of the GaN by CL emission from the AlGaN layer.

With increasing energy, the electron beam will excite an increasing proportion of material near and at the interface between the AlGaN and underlying GaN layer. We might expect to observe emission associated with defects due to this interface. We attribute the peak observed at ≈ 3.59 eV to such defect emission, the intensity of this emission increasing with increasing electron beam energy as the electron beam excites higher proportions of defected material.

We propose that the shoulder on the high energy side of the GaN bound exciton band at ≈ 3.52 eV is due to an AlGaN layer with low aluminium concentration (≈ 2%) [2] at the interface between the AlGaN and GaN layers. We base this conclusion on the fact that this shoulder appears at the same electron beam energy as the GaN bound exciton band and must therefore occur at a similar depth in the sample as the top of the GaN layer. The intensity of the shoulder while similar to that of the GaN bound exciton band at 8 keV, drops to less than half the intensity of the GaN emission band at 16 keV. This reduction in relative intensity of the shoulder may be due to the absence of the source of this emission as the electron beam probes deeper into the sample, i.e., the shoulder emission at ≈ 3.52 eV is due to a thin sandwich layer.

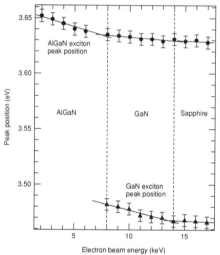

Figure 3. *Shift of the peak of the AlGaN and GaN exciton peaks as a function of electron beam energy.*

To understand the origin of the observed red shift of the AlGaN bound exciton band with increasing energy, we plot its peak position as a function of electron beam energy in figure 3. We also plot the dependence of peak position of the GaN bound exciton band. Figure 3 shows that the peak of the GaN bound exciton band shifts to the red by 15 meV as the electron beam energy is increased from 8 to 14 keV. These electron beam energies correspond to penetration depths of approximately 480 nm and 1.2 μm respectively, i.e., from the approximate location of the AlGaN/GaN interface to the GaN/sapphire substrate interface. We attribute this red shift to a relaxation towards the sapphire substrate of compressive strain in the GaN layer, where the strain is due to the overlying AlGaN layer. From the work of Kim et al [3] and Edwards et al [4], we estimate the change in stress across the GaN layer to be of order 4 kbar, this corresponds to a change in strain of approximately 0.15%. The AlGaN peak is seen to red shift by 20 meV as the electron beam energy is increased from 2 to 8 keV. These electron beam energies correspond to the electron beam penetrating from 50 nm to 480 nm into the layer, i.e., from close to the surface of the AlGaN layer to the AlGaN/GaN interface. As the electron beam energy increases from 8 keV to 16 keV the AlGaN peak continues to red shift as energy is deposited deeper into the sample. A total shift of 22 meV is observed. In this case there is more than one possible explanation for the red shift of the peak. It may be attributed to the relaxation towards the surface of the AlGaN layer of the tensile strain to which the underlying GaN subjects the AlGaN layer. Alternatively, the shift may be due to an increase in the aluminium concentration. A red shift of 22 meV from ≈ 3.65 eV corresponds to a change in aluminium mole fraction from ≈ 8% at the surface of the sample to ≈ 7% at the interface between the AlGaN and GaN epilayers [2]. It could be also due to a combination of a variation of strain and aluminium concentration. However, without an independent method of measuring either the strain or the aluminium concentration with a resolution of order 10 nm we cannot come to a definitive conclusion.

A variable energy electron beam has also been used to investigate an InGaN epilayer (≈ 400 nm thick) grown on an epilayer of GaN (1.0 μm), which in turn was grown on a sapphire substrate [5]. Figure 4 shows the peak position of the InGaN

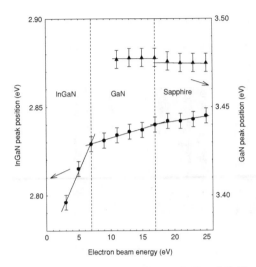

Figure 4. *Shift of the peak of the InGaN and GaN emission peaks as a function of electron beam energy*

emission band as a function of electron beam energy. We also plot the dependence of peak position of the GaN emission band. In this case there is no significant shift of the GaN peak position therefore no change in strain is detectable. The InGaN peak is seen to blue shift by 33 meV as the electron beam energy is increased from 3 to 7 keV. No beam current dependence of this peak shift is observed. These electron beam energies correspond to the electron beam penetrating from 90 nm to 350 nm into the layer, i.e., from close to the surface of the InGaN layer to the neighbourhood of the InGaN/GaN interface. As the electron beam energy increases from 7 keV to 25 keV the InGaN peak continues to blue shift as more energy is deposited deeper into the sample. A total shift of 50 meV is observed. As for the AlGaN epilayer discussed previously, there is more than one possible explanation for the observed blue shift. It may be attributed to the relaxation towards the surface of the InGaN layer of the compressive strain to which the underlying GaN subjects the InGaN layer. Alternatively, the shift may be due to an increase in the indium concentration towards the surface of the epilayer. A blue shift of 50 meV from ≈ 2.8 eV corresponds to a change in indium mole fraction from ≈ 15% at the surface of the sample to ≈ 14% at the interface between the InGaN and GaN epilayers [6]. We have revised our values for the indium mole fraction from those published in [5] in line with the results discussed in [6]. The shift could also be due to a combination of a variation of strain and indium concentration. In this case we suggest that a change in indium concentration may indeed play a role. In [5] we attributed the shift to the "compositional pulling effect" as reported by Shimuzu et al (1997) [7] and Hiramatsu et al (1997) [8]. That is the indium mole fraction is observed to be low during the initial stages of InGaN growth on GaN epilayers, but increases with increasing film thickness. Again, as previously discussed for the AlGaN epilayer, without an independent method for measuring either the strain or the indium concentration with a resolution of order 10 nm we cannot come to a definitive conclusion as to the origin of the shift of the InGaN peak.

EBSD patterns from GaN epilayers grown on misoriented sapphire substrates

EBSD is a diffraction technique that allows crystallographic information to be extracted from samples in a scanning electron microscope [9]. In particular it allows the measurement of strain with a lateral spatial resolution of order 20 nm [10]. We present here some preliminary measurements from two silicon-doped 3 μm thick GaN epilayers grown on an on-axis (0001) sapphire substrate and a sapphire substrate misoriented by

$10°$ toward the m-plane (10 $\bar{1}$ 0). Figures 5 (a) and (b) show the EBSD patterns from the on-axis GaN epilayer and the $10°$ epilayer respectively.

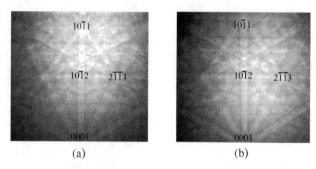

(a)   (b)

Figure 5. *EBSD patterns from GaN epilayers grown (a) on and (b) $10°$ off-axis respectively.*

A compression of the pattern towards the [10$\bar{1}$1] zone axis (pole) from the [0001] zone axis is observed in figure 5 (a) when compared to figure 5 (b). We tentatively attribute this shift to a difference of the in-plane compressive strain between the 2 samples where the $10°$ epilayer is under less compressive strain than the on-axis epilayer [11]. This result is consistent with those reported by Edwards et al [4] that GaN epilayers grown on vicinal substrates are less strained.

In the future we plan to quantify strain measurements using the EBSD technique. This will enable us to map the strain in GaN epilayers and will also allow us to make independent measurements of the strain in alloy samples such as the InGaN and AlGaN epilayers described in this paper. We may therefore come to definitive conclusions as to the origin of the observed shifts in the CL emission bands from these materials.

REFERENCES

1. E. Napchan and D B Holt, Inst. Phys. Conf. Ser. **87**, eds A. G. Cullis et al., (IOP, 1987) pp. 733.
2. G. Steude et al, *MRS Internet J. Nitride Semicond. Res.* **4S1**, G3.26 (1999).
3. S. Kim, et al, *Appl. Phys. Lett.,* **67** 380 (1995).
4. N. V. Edward et al, *Appl. Phys. Lett.,* **73** 2808 (1998).
5. C. Trager-Cowan, et al, *Mat. Res. Soc. Symp. Proc.,* **482**, 715 (1998).
6. K. P. O'Donnell et al, this conference reference number W11.26
7. M. Shimuzu et al, Sol. St. Electr., **41** 145 (1997).
8. K. Hiramatsu et al, *MRS Internet J. Nitride Semicond. Res.,* **2** 6 (1997).
9. D. J. Dingley et al, *Atlas of Backscattering Kikuchi Diffraction Patterns,* (IOP, 1995).
10. D. J. Dingley and D. P. Field, *Materials Science and Technology,* **13** 69 (1997).
11. K. Z. Troost et al, *Appl. Phys. Lett.,* **62** 1110 (1993).

Growth—MOCVD, HVPE, Bulk

Mat. Res. Soc. Symp. Vol. 595 © 2000 Materials Research Society

MOVPE Growth of Quaternary (Al,Ga,In)N for UV Optoelectronics

Jung Han,[*] Jeffrey J. Figiel, Gary A. Petersen, Samuel M. Myers, Mary H. Crawford, Michael A. Banas, and Sean J. Hearne
Sandia National Laboratories, Albuquerque, NM 87185

ABSTRACT

We report the growth and characterization of quaternary AlGaInN. A combination of photoluminescence (PL), high-resolution x-ray diffraction (XRD), and Rutherford backscattering spectrometry (RBS) characterizations enables us to explore the contours of constant- PL peak energy and lattice parameter as functions of the quaternary compositions. The observation of room temperature PL emission at 351nm (with 20% Al and 5% In) renders initial evidence that the quaternary could be used to provide confinement for GaInN (and possibly GaN). AlGaInN/GaInN MQW heterostructures have been grown; both XRD and PL measurements suggest the possibility of incorporating this quaternary into optoelectronic devices.

INTRODUCTION

Development of nitride-based ultraviolet (UV) emitters has thus far been overshadowed by the intense effort in the InGaN-based visible light emitters.[1] The potential usage of a compact UV light source ranges from energy-efficient indoor lighting (as a white-light source), free space satellite communication, compact chemical, environmental, and biological sensing devices, applications in medical diagnosis, treatment, and possibly optical storage. With all the rewarding promises, however, the development of UV emitters faces many challenging issues. Preliminary work in the AlGaN/GaN quantum well (QW) UV emitters[2] suggested that the use of a binary GaN active region results in a low optical output power. Furthermore, mismatch-induced tensile stress was observed during growth of AlGaN heterostructures on thick GaN templates for UV devices,[3] leading to relaxation through crack generation. Quaternary AlGaInN compound semiconductors are expected to enclose a finite (non-zero) area on the plot of energy gap versus lattice constant. In principle the employment of quaternary compounds should render flexibility in tailoring bandgap profile while maintaining lattice matching and structural integrity. In this paper we will summarize our investigation on the issues of optical efficiency and strain control using quaternary AlGaInN compound.

EXPERIMENTAL DETAILS

The MOVPE growth is carried out in a vertical rotating-disc reactor. All of the quaternary epilayers and multiple quantum well (MQW) structures were grown on 1-μm GaN epilayers at 1050°C using a standard two-step nucleation procedure (on sapphire) with low-temperature GaN grown at 550 °C.[4] The growth temperature of the quaternary

---

[*] Electronic mail: jhan@sandia.gov

AlGaInN was varied between 750 and 820 °C. The reactor pressure was held constant at 200 Torr. The NH_3 and N_2 flows were set at 6 l/min each. An additional flow of H_2 (~400 cm^3/min) was employed as a carrier gas. Trimethylgallium, triethylaluminum, and trimethylindium were employed as metalorganic precursors.

Given the scarcity of reliable information concerning the quaternary, it is imperative to accurately determine the concentrations of the constituent elements. Concentrations of In and Al in the films were measured using Rutherford backscattering spectrometry (RBS)[5] with a 2.5 MeV $^4He^+$ ion beam. Ion-channeling effects in the RBS spectra were randomized by tilting the sample 10° from normal to the analysis beam and continuously rotating the sample about its normal during the analysis. Al and In concentrations were then extracted from the spectra using the SIMNRA (ver. 4.4) simulation program.[6] Figure 1 shows a representative RBS spectrum from AlGaInN. The solid line is the fitted simulation, while the dashed line is a simulation for the same In concentration with no Al. The inset shows a diagram of the measured film on GaN, along with the fitted concentration values for In and Al. The large backscattering yield from In, coupled with the accurately known nuclear cross section, results in a relatively high accuracy for the determination of the In concentration. We estimate an uncertainty of ±0.002 at an In concentration of 0.01, and ±0.005 for a concentration of 0.10. Because the backscattering yield from Al is much smaller, the Al concentration was determined less directly by measuring a reduction, or dip, in the yield from Ga, due to the presence of the Al and In in the layer. An uncertainty of ±0.01 at an Al concentration of 0.10, and ±0.02 for a concentration of 0.20 was estimated for the cited data.

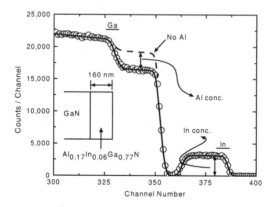

Figure 1. An RBS spectrum from an AlGaInN sample. The solid line is a simulated fit.

A high-resolution triple-axis x-ray diffractometer (Philips X'Pert System) was employed to assess the structural quality and determine both in-plane (a) and out-of-plane (c) lattice parameters using symmetric (0002) and asymmetric (202-4) (using grazing-incidence) diffractions. Photoluminescence (PL) was measured using a HeCd laser (325 nm) as the excitation source in conjunction with a 0.3 meter spectrometer and a UV enhanced CCD detector. Typical excitation level is around 30 W/cm^2. Real time wafer curvature measurements were performed with a multi-beam optical stress sensor (MOSS)[7] modified for use on our reactor. To determine the wafer curvature, the divergence of an array of initially parallel laser beams is measured on a CCD camera after reflection of the array from the film/substrate surface. Changes in wafer curvature induce a proportional change in the beam spacing on the camera. This technique provides a direct measurement of the stress-thickness product during MOVPE of GaN.[8]

In attempting to extract band gap energy from a PL measurement for the AlGaInN quaternary system, one is reminded that the determination of band gap energy for GaInN has been a subject of much debate.[9] Indium-related compositional fluctuation is expected to contribute to a Stokes-like shift with a magnitude of less than 100 meV.[10] Field-induced separation of electrons and holes (due to piezo- and spontaneous polarization) causes a red shift (up to 300 meV) of PL emission energy in quantum well structures[11] but should not be of significance in epilayers thicker than 50 nm. Finally, the existence of a compressive or tensile biaxial strain, due to a combination of lattice mismatch[12] and thermal mismatch,[8] introduces a blue or red shift, respectively, to the band gap energy as determined by the deformation potentials.[13] Given that the In fraction is limited to 10% and the pseudomorphic strain in the quaternary does not exceed 1% in compression, we estimate (using published deformation potentials[14]) that the difference between PL emission energy and band gap to be less than 100 meV in the epilayers reported here.

RESULTS AND DISCUSSION

GaN and GaN:In Active region Study

In our earlier work we have shown an AlGaN/GaN QW LED emitting at a wavelength of 354 nm.[2] The power output of the prototype UV emitter was around 13 μW at a current level of 20 mA, which is about 2 orders of magnitude lower than the commercial high-brightness blue LEDs. A series of test samples, consisted of 0.5 μm of GaN test layers grown on a standard GaN epilayer (of 1 μm thick), were grown to optimize the optical quality of the active region. Photoluminescence (PL) at room temperature (RT) was employed to measure the optical quality. Among the growth parameters investigated, reactor pressure was found to have a distinct effect on the PL emission intensity. As the growth pressure was increased from 80 Torr (the pressure employed in the previous LEDs and chosen to mitigate the gas-phase reaction during growth of AlGaN) to 250 Torr, integrated PL intensity increased by more than 10 times and peaked at around 200 Torr. The mechanism responsible for the observed enhancement of GaN emission is not known. One might speculate that a reduction of carbon incorporation, due to an increase of the partial pressures of both NH_3 and H_2, reduces the non-radiative recombination paths.[15]

Recently it has been reported by several groups[16,17,18] that the addition of a small amount of indium could lead to an enhancement in PL emission efficiency. Mukai et al.[16] demonstrated an GaN:In UV LED with an indium content of around 2% and optical power output as high as 5 mW. A series of GaN:In samples (0.2 μm thick on 1 μm GaN epilayers) were grown in which the indium fraction was increased from 0 to approximately 4% as determined by x-ray and RBS. Figure 1 shows the PL spectrum of the GaN:In layers with the spectrum from an "optimized" GaN layer as a reference (the bottom trace peaking at 364 nm). At the expense of a shift of emission peak toward longer wavelengths (~380 nm), we attained an increase of more than one order of magnitude in integrated light intensity. Two explanations have been offered to explain the observed enhancement in light emission: carrier localization due to indium incorporation [16] or the suppression of non-radiative recombination defects in the presence of indium. [17] Details of the optical investigation will be reported in a separate article in this proceeding. [19]

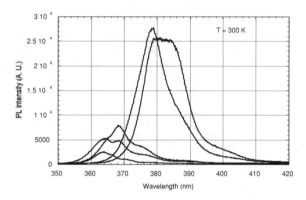

Figure 2. PL of GaInN epilayers with various amount of Indium. The integrated intensity increases by more than 10 times from GaN to 4% GaInN.

Quaternary AlGaInN

Only limited information has been published concerning the AlGaInN quaternary system. Matsuoka[20] predicted the presence of an unstable mixing region (spinodal phase separation) in wurtzite AlGaInN as the indium content increases. The challenges associated with the growth of AlGaInN, primarily the selection of conditions necessary to enable surface diffusion of Al and Ga species while at the same time preventing the surface re-evaporation of In, are partially manifested by the scarcity of experimental data. McIntosh *et al.*[21] and Yamaguchi *et al.*[22] reported the physical incorporation of Al and In into GaN by MOVPE. Aiming at proof-of-concept demonstrations of material synthesis, little information was provided in either work concerning the bandgap versus alloy compositions. The functional feasibility of AlGaInN as a confinement barrier also remained ambiguous in these studies since PL emission (energies) from samples investigated did not exceed the emission (energy) of GaN (~363 nm/ 3.42 eV)). The concern was reiterated by Peng *et al.*[23] in stating, based on optical absorption measurements of sputtered polycrystalline films, that AlInN and AlGaInN do not seem to provide good confinement (to GaInN) due to a very strong bowing effect. We have shown from in-situ stress measurement that the use of high Al-fraction (or thick) AlGaN barriers on GaN templates, required for electrical and optical confinement in UV devices, leads to a buildup of tensile stress and subsequently the occurrence of cracking. [3] As discussed in the above section, introducing indium into GaN greatly enhances the optical efficiency but at the cost of an increase in emission wavelength. Investigation of quaternary AlGaInN was initiated by the two constraints encountered in the use of ternaries AlGaN and GaInN as barriers and wells, respectively. It is expected that AlGaInN would render flexibility in controlling the mismatch strain and bandgap profile.

Figure 3 shows (0002) $2\theta-\omega$ x-ray diffraction (XRD) scans (with an analyzer-crystal detector) for several AlGaInN (a, c-f) and GaInN (b) epilayer samples. Thickness of the samples varies from 0.14 to 0.20 μm, as indicated by the change of the period of the diffraction fringes. (Thin film thickness derived independently from x-ray fringes and from RBS simulations differed by less than 5% in most cases.) By adjusting the relative Al and In compositions, one observes a change of the out-of-plane lattice parameter from being larger (top trace) to smaller (bottom trace) than that of GaN. Pseudomorphic growth is verified by measuring the in-plane lattice parameter (*a*) of the epilayers. Unlike the case in zincblende semiconductor heterostructure, the appearance of an overlap between the (0002) diffraction peak from the quaternary and the peak from GaN (a condition between scans (d) and (e)) does not necessarily imply a lattice-matched strain-free epitaxy as the in-plane and out-of-plane lattice parameters of AlN, GaN, and InN do not vary in a linear manner. Using the published lattice parameters,[12] exact match of the *a* lattice constant of AlInN to GaN should occur at an Al fraction of 18% while it takes approximately 28% Al in AlInN to match the out-of-plane (*c*) lattice parameter of GaN. A quaternary compound with a *c* lattice parameter matched to GaN is speculated to be under biaxial compression during growth.

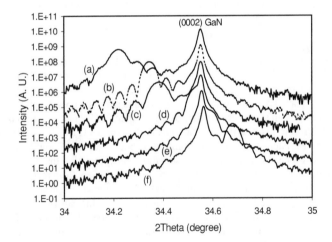

Figure 3. *(0002) $2\theta-\omega$ x-ray diffraction of AlGaInN (solid traces) and GaInN (dashed line) epilayers showing a range of lattice parameters bracketing the GaN lattice parameter. The (0002) GaN diffraction peaks move progressively toward higher-angle side as the lattice parameters of the upper alloy layers reduce.*

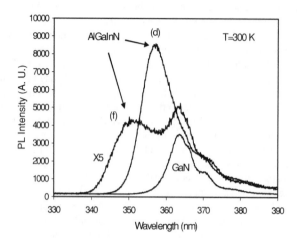

Figure 4. *Room-temperature photoluminescence from a GaN (solid line) epilayer and two AlGaInN epilayers (samples d and f).*

Optical properties of the AlGaInN samples (d and f in Figure 3) were investigated with RT PL (Figure 4). The PL from a "standard" GaN epilayer was included for comparison. It is clear from Figure 4 that the AlGaInN epilayer emits with energy higher than that of GaN. The $Al_{0.14}Ga_{0.82}In_{0.04}N$ epilayer (sample (d) in Figures 3 and 4) gives the indication that one can synthesize a wider bandgap (than GaN) quaternary AlGaInN compound yet with an in-plane lattice constant *larger* than that of GaN. It is worth noting that the integrated PL intensity from this quaternary is about three times brighter than that from GaN. There has been much empirical observation and speculation that the presence of indium in GaInN ternary layers seems to preserve excited carriers from recombining at non-radiative centers.[24] Mukai *et al.*[16] reported, in the development of ultra-violet LEDs, a rapid reduction of optical efficiency as the In content in the GaInN active region decreased below a certain value. The implication is that the carrier-localization effect,[24] possibly due to the presence of a compositional fluctuation of indium, could make AlGaInN an alternative (to GaN or AlGaN) as a short-wavelength ($\lambda < 360$nm) light-emitting medium

For the design and implementation of quaternary heterostructure devices, it is helpful to superimpose contours of constant lattice parameter and "bandgap" (or more precisely, PL peak) energy as functions of the alloy compositions. A series of AlGaInN, GaInN, and AlGaN epilayers have been grown and the characterizations are summarized in Figure 5. $Al_yGa_{1-x-y}In_xN$ is described in this work by a right-angle triangle with AlGaN and GaInN forming the two sides around the right-angle vortex. (The hypotenuse corresponds to the ternary AlInN, not shown in Figure 5.) The location of any $Al_yGa_{1-x-y}In_xN$ quaternary is uniquely specified by the In (x) and Al (y). The triangles, diamonds, and circles represent the data points for AlGaN, GaInN, and AlGaInN, respectively. The compositions of the ternaries were determined primarily by x-ray diffraction and independently confirmed by RBS, the quaternary compositions were determined by RBS. The solid line with a slope of around 4.4 defines the quaternary alloys that are lattice matched to GaN; one can construct an array of parallel lines for lattice-matched growth with different in-plane lattice parameters. (Vegard's law predicts that the in-plane lattice parameter will remain unchanged when the increments of Al (ΔAl) and In (ΔIn) maintain a ratio of 4.44.) Quaternary alloys located to the lower right half of the straight line are expected to be under compression on a GaN template. The distance between a given data point on this plot and the solid line gives a qualitative measure of mismatch-induced strain.

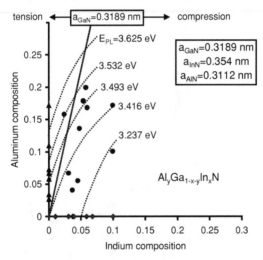

Figure 5. *Contour plot of constant (in-plane) lattice parameter and emission energy versus Al and In compositions for AlGaInN epilayers.*

PL investigation of the near band-edge emission is summarized by the dashed curves (denoting contours of constant PL peak energy) in Figure 5. The dashed contours are derived based on two assumptions: i) all the data points should be fitted by contours with nearly identical forms, and ii) the contours should be consistent with the available (theoretical[25] and experimental[26]) bowing information of ternary AlInN. Along each dashed line, the emission wavelength remains constant while the in-plane lattice parameter (and therefore strain) can be adjusted by varying the Al and In fractions. Since the In-related Stokes-shift tends to increase with an increasing In composition, the slopes of the contours for the actual bandgap energy might be less steep.

AlGaInN/GaInN Multiple Quantum Wells

To demonstrate the possibility of heterostructures incorporating quaternary AlGaInN, we have grown multiple quantum well (MQW) structures consisting of 10 periods of AlGaInN barriers (Al~14%, In~4%, ~100Å) and GaInN wells (In~4%, ~45Å).

Figure 6 shows the (0002) 2θ–ω XRD from the MQW structure. The presence of satellite peaks and other interference fringes is indicative of structural order and coherency (along the growth direction). Cross-sectional TEM images (Figure 7) indicated that the AlGaInN/GaInN interfaces are smooth and coherent on the atomic scale. One also notices from Figure 6 that the AlGaInN/GaInN MQW has a zero-order peak that corresponds to the lattice constant of a 2.0% GaInN alloy. In-situ stress/strain monitoring was conducted during the growth of a similar MQW structure and is shown in Figure 8. While the GaN layer is under tension[8] during growth (0.40 GPa), the AlGaInN/GaInN MQW region is under compression during growth (~0.47±0.07 GPa). Assuming a compressive thermal mismatch stress of 0.2 GPa from 1050°C to 800°C, one can estimate an average compressive mismatch strain of 0.15%, corresponding to the lattice constant of a 1.4% GaInN epilayer.

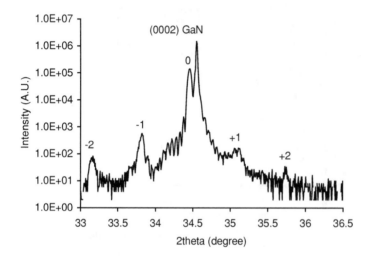

Figure 6. *(0002) 2θ-ω x-ray diffraction of an AlGaInN (100Å)/GaInN (45Å) multiple (10) quantum wells.*

Optical quality of the AlGaInN/GaN MQWs was investigated using temperature-dependent PL. It is generally assumed that the non-radiative recombination mechanisms are thermally activated and the decay of integrated PL intensity from low temperature to room temperature renders a quantative measurement of the nonradiative processes.[27,28] Figure 9 shows the temperature-dependent PL with a plot of the integrated intensity in the inset. A reduction of 4.5 times of the integrated PL intensity from 4K to RT is comparable to the best reported for GaInN MQWs emitting at around 450 nm.[29] This also implies that an upper limit to the internal quantum efficiency of 20% is obtained at RT.

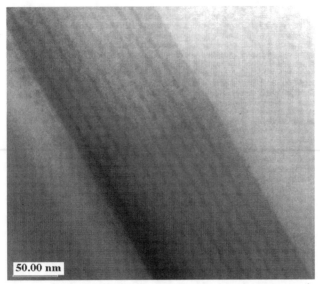

Figure 7. *Cross-sectional TEM image of an AlGaInN (100Å)/GaInN (45Å) multiple (10) quantum wells.*

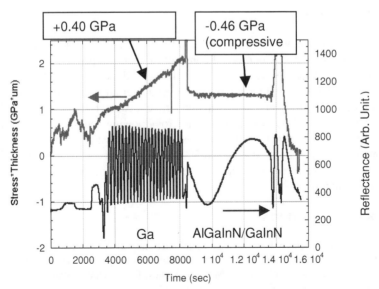

Figure 8. *In-situ stress measurement of during growth of an MQW LED.*

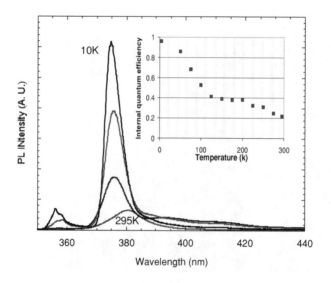

Figure 9. *Temperature-dependent PL (at 10, 75, 150, and 295 K with decreasing intensity) suggesting an internal quantum efficiency of around 20% from the AlGaInN/GaInN MQWs.*

CONCLUSIONS

In conclusion, we reported the MOVPE growth and characterization of quaternary AlGaInN. Contours of constant PL peak energy and lattice parameter of AlGaInN were projected onto a two-dimensional plot. We confirmed that the employment of this novel quaternary compound could facilitate the tailoring of energy gap and lattice mismatch for design flexibility. In spite of the previous speculation that strong bowing could hinder the usage as a short-wavelength light emitter, we observed PL emission as short as 351 nm from a layer with about 20% Al and 5% In. The possibility of replacing AlGaN as an alternative barrier material was demonstrated through an AlGaInN/GaInN MQW structure.

ACKNOWLEDGMENTS

The authors acknowledge stimulating interactions with J. Y. Tsao, A. F. Wright, S. R. Lee, and H. Amano (Meijo University). Sandia is a multiprogram laboratory, operated by Sandia Corporation, a Lockheed Martin company, for the United States Department of Energy, under contract DE-AC04-94AL8500.

REFERENCES

[1] For a review, see S. Nakamura and G. Fasol, *The Blue Laser Diode*, Springer-Verlag, Berlin (1997).

[2] J. Han, M. H. Crawford, R. J. Shul, J. J. Figiel, M. Banas, L. Zhang, Y. K. Song, H. Zhou, and A. V. Nurmikko, Appl. Phys. Lett, 73, 1688 (1998)

[3] J. Han, M. H. Crawford, R. J. Shul, S. J. Hearne, E. Chason, J. J. Figiel, and M. Banas, MRS Internet J. Nitride Semicond. Res. 4S1 G7.7 (1999).

[4] J. Han, T. –B. Ng, R. M. Biefeld, M. H. Crawford, and D. M. Follstaedt: Appl. Phys. Lett. **71** (1997) 3114.

[5] W-K. Chu, J. W. Mayer, and M.-A. Nicolet: *Backscattering Spectrometry* (Academic, New York, 1978).

[6] M. Mayer: *SIMNRA User's Guide*, Technical Report IPP 9/113 (Max Planck-Institut fur Plasmaphysik, Garching, Germany, 1997).

[7] C. Taylor, D. Barlett, E. Chason, J. A. Floro, Ind. Physicist 4, 25 (1998)

[8] S. Hearne, E. Chason, J. Han, J. A. Floro, J. Hunter, J. J. Figiel: Appl. Phys. Lett. **74** (1999) 356.

[9] T. Takeuchi, H. Takeuchi, S. Sota, H. Sakai, H. Amano, and I. Akasaki: Jpn. J. Appl. Phys. **36** (1997) L177; S. Nakamura and T. Mukai: J. Vac. Sci. Tech. A **13** (1995) 705.

[10] S. Chichibu, T. Azuhata, T. Sota, and S. Nakamura: Appl. Phys. Lett. **70** (1997) 2822; S. F. Chichibu, A. C. Abare, M. P. Mack, M. S. Minsky, T. Deguchi, D. Cohen, P. Kozodoy, S. B. Fleischer, S. Keller, J. S. Speck, J. E. Bowers, E. Hu, U. K. Mishra, L. A. Coldren, S. P. DenBaars, K. Wada, T. Sota, and S. Nakamura: Mat. Sci. Eng. **B59** (1999) 298.

[11] S. Chichibu, T. Deguchi, T. Sota, K. Wada, and S. Nakamura: Mat. Res. Soc. Symp. Proc. **482** (1998) 613.

[12] O. Ambacher, J. Phys. D: Appl. Phys. 31, 2653 (1998).

[13] C. Kisielowski, J. Kruger, S. Ruvimov, T. Suski, J. W. Ager III, E. Jones, Z. Liliental-Weber, M. Rubin, E. R. Weber, M. D. Bremser, and R. F. Davis: Phys. Rev. **B 54** (1996) 17745.

[14] S. Chichibu, A. Shikanai, T. Azuhata, T. Sota, A. Kuramata, K. Horino, and S. Nakamura, Appl. Phys. Lett. 68, 3766 (1996)

[15] S. M. Bedair, *Gallium Nitride (GaN) I*, Semiconductors and Semimetals, v50, p127 (1998), Academic Press, San Diego

[16] T. Mukai, D. Morita, and S. Nakamura, J. Cryst. Growth 189/190, 778 (1998)

[17] Y. Narukawa, S. Saijou, Y. Kawakami, S. Fujita, T. Mukai, and S. Nakamura, Appl. Phys. Lett. 74, 558 (1999)

[18] X. Shen and Y. Aoyagi, Jpn. J. Appl. Phys. 38, L14 (1999).

[19] M. H. Crawford et al., Mat. Res. Soc. Symp.Proc. for Fall MRS Meetings, 1999.

[20] T. Matsuoka: Appl. Phys. Lett. **71** (1997) 105.

[21] F. G. McIntosh, K. S. Boutros, J. C. Roberts, S. M. Bedair, E. L. Piner, and N. A. El-Masry: Appl. Phys. Lett. **68** (1996) 40; S. M. Bedair, F. G. McIntosh, J. C. Roberts, E. L. Piner, K. S. boutros, N. A. El-Masry: J. Cryst. Growth **178** (1997) 32.

[22] S. Yamaguchi, M. Kariya, S. Nitta, H. Kato, T. Takeuchi, C. Wetzel, H. Amano, and I. Akasaki: J. Cryst. Growth **195** (1998) 309.

[23] T. Peng, J. Piprek, G. Qiu, J. O. Olowolafe, K. M. Unruh, C. P. Swann, and E. F. Schubert: Appl. Phys. Lett. **71** (1997) 2439.

[24] S. F. Chichibu, H. Marchand, M. S. Minsky, S. Keller, P. T. Fini, J. P. Ibbetson, S. B. Fleischer, J. S. Speck, J. E. Bowers, E. Hu, U. K. Mishra, S. P. DenBaars, T. Deguchi, T. Sota, and S. Nakamura: Appl. Phys. Lett. **74** (1999) 1460.

[25] A. F. Wright and J. S. Nelson: Appl. Phys. Lett. **66** (1995) 3051; A. F. Wright and J. S. Nelson: Appl. Phys. Lett. **66** (1995) 3465.

[26] K. S. Kim, A. Saxler, P. Kung, M. Razeghi, and K. Y. Lim: Appl. Phys. Lett. **71** (1997) 800.

[27] M. Leroux, N. Grandjean, B. Beaumont, G. Nataf, F. Semon, J. Massies, and P. Gibart, J. Appl. Phys. 86, 3721 (1999).

[28] Y. Narukawa, Y. Kawakami, S. Fujita, and S. Fujita, Phys. Rev. B 55, R1938 (1997).

[29] S. F. Chichibu, private communication (1999).

Mat. Res. Soc. Symp. Vol. 595 © 2000 Materials Research Society

Homo-epitaxial growth on misoriented GaN substrates by MOCVD

A.R.A. Zauner[1], J.J. Schermer[1], W.J.P. van Enckevort[1], V. Kirilyuk[1], J.L. Weyher[1,2], I. Grzegory[2], P.R. Hageman[1], and P.K. Larsen[1]

[1]Research Institute for Materials, University of Nijmegen, Toernooiveld, 6525ED Nijmegen, The Netherlands
[2]High Pressure Research Center, Polish Academy of Sciences, Sokolowska 29/37, 01-142 Warsaw, Poland

ABSTRACT

The N-side of GaN single crystals with off-angle orientations of 0°, 2°, and 4° towards the $[10\bar{1}0]$ direction was used as a substrate for homo-epitaxial MOCVD growth. The highest misorientation resulted in a reduction of the density of grown hillocks by almost two orders of magnitude as compared with homo-epitaxial films grown on the exact $(000\bar{1})$ surface. The features still found on the 4° misoriented sample after growth can be explained by a model involving the interaction of steps, introduced by the misorientation and the hexagonal hillocks during the growth process.

INTRODUCTION

Metalorganic Chemical Vapour Deposition (MOCVD) growth of Gallium Nitride (GaN) in the $[000\bar{1}]$ direction is associated with the formation of surface defects, such as hexagonal hillocks [1,2,3]. The inversion domain, located at the centre of the point-topped pyramids, apparently causes hillock formation as a result of the higher growth rate of this defect compared with the growth rate of the surrounding matrix [3]. Also for homo-epitaxial growth on the N-side of GaN substrates formation of hexagonal pyramids is observed [4,5].

Homo-epitaxial growth in the $[000\bar{1}]$ direction has the advantage that the N-side of GaN single crystals can be mechano-chemically polished to obtain epi-ready substrates [6], while the Ga-side (for growth in the [0001] direction) can only be mechanically polished, so that reactive ion etching is needed to prepare epi-ready substrates [7].

For device applications smooth surfaces are required, therefore the formation of hillocks should be avoided. A common way to avoid growth features on the surface is the use of misoriented substrates [8]. In the present work the surface morphology of homo-epitaxial GaN layers is studied for different off-angle orientations from the exact $[000\bar{1}]$ direction. It is found that the formation of hexagonal pyramids can be strongly suppressed by the use of a sufficiently large misorientation, resulting in much smoother layers.

EXPERIMENTAL

GaN single crystals [9] were used as substrates for MOCVD growth. The $(000\bar{1})$ substrate surfaces were mechano-chemically polished [6] to obtain off-angle orientations of 0°, 2°, and 4° towards the $[10\bar{1}0]$ direction. The misorientation was confirmed by X-ray diffraction analysis.

Homo-epitaxial GaN growth was performed at a temperature of 1040°C and a pressure of 50 mbar using trimethylgallium (TMG) and ammonia (NH_3) as precursors, and hydrogen (H_2) as carrier gas. The substrate crystals were heated to growth temperature under a NH_3 gas flow diluted in nitrogen (N_2). Growth was performed with a V/III molar ratio of about 1700 with a total flow of about 5 slm.

The surfaces of the homo-epitaxial GaN layers were investigated by optical differential interference contrast microscopy (DICM), a technique very sensitive for detecting local differences in surface slope, which are visualised by different shades of grey in the figures.

RESULTS AND DISCUSSION

Optical examination with DICM of layers grown on $(000\bar{1})$ substrates without misorientation reveals a large number of hexagonal growth hillocks covering almost the entire surface. As shown in figure 1a the hexagonal base of the hillocks is 10-50 μm in size. The majority of the hillocks are regular point-topped pyramids, although some of them are macroscopically flat-topped or disrupted. The highest hillocks, with steeper side facets, show an increased contrast. Similar hillock morphology has been observed for hetero-epitaxial GaN layers grown in the $[000\bar{1}]$ direction [3,10]. Faster growing inversion domains of Ga-polarity in a matrix of N-polarity lead to the formation of hillocks [3]. Also for homo-epitaxial growth on the N-side hexagonal pyramids were observed [4,5].

The virtual $(000\bar{1})$ plane, of macroscopically flat-topped pyramids, is faceted due to steps generated by a step source located at the centre of the hexagon. These steps originate from dislocations at the centre of such hillocks, as was concluded from the occurrence of interlaced spirals with atomic height steps [11].

For the sample with a 2° off-angle orientation towards the $[10\bar{1}0]$ direction the density of hillocks at the surface is significantly lower compared to the exactly oriented samples (see table I). The misorientation induced step flow started to overgrow part of

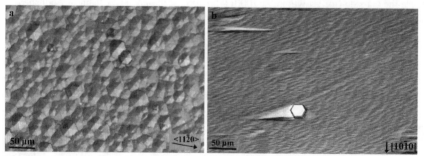

Figure 1. *DICM image of a homo-epitaxial GaN layer grown on the N-side $(000\bar{1})$ of a GaN single crystal (a)without misorientation and (b) 4° misorientation towards the $[10\bar{1}0]$ direction.*

Table I. *The dependence of the hillock density on the misorientation of the GaN substrates.*

misorientation (°)	hillock density (cm^{-2})
0	1.2E+05
2	2.8E+04
4	1.3E+03

the hexagons. However, for the majority of the features on the surface the source of growth is still present. The step flow resulting from the 2° misorientation appeared to be insufficient to overgrow the centres of the hillocks.

The estimated hillock density for the 4° misoriented specimen is nearly two orders of magnitude lower as compared with specimens of exact orientation (see table I). Obviously the step flow resulting from the misorientation, which is doubled compared to the 2° off-angle sample discussed above, is sufficient to overgrow the centres of the majority of the hexagons.

Although, for 4° misorientation most hexagonal features disappeared from the surface (figure 1b), there is still a number of partial and complete hexagonal hillocks that can be recognised. A number of these features is shown in figure 2. Figure 2a shows a kind of plateau with a partly overgrown hexagonal hillock near its centre, dividing the plateau in two parts. The two regions, which border on the edges of the hillock, are not equally sized. From the difference in size of those two areas it can be concluded that the misorientation is not exactly in the [10$\bar{1}$0] direction. The plateau ends abruptly in a steep step bunch perpendicular to the [10$\bar{1}$0] direction. In figure 2b a part of a hexagon can still be recognised. Like the hillock shown in figure 2a, the two side areas that border on the hexagon are not equal in size and the plateau terminates in the [10$\bar{1}$0] direction resulting in a sharp, steep edge. For the feature shown in figure 2c a discontinuity in slope of the plateau itself can be recognised at the place where the plateau starts to narrow. On the plateaus of the surface features of figure 2a-2c steps sloping towards the <11$\bar{2}$0> directions can be seen.

A plausible explanation for the occurrence of the surface features shown in figure 2

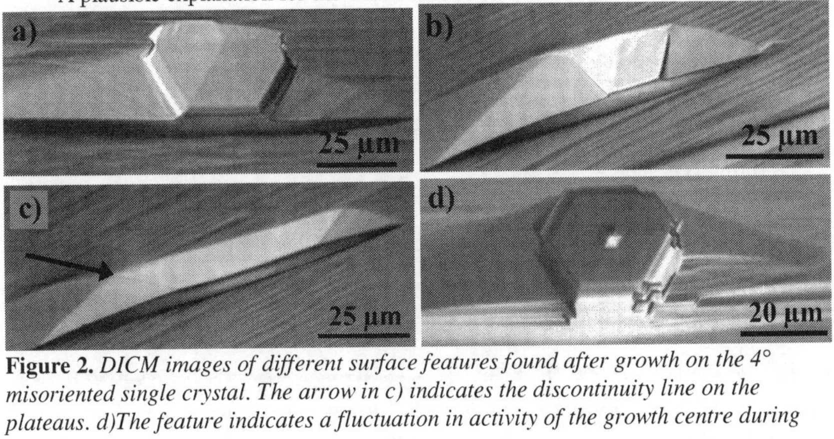

Figure 2. *DICM images of different surface features found after growth on the 4° misoriented single crystal. The arrow in c) indicates the discontinuity line on the plateaus. d)The feature indicates a fluctuation in activity of the growth centre during growth.*

can be given. The sample was mechano-chemically polished to a 4° misorientation, thereby introducing steps approximately parallel to [$1\bar{1}20$], over the sample. At the beginning of growth shallow as well as steep hexagonal hillocks tend to be formed. However, only steep hexagonal hillocks, which are formed by a more active step source, survive the propagation of the misorientation step train and will be present after a certain period of growth (see figure 2). Less steep hillocks are overgrown in an early stage of the process or cannot be formed at all. In figure 1a it can be seen that the height of the hillocks varies over the sample. About 3% of the hillocks presented in figure 1a show an enhanced contrast which indicates steeper hillocks. This percentage is in agreement with the previous observation of the reduced hillock density by almost two orders of magnitude for the 4° misoriented sample as compared with the samples without misorientation.

In addition, the activity of many individual hillocks decreases or fluctuates during growth (see figure 2d). If the activity of the step source at the hillock centre decreases vertical growth slows down but the base of the hillock still expands. The time dependent overgrowth of a steep hillock of which the vertical growth is suddenly almost stopped is represented schematically in figure 3.

In its steep period as well as in its first time of slow vertical growth the slowest advancing steps, in the <$10\bar{1}0$> directions, bound the hillock (figure 3a). The steps introduced by the misorientation are more or less in the [$10\bar{1}0$] direction, which means that these steps also move slowly. Steps in other directions move considerably faster. Steps from the left and right hand side of the hillock form a re-entrant corner with the steps induced by the misorientation (see figure 3). Due to an increased effective supersaturation near this region the step velocity increases and the re-entrant corner becomes rounded [12]. Now a whole range of step orientations is created, and steps in the fast moving <$11\bar{2}0$> directions are selected to propagate over the plateaus.

Figure 3b represents a hillock of which the step source has already been inactive for some time at the moment that the misorientation-induced steps just start to overgrow the flat top of the hillock. The steps on the plateau, left and right from the hillock, move faster than the steps introduced by the misorientation. During continued growth the hexagon-centre will be overgrown (figure 3d-3e).

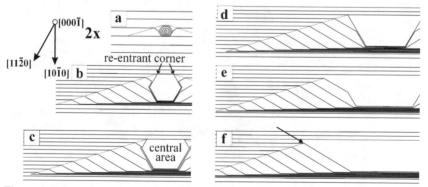

Figure 3. *Schematic representation of the time-dependent development and subsequent overgrowth of a hexagonal hillock. Image a) is two times enlarged. The arrow in f) indicates the discontinuity line on the plateau, cf. figure 2c.*

From the moment that the hillock is half overgrown, and no steps opposite to the misorientation-steps exist (figure 3d), the position of the re-entrant corner moves along the intersection line of the remaining hillock side facet and the plateau during prolonged growth (figure 3e) and will finally evanesce (figure 3f). The boundary between the stepped plateau areas and the central region of the plateau, which is more or less parallel to the $(000\bar{1})$ plane, now corresponds with the last fast $<11\bar{2}0>$ step emitted from the vanishing hillock, cf. figure 2c.

From the mechanism of hillock overgrowth shown in figure 3 it becomes also clear that as long as the hillock, or part of it, is present on the surface the plateaus next to the hillock become wider along $<10\bar{1}0>$. Since the plateaus are formed from steps emerging from the re-entrant corner, the plateaus were narrower at earlier times of growth when the hillocks were smaller. This explains the triangular shape of both plateaus parts adjacent to the hillocks and the fact that the hillock is more or less located at the centre of the plateau.

During continued growth and assuming that the steps at the steep edge of the plateau propagate as fast as the misorientation steps, the plateau expands in the left and right directions, only due to expansion of the flat central area. Apart from translation the stepped triangular plateau regions do not change in time (figure 3).

The moment at which the situation represented in figure 3b occurs during the growth process depends on the time-dependent activity of the growth source of the hillock during the earlier stage of the growth run, which is different for the individual hillocks. Therefore, different stages of feature development appear side by side on the same layer after a certain period of growth as shown in figure 2. A hillock of which the step source has become inactive at a very early stage of growth will be overgrown much faster, i.e. the time between the different stages of the overgrowth process presented in figure 3 will be much shorter. Overgrowth of the hillock will also be faster for higher misorientations towards the $[10\bar{1}0]$ direction and same step source activity of the hillock. Overgrowth of the remaining part of the hillock can only be realised if steps introduced by the misorientation move faster than the steps of the remainder of the hillock.

Steps in the $<11\bar{2}0>$ directions move 2.5 times as fast as steps in the $<10\bar{1}0>$ directions, as is determined by measuring the distances between the small step bunches on both the misoriented surface and the plateau. Therefore, a misorientation towards the $[11\bar{2}0]$ direction should be chosen in order to overgrow the surface features completely.

CONCLUSIONS

During MOCVD growth of GaN films on the N-side of exactly oriented GaN single crystal substrates, hexagonal pyramids are formed. In this study it is found that the formation of these pyramids can be largely suppressed by the use of substrates with a slight misorientation towards the $[10\bar{1}0]$ direction. For a substrate misorientation of 4° the density of the elevations on the homo-epitaxial film is reduced by almost two orders of magnitude as compared with exactly oriented GaN substrates. The morphologies of those features that persist on the 4° off-angle sample during growth can be explained by a model involving the interaction of steps introduced by the misorientation with steps originating from the hillocks.

It is found that the step velocity is dependent on the step orientation: steps in the $<11\bar{2}0>$ directions move 2.5 times as fast as steps in the $<10\bar{1}0>$ directions.

Considerations following from the model indicate that for a further improvement of the surface morphology by reduction of the hillock density either substrates with larger misorientations or, since the step velocity is highly anisotropic, with misorientations towards the $[11\bar{2}0]$ direction are needed.

ACKNOWLEDGEMENTS

This work was financially supported by the Dutch Technology Foundation (STW). JLW wishes to thank for the grant of NATO Scientific Affair Division.

REFERENCES

[1] M. Seelmann-Eggebert, J.L. Weyher, H. Obloh, H. Zimmermann, A. Rar, and S. Porowski, Appl. Phys. Lett 71 (1997) 2635.

[2] F.A. Ponce, D.P. Bour, W.T. Young, M. Saunders, and J.W. Steeds, Appl.Phys.Lett. 69 (1996) 337.

[3] J.L. Rouvière, M. Arlery, R. Niebuhr, K.H. Bachem, and O. Briot, Materals Science and Engineering B43 (1997) 161.

[4] J.L. Weyher, P.D. Brown, A.R.A. Zauner, S. Müller, C.B. Boothroyd, D.T. Foord, P.R. Hageman, C.J. Humpreys, P.K. Larsen, I. Grzegory, and S. Porowski, J. Crystal Growth 204 (1999) 419.

[5] P.D. Brown, J.L. Weyher, C.B. Boothroyd, D.T. Foord, A.R.A. Zauner, P.R. Hageman, P.K. Larsen, M. Bockowski, and C.J. Humphreys. XI MSM Conf. proceedings, 1999, in press.

[6] J.L. Weyher, S. Müller, I. Grzegory, and S. Porowski, J. Crystal Growth 182 (1997) 17.

[7] M. Schauler, F. Eberhard, C. Kirchner, V. Schwegler, A. Pelzmann, M. Kamp, K.J. Ebeling, F. Bertram, T. Riemann, J. Christen, P. Prystawko, M. Leszczynski, I. Grzegory, and S. Porowski, Appl. Phys. Lett. 74 (1999) 1123.

[8] W.J.P. van Enckevort, G. Janssen, W. Vollenberg, and L.J. Giling, J. Crystal Growth 148 (1995) 365.

[9] S. Porowski Mater. Science Eng. B44 (1997) 407.

[10] L.T. Romano and T.H. Myers, Appl. Phys. Lett. 71 (1997) 3486.

[11] G. Nowak, K. Pakula, I. Grzegory, J.L. Weyher, and S. Porowski, Phys. Stat. Sol. (in press).

[12] B. Van der Hoek, J.P. Van der Eerden, and P. Bennema, J. Crystal Growth 56 (1982) 108.

Mat. Res. Soc. Symp. Vol. 595 © 2000 Materials Research Society

AlN Wafers Fabricated by Hydride Vapor Phase Epitaxy

Andrey Nikolaev[1], Irina Nikitina[1], Andrey Zubrilov[1], Marina Mynbaeva[1],
Yuriy Melnik[2] and Vladimir Dmitriev[2],
[1] Ioffe Institute and Crystal Growth Research Center,
St. Petersburg, 194021 Russia
[2] TDI, Inc.,
Gaithersburg, MD 20877, USA.

ABSTRACT

We report on AlN wafers fabricated by hydride vapor phase epitaxy (HVPE). AlN thick layers were grown on Si substrates by HVPE. Growth rate was up to 60 microns per hour. After the growth of AlN layers, initial substrates were removed resulting in free-standing AlN wafers. The maximum thickness of AlN layer was about 1 mm. AlN free-standing single crystal wafers with a thickness ranging from 0.05 to 0.8 mm were studied by x-ray diffraction, transmission electron microscopy, optical absorption, and cathodoluminescence.

INTRODUCTION

AlN substrates are needed for high-power microwave devices based on group III nitride heterostructures and UV photodetectors. High-performance GaN-based microwave transistors and solar-blind photodetectors have already been demonstrated on sapphire and SiC substrates. AlN substrates are expected to improve device characteristics further due to close lattice- and thermal-match with GaN-based device structures, high thermal conductivity and good insulating properties. High quality AlN materials have been grown in thin layer form, but bulk growth of AlN is much more difficult technological task. Main methods to grow bulk AlN are sublimation [1] and growth from a melt [2]. Experimental AlN crystals of about 15 mm in diameter have been reported for laboratory development [3]; commercial AlN wafers are not available.

Recently, we have demonstrated high quality AlN materials grown by the HVPE method [4]. AlN layers about 0.5 μm thick were grown on 6H-SiC substrates. Cathodoluminescence of AlN was measured at room temperature revealing edge peak at photon energy of 5.9 eV. The minimum value of the full width at a half maximum (FWHM) of ω-scan x-ray rocking curve was about 120 arc sec. The value of the specific resistivity of AlN was found to be 10^{13} Ohm cm at 300 K and 10^8 Ohm cm at 700 K.

In this paper, we report on HVPE growth of thick (up to 1 mm) AlN layers.

EXPERIMENTAL PROCEDURE AND RESULTS

AlN deposition experiments were carried out on the HVPE growth machine equipped with horizontal open-flow hot wall reactor and two-zone resistively heated furnace (Fig. 1). One zone is the source zone and another is the growth zone. The maximum growth zone temperature is 1200°C. Substrate temperature for AlN was varied from 900°C to 1200°C. In our HVPE processes, Ar serves as a carrier gas. Ammonia and HCl are supplied from gas tanks. A boat containing metallic Al was placed in Al source tube. Substrates were placed on quarts pedestal in the growth zone of the reactor. AlN layers were grown using a reaction between aluminum chloride and ammonia. Aluminum chloride gas was formed by reaction between metallic Al and HCl, and was transported from the source zone to the growth zone by Ar flow.

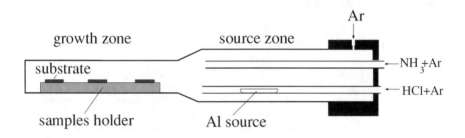

Figure 1. *Schematic view of HVPE growth reactor.*

AlN layers up to 1 mm thick were grown by HVPE technique described above. Sapphire, SiC, and Si wafers were used as substrates. AlN growth rate was controlled in the range from 0.1 to 1 μm/min. After the growth, Si substrate was removed by chemical etching. Free-standing AlN wafers up to 3x3 cm^2 in size were obtained (Fig. 2). The size of the wafer was equal to the size of the initial Si substrate. These are the largest AlN wafers reported to date. Results of material characterization of these AlN wafers are given bellow.

Figure 2. *Photo of free-standing AlN wafer.*

Fabricated AlN free-standing wafers were studied in terms of surface morphology, crystal structure, and optical properties. Top (as-grown) surface of AlN wafers contained growth pyramids (Fig. 3a). The back surface (former Si substrate – AlN interface) is flat and shiny. Usually, the back surface contained cracks (Fig. 3b). In order to use as-grown surface of AlN wafers for subsequent AlN or GaN epitaxy, this surface has to be polished.

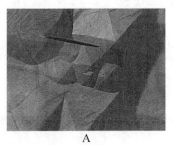
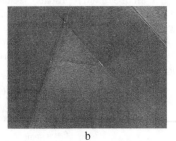

A   b

Figure 3. *Surface morphology of AlN free-standing wafer:*
a)   *As-grown surface;*
b)   *Former interface.*

Crystal structure of free-standing AlN wafers was measured by x-ray diffraction. It was found that AlN has single crystal structure. X-ray diffraction scan measured in the range from 20 to 80 degrees showed only (0002) and (0004) AlN reflections indicating that the grown material has wurtzite structure and that the grown surface has (0001) orientation. X-ray rocking curves measured in ω–2Θ- and ω-scanning geometries had the FWHM values of about 100 arc sec and about 40 arc min, respectively (Fig. 4). Transmission electron microscopy study, which was done at Erlangen University, indicated that dislocation density in AlN wafers decreases from 10^9 – 10^{10} cm^{-2} in the former interface region down to 10^8 cm^{-2} in the surface region [5].

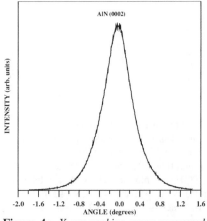
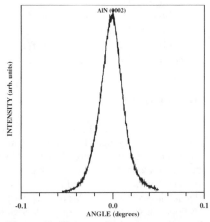

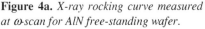

Figure 4a. *X-ray rocking curve measured at ω-scan for AlN free-standing wafer.*

Figure 4b. *X-ray rocking curve measured at ω-2Θ–scan for AlN free-standing wafer.*

Optical properties of grown AlN was studied using cathodoluminescence and optical transmission (Fig. 5). Cathodoluminescence spectra contained two peaks, one at ~375 nm (~3.3 eV) and another at ~207 nm (~5.98 eV). The most intense peak at a wavelength of 375 nm is probably related to oxygen impurity. The edge peak at 207 nm is related to near band-gap optical transitions. The nature of both peaks must be investigated in a future.

We would like to mention that the fabricated AlN wafers were used as seeds for AlN sublimation bulk growth by sublimation [6] and no cracks have been observed in the bulk AlN materials.

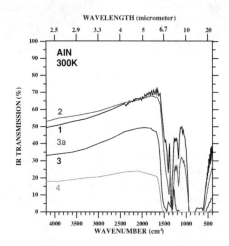

Fig. 5. *IR transmission spectra (300K) for free standing AlN samples:*
1 - thickness is 0.05 mm, $\alpha = 17$ cm^{-1};
2 - thickness is 0.08 mm, $\alpha = 20$cm^{-1};
3 - thickness is 0.3 mm, $\alpha = 15$ cm^{-1}.
3a - thickness is 0.29 mm, $\alpha = 20$ cm^{-1}.
4 - thickness is 0.49 mm, $\alpha = 22$ cm^{-1}.

CONCLUSIONS

AlN wafers up to 3x3 cm^2 in size were fabricated using HVPE technique. Thickness of AlN ranged from 0.05 to 1 mm. Lateral size of AlN wafers was equal to the size of initial substrates. Free-standing AlN wafers were fabricated employing Si substrate removal by chemical etching. It was found that AlN wafers are single crystals with 2H structure. Fabricated AlN wafers had (0001) orientation. The wafers exhibited edge luminescence at 5.9 eV and deep level luminescence peaking at 3.3eV (4 K). Future development will be focused on the fabrication of 2 inch diameter AlN substrates for group III nitride electronic devices.

ACKNOWLEDGMENTS

We thank M. Albrecht and H. Strunk for TEM measurements, A. Davydov for x-ray measurements and J. Freitas for cathodoluminescence study of AlN samples. Work at Ioffe Institute and CGRC was partly supported by INTAS programs # 96-1031 and #96-2131. Work at TDI was partially supported by ONR (contract monitor Colin Wood).

REFERENCES

1. G.A. Slack and T. McNelly, *J.Cryst.Growth*, **34,** 263 (1976).
2. C. Dugger, *Mat.Res.Bull.* **9**(3), 331 (1974).
3. L. Schowalter, ONR Progress Seminar on Bulk Nitride Semiconductor Growth and Characterization, Naval Research Laboratory, Washington DC, November, 1999.
4. Yu.V. Melnik, A.E. Nikolaev, S.I. Stepanov, I.P. Nikitina, A.I. Babanin, N.I. Kuznetsov, V.A. Dmitriev, in Proc.: 2nd European Conference on Silicon Carbide and Related Materials, ECSCRM'98, Montpellier, France, September 2-4, 1998, P.199-200.
5. M. Albrecht , H.P. Strunk, I.P. Nikitina, A.E. Nikolaev, Yu.V. Melnik, K. Vassilevski, V.A. Dmitriev, *this volume*.
6. W.L. Sarney , L. Salamanca-Riba, R.D. Vispute, T. Hossain, P. Zhou, H.N. Jayatirtha, M. Spencer, V. Dmitriev, Y. Melnik, A, Nikolaev and K.A. Jones, *this volume*.

Mat. Res. Soc. Symp. Vol. 595 © 2000 Materials Research Society

GaN 20-mm Diameter Ingots
Grown from Melt-Solution by Seeded Technique

V.A. Sukhoveyev[1], V.A. Ivantsov[2,3], I.P. Nikitina[3], A.I. Babanin[3], A.Y. Polyakov[4], A.V. Govorkov[4], N.B. Smirnov[4], M.G. Mil'vidskii[4], and V.A. Dmitriev[2]
[1]Crystal Growth Research Center, 29 Ligovsky Pr., 193036 St. Petersburg, Russian Federation
[2]TDI, Inc., 8660 Dakota Dr., Gaithersburg, MD 20877, U.S.A.
[3]Ioffe Institute, 26 Polytechnicheskaya Str., 194021 St. Petersburg, Russian Federation
[4]Institute of Rare Metals, 5 B.Tolmachevsky, 109017 Moscow, Russian Federation

ABSTRACT

In this paper, we describe the seeded growth of ~20 mm diameter 15 mm long GaN ingots from the melt-solution. This is the first successful attempt to conduct growth of GaN boule-crystals. GaN ingots were grown from Ga-based melt in the temperature range of 800-1000°C at less than 2 atm ambient pressure. Growth was performed at ~2 mm/hr growth rate. X-ray diffraction revealed polycrystalline structure of the ingots. Homoepitaxial GaN layers were deposited by HVPE technique on the substrates, which were fabricated from the grown GaN ingots.

INTRODUCTION

The lack of GaN substrates limits the performance of GaN-based devices including light emitters and microwave power transistors. No doubts, only GaN itself can meet the requirements on perfect substrates for advanced GaN-based electronics.

Recently, we have demonstrated that high quality bulk GaN single crystals can be grown from the liquid phase at reasonably low temperatures (<1100°C) and pressures (<2 atm) [1]. The crystals were grown via spontaneous nucleation on the melt surface. It was found that crystals possessed so fine optical properties that for the first time room temperature stimulated emission was registered from the bulk GaN material. Measured in wide temperature range, optical bandgap of these crystals was occurred closer to that of epitaxial material than to bandgap of bulk GaN grown at high pressure [2, 3]. In this paper, we present recently obtained results on the seeded growth from the liquid phase of GaN boules.

EXPERIMENTAL EQUIPMENT AND PROCEDURE

Growth experiments were conducted in a vertical 300-mm diameter water-cooled chamber. The chamber was made of stainless steel and equipped with viewing and vacuum ports. Quartz viewports gave a chance of *in situ* observation of the growing crystal during the growth run. A 5 MHz rf-generator supplied up to 10 kW power for heating of a graphite susceptor. Linear and rotation motion of the seed was activated using manual and/or motorized actuators.

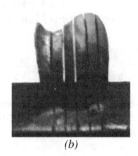

| (a) | (b) |

Figure 1. *As-grown ~20-mm diameter 15-mm in length GaN ingot (a). The same ingot after etching in HF+HNO₃ mixture and sawing in ~2 mm thick wafers (b).*

Prior to the growth run, the hot zone was degassed at 1100°C down to 10^{-5} Torr residual pressure. Then, charge materials were put in a graphite crucible and chamber was evacuated as soon as possible to reduce an inevitable contamination of the growth zone from the atmosphere. After the evacuation, the chamber was filled with nitrogen-rich gas mixture at a total pressure not exceeding 2 atm. Meanwhile, temperature of the heater was rised to a growth temperature chosen from 850-1000°C range. The started temperature was maintained constant for the launch period within an accuracy of ±0.5°C and decreased after that down to 800°C by the end of the growth run. Growth run lasted no longer than 10 hours.

The growth of GaN boule was initiated on the graphite rod that was used as a seed. Normal growth rate was piloted by the velocity of seed linear motion that was settled at about 2 μm/hr. As a result of the growth experiments, GaN ingots up to 21 mm in diameter and 15 mm in length were grown (figure 1). The ingots were sliced in approximately 2 mm thick wafers. In order to evaluate these wafers, homoepitaxial GaN layers were grown by hydride vapor phase epitaxy (HVPE).

The composition analysis of the wafers was carried out in scanning Auger electron spectroscopy–secondary ion mass-spectrometry (AES-SIMS) homemade analytical setup with the minimum back pressure of 10^{-11} Torr. For the SIMS measurements, secondary ion complexes are mass-filtered and focused by an objective system, which allows independent adjustment of ion beam diameter and density. The area of scanning zone is 500×500 μm^2 and ion current ranges from 0.2 to 20 mA/cm^2. High transmission secondary ions optics with a narrow-band pass energy provides mass resolution of the quadruple filter (*3M*) with a dynamic range of six orders of magnitude. It is a main feature of the setup that, due to the weak draw field of secondary ions, simultaneous AES and SIMS analysis are possible. For the AES measurements, cylindrical mirror analyzers with energy resolution of 0.5 % and a coaxial electron gun with an energy of 3 keV and an electron beam size of 5 μm were used. Electron beam is matched with the center of the crater being ion etched. The specific procedure of surface cleaning during the measurements was developed to reduce oxygen reabsorption effect. Monitoring the signals of Ga_{LMM} and N_{KLL} transitions at *1066 eV* and *376 eV*, respectively, the composition of bulk GaN was measured.

To characterize the structure of the ingots and HVPE grown homoepitaxial layers, x-ray diffraction (XRD) measurements were done in ω-2θ scanning geometry. As a reference crystal in double-axis XRD setup, high quality Lely 6H-SiC crystal with lattice parameters of $a=3.0817$ Å and $c=15.1183$ Å was used.

Room temperature micro-cathodoluminescence (micro-CL or SEM-CL) measurements in scanning electron microscope (SEM) were employed together with routine CL measurements at ~90 K for optical characterization of GaN wafers and epilayers. The SEM-CL study was conducted at electron beam current of 10^{-8} A and beam diameter of 1000 Å.

RESULTS

Several GaN ingots having the diameters from 16 to 21 mm and lengths from 10 to 15 mm were successfully grown. The AES spectrum features gallium and nitrogen related peaks typical for GaN. The oxygen (O_{KLL} at 504 eV) and carbon (C_{KLL} at 263 eV) intense peaks were believed to appear due to the surface oxidation and contamination. Apart from carbon and oxygen, there was not any other element being detected within the limits of instrumental resolution. The sample was sputtered for 20 sec by Ar^+ beam in order to remove surface contamination (figure 2, left). Oxygen and carbon peaks still presented in the spectrum, but intensities of them significantly reduced. To detail the distribution of the impurities in the bulk of the crystal, a depth-profile SIMS measurements were performed (figure 2, right). The traces of oxygen and carbon, which were detected in the bulk of GaN, most probably are associated with grain boundaries, since microscopical observations made after chemical etching showed that sliced wafers are composed of separate micron-size crystallites having pyramidal and platelet shapes.

Secondary electron microscopical image and SEM-CL map of GaN wafer are shown in figure 3. They showed as small as 1-3 μm size grains with the boundaries exhibiting strong signal in yellow luminescence collection mode.

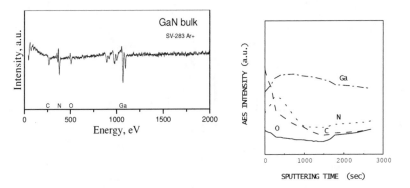

Figure 2. *Auger electron spectrum (left) and secondary ion depth-profile (right) taken from the bulk GaN polycrystalline sample after Ar^+-beam sputtering. Peak intensity variation of N and Ga signals are likely due to "sputtering effect".*

(a)

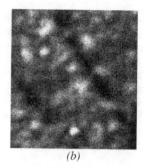
(b)

Figure 3. *Secondary electron microscopy image (a), and SEM cathodoluminescence map (b) of the wafer that was sliced from the GaN ingot (RT, 1000ˣ).*

SEM-CL spectra measured at both room temperature (RT) and 95 K detected strong yellow luminescence and weak exciton-related lines (figure 4). Donor-acceptor pair recombination did not observed.

The XRD measurements showed the presence of (10$\underline{1}$1), (10$\underline{1}$0) and (0002) reflexes (figure 5, solid line). The highest intensity of the (10$\underline{1}$1) peak indicates that crystallites have preferred orientation and this orientation corresponds to major growth direction that is normal to (10$\underline{1}$1) crystallography plane.

In order to estimate the potential of the developed technique for GaN substrate manufacturing, preliminary HVPE experiments were performed on GaN substrates, which were prepared from the grown ingots. To remove residua of the melt, the wafers were etched in HF+HNO$_3$ mixture and cleansed in organic solvents and deionized water. The wafers were polished and lapped to finish the surface of the wafers for subsequent

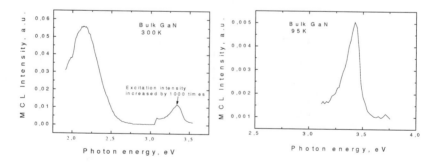

Figure 4. *Cathodoluminescent spectra taken from a GaN wafer at room temperature (left) and edge CL peak measured at 95 K (right).*

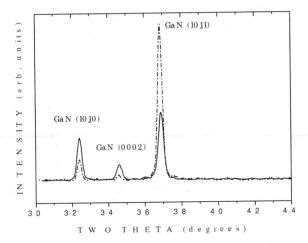

Figure 5. *X-ray diffraction spectra for GaN substrate (solid line) and deposited homoepitaxial GaN layer (dash-dot line).*

epitaxy. About 3-μm thick GaN layer was successfully grown using routine HVPE procedure developed for GaN epitaxy on SiC and sapphire substrates [4]. Epitaxial layer was free of cracks. Results of the XRD measurements are shown in Figure 5 (dash-dot line). It is clearly seen that (10$\underline{1}$1) reflex becomes strongly dominant reflex of the epilayer spectrum. It seems to be reasonable to assume that, if this tendency will be preserved, thicker HVPE layer could be conversed to single-crystalline material.

CONCLUSION

GaN ingots were grown from the Ga-based melt in the temperature range of 800-1000°C at less than 2 atm ambient pressure. Growth was performed on the graphite seed at ~2 mm/hr growth rate. Grown ingots were sliced in ~2 mm thick wafers, which were characterized on structural and optical properties. X-ray diffraction analysis revealed polycrystalline structure of the wafers with the [10$\underline{1}$1] preferred orientation of the composing crystallites.

GaN 3 μm thick homoepitaxial layer deposited on GaN substrate fabricated from the ingot was stress-free and displayed improved structural and optical properties in comparison with the substrate.

Presented results show that large volume GaN ingots can be grown from liquid phase using seeded technique. We believe that these results open the opportunity to develop seeded technique for bulk growth of bulk GaN crystals. Future research will be directed to single crystal growth of bulk GaN from liquid phase.

ACKNOWLEDGMENTS

The authors greatly acknowledge Dr. M. Spencer for extremely helpful discussions. The research at the CGRC and Ioffe Institute was partially funded by the Russian Fund for Fundamental Research (Project No. 97-02-18057). Work at TDI, Inc. was supported by ONR Contract N00014-99-C-0346 (Program Monitor Dr. Colin Wood).

REFERENCES

1. V.A. Ivantsov, V.A. Sukhoveyev, and V.A. Dmitriev, *Mat. Res. Soc. Symp. Proc.*, **468**, 143 (1997).
2. V. Sukhoveyev, V. Ivantsov, A. Zubrilov, V. Nikolaev, I. Nikitina, V. Bougrov, D. Tsvetkov, and V. Dmitriev, *Materials Science Forum*, **264-268**, 1331 (1998).
3. H. Teisseyere, P. Perlin, T. Suski, I. Grzegory, S. Porowski, J. Jun, A. Pietraszko, and T.D. Moustakas, *J. Appl. Phys.*, **76**, 2429 (1994).
4. Yu.V. Melnik, I.P. Nikitina, A.S. Zubrilov, A.A. Sitnikova, Yu.G. Musikhin, and V.A. Dmitriev, *Inst. Phys. Conf. Ser.*, **142**, 863 (1996).

Mat. Res. Soc. Symp. Vol. 595 © 2000 Materials Research Society

Preparation and Characterization of Single-crystal Aluminum Nitride Substrates

Leo J. Schowalter,[*,**] J. Carlos Rojo,[**] , Nikolai Yakolev[*], Yuriy Shusterman,[*] Katherine Dovidenko,[*,+] Rungjun Wang,[*] Ishwara Bhat,[*] and Glen A. Slack[*,**]

[*]Rensselaer Polytechnic Institute
Troy, NY 12180

[**]Crystal IS, Inc.
Latham, NY 12110

ABSTRACT

Large (up to 10mm diameter) aluminum nitride (AlN) boules have been grown by the sublimation-recondensation method to study the preparation of high-quality single crystal substrates. The growth mechanism of the boules has been studied using AFM. It has been determined that large single crystal grains in those boules grow with a density of screw dislocations below 5×10^4 cm^{-3} while edge dislocations are at lower density (none were observed). High-quality AlN single crystal substrates for epitaxial growth have been prepared and characterized using Chemical Mechanical Polishing (CMP) and AFM imaging, respectively. Also, the differential etching effect of KOH solutions on the N and Al-terminated faces of AlN on vicinal c-faces has been investigated. In order to identify the N or Al-terminated face, convergent beam electron diffraction has been used.

INTRODUCTION

Single crystal AlN is attractive for the fabrication of substrates for III-nitride epitaxial growth. Applications of wide-bandgap and high-temperature semiconductors include the development of blue/UV solid-state charge injection lasers, UV optical sources and detectors, high power microwave devices, high power switches, and high temperature applications. Enhanced III-nitride epitaxy on AlN substrates, due to smaller lattice mismatch, smaller thermal expansion mismatch and good chemical compatibility, may allow substantial improvement in the ability to manufacture these devices. In addition, very low defect material is needed for developing solid-state low noise and power devices for use in power supplies, communications, fire control, surveillance, and multifunctional RF systems using the nitride, wide-bandgap semiconductors. Foreign substrate induced defects have also been implicated in the performance of high power microwave devices. Substrates of AlN are also attractive for surface acoustic wave (SAW) applications, which would take advantage of the very high sound velocity in AlN. This would have application in high power, high frequency amplifiers for wireless communication base stations. The high thermal conductivity and electrically insulating properties of AlN are also attractive for high performance substrates where heat sinking is a key issue.

[+] Currently at the University at Albany, NY 12222

For III-nitride epitaxial growth, the preparation of nearly-atomically flat surface is required. Due to the extreme hardness of AlN, we have found that epitaxial-growth quality surfaces can be only achieved by using chemical-mechanical polishing (CMP) methods. The determination of the etching properties of single crystal substrates is of great importance for epitaxial growth and device preparation. We have studied the effects of KOH solution on the both the N and Al-terminated surfaces along the c-axis crystallographic direction on high-quality single crystal substrates. Also, we discuss a TEM-based method to unambiguously identify the N and Al-terminated faces along the c-axis.

CRYSTAL GROWTH OF BULK ALN

Several (see Fig. 1) AlN boules of up to 10mm of diameter have been produced at Crystal IS facilities using a sublimation-recondensation technique first developed for this material by Slack and McNelly [1], which demonstrated crystal growth rates up to 0.3mm/hr. AFM imaging of the as-grown facets reveals interesting information about the growth mechanism of AlN. The atomic arrangement of the facets on the growing front of the crystal, shown in figure 2, corresponds to the characteristic structure of c-face fronts of hexagonal wurzite-type crystals. The 0.25nm-high monolayer steps consist of segments of triangles that form 60° angles at their corners. The triangles on each successive monolayer are rotated 60° or 180° with respect to the preceding one. A step going in [$1\bar{1}00$] direction has only 1 broken bond per edge atom in one layer. A parallel step in the next layer would have 2 broken bonds per edge atom; a step having again 1 broken bond per edge atom in the latter layer runs at 60° angle with respect to the [$1\bar{1}00$] step in the former layer. The straight segments of the steps indicate that the migration length of atoms along the steps is far more than a micron. The diffusion length of the atoms on the terraces must be even higher than that. We found that the origin of the step flow results from screw dislocations (see figure 3) which intersect the growing surface. The density of screw dislocations has been estimated to be ~5·10^4 cm^2. It is seen that the apex starts from the double step, hence the normal component of the Burgers vector is c = 0.498 nm. The step edge has the shape of the Archimede spiral without distinct straight segments. This means that the crystal growth is governed not only by the diffusion but also by the strain field of the dislocation, i.e., positions of an atom at the same distance of the dislocation are equally favorable without respect to the number of broken bonds.

Figure1. *Example of AlN boule grown at Crystal IS.*

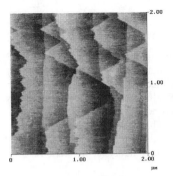

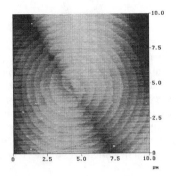

Figure 2. AFM image of the step
flow growth mechanism of (0001)
facets on the as-grown surface

Figure 3. AFM image of a screw
dislocation on a (0001) facet on
the as-grown surface of AlN

EPITAXIAL GROWTH ON A-FACE SUBSTRATES

Boules of AlN have been cut to obtain several c-face and a-face substrates. Due to the extreme hardness of AlN, preparing high quality surfaces by common mechanical polishing process is both time-consuming and difficult. Mechanical polishing using diamond powder was not successful in obtaining an adequate surface for epitaxial growth as reported in a previous work [2]. Chemical-Mechanical Polishing (CMP) was, however, very successful in achieving epitaxial-growth-quality surfaces. Examination of the substrate by AFM showed a nearly atomically flat surface with monolayer steps and with all evidence of mechanical damage (sawing damage, scratches, etc.) removed. Rutherford backscattering/ion channeling (RBS) spectra were used to determine the crystal quality of the substrate surface and the grown epitaxial layers. The minimum yield χ_{min} is used as a measure of the crystal quality and is defined as the ratio between the backscattered ion yield when the incident beam is aligned along a particular crystallographic axis and the backscattered ion yield when the incident beam is aligned randomly (along some non-crystallographic direction). The χ_{min} will increase with increasing defect density. We have measured χ_{min} in the spectral window corresponding to ion scattering from immediately below the sample surface peak. The results (see figure 4) show a χ_{min}'s of 1.4% along the $[11\overline{2}0]$ axis of one of our AlN substrate after CMP. This value is very close to the theoretical minimum indicating that the CMP process is successful in removing mechanical damage near the surface of the substrate and was consistent with the AFM results. After growing a 0.7-µm-thick AlN and a 1-µm-thick $Al_{0.5}Ga_{0.5}N$ epitaxial layer, RBS spectra were also used to determine the crystal quality of the epitaxial layers. Details on the epitaxial growth process and the ion-channeling results can be found in Ref. (Schowalter Tucson). For the 0.7-µm-thick homoepitaxial layer of AlN, the χ_{min} remained 1.5% again indicating excellent crystal quality. The χ_{min} measured for the $Al_{0.5}Ga_{0.5}N$ layer was 2.2%, which is still excellent considering that no attempt was made to optimize the growth parameters and that no buffer layer was needed to achieve them.

DETERMINATION OF THE ETCHING EFFECTS ON VICINAL C-FACES

Several studies performed on polycrystalline AlN have shown high etch rates obtain by using KOH based solutions. However, very little work has been done on etching of AlN single crystals. Recently, Mileham et al. [3] reported the first etching study on AlN single crystal grown on Al$_2$O$_3$. It was found that the etching rate depends critically on the crystalline quality and that the etching process is much faster at the grain boundary or wherever the density of defects is higher. Nevertheless, those crystal layers were defective and they do not report any orientation-dependent etching effect. The production of large, high-quality AlN single crystal substrates has made it possible to perform a more detailed study on the effect of KOH-based solutions on this material and to distinguish the differences of the etching process on different crystal orientations.

During the preparation of a vicinal c-face substrate (cut 20° off axis), we observed that cleaning the substrate with a phosphoric/sulfuric acid mixture gave different results for the two sides of the substrate. One side, which turned out to be the N-terminated face, became very rough surface. When a substrate with the same orientation was submerged in 1:2.5 KOH:water (by weight) solution for 3 minutes, AFM observation of the sample revealed the same unequal result for each side of the substrate. However, the KOH solution resulted in a slower etching rate than that produced by the phosphoric acid. AFM imaging (see figure 5) revealed a surface consisting of pyramids with a height between 0.5 and 1.5 μm high. This surface cannot be considered adequate for epitaxial growth. However, on the side that turned out to be the Al-terminated face the etching rate was unobservable resulting in a flatter surface.

STUDY OF THE SURFACE POLARITY BY TEM

In order to identify the atomic termination of the side that resulted in a larger etching effect on the vicinal c-face of our substrate, a TEM-based method was used. Transmission electron microscopy (TEM) was carried out using JEOL 2010 and JEOL 200CX operated at 200 kV. We used a conventional TEM sample preparation technique consisting of mechanical polishing followed by Ar$^+$ ion milling of the specimen. The MacTempas$^©$ program was used for calculations of the diffracted beam intensities. Large, electron-transparent areas of the crystal were studied by TEM using different sets of two-beam, bright-field/dark-field conditions. The imaging was done close to [11$\bar{2}$0]

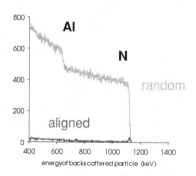

Figure 4. *Ion-channeling results obtained on the a-face after CMP process*

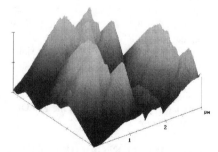

Figure 5. *AFM image on a 2x2 μm^2 area of the N-terminated c-face of AlN after KOH etching. Each height division represents 0.5 μm*

zone of AlN. No structural defects such as stacking faults or dislocations were observed over the studied area of about 80 square microns of the sample. A typical TEM image of the crystal taken under g=10$\bar{1}$0 bright field conditions is shown in figure 6. Selected area electron diffraction patterns were taken from large (about 1.5 micron in diameter) and small (about 0.1 micron) areas of the AlN sample. No streaking of the diffraction spots or other indications of the possible defects in the sample was observed. An example of the [11$\bar{2}$0] AlN zone axis diffraction pattern taken from the area of 1.5 micron in diameter is shown in figure 7.

We have studied the polarity of the crystal surfaces using the method described by Dovidenko et al. [4]. The method is based on the specific dependence of diffracted beam intensity vs. specimen thickness for the diffracted beams revealing the non-centrosymmetric nature of the crystal structure. We have carried out calculation of the intensities of different diffracted beams for the [11$\bar{2}$0] zone of wurtzite AlN crystal structure using multi-slice image simulations. For our calculations, we assumed the positive [0001] direction of the structure to go from Al to N atom. The plots in figure 8 show the calculated intensities vs. crystal thickness for 0002, 000$\bar{2}$, 10$\bar{1}$2 and 10$\bar{1}$$\bar{2}$ AlN diffracted beams. It is seen from the plot that, for the sample thickness in the range of 10 nm, the intensity of the 0002 beam is 2-6 times higher than the intensity of the 000$\bar{2}$ beam. This is due to the non-centrosymmetric nature of the crystal structure and the non-equivalency of the [0001] and [000$\bar{1}$] directions. Similar calculations were done for different diffracted beams belonging to the AlN [11$\bar{2}$0] zone. To compare the intensities of the experimental diffracted beams and thus determine the polarity of the crystal, we obtained the convergent beam electron diffraction (CBED) patterns from the thin regions of the specimen. Since the intensity of the beams is a function of specimen thickness, we estimated the thickness by doing high-resolution TEM at the points where CBEDs were taken and comparing our high-resolution images with the calculated ones. The experimental CBED pattern shown in figure 9 was obtained from a region of the

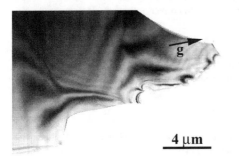

4 μm

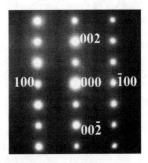

Figure 6. Typical TEM image of the crystal taken under g=10$\bar{1}$0 bright field conditions, the direction of g vector is shown by arrow.

Figure 7. Selected area electron diffraction pattern of the AlN crystal taken from the area of about 1.5 μm in diameter ([11$\bar{2}$0] zone axis).

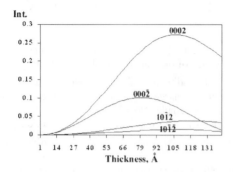

Figure 8. *Calculated intensities (arbitrary units) of the 0002, 000$\bar{2}$, 10$\bar{1}$2 and 10$\bar{1}\bar{2}$ diffracted beams vs. sample thickness (Å) for [11$\bar{2}$0] zone of wurtzite AlN crystal structure.*

Figure 9. *Convergent beam electron diffraction pattern taken from the thin (about 100 Å) region of the AlN crystal in [11$\bar{2}$0] zone axis.*

specimen which about 12 nm thick. The intensity of one of {0002}-type diffraction discs is significantly higher than that of the opposite one, the same is true for the {10$\bar{1}$2}-type of reflections. This allows the unambiguous indexing of the pattern and thus the determination of the absolute polarity of the crystal. The surface of the crystal, which was most affected by KOH etching, was found to be Al-terminated.

CONCLUSIONS

We have demonstrated the growth of large AlN boules with low-dislocation-density, single-crystal grains. High quality a-face AlN substrates have been prepared using a chemical-mechanical polishing technique. High quality epitaxial layers of AlN and AlGaN have been grown on substrates of this orientation without the need for buffer layers. Finally, we have reported the different morphology and behavior after KOH and phosphoric acid etching on the N and Al-terminated faces along the c-axis. The identification of these faces has been carried out with convergent beam electron diffraction.

ACKNOWLEDGEMENTS – Financial support from ONR, BMDO, and the US Navy is acknowledged as well as the support of our contract monitors Dr. C.E.C. Wood (ONR), Dr. C. Litton (AFRL), and T. Groshens (Naval Surface Warfare Center).

REFERENCES

1. G.A. Slack and T.F. McNelly, *J. Crystal Growth* **34**, 263 (1976); and G.A. Slack and T.F. McNelly, *J. Crystal Growth* **42**, 560 (1977).
2. J. Schowalter, J.C, Rojo, G.A. Slack, Y. Shusterman, R. Wang, I. Bhat, and G. Arunmozhi, *J. Crystal Growth* (accepted 99).
3. J.R. Mileham, S.J. Pearton, C.R. Abernathy, J.D. MacKenzie, R. J. Shul, S.P. Kilcoyne. *Appl. Phys. Lett.* **67**, 1119 (1995).
4. K. Dovidenko, S. Oktyabrsky, J. Narayan, and M. Razeghi, *MRS Internet J. Nitride Semicond. Res.* 4S1, G6.46 (1999).

Mat. Res. Soc. Symp. Vol. 595 © 2000 Materials Research Society

GROWTH OF CRACK-FREE THICK AlGaN LAYER AND ITS APPLICATION TO GaN-BASED LASER DIODE

I. Akasaki[1,2], S. Kamiyama[2], T. Detchprohm[2], T. Takeuchi[1], and H. Amano[1,2]

[1] Department of Electrical and Electronic Engineering, Meijo University,
1-501 Shiogamaguchi, Tempaku-ku, Nagoya 468-8502, Japan
[2] High-Tech Research center, Meijo University,
1-501 Shiogamaguchi, Tempaku-ku, Nagoya 468-8502, Japan

ABSTRACT

In the field of group-III nitrides, hetero-epitaxial growth has been one of the most important key technologies. A thick layer of AlGaN alloy with higher AlN molar fraction is difficult to grow on sapphire substrate, because the alloy layer is easily cracked. It is thought that one cause of generating cracks is a large lattice mismatch between an AlGaN and a GaN, when AlGaN is grown on the underlying GaN layer. We have achieved crack-free $Al_{0.07}Ga_{0.93}N$ layer with the thickness of more than $1\mu m$ using underlying $Al_{0.05}Ga_{0.95}N$ layer. The underlying $Al_{0.05}Ga_{0.95}N$ layer was grown directly on sapphire by using the low-temperature-deposited buffer layer (LT-buffer layer). Since a lattice mismatch between the underlying $Al_{0.05}Ga_{0.95}N$ layer and upper $Al_{0.07}Ga_{0.93}N$ layer is relatively small, the generation of cracks is thought to be suppressed. This technology is applied to a GaN-based laser diode structure, in which thick n-$Al_{0.07}Ga_{0.93}N$ cladding layer grown on the $Al_{0.05}Ga_{0.95}N$ layer, improves optical confinement and single-robe far field pattern in vertical direction.

INTRODUCTION

GaN-based semiconductor lasers are promising for light sources of high-density optical data storage systems, because of their short wavelength of around 400nm. However, we have worked with much effort for a long time in order to obtain a high quality crystal, because of the lack of substrates lattice-matched to this material system. In 1986, on the growth method using a LT-buffer layer[1] on sapphire substrate, which has a large lattice-mismatch of about 16%, GaN film with an excellent flatness and crystallinity was successfully obtained. Further, p-type low resistance crystal was realized by Mg-doping and electron-beam irradiation [2], or anneal process [3]. After such progress, high brightness blue and green light emitting diodes (LEDs) have been realized [4],[5]. More recently, several groups have achieved continuous-wave (CW) operation of violet-blue laser diodes(LD)[6]-[9]. It is thought that the GaN-based LD for the practical use will be commercially available in the near future.

However, the further improvements concerning with a crystalline quality and device performance are required against the practical use. In general, high quality GaN layer is grown on LT-buffer layer / sapphire substrate. When the AlGaN layers, commonly used as cladding layers in LDs, are grown on the GaN layer, the crack is often observed in the surface. This is thought to be due to the difference of lattice constants between GaN and AlGaN. Therefore, the thickness and AlN molar fraction of AlGaN layer, which are considerably important for the device performance, are greatly limited in order to avoid the crack generation.

The beam profiles of GaN-based diode laser in the direction perpendicular to the junction (vertical mode) are tend to be composed of multiple spots, which can be seen in the far field pattern [10],[11]. A poor optical confinement causes optical leakage from a waveguide layer to the underlying layer, such as an underlying n-GaN base layer. In order to suppress this optical leakage, the thickness of an AlGaN cladding layer should be much thicker than that usually embedded in a conventional structure. However it is quite difficult to grow crack-free thick AlGaN directly on GaN as mentioned above. Thus, the suppression of the crack in AlGaN is important subject for the improvement of device performance.

In this report, we demonstrate a crack-free $Al_{0.07}Ga_{0.93}N$ layer with the thickness of more than 1μm using underlying AlGaN layer grown directly on LT-buffer layer. We confirmed that the crystalline quality of AlGaN just on the LT-buffer layer is comparable as that of GaN. We also achieved a single-robe far field pattern in the LD with the thick and crack-free n-AlGaN layer as a cladding layer.

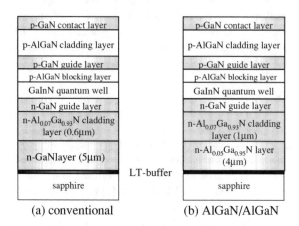

(a) conventional   (b) AlGaN/AlGaN

Figure 1: Schematic of laser structure with AlGaN/LT-buffer/sapphire.

CRYSTAL GROWTH

In a conventional laser structure, whose schematic diagram is shown in Fig. 1 (a), a n-GaN layer grown just on the LT-buffer layer is used as a base layer for laser structure, as well as a n-contact layer. However, this layer might cause the crack in n-AlGaN cladding layer grown on the layer, because of the tensile strain from the difference of lattice constants between GaN and AlGaN. We applied a new approach to overcome the problem that n-AlGaN is used as the base layer shown in Fig. 1 (b). We investigated a crystalline quality of n-$Al_{0.05}Ga_{0.95}N$ layer directly grown on the LT-buffer layer / sapphire substrate, using X-ray diffraction measurement and transmission electron microscopy (TEM). The FWHM of X-ray rocking curve was about 400 arcsec, and dislocation density from TEM image was $4 \times 10^9 cm^{-2}$, which values are comparable to those of GaN grown on LT-buffer layer. From these results, the n-$Al_{0.03}Ga_{0.97}N$ layer directly grown on the LT-buffer layer has a capability for the base layer of GaN-based laser diodes.

Laser structure was grown on c-face sapphire substrates by low-pressure metalorganic vapor phase epitaxy. In the new structure shown in Fig. 1 (b), instead of underlying n-GaN layer, a 4 µm thick n-$Al_{0.05}Ga_{0.95}N$ layer was directly grown on LT-buffer layer. After the growth of the underlying n-$Al_{0.05}Ga_{0.95}N$ layer, the other layers of laser structure were grown, which are 1µm n-$Al_{0.07}Ga_{0.93}N$ cladding layer, 0.1µm n-GaN

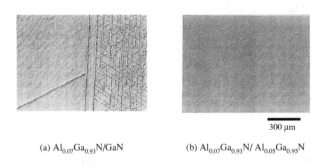

300 µm

(a) $Al_{0.07}Ga_{0.93}N/GaN$ (b) $Al_{0.07}Ga_{0.93}N/ Al_{0.05}Ga_{0.95}N$

Figure 2: Surface morphologies of the laser structures

guiding layer, 3 nm $Ga_{0.9}In_{0.1}N$ / 10 nm Si-doped $Ga_{0.98}In_{0.02}N$ 3QWs active layer with 15nm $Al_{0.15}Ga_{0.85}N$ electron blocking layer, 0.1µm p-GaN guiding layer, 0.5µm p-$Al_{0.07}Ga_{0.93}N$ layer, and 0.1µm p-GaN contact layer.

The surface morphologies of the conventional and the new laser structures are shown in Fig.2. The conventional laser structure has many cracks due to the lattice mismatch between GaN and $Al_{0.07}Ga_{0.93}N$, while the new structure does not have any crack. From the X-ray diffraction measurement, we have obtained the tensile strain in n-$Al_{0.07}Ga_{0.93}N$ layer of new structure is about 0.06%, which value is almost the halt in the case of conventional structure. This lower tensile strain could contribute to few cracks in the wafer irrespective of thicker n-$Al_{0.07}Ga_{0.93}N$ layers.

DEVICE PERFORMANCE

Using this laser wafer, we fabricated the 2-μm-wide ridge waveguide devices with 500-μm-long cavity. The laser facets were formed by cleaving along (1-100) plane of sapphire. High reflection coating was performed on the facets (front:80%, rear: 90%).

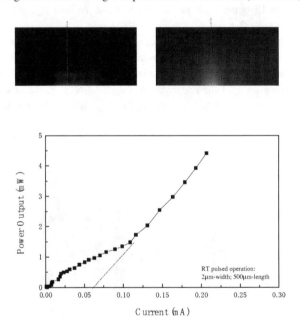

Figure 3: L-I Characteristic under RT Pulsed Condition

Figure 3 shows L-I characteristics of the new type laser. The measurements were pursued at room temperature under pulsed condition with a duty ration of 0.05 %. The threshold current of this device is as low as 65mA, which corresponds to threshold current density of 6.5 kA/cm^2. This value is equal to or better than that of conventional one in our experiments, indicating that the material quality of layers on underlying AlGaN layer / LT-buffer / sapphire is high enough in terms of device performances. The vertical profiles of far field pattern for the conventional and the new type lasers are shown in Fig. 4. The new type laser clearly shows single peak in far field pattern, while the conventional one shows multi peak in far field pattern as other groups already reported.[10],[11] These results indicate that the thick AlGaN cladding layer can successfully suppress the optical leakage from the waveguide region to the other regions. We also plotted the theoretical profiles of far field patterns calculated. The calculated results are good agreement with the experimental results.

CONCLUSION

In summary, we have realized thick layer of AlGaN free from cracks, using AlGaN / LT-buffer layer structure. Applying the structure to laser diode, we have demonstrated a low threshold current operation and a single spot near field and far field patterns in the direction perpendicular to the junction. A $1\mu m$ $Al_{0.07}Ga_{0.93}N$ cladding layer attributes to suppress the optical leakage and results in a improvement of optical confinement. This structure should be the key technology for practical use of high-performance GaN-based laser diode.

ACKNOWLEDGEMENT

This work was partly supported by the JSPS Research for the Future Program in the Area of Atomic Scale Surface and Interface Dynamics under the project of "Dynamic Process and Control of the Buffer Layer at the Interface in a Highly-Mismatched System" and Ministry of Education, Science, Sports and Culture of Japan (High-Tech Research Center Project and contract nos. 11450131).

REFERENCES

1. H. Amano, N. Sawaki, I. Akasaki, and T. Toyoda, Appl. Phys. Lett. 48, 353 (1986).
2. H. Amano, M. Kito, K. Hiramatsu, and I. Akasaki, Jpn. J. Appl. Phys. 28, L2112 (1989).
3. S. Nakamura, N. Iwata, M. Senoh, and T. Mukai, Jpn. J. Appl. Phys. 31,1258 (1992).
4. I. Akasaki and H. Amano, Mater, Res. Soc. Fall Meeting Proc. 1991, Boston, 383.
5. S. Nakamura, T. Mukai, and M. Senoh, Appl. Phys. Lett. 64, 1687 (1994).
6. S. Nakamura, M. Senoh, S. Nagahama, N. Iwasa, T. Yamada, T. Matsushita, Y. Sugimoto, H. Kiyoku, Appl. Phys. Lett., 69, 4056 (1996).
7. T. Kobayashi, F. Nakamura, K. Nagahama, T. Tojyo, H. Nakajima, T. Asatsuma, H. Kawai, M. Ikeda, Electron. Lett. 34, 1494 (1998).
8. A. Kuramata, S. I. Kubota, R. Soejima, K. Domen, K. Horino, T. Tanahashi, Jpn. J. Appl. Phys. 37, L1373 (1998).
9. M. Kuramoto, C. Sasaoka, Y. Hisanaga, A. Kimura, A. Yamaguchi, H. Sunakawa, N. Kuroda, M. Nido, A. Usui, and M. Mizuta, Jpn. J. Appl. Phys. 38, L184 (1999).

10. D. Hofsteller, D.P. Bour, R.L. Thornton, and N.M. Johnson, Appl. Phys. Lett. 70, 1650 (1997).
11. S. Nakamura, Mat. Sci. Eng. B50, 277 (1997)

Growth—MBE, Cubic GaN, GaAsN, Si, Substrates

Mat. Res. Soc. Symp. Vol. 595 © 2000 Materials Research Society

High-Quality AlGaN/GaN Grown on Sapphire by Gas-Source Molecular Beam Epitaxy using a Thin Low-Temperature AlN Layer

M. J. Jurkovic, L.K. Li, B. Turk, W. I. Wang
Department of Electrical Engineering, Columbia University,
New York, NY, 10027

S. Syed, D. Simonian, H. L. Stormer
Department of Physics, Columbia University,
New York, NY, 10027

ABSTRACT

Growth of high-quality AlGaN/GaN heterostructures on sapphire by ammonia gas-source molecular beam epitaxy is reported. Incorporation of a thin AlN layer grown at low temperature within the GaN buffer is shown to result in enhanced electrical and structural characteristics for subsequently grown heterostructures. AlGaN/GaN structures exhibiting reduced background doping and enhanced Hall mobilities (2100, 10310 and 12200 cm^2/Vs with carrier sheet densities of 6.1 x 10^{12} cm^{-2}, 6.0 x 10^{12} cm^{-2}, and 5.8 x 10^{12} cm^{-2} at 300 K, 77 K, and 0.3 K, respectively) correlate with dislocation filtering in the thin AlN layer. Magnetotransport measurements at 0.3 K reveal well-resolved Shubnikov-de Haas oscillations starting at 3 T.

INTRODUCTION

A unique combination of electrical, thermal, chemical, and structural characteristics render the AlGaN/GaN material system extremely suitable for application in high-speed, high-power electronic applications. Attractive properties include: a wide bandgap (3.4 for GaN, 6.2 eV for AlN), large breakdown field (~3 MV/cm), high peak and saturation electron velocities (~3 x 10^7 cm/s), robust structural and thermal stability, and chemical inertness. The large conduction band offset at the AlGaN/GaN interface and large spontaneous and piezoelectric polarization fields in the strained heterostructure enable the formation of a high-density two-dimensional electron gas (2DEG) on the order of 10^{13} cm^{-2} in AlGaN/GaN heterostructures grown in the (0001) direction. The Ga- or A-face is uniquely identified by reflection high-electron energy diffraction (RHEED) patterns exhibiting 2x2, 5x5, or 6x4 surface reconstructions as opposed to the N- or N-face which exhibit 1x1, 3x3, 6x6 and c(6x12) patterns [1]. Thus, the same role fullfilled by sheet-doping in modulation-doped FET's (MODFET's) is achieved in the AlGaN/GaN system without the mobility-limiting scattering mechanisms present in intentionally-doped structures. Consequently, piezoelectrically-doped high electron mobility transistor (HEMT) structures exhibiting high-speed, high-power operation and which can be subject to adverse environmental conditions can be achieved utilizing the AlGaN/GaN system.

AlGaN/GaN HEMT's which exhibit promising high-frequency, high-temperature, and high-power characteristics have been achieved [2-5] in spite of the lack of a lattice-matched substrate. Enhanced device characteristics for HEMT's grown on 6H-SiC substrates have been partially attributed to the higher thermal conductivity and closer lattice match to GaN (3.4%) as compared to that for sapphire (13.8%). AlGaN/GaN heterostructures exhibiting mobilities of 2000 cm^2/Vs, 9000 cm^2/Vs, and 11000 cm^2/Vs with sheet densities of 1 x10^{13} cm^{-2}, 1 x10^{13} cm^{-2}, 7 x10^{12} cm^{-2} at 300 K, 77 K, and 4.2 K, respectively, have been grown on conducting 6H-SiC substrates by metalorganic chemical vapor deposition (MOCVD) [6]. The highest attained mobility for similar structures grown on sapphire by MOCVD is 10300 cm^2/Vs with an electron sheet density of 6.2 x 10^{12} cm^{-2} at 1.5 K in spite of the large lattice mismatch [7]. Recently, lattice-matched homoepitaxy has been demonstrated by the use of a GaN template grown on sapphire by MOCVD for a plasma-assisted molecular beam epitaxy (PA-MBE) growth process, resulting in mobilities as high as 1150 cm^2/Vs, 24000 cm^2/Vs, and 51700 cm^2/Vs with sheet electron densities of 1.4 x 10^{13} cm^{-2}, 2.5 x 10^{12} cm^{-2}, 2.2 x 10^{12} cm^{-2}, at 300 K, 77 K, and 13 K, respectively, although the template layer was shown to exhibit a high background doping [8, 9].

In contrast, AlGaN/GaN heterostructures grown directly on highly mismatched (0001)-oriented sapphire substrates have also been achieved by molecular beam epitaxy (MBE) [10, 11], resulting in much lower mobilities: 1211 cm^2/Vs, and 5660 cm^2/Vs with electron sheet densities of 4.9 x 10^{12} cm^{-2} and 5 x 10^{12} cm^{-2} at 300 K and 77 K, respectively. However, for organometallic vapor phase epitaxial (OMVPE) growth of GaN on sapphire, the insertion of one or more low-temperature-grown (LT) -AlN or LT-GaN interlayers within the high-temperature-grown GaN has recently been shown to result in a reduction in threading dislocation density for subsequently grown epilayers [12]. More specifically, a defect filtering process was observed as a large portion of the stress-induced threading dislocations originating from the GaN/sapphire interface were terminated at the LT-AlN and LT-GaN interlayers, resulting in improved quality for the epilayers grown on the buffer structure. Although experimental data indicates that higher quality growth can be achieved on 6H-SiC, it remains highly desirable to obtain high-quality growth of GaN-based heterostructures on sapphire due to substrate availability, low cost, and high resistivity.

In this work, we report on the incorporation of a single LT-AlN interlayer within the GaN buffer of a AlGaN/GaN HEMT structure grown directly on sapphire (0001) by ammonia gas source MBE (GS-MBE), resulting in a significant enhancement in the structural and electrical characteristics of the subsequently grown heterostructure. AlGaN/GaN structures exhibiting reduced background doping and Hall mobilities of 2100, 10310 and 12200 cm^2/Vs with carrier sheet densities of 6.1 x 10^{12}, 6.0 x 10^{12}, and 5.8 x 10^{12} cm^{-2} at 300 K, 77 K, and 0.3 K, respectively, confirm the effectiveness of the buffer layer structure. The existence of a high-density two-dimensional electron gas is verified by magetotransport measurements performed at 0.3 K exhibiting Shubnikov-de Haas oscillations for fields as low as 3 T and a negatively-sloped magnetoresistance, indicating a low background doping in the buffer structure. Finally, a mobility of 2210 cm^2/Vs (10360 cm^2/Vs) with a sheet charge density 5.5 x 10^{12} cm^{-2} (5.9 x 10^{12} cm^{-2}) for the two-dimensional electron gas at room temperature (77 K) is calculated based on a two-layer conduction model. The results demonstrate that high-quality AlGaN/GaN HEMT structures can be grown by GS-MBE directly on sapphire in a single growth process by the incorporation of a single LT-AlN interlayer within the GaN buffer.

EXPERIMENT

AlGaN/GaN HEMT samples were grown on 2''-diameter, basal-plane sapphire substrates by GS-MBE in a Varian GEN II MBE system equipped with an RF plasma source (SVT Associates). Conventional Knudsen effusion cells were used as the Ga and Al sources while high-purity ammonia gas was used as the nitrogen source. After degreasing, the sapphire substrates were loaded into the MBE system and outgassed at 900 °C for 30 minutes. All epitaxial layers were unintentionally doped and, with the exception of the AlN buffer layers, grown at 810 °C.

The heteroepitaxial growth was achieving utilizing a two-step growth process [13-16]. First, nitridation of the surface was performed at a substrate temperature of 620 °C, followed by low-temperature growth (450 °C) of a thin (20-30 nm) AlN nucleation layer and an anneal at 810 °C. Second, growth proceeded with the epitaxy of a 3μm undoped-GaN buffer. Growth of the GaN buffer at 810 °C was interrupted after 600-700 nm in order to insert a thin, low-temperature (450 °C) AlN layer. The thickness of the AlN layer was varied from 20 nm to 60 nm for a set of samples for which an optimum thickness of 30nm was determined from subsequent van der Pauw Hall characterization.

A 30 nm AlGaN barrier followed growth of the GaN buffer. The Al composition for the barriers of the samples reported was determined by x-ray diffraction (XRD) measurements to be close to 20%. In addition, reference samples, consisting of the buffer layer structure both with and without the low-temperature AlN insertion layer, were grown in order to enable estimation of the carrier concentration and mobility of the 2DEG utilizing a two-layer conduction model.

Transport characterization was performed upon prepared samples at room temperature and at 77 K by van der Pauw Hall measurement. The existence of the piezoelectrically-induced 2DEG at the AlGaN/GaN interface was confirmed by well-resolved Shubnikov-de Haas oscillations observed at 3 T and above during magnetotransport measurements performed at 0.3 K. Also, RHEED *in situ* monitoring and atomic force microscopy analysis were employed in order to characterize the structural properties of the epitaxial films while photoluminescence (PL) measurements were performed at room temperature and at 10 K.

RESULTS AND DISCUSSION

A marked difference in PL spectra for AlGaN/GaN HEMT structures grown with and without the incorporation of the LT-AlN layer within the GaN buffer structure is shown in Figure 1. The spectra were obtained at 10 K. The dominant emission peak, located at 3.490 eV, for HEMT samples which incorporate the LT-AlN layer is attributed to transitions associated with free B excitons (FE$_B$) [17] while that for the samples grown without the AlN layer is ascribed to neutral-bound-excitons with a transition energy of 3.476 eV. Also, emission peaks at 3.494 eV, 3.503 eV, and 3.512 eV, (the latter two of which are associated with the band-to-band transitions to A and B valence bands, respectively [18]) are readily apparent in the LT-AlN samples while the non-LT-AlN samples exhibited less defined peaks at 3.480 eV and 3.499 eV. The highly-resolved A and B valence band peaks exhibited by the LT-AlN samples correspond with high-quality epitaxial growth, in contrast to the non-LT-AlN samples.

The RHEED patterns for all samples were observed to be relatively streaky. However, enhanced sharpness in the RHEED patterns became apparent during epitaxial growth of the GaN buffer and subsequent layers immediately upon growth of the LT-AlN layer. More specifically, the minor spots and facets observed prior to the LT-AlN layer vanished. In addition, analysis of the samples by atomic force microscopy revealed that the average grain-size for AlGaN/GaN heterostructures grown on conventional GaN

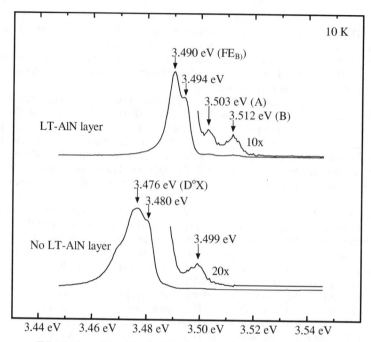

PL Intensity (Arb. Units) vs. Photon Energy

Figure 1. Band edge region of the photoluminescence spectra for $Al_{0.2}Ga_{0.8}N/GaN$ samples grown with and without a single LT-AlN interlayer within the GaN buffer.

buffers and those grown on buffers which incorporate the LT-AlN layer, were 200-500 nm and 1.2-1.5 μm, respectively. The line width (FWHM) of x-ray diffraction rocking curve for the heterostructure grown with the LT-AlN interlayer was 100 arc-sec.

The above results obtained from structural and optical characterization indicate a significant enhancement in the crystallinity of the epitaxial layers with the incorporation of the LT-AlN layer and correspond well with the reduced etch pit and threading dislocation density, and superior x-ray diffraction and PL linewidths reported for GaN films grown by OMVPE incorporating single and multiple LT-AlN and LT-GaN interlayers [12]. The enhanced resolution of the PL peaks, increased sharpness of the RHEED patterns, and reduced density of grain boundaries observed in the present work correlate with enhanced electrical characteristics, as confirmed by van der Pauw and magnetoresistance measurements.

The presence of a 2DEG at the strained AlGaN/GaN interface was confirmed by strong and well-resolved Shubnikov-de Haas (SdH) oscillations observed in the magnetoresistance (R_{XX}) measurements at 0.3 K shown in Figure 2. The low onset of oscillation, 3 T, indicates a relatively small amount of disorder in the structures while the decreasing minima with increasing magnetic field correlates with minor sub-channel conduction due to a low background doping in the buffer. The electron mobility at 0.3 K was found to be 12200 cm^2/Vs while an electron sheet density of 5.8×10^{12} cm^{-2} was determined from the SdH characteristic. Also, distortion of the SdH characteristic was

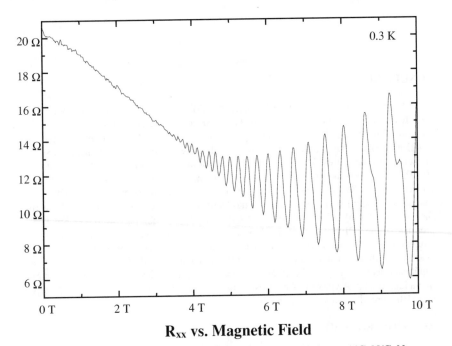

R_{xx} vs. Magnetic Field

Figure 2. Magnetoresistance, R_{xx}, vs. magnetic field at 0.3 K for an AlGaN/GaN heterostructure grown on sapphire by GS-MBE using a buffer which incorporates a LT-ALN interlayer. Shubnikov-de Haas oscillation starts at 3 T.

observed above 7 T in the form of the appearance of secondary minima near the SdH maxima. These features may be attributed to spin-splitting of the Landau levels at higher magnetic fields, a phenomenon that is thought to be observable only in crystals of exceptional quality. These additional SdH features at high magnetic fields have also been recently reported for AlGaN/GaN and GaN structures exhibiting high mobilities achieved by other means [7, 8].

The electron mobilities for AlGaN/GaN heterostructures grown on GaN buffers incorporating a single LT-AlN interlayer were determined by van der Pauw Hall measurements to be 2100 and 10310 cm^2/Vs with carrier sheet densities of 6.1 x 10^{12} cm^{-2}, 6.0 x 10^{12} cm^{-2}, at 300 K and 77 K, respectively. A reference buffer sample incorporating the LT-AlN interlayer was found to exhibit a reduced background electron density of 4.7 x 10^{16} cm^{-3} (4.3 x 10^{16} cm^{-3}) at 300 K (77 K) as compared to the density, 1.1 x 10^{17} cm^{-3} (5.7 x 10^{16} cm^{-3}), of a buffer sample grown without the LT-AlN interlayer. A mobility of 2210 cm^2/Vs (10360 cm^2/Vs) with a sheet charge density of 5.5 x 10^{12} cm^{-2} (5.9 x 10^{12} cm^{-2}) for the two-dimensional electron gas at room temperature (77 K) is calculated based on a two-layer conduction model using the values of electron sheet density and mobility for the reference GaN buffer sample grown with the LT-AlN interlayer and those obtained above by van de Pauw Hall measurement for the AlGaN/GaN heterostructure. The room temperature mobility value for the 2DEG is the highest among the methods reported for epitaxial growth of AlGaN/GaN heterostructures, including plasma-assisted-MBE on MOCVD-grown GaN/sapphire templates, as well as MOCVD or OMVPE on sapphire, 4H-SiC and 6H-SiC [6-11].

CONCLUSION

In conclusion, high quality AlGaN/GaN HEMT structures have been grown directly on sapphire (0001) substrates by GS-MBE in a single growth process. Incorporation of a single LT-AlN interlayer within the GaN buffer is shown to result in significant enhancement in the structural, optical, and electrical transport characteristics of such heterostructures. More specifically, measured electron mobilities were 2100, 10310, and 12200 cm^2/Vs with carrier sheet densities of 6.1 x 10^{12}, 6.0 x 10^{12}, and 5.8 x 10^{12} cm^{-2} at 300 K, 77 K, and 0.3 K, respectively. In addition, mobility of the 2DEG was calculated using a two-layer conduction model to be 2210 cm^2/Vs (10360 cm^2/Vs) with a sheet charge density 5.5 x 10^{12} cm^{-2} (5.9 x 10^{12} cm^{-2}) at room temperature (77 K). The presence of the 2DEG was confirmed by well-resolved SdH oscillations starting at 3 T. A reduction in background doping was observed upon incorporation of the LT-ALN interlayer within the GaN buffer while PL measurements and AFM analysis confirmed a significant decrease in grain boundary and point defect densities. The results demonstrate that a high quality AlGaN/GaN heterostructure can be achieved in a single growth step by GS-MBE directly on a sapphire substrate.

ACKNOWLEDGEMENTS

This work was supported in part by ONR, monitored by Dr. C. E. C. Wood.

REFERENCES

1. A. R. Smith, R. M. Feenstra, D. W. Greve, M.-S. Shin, M. Skowronski, J. Neugebauer, J. E. Northrup, *Appl. Phys. Lett.*, **72**, 2114 (1998).

2. M. A. Khan, Q. Chen, M. S. Shur, B. T. Dermott, J. A. Higgins, J. Burm, W. Schaff, L. F. Eastman, *Electron. Lett.*, **32**, 357, (1996).

3. Y.-F. Wu, B. P. Keller, P. Fini, S. Keller, T. J. Jenkins, L. T. Kehias, S. P. Denbaars, U. K. Mishra, *Electron Dev. Lett.*, **19**, 50 (1998).

4. R. Li, S. J. Cai, L. Wong, Y. Chen, K. L. Wang, R. P. Smith, S. C. Martin, K. S. Boutros, J. M. Redwing, *Electron Dev. Lett.*, **20**, 323 (1999).

5. S. T. Sheppard, K. Doverspike, W. L. Pribble, S. T. Allen, J. W. Palmour, L. T. Kehias, T. J. Jenkins, *Electron Dev. Lett.*, **20**, 161 (1999).

6. R. Gaska, M.S. Shur, A. D. Bykhovski, A. O. Orlov, G. L. Snider, *Appl. Phys. Lett*, **74**, 287 (1999).

7. T. Wang, Y. Ohno, M. Lachab, D. Nakagawa, T. Shirahama, S. Sakai, H. Ohno, *Appl. Phys. Lett.*, **74**, 3531 (1999).

8. C. R. Elsass, I. P. Smorchkova, B. Heying, E. Haus, P. Fini, K. Maranowski, J. P. Ibbeston, S. Keller, P. M. Petroff, S. P. Denbaars, U. K. Mishra, J. S. Speck, *Appl. Phys. Lett.*, **74**, 3528 (1999).

8. I. P. Smorchkova, C. R. Elsass, J. P. Ibbeston, R. Vetury, B. Heying, P. Fini, E. Haus, S. P. Denbaars, J. S. Speck, U. K. Mishra, *J. Appl. Phys.*, **86**, 4520 (1999).

10. L. K. Li, J. Alperin, W. I. Wang, D. C. Look, D. C. Reynolds, *J. Vac. Sci. Technol. B*, **16**, 1275 (1998).

11. J. B. Webb, H. Tang, S. Rolfe, J. A. Bardwell, *Appl. Phys. Lett.*, **75**, 953 (1999).

12. H. Amano, M. Iwaya, T. Kashima, M. Katsuragawa, I. Akasaki, J. Han, S. Hearne, J. A. Floro, E. Chason, J. Figiel, *Jpn. Journ. Appl. Phys., Part 2*, **37**, L1540 (1998).

13. W. I. Wang, *Appl. Phys. Lett.*, **44**, 1149 (1984).

14. J. S. Harris, S. M. Koch, S. J. Rosner, *Mater. Res. Soc. Symp. Proc.*, **91**, 3 (1987).

15. H. Kroemer, *J. Crystal Growth*, **81**, 193 (1987).

16. H. Amano, N. Sawaki, I. Akasaki, Y. Toyoda, *Appl. Phys. Lett.*, **48**, 353 (1986).

17. D. C. Renolds, D. C. Look, *J. Appl. Phys.*, **80**, 594 (1996).

18. K. C. Zeng, J. Y. Lin, H. X. Jiang, W. Yang, *Appl. Phys. Lett.*, **74**, 3821 (1999).

Mat. Res. Soc. Symp. Vol. 595 © 2000 Materials Research Society

High Quality AlN and GaN Grown on Si(111) by Gas Source Molecular Beam Epitaxy with Ammonia.

Sergey A. Nikishin, Nikolai N. Faleev, Vladimir G. Antipov, Sebastien Francoeur, Luis Grave de Peralta, George A. Seryogin, Mark Holtz[1], Tat'yana I. Prokofyeva[1], S. N. G. Chu[2], Andrei S. Zubrilov[3], Vyacheslav A. Elyukhin[3], Irina P. Nikitina[3], Andrei Nikolaev[3], Yuriy Melnik[4], Vladimir Dmitriev[4] , and Henryk Temkin
Dept of Electrical Engineering, Texas Tech University, Lubbock, TX 79409, U. S. A.
[1]Dept of Physics, Texas Tech University, Lubbock, TX 79409, U. S. A.
[2]Lucent/Bell Labs, Murray Hill, NJ 07974, U. S. A.
[3]Ioffe Physical-Technical Institute, St.Petersburg, 194021, Russia
[4]TDI, Inc., Gathersburg, MD 20877, U. S. A.

ABSTRACT

We describe the growth of high quality AlN and GaN on Si(111) by gas source molecular beam epitaxy (GSMBE) with ammonia (NH_3). The initial nucleation (at 1130-1190K) of an AlN monolayer with full substrate coverage resulted in a very rapid transition to two-dimensional (2D) growth mode of AlN. The rapid transition to the 2D growth mode of AlN is essential for the subsequent growth of high quality GaN, and complete elimination of cracking in thick (> 2 μm) GaN layers. We show, using Raman scattering (RS) and photoluminescence (PL) measurements, that the tensile stress in the GaN is due to thermal expansion mismatch, is below the ultimate strength of breaking of GaN, and produces a sizable shift in the bandgap. We show that the GSMBE AlN and GaN layers grown on Si can be used as a substrate for subsequent deposition of thick AlN and GaN layers by hydride vapor phase epitaxy (HVPE).

INTRODUCTION

There have been numerous recent attempts to prepare AlGaN-based heterostructures on Si (111) substrates [1-4]. The achievements of plasma assisted MBE (PAMBE) of AlN and GaN on Si(111) have been summarized in [5, 6]. GSMBE, with direct decomposition of NH_3 on the substrate surface,[7] can also be used to grow high quality AlN [8] and GaN [7, 9-11]. In the present paper we describe growth of AlN and GaN by GSMBE with NH_3 on Si(111) substrates.

We use *in situ* reflection high-energy electron diffraction (RHEED), low energy electron diffraction (LEED) and *ex situ* triple-crystal x-ray diffraction (XRD), transmission electron microscopy (TEM), RS, and PL to study structure and optical properties. XRD results from AlN and GaN layers, grown by HVPE on the GSMBE AlN and GaN layers are also reported.

EXPERIMENTAL DETAILS AND DISCUSSION

All growth experiments were carried out on 2 and 3-in oriented Si (111) substrates prepared by wet chemical etching [12]. The procedure used results in a hydrogen terminated Si surface. Substrates prepared by this process showed the (7×7) surface structure, observed by RHEED, after heating to 920 K for few minutes. The well known

Si(111) surface reconstruction transition, from (7×7) to (1×1), which occurs at T_t=1100 K [13, 14], provided a convenient temperature calibration point [9].

It is well known that once the growth chamber is used for the growth of nitrides, the residual nitrogen at the temperature range of 1100-1200 K induces formation of Si-N bonds on the substrate surface [4-6, 15]. The exposition of Si under the active nitrogen flux during PAMBE results in formation of thick SiN layers (down to ~1.5 nm) [16]. The formation of amorphous SiN islands leads to disordered growth of AlN and GaN on Si [6, 17]. Different initial treatments have been tried during MBE to eliminate SiN formation. These are mostly based on the use so-called Al/Si γ-phase as starting surface for AlN growth [5, 6, 9, 18, 19]. However, above 1100K Al/Si γ-phase does not completely protect the Si surface. Approximately 34% of the total substrate area remains bare Si [5] when the AlN growth is initiated. This situation deteriorates at higher temperatures (> 1190 K) where the best AlN crystal perfection can be achieved [6, 20].

The background pressure of NH_3 in our Riber-32 was ~ 5×10^{-8} torr. Since the efficiency of NH_3 decomposition at the 1100 K is less than a few percent [21], we estimate the background beam-equivalent pressure (BEP) of the potentially active NH_3 ~ 10^{-10} torr. We have shown [8, 22] that under these conditions the formation of Al/Si γ-phase is inhibited by SiN_x or SiNH [23] islands. These islands form due to the presence of active nitrogen from the background ammonia in the growth chamber.

In order to control the process of silicon nitridation by background NH_3, high growth temperature (1130-1190 K), combined with carefully controlled alternating pulses of Al and NH_3 [24], have been employed [8, 10, 22]. The onset of epitaxy thus suppresses formation of SiN_x and results in very rapid transition to 2D growth mode of AlN [8]. But, the *in situ* RHEED control of Si nitridation by background NH_3 is a very complicated task. Figure 1 (a) shows a typical RHEED pattern resulting from partial nitridation of Si (111) surface by background NH_3, which happens at temperatures above 920 K. To confirm that this pattern is related to the N-Si bond formation, the surface was exposed to higher NH_3 flux (up to BEP ~10^{-6} torr) and for different durations. The symmetry of the pattern does not change with exposure time up to 2 min. (920-1200 K, NH_3 BEP ~10^{-6} torr). Beyond 2 min., discrete RHEED reflections disappear suggesting full surface coverage SiN_x. The pattern in figure 1 (a) also did not change with temperature (down to 300 K) or with different electron energy of ~ 5-9 keV.

In contrast, LEED measurements provide important details about the Si surface due to the shallower region it probes. Figure 2 shows room temperature LEED patterns of Si

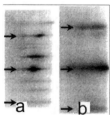

Figure 1. *RHEED patterns along the Si [110} azimuth: (a) partially nitrided Si (111). Position of the (01), (00), and (-01) reflection from the (1x1) structure are indicated from top to the bottom; the extra reflection arise from partial nitridation with NH_3; (b) (1x1) surface structure of the first monolayer of AlN.*

(111) surface after NH_3 exposure at AlN growth temperature. The pattern, corresponding to an ordered surface, only arises after exposure under background NH_3 for the 1–15 min. or under NH_3 flux with BEP~10^{-6}torr for 1-2 s. We observe extra reflections in the LEED images at different electron energies (20-110 eV), i.e., over the range most sensitive to surface structure. The extra reflections vanish under high or prolonged NH_3 exposure. The presence of the extra reflections in the LEED pattern signal a well ordered surface of the Si substrate, which is found to be best suited for high quality AlN growth.

We grew AlN at 1160 ± 30 K. The formation of the (1x1) surface structure could be seen both by RHEED (figure 1 (b)) and LEED after deposition of 2-3 nm of AlN. The growth was started by turning on both Al and NH_3 after deposition at least one monolayer of AlN. Growth rates of 90-400 nm/h were obtained. A completed 2D growth mode could be seen, by RHEED, after about 5-7 nm. The rapid transition to 2D growth is of interest for several reasons. First, it results in complete relaxation of AlN [25] by the formation of a coincidence lattice at AlN/Si interface [19]. Second, the rapid transition results in a decrease in both the defect density [8, 22] and in the strain energy of AlN layer [26]. From RHEED images we have found the $\{b_{Si(111)}=(a_{Si}/\sqrt{2})\}/a_{AlN(0001)}$ ratio of ~1.233 [8]. This compares to 1.235 obtained from standard values of lattice constants of Si (a_{Si}=0.542 nm) and AlN (a_{AlN}=0.311 nm). Thus the condition 4:5 along [110] azimuth of Si is evident. A similar effect of matching has been observed for c-GaN/GaAs(001) interface [27].

The growth of GaN was started after AlN buffer layer completed the transition to the 2D growth mode. GaN layers were grown at a temperature of 1000±30 K and growth rates of 0.4-1.5 μm/h. Formation of a (2x2) surface structure could be seen by RHEED after deposition of 20-100 nm of GaN. The fastest 3D-2D transition of growth mode was reached at the highest growth rate, possibly due to the high value of lateral growth rate reported for GaN [28]. We also prepared structures incorporating several (1-4) short period AlN/GaN and AlGaN/GaN superlattices (SLs). The effectiveness of such SLs, in controlling defect propagation from the AlN/Si and AlN/GaN interfaces, is demonstrated in the TEM cross section of figure 3. We stress that the third GaN layer and SL (figure 3) were grown in a 2D growth mode with (2x2) surface structure. Recently, similar SLs has been used to improve crystal quality of GaN grown on Si(001) by metal-organic chemical vapor deposition (MOCVD) [29].

From XRD measurements the full width at half maximum (FWHM) of the (0002) AlN peak in triple-crystal ω-2θ scan was less than 100 arc sec. This is the best value reported for ~200 nm thick AlN grown on Si(111). In GaN samples 1.5-2 μm thick the FWHM of the (0002) peak was as narrow as 14 arc sec. The GaN layers were free of cracks.

The layers of GaN grown on silicon are known to crack during the cool-down cycle

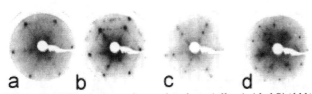

Figure 2. *LEED patterns of an ordered partially nitrided Si (111) surface at different energy of electron beam: (a) E_e=23 eV; (b) E_e=35 eV; (c) E_e=53 eV; (d) E_e=108 eV. All the RHEED and LEED patterns are displayed as negative images for clearness.*

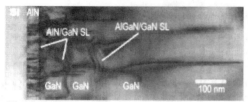

Figure 3. *TEM cross-section of GaN/(buffer layer)/Si structure.*

(see references in [8]). Cracking is believed to occur due to the thermal expansion coefficient mismatch between GaN and Si. Based on the result present both in this article and [10, 19, 22, 24-26], we can assume that cracking of GaN can be completely eliminated by assuring uniform growth of AlN at the Si-N-Al interface. From Raman measurements [10] we found the value of a residual tensile stress of < 0.2 GPa in our GaN with the thickness of 0.4-2.3 μm. This low stress is important because it is well below the ultimate strength of breaking of 0.4 GPa, under tensile stress, for GaN [31].

PL spectra at temperatures ranging from 77K to 495K were excited by a 2 kW pulse (10ns, 100Hz) nitrogen laser. Above 100 K, the PL spectra are dominated by the free exciton (FE) recombination [32, 33]. The red shift of the FE peak, due to the bandgap shrinkage, as well as usual quenching were observed with increasing temperature. The room temperature PL spectra of GaN show a FE peak position at 3.408 eV with the FWHM of less than 40 meV. The temperature dependence of the bandgap energy calculated from PL data for our GaN/Si crystals is presented in figure 4. Experimental data were fitted with both Passler's model [34] and Varshni's empirical approximation [35]. The PL results will be reported in detail in a later paper [36]. Also presented in figure 4 is the temperature dependence of the bandgap energy for a GaN layer grown by MOCVD on 6H-SiC substrate [37] and for a stress-free bulk GaN crystal grown HVPE [38]. Comparison of the $E_g(T)$ dependence for GaN grown on Si(111) and on SiC

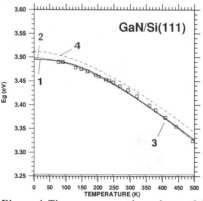

Figure 4. *The temperature dependence of the bandgap. Squares – this work; 1 - Passler's model [34] fit; 2 – Varshni's model [35] fit; 3 – GaN layer grown by MOCVD on SiC (recalculated data from [37]; 4 – Strain free bulk GaN grown by HVPE [38].*

substrates, with that of bulk GaN indicates residual tensile mechanical stresses in epilayers on both substrates. For GaN/SiC this has a value of about 0.5-1.0 GPa, which is primarily due to differences in values of the thermal expansion coefficients of GaN and SiC. This is at twice as large as what we see in our GaN/Si samples. That is why the $E_g(T)$ dependence is closer to that for strain free bulk GaN as shown in figure 4. Our estimation of the tensile stresses in our GaN/Si samples, based on the PL measurements, ranges from 0.08 to 0.3 GPa. These values are in a good agreement with the results of our Raman measurements.

The AlN and GaN layers grown by MBE on Si were used as substrate materials for growth of thick AlN and GaN layers by HVPE. The FWHM of the AlN (0002) ω-2θ-scan peak in the structure AlN(50 μm thick, HVPE)/GaN-AlN(MBE)/Si was 93 arcsec. For a 50 μm thick AlN layer grown by HVPE directly on Si, a FWHM value of 61 arcsec was obtained. FWHM of (0002) GaN peak in the structure GaN(1.5 μm thick, HVPE)/GaN-AlN(MBE)/Si at ω-2θ-scan was 48 arcsec. The c-axis lattice constant of HVPE GaN was higher than that of MBE grown GaN. We are currently performing experiments to understand this phenomenon.

CONCLUSIONS

This study shows how to grow high quality of GaN and AlN by GSMBE with NH_3. We can conclude that cracking of GaN grown on Si (111) can be completely eliminated by assuring uniform growth of AlN at the Si-N-Al interface.

ACKNOWLEDGMENTS

Work at TTU is supported by DARPA (monitored by Dr. Robert F. Leheny), AFOSR (monitored by Major Dan Johnstone) and the J. F Maddox Foundation.

REFERENCES

1. K. S. Stevens, M. Kinniburgh, and R. Beresford, Appl. Phys. Lett. **66**, 3518 (1995).
2. S. Guha, N. A. Bojarczuk, Appl. Phys. Lett. **72**, 415 (1998).
3. D. Kuksenkov, H. Temkin, R. Gaska, and J. W. Yang, IEEE Electron. Device Lett., **19**, 222 (1998).
4. H. Marchand, N. Zhang, L. Zhao, Y. Golan, S. J. Rosner, G. Girolami, Paul T. Fini, J. P. Ibbetson, S. Keller, Steven DenBaars, J. S. Speck, and U. K. Mishra, MRS Internet J. Nitride Semicond. Res. **4**, 2 (1999) and references therein.
5. K. Yasutake, a. Takeuchi, H. Kakiuchi, and K. Yoshii, J. Vac. Sci. Technol., **A 16**, 2140 (1998).
6. E. Calleja, M. A. Sánchez-García, F. J. Sánchez, F. Calle, F. B. Naranjo, E. Muñoz, S. I. Molina, A. M. Sánchez, F. J. Pacheco, R. García, J. Cryst. Growth, **201/202**, 296 (1999) and references therein.
7. M. Godlewski, J. P. Bergman, B. Monemar, U. Rossner, A. Barski, Appl. Phys. Lett. **69**, 2089 (1996).
8. S. A. Nikishin, V. G. Antipov, S. Francoeur, N. N. Faleev, G. A. Seryogin, V. A. Elyukhin, H. Temkin, T. I. Prokofyeva, M. Holtz, A. Konkar and S. Zollner, Appl. Phys. Lett., **75**, 484 (1999).
9. E. S. Hellman, D. N. E. Buchanan, and C. H. Chen, MRS Internet J. Nitride

Semicond. Res. **3**, 43 (1998).

10. S. A. Nikishin, N. N. Faleev, V. G. Antipov, S. Francoeur, L. Grave de Peralta, G. A. Seryogin, H. Temkin, T. I. Prokofyeva, M. Holtz, and S. N. G. Chu, Appl. Phys. Lett. **75**, 2073 (1999).
11. N.-E. Lee, R. C. Powell, Y.-W. Kim, and J. E. Greene, J. Vac. Sci Thechnol, **A 13**, 2293 (1995).
12. V. G. Antipov, S. A. Nikishin, and D. V. Sinyavskii, Tech. Phys. Lett. **17**, 45 (1991).
13. J. J. Lander, Surf. Sci., **1**, 125 (1964).
14. N. Osakabe, Y. Tanishiro, K. Yagi, and G. Honjo, Surf. Sci., **109**, 353 (1981).
15. V. G. Antipov, S. A. Nikishin, A. S. Zubrilov, D. V. Tsvetkov, and V. P. Ulin, *Proceedings of the Sixth International Conference on silicon Carbide and Related Materials,* Kyoto, Japan, 18-21 Sept. 1995, Inst. Phys. Conf. Ser. No 142 (IOP, London, 1996), Vol. XXV+1120, p.847 and references therein.
16. Y. Nakada, I. Aksenov, and H. Okumura, Appl. Phys. Lett., **73**, 827 (1998).
17. A. Ohtani, K. S. Stevens, and R. Beresford, Appl. Phys. Lett., **65**, 61 (1994).
18. A. Bourret, A. Barski, J. L. Rouvière, G. Renaud, and A. Barbier, J. Appl. Phys. Lett. **83**, 2003 (1998).
19. H. P. D. Schenk, G. D. Kipshidze, V. B. Lebedev, S. Shokhovets, R. Goldhahn, J. Ktäußlich, A. Fissel, Wo. Richter, J. Cryst. Growth, **201/202**, 359 (1999);
20. E. Calleja, M. A. Sánchez-García, E. Monroy, F. J. Sánchez, E. Muñoz, A. Sanz-Hervás, C. Villar, and M. Aguilar, J. Appl. Phys. **82**, 4681 (1997).
21. R. C. Powell, N.-E. Lee, and J. E. Greene, Appl. Phys. Lett., **60**, 2505 (1992).
22. S. A. Nikishin, N. N. Faleev, H. Temkin, and S. N. G. Chu, *State-of-the-Art Program on Compound Semiconductors (SOTAPOCS XXXI), Honolulu, Hawaii, 1999, October 17-22,* Electrochemical Society Proceedings Volume 99-17, p. 238 (1999).
23. *Silicon Nitride in Electronics,* edited by A. V. Rzhanov; Novosibirsk, Nauka (1982).
24. S. A. Nikishin, and H. Temkin, unpublished.
25. A. Bourret, A. Barski, J. L. Rouvière, G. Renaud, A. Barbier, J. Appl. Phys. **83**, 2003, (1998).
26. G. D. Kipshidze, H. P. Schenk, A. Fissel, U. Kaiser, J. Schulze, Wo. Richter, M. Weihnacht, R. Kunze, J. Ktäusslich, Semiconductors, **33**, 1241 (1999)
27. A. Trampert, O. Brandt, H. Yang, and K. H. Ploog, Appl. Phys. Lett., **70**, 583 (1997).
28. T. D. Moustakas, Mater. Res. Soc. Symp. Proc. **395**, 111 (1996).
29. X.Zhang, S.-J.Chua, P.Li, K.-B.Chong, Z.-C.Feng, Appl. Phys. Lett. **74**, 1984 (1999).
30. H. Tang and J. B. Webb, Appl. Phys. Lett., **74**, 2373 (1999).
31. Semiconductors and Semimetals, Vol. 57, Gallium Nitride (GaN) II, edited by J. I. Pankove and T. D. Moustakas (Academic Press, San Diego, 1999), p.294.
32. B.Monemar, J.P.Bergman, I.A.Buyanova, W.Li, H.Amano, I.Akasaki. MRS Internet Journal: Nitride Semiconductor Research **1**, 2 (1996).
33. M. Smith, G.D.Chen, J.K.Lin, H.X.Jiang, M.Asif Khan, C.J.Sun, Q.Chen, J.W.Yang, J.Appl. Phys.**79**, 7001 (1996).
34. R. Passler, Phys. Stat. Sol.(b), **200**, 155 (1997).
35. Y. P. Varshni, Physica, **34**, 149 (1967).
36. A. S. Zubrilov, S. A. Nikishin, and H. Temkin (unpublished).
37. A. S. Zubrilov, V. I. Nikolaev, D. V. Tsvetkov, V. A. Dmitriev, K. G. Irvine, J. A. Edmond, and C. H. Carter, Jr. Appl. Phys. Lett. **67**, 533 (1995).
38. A. S. Zubrilov, Yu. V. Melnik, A. E. Nikolaev, M. A. Jakobson, D. K. Nelson, V. A. Dmitriev, Semiconductors **33**, 1067 (1999).

Mat. Res. Soc. Symp. Vol. 595 © 2000 Materials Research Society

MBE Growth of Nitride-Arsenide Materials for long Wavelength Opto-electronics

Sylvia G. Spruytte, Christopher W. Coldren, Ann F. Marshall[1], Michael C. Larson[2], James S. Harris
Solid State and Photonics Lab, Stanford University,
Stanford, CA 94305, U.S.A.
[1]Center for Materials Research, Stanford University,
Stanford, CA 94305, U.S.A.
[2]Lawrence Livermore National Laboratory, P.O. Box 808, L-222,
Livermore, CA. 94551, U.S.A.

ABSTRACT

Nitride-Arsenide materials were grown by molecular beam epitaxy (MBE) using a radio frequency (rf) nitrogen plasma. The plasma conditions that maximize the amount of atomic nitrogen versus molecular nitrogen were determined using the emission spectrum of the plasma. Under constant plasma source conditions and varying group III flux, the nitrogen concentration in the film is inversely proportional to the group III flux (i. e. the nitrogen sticking coefficient is unity). The relationship between nitrogen concentration in the film and lattice parameter of the film is not linear for nitrogen concentrations above 2.9 mole % GaN, indicating that some nitrogen is incorporated on other locations than the group V lattice sites. For films with these higher nitrogen concentrations, XPS indicates that the nitrogen exists in two configurations: a Gallium-Nitrogen bond and another type of nitrogen complex in which nitrogen is less strongly bonded to Gallium atoms. Annealing removes this nitrogen complex and allows some of the nitrogen to diffuse out of the film. Annealing also improves the crystal quality of GaAsN quantum wells.

INTRODUCTION

Group III-Nitride-Arsenides are promising materials for 1.3μm and 1.55μm telecommunications optoelectronic devices grown on GaAs substrates[1,2,3]. The role of nitrogen is two fold: the nitrogen causes the bulk bandgap to decrease dramatically and the smaller lattice constant of GaN results in less strain in InGaNAs compared to InGaAs. However, the growth of such nitride-arsenides is complicated by the difficulty of generating a reactive nitrogen source and the divergent properties of nitride and arsenide materials. The luminescence properties of InGaNAs deteriorate rapidly with increasing nitrogen concentration[4]. It is common to increase the luminescence efficiency of these InGaNAs quantum wells by a short high temperature anneal[3].

In the present study, growth of nitride-arsenides was performed by elemental source MBE using a rf nitrogen plasma. We have investigated the incorporation of nitrogen using X-Ray photoelectron spectroscopy (XPS), secondary ion mass spectroscopy (SIMS), high resolution X-ray diffraction (HRXRD), and electron microprobe. The effect of thermal annealing on crystal quality and nitrogen incorporation was studied using XPS and HRXRD.

DETAILS OF MBE NITRIDE-ARSENIDE GROWTH

The growth of Nitride-Arsenides was performed in a Varian Gen II system by elemental source MBE. Group III fluxes are provided by thermal effusion cells, dimeric arsenic is provided by a thermal cracker, and reactive nitrogen is provided by an rf plasma cell. The plasma cell is operated at 250-350 W with a nitrogen gas flow of 0.1-0.5 sccm. Typical growth rates are 0.4-4 Å/s. These conditions allow to add up to 5 atomic percent nitrogen to III-V materials.

To minimize ion damage to the surface of the growing film, an rf plasma is used to generate atomic nitrogen[5]. The exit aperture of this source was custom designed so that the plasma can operate with low nitrogen flow rates without compromising the uniformity of the generated atomic nitrogen. The plasma conditions that maximize the amount of atomic nitrogen versus molecular nitrogen were determined using the emission spectrum of the plasma (see Figure 1). The intensity of the first set of bands at approximately 550, 580, and 650 nm is a measure of the amount of molecular nitrogen; the intensity in the bands at 740, 820 and 870 nm is proportional to the amount of atomic nitrogen present in the plasma[6]. Both the ratio of atomic nitrogen versus molecular nitrogen and the total amount of atomic nitrogen created, increase with increasing plasma power. For our plasma source design, the plasma power is more effective than the nitrogen flow rate to control the generation of atomic nitrogen. We optimized the start-up procedure of the plasma to maximize the generation of atomic nitrogen.

We have verified with SIMS that the impurity concentration (H, O, C and B) in our films is below $2.10^{17}/cm^3$.

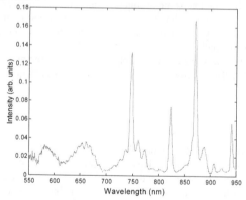

Figure 1. *Emission spectrum of the rf nitrogen plasma for an nitrogen supply of 0.25 sccm and power of 300 W.*

STUDY OF NITROGEN INCORPORATION

For the plasma operating with 300 W power and 0.25 sccm nitrogen flow, the nitrogen concentration in the film is dependent on the gallium arsenide (GaAs) growth rate as shown in Figure 2. The nitrogen concentration is inversely proportional to the GaAs growth rate because the atomic nitrogen is so reactive that all the supplied atomic nitrogen is consumed to form GaNAs. N_2 formation is limited because the generated amount of atomic nitrogen is small compared to the As_2 and Gallium overpressures in the MBE system[7]. As the films are grown under an As_2 pressure of twenty times the gallium pressure, the GaAs growth rate is only dependent on the Gallium flux. Therefore, the nitrogen concentration of the InGaNAs films can be controlled solely by the group III flux. This indicates that the InGaNAs system might have some advantages in terms of yield and reproducibility compared to the arsenide-phosphide system where the group V flux control is very critical and strongly temperature dependent[8,9].

We have observed that the relation between lattice parameter and nitrogen concentration in GaNAs films is not linear for nitrogen concentration above 2.9 mol % GaN (see Figure 3). This indicates that at high nitrogen concentration, some nitrogen is being incorporated on other locations than the group V lattice sites. The fact that nitrogen is such a small atom compared to gallium and arsenic makes this incorporation on other sites than the group V lattice sites more likely.

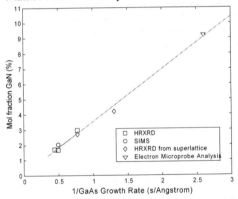

Figure 2. *Concentration of nitrogen in GaNAs films as function of the GaAs growth rate. The nitrogen concentration was measured by HRXRD, SIMS, and electron microprobe analysis. The nitrogen plasma was operated with a power of 300 W and a nitrogen flow of 0.25 sccm.*

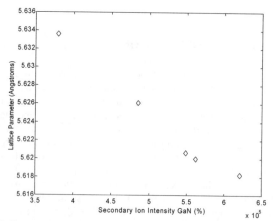

Figure 3. *Lattice parameter of different GaNAs films as function of nitrogen concentration The lattice parameter of the films was determined using HRXRD in a Philips Materials Research Diffractometer. The nitrogen concentration was measured by SIMS in an CAMECA IMS 4.5f instrument.*

XPS analysis was done to confirm this hypothesis. The N(1s) spectrum for a $GaN_{0.06}As_{0.94}$ film covered with GaAs cap (see Figure 4) indicates that the nitrogen exists in two configurations: a Ga-N bond and another nitrogen-complex in which N is less strongly bonded to Ga atoms. As the GaNAs layer is buried, surface contamination should have no influence on the shape of the nitrogen peak. Rapid thermal annealing (30 seconds at 775 °C) removes this nitrogen complex but also results in out-diffusion of the nitrogen.

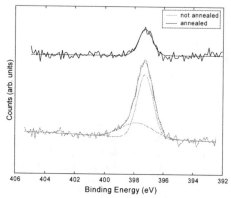

Figure 4. *XPS signal from N(1s) peak from an annealed and an not annealed $GaN_{0.06}As_{0.94}$ film measured in an PHI 5800 instrument. The Gallium Auger peak around 398 eV was removed by dividing the obtained spectrum by the spectrum of a GaAs reference sample. Annealing was done for 30 seconds at 775 °C in an AG heatpulse 310.*

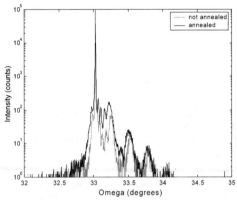

Figure 5. *HRXRD rocking curve from an annealed and not annealed 6 period 70 Å GaN$_{0.05}$As$_{0.95}$/130 Å GaAs superlattice. Annealing was done for 10 seconds at 775 °C in an AG heatpulse 310.*

Rapid thermal annealing and the resulting disappearance of this nitrogen complex also improves of the crystal quality of GaN$_{0.05}$As$_{0.95}$ quantum wells as indicated by the increased X-ray intensity diffracted from an annealed 6 period 70 Å GaN$_{0.05}$As$_{0.95}$/130 Å GaAs superlattice (see Figure 5). The out-diffusion of the nitrogen is responsible for the slight shift of the peaks.

CONCLUSION

Growth of Nitride-Arsenides was performed by MBE using an rf plasma to generate atomic nitrogen. We have demonstrated that the nitrogen concentration can be controlled accurately by the group III growthrate as the nitrogen concentration is inversely proportional to the group III growth rate. Nitrogen exists in two binding configurations in as grown Nitride-Arsenides. We have investigated the effect of anneal on crystal quality and nitrogen binding configuration using HRXRD and XPS.

ACKNOWLEDGEMENTS

This research was supported by DARPA and ARO through contract DAAG55-98-1-0437.

The authors would like to thank Drs. P. Van Lierde and T. Schuerlein from Charles Evans and Associates for the SIMS and XPS measurements. They would also like to thank the Center for Materials Research at Stanford University and in particular B. Jones and H. Kirby for the use of the electron microprobe and the Phillips MRD.

REFERENCES

[1] C.W. Coldren, M.C. Larson, S.G. Spruytte, and J.S. Harris, *Electron. Lett.,* to be published.

[2] Nakahara, K., Kondow, M., Kitatani, T., Larson, M.C., Uomi, K., *IEEE Photonics Technol. Lett.*, **10**, 487 (1998).

[3] C.W. Coldren, S.G. Spruytte, A.F. Marshall, J.S. Harris, M.C. Larson, *J. Vac. Sci. Technol.*, to be published.

[4] C.W. Coldren, S.G. Spruytte, A.F. Marshall, J.S. Harris, M.C. Larson, presented at the 18th Noth American Conference on Molecular Beam Epitaxy.

[5] V. Kirchner, H. Heinke, U. Birkle, S. Einfeldt, D. Hommel, H. Selke, and P.L. Ryder, *Phys. Rev. B*, **58**, 15749 (1998).

[6] R. M. Park, *J. Vac. Sci. Technol.* A, **10**,701 (1992).

[7] Y. Qui, S. A. Nikishin, H. Temkin, V. A. Elyukhin, and Y. A. Kurdriavtsev, *Appl. Phys. Lett.*, 70, 2831 (1997).

[8] R. R. LaPierre, B. J. Robinson, and D. A. Thompson, *J. Appl. Phys.,* **79**, 3021 (1996).

[9] T. L. Lee, J. S. Liu, H. H. Lin, *J. Electron. Mater.*, **25**, 1469 (1996).

Theory Doping

Mat. Res. Soc. Symp. Vol. 595 © 2000 Materials Research Society

Structural and Electronic Properties of Line Defects in GaN

Joachim Elsner[1,2], Alexander Th. Blumenau[1,2], Thomas Frauenheim[1],
Robert Jones[2], Malcolm I. Heggie[3]

[1]Fachbereich Physik, Universität Paderborn, D-33095 Paderborn, Germany
[2]Semiconductor Physics Group, University of Exeter, Exeter, EX4 4QL, UK
[3]CPES, University of Sussex, Falmer, Brighton, BN1 9QJ, UK

ABSTRACT

We present density-functional theory based studies for several types of line defects in both hexagonal and cubic GaN. {10-10} type surfaces play an important role in hexagonal GaN since similar configurations occur at open-core screw dislocations and nanopipes as well as at the core of threading edge dislocations. Except for full-core screw dislocations which possess heavily strained bonds all investigated stoichiometric extended defects in hexagonal GaN do not induce deep acceptor states in the band-gap and thus cannot be responsible for the yellow luminescence. However, electrically active point defects in particular gallium vacancies and oxygen related defect complexes are found to be trapped at the stress field of the dislocations. Preliminary calculations for cubic GaN find the ideal stoichiometric 60°-dislocations to be electrically active. As in hexagonal material, vacancies and impurities like oxygen are likely to be trapped at the dislocation core.

INTRODUCTION

GaN plays an important role in today's optoelectronics. This is mainly due to its large band gap (3.4 eV for hexagonal GaN) which makes blue light emission possible. Many devices like blue light emitting diodes or lasers are based on hexagonal GaN (see [1] for an overview), although recently the first light emitting diodes using cubic GaN have been constructed [2,3].

Electronic states in the band gap induced by extended defects can significantly alter the optical performance. This fact becomes extremely important in laser devices, where parasitic components in the emission spectrum are highly undesirable. Moreover, point defects could be trapped in the stress field of extended defects giving rise to charge accumulation in their vicinity. The resulting electrostatic field can lead to electron scattering which will severely affect the electron mobility (see Look and Sizelove [4] for a recent model in hexagonal GaN). Therefore, there is considerable interest in understanding the microstructure of extended defects in GaN and their interaction with point defects.

In this paper we discuss a variety of line defects in hexagonal GaN. We discuss the line defects in their pure form as well as the interaction with point defects which are likely to be trapped in the stress field. In addition first results on dislocations in cubic GaN are presented.

For our calculations we used two different methods to obtain geometries, energetics and electrical properties of the examined structures: AIMPRO, an ab initio local density-functional (LDF) pseudopotential method and SCC-DFTB, a self-consistent charge density-functional tight-binding method. On the one hand the AIMPRO method allows accurate determination of the electronic structure. The SCC-DFTB method on the other hand can be applied to larger supercells and thus makes possible the calculation of formation energies of extended defects. Details of both methods and their application to GaN and oxygen related defect complexes in GaN have been given previously [5-7].

DISLOCATIONS IN HEXAGONAL GaN

Threading Screw Dislocations

Threading screw dislocations in hexagonal material have a Burgers vector parallel to the dislocation line [0001]. The smallest screw dislocations have thus elementary Burgers vectors ±c. Since they nucleate in the early stages of growth at the sapphire interface and thread to the surface of the crystallites, screw dislocations are believed to arise from the collisions of islands during growth [8]. At a screw dislocation the surface is rough and has a high energy which favours the nucleation of islands. They are thus vital for the growth process.

We consider first a screw dislocation with a full core [5]. This type has been observed by Xin et al. [9] using high resolution Z-contrast imaging. Both the AMIPRO and the SCC-DFTB method found heavily distorted bond lengths (by as much as 0.4 Å) yielding deep gap states ranging from 0.9 to 1.6 eV above the valence band maximum, VBM, and shallow gap states at 0.2 eV below the conduction band minimum, CBM. An analysis of these gap states revealed that the states above the VBM are localised on N core atoms, whereas the states below CBM are localised on core atoms but have mixed Ga and N character. Therefore the full-core screw dislocation is electrically active and could act as a non-radiative centre [5]. Similarly one could expect that dislocations of mixed type would also have deep states in the gap as a result of the distortion arising from their screw component. Indeed atomic force microscopy in combination with CL imaging has shown that threading dislocations with a screw component act as non-radiative combination sites [10].

A calculation in a supercell containing a full-core screw dipole consisting of two dislocations with [0001] and -[0001], which are symmetrically equivalent, gave a high line energy of 4.88 eV/Å which is due to the strong distortion of the bonds of the core atoms.

Similar calculations were then carried out with the hexagonal core of the screw dislocation removed giving a core with a narrow opening of approximately 7.2 Å (see Fig. 1). The atoms on the walls adopt three fold coordinations similar to those found on the (10-10) surface. Thus Ga (N) atoms develop sp^2 (p^3) hybridisations which lower the surface energy and clear the gap of deep states. We found shallow gap states which are induced by the distortion arising from the Burgers vector. This was verified by comparison with the undistorted (10-10) surface which shows no shallow states.

The distortion in the open-core screw dislocation (maximum of 0.2 Å) is significantly less than that in the full-core screw dislocation. It is therefore not surprising that the calculated line energy of 4.55 eV/Å is lower than the line energy of the full-core screw dislocation. The energy required to form the surface at the wall is compensated by the energy gained by reducing the strain. However, a further opening gave a higher line energy and we conclude that the equilibrium diameter is approximately 7.2 Å. This opening has also been observed by Liliental-Weber *et al.* [11] who found some of the screw dislocations to have holes which are three atomic rows wide (see Fig. 1).

A theoretical approach to predict the opening of a screw dislocation was deduced by Frank [12]. By balancing the elastic dislocation strain energy released by the formation of a hollow core against the energy of the resulting free surfaces, he showed that, for isotropic linear elasticity and a cylindrical core, the equilibrium core radius is

$$r_{eq} = \frac{\mu b^2}{8\pi^2 \gamma},\tag{1}$$

where γ is the surface energy, μ is the shear modulus and b is the Burgers vector. For a rough estimate of the equilibrium radius, we use the theoretical value for the surface energy of {10-10} facets which we found to be $\gamma = 121$ meV/Å^2. Taking $\mu = 8 \cdot 10^{10}$ Nm^{-2} as an upper limit and $b = 0.5$ nm for the Burgers vector of an elementary screw dislocation yields 0.2 nm as the equilibrium radius. It is unlikely, that isotropic elasticity theory can describe the severely distorted full core and give precise quantitative values. Nevertheless our calculated value of 7.2 Å is reasonably close.

Oxygen and the Formation of Nanopipes

Nanopipes in hexagonal GaN thread along the **c**-axis and have hexagonal cross sections, i.e. they are enclosed by {10-10} type walls (see Fig. 2). The first suggestion was that they were the manifestation of screw dislocations with empty cores as discussed long ago by Frank [12]. However, as shown above neither *ab initio* calculations nor Frank´s theorem support the idea that in GaN the core of a screw dislocation with Burgers vector equal to **c** is open with a large diameter. Liliental-Weber *et al.* [13] found

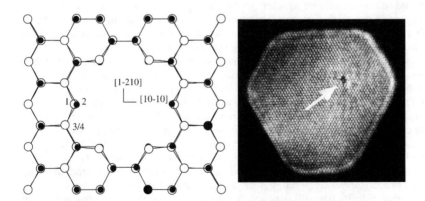

Figure 1 Left: *Top view (in [0001]) of the relaxed core of the open-core screw dislocation. The three fold coordinated atoms 1 (Ga) and 2 (N) adopt a hybridisation similar to the (10-10) surface atoms.* **Right:** *TEM image of a nanopipe containing a dislocation with a screw component. During growth the nanopipe closes leaving the dislocation with an opening of three rows (approx. 8 Å) wide. Z. Liliental-Weber [11].*

the diameters and densities of nanotubes to be increased in the presence of impurities, e.g. O, Mg, In and Si, and argued that these impurities decorate the {10-10} walls of the nanotubes inhibiting overgrowth. O being the main source of unintentional doping in GaN, we will now discuss how O can cause the formation of nanopipes.

There is experimental evidence that oxygen acts as a donor in bulk GaN [14] and total energy calculations show that O sits on a N site [15]. Since the internal surfaces of screw dislocations are very similar to those of the low energy (10-10) surface, we investigated [16] the likely surface sites for oxygen replacing N atoms. We found that the energy of a neutral O_N defect is 0.8 eV lower at the relaxed (10-10) surface. This shows that there is a tendency for O to segregate to the surface. The added oxygen has an additional electron occupying a state near the CBM. The defect has therefore a high energy and would attract acceptors resulting in a neutral complex. One possible acceptor, other than added dopants, would be a gallium vacancy (V_{Ga}) which acts as a triple acceptor and has been calculated to have a low formation energy in n-type GaN [17,18]. Consequently, we suppose that the surface oxygen concentration could be sufficiently large, and the oxygen atoms sufficiently mobile, that the three N neighbours of V_{Ga} at the (10-10) surface are replaced by O forming the V_{Ga}–$(O_N)_3$ defect.

Our calculations [16] showed that V_{Ga}–$(O_N)_3$ is more stable at the surface than in the bulk by 2.15 eV. Two O neighbours of the surface vacancy lie below the surface and each is bonded to three Ga neighbours, but the surface O is bonded to only two subsurface Ga atoms in a normal oxygen bridge site. The defect is electrically inactive with the O atoms passivating the vacancy in the same way as VH_4 in Si.

The question then arises as to the influence of the defect on the growth of the material. Growth over the defect must proceed by adding a Ga atom to the vacant site but this leaves three electrons in shallow levels near the conduction band resulting in a very high energy. This suggests that the defect can stabilise the surface and thus inhibit growth. From this we can conclude that such defects lead to the formation of nanopipes if we assume that during growth of the epilayers, either nanopipes with very large radii are formed which gradually shrink when their surfaces grow out, or there is a rapid drift of oxygen to a pre-existing nanopipe. In either case the concentration of oxygen and V_{Ga}–$(O_N)_3$ defects increases at the walls of the nanopipe. The maximum concentration of this defect would be reached if 50 % (100 %) of the first (second) layer N atoms were replaced by O and further growth then would be prevented. It is, however, likely that far less than the maximum concentration is necessary to stabilise the surface and make further shrinkage of the nanopipe impossible. Provided oxygen could diffuse to the surface fast enough, the diameter and density of the holes would be related to the initial density of oxygen atoms in the bulk. This model requires that the walls of the nanopipe are coated with oxygen although the initial stages of formation of the pipe are obscure.

In conclusion, we have shown that oxygen tends to segregate to the (10-10) surface and forms stable and chemically inert V_{Ga}–$(O_N)_3$ defects. These defects increase in concentration when the internal surfaces grow out. When a critical concentration of the order of a monolayer is reached, further growth is prevented. This model leads to nanopipes with (10-10) walls coated with GaO and supports the suggestions of Liliental-Weber *et al.* that nanopipes are linked to the presence of impurities [13].

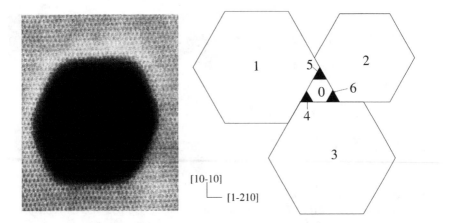

Figure 2 Left: *High resolution Z-contrast image along [0001] of a nanopipe. Y. Xin unpublished.* **Right:** *Suggested mechanism for the formation of a nanopipe (area No. 0). Three hexagons (No. 1,2,3) are growing together. As the surface to bulk ratio at edges (No. 4,5,6) is very large, they grow out quickly leaving a nanopipe (area No. 0) with {10-10} type facets.*

Threading Edge Dislocations

Pure edge dislocations lie on {10-10} planes and $\mathbf{b} = \mathbf{a} = [1\text{-}210]/3$ is their Burgers vector. They are a dominant species of dislocation, occurring at extremely high densities of approximately $10^8 - 10^{11}$ cm^{-2} in hexagonal GaN grown by MOCVD on (0001) sapphire and in analogy to screw dislocations are thought to arise from the collisions of islands during growth [8].

The relaxed core of the threading edge dislocation is shown in Fig. 3. The corresponding bond lengths and bond angles of the most distorted atoms are given in Table I. With respect to the perfect lattice the distance between columns (1/2) and (3/4) [and the equivalent on the right] is 9 % contracted while the distance between columns (9/10) and (7/8) [and the equivalent on the right] is 13 % stretched. This atomic geometry for the threading edge dislocation has recently been confirmed by Xin *et al.* [9] using atomic resolution Z-contrast imaging. Consistent with our calculation they determined a contraction (stretching) of 15±10 % of the distances between the columns at the dislocation core. Our calculations show that in a manner identical to the (10-10) surface, the three-fold coordinated Ga (N) atoms (no. 1 and 2 in Fig. 3) relax towards sp^2 (p^3) leading to empty Ga dangling bonds pushed towards the CBM, and filled lone pairs on N atoms lying near the VBM. Thus using the SCC-DFTB method we find full-core threading edge dislocations to induce no deep states into the band gap and conclude that in their pure form they should not be charged. Concerning the electrical properties of full-core edge dislocations it should be noted that using plane-wave scf calculations Wright *et al.*. [19] found the full core edge dislocation to possess states below the CBM (0.75 eV). However, recent Monte Carlo calculations for different background dopant densities based on scf plane-wave energy calculations by Wright and co-workers [20] have shown

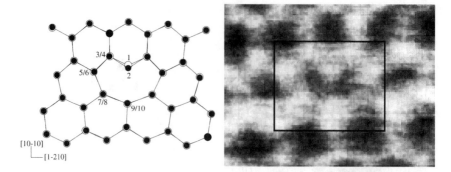

Figure 3 Left: *Top view (along [0001]) of the relaxed core of the threading edge dislocation (**b**=1/3[1-210]). The three fold coordinated atoms 1 (Ga) and 2 (N) adopt a hybridisation similar to the (10-10) surface atoms. The distance between columns (1/2) and (3/4) is contracted by 9 % while the distance between columns (7/8) and (9/10) is stretched by 13 %.* **Right:** *High resolution Z-contrast image of a threading edge dislocation looking down [0001]. The bright dots are atomic columns of alternating Ga and N atoms. The dislocation core is shown in the boxed region. Y. Xin et al. [9].*

no significant charge accumulation at the full-core threading edge dislocation even in heavily n-doped material.

From a supercell calculation, we obtain a line energy of 2.19 eV/Å for the threading edge dislocation. We note that this line energy is considerably lower than the one found for the screw dislocation with a narrow opening. This can be interpreted by noting that the edge dislocation has a smaller number of three fold coordinated atoms than the open-core screw dislocation as well as a smaller elastic strain energy. The latter arises from the smaller Burgers vector and from the smaller energy factor $k = (1-v)^{-1}$ for edge dislocations where v is the Poissons ratio (for screw dislocations k is equal to 1).

In analogy to the open-core screw dislocations we have investigated whether the energy of the threading edge dislocation could be lowered by removing the most distorted

Table I. Bond lengths, min-max (average) in Å and bond angles, min-max (average) in ° for the most distorted atoms at the core of the threading edge dislocation **b**=1/3[1-210]). Atom numbers refer to Fig. 3.

Atom	bond lengths	bond angles
1 (Ga$_{3\,x\,coord.}$)	1.85 – 1.86 (1.85)	112 – 118 (116)
2 (N$_{3\,x\,coord.}$)	1.88 – 1.89 (1.86)	106 – 107
3/4 (Ga/N$_{4\,x\,coord.}$)	1.86 – 1.95 (1.91)	97 – 119
5/6 (Ga/N$_{4\,x\,coord.}$)	1.92 – 2.04 (1.97)	100 – 129
7/8 (Ga/N$_{4\,x\,coord.}$)	1.94 – 2.21 (2.06)	94 – 125
9/10 (Ga/N$_{4\,x\,coord.}$)	1.95 – 2.21 (2.11)	100 – 122

core atoms (see Fig. 3). However, removal of either the columns of atoms (9,10), or the columns (1,2), (3,4), (5,6), (7,8) and their equivalents on the right, leads to considerably higher line energies. This implies that, in contrast with screw dislocations, which as discussed above can exist with a variety of cores, the threading edge dislocations should exist with a full core. Recent experimental results (Z-contrast imaging and EELS) by Xin *et al.* [21] show that the vast majority of edge dislocations have indeed a full core independent of the type of doping.

Deep Acceptors Trapped at Threading Edge Dislocations

The $V_{Ga} -(O_N)_3$ defect considered above in the context of the formation of nanotubes is electrically inactive at a (10-10) surface site but defects like $V_{Ga} -(O_N)_2$ and $V_{Ga} -O_N$ act as single and double acceptors, respectively. If these were trapped in the strain field of a dislocation, then we would expect the dislocation to appear electrically active [7]. We expect the energy of $V_{Ga} -(O_N)_n$ to be much lower at the dislocation core than in the bulk. This is because a neighbouring pair of three-fold coordinated Ga and N atoms are removed from the core and an O atom is inserted into the N site. The oxygen atom then lies in bridge site between two Ga atoms in a normal bonding configuration. Also other positions near to the dislocation core are found to be energetically favourable. Indeed, assuming Ga-rich growth conditions, O in equilibrium with Ga_2O_3 and n-type material our calculations find the resulting formation energies of these defects at the dislocation core and in the dislocation stress field to be significantly lower than in bulk material (see [7] for more details). Thus the core of the dislocation will spontaneously oxidise if oxygen is mobile. In any event, we anticipate that electrically active donor and acceptor pairs will be trapped at the core and in the dislocation stress field possibly giving rise to a negatively charged dislocation line in n-type material. This is in agreement with temperature-dependent Hall-effect measurements [4].

DISLOCATIONS IN CUBIC GaN

The 60°-Shuffle-Dislocation

We now investigate the properties of dislocations in cubic GaN. Our results are limited to possible models for 60°-shuffle-dislocations. Indeed, 60°-dislocations have recently been observed in MBE grown GaN on (001) GaAs by TEM [22].

60°-dislocations are dislocations of mixed edge and screw character with a Burgers vector inclined at an angle of 60° to the direction of the dislocation line. In zincblende lattices we distinguish the glide and the shuffle type. Here only the latter is discussed.

Figure 4 (left) shows the projection of the relaxed 60°-shuffle-dislocation with nitrogen core. The dislocation lies along [1-10] thus normal to the projection plane. Bond lengths and bond angles of the three-fold coordinated nitrogen core atom (atom (1) in Fig. 4 (left)) are given in Table II. With respect to the prefect lattice the bonds between the core atom (1) and its next gallium neighbours (2/3) are 7 % contracted while those between (4), (5), (6), (7) and (8) are 7 % stretched. Although the three-fold coordinated nitrogen core atoms (1) relax along -[111] they are still sp^3 hybridised. We therefore

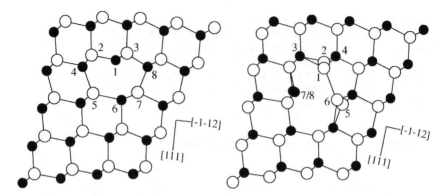

Figure 4 *Left: 60°-shuffle-dislocation with nitrogen core.* **Right:** *60°-shuffle-dislocation with gallium core.*

expect the dangling bonds of those atoms to induce deep gap states. Indeed our calculations show acceptor-like states giving rise to a partially filled band 0.3 eV above the VBM. This band is approximately 0.3 eV wide.

The relaxed core of the gallium core 60°-shuffle-dislocation is shown in Fig. 4 (right). The dislocation lies along [1-10] thus normal to the projection plane. The corresponding bond angles of all three-fold coordinated atoms are given in Table III. Unlike the nitrogen core the gallium core shows a reconstruction where Ga – N bonds are broken (between columns (5/6) and (7/8)) and the whole core structure is heavily distorted. Of the three-fold coordinated atoms every second atom along the dislocation line ([1-10] direction) is distorted inwards (see columns (1/2), (5/6) and (7/8) in Fig. 4). Table IV lists the distances between the three-fold atoms. As for the nitrogen core we find a partially filled band, this time 0.4 eV above the VBM and 0.3 eV wide. Additionally a very narrow band exists at mid-gap position..

Up to now the lack of detailed experimental TEM or Z-contrast data on dislocations in cubic GaN prevents a direct comparison and evaluation of our models.

Table II. Bond lengths, min-max (average) in Å and bond angles, min-max (average) in ° for the nitrogen core atom of the 60°-shuffle-dislocation. Atom numbers refer to Fig. 4 (left).

Atom	bond lengths	bond angles
1 (N$_{3\,\text{x coord.}}$)	1.81 – 1.84 (1.82)	105 – 122 (114)

Table III. Bond angles, min-max (average) in ° for all three-fold coordinated atoms of the 60°-shuffle-dislocation with gallium core. Atom numbers refer to Fig. 4 (right).

Atom	bond angles
1 ($Ga_{3\ x\ coord.}$)	99 – 110 (105)
2 ($Ga_{3\ x\ coord.}$)	110 – 121 (116)
5 ($Ga_{3\ x\ coord.}$)	112 – 124 (117)
6 ($Ga_{3\ x\ coord.}$)	100 – 110 (104)
7 ($N_{3\ x\ coord.}$)	108 – 121 (115)
8 ($N_{3\ x\ coord.}$)	105 – 117 (112)

Table IV. Distances in Å between the three-fold coordinated atoms of the 60°-shuffle-dislocation with gallium core. Atom numbers refer to Fig. 4 (right).

Atoms	distance
1 (Ga) – 6 (Ga)	3.33
2 (Ga) – 5 (Ga)	3.66
6 (Ga) – 8 (N)	3.33
5 (Ga) – 7 (N)	3.66

Interaction of Impurities with the Nitrogen Core 60°-Shuffle-Dislocation

We investigated the interaction of impurities like the nitrogen vacancy, oxygen and carbon with the nitrogen core 60°-shuffle-dislocation. The calculations show that O_N, V_N and C_N are stable in the dislocation core (at position (1) in Fig. 4 (left)). The formation energies of the neutral charge states of V_N and O_N in the dislocation core are lower than in the bulk (4.4 and 2.2 eV respectively). For C_N no difference in formation energy was found. This means V_N and O_N are bound to the core and C_N is not. We thus conclude that oxygen and nitrogen vacancies are likely to be accumulated at the nitrogen core of the 60°-shuffle-dislocation.

To learn about the effects of oxygen on the electronic structure, we increased the oxygen concentration from ¼ to ¾ of all core atoms at position (1) and found no significant change in the energetic position or width of the dislocation-induced band. However, oxygen seems to be able to passivate the electrical activity by filling up the dislocation band at least in n-type material.

To get deeper insights here, further investigations have to include different charge states of the defects as well as charge accumulations at the dislocations themselves.

SUMMARY

We have presented calculations for a variety of line defects threading along the **c**-axis in hexagonal GaN. We found full-core screw dislocations to have a large distortion of the bonds at the dislocation core resulting in deep states in the band gap. Open-core screw dislocations and in particular threading edge dislocations, which occur at very high densities, have a core structure similar to (10-10) surfaces and therefore do not induce deep acceptor states in their impurity-free form.

Oxygen-related defect complexes, some of which are electrically active, are found to possess very low formation energies at the core and in the stress field of threading edge dislocations in hexagonal material. One specific oxygen related defect complex, $V_{Ga} - (O_N)_3$ is believed to be responsible for the formation of nanopipes in hexagonal GaN.

In cubic GaN we found the 60°-shuffle-dislocation to be electrically active. Similarly to dislocations in hexagonal material the 60°-shuffle-dislocation is likely to trap point defects and impurities which could alter the electrical properties.

AKNOWLEDGEMENTS

The authors would like to thank A.F. Wright, A. Mühlig, Z. Liliental-Weber and Y. Xin for useful and stimulating discussions.

REFERENCES

1. S. Nakamura and G. Fasol, *The Blue Laser Diode*, Springer Verlag, Berlin 1997.
2. H. Yang, L.X. Zheng, J.B. Li, X.J. Wang, D.P. Xu, Y.T. Wang and X.W. Hu, *Appl. Phys. Lett.*, **74**, 2498 (1999).
3. D.J. As and K. Lischka, *priv. comm.* (1999).
4. D.C. Look and J.R. Sizelove, *Phys. Rev. Lett.*, **82**, 1237 (1999).
5. J. Elsner, R. Jones, P.K. Sitch, V.D. Porezag, M. Elstner, Th. Frauenheim, M.I. Heggie, S. Öberg and P.R. Briddon, *Phys. Rev. Lett.*, **79**, 3672 (1997).
6. M. Elstner, D. Porezag, G. Jungnickel, J. Elsner, M. Haugk, Th. Frauenheim, S. Shuhai and G. Seifert, *Phys. Rev.*, **B 58**, 7260 (1998).
7. J. Elsner, R. Jones, M. Haugk, Th. Frauenheim, M.I. Heggie, S. Öberg and P. R. Briddon, *Phys. Rev.*, **B 58**, 12571 (1998).
8. X.J. Ning, F.R. Chien and P. Pirouz, *J. Mater. Res.*, **11**, 580 (1996).
9. Y. Xin, S. J. Pennycook, N. D. Browning, P. D. Nelist, S. Sivananthan, F. Omnes, B. Beaumont, J-P. Faurie and P. Gibart, *Appl. Phys. Lett.*, **72**, 2680 (1998).
10. S.J. Rosner, E.C. Carr, M.J. Ludowise, G. Girlami and H.I. Erikson, *Appl. Phys. Lett.*, **70**, 420 (1997).
11. Z. Liliental-Weber, *priv. comm. at the EDS* (1998).
12. F.C. Frank, *Acta. Crys.*, **4**, 497 (1951).
13. Z. Liliental-Weber, Y. Chen, S. Ruvimov and J. Washburn, *Phys. Rev. Lett .*, **79**, 2835 (1997).
14. C. Wetzel, T Suski, J.W. Ager, E.R. Weber, E.E. Haller, S. Fischer, B.K. Meyer, R.J. Molnar and P. Perlin, *Phys. Rev. Lett.*, **78**, 3923 (1997).
15. J. Neugebauer and C.G. Van de Walle, *Festkörperprobleme*, **35**, 25 (1996).
16. J. Elsner, R. Jones, M. Haugk, R. Gutierrez, Th. Frauenheim, M.I. Heggie, S. Öberg and P.R. Briddon, *Appl. Phys. Lett.*, **73**, 3530 (1998).
17. J. Neugebauer and C.G. Van de Walle, *Appl. Phys. Lett.*, **69**, 503 (1996).
18. P. Bougulawski, E.L. Briggs and J. Bernholc, *Phys. Rev.*, **B 51**, R17255 (1995).
19. A.F. Wright and U. Gossner, *Appl. Phys. Lett.*, **73**, 2751 (1998).
20. K. Leung, A.F. Wright and E.B. Stechel, *Appl. Phys. Lett.*, **74**, 2495 (1999).
21. Y. Xin, E.M. James, I. Arslan, S. Sivananthan, N.D. Browning, S.J. Pennycook, F. Omnes, B. Beaumont, J-P. Faurie and P. Gibart, *submitted to Appl. Phys. Lett.* 1999
22. S. Kaiser, *priv. comm.* (1999).

Mat. Res. Soc. Symp. Vol. 595 © 2000 Materials Research Society

Simulation of H Behavior in p-GaN(Mg) at Elevated Temperatures

S. M. Myers, A. F. Wright, G. A. Petersen, C. H. Seager, M. H. Crawford, W. R. Wampler, and J. Han
Sandia National Laboratories, Albuquerque, NM 87185-1056

ABSTRACT

The behavior of H in p-GaN(Mg) at temperatures >400°C is modeled by using energies and vibration frequencies from density-functional theory to parameterize transport and reaction equations. Predictions agree semiquantitatively with experiment for the solubility, uptake, and release of the H when account is taken of a surface barrier.

INTRODUCTION

Hydrogen is introduced into GaN during growth by metal-organic chemical vapor deposition (MOCVD) and subsequent device processing [1]. This impurity greatly affects electrical properties, notably in p-type GaN doped with Mg where it can reduce the carrier concentration by orders of magnitude [2]. Application of density-functional theory to the zincblende[3] and wurtzite [4] forms of GaN has indicated that dissociated H in interstitial solution can assume positive, neutral, and negative charge states, with the neutral species being less stable than one or the other of the charged states for all Fermi energies. Hydrogen is predicted to form a bound neutral complex with Mg, and a local vibrational mode ascribed to this complex has been observed [5].

We are developing a unified mathematical description of the diffusion, reactions, uptake, and release of H in GaN at the elevated temperatures of growth and processing. Our treatment is based on zero-temperature energies from density functional theory. Objectives are to facilitate comparison of theory with experiment and, ultimately, to describe quantitatively H behavior pertinent to device processing. Herein we summarize work relating to p-type GaN(Mg), with details to follow in a more comprehensive article.

THEORETICAL FORMALISM AND PARAMETER EVALUATION

The diffusion flux of H in GaN along the c-axis direction is given by

$$\Phi_{\text{dif}} = -D^0 \frac{\partial}{\partial x}\left[H^0\right] - D^+ \frac{\partial}{\partial x}\left[H^+\right] - D^- \frac{\partial}{\partial x}\left[H^-\right]$$
$$- \left(D^+\left[H^+\right] - D^-\left[H^-\right]\right)\left(\frac{1}{kT}\right)\frac{\partial}{\partial x}\left(E_f - E_v\right) \tag{1}$$

where brackets denote concentration, the quantities D are diffusion coefficients, E_f is the Fermi energy, and E_v is the energy of the valence-band edge. The final term on the right describes driven migration due to spatial variation of the Fermi level. We assume rapid local equilibration among the H states, which is believed appropriate for the temperatures >400°C of concern herein. One then has [6]

$$\frac{\left[H^+\right]}{\left[H^0\right]} = \exp\left(\frac{G_H^{+\backslash 0} - E_f}{kT}\right) \tag{2}$$

$$\frac{\left[H^0\right]}{\left[H^-\right]} = \exp\left(\frac{G_H^{0\backslash -} - E_f}{kT}\right) \tag{3}$$

where the quantities G_H are bandgap levels associated with the electronic transitions and are defined to include degeneracy. Populations relating to the Mg acceptor are given by

$$\frac{N_{GaN}\left[MgH\right]}{\left[Mg^-\right]\left[H^+\right]} = \exp\left(\frac{G_{MgH}}{kT}\right) \tag{4}$$

$$\frac{\left[Mg^0\right]}{\left[Mg^-\right]} = \exp\left(\frac{G_{Mg}^{0\backslash -} - E_f}{kT}\right) \tag{5}$$

where N_{GaN} is the formula-unit density of GaN, MgH is the neutral complex, and G_{MgH} is the nonconfigurational part of the change in Gibbs free energy arising from dissociation of the MgH complex. The hole concentration is

$$\left[h^+\right] = \int_{-\infty}^{E_v} N_{val}(E)\, F_h(E)dE \tag{6}$$

where N_{val} is the density of states in the valence band and F_h is the Fermi distribution function for holes. The electron concentration is given by an analogous integral of the conduction band. The description is completed by imposing charge neutrality. Equilibrium with external H_2 gas is described by the solubility equation

$$\left[H^0\right]^{eq} = N_{GaN}\left(P_{gas}^*\right)^{1/2} \exp\left(-\frac{G_{sol}^0}{kT}\right) \tag{7}$$

where P_{gas}^* is the fugacity of the gas. This relationship is expressed in terms of H^0 to avoid a dependence on Fermi level. The system of equations is solved numerically.

Parameters in Eqs. (1)-(7) are evaluated using zero-temperature results from density functional theory for wurtzite GaN. The theoretical procedure and some results have been reported previously [4]. Using the H atom in vacuum as the reference state, we find an energy of +0.34 eV for interstitial H^0, (-2.67 eV + E_f - E_v) for H^+, and (+1.52 eV - E_f + E_v) for H^-. The formation of the neutral Mg-H complex from Mg^- and H^+ has a binding energy of 0.70 eV. The ionization level for the Mg acceptor is taken from experimental studies to be E_v+0.16 eV [7]. The diffusion activation energies from density functional theory are 0.5 eV for H^0, 0.7 eV for H^+, and 1.6 eV for H^-; the diffusion prefactors are set equal to 0.001 cm^2/s, which is representative for interstitials. The densities of states for the valence and conduction bands are evaluated from density functional theory.

The quantities G in Eqs. (1)-(7) differ between 0 K, where the results of density functional theory apply directly, and the elevated temperatures of interest. Causes include the temperature-dependent thermodynamics of H$_2$ gas and the enthalpy and entropy arising from H vibrations in the semiconductor. We treat these effects for deuterium (D), the isotope used in our experiments. The Gibbs free energy of D$_2$ gas is taken from the literature [8]. Vibrational effects are included for the dominant D states in p-type GaN(Mg), D$^+$ and the Mg-D complex. Density functional theory reveals two nearly degenerate energy minima of D$^+$, one at the N antibonding site oriented transverse to the c-axis, and the other at the c-axis bond-center position. Computed vibrational rates for D at the antibonding site are 2100 cm^{-1} (stretch) and 599 cm^{-1} (wag), while for the bond-center site we find 2418 cm^{-1} (stretch) and 355 cm^{-1} (wag). The Mg-D complex has a single stable configuration with the D at a N antibonding site diagonally opposite from a Mg neighbor along a bond direction transverse to the c-axis. The vibrations of this D are taken to be the same as for the N antibonding site without Mg. The vibrational enthalpy and entropy are computed from these results in the harmonic approximation [9].

PREDICTIONS AND COMPARISON WITH EXPERIMENT

Figure 1 shows the solubility of D in GaN(Mg) at 700°C as predicted by the theoretical model. Also given are experimental results for MOCVD GaN(Mg) on sapphire that was first activated by vacuum annealing for 1 hour at 900°C and then heated in D$_2$ gas at 700°C. The D concentration was measured using the nuclear reaction ^{2}D(^{3}He,p)^{4}He [10]. That the introduced D behaved as expected was evidenced by infrared spectroscopy, which exhibited the absorption near 2320 cm^{-1} ascribed to the Mg-D complex [5], and also by electrical measurements, which showed an increase in

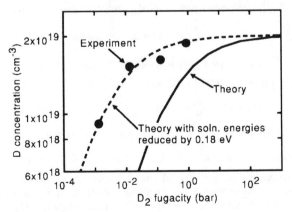

Figure 1. *Solubility of D in GaN(Mg) at 700°C as a function of D$_2$ fugacity.*

room-temperature resistivity from 3 to 14,000 Ω-cm upon D charging. In the model calculation, the concentration of electrically active Mg acceptors is 2.4×10^{19} cm^{-3}, somewhat below the total Mg concentration of $\sim5\times10^{19}$ cm^{-3}, and compensating shallow donors are included at a concentration of 0.4×10^{19} cm^{-3}. These plausible choices produce agreement with the observed saturation of D uptake at $\sim2\times10^{19}$ cm^{-3} and with the room-temperature hole concentration of $\sim4\times10^{17}$ cm^{-3} measured before D introduction. Other parameters are evaluated as discussed in the preceding section.

The horizontal displacement between theory and experiment in Fig. 1 provides a test of the mechanistic assumptions and computed H energies of the theoretical model. Interpretation is facilitated by applying a single offset to all of the D energies in the solid relative to D in vacuum so as to produce agreement with experiment. The result, given by the dashed line in Fig. 1, is achieved by reducing the energies in the solid by 0.18 eV. This shift is believed to be less than the combined theoretical uncertainties.

We also used the theoretical model to treat thermal release of D from GaN(Mg), initially assuming no surface barrier. This yielded a much greater release rate than observed in our experiments; during vacuum annealing at 700°C, the measured release rate exceeded predictions by 6 orders of magnitude. Similarly large disparities were found for uptake from D_2 gas. This leads us to propose a surface permeation barrier.

The barrier is provisionally modeled in the conventional manner by assuming equilibrium between solution and surface states so that the rate-determining step is desorption with H_2 formation for release and dissociative adsorption for uptake [11]. The D atomic flux through the surface is then given by

$$\Phi_{sur} = \Phi_0^{in} P_{gas}^* \left(1-\theta_{sur}\right)^2 \exp\left(-\frac{E_{ads}}{kT}\right) - \Phi_0^{out} \left(\theta_{sur}\right)^2 \exp\left(-\frac{E_{des}}{kT}\right) \qquad (8)$$

$$\frac{\theta_{sur}}{1-\theta_{sur}} = \frac{\left[H^0\right]_{x\rightarrow0}}{N_{GaN}} \exp\left(-\frac{E_{sur}^0}{kT}\right) \qquad (9)$$

where Φ_0^{in} and Φ_0^{out} are constant prefactors, θ_{sur} is the fractional occupation of surface sites, and $\left[H^0\right]_{x\rightarrow0}$ is the solution concentration near the surface. The first, uptake term on the right is parameterized using two measured uptake D fluxes, $\sim3\times10^{11}$ cm^{-2}s^{-1} for 700°C and a D_2 pressure of 0.013 bar, and $\sim3\times10^{11}$ cm^{-2}s^{-1} for 500°C and 0.88 bar. This yields $E_{ads} = 1.36$ eV and $\Phi_0^{in} = 2.3\times10^{20}$ cm^{-2}s^{-1}. In the second, release term, the barrier E_{des} is estimated to be 1.8 eV from surface desorption studies [12]; E_{sur}^0 and Φ_0^{out} are then evaluated by requiring that Eq. (7) hold when $\Phi_{sur} = 0$ in Eq. (8).

In Fig. 2, the predictions of the model with surface barrier included are compared with measurements of isothermal release into vacuum at 800°C. The second-order release kinetics of Eq. (8) imply that, after the residual concentration of D has become small enough to have little influence on the Fermi level, the reciprocal of retained D concentration should vary nearly linearly with time. The experimental data are seen in

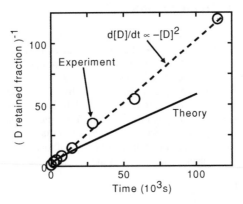

Figure 2. *Isothermal D release from GaN(Mg) at 800ºC.*

the figure to exhibit this property. Since few alternative rate-determining processes would show such kinetics, the finding evidences the presence of the barrier. Moreover, the predicted and measured release rates differ by less than a factor of 2.

Figure 3 shows predictions of the theoretical model for D release from initially saturated GaN(Mg) during isothermal annealing and the resultant increase in hole concentration at 300 K. Release data for 700 and 800ºC are included and agree semiquantitatively with the model. Comparison can also be made with published data for electrical activation of MOCVD GaN(Mg) during 20-minute isochronal anneals at 100ºC intervals [13]. Our model simulation yields 36% recovery of the hole concentration at 300 K after the anneal at 700ºC, whereas the measured room-temperature conductivity rose to ~1/2 of its ultimate value at this temperature.

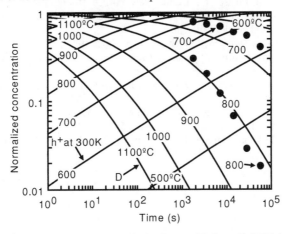

Figure 3. *Predicted isothermal release of D from GaN(Mg) and associated increase of hole concentration at 300 K.*

CONCLUSIONS AND IMPLICATIONS

The semiquantitative agreement found in this study between predicted and measured solubilities of D in GaN(Mg) supports the H energies obtained from density functional theory. Our combined theoretical and experimental results for uptake and release indicate a surface barrier with second-order release kinetics.

In our view, the modeling approach presented here has the potential ultimately for quantitative prediction of H behavior at temperatures >400°C where local equilibrium among H states is expected. Remaining areas for refinement include the surface barrier, whose properties are incompletely known. At lower temperatures where the activated transitions between H states are not rapid, the assumption of local equilibration should give way to an explicit treatment of the reaction rates.

ACKNOWLEDGMENT

This work was supported by the U. S. Dept. of Energy under Contract DE-AC04-94AL85000, primarily under the auspices of the Office of Basic Energy Sciences.

REFERENCES

1. S. J. Pearton, J. C. Zolper, R. J. Shul, and F. Ren, *J. Appl. Phys.*, **86**, 1 (1999).
2. S. Nakamura, N. Iwasa, M. Senoh, and T. Mukai, *Jpn. J. Appl. Phys.*, **31**, 1258 (1992).
3. J. Neugebauer and C. G. Van de Walle, *Phys. Rev. Lett.*, **75**, 4452 (1995).
4. A. F. Wright, *Phys. Rev. B*, **60**, R5101 (1999).
5. W. Götz, N. M. Johnson, D. P. Bour, M. D. McCluskey, and E. E. Haller, *Appl. Phys. Lett.*, **69**, 3725 (1996).
6. C. Herring and N. M. Johnson, in *Hydrogen in Semiconductors,* edited by J. I. Pankove and N. M. Johnson (Academic, New York, 1991) pp. 234-235.
7. W. Götz, N. M. Johnson, J. Walker, D. P. Bour, and R. A. Street, *Appl. Phys. Lett.*, **68**, 667 (1996), and citations therein.
8. *J. Phys. Chem. Ref. Data*, Vol. 14, Suppl. 1, p. 1002 (1985).
9. T. L. Hill, *An Introduction to Statistical Thermodynamics* (Dover, New York, 1986) pp. 86-93.
10. S. M. Myers, G. R. Caskey, Jr., D. E. Rawl, Jr., and R. D. Sisson, Jr., *Metall. Trans. A*, **14**, 2261 (1983).
11. F. C. Tompkins, *Chemisorption of Gases on Metals* (Academic, New York, 1978).
12. R. Shekhar and K. F. Jensen, *Surf. Sci.*, **381**, L581 (1997).
13. S. Nakamura, T. Mukai, M. Senoh, and N. Iwasa, *Jpn. J. Appl. Phys.*, **31**, L139 (1992).

Mat. Res. Soc. Symp. Vol. 595 © 2000 Materials Research Society

Mg SEGREGATION, DIFFICULTIES OF P-DOPING IN GaN

Z. LILIENTAL-WEBER, M. BENAMARA, W. SWIDER and J. WASHBURN, I. GRZEGORY[#] and S. POROWSKI[#], R.D. DUPUIS[*] and C.J. EITING[*].

Materials Science Division, Lawrence Berkeley National Laboratory, Berkeley CA 94720, 62/203
[#] High Pressure Research Center "Unipress," Polish Academy of Sciences, Warsaw, Poland
[*] The University of Texas at Austin, Microelectronics Research Center, PRC/MER 1.606D-R9900, Austin TX 78712-1100 USA

ABSTRACT

Transmission electron microscopy has been used to study defects formed in Mg-doped GaN crystals. Three types of crystals have been studied: bulk crystals grown by a high pressure and high temperature process with Mg added to the Ga solution and two types of crystals grown by metal-organic chemical vapor deposition (MOCVD) where Mg was either delta-doped or continuously doped. Spontaneous ordering was observed in bulk crystals. The ordering consists of Mg rich planar defects on basal planes separated by 10.4 nm and occurs only for growth in the N to Ga polar direction ($000\bar{1}$ N polarity). These planar defects exhibit the characteristics of stacking faults with a shift vector of a $1/3$ [$1\bar{1}00$] +c/2 but some other features identify these defects as inversion domains. Different type of defects were formed on the opposite site of the crystal (Ga to N polar direction), where the growth rate is also an order of magnitude faster compared to the growth with N-polarity. These defects are three-dimensional: pyramidal and rectangular, empty inside with Mg segregation on internal surfaces. The same types of defects seen for the two growth polarities in the bulk crystals were also observed in the MOCVD grown GaN samples with Mg delta doping, but were not observed in the crystals where Mg was added continuously.

INTRODUCTION

Obtaining efficient p-type doping has been a continual challenge in GaN technology. GaN can easily be grown with n-conductivity but obtaining p-doping is more difficult. The most commonly used p-dopant is Mg. Despite the success in using Mg as p-dopant to fabricate light emitting diodes (LEDs) and lasers [1], many aspects of Mg-doping in GaN are still not understood. We report here evidence that, under certain growth conditions, Mg segregates and causes formation of several different types of defects depending on the details of the growth condition and on the growth polarity.

EXPERIMENTAL

Three different types of samples were studied: bulk samples, MOCVD samples with Mg delta doping and MOCVD samples where Mg was added continuously. The bulk Mg doped GaN crystals were grown by the High Nitrogen Pressure Method from a solution of liquid gallium containing 0.1-0.5 at.% Mg [2]. P-type conductivity was not achieved in these crystals, all were highly resistive.

The Mg delta-doped structure was grown at 1030°C by MOCVD at 200 Torr in a hydrogen (H_2) ambient using trimethylgallium (TMGa) and ammonia (NH_3). Cp_2Mg was used as the Mg precursor. The superstructure of a 130-period GaN/Mg delta doped layer consisted of 104 Å-thick layers of GaN, each followed by a 15 second exposure of Cp_2Mg. During the Cp_2Mg exposure, the TMGa was vented, but the NH_3 and H_2 remained flowing into the chamber. At the end of the growth samples were in`-`situ annealed at 850°`C for 10 min in order to dissociate Mg-H complexes and activate the Mg atoms.

The p-i-n structures were also grown by MOCVD at the same temperature (1030°C) but with continuous Cp_2Mg exposure and similar growth rate (5.5 Å/s) compared to (5.3Å/s) used for the delta doping. The same annealing at 850°C for 10 min was also performed. All crystals have been studied using transmission electron microscopy (TEM). Dopant concentration and impurity levels were determined using Secondary Ion Mass Spectrometry (SIMS) and Energy Dispersive x-ray spectroscopy (EDX). Cross-section samples were prepared along the [1$\overline{1}$00] direction for Convergent Beam Electron Diffraction (CBED) in order to determine crystal polarity and along the [11$\overline{2}$0] direction for high resolution cross-section studies (HREM).

RESULTS

Table I: Dopant and impurity concentration based on SIMS studies

Sample	Mg (cm^{-3})	O (cm^{-3})	C (cm^{-3})	Si (cm^{-3})
B	6e19	2.5e19	1.5e17	4e16
M	5e19	3e19	5e17	1e16 -1e17
Ba	4-6e^{19}	7e^{19}	1.5e18	1-2e19
P	3e19	<3e18	2-10e17	1-10 e16
MOCVD	1.5-3e19	5e16	5e16	
MOCVD	3e19	5e16	5e16	

Fig.1a shows a high resolution image of cross-sectional TEM micrograph from the the bulk sample B together with a ball model (Fig. 1b) indicating that this part of crystal was grown with [000$\overline{1}$] N polarity. This figure shows the presence of planar defects separated from each other by equal distances about 10.4 nanometers (20 c-where c is the lattice parameter). Selective area diffraction patterns (SAD) show that the formation of these regularly spaced planar defects leads to additional diffraction spots dividing the (0001) reciprocal distance into 20 equal parts (Figs.1c, d). Our earlier studies of these defects show that these monolayers have a 1/3 [1$\overline{1}$00]+c/2 displacement vector [3-5] characteristic of a stacking fault. However, a splitting of the (0001) and (0003) reflections (Fig.1e) indicates that these defects contain inversion domain boundaries (IDBs). This was also confirmed by a multi-beam dark field image obtained from the [11$\overline{2}$0] zone axis. By placing either (0001) or (000$\overline{1}$) diffracted beams on the zone axis with a small objective aperture reverse contrast on the defects was obtained. This confirms that these defects contain inversion domain boundaries (IDBs).

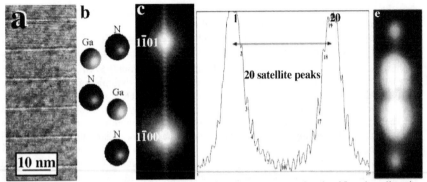

Fig. 1. (a) High magnification of sub-surface area, grown in the N polar direction indicated by ball model shown in (b), showing periodic arrangement of planar defects; (c) fraction of the diffraction pattern between $(1\bar{1}00)$ and $(1\bar{1}01)$ diffraction spots showing formation of satellite spots due to formation of ordered structure; (d) intensity distribution through this pattern showing division of (0001) reciprocal distance into 20 equal spacings; (e) A splitting of (0001) reflection.

EDX analysis [4] with an electron beam size of the order of 1 nm which was placed either on the defect or in the area between the defects confirmed that these planar defects are Mg rich. The ordering close to the N polarity surface was not observed in all studied crystals, such as Ba and P (Table I). In some of them only partial ordering took place and thick layers of regular hexagonal material were present between ordered areas (not shown for lack of space). Some specimens had no ordering at all. This suggests that the ordered structures form only for certain critical growth conditions.

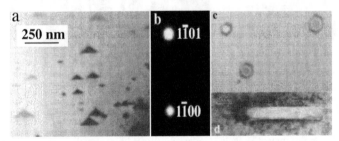

Fig. 2. (a) Tetrahedral-shaped defects formed in bulk GaN crystals grown with Ga polarity, (b) Fraction of the diffraction pattern between $(1\bar{1}00)$ and $(1\bar{1}01)$ diffraction spots obtained from this area of the crystal. Note lack of satellite spots; (c) Defects observed in plan-view configuration indicating that they are empty inside; (d) Another type of hollow defect (in cross-section) formed for growth with Ga polarity.

This opposite side of the bulk crystals (Ga-polarity) had a completely different defect structure as shown in (Fig. 2a). They appear in [11$\bar{2}$0] cross-section TEM micrographs as triangular features with a base on (0001) c-planes and six $\{11\bar{2}2\}$ side facets. All these triangles were oriented in a direction with the base closest and parallel to the sample surface with Ga-polarity, e.g. from the triangle tip to the base a long bond direction along the c-axis is from Ga to N. No additional diffraction spots (or satellite spots) were observed on this side of the crystal (Fig. 2b).

These triangular defects were observed in many of the bulk samples. The dimensions of the largest defects are in the range 100 nm (measured length of their bases) and the smallest are about 3-5 nm. The density of these defects is about $2.5 \times 10^9 \text{cm}^{-2}$. Studies in the plan-view configuration confirm that these defects are pyramids and that they are empty inside (Fig. 2c). CBED studies of these defects indicate that some reconstruction exist on the internal surfaces of these pinholes most probably caused by Mg segregation [5]. There were also crystals like M and Ba (see Table I) where these defects were not observed at all. Especially for crystal Ba in which there were no structural defects at all despite the fact that Mg concentration was comparable to those where ordering was present (B and M) and where highest concentration of oxygen was observed. A second type of hollow defect (empty inside) was also observed in cross-section samples, a rectangular defect delineated by a cubic layer on top and bottom basal plane of internal surfaces (Fig. 2d). The different types of defects observed on the crystal sides grown with N and Ga-polarity is likely to be associated with different surface reconstructions and different positions within the unit cell where the Mg atoms are incorporated.

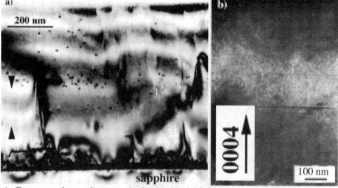

Fig. 3. (a) Cross-section micrograph from the MOCVD grown GaN sample with Mg-delta doping. Note high density of triangular and also rectangular defects typical for growth with Ga polarity. The arrows show the area where planar defects are formed typical for N polarity. Their higher magnification of the area with planar defects is shown in (b).

These different types of defects (planar and three-dimensional) formed for growth with different polarities were observed initially in bulk crystals doped with Mg [3-5], where many crystals are grown during a single growth run, and the precise growth condition for each particular crystal is rather indeterminate. Therefore, experiments were also carried on heteroepitaxial samples grown by MOCVD, where the growth conditions can be controlled for each crystal.

TEM studies on cross-section Mg-delta doped samples also show both types of defects formed for the two opposite polarities in bulk GaN:Mg, e.g. planar defects (polytypoids) which were characteristic for growth with N polarity and pyramidal and rectangular pinholes observed in the bulk GaN:Mg grown with Ga polarity (Fig. 3). This observation clearly suggests that crystal growth polarity is changing due to Mg doping in agreement with earlier studies [6].

Fig. 3a shows that growth of about 150 nm of GaN:Mg above the buffer layer did not have any visible change in defect arrangement compared to undoped samples [3]. However, in the following layer, about 200 nm thick, planar defects, like those in GaN:Mg bulk crystals grown with N-polarity was observed (Fig. 3b). In the layer following these planar defects a high density ($\sim 10^{10}$ cm^{-2}) of triangular and rectangular defects like those observed in bulk crystals grown with Ga polarity was observed. SIMS analysis shows (Fig. 4a) much higher Mg concentration (4×10^{19} cm^{-3}) present in the areas where ordering took place. At the layer thickness at which the hollow defects (rectangular and triangular) were formed Mg concentration dropped to 2×10^{19} atoms/cm^3 and stayed almost constant with a slow increase to reach a concentration of 4×10^{19} cm^{-3} at the sample surface. Impurity levels in this sample were very low showing carbon and oxygen concentrations on the level of 4-5×10^{16} cm^{-3} (three orders of magnitude lower than in the bulk samples), therefore, defects which were formed in these crystals must be caused by the Mg presence rather than by oxygen impurity level.

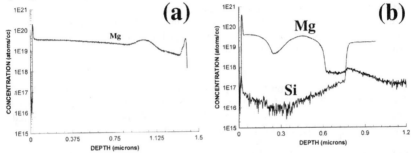

Fig. 4. (a) SIMS measurements obtained from a MOCVD sample with Mg delta doping. A higher Mg concentration was measured in the area where planar defects were formed, similar to those observed in the bulk GaN:Mg, (b) Similar analysis for the sample with continuous Mg doping where formation of structural defects was not observed despite the comparable Mg concentration.

However, in a sample where Mg was supplied continuously, despite similar Mg concentration (Fig. 4b), similar impurity levels and practically the same growth rate, formation of the previously described planar and/or three-dimensional defects was not observed. Therefore, it is not surprising that not all bulk samples showed ordering, since some deviations in composition, temperature or growth rate are possible from crystal to crystal growing in different parts of the container.

SUMMARY

Ordering observed in GaN:Mg for growth with N-polarity appears to be similar to the polytypoids formed in AlN rich in oxygen, or in Mg-Al-N-O-Si compounds [7-8]. Different types of polytypoids were observed depending on the M/X ratio (M-metal and X-nonmetal). In all these samples oxygen concentration was at a level close to 1 at. % or higher [7-8]. In our bulk samples oxygen concentration was high, but only in the range of 5×10^{19} cm^{-3}. However, in the samples with Mg delta doping oxygen concentration was only at the level of 5×10^{16} cm^{-3}, therefore, the polytypoids which are formed in our GaN:Mg samples are probably not oxygen related. It is anticipated that they must be caused most by the Mg presence, which was confirmed by EDX studies [4]. Our experimental observations indicate that planar defects formed for growth with N polarity show a shift of 1/3 [1 $\bar{1}$ 00] +c/2 and contain an inversion. Different Mg behavior for N- and Ga-polarity can be expected based on the theoretical work of Bungaro et al [9], but ordering was a rather unexpected finding. Segregation of Mg was expected for N-polarity, but the present studies show that for specific growth conditions Mg segregation can also occur for Ga polarity. The presence of Mg also leads to formation of hollow defects in the form of pyramids and rectangular defects with surface reconstruction due to Mg segregation. This indicates that Mg segregated on particular planes tends to prevent further growth. These three dimensional defects were observed in the majority of bulk GaN samples but also in GaN:Mg delta doped samples grown by MOCVD.

An extremely interesting observation in this study, that deserves further investigation, is that in the MOCVD layers grown under similar conditions and with similar Mg concentration, but where Mg is introduced continuously rather than using the delta doping process, none of the defects were formed. Since growth of MOCVD layers can be better controlled than bulk growth this can lead to further understanding of the behavior of Mg in GaN structures. Optimum growth conditions may be identified which result in active dopant as needed for p-type conductivity but which avoid formation of defects. Formation of these structural defects for Mg doped samples helps to explaining why consistent p-doping has been rather difficult.

ACKNOWLEDGMENT

This work was supported by U.S. Department of Energy, under the Contract No. DE-AC03-76SF00098. The use of the facility at the NCEM at LBNL is greatly appreciated. Work at UT-Austin was supported by the ONR under N00014-95-1-1302 (monitored by Dr. J. C. Zolper) and the NSF under Grant CHE-89-20120.

REFERENCES:

1. S. Nakamura, M. Senoh, and T. Mukai, Jpn. J. Appl. Phys. 31 (1991) L1708; S. Nakamura, Paper Plenary 1, presented at the 24th International Symposium on Compound Semiconductors, San Diego CA, 8-11 September 1997.
2. S. Porowski, M. Bockowski, B. Lucznik, I. Grzegory, M. Wroblewski, H. Teisseyre, M. Leszczynski, E. Litwin-Staszewska, T. Suski, P. Trautman, K. Pakula and J.M. Baranowski, Acta Physica Polonica A 92 (1997) 958.
3. Z. Liliental-Weber, M. Benamara, J. Washburn, I. Grzegory, and S. Porowski, Materials Res. Soc. Symp.Proc., vol. 572, 363 (1999).
4. Z. Liliental-Weber, M. Benamara, J. Washburn, I. Grzegory, and S. Porowski, Phys. Rev. Lett. 83, 2370 (1999).
5. Liliental-Weber, M. Benamara, W. Swider, J. Washburn, I. Grzegory, and Z S.Porowski, D.J.H. Lambert, R.D. Dupuis, and C.J. Eiting, Physica B Vol. 273, 124 (1999).
6. V. Ramachandran, R.M. Feenstra, W.L. Sarney, L. Salamanca-Riba, J.E. Northrup, L.T. Romano, and D.W. Greve, Appl. Phys. Lett. 75, 808 (1999).
7. K.H. Jack, J. Mat. Sci. 11 (1976) 1135.
8. Y.Yan, M. Terauchi and M. Tanaka, Phil. Mag. A 77 (1998) 1027.
9. Bungaro, K. Rapcewicz, and J. Bernholc, Phys. Rev. B 59 (1999) 9771.

Mat. Res. Soc. Symp. Vol. 595 © 2000 Materials Research Society

Optical Activation Behavior of Ion Implanted Acceptor Species in GaN

B.J. Skromme and G.L. Martinez
Department of Electrical Engineering and Center for Solid State Electronics Research,
Arizona State University, Tempe, AZ 85287-5706

ABSTRACT

Ion implantation is used to investigate the spectroscopic properties of Mg, Be, and C acceptors in GaN. Activation of these species is studied using low temperature photoluminescence (PL). Low dose implants into high quality undoped hydride vapor phase epitaxial (HVPE) material are used in conjunction with high temperature (1300 °C) rapid thermal anneals to obtain high quality spectra. Dramatic, dose-dependent evidence of Mg acceptor activation is observed without any co-implants, including a strong, sharp neutral Mg acceptor-bound exciton and strong donor-acceptor pair peaks. Variable temperature measurements reveal a band-to-acceptor transition, whose energy yields an optical binding energy of 224 meV. Be and C implants yield only slight evidence of shallow acceptor-related features and produce dose-correlated 2.2 eV PL, attributed to residual implantation damage. Their poor optical activation is tentatively attributed to insufficient vacancy production by these lighter ions. Clear evidence is obtained for donor-Zn acceptor pair and acceptor-bound exciton peaks in Zn-doped HVPE material.

INTRODUCTION

The properties of shallow acceptor dopants in GaN are important in various device applications requiring p-type doping, such as lasers, light-emitting diodes, photodiodes, junction field-effect transistors, thyristors, bipolar and heterojunction bipolar transistors, and high-voltage junction termination extension in high power devices. The spectroscopic properties and binding energies of many of these acceptors have been characterized only to a limited degree. Ion implantation is a method which permits the easy, controlled introduction of virtually any dopant species (or combination thereof) without complications due to unintentional passivating agents such as H. It is therefore potentially highly useful to survey the properties of potential dopant species. Such a broad survey of 35 elements was performed in pioneering work by Pankove and Hutchby [1]. However, the implantation doses used were very heavy, in order to overcome the high residual background doping common at that time, and the resulting spectra were therefore quite broad. Also, the annealing temperature used (1050 °C) was significantly lower than that now known to be needed to remove damage effectively and activate dopants effectively in GaN (at least 1300 °C) [2-4]. Moreover, NH_3 was used as the annealing ambient, which is now known to lead to H passivation of various dopants [5], and may have strongly affected the spectra. Also, the photoluminescence (PL) measurements were performed at 77 K, which does not yield the full range of information available at liquid He temperatures, due to thermal delocalization of excitons and thermal phonon broadening. Subsequent PL studies of specific implanted impurities have been reported [6,7], but typically employed high dose implantations and sometimes low annealing temperatures. High quality PL spectra typically require low doping levels to avoid the effects of inhomogeneous Stark effect broadening and impurity interactions, which lead to spectral broadening.

The present study aims to demonstrate the feasibility of using low dose implantations of selected impurities into high purity starting material to obtain high quality PL spectra of the resulting localized exciton and impurity states. Annealing is performed under conditions known to recover the damage due to the implantation without degrading the material [2,3]. This approach also permits us to assess the activation behavior of implanted dopants using optical methods. Electrical characterization of acceptor activation is difficult because of the inefficient ionization of most acceptors in GaN at room temperature. Our initial study employs the most common acceptor dopant, Mg, in addition to two potentially shallower alternative acceptors, Be and C. We find that the latter two impurities do not activate efficiently without a co-implantation under our conditions, but that Mg does. We also investigate the spectroscopic properties of Zn acceptors, using HVPE material doped with Zn during growth.

EXPERIMENTAL APPROACH

The starting material for the implantations is a 20 μm thick heteroepitaxial GaN layer grown on sapphire(0001) by hydride vapor phase epitaxy (HVPE). This unintentionally doped layer has a net residual doping level of $n=5.1 \times 10^{16}$ cm^{-3} and an electron mobility $\mu_n = 680$ cm^2/Vs. The full width at half maximum (FWHM) of the neutral donor-bound exciton (D$^\circ$,X) peak in the low temperature PL spectrum of this sample is only 1.8 meV, confirming its good quality. Moreover, the only shallow donor-acceptor pair (D$^\circ$-A$^\circ$) PL peak in this sample, at 3.27 eV, is only barely detectable in comparison to the 2LO phonon replica of the (D$^\circ$,X) peak. The residual yellow and red PL bands at 2.2 and 1.6 eV, respectively, are similarly very weak, having maximum heights at least 5000× smaller than that of the dominant (D$^\circ$,X) peak. This material is therefore ideal for implantation studies of acceptors, since even very weak peaks due to low dose implants will be readily visible above the residual background peaks.

Pieces of the above wafer were implanted with ^{24}Mg, ^{9}Be, or ^{12}C ions, using four different energies in each case to achieve a roughly flat profile from about 20 to 430 nm. The average concentration was set to be 10^{15}, 10^{16}, and 10^{17} cm^{-3} in different pieces for each impurity. The samples were then encapsulated with a ~50-70 nm thick layer of AlON, which was reactively sputtered using a high purity Al target, Ar, and N$_2$. The samples were heated to 500 °C during the sputtering process. Rapid post-implantation thermal annealing was then performed for 8 s at 1300 °C in flowing N$_2$.

Photoluminescence (PL) spectra were then recorded using excitation provided by a He-Cd laser at 325 nm or by an Ar$^+$ laser at 305.5 nm. Details of the measurement system are given elsewhere [8]. All spectra are corrected for the spectral response of the measurement system.

PHOTOLUMINESCENCE OF Mg, Be, AND C IMPLANTED GaN

Low temperature PL spectra as a function of implantation dose are shown in Figure 1 for the case of Mg implantation. The highest energy peaks involve donor and acceptor-bound excitons and are discussed below. Even at the lowest dose (10^{15} cm^{-3}), a clear (D$^\circ$-A$^\circ$) peak at 3.270 eV and its LO and acoustic phonon replicas become visible, although they are more than 100× weaker than the original (D$^\circ$,X) peak. In contrast, unimplanted control samples annealed under the same conditions show essentially no changes in their spectra (not shown), except for an overall increase in the absolute intensity of the spectrum and a slight broadening of the (D$^\circ$,X) peak to 3.6 meV FWHM. The new peaks can therefore be assigned unambiguously to the implanted Mg atoms (they did not occur at similar doses for Be and C, as discussed below). Moreover, a clear dose dependence is observed; the (D$^\circ$-A$^\circ$) peak grows steadily stronger compared to the excitonic peaks with increasing dose (note the various vertical scale expansion factors). It also becomes broader due to the increasing impurity content. In the deeper portions of the spectrum (note the broken axis), a new green PL band becomes prominent at 2.35 eV, together with a weaker red band around 1.73 eV (again note the much reduced vertical scale expansion factor at the highest dose). It is not clear from our data if either deep band is uniquely associated with Mg impurities, since they might simply originate from defects whose formation is stimulated by moving the Fermi level close to the valence band.

Study of the Mg-related features is facilitated by the narrow linewidths in these spectra and by the absence of features which may arise from the presence of H [9]. The (D$^\circ$-A$^\circ$) peak shifts to higher energy with increasing excitation intensity at a rate of a few meV per decade of intensity and broadens (not shown). This is the normally expected saturation behavior expected for this type of transition, due the Coulomb interaction in the final state. The temperature dependence of the spectra is shown in Figure 2 for the lowest implantation dose, since that sample shows the best-resolved peaks as a function of temperature. As the temperature increases, the (D^o,X) peak quenches in favor of A free exciton (X_A) recombination, and shifts to lower energy as the band gap is reduced. The $(D^o\text{-}A^o)$ peak quenches in favor of a distinct, higher energy peak located, e.g., at 3.284 eV at 50 K. This peak grows to dominate the spectrum

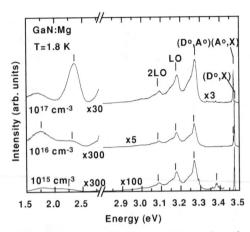

Figure 1. Overall PL spectra of Mg-implanted HVPE GaN as a function of dose.

at the highest temperature and becomes broader. We therefore attribute it to conduction band-to-acceptor $(e\text{-}A^o)$ transitions involving the Mg acceptor level, which become dominant as the shallow donors thermally ionize into the conduction band. The X_A position at 50 K is 3.479 eV. Adding the previously determined value of the free exciton binding energy to this value (26.4 meV [8]) yields a band gap at this temperature (including any effects of strain) of 3.505 eV. The $(e\text{-}A^o)$ peak position then yields a Mg acceptor binding energy of 224 meV (accounting for the kT/2 kinetic energy of the free electrons), in good agreement with our prior determination [9].

The excitonic portions of the PL spectra in Figure 1 are shown in more detail in Figure 3, to illustrate their dependence on Mg dose. A simple reflectance spectrum of the 10^{16} cm^{-3} sample is also included, to aid in identifying the A and B free exciton features. Similar reflectance spectra were recorded on each sample, and some PL spectra in Figure 3 have been rigidly shifted by up to 2 meV to align the X_A peaks, to compensate for minor strain variations and facilitate comparisons. In the lowest dose sample, the (D^o,X)

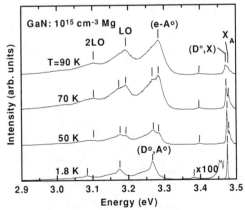

Figure 2. Temperature dependence of the PL spectra of a lightly Mg-implanted sample.

peak at 3.4760 eV remains dominant as it was in the starting material. A very weak neutral acceptor-bound exciton (A^o,X) peak is however also observed at 3.4704 eV, which was absent in the starting material. This peak is even more clearly evidenced in the LO phonon replicas of the bound excitons, where its phonon replica is comparable in strength to that of the (D^o,X) peak at 3.384 eV. This effect is due to the characteristically much stronger LO phonon coupling strength of the (A^o,X) peak compared to the (D^o,X) peak [9]. As the dose is increased, the (A^o,X) peak increases in intensity relative to the (D^o,X) peak and eventually dominates it. The (A^o,X) peak is quite sharp (2.6 meV FWHM), enabling an accurate determination of the localization energy of the Mg acceptor-bound exciton as about 12.2 meV. This value is in reasonable agreement with the ~11 meV value derived from homoepitaxial Mg-doped material [10], although it is smaller than the ~14-16 meV value we derived earlier from MBE GaN:Mg with much broader excitonic linewidths [9]. The present value should be more accurate.

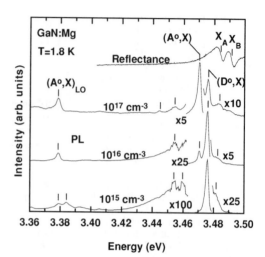

Figure 3. Excitonic portion of the PL spectra of Fig. 1 on an expanded scale, together a simple reflectance spectrum of the 10^{16} cm^{-3} dose sample.

The behavior of the Be and C implanted samples is dramatically different. The overall PL spectra are shown as a function of implantation dose for the C case in Figure 4. No (A°, X) peak whatsoever or its LO phonon replica can be detected at any of the three doses. Moreover, only a weak $(D^\circ - A^\circ)$ peak (note the scale expansion factors), possibly involving C acceptors, can be detected at the highest dose. The activation efficiency is estimated to be least 50× worse than that of Mg. Judging from the $(D^\circ - A^\circ)$ peak position, the C acceptor level may be about 15 meV shallower than that of Mg, but data on more clearly activated samples will be necessary to obtain a reliable value. In the deep portions of the spectrum, the 2.2 eV yellow PL band is strongly enhanced with increasing dose. This effect might be taken as evidence that the yellow band directly involves C impurities, as argued previously [11]. However, we observe an identical effect in Be implanted material, as shown in Figure 5. It therefore appears much more likely that the yellow PL band is somehow enhanced by residual implantation damage, as observed by

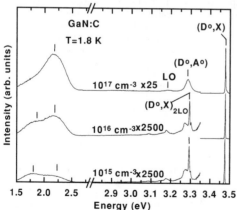

Figure 4. Overall PL spectra of C-implanted HVPE GaN as a function of dose.

Pankove and Hutchby [1]. This band is notably absent, however, in the Mg-implanted material, which might be due to the change in Fermi level resulting from the effective *p*-type doping with Mg. The Fermi level position may influence the formation of this defect [12].

The data as a function of Be implantation dose in Figure 5 are in all respects quite similar to that of the C-implanted case in Figure 4. Again a slightly shallower, and less strongly phonon-coupled $(D^\circ - A^\circ)$ peak is introduced, but only weakly and at the highest dose. We see no evidence for a much shallower, ~3.385 eV $(D^\circ - A^\circ)$ peak reported previously in MBE GaN:Be [13,14]. Our $(D^\circ - A^\circ)$ peak is more similar to the deeper one observed

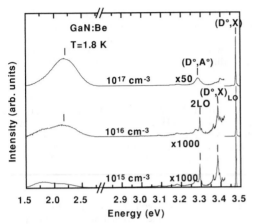

Figure 5. *Overall PL spectra of Be-implanted HVPE GaN as a function of dose.*

in MBE GaN:Be by Salvador et al. [15].

Several possible explanations could be advanced to explain the poor optical activation of Be and C in our samples, compared to that of Mg. It could be assumed that these elements simply do not produce shallow acceptor levels in GaN, or that they are naturally reluctant to incorporate substitutionally to do so. In particular, Be might be expected to form interstitial donors (although we see no direct spectroscopic evidence of new donor states). We believe, however, that the problem could be linked to the lighter mass of these elements compared to that of Mg. They produce less damage when implanted, and fewer vacancies are therefore created for them to substitute into. Further experiments are planned to co-implant inert elements of mass similar to Mg (such as Al) together with Be and/or C. If shallow acceptor levels do not appear under those conditions, the problem must be linked to the fundamental behavior of these species in GaN.

PROPERTIES OF Zn ACCEPTOR LEVELS IN Zn-DOPED HVPE MATERIAL

We also investigated the PL properties of a Zn-doped HVPE layer to determine the properties of Zn acceptors in GaN. The structure consisted of a 17 μm thick undoped GaN buffer layer on a sapphire(0001) substrate, followed by a 1.5 μm thick layer doped with Zn to a concentration of roughly 1×10^{18} cm^{-3}. The sample was grown in the same reactor used to produce our undoped wafer, so the observed new features can be confidently attributed to Zn.

The resulting PL spectra are shown in Fig. 6 as a function of excitation intensity. In addition to the usual residual (D°,X) peak at 3.4774 eV, we observe a prominent new acceptor-bound exciton at 3.4595 eV, with a localization energy with respect to the A free exciton of 24.5 meV. This value is in good agreement with previous data published by Amano et al. [16]. The (A°,X) peak is accompanied by prominent LO phonon replicas as well as a lower energy satellite feature at 3.444 eV of undetermined origin. A broad, structured emission extending from ~2.6 to 3.1 eV (peaking at ~2.92 eV) is also observed as seen previously [17]. We clearly resolve the no-

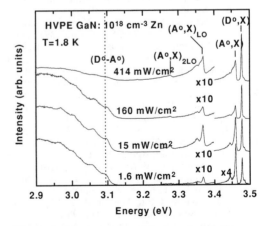

Figure 6. *PL spectra as a function of excitation intensity for HVPE GaN doped with 10^{18} cm^{-3} Zn during growth.*

phonon feature of this structure at around 3.1 eV, together with a series of LO and probably acoustic phonon replicas with strong coupling strengths. Most significantly, we observe a clear shift of the no-phonon peak to higher energy with increasing excitation intensity, positively identifying it as a $(D^\circ\text{-}A^\circ)$ transition. Comparing the $(D^\circ\text{-}A^\circ)$ peak position to that of Mg, we estimate the binding energy of the Zn acceptor as 400 meV.

CONCLUSIONS

We have demonstrated that low dose ion implantation into high purity material, coupled with suitably high temperature annealing in a H-free ambient, is a highly suitable method to study the spectroscopic properties of impurities in GaN. Successful study of low mass impurities may however require either co-implantation with heavier elements or annealing at even higher temperatures. Both types of experiment are planned.

ACKNOWLEDGMENTS

This work was supported by the Materials Research Science and Engineering Center at ASU, under Grant No. DMR 96-32635 from the National Science Foundation. We very gratefully acknowledge R. Molnar of MIT Lincoln Laboratories for providing the undoped and Zn-doped HVPE material used in this work.

REFERENCES

1. J.I. Pankove and J.A. Hutchby, J. Appl. Phys. **47**, 5387 (1976).
2. H.H. Tan, J.S. Williams, J. Zou, D.J.H. Cockayne, S.J. Pearton, J.C. Zolper, and R.A. Stall, Appl. Phys. Lett. **72**, 1190 (1998).
3. X.A. Cao, C.R. Abernathy, R.K. Singh, S.J. Pearton, M. Fu, V. Sarvepalli, J.A. Sekhar, J.C. Zolper, D.J. Rieger, J. Han, T.J. Drummond, R.J. Shul, and R.G. Wilson, Appl. Phys. Lett. **73**, 229 (1998) and references therein.
4. T. Suski, J. Jun, M. Leszczynski, H. Teisseyre, S. Strite, A. Rockett, A. Pelzmann, M. Kamp, and K.J. Ebeling, J. Appl. Phys. **84**, 1155 (1998).
5. S. Nakamura, N. Iwasa, M.Senoh, and T. Mukai, Jpn. J. Appl. Phys. **31**, 1258 (1992).
6. S. Strite, P.W. Epperlein, A. Dommann, A. Rockett, and R.F. Broom, Mater. Res. Soc. Symp. Proc. **395**, 795 (1996).
7. E. Silkowski, Y.K. Yeo, R.L. Hengehold, M.A. Khan, T. Lei, K. Evans, and C. Cerny, Mater. Res. Soc. Symp. Proc. **395**, 813 (1996).
8. B.J. Skromme, H. Zhao, B. Goldenberg, H.S. Kong, M.T. Leonard, G.E. Bulman, C.R. Abernathy, and S.J. Pearton, Mater. Res. Soc. Symp. Proc. **449**, 713 (1996).
9. M.A.L. Johnson, Z. Yu, C. Boney, W.C. Hughes, J.W. Cook, Jr., J.F. Schetzina, H. Zhao, B.J. Skromme, and J.A. Edmond, Mater. Res. Soc. Symp. Proc. **449**, (1996).
10. M. Godlewski, A. Wysmolek, K. Pakula, J.M. Baranowski, I. Grzegory, J. Jun, S. Porowski, J.P. Bergman, and B. Monemar, in *Proc. Internat. Sympos. Blue Laser & Light Emitting Diodes*, p. 356 (1996).
11. R. Zhang and T.F. Kuech, Mater. Res. Sympos. Proc. **482**, 709 (1998).
12. K. Saarinen, J. Nissilä, P. Hautojärvi, J. Likonen, T. Suski, I. Grzegory, B. Lucznik, and S. Porowski, Appl. Phys. Lett. **75**, 2441 (1999).
13. D.J. Dewsnip, A.V. Andrianov, I. Harrison, J.W. Orton, D.E. Lacklison, G.B. Ren, S.E. Hooper, T.S. Cheng, and C.T. Foxon, Semicond. Sci. Technol. **13**, 500 (1998).
14. F.J. Sánchez, F. Calle, M.A. Sanchez-Garcia, E. Calleja, E. Muñoz, C.H. Molloy, D.J. Somerford, F.K. Koschnick, K. Michael, and J.-M. Spaeth, MRS Internet J. Nitride Semicond. Res. **3**, 19 (1998).
15. A. Salvador, W. Kim, Ö. Aktas, A. Botchkarev, Z. Fan, and H. Morkoç, Appl. Phys. Lett. **69**, 2692 (1996).
16. H. Amano, K. Hiramatsu, and I. Akasaki, Jpn. J. Appl. Phys. **27**, L1384 (1988).
17. M. Ilegems, R. Dingle, and R.A. Logan, J. Appl. Phys. **43**, 3797 (1972).

Contact, Point Defects, Processing

Mat. Res. Soc. Symp. Vol. 595 © 2000 Materials Research Society

Characteristics of Ti/Pt/Au Ohmic Contacts on *p*-type GaN/Al$_x$Ga$_{1-x}$N Superlattices

L. Zhou, F. Khan, A.T. Ping, A. Osinski[1] and I. Adesida
Microelectronics Laboratory,
Department of Electrical and Computer Engineering,
Univ. of Illinois, Urbana-Champaign, IL 61801, U.S.A.
[1]NZ Applied Technologies, Woburn, MA 01801, U.S.A.

ABSTRACT

Ti/Pt/Au metallization on p-type GaN/Al$_x$Ga$_{1-x}$N (x=0.10 and 0.20) superlattices (SL) were investigated as ohmic contacts. Current-voltage and specific contact resistance measurements indicate enhanced p-type doping in the superlattice structures compared to that in GaN. Ti/Pt/Au is shown to be an effective ohmic metallization scheme on p-type GaN/Al$_x$Ga$_{1-x}$N superlattices. A specific contact resistance of R$_c$ = 4.6x10^{-4} Ω-cm^2 is achieved for unalloyed Ti/Pt/Au on GaN/Al$_{0.2}$Ga$_{0.8}$N SL. This is reduced to 1.3x10^{-4} Ω-cm^2 after annealing for 5 minutes at 300 °C.

INTRODUCTION

Ohmic contacts with very low contact resistances are essential for improving the performance of optoelectronic devices fabricated on GaN-based materials. For n-type GaN, metallization schemes containing Ti have been very effective in lowering the specific contact resistance (R$_c$) to less than 10^{-5} Ω-cm^2 [1-3]. In contrast, because of the wide bandgap and the relatively deep acceptor level of the commonly used Mg dopant, it has been very difficult to find ohmic contacts with R$_c$ less than 10^{-3} Ω-cm^2 for p-GaN [4]. So far, only three ohmic metallization and processing schemes on p-GaN have been reported to yield R$_c$ values less than 10^{-4} Ω-cm^2. One of these schemes requires a post-metallization anneal of 20 minutes at 800 °C [5], while another requires a boiling aqua regia surface etch prior to metal deposition [6, 7]. The third method reduces the resistance of Ni/Au contacts by oxidizing the metallization at 500 °C for 10 minutes and form a thin NiO layer[8]. This work aims at finding a low-resistance ohmic contact on p-GaN which requires less aggressive treatments, thereby making the metallization process more compatible with device processing.

Schubert *et al.* proposed that the overall hole concentration in p-type GaN can be increased through the generation of valence band edge oscillations in a superlattice[9]. This technique improves the activation efficiency of deep acceptors by allowing them to ionize when tunneling occurs from the larger bandgap material into the adjacent material with a narrower bandgap. Significantly enhanced p-type doping efficiency in GaN/Al$_x$Ga$_{1-x}$N short-period superlattices (SL) has been reported recently [10-11]. In this paper, we demonstrate excellent ohmic contacts on GaN/Al$_x$Ga$_{1-x}$N SL with a Ti/Pt/Au contact scheme without any aggressive surface treatment and post-metallization anneal. The results are compared to that obtained from the same metallization on p-GaN.

EXPERIMENTAL DETAILS

The superlattice layers used in this study were grown by MBE on sapphire and consisted of 20 periods each of Mg-doped GaN (10 nm) and Al$_x$Ga$_{1-x}$N (10 nm). Samples with two different aluminum concentrations, x=0.1 and 0.2, were investigated. Both samples exhibited a mobility of ~1 cm^{2}V^{-1}s^{-1} from room temperature Hall measurements. The reference p-GaN layer used in this study has a 0.3 μm doped layer with a nominal bulk carrier concentration of 2.5x10^{17} cm^{-3} and a mobility of 9 cm^{2}V^{-1}s^{-1} as obtained from Hall measurements at room temperature. Contact resistances were determined using the linear transfer length method (TLM). The structures used for the TLM measurements were fabricated by first patterning the p-GaN and SL samples with 100 x 650 μm mesas. The pattern was then transferred into the samples using reactive ion etching, which electrically isolated the mesas. Rectangular pads were then patterned. Prior to transferring the samples into the evaporation chamber, the surfaces were cleaned in O$_2$ plasma, followed by dips in a dilute HCl:DI (1:2) solution. Electron beam evaporation was used to deposit Ti (15 nm) and Pt (50nm), while Au (80nm) was thermally evaporated as the last capping layer. Post-deposition heat treatment was carried out while flowing N$_2$ at 1 atmosphere in a rapid thermal annealing (RTA) system. The contact characteristics were studied using current-voltage (I-V) and four-probe linear TLM techniques at room temperature. For accurate specific contact resistance determination, actual pad distances were determined using scanning electron microscopy (SEM) after metal lift-off.

RESULTS AND DISCUSSIONS

There are several reasons for adopting Ti/Pt/Au metallization for this study. Based on the Schottky-Mott model of metal-semiconductor contacts, the metal-semiconductor barrier height ϕ_b is related to the work function of the metal ϕ_m and the electron affinity of a *p*-type semiconductor χ_s by $\phi_b=(\chi_s+E_g)-\phi_m$, provided that there is no Fermi level pinning at the M-S interface. Therefore, metals with high work functions are required to reduce the M-S barrier height for wide band-gap semiconductors such as GaN. Pt, with a ϕ_m of 5.65 eV, has the highest work function of all metals and is chemically stable. However, experiments have shown that Pt alone suffers from poor adhesion on GaN and therefore, a thin Ti layer is used to improve adhesion as well as getter surface contamination or oxides. The excellent ohmic contacts obtained here for Ti/Pt/Au shows that Ti has been effective in these roles.

Figure 1 shows the current-voltage characteristics of as-deposited Ti/Pt/Au contacts on all three samples. The as-deposited contact is rectifying on the reference p-type GaN. In contrast, due to the enhanced doping efficiencies, the

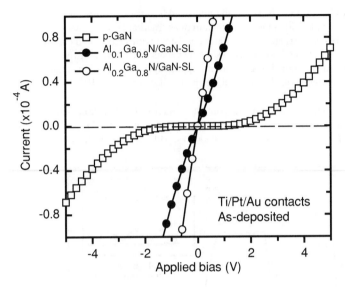

Figure 1. Current-voltage characteristics of as-deposited Ti/Pt/Au contacts on p-Type GaN/Al$_x$Ga$_{1-x}$N superlattices (x=0.1 and 0.2)

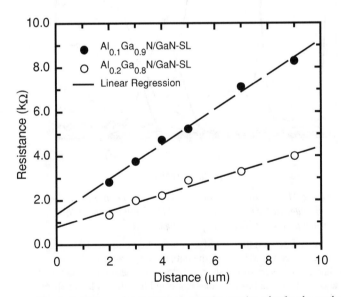

Figure 2. Measured resistance vs. contact spacing plot for the as-deposited Ti/Pt/Au contacts on p-type GaN/Al$_x$Ga$_{1-x}$N superlattices (x=0.1 and 0.2). Dotted lines are least-square best fits to the different data sets.

I-V curves for GaN/Al$_x$Ga$_{1-x}$N SL exhibit improved linearity. Figure 1 also reveals that the sample with a higher aluminum mole fraction (x = 0.2) was more ohmic in its I-V characteristics. This is consistent with a calculated carrier concentration of 3x10^{18} cm^{-3} for the SL (x = 0.1) versus 5x10^{18} cm^{-3} for the other (x = 0.2). An additional mechanism in support of this observation is that the total polarization charge should be higher in the SL with x=0.2 [10].

The contact resistivity (ρ_c) and specific contact resistance (R$_c$) were calculated from the measured resistance vs. contact pad spacing data using the linear TLM method. Figure 2 plots the measured resistance vs. pad distance for the as-

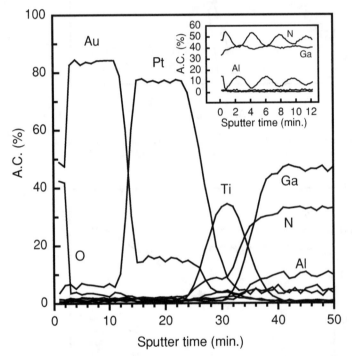

Figure 3 AES depth profile of Ti /Pt /Au contacts after annealing for 5 minutes at 300°C. Inset shows AES depth profile of the as-grown p-type GaN/Al$_{0.2}$Ga$_{0.8}$N sample.

deposited Ti/Pt/Au contacts on the two SL samples. Least-square linear regression lines were used to extrapolate the data to the y-axis. The slope of the regression line is a measure of the sheet resistance of the semiconductor. It is clear that the SL with x=0.2 has a smaller slope and therefore a lower sheet resistance, again due to the increased carrier concentration coming from its higher Al mole fraction. ρ_c and R$_c$ values can be extracted from the slope of the line, x- and y-intercepts and the pad width. These results are summarized in Table 1. For contacts on semiconductors with N$_A$>10^{17}cm^{-3}, the tunneling process is expected to predominate. In this case, $R_c \propto \exp[m(N_A)^{-1/2}]$, where m is a parameter related to the contact barrier height and hole effective mass, and N_A, which is the effective carrier concentration. It is clear from this relationship that R_c will be lower for the SL with x = 0.2, and R_c of both SL should be much lower than that of the p-GaN, which is consistent with the results shown in Figure 1 and Table 1.

Table 1: Summary of results for as-deposited and annealed (at 300 $^{\circ}$C for 5 minutes) Ti/Pt/Au contacts on p-type GaN and GaN/Al$_x$Ga$_{1-x}$N superlattices (x=0.1 and 0.2).

Semiconductor Layer Structure	Contact Resistivity (Ω-mm)		Specific Contact Resistance (Ω-cm^2)	
	as-dep	annealed	as-dep	annealed
p-GaN	Schottky	~200	Schottky	~6×10^{-3}
GaN/Al$_{0.1}$Ga$_{0.9}$N SL	72	62	6.6×10^{-4}	4.7×10^{-4}
GaN/Al$_{0.2}$Ga$_{0.8}$N SL	41	24	4.6×10^{-4}	1.3×10^{-4}

Preliminary results on the effectiveness of Ti/Pt/Au contacts on p-type GaN and Al$_x$Ga$_{1-x}$N alloys after annealing are also included in Table 1. Contacts annealed at 300 $^{\circ}$C for 5 minutes showed clear improvement. Figure 3 shows an Auger electron spectroscopy (AES) depth profile of the contact metallization after this treatment. Annealing at temperatures higher than 400 $^{\circ}$C resulted in rapid degradation of the contacts. The degradation mechanism is currently still under investigation.

CONCLUSIONS

This study presented the electrical characteristics of Ti/Pt/Au contact on p-type GaN/Al$_x$Ga$_{1-x}$N superlattices. The results indicate effective increase in carrier concentration by using superlattice structures. Ti/Pt/Au has been demonstrated as an effective ohmic metallization scheme on p-type GaN/Al$_x$Ga$_{1-x}$N superlattices, which achieved a specific contact resistance of 1.3x10^{-4} Ω-cm^2 on GaN/Al$_{0.2}$Ga$_{0.8}$N superlattice after annealing for 5 minutes at 300 $^{\circ}$C.

Acknowledgements: The materials work at NZ Applied Technologies was supported by BMDO contract N00014-99-M-0277 (Dr. J. Zolper). The work at the University of Illinois was supported by DARPA Grant DAAD19-99-1-0011 (Dr. N. El-Masry and Dr. E. Martinez) and NSF Grant ECS 95-21671.

REFERENCES

1. Fan, Z.F., Mohammad, S.N., Kim, W., Aktas, O., Botchkarev, A.E. and Morkoc, H.: "Very low resistance multilayer ohmic contacts to n-GaN", Appl. Phys. Lett., 1996, **68**, 12, pp.1672-1674

2. Schmitz, A.C., Ping, A.T., Khan, M.A., Chen, Q., Yang, J.W. and Adesida, I.: "Metal contacts to n-type GaN", J. Electron. Mat., 1998, **27**, 4, pp.255-260

3. Smith, L.L., Davis, R.F., Liu, R.J., Kim, M.J. and Carpenter, R.W.: "Microstructure, electrical properties and thermal stability of Ti-based ohmic contacts to n-GaN", J. Mater. Res., 1999, **14**, 3, pp. 1032-1038

4. Liu, Q. Z. and Lau, S. S. : "A review of the metal-GaN contact technology", Solid State Electronics, 1998, **42**, (5), pp.677-691

5. Suzuki, M., Kawakami, T., Arai, T., Kobayashi, S., Koide, Y., Uemura, T., Shibata, N. and Murakami, M.: "Low-resistance Ta/Ti ohmic contacts for p-type GaN", Appl. Phys. Lett., **74**, 2, pp.275-277

6. Kim, J.K., Lee, J.L., Lee, J.W., Shin, H.E., Park, Y.J. and Kim, T.: "Low resistance Pd/Au ohmic contacts to p-type GaN using surface treatment", Appl. Phys. Lett., 1998, **73**, 20, pp.2953-2955.

7. Kim, J.K., Lee, J.L., Lee, J.W., Park, Y.J. and Kim, T.: "Low transparent Pt ohmic contact to p-type GaN by surface treatment using aqua regia", Appl. Phys. Lett., 1997, **70**, 10, pp. 1275-1277.

8. Ho, J.K., Jong, C.S., Chiu, C.C., Huang, C.N., Chen, C.Y. and Shih K.K.: "Low-resistance ohmic contacts to p-type GaN", Appl. Phys. Lett., **74**, 9, pp.1275-1277.

9. Schubert, E.F., Grieshaber, W. and Goepfert, I.D.: "Enhancement of deep acceptor activation in semiconductors by superlattice doping", Appl. Phys. Lett., 1996, **69**, pp.3737-3739

10. Kozodoy, P., Hansen, M., DenBaars, S. P. and Mishra, U.K.: "Enhanced Mg doping efficiency in $Al_{0.2}Ga_{0.8}N$/GaN superlattices", Appl. Phys. Lett., 1999, **74**, 24, pp.3681 – 3683

11. Goepfert, I.D., Schubert, E.F., Osinski, A. and Norris, P.E.: "Demonstration of efficient p-type doping in $Al_xGa_{1-x}N$/GaN superlattice structures", Elect. Lett., 1999, **35**, 13, pp. 3288-3290

Mat. Res. Soc. Symp. Vol. 595 © 2000 Materials Research Society

HIGH QUALITY NON-ALLOYED Pt OHMIC CONTACTS TO *P*-TYPE GaN USING TWO-STEP SURFACE TREATMENT

Ja-Soon Jang, Seong-Ju Park, and Tae-Yeon Seong*
Department of Materials Science and Engineering, Kwangju Institute of Science and Technology (K-JIST), Kwangju 500-712, Korea, *e-mail: tyseong@kjist.ac.kr

ABSTRACT

Two-step surface-treatment is introduced to obtain low resistance Pt contacts to *p*-type GaN. The first step is performed after the mesa etching process using buffered oxide etch (BOE) and ammonium sulfide $[(NH_4)_2S_x]$. This is followed by the second step using BOE. The Pt contact, which was treated sequentially using ultrasonically boiled BOE (10 min) and boiled $(NH_4)_2S_x$ (10 min), produces a specific contact resistance of 3.0 $(\pm 3.8) \times 10^{-5}$ Ωcm^2. However, the contact, that was simply BOE-treated, yields 3.1 $(\pm 1.1) \times 10^{-2}$ Ωcm^2. This indicates that the two-step surface treatment is promising technique for obtaining high quality ohmic contacts to *p*-GaN. Investigation of the electronic transport mechanisms using current-voltage-temperature (I-V-T) data indicates that thermionic field emission is dominant in the surface-treated Pt contacts.

INTRODUCTION

GaN and III-V nitride layers have been extensively investigated, since the realisation of short wavelength light emitting diodes (LEDs) and laser diodes (LDs)[1,2] and the demonstration of metal-semiconductor field effect transistors (MESFETs)[3] and heterojunction bipolar transistors (HBTs).[4] Low resistance and thermally stable ohmic contacts are crucial for improving such device performance. However, there are some obstacles, such as difficulty in increasing *p*-GaN near-surface carrier concentrations and the absence of metals having work function larger than that of *p*-GaN (sum of bandgap of 3.4 eV and electron affinity of 3.3 eV),[5] which make it difficult to achieve low resistance ohmic contacts to *p*-GaN. Jang et al.[6] investigating ohmic contacts to *p*-GaN using Ni/Pt/Au metallisation schemes, showed that the metal contact was ohmic with a contact resistance of 2.1×10^{-2} Ωcm^2 when annealed at 500 °C for 30 s in a flowing Ar atmosphere. Mori et al.,[7] investigating ohmic contacts on *p*-GaN using Pt, Ni, Au, and Ti single layers, showed that the as-deposited Pt contact was ohmic with a specific contact resistance of 1.3×10^{-2} Ωcm^2. Cao et al.,[8] investigating thermal stability of W and WSi_x contacts on *p*-GaN, reported a specific contact resistance of $\sim 10^{-2}$ Ωcm^2 for the 300 °C annealed WSi_x.

To achieve low resistance ohmic contacts to *p*-GaN, surface treatments using the solutions of KOH and HNO_3:HCl (1:3) have been performed.[9,10] Lee et al.[9] employed KOH to modify surface conditions and showed that for Pd/Au contacts, the surface treatment leads to a decrease in the specific contact resistance up to 7.1×10^{-3} Ωcm^2. Kim et al.[10] used $HNO_3:HCl$ (1:3) to modify surface conditions and showed that for Pd/Au ohmic contacts to p-GaN, the surface modification results in a specific contact resistance of 4.1×10^{-4} Ωcm^2. They attributed the low resistance to the removal of a native oxide layer that inhibits hole transport from the metal to p-GaN.

In this paper, we report on the formation of low resistance Pt contacts to p-GaN by two-step surface treatment technique using buffered oxide etch (BOE) and ammonium sulfide [$(NH_4)_2S_x$]. It is shown that specific contact resistances and Schottky barrier heights depend sensitively on the surface-treated conditions. In addition, the electronic transport mechanisms for the surface-treated Pt contacts are described and discussed.

EXPERIMENTAL PROCEDURE

Metalorganic chemical vapor deposition (Emcore DGaN125^{TM}) was used to grow 1-μm-thick p-GaN:Mg (n_a = 1.5-3×10^{17} cm^{-3}) on (0001) sapphire substrates. The GaN layer was ultrasonically degreased in trichloroethylene, acetone, methanol, and ethanol, and rinsed in deionised (DI) water for 5 min. Prior to the fabrication of TLM patterns, mesa structures were patterned by inductively coupled plasma etching (Oxford Plasma 100) using $Cl_2/Ar/H_2$. The first-step surface treatment was performed after the mesa etching process. The mesa-patterned layers were chemically treated by three different conditions: (i) not-treated (termed here 'A-treated'); (ii) ultrasonically boiled in BOE solution for 10 min ('B'); (iii) first ultrasonically boiled in BOE for 10 min and then boiled in $(NH_4)_2S_x$ for 10 min ('C'). After the first-step treatment, TLM patterns were defined by photolithographic technique. The size of the pads was 100×200 μm^2 and the spacing between the pads was 5, 10, 15, 20, 25, and 35 μm. After the TLM patterning, the second-step treatment was performed. All the TLM-patterned layers were dipped into BOE for 30 s. Metallisation patterns were defined using lift-off technique. The samples were then rinsed in DI water, blown dry by N_2, and immediately loaded into an electron beam evaporation chamber (PLS 500). The thickness of the Pt films was 25 nm. Current-voltage (I-V) data were measured at room temperature using a parameter analyzer (HP 4155A) and Schottky barrier heights (SBHs, ϕ_b) were calculated using the I-V method. Electronic transport mechanisms were investigated by I-V-T data. X-ray photoemission spectroscopy (XPS) and Auger electron spectroscopy (AES) were used to investigate the variously treated surfaces of p-GaN.

RESULTS AND DISCUSSION

Figure 1 shows the I-V characteristics of Pt contacts on the various surface-treated p-GaN. The A-treated Pt contact reveals nonlinear I-V behaviour. However, the B- and C-treated contacts show near linear and linear characteristics, respectively. Specific contact resistances (R_{sc}) were determined from a plot of the measured resistances versus the spacings between the TLM pads. The least square method was used to fit a straight line to the experimental data. R_{sc} was determined to be 3.1 ($\pm$1.1)$\times10^{-2}$ for the A-treated sample, 2.2 ($\pm$1.6)$\times10^{-3}$ for the B-treated sample, and 3.0 ($\pm$3.8)$\times10^{-5}$ Ωcm^2 for the C-treated sample. It is noteworthy that the C-treatment results in a dramatic reduction (by about three orders of magnitude) in R_{sc} as compared to that of the A-treatment. The result of the A-treated sample is comparable to that reported by Mori et al.[7]

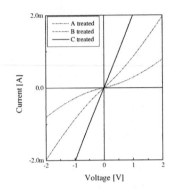 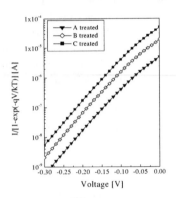

Figure 1. I-V characteristics of the variously surface-treated Pt contacts on p-GaN.

Figure 2. Plot of I/[1-exp(-qV/kT)] vs V for the variously surface-treated samples.

In order to investigate the electronic transport mechanisms, the effective SBHs and the relation between R_{sc} and temperature were calculated by the I-V and I-V-T measurements, respectively. The I-V relation is given by [11]

$$I = I_o \exp(qV/nkT) \, [1 - \exp(-qV/kT)]$$
$$I_o = AA^{**}T^2 \exp(-\phi_b/kT)$$

The value of A^{**} was calculated to be 104 A cm^{-2} k^{-2} assuming the effective hole mass (m_h^*) of 0.8 m_e for p-GaN.[12] It is known that the measured SBHs (ϕ_b) are not significantly affected by the variation of A^{**}.[13] Thus, the value of 104 A cm^{-2} k^{-2} was used for A^{**} to calculate SBHs. The effective SBHs were determined at zero voltage as shown in Fig. 2. The SBH was 0.49 ($\pm$0.01) eV for the A-treatment, 0.46 ($\pm$0.01) eV for the B-treatment, and 0.43 ($\pm$0.015) eV for the C-treatment. This indicates that the reduction of R_{sc} can be attributed to the decrease in the SBHs.

According to the electronic transport theory on metal-semiconductor contacts,[11] in principal, there are three mechanisms: thermionic emission (TE), thermionic field emission (TFE), and field emission (FE), which dominate the carrier flow. Relation between R_{sc}, E_{oo}, and SBH can be given by

$$Rsc \propto \exp(q\phi_b/kT) \text{ for TE } (E_{oo}/kT \ll 1) \quad (1)$$
$$Rsc \propto \exp[q\phi_b/(E_{oo} \coth(E_{oo}/kT))] \text{ for TFE } (E_{oo}/kT \sim 1) \quad (2)$$
$$Rsc \propto \exp(q\phi_b/E_{oo}) \text{ for FE } (E_{oo}/kT \gg 1) \quad (3)$$
$$E_{00} = hq/4\pi \, [Na/m_h^*\varepsilon]^{1/2}: \text{ tunneling parameter} \quad (4)$$

The temperature dependence of R_{sc} was obtained using the equations (1) – (4). (In these calculations, the SBHs and carrier concentrations of the contacts obtained by the I-V and Hall measurements were used as initial values.) Detailed results about the depth dependence of the effective carrier concentrations and the temperature dependence of tunneling parameters and effective SBHs will be published elsewhere.[14] Figure 3 shows the relation between R_{sc} and temperature. The dotted lines indicate the values that are theoretically calculated using the equations (1) – (4). For the A-treated sample, the measured specific contact resistance decreases with increasing temperature. This is in agreement with the calculated result, indicating that the dominant transport mechanism is TE. For the B- and C-treated samples, however, R_{sc} remains virtually unchanged over the given temperatures. The experimental results are consistent with those calculated using the approximation equations (2) and (3). This shows that TFE is dominant in the B-treated sample, whereas FE dominates in the C-treated sample. Furthermore, the SBHs of the B- and C-samples were calculated from the theoretical I-V characteristics for TFE and FE.[15] The calculations showed that the SBH is 0.46 eV for the B-treatment and 0.42 eV for the C-treatment. It is noteworthy that these values are comparable to those obtained by the I-V method for TE.

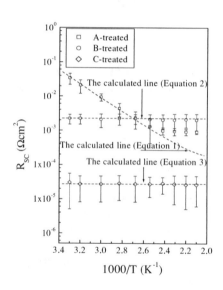

XPS and AES were employed to investigate the variously treated surfaces of p-GaN. Figure 4 (a) shows that the intensity of oxygen peak (O1s) varies with surface-treated conditions. This is consistent with the results of AES spectra, as shown in Fig. 4(b). It is noteworthy that no changes in the Ga 2p and the N1s peaks were observed (not shown). According to the metal-semiconductor band theory,[11,16] the effective SBHs can be influenced by the presence

Figure 3. Plot of R_{sc} vs temperature for the various surface-treated samples using I-V-T measurements.

of an oxide layer (with a thickness of δ) at the Pt/p-GaN interface.

$$q\phi_b = q\phi_{bo} + 4\pi kT/h\,(2m\chi)^{1/2}\,\delta \qquad (5)$$

where χ is the mean tunneling barrier for carrier injection from metal to p-GaN and m is the mean tunneling effective mass of carriers. This indicates that the removal of native oxide can contribute to the reduction of the SBHs of the surface-treated contacts. Thus, the complete removal of a native oxide layer with a thickness of ~2.5 nm is expected to result in reduction in R_{sc} by a factor of 15 – 18.

Based upon the results obtained using the I-V, I-V-T, XPS and AES measurements, the ohmic behaviour of the surface-treated contacts could be explained as follows. First, the reduction in the contact resistance can be attributed to the effective removal of the native oxide on the surface.[10] Second, the surface treatments could result in an increase in the carrier concentration near the surface of the p-GaN layers.[14] In this work, Hall measurements were made of the A- and C-treated GaN layers before the metal deposition. Indeed, it was shown that the C-treatment leads to an increase (by about a factor of 10) in the carrier concentration as compared to the A-treatment. This is indicative of the contact systems that may consist of metal/p^+-GaN /p-GaN structures for the B- and C-treated samples, making it possible for carriers to tunnel through the barriers. Third, the improvement in the contact resistance may be associated with an increase in the contact area between the metal and the p-GaN layer,[17,18] since the surface treatment may cause the roughening of the layer surface. Therefore, we suggest that the improvement in the specific contact resistances of the surface-treated contacts could be due to either the removal of the native oxide, an increase in the carrier concentration, an increase in the contact area, or their combined effects.

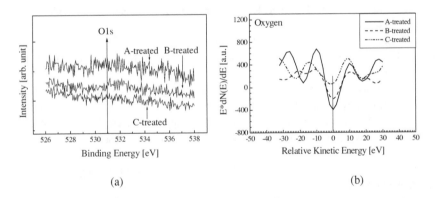

(a)

(b)

Figure 4. (a) XPS and (b) AES spectra of oxygen for the various surface-treated samples (A, B and C). Comparison clearly shows that native oxide on the C-treated sample is more effectively removed than the others.

SUMMARY

The effect of the surface treatment on the ohmic behaviour of the Pt contacts to p-GaN:Mg (1.5~3×10^{17} cm^{-3}) was investigated. Prior to the metal deposition, the two-step surface treatment was performed to modify the surface structures of the p-GaN layers: first, the layers were treated using boiled BOE and $(NH_4)_2S_x$ solutions; second, all the layers were then dipped into BOE. The measurements showed that specific contact resistance was highly sensitive to the surface-treated conditions. The A-treated sample produced 3.1 (± 1.1)$\times 10^{-2}$ Ωcm^2, while the BOE/$(NH_4)_2S_x$-treated contact yielded 3.0 (± 3.8)$\times 10^{-5}$ Ωcm^2. Ohmic behaviour could be due to the effective removal of native oxide on p-GaN, the increase in carrier concentration at near surface, and the increase in the contact area.

ACKNOWLEDGMENT

This work was supported by Korea Ministry of Information and Communications (C1-98-0929) and Critical Technology 21 (Development of Power Semiconductors) (98-N5-01-01-A-08).

REFERENCES

1. S. Nakamura, T. Mukai, and M. Senoh, J. Appl. Phys. **76**, 8189 (1994).
2. S. Nakamura, M. Senoh, N. Iwasa, T. Yamada, T. Matsushita, H. Kiyoku, and Y. Sugimoto, Jpn. J. Appl. Phys. **35**, L217 (1996).
3. M.A. Khan, J.N. Kuznia, A.R. Bhattarai and D.T. Olson, Appl. Phys. Lett. 62 1786 (1993).
4. J. Pankove, S.S. Chang, H.C. Lee, R.J. Molnar, T.D. Mustakas and B. Van Zeghbroeck, IEDM **94**-389 (1994).
5. V. M. Bermudez, J. Appl. Phys. **80**, 1190 (1996).
6. J. S. Jang, K. H. Park, H. K. Jang, H. G. Kim, and S. J. Park, J. Vac. Sci. Technol. B **16**, 3105 (1998).
7. T. Mori, T. Kozawa, T. Ohwaki, Y. Taga, S. Nagai, S. Yamasaki, S. Asami, N. Shibata, and M. Koike, Appl. Phys. Lett. **69**, 3537 (1996).
8. X. A. Cao, S. J. Pearton, F. Ren, and J. R. Lothian, Appl. Phys. Lett. **73**, 942 (1998).
9. J. L. Lee, J. K. Kim, J. W. Lee, Y. J. Park, and T. I Kim, Sol. Stat. Electron. **43**, 435 (1999).
10. J. K Kim, J. L. Lee, J. W. Lee, H. E. Shin, Y. J. Park, and T. I Kim, Appl. Phys. Lett. **73**, 2953 (1998).
11. E.H. Rhoderick and R.H. Williams, *Metal-Semiconductor Contacts* (Clarendon, Oxford 1988).
12. J.I. Pankove, S. Bloom, and G. Harbeke, RCA Rev. **36**, 163 (1975).
13. J. K. Sheu, Y. K. Su, G. C. Chi, M. J. Jou, and C. M. Chang, Appl. Phys. Lett. **72**, 3317 (1998).
14. J. S. Jang, and T. Y. Seong (unpublished).
15. A.Y.C. Yu, Solid State Electron. **13**, 239 (1970)
16. K. Hattori and Y. Izumi, J. Appl. Phys. **53**, 6906 (1982).
17. J. S. Jang, I. S. Chang, H. K. Kim, T-Y. Seong, S. H. Lee, and S. J. Park, Appl. Phys. Lett. **74**, 70 (1999).
18. J. S. Jang, S. J. Park, and T. Y. Seong, J. Vac. Sci. Technol.B 1999 (in press).

Mat. Res. Soc. Symp. Vol. 595 © 2000 Materials Research Society

Electrical Measurements in GaN: Point Defects and Dislocations

David C. Look, Zhaoqiang Fang, and Laura Polenta[1]

Semiconductor Research Center, Wright State University, Dayton, OH 45435

[1]INFM and Dipartimento di Fisica, University of Bologna, I-40126 Bologna, Italy

ABSTRACT

Defects can be conveniently categorized into three types: point, line, and areal. In GaN, the important point defects are vacancies and interstitials; the line defects are threading dislocations; and the areal defects are stacking faults. We have used electron irradiation to produce point defects, and temperature-dependent Hall-effect (TDH) and deep level transient spectroscopy (DLTS) measurements to study them. The TDH investigation has identified two point defects, an 0.06-eV donor and a deep acceptor, thought to be the N vacancy and interstitial, respectively. The DLTS study has found two point-defect electron traps, at 0.06 eV and 0.9 eV, respectively; the 0.06-eV trap actually has two components, with different capture kinetics. With respect to line defects, the DLTS spectrum in as-grown GaN includes an 0.45-eV electron trap, which has the characteristics of a dislocation, and the TDH measurements show that threading-edge dislocations are acceptor-like in n-type GaN. Finally, in samples grown by the hydride vapor phase technique, TDH measurements indicate a strongly n-type region at the GaN/Al_2O_3 interface, which may be associated with stacking faults. All of the defects discussed above can have an influence on the dc and/or ac conductivity of GaN.

INTRODUCTION

Because GaN is a wide-bandgap semiconductor, it is commonly assumed that large quantities of point defects will be present, because of self-compensation affects. For example, the first GaN, grown some thirty years ago, was strongly n-type, and the natural assumption was that the relevant donors were associated with the N vacancy V_N [1]. Only recently has this idea been shown not to be true, at least for the best, present-day GaN epitaxial layers [2]. In fact, accurate first-principles theoretical calculations indeed suggest a low concentration of V_N centers in n-type GaN, but a much higher concentration of Ga vacancy V_{Ga} defects [3]. However, the theoretical calculations assume equilibrium growth conditions, which may not be true, especially for the molecular-beam epitaxial (MBE) process. In fact, at the present time, MBE-grown GaN layers contain higher concentrations of donors, acceptors, and traps, than layers grown by metal organic chemical vapor deposition (MOCVD), or hydride vapor phase epitaxy (HVPE). The question is whether those donors, etc., are associated with defects, or impurities, or both. With regard to line defects, i.e., dislocations, the situation is somewhat clearer. It is known that dislocations act as nonradiative recombination centers [4], and affect the performance of GaN-based light emitters [5]. More recently, it has been shown that they are also strong scattering centers, especially at the typical concentrations found in mismatched GaN/Al$_2$O$_3$ layers [6]. Finally, there is evidence from deep-level transient spectroscopy (DLTS) experiments, that dislocations act as traps [7]. The V$_N$ center also is a trap [8].

Besides point defects and dislocations, there are two-dimensional defects, such as stacking faults, which we will classify as areal defects [9,10]. Recent theory suggests that stacking faults in GaN are not electrically active [10]; however, it is also known that the heavily-faulted interface region between HVPE GaN and the Al$_2$O$_3$ substrate is strongly n-type [11], possibly due to donor-like defects, although impurities cannot be ruled out. Thus, point, line, and areal defects all may play important roles in determining the concentrations of donors, acceptors, traps, and recombination centers in GaN.

LINE DEFECTS: DISLOCATIONS

In GaN layers with high dislocation densities, the mobilities are usually low [12]. However, until recently, it was not clear whether or not dislocations had a direct or indirect effect on mobility. For example, one of the two layers shown in figure 1 has a much higher mobility than the other, and also a much lower dislocation density [4 x 10^8 cm^{-2} vs. 2 x 10^{10} cm^{-2}, respectively, as measured by transmission electron microscopy (TEM)]. But there are also other differences in the samples, which might be expected to have an even greater effect on the mobilities. To quantify the scattering effects of dislocations, we recently showed that the scattering rate for a line charge, such as a

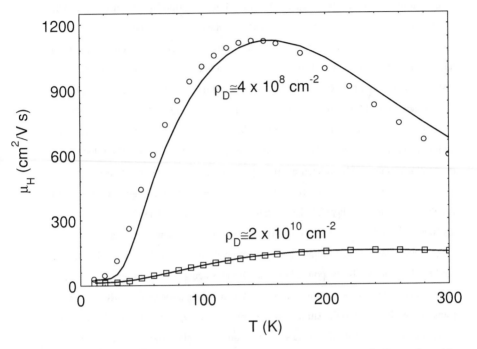

Figure 1. *Hall mobility as a function of temperature for two MOCVD GaN samples with different dislocation densities (ρ_D's). The solid lines are theoretical fits.*

dislocation, is given by the inverse of the following relaxation time [6]:

$$\tau_{dis}(k) = \frac{\hbar^3 \varepsilon_0^2 c^2}{N_{dis} m^* f^2 e^4} \frac{\left(1 + 4\lambda^2 k_\perp^2\right)^{3/2}}{\lambda^4} \tag{1}$$

where c is the c-lattice distance (5.185 Å), f/c is the linear charge density, λ is a screening parameter, $k_\perp$ is the wave vector in a direction perpendicular to the c axis, and the other symbols have their usual meanings. Since $\lambda \propto n^{1/2}$, it can be seen that dislocations are strongly screened by free electrons, and thus affect mobility more in the better (lower-n) samples. By adding the scattering rates of other important scattering mechanisms, and solving the Boltzmann transport equation to determine mobility, data sets such as those shown in figure 1 can be fitted, and parameters such as N_{dis} determined from the best fit [6]. The agreement shown in figure 1 is excellent, especially considering that the only fitting parameters, other than N_{dis}, are the donor concentration N_D and energy E_D. With an assumption of one negative charge per c-lattice distance (f=1), the fitted dislocation densities are 4.2 x 10^8 and 2.3 x 10^{10} cm^{-2}, respectively, in very good agreement with the TEM results. First-principles theory suggests that a likely configuration for the dislocation core in n-type GaN includes vacancies at the Ga sites [13]. It is expected that Ga vacancies could hold up to three negative charges per site (f=3); however, a theoretical analysis which includes electron-electron repulsion [14] leads to a value of about one charge per site (f=1), in agreement with the Hall results [6]. Note that such dislocations can strongly compensate the layer; e.g., 2 x 10^{10} dislocations per cm^2 will compensate (2 x 10^{10})/(5.185 x 10^{-8})=4 x 10^{17} electrons per cm^3, a higher value than that found in some high-quality GaN layers these days. Dislocation densities less than 10^8 cm^{-2} will have little effect on the mobility of present-day samples, but may become important again as impurity and defect levels are eventually reduced to the 10^{15}- cm^{-3} level. It should also be pointed out that, besides being strong scattering centers, there is good evidence that dislocations can also act as recombination centers (RCs) [4] and traps [7]. With regard to the latter, consider trap C1, shown in the data of figure 2; pulse- filling studies verify that this trap has unusual capture kinetics, more suggestive of a line charge (dislocation) than a point charge. Note that trap and RC effects often begin to show up at far lower defect concentrations than those found to be important for compensation and scattering. Thus, efforts to rid GaN of threading dislocations should be vigorously pursued.

AREAL DEFECTS: STACKING FAULTS

Type I stacking faults (one violation of the stacking rule) in GaN have a small formation energy (10 meV per unit-cell area), and thus are expected to be abundant [10]. Indeed, they are quite common in GaN, especially near the film/substrate interfaces [9]. In HVPE-grown GaN, especially, a region of about 2000-Å thickness near the GaN/Al$_2$O$_3$ interface is heavily faulted, and, interestingly, this same region is strongly n- type, often with a sheet density of about 10^{15} cm^{-2} or greater [11]. There is some disagreement in the literature whether or not stacking faults (SFs) should have electronic levels in the band gap; one theoretical calculation finds a level at E_V + 0.13 eV [15],

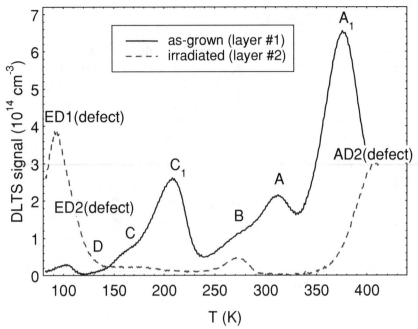

Figure 2. *DLTS spectra for an as-grown MBE GaN layer (solid line), and an irradiated MOCVD GaN layer (dashed line).*

while another finds no levels at all [10]. Neither of these scenarios can explain the highly degenerate n-type region, so that the origin of these shallow donors remains a mystery. It is possible that other donor-type defects, such as N vacancies, are present along with the SFs, or there may be a high concentration of impurity donors, such as O_N. In any case, this interface region has a strong effect on the measured electrical properties, as seen in figure 3. In this figure, the dashed line represents the true (bulk) mobility, determined by analyzing the Hall data with a two-layer conduction model. The carrier concentration (not shown) is affected even more, and an analysis of the raw (uncorrected) concentration data, without accounting for the interface region, leads to very inaccurate donor energies [11]. Also, any devices fabricated on such layers could experience severe current shunting. The elimination of these interface donors, whether or not they result directly from defects, would be highly desirable.

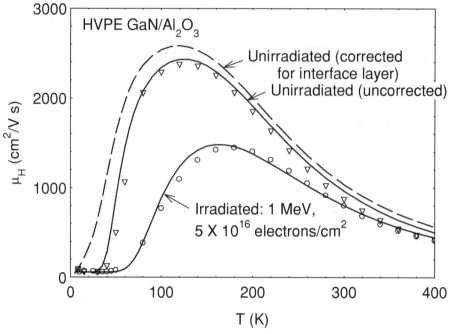

Figure 3. *Hall mobility as a function of temperature for unirradiated and irradiated HVPE GaN layers. Also shown (dashed line) is the mobility of the unirradiated sample corrected for a degenerate interface layer.*

POINT DEFECTS: VACANCIES AND INTERSTITIALS

Although point defects have longed been blamed for many of the undesirable properties of GaN materials, it is only in recent years that some of the true facts have become known. To be able to accurately investigate defects, it must be possible to produce them in a controlled manner, and this process can be carried out with a high-energy electron beam, which knocks out atoms, but does not create massive damage and large defect complexes [16]. Figure 3 shows the typical mobility degradation due to irradiation with 1-MeV electrons to a fluence of 5×10^{16} electrons per cm^2. The sample here is a 60-μm-thick, HVPE, GaN/Al$_2$O$_3$ layer, which, before irradiation, exhibited a "world's record" mobility of 950 cm^2/V-s at room temperature. Theoretical fits of the μ-vs.-T data (figure 3), and n-vs.-T data (not shown), indicate that each 1-MeV electron per cm^2 produces about one 0.06-eV donor, and one deeper acceptor per cm^3, and the interpretation, from several considerations, is that the donor is the N vacancy, and the acceptor, the N interstitial [2]. An immediate conclusion from these assignments is that the *residual* donor in this material, and in most other state-of-the-art GaN layers, is not V_N, because the residual donor energy in such layers is typically only about 0.02 eV, not 0.06 eV. The usual impurity candidates suggested for residual donors are O_N and Si_{Ga}, and, indeed, each of these would be expected to have an energy of about 0.02 eV at the 10^{17}-cm^{-3} level. Unfortunately, however, it is difficult to distinguish between these two impurities, because it is not easy at the present time to accurately determine concentrations of O and Si (or anything else) in the low-10^{17}-cm^{-3} range. For some samples having very high free-electron concentrations, say higher than mid-10^{18} cm^{-3}, there is evidence that impurities cannot account for all of the donors; in such cases, V_N donors may also be important, but verification is difficult because the donor-energy fingerprint (i.e., 0.06 eV) will be greatly modified by screening effects in high-concentration samples.

We now consider the use of DLTS to study point defects created by irradiation. The dashed line in figure 4 shows the *change* in the DLTS spectrum of a 4-μm-thick MOCVD GaN/Al$_2$O$_3$ layer after a 1-MeV electron irradiation to a fluence of 6 x 10^{15} cm^{-2}. The only trap produced by the irradiation (in this DLTS temperature range) is the one designated E, with a peak near 115 K. It turns out that trap E is produced by several types of irradiation, including electrons [8], protons [17], He ions [7], and sputtered Au atoms [18], and thus must be a rather common point defect. The usual Arrhenius analysis of E gives energies ranging from 0.13 to 0.20 eV, depending on the type of irradiation [7,8,17,18]; however, very recently it has been shown that trap E, or at least the version produced by 1-MeV electrons, really consists of two traps, ED1 and ED2, as shown in figure 5 [19]. (Note that trap D in this spectrum existed before the irradiation.) Both ED1 and ED2 have thermal energies of 0.06 eV, identical to that of the donor found from the TDH measurements; however, they have different capture cross sections, with that of ED2 being temperature dependent. It is likely that ED1 and ED2 are both related to V_N, but that, in fact, they are different complexes of V_N. For example, one of them could be V_N-N_I, and the other, V_N-N_{Ga}-Ga_N-N_I. That is, from a simple theoretical model, presented in reference 20, each of these "chain" defect complexes could be produced by 1-MeV

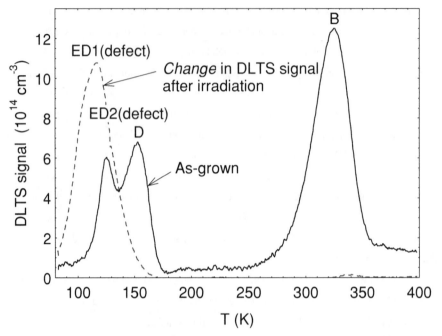

Figure 4. *DLTS spectra for an as-grown MBE GaN layer (solid line), and an irradiated MOCVD GaN layer (dashed line). The spectrum for the irradiated sample actually represents only the change due to irradiation.*

electrons. At a low enough irradiation energy, it should be possible to produce only the V_N-N_I defect, if it is stable at room temperature, and the presence of only one defect would simplify the analysis. At present, an "exact" analysis, such as that shown in figure 5, is absolutely essential to understand the important relationship between the irradiation-produced DLTS and TDH centers.

The practical importance of trap E becomes obvious when comparing with the DLTS spectrum of an as-grown MBE GaN/Al$_2$O$_3$ layer, also shown in figure 4. Note that the as-grown layer contains trap ED2, at a reasonably high concentration. There is independent evidence that trap ED2 (called E_1 in reference 21) is reduced by using higher N fluxes in the growth; this information is consistent with the V_N nature of trap E.

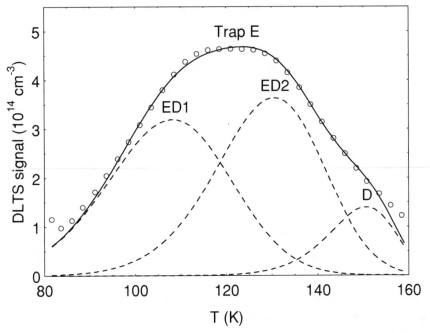

Figure 5. *DLTS spectrum of Trap E. The solid line is the composite signal resulting from the three simulated traps shown as dashed lines.*

By going to higher DLTS temperatures, another irradiation-induced trap appears, trap AD2, as shown in figure 2. The energy of trap AD2 is about 0.9 eV, close to that of a photoluminescence band which appears in electron-irradiated GaN, and which has tentatively been assigned to a Ga_I-X complex [22]. Interestingly, the as-grown MBE GaN layer shown in figure 2 displays trap A_1, which also has an energy of about 0.9 eV, and may be related to the defect trap AD2. (Trap A_1 also appears in MOCVD and HVPE layers.) Traps ED2 and AD2 may be expected to influence GaN devices, because these traps can reach levels of 10^{15} cm^{-3} or higher in some samples.

Another point defect which has been studied in as-grown GaN is the Ga vacancy, V_{Ga}. Theory suggests that V_{Ga} centers should be prevalent in n-type GaN [3], and indeed, they have been identified by positron annihilation experiments [23]. It is a mystery as to why the V_{Ga} defect has not yet been found in irradiated GaN, but the V_{Ga}-Ga_I Frenkel pairs possibly recombine as soon as they are created, in n-type material, due to the strong coulomb attraction between V_{Ga}^{3-} and Ga_I^+.

The results presented here show that point defects in GaN can be donors, acceptors, traps, and recombination centers, and that they exist in as-grown GaN layers. Although not specifically discussed in this paper, it is also important to note that defects will be produced in a radiation environment, such as that associated with the Van Allen belt. Future tasks include showing how point defects may affect particular types of devices, and then, how to get rid of them. This program is likely to engage the GaN community for some time to come.

SUMMARY

Three types of defects, point, line, and areal, have been found in n-type GaN, and can influence the electrical properties of the material. Line defects, namely threading dislocations, have been shown to act as acceptors, recombination centers, scattering centers, and possibly traps, while areal defects, stacking faults in this case, induce a highly conductive interface region in HVPE GaN/Al_2O_3 layers. We have employed high-energy electron irradiation to create point defects, and have used TDH and DLTS to show that at least one of them is a donor, and that two of them are traps. We have also shown that very similar traps exist in as-grown GaN layers, and may have an effect on GaN devices.

ACKNOWLEDGMENTS

We wish to thank H. Morkoç and R. Molnar for samples; S. Keller for raw, Hall data; and J.W. Hemsky for electron irradiations. D.C.L. and Z-Q. F. were supported under U.S. Air Force Contract F33615-95-C-1619 and U.S. Navy Grant N00014-99-1-1067. LP was supported by INFM (Italian National Institute for Physics of Matter). Also, partial support was received from the Air Force Office of Scientific Research. Most of the work of D.C.L. was performed at the Air Force Research Laboratory, Wright-Patterson Air Force Base, Ohio.

REFERENCES

1. H. P. Maruska and J.J. Tietjen, Appl. Phys. Lett. **15**, 327 (1969).

2. D.C. Look, D.C. Reynolds, J.W. Hemsky, J.R. Sizelove, R.L. Jones, and R.J. Molnar, Phys. Rev. Lett. **79**, 2273 (1997).

3. J. Neugebauer and C. G. Van de Walle, Phys. Rev. B **50**, 8067 (1994).

4. S. Evoy, H.G. Craighead, S. Keller, U.K. Mishra, and S.P. DenBaars, J. Vac. Sci. Technol. B **17**, 29 (1999).

5. T. Mukai, K. Takekawa, and S. Nakamura, Jpn. J. Appl. Phys. **37**, L839 (1998).

6. D.C. Look and J.R. Sizelove, Phys. Rev. Lett. **82**, 1237 (1999).

7. F.D. Auret, S.A. Goodman, F.K. Koschnick, J-M. Spaeth, B.Beaumont, and P.Gibart, Appl.Phys. Lett. **73**, 3745 (1998).

8. Z-Q. Fang, J.W. Hemsky, D.C. Look, and M.P. Mack, Appl. Phys. Lett. **72**, 448(1998).

9. L.T. Romano, B.S. Krusor, and R.J. Molnar, Appl. Phys. Lett. **71**, 2283 (1997).

10. C. Stampfl and C.G. Van de Walle, Phys. Rev. B **57**, R15052 (1998).

11. D.C. Look and R.J. Molnar, Appl. Phys. Lett. **70**, 3377 (1997).

12. S. Keller, B.P. Keller, Y-F. Wu, B. Heying, D. Kapolnek, J. S. Speck, U. K. Mishra, and S. P. DenBaars, Appl. Phys. Lett. **68**, 1525 (1996).

13. A.F. Wright and U. Grossner, Appl. Phys. Lett. **73**, 2751 (1998).

14. K. Leung, A.F. Wright, and E.B. Stechel, Appl. Phys. Lett. **74**, 2495 (1999).

15. Z.Z. Bandić, T.C. McGill, and Z. Ikonić, Phys. Rev. B **56**, 3564 (1997).

16. F. Agullo-Lopez, C.R.A. Catlow, and P.D. Townsend, *Point Defects in Materials*, (Academic, New York, 1988).

17. F.D. Auret, S.A. Goodman, F.K. Koschnick, J-M. Spaeth, B.Beaumont, and P.Gibart, Appl.Phys. Lett. **74**, 407 (1999).

18. F.D. Auret, S.A. Goodman, F.K. Koschnick, J-M. Spaeth, B.Beaumont, and P.Gibart, Appl.Phys. Lett. **74**, 2173 (1999).

19. L. Polenta, Z-Q. Fang, and D.C. Look, 1999 (unpublished).

20. D.C. Look, J.W. Hemsky, and J.R. Sizelove, Phys. Rev. Lett. **82**, 2552 (1999).

21. Z-Q. Fang, D.C. Look, W. Kim, Z. Fan, A. Botchkarev, and H. Morkoç, Appl. Phys. Lett. **72**, 2277 (1999).

22. M. Linde, S.J. Uftring, G.D. Watkins, V. Härle, and F. Scholz, Phys. Rev. B **55**, R10177 (1997).

23. K. Saarinen, J. Nissilä, P. Hautojärvi, J. Likonen, T. Suski, I. Grzegory, B. Lucznik, and S. Porowski, Appl. Phys. Lett. **75**, 2441 (1999).

Mat. Res. Soc. Symp. Vol. 595 © 2000 Materials Research Society

PROPERTIES AND EFFECTS OF HYDROGEN IN GaN

S.J. Pearton [1], H. Cho [1], F. Ren [2], J.-I. Chyi [3], J. Han [4], R.G. Wilson [5]

[1] Department of Materials Science and Engineering, University of Florida, Gainesville FL 32611, USA
[2] Department of Chemical Engineering, University of Florida, Gainesville FL 32611, USA
[3] Department of Electrical Engineering, National Central University, Chung-Li 32054, Taiwan
[4] Sandia National Laboratories, Albuquerque NM 87185, USA
[5] Consultant, Stevenson Ranch, CA 91381, USA

ABSTRACT

The status of understanding of the behavior of hydrogen in GaN and related materials is reviewed. In particular, we discuss the amount of residual hydrogen in MOCVD-grown device structures such as heterojunction bipolar transistors, thyristors and p-i-n diodes intended for high power, high temperature applications. In these structures, the residual hydrogen originating from the growth precursors decorates Mg-doped layers and AlGaN/GaN interfaces. There is a significant difference in the diffusion characteristics and thermal stability of implanted hydrogen between n- and p-GaN, due to the stronger affinity of hydrogen to pair with acceptor dopants and possibly to the difference in H_2 formation probability.

INTRODUCTION

It is our experience that atomic hydrogen readily permeates into GaN during many different device processing steps. The indiffusion is likely to be strongly enhanced by the presence of the high defect density in heteroepitaxial material. Examples of the processes in which hydrogen is incorporated into GaN at low temperatures (< 300 °C) include plasma enhanced chemical vapor deposition of dielectrics (the source of the hydrogen can be the plasma chemistry or erosion of the photoresist mask), boiling in solvents (including water), annealing under H_2 or NH_3 ambients and wet etching in acid or base solutions (e.g. NaOH, KOH). Secondary Ion Mass Spectrometry (SIMS) profiling after processing with deuterated chemicals to enhance the detection sensitivity show that hydrogen can diffuse into GaN at temperatures as low as 80 °C. The main effect of the hydrogen is passivation (electrical deactivation) of Mg acceptors in p-GaN through formation of neutral Mg-H complexes, which can be dissociated through either minority carrier (electron) injection or simple thermal annealing. There is evidence that all of the acceptor species in GaN, namely Mg, C, Ca and Cd are found to form complexes with hydrogen.

In this paper we will summarize what is known about the effects of hydrogen in GaN and discuss new results on the incorporation of hydrogen in electronic device structures.

PRESENT STATE OF KNOWLEDGE

The following information is firmly established for the behavior of hydrogen in GaN.[1-26]

1. GaN(Mg) grown by atmospheric pressure Metal Organic Chemical Vapor Deposition is highly resistive due to formation of neutral Mg-H complexes.[18,21,22] The Mg can be activated (i.e. produce p-type conductivity) by post-grown annealing at $\geq$ 700 °C in N_2[22], low energy electron-beam irradiation near room temperature[18] or by forward biasing of a p-n junction to inject minority carriers (electrons).[27]

2. Thermal annealing in NH_3 at $\geq$ 400 °C causes the GaN(Mg) to revert to a highly resistive (> 10^6 $\Omega\cdot$cm) state.[21,22]

3. GaN(Mg) grown by low-pressure MOCVD, Reactive Molecular Beam Epitaxy (solid Ga, NH_3 gas) or by plasma-assisted MBE (solid Ga, plasma N_2) can be p-type without annealing. This is presumably due to the lower H_2 fluxes during growth, and the vacuum anneal undergone by RMBE material which is typically cooled without the plasma on.[28]

4. The residual hydrogen concentration in all GaN(Mg) tracks the acceptor concentration.[12]

5. All acceptor species show pairing with hydrogen (i.e. Mg, Zn, C, Ca, Cd).[13,18,21,29-31]

6. Donor dopant-hydrogen complexes have not yet been detected, and hydrogen concentrations are generally much lower in n-type GaN than in p-type material.[32]

7. Hydrogen enters GaN during many different device processing steps, even at 100 °C or lower.[11]

Information that is not as firmly established includes the following:

1. The energy levels and charge states (H°, H^+, H^-) of hydrogen in GaN.

2. The existence of hydrogen molecules or larger clusters.

3. The role of hydrogen in facilitating p-type doping. Van Vechten[26] has suggested that hydrogen compensates native defects, making p-type doping possible, while Neugebauer and Van der Walle[14,19] built on an earlier idea of Neumark's[33] from II-VI compounds that the presence of hydrogen increases Mg solubility and decreases compensating native vacancy defect concentration in GaN.

4. The role of line and point defects in enhancing hydrogen diffusivity (this will be answered by a comparison of its incorporation in the new epitaxial lateral overgrowth material with conventional heteroepitaxial GaN).

5. Hydrogen solubility – as with other semiconductors, the apparent solubility at low temperatures ($\leq$ 500 °C) is likely to be dominated by the concentration of sites to which the hydrogen can be bonded, i.e. dopants, defects and impurities.[34]

In the following sections we will cover the incorporation of hydrogen during growth and processing, its thermal stability in GaN and some of the complexes it forms with dopants and defects.

HYDROGEN IN AS-GROWN GaN

As mentioned above, the hydrogen concentration in p-GaN tracks the Mg concentration, and is correlated with the active Mg, not the total Mg density. Since Mg has a relatively deep ionization level (160-170 meV), the hole concentration at room temperature is typically only a percent or so of the total Mg in the crystal. We have found typical hydrogen concentrations in the 10^{18}-10^{19} range in virtually all GaN grown by gas-phase techniques; in some samples doped with Si, we have seen the hydrogen present at higher concentrations (10^{20} cm^{-3}), similar to that of the Si (Figure 1). This suggests the possibility of (Si-H) pair formation, but more work must be done to confirm this idea. It is a general result in semiconductors that acceptor passivation by hydrogen is more thermally stable than donor passivation and therefore is more readily observed experimentally.[34]

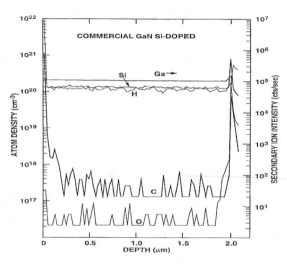

Figure 1. SIMS profiles of H and other background impurities in MOCVD-grown, Si-doped GaN.

For the case of p-GaN, the hydrogen concentration is usually the same as the active Mg (i.e. ~ 1% of 10^{20}cm^{-3}) after growth. A 700 °C, 20 min post-growth anneal in N_2 only reduces the hydrogen concentration by a factor of 2 or 3, but this is enough to have strong p-type conductivity. An interesting question is whether the hydrogen is present in the GaN at the growth temperature (where it would be unbound and extremely diffusive) and then settles on the Mg during cooldown, or whether it actually enters the GaN from surface cracking of NH_3 (to N_2, NH_x and H) during cooldown after growth (probably in the temperature range 300-600 °C). The latter is the passivation mechanism in p-type InP and GaAs grown by MOCVD, where hydrogen is found to deactivate 10-60% of Zn or C acceptors.

The annealing ambient may also affect the apparent thermal stability of the passivation, with reactivation of passivated Mg occuring at lower temperatures for N_2 anneals. This is also a familiar phenomenon from other semiconductors. Since reactivation of the acceptor only requires dissociation of the Mg-H complex and a short range diffusion of the hydrogen away from the acceptor, it is also found that the apparent reactivation temperature depends therefore on the doping concentration (and therefore the associated hydrogen concentrations) and the layer structure and thickness (e.g. heteroepitaxial materials cladding the p-GaN layer may serve to retard hydrogen out-diffusion).

In contrast to other semiconductors in which acceptors are passivated by atomic hydrogen occupying a bond-centered position between the dopant and a neighboring lattice atom, in GaN the ionicity of the bonds means there is no local maximum in the charge density at the bond center. Both theoretical[14] and experimental[23] evidence suggests that in GaN(Mg) the hydrogen attaches to a nitrogen atom in an antibonding orientation. The predicted stretch frequency of hydrogen in the Mg-H complex is 3360 cm^{-1},[14] similar to that in NH_3 molecules (3441 cm^{-1}). Experimentally, a frequency of 3125 cm^{-1} has been reported.[23]

HYDROGEN IN AS-GROWN DEVICE STRUCTURES

There is a lot of interest in the development of GaN-based high power, high temperature electronics for power switching and microwave applications. One potential device for ultra-high power switching is a thyristor, in either the npnp or pnpn configurations. Figure 2 shows a SIMS profile of Mg and H in an MOCVD-grown npnp thyristor structure. Note how the residual hydrogen tracks the Mg. Since the Mg-doped layer at ~ 5.5 μm depth was the first to be grown, it contains less hydrogen than the Mg-doped region nearer to the surface which has spent less time at the growth temperature of 1040 °C. The source of the hydrogen is either the precursors, $(CH_3)_3Ga$

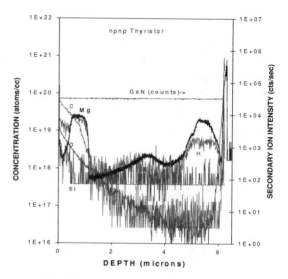

Figure 2. SIMS profiles of H and other background impurities in an MOCVD-grown npnp thyristor structure.

and AsH_3, or the carrier gas, which is generally H_2. Sugiura et.al.[17] found that if they employed a N_2 carrier gas, they could get p-GaN without post-growth annealing. They also found that using a mixed H_2/N_2 carrier gas reduced the p-type doping level in as-grown samples. These results implicate the H_2 carrier gas as the main source of hydrogen for passivation of the Mg acceptors, but previous experience in MOCVD growth of highly doped p-GaAs suggests that both the $(CH_3)_3Ga$ and AsH_3 are likely to also play a role. In particular, the cool down in AsH_3 ambient after growth is likely a source of hydrogen for passivation.

Schottky and p-i-n diodes are employed as high-voltage rectifiers in power switching application. To suppress voltage transients when current is switched to inductive loads such as electric motors, these diodes are placed across the switching transistors. The advantage of simple metal-semiconductor diodes relative to p-n junction diodes is the faster turn-off because of the absence of minority carrier storage effects and lower power dissipation during switching. Wide bandgap semiconductors such as GaN offer additional advantages for fabrication of diode rectifiers, including much higher breakdown voltages and operating temperatures. There is much interest in developing advanced switching devices and control circuits for CW and pulsed electrical sub-systems in emerging hybrid-electric and all-electric vehicles, more-electric airplanes and naval ships and for improved transmission, distribution and quality of electric power in the utilities industry. Eventually one would like to reach target goals of 25 kV stand-off voltage, 2 kA or higher conducting current, forward drop less than 2% of the rated voltage and maximum operating frequency of 50 kHz.

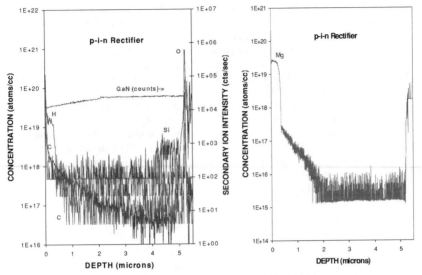

Figure 3. SIMS profiles of H and Si (left) and Mg (right) in an as-grown, MOCVD p-i-n rectifier structure.

Figure 3 shows a SIMS profile of H and other background impurities (along with intentional Si doping) in an MOCVD-grown p-i-n diode structure (left), together with the Mg profile in the structure (right). Notice once again that the H decorates the Mg due to formation of the neutral $(Mg-H)^{\circ}$ complexes. About 70-80% of the Mg atoms have hydrogen attached.

To create a Schottky rectifier with high breakdown voltage, one needs a thick, very pure GaN depletion layer. Figure 4 shows SIMS profile of H and other background impurities in a 2 μm thick, high resistivity ($10^7 \Omega \cdot cm$) GaN layer grown by MOCVD. The reverse breakdown voltage of simple Schottky rectifiers fabricated on this material was > 2 kV, a record for GaN. Notice that in this material the hydrogen concentration is at the detection sensitivity of the SIMS apparatus. The amount of hydrogen present in GaN after cooldown from the growth temperature will

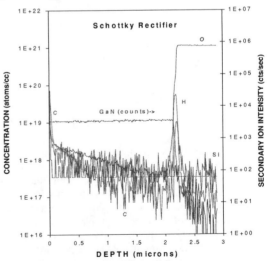

Figure 4. SIMS profiles of H and other background impurities in as-grown, MOCVD Schottky rectifier structure.

depend on the number of sites to which it can bond, including dopants and point and line defects. In the absence of p-type doping, it is clear that the number of these sites is $\leq$ 8×10^{17} cm^{-3} under our growth conditions.

Another important device for power microwave applications is the heterojunction bipolar transistor (HBTs). Several groups have reported GaN/AlGaN HBTs with gains less than 10 at room temperature. Residual hydrogen passivation of the Mg acceptors in the base of such devices would create instabilities in the apparent gain during operation, since minority carrier injection would lead to time-dependent dopant reactivation, i.e. the gain would show an initial exponentially decreasing current gain, which might be interpreted as a reliability problem in the absence of an understanding of the effects of hydrogen. Figure 5 shows SIMS profiles of the Mg (left) and H (right) in an as-grown GaN/AlGaN HBT structure. The hydrogen is found to decorate both the GaN(Mg) base layer at ~ 0.4 μm depth and also the AlGaN emitter layer, which is a region of additional strain within the structure.

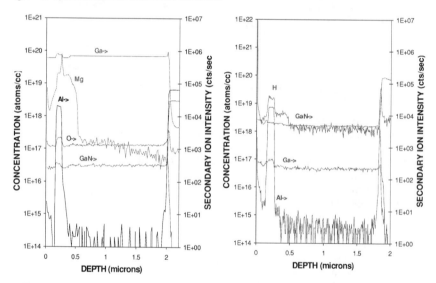

Figure 5. SIMS profiles of Mg (left) and H (right) in as-grown MOCVD GaN/AlGaN HBT structure.

Once the device structure is grown and subsequently annealed to minimize the amount of residual hydrogen, there can still be problems associated with indiffusion during device fabrication. Atomic hydrogen is found to be unintentionally incorporated into the nitrides during many processing steps, including boiling in water, dry etching, wet etching, chemical vapor deposition of dielectrics and annealing in H_2 or NH_3. This shows that most of wet chemical or plasma processes involved in device processing are capable of causing passivation of dopants due to hydrogen incorporation in GaN and related alloys. SIMS profiling of GaN exposed to hydrogen-containing gases or chemicals shows that the hydrogen can diffuse into these layers at temperatures as low as 80 °C. Even though reactivation of dopants can be achieved by relatively low temperature thermal annealing after a given process step, the hydrogen may be cause similar problems after the subsequent processing so that one should be aware of its effects.

DIFFUSION OF H IN IMPLANTED OR PLASMA-TREATED GaN

Implantation of protons is commonly employed for inter-device isolation of electronic and photonic devices and for current guiding in various laser diode structures. Figure 6 shows SIMS profiles of ^2H implanted into n-GaN, as a function of the subsequent annealing temperature under a N_2 ambient. There is no redistribution of the deuterium until ≥ 800 °C, while at higher temperatures the remaining ^2H decorates the implant damage profile which is slightly closer to the surface than the atomic profile. Note that even after a 1200 °C anneal there is a significant concentration of deuterium remaining at the

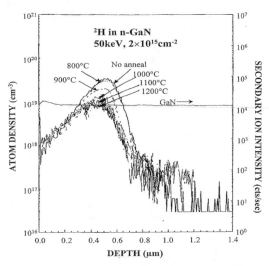

Figure 6. SIMS profiles of ^2H implanted into n-GaN at a dose of 2×10^{15} cm^{-2} and 50 keV, as a function of subsequent annealing temperature (10 secs. duration).

peak of the damage profile. It appears that after high temperature annealing the deuterium is either trapped at defects or escapes the crystal.

By sharp contrast, the redistribution of implanted ^2H in p-GaN is more complicated. Figure 7 shows SIMS profiles of the deuterium as a function of annealing temperature. The redistribution begins at ≥ 500 °C and at higher temperatures there is decoration of the residual implant damage and the GaN/Al$_2$O$_3$ interface, in addition to the formation of a plateau region at ~ 0.8 μm where Mg-H complexes are the dominant species. These results show that there is competition for trapping of deuterium between different sites, including strain and defects at interfaces, Mg dopants in the GaN, damage-related defects associated with the implant process and finally the surface of the GaN, through which the deuterium can evolve.

An example of the indiffusion of deuterium into nitrides (in this case, AlN) as a result of exposure to a ^2H plasma at low temperatures (200 or 250 °C) is shown in Figure 8. The deuterium has saturated the 1 μm thick layer after the 250 °C exposure. This gives a rough estimate for the diffusivity as 10^{-12} cm^2·sec^{-1} at 250 °C. Similar values are obtained for deuterium in GaN. It is likely that the diffusivity in defect-free (bulk or homoepitaxial) nitrides is considerably lower than these values.

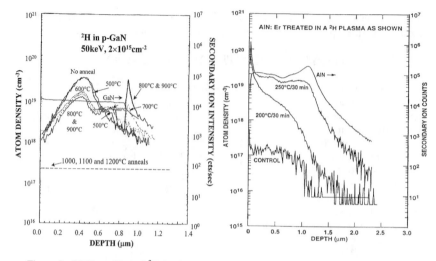

Figure 7. SIMS profiles of ^{2}H implanted into p-GaN at a dose of 2×10^{15} cm^{-2} and 50 keV, as a function of subsequent annealing temperature (10 secs. duration).

Figure 8. SIMS profiles of ^{2}H in AlN exposed to deuterium plasmas at 200 °C or 250 °C for 30 mins.

SUMMARY AND CONCLUSIONS

Hydrogen is found to bond to regions of strain in GaN-based structures, such as interfaces or defects, as well as acceptor dopants. There are many potential sources of hydrogen during growth and processing of nitrides so that one needs to be vigilant for its effects, i.e. a reduction in hole concentration with an associated increase in hole mobility. It is expected that unintentional hydrogen incorporation will play a role in GaN electronic devices which contain p-n junctions, leading to high p-contact resistances and time-dependent current gains if precautions are not taken to remove the hydrogen.

ACKNOWLEDGMENTS

The work at UF is partially supported by grants from DARPA/EPRI, MDA 972-98-1-0006, monitored by ONR (J.C. Zolper) and NSF 97-32865 (L. Hess). The work at NCU is partially supported by a grant from the National Science Council of R.O.C. under contract No. NSC-88-2215-E-008-012. Sandia is a multiprogram laboratory operated by Sandia Corporation, a Lockhead-Martin Company, for the US Department of Energy under contract No. DEAC04-94-AL85000. The work of RGW is partially supported by ARL (J.M. Zavada).

REFERENCES

1. see for example, S.J. Pearton, J.C. Zolper, R.J. Shul and F. Ren, "GaN: Defects, Processing and Devices", J. Appl. Phys. **78** R1 (1999).

2. C.G. Van de Walle, "Theory of Hydrogen is Semiconductors", Mat. Res. Soc. Symp. Proc. **513** 55 (1998).
3. S.M. Myers, T.J. Headly, C.R. Hills, J. Han. G.A. Petersen, C.H. Seager, W.R. Wampler, "The behavior of Ion-Implanted Hydrogen in GaN", MRS Internet J. Nitride Semicond. Res. **4S1** G5.8 (1999).
4. W.R. Wampler and S.M. Myers, "Ion Channeling Analysis of GaN Implanted with Deuterium", MRS Internet J. Nitride Semicond. Res. **4S1** G3.73 (1999).
5. M.G. Weinstein, C.Y. Song, M. Stavola, S.J. Pearton, R.G. Wilson, R.J. Shul, K.P. Killeen and M.J. Ludowise, "H Decorated Lattice Defects in Proton-Implanted GaN", Appl. Phys. Lett. **72** 1703 (1998).
6. M.G. Weinstein, M. Stavola, C.Y. Song, C. Bozdog, H. Przbylinska, G.D. Watkins, S.J. Pearton and R.G. Wilson, "Spectroscopy of Proton-Implanted GaN", MRS Internet J. Nitride Semicond. Res. **4S1** G5.9 (1999).
7. J.M. Zavada, R.G. Wilson, C.R. Abernathy and S.J. Pearton, "Hydrogenation of GaN, AlN and InN", Appl. Phys. Lett. **64** 2724 (1994).
8. R.G. Wilson, S.J. Pearton, C.R. Abernathy and J.M. Zavada, "Outdiffusion of deuterium from GaN, AlN and InN", J. Vac. Sci. Technol. A **13** 719 (1995).
9. M. Miyachi, T. Tanaka, Y. Kimura and H. Ota, "The Activation of Mg in GaN by Annealing with Minority Carrier Injection", Appl. Phys. Lett. **72** 1101 (1998).
10. V.J.B. Torres, S. Oberg and R. Jones, "Theoretical Studies of Hydrogen Passivated Substitutional Mg Acceptor in Wurzite GaN", MRS Internet J. Nitride Semicond. Res. **2** 35 (1997).
11. S.J. Pearton, C.R. Abernathy, C.B. Vartuli, J.W. Lee, J.D. MacKenzie, R.G. Wilson, R.J. Shul, F. Ren and J.M. Zavada, "Unintentional Hydrogenation of GaN and Related Alloys During Processing", J. Vac. Sci. Technol. A **14** 831 (1996).
12. Y. Ohba and A. Hatano, "Mg Doping and H Incorporation in GaN MOCVD", Jpn. J. Appl. Phys. **33** L1367 (1994).
13. J.W. Lee, S.J. Pearton, J.C. Zolper and R.A. Stall, "Hydrogen Passivation of Ca Acceptors in GaN", Appl. Phys. Lett. **68** 2102 (1996).
14. J. Neugebauer and C.G. Van de Walle, "Role of H in Doping of GaN", Appl. Phys. Lett. **68** 1829 (1996).
15. S.M. Myers, J. Han, T.J. Headly, C.R. Hills, G.A. Petersen, C.H. Seager, W.R. Wampler and A.F. Wright, "Behavior of Ion-Implanted Hydrogen in GaN at Concentrations ≥ 1 at.%", Phys. Rev. B (in press).
16. H. Harima, T. Inoue, S. Nakashima, M. Ishida and M. Taneya, "Local Vibrational Modes as a Probe of Activation Process in p-type GaN", Appl. Phys. Lett. **75** 1383 (1999).
17. L. Sugiura, M. Suzuki and J. Nishino, "P-type Conduction in As-Grown Mg-Doped GaN Grown by MOCVD", Appl. Phys. Lett. **72** 1748 (1998).
18. H. Amano, M. Kito, K. Hiramatsu and I. Akasaki, Jap. J. Appl. Phys. **28** L112 (1989).
19. C.G. Van de Walle, "Interaction of Hydrogen with Native Defects in GaN", Phys. Rev. B **56** R10020 (1997).
20. S.K. Estreicher and D.M. Maric, "Theoretical Study of H in Cubic GaN", Mat. Res. Soc. Symp. Proc. **423** 613 (1996).
21. S. Nakamura, N. Iwasa, M. Senoh and T. Mukai, "Hole Compensation Mechanism of p-GaN Films", Jpn. J. Appl. Phys. **31** 1258 (1992).

22. S. Nakamura, T. Mukai, M. Senoh and N. Iwasa, "Thermal Annealing Effects on p-type, Mg-Doped GaN Films", Jpn. J. Appl. Phys. **31** L139 (1992).
23. W. Gatz, N.H. Johnson, D.P. Bour, M.D. McCluskey and E.E. Haller, "Local Vibrational Modes of the Mg-H Acceptor Complex in GaN", Appl. Phys. Lett. **69** 3725 (1996).
24. A. Bosin, V. Fiorentini and D. Vanderbilt, "H, Acceptors, and H-Acceptors Complexes in GaN", Mat. Res. Soc. Symp. Proc. **395** 503 (1996).
25. Y. Okamoto, M. Saito and A. Oshiyama, "First Principles Calculations on Mg and Mg-H in GaN", Jap. J. Appl. Phys. **35** L807 (1996).
26. J.A. Van Vechten, J.D. Zook, R.D. Horning and B. Goldenberg, "Detecting Compensation in Wide Bandgap Semiconductors by Growing in H That is Removed by Low Temperature De-Ionizing Radiation", Jpn. J. Appl. Phys. **31** 3662 (1992).
27. S.J. Pearton, J.W. Lee and C. Yuan, "Minority Carrier Enhanced Passivation of H-Passivated Mg in GaN", Appl. Phys. Lett. **68** 2690 (1996).
28. see for example G. Popovici and H. Morhoe, "Growth and Doping of Defects in III-Nitrides", In GaN and Related Materials II, ed. S.J. Pearton (Gordon, Breach, NY, 1999).
29. H. Amano, I. Akasaki, T. Kozawa, N. Sawaki, K. Ikeda and Y. Ishii, "Doping of GaN with Zn", J. Lumin. **4** 121 (1988).
30. S.J. Pearton, C.R. Abernathy and F. Ren, "Electrical Passivation in H-Plasma-Exposed GaN", Electron. Lett. **30** 527 (1994).
31. A. Burchaid, M. Deicher, D. Forkel-Wirth, E.E. Haller, R. Magerle, A. Prospero and R. Stotzler, "First Microscopic Observation of Cd-H Pairs in GaN", Mat. Res. Soc. Symp. Proc. **449** 961 (1997).
32. N.M. Johnson, W. Gotz, J. Neugebauer and C.G. Van de Walle, "Hydrogen in GaN", Mat. Res. Soc. Symp. Proc. **395** 723 (1996).
33. C.F. Neumark, "Defects in Wide Bandgap II-VI Crystals", Mat. Sci. Eng. R **21** 1 (1997).
34. S.J. Pearton, J.W. Corbett and M. Stavola, Hydrogen in Crystalline Semiconductors (Springer-Verlag, Berlin 1992).
35. G. R. Antell, A.T.R. Briggs, B.P. Butler, S.A. Kitching, J.P. Stagg, A. Chew and D.E. Sykes, "Passivation of Zn Acceptors in InP by Atomic Hydrogen Coming from AsH$_3$ during MOVPE", Appl. Phys. Lett. **53** 758 (1988).

Mat. Res. Soc. Symp. Vol. 595 © 2000 Materials Research Society

Lattice Location of Deuterium in Plasma and Gas Charged Mg Doped GaN

W. R. Wampler, J. C. Barbour, C. H. Seager, S. M. Myers, A. F. Wright and J. Han
Sandia National Laboratories, Albuquerque, NM 87185-1056

ABSTRACT

We have used ion channeling to examine the lattice configuration of deuterium in Mg doped GaN grown by MOCVD. The deuterium is introduced by exposure to gas phase or ECR plasmas. A density functional approach including lattice relaxation, was used to calculate total energies for various locations and charge states of hydrogen in the wurtzite Mg doped GaN lattice. Results of channeling measurements are compared with channeling simulations for hydrogen at lattice locations predicted by density functional theory.

INTRODUCTION

Mg doped GaN grown by MOCVD has high resistivity due to passivation of acceptors by hydrogen incorporated during growth [1]. Low resistivity p-type material required for devices can be obtained through post-growth activation by thermal annealing [2, 3], low energy electron beam irradiation (LEEBI) [4, 5] or minority carrier injection [6, 7]. Mechanisms for these activation treatments are believed to involve thermal or electronically induced dissociation of hydrogen from acceptors leading to loss of hydrogen from the material, or to a change in atomic configuration or charge state of the hydrogen. In this study we investigate the atomic configuration of hydrogen in GaN. We use density functional theory to predict the lattice configuration for various states of hydrogen in wurtzite Mg doped GaN, and we use ion channeling [8] to experimentally examine the hydrogen lattice location.

EXPERIMENT

Wurtzite GaN:Mg films with (0001) orientation and thickness in the range of 1.4 - 2.3 μm were grown epitaxially by MOCVD on c-oriented sapphire substrates as described elsewhere [9]. The GaN:Mg was doped with Mg at concentrations in the range from 5 to 7×10^{19}/cm^3 as determined by SIMS.

Samples were passivated by exposure to D_2 gas at 88 kPa and 700° C for four hours. Conductivity measurements showed this treatment yielded high resistivity passivated films [3]. Other samples were exposed to deuterium ECR plasmas for one hour. The sample temperature during ECR plasma exposure was chosen to be 600°C based on a previous study [10] which showed deuterium incorporation into p-type GaN at 600°C but not at 400°C from exposure to remote plasma. During ECR plasma exposure the incident D has about 30 eV of kinetic energy which is high enough to penetrate a few lattice spacings but too low to create bulk defects by atomic collisions [11].

Prior to gas or plasma exposures the samples were vacuum annealed at 900°C for one hour to remove hydrogen present from growth. After deuterium exposures, the concentration and depth distribution of deuterium in the films was characterized by D(^{3}He,p)α nuclear reaction analysis (NRA) [12] and by secondary ion mass spectroscopy. The D concentrations were uniform throughout the Mg doped region of the films and are given in the legends of figures 1 and 2. In the gas charged sample the D concentration is approximately half the Mg concentration, whereas in the two plasma charged samples the D concentration is higher than the Mg concentration, which indicates that the chemical potential for D from the plasma charging is much higher than from the gas charging.

Transmission infrared absorption spectra were also measured in gas and plasma charged samples. In both cases similar absorption peaks were seen at 2320 cm^{-1} (shown in figures 1 and 2) which were not present prior to deuterium exposure. The increase in energy to 3120 cm^{-1} seen for this absorption peak in similar samples exposed to hydrogen shows that the absorption peak is due to stretch vibrations of H or D bound to nitrogen. This is also consistent with our observation that the 2320 cm^{-1} absorption peak disappears when D is removed from the samples by vacuum annealing. The fact that the peak area is similar for the gas and plasma charged samples while the D content is much larger in the plasma charged sample shows that at least in the plasma charged sample much of the D is in an IR inactive state. Similar IR absorption peaks were reported previously for GaN:Mg deuterated with a remote plasma system [13] and were ascribed to D at an anti-bonding position bound to nitrogen neighboring the Mg acceptor.

Ion channeling studies of D were done by counting protons from the D(^{3}He,p)α nuclear reaction using an incident analysis beam of 850 keV ^{3}He$^+$ ions. This gives counts from D to depths of about 1 µm. The analysis beam size was 1x1 mm. Channeling measurements were done with the samples at room temperature. The proton yield was measured as a function of angle between the analyzing ion beam and the crystallographic c-axis of the samples. When the analysis beam is aligned along the c-axis, channeling reduces the ion flux near the rows of host atoms and increases the flux near the center of the open channels, which causes a dip in the NRA yield if the D is near the host atom rows, or conversely, to a peak in yield if the D is near the center of the channel.

Figure 1 shows the measured NRA yield normalized to the off-axis or random yield versus the angle between the analysis beam direction and the c-axis for the gas charged sample. The solid circles show measurements taken beginning on axis and stepping progressively farther from the axis, using an analysis ion beam dose of 1 microcoulomb at each angular position. The open circles show a repeat angular scan at the same location on the sample. The first scan shows a dip with halfwidth almost 1 degree and a small narrow central peak. In the second scan the dip is gone but a peak remains. The result that the on-axis yield is higher for the second scan than for the first scan shows that the analysis beam used for the first angular scan has caused a change in lattice location of some of the deuterium.

Figure 2 shows the NRA channeling yield for the two plasma charged samples. The plasma charged sample with lower D concentration showed a broad dip similar to the gas charged sample whereas the plasma charged sample with more deuterium showed little variation of yield with angle. These measurements were made using 0.5 and 0.25 microcoulomb at each angular position for the low-D and high-D samples respectively. Repeat angular scans at the same location as the first scan were also done on the plasma charged samples to test for effects of analysis beam dose on channeling. These repeat scans showed little variation in yield with angle, i.e. the broad dip seen in the first scan on the low-D plasma charged sample was no longer present. Based on measurements of yield versus analysis beam dose at fixed on-axis angular position on the low-D plasma charged sample, we infer that the on-axis yield for the first scan was not significantly influenced by the analysis beam.

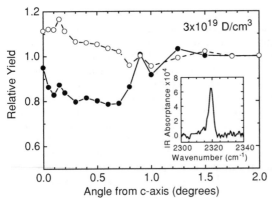

Figure 1 Effect of channeling on NRA yield for a GaN:Mg sample after exposure to deuterium gas. Filled circles show the initial measurement. Open circles are a repeat measurement at the same position. Inset shows IR absorption peak before ion beam exposure, due to N-D stretch local vibration.

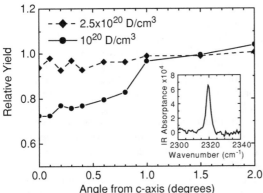

Figure 2 Effect of channeling on NRA yield for two GaN:Mg samples after exposure to deuterium ECR plasma. Inset shows IR absorption peak due to N-D stretch local vibration.

EFFECTS OF ANALYSIS ION BEAM

Next we consider two mechanisms by which the analysis beam might cause a change in D lattice location and thus affect the measured channeling yields. Collisions with lattice atoms produce vacancies and interstitial defects. Using the TRIM Monte Carlo particle transport code [14] we estimate the number of displacements produced by one microcoulomb of 850 keV He^3 on an area of 1 mm^2 to be ~ 10^{20} displacements/cm^3. This is comparable to the concentrations of Mg and D in the samples. If the D or the defects are mobile at room temperature, D-defect complexes might form which would change the lattice location of the D. However, when the analysis beam is aligned along the c-axis, channeling reduces the ion beam flux at the GaN lattice sites, and thus also the number of displacement events by about two orders of magnitude. Therefore we expect that with the beam near the c-axis on a fresh spot the number of displacements produced by the first few microcoulombs should be small compared to the concentration of D and Mg in the sample and is therefore not likely to have a large effect on the location of the D in the lattice.

Free electrons and holes produced by the analysis ion beam might also cause a change in lattice location of hydrogen. Mg acceptors in GaN passivated by hydrogen are observed to be activated by electron beam irradiation [4, 5]. While mechanisms for this activation are not known, a change in lattice location of hydrogen initially bound at the acceptor site, induced by a change in charge state, could be involved. Acceptor activation by electron-hole production should be similar for ion and electron irradiation. The energy deposition required to activate Mg acceptors in GaN by 15 keV electron irradiation is about 10^{21} eV/cm^3 [4, 5] which will produce about 10^{20} /cm^3 electron-hole pairs in GaN [15]. The channeling analysis ion beam dose required to generate this number of electron- hole pairs is estimated, using the known stopping power of helium ions [14], to be 0.3 nanocoulomb. Since this dose is much less than the dose used for our channeling measurements we expect acceptor activation by electron-hole production to occur very early in the channeling measurement. We therefore interpret our channeling results to represent material in which some or all of the Mg acceptors are not passivated or compensated by deuterium. The He stopping power and hence also electron-hole production is only slightly affected by channeling and will therefore not be very different for on versus off-axis alignment.

DISCUSSION

A density functional approach within the local density approximation, was used to calculate total energies, including effects of lattice relaxation, for various locations and charge states of hydrogen in the wurtzite GaN lattice. The minimum energy configuration and energy barriers for diffusion depend on the H charge state. In the undoped lattice the most stable sites for H^0 and H^- are near the center of the trigonal channel along the c-axis whereas H^+ prefers sites nearer the nitrogen [16]. In wurtzite GaN there are two types of antibonding (AB) and bond-centered (BC) sites which we refer to as $AB_{N\parallel}$ and $BC_{N\parallel}$, when the N-H$^+$ direction is parallel to the c axis, and $AB_{N\perp}$ and $BC_{N\perp}$ when the N-H$^+$ direction is roughly perpendicular to the c axis. In undoped GaN the lowest energy sites for H^+ are the $AB_{N\perp}$ and $BC_{N\parallel}$ sites which have about the same formation energy. In p-type material the H^+ charge state has the lowest energy so one might expect most of the H to occupy $AB_{N\perp}$ sites since there are three of these for each $BC_{N\parallel}$ site. However, several additional effects may influence the location of H in p-type GaN. First, the energy of H^+ is lower when it is near a Mg atom. The minimum energy site for H^+ near the Mg is the $AB_{Mg-N\perp}$ site of nitrogen atoms neighboring the Mg atom, which is 0.056 nm from the center of the trigonal channel (indicated as site AB in the inset diagram in figure 3). Thus, a passivated neutral acceptor-hydrogen complex forms with a binding energy of about 0.7 eV. The energies for H^+ at other sites near Mg were 0.2 eV or more higher in energy than the $AB_{Mg-N\perp}$ site. Density functional theory also predicts that H_2 may form at the center of the trigonal channels [16] if the Mg is fully passivated or if the H concentration is so high that H becomes the dominant donor/acceptor. In our experiments, the high D concentrations from plasma charging could lead to formation of interstitial D_2.

We have carried out computer simulations of the yield versus angle for various D locations in the GaN lattice. These simulations were done using a statistical equilibrium continuum (SEC) model [17] modified for the case of channeling along the c axis in wurtzite GaN. Our calculations use Doyle Turner potentials [18] for the GaN lattice with 24 rows. The model includes dechanneling due to thermal vibration of the host atoms. RMS vibrational amplitudes of 0.00735 nm for Ga and 0.00806 nm for N were used [19]. The SEC model gave good agreement with the observed channeling dip for 2 MeV ^{4}He backscattered from host lattice Ga as shown in figure 3, providing an important validation of the model. Figure 3 shows the NRA yield predicted by the SEC model for various

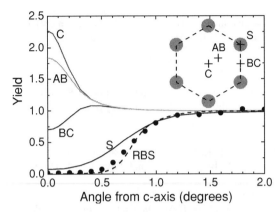

Figure 3 Solid curves show simulated NRA channeling yields for the various D lattice locations indicated in the inset. The dots and dashed curve show measured and simulated yields for 2MeV He^4 scattered from host Ga atoms.

locations of D in the channel as indicated in the inset diagram. Curve C is for D at the channel center, curve S is for D in-line with the host atom rows which includes BC$_{N\parallel}$ sites. Curve AB is for D 0.056 nm from the channel center, corresponding to the AB$_{Mg-N\perp}$ site. Also shown is the yield for D midway between the Ga and N atoms at the channel edge. D vibrational amplitudes used in the simulations (0.026 nm for BC$_{N\parallel}$ and 0.018 nm for AB$_{N\perp}$ and other sites) were estimated assuming a harmonic oscillator model with vibrational frequencies from the density functional calculations [3]. The D vibrational amplitude is mainly due to the low frequency bending modes. Except for the S site, changes in D vibrational amplitude by factors of 2 do not significantly change the calculated yield curves. The channeling simulations show that D near the channel center (which includes AB$_{N\perp}$ sites) gives a narrow peak in the channeling yield and D along the host atom rows gives a broad dip. D at the channel edge midway between host atom rows gives a dip which is narrower and shallower.

The channeling results for the gas charged sample show that the D occupies at least two different sites. The narrow peak shows that 10-20% of the D is near the center of the trigonal channel. This peak is consistent with D at the AB$_{Mg-N\perp}$ site predicted by density functional theory for D$^+$ passivating Mg acceptors, but could also be from neutral D^0 or interstitial D$_2$ which are also predicted to occupy sites near the channel center. In addition to the narrow peak there is a broad dip in channeling yield which shows that ~30% of the D is at a different site near the rows of host atoms along the c direction. This may be D in the activated state whose configuration is not known. D is expected to be in this activated state during the channeling measurements due to electronic excitation by the analysis ion beam.

The two plasma charged samples have D concentrations exceeding the Mg concentration. Solubility calculations show that such high D concentrations indicate very high D chemical potentials [3]. One of the plasma charged samples shows a broad channeling dip similar to the gas charged sample, which may again be due to D in the activated state. This broad dip is not seen in the second plasma charged sample with the highest D concentration. The absence of channeling features in this sample indicates the D is incoherent with the GaN lattice, possibly due to precipitation or formation of extended defects induced by the very high D chemical potential. The absence of a narrow peak in both plasma charged samples shows that little of the D is localized near the channel center which rules out the presence of significant amounts of neutral D^0, D$_2$ or

D^+ passivating Mg. Finally, we observe that high off-axis ion beam doses cause a change in D location presumably due to production of lattice defects by collisional displacements.

ACKNOWLEDGMENTS
This work was supported by the U.S. Department of Energy under contract DE-AC04-94AL85000, primarily under the auspices of the Office of Basic Energy Sciences.

REFERENCES
1   S. J. Pearton, J. C. Zolper, R. J. Shul and F. Ren, J. Appl. Phys. **86** (1999) 1.
2   S. Nakamura, T. Mukai, M. Senoh and N. Iwasa, Jpn. J. Appl. Phys. Part 2, **31** (1992) L139.
3   S. M. Myers, et. al. These proceedings.
4   M. Inamori, H. Sakai, T. Tanaka, H. Amano and I. Akasaki, Jpn. J. Appl. Phys. Part 1, **34** (1995) 1190.
5   X. Li and J.J.Coleman, Appl. Phys. Lett. **69** (1996) 1605.
6   S.J.Pearton, J.W.Lee and C. Yuan, Appl. Phys. Lett. **68** (1996) 2690.
7   M. Miyachi, T. Tanaka, Y. Kimura and H. Ota, Appl. Phys. Lett. **72** (1998) 1101.
8   L. C. Feldman, J. W. Mayer and S. T. Picraux, *Materials Analysis by Ion Channeling*, (Academic, New York, 1982), pp. 88-135.
9   J. Han, T. B. Ng, R. M. Biefeld, M. H. Crawford, and D. M. Follstaedt. Appl. Phys Lett. **71,** 3114 (1997).
10   W. Götz, N.M.Johnson, J.Walker, D.P.Bour, H.Amano and I. Akasaki, Appl. Phys. Lett. **67** (1995) 2666.
11   C.A. Outten, J.C. Barbour and W.R. Wampler, J. Vac. Sci. Technol. **A9** (1991) 717.
12   S.M.Myers, G.R.Caskey, D.E.Rawl, and R.D.Sisson, Metall. Trans. **14A** (1983) 2261.
13   W. Goetz, N.M. Johnson., D.P. Bour, M.D. McCluskey and E.E. Haller, Appl. Phys. Lett. **69** (1996) 3725.
14   J.F.Ziegler, J.P.Biersack and U.L.Littmark, *The stopping and Range of Ions in Matter*, Vol. 1, Pergamon Press (1985) New York.
15   C.A.Klein, J. Appl. Phys. **39** (1968) 2029.
16   A.F. Wright, Phys. Rev. **B60**, (1999) 5101.
17   B. Bech-Nielsen, Phys. Rev. **B37**, 6353 (1988).
18   P. A. Doyle and P. S. Turner, Acta Crystalogr. Sect.A **24**, 390 (1968).
19   A. Yoshiasa, K. Koto, H. Maeda and T. Ishii, Jpn. J. Appl. Phys. **36**, 781 (1997).

Mat. Res. Soc. Symp. Vol. 595 © 2000 Materials Research Society

Surface Conversion Effects in Plasma-Damaged p-GaN

X.A. Cao[1], S.J. Pearton[1], G.T. Dang[2], A.P. Zhang[2], F. Ren[2], R.J. Shul[3], L. Zhang[3], R. Hickman[4] and J.M. Van Hove[4]

[1] Dept. Materials Science and Engineering
University of Florida, Gainesville, FL 32611
[2] Dept. Chemical Engineering
University of Florida, Gainesville, FL 32611
[3] Sandia National Laboratories, Albuquerque, NM 87185
[4] SVT Associates, Eden Prairie, MN 55344

ABSTRACT

The near-surface (400-500Å) of p-GaN exposed to high density plasmas is found to become more compensated through the introduction of shallow donors. At high ion fluxes or ion energies there can be type-conversion of this surface region. Two different methods for removal of the damaged surface were investigated; wet etching in KOH, which produced self-limiting etch depths or thermal annealing under N_2 which largely restored the initial electrical properties.

INTRODUCTION

Very little work has been performed to understand the effect of plasma damage on p-GaN. Shul et.al. [1] reported that the sheet resistance of p-GaN increased upon exposure to Inductively Coupled Plasmas (ICP) of pure Ar. The increases were almost linearly dependent on ion energy (90% increase at −350 eV), but weakly dependent on ion flux (~25% increase at ~10^{17} ions·cm^{-2}·s^{-1}).

Most past work in this area has focused on n-type material. The sheet resistances of GaN [2], InGaN [3], InAlN [3] and InN [3] samples were found to increase in proportion to ion flux and ion energy in an Electron Cyclotron Resonance (ECR) Ar plasma. Ren et.al. [4,5] examined the effect of ECR BCl_3/N_2 and CH_4/H_2 plasmas on the electrical performance of InAlN and GaN channel field effect transistors. They found that hydrogen passivation of the Si doping in the channel may occur if H_2 is a part of the plasma chemistry and that preferential loss of N_2 degraded the rectifying properties of Schottky contacts deposited on plasma-exposed surface. Ping et.al. [6,7] found more degradation in GaN Schottky contacts exposed to Ar plasmas relative to $SiCl_4$ exposure, which would be expected on the basis of the faster etch rate with the latter and hence improved damage removal. In general it is found the ion damage tends to increase the n-type doping level at the surface, most likely through preferential loss of N_2, and that the bandedge photoluminescence intensity decreases through introduction of non-radiative levels [8-12].

Understanding the effects of plasma-induced damage in GaN has become more important as the interest in electronic devices for high temperature, high power applications has increased. One clear example is in the fabrication of GaN/AlGaN heterojunction bipolar transistors (HBTs) [13-15] or GaN bipolar junction transistors (BJTs) [16] where it is necessary to etch down to both a p-type base layer and an n-type subcollector layer. Energetic ion damage may result in increased surface and bulk leakage currents and changes in the electrical properties of the near-surface region through a change in Ga/N stoichiometry [1]. In other compound semiconductor systems it is often possible to remove plasma-damaged regions using slow wet chemical etching [17]. In the GaN system, much less is known about the electrical effects of dry etch damage, and its subsequent removal by wet etching or annealing.

We describe the results of experiments in which p-GaN was exposed to Inductively Coupled Plasmas (ICP) of Cl_2/Ar, H_2 or Ar. The changes in electrical properties were measured by diode breakdown voltage.

EXPERIMENTAL

A schematic of the final Schottky diode structures is shown in Figure 1. Plasma exposures were performed with the contacts in place. The layer structure consisted of 1μm of undoped GaN (n-5×10^{16} cm^{-3}) grown on a c-plane Al_2O_3 substrate, followed by 0.3μm of Mg doped (p-10^{17} cm^{-3}) GaN. The samples were grown by rf plasma-assisted Molecular Beam Epitaxy [18]. Ohmic contacts were formed with Ni/Au deposited by e-beam evaporation, followed by lift-off and annealing at 750°C. The GaN surface was then exposed for 1 min to ICP Cl_2/Ar, H_2 or Ar plasmas in a Plasma-Therm 790 System. The 2MHz ICP source power was varied from 300-1400 W, while the 13.56 MHz rf chuck power was varied from 20-250 W. The former parameter controls ion flux incident on the sample, while the latter controls the average ion energy. Prior to deposition of 250μm diameter Ti/Pt/Au contacts through a stencil mask, the plasma exposed surfaces were either annealed under N_2 in a rapid thermal annealing system, or immersed in boiling NaOH solutions to remove part of the surface. As reported previously it is possible to etch damaged GaN in a self-limiting fashion in hot alkali or acid solutions [19-21]. The current-voltage (I-V) characteristics of the diodes were recorded on an HP 4145A parameter analyzer.

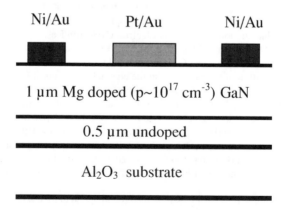

Figure 1. Schematic of completed p-GaN Schottky diode.

RESULTS AND DISCUSSION

We observed an increase in reverse breakdown voltage (V_B) on the p-GaN, whose magnitude was dependent on both ion energy and ion flux. The increase in breakdown voltage on the p-GaN is due to a decrease in hole concentration in the near-surface region through the creation of shallow donor states. The key question is whether there is actually conversion to an n-type surface under any of the plasma conditions. Figure 2 shows the forward turn-on characteristics of the p-GaN diodes exposed to different source power Ar discharges at low source power (300 W). The turn-on voltage remains close to that of the unexposed control sample. However there is a clear increase in the turn-on voltage at higher source powers, and in fact at ≥750 W the characteristics are those of an n-p junction [22]. Under these conditions the concentration of plasma-induced shallow donors exceeds the hole concentration and there is surface conversion. In other words the metal-p GaN diode has become a metal-n GaN-p GaN junction. We always find that plasma exposed GaN surfaces are N_2-deficient relative to their unexposed state [1,3,4,5,10], and therefore the obvious conclusion is nitrogen vacancies create shallow donor levels. This is consistent with thermal annealing experiments in which N_2 loss from the surface produced increased n-type conduction [23,24].

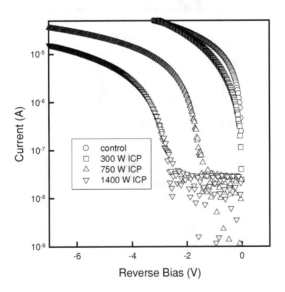

Figure 2. Forward turn-on characteristics of diodes exposed to ICP Ar discharges (150 W rf chuck power) at different ICP source powers prior to deposition of the Ti/Pt/Au contact.

The influence of rf chuck power on the diode I-V characteristics is shown in Figure 3 for both H_2 and Ar discharges at fixed source power (500 W). A similar trend is observed as for the source power experiments, namely the reverse breakdown voltage increases, consistent with a reduction in p-doping level near the GaN surface.

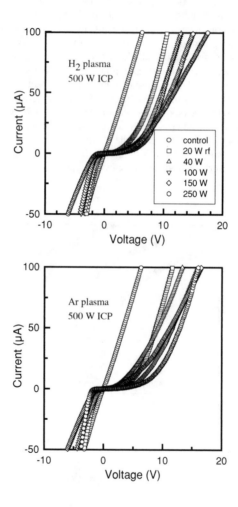

Figure 3. I-V characteristics from samples exposed to either H_2 (top) or Ar (bottom) ICP discharges (500 W source power) as a function of rf chuck power prior to deposition of the Ti/Pt/Au contact. The breakdown voltage increases with the rf chuck power employed during the plasma exposure.

Figure 4 plots breakdown voltage and dc chuck self-bias as a function of the applied rf chuck power. The breakdown voltage initially increases rapidly with ion energy (the self bias plus ~25 V plasma potential) and saturates above ~100 W probably due to the fact that sputtering yield increases and some of the damaged region is removed. Note that there are very large changes in breakdown voltage even for low ion energies, emphasizing the need to carefully control both flux and energy. We should also point out that our experiments represent worse-case scenarios because with real etching plasma chemistries such as Cl_2/Ar, the damaged region would be much shallower due to the much higher etch rate. As an example, the sputter rate of GaN in a 300 W source power, 40 W rf chuck power Ar ICP discharge in ~40Å·min^{-1}, while the etch rate in a Cl_2/Ar discharge under the same conditions is ~1100Å·min^{-1}.

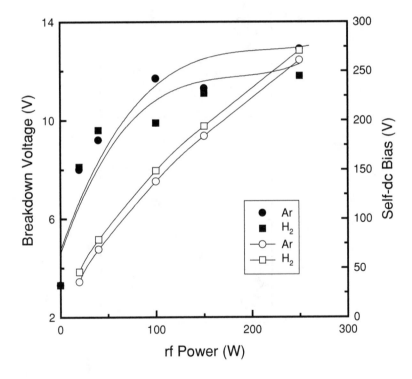

Figure 4. Variation of diode breakdown voltage in samples exposed to H_2 or Ar ICP discharges (500 W source power) at different rf chuck powers prior to deposition of the Ti/Pt/Au contact (solid symbols). The dc chuck self-bias during plasma exposure is also shown (open symbols).

Figure 5 shows I-V characteristics from samples that were wet etched to various depths in NaOH solutions after exposure to either Cl$_2$/Ar or Ar discharges (500 W source power, 150 W rf chuck power, 1 min). For these plasma conditions we did not observe type conversion of the surface. However, we find that the damaged GaN can be effectively removed by immersion in hot NaOH, without the need for photo- or electrochemical assistance of the etching. The V$_B$ values increase on p-GaN after plasma exposure due to introduction of shallow donor states that reduce the net acceptor concentration.

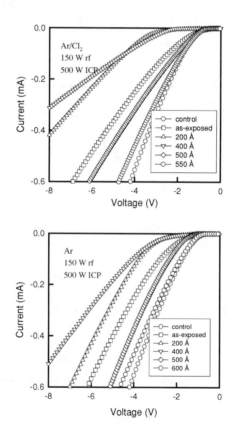

Figure 5. I-V characteristics from p-GaN samples exposed to ICP Cl$_2$/Ar (top) or Ar (bottom) discharges (500 W source power, 150 W rf chuck power) and wet etched in boiling NaOH to different depths prior to deposition of the rectifying contact.

Figure 6 shows the wet etch depth in plasma damaged p-GaN as a function of etching time. The etch depth saturates at depths of 500-600Å, consistent with the electrical data. It has previously been shown by Kim et.al. [19] that the wet etch depth on thermally- or ion-damaged GaN was self-limiting. This is most likely a result of the fact that defective or broken bonds in the material are readily attacked by the acid or base, whereas in undamaged GaN the etch rate is negligible. I-V characteristics were recorded from samples exposed to 750 W source power, 150 W rf chuck power (-160 V dc chuck bias) Ar discharges and subsequently wet etched to different depths using 0.1 M NaOH solutions before deposition of the Ti/Pt/Au contact. Figure 7 shows the effect of the amount of material removed on the diode breakdown voltage. Within the experimental error of ±12%, the initial breakdown voltage is reestablished in the range 400-450Å. This is consistent with the depth obtained from the etch rate experiments described above. These values are also consistent with the damage depths we established in n-GaN diodes exposed to similar plasma conditions [25].

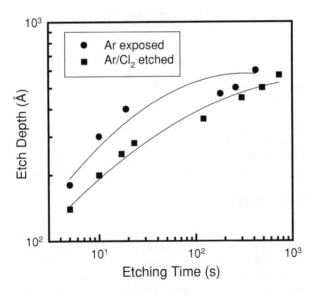

Figure 6. Wet etch depth versus etch time in boiling 0.1M NaOH solutions for plasma damaged p-GaN.

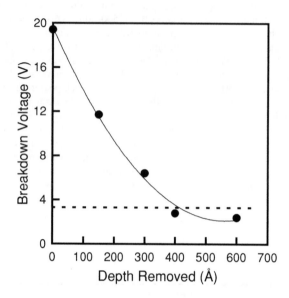

Figure 7. Variation of V_B with depth of plasma exposed p-GaN removed by wet etching prior to deposition of the rectifying contact. The dashed line shows the control value in unexposed p-GaN diodes.

We also performed annealing experiments on the plasma-damaged p-GaN at temperatures up to 900°C under a N_2 ambient. At this temperature the V_B values were returned to within 10% of their control values. Annealing at higher temperatures degraded the electrical properties, most likely due to nitrogen loss for the surface.

SUMMARY AND CONCLUSIONS

ICP plasma exposure of p-GaN produced large increases in reverse breadown voltage, independent of the plasma chemistry employed. The results are consistent with creation of nitrogen vacancy-related shallow donors in the near-surface region (<600Å). Although not discussed here, the damage was found to accumulate in this region even for very short plasma exposure times (3-4 secs). Two different methods for restoring V_B were investigated. Annealing at 900°C essentially returned V_B to its original, control value, whereas wet etch removal of the damaged surface in boiling NaOH prior to deposition of the Schottky contact was also effective in restoring the initial electrical properties.

ACKNOWLEDGMENTS

The work at UF is partially supported by a DARPA/EPRI grant, No. MDA972-98-1-0006 (D. Radack and J. Melcher) monitored by ONR (J.C. Zolper) and a NSF grant DMR97-32865 (L.D. Hess). Sandia is a multiprogram laboratory operated by Sandia Corporation, a Lockheed-Martin Company, for the US Department of Energy under contract No. DEAC04-94-AC-85000.

REFERENCES

1. R.J. Shul, L. Zhang, A.G. Baca, C.G. Willison, J. Han, S.J. Pearton, F. Ren, J.C. Zolper and L.F. Lester, "High Density Plasma-Induced Etch Damage in GaN," Mat. Res. Soc. Symp. Proc. Vol. 573, 271 (1999).
2. C.R. Eddy, Jr. and B. Molnar, "The Effect of H2-Based Etching on GaN," Mat. Res. Soc. Symp. Proc. 395, 745 (1996); "Plasma Etch-Induced Conduction Changes in GaN," J. Electron. Mater. 28, 314 (1999).
3. S.J. Pearton, J.W. Lee, J.D. MacKenzie, C.R. Abernathy and R.J. Shul, "Dry Etch Damage in InN, InGaN and InAlN," Appl. Phys. Lett. 67, 2329 (1995).
4. F. Ren, J.R. Lothian, S.J. Pearton, C.R. Abernathy, C.B. Vartuli, J.D. MacKenzie, R.G. Wilson and R.F. Karlicek, "Effects of Dry Etching of Surface Properties of III-Nitrides," J. Electron. Mater. 26, 1287 (1997).
5. F. Ren, J.R. Lothian, Y.K. Chen, J.D. MacKenzie, S.M. Donovan, C.B. Vartuli, C.R. Abernathy, J.W. Lee and S.J. Pearton, "Effect of BCl3 Dry Etching on AlInN Surface Properties," J. Electrochem. Soc. 143, 1217 (1996).
6. A.T. Ping, A.C. Schmitz, I. Adesida, M.A. Khan, A. Chen and Y.W. Yang, "Characterization of RIE Damage to GaN using Schottky Diodes," J. Electron. Mater. 26, 266 (1997).
7. A.T. Ping, Q. Chen, J.W. Yang, M.A. Khan and I. Adesida, "The Effects of RIE Damage on Ohmic Contacts to n-GaN," J. Electron. Mater. 27, 261 (1998).
8. J.Y. Chen, C.J. Pan and G.C. Chi, "Electrical and Optical Changes in the Near-Surface of GaN," Solid-State Electron. 43, 649 (1999).
9. F. Ren, J.R. Lothian, Y.K. Chen, R. Karlicek, L. Tran, M. Schurmann, R. Stall, J.W. Lee and S.J. Pearton, "Recessed Gate GaN FET," Solid-State Electron. 41, 1819 (1997).
10. R.J. Shul, J.C. Zolper, M.H. Crawford, R.J. Hickman, R.D. Briggs, S.J. Pearton, J.W. Lee, R. Karlicek, C. Tran, M. Schurmann, C. Constantine and C. Barratt, "Plasma-Induced Damage in GaN," Proc. Electrochem. Soc. 96-15, 232 (1996).
11. A.S. Usikov, W.L. Lundin, U.I. Ushakov, B.V. Pushnyi, N.N. Schmidt, Y.Y. Zadiranov and T.V. Shubtra, "Electrical and Optical Properties of GaN After Dry Etching," Proc. Electrochem. Soc. 97-14, 57 (1998).
12. K. Satore, A. Matsutani, T. Shirasawa, M. Mori, T. Honda, T. Sakaguchi, F. Kagana and K. Iga, "RIBE of GaN," Mat. Res. Soc. Symp. Proc. 449, 1029 (1997).
13. L.S. McCarthy, P. Kozodoy, S.D. DenBaars, M. Rodwell and U.K. Mishra, 25th Int. Symp. Compound Semicond., Oct. 1998, Nara, Japan.
14. F. Ren, C.R. Abernathy, J.M. Van Hove, P.P. Chow, R. Hickman, J.J. Klaassen, R.F. Kopf, H. Cho, K.B. Jung, R.G. Wilson, J. Han, R.J. Shul, A.G. Baca and S.J. Pearton, "300°C GaN HBT," MRS Internet J. Nitride Semicond. Res. 3, 41 (1998).

15. J. Han, A.G. Baca, R.J. Shul, C.G. Willison, L. Zhang, F. Ren, A.P. Zhang, G.T. Dang, S.M. Donovan, X.A. Cao, H. Cho, K.B. Jung, C.R. Abernathy, S.J. Pearton and R.G. Wilson, "Growth and Fabrication of GaN HBT," Appl. Phys. Lett. 74, 2702 (1999).

16. S. Yoshida and J. Suzuki, "High Temperature Reliability of GaN MESFET and BJT," J. Appl. Phys. 85, 7931 (1999).

17. S.J. Pearton, "RIE of Compound Semiconductors," Int. J. Mod. Phys. B8, 1781 (1994).

18. J.M. Van Hove, R. Hickman, J.J. Klaassen, P.P. Chow and P.P. Ruden, "High Quality GaN Grown by rf-MBE," Appl. Phys. Lett. 70, 282 (1997).

19. B.J. Kim, J.W. Lee, H.S. Park and T.I. Kim, "Wet Etching of GaN Grown by OMVPE," J. Electron. Mater. 27, L32 (1998).

20. S.A. Stocker, E.F. Schubert and J.M. Redwing, "Crystallographic Wet Etching of GaN," Appl. Phys. Lett. 73, 2345 (1998).

21. J.-L. Lee, J.K Kim, J.W. Lee, Y.J. Park and T. Kim, "Effect of Surface Treatment by KOH on p-Ohmic Contacts to GaN," Solid-State Electron. 43, 435 (1999).

22. S.M. Sze, Physics of Semiconductor Devices (Wiley-Interscience, NY, 1981).

23. J.C. Zolper, D.J. Rieger, A.G. Baca, S.J. Pearton, J.W. Lee and R.A. Stall, "Sputtered AlN for Annealing GaN," Appl. Phys. Lett. 69, 538 (1996).

24. J. Brown, J. Ramer, K. Zhang, L.F. Lester, S.D. Hersee and J.C. Zolper, "ECR Etching of GaN With and Without H_2," Mat. Res. Soc. Symp. Proc. Vol. 395, 702 (1996).

25. X.A. Cao, H. Cho, S.J. Pearton, G.T. Dang, A. Zhang, F. Ren, R.J. Shul, L. Zhang, F. Ren, R.J. Shul, L. Zhang, A.G. Baca, R. Hickman and J.M. Van Hove, "Depth and Thermal Stability of Dry Etch Damage in GaN Schottky Diodes," Appl. Phys. Lett. 75, 232 (1999).

Mat. Res. Soc. Symp. Vol. 595 © 2000 Materials Research Society

ZIRCONIUM MEDIATED HYDROGEN OUTDIFFUSION FROM p-GaN

E.Kaminska[1], A.Piotrowska[1], A. Barcz[1,3], J.Jasinski[2], M. Zielinski[3], K.Golaszewska[1], R.F.Davis[4], E. Goldys[5], K. Tomsia[5]

[1]Institute of Electron Technology, Al.Lotnikow 46, Warsaw, Poland, eliana@ite.waw.pl.
[2]Institute of Experimental Physics, Warsaw University, Warsaw, Poland
[3]Institute of Physics PAS, Warsaw, Poland,
[4]Department of Material Science and Eng., NCSU, Raleigh, NC 27695-7907, USA
[5]SMPCE, Macquarie University, Sydney NSW, Australia

ABSTRACT

We have shown that Zr-based metallization can effectively remove hydrogen from the p-type GaN subsurface, which eventually leads to the formation of an ohmic contact. As the release of hydrogen starts at ~900°C, the thermal stability of the contact system is of particular importance. The remarkable thermal behavior of the ZrN/ZrB_2 metallization is associated to the microstructure of each individual Zr-based compound, as well as to the interfacial crystalline accommodation.

INTRODUCTION

The role of hydrogen in the passivation and activation of dopants in MOCVD grown GaN is commonly recognized, however, experimental knowledge on its behavior remains still incomplete [1]. Recent interest in GaN-based devices has stimulated studies on p-type doping issues [2, 3] as well as the search for novel ohmic contacts. In particular, Murakami [4] recommended metals with high binding energies to hydrogen for making ohmic contacts to p-type GaN, and this idea has been experimentally verified with the use of Ta/Ti contacts [5]. These contacts provided the lowest reported contact resistivity values ($3x10^{-5}\Omega cm^2$ for hole concentration of $7x10^{17}cm^{-3}$), but they quickly degraded.

The accumulation of hydrogen in areas of high defect density has been considered [6], it is therefore very likely that hydrogen could be gettered in the superficial layer of p-GaN. In consequence, the subsurface volume may be characterized by a free carrier concentration lower than the bulk. The present study explores the problem of the accumulation of hydrogen in the near-surface region of p-GaN and the possibility of its removal by using ZrN/ZrB_2 metallization and annealing. Zirconium possesses one of the highest absorptive capabilities for hydrogen of metal hydride systems, while ZrN and ZrB_2 are distinguished for their exceptional low resistivities and high melting points. The content of hydrogen was quantified using secondary ion mass spectrometry (SIMS), while the stability of Zr-based caps on GaN under annealing has been examined using transmission electron microscopy (TEM) together with SIMS.

EXPERIMENTAL PROCEDURE

The samples used in this study were (0001) oriented, Mg-doped GaN epilayers, 0.5-2 μm thick, grown via MOCVD on undoped GaN, with AlN buffer films predeposited on 6H-SiC or sapphire substrates. The hole concentration was p = 1-5*10^{17}cm^{-3}. Prior to insertion into the deposition chamber, the surface of GaN was etched in buffered HF, and dipped in NH$_4$OH:H$_2$O (1:10).

Thin films of Zr-N and Zr-B were deposited by DC magnetron sputtering in Ar discharge, from ZrN(99.5%) and ZrB$_2$ (99.5%) targets, respectively. The process parameters were first optimized with regard to the stoichiometric composition of deposited films. Finally, a bilayer ZrN/ZrB$_2$ metallization of a thickness (100nm)/(100nm), sequentially deposited without breaking the vacuum, was used throughout this study. Heat treatments were carried out in a rapid thermal annealer (RTA), in flowing N$_2$, at temperatures in the range 700-1150^{0}C for time 30 s. to 5 min. During RTA the samples were protected by a piece of GaN/Al$_2$O$_3$ as a proximity cap.

The evolution of the microstructure of Zr-based films deposited on GaN, under annealing, was investigated by cross-sectional transmission electron microscopy (XTEM) and high resolution imaging (HREM) in JEOL 2000EX microscope. SIMS profiling was performed with a Cameca 6F instrument using either oxygen or cesium primary beam.

The electrical characterization involved measuring the current-voltage characteristics and the specific contact resistance using circular transmission line method. A complementary study of electrical properties of Zr-based contacts to n-type GaN (n = 1*10^{17}cm^{-3}) has also been carried out.

RESULTS

Microstructure and thermal stability of GaN/ZrN/ZrB$_2$ system

The as-deposited ZrN and ZrB$_2$ films exhibit golden-yellow and pale-yellow color, respectively, characteristic of stoichiometric compounds. XTEM micrographs and corresponding SIMS profiles of as-deposited and annealed GaN/ZrN/ZrB$_2$ structures are given in Fig. 1 and 2. The initial morphology of ZrN film is amorphous with islands of fine-grain polycrystalline material. ZrB$_2$ film is amorphous. Annealing at 800^{0}C causes partial crystallization of ZrN into elongated grains, whereas the microstructure of ZrB$_2$ film does not change noticeably. The final morphology, after annealing at 1100^{0}C, of ZrN film is columnar, while the ZrB$_2$ top layer reveals a fine-grained structure.

SIMS depth profiles of ^{11}B$^+$, ^{14}N$^+$, ^{90}Zr$^+$, ^{90}Zr^{14}N and ^{69}Ga$^+$, measured using O$_2$$^+$ primary beam, has been chosen to illustrate the in-depth structure of GaN/ZrN/ZrB$_2$ contacts. In Fig.2.a., a relatively high value of ^{90}Zr^{14}N signal in ZrB$_2$ layer is a result of ^{90}Zr^{14}N and ^{94}Zr^{10}B mass interference. Similarly, the high intensity of ^{90}Zr in GaN layer is probably due to the interference with gallium-nitrogen cluster – the 5 orders of magnitude decrease of the ^{90}Zr^{14}N profile in GaN near the interface region indicates that Zr does not penetrate significantly into GaN.

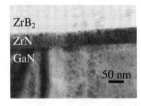 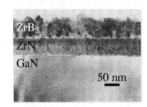 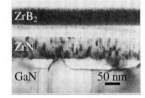

Figure 1. XTEM micrographs of GaN/ZrN/ZrB$_2$ contact: a) as-deposited, b) annealed at 800^{0}C, 5 min.; c) annealed at 1100^{0}C, 30 s.

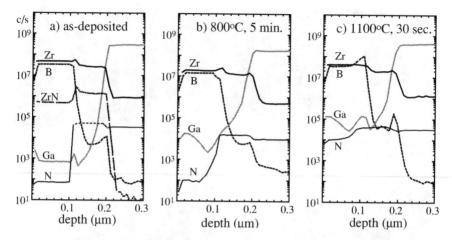

Figure 2. SIMS profiles for GaN/ZrN/ZrB$_2$ contact: a) as-deposited, b) annealed at 800^{0}C, 5 min., c) annealed at 1100^{0}C, 30 s.

SIMS depth profiles of contacts unprocessed and heat treated at 800^{0}C indicate that GaN/ZrN/ZrB$_2$ system remains stable. In contacts subjected to annealing at 1100^{0}C, however, the Ga signal in Zr-containing films increases about two orders of magnitude, suggesting that the interaction at the GaN/ZrN interface has taken place.

In order to evaluate the extent of this interaction, in Fig.3 are shown HREM images of GaN/ZrN interfaces after annealing at 800^{0}C and 1100^{0}C. While the interface of the contact processed at 800^{0}C is sharp and abrupt, regular intrusions form at the GaN/ZrN interface annealed at 1100^{0}C. High-resolution image and electron diffraction pattern revealed evidence for the formation of a new phase, lattice-matched to GaN. This phase, composed of Ga, Zr and N, as indicated by EDX analysis, could not be identified as yet.

Hydrogen in GaN/ZrN/ZrB$_2$ system

The behavior of hydrogen in GaN/ZrN/ZrB$_2$ under annealing was studied with SIMS. Since Zr has a high affinity for hydrogen, it is obvious that some hydrogen is trapped in Zr-based metallization during the sputter-deposition. In Table I are shown relative H$^-$ yields, with respect to Zr$^-$ ions, in ZrN and ZrB$_2$ films as a function of annealing temperature. A noticeable evolution of hydrogen starts at 750^{0}C from ZrB$_2$ layer. After heat treatment at 850^{0}C the Zr-based metallization is practically free of hydrogen.

Table I. Relative H$^-$ yields with respect to Zr$^-$ in ZrN and ZrB$_2$ films after annealing.

material/ annealing	as-deposited	750^{0}C, 5 min.	800^{0}C, 5 min.	850^{0}C, 5 min.	1100^{0}C, 30 s.
ZrB$_2$	3.4	0.4	0.4	0.3	0.1
ZrN	5.0	2.0	3.0	0.1	0.1

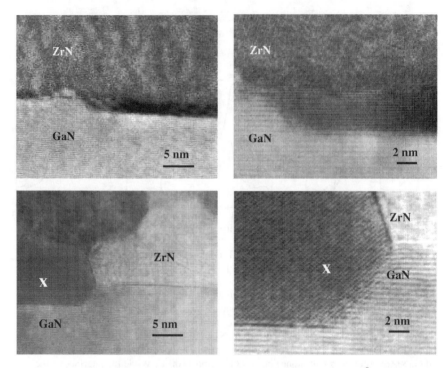

Figure 3. HREM images of GaN/ZrN interface after annealing at: a) 800^0C, 5 min., b) 1100^0C, 30 s.

Mg and hydrogen depth profiles in GaN are presented in Fig.4. Concentrations of hydrogen and Mg in GaN were evaluated using relative sensitivity factors from ref. [7]. While the level of Mg in the entire layer of p-type GaN remains constant, a noticeably higher concentration of hydrogen in the near-surface region is clearly seen. For comparison, we have also analyzed the profile of hydrogen in n-GaN, confirming the absence of this impurity in n-type MOCVD-grown material with the detection limit $\sim 10^{17}$cm^{-3}. Heat treatment at 800^0C, i.e. at temperature effective for activation of Mg as a dopant [8], left the amount of hydrogen in p-type material practically unchanged. The process of evolution of hydrogen from the subcontact region starts at about 900^0C, and after annealing at 1100^0C the level of this impurity in the subsurface layer of p-GaN is substantially reduced, as demonstrated in Fig. 4.b.

Electrical properties of GaN/ZrN/ZrB$_2$ contacts

The release of hydrogen from p-GaN subsurface region perfectly correlates with the electrical properties of GaN/ZrN/ZrB$_2$ contacts. Contacts annealed below 900^0C are non-linear and highly resistive. After heat treatment at 900^0C for 30 sec. they become linear. The minimum specific contact resistance of $1*10^{-4}$ Ωcm^{-3} (p = $5*10^{17}$cm^{-3}) was observed for contacts annealed at 1000^0C. Annealing at 1100^0C causes the increase of

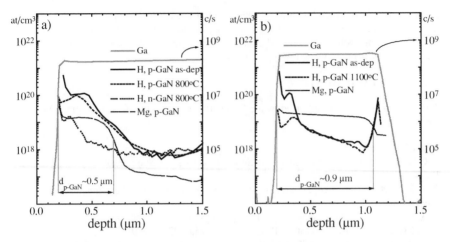

Figure 4. Magnesium and hydrogen depth profiles in p-GaN annealed under ZrN/ZrB$_2$: a) as-deposited and annealed at 800^{0}C, 5 min., b) as-deposited and annealed at 1100^{0}C, 30 s.

the contact resistivity. Ohmic contacts formed at 1000^{0}C were subjected to accelerated life-time testing at 600^{0}C for 8 hours in N$_2$ flow. Both, the low value of specific contact resistance and the surface morphology were preserved throughout the aging.

ZrN/ZrB$_2$ contacts, when deposited on n-type GaN become ohmic after heat treatment at 1100^{0}C for 30 sec. with a resistivity of $6*10^{-4} \Omega cm^{-3}$ (n = $1*10^{17}cm^{-3}$).

DISCUSSION AND CONCLUSIONS

We have shown that Zr-based metallization can effectively remove hydrogen from the p-type GaN subsurface which eventually leads to the formation of an ohmic contact. As the release of hydrogen starts at temperatures as high as 900^{0}C, the thermal stability of the contact system is of particular importance. The thermal behavior of the ZrN/ZrB$_2$ metallization is associated to the microstructure of each individual Zr-based compound, as well as to the interfacial crystalline accommodation. Amorphous transition metal nitrides have been previously proven as effective diffusion barriers in integrated circuits technology [9], for thermal treatments up to 600^{0}C. In this work we demonstrate that by properly designing the configuration of the metallization system, the thermal stability of Zr-based compounds can be enhanced well above their crystallization temperature. The use of ZrB$_2$ layer on top of ZrN film effectively encapsulates the system, preventing the decomposition of the surface of GaN under annealing, even in case when the microstructure of ZrN film becomes columnar with grain boundaries providing the path for fast diffusion. On the other hand, ZrN interlayer between GaN and ZrB$_2$ film impedes the recrystallization of the latter.

Whereas Zr-based metallization enables hydrogen to outdiffuse from p-type subcontact region, the resistivity of ohmic contact increases, when the thermal processing is conducted at temperature ~1100^{0}C, Our experiments clearly demonstrate that this coincides with a reaction at the contact interface. Thus the degradation of ohmic contact properties can be explained in terms of the decomposition of GaN and formation of nitrogen vacancies, being donors that compensate Mg acceptor centers. This would also explain the ohmic behavior, observed after heat treatment at 1100^0C, when the same metallization scheme is applied to n-type GaN.

ACKNOWLEDGEMENTS

This work was partially supported by the Committee for Scientific Research, grant No. PBZ 28.11/P9.

REFERENCES
1. C.G. Van de Walle, N.M.Johnson, in Semiconductors and Semimetals 57 (Academic Press, 1999), Chap.4.
2. S. J. Pearton, J. W. Lee, in Semiconductors and Semimetals 61 (Academic Press, 1999), Chap.10.
3. J. Neugebauer, C.G. Van de Walle, J. Appl. Phys. 85, 3003 (1999).
4. M.Murakami, Y.Koide, Critical Rev. Sol. State Mat. 23, 1 (1998).
5. M. Suzuki, T. Kawakami, T. Arai, S. Kobayashi, Y. Koide, T. Uemura, N. Shibata, M. Murakami, Appl. Phys. Lett. 74, 275 (1999).
6. S.J. Pearton, S. Bendi, K. S. Jones, V. Krishnamoorthy, R. G. Wilson, R. F. Karlicek, Jr., R. A. Stall, Appl. Phys. Lett. 69, 1879 (1996).
7. J. W. Erickson, Y. Gao, R. G. Wilson, Mat. Res. Soc. Symp. Proc. Vol.395, 363 (1996).
8. W. Gotz, N. M. Johnson, J. Walker, D. P. Bour, R. A. Street, Appl. Phys. Lett. 68, 667 (1996).
9. L. Krusin-Elbaum, M. Wittmer, C-Y. Ting, J.J. Cuomo, Thin Solid Films 104, 81 (1983).

Devices, Optical Characterization,
Processing, Contacts, Defects

Mat. Res. Soc. Symp. Vol. 595 © 2000 Materials Research Society

A Comparative Study Of GaN Diodes Grown by MBE on Sapphire and HVPE-GaN /Sapphire Substrates.

Anand V. Sampath, Mira Misra, Kshitij Seth, Yuri. Fedyunin, Hock M. Ng[*], Eleftherios Iliopoulos, Zeev Feit[1], Theodore .D. Moustakas

ECE Department, Boston University, 8. St. Mary's St. Boston, MA 02215
[1] Photonics Center, Boston University, 8. St. Mary's St. Boston, MA 02215

ABSTRACT

In this paper we report on the fabrication and characterization of GaN diodes (Schottky and p-n junctions) grown by plasma assisted MBE. We observed that Schottky diodes improve both in reverse as well as forward bias when deposited on 5 μm thick HVPE n$^+$-GaN/sapphire instead of bare sapphire substrates. These improvements are attributed to the reduction of disloctions in the MBE homoepitaxially grown GaN. Similar benefits are observed in the reverse bias of the p-n junctions which according to EBIC measurements are attributed to the reduction of etch pits in the MBE grown p-GaN.

INTRODUCTION

GaN photodiodes have recently been reported by a number of laboratories [1-3]. Some studies suggest that threading dislocations in these devices act as leakage paths resulting in increased dark current [4-6]. The Santa-Barbara group has demonstrated that p-n junctions fabricated by the deposition of p-GaN by the MOCVD method on epitaxially laterally overgrown (ELO) GaN stripes show orders of magnitude reduction in the reverse-bias dark current [7]. However, such devices are small and impractical as photodetectors because of the narrow width of the ELO-stripes (~ 8 μm).An alternative approach is to use the HVPE method for the growth of GaN substrates. Due to its high growth rates, this method is capable of producing thick GaN films with significantly reduced concentration in threading defects at the free surface [8]. Furthermore, significantly broader ELO GaN stripes can be produced by this method [9]. Also, Smith et. al. have recently demonstrated that photoconductive detectors grown on HVPE GaN/ sapphire have improved optical response and sharpness than those grown directly on sapphire [10].

In this paper we report on the fabrication and characterization of GaN Schottky diodes grown by molecular beam epitaxy on c-plane sapphire as well as 5-6 μm thick HVPE grown GaN films. Furthermore, the fabrication and characterization of p-n junctions formed by depositing p-GaN by MBE on n$^+$ GaN grown by HVPE is presented.

---

[*] Presently at Lucent Technology, Bell Labs

EXPERIMENTAL METHODS

The 5 –6 µm thick n⁺- GaN /sapphire substrates employed in this study were grown by the HVPE method. Briefly, the films were grown in a horizontal quartz reactor by flowing HCl over Ga melt that produces GaCl. Using N_2 as a carrier gas, the GaCl is transported to the substrate where it reacts with NH_3 and forms GaN. The films grown by this method are typically unintentionally doped n-type with carrier concentration between $5x10^{18}$ to $1x10^{19}$ cm⁻³. The density of dislocations in these relatively thin GaN substrates was estimated from photoelectrochemical etching to be on the order of ~ 10^9 cm⁻².

The active layers for both types of diodes were grown by plasma assisted MBE. Active nitrogen for the MBE growth was produced from molecular nitrogen by a microwave plasma-assisted electron cyclotron resonance source. The active layers (n⁺/n⁻, or p) were grown at 700- 800 °C using procedures described in our recent papers [11]. The films were doped n-type with silicon by varying the cell temperature from 1025 to 1100 °C for the n⁻ and n⁺ layers respectively. The p-type films were doped with Mg by varying the cell temperature from 300 – 350 °C. For the devices grown directly on (0001) sapphire, the substrates were first subjected to a nitridation step at 800 °C (conversion of the surface of Al_2O_3 to AlN) using procedures developed first in our laboratory [12]. This step was followed by the deposition of a 20-30 nm thick AlN buffer grown at 700-750 °C. The growth of the active layers on the HVPE grown GaN/sapphire proceeded directly without any plasma pretreatment.

All the diodes were fabricated using a similar process. First a 300 µm diameter mesa was defined by RIE using Cl_2 gas. Etching proceeded at a flow rate of 20 sccms, a pressure of 13 mTorr and 100W RF power (-350 V DC bias). The etch mask consisted of a thin e-beam deposited SiO_2 layer followed by AZ4620 photoresist. Subsequently the structures were rapid thermally annealed in forming gas at 700°C for 20sec prior to metalization. The e-beam deposited contact to the n⁺- GaN film consists of a combination of Ti/Al/Ni/Au at thicknesses of 20 nm/50nm/15nm/400 nm respectively. An e-beam evaporated Ni/Pt/Au stack at thicknesses of 20nm/100nm/200nm was used as the Schottky contact to the n⁻ - GaN layer and a Ni/Au stack of 2 nm\20 nm was used as the ohmic contact to the p-GaN layer. The metal contacts were defined using standard lift-off techniques. The p-GaN ohmic contact was annealed in air at 500 °C for 10 mins to improve ohmicity as described by Ho and coworkers [16].

The diodes IV characteristics were measured using an HP4155A Semiconductor Parameter Analyzer and a Karl Suss PM5 probe station. The CV characteristics were measured using a HP4275 Muli-frequency LCR meter. All measurements were performed at 1 MHz frequency and 130 mV test signal.

RESULTS AND DISCUSSION

A. Schottky Diodes

Typical I-V characteristics for the two types of Schottky diodes are shown in Figure 1. The ideality factor and reverse saturation current for the homoepitaxially and heteroepitaxially grown diodes are $1.4x10^{-9}$ A/cm² and $1.2x10^{-9}$ A/cm² respectively. The C-V measurements demonstrate that both types of Schottky diodes have a free donor

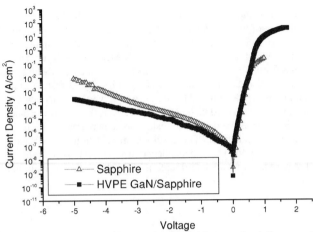

Figure 1. I-V Characteristics of homoepitaxial and heteroepitaxially grown Schottky diodes

carrier concentration of $6-9 \times 10^{16}$ cm^{-3} as well as barrier height of .9-1 V. The 300 μm diameter diode fabricated on HVPE-GaN has leakage current density of 2.75×10^{-4} A/cm^2 at –5V reverse bias, which is over 1 order of magnitude lower than the heteroepitaxially grown one. We attribute this improvement to the reduction of threading defects in the homoepitaxially grown device. This conclusion is consistent with the results of both Kozodoy et. al [7], who investigated the quality of GaN p-n junctions grown on epitaxially overgrown GaN/sapphire substrates as well as Shiojima et. al. [5], who investigated the performance of Schottky diodes grown on sapphire with varying Si doping.

The forward bias characteristics of the diodes are compared in Figure 2 employing a modification to the ideal diode equation that accounts for the series resistance of the device [13]. Namely, $I = I_o \ast \exp(q(V - I \ast R_s)/nkT)$, where I is the diode

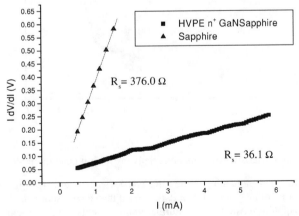

Figure 2. Plot of I*dV/dI vs. I for both the homoepitaxially and heteroepitaxially grown Schottky diodes. The slope of the plot is the forward series resistance (R_S) of the diode.

current, V is the applied voltage and R_s is the series resistance of the diode.

The series resistance of the two devices differ by about one order of magnitude. We believe that the larger resistance of the heteroepitaxial device has two origins. One is due to the larger series and contact resistance of the n^+-GaN MBE grown layer. The second contribution comes from the reduction of the active area in the n^- GaN layer due to the higher concentration of dislocations in this device. Specifically, as discussed in another paper presented in this symposium [14], the transport in vertical GaN devices does not suffer from scattering by charged dislocations. However, the series resistance of the device is expected to increase as the density of dislocations increases due to the reduction of the active area for conduction. This results from depletion around each dislocation. A quantitative description of the dependence of series resistance in the forward direction will require an accurate determination of the dislocation density in the sample.

B. p-n Diodes

The IV characteristics of a diode formed by depositing p-GaN (5×10^{17} cm^{-3}) by MBE on 5 μm thick HVPE n^+- GaN is shown in Figure 3. A dark current of 10^{-8} A/cm^2 was measured at –2V for these devices. The dark current seen in these diodes is attributed

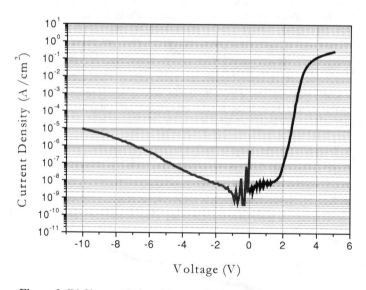

Figure 3. IV Characteristics of homoepitaxially grown p-n junction diode.

to the density of dislocations remaining in the film. The devices discussed in Figure 3 show more than an order of magnitude lower dark current than a corresponding device grown by MBE on sapphire substrates [15]. Further improvements should be achieved with the use of thicker HVPE n-GaN films since the dislocation density in such films

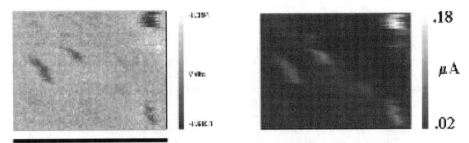

2.5 μm

SEM Image   EBIC Image

Figure 4. SEM (left) and EBIC (right) image of a p-n junction diode. The diode shows larger recombination current at defects revealed in the SEM image.

have been shown to decrease monotonically with film thickness [8]. Applying the ideal diode equation to these diodes, the ideality constant was found to be ~3.5.

To investigate this hypothesis we have performed electron beam induced current (EBIC) studies. For these measurements diodes with 290 μm diameter ring contact was used which allows the electron beam to directly probe the p-GaN surface. The SEM and EBIC images of such a device are shown in Figure 4. We notice that there is a correspondence between the pits in the p-GaN layer shown in the SEM image and the amount of current conducting through those areas in the EBIC image. Thus, we believe that this result unambiguously correlates the leakage current under reverse bias with the number of dislocations in the material.

CONCLUSIONS

High quality GaN Schottky diodes have been fabricated on both c-plane sapphire as well as HVPE GaN on sapphire substrates. The homoepitaxially grown diodes are both more conductive under forward bias (37 Ω), as well as less leaky under reverse bias (2.75x10^{-4} A/cm^2 at –5V) than the heteroepitaxially grown devices. We attribute this to the reduction of dislocations in the homoepitaxially grown MBE GaN layers. Low leakage p-n junction diodes (5x10^{-7} A/cm^2 at –5V) have also been fabricated by depositing p-GaN by MBE on HVPE GaN on sapphire substrates. EBIC measurements in such devices correlate the leakage current to threading defects. These devices have poor forward bias characteristics which we attribute to poor p-type contacts.

ACKNOWLEDGEMENTS

The authors are grateful to Dr. Phillip. A. Lamarre and Mr. Dave Parker, Lockheed-Martin Infrared Imaging Sytems, for their assistance with photolithography and C-V measurements. We also thank Dr. Shawn Burke, Dr. Peter McDonald and the Photonics Systems Laboratory for use of their equipment as well as Mr. Paul Mak for his assistance in device fabrication. The authors are grateful to Ms. Worthen and Clariant Corporation for assistance with their photoresist products.The device processing for this work was done in the Optoelectronics Processing Facility at the Boston University Photonics Center. This work was supported by ONR under grant # N00014-99-1-0309, monitored by Dr. Y.S. Park.

REFERENCES

[1] C.J. Collins, T. Li, A.L. Beck, R. D. Dupuis, J.C. Campbell, J.C. Carrano, M.J. Schurman, I.A. Ferguson, *Appl. Phys. Lett,* 75, 2138, (1999)

[2] G.Y. Xu, A. Salvador, W. Kim, Z.Fan, C. Lu, H. Tang, H. Morkoc, *Appl. Phys. Lett,* 71,15, (1997)

[3] D. Walker, A. Saxler, P. Kung, X. Zhang, M. Hamilton, J. Diaz, M. Razeghi, *Appl. Phys. Lett,* 72, 3303, (1998)

[4] D.V.Kuksenkov, H. Temkin, A. Osinsky, R. Gaska, M.A. Khan, *Appl. Phys. Lett,* 72, 1365, (1998)

[5] Kenji Shiojima, Jerry M. Woodall, Christopher J. Eiting, Paul A. Grudowski, Russell D. Dupuis, *J. Vac Sci Technol B.* 17(5), 1999 2030

[6] J.T. Torvik, J.I Pankove, S. Nakamura, I. Grzegory, S. Porowski, *J. Appl. Phys,* 86, 4588 (1999)

[7] P. Kozodoy, J.P. Ibbetson, H. Marchand, P.T. Fini, S. Keller, J.S. Speck, S.P DenBaars, U.K Mishra, *Appl. Phys. Lett,* 73, 975, (1998)

[8] R.P Vaudo, V.M. Phanse, ECS Proceedings Vol 98-18, edited by T.D. Moustakas, S.E. Mohney and S.J. Pearton, 79 (1999)

[9] Dharanipal Doppalapudi, Kyung J. Nam, Anand V. Sampath, Rajminder Singh, Hock Min Ng, S.N. Basu, Theodore D. Moustakas, III-V Nitride Materials and Processes, T.D. Moustakas, S. Mohney, S. Pearton Eds, ECS Proc Vol. 98-18, 87 (1999)

[10] G. M. Smith, J. M. Redwing, R. P. Vaudo, E. M. Ross, J. S. Flynn, V. M. Phanse, *Appl. Phys. Lett,* 75, 25, (1999)

[11] Hock Min Ng, Dharanipal Doppalapudi, Dimitris Korakakis, Rajminder Singh ,Theodore D. Moustakas, *J. Cryst. Growth,* 189, 349, (1998)

[12] T.D. Moustakas, T. Lei, R.J. Molnar, Physica B, 185, 36 (1993)

[13] Dieter K. Schroder, "Semiconductor Material and Device Characterization", 2nd edition, 205, John Wiley and Sons, New York, (1998)

[14] Mira Misra, Anand V. Sampath, Theodore D. Moustakas, Fall MRS Meeting, 1999

[15] Peter P. Chow, Jody J. Klaasen, James V. VanHove, Andrew Wowchak, Christina Polley, David King, SPIE- Optoelectronics 2000 Meeting, 2000

[16] Jin-Kuo Ho, Charng-Shyang Jong, Chien C. Chiu, Chao-Nien Huang, and Kwang-Kuo Shih, Li-Chien Chen, Fu-Rong Chen, and Ji-Jung Kai, *J. Appl. Phys,* 86, 4491 (1999)

Mat. Res. Soc. Symp. Vol. 595 © 2000 Materials Research Society

VERTICAL TRANSPORT PROPERTIES OF GaN SCHOTTKY DIODES GROWN BY MOLECULAR BEAM EPITAXY

M. Misra, A.V. Sampath and T.D. Moustakas
Department of Electrical Engineering and the Photonics Center, Boston University, MA
02215

ABSTRACT

Lateral and vertical electron transport parameters were investigated in lightly doped n-GaN films, grown by MBE. Diodes were fabricated by forming Schottky barriers on n⁻-GaN films using a mesa-etched vertical geometry. Doping concentrations and barrier heights were determined, from C-V measurements, to be $8\text{-}9 \times 10^{16}$ cm^{-3} and 0.95-1.0 eV respectively. Reverse saturation current densities were measured to be in the $1\text{-}10 \times 10^{-9}$ A/cm^2 range. Using the diffusion theory of Schottky barriers, vertical mobility values were determined to be 950 $cm^2/V\text{-}s$. Lateral mobility in films grown under similar conditions was determined by Hall effect measurements to be 150-200 $cm^2/V\text{-}s$. The significant increase in mobility for vertical transport is attributed to reduction in electron scattering by charged dislocations.

INTRODUCTION

Lateral transport in lightly doped n-GaN thin films has been extensively studied, most often by using Hall effect measurements. These studies indicate that electron transport in the lateral direction is dominated by scattering from charged dislocations, which serves to reduce the electron mobility.[1-2] Based on this model, transport in the vertical direction should be relatively unaffected by the presence of the threading dislocations and electron mobility should be significantly increased. Since a large number of GaN devices, including lasers, LEDs, detectors and bipolar junction transistors [3-5], have vertical geometry, it is important to investigate electron mobility in this direction to better understand and engineer device performance.

In this paper we have used Schottky barrier diodes, fabricated with a vertical geometry on lightly doped n-GaN films grown by ECR-MBE, to investigate electron mobility for vertical transport. The diodes were evaluated by I-V and C-V measurements to determine diode ideality, reverse saturation current, doping concentration and barrier height. Using the diffusion model for current transport in Schottky barriers, mobility for vertical transport was determined. The results are compared with lateral electron mobility determined by Hall Effect measurements.

EXPERIMENTAL METHODS

The GaN films were grown on c-plane sapphire, by plasma-assisted molecular beam epitaxy, following procedures described earlier.[5] In this paper we present only a brief description. Prior to introduction into the growth chamber, the substrates were subjected to solvent degreasing. The Al_2O_3 surface was converted to AlN by exposing the substrate, held at 800 °C, to nitrogen plasma using an electron cyclotron resonance (ECR) microwave plasma source. This step was followed by the deposition of 1000 Å

AlN buffer. Two types of devices (Figure 1) were investigated. In the first, 2.25 μm thick n^+ GaN ($2x10^{18}$ cm^{-3}) was deposited first, followed by a n$^-$ GaN ($1x10^{17}$ cm^{-3}) of about the same thickness.The second device consists of 4 μm thick n- GaN ($1x10^{17}$ cm^{-3}). All the layers were grown under Ga rich conditions. RHEED patterns, taken during growth of the n-GaN layers, showed 2x2 surface reconstruction, indicating that the films were grown with the Ga-polarity.[6]

Circular mesas, approximately 3 μm deep and 300 μm in diameter, were defined on both samples by reactive ion etching, using Cl$_2$ gas. Ohmic contacts were defined on the bottom of the mesa by electron beam deposition of 200/1000/200/2500Å Ti/Al/Ni/Au metal multilayers. The Ohmic contacts were rapid thermal annealed at 750°C for 60 seconds, prior to deposition of the Schottky contacts. Schottky contacts were formed at the top of the mesa by patterning 200 μm dots by photolithography and depositing a 200/1000/3000Å Ni/Pt/Au metal multilayer stack by electron beam evaporation. Some of the diodes were mounted on transistor headers and Au wires were bonded to the contacts by thermosonic bonding. Figure 1 shows a schematic illustration of the two types of samples reported in this paper.

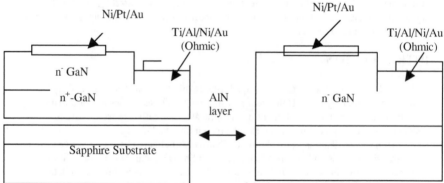

Figure 1: Schematic illustration of the two types of devices investigated.

Hall Effect measurements were performed on the n-GaN films by the Van der Pauw method, using indium contacts in the clover leaf configuration. Current–voltage characteristics were measured using the HP 4155A semiconductor parameter analyzer. Capacitance-voltage measurements were performed using the HP 4275A LCR measurement system. For some of the type II devices, Ohmic contacts were deposited at the top and bottom of the mesas, and their I-V characteristics were evaluated.

EXPERIMENTAL RESULTS

The current voltage characteristics for the structures with Ohmic contacts at both ends are shown in Figure 2. The resistance of the devices was measured to be in the 3-10 Ω range. In order to investigate the origin of this resistance, I-V measurements were performed as a function of temperature. The resistance was found to slightly increase with temperature, while our evidence from planar Hall effect measurements indicate that the resistance decreases with temperature. This suggests that the measured resistance in the vertical direction using two ohmic contacts is dominated by the contacts rather than the bulk GaN resistance. From the data of Fig. 2, the specific contact resistivity was estimated to be on the order of $10^{-3} \, \Omega*cm^2$. Such values of contact resistivity for lightly doped GaN layers are quite typical [7]. This observation has important implications for vertical devices with high-mobility layers because it indicates that in order to obtain the maximum current carrying capacity of the active layer, the Ohmic contact resistance must be brought down by several orders of magnitude.

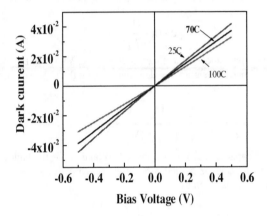

Figure 2: I-V characteristics of veritical GaN resistors measured as described in the text at different temperatures.

Typical current voltage characteristics of the two types of Schottky barrier diodes investigated are shown in Figure 3.

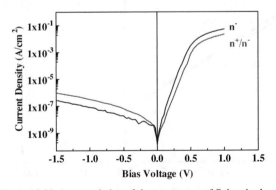

Figure 3: Typical I-V characteristics of the two types of Schottky barrier diodes

In the forward direction the data can be fitted to Eq. 1. where J_o is the saturation current density and n is the ideality factor. Applying Eq. 1 to the data of Figure 3, we find that the diodes are practically ideal (n=1.15-1.2) and the value of the reverse saturation current densities for the n^+/n^- and n^--GaN samples are 1.08×10^{-9} A/cm^2 and 1.0×10^{-8} A/cm^2 respectively. A number of devices were probed on each wafer, and found to give very good uniformity over an area approximately equal to a quarter of a 2" wafer.

$$J = J_0 \exp(\frac{qV}{nkT}) \tag{1}$$

Results of the capacitance-voltage measurements are shown in Fig. 4. The doping concentration and barrier height were determined from the C-V measurements, using the following equation [8]:

$$\frac{1}{C^2} = \frac{2}{A_e^2}\left(\frac{(V_{bi} - V)}{q\varepsilon_s N_d}\right) \tag{2}$$

where C is the measured capacitance, V is the applied voltage, V_{bi} is the built-in voltage, ε_s is the dielectric constant, N_d is the doping concentration and A_e is the geometric area of the device. The barrier height is given by

$$\phi = \left(V_{bi} + \left(\frac{kT}{q}\right)\ln\left(\frac{N_c}{N_d}\right)\right) \tag{3}$$

where N_c is the effective density of states in the conduction band.

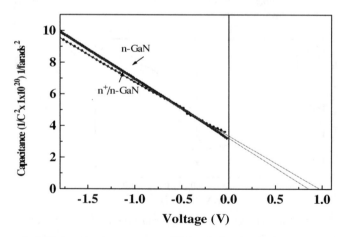

Figure 4: C-V plots for the two types of Schottky barrier diode structures investigated.

The doping concentration and barrier height obtained by fitting the data of Fig. 4 to Eqs. (2) and (3), were determined to be 9×10^{16} cm^{-3} and 1.0 V for the n$^+$/n$^-$-GaN diode, 8.4×10^{16} cm^{-3} and 0.95 V for the n$^-$-GaN diode. These results are in general agreement with those reported earlier. [9-10] A number of devices, fabricated on the same wafer were tested and were found to give similar values of barrier heights, doping concentration and reverse saturation current densities.

DISCUSSION

The carrier concentration in the two types of samples was determined by measuring the Hall effect on the n$^-$-GaN films as well as by measuring the capacitance voltage characteristics of both types of Schottky barrier diodes.

From the Hall effect measurements, the carrier concentration of the n$^-$-GaN film was determined to be 1.4×10^{17} cm^{-3} and the lateral mobility (μ_l) was 160 cm^2/V-s. Although the dislocation densities in these particular samples have not been measured, from measurements on samples grown under similar conditions, the threading dislocation density is estimated to be in the range of $5\text{-}10 \times 10^9$ cm^{-2}. As discussed previously, the lateral electron mobility in our samples attains its highest value at 300-400 cm^2/V-s at carrier concentrations of $3\text{-}5 \times 10^{17}$ cm^{-3} and then it reduces at lower carrier concentrations. This result was accounted for by scattering from negatively charged dislocations [1,2].

In order to investigate the electron mobility for vertical transport, the diffusion theory of Schottky barriers, which expresses the relation between the reverse saturation current as a function of barrier height and the diffusion coefficient of electrons in the space charge region was employed. According to this model, the reverse saturation current density is given by the relation [11]

$$ J_0 = \left\{ qN_c \mu_v \left[\frac{2qV_{bi}N_d}{\varepsilon_s} \right]^{1/2} \right\} \exp\left(\frac{-q\phi}{kT} \right) \tag{4} $$

where μ_v is the electron mobility in the vertical direction. Using the values of J_0, N_d, ϕ and V_{bi} in Eq. (4), we obtain a vertical mobility of μ_v=932 cm^2/V-s for the n$^+$/n$^-$GaN and μ_v =949 cm^2/V-s for the n$^-$-GaN film.

In the calculation of the vertical mobility, we have assumed that the active area is equal to the geometric area of the device. A reduction of the active area (up to one-third for dislocation density of 5×10^9 cm^{-2} and doping concentrations of 1×10^{17} cm^{-3}) is expected due to depletion around each charged dislocation. Specifically, considering the active area to be two- thirds the geometric area, increases the doping concentration and reverse saturation current density to be 1.9×10^{17} cm^{-3}, 1.62×10^{-9} A/cm^2 respectively for the n$^+$/n$^-$ device. This leads to a vertical mobility of 1070 cm^2/V-s for the same device. These results are in good agreement with Monte-Carlo calculations for electron mobility in GaN thin films.[2]

CONCLUSIONS

Investigation of lateral and vertical transport in n$^-$-GaN films indicates that for samples doped at the level of 1×10^{17} cm^{-3}, the vertical mobility is more than six times higher than the lateral mobility. This result is attributed to the reduction of electron scattering by charged dislocations during vertical transport.

ACKNOWLEDGEMENTS

We wish to thank Prof. C.R Eddy and Dr. N.G. Weimann for useful discussions. We are grateful to Dr. P.A. Lamarre and Mr. D. Parker of Lockheed-Martin for help with device fabrication and C-V measurements and Mr. P. Mak of the Photonics Center at Boston University for help with packaging of the devices. The authors are grateful to Ms. Worthen and Clariant Corporation for assistance with their photoresist products. The device processing for this work was done in the Optoelectronics Processing Facility at the Boston University Photonics Center. This work was supported by ONR under grant # N00014-99-1-0453, monitored by Dr. J. C. Zolper.

REFERENCES

1. H.M. Ng, D. Doppalapudi, T.D. Moustakas, N.G. Weimann and L.F. Eastman, Appl. Phys. Lett., 73, 821, (1998)
2. N.G. Weimann, L.F. Eastman, D. Doppalapudi, H.M. Ng and T.D. Moustakas, J. Appl. Phys. 83, 3656, (1998)
3. S. Nakamura, in Gallium Nitride I edited by J. Pankove, T.D. Moustakas,, Semiconductors and Semimetals, Vol.50, Chapter 14, 431, Academic Press, New York, 1998
4. M.S. Shur and M. Asif Khan, in Gallium Nitride I edited by J. Pankove, T.D. Moustakas,, Semiconductors and Semimetals, Vol.57, Chapter 10, 407, Academic Press, New York, 1998
5. T.D. Moustakas, in Gallium Nitride I edited by J. Pankove, T.D. Moustakas,, Semiconductors and Semimetals, Vol.57, Chapter 2, Academic Press, New York, 1998
6. A.R. Smith, R.M. Feenstra, D.W. Greve, M-S Shin, M. Skowronski, J. Neugebauer, J.E. Northrup, Appl. Phys. Lett., 72, 2114, (1998)
7. A.V. Sampath, M.S. Thesis, Boston University, (1996)
8. S.M. Sze, "Physics of Semiconductor Devices", 2nd edition, 258, John Wiley and Sons, New York, (1981)
9. A.C. Schmitz, A.T. Ping, M. Asif Khan, Q. Chen, J.W. Yang, I. Adesida, Semicond. Sci. Technol. 11, 1464, (1996)
10. L. Wang, M.I. Nathan, T.H. Lim, M.A. Khan, Q. Chen, Appl. Phys. Lett., 68, 1267, (1996)
11. E. Spenke, Electronic Semiconductors, (McGraw-Hill, New York, 1958), 84

Mat. Res. Soc. Symp. Vol. 595 © 2000 Materials Research Society

Pulsed-laser-deposited AlN films for high-temperature SiC MIS devices

R. D. Vispute, A. Patel, K. Baynes, B. Ming, R. P. Sharma, and T. Venkatesan
CSR, Department of Physics, University of Maryland, College Park, MD 20742.

C. J. Scozzie, A. Lelis, T. Zheleva, and K. A. Jones
United States Army Research Laboratory, Adelphi, MD 20783.

ABSTRACT

We report on the fabrication of device-quality AlN heterostructures grown on SiC for high-temperature electronic devices. The AlN films were grown by pulsed laser deposition (PLD) at substrate temperatures ranging from 25 °C (room temperature) to 1000 °C. The as-grown films were investigated using x-ray diffraction, Rutherford backscattering specttroscopy, ion channeling, atomic force microscopy, and transmission electron microscopy. The AlN films grown above 700 °C were highly c-axis oriented with rocking curve FWHM of 5 to 6 arc-min. The ion channeling minimum yields near the surface region for the AlN films were ~2 to 4%, indicating their high degree of crystallinity. TEM studies indicated that AlN films were epitaxial and single crystalline in nature with a large number of stacking faults as a results of lattice mismatch and growth induced defects. The surface roughness for the films was about 0.5 nm, which is close to the unit cell height of the AlN. Epitaxial TiN ohmic contacts were also developed on SiC, GaN, and AlN by in-situ PLD. Epitaxial TiN/AlN/SiC MIS capacitors with gate areas of $4 * 10^{-4}$ cm^2 were fabricated, and high-temperature current-voltage (I-V) characteristics were studied up to 450 °C. We have measured leakage current densities of low 10^{-8} A/cm^2 at room temperature, and have mid 10^{-3} A/cm^2 at 450°C under a field of 2 MV/cm.

INTRODUCTION

Wide Band Gap (WBG) semiconductors are important for the development of high-temperature and high-power electronics. For these applications, SiC, GaN, and their heterostructures have been identified as potential candidates due to their wide band gaps, desirable electronic and optical properties, thermal and chemical stability, doping capabilities and the possibility of their use in fabricating high quality heterostructures and devices [1-2]. The most significant challenge in this area is the fabrication of high quality metal-oxide-semiconductor structures. High quality thin dielectric films having low leakage currents at temperatures up to 450°C under operating fields of about 1 MV/cm are necessary for the high temperature electronics. Thin films of AlN seem to be promising for these applications compared to other materials such as SiO$_2$. An additional advantage of using AlN on SiC or GaN is the very small lattice mismatch (~5%). Although AlN has been investigated previously [3-5], its full potential as a dielectric material has not been successfully demonstrated due to many problems associated with the defects such as columnar structures, threading dislocations, and low angle grain boundaries. These defects are responsible for leakage currents at room temperature as well as at elevated temperatures.

In general, the lattice and structural mismatches and the film-substrate interfaces are the dominating parameters governing the microstructure of the AlN films and hence their properties. Conventional growth techniques such as chemical vapor deposition

(CVD) and molecular beam epitaxy (MBE) are based on the equilibrium growth conditions and produce columnar structures in the III-V nitrides on SiC and sapphire. A highly non-equilibrium growth process is desirable to circumvent these growth problems. For example, energetic processes involved in the non-equilibrium growth conditions can be utilized to form stacking faults in the plane of the interface to manage the lattice mismatch induced strain rather than the formation of vertical threading dislocations, or columnar structures in the films. Such energetic and non-equilibrium growth conditions can be achieved, for example, in Pulsed Laser Deposition (PLD) [6]. Such a process can be utilized in the early stages of growth of the dielectric layer without the formation of columnar structures or vertical grain boundaries. In this paper, we highlight the growth of AlN films by PLD, their characterization, and their application for the development of high-temperature SiC based thyristors. Our major objective is to develop and fabricate thyristors based on SiC, with integration of other wide band gap materials suitable for high-temperature operation (300-500 °C). In this context, we have studied the suitability of the PLD technique for the fabrication of AlN thin films for encapsulation, passivation, and as a dielectric layer on SiC.

EXPERIMENTAL

The schematic of the PLD process is shown in Fig. 1. A KrF excimer laser ($\lambda = 248$ nm, $\tau = 25$ ns) was used for the ablation of a polycrystalline, stoichiometric AlN target (99.99 purity) at an energy density of ~1 J/cm^2. Upon laser absorption by the target surface, a strong plasma plume is produced as shown in Figure 1. The laser-induced plasma consists of atoms, molecules, excited atomic and molecular species, and clusters. The laser ablated species are allowed to condense on to a substrate kept at suitable temperature. The NH$_3$ background gas pressure in our experiment was varied from 10^{-6} to 10^{-3} Torr. The deposition rate and the film thickness were controlled by the pulse repetition rate (5 to 10 Hz) and total deposition time (30 to 60 min.). The PLD films were characterized by four-circle x-ray diffraction (XRD), atomic force microscopy (AFM), UV-visible spectroscopy, Rutherford backscattering spectrometry (RBS) and ion channeling, transmission electron microscopy (TEM), and electrical transport (I-V) measurements. TEM studies were performed using JEOL 2010 system operated at 200KV.

RESULTS AND DISCUSSION

The crucial parameters in the PLD of AlN dielectric films are the deposition temperature and laser fluence. We have studied the dependence of these parameters on the crystalline quality, surface morphology, and the optical and electrical properties of the III-V nitride films [6]. We found that AlN grows epitaxially on sapphire, SiC, and GaN at a substrate temperature as low as 600 °C. However, the crystalline quality of these films improves with an increase in the substrate temperature. High epitaxial quality was obtained when the films were grown under a NH$_3$ background gas pressure of 5×10^{-5} Torr and a substrate temperature of 850 to 1000 °C. The XRD θ–2θ angular scans of a 2000-5000 Å thick film clearly show only a {000l} family of the planes of wurtzite-AlN with full-width-at-half-maximum (FWHM) of the rocking curve (ω) for the (0002) peak of about 5 to 7 arc-minutes. The quantitative analysis of the crystalline quality, composition, and interface structure of the III-V nitride films was carried out by RBS and ion-channeling techniques. The ratio of the RBS yield with the He$^+$ beam incident along [0001] (channeled) to that of a random direction, respectively, (χ_{min}), reflects the epitaxial quality of the film. Figures 2 show the aligned and random backscattering spectra for the AlN films. The χ_{min} near the surface region of the films is ~3%, indicating a high degree of crystallinity. This χ_{min} value is close to that of high quality MOCVD and MBE films.

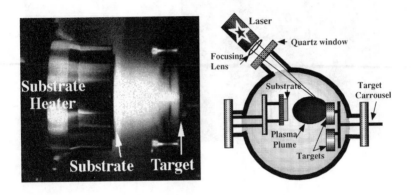

Figure 1. Laser induced plasma plume of III- nitrides and schematic of the PLD.

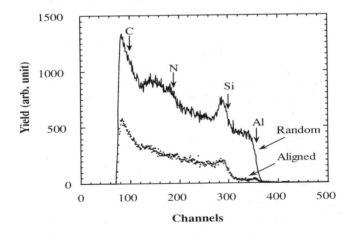

Figure 2. RBS and ion channeling spectra of PLD AlN dielectric film on 6H-SiC (0001).

measured by AFM for AlN (Figure 3) was ~ 0.5 to 2 nm. The mean surface roughness of 0.5 nm obtained for the AlN film grown at 1000°C is essentially the unit cell height of the AlN. The high vacuum conditions have also enabled us to grow pinhole free, uniform films with low lattice stress which are crucial factor for the encapsulation of ion implanted SiC and the fabrication of device quality dielectric layers.

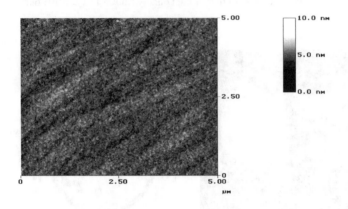

Figure 3. AFM of PLD AlN thin film deposited on SiC.

Microstructural characteristics, epitaxy, and interfaces were characterized using high-resolution transmission electron microscopy. TEM studies indicated that the PLD AlN films grown above 700 °C were epitaxial and single crystalline. As-grown films on 6H-SiC are essentially single crystalline of good crystallographic quality as revealed by selective area diffraction (SAD) pattern (not shown) and the contrast from the micrographs in imaging mode. Observation in low magnification of the AlN film also showed uniform contrast indicating high crystalline quality of the film. Figure 3 shows high-resolution TEM lattice image of AlN/SiC interface. In the HRTEM investigation, we observed numerous stacking faults than the threading dislocations. This is in contrast to the columnar structures usually observed for the III-V nitrides grown by CVD and MBE. At the present juncture, we feel that the growth process in the pulsed laser deposition of AlN is away from equilibrium conditions due to high kinetic energy and instantaneously high arrival rate of the laser ablated particles. Note also that the interface with the 6H-SiC substrate is clean, smooth, and shows no indication of amorphous phase.

Epitaxial TiN gate electrodes with low resistivity were fabricated using pulsed laser deposition of TiN at 600C [22]. The TiN/AlN/SiC MIS capacitors with gate areas of $4x10^{-4}$ cm^2 were patterned using ion milling, and high-temperature current-voltage (I-V) characteristics and were studied up to 450 °C. The thickness of the AlN films used as a dielectrics was varied from 200 nm to 600 nm. A typical substrate for these studies was research grade Si face-6H-SiC, 3.5° off and highly doped ($4x10^{17}$cm^{-3}) n-type. We measured leakage current densities of around 10^{-8} A/cm^2 at room temperature and of around $7*10^{-3}$ A/cm^2 at 450 °C for a 1.7 MV/cm field, as shown in Fig. 4. These leakage currents are orders of magnitude lower than values that have ever been previously reported for AlN thin films. The mechanism of the leakage currents in PLD AlN films at high temperature is discussed in our earlier papers [7-8]. It is now interesting to note that the microstructure of the PLD AlN described earlier is responsible for the low leakage current densities.

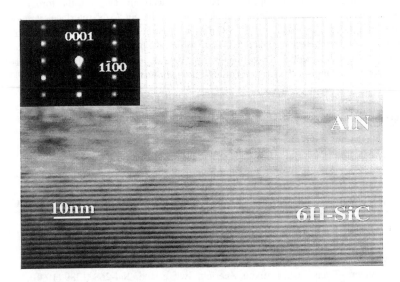

Figure 4. HRTEM of PLD AlN dielectric film on 6H-SiC(0001) substrate.

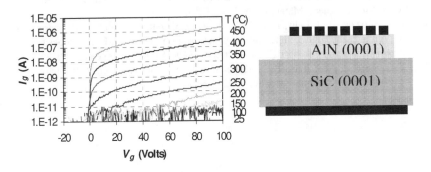

Figure 5. I-V characteristics and schematic of PLD TiN/AlN/SiC MIS capacitors

We have shown that high-quality thin films of AlN can be fabricated by PLD at substrate temperatures (750 to 850 °C) which are lower than those employed in (MOCVD) (1000 to 1100 °C), an alternate growth method. These films show a high degree of crystallinity and good optical properties. Most importantly, the PLD AlN films show leakage current densities below 10E-08 A/cm^2 at room temperature and about 5x10

[3] A/cm^2 at 450°C under 2 MV/cm accumulation field. Their properties of PLD AlN are excellent for its applicability as a dielectric for SiC thyristors. The interfaces between the AlN and SiC were sharp. PLD is also used successfully for epitaxial growth of TiN ohmic contacts. It is demonstrated that the PLD AlN films on SiC substrates do not have the usual columnar structure that allows grain boundaries to propagate parallel to the c-axis. Rather, the mismatch strain appears to be mostly accommodated by the formation of stacking faults. Since these defects are parallel to the SiC/AlN interface, it may be possible to fabricate high quality MIS SiC device structures. After the growth of a critical thickness of the dielectric layer without forming columnar defects, conventional deposition processes can be utilized for the large area depositions, integration, and in-situ multilayer processing for device fabrications and manufacturing.

Acknowledgements: Funding support from ARL, Adelphi, MD, is acknowledged.

References:

[1] S. Nakamura and G. Fasol, *The Blue Laser Diode* (Springer, Berlin, 1997).

[2] S. Strite and H. Morkoc, J. Vac. Sci. Technol. B **10**, 1237 (1992).

[3] C. M. Zetterling, K. Wongchotigul, M. G. Spencer, C. I. Harris, S. S. Wong, and M. Ostling, Mater. Res. Soc. Symp. Proc. **423**, 667 (1996).

[4] C. M. Zetterling, M. Ostling, K. Wongchotigul, M. G. Spencer, X. Tang, C. I. Harris, N. Nordell, and S. S. Wong, J. Appl. Phys. **82**, 2990 (1997).

[5] T. Ouisse, H. P. D. Schenk, S. Karmann, and U. Kaiser, in *Proceedings of the International Conference on Silicon Carbide III-Nitrides and Related Materials-1997*. Materials Science Forum, edited by G. Pensl, H. Morkoc, B. Monemar, and E. Janzen ~Trans Tech, Uetikon, Zuerich, Switzerland. **264-268**, 1389 (1998).

[6] R. D. Vispute, S. Choopun, R. Enck, A. Patel, V. Talyansky, R. P. Sharma, T. Venkatesan, W.L. Sarney, L. Salamanca-Riba, S.N. Andronescu, and A. A. Iliadis, K. A. Jones, J. Elect. Materials, **28** 275 (1999).

[7] C. J. Scozzie, A. J. Lelis, B. F. McLean, R. D. Vispute, A. Patel, R. P. Sharma, and T. Venkatesan."High Temperature Characterization of Pulsed-Laser-Deposited AlN on 6H-SiC from 25 to 450°C." submitted to ISCRM 99 Conference (1999).

[8] C. J. Scozzie, A. J. Lelis, B. F. McLean, R. D. Vispute, A. Patel, R. P. Sharma, and T. Venkatesan, J. Appl. Phys. **86**, 4052 (1999).

Mat. Res. Soc. Symp. Vol. 595 © 2000 Materials Research Society

FABRICATION AND CHARACTERIZATION OF METAL-FERROELECTRIC-GAN STRUCTURES

W.P.Li, R.Zhang, J.Yin, X.H.Liu, Y.G.Zhou, B.Shen, P.Chen, Z.Z. Chen, Y.Shi, R.L.Jiang, Z.G.Liu, Y.D.Zheng
National Laboratory of Solid State Microstructure and Department of Physics, Nanjing University, Nanjing 210093, China
Z.C.Huang
Raytheon ITSS, 4500 Forbes Boulevard, Maryland 20771

ABSTRACT

GaN-based metal-ferroelectric-semiconductor (MFS) structure has been fabricated by using ferroelectric $Pb(Zr_{0.53}Ti_{0.47})O_3$ (PZT) instead of conventional oxides as gate insulators. The GaN and PZT films in the MFS structures have been characterized by various methods such as photoluminescence (PL), wide-angle X-ray diffraction (XRD) and high-resolution X-ray diffraction (HRXRD). The Electric properties of GaN MFS structure with different oxide thickness have been characterized by high-frequency C-V measurement. When the PZT films are as thick as 1 μm, the GaN active layers can approach inversion under the bias of 15V, which can not be observed in the traditional GaN MOS structures. When the PZT films are about 100 nm, the MFS structures can approach inversion just under 5V. All the marked improvements of C-V behaviors in GaN MFS structures are mainly attributed to the high dielectric constant and large polarization of the ferroelectric gate oxide.

I. INTRODUCTION

Recently, the semiconductor gallium nitride (GaN) has been recognized for several decades for their potential and, recently, commercial viability in wide band optoelectronic device applications [1]. The electronic devices that can be used for high power and high temperature applications have also been fabricated, among which GaN-based MOSFETs are actively studied [2]. The traditional GaN MOS structures fabricated with conventional oxides, like SiO_2 [3,4], Si_3N_4 [3], $Ga_2O_3(Gd_2O_3)$ [4], as insulators have to be improved. One important reason is that large applied voltage, which is incompatible with most other electronic devices, is imposed on all the previous GaN MOS structures. In our work, ferroelectric oxides have been used in GaN MOS structures to address this problem because of large polarization provided by ferroelectric and the high dielectric constant of ferroelectric gate.

Since the field-effect transistors (MFSFETs) was first proposed by Wu [5, 6], it has been extensively studied because of their applications in non-volatility and high speed memories and integration circuits [7-10]. Nowadays, the semiconductor used in metal-ferroelectric-semiconductor (MFS) structures usually is Si. The instability of the ferroelectric/Si interface due to interdiffusion between ferroelectrics and Si substrates has still impeded the development of the novel device [8]. Consequently, many kinds of buffer layers have been deposited between ferroelectrics and Si substrates to prevent the behavior of interdiffusion [11-13], which also decreases the control of ferroelectric polarization on the potential of Si surface. GaN is so stable that it can work without weight loss at 1000 ^0C[14]. GaN MFS structure is a good candidate to develop ferroelectric/semiconductor interface and MFSFETs with high-temperature stability.

II. EXPERIMENTS

In the GaN MFS structure developed in this work, n-type GaN active layer is grown by light radiant heating low-pressure metalorganic chemical vapor deposition (LRH-LP-MOCVD) on the (0001) oriented sapphire (Al_2O_3) substrate which is pre-cleaned by organic solvents and $H_2SO_4 : H_3PO_3(3:1)$ solutions [15,16]. Before the epitaxial growth of GaN layer, a thin GaN buffer (about 30nm) was deposited at 520 ^{0}C on sapphire substrate. Then the GaN active layer ($n \sim 10^{17} cm^{-3}$) was deposited at 1040 ^{0}C. Afterwards, the ferroelectric $Pb(Zr_{0.53}Ti_{0.47})O_3$ (PZT) films with different thickness have been deposited directly on GaN films by pulsed laser deposition (PLD). The experimental condition is showed in the Table 1.

After the deposition of PZT, the samples are annealed in-situ in the chamber with 0.5 atm. O_2 for 30 minutes. In the following process, very thin SiO_2 films are deposited on thin PZT films (about 130 nm) by plasma enhanced chemical vapor deposition (PECVD) to decrease the leakage current. Then the top electrodes (about 200 nm) are fabricated by magnetron sputtering and patterned with a shadow mask of holes 0.2mm in diameter, while the bottom electrodes are contacted with aluminum (Al) from the edge of GaN top surface. Finally, the whole samples are annealed at the temperature of 600 ^{0}C after which excellent ohmic contacts on GaN are formed.

Table I. The experimental condition in the deposition
of PZT by pulsed laser deposition (PLD)

Item	Condition
Target	ZrO_2, TiO_2 and PbO
Substrate	n-type GaN(0001)
Substrate temperature	750 ^{0}C
Deposition time	5~25 minutes
Laser wavelength	248 nm (KrF excimer)
Laser frequency	5 Hz
Laser power	2.5J/cm^2
ambient	O_2 (20 Pa)

III. DISCUSSION

The GaN films deposited by LRH-LP-MOCVD have been characterized by photoluminescence (PL) and high-resolution x-ray diffraction (HRXRD) methods. In the PL spectra, strong band-edge luminescence has been observed and undesirable YL peaks do not exist. The full width at half maximum (FWHM) of GaN in the HRXRD pattern is 8.6 min.

The PZT films directly on GaN (0001) are characterized by X-ray diffraction (XRD). The results show that the PZT films are deposited along the orientations of <100>, <110>, <211>

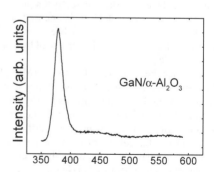

Figure 1 A typical PL spectrum of GaN on sapphire by LRH-LP-MOCVD.

and <111>, among which the <110> peak is the strongest. No undesirable peak of pyrochlore phase appears in the XRD pattern, which shows that PZT films are well crystallized with perovskite structure.

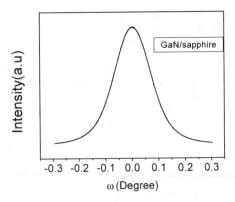

Figure 2 A typical HRXRD spectrum GaN on sapphire grown by LRH-LP-MOCVD.

The high-frequency capacitance-voltage (C-V) measurement is an important characterization method of MOS structures. When the applied bias are large enough to make the GaN surface approach depletion, the bias voltage (V_b) has been distributed between gate insulator (V_i) and depletion layer (V_d):

$$V_b = V_i + V_d \qquad (1)$$

The GaN MOS structures under the bias can be expressed as the two capacitors, the insulator capacitor (C_i) and depletion capacitor (C_d), in series. The voltage and capacitance distributions have the following relationship:

$$\frac{V_d}{V_i} = \frac{C_i}{C_d} \qquad (2)$$

Therefore, the bias voltage can be expressed into the following form:

$$V_b = (1 + {C_d}\big/{C_i})V_d \qquad (3)$$

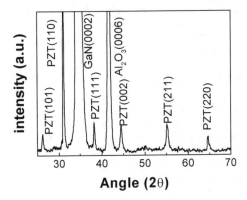

Figure 3. An X-ray spectrum of PZT directly on GaN (0001) by pulsed laser deposition.

As to a specific semiconductor sample, the V_d is a constant. Therefore, in order to decrease the bias voltage (V_b), we have to increase the insulator capacitance (C_i). In traditional GaN MOS structures with SiO_2 or Si_3N_4 as gate oxides, the effort made to decrease the large applied bias has focused on reducing the thickness of gate oxides, which results in the increase of C_d. However, these approaches to decrease the bias voltage are not desirable. It is very difficult to allow GaN to approach inversion under bias of 15 volts although the oxide insulators are thinner than 60nm [4].

In GaN MFS structures, the insulator capacitance (C_i) has been greatly increased because of the high dielectric constant of ferroelectrics. By using the ferroelectric oxides as gate insulators, we need not reduce the oxide thickness to increase the capacitance.

Moreover, in the MFS structures, there are polarization fields (on the order of 10^6 V/cm) much larger than external applied field (on the order of $10^{-4} \sim 10^{-5}$ V/cm) at the ferroelectric/semiconductor interfaces. Figure 4 shows the C-V behavior of the GaN MFS structures with 1 μ PZT films. We can find that the GaN MFS structures with thick gate oxides can approach inversion under the bias of 15 volts.

The GaN MFS structures with 100 nm PZT films have also been characterized by high-frequency C-V method. In Figure 5, we can find the applied bias has been decreased sharply. The GaN active layer can approach inversion just under 5V, which can satisfy the practical need of the GaN MOS structures.

IV. CONCLUSION

In our work, GaN-based metal-ferroelectric-semiconductor (MFS) structures have been fabricated by using the ferroelectric oxide $Pb(Zr_{0.53}Ti_{0.47})O_3$ as gate insulators. The electrical properties of GaN MFS structures with different oxide thickness have been studied by the high-frequency C-V measurement. Due to the high dielectric constant and large polarization field of PZT films, the large applied voltage on conventional GaN MIS structures has been decreased sharply with comparison to that of traditional GaN MOS structures.

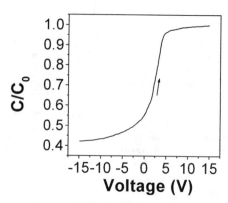

Figure 4 The C-V behaviors of GaN MFS structures with the 1 μ PZT films. The structures can approach reversion under 15 V, which is better than the C-V behaviors of GaN MOS structures with the same thick SiO_2 films.

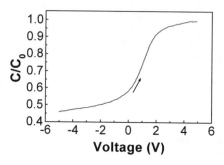

Figure 5 The C-V behaviors of GaN MFS structures with the 100nm PZT films. The structures can approach inversion just under small bias.

REFERENCES

1. T. Egawa, T. Jimbo, and M. Umeno, J. Appl. Phys. **82**, 5816 (1997).
2. F. Ren, M. Hong, S. N. G. Chu, M. A. Marcus, M. J. Schurman, A. Baca. Appl. Phys. Lett. **73**. 3893. (1998).

3. S. Arulkmaran, T. Jimbo, H. Ishikawa, S. Arulkumaran, T. Egawa. Appl. Phys. Lett. **73**. 809. (1998)
4. H. C. Casey. Jr., G. Gfoutain, R. G. Alley, B. P. Keller, and Steven P.DenBarrs. Appl. Phys. Lett. **68**. 1850. (1996).
5. S. Y. Wu, IEEE Trans. Electron Dev. **ED-21**, 499 (1973).
6. S. Y. Wu, Ferroelectrics **11**, 379 (1976).
7. Timonthy A. Rost, He Lin, and Thomas A. Rabson Appl. Phys. Lett. **59**, 3654(1991).
8. Marin Alexe. Appl. Phys. Lett. **72**, 2283. (1998)
9. Junji Senzaki, Koji Kurihara, Naoki Namura, Osamu Mitsunaga, Yoshitaka Iwasaki and Tomo Ueno. Jpn. J. Appl. Phys. **37**, 5150. (1998)
10. Takeshi Kijima and Hironori Matsunaga, Jpn. J. Appl. Phys. **37** (1998) 5171.
11. Junji Senzaki, Koji Kurihara, Naoki Nomura, Osamu Mitsunaga, Yoshitaka Iwasaki and Tomo Ueno, Jpn. J. Appl. Phys. **37** (1998) 5171.
12. Takeshi Kijika, Sakiko Satoh, Hironori Matsunaga and Masatoshi Koba, Jpn. J. Appl. Phys. **35** (1995) 1246.
13. Joon Lee, Young-Chul Choi, and Byung Lee, Jpn. J. Appl. Phys. **36** (1997) 3644
14. Y. Morimoto. J. Electrochem. Soc. **121**, 1381. (1974)
15. B. Shen, Y. G. Zhou, P. Chen, Z. Z. Chen, L. Zang, R. Zhang, Y. Shi. Y. D. Zheng. Appl. Phys. A **68**, 593. (1999)
16. Y. G. Zhou, B. Shen, Z. Z. Chen, P. Chen, R. Zhang, Y. Shi, Y. D. Zheng. Chinese Journal of Semiconductor **20**, 147. (1999)

Mat. Res. Soc. Symp. Vol. 595 © 2000 Materials Research Society

Growth and Characterization of Piezoelectrically Enhanced Acceptor-Type AlGaN/GaN Heterostructures

A. Michel, D. Hanser*, R.F. Davis*
Dept. Chemical Engineering, *Dept. Materials Science and Engineering
North Carolina State University, Raleigh, NC 27695

D. Qiao, S.S. Lau, L.S. Yu, W. Sun, P. Asbeck
Dept. Electrical and Computer Engineering
University of California San Diego, LaJolla, CA 92093-0407

ABSTRACT

Acceptor (Mg)-doped AlGaN/GaN heterostructures were grown via MOVPE and compared to similarly doped GaN standard films grown in the same reactor. Chemical analysis of the films, via secondary ion mass spectrometry (SIMS), revealed comparable Mg concentrations of ~$2x10^{19}$ atoms/cm^3 in all films. The Mg-doped GaN standard sample had a sheet conductance of 7-μS compared to a sheet conductance of 20-μS for an AlGaN/GaN heterostructure. The sheet conductance of the AlGaN/GaN heterostructures was higher due to piezoelectric acceptor doping and modulation doping effects in addition to conventional Mg acceptor doping.

INTRODUCTION

Heterostructures of AlGaN and GaN are being extensively investigated and employed in the fabrication of high electron mobility transistors (HEMT) and heterojunction bipolar transistors (HBT) for high-frequency and high-power applications including microwave amplifiers and compact and efficient power supplies. Prior HEMT research has confirmed the existence of a two-dimensional electron gas (2DEG) at the AlGaN/GaN heterointerface [1]. Strong piezoelectric effects induce a 2D-gas at the AlGaN/GaN interface without doping. Electron sheet concentrations as high as $3X10^{13}$ cm^{-2} have been reported [2], [3].

The development of GaN-based HBTs has received recent attention for high power/switching devices. The intrinsic high breakdown electric field of GaN should permit a GaN collector drift region that can be either much thinner and/or doped much higher relative to other semiconductors with smaller breakdown strengths (about a factor 30 compared to Si) [4]. Recent reports regarding the electrical properties of GaN/AlGaN HBTs indicate that the resistance in the extrinsic base region is too high to obtain satisfactory common-emitter characteristics [5]. It is desirable to have a p-type base region and the development of nitride-based HBTs centers on developing a more conductive base layer.

Acceptor-type doping of GaN materials has been problematic due in large part to passivation of Mg dopants by H. It has been established that H enhances the incorporation of dopants such as Mg, but the Mg-H bonds must be broken by post-growth annealing to achieve activation [6], [7]. However, the large piezoelectric effects from AlGaN/GaN heterostructures can potentially contribute significant conductivity in the base material relative to conventional Mg-doping.

This paper examines the extent of piezoelectric acceptor doping and modulation doping and compares these results with conventional Mg acceptor doping in AlGaN/GaN system. Material characterization including the Mg concentration profile and AlGaN/GaN interface, as determined by secondary electron mass spectrometry (SIMS), are discussed in relation to the sheet conductivity of the films.

EXPERIMENTAL PROCEDURE

Each heterostructure consisted of a GaN (0001) film grown at 1000°C on an $Al_xGa_{1-x}N$ (0001) layer. The latter was deposited at 1020°C on an AlN (0001) buffer layer previously deposited at 1100°C on an on-axis 6H-SiC (0001) substrate. Figure 1 shows the doping and Al profiles of the GaN and $Al_xGa_{1-x}N$ layers in selected samples.

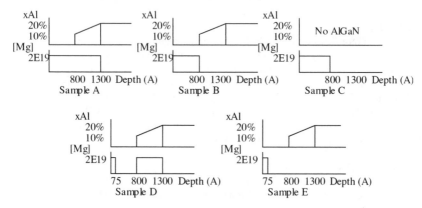

Figure 1. Doping and Al profiles in selected GaN/$Al_xGa_{1-x}N$ heterostructures. The zero point of the depth scale is the top of the GaN layer.

The AlN buffer layer and all subsequent films were grown in a cold-wall, vertical, pancake-style, RF inductively heated metalorganic vapor phase epitaxy (MOVPE) system. Ammonia (NH_3), triethylaluminum (TEA) and triethylgallium (TEG) were used as precursors. Bis-cyclopentadienyl-magnesium (Cp_2Mg) was employed for the p-type doping. High-purity H_2 was used as both the carrier and the diluent gas. After cooling to room temperature the samples were annealed at 800°C in N_2 to activate the Mg acceptors. Additional details of the growth experiments in the NCSU reactor have been previously reported [8].

Sample A contains all doping contributions: piezoelectric (PZ), modulation doping (MD) and conventional acceptors (ACC). Sample B lacks the modulation doping component. Sample C is an Mg-doped GaN film and is the control sample. It has no PZ charge or modulation doping. Sample D has doping from MD and PZ. Sample E has only PZ doping.

The Mg concentration profiles and the Al concentrations were determined using Secondary Ion Mass Spectrometry (Cameca IMS-6f) having a 100nA 10keV O_2^+ primary beam. Sputtering rates and Mg sensitivity factors for varying Al concentrations have been previously determined [9]. Analysis of the AlGaN/GaN interface was conducted by comparing the Al and Ga signals. Electrical measurements of the films were conducted using CV and Hall measurements (van der Pauw configuration). The Au(400A)/Ni(600A) ohmic contacts were produced by evaporization of the individual metals and subsequently alloying at 650ºC for 20 minutes. The sheet resistance and the Hall mobility were determined for each layer.

RESULTS AND DISCUSSION
      Magnesium depth profiles determined in Samples A-E are shown in Figure 2.

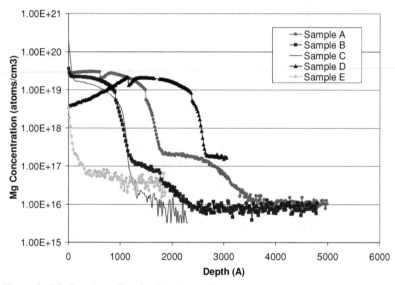

Figure 2. Mg Depth profiles for Samples A-E

The Mg concentration in the GaN and the $Al_{0.1}Ga_{0.9}N$ layers ~ 2×10^{19} atoms/cm^3. A difficulty with Mg doping is the memory effect in the reactor due to the adsorption of the metal-organic precursor [10]. This leads to slightly higher conductivity from Sample D than expected from a more ideal profile, but does not affect the analysis significantly. The contribution of the polarization to the effective doping contains both a part associated with spontaneous polarization, and a part associated with strain (piezoelectric contribution). The abruptness of the interface affects the distribution of the effective charge (although not its total magnitude, provided there is no lattice relaxation). A total polarization charge as high as ps~1×10^{13} cm^{-2} is expected at an $Al_{0.25}Ga_{0.75}N$/GaN interface [3]. The sharpness of the AlGaN/GaN interface in samples investigated in this research is revealed in the SIMS plots of the Al/Ga signal ratio as a function of film depth. The AlGaN/GaN interface composition varies within a thickness of 300Å.

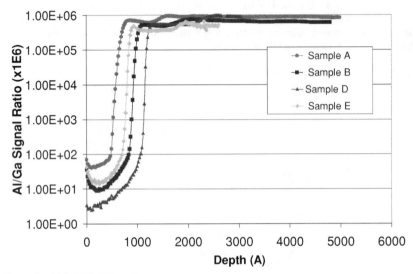

Figure 3. AlGaN/GaN interface

     The sheet conductivities of the films with the various doping contributions are presented in Figure 4. The conductivity of Sample E could not be measured.

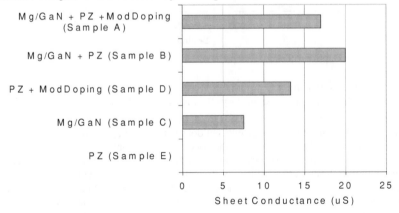

Figure 4. Sheet conductance for Samples A-D.

All samples were more conductive than the Mg-doped GaN standard (Sample C). As all samples contained essentially equal Mg concentrations, as shown in Figure 3, the electrical data indicate that interfacial strain and modulated doping were effective in increasing the conductivity of the base material. The magnitude of the observed conductivity increase is in good accord with what would be expected on the basis of estimated polarization and modulation doping effects. For example, an added hole density of the order of 1×10^{13} cm^{-2}, with a mobility of 10 cm^2/Vsec, would provide an extra component of conductivity of 16 uS.

In our experiments, the GaN layer thicknesses and dopings were chosen to be of the same order as those that could be used in the base layer of an HBT. The AlGaN layers could be used as the emitter layer of the HBT, provided the structure is fabricated with the collector on the wafer surface, and the emitter is grown below the base ("collector-up" device). By contrast, the HBT structures reported to date use an AlGaN emitter on the wafer surface, on top of the GaN base ("emitter-up" device). In these structures (with (0001) films, or Ga-face growth direction), the polarization doping contributions from the emitter-base AlGaN/GaN interface provide a donor-like charge, which compensates the Mg-doping contribution, rather than adding to it.

CONCLUSIONS

Acceptor (Mg)-doped Al$_x$Ga$_{1-x}$N/GaN heterostructures and a similarly doped GaN standard layer have been grown via MOVPE and the sheet conductance determined from CV measurements. Secondary ion mass spectroscopy shows Mg concentrations in the films of ~2×10^{19} cm^{-3}. The Mg concentration in all the films was essentially uniform; however, each film possessed a different conductivity depending on the presence and the magnitude of piezoelectric charges and the modulated doping. Additionally SIMS analysis of each film showed a sharp AlGaN/GaN interface (~300 Å) that enhances both the interfacial strain and the piezoelectric charge. Sheet conductance measurements showed that all Al$_x$Ga$_{1-x}$N/GaN heterostructures exhibited greater conductivity than the pure GaN sample. This strongly indicates that piezoelectric doping and modulated doping increase the base material conductivity over that achieved by conventional acceptor doping.

Acknowledgement:
The authors acknowledge Cree Research, Inc. for the SiC wafers. This work is supported by the Office of Naval Research under contracts N0W14-98-1-0654 (John Zolper, monitor).

REFERENCES

[1] A. Bykhovski, B. Gelmont, M. Shur, J. Appl. Phys. **74**, 6734 (1993).

[2] P. M. Asbeck, E.T. Yu, S.S. Lau, G.J. Sullivan, J. VanHove, J.M. Redwing, Electron. Lett. **33**, 1230 (1997).

[3] E. T. Yu, G.J. Sullivan, P.M. Asbeck, C.D. Wang, D. Qiao, S. Lau, Appl. Phys. Lett. **71**, 2794 (1997).

[4] H. Morkoc, R. Cingolani, W. Lambrecht, B. Gil, H.-X. Jiang, J. Lin, D. Pavlidis, K. Shenai, MRS Internet J. Nitride Semicond. Res. **4S1**, G1.2 (1999).

[5] J. Han, A.G. Baca, R.J. Shul, C.G. Willison, L. Zhang, F. Ren, A.P. Zhang, G.T. Dang, S.M. Donovan, X.A. Cao, H. Cho, K.B. Jung, C.R. Abernathy, S.J. Pearton, R.G. Wilson, Appl. Phys. Lett. **74**, 2702 (1999).

[6] S. Nakamura, T. Mukai, M. Senoh, Appl. Phys. Lett. **64**, 1687 (1994).

[7] F. A. Reboredo, S.T. Pantelides, MRS Internet J. Nitride Semicond. Res. **4S1**, G5.3 (1999).

[8] W. Weeks, M. Bremser, K. Alley, E. Carlson, Appl. Phys. Lett. **67**, 401 (1995).

9   D. P. Griffis, R. Loesing, D.A. Ricks, M.D. Bremser, R.F. Davis, in *Quantitative Analysis of C, O, Si and Mg Impurities in AlGaN*, Orlando, FL, 1997 (John Wiley and Sons), p. 201-204.

10   Y. Ohba, A. Hatano, J. Cryst. Growth **145**, 214 (1994).

Mat. Res. Soc. Symp. Vol. 595 © 2000 Materials Research Society

Low-Frequency Noise in SiO_2/AlGaN/GaN Heterostructures on SiC and Sapphire Substrates

N. Pala[1], R. Gaska[1], M. Shur[1], J. W. Yang[2] and M. Asif Khan[2]

[1]*Department of ECSE, Rensselaer Polytechnic Institute, Troy, New York 12180, USA*
[2]*Department of ECE, University of South Carolina, Columbia, South Carolina 29208, USA*

ABSTRACT

The low-frequency noise in GaN-based Metal-Oxide-Semiconductor Heterostructure Field Effect Transistors (MOS-HFETs) and HFETs on sapphire and n-SiC substrates were studied. Hooge parameter at zero gate bias was calculated about 8 x 10^{-4} for both types of the devices. The AlGaN/GaN MOS-HFETs exhibited extremely low gate leakage current and much lower noise at both positive and negative gate biases. These features demonstrate the high quality of the SiO_2/AlGaN heterointerface and feasibility of this technology for high-power microwave transmitter and high-power, high-temperature switches.

INTRODUCTION

Heterostructure Field Effect Transistors (HFETs) based on AlGaN/GaN material system are expected to offer superior performance in microwave and optical communication systems [1,2]. AlGaN/GaN HFETs grown on sapphire and silicon carbide substrates with impressive DC and microwave characteristics demonstrated [3,4].It is important to know the low frequency noise since it is the limiting figure for all kinds of HEMTs and MOSFETs. Especially when these devices are used as oscillators or mixers, low-frequency noise limits the phase-noise characteristics and degenerates the performance of the electronic system. It was shown that the level of the low frequency noise strongly depends on the gate leakage current I_g even at $I_g/I_d < 10^{-4}$-10^{-5}. (I_d is the drain current) [5,6]. Recently, novel AlGaN/GaN Metal-Oxide-Semiconductor Heterostructure Field Effect Transistors (MOS-HFETs) with high-quality SiO_2/AlGaN interfaces have been demonstrated [7]. The devices had output characteristics similar to AlGaN/GaN HFETs. However, the introduction of SiO_2 reduced the gate leakage by approximately six orders of magnitude, which is extremely important for high-power and low noise applications.

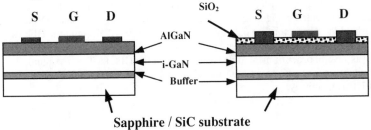

Sapphire / SiC substrate

Fig. 1. Schematics of AlGaN/GaN HFET and MOS-HFET structures.

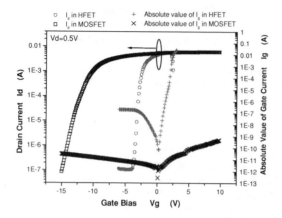

Fig. 2. Comperative input characteristics of the MOS-HFET and base line HFET.

DEVICE PROPERITIES

Figure 1 shows the schematic device structures of the MOS-HFETs and the base line HFETs. The AlGaN/GaN heterostructures were grown by Metal Organic Chemical Vapor Deposition (MOCVD) on (0001) sapphire and conducting 6H-SiC substrates. A 50 nm AlN buffer layer was first grown on the substrate followed by the deposition of a 1.5 and 0.5 μm semi-insulating GaN layer for sapphire and SiC, respectively. The heterostructures were capped with a 30-nm $Al_{0.2}Ga_{0.8}N$ barrier layer, which was doped with silicon to approximately to 2×10^{18} cm^{-3}. Both MOS-HFET and HFET devices were fabricated on the same wafer having the electron sheet density and the Hall mobility close to 9×10^{12} cm^{-2} and 1000 cm^2/Vs, respectively. For MOS-HFET structures, a 100 nm SiO_2 layer was deposited on the AlGaN/GaN heterostructure using plasma enhanced chemical vapor deposition (PECVD) on one half of the wafer. We studied 100 μm wide transistors with 10 μm gate length and 30 μm source-drain spacing. The large gate area and thick SiO_2 were chosen in order to study the gate leakage effects on the low frequency noise in both types of devices.

The input current-voltage characteristics of the MOS-HFET and HFET on sapphire are presented in Figure 2. The threshold voltages were -15V and -3V, and subthreshold ideality factors are 6.41 and 4.54 for the MOS-HFETs and the HFETs, respectively. The maximum gate current in the MOS-HFET was on the order of 10 pA, which was several order of magnitude smaller than in the HFET . The maximum transconductance, g_m = 26.5 mS/mm and 45.5 mS/mm were measured for MOS-HFET and HFET, respectively.

LOW FREQUENCY NOISE MEASUREMENT

Figure 3 shows the frequency dependencies of the relative Spectral Noise Density (SND) in MOS-HFET and HFET structures under different biasing conditions. The figure clearly reveals that the low-frequency noise in both MOS-HFET and HFET is a typical 1/f noise for all bias conditions. The noise level in different materials is usually characterized by the dimensionless Hooge parameter, α [8]:

$$\alpha = \frac{S_I}{I^2} fN \qquad (1)$$

where f is the frequency, N is the total number of carriers in the channel between source and drain. Note that we used the the total number of carriers in the channel between source and drain not only beneath the gate since there was no significant effect of the gate with 0V bias on the channel. The latter case would give an α parameter about 10^{-4} which is comparable to the lowest reported ones. The devices exhibited the same noise level at the gate bias $V_g = 0$ V resulting the Hooge parameter $\alpha = 8\times 10^{-4}$. This value of α is somewhat less than the reported values in Ref. [6,9,10] for HFETs grown on sapphire. Lower values for α were also reported [11,12].

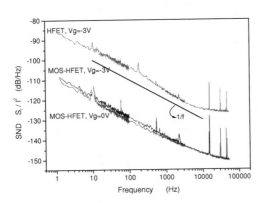

Fig. 3 Frequency dependence of Spectral Noise Density (SND) in MOS-HFET and HFET structures on sapphire. $V_d = 0.5$V S/D=30 µm

The Figure 4 compares the gate bias dependence of the SND at the frequency $f = 200$ Hz and the gate current in MOS-HFET and HFET. A similar level of the low-frequency noise at zero gate bias in both types of structures indicates that no traps or defects, which might degrade the device performance, were introduced by oxidation. MOS-HFETs and HFETs demonstrated rather different noise characteristics under applied gate bias. HFETs on the sapphire and SiC substrates exhibited the similar noise-gate bias dependence except for the somehow higher value for HFET on SiC at 0V gate bias. The sharp increase in noise density in HFET was attributed to the high gate leakage current at high gate biases as in Ref. [5,6,9]. Some authors explained the similar gate bias dependence of the noise by the screening effect [9]. However, as it is seen from Figure 4, the noise in HFET increases not only for negative gate voltage V_g but also for positive V_g. That increase can not be explained by the screening effect. It should be also mentioned that HFETs on the same substrate with identical current voltage characteristics but with a higher level of the gate leakage current I_g had a higher noise.

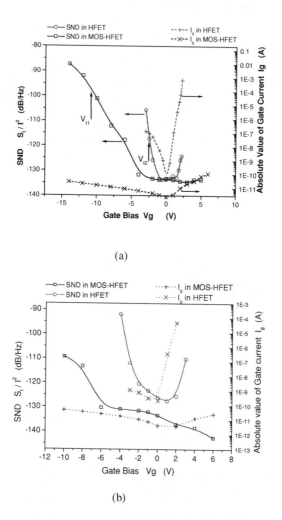

(a)

(b)

Fig. 4. . Gate bias dependence of the noise density at $f = 200$ Hz and the gate current in MOS-HFET and HFET on (a) sapphire (b) n-SiC. $V_d = 0.5$ V. Also shown the threshold voltages V_{t1} and V_{t2}, for MOS-HFET and HFET, respectively.

MOS-HFET structures had an extremely low gate leakage current and practically no gate voltage dependence for small values of the gate bias, V_g (see Fig. 4.). However for the $V_g < -5V$ the noise in the MOS-HFET increases sharply. On the other hand, as seen in Fig. 4b, the MOS-HFET structures on SiC presented almost the same noise level with their counterparts on sapphire in 0 V gate bias but the noise continued decreasing slightly with increasing positive bias. Further experiments are needed in order to establish the nature of the noise dependence on the gate bias and/or on the drain current.

We also measured the dependence of SND on sheet electron density in the heterostructures grown on 6H-SiC using the conducting substrate as a back-gate. This allowed us to modulate the electron density in 2D gas at AlGaN/GaN heterointerface. The obtained results in Figure 5 demonstrate that the substrate bias may reduce SND by more than order of magnitude.

CONCLUSION

In conclusion, we present the comparative low-frequency noise analysis in SiO_2/AlGaN/GaN MOSFETs and base line HFETs. The low frequency noise in these MOSFETs with extremely low gate leakage current is less than HFETs at both positive and negative gate biases. That makes the novel MOSFET structure a promising device for high frequency, low noise applications.

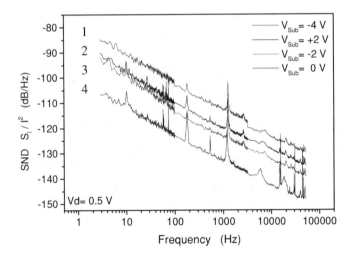

Fig. 5 Variation of the SND with the modulation of the substrate voltage V_{Sub} applied on the n-SiC .1)V_{Sub} = -4V, 2) V_{Sub} = +2V, 3) V_{Sub} = -2V, 4) V_{Sub} = 0V

ACKNOWLEDGEMENTS

This work was supported by the Office of Naval Research. Project monitor is Dr. J. Zolper. The authors are grateful to Professor M. E. Levinshtein and S. Rumyantsev for valuable discussions.

REFERENCES:

1. S.J. Pearton and C. Kuo, MRS Bulletin, February 1997, 17-19.
2. S. Nakamura, MRS Bulletin, February 1997, 29-35.
3. R. Gaska, Q. Chen, J. Yang, A. Osinsky, M.A. Khan, M. Shur, IEEE Elect. Dev. Lett. Vol. 18, pp. 492-494, Oct. 1997.
4. A. T. Ping , Q. Chen, J. W. Yang, M.A. Khan and I. Adesida, IEDM Tech. Dig. 1997, pp. 561.-564
5. M. E. Levinshtein, S. L. Rumyantsev, R. Gaska, J. W. Yang, M. S. Shur "AlGaN/GaN high electron mobility field effect transistors with low 1/f noise Appl. Phys. Lett, v.73 N8, pp. 1089-1091 (1998)
6. S. Rumyantsev, M. E. Levinshtein, R. Gaska, M. S. Shur, J. W. Yang, and M. A. Khan "Low-frequency noise in AlGaN/GaN HFEts on SiC and Saphire substrates" Journ. Appl. Phys. Submitted for publication
7. M. Asif Khan, X. Hu, G. Simin, A. Lunev, J.Yang, R.Gaska, M.S. Shur, "AlGaN/GaN Metal-Oxide-Semiconductor Field effect Transistor", IEEE El. Dev. Lett. in print
8. F. N. Hooge, IEEE Trans. Elect. Dev. Vol. 41, No. 11, p. 1926, (1994)
9. S. Rumyantsev, M. E. Levinshtein, R. Gaska, M. S. Shur, A. Khan, J. W. Yang, G. Simin, A. Ping and T. Adesida, "Low 1/f noise in AlGaN/GaN HFETs on SiC substrates", Abstracts of 3rd Int. Conf. On Nitride Semiconductors (ICNS3), Montpellier, France, July-04-july 09, 1999, pp. 125-126.
10. D. V. Kuksenkov, H. Temkin, R. Gaska, J. W. Yang, "Low-frequency noise in AlGaN/GaN Heterostructure Field Effect Transistors", IEEE Electron Device Letters, 19 (7), pp. 222-224 (1998)
11. J. A. Garrido, F. Calle, E. Munoz, I. Izpura, J.L. Sanchez-Rojas, R. Li, and K.L Wang, "Low frequency noise and screening effects in AlGaN/GaN HEMTs, El. Lett. v. 34, no. 24, pp. 2357-2359, (1998)
12. A. Balandin, S.V. Morozov, S. Cai, R. Li, K.L. Wang, G. Wijertane, C. R Viswanathan, "Low flicker-noise GaN/AlGaN heterostructure field effect transistors for microwave communications", IEEE, Trans. Microwave Theory and Tech. Vol. 47, No. 8, p. 1413, (1999).

Mat. Res. Soc. Symp. Vol. 595 © 2000 Materials Research Society

Electrical transport of an AlGaN/GaN two-dimensional electron gas

A. Saxler,[a] P. Debray,[a,b] R. Perrin,[a] S. Elhamri,[a,c] W. C. Mitchel,[a] C.R. Elsass,[d] I.P. Smorchkova,[d] B. Heying,[d] E. Haus,[d] P. Fini,[d] J.P. Ibbetson,[d] S. Keller,[d] P.M. Petroff,[d] S.P. DenBaars,[d] U.K. Mishra,[d] and J.S. Speck[d]

[a]Air Force Research Laboratory, Materials and Manufacturing Directorate, AFRL/MLPO, Wright-Patterson AFB, Ohio 45433-7707
[b]Permanent address: Service de Physique de l'Etat Condensé, Centre d'Etudes de Saclay, F-91191 Gif-sur-Yvette Cedex, France.
[c]Permanent address: Department of Physics, University of Dayton, Dayton, OH 45469.
[d]College of Engineering, University of California, Santa Barbara, CA 93106

ABSTRACT

An $Al_xGa_{1-x}N/GaN$ two-dimensional electron gas structure with x = 0.13 deposited by molecular beam epitaxy on a GaN layer grown by organometallic vapor phase epitaxy on a sapphire substrate was characterized. Hall effect measurements gave a sheet electron concentration of $5.1x10^{12}$ cm^{-2} and a mobility of $1.9 x 10^4$ cm^2/Vs at 10 K. Mobility spectrum analysis showed single-carrier transport and negligible parallel conduction at low temperatures. The sheet carrier concentrations determined from Shubnikov-de Haas magnetoresistance oscillations were in good agreement with the Hall data. The electron effective mass was determined to be 0.215 ± 0.006 m_0 based on the temperature dependence of the amplitude of Shubnikov-de Haas oscillations. The quantum lifetime was about one-fifth of the transport lifetime of $2.3 x 10^{-12}$ s.

INTRODUCTION

Many applications exist for the III-Nitrides in high-power and high-temperature electronics,[1-3] solar-blind ultraviolet photodetectors,[4] and blue and ultraviolet light emitting and laser diodes.[5] A structure of particular interest for the electronic devices is the AlGaN/GaN two-dimensional electron gas (2DEG) for use in high electron mobility transistors (HEMTs). In this paper, we study in detail the electrical transport properties of a high-mobility AlGaN/GaN 2DEG. Temperature and magnetic field dependent Hall effect measurements are used to study the basic transport properties. Temperature dependent Shubnikov – de Haas (SdH) measurements permitted the extraction of the electron effective mass in GaN. Determination of this constant is important for use in device modeling, but there is considerable scatter in the reported data for the electron effective mass which ranges from 0.18-0.23 m_0.[6-9] Two previous reports of the effective mass using SdH measurements for AlGaN/GaN 2DEGs have been made.[6,7] The SdH measurement for a structure grown on a SiC substrate[6] yielded a significantly different result than that measured by other techniques. In this paper, we will use the SdH measurements to estimate the electron effective mass, scattering time, and carrier concentration and compare the results to those obtained by the Hall effect measurements.

EXPERIMENT

An $Al_xGa_{1-x}N/GaN$ two-dimensional electron gas structure was deposited by molecular beam epitaxy on a 2-3 µm thick GaN layer grown by organometallic vapor phase epitaxy on a sapphire substrate. The $Al_xGa_{1-x}N$ was approximately 50 nm thick. The details of the growth have been reported previously.[10]

Hall effect measurements were performed over a temperature range of $10 - 300$ K and a magnetic field range of $0 - 2$ T. Both the Hall effect and the SdH measurements were taken in the four-probe van der Pauw configuration using annealed Ti/Al contacts placed at the four corners of a square.

Measurements of SdH magnetoresistance oscillations were performed using a low ac bias current to avoid electron heating. A magnetic field of $0 - 9$ T was used. The temperature was controlled from $1.2 - 4.2$ K by immersing the sample in liquid helium and setting the pressure. A calibrated thermometer was then used to measure the temperature.

RESULTS AND DISCUSSION

As seen in Fig. 1, the sheet electron concentration and mobility measured by the Hall effect are nearly independent of temperature below about 100 K. This behavior is typical of two dimensional electron gas structures. At 10 K, the electron concentration, n_H, was 5.06×10^{12} cm^{-2}, and the Hall mobility, μ_H, was 1.91×10^4 cm^2/Vs. The transport scattering time, $\tau_c = \mu_H m^*/e$, where m^* is the effective mass and e is the electron charge, was found to be 2.34 ps.

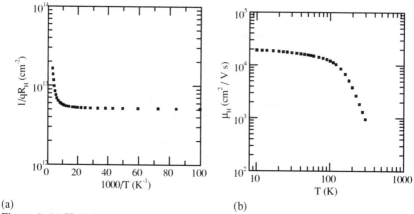

(a) (b)

Figure 1. (a) Hall sheet carrier concentration n_H as a function of inverse temperature. (b) Hall mobility μ_H as a function of temperature.

In order to determine if there was any significant parallel conduction at low temperature, magnetic field dependent Hall effect measurements were taken. The data taken at 10 K is plotted in Fig. 2 (a) along with the fit as in the analysis of Kim et al.[11] The reduced conductivities, X and Y, are σ_{xx} and $2\sigma_{xy}$ respectively divided by the zero field conductivity. When the reduced conductivities are plotted in this manner and only a single electron mobility μ is present, at a magnetic field $B=1/\mu$, X passes through 0.5 and Y peaks with a magnitude of 1. It is clear from this data that there is only one single-carrier conduction path at 10 K. The single carrier fit to this data gives an electron concentration of 5.08 x 10^{12} cm^{-2} and a mobility of 1.90 x 10^4 cm^2/Vs, in very good agreement with the single magnetic field Hall taken at 0.5 T. The mobility spectrum shown in Fig. 2(b) was obtained using the magnetic field dependent data of Fig. 2(a) processed with software from Lake Shore Cryotronics using the quantitative mobility spectrum analysis technique.[12] This plot also illustrates that there is a single dominant high mobility channel. Fig. 2(c) shows the reduced conductivity at a higher temperature of 160K. From this plot it is apparent that there is more than one conduction path in the sample since Y does not reach −1. A two-carrier fit gives n_1=5.0 x 10^{12} cm^{-2} , μ_1 = 0.82 x 10^4 cm^2/Vs, $n_2 \sim 3$ x 10^{13} cm^{-2} , and $\mu_2 \sim 2$ x 10^2 cm^2/Vs and is plotted as a line. The mobility spectrum also shows a high mobility channel and a lower mobility channel with significantly lower conductivity. The limited magnetic field range is responsible for reduced certainty in determining the lower mobility. The second lower mobility channel can be interpreted as the bulk-like unintentionally-doped GaN layer which freezes out at low temperatures.

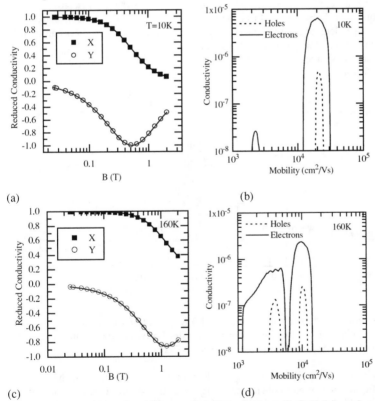

(a)   (b)   (c)   (d)

Figure 2. Reduced conductivity as a function of magnetic field (a), and mobility spectrum (b) at 10 K showing that a single-carrier high mobility channel dominates the conduction; and 160 K (c),(d) showing the presence of a second lower mobility channel .

The SdH oscillations in the magnetoresistance $R_{xx}(B)$ were measured at 1.3 K along two mutually perpendicular sides (labeled R1 and R2) of the square van der Pauw sample. A "beating" effect was observed in the amplitude of the R1, as would be expected from two interfering oscillations of slightly different frequencies. Assuming only one subband is occupied and neglecting the contribution of higher harmonics, the oscillatory part $\Delta\rho_{xx}$ of the magnetoresistivity can be expressed as:[13]

$$\frac{\Delta\rho_{xx}}{2\rho_0} = 2\frac{\chi}{\sinh(\chi)}\exp\left(\frac{-\pi}{\omega_c\tau_q}\right)\cos\left(\frac{2\pi\varepsilon}{\hbar\omega_c} - \pi\right) \tag{2}$$

where $\chi = 2\pi^2 k_B T / \hbar\omega_c$, $\omega_c = eB/m^*$, $\varepsilon = \pi\hbar^2 n/m^*$ is the Fermi energy, τ_q is the quantum scattering time, B is the magnetic field, k_B is the Boltzmann constant, T is the absolute temperature, k_B is the reduced Plank constant, and n is the sheet electron concentration of the 2DEG. From this equation and the observed magnetic field and temperature dependent magnetoresistance data we will extract the carrier concentration, the effective mass, and the quantum scattering time. This is possible since the magnetoresistance is directly proportional to the magnetoresistivity through a sample geometrical factor.

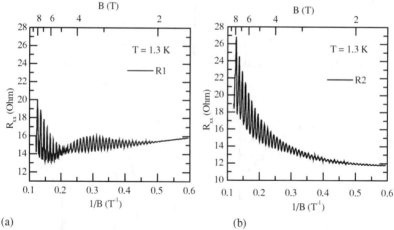

(a)   (b)

Figure 3. Shubnikov-de Haas oscillations at 1.3 K from two mutually perpendicular sides (a) R1 and (b) R2 of the van der Pauw sample. Notice the attenuation of R1 amplitudes due to a "beating" effect.

Since the observed SdH oscillations do not show the presence of higher harmonics, Eq. 6 can be used to obtain the carrier concentration from the frequency of the oscillations in 1/B. Fig. 4(a) shows the inverse magnetic field at which the peak maxima occur. The sheet carrier concentrations for the R1 and R2 oscillations were found to be 4.77 and 4.94 x 10^{12} cm^{-2}, respectively. The corresponding drift mobilities determined using the observed value of ρ_0 were 2.03 and 1.96 x 10^4 cm^2/Vs. Interestingly, the frequency difference determined from the beating period of the R1 oscillations is found to correspond to an electron concentration difference of about 0.18 x 10^{12} cm^{-2}. The sheet carrier concentrations obtained from the SdH oscillations are slightly lower than that of the Hall value which may indicate the presence of a very small parallel conduction. The background rise of the R2 oscillations (Fig. 8) indicates field-induced localization of some donor impurities, consistent with small parallel conduction. The small anisotropies observed in the carrier concentration and mobility may result from a slight inhomogeneity in the sample, especially because in the van der Pauw configuration the current paths are not necessarily linear. The observation of the beating effect may also result from this inhomogeneity.

The effective mass can be extracted by measuring the amplitude of the SdH oscillations as a function of temperature. In Fig. 4(b), the natural logarithm of the amplitude A divided by temperature T is plotted as a function of the temperature. The data was fit simultaneously for two different resistance configurations (labeled R1 and R2) and two different magnetic fields (3 T and 4 T). The value obtained from the fit was 0.215±0.006 m_0 which is in good agreement with recently reported values obtained from other techniques such as cyclotron resonance.[7-9]

Fig. 4(c) is a "Dingle plot" of the R2 oscillation amplitudes (Fig. 8). The quantum relaxation time τ_q was obtained by fitting the data. Since the amplitudes of R1 oscillations are attenuated by the beating effect, they were not used. The rather poor fit may indicate that some beating is also present in the R2 oscillations. The fit yielded a value of about 0.5 ps for the quantum scattering time, which is significantly lower than the transport scattering time τ_p of 2.34 ps obtained in the previous section. Note that τ_q is given by the total scattering rate, whereas τ_p is weighted by the scattering angle in such a way so as to minimize low-angle scattering. For the 2DEG at the heterointerface, we expect the dominant scattering to be due to long-range potentials associated with donor impurities outside the well which produce predominantly small-angle scattering. A value of τ_p higher than τ_q is therefore expected and has often been observed.[20]

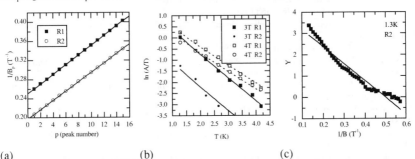

(a)   (b)   (c)

Figure 4. (a) Peak positions in reciprocal magnetic field of SdH oscillations and fit to obtain the carrier concentrations. (b) Fitting for the effective mass for two magnetic field values and two resistivity configurations simultaneously gives m* = 0.215±0.006 m$_0$. (c) Dingle plot for determination of scattering time by fitting the field dependent amplitude.

CONCLUSIONS

An $Al_xGa_{1-x}N/GaN$ two-dimensional electron gas structure with x = 0.13 was deposited by molecular beam epitaxy on a GaN layer grown by organometallic vapor phase epitaxy on a sapphire substrate. Mobility spectrum analysis showed negligible parallel conduction at low temperatures. The sheet carrier concentrations determined from SdH magnetoresistance oscillations were in good agreement with the Hall data. The uncorrected electron effective mass was determined to be 0.215±0.006 based on the temperature dependent amplitude of the SdH oscillations. The quantum lifetime was found to be about one-fifth of the transport lifetime of 2.3 x 10^{-12} s. We have observed small anisotropies in the electronic properties of the 2DEG at the AlGaN/GaN heterointerface. We currently attribute this anisotropy to a slight inhomogeneity in the sample.

ACKNOWLEDGMENTS

The work at UCSB was supported in part by the MRL Central Facilities supported by the NSF under Award No. DMR-9123048. The authors also gratefully acknowledge G. Landis, S. Davidson, A. C. Gossard, J. English, the ONR IMPACT MURI Center at UCSB (C. Wood and J. Zolper contract monitors), AFOSR (G. Witt and D. Johnstone), the Center for Quantized Electronic Structures, HRL Laboratories, the UC-MICRO program, and the NDSEG Fellowship Program.

REFERENCES

[1] O. Aktas, Z. F. Fan, S. N. Mohammed, A. E. Botchkarev, and H. Morkoç, Appl. Phys. Lett. **69**, 3872 (1996).

[2] U.K. Mishra, Y.-F. Wu, B.P. Keller, S. Keller and S.P. DenBaars, *IEEE Trans. on Microwave Theory and Tech.*, **46, 756** (1998).

[3] L. McCarthy, P. Kozodoy, M. Rodwell, S. DenBaars, and U. Mishra, *Compound Semiconductor* **4**(8), 16 (1998).

[4] M. Razeghi and A. Rogalski, J. Appl. Phys. **79**, 7433 (1996).

[5] S. Nakamura, M. Seno, S. Nagahama, N. Iwasa, T. Yamada, T. Matsushita, Y. Sugimoto, and H. Kiyoku, Appl. Phys. Lett. **69**, 4056 (1996).

[6] S. Elhamri, R. S. Newrock, D. B. Mast, M. Ahoujja, W. C. Mitchel, J. M. Redwing, M. A. Tischler, and J. S. Flynn, Phys. Rev. B **57**, 1374 (1998).

[7] L. W. Wong, S. J. Cai, R. Li, K. Wang, H. W. Jiang, and M. Chen, Appl. Phys. Lett. 73, 1391 (1998).

[8] J. S. Im, A. Moritz, F. Steuber, V. Härle, F. Scolz, and A. Hangleiter, Appl. Phys. Lett. **70**, 631 (1997).

[9] S. W. King, C. Ronning, R. F. Davis, M. C. Benjamin, and R. J. Nemanich, J. Appl. Phys. **84**, 2086 (1998).

[10] C.R. Elsass, I.P. Smorchkova, B. Heying, E. Haus, P. Fini, K. Maranowski, J.P. Ibbetson, S. Keller, P.M. Petroff, S.P. DenBaars, U.K. Mishra, and J.S. Speck, Appl. Phys. Lett. **74**, 3528 (1999).

[11] J. S. Kim, D. G. Seiler, and W. F. Tseng, J. Appl. Phys. **73**, 8324 (1993).

[12] I. Vurgaftman, J. R. Meyer, C. A. Hoffman, D. Redfern, J. Antoszewski, L. Faraone, and J. R. Lindemuth, J. Appl. Phys. **84**, 4966 (1998).

[13] P. T. Coleridge, R. Stoner, and R. Fletcher, Phys. Rev. B **39**, 1120 (1989); A. Isihara and L. Smrcka, J. Phys. C **19**, 6777 (1986).

Mat. Res. Soc. Symp. Vol. 595 © 2000 Materials Research Society

CORRELATION BETWEEN SHEET CARRIER DENSITY-MOBILITY PRODUCT AND PERSISTENT PHOTOCONDUCTIVITY IN ALGAN/GAN MODULATION DOPED HETEROSTRUCTURES

J. Z. Li, J. Li, J. Y. Lin, and H. X. Jiang
Department of Physics, Kansas State University
Manhattan, KS 66506-2601

ABSTRACT

High quality $Al_{0.25}Ga_{0.75}N$/GaN modulation-doped heterojunction field-effect transistor (MOD-HFET) structures grown on sapphire substrates with high sheet carrier density and mobility products ($n_s\mu > 10^{16}$/Vs at room temperature) have been grown by metal organic chemical vapor deposition (MOCVD). The optimized structures were achieved by varying structural parameters, including the AlGaN spacer layer thickness, the Si-doped AlGaN barrier layer thickness, the Si-doping concentration, and the growth pressure. In these structures, the persistent photoconductivity (PPC) effect associated with the two-dimensional electron gas (2DEG) system was invariantly observed. As a consequence, the characteristic parameters of the 2DEG were sensitive to light and the sensitivity was associated with permanent photoinduced increases in the 2DEG carrier mobility (μ) and sheet carrier density (n_s). However, we observed that the magnitude of the PPC and hence the photoinduced instability associated with these heterostructures were a strong function of only one parameter, the product of n_s and μ, which is the most important parameter for the HFET device design. For a fixed excitation photon dose, the ratio of the low temperature PPC to the dark conductivity level was observed to decrease from 200% to 3% as the $n_s\mu$ (300 K) product was increased from 0.048 x 10^{16}/Vs to 1.4 x 10^{16}/Vs. Based on our studies, we suggest that the magnitude of the low temperature PPC can be used as a sensitive probe for monitoring the electronic quality of the AlGaN/GaN HFET structures.

INTRODUCTION

Recent progresses in III-nitride material growth and device processing have greatly extended their applications in the area of electronic as well as optoelectronic devices [1,2]. For electronic device applications, $Al_xGa_{1-x}N$/GaN heterojunction field-effect transistors (HFETs) have shown great promises in microwave and millimeter-wave electronic device applications [3-5]. However, the performance of AlGaN/GaN HFETs still falls far from that of the theoretical prediction [6]. Further improvements in AlGaN/GaN heterojunction material quality as well as in structural design are needed. Routine but powerful material and device characterization methods must be established.

In this work, $Al_{0.25}Ga_{0.75}N$/GaN modulation-doped heterojunction field-effect transistor (MOD-HFET) structures grown on sapphire substrates have been produced by our metal organic chemical vapor deposition (MOCVD) system and characterized by Hall and persistent photoconductivity measurements. The persistent photoconductivity (PPC) effect, which has been observed previously in GaN materials by several groups [7-11], was universally presented in these AlGaN/GaN HFET structures, indicating the presence of charge trapping effects. However, the magnitude of the PPC was observed to be a strong function of only one parameter, the product of sheet carrier density (n_s) and the two-dimensional electron gas mobility (μ). Since $n_s\mu$ is the most important intrinsic material parameter for the HFET structural design, the magnitude of PPC in turn can be utilized as a sensitive probe for monitoring the electronic qualities of the AlGaN/GaN MOD-HFET structures.

EXPERIMENTAL DETAILS

The inset of Fig. 1 is a schematic diagram showing the generic structure of samples used in this work which consisted of a 1.3 μm highly insulating GaN epilayer followed by an $Al_{0.25}Ga_{0.75}N$ spacer layer and finally a Si-doped $Al_{0.25}Ga_{0.75}N$ layer. The structures were deposited over basal plane sapphire substrates using a variable pressure MOCVD system at a growth temperature of 1050 °C. A total of seven samples with varying growth or structural parameters were studied. For the PPC measurements, a 1.5 V bias was supplied to the sample, a Hg lamp was used as an excitation source, and the conductivity was monitored through current (I_{ppc}) by using an electrometer. To ensure that each set of data obtained under different temperatures have the same initial conditions, the system was always heated up to 300 K, then cooled down in darkness to the desired measurement temperatures. The excitation intensity and buildup time span are fixed for different temperatures and samples. Fig. 1 illustrates the temperature variations of the sheet carrier density and mobility for one of our optimized structures. The values of the sheet carrier density and mobility products ($n_s\mu$) at different temperatures shown in Fig. 1 are among those highest values reported for the AlGaN/GaN HFET structures grown on sapphire substrates [12-14].

For all seven structures investigated here, the conductivity of the 2DEG channel is enhanced after exposure to light. Moreover, the light enhanced conductivity persists for a

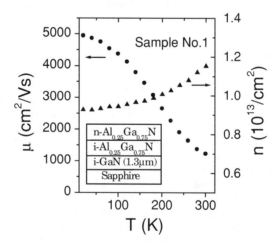

Fig. 1   2DEG sheet carrier density (n_s) and mobility (μ) versus temperature, T, measured in a dark state for one of our optimized $Al_{0.2}Ga_{0.8}N$/GaN MOD-HFET structures. The inset shows the generic structures for the samples used in this study.

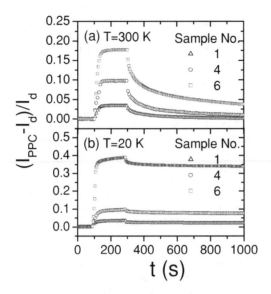

Fig. 2   Buildup and decay kinetics of PPC associated with the 2DEG system in three Al$_{0.25}$Ga$_{0.75}$N/GaN MOD-HFET structures measured at two representative temperatures, (a) T = 300 K and (b) T = 20 K. Here I_{ppc} denotes the persistent photocurrent measured at time t and I_d the initial dark current levels.

long period of time after the removal of excitation source, an effect which is referred to as PPC. In Fig. 2, we present the PPC results obtained for three representative samples measured at two different temperatures. We can see that the decay time constants of the low temperature PPC are very long. By formulating the PPC buildup and decay kinetics in an AlGaN/GaN heterostructure, we have previously attributed the PPC effect to the photoionization of deep level centers in AlGaN barrier [11]. It is well known from the earlier works on the AlGaAs/GaAs MODFETs that the PPC effects are detrimental to their performance [15]. The effects include sensitivity to light, a shift of the threshold voltage, and collapse of the drain I-V characteristics, all of which reduced the usefulness of the AlGaAs/GaAs MODFET for integrated circuits. Interestingly, for AlGaN/GaN HFET structures studied here, we observe that the magnitude of PPC and hence the device instabilities can be minimized by varying the growth conditions as well as structural parameters. Table 1 summarizes our Hall and PPC measurement results for all seven structures with different structural and growth parameters. As shown in Table 1, the magnitude of PPC or the photoinduced conductivity enhancement $(I_{ppc} - I_d)$ over its dark level (I_d) is about 200% in sample #7, but is negligibly small (only about 3%) in sample #1.

The general trends shown in Table 1 are that samples possess higher mobilities as well as higher sheet carrier densities exhibit reduced PPC, however, with a few exceptions (e.g., samples #4 & #5). By carefully inspecting the results summarized in Table 1, we can clearly see that the magnitude of PPC has a systematic dependence only on the product of the 2DEG sheet carrier density and mobility, i.e., $n_s\mu$, the most important intrinsic material parameter for the HFET device design. In Fig. 3, we have re-plotted the the the magnitude of PPC versus $n_s\mu$ measured at 20 K (a) and 300 K (b). At both temperatures, the magnitude of PPC decreases monotonously with an increase of $n_s\mu$, follows the relationship of

$$R_{ppc} = A\,(n_s\mu)^{-\alpha}.$$

Here the magnitude of PPC (R_{ppc}) is defined as $[(I_{ppc} - I_d)/I_d]$ and A and α are two constants. As illustrated in Table 1, the better structures (i.e., larger values of $n_s\mu$) were achieved by varying (i) the AlGaN space layer thickness, (ii) the Si-doped AlGaN layer thickness, (iii) the doping levels in the Si-doped AlGaN layer, and (iv) the growth pressure. The electronic qualities (or $n_s\mu$ values) of these HFET structures can be further improved slightly by adjusting the four parameters described above until the magnitude of PPC further reduces to zero. However, we speculate that further dramatic enhancements in AlGaN/GaN heterostructure material quality require further enhancements in epitaxial film qualities of both the underlying i-GaN and the top AlGaN epilayers.

Table 1. Hall-effect and PPC measurement results of the seven $Al_{0.25}Ga_{0.75}N/GaN$ MOD-HFET structures obtained at 20 K and 300 K, together with the structural parameters (spacer and Si-doped AlGaN layer thicknesses), the relative Si-doping levels (SiH_4 flow rate), and the growth pressure. All structures were grown at 1050 ^0C in a variable pressure MOCVD. Here I_{ppc} denotes the buildup levels of the persistent currents for a fixed excitation intensity and buildup time span and I_d the initial dark current levels.

Sample No.	μ (cm^2/Vs) 20 K/300 K	n_s $(10^{13}/cm^2)$ 20 K/300 K	$n_s\mu$ $(10^{16}/Vs)$ 20 K/300 K	PPC ratio $[(I_{PPC}-I_d)/I_d]\times100\%$ 20 K/300 K	i-AlGaN/n-GaN Thickness (nm)	SiH_4 (sccm)	Pressure (torr)
1	4950 / 1230	0.93 / 1.15	4.95 / 1.42	3.31 / 3.42	6/25	3	100
2	3760 / 870	0.76 / 1.37	2.86 / 1.19	5.84 / 5.32	8/25	1	150
3	2920 / 884	0.78 / 1.13	2.28 / 1.00	7.10 / 6.75	4/25	1	150
4	2150 / 573	0.84 / 0.92	1.81 / 0.53	8.34 / 11.15	6/25	1	150
5	1620 / 703	0.69 / 0.74	1.13 / 0.52	37.35 / 14.34	6/25	1	100
6	600 / 280	0.99 / 1.16	0.59 / 0.33	39.18 / 18.27	6/25	5	100
7	2800 / 485	0.11 / 0.09	0.31 / 0.048	200.35 / 300.24	25/25	0	77

The strong PPC effect observed in several samples here is associated with both the photoinduced increases in 2DEG sheet carrier density and mobility. In these structures, the sheet carrier densities at fixed temperatures in a single sample can be continuously varied by varying the excitation photon dose, while the increases in sheet carrier densities also result in enhancements in 2DEG mobilities. Such behaviors have been seen previously in an AlGaN/GaN HFET structure [11], which clearly demonstrates that one of the possible ways to enhance the 2DEG mobilities is to increase the sheet carrier densities in the 2DEG channel region. In desired structures with minimal PPC effects, enhanced mobilities can be accomplished by barrier or channel doping. However, at much greater sheet carrier densities, the population of the higher-lying subbands could limit the overall mobility of the structure due to intersubband scattering as well as the

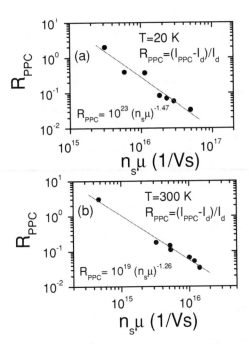

Fig. 3 The magnitude of PPC, R_{ppc}, or the photoinduced conductivity enhancement for a fixed excitation intensity and buildup time span $(I_{ppc} - I_d)$ over its dark level (I_d) as a function of the product of the sheet carrier density and mobility, $n_s\mu$, obtained from seven different HFET structures. The solid lines are the least squares fits to experimental data with $R_{ppc} = A(n_s\mu)^{-\alpha}$. The fitted values of α (A) are 1.47 (10^{23}) and 1.26 (10^{19}) at 20 K and 300 K, respectively.

loss of the true two-dimensional behavior due to a virtual continuum of bands being populated [12,16,17]. Thus a trade-off between these effects must be considered.

In summary, we have shown in this work that the most important intrinsic material parameter for the AlGaN/GaN HFET device design, namely the sheet carrier and mobility product $(n_s\mu)$ is strongly correlated with the magnitude of PPC in these structures. As a result, the magnitude of PPC can be utilized as a sensitive probe for monitoring the electronic qualities of the AlGaN/GaN MOD-HFET structures. We observed that the magnitude of the PPC and hence the photoinduced instability associated with these heterostructures decreases rapidly with an increase of $n_s\mu$, according to $R_{ppc} \propto (n_s\mu)^{-\alpha}$, with α varying from 1.26 to 1.47 in the temperature range of 20 to 300 K.

ACKNOLEGEMENTS

This research is supported by BMDO (Dr. Kepi Wu) monitored by USASSDC, NSF (DMR-99502431), and DOE(DE-FG03-96ER45604).

REFERENCES:

1. H. Morkoc, S. Strite, G. B. Gao, M. E. Lin, B. Sverdlov, and M. Burns, J.Appl.Phys.**76**, 1363 (1994).
2. S. N. Mohammad, A. A. Salvador, and H. Morkoc, Proc. IEEE **83**, 1306 (1995).
3. M. A. Khan, M. S. Shur, J. N. Kuznia, Q. Chen, J. Burn, and W. Schaff, Appl. Phys. Lett. **66**, 1083 (1994).
4. L. Eastman, K. Chu, W. Schaff, M. Murphy, N. G. Weimann, and T. Eustis, MRS internet J. Nitride Semicond. Res. **2**, 17, (1997).
5. Y. F. Wu, B. P. Keller, P. Fini, S. Keller, T. J. Jenkins, L. T. Kehias, and S. P. DenBaars, IEEE Electron. Device Lett. **19**, 50 (1998).
6. B. Gelmont, K. S. Kim, and M. Shur, J. Appl. Phys. **74**, 1818 (1993).
7. C. Johnson, J. Y. Lin, H. X. Jiang, M. Asif Khan, and C. J. Sun, Appl. Phys. Lett. **68**, 1808 (1996).
8. G. Beadie, W. S. Rabinovich, A. E. Wickenden, D. D. Koleske, S. C. Binari, and J. A. Freitsa, Jr., Appl. Phys. Lett. **71**, 1092 (1997).
9. C. H. Qiu and J. I. Pankove, Appl. Phys. Lett. **70**, 1983 (1997).
10. M. T. Hirsch, A. Wolk, W. Walukiewicz, and E. E. Haller, Appl. Phys. Lett. **71**, 1098 (1997).
11. J. Z. Li, J. Y. Lin, H. X. Jiang, M. Asif Khan, and Q. Chen, J. Appl. Phys. **82**, 1227 (1997). J. Vac. Sci. Technol. **B15**, 1117 (1997).
12. R. Gaska, M. S. Shur, A. D. Bykhovski, A. O. Orlov, and G. L. Snider, Appl. Phys. Lett. **74**, 287 (1999).
13. T. Wang, Y. Ohno, M. Lachab, D. Nakagawa, T. Shirahama, S. Sakai, and H. Ohno, Appl. Phys. Lett. **74**, 3531 (1999).
14. C. R. Elsass, I. P. Smorchkova, B. Heying, E. Haus, P. Fini, K. Maranowski, J. P. Ibbetson, S. Keller, P. M. Petroff, S. P. DenBaars, U. K. Mishra, and J. S. Speck, Appl. Phys. Lett. **74**, 3528 (1999).
15. P. M. Mooney, J. Appl. Phys. **67** R1 (1990).
16. X. Z. Dang, P. M. Asbeck, E. T. Yu, G. J. Sullivan, M. Y. Chen, B. T. McDermott, K. S. Bouttros, and J. M. Redwing, Appl. Phys. Lett. **74**, 3890 (1999).
17. L. Hsu and W. Walukiewicz, Phys. Rev. **B56**, 1520 (1999).

Mat. Res. Soc. Symp. Vol. 595 © 2000 Materials Research Society

Full Band Monte Carlo Comparison of Wurtzite and Zincblende Phase GaN MESFETs

Maziar Farahmand and Kevin F. Brennan
School of Electrical and Computer Engineering
777 Atlantic Dr.
Georgia Tech
Atlanta, GA 30332-0250 U.S.A.

ABSTRACT

The output characteristics, cutoff frequency, breakdown voltage and the transconductance of wurtzite and zincblende phase GaN MESFETs have been calculated using a self-consistent, full band Monte Carlo simulation. It is found that the calculated breakdown voltage for the wurtzite device is considerably higher than that calculated for a comparable GaN zincblende phase device. The zincblende device is calculated to have a higher transconductance and cutoff frequency than the wurtzite device. The higher breakdown voltage of the wurtzite phase device is attributed to the higher density of electronic states for this phase compared to the zincblende phase. The higher cutoff frequency and transconductance of the zincblende phase GaN device is attributed to more appreciable electron velocity overshoot for this phase compared to that for the wurtzite phase. The maximum cutoff frequency and transconductance of a 0.1 μm gate-length zincblende GaN MESFET are calculated to be 220GHz and 210 mS/mm, respectively. The corresponding quantities for the wurtzite GaN device are calculated to be 160GHz and 158 mS/mm.

INTRODUCTION

The particular features that make the III-nitrides and SiC attractive for high frequency, high power device applications are their relatively high breakdown voltage, high saturation electron drift velocity, low dielectric constants, and for SiC, high thermal conductivity [1-3].

To the authors' knowledge, only a few other theoretical studies of the behavior of GaN based FETs have been made [4-7]. The theoretical study of GaN based devices is frustrated to some extent by the paucity of reliable transport parameters. As such, numerical studies using advanced drift diffusion and hydrodynamic simulation tools [8] that require extensive parameterization are, at present, challenging. Alternatively, less parameterized models, such as the full band ensemble Monte Carlo technique that can proceed relatively independently of experiment, are presently more suitable for studying transport in GaN and in modeling GaN based devices [9]. It has been shown that Monte Carlo simulations using an analytical band structure approximation are generally insufficiently detailed to properly describe high field transport dynamics [10]. Inclusion of the full details of the band structure is necessary to provide an accurate accounting of the breakdown properties of a device. For these reasons, we employ a full band ensemble Monte Carlo simulation to examine the relative performance characteristics of wurtzite and zincblende phase GaN MESFETs operated near breakdown. The model is summarized in the next section. In section III, the calculated results are presented and conclusions are drawn in Section IV.

MODEL DESCRIPTION

The full band Monte Carlo technique relies on the knowledge of fairly well known parameters i.e., lattice constants, energy gaps, phonon energies, dielectric constants, etc. From these quantities a reasonably accurate description of the band structure, impact ionization transition rate and phonon scattering rates can be obtained, enabling study of the basic transport properties of the material and its related devices. The full band Monte Carlo simulator can thus be developed without extensive experimental knowledge of the transport parameters of a material. Though the Monte Carlo model suffers from some basic uncertainties, these mainly being in the details of the deformation potentials and the concomitant high energy scattering rates, it is nonetheless presently the most reliable simulation tool for investigating high field transport.

The full details of our Monte Carlo model have been presented elsewhere [9]. In addition, the details of how our Monte Carlo model has been merged with a Poisson solver to enable device simulation have also been discussed elsewhere [6]. For brevity we will not repeat these details here.

This study is comparative in nature. The study is not meant to project the ultimate limits of potential performance of GaN MESFETs. Instead, it is our goal to examine devices with the same geometry and doping concentrations but made using the two different polytypes of GaN, wurtzite and zincblende, to provide some basis of comparison of how the device performance varies, if at all, with choice of polytype. The geometry and doping concentrations for the device structure used in this study are shown in Figure 1. The very small dimensions of the device are chosen mainly to manage the computational demands of the simulator. All of the calculations are made assuming an ambient temperature of 300K. The dopants are assumed to be fully ionized and no doping compensation is present.

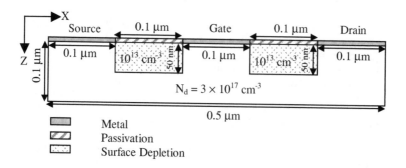

Figure 1. Cross section of the MESFET.

In our previous theoretical studies of the transport properties of the III-nitrides [11-14] we have encountered new physical effects which influence the transport calculations in non-cubic symmetry semiconductors [15]. Perhaps the most important effect encountered in modeling non-cubic symmetry semiconductors is that of band intersections. In materials with highly complex band structures, such as wurtzite phase GaN, numerous band intersection points occur. Proper treatment of these points is necessary to render an accurate description of the transport dynamics. We have developed two approaches to treating transport near these points. The first approach is based on the evaluation of the overlap integral of the cell periodic part of the wave function between the initial and possible final states of the intersecting bands. The final state is selected stochastically based on the relative probabilities given by the overlap integrals [12]. The second approach is based on a velocity continuation test performed at band intersections [6]. As in the case for the overlap integral test, the final state is selected stochastically but with the difference that the choice is based on the requirement of velocity continuation during the drift. Good agreement between the two models has been observed. It should be noted however that neither one of these techniques definitively describes the transport physics at band intersections. Instead, these approaches are only initial attempts at modeling the effects of band intersections and as such provide only a first order approximation of multiband transport effects. A more complete study of the physics of transport near band intersections is currently underway.

In the present study we adopt the velocity continuation approach for treating band intersection points. Though we currently believe that the overlap integral method is more fundamentally sound, that technique presently requires more computational resources. The device simulator used here, when band intersection effects are neglected as can be done in the zincblende phase GaN calculations, already highly stresses our computational capabilities. Therefore, when band intersection effects must be included, as is necessary in wurtzite phase GaN, computational efficiency becomes of the highest priority. Though we cannot presently be certain as to how accurate the velocity continuation test is, its observed empirical agreement with the overlap integral test gives some confidence that this model is acceptable, at least to first order. In any event, the present limitations of our computational resources precludes doing much else.

CALCULATED RESULTS

The calculated drain current, I_d, versus drain-source voltage, V_{ds}, for gate-source voltages, V_{gs}, varying from -0.1V to -5.1V in 1V steps for the zincblende phase and the wurtzite phase devices are shown in Figure 2. The Schottky barrier potential height was included in V_{gs}. For each bias point, the drain current was calculated under two conditions. These were with and without the presence of impact ionization. This was done in order to calculate the breakdown voltage[7].

As it can be observed from Figure 2, the breakdown voltages are significantly higher in the wurtzite phase device than in the zincblende phase device. The higher breakdown voltage of the wurtzite device is attributed to the higher density of electronic states in the wurtzite phase than in the zincblende phase GaN. As a result of the greater electronic density of states, the electrons are, on the average, cooler in the wurtzite phase than in the zincblende phase under the same biasing conditions.

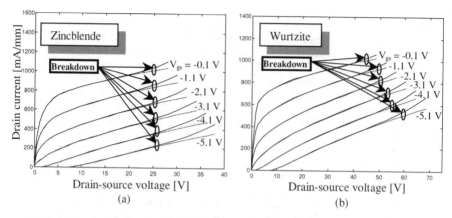

Figure 2. Output characteristics for the zincblende phase device (a) and the wurtzite phase device (b). The gate-source voltage, V_{gs} includes the Schottky barrier height. The drain currents have been calculated with impact ionization (solid lines) and without impact ionization (dashed lines). The ovals show the breakdown locations on the I-V curves.

The current-voltage characteristics shown in Figure 2, exhibit a finite slope even at the lowest gate-source voltage reported which is mainly a consequence of the very small gate length examined. The late turn-on observed for the curves with the most negative gate-source voltages is due to the potential barrier under the gate which blocks the source electrons from reaching the drain contact. This potential barrier limits the drain-source current until a sufficiently large drain-source bias is applied, which lifts the barrier to allow an appreciable current flow.

The current gain cutoff frequency, f_T was calculated from the following expression,

$$f_T = \frac{1}{2\pi} \frac{g_m}{C_g} \qquad (1)$$

where g_m is the transconductance and C_g is the gate capacitance obtained by the incremental charge partitioning method [16]. The calculated gate capacitance was 1.519 pF/cm for the zincblende device and 1.572 pF/cm for the wurtzite device. The calculated maximum cutoff frequency and transconductance for the zincblende device were calculated to be $f_T = 220$ GHz and $g_m = 210$ mS/mm which are higher than the cutoff frequency of $f_T = 160$ GHz, and transconductance of $g_m = 158$ mS/mm predicted for an otherwise equivalent wurtzite device. The higher transconductance and cutoff frequency of the zincblende phase GaN MESFET can be explained as follows. Due to the short device length, the electrons within the device experience velocity overshoot. The velocity overshoot is to first order, proportional to the inverse effective mass. The electrons have a lower effective mass in the zincblende phase than the wurtzite phase and consequently, the electron velocity overshoot in the zincblende phase is higher than that in the wurtzite phase. The higher magnitude of velocity overshoot in the zincblende phase as compared to the wurtzite phase has also been observed by Foutz et al. [17]. It is hypothesized that the calculated higher cutoff frequency and transconductance for the zincblende phase device is due to the higher overshoot velocity in zincblende phase GaN material.

Conclusions

In this paper, the first full band comparison of the breakdown and high frequency properties of both zincblende and wurtzite phase GaN MESFETs was presented. The calculations were made using a full band, ensemble Monte Carlo simulation including a numerical formulation of the impact ionization transition rates. It was found that the drain current in both phases of the GaN MESFET increases gradually with increasing drain voltage after the onset of breakdown.

It was further found that the breakdown voltage of the zincblende phase GaN device is less than the breakdown voltage of an otherwise equivalent wurtzite phase device. This is due to the lower density of states in zincblende phase GaN, which results in a warmer electron distribution than in the wurtzite phase GaN under similar field conditions. The higher breakdown voltage of the wurtzite phase device results in higher maximum output power.

The cutoff frequency and transconductance of the zincblende phase device was found to be higher than that of the wurtzite phase device. The difference in the calculated cutoff frequencies was attributed to the higher electron overshoot velocity in the zincblende phase. The maximum transconductance and cutoff frequency were calculated from the Monte Carlo simulation to be 210 mS/mm and 220 GHz for the zincblende phase MESFET, and 158 mS/mm and 160 GHz for the wurtzite phase MESFETs respectively.

It should be noted that due to the basic uncertainties in the band structure, phonon scattering rates, and ionization coefficients, the calculations presented herein serve only as a general guide to the expected performance measures of GaN MESFET devices. It should be further noted that the breakdown properties predicted here rely on previous theoretical estimates of the impact ionization coefficients that have a high degree of uncertainty and have yet to be experimentally verified. Nevertheless, on the basis of these calculations it can be expected that very high frequency performance at high breakdown voltage for GaN based MESFET structures can be obtained. In addition, it is also reasonably certain that a zincblende phase GaN MESFET structure will have a higher cutoff frequency and transconductance than an otherwise identical wurtzite phase GaN MESFET device. On the other hand, the GaN wurtzite phase device will have a higher breakdown voltage than the GaN zincblende phase device.

ACKNOWLEDGEMENTS

This work was sponsored by the National Science Foundation through grant ECS-9811366. Additional support was received by the Office of Naval Research through contract E21-K09, by the Office of Naval Research through subcontract E21-K69 made to Georgia Tech through the UCSB MURI program. Usage of the Intel Corp. Advanced Platform Computer Cluser at Georgia Tech is gratefully acknowledged.

REFERENCES

1. C. E. Weitzel, Inst. Phys. Conf. Ser. No. 142:Chapter 4, 765 (1996).
2. E. R. Brown, Solid-State Electron., **42**, 2119 (1998).
3. K. Shenai, R. S. Scott, and B. J. Baliga, IEEE Trans. Electron Devices, **36**, 1811 (1989).
4. F. Schwierz, M. Kittler, H. Forster, and D. Schipinski, Diamond and Rel. Mater., **6**, 1512 (1997).
5. F. Dessenne, D. Cichocka, P. Desplanques, and R. Fauquembergue, Mat. Sci. Eng., **B50**, 315 (1997).
6. M. Farahmand and K. F. Brennan, IEEE Trans. Electron Devices, **46**, 1319 (1999).
7. M. Farahmand and K. F. Brennan, submitted to IEEE Trans. Electron Devices, 1999 (unpublished).
8. A. W. Smith and K. F. Brennan, Prog. Quantum Electr., **21**, 293 (1998).
9. K. F Brennan., J. Kolnik, I.H. Oguzman, E. Bellotti, M. Farahmand, P. P. Ruden, R. Wang, and J. D. Albrecht, in *GaN and Related Materials II-Volume 7*, edited by S. J. Pearton (Gordon and Breach, Amsterdam, 2000) pp. 305-359.
10. M. V. Fischetti and S. E. Laux, Phys. Rev. B, **38**, 9721 (1988).
11. J. Kolnik, I. H. Oguzman, K. F. Brennan, R. Wang, and P. P. Ruden, J. Appl. Phys., **81**, 726 (1997).
12. I. H. Oguzman, E. Bellotti, K. F. Brennan, J. Kolnik, R. Wang, and P. P. Ruden, J. Appl. Phys., **81**, 7827 (1997).
13. E. Bellotti, B. K. Doshi, K. F. Brennan, J. D. Albrecht, and P. P. Ruden, J. Appl. Phys., **85**, 916 (1999).
14. I. H. Oguzman, J. Kolnik, K. F. Brennan, R. Wang, T.-N. Fang, and P. P. Ruden, J. Appl. Phys., **80**, 4429 (1996).
15. K. F. Brennan, E. Bellotti, M. Farahmand, H-E. Nilsson, P. P. Ruden and Y. Zhang, submitted to IEEE Trans. Electron Dev., Special issue on Computational Electronics, 1999 (unpublished).
16. S. E. Laux, IEEE Trans. Electron Dev., **32**, 2028 (1985).
17. B. E. Foutz, L. F. Eastman, U. V. Bhapkar, and M. S. Shur, Appl. Phys. Lett., **70**, 2849 (1997).

Mat. Res. Soc. Symp. Vol. 595 © 2000 Materials Research Society

New Materials-Theory-Based Model for Output Characteristics of AlGaN/GaN Heterostructure Field Effect Transistors

J.D. Albrecht[*†], P.P. Ruden[*], and M.G. Ancona[†]

[*]Department of Electrical and Computer Engineering, University of Minnesota, Minneapolis, MN 55455, U.S.A.

[†]Electronic Science and Technology Division, U.S. Naval Research Laboratory, Washington, DC 20375, U.S.A.

ABSTRACT

A new model is used to examine the DC output characteristics of AlGaN/GaN heterostructure field effect transistors. The model is based on the charge-control/gradual-channel approximation and takes into account the non-linear current vs. voltage characteristics of the ungated AlGaN/GaN heterostructure channel regions. The model also includes thermal effects associated with device self-heating. For the power dissipation levels considered for many applications, the thermal degradation of the carrier drift velocity is shown to cause a negative output conductance in saturation. The temperature is incorporated self-consistently into the model through the field and temperature dependent mobility obtained from Monte Carlo transport simulations for electron transport in GaN. Calculated results presented for the DC output characteristics of several AlGaN/GaN field effect transistors show a strong dependence on the thermal properties of the substrate material. The substrate materials considered in this work are sapphire, SiC, AlN, and GaN.

INTRODUCTION

The continuing improvement of group III-nitride compound semiconductor material quality has led to the demonstration of viable AlGaN/GaN heterostructure field effect transistors (HFETs) with widths up to 1mm[1], suitable for high power amplifiers. For these applications self-heating of the device is important and, since the heat is conducted out of the active region through the substrate, the thermal properties of the substrates play a significant role in the device performance. The traditional sapphire substrate has shown itself to be inadequate and recent GaN-based power amplifier devices[2] have explored new substrate materials. Alternative substrates include SiC, GaN, and AlN, which offer enhanced thermal conductivities. Power densities as large as 6.9 W/mm at 10GHz have been demonstrated for GaN-based devices on SiC[3]. Here, we present a comparison of the DC output characteristics of AlGaN/GaN HFETs on sapphire, SiC, AlN, and GaN.

Monte Carlo transport calculations for electrons in GaN have yielded results for the degradation of the electron velocity with increasing temperature[4]. An HFET model in the framework of the charge-control/gradual-channel approximation and based on a parameterization of the calculated transport properties has been developed. Initial simplified versions of this model have been applied successfully to the analysis of DC current vs. voltage characteristics of gated[5] and ungated[6] AlGaN/GaN heterostructure devices fabricated on sapphire substrates. The model will be summarized only briefly here. A detailed description will be published separately[7].

The thermal properties of the substrates considered are included via temperature-dependent thermal impedances obtained from independent simulations of the two-dimensional heat flow. Dissipation of heat generated in the channel region occurs through conduction through the substrate, which is coupled to a 300K heatsink. Although the thermal impedance is determined for a particular lithographic design, the qualitative effects on the device output characteristics of the different substrate thermal properties are meaningful for many device designs with the exception of those in which heatsinking is accomplished by thermal contacts to the top surface or where the substrate material has been partially removed after processing.

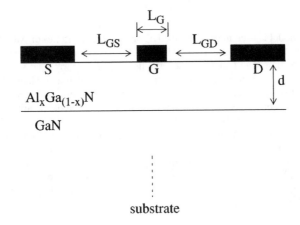

Figure 1: *Schematic HFET structure.*

MODEL DESCRIPTION

Materials-theory-based device models are an extension of calculations of fundamental material properties such as pseudopotential electronic bandstructure calculations[8] and Monte Carlo simulations of carrier transport[4,9] over wide ranges of applied electric field and lattice temperature. Through the parameterization of the electron transport properties, the HFET device model is fully self-contained and does not rely on external input parameters to describe the output conductance or series resistances associated with the ungated portions of the channel. The materials-theory-based methodology permits an investigation of extrinsic effects on device performance, e.g. the effects of different substrate thermal properties, in a natural way.

Lying at the core of the HFET model are the semi-analytic charge-control/gradual–channel approximation models for ungated and gated structures developed in refs. 5 and 6, respectively. These models yield implicit device equations for the steady-state current in terms of the charge distribution along the channel of the device structure shown schematically in Figure 1. The two-dimensional channel carrier concentration, n, as a function of channel position, x, is calculated from the potential difference between the surface (denoted by subscript s) and channel (ch) potentials by

$$n(x) = \frac{\varepsilon}{q(d + \Delta d)} \left[\phi_s(x) - \phi_{ch}(x) - V_T \right] \tag{1}$$

where q is the electron charge, ε is the appropriate dielectric constant, and V_T is the threshold voltage. The AlGaN barrier layer thickness is denoted by d. The spatial extent of the two-dimensional electron gas, Δd, is calculated form a self-consistent solution of the coupled, one-dimensional Schroedinger and Poisson equations and verified by capacitance measurements on similar structures[5,6]. For the carrier concentrations relevant here, $\Delta d \sim 30$A. The threshold voltage and barrier layer thickness chosen in this work are -4.0V and 200Å, respectively. The surface potential is taken as constant under the gate and linearly varying between the source and drain contacts and the gate.

The DC drain current calculation involves matching the channel carrier concentration at channel positions directly beneath the edges of the gate metal. The channel potential boundary conditions are related to the applied voltage, V_{DS}, and the voltage drops across the drain and source contact resistances. Finally, the current, normalized to the gate width is calculated from

$$I_D = qn(x)\mu(T, \frac{dV_{ch}}{dx}) \cdot \frac{dV_{ch}}{dx} - qD_n(T)\frac{dn}{dx} \tag{2}$$

and the appropriate boundary conditions. The carrier diffusivity is obtained from the temperature dependent low-field mobility using the classical Einstein relation. The room temperature mobility at low field is 1100 cm^{2}V^{-1}sec^{-1}. Drain and source contact resistances equal to 0.5Ωmm are assumed[5]. The channel carrier concentration in the ungated access regions between the gate and the source and drain contacts is non-uniform. Consequently, these regions give rise to non-linear resistances as has been demonstrated experimentally and theoretically previously[6].

substrate	Z_o (Kmm/W)	T_o (K)
Al$_2$O$_3$	25.0	1000
GaN	10.7	750
AlN	8.9	1000
SiC	4.9	470

Table I: *Structural thermal impedance parameters obtained from two-dimensional heat flow simulations.*

Heat dissipation is treated self-consistently using a temperature-dependent thermal impedance. The channel is assumed to be separated from the substrate by a 3μm thick GaN buffer layer. For convenience of comparison, the substrates are all assumed to be

330μm thick. The temperature dependence of the thermal conductivities of the constituent materials has been taken into account using

$$(T - 300K) = Z_{th} \cdot I_D \cdot V_{DS}$$
$$Z_{th}(T) = Z_o(1 + \frac{T - 300K}{T_o}) \qquad (3)$$

where Table I gives the calculated room temperature thermal impedances, Z_o, for the various substrates considered, together with a characteristic temperature, T_o, that describes the temperature dependence of the thermal impedance. The temperature of the heatsink is 300K. The temperature dependence of the electron drift velocity vs. electric field characteristics of GaN is taken into account parametrically[4].

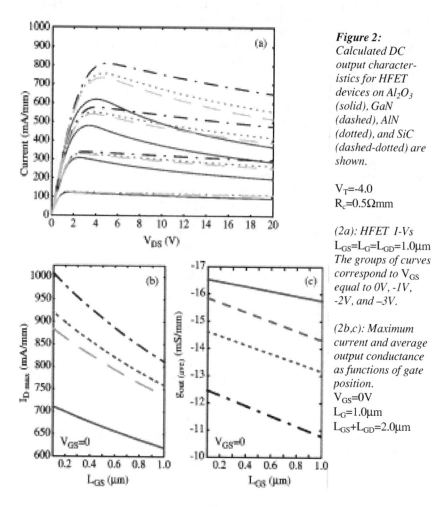

Figure 2:
Calculated DC output character-istics for HFET devices on Al₂O₃ (solid), GaN (dashed), AlN (dotted), and SiC (dashed-dotted) are shown.

V_T=-4.0
R_c=0.5Ωmm

(2a): HFET I-Vs
$L_{GS}=L_G=L_{GD}=1.0\mu m$
The groups of curves correspond to V_{GS} equal to 0V, -1V, -2V, and –3V.

(2b,c): Maximum current and average output conductance as functions of gate position.
$V_{GS}=0V$
$L_G=1.0\mu m$
$L_{GS}+L_{GD}=2.0\mu m$

RESULTS AND DISCUSSION

Results for three device designs are shown in Figure 2. Each device design was assumed to be fabricated on the four different substrates mentioned above and characterized by the thermal impedances listed in Table 1. Referring to the schematic HFET in Figure 1, the first device has $L_G=L_{GS}=L_{GD}=1.0\mu m$. The current vs. voltage characteristics for this device are shown in Figure 2(a). The effects of the substrate thermal properties are evident from the variation in the maximum currents obtained for $V_{GS}=0$. Due to the high thermal impedance, the negative output conductance is a much stronger effect for the device on sapphire than for devices on the other substrates. The calculated results for devices fabricated on sapphire substrates are consistent with measurements[7]. For small applied bias, the thermal effects are negligible and therefore, the current vs. voltage characteristics of all devices shown in Figure 2(a) are similar in that range.

In Figures 2(b) and 2(c), the effects of the gate position are examined. For a device of the same gate length as in Figure 2(a), the gate is moved closer to the source such that $L_G=1.0\mu m$ and $L_{GS}+L_{GD}=2.0\mu m$. The resulting output characteristics have features associated with the carriers experiencing decreasing electric fields under the gate region closest to the drain due to the increasing L_{GD}. This effect is characterized by increased maximum device currents as shown in Figure 2(b). The maximum current increases as the gate is moved closer to the source. A larger knee voltage is required to obtain the maximum current because the electric field under the gate is smaller for a given applied voltage than in the case of the device in Figure 2(a). The combined effects of the increase in device current and the shifting of the current maxima to larger V_{DS} is shown in figure 2(c) as an average output conductance defined by the current drop from its maximum to $V_{DS}=20V$, such that $g_{out(ave)} =$

$$(I_{V_{DS}=20V} - I_{max})/(20V - V_{knee}).$$

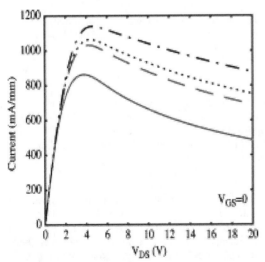

Figure 3: *Calculated current vs. voltage characteristics for a scaled HFET device on Al₂O₃ (solid), GaN (dashed), AlN (dotted), and SiC (dashed-dotted) are shown. The curves correspond to $V_{GS}=0V$. $V_T=-4.0$, $R_c=0.5\Omega mm$, $L_{GS}=L_G=L_{GD}=0.5\mu m$.*

In Figure 3, a scaled device is examined. In this case, $L_G=0.5\mu m$ and $L_{GS}=L_{GD}=0.5\mu m$. The contact resist-ances are held constant. A clear increase in output current accompanies the scaling and the trans-conductance has improved. The maximum current output has increased for the device with the best heatsinking (SiC substrate) from 812mA/mm to 1180mA/mm. For the device on sapphire a similar increase in the maximum current from 615mA/mm to 860mA/mm is seen. Clearly, with the increase in power the thermal impedance plays an even greater role.

CONCLUSIONS

We have presented calculated current vs. voltage characteristics for a series of GaN-based HFETs on a variety of substrates, including sapphire, SiC, GaN, and AlN. Our device model is based on the field and temperature dependent electron velocity for GaN obtained by Monte Carlo transport simulations. The temperature is included self-consistently via a thermal impedance that takes into account the temperature dependence of the thermal conductivity of the substrate and the epitaxial materials. We demonstrate that the thermal properties of the substrate play a significant role in limiting the HFET performance under high-power conditions. In each of our simulations, the effects of improved heatsinking are shown to increase the DC output current and to decrease the negative output conductance observed for GaN HFETs.

In addition, our investigation included the effects of contact spacing as a second limiting factor for the device performance. By examining device geometries with the gate positioned closer to the source and a scaled device, it is shown that significant gains in the maximum current and transconductance can be achieved with these design improvements.

ACKNOWLEDGEMENT

This work was supported in part by the National Science Foundation, the Office of Naval Research, the Minnesota Supercomputer Institute, and the National Research Council.

REFERENCES

[1] N.X. Nguyen, C. Nguyen, and D.E. Grider, *Electr. Lett* **34**, 811 (1998).
[2] J.J. Xu, Y.-F. Wu, S. Keller, G. Parish, S. Heikman, B.J. Thibeault, U.K. Mishra, and R. A. York, *IEEE Microwave and Guided Wave Letters.* **9**, 277 (1999).
[3] S.T. Sheppard, K. Doverspike, W.L. Pribble, S.T. Allen, J.W. Palmour, L.T. Kehias, and T.J. Jenkins, *IEEE Electr. Dev. Lett.* **20**, 161 (1999).
[4] J.D. Albrecht, R.P. Wang, P.P. Ruden, M. Farahmand, and K.F. Brennan, *J. Appl. Physics* **83**, 4777 (1998).
[5] P.P. Ruden, J.D. Albrecht, A. Sutandi, S.C. Binari, K. Ikossi-Anastasiou, M.G. Ancona, D.D. Koleske, and A.E. Wickenden, *MRS Int. J. Nit. Sem. Res.* **4S1**, GG6.35 (1999).
[6] J.D. Albrecht, P.P. Ruden, S.C. Binari, K. Ikossi-Anastasiou, M.G. Ancona, R.L. Henry, D.D. Koleske, and A.E. Wickenden, *Mat. Res. Soc. Symp. Proc.* **572**, 489 (1999).
[7] J.D. Albrecht, P.P. Ruden, M.G. Ancona, and S.C. Binari, submitted for publication.
[8] R. Wang, P.P. Ruden, J. Kolnik, I. Oguzman, and K.F. Brennan, *J. Phys. Chem. Sol.* **58**, 913 (1997).
[9] J.D. Albrecht, R.P. Wang, P.P. Ruden, M. Farahmand, E. Bellotti, and K.F. Brennan, *Mat. Res. Soc. Symp. Proc.* **482**, 815 (1998).

Mat. Res. Soc. Symp. Vol. 595 © 2000 Materials Research Society

High-Gain, High-Speed ZnO MSM Ultraviolet Photodetectors

H. Shen[1], M. Wraback[1], C. R. Gorla[2], S. Liang[2], N. Emanetoglu[2], Y. Liu[2], and Y. Lu[2]
[1]US Army Research Laboratory, Sensors and Electron Devices Directorate,
AMSRL-SE-EM, Adelphi, MD 20783
[2]Rutgers University, Department of Electrical and Computer Engineering, Piscataway, NJ
08854

ABSTRACT

High quality zinc oxide (ZnO) films were epitaxially grown on R-plane sapphire substrates by metalorganic chemical vapor deposition at temperatures in the range 350-600°C. In-situ nitrogen compensation doping was performed using NH_3. The metal-semiconductor-metal ultraviolet-sensitive photodetectors were fabricated on nitrogen-compensation-doped epitaxial ZnO films. The photoresponsivity of these devices exhibits a linear dependence upon bias voltage up to 10 V, with a photoresponsivity of 400 A/W at 5 V. The rise and fall times are 1 and 1.5 μs, respectively.

INTRODUCTION

ZnO has found many electrical and optical applications[1-10] because it is a wide band gap semiconductor with a strong piezoelectric effect. However, little work has been done on ZnO UV photodetectors. Early studies[11,12] indicated that ZnO detectors exhibit large photoconductive gain. However, this photoconductive gain is related to a slow chemical process. On the other hand, H. Fabricius et. al[13] reported a high speed, ultraviolet-sensitive photodiode using Au as a Schottky barrier on a thin sputtered layer of polycrystalline ZnO. A rise time of 20μs and a decay time of 30μs were observed. Unfortunately, the quantum efficiency for these devices was low (on the order of 1%) and the photoresponsivity was small (on the order of 3mA/W).

In this paper, we report high-gain, high-speed ZnO MSM ultraviolet photodetectors using high quality epitaxial ZnO films grown on R-plane sapphire. In comparison with their polycrystalline counterpart, the epitaxial ZnO detectors exhibit a higher photoresponsivity and a faster photoresponse time.

DEVICE FABRICATION

ZnO thin films were grown on R-sapphire. A rotating disc vertical flow MOCVD reactor was used. Diethylzinc (DEZn) and oxygen were used as the Zn metalorganic source and oxidizer respectively. Detailed descriptions on the growth system and the growth process have been reported elsewhere.[14,15] In-situ nitrogen compensation doping (p-type) from NH_3 was used to reduce the electron concentration in undoped ZnO.

MSM structures were fabricated on nitrogen-compensation-doped ZnO films. Al (200 nm thick) was used as the contact metal because of its low electronegativity (of 1.5)[16] . Circular patterns were used for contact electrodes. The spacing of the two electrodes was varied from 2μm to 16μm. The ZnO surface was cleaned using a standard ultrasonic solvent cleaning process. E-beam metallization, photolithography, and wet chemical etching techniques were used to generate the electrode pattern.

EXPERIMENTAL RESULTS

Figure 1 shows the effect of electrical resistivity on growth temperature. The resistivity decreases as the growth temperature increases. A highly resistive film was obtained at 390°C. Undoped ZnO generally exhibits n-type behavior due to the presence of oxygen vacancies or zinc interstitials. Incorporation of nitrogen reduces the electron carrier concentration. Although a higher growth temperature is preferred to crack NH_3, it actually results in lower resistivities. This is due to an increase in ZnO non-stoichiometry (Zn interstitials or O vacancies) at higher growth temperatures.

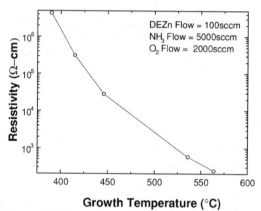

Figure 1 Effect of MOCVD growth temperature on the electrical resistivity of N-doped ZnO films

The solid line in Figure 2 shows the current-voltage (I-V) characteristics of the detector. Low dark current of 450 nA is achieved at 5V due to high resistivity. The dashed line in Figure 2 shows I-V characteristics under 6.4nW, 365nm UV photo-illumination. The linear I-V relations under both forward and reverse bias indicated the presence of an ohmic

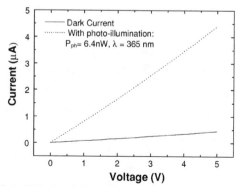

Figure 2 Dark (solid line) and photo-illuminated (dashed line) I-V

metal-semiconductor contact and that the device is operating in photoconductive mode.

The spectral photoresponse of the MSM photodetectors was measured using a Xe lamp and monochromator combination. A Newport 1830-C Optical Power Meter was employed to calibrate the light beam intensity. Figure 3 shows the spectral response of a ZnO photoconductive detector. A sharp cut-off near 373 nm is observed. The photoresponse drops by more than two orders of magnitude across the cut-off wavelength within 15 nm of the band edge. A photoresponsivity of about 400A/W is obtained under 5 V bias voltage at wavelengths from 373 nm to 300 nm.

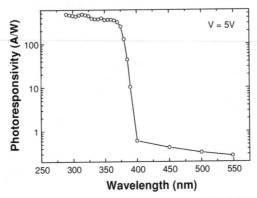

Figure 3 Spectral response of a ZnO MSM photodetector under 5 V bias.

Figure 4 shows the photoresponse as a function of bias voltage from 1mV to 5V at 6.4nW, 365nm UV photo-illumination. A linear relation indicates that the device is operating in constant mobility regime. Assuming a carrier lifetime of 1 μs (discussed below), we find an electron mobility of ~ 100 cm^2/V*s.

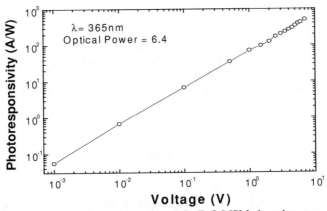

Figure 4 Photoresponsivity vs. bias voltage of a ZnO MSM photodetector.

We have also measured the photoresponse speed of the detector. The optical excitation source was a Ti:Sapphire regenerative amplifier-pumped optical parametric amplifier, which produces visible ultrashort pulses at a 175 kHz repetition rate. These pulses were compressed to less than 100 fs and frequency doubled in barium borate (BBO) to obtain ultraviolet pulses tunable between 300 and 375 nm. Neutral density filters were used to control the optical power on the detector. The optical energy on the detector was about 5.6fJ per pulse. The signal from the detector was recorded by a digital scope with a time resolution better than 1 ns. Figure 5 shows the response of the ZnO detector under 5 V bias. The rise time is about 1 μs, while the fall time is about 1.5 μs. Because the optical pulse used in our measurement was less than 100fs, we used the

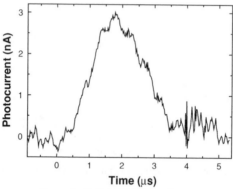

Figure 5 Photocurrent vs. time of a ZnO MSM photodetector under 100fs pulsed excitation.

average power at the frequency of 175kHz to calculate the gain of the device and found that it is about 10.

DISCUSSION

Early studies[11,12] indicated that the photoresponse of ZnO consisted of two parts, a rapid and reproducible solid state process [$h\nu \rightarrow h^+ + e^-$], where e and h represent electron and hole respectively, and a slow process with a large response. The latter is mainly due to a two step process: (a) oxygen adsorption under dark conditions by trapping of an electron,

$$O_2 + e^- \rightarrow O_2^-$$

and (b) photodesorption of O_2 by capturing a photo-generated hole,

$$O_2^- + h^+ \rightarrow O_2$$

We believe that the fast solid state photoresponse in our devices is enhanced due to the higher mobility and longer lifetime of our improved ZnO film. The suppression of the slow O_2 related process may be due to the reduction of electron concentration in our nitrogen compensation doped ZnO films. We notice that the slow process is usually associated with polycrystalline ZnO films deposited on glass substrates.[17-19] Polycrystalline ZnO films have a large number of grain boundaries. Therefore, the suppression of the slow process in our devices may also be attributed to the reduction of the grain boundaries in our high quality epitaxial ZnO film.

CONCLUSION

We have demonstrated high quality ZnO epitaxial films MOCVD grown on R-plane sapphire. MSM UV-sensitive photodetectors have been successfully fabricated on nitrogen-compensation-doped ZnO epitaxial films. The low frequency photoresponsivity is on the order of 400A/W at 5V bias. The minority lifetime of the devices is estimated to be in the 1 µs range. The rise time and fall time are about 1 µs, and 1.5 µs, respectively.

ACKNOWLEDGEMENT

This work is partially supported by a Rutgers University SROA grant, and by the BMDO Innovative Science and Technology Office under contract DASG60-98-M-0067 (Monitored by Mr. Esam Gad at U.S. Army Space and Missile Defense Command).

REFERENCES

1. J. Hu and R. G. Gordon, *J. Electrochem. Soc.* **139**, 2014 (1992).
2. K. L. Chopra, S. Major and D. K. Pandya, *Thin Solid Film* **102**, 71 (1984).
3. N. W. Emanetoglu, S. Liang, C. Gorla, Y. Lu, S. Jen, and R. Subramanian, 1997 *IEEE Ultrasonics Symposium*, p. 195.
4. Y. Makishima, K. Hashimoto and M. Yamaguchi, *Jpn. J. Appl. Phys.* Part 1, **33**, 2998 (1994).
5. M. Wu, A. Azuma, T. Shiosaki and A. Kawabata, *J. Appl. Phys.* **62**, 2482 (1987).
6. A. Barker, S. Crowther, D. Rees, *Sensors and Actuators* A **58**, 229 (1997).
7. D. Peregol, E. Pic and J. Plantier, *J. Appl. Phys.* **62**, 2563 (1987).
8. R. G. Heideman, P. V. Lambeck, J. G. E. Gardeniers, *Optical Materials* **4**, 741 (1995).
9. F. Hamdani, D. J. Smith, H. Tang, W. Kim, A. Salvador, A. E. Botcharev, J. M. Gibson, A. Y. Polyakov, M. Skowronski, and H. Morkoc, *J. Appl. Phys.* **83**, 983 (1998).
10. M. A. L. Johnson, S. Fujita, W. H. Rowland, W. C. Hughes, J. W. Cook, and F. Schetzina, *J. Electron. Mater.* **25**, 855 (1996).
11. Erich Mollwo, *Photoconductivity Conf.* (Wiley, New York), p. 509, (1956).
12. P. H. Miller, Jr., *Photoconductivity Conf.* (Wiley, New York), p. 287, (1956).
13. H. Fabricius, T. Skettrup and P. Bisgaard, Appl. Optics, **25**, 2764 (1986).
14. S. Liang, C. R. Gorla, N. Emanetoglu, Y. Liu, W. E. Mayo and Y. Lu, *J. Electron. Mater.*, **27**, L72 (1998).
15. C. R. Gorla, N. W. Emanetoglu, S. Liang, W. E. Mayo, Y. Lu, M. Wraback and H. Shen, *J. Appl. Phys.*, **85**, 2595 (1999).
16. C. A. Mead, *Pyhs. Lett,* **18**, 218 (1965).

17. Y. Takahashi, M. Kanamori, A. Kondoh, H. Minoura and Y. Ohya, *Jpn. J. Appl. Phys.* **33**, 6611 (1994).
18. D. H. Zhang and D. E. Brodie, *Thin Solid Films* **261**, 334 (1995).
19. D. H. Zhang, *J. Phys. D Appl. Phys.* **28,** 1273 (1995).

Mat. Res. Soc. Symp. Vol. 595 © 2000 Materials Research Society

Temperature Distribution in InGaN-MQW LEDs under Operation

**Veit Schwegler[1], Matthias Seyboth[1], Sven Schad[1],Marcus Scherer,[1]
Cristoph Kirchner[1], Markus Kamp[1], Ulrich Stempfle[2], Wolfgang Limmer[2], and
Rolf Sauer[2]**
[1]Dept. of Optoelectronics, University of Ulm, 89069 Ulm, Germany
[2]Dept. of Semiconductor Physics, University of Ulm , 89069 Ulm, Germany

ABSTRACT

The temperature distribution in InGaN-MQW light emitting diodes was examined during operation with spatially resolved micro-Raman and micro-Electroluminescence measurements. The experimental results were compared to finite element simulations. A good agreement between the different experimental and calculated data is found. Maximum operation temperatures up to 140 °C at a moderate forward currents of 30 mA are revealed by all three independent methods. Influences of substrate thickness, different substrates, and even bond-wires are shown.

INTRODUCTION

Generation of heat by ohmic losses limits performance and lifetime of light-emitting diodes (LEDs). Due to the high bandgap voltages GaN-based devices are subject of especially severe ohmic heating. Increased temperature reduces quantum efficiency and enhances diffusion of impurities (e.g. dopands, contact metals) as well as migration of dislocations [1].

UV-emitting InGaN-MQW LEDs, grown by low pressure MOVPE on sapphire substrates, are investigated regarding their junction temperature at different injection currents. Temperature distributions are determined by E2 phonon scattering (Raman), spatially resolved μ-EL, and finite element simulations. Influences of substrate thickness (sapphire) and material (sapphire/SiC) on device temperatures are studied.

EXPERIMENTAL

Our UV-LEDs show narrow single peak emission at about 410 nm. A buffer layer of 2 μm undoped GaN is followed by 1 μm Si-doped GaN. The active region consists of a 4 x InGaN/GaN (1.5 nm/5.5 nm) quantum well structure followed by a Mg-doped $Al_{0.08}Ga_{0.92}N$ electron barrier and 300 nm p-GaN cladding layer. In the Mg-doped GaN we have free carrier concentrations of $p=2-3 \times 10^{17}$ cm^{-3} in the Si-doped GaN of $n=5 \times 10^{18}$ cm^{-3}. A circular mesa is structured by chemically assisted ion beam etching. Contacts consist of an evaporated Ni/Au metallization with an inner circular area as p-contact and a surrounding n-contact. These devices have series resistances of approximately 30 Ω. For measurements they are diced into pieces with a size of some 800 μm x 1800 μm and mounted on teflon or metal heat sinks.

Operation of these LEDs on a temperature controlled heat sink displays impressively the problems aligned with temperature creation: Output power decreases exponentially with increased heat sink temperature (Figure. 1), since the internal quantum

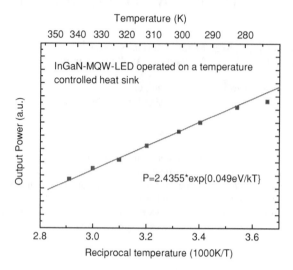

Figure 1: *Temperature dependence of output power showing the almost exponential decrease of light output with increased temperature.*

efficiency is depending almost exponentially on temperature according to equation (1) approximated for higher temperatures.

$$\eta_i = \left(1 + (\frac{\tau_t p N_t}{\tau_r n N_r}) \exp(-\frac{E_t - E_r}{k_B T})\right)^{-1} \quad (1)$$

with $\eta_i, \tau_i, n, p, N_i, E_i$ being the internal quantum efficiency, the lifetime, hole and electron concentration, concentrations and energies of radiative and non-radiative recombination mechanisms. The indices t and r account for trap and radiative related transitions.

The temperature of a LED can be derived from the emission spectrum with a fit using several fitting parameters [2], it is also possible to exploit only the short wavelength slope near the maximum of the EL spectrum for temperature determination according to reference [3].

Full fits on μ-EL spectra are applied to get a spatially resolved picture of the temperature distribution. The backside emission of a LED was therefore spectrally and spatially mapped through the sapphire substrate. Figure 2 shows the integral EL intensity and wavelength distribution of a LED driven with a current density of 91 A/cm². The elevated current density beneath the p-contact is reflected by an increased temperature. A heat drain effect by the bondwire can be seen in a reduced redshift. Details are published in [4].

Temperature determination with Raman measurements makes use of the phonon shift. From the relative shift of the GaN E_2-phonon frequency of a LED under operational current with respect to the frequency at zero current, the temperature of the device can be

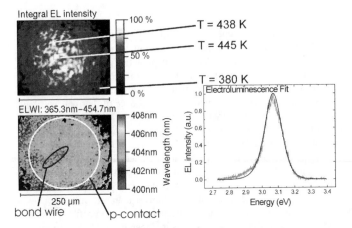

Figure 2. Temperature distribution derived at different points from μ-EL (current density j=320 A/cm²), the redshift beneath the p-contact is due to increased temperature resulting from higher current density, the reduced redshift beneath the bond wire is due to heat drain effect. For an extended discussion of measuring method and results cf. [4]

computed. A detailed description of this measurement method is given in reference [5]. A spot diameter of 0.7 μm permits high spatial resolution.

Results of Raman measurements are depicted in Figure 3 in comparison with data obtained from high energy slope fits of EL spectra. In good agreement, both methods reveal temperatures of about 415 K already at moderate currents of 30 mA.

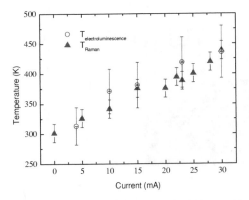

Figure 3. Temperature of LEDs under operation, LEDs mounted on teflon heat sink. Comparison of temperatures derived from EL and Raman measurements. In good agreement, both methods yield temperatures as high as 415 K even at moderate currents of 30 mA.

SIMULATION

The finite element calculations for temperature distributions in LEDs are done using ANSYS, a well established commercial simulation program which is widely used for simulations of mechanical and electrical problems. Its capabilities include a combined thermal-electrical solver employing all linear effects and non-linear thermal transport properties. The simulations include the material parameters of GaN, InGaN, AlGaN, Al_2O_3, SiC and Ni/Au as contact materials. Temperature dependence of thermal conductivity, carrier concentration (particularly thermally depending Mg acceptor ionization), carrier mobility, and contribution to thermal conductivity from free carriers have been incorporated into the model. To account for our particular problem the geometry and electrical IV-characteristics of the LEDs are included. There are no fitting parameters within the simulation. Besides the mentioned material and device parameters the only input is the backside temperature serving as a boundary condition. This temperature has been measured with a thermal probe. The solver than returns a 3-dimensional distribution (with radial symmetry) of electrical and thermal profiles present in the LEDs under various injection conditions as well as in dependence of the boundary conditions, i.e. thermal contact and properties of the base to which the LEDs are mounted. Figure 4 shows a cross section of such a simulated temperature distribution. Surface temperatures drawn of simulations were compared to Raman data (Figure 5). A refined simulation is improved by implementing heat drain by bond wires. A good match of both methods could be obtained.

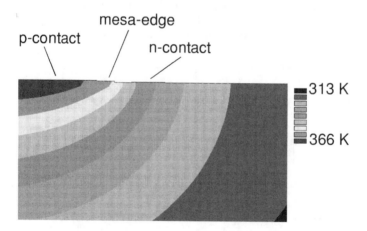

Figure 4. Cross section of temperature distribution in a LED under operational current of 63 mA mounted on metal heat sink obtained with finite element calculation. The elevated temperature beneath the p-contact is clearly visible.

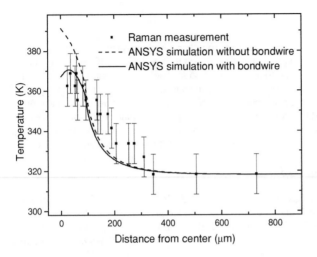

Figure 5. *Comparison of spatially resolved temperature profiles obtained from Raman measurements and finite element calculations. Operation conditions as in Figure 4. The data show good agreement of both methods. The impact of a bond-wire on device temperature is demonstrated.*

Considering the good agreement between simulation and experiment, we used ANSYS to calculate the influence of substrate thickness and material on temperatures in LEDs. Results are compiled in Figure 6. The simulations show strong impacts both of substrate thickness and material. Especially the improved heat transport through SiC substrates is clearly visible.

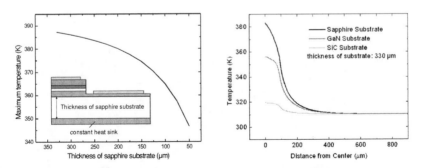

Figure 6. *Device simulations with varied substrate thickness or material show the impact of the thermal conductivity of the substrate. Advantages of SiC are demonstrated.*

CONCLUSIONS

Using Raman and EL measurements together with finite element calculations we obtained a coherent picture on temperature distribution in InGaN-MQW LEDs. The different approaches reveal good agreement both in absolute temperature and in spatial distribution. The operating current causes significant heating limiting the output power and lifetime of devices. Even for moderate currents our devices show strong heating. Most heat is generated beneath the p-contact. Temperature drain effects of bond wires can be detected and simulated. The good agreement of simulation and experiment allows the employment of simulation as a versatile tool for further improvement of devices.

ACKNOWLEDGEMENTS

The authors would like to thank P. Fischer, J. Christen, and M. Zacharias (Univ. Magdeburg) for the μ-EL measurements. Also we are grateful to A.E. Yunovich for the valuable discussions and assistance on fits of EL spectra. Financial support by German Ministry of Science and Education (BMBF) under Contract No. 01 BS 802 is gratefully acknowledged.

REFERENCES

1. L. Sugiura, J. Appl. Phys. **81**, 1633 (1997).
2. K.G. Zolina, V. E. Kudryashov, A.N. Turkin, and A.E. Yunovich, Semiconductors **31**, 901 (1997).
3. A.E. Yunovich, V.E. Kudryashov, A.N. Turkin, A.N. Kovalev, F.I. Manyakhin, MRS Internet J. Nitride Semic. Res. **4S1**, G6.29 (1999).
4. P. Fischer, J. Christen, M. Zacharias, V. Schwegler, C. Kirchner, and M. Kamp, Appl. Phys. Lett. **75**,. 3440 (1999).
5. A. Link, K. Bitzer, W. Limmer, R. Sauer, C. Kirchner, V. Schwegler, M. Kamp, D.G. Ebling, and K.W. Benz, J. Appl. Phys. **86**, 6256 (1999).

Mat. Res. Soc. Symp. Vol. 595 © 2000 Materials Research Society

Electron Beam Pumping in Nitride Vertical Cavities with GaN/ $Al_{0.25}$ $Ga_{0.75}$ N Bragg Reflectors

H. Klausing, J. Aderhold, F. Fedler, D. Mistele, J. Stemmer, O. Semchinova, and J. Graul ; J. Dänhardt[1] and S. Panzer[1]
Laboratorium für Informationstechnologie, Universität Hannover,
30167 Hannover, Germany
[1]Fraunhofer Institut für Elektronenstrahl- und Plasmatechnik,
01277 Dresden, Germany

ABSTRACT

Electron beam pumped surface emitting lasers are of great interest for a variety of applications, such as Laser Cathode Ray Tubes (LCRT) in projection display technology or high power UV light sources for photolithography.
Two distributed Bragg reflector (DBR) samples were grown by plasma assisted molecular beam epitaxy (PAMBE). The active regions of the samples are a GaN:Si bulk layer and a multihetero (MH) structure, respectively.
Also, a separately grown single DBR stack was studied to find optical transmission and reflection properties which were compared to transfer matrix simulations.
Scanning electron beam pumping at 80 K with an excitation energy of 40 keV at varying beam currents revealed luminescence emission maxima located at about 3.45 eV for the sample with the MH structure active region. Optical modes appeared for excitation powers greater than 0.85 MW/ cm^2 . Further increasing the excitation power density the number of modes increased and a broadening and redshift of the luminescence spectrum could be observed.
Based on our experimental results, we discuss the dependence of optical parameters of the nitride vertical cavity and sample surface reactions on primary electron beam power.

INTRODUCTION

Great progress has been achieved in the development of III-V nitride semiconductor light emitting diodes (LEDs) and room temperature continuous wave (cw) laser diodes (LDs) operating in the blue-green to ultraviolet (UV) spectral range. With the success of commercial production injection current LDs, based on InGaN/GaN/AlGaN compounds, focus has shifted towards the development of vertical cavity surface emitting lasers (VCSEL), which should be applicable in optical storage technology and UV light source photolithography.
Recent examples are vertical cavities consisting of a GaN/AlGaN DBR as bottom reflector and a dielectric DBR as top reflector [1], GaN/AlGaN DBRs [2,3] or dielectric mirrors [4] on both sides of the active medium, and a high gain quantum dot-like active medium with one GaN/AlGaN DBR only [5].
Stimulated emission in the blue and near UV spectral range of the just mentioned VCSEL structures has been accomplished by optical pumping. Another excitation method is electron beam pumping (EBP), which could be applied in projection display technology. The excitation process within semiconductor laser structures induced by accelerated electrons is not well understood and differs in many aspects from the better known

photonic excitation. The characteristics of the EBP emission spectrum depend on the electron beam intensity distribution, excitation time, penetration depth, etc.

Realization of VCSELs relies on the fabrication of highly reflective mirrors, which could be produced by depositing suitable dielectric multilayers [4], resulting in several after growth processing steps. This can be avoided by integration of distributed Bragg reflector (DBR) stacks (made from multiple GaN/AlGaN/AlN layers) directly into the growth process, which also allows doping necessary for current injection. Many researchers have investigated DBR quarter wave stacks and achieved reflectivities of up to 96-97% at wavelengths in the blue-green and near UV spectral range[6-9].

We particularly focus attention on the two GaN/AlGaN DBR vertical cavity test structures.

EXPERIMENT

Our four DBR samples were grown by plasma assisted molecular beam epitaxy on sapphire (0001). On the first sample (V54) growth started with an initial 18 nm AlN buffer layer followed by a 45 nm GaN spacer layer. Then the deposition of the structure under study began with first a 15.5 period GaN/ $Al_{0.25}Ga_{0.75}N$ DBR stack, secondly the 1.2 µm GaN:Si active region. Growth concluded with the deposition of another identical DBR and a 100 nm Al cap layer. The second sample (V60) has a GaN buffer layer and a MH structure active region of 8.5 periods GaN/$Al_{0.08}Ga_{0.92}N$, but is otherwise identical. The remaining two samples each consist of a single DBR stack (10.5 periods of GaN/$Al_{0.25}Ga_{0.75}N$), one containing an $Al_{0.1}Ga_{0.9}N$ buffer layer (V51), the other a GaN buffer layer, followed by a 45 nm GaN spacer layer (V52).

Reflection and transmission measurements done with the latter single DBR samples were compared to transfer matrix simulations. The extracted data was used to estimate the reflectivity of the lower and upper DBR stacks, including the Al layer of sample V60. Both vertical cavity samples were pumped at $T = 80$ K with a cw electron beam of $E_e =$ 40 keV. The beam velocity v_{sc} was tunable from 10 to 480 m/s and the beam current I_e was varied between 10 and 300 µA at a constant electron beam cross-section d_e of about 25 ± 1 µm as measured by SEM.

Fig.1 shows the optical path of the experimental setup. Light emitted by the sample through the sapphire was spectrally analyzed by a double monochromator in conjunction with a photomultiplier tube (PMT). Our setup had a 0.2 nm spectral resolution at 500 nm, a repeatability of ± 0.2 nm, and an accuracy of ± 0.4 nm. Finally the samples were subjected to profilometry and Auger electron spectroscopy (AES) in order to gain information about the interaction of the electron beam and the sample surface.

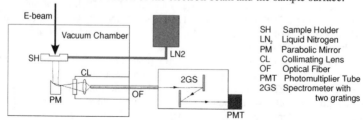

SH   Sample Holder
LN₂   Liquid Nitrogen
PM   Parabolic Mirror
CL   Collimating Lens
OF   Optical Fiber
PMT   Photomultiplier Tube
2GS   Spectrometer with two gratings

Figure 1: *Experimental setup for detection of surface emission*

RESULTS

Shown in Fig. 2 (left) are spectra of sample V54, collected with e-beam currents of I_e=100 μA and I_e=200 μA. Optical modes appear within the spectral range of 3.2 to 3.48 eV. Additional scans proved that there is no increment of intensity in any other range. Optical modes of relatively high intensity accumulate between 3.33 and 3.45 eV. The visibility V ($V=I_{max}/I_{min}$, where I_{max} and I_{min} are maximum and minimum intensities) is bigger for I_e=100 μA excitation current than for I_e=200 μA. Furthermore, two broad spontaneous emission peaks occur at about 3.395 and 3.45 eV. The apparent optical modes in the emission spectra of sample V60 (MH active region) (Fig. 3, right) differ from the ones of sample V54. Applying current densities of 16 A/cm^2 (I_e=80 μA) and above optical modes can be observed. The higher the excitation current densities are, the higher the number of modes and the broader and more intense the spontaneous emission envelope becomes. Increasing the current density from 2 to 16 A/cm^2 results in a 4 meV redshift of the emission envelope maximum. The redshift between 16 and 41 A/cm^2 (I_e=80 μA and I_e=200 μA) is 14.7 meV.

Above an excitation current density of 41 A/cm^2 the spontaneous emission intensity increases in the lower energy tail, but it breaks down in the range near 3.45 eV, where optical modes become more intensive, which increases their visibility.

We suggest that the observed modes result from multiple reflections of amplified spontaneous emission (ASE) between the ends of the Fabry-Perot cavity formed by the DBRs. This effect has been previously discussed [10] in the case of GaAs lasers, and has also been reported in the context of meandering off-axis rays in VCSELs [11], which by ASE in transverse direction cause an increase of the laser threshold value.

The energy transfer from the emission envelope into a number of optical modes is evidenced by increased mode visibility and diminished envelope intensity around 3.4 eV (Fig. 2, right). Binet *et al.* [12] ascribed the appearance of multiple modes in emission spectra, obtained by optical pumping of GaN/InGaN heterostructures to stimulated emission in the etched lateral Fabry-Perot cavity. Assuming maximum quantum efficiency of the radiative transition the maximum energy efficiency of an EBP semiconductor laser is 33%. Only a third of the implanted average energy $\langle E \rangle = 3$ Eg can be transformed into photons by radiative recombination [11].

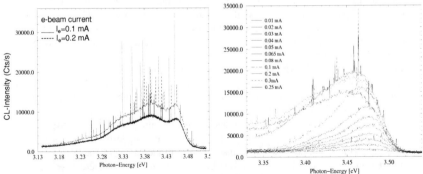

Figure 2: *Emission spectra of sample V54 (GaN: Si active region) (left) and the emission spectra of sample V60 (GaN/Al$_{0.08}$Ga$_{0.92}$N active region) (right).*

We attribute the redshift of the spontaneous emission background maximum of sample V60 to bandgap renormalization. The bandgap shrinkage results from manybody interaction, when the density of the electron-hole plasma (EHP) increases [11,13,14]. At high nonequilibrium carrier densities, the free carriers begin to screen the deformation potential of the electron-phonon interaction [11]. Consequently, the thermalisation time τ is increased, i.e. the transfer of energy to the lattice is reduced. Therefore, the temperature of the EHP rises relatively to the lattice temperature (The distribution of the kinetic energy of hot carriers by Coulomb interaction determines the effective temperature). The heating of the quasiequilibrium EHP is a limiting factor for the efficiency of radiative recombination. If the lifetime τ becomes of the order of the spontaneous lifetime τ_{sp}, i.e. the lifetime of the EHP, the threshold power for lasing is increased and population inversion can be prevented.

The excited region of the sample also experiences a quasiadiabatic heating [13,15], if the excitation time is smaller than the time the temperature distribution needs to expand over the e-beam cross-section. This condition is fulfilled in our case, so a relative, local temperature rise of several hundred Kelvins also contributes to the bangap shrinkage. In our case it is not evidenced which effect has more influence on the emission redshift nor whether the contributions are of the same order.

The reduced visibility $V=I_{max}/I_{min}$ with higher excitation current densities (Fig. 2, right) can be attributed to an effect which strongly influences the propagation of the electromagnetic field within the cavity: the distribution of the nonequilibrium carrier density in the active region, generated by EBP, is inhomogeneous. This implicates a strong inhomogeneity in the gain distribution along the cavity axis. The gain in longitudinally pumped vertical cavities depends on the effective penetration depth [15] as shown in equation 1

$$x_0 \approx \frac{1}{3} \cdot 2.1 \cdot 10^{-12} \frac{U_a^2}{\rho} \; [cm] \qquad 10\,keV \le e \cdot U_a \le 100\,keV, \qquad (1)$$

where U_a is the acceleration voltage. Consequently, the inhomogeneous carrier distribution influences the emission efficiency and the laser threshold as follows [11]: the carrier density in the most excited, active region is considarably higher than in the passive part. A carrier-density induced bandgap shrinkage occurs in the highly excited region and shifts the gain profile towards longer wavelenghts. This reduces the resonant absorption in the passive part, so that the active transition is out of resonance with the reverse transition. If the pumping becomes larger, the density of free carriers in the weakly excited part increases because of the reabsorption of radiation. This leads to an equalisation of the gain profile along the cavity axis and enhances the losses for the propagating field, unless the optical field is below the saturation value (in that case the passive region becomes transparent).

In contrast to sample V54 with GaN bulk active region, the use of MH structures (V60) minimizes this effect, because the bandgap of the passive region is slightly larger than in the active region. Therefore, the use of quantum wells or quantum dots as active part of the cavity is to be preferred for prevention of gain deterioration.

The deterioration of gain in EBP semiconductor lasers is also attributed to ASE propagating in the transverse direction [16]. The transverse single-pass gain is several times larger than the longitudinal single-pass gain. With increasing diameter of the excited region the threshold value for lasing in longituginal direction increases, because the ASE along the transverse direction of the cavity depletes the longitudinal gain.

These three-dimensional spatial and spectral gain inhomogeneities [16] may also have caused the relative minimum in the spontaneous emission spectrum of sample V54 (Fig. 2, left), located at about 3.43 eV. We assume that the peaks at 3.395 and 3.45 eV indicate recombination of an EHP and phonon-assisted exciton decay at higher photon energy. This has also been observed for optically pumped GaN epilayers [17], where a further contribution of band-to-band transitions to the high energy emission peak is assigned.

The results of the reflection measurements of V51 and V52 are shown in Fig. 3 (left). The spectrum of V51 reveals a maximum reflectivity of 43% at 364 nm. Sample V52 shows a maximum reflectivity of 43% at 387 nm. The peak located at 364 nm reveals a reflectivity of 34%. Transfer-matrix simulations were done with dispersion data from reflection and transmission measurement on single GaN and AlGaN layers. The obtained values for the reflection maxima are 56% at 380 nm (V52) and 48% at 365 nm (V51). The experimental values are 23% and 10% smaller than the simulated values, respectively. We ascribe this to the dispersion data, where losses, probably due to the buffer layer, could not be specified.

Preliminary X-ray diffraction (XRD) $\omega/2\Theta$ -scans hint at a higher than expected AlN content of the quarter wave layers.

Taking into account the lower experimental values of reflectivity, we can estimate the reflectivity of the lower and the upper DBR stacks of V54 and V60. Therefore we simulated the spectra for the bottom and top stack of V60 (Fig. 3, right) with and without the Al cap layer. A maximum reflectivity of 67% results from a single 15.5 period DBR stack. Adding the Al cap layer, the reflectivity of the overall reflector amounts to 90%. In addition, the Al cap layer reduces electron backscattering and prevents supercharging. If we reduce the values for reflectivity by a factor of 0.75, considering the discrepancy between sample and simulation (V51, V52), we obtain 68% for the upper reflector and 50% for the lower reflector.

Apart from the increase in reflectvity of the upper reflector, induced by the Al cap layer, the peaks also broaden. Broadening enables the propagation of mutiple modes, which is observed in our spectra.

As the threshold condition gives the cavity gain necessary to overcome the cavity loss

$$\Gamma \cdot g_{th} = \alpha_{loss} - \frac{1}{2L} \ln(r_b \cdot r_t) \qquad\qquad (2)$$

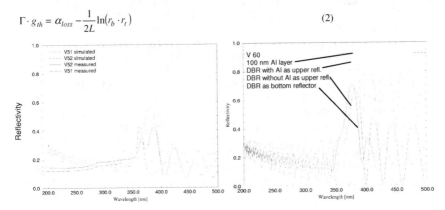

Figure 3: *Reflectivity vs wavelength of the single DBR stacks V51 and V52 with simulations (left) and reflectivity of upper and lower DBR of sample V60 (right).*

for any photon energy [14], one obtains a threshold gain g_{th} of about 2250 cm^{-1} for V54 and 4240 cm^{-1} for V60, inserting reflectivity coefficients $r_b = 0.7$ and $r_t = 0.82$, cavity length $L = 1.2$ μm, a confinement factor $\Gamma = 1$ ($\Gamma = 0.53$) in the case of V54 (V60) and neglecting any internal losses α_{loss} (free carrier absorption, absorption by the DBRs etc.). AES and SEM measurements show that the accelerated electrons induce a reaction on the surface of the Al cap layer with carbon compounds, which reside in the vacuum chamber at a pressure of ~2×10^{-5} mbar. Electrons with low kinetic energy induce a polymerisation of carbon compounds [15] as can be seen in the outer range of the e-beam cross-section. At the center of the e-beam cross-section, the molecular chains fractionize and suppress polymerisation. This is confirmed by profilometry and AES measurements. The Al cap layer did not alter, so we can exclude a decrease of reflectivity of the upper reflector.

CONCLUSIONS

Electron beam pumped vertical cavities with active regions made of a GaN epilayer and a MH structure have shown multimode emission. The optical modes indicate multiple reflections of ASE within the cavity, which is confined by 15.5 period GaN/ Al$_{0.25}$ Ga$_{0.75}$ N DBRs with reflectivities of about 70% and 50%, respectively. The use of a multiquantumwell (MQW) or quantum dot active region supposedly prevents gain deterioration, which limits the emission efficiency, and therefore reduces the threshold for lasing.
We conclude that fabrication of DBRs based on the AlGaN/AlN-system is a promising way to realize highly reflective mirrors with low absorption in the near UV spectral range.

REFERENCES

1. T. Someya and Y. Arakawa, *Jpn. J. Appl. Phys.* **37**, L 1424-1426, (1998).
2. J. Redwing and J. Flynn, *Appl. Phys. Lett.* **69**(1), 1, (1996).
3. N. Anderson, J. M. Redwing, S. Flynn, *Jpn. J. Appl. Phys.* **38**, 4794-4795, (1999).
4. Y.-K. Song and C. Carter-Coman, *Appl. Phys. Lett.* **74**(23), 3441, (1999).
5. I. L. Krestnikov and N. N. Ledentsov, *Appl. Phys. Lett.* **75**(9), 1192, (1999).
6. H. Ng and T. Moustakas, *Appl. Phys. Lett.* **74**(7), 1036, (1999).
7. M. Khan and D. Olsen, *Appl. Phys. Lett.* **59**(12), 1449, (1991).
8. R. Langer and L. Dang, *Appl. Phys. Lett.* **74**(24), 3610, (1999).
9. T. Someya and Y. Arakawa, *Appl. Phys. Lett.* **73**(25), 3653, (1998).
10. M. I. Nathan, *Phys. Rev. Lett.* **11**(4), 152–154, (1963).
11. O. Bogdankevich, *Sov. J. Quantum Electron.* 24(12), 1031–1053, (1994).
12. Binet and O. Briot, *Mat. Sc. & Eng.* **B50**, 183–187, (1997).
13. Nasibov and E. Shemchuk, *Sov. J. Quantum Electron.* **8**(9), 1082–1085 (1978).
14. J. Singh, *Semiconductor Optoelectronics*, McGraw-Hill, New York (1995).
15. S. Schiller, U. Heisig, and S. Panzer, *Elektronenstrahltechnologie*, Wissenschaftl. Verlagsgesellschaft mbH, Stuttgart, 1977.
16. Khurgin and D. A. Davids, *Optical and Quantum Electronics* **25**, 451–465 (1993).
17. J.-Chr. Holst and A. Hoffmann, *MRS Internet J. Nitride Semicond. Res.* **2**, Art. 25 (1997).

Mat. Res. Soc. Symp. Vol. 595 © 2000 Materials Research Society

Microstructure-based lasing in GaN/AlGaN
separate confinement heterostructures

S. Bidnyk, J. B. Lam, B. D. Little, G. H. Gainer, Y. H. Kwon, J. J. Song,
G. E. Bulman[1], and H. S. Kong[1]
Center for Laser and Photonics Research and Department of Physics,
Oklahoma State University, Stillwater, OK 74078, U.S.A.
[1]Cree Research, Inc., Durham, NC 27713, U.S.A.

ABSTRACT

We report on an experimental study of microstructure-based lasing in an optically pumped GaN/AlGaN separate confinement heterostructure (SCH). We achieved low-threshold ultra-violet lasing in optically pumped GaN/AlGaN separate confinement heterostructures over a wide temperature range. The spacing, directionality, and far-field patterns of the lasing modes are shown to be the result of microcavities that were naturally formed in the structures due to strain relaxation. The temperature sensitivity of the lasing wavelength was found to be twice as low as that of bulk-like GaN films. Based on these results, we discuss possibilities for the development of ultra-violet laser diodes with increased temperature stability of the emission wavelength.

INTRODUCTION

Recent progress in the development of InGaN-based lasing structures was highlighted by Nichia's announcement of "engineering sample shipments" of their blue laser diode (LD) [1]. However, due to the lack of ideal substrates for the growth of thin film nitrides, a large number of dislocations and cracks are naturally formed in the epitaxial layer to alleviate the lattice mismatch and the strain of postgrowth cooling. This does not always negatively affect lasing characteristics but sometimes introduces interesting lasing properties due to self-formed high-finesse microcavities [2] that could be utilized for the development of near- and deep-UV LDs.

In this work we describe the results of optical pumping experiments on GaN/AlGaN separate confinement heterostructures in the presence of a large number of naturally formed microcavities. We achieved ultra-violet lasing in these structures over a wide temperature range. The spacing, directionality, and far-field patterns of the lasing modes were correlated to the geometry of microcavities. The temperature sensitivity of the lasing wavelength was measured and compared to that of bulk-like GaN films.

EXPERIMENTAL DETAILS

The GaN/AlGaN separate confinement heterostructure (SCH) samples used in this work were grown by metalorganic chemical vapor deposition on 6H-SiC (0001) substrates with ~3-μm-thick GaN epilayers deposited prior to the growth of the SCH region. The SCH sample under discussion has a 150-Å-thick GaN active layer, surrounded by 1000-Å-thick $Al_{0.05}Ga_{0.95}N$ cladding layers and 2500-Å-thick $Al_{0.10}Ga_{0.90}N$ waveguide layers symmetrically located on each side. For the purpose of comparison, we also studied a 4.2-μm-thick GaN epilayer grown on (0001) 6H-SiC. The samples were mounted on a copper heat sink attached to a wide temperature range cryostat. Conventional photoluminescence (PL) spectra were measured in the back-scattering geometry using a frequency-doubled Ar$^+$ laser (244 nm) as the excitation source. In order to study the lasing phenomena, a tunable dye laser pumped by a frequency-doubled, injection-seeded Nd:YAG laser was used as the primary optical pumping source. The excitation beam was focused to a line on the sample using a cylindrical lens. The emission from the edge of the sample was coupled into a 1-m spectrometer with a side-mounted optical multi-channel analyzer and photomultiplier tube. Special precautions were taken to avoid fluctuations in sample position due to the thermal expansion of the mounting system. This allowed us to spatially "pin" the sample and obtain lasing modes from a single microcavity over the entire temperature range studied.

DISCUSSION

In order to evaluate the effects of cracks on the optical properties, we cleaved our samples into submillimeter-wide bars (note that when GaN is grown on SiC, it can be easily cleaved along the ($11\overline{2}0$) direction [3]). Before cleaving, the samples did not exhibit any noticeable defects on the surface. After the cleaving process, however, cracks were observed along all three cleave planes associated with a hexagonal crystal structure, with the majority running parallel to the length of the bar, as shown in Figure 1.

Figure 1. *A picture of the sample surface of a GaN/AlGaN separate confinement heterostructure after cleaving. Before cleaving, the sample exhibited no noticeable defects on the surface. After the cleaving process, however, cracks can be seen running parallel to the length of the bar.*

When the sample was excited above the lasing threshold, high-finesse cavity modes were observed. Typical emission spectra at pump densities above and below the lasing threshold are depicted in Figure 2. A series of equally spaced and strongly polarized (TE:TM ≥ 300: 1) lasing modes with full width at half maximum of ~3 Å appears on the low energy side of the GaN-active-region peak. We note, however, that when the spontaneous emission was collected from the sample edge instead of the surface, the lasing modes appeared on the higher energy side of the spontaneous emission peak [4]. This phenomenon is due to strong re-absorption that introduces a shift of several nanometers to the spontaneous emission peak. Lasing in our samples occurs in the wavelength range of 360 to 364 nm at room temperature, which is much deeper in the UV than for InGaN/GaN-based structures. We were able to correlate the spacing between the modes in Figure 2 to the distance between the cracks depicted in Figure 1. Assuming that the unity round-trip condition is satisfied and there is no loss due to absorption in the GaN layer, the threshold gain can be estimated from [5]:

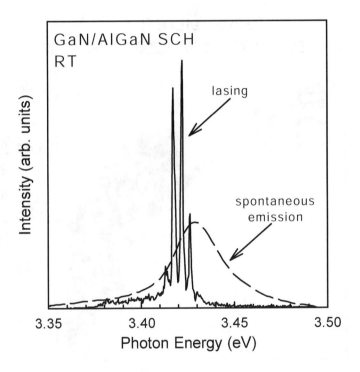

Figure 2. *Lasing and spontaneous emission in the GaN/AlGaN separate confinement heterostructure at room temperature. Lasing is believed to be of microcavity origin.*

$$g_{th} > \frac{\ln(R^{-2})}{2L},$$

(1)

where R is the mirror reflectivity (20% for a GaN-air interface) and L is the distance between the cracks. Alternatively, L could be extracted through an analysis of the mode spacing $\Delta\nu$ through the relation $\Delta\nu = c/(2nL)$. We obtained a gain value of ~500 cm^{-1} (corresponding to a threshold pump density of 105 kW/cm^2), which is on the same order of magnitude as the near-threshold gain values measured in GaN by the variable stripe technique (see Refs. [6,7] for example). We note, however, that due to the limited quality of GaN and AlGaN layers, the photon flux inside a microcavity encounters additional losses associated with absorption and scattering of light. In view of this, the value for gain calculated through Eq. 1 represents the lower limit for the actual gain at threshold.

It was found that by pumping different areas along the length of the bar, the emission exhibited varying degrees of cavity finesse. The finesse is directly linked to the end mirror reflectivities and internal losses. Thus the different degree of finesse is

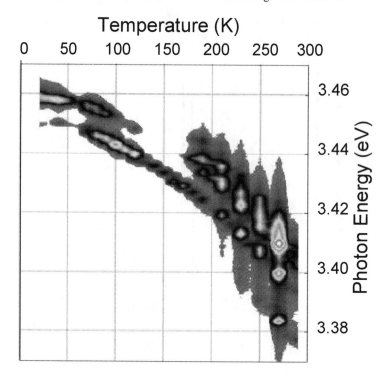

Figure 3. *Lasing modes as a function of temperature for a single microcavity. This image was obtained by consistently taking lasing spectra with an optical multi-channel analyzer at $1.3 \times I_{th}(T)$ while the temperature was gradually varied from 20 K to 300 K.*

presumably due to the presence of cavities of varying quality formed by parallel cracks running through the active layer. Areas exhibiting high finesse consistently had a narrower far-field pattern than those exhibiting low finesse possibly indicating larger scattering loss.

We further note that the high finesse emission exits the sample at an angle of $\phi \sim 18°$ rather than parallel to the sample surface. In order to understand this phenomenon, the samples were examined in cross section using a scanning electron microscope. A large number of cracks were noticed lying at small angles ($\alpha \sim 7°$) to the c axis. Taking into account the refractive index of GaN, we were able to correlate the geometry of the cracks to the emission angle through Snell's law:

$$\sin \phi = n \sin \alpha . \tag{2}$$

A dramatic decrease in the lasing threshold of our SCH structure in comparison to a bulk-like GaN epilayer was observed over the entire temperature range studied. For one of the SCH samples, the lasing threshold was measured to be as low as 65 kW/cm^2 at room temperature. We made a direct comparison of threshold carrier densities in GaN/AlGaN SCHs and bulk-like GaN epilayers and concluded that lasing in SCHs occurs at carrier densities $< 5\times10^{17}$ cm^{-3}, which is considerably lower than that required for the formation of an electron-hole plasma. By further examining the relative energy position between the spontaneous emission and lasing modes we concluded that the exciton-exciton scattering mechanism remains dominant in SCH lasing structures even at room temperature. The details of carrier density calculations and photon energy analysis are given elsewhere [8].

The temperature dependence of the energy position of the lasing modes was also studied. To avoid spatial displacement of the sample due to a non-zero temperature expansion coefficient of the cryostat system, we chose to mount the cryostat horizontally so that the rod expansions occurred in the direction parallel to the direction of the laser beam. This allowed us to obtain lasing from the same microcavity over the entire temperature range. We also used the optical multi-channel analyzer to consistently acquire spectra at $1.3\times I_{th}$ for each temperature. Figure 3 shows a plot of mode energy position versus temperature. The gray pallet corresponds to different emission intensities. We note that the energy position of the gain maximum in the GaN/AlGaN SCH changes by about 50 meV when the temperature is varied from 10 to 300 K, which is twice smaller than that expected from a bulk-like GaN epilayer [9]. Such a reduced temperature sensitivity of the laser emission suggests that the GaN/AlGaN SCH could be well suited for UV applications that require increased temperature stability of the emission wavelength.

CONCLUSIONS

We achieved ultra-violet lasing in optically pumped GaN/AlGaN separate confinement heterostructures over a wide temperature range. The spacing, directionality, and far-field patterns of the lasing modes were shown to be the result of microcavities that were naturally formed in the lasing structures due to strain relaxation. The temperature sensitivity of the lasing wavelength was found to be twice lower than that of bulk-like GaN films, indicating that the separate confinement heterostructures could be used in applications that require increased temperature stability of the emission wavelength.

ACKNOWLEDGEMENTS

The authors would like to thank ONR, DARPA, AFOSR, and ARO for financial support of this research project.

REFERENCES

1. *See homepage of Nichia America Corporation at* http://www.nichia.com.
2. S. Bidnyk, Ph.D. Dissertation, Oklahoma State University (1999).
3. B. D. Cullity, *"Elements of X-ray diffraction,"* 2nd edition (Addison-Wesley Publishing company, Reading, Massachusetts, 1978), 502.
4. T. J. Schmidt, X. H. Yang, W. Shan, J. J. Song , A. Salvador, W. Kim, Ö. Aktas, A. Botchkarev, and H. Morkoç, Appl. Phys. Lett. **68**, 1820 (1996).
5. D. M. Bagnall and K. P. O'Donnell, Appl. Phys. Lett. **68**, 3197 (1996).
6. D. Wiesmann, I. Brener, L. Pfeiffer, M. A. Khan, and C. J. Sun, Appl. Phys. Lett. **69**, 3384 (1996).
7. R. Dingle, K. L. Shaklee, R. F. Leheny, and R. B. Zetterstrom, Appl. Phys. Lett. **19**, 5 (1971).
8. S. Bidnyk, J. B. Lam, B. D. Little, Y. H. Kwon, J. J. Song, G. E. Bulman, H. S. Kong, T. J. Schmidt, Appl. Phys. Lett. **75**, 3905 (1999).
9. S. Bidnyk, T. J. Schmidt, B. D. Little, and J. J. Song, Appl. Phys. Lett. **74**, 1 (1999).

Mat. Res. Soc. Symp. Vol. 595 © 2000 Materials Research Society

Dependence of aging on inhomogeneities in InGaN/AlGaN/GaN light-emitting diodes

V.E.Kudryashov, S.S.Mamakin, A.N.Turkin, <u>A.E.Yunovich</u>,
M.V.Lomonosov Moscow State University,
Department of Physics,
119899 Moscow, Russia
E-mail: yunovich@scon175.phys.msu.su.
A.N.Kovalev, F.I.Manyakhin,
Moscow Institute of Steel and Alloys,
Leninski Prospect 4, Moscow 117235, Russia

ABSTRACT

Changes of properties of green LEDs based on $In_xGa_{1-x}N/Al_yGa_{1-y}/GaN$ heterostructures were studied during $150 \div 200$ hours at currents J = $30 \div 80$ мА. The radiation intensity at low currents ($0.1 \div 1$ mA) is quite sensitive to such an aging, it falls down $10 \div 100$ times. Quantum efficiency and spectral parameters at normal currents (J ≈ 10 mA) change non-monotonically during aging, some degradation is observed after 168 hours. The degradation is observed also after a short (< 1 min) period of reverse current. These phenomena are discussed in terms of under threshold defect's formation and their migration in the space charge region of p-n-heterojunction. Potential fluctuations in the space charge region are quite sensitive to this process.

INTRODUCTION

Problems of aging and degradation of GaN-based devices were discussed during last years in papers [1-6]. Lifetimes of GaN- based light-emitting diodes (LEDs) at normal current conditions may be of 10^5-10^6 hours. That is why it is essential to evaluate these lifetimes using short periods of working at high currents. It was shown that quantum efficiency of LEDs with single quantum wells (of Nichia Chemical) at relatively high currents (J = 80 мА) can grow up during ≈100 hours and then gradually fall down during a working time of ≈1000 hours [4-6].

It was interesting to study a process of aging and degradation in analogous conditions on LEDs of various laboratories. We have shown in a recent paper [7] that minor differences in quantum efficiency (≈10% at working currents) of green LEDs based on $In_xGa_{1-x}N/Al_yGa_{1-y}/GaN$ heterostructures with multiple quantum wells (MQW) (of Hewlett-Packard laboratories [8, 9]) are connected with large differences of intensity and quantum efficiency dependence on current. These differences are caused by different distribution of effective charges in the space charge regions and role of tunnel component of currents at low voltages. In this work we study processes of aging and degradation of these LEDs at relatively high currents at room temperature.

EXPERIMENTAL

Green LEDs based on $In_xGa_{1-x}N/Al_yGa_{1-y}/GaN$ heterostructures with MQWs [7-9] were studied. They were divided on three groups (Q, P, N) of 20 samples according to a difference in quantum efficiencies (±10%) at normal currents (J = 10 mA). Properties of these groups of LEDs were studied in a wide range of J (10^{-7} – $3 \cdot 10^{-1}$ A) in [7].

Parameters of 3 LEDs chousen for careful measurements are given in Table 1. Spectral maxima ($\hbar\omega_{max}$), power efficiencies $\eta\%$, the hole concentration p on the p- side of the junction, a width of the space charge region w are given in the Table 1 with parameters describing the exponential tails of the main spectral band. The long wavelength side parameter is E_0; the short wavelength one is $E_1 = m\cdot kT$. The distributions of charges in the space charge regions N_A^- (w) and p-values for Q, N and P LEDs derived from dynamic capacitence measurements see Fig. 3 in [7]. The LEDs Q and P had more compensated space charge regions; the LED N had the lowest value of the width of the space charge region and clearly pronounced tunnel radiation band with $\hbar\omega_{max} \approx eU$.

Table I. Parameters of LEDs before aging.

Parameter	Q	N	P
P, $\times 10^{17}$cm^{-3}	6	3	2
w, nm	200	120	130
$\eta\%$(10 mA)	2.27	2.41	2.57
$\eta\%$(0.1 mA)	1.65	0.006	0.001
$\hbar\omega_{max}$ (10 mA), eV	2.485	2.486	2.500
W, mW (10 mA)	0.84	0.7	0.63
E_0, meV	56.7	59.1	56.1
E_1=mkT, meV	33.2	31.8	33.6

The first experiment of LED's aging was performed in the same conditions as in [4-6] that is at J = 80 mA. The LEDs at this current strongly degraded during 24 hours. That is why the conditions were choused by gradual changing of working currents in the range J = 30 ÷ 60 mA during 168 hours (see Table 2). Spectra and quantum efficiency of the LEDs were measured in the range J = 10 µA ÷ 60 mA each 24 hours.

Table II. Conditions of aging.

Time, hours	Current, mA	Voltage (N), V	Voltage (P), V	Voltage (Q), V
0-24	30	3.53	3.71	3.56
24-48	40	3.76	3.95	3.71
48-72	40	3.68	3.94	3.70
Reverse current, <1 min	-1 (V ≈ -10 V)			
72-96	40	4.2	3.93	3.72
96-120	40	3.63	3.90	4.12
120-144	50	3.76	4.29	
144-168	60	3.93	4.47	3.90

EXPERIMENTAL RESULTS

Spectra of two LEDs (N and P) before and after aging during 144 hours are given in Figs. 1 (a, b, c, d). Spectral maxima at J > 1 mA shift with current in the range $\hbar\omega_{max} \approx 2.36 \div 2.50$ eV. The tunnel band at low currents had maxima according to the applied voltage: $\hbar\omega_{max} \approx eU = 1.92 \div 2.13$ eV. It is essential that we could trace changes of spectral form during aging. The exponents of spectral tails (E_0 and $E_1 = m \cdot kT$, see a discussion lower) change during the aging. Spectral intensities at low currents fell down during aging. This is more pronounced for the LED P (the lowest current at which the spectra could be measured was 60 μA before aging and 1 mA after aging).

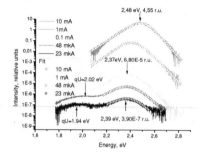

Figure 1a. Spectra of LED N and their approximations before degradation.

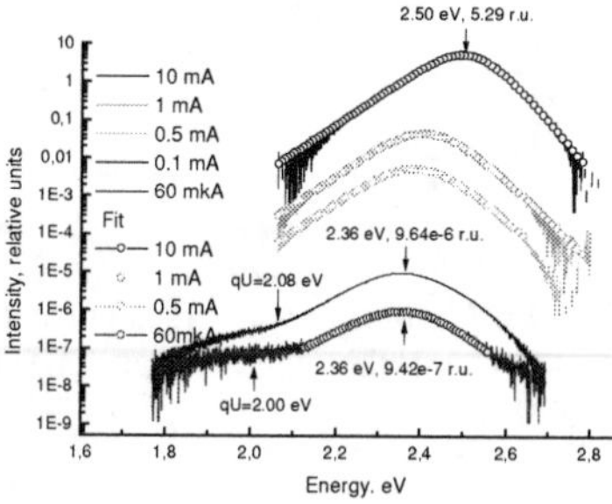

Figure 1c. Spectra of LED P and their approximations before degradation.

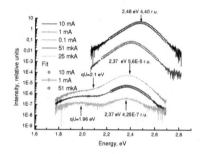

Figure 1b. Spectra of LED N and their approximations after degradation.

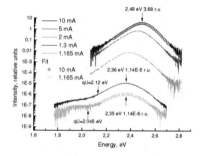

Figure 1d. Spectra of LED P and their approximations after degradation.

Changes of LED's integrated external radiation power versus current are shown during aging in Figs. 2 (a, b, c). The integral intensity at normal current (J = 10 mA) for LED Q fell down of 15%, for P fell down of 2%, it has grown up for LED N of 3%. But the integral intensity at low currents fell down 10÷100 times.

These changes correlate with growing of nonradiative leakage currents at low voltages, see current-voltage characteristics J(V) after aging in Fig. 3. Before aging the J(V) curves differed slightly. The J(V) curve of a not aged LED Q is shown for a comparison.

It is to be noted that the process of aging was step like. We could see at some moment a jump of light intensity and a jump of voltage on the LED at a constant current. The light intensity changed non-monotonically: sometimes jumps were positive, sometimes negative. During the last periods of aging the intensity and voltage were unstable: sometimes properties of LEDs recovered to previous measurements.

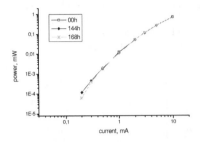

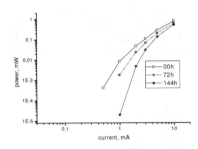

Figure 2a. Output power versus current in dependence on the degradation time for LED N.

Figure 2b. Output power versus current in dependence on the degradation time for LED P.

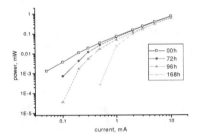

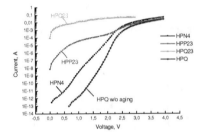

Figure 2c. Output power versus current in dependence on the degradation time for LED Q.

Figure 3. Current - voltage characteristics after aging (for LEDs N, P, Q) and without aging (for LED Q).

DISCUSSION

The spectral changes were analyzed by a phenomenological equation with three adjustable parameters: $\hbar\omega_{max}$, E_0 и $E_1 = m \cdot kT$, which describes the spectra with a good accuracy [7]:

$$I(\hbar\omega) \sim [1 + \exp(- (\hbar\omega - E_g^{eff})/E_0)]^{-1} \cdot [1 + \exp(\hbar\omega - \hbar\omega_{max})/ E_1]^{-1}$$

$$E_g^{eff} = \hbar\omega_{max} + E_0; E_1 = m \cdot kT. \qquad (1)$$

This approximation was used to determine electron temperatures in the active region at aging currents (see details in [4, 10]). The m values were adjusted at low currents ($J = 5 \div 10$ mA) and changes of E_1 then were accounted to changes of T. Temperatures T at $J = 50 \div 60$ mA were of $330 \div 350$ K.

Changes of spectral parameters at $J = 10$ mA during the aging process are shown in Figs. 4 (a, b, c). Values of E_0 were changing in the range $55 \div 61$ meV, $E_1 = m \cdot kT$ – in the range $30 \div 36$ meV (range of m = 1.2 ÷ 1.4). These changes were of the decimal order higher then the accuracy of fitting the parameters ($\delta E_0 \approx 0.4$ meV, $\delta E_1 \approx 0.1$ meV). A distinct change of parameters is seen after a short period of reverse current (at 72 hours point, $J < 1$ mA, $V \approx -10$ V). After this procedure spectra changed with aging more quickly.

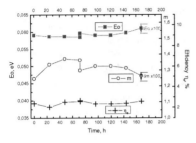

4a.

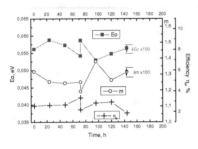

4b.

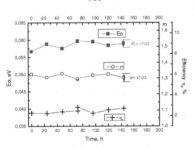

4c.

Figure 4. Dependence of spectral parameters and efficiency of LEDs N (a), P (b), Q (c) on the degradation time.

It is to be noted that parameter of the low energy tails E_0 may be connected with microscopic potential fluctuations in the active region [7]. It has grown after the aging process. However there were small non- monotonic jumps of E_0 during the aging. Spectra of green GaN- LEDs with MQWs (from Toyoda Gosei) had on the high-energy side an additional band of low intensity [11]. This was described due to macroscopic potential fluctuation. Values of m in the fitting procedure were growing due to the additional band and changes of m were understood due to these fluctuations. We may understand changes of m during the aging in our case also due to some changes of these fluctuations.

It is necessary to explain why some defects can be formed at comparatively low temperatures 330÷350 K during degradation. We will use a model of under threshold defect's creation by hot electrons in heterostructures with quantum wells and defects migration in the space charge region. This model was applied for GaN LED's structures in [4, 6]. Electrons are injected from n-GaN side into the active region with quantum wells. Electron energy over the bottom of conduction band is >> kT if the mean free path l_{fp} is higher than the width of the active layer.

Some hot electrons at high currents do not recombine in the active region and enter compensated layer on the p-side of the active region. If the width of the compensated layer is higher than the mean free path, electrons with an electron temperature T_e can create defects with a probability $w \sim \exp(-E_D/kT_e)$, E_D being an energy threshold for a point defect creation [12]. In such a model it is understandable why the degradation is so sensitive to a reverse current at voltages sufficient for impact ionization ($V \approx$ -10 V). It is understandable why a process of aging is sensitive to the voltage $eV > E_g$ on the structure (see Table II). It is understandable why LEDs P and Q (with a higher width of the compensated layer) degrade more quickly than LED N. Concentration of various complex defects in the compensated layers is greater than in non-compensated ones. Creation of defects by hot electrons is also more probable near the existing defects.

Created defects may migrate in the space charge region and in some cases form conductive channels, for example along dislocations. These channels will shunt the compensated region or p-n junction. The shunting channels will work as nonraditive paths of the current at low voltages. That may describe the aging in our experiments.

One of the main problems in growing of GaN structures is p-type doping. A distribution of charged acceptors N_A^+ on the p-side of LED's from various origins was compared in [7]. Capacitance-voltage measurements had shown that Nichia's green LEDs have higher N_A^+ ($\approx 5 \cdot 10^{17}$ cm^{-3}) and lower values of the space charge region width w (≈ 80 nm, that is less compensation). This may be a cause of longer times of aging (100-1000 hours [4-6]) in those samples than in samples studied in this paper.

CONCLUSIONS

Aging of green LEDs based on InGaN/AlGaN/GaN heterostructures with multiple quantum wells at currents J = 30 ÷ 60 mA can lead to remarkable changes of quantum efficiency at low currents during 100-200 hours. Spectral form is sensitive to the aging and it is possible to study this process by changes of exponential spectral tails.

Minor differences in LED's parameters before aging may cause remarkable different behavior during aging. The LEDs with more compensated space charge region near the active MQW layer are more sensitive to the aging process. Quantum efficiency and spectral parameters of the LEDs change non-monotonically during aging.

Mechanism of defect's creation by hot electrons entering the compensated layer and migration of these defects forming shunting paths for the current is proposed to explain experimental results.

ACKNOWLEDGEMENTS
Authors thank Dr. Paul Martin (HP Labs.) for sending LEDs to MSU.

REFERENCES

1. M.Osinski, P.Perlin, P.G.Eliseev, G.Liu, D.Barton, *MRS Symp. Proc.*, **449**, 1179-1183 (1997).
2. T.Egawa, H.Ishikawa, T.Jimbo, M.Umeno. *Appl. Physics Letters* **69**(6), 830 (1996).
3. M.Osinski, D.Barton. *Journal of the Korean Physical Societ,* **30**, S13 (1997).
4. A.N.Kovalev, F.I.Manyakhin, A.E.Yunovich. *MRS Internet J.of Nitride Semic. Res.*, **3**, 52 (1998). http:\\nsr.mij.mrs.org\3\52.
5. A.E.Yunovich, A.N.Kovalev, V.E.Kudryashov, F.I.Manyakhin, A.N.Turkin. *Mat. Res. Soc. Symp. Proc.*, **482**, 1041 (1998).
6. A.N.Kovalev, F.I.Manyakhin, V.E.Kudryashov, A.N.Turkin, A.E.Yunovich. *Semiconductors*, **33**, 2, (1999).
7. A.E.Yunovich, V.E.Kudryashov, S.S.Mamakin, A.N.Turkin, A.N.Kovalev, F.I.Manyakhin, *Physica status solidi (a)*, **176**, N1, p.125 (1999).
8. P.S.Martin. *Abstr. of The Third Eur. GaN Workshop (EGW3)*, 36-O, 36, (1998).
9. S.D.Lester, M.J.Ludowise, K.P.Killeen, B.H.Perez, J.N.Miller, S.J.Rosner. *Proc. of The Second Intern. Conf. On Nitride Semiconductors (ICNS'97)*, F3-5, 510 (1997).
10. V.Schwegler, C.Kirchner, M.Kamp, K.Ebeling, V.E.Kudryashov, A.N.Turkin, A.E.Yunovich, A.Link, W.Limmer. *Phys. Stat. Sol.(a)*, **176**, N1, p.783-786 (1999).
11. V.E.Kudryashov, A.N.Turkin, A.E.Yunovich, A.N.Kovalev, F.I.Manyakhin. *Semiconductors*, **33** (4), p.429-435 (1999).
12. F.I.Manyakhin. *Izvestiya VUZ'ov, Ser. Mater. Electron. Techn.*, N 4, p. 56-60 (1998) (in Russian).

Mat. Res. Soc. Symp. Vol. 595 © 2000 Materials Research Society

Optical Spectroscopy and Composition of InGaN

K.P. O'Donnell[1], **R.W. Martin**[1], **M.E. White**[1], **K. Jacobs**[2], **W. Van der Stricht**[2],
P. Demeester[2], **A. Vantomme**[3], **M.F. Wu**[3] and **J.F.W. Mosselmans**[4]

[1]Dept. of Physics and Applied Physics, Strathclyde University, Glasgow, G4 0NG, U.K.
[2]Department of Information Technology, University of Ghent, Ghent 7500, Belgium.
[3]Inst. Kern- en Stralingsfysica, Univ. of Leuven, B-3001 Leuven, Belgium.
[4]CLRC, Daresbury Laboratories, Warrington WA4 4AD, England, U.K..

ABSTRACT

Commercial light emitting devices (LEDs) containing InGaN layers offer unrivalled performance in the violet (~400 nm), blue (~450 nm) and green (~520 nm) spectral regions. Nichia Chemicals Company has also produced amber InGaN LEDs with peak output near 590 nm. Here, we predict, on purely theoretical grounds, a surprisingly high limiting value of 1020 nm (peak) for InGaN intrinsic emission. We partly confirm this prediction by spectroscopic measurements of samples with photoluminescence (PL) peaks between 370 nm and 980 nm. In addition, we have measured the indium content of a range of light-emitting layers, using Rutherford Backscattering Spectrometry (RBS), Extended X-Ray Absorption Fine Structure (EXAFS) and Energy Dispersive X-Ray Analysis (EDX). The PL peak energy is found to depend linearly on the indium fraction: violet-emitting layers have an indium content of ~8%, blue layers ~16% and green layers ~25%. A linear extrapolation to the limit set by the Stokes' shift prediction, mentioned earlier, yields a limiting indium concentration of only ~52%. The profound impact of these results on future extensions of nitride technology and current theoretical models of InGaN is briefly discussed.

INTRODUCTION

The extension of nitride technology to the full visible spectrum, 400 nm to 700 nm, is a technical challenge that depends upon the successful incorporation of increasing amounts of indium nitride into gallium nitride while maintaining high fluorescence efficiency in the solid solution, $In_xGa_{1-x}N = x(InN) + (1-x)GaN$. Of the commercial suppliers, only Nichia Chemicals Company has presented nitride LEDs that emit efficiently in the amber (~590 nm) and, recently, the red (~675 nm) spectral regions [1].

Some time ago, the Strathclyde group reported photoluminescence (PL) emission peaks up to 650 nm in epilayers grown at low temperatures [2]. It seemed reasonable to expect that the limit to the 'redshifting' of nitride technology would be set by the band gap of indium nitride (which, at 1.89 eV [3], is equivalent to a wavelength of 656 nm) but here we show, both theoretically and practically, that the limit, at 1.21 eV (1020 nm), is actually well below the InN band gap. On the basis of our Stokes' shift model, the peak photon energy of the dominant luminescence band in InGaN samples is shown to be linearly related to the alloy band gap [4]. Denoting these optical energies by E_p and E_g, respectively, we find, from our own spectroscopic results and a survey of the relevant literature, that:

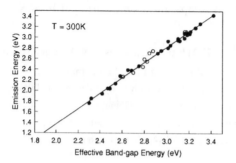

Figure 1. The linear relation between optical energies of InGaN layers (●) and diodes (o).

$$E_p = (1.45\ E_g - 1.54)\ eV \qquad\qquad (1)$$

as shown in Figure 1. The PL energy is always lower than the band gap of a given alloy by virtue of the Stokes' shift, which increases as the band gap decreases [4]. We assume that a *natural limit* to the Stokes' shift exists, since the alloy band gap can not have a value lower than the band gap of pure InN. Substituting $E_g = 1.89$ eV into equation (1) yields the limiting PL peak energy of 1.21 eV (a photon wavelength of 1020 nm). Further, by extrapolating the measured linear correlation between PL emission and indium content, described below, to the same limit, we discover a limiting indium content of only 52%.

EXPERIMENTAL TECHNIQUES

The basic experimental procedure in this work is the systematic comparison of a set of optical measurements on laterally inhomogeneous samples [2], with measurements of the local indium content, measured *at the same locations* on the same samples. Maps of PL peak wavelength, such as that shown in Figure 2, are built up by analysing a set of conventional PL spectra taken under identical conditions from a grid of points on the surface of samples about 5 mm square. Each spectrum represents total emission from an area, defined by the laser excitation spot, which is typically less than 0.1 mm wide.

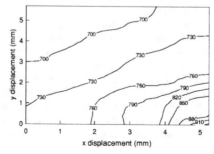

Figure 2. PL peak wavelength map of a laterally inhomogeneous InGaN layer

We will compare in the following work the results of several techniques that allow completely independent estimates of the indium content of our films. Rutherford Backscattering Spectrometry [5], RBS, is a nuclear measurement technique that is capable of determining both the composition and thickness of layers under investigation, independent of sample strain, etc. Extended X-Ray Absorption Fine Structure [6], EXAFS, is due to the self-interference of core photo-electrons, excited out of a particular atom in a solid by X-ray absorption above a characteristic edge in the spectrum, when they are subsequently scattered by the atom's neighbours. EXAFS is a probe of the average local atomic structure in the vicinity of a target atomic species. Its special application to the measurement of alloy composition is described here. Energy-dispersive X-ray analysis [7], EDX, complementary to electron microscopy, identifies and counts atoms through the detection of characteristic x-ray emission lines.

All of the samples in our study were produced as InGaN (~0.25 μm) – GaN (>2 μm) bilayers on sapphire in a Thomas Swan rotating-disc reactor specifically designed and adapted for growth of nitrides [8]. While PL mapping of InGaN layers grown at the University of Ghent during the last three years has provided a large number (thousands) of individual spectra, indicative of a very wide range of indium content, we have not yet examined the full range of samples available to us by all of the assay techniques that are mentioned above. There are two main reasons for this failure. At present, both RBS and EXAFS are somewhat limited in terms of achievable spatial resolution, compared to optical experiments; values measured by these techniques must represent averages taken over a few square mm of sample surface. The techniques are also rather expensive in terms of time and resource. Absorption data from the film shown in Fig. 2 confirm the trend in Fig. 1, but were not accessible for the regions emitting at the longest wavelengths, where the composition is changing rapidly.

EXPERIMENTAL RESULTS AND ANALYSIS

The large range of PL spectral energies shown by InGaN layers is illustrated in Figure 3, with excitation power <50 mW at either 325 or 350 nm. The PL peak energies range from values, at room temperature, similar to that of pure GaN, near 3.41 eV, down to a value, a literature record so far, of 1.27 eV [9]. It should be emphasised that the InGaN spectra form a *continuous set*: all intermediate values of peak energy are represented in our samples.

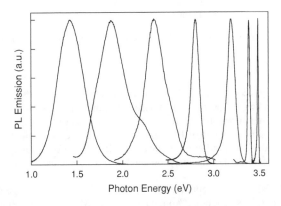

Figure 3. Representative 10 K PL spectra of a range of InGaN samples.

It is very obvious from Figure 3 that PL spectra of InGaN layers increase in line width as the peak shifts to the red: it has been shown that the magnitude of the line width is proportional to that of the peak's deviation from the band edge of GaN [10]. At the same time the PL peak intensity decreases [2].

To save space, Figure 4 collects and compares measurements of indium content, obtained by diverse techniques, with the equivalent PL peak energies. It is clear that there is a linear relationship between the two, and that the different measurement techniques show a high degree of mutual agreement if the PL peak energy is used as a benchmark. The relationship between PL peak energy and indium fraction, x, obtained from fitting a straight line to all of the data presented in Figure 4, is:

$$E_p = (3.45 - 4.3x) \text{ eV} \qquad (2)$$

Violet-emitting layers have an indium content of ~8%, blue layers ~16% and green layers ~25%. Clearly, the combination of equations (1) and (2) yields a similar linear relationship between optical band gap (as defined in [4]) and indium fraction for 0<x<0.4:

$$E_g = (3.44 - 3.0x) \text{ eV} \qquad (3)$$

The value of x obtained by setting E_g equal to 1.89 eV, the band gap energy of pure InN, in equation (3), is surprisingly only 0.52.

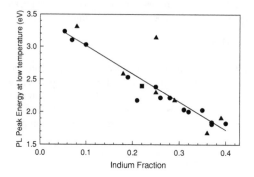

Figure 4: PL peak energies vs. In fraction, according to RBS(•), EXAFS(▲), EDX(■).

DISCUSSION

The range of possible emission wavelengths for InGaN-based LEDs extends well above the false ceiling set by the band gap of pure indium nitride (656 nm). Similar 'sub-bandgap' emission is observed in GaN quantum dots[11]. If we consider the peak of the emission band alone, theory has suggested a limiting value of 1020 nm, while we have experimentally reached a value close to one micron, albeit in a rather exceptional sample, which may be unique at the present time. The long-wavelength tail of the emission band of this sample extends very far into the infra red.

These results mark InGaN as an exceptional alloy in terms of the extreme range in wavelength of its intrinsic emissions, extending from the band edge of GaN, near 360 nm, to well above 1200 nm. However, it should be noted that we are dealing here not with bulk samples but with quite thin layers that are grown on top of thick accommodation layers of GaN in order to reduce lattice mismatch with an underlying sapphire substrate. While the strain state of the emitting layers is not our concern here, it may have a decisive effect on the optical properties, through the separation of charge carriers in a strain-induced internal electric field [12]. Indeed the action of these intense electric fields within In-rich quantum dots is the most likely explanation for the long wavelength emission.

A misunderstanding of the differences that may exist between the properties of bulk samples and thin layers has also, in our opinion, dogged recent accounts in the literature of the functional dependence of the band gap of InGaN alloys on the indium content. There is no empirical support for the use of a "bowing" parameter to describe the dependences of E_p and E_g on x, the indium content of our samples, which are clearly linear. Our samples are not alloys in a simple sense. On the other hand, some matters, revealed by the present set of investigations, deserve serious consideration. Figure 5 summarises a survey of recent literature that has claimed to examine the $E_g(x)$ or $E_p(x)$ relationship. There is a substantial measure of agreement between our findings and other recent work, although our data set is more extensive than any other in terms of the range of indium contents encountered. Our results disagree substantially with those of earlier literature, for reasons to be discussed elsewhere.

It is bad practice and quite dangerous to extrapolate data fits beyond their legitimate range, for example by including the 'trivial' point $(1, E_g(InN))$ on a graph of optical energies as a function of indium fraction. The extrapolation that we proposed earlier may also be hard to justify, but is less likely to be erroneous. Our conjecture depends upon the continuation of two linear trends, represented by equations (1) and (2), up to the limit set by the bandgap of InN. While we can not be certain of this trend's continuing beyond x = 0.4, we have presented evidence to extend the observation of a linear behaviour down to much lower photon peak energies than any described previously. A deviation from linearity beyond the point reached by the current set of experiments is possible but would result in very extraordinary behaviour of $E_p(x)$.

In physical terms, our conjecture is equivalent to neglecting the effect of the internal electric field on the optical band gap, as determined from the absorption spectrum. We suppose, as a zero–order approximation, that the Stokes'shift is due entirely to the relaxation of photo-excited carriers prior to their recombination.

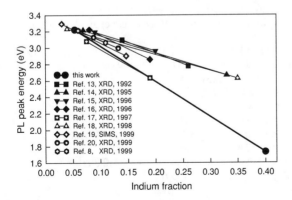

Figure 5. Comparison of this work with results from recent literature.

Finally, we question the meaning of the value estimated for the limiting concentration: is its closeness to one half significant? To recap, this is the extrapolated indium fraction, for an InGaN-on-GaN bilayer, at which we expect the measured optical band gap to reach a value equal to that of bulk InN.

CONCLUSIONS

A prediction, on the basis of established theory, that InGaN layers of high indium content are capable of infrared light emission, has been substantially verified by spectroscopic investigations. The theoretical limit is a peak at 1020 nm while the experimental one, only slightly less, at 980 nm, may be exceeded in new samples.

The indium content of several InGaN layers has been measured by a number of different techniques: RBS, EXAFS and EDX, with closely comparable results. The variation of optical energies with indium content is found to be linear, in the range $0<x<0.4$, with a slope energy of 4.3 eV for emission and 3.0 eV for absorption, in fair agreement with some recent work, but in disagreement with some earlier literature.

The projected limiting indium concentration for a layer emitting with a peak wavelength of 1020 nm is ~52%. The physical meaning of this limit is at present unknown.

REFERENCES

1.   T. Mukai, M. Yamada, and S. Nakamura, Jpn. J. Appl. Phys **38**, 3976 (1999).
2.   K. P. O'Donnell et al, Appl. Phys. Lett. **73**, 3273 (1998).
3.   B. Gil and M. Leroux, Properties of InN, in EMIS Datareview **23** (1999).
4.   R.W. Martin, P.G. Middleton and K.P. O'Donnell, Appl. Phys. Lett. **74**, 263 (1999).
5.   K. P. O'Donnell et al, phys. stat. sol (b) **216**, 171 (1999).
6.   K. P. O'Donnell et al, phys. stat. sol (b) **216**, 151 (1999).
7.   K. P. O'Donnell et al, Mat. Sci. Eng. **B59**, 288 (1999).
8.   W. Van Der Stricht, Ph.D. Thesis, University of Ghent (1999, unpublished).
9.   K. P. O'Donnell et al, phys. stat. sol (b) **216**, 141 (1999).
10. K. P. O'Donnell et al, Appl. Phys. Lett. **70**, 1843 (1997).
11. F. Widmann et al, Phys. Rev. **B58**, 15989 (1999); B. Danilano et al, Appl. Phys. Lett. **75**, 962 (1999).
12. K. P. O'Donnell, R.W. Martin and P.G. Middleton, Phys. Rev. Lett. **82**, 237 (1999).
13. S. Nakamura and T. Mukai, Jpn. J. Appl. Phys. **31**, 1457 (1992).
14. S. Nakamura, J. Vac. Sci. Technol. A **13(3)**, 705 (1995).
15. S. Keller et al, Appl. Phys. Lett. **68**, 3147 (1996).
16. W. Shan et al, Appl. Phys. Lett. **69**, 3315 (1996).
17. T. Takeuchi et al, Jpn. J. Appl. Phys. **36**, 177 (1997).
18. Collection of early data from Figure 16 of S. M. Bedair, *Gallium Nitride I*, Semiconductors and Semimetals, **50**, 147 (1998).
19. J. Wagner et al, MRS Internet J. Nitride Semicond. Res. **4S1**, G2.8 (1999).
20. Y.-H. Kwon et al, Appl. Phys. Lett. **75**, 2545 (1999).

Mat. Res. Soc. Symp. Vol. 595 © 2000 Materials Research Society

Sign of the piezoelectric field in asymmetric GaInN/AlGaN/GaN single and double quantum wells on SiC

Jin Seo Im and A. Hangleiter
Institut für Technische Physik, Technische Universität Braunschweig, Mendelssohnstr. 2, D-38106 Braunschweig, Germany

J. Off and F. Scholz
4. Physikalisches Institut, Universität Stuttgart, Pfaffenwaldring 57, D-70550 Stuttgart, Germany

ABSTRACT

We study both GaInN/GaN/AlGaN quantum wells with an asymmetric barrier structure grown on SiC substrate and GaN/AlGaN asymmetric double quantum well (ADQW) structures. In the first case, a time-resolved study reveals an enhanced oscillator strength when the AlGaN barrier is on top of the GaInN quantum well. In comparison to our previous study of the same structure grown on sapphire, we find that the sign of the field is the same in both cases: the field points towards the substrate. In the case of ADQW, we observed not only intrawell transitions of both a 4 nm and a 2 nm QW separated by a 2.5 nm AlGaN barrier but also an interwell transition between the two QWs in the photoluminescence. The lifetimes and emission energies of the transitions can be well explained by the existence of the piezoelectric field built in the QWs.

INTRODUCTION

GaInN/AlGaN/GaN-based quantum wells (QWs) have played a key role in the rapid development of short-wavelength light emitters [1]. To explain the puzzling optical properties of the quantum wells, the piezoelectric field effect has been recently discussed actively [2,3,4,5]. In this work, we explore the effect in more detail studying two structures designed specially: first, asymmetric barrier structures, i.e., a GaInN quantum well sandwiched between an AlGaN and a GaN layer, and secondly, GaN/AlGaN asymmetric double quantum well structures. The asymmetry introduced in hese structures exhibits clearly the dominating influence of the piezoelectric field on the optical transitions in the quantum wells.

EXPERIMENTAL

Our samples were grown on (0001)-oriented SiC substrates using low-pressure metalorganic vapor phase epitaxy (LP-MOVPE). The GaInN layers were grown below temperature of 800°C with N_2 as a carrier gas. The growth temperature of GaN and AlGaN layers was 1000°C. Reciprocal space mapping of X-ray diffraction intensity shows that GaInN grown on GaN buffers is coherently strained up to thickness of some 100 nm [6]. A nominally undoped 7 nm GaInN QW is sandwiched between asymmetric barrier layers, which consist of a 300 nm GaN buffer layer, a 60 nm GaN cap layer, and a 20 nm AlGaN layer below or above the quantum well (see Fig. 1). The asymmetric double quantum well structure consists of a 2 nm and a 4 nm GaN QW, which are grown in succession on a 700 nm AlGaN buffer layer and separated by a 2.5 nm AlGaN layer. Time-resolved photoluminescence (TRPL) spectroscopy with resonant excitation of the quantum wells was performed at 5 K using a setup already described elsewhere [3].

Fig. 1. *Schematic pictures of sample structure of GaInN/GaN quantum wells with an AlGaN barrier.*

RESULTS

Asymmetric barrier structure on SiC substrate

In this section, we study GaInN/GaN QWs grown on SiC substrates with an additional AlGaN barrier above or below the quantum well. Fig. 2 shows low-temperature photoluminescence spectra of two QWs with asymmetric structure and a simple QW without the AlGaN barrier. The simple QW has an emission maximum at 3.08 eV and a phonon replica is recognizable. The sample with an AlGaN barrier below the quantum well exhibits no emission correlated with the GaInN layer, but shows a broad emission band around 2.67 eV emitted by the SiC substrate. In contrast, the sample with an AlGaN barrier above the quantum well shows an emission, and its maximum lies higher than that of the simple QW. A more detailed picture is given by decay traces at the respective luminescence maxima (see Fig. 3). We find that the luminescence intensity of the sample with an AlGaN barrier above the quantum well decays much faster than that of a simple quantum well without an AlGaN layer. At long delay time, we obtain a decay time of 25 ns for the former and 100 ns for the latter.

The origin of the difference induced by the introduction of the additional AlGaN barrier can be well explained by the existence of a piezoelectric field in the quantum well: if there were no electric field in the quantum well, the quantum wells with asymmetric structure would show identical optical

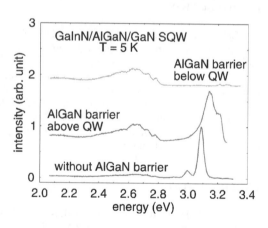

Fig. 2. *Time-integrated low-temperature photoluminescence spectra of GaInN/GaN quantum wells with and without an additional AlGaN barrier.*

properties. In the presence of a piezoelectric field, the additional AlGaN barrier leads to increased electron confinement and oscillator strength when it is placed where electrons are pushed by the electric field. If the AlGaN layer is grown on the other side of the quantum well, we expect a decrease of electron confinement and oscillator strength, leading to quenching of the optical transitions.

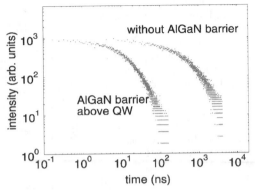

Fig. 3. *Photoluminescence decay of GaInN/GaN quantum wells without and with an additional AlGaN barrier above the quantum well.*

The sample with an AlGaN barrier on top of the quantum well exhibits a decreased lifetime compared with the simple quantum well, which indicates increased oscillator strength in this structure. In addition, the energetic position shifts toward higher energy due to the increased confinement energy. Therefore, our experimental results indicate directly that the piezoelectric field points towards the substrate.

It should be noticed that the additional AlGaN layer is grown at the same growth temperature of GaN layers to avoid a different interface quality among samples due to different growth temperatures. The interface effect therefore is not large enough to cause the observed difference of optical properties among samples.

It is interesting to compare this result with our previous work on the same structure grown on a sapphire substrate. The sample with an AlGaN barrier below the QW grown on the sapphire substrate exhibited a transition that is not quenched, but decays more slowly by 2-3 orders of magnitude than that with an AlGaN barrier above QW, indicating the same results of decreased electron confinement [5]. And we observed that an additional AlGaN barrier above the quantum well enhances the electron confinement in the sample grown on the sapphire substrate. This leads to the conclusion that the direction of the field is not dependent on the choice of SiC or sapphire substrate for structures grown by MOVPE.

We consider now the crystallographic polarity of our samples in relation to the piezoelectric polarity. Principally, there is no conclusive relationship between the crystallographic and piezoelectric polarity. But, guided by the theoretical sign of piezoelectric coefficient [7], the direction of the piezoelectric field in GaInN/GaN QWs is opposite to the [0001]-direction, which indicates that our samples are grown in the [0001]-direction, i.e., with Ga-face polarity. Therefore, the crystallographic polarity is also independent of the choice SiC or sapphire substrate, as is the piezoelectric polarity.

Asymmetric double quantum well

On the basis of the results in the previous section, we study an asymmetric GaN/AlGaN double quantum well grown on a SiC substrate. The two 2 nm and 4 nm

GaN quantum wells are separated by a 2.5 nm AlGaN barrier layer. Low-temperature spectra of the sample are summarized Fig. 4. To start with the time-integrated spectrum (dotted curve), we find a main emission line at 3.44 eV. This line is neighbored by two lines at 3.34 eV and 3.56 eV. It is interesting to notice that the energy differences between the main line and the two lines are similar. Time-resolved measurements reveal a more detailed picture. Fig. 4 shows normalized photoluminescence time-resolved spectra (solid curves) at increasing delay times. One can clearly distinguish three peaks again

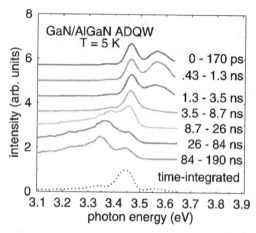

Fig. 4. Time-integrated and time-resolved photoluminescence spectra of a GaN/AlGaN asymmetric double quantum well.

dominating the spectrum for different delay times. Within approximately 8.7 ns after excitation, the high-energy peak intensity vanishes almost completely and the middle-energy peak dominates the spectrum. With further evolution in time, the low-energy peak takes over the maximum position.

A comparison of the decay times of the emission lines is given in Fig. 5. The high-energy peak intensity decays with a lifetime of about 0.3 ns and the middle-energy peak shows a rather increased lifetime of 4 ns. The decay time of the low-energy emission line is dramatically increased up to a time scale in the microsecond range.

Let us now discuss the origin of these peaks. The high- and middle-energy lines can be interpreted as intrawell transitions in the 2 nm and 4 nm quantum well, respectively. The difference of the energy position and the lifetime of these lines is mainly induced by the piezoelectric field in the strained GaN quantum wells as observed earlier [3]. To explain the low-energy line, we take a close look at the band diagram depicted schematically in Fig. 6. The AlGaN barriers are assumed to be unstrained and have no piezoelectric field, while the GaN QWs are under a biaxial

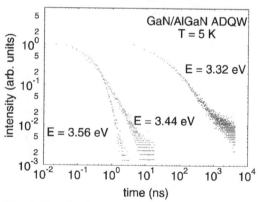

Fig. 5. Photoluminescence decay of a GaN/AlGaN asymmetric double quantum well.

W11.28.4

compressive strain, which induces a piezoelectric field F_{piezo}. In thermal equilibrium, the global band bending due to background doping gives rise to a depletion region in the AlGaN barriers below the GaN QWs and electrons accumulating in the topmost QW [5]. The resulting space charges partially screen the field in both QWs in an almost homogeneous manner. Since the barrier layer between the two QWs is much thinner than the typical Debye length of several 10 nm, the barrier layer does not carry any significant space charge. Nevertheless, the screening field F_{scr} induces a reverse electric field in the AlGaN barrier. On the other hand, a reduced effective electric field

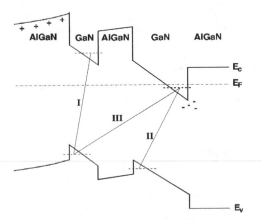

Fig. 6. *Schematic band diagram of an asymmetric GaN/AlGaN double quantum well with the piezoelectric field. Due to Fermi level alignment, the conduction bands of the AlGaN buffer layers are bent upward, and electrons are depleted.*

$F_{eff} = F_{piezo} - F_{scr}$ is built in both GaN QWs.

The effective field F_{eff} affects the energies of the intrawell transitions in the 2 nm and 4 nm QWs, which are labeled with I and II, respectively. The transition energy of QWs which are thicker than 2 nm varies approximately linearly with the well width [3]. Within this approximation, the energy difference between I and II is given as $\Delta E_{I,II} = F_{eff} \times \Delta L$, where ΔL is the well width difference of both GaN QWs. On the other hand, we can express the energy difference of the interwell (III) and the intrawell (II) transition as $\Delta E_{II,III} = F_{eff} \times L_I - F_B \times L_B$, where L_I and L_B is the width of the 2 nm GaN QW and of the AlGaN barrier layer, respectively.

Our photoluminescence measurements result in $\Delta E_{I,II}$ of 120 meV and $\Delta E_{II,III}$ of 100 meV. A comparison of this observation to the model indicates that the electric field F_B, which is induced by a screening field F_{scr} in the AlGaN barrier, is about 10 % of the piezoelectric field F_{piezo}.

Furthermore, the electron-hole separation for the interwell transition (III) can be approximated to the electron-hole separation in a 8.5 nm quantum well. This results in a small overlap of electron and hole wave functions. The lifetime in a microsecond range of the low-energy emission complies well with this expectation. Both the energetic position and the lifetime of the low-energy line can be well understood by the model considering an electric field in the double quantum wells.

The intrinsic electric field has two origins: first, a strain-induced piezoelectric field, and secondly, a spontaneous polarization. Especially, theoretical work predicted that spontaneous polarization in GaN/AlGaN QWs plays a more significant role compared to GaInN/GaN QWs [7]. Our experiments are sensitive to the total field, but cannot distinguish between the piezoelectric and spontaneous contributions.

It should be noticed that the appearance of the interwell transition depends crucially on the redistribution of excess carriers. Without this redistribution each quantum well would be semi-isolated, and only the intrawell transitions would be observed. The TRPL shows that the intrawell transition in the 2 nm QW disappears within about 8 ns. This indicates that the excess electrons in the 2 nm QW escape to the 4 nm QW or recombine with holes on this time scale. In this context, we can expect that the change of the position of the two quantum wells causes a different distribution of excess carriers in the QWs. In this case, the excess electrons in the 4 nm QW escape to the 2 nm QW, which has an effect on the interwell transition of the 4 nm QW.

SUMMARY

A time-resolved study on GaInN/GaN QW with an asymmetric barrier on SiC substrate reveals an enhanced oscillator strength when the AlGaN barrier is on top of the GaInN quantum well. In comparison to our previous work on a structure grown on sapphire, we find that the sign of the field is the same in both cases: the field points toward substrate. In the case of GaN/AlGaN ADQW, we observed an inter- and two intra-well transitions in the photoluminescence, and their lifetimes and emission energies confirm that these transitions are strongly influenced by the piezoelectric field.

ACKNOWLEDGMENTS

Research supported by Deutsche Forschungsgemeinschaft (DFG) under contract No. Ha. 1670/10. One of the authors (J.S.I.) gratefully acknowledges the support of the Deutscher Akademischer Austauschdienst (DAAD).

REFERENCES

1. S. Nakamura and G. Fasol, *The Blue Laser Diode* (Springer, Berlin, 1997).
2. T. Takeuchi, S. Sota, M. Katsuragawa, M. Komori, H. Takeuchi, H. Amano, and I. Akasaki, Jpn. J. Appl. Phys. **36**, L382 (1997).
3. J. S. Im, H. Kollmer, J. Off, A. Sohmer, F. Scholz, and A. Hangleiter, Phys. Rev. B **57**, R9435 (1998).
4. T. Takeuchi, C. Wetzel, S. Yamaguchi, H. Sakai, H. Amano, and I. Akasaki, Appl. Phys. Lett. **73**, 1691 (1998).
5. J. S. Im, H. Kollmer, O. Gfrörer, J. Off, F. Scholz, and A. Hangleiter, MRS Internet J. Nitride Semicond. Res. **4S1**, G6.20 (1999).
6. J. Off, A. Kniest, C. Vorbeck, F. Scholz, and O. Ambacher, J. Crystal Growth **195**, 286 (1998).
7. F. Bernardini, V. Fiorentini, and D. Vanderbilt, Phys. Rev. B **56**, R10024 (1997).

Mat. Res. Soc. Symp. Vol. 595 © 2000 Materials Research Society

THE FORMATION OF IN-RICH REGIONS AT THE PERPHERY OF THE INVERTED HEXAHONAL PITS OF InGaN THIN-FILMS GROWN BY METALORGANIC VAPOR PHASE EPITAXY

P. Li[1], S. J. Chua[1,2], M. Hao[2], W. Wang[2], X. Zhang[1] T. Sugahara[3] and S. Sakai[3]

[1]Center for Opto-electronics
Dept. of Electrical Engineering
National University of Singapore
SINGAPORE 119260

[2]Institute of Materials Research and Engineering
SINGAPORE

[3]Dept. of Electrical and Electronics Engineering,
The University of Tokushima,
2-1 Minami-Josanjima 770-8506,
Japan

ABSTRACT

InGaN thin films were grown by low-pressure metalorganic-vapor-phase-epitaxy (MOVPE) and characterized by cathodoluminescence (CL) and scanning electron microscopy (SEM). SEM images showed that InGaN samples have inverted hexagonal pits which are formed by the In segregation on the $(10\bar{1}1)$ surfaces. Room temperature CL at the wavelengths corresponding to the GaN band edge, the In-poor and In-rich regions showed that the In–rich regions formed at the periphery of the hexagonal pits.

INTRODUCTION

The development of blue LEDs and laser diodes has attracted considerable research activities on the growth of GaN based III-V nitrides. The band gap of InGaN can be varied over nearly the whole spectral range from near UV to red, so it is usually used in active regions of these devices.

The growth of $In_xGa_{1-x}N$ alloys is extremely difficult mostly due to the trade-off between the epilayer quality and the amount of InN incorporation into the alloy. Growth at the temperatures of approximately 800^0C typically results in high crystalline quality but the amount of InN in the solid is limited to low values because of the high volatility of In. Lowering the growth temperature results in an increase in the In content at the cost of reduced crystalline quality. The difference in lattice constant and thermal stability between the two constituents, InN and GaN, also complicates the growth of $In_xGa_{1-x}N$. The lattice mismatch can lead to a miscibility gap[1], which causes fluctuations of In content across the film. Singh and co-workers [2] provided strong evidence of phase separation in InGaN thick films grown by MBE. El-Masry *et al* [3]reported phase separation in thick InGaN films grown by MOVPE. Tran *et al* [4] showed that when phase-separation occurs, InGaN clusters with different indium compositions can coexist and that the high brightness in blue and green LEDs is due to the radiative recombination in the In-rich InGaN clusters.

The mechanism of phase-separation is still not clear. Wu *et al* [5] reported that the V-defects initiate at the threading dislocations in one of the quantum wells in a multiple quantum well (MQW) stack and found that V-defect is correlated with the localized excitonic recombination centers that give rise to a long-wavelength shoulder in photoluminescence (PL) and cathodoluminescence (CL) spectra.

Sugahara and Sakai [6] discussed the role of dislocation in InGaN phase separation. They showed that the dislocations in InGaN act as nonradiative recombination centers and confirmed that the phase separation in InGaN is caused by the spiral growth due to mixed dislocations. They demonstrated that dislocations with the screw component favor the formation of In-rich regions. The extra combinations of the dangling bonds in the dislocated areas can prevent the evaporation of InN during the InGaN growth.

EXPERIMENTAL

The four samples (245K1, 245K1A, 245K1B, 245K1C) of $In_xGa_{1-x}N$ films were grown by MOVPE (Emcore D125) on (0001) sapphire substrates. MOVPE was conducted using TMGa, TMIn and NH_3 as precursors. A 2μm thick undoped bulk GaN was first grown on the 250Å thick GaN buffer layer. The growth temperature was 530^0C and 1050^0C for the GaN buffer and bulk layer, respectively. After deposition of the GaN bulk layer, the growth temperature was lowered down to about 700^0C for the deposition of InGaN. The InGaN layer thickness was about 500Å. There was no cap layer on the top of InGaN. H_2 and N_2 were used as carrier gases for the growth of GaN and InGaN, respectively. The TMIn/TMGa molar ratios were 0.8789, 0.89, 0.8935, and 0.8935, respectively for the four samples. The samples were analyzed by cathodoluminescence (CL), high–resolution X-ray diffraction (HRXRD) and scanning electron microscopy (SEM) at room temperature (RT). CL measurements were performed with an acceleration voltage of 5 or 15kV using a JOEL 6400 SEM equipped with an Oxford Mono CL2.

RESULTS

Cathodoluminescence Spectra and Mapping

Figure 1 shows the room temperature CL spectra of the four samples (245k1, 245k1A, 245k1B, 245k1C) by scanning the incident electron beam in relatively wide area ($10.6x8.1μm^2$). The CL spectra of sample 245k1A, and 245k1C exhibit two peaks. The peaks with longer wavelengths were attributed to band-edge related transitions in the In-rich regions. The In–contents of the In-poor regions determined by HRXRD are 16.19%, 16.42%, 19.53% and 20.02%, respectively. In addition to the In-poor regions, In-rich regions with In-contents as high as 51.3% have been found for the four samples.

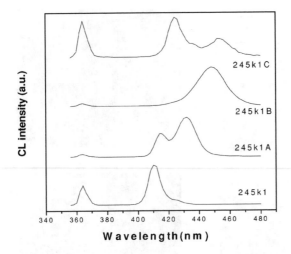

Fig.1. Cathodoluminescence spectra of the four samples, 245k1,245k1A,245k1B, 245k1C at RT.

Figure 2 shows plan-view SEM and CL images taken at wavelengths corresponding to the GaN band edge (364nm), InGaN In-poor regions (410nm, 414nm, 424nm, and 424nm for sample 245k1, 245k1A, 245k1B, and 245k1C, respectively), and In-rich regions (424nm, 432nm, 448nm, and 454nm for sample 245k1, 245k1A, 245k1B, and 245k1C, respectively).

In the SEM images, all the samples have "inverted hexagonal pits" (IHPs) but with different size and density. Samples grown under higher TMIn/TMGa molar ratios tend to have cathodoluminescence with longer wavelengths and larger IHPs. Sample 245k1C has the largest IHPs with a diameter of approximately 1 μm.

For any one of the four samples, the GaN band-edge CL at 364nm is not uniform. The dark spots correspond to non-radiative recombination centers which were reported to be related with dislocations [6]. The CL from the In-poor regions is not uniform either, which demonstrates the non-homogeneous nature of hetero-epitaxially grown InGaN. The CL from In-rich region comes *only* from the periphery of the IHPs. The situation is most clearly shown in the CL at 454nm for the sample 245k1C where the CL from In-rich regions is associated with the periphery of the two large IHPs.

One may argue that the longer wavelength CL comes from the defects which are captured by dislocations. To further clarify this, we performed temperature-dependent photoluminescence (PL) [7] and excitation-power-dependent PL as well as time-resolved PL (TRPL) [8]. The PL of the four samples have long wavelength peaks or shoulders, the temperature–dependent behavior of the long wavelength peak is S-shaped-like which was reported by Cho et al [9]. The TRPL demonstrated nano-second-order lifetime of the long wavelength peak. This fact excludes the possibility of donor-acceptor-pair transition (DAP) which usually has a longer lifetime in the micro-second order [9]. Based on the above PL experiments and HRXRD, we can confirm that In-rich regions occur in all of our four samples.

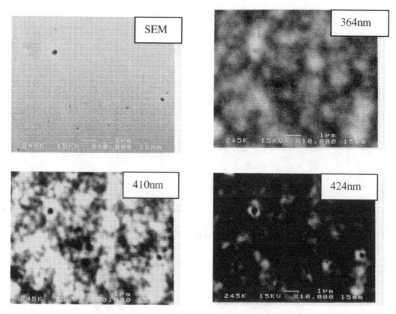

Fig.2-1. Top-view SEM and CL of the sample 245K1

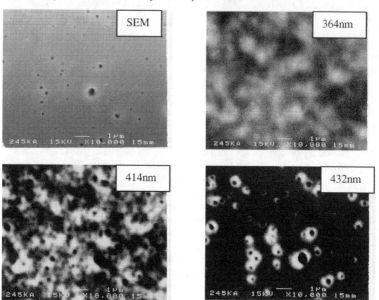

Fig.2-2. Top-view SEM and CL of the sample 245K1A

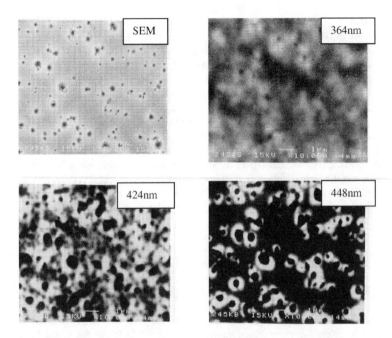

Fig.2-3. Top-view SEM and CL of the sample 245K1B

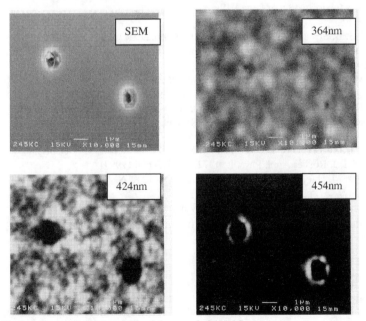

Fig.2-4. Top-view SEM and CL of the sample 245K1C

DISCUSSIONS

The formation of IHPs was investigated theoretically by Northrup [10]. A strong preference for In surface segregation and occupation on a specific surface (10$\bar{1}$1) was demonstrated using first-principle calculations. His calculations indicate that in the absence of In, the equilibrium pit size is on the order of a few nanometers. However, the segregation of In on (10$\bar{1}$1) surfaces can provide a strong driving force to reduce the surface energy and thereby allow the formation of large IHPs. In fact, In behaves as a differential surfacant, reducing the energy of the (10$\bar{1}$1) surface relative to (0001) surface. This effect promotes the formation of larger IHPs. In our samples, these IHPs as large as 1μm in diameter were found. The strong CL with longer wavelengths at room temperature from the regions near IHPs can be attributed to the quantum-confinement enhanced effect in the form of quantum well or quantum dots with high In content near IHPs. The formation of In-rich region near IHPs can be explained by carefully examining the role of dislocation in the growth of InGaN: the extra dangling bonds in dislocated area can prevent the evaporation of InN during the InGaN growth [7]. Therefore, the dislocations can become the triggering centers for In-rich regions.

It is worthwhile to distinguish the phase-separation (bulk diffusion) process with the segregation that may have occurred during the growth. In this paper, the latter: the In segregation on the growing surface and the preference for InN bonds on (10$\bar{1}$1) can explain the formation of the In-rich regions at the periphery of IHPs.

Conclusions

In conclusions, RT CL at wavelengths corresponding to the GaN near band egde, the In-poor and In-rich regions were performed on the InGaN samples. It was found that the In-rich regions formed at the periphery of the IHPs. The segregation of In on the (10$\bar{1}$1) surfaces allows the formation of large IHPs.

REFERENCES

1. I. Ho and G. B. Stringfellow, Appl. Phys. Lett. **69**, pp.2701, (1996).
2. R. Singh, D. Doppalapudi, T. D.Moustakas, and L. T. Romano, Appl. Phys. Lett. **70**, pp.1089, (1997).
3.N. A. El-Masry, E. L. Piner, S. X. Liu, and S. M. Bedair ,Appl. Phys. Lett., **72**, pp.40, (1998).
4. C. Tran, R. Karlicek, M. Schurman, V. Merai, A. Osinsky, Y. Li, I. Eliashevich, M. Brown, J. Nering, T. DiCarlo, I. Ferguson and R. Stall, 2nd Inter. Symp. On Blue Laser and LEDs, Chiba, sept.-29, Oct2. 1998, pp246-249
5. H. Wu, C. R. Elsass, A. Abare, M. Mack, S. Keller, P. M. Petroff, S. P. DenBaars, J. S. Speck, and S. J. Rosner Appl. Phys. Lett **72**, pp.692-694,(1997).
6. T. Sugahara, M. Hao, T. Wang, D. Nakagawa, Y. Naoi, K. Nishino and S. Sakai Jpn.J.Appl. Phys., **37**, pp.1195, (1998).
7. P. Li, S. J.Chua, G. Li, W.Wang, X.C. Wang, and Y. P. Guo, The Third International Conference on Nitride Semiconductor, Montpellier, France, July 4-july 9, 1999.
8. S.J. Chua, and G . Li to be published.
9.Yong-Hoon Cho, B.D. Little, G. H. Gainer, J. J. Song, S. Keller, U. K. Mishra, and S. P. Denbaars, Section G, GaN and related Materials, MRS Fall meeting 1998.
10. J. E. Northup, L. T. Romano and J. Neugebauer, Appl.Phys. Lett.,**74**,2319 (1999).

Mat. Res. Soc. Symp. Vol. 595 © 2000 Materials Research Society

Emission Enhancement of GaN/AlGaN Single-Quantum-Wells Due to Screening of Piezoelectric Field

A. Kinoshita[1 2], H. Hirayama[1], P. Riblet[1], M. Ainoya[1 2], A. Hirata[2] and Y. Aoyagi[1]
[1] The Institute of Physical and Chemical Research (RIKEN),
Hirosawa 2-1, Wako-shi, Saitama, 351-0198, Japan
[2] Department of Chemical Engineering, Waseda University,
Okubo 3-4-1, Shinjuku-ku, Tokyo, 169-8555, Japan

ABSTRACT

Photoluminescence (PL) enhancement due to the screening of piezoelectric field induced by Si-doping is systematically studied in GaN/AlGaN quantum wells (QWs) fabricated by metal organic vapor-phase-epitaxy (MOVPE). The PL enhancement ratio of QWs for Si-doped directly into the wells was much larger than that for doped only into the barrier layers. This result shows that the crystal quality of the quantum well is not so damaged by heavy Si-doping, which is different from the cases of GaAs or InP material systems. The PL intensity enhancement ratio was especially large for thick wells. The typical value of the enhancement ratio was 30 times for a 5 nm-thick single QW. The optimum Si-doping concentration was approximately 4×10^{18} cm^{-3}. From the well width dependence of the PL enhancement ratio and PL peak shift under high excitation conditions, we determined that the dominant effect inducing the PL enhancement is screening of piezoelectric field in the QWs. These results indicate that Si-doping is very effective for the application of GaN/AlGaN QWs to optical devices.

INTRODUCTION

GaN and AlGaN are currently of great interest for application to ultraviolet (UV) laser diodes (LDs) and light-emitting diodes (LEDs) [1,2], because the wide band gap direct transition can be adjusted between 6.2eV (AlN) and 3.4eV (GaN). Recently, UV LDs or LEDs are expected to realize the large capacity optical memories, new chemical processes, medical applications and disposal facilities to industrial wastewater [3].

Some research groups reported on the approaches to the realization of UV optical devices using GaN or AlGaN materials [4,5]. We have already achieved 333 nm current injection emission from $Al_{0.03}Ga_{0.97}N/Al_{0.25}Ga_{0.75}N$ quantum wells (QWs) and 234 nm efficient photoluminescence from AlN/AlGaN QWs. However there are some large problems preventing them from achieving UV optical devices. The most serious problems are difficulty to obtain efficient UV emission from AlGaN QWs, in contrast to InGaN QWs, and current injection through high Al content (more than 30%) AlGaN alloy. The emission enhancement in GaN/AlGaN QWs using Si- or In- [6] doping have been reported, and some mechanisms of the emission enhancement were discussed in previous papers [7-9]. However, the dominant mechanism of them was not yet clarified.

In this paper, we systematically investigate the PL intensity enhancement induced by Si-doping into GaN/AlGaN single (S) and multi quantum well (MQW) structures and reveal that the dominant effect is due to screening of piezoelectric field in the QWs. At first, the PL intensity enhancement due to Si-doping is compared between well doping

MQWs and barrier doping ones. Then, the dependencies of Si-doping concentration and well thickness on the PL enhancement for the SQWs are investigated. Finally, the maximum blue shift by changing the PL excitation power is compared between Si-doped and undoped SQWs.

SAMPLE PREPARATION

The samples were grown at 76 Torr on the Si faces of on-axis 6H-SiC(0001) substrates by a conventional horizontal-type metal organic vapor phase epitaxy (MOVPE) system. The substrates were loaded onto a SiC-coated graphite susceptor inductively heated by RF. As precursors, ammonia (NH_3), tetraethylsilane (TESi), trimethylaluminum (TMAl), and trimethylgallium (TMGa) were used with H_2 or N_2 as carrier gas. Typical gas flows were 2 standard (at 760 Torr) liters per minute (SLM), 2 SLM, and 2 SLM for NH_3, H_2, and N_2, respectively. The molar fluxes of TMGa and TMAl for AlGaN growth were 38 and 4.2 μmol/min, respectively. The substrate temperature during growth was measured with a thermocouple located at the substrate susceptor.

In order to achieve a surface suitable for growth of GaN QWs, first an approximately 500 nm-thick $Al_{0.15}Ga_{0.85}N$ buffer layers following extremely thin (~ 5 nm) AlN layers were deposited on a 6H-SiC substrate at 1100°C. Then 5 layers of MQW or SQW of GaN/AlGaN was grown at 1080°C. The well widths were ranging from 2 to 5 nm. The barrier width was 6 nm. Si was doped to the well layer(s) or barrier layers, and the doping concentration was ranging up to 9×10^{18} cm^{-3}. Finally, a 25-nm-thick $Al_{0.15}Ga_{0.85}N$ cap layer was grown at the same temperature as used for the QWs. The thickness of each layer was estimated based on the growth rate of equivalent bulk samples. The full width at half maximum of the X-ray rocking curve of bulk GaN and AlGaN sample were 95 sec and 122 sec, respectively. All of the PL measurement was performed at 77 K under He-Cd laser excitation except for the last measurement (Fig. 5).

MEASUREMENT RESULTS AND DISCUSSION

At first, the well width dependence of PL properties of undoped MQWs was investigated in order to find the optimized well width for PL intensity. Figure 1 shows the PL spectra of undoped GaN/$Al_{0.16}Ga_{0.84}N$ 5MQWs with various well widths. The intense single peak emission from QW regions were observed. The peak wavelength was shifted from 335 nm to 339 nm when the well width was varied from 2.5 to 4.5 nm. The measured PL peak energy agrees well with the theoretical value calculated taking into account piezoelectric field in the well. The emission around at 332 nm is from barrier layers. The phonon-replica peak was observed at lower energy side of the QW's peak indicating the high crystal quality of the QWs. The PL intensity heavily depends on the well width.

The reason for the immediate PL intensity reduction for thick MQWs is explained by the separation of electron and hole wave functions induced by the large piezoelectric field in the QWs [8]. For thin MQWs, the broadening of electron wave function and an increase of nonradiative transition at heterointerfaces induces the reduction of PL intensity. The

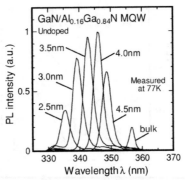

Figure 1. PL spectra of undoped GaN/Al$_{0.16}$Ga$_{0.84}$N 5MQWs with various well width measured at 77K. The PL peak energy and intensity strongly depend on the well width.

almost the same well width dependence.

Then well doping and barrier doping samples were compared with respect to the PL intensity enhancement ratio in GaN/Al$_{0.16}$Ga$_{0.84}$N five layer MQW structures. Figure 2 shows the PL enhancement ratio compared with undoped one as a function of Si-doping concentration for well doping and barrier doping samples. The well width was 4 nm, which was the optimum value for undoped case. Si-doping concentration was varied ranging from 0 to 9×10^{18} cm^{-3}. The enhancement ratio strongly depends on the Si-doping concentration. The larger value of enhancement ratio was obtained for well doping cases. The Si-doping concentration, which gives the maximum enhancement ratio, was smaller for well doping. The typical value of the enhancement ratio was seven times for well doping of 3×10^{18} cm^{-3}, and three times for barrier doping of 5×10^{18} cm^{-3}. For both series of samples, the enhancement ratio increases with an increase of doping concentration at the beginning, then reduces for higher doping concentration. The mechanism of PL intensity enhancement is based on the screening of the piezoelectric field in the wells that

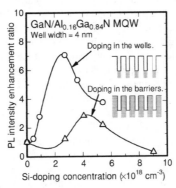

Figure 2. PL intensity enhancement ratio for Si-doped GaN/Al$_{0.16}$Ga$_{0.84}$N 5MQWs as a function of Si-doping concentration. Si is doped in the wells or barriers. The well and barrier width were 4nm and 6nm, respectivery.

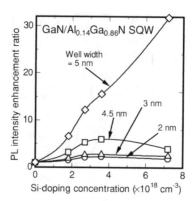

Figure 3. *PL intensity enhancement ratio for Si-doped GaN/Al$_{0.14}$Ga$_{0.86}$N SQWs with various well width as a function of Si-doping concentration.*

is to be mentioned in detail later. The enhancement ratio reduction at high doping concentration is induced by the deterioration of the crystal quality.

We discuss the mechanism of PL intensity enhancement due to Si-doping. Some previous works reported about the influence of Si-doping in bulk GaN [9] or barrier layers of InGaN MQWs [8], but there is no report about the comparison of well doping and barrier doping. In this work, the well doping is shown to be more efficient for GaN system, in contrast to the modulation doping (barrier doping) method which was widely applied for GaAs or InP systems. This is because the influence of the deterioration in GaN system is smaller than that in the other materials. This result seems to be notable as evidence for the peculiarity of this material.

There are two effects of Si-doping considered to induce the PL enhancement. One is screening of piezoelectric field in the wells, and the other is improvement of crystal quality. The screening effect is determined by an increase of carrier concentration and also a lattice stress relaxation of the well [11]. The effective carrier concentration in the well is larger for well doping case. For barrier doping case, the carrier injected from the barrier layers causes the same effect [8]. In the GaN/AlGaN QWs, the well is expected to be compressively strained. In our experiment the value of compressive strain in the well is estimated to be approximately 0.64% simply from the lattice constant of GaN and Al$_{0.14}$Ga$_{0.86}$N. By introducing Si-doping in GaN, the compressive strain is expected to be slightly relaxed [11], because Si in GaN shortens the lattice constant along c-axis. However, the value of the stress relaxation by Si-doping is reported to be small in case doping level is 10^{18}-10^{19} cm^{-2} (for example 0.027% for Si-doped GaN on sapphire [11] or negligible value for Si-doped GaN/InGaN QWs [12]). Consequently, the carrier increasing is considered to be dominant for the screening of piezoelectric field.

On the other hand, improvement of crystal quality due to Si-doping causes the reduction of luminescence killers [9]. The reduction of defects, which is reported by S. Ruvimov et al., is expected to occur for every Si-doped sample in this experiment [11]. In our experiments, no drastic difference of the line width of PL spectra between well doping and barrier doping was observed. Hence, the screening of the piezoelectric field in the wells is considered to be the dominant mechanism of the PL intensity enhancement.

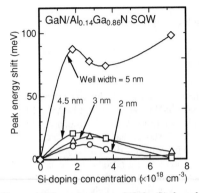

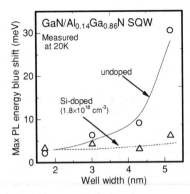

Figure 4. PL peak energy shift for Si-doped GaN/Al$_{0.14}$Ga$_{0.86}$N SQWs with various well width as a function of Si-doping concentration.

Figure 5. PL peak energy difference between high and low excitation conditions for undoped and Si-doped GaN/Al$_{0.14}$Ga$_{0.86}$N SQWs as a function of well width. Si-doping concentration was 1.8×10^{18} cm^{-3}.

Figure 3 shows the PL intensity enhancement ratio of Si-doped GaN/Al$_{0.10}$Ga$_{0.90}$N SQWs for various well widths as a function of Si-doping concentration. SQW structures were used in order to minimize the influence of well width fluctuation and the piezoelectric field of barrier layers. Si was doped only in the well. The well width and Si-doping concentration was varied from 2 to 5 nm, and from 0 to 8×10^{18} cm^{-3}, respectively. The maximum value of the enhancement ratio increased with the increase of well width. The typical value was 30 times for 5 nm-thick SQW of 8×10^{18} cm^{-3} Si-doped. For every well width, the enhancement ratio saturates at high doping concentration. The Si-doping concentration at the saturated region is higher for thick well. This is because for thick QW the piezoelectric field is strong and then high doping was needed to screen the field. These results fairly explain that screening of the piezoelectric field in the well induces the PL intensity enhancement.

Figure 4 shows the PL peak shift for Si-doped SQWs from that obtained for the undoped sample as a function of Si-doping concentration. The PL peak energy shows a large blue shift at low Si-doping concentration, and then a gradual red shift. The amount of the blue shift was larger for thick SQWs. The blue shift is considered due to the screening of the piezoelectric field in the well. The amount of blue shift becomes large for a thick QW, because the piezoelectric field becomes stronger. On the other hand, the PL peak red shift at high Si-doping concentration more than 3×10^{18} cm^{-3} was considered to be induced by the band gap narrowing due to Si-doping.

Figure 5 shows the maximum PL peak shift from low to high excitation conditions as a function of well width for Si-doped and undoped SQWs. The well width was varied from 2 nm to 5 nm. The PL measurement was performed at 20K. The excitation lasers used were He-Cd laser (325 nm, approximately 3 W/cm^2) and Xe-Cl excimer laser (308 nm, approximately 200 kW/cm^2) for low and high excitation conditions, respectively. For Si-doped sample, the maximum blue shift was independent of well width. On the other hand, for undoped sample, the maximum blue shift increased with the increase of the well width. This shows that the piezoelectric field in the Si-doped SQW is screened even under low excitation conditions by electron charge induced by Si-doping.

CONCLUSIONS

In conclusion, we systematically study the PL enhancement due to the screening of piezoelectric field induced by Si-doping in GaN/AlGaN QWs fabricated by MOVPE. The PL enhancement ratio of QWs for Si-doped directly into the wells was much larger than that for doped only into the barrier layers. This result shows that the crystal quality of the quantum well is not so damaged by heavy Si-doping, which is different from the cases of GaAs or InP material systems. The PL intensity enhancement ratio was especially large for thick wells. The typical value of the enhancement ratio was 30 times for a 5 nm-thick single QW. The optimum Si-doping concentration was approximately 4×10^{18} cm^{-3}. From the well width dependence of the PL enhancement ratio and PL peak shift under high excitation conditions, we determined that the dominant effect inducing the PL enhancement is screening of the piezoelectric field in the QWs. These results indicate that Si-doping is very effective for the application of GaN/AlGaN QWs to optical devices.

REFERENCES

1. S. Nakamura and G. Fasol, The Blue Laser Diode (Springer, Berlin, 1997).
2. J. Han, M. H. Crawford, R. J. Shul, J. J. Figiel, M. Banas, L. Zhang, Y. K. Song, H. Zhou and A. V. Nurmikko, Appl. Phys. Lett. **73,** 1688 (1998).
3. T. Ogita, H. Hatta, S. Nishimoto and T. Kagiya, J. Jpn. Chem. Eng. Soc. **5,** 970 (1985)
4. P. Kozodoy, M. Hansen, S. P. DenBaars and U. K. Mishra, Appl. Phys. Lett. **74,** 3681 (1999).
5. T. Mukai, H. Narimatsu and S. Nakamura, J. Cryst. Growth **189/190,** 778 (1998).
6. X. Q. Shen and Y. Aoyagi, Jpn. J. Appl. Phys. **38,** L 14 (1999).
7. Han and M. H. Crawford, Extended Abstracts of the 1999 Int. Conf. on Solid. State Devices and Materials, Tokyo, 1999 (The Japan Society of Applied Physics, Tokyo, 1999) pp. 46.
8. S. Chichibu, D. A. Cohen, M. P. Mack, A. C. Abare, P. Kozodoy, M. Minsky, S. Fleischer, S. Keller, J. E. Bowers, U. K. Mishra, L. A. Coldren, D. R. Clarke and S. P. DenBaars, Appl. Phys. Lett. **73,** 496 (1998).
9. E. F. Schubert, I. D. Goepfert, W. Grieshaber and J. M. Redwing, Appl. Phys. Lett. **71,** 921 (1997).
10. I. H. Lee, I. H. Choi, C. R. Lee and S. K. Noh, Appl. Phys. Lett. **71,** 1359 (1997).
11. S. Ruvimov, Z. L. Weber, T. Suski, J. W. Ager III, J. Washbum, J. Krueger, C. Kisielowski, E. R. Weber, H. Amano and I. Akasaki, Appl. Phys. Lett. **69,** 990 (1996).
12. Y. H. Cho, J. J. Song, S. Keller, M. S. Minsky, E. Hu, U. K. Mishra, and S. P. DenBaars, Appl. Phys. Lett. **73,** 1128 (1998).

Mat. Res. Soc. Symp. Vol. 595 © 2000 Materials Research Society

Correlation between structural properties and optical amplification in InGaN/GaN heterostructures grown by molecular beam epitaxy

A. Kaschner[1], J. Holst[1], U. von Gfug[1], A. Hoffmann[1], F. Bertram[2], T. Riemann[2], D. Rudloff[2], P. Fischer[2], J. Christen[2], R. Averbeck[3], and H. Riechert[3]

[1]Institut für Festkörperphysik, TU Berlin, Hardenbergstrasse 36, 10623 Berlin, Germany
[2]Institut für Experimentelle Physik, Otto-von-Guericke-Universität, PO Box 4120, 39016 Magdeburg, Germany
[3]Infineon Technologies, Corporate Research, CPR 7, 81730 München , Germany

Abstract

We comprehensively studied InGaN/GaN heterostructures grown by molecular beam epitaxy (MBE) using a variety of methods of optical spectroscopy, such as cathodoluminescence microscopy (CL), time-integrated and time-resolved photoluminescence. To correlate the fluctuations in emission wavelength with values for the optical amplification we performed gain measurements in edge-stripe geometry. The lateral homogeneity can be drastically improved using a template of GaN grown on the sapphire substrate by metal-organic vapor phase epitaxy (MOVPE). Gain values up to 62 cm^{-1} were found in samples with low indium fluctuations, which is comparable to values for high-quality InGaN/GaN heterostructures grown by MOVPE.

Introduction

Heterostructures of compound group-III nitride-semiconductors are of great importance for the rapidly increasing market of optoelectronical devices in the blue spectral range [1]. While most articles in this field report on investigation of samples grown by metal-organic vapor phase techniques [2-4] we focus here on InGaN grown by MBE. It is known that fluctuations in the Indium concentration as well as the miscibility gap [5], are major issues concerning the optical properties of the InGaN alloy system. Recently, the formation of InN quantum dots [6] or InGaN with potential fluctuations [7] was suggested to be the origin of the luminescence in the material. The degree of Indium fluctuations in MBE-grown heterostructures can drastically be reduced by using a template of MOVPE-GaN on the sapphire substrate, which leads to improved optical properties and gain characteristics. Due to the high Indium desorption rate it is necessary to grow at temperatures typically below 700°C during MBE growth, which is far away from the thermodynamic equilibrium. However, we show that high-quality InGaN/GaN heterostructures can be grown under these conditions exhibiting gain values up to 62 cm^{-1} at 7 K.

Experimental Details

Time-integrated photoluminescence experiments at different temperatures were performed using the 325 nm line of a He-Cd Laser. For time-resolved measurements a single photon counting setup was used with a 50 ps FWHM response to the laser pulse. A frequency-doubled dye laser pumped by the third harmonic of a Nd:YAG was used for excitation. The photoluminescence signal was analyzed in a 0.35 m subtractive double spectrometer and detected by a microchannel plate photomultiplier. The setup for the CL microscopy experiments is described in Ref. [8]. For the time integrated high-excitation investigations we used a dye laser pumped by an excimer laser, providing pulses with a duration of 15 ns at a rate of 30 Hz and a total energy of up to 20 µJ at 340 nm. The samples were mounted in a bath cryostat at 1.8 K. Gain measurements were performed using the variable-stripe-length method [9].

In this paper we focus on four typical InGaN/GaN heterostructures which are chosen from a variety of samples. All samples are grown on sapphire, followed by a 1.8 µm GaN (MOVPE) layer for samples A-C and capped by a 30 nm GaN layer. Samples A and B are double heterostructures including a 40 nm thick InGaN layer, with an Indium concentration of 10% and 13.7%, respectively, as determined by XRD [10]. Sample C is a 10x5 nm InGaN multiple-quantum well (MQW) with 4 nm GaN barriers. The growth conditions for the InGaN wells were the same as for sample B. Finally, sample D contains an InGaN layer of 120 nm thickness and 21% Indium not grown on a MOVPE-GaN template.

Results

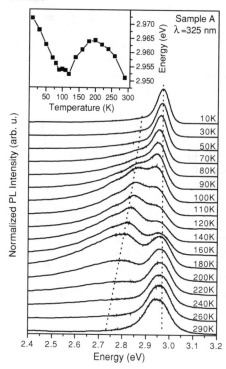

Fig. 1: PL spectra of sample A at different temperatures. The inset shows the position of the high-energy peak vs. temperature.

Fig. 1 shows the temperature dependence of the photoluminescence of sample A. One can distinguish between two different PL peaks, which change their relative intensities with temperature. At very low temperatures (10 K-70 K) and at room temperature the high-energy peak dominates the spectra. Furthermore this luminescence exhibits a "S-shaped" emission position shift with temperature as can be seen in the inset of Fig. 1. This behavior (redshift-blueshift-redshift) was first explained by Cho et al. [11] in terms of inhomogeneity and carrier localization in the InGaN. Recently, it has been shown that the "S-shape" behavior becomes less pronounced with increasing excitation power which can also be understood in terms of local potential fluctuations in the InGaN [12]. We think that thermalization and carrier freeze out in potential fluctuations leads to this unexpected temperature behavior. In this meaning the high-energy peaks is not only one emission line, but the spatially integrated luminescence of locally different recombination energies, which mirrors the distribution of the potential fluctuations.

We assign the low-energy peak being dominant between 90 K and 180 K as

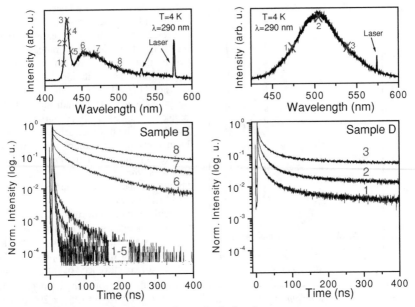

recombination from electronic states deeper in the band gap.

Fig. 2: Transients of sample B and D at different spectral positions (lower part) as indicated in the PL spectra (upper part).

To further investigate the underlying recombination processes we performed time-resolved PL experiments. Fig. 2 shows the result for the samples B (left hand side) and D (right hand side). In the upper part the PL spectra at 4 K are depicted and in the lower part the transients taken at the indicated spectral positions. In principle the PL spectrum of

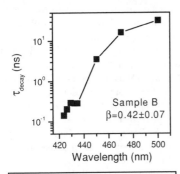

Fig. 3: Decay time as function of the emission wavelength for sample B.

sample B is similar to those of sample A and C, all grown on the MOVPE-GaN template. In contrast, the sample D exhibits only one broad luminescence structure. The temporal behavior of the luminescence is a multiple exponential decay for any spectral position and for all the samples. Pophristic et al. [13] claimed that this decay behavior can be described by $\exp[-(t/\tau)^{\beta}]$, the so-called stretched exponential. With this function it is possible to describe the recombination dynamics in heavily disordered systems. In Fig. 3 the decay times as determined by this model are depicted for different spectral positions of sample B. The general trend -also for sample A and C- is an increase in decay time with the detection wavelength. We find decay times of 150-300 ps

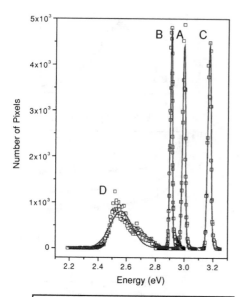

Fig. 4: CL histogramms of all four samples and the related Gaussian fit.

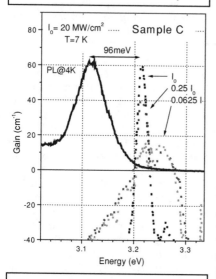

Fig. 5: Gain spectra of sample C at 7 K for different excitation densities. For comparison a low-excitation PL is shown.

and 2-32 ns for the high-energy and low-energy PL peak, respectively. The luminescence of sample D exhibits only long decay times in the ns range. It was shown that the decay times strongly depend on the excitation energy and intensity [14]. However, our results for the luminescence decay indicate that the samples under investigation are disordered. The disorder arises from local fluctuations in the Indium concentration leading to potential fluctuations. This is in line with the considerations about the thermodynamic growth conditions in the introductory part.

To achieve more information about the potential fluctuations we use the results of the CL measurements. Fig. 4 shows the histogramms of the four samples, i.e. the number of occurrence of an emission energy in the CL wavelength image of an area of 7 µm x 10 µm (256 x 200 pixels) is depicted. For samples A-C we find relatively narrow Gaussian distributions, whereas for the sample D the plotted histogramm is much broader and non-Gaussian. A Gaussian distribution indicates that the emission energies are random distributed on a lateral scale much smaller than the spatial resolution of the CL measurements (45 nm). For sample D the fluctuations are on a much larger scale, i.e. in the order of the spatial resolution or above. The difference between sample D and the others can not be explained by the higher Indium content of sample B, since alloy broadening following $[x(1-x)]^{1/2}$ can not explain the much broader distribution with a change in line shape. We think that the improvement from sample D to sample A-C is a result of better adjusted growth conditions and the use of the MOVPE-GaN templates.

In the following we will discuss how the degree of fluctuations affects the optical amplification. Fig. 5 shows the optical gain of sample C as function of the energy for different excitation energy. For comparison the low-intensity excitation PL is plotted. A gain structure between 3.2 eV and 3.3 eV appears for the lowest excitation. For higher power densities a sharp structure centered around 3.22 eV strongly increases having a maximum gain value of 62 cm^{-1}. This is assigned to the onset of lasing in the structure. The lasing peak is 96 meV higher in energy than the PL maximum and can be explained by the filling of states due to very high excitation densities during the gain measurements. The correlation between optical gain and structural properties is given in Table 1.

Sample	Gaussian Fit of CL Histogramms		Gain (7 K)	
	E_{center} (eV)	σ	E_{gain} (eV)	g_{max} (cm^{-1})
A	2.996	0.009745	3.131	37
B	2.908	0.007895	2.899	10
C	3.171	0.011265	3.213	62
D	2.561	0.0805	-	-

Table 1: Giving the results of the Gaussian fit of the CL histogramms (51200 data points) and the gain values at 7 K. The σ is the standard deviation.

We find that the samples A-C which exhibit a Gaussian CL histogramm show material gain at 7 K. In contrast, sample D (grown without MOVPE-GaN template) has a broad non-Gaussian luminescence distribution and shows no optical amplification. The highest gain value is found for sample C, i.e. the 10x5 nm MQW. We suggest that this is a result of confinement effects. On the other hand, the MQW has the broadest CL distribution of the three high-quality samples, which originates from Indium fluctuations but also from interface roughness [15]. However, on the small data basis of three samples we can not make a general statement on the dependence of g_{max} on σ for the high-quality MBE samples.

Conclusions

We have investigated the optical and structural properties of InGaN/GaN heterostructures grown by MBE using time-resolved PL, CL microscopy and gain spectroscopy. For all the samples we found evidence for potential fluctuations. Particularly, the recombination dynamics resulting in transients, which can be fitted by stretched exponential decays, and the CL measurements are explained with this model. We found a Gaussian CL distribution and optical amplification for the samples grown on a MOVPE-GaN template. The best sample, a 10x5 nm MQW, shows gain values up to 62 cm^{-1} at 7 K and 10 cm^{-1} at 150 K. Further improvement in growth conditions is expected to lead to optical amplification at RT, which is the requirement for laser fabrication.

Acknowledgments

The authors acknowledge the supply of the MOVPE-GaN templates on sapphire by R. Handschuh and B. Hahn (Osram Opto Semiconductors). A. Kaschner is gratefully for an Ernst-von-Siemens-scholarship. Work at Siemens/Infineon was partly supported by the German Ministry of Education and Research (BMBF).

References

[1]   S. Nakamura and G. Fasol: *The Blue Laser Diode* (Springer, Berlin, 1997).

[2]   Y.-H Kwon, G. H. Gainer, S. Bidnyk, Y.-H. Cho, J. J. Song, M. Hansen, and S. P. DenBaars, Appl. Phys. Lett. **75**, 2545 (1999).

[3]   T. Wang, D. Nakagawa, M. Lachab, T. Sugahara, and S. Sakai, Appl. Phys. Lett. **74**, 3128 (1999).

[4]   H. Kollmer, J. S. Im, S. Heppel, J. Off, F. Scholz, and A. Hangleiter, Appl. Phys. Lett. **74**, 82 (1999).

[5]   I.-H. Ho and G. B. Stringfellow, Appl. Phys. Lett. **69**, 2701 (1996).

[6]   K. P. O'Donnell, R. W. Martin, and P. G. Middleton, Phys. Rev. Lett. **82**, 237 (1999).

[7]   R. Averbeck, A. Graber, H. Tews, D. Bernklau, U. Barnhöfer, and H. Riechert, SPIE Vol. **3279**, 28 (1998).

[8]   F. Bertram, T. Riemann, J. Christen, A. Kaschner, A. Hoffmann, C. Thomsen, K. Hiramatsu, T. Shibata, and N. Sawaki; Appl. Phys. Lett. **74**, 359 (1999).

[9]   K. L. Shaklee, R. E. Nahory, and R. F. Leheny, J. Lum. **7**, 284 (1973).

[10] M. Schuster, P. O. Gervais, B. Jobst, W. Hösler, R. Averbeck, H. Riechert, A Iberl, and R. Stömmer, J. Phys. D Appl. Phys. **32**, 56 (1999).

[11] Y.-H. Cho, G. H. Gainer, A. J. Fischer, J. J. Song, S. Keller, U. K. Mishra, and S. P. DenBaars, Appl. Phys. Lett. **73**, 1370 (1998).

[12] J. Christen, T. Riemann, P. Fischer, J. Holst, A. Hoffmann, and M. Heuken, MRS Fall Meeting Boston 1999.

[13] M. Pophristic, F. H. Long, C. Tran, I. T. Ferguson, and R. F. Karlicek, J. Appl. Phys. **86**, 1114 (1999).

[14] M. Pophristic, F. H. Long, C. Tran, R. F. Karlicek, Z. C. Feng, and I. T. Ferguson, Appl. Phys. Lett. **73**, 815 (1998).

[15] Y.-H. Cho, J. J. Song, S. Keller, U. K. Mishra, and S. P. DenBaars,. Appl. Phys. Lett. **73**, 3181 (1998).

Mat. Res. Soc. Symp. Vol. 595 © 2000 Materials Research Society

Optical Properties of AlGaN Quantum Well Structures

Hideki Hirayama[1], Yasushi Enomoto[12], Atsuhiro Kinoshita[12], Akira Hirata[2] and Yoshinobu Aoyagi[1]
[1] The Institute of Physical and Chemical Research (RIKEN),
2-1, Hirosawa, Wako-shi, Saitama, 351-0198, Japan, hirayama@postman.riken.go.jp
[2] Department of Chemical Engineering, Waseda University,
3-4-1, Okubo, Shinjuku-ku, Tokyo, 169-8555, Japan

ABSTRACT

We demonstrate 230-250 nm efficient ultraviolet (UV) photoluminescence (PL) from AlN(AlGaN)/AlGaN multi-quantum-wells (MQWs) fabricated by metal-organic vapor-phase-epitaxy (MOVPE). Firstly, we show the PL properties of high Al content AlGaN bulk (Al content: 85-95%) emitting from near band-edge. We systematically investigated the PL properties of AlGaN-MQWs consisting of wide bandgap AlGaN (Al content: 53-100%) barrier. We obtained efficient PL emission of 234 and 245 nm from AlN/$Al_{0.18}Ga_{0.82}N$ and $Al_{0.8}Ga_{0.2}N$/$Al_{0.18}Ga_{0.82}N$ MQWs, respectively, at 77 K. The optimum value of well thickness was approximately 1.5 nm. The emission from the AlGaN MQWs were several tens of times stronger than that of bulk AlGaN. We found that the most efficient PL is obtained at around 240 nm from AlGaN MQWs with $Al_{0.8}Ga_{0.2}N$ barriers. Also, we found that the PL from AlGaN MQW is as efficient as that of InGaN QWs at 77 K.

INTRODUCTION

AlN or AlGaN is very attractive material for the application to ultraviolet (UV) laser diodes (LDs), light-emitting diodes (LEDs) or photo detectors, because the direct transition energy can be adjusted between 6.2 eV (AlN) and 3.4 eV (GaN). UV LDs or LED are expected to realize the large capacity optical memories or compact UV measurement systems. Recently, the efficient UV light sources are expected in the field of chemical industry or medical applications.

However, there are some large problems preventing us from achieving UV optical devices. The most serious problems are difficulty to obtain efficient UV emission from AlGaN quantum wells (QWs), in contrast to InGaN QWs, and hole injection through high Al content (more than 30%) AlGaN alloy. Recently some research groups reported on the approaches to realize UV optical devices using GaN or AlGaN materials [1,2]. We have already achieved 333 nm current injection emission from $Al_{0.03}Ga_{0.97}N$/$Al_{0.25}Ga_{0.75}N$ QWs using Mg-doped AlGaN/GaN super-lattice hole conducting region [3].

The purpose of this work is to obtain efficient UV emission as well as blue emission already obtained from InGaN QWs. Last year, we have obtained 280 nm intense photoluminescence (PL) emission from AlGaN QWs at 77 K [4]. Also, we reported on the PL enhancement in GaN/AlGaN QWs induced by Si-doping due to the compensation of large piezoelectric field in the well [5].

In this work, we demonstrate the first intense UV PL around 230 nm from AlGaN multi-quantum-wells (MQWs) consisting of high Al content AlGaN or AlN barriers. We obtained the growth condition of high Al content AlGaN. We obtained a single peak spectrum from high Al content AlGaN bulk (Al content up to 80%) emitting from the near band-edge. Then we fabricated AlGaN-MQWs consisting of wide bandgap AlGaN (Al content: 53-100%) barriers. The well thickness dependence of PL properties is systematically investigated. We show efficient emission around 230 nm obtained from the AlN/AlGaN MQWs. Finally, the PL intensities are compared between AlGaN MQWs with various Al content of AlGaN barriers, and InGaN and GaN QWs.

EXPERIMENTS AND DISCUSSIONS

The samples were grown at 76 Torr on the Si-face of an on-axis 6H-SiC(0001) substrate, by a conventional horizontal-type MOVPE system. As precursors ammonia (NH_3), trimethylaluminum (TMAl), and trimethylgallium (TMGa) were used with H_2 as carrier gas. N_2 gas was independently supplied by a separate line in order to control the gas flow. Typical gas flows were 2 standard liters per minute (SLM), 2 SLM, and 0.5 SLM for NH_3, H_2, and N_2, respectively. The molar fluxes of TMGa and TMAl for the growth of $Al_xGa_{1-x}N$ (x=0.11-1) were 38 and 2.6-45 µmol/min, respectively. At this condition, the growth rate of $Al_{0.11}Ga_{0.89}N$, $Al_{0.40}Ga_{0.60}N$ and AlN were approximately 2.4, 1.0 and 0.4 µm/h, respectively. The substrate temperature measured with a thermocouple located at the substrate susceptor during the growth was 1140 °C for all layer. All samples were undoped.

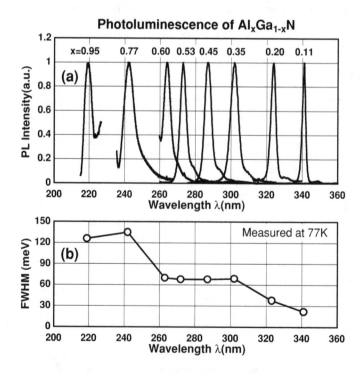

Figure 1. PL spectra of $Al_xGa_{1-x}N$ (x=0.11-0.95) films grown on 6H-SiC measured at 77 K.

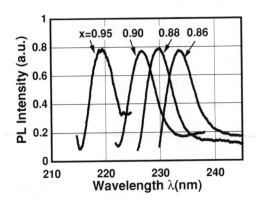

Figure 2. PL spectra of high Al content $Al_xGa_{1-x}N$ (x=0.86-0.95) films measured at 77 K.

At first, we show the optical properties of high Al content AlGaN alloy. Figure 1 shows (a) the PL spectra and (b) full width at half maximum (FWHM) of the PL peak of AlGaN films measured at 77 K. The AlGaN alloy was grown directly on a very thin (~5 nm) AlN layer deposited on SiC. Figure 2 also shows the PL spectra of high Al content AlGaN (Al content is 85-95%). The thickness of AlGaN film was approximately 250 and 400 nm for AlN and $Al_{0.11}Ga_{0.89}N$, respectively. As seen in Fig. 1 and 2, single peak spectra were obtained for Al contents of 0.11-0.95 emitting from near band edge. The yellow emission around 500-550 nm was negligible even for high Al content AlGaN. The phonon-replica peaks are seen at the low energy side of each spectra for Al content of 0.11-0.53, confirms the good crystal quality of the AlGaN. The typical value of FWHM of the spectrum was approximately 20, 65 and 120 meV for the Al content of 10-12%, 35-60% and 70-95%, respectively. The increased FWHM observed for the Al content of 35-60% may be due to the layer composition fluctuation during the growth. The large FWHM (120 meV) observed for high Al content AlGaN indicates that at this moment the crystal quality of AlN is not so good compared with that of GaN.

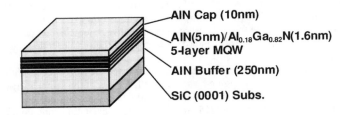

Figure 3. Schematic layer structure of fabricated $AlN/Al_{0.18}Ga_{0.82}N$ MQW sample.

Table I. Material and the thickness of buffer, barrier, well layers, and PL peak wavelength range of each series of AlGaN MQWs.

Sample Series	Structure	Buffer (Thickness)	Barrier (Thickness)	Well (Thickness)	Peak Wavelength
(a)	5-layer MQW	AlN (250 nm)	AlN (5 nm)	$Al_{0.18}Ga_{0.82}N$ (1.2-3.3 nm)	229-285 nm
(b)	5-layer MQW	$Al_{0.8}Ga_{0.2}N$ (250 nm)	$Al_{0.8}Ga_{0.2}N$ (5 nm)	$Al_{0.18}Ga_{0.82}N$ (0-3.3 nm)	238-288 nm
(c)	5-layer MQW	$Al_{0.7}Ga_{0.3}N$ (300 nm)	$Al_{0.7}Ga_{0.3}N$ (5 nm)	$Al_{0.12}Ga_{0.88}N$ (1.4-2.7 nm)	255-303 nm
(d)	5-layer MQW	$Al_{0.53}Ga_{0.47}N$ (400 nm)	$Al_{0.53}Ga_{0.47}N$ (5 nm)	$Al_{0.11}Ga_{0.89}N$ (0-6.7 nm)	272-343 nm

Then, we fabricated four series of AlGaN MQW samples, consisting of different Al content AlGaN barriers. Figure 3 shows an example of schematic layer structure. Table I summarizes the material and the thickness of buffer, barrier, well layers, and PL peak wavelength range of each series of AlGaN MQWs. In order to achieve a flat surface suitable for the growth of AlGaN quantum wells, an approximately 250-400 nm thick AlN(AlGaN) buffer layer followed by a very thin AlN layer was deposited. We confirmed a step-flow grown surface by atomic force microscopy (AFM) for the sample series (a), (b), and (c). After that, a five-layer MQW structure consisting of several-nm thick $Al_xGa_{1-x}N$ wells (x=0.11-0.18) and 5 nm-thick $Al_yGa_{1-y}N$ barriers (y=0.53-1), and 10nm-thick $Al_yGa_{1-y}N$ cap (y=0.53-1) were grown. The well or barrier thickness is estimated simply from the growth rate of bulk material.

Figure 4 shows PL spectra of (a) $AlN/Al_{0.18}Ga_{0.82}N$ and (b) $Al_{0.80}Ga_{0.20}N/Al_{0.18}Ga_{0.82}N$ five-layer MQWs for various well thickness, excited with Xe-lamp light source (215 nm) measured at 77 K. The well thickness was varied ranging 1.2-3.3 nm and 0-3.3 nm for sample series (a) and (b), respectively. We obtained single-peak intense PL emission from each MQWs. The most efficient emission was obtained at the wavelength of 234 and 245 nm from $AlN/Al_{0.18}Ga_{0.82}N$ and $Al_{0.8}Ga_{0.2}N/Al_{0.18}Ga_{0.82}N$ MQWs, respectively. The optimum value of well thickness was approximately 1.5 nm. The PL intensity of the MQWs were several tens of times larger than that of AlGaN bulk.

The quantized level shift is obviously observed. The PL intensity heavily depends on the well thickness. The rapid reduction of the PL intensity with the increase of the well thickness may be caused by a reduction of the radiative recombination probability due to the large piezoelectric field in the well [6]. The reduction of the emission intensity for thin well may be mainly due to the increase of nonradiative recombination on the hetero-interfaces. For the sample series (c) and (d), we observed the similar tendency obtained for (a) and (b).

Finally, the PL intensities are compared between AlGaN MQWs with various Al content of AlGaN barriers. Figure 5 shows the PL spectra of the samples (a)-(d) shown in Table I for optimized QW thickness measured at 77 K. All samples were excited with Xe-lamp light source (215 nm) with the same excitation condition. From Fig. 5, we found that the most efficient emission is obtained around 245 nm for AlGaN MQW systems and that the optimum Al content for AlGaN barrier is 80%. The PL intensity of QW is strongly depending on the buffer conditions. More efficient emission is expected from AlN/AlGaN QWs by improving the crystal quality of AlN buffer and barriers. We have also found that the PL intensity of AlGaN QWs is as strong as that of the $In_{0.02}Ga_{0.98}N/In_{0.20}Ga_{0.80}N$ QW and much stronger than that of $Al_{0.12}Ga_{0.88}N/GaN$ MQWs at 77 K [4]. However, at room temperature, the emission from AlGaN and GaN QWs are much weaker in comparison with InGaN QWs. The next subject is to obtain efficient UV emission at room temperature.

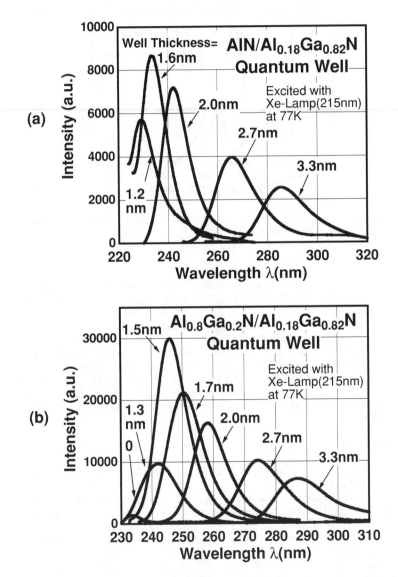

Figure 4. PL spectra of (a) $AlN/Al_{0.18}Ga_{0.82}N$ and (b) $Al_{0.80}Ga_{0.20}N/Al_{0.18}Ga_{0.82}N$ 5-layer MQWs for various well thickness.

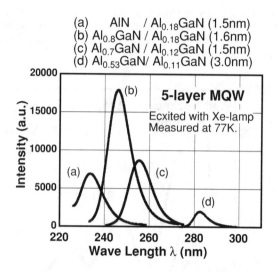

Figure 5. Comparison of PL intensity at 77 K between AlGaN 5-layer MQWs with different Al content AlGaN barriers.

CONCLUSION

We demonstrated 230-250 nm efficient UV PL from AlN(AlGaN)/AlGaN MQWs fabricated by MOVPE. We showed PL spectra of high Al content AlGaN bulk (Al content: 85-95%) emitting from near band-edge. We systematically investigated the PL properties of AlGaN-MQWs consisting of wide bandgap AlGaN (Al content: 53-100%) barrier. We obtained efficient PL emission of 234 and 245 nm from AlN/$Al_{0.18}Ga_{0.82}$N and $Al_{0.8}Ga_{0.2}$N/$Al_{0.18}Ga_{0.82}$N MQWs, respectively, at 77 K. The optimum value of well thickness was approximately 1.5 nm. The emission from the AlGaN MQWs were several tens of times larger than that of bulk AlGaN. We found that the most efficient PL is obtained at around 240 nm from AlGaN MQWs with $Al_{0.8}Ga_{0.2}$N barrier. Also, we found that the PL from AlGaN MQW is as efficient as that of InGaN QWs at 77 K.

REFERENCES

1. Peter Kozodoy, Monica Hansen, S. P. DenBaars and U. K. Mishra, *Appl. Phys. Lett.* **74,** 3681 (1999).
2. T. Mukai, H. Narimatsu and S. Nakamura, *J. Crystal Growth* **189/190,** 778 (1998).
3. A. Kinoshita, H. Hirayama, M. Ainoya, A. Hirata and Y. Aoyagi, to be submitted.
4. H. Hirayama and Y. Aoyagi, *Mater. Res. Soc. Proc.* **537,** G3.74 (1999).
5. A. Kinoshita, H. Hirayama, A. Hirata and Y. Aoyagi, *International Conference on Solid State Device and Materials (SSDM)*, **C-4-2,** Tokyo (1999).
6. S. Chichibu, D. A. Cohen, M. P. Mack, A. C. Abare, P. Kozodoy, M. Minsky, S. Fleischer, S. Keller, J. E. Bowers, U. K. Mishra, L. A. Coldren, D. R. Clarke and S. P. DenBaars, *Appl. Phys. Lett.* **73,** 496 (1998).

Mat. Res. Soc. Symp. Vol. 595 © 2000 Materials Research Society

Characterization of InGaN Quantum Wells Grown by Molecular Beam Epitaxy (MBE) Using Ammonia as the Nitrogen Source

F. Semendy, L.K. Li [1], M.J. Jurkovic [1], and W.I.Wang [1]

AMSRL-SE-EM, SEDD, Army Research Laboratory, Adelphi, MD 20783

[1]Electrical Engineering Department, Columbia University, New York, NY 10027

ABSTRACT

Single quantum well InGaN was grown by molecular beam epitaxy with ammonia as the nitrogen source. The samples were grown on (0001) sapphire substrates. The photoluminescence (PL) intensity of InGaN quantum wells showed band-edge emissions at 2.71eV at low temperature (10 K). PL was investigated as a function of excitation intensity and temperature. The relationship between PL intensity and excitation intensity, as well as the relationship between PL intensity and lattice temperature was studied. Also studied was the combined effect of temperature and intensity variation. Detailed results are reported here.

INTRODUCTION

III-V GaN based semiconductors have recently attracted attention for their potential applications as multicolor light emitters, solar-blind ultraviolet detectors, and high power/high-temperature electronics. In $_x$Ga$_{1-x}$N alloy systems and related heterostructures such as quantum wells (QWs) are especially attracting attention because they are used as active layers for high-brightness blue, green, and yellow light emitting diodes (LEDs) [1,2] and cw blue laser diodes (LDs) [3]. A large number of studies have been reported on the optical properties of InGaN epilayers and InGaN/GaN QW structures particularly pressure dependent PL studies [4] on InGaN/GaN multiple quantum wells , time resolved PL studies of InGaN/GaN single quantum wells at room temperature [5], determination of PL mechanism in InGaN quantum wells [6] and temperature dependent PL line shapes in InGaN [7]. Despite the commercial success of InGaN based LEDs, their optical emission properties are not completely understood.

EXPERIMENT

The InGaN SQW structures with varying thickness examined in this study were grown by molecular beam epitaxy (MBE) with ammonia used as the nitrogen source. The samples were grown on (0001) sapphire substrates. Before the growth, the sapphire substrate was cleaned by an SVT Associates RF nitrogen source at 800°C for 10 minutes. We used low substrate temperature (<650 °C) for InGaN to avoid the evaporation of In from the growing surface. The structure involved are the substrate

(0001) sapphire, a GaN buffer layer (1.5 µm), InGaN quantum well (varying thickness - 30, 60 Å), and a GaN capping layer of thickness 300 Å as schematically illustrated in Fig.1. We measured PL using the 325 nm line of a cw He-Cd laser operating at 50 mW. Variable intensity filter was used to change the laser intensity and the optical power was measured using Newport Power meter model 1815C. This intensity variation study was conducted at room temperature and varying temperatures starting from 10 K. We also conducted temperature-dependent and intensity-dependent studies as well as combined temperature and intensity dependent studies between 10 and 300 K. PL measurements were done with a Spex model 1404 with 0.85 m double-path spectrophotometer with the detector being a Hamamatsu water cooled GaAs photo-multiplier tube. A Janis cryogenic cooler was used to cool the sample. For all the experimental studies, one InGaN specimen was used for consistency.

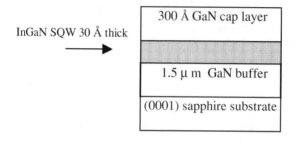

FIG. 1. Layer structure of InGaN single quantum well grown by molecular beam epitaxy with ammonia used as the nitrogen source. Note that the well and film regions are not drawn to scale.

RESULTS AND DISCUSSION

We first confirmed that the samples used for the optical measurements were of good quality throughout by performing PL measurements at different positions on the sample. Experiments were performed for varying temperatures and laser intensities. Figure 2 shows a transition energy peak at 2.71 eV for the $In_{0.3}Ga_{0.7}N$ SQW structure with a well width of 30 Å. As can be seen from the figure, the PL spectrum is a strong and narrow well-defined peak with a FWHM value of 125 meV, indicating the quality of the SQW. Shown in Figure 3 is a series of PL spectra of $In_{0.3}Ga_{0.7}N$ from 10 to 60 K. At 10 K, near band gap emission manifests as the main peak at 2.71 eV.

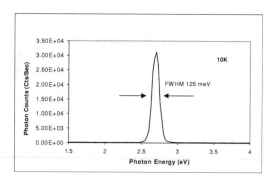

FIG. 2. Low temperature photoluminescence spectra for the 30 Å
In $_{0.3}$Ga $_{0.7}$N /GaN single quantum well structure grown at 650 °C by RF
plasma source.

As temperature goes up above T>60 K (not shown here) the carriers are becoming
thermalized while below T<60 K the carriers are not thermalized, presumably as a
result of localization. Also upto 60 K the spectra are shifted to the blue region as can
be observed from the figure. Beyond that temperature the peaks shift toward the red.
Figure 4 is the plot of energy peak positions (eV) of In $_{0.3}$Ga $_{0.7}$N SQW against
temperature. In this graph we see that at lower temperature (<60 K), bandgap energy
E_g increases shifting to blue. This can be explained by non-thermal equilibrium
distribution of carriers in traps with energies lower energies below E_g. The magnitude
of the blue shift is about 20 meV for the temperature range of 10 to 60 K. Beyond 60
K the band gap energy shifts lower indicating the red shift. The full width half-
maximum (FWHM) value of each peak widens and peak energy value goes down as
the temperature goes up, indicating the thermal effect on the excitons. Rough
calculations indicate that these FWHM changes are almost linear with temperature.

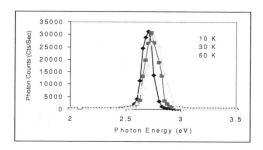

FIG. 3. PL of In $_{0.3}$Ga $_{0.7}$ N SQW at various temperatures.

Low temperature (10 K) PL studies on In $_{0.3}$Ga $_{0.7}$N SQW were done for various pump powers. The PL output at peak wavelength of 2.71 eV decreased almost linearly with decreasing laser excitation power. In all experiments the sample was maintained at varying temperatures in the cryogenic cooler. Experiments were performed for various pump powers (30, 15, 8, 4, 2, 1 mW) of the HeCd laser for each temperature of interest. The FWHM values of the spectrum increase as the pumping power goes down, indicating a spectral broadening.

Similar experiments were performed at higher temperatures for the same pumping power, as shown in Figure 5. As can be seen from the plot, even at 10 K nonlinearity can be observed. However, the 200 and 300 K experimental values are noticeably nonlinear.

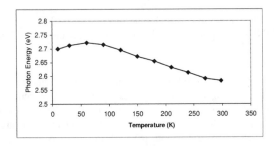

FIG. 4. PL emission peak positions versus temperature for In $_{0.3}$Ga $_{0.7}$ N SQW.

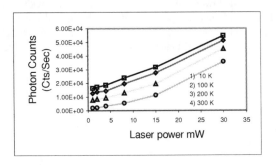

FIG .5. PL intensity as a function of laser excitation intensity for various temperatures.

CONCLUSIONS

In summary, we have grown InGaN single quantum well structures using ammonia and an RF plasma nitrogen source. These samples were used for a detailed PL study, including intensity and temperature variation studies. InGaN quantum wells showed band-edge emissions at 2.71 eV at low temperatures (10 K) which shifted to different values at higher temperatures. At lower temperature, a blue shift in the PL energy peak and at higher temperatures a classical red shift occurred caused by variations in inter-band transitions and thermalization. In the intensity variation study, it was found that the PL intensity increases with increasing laser excitation power. We observed for the first time in these materials that when both excitation intensity and lattice temperature are varied, there is noticeable nonlinear increase in the PL intensity of free excitons above 200 K. This could be explained as due to a strong dependence of the excitation related population in the specimen because of the excitation intensity and the lattice temperature for the sample. However, further modeling and studies are required and hence being pursued for a full explanation of this phenomenon.

REFERENCES

1. S. Nakamura, M. Senoh, N. Iwasa, and S.-I. Nagahama, Jpn. J. Appl. Phys. **34**, L797 (1995).
2. S. Nakamura, M. Senoh, N. Iwasa, and S.-I. Nagahama, Appl. Phys. Lett. **67**, 186 (1995).
3. S. Nakamura, M. Senoh, S. Nagahama, N. Iwasa, T. Matsushita, Y. Sugimoto, and H. Kiyoku, Appl. Phys. Lett. **70**, 868 (1997).
4. W. Shan, P. Perkin, J. W. Ager III, W. Walukiewicz, E.E.Haller, M. D. McCluskey , N. M. Johnson, and D. P. Bour, Appl. Phys. Lett. **73**, 1613 (1998).
5. C. K. Sun, T. L. Chieu, S. Keller, G. Wang, M. S. Minsky, S. P. DenBaars,. and J E. Bowers, Appl. Phys. Lett. **71**, 425 (1997).
6. P.Riblet, H. Hirayama, A. Kinoshita, A. Hirata, T. Sugano, and Y. Aoyagi, Appl. Phys. Lett. **75**, 2241 (1998).
7. K. L. Teo, J. S. Cotton, P.Y. Yu, E. R. Weber, M. F. Liu, K.Uchida,H. Tokunaga, N. Akutsu, and K. Matsumoto, Appl. Phys. Lett. **73**, 1697 (1998).

Mat. Res. Soc. Symp. Vol. 595 © 2000 Materials Research Society

Comparative study of structural properties and photoluminescence in InGaN layers with a high In content

A. Vantomme[1], M.F. Wu[1], S. Hogg[1], G. Langouche[1], K. Jacobs[2], I. Moerman[2], M.E. White[3], K.P. O'Donnell[3], L. Nistor[4,5,6], J. Van Landuyt[5] and H. Bender[6]

[1]Inst. Kern- en Stralingsfysica, Univ. of Leuven, B-3001 Leuven, Belgium
[2]Dept. of Information Technology, Univ. of Gent-IMEC, B-9000 Gent, Belgium
[3]Dept. of Physics and Applied Physics, Univ. of Strathclyde, Glasgow, U.K.
[4]Institute of Atomic Physics, Bucharest, Romania
[5]EMAT, Univ. of Antwerp (RUCA), B-2020 Antwerp, Belgium
[6]IMEC, B-3001 Leuven, Belgium

ABSTRACT

Rutherford backscattering and channeling spectrometry (RBS), photoluminescence (PL) spectroscopy and transmission electron microscopy (TEM) have been used to investigate macroscopic and microscopic segregation in MOCVD grown InGaN layers. The PL peak energy and In content (measured by RBS) were mapped at a large number of distinct points on the samples. An indium concentration of 40%, the highest measured in this work, corresponds to a PL peak of 710 nm, strongly suggesting that the light-emitting regions of the sample are very indium-rich compared to the average measured by RBS. Cross-sectional TEM observations show distinctive layering of the InGaN films. The TEM study further reveals that these layers consist of amorphous pyramidal contrast features with sizes of order 10 nm. The composition of these specific contrast features is shown to be In-rich compared to the nitride matrix.

INTRODUCTION

InGaN compounds have been used successfully for the fabrication of highly efficient blue, green, amber and red light emitting diodes. The wavelength of the emitted light of these diodes has been correlated with the overall In-content of the ternary nitride [1]. However, from electro- and photoluminescence (PL) experiments, it was suggested that the luminescence originates from In-rich inclusions, the *size* of which determines the wavelength of the emitted light [2]. The size of efficient quantum dots is expected to be of the order of nm [3].

Several groups have tried to find direct evidence of phase separation in $In_xGa_{1-x}N$ layers. In particular, due to their small sizes, the quest for quantum dots is an challenging issue. Recently, evidence was found for segregation of $In_xGa_{1-x}N$ multiple quantum wells (MQW's) and single layers [4-6]. In all of these studies, phase segregation was only observed for In-contents exceeding ~0.30, and it was generally evidenced from extra peaks in X-ray diffraction (XRD) patterns [4-6], from extra spots in the transmission electron diffraction patterns [5], or from energy dispersive X-ray (EDX) analysis [6]. These findings are in agreement with the calculations of Ho and Stringfellow [7], who calculated the critical temperature T_c above which the InN-GaN system is completely miscible. This temperature, 2457 K, by far exceeds the typical InGaN growth temperature at which a maximum In solubility in GaN of less than 6 % is predicted. MQW's were shown to be much more stable than thick single layers, separation only occurring after annealing at 950°C [6]. This enhanced stability is due to the high elastic strain experienced in thin heterostructures and quantum wells, which results in a significant decrease of the critical temperature for single phase stability [8]. In all of the aforementioned studies, typical precipitate sizes of (several) tens of nanometer were reported, hence too large to result in efficient luminescence [3].

In this paper, we present a correlation between the PL-energy of single $In_xGa_{1-x}N$ thick films and their *overall* and *local* In-content. In particular, we will focus on the effect of local variations of the composition on the nanoscale, resulting in a phase segregation of the ternary nitride.

EXPERIMENTAL

Two $In_xGa_{1-x}N$ layers, each with a laterally varying composition, were grown at 780°C by metalorganic chemical vapour deposition on a thick GaN buffer / Al_2O_3 substrate. The thickness of the layers is about 250 nm. The exact In-content and thickness, as well as the crystalline quality, was mapped by Rutherford backscattering and channeling spectrometry (RBS/C), using a beam spot of approximately 1 mm^2. The crystallinity, azimuthal orientation and the phases present in the sample were further monitored with low (Rigaku system with a rotating anode) and high (Bruker D8 discover) resolution XRD in θ–2θ geometry. PL mapping was performed at low temperature (< 30 K) using an Ar^+ laser with a spot diameter of 100 μm. Samples for transmission electron microscopy (TEM) were prepared in cross sections and studied with a Philips CM 30 FEG electron microscope, operating at 300 kV. Electron dispersive X-ray analysis (EDX) was performed with a Link instrument on a Philips CM 20 microscope.

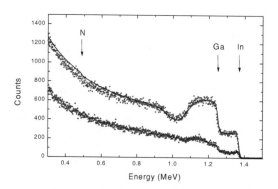

Figure 1. *Random (o), aligned (Δ) and simulated (solid line) RBS spectra of the* $In_{0.25}Ga_{0.75}N/GaN/Al_2O_3(0001)$ *layer. All TEM results shown further have been obtained from this part of the specimen.*

RESULTS AND DISCUSSION

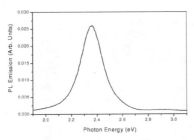

Figure 2. *Low temperature PL spectrum of the same sample as shown in Fig. 1.*

Using RBS, the exact In-fraction of the $In_xGa_{1-x}N$ layers was mapped at a large number of distinct points on the samples, x-values in the range from 0.2 to 0.4 were found [9]. Figure 1 shows the random and aligned RBS spectra of a particular area of the sample which will be discussed in detail below. From the random spectrum, a composition of $In_{0.25}Ga_{0.75}N$ and a thickness of 290 nm are deduced. The spectrum measured with the incoming beam aligned along the <0001> direction indicates that the minimum yield χ_{min} of the $In_{0.25}Ga_{0.75}N$ layer is 20 % (this is the ratio of the backscattering yields of the aligned and random spectra, and is a measure for the crystalline quality of the layer). The value of χ_{min} is found to increase with the In-content

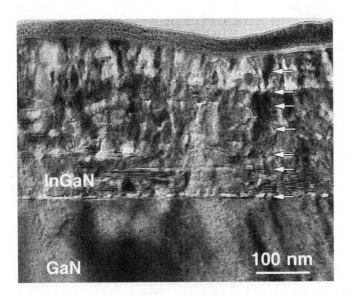

Figure 3. *TEM image of a cross section of the InGaN layer. Rows of pyramidal-shaped inclusions of brighter contrast, parallel to the interface with the GaN buffer layer are marked with arrows.*

(i.e. increasing lattice mismatch with respect to the GaN buffer layer), indicating a deteriorating crystallinity.

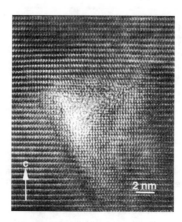

Figure 4. *HREM image along the [01$\underline{1}$0] zone axis showing structural details of a pyramidal-shaped inclusion.*

PL measurements revealed a direct, linear relationship between PL peak energy and In content [9]. Taking into account the linear variation of the PL peak with band gap [10], this result implies that band bowing is absent. The indium content measured by RBS is systematically lower than that expected from previous estimates based on a comparison of PL with X-ray diffraction [11]. As an example, an indium concentration of 40%, the highest measured in this work, corresponds to a PL peak of 710 nm, strongly suggesting that the light-emitting regions of the sample are very indium-rich compared to the overall In content of the nitride, as measured by RBS. On the other hand, the region with an In-concentration of 25 % results in a PL peak energy of 525 nm (Figure 2).

In an attempt to find direct evidence for the phase segregation suggested by the PL measurements, XRD was performed in θ–2θ geometry, using a detector angle between 10° and 120°. In none of the low resolution/high intensity diffraction patterns could any trace of In-rich phases be found. However, preliminary high resolution measurements (which also allow us to probe a much smaller – hence perhaps more uniform – part of the inhomogeneous layer) give evidence of a small fraction of $In_xGa_{1-x}N$, with a composition close to x = 0.50. A detailed high-resolution investigation of samples with varying In compositions will be indispensable to understand the mechanism of this phase segragation, and in particular to elucidate whether the value of x = 0.50 gives evidence of ordering in the ternary system.

On the other hand, transmission electron microscopy of the 290 nm thick $In_{0.25}Ga_{0.75}N$ layer reveals the presence of 10-20 nm islands, aligned at the interface with the GaN buffer layer, and parallel to it at different depths in the InGaN layer. Figure 3 is a cross sectional transmission electron microscopy (TEM) image showing the InGaN layer on the top of the GaN buffer layer. Besides planar defects and strain contrast, the image reveals some pyramidal-shaped zones of a different brighter contrast. Seven such rows of pyramidal islands are visible (indicated by arrows in Fig. 3), the first one lying

near the interface with the GaN buffer layer. The distance between the rows of pyramidal
inclusions in the bulk of the film is about 30-40 nm.

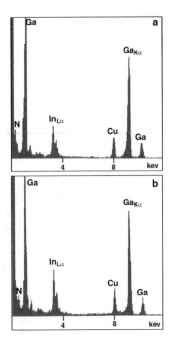

*Figure 5. EDX spectra from the matrix (a) and from a pyramidal-shaped inclusion
(b) showing an increased In concentration in the latter.*

In an attempt to obtain more detailed structural information about these precipitates
high resolution electron microscopy (HREM) has been performed. Figure 4 shows a
HREM image along the [01$\bar{1}$0] zone axis of a region in the InGaN layer containing a
pyramidal shaped inclusion. It is clear that the material in these precipitates is
amorphous, which rules out a direct correlation to the In-rich phase observed by X-ray
diffraction.

To compare the composition of the pyramidal precipitates to that of the
surrounding matrix, energy dispersive X-ray (EDX) analysis was performed. The count
rate was maintained constant for the different measurements, so that the absolute
intensities of the peaks can be compared. The spot size of the electron beam was 15 nm.
Care was taken to analyse neighbouring regions of the islands and the matrix, so that the
sample thickness could be considered constant from one zone to the other. In this way
one can presume that the same volume of specimen was analysed in the islands and in the
matrix. The EDX results (Figure 5) show that the amorphous pyramidal-shaped
inclusions are In-rich as compared to the surrounding matrix. The relative ratio of the
InL_α to GaK_α peak intensities, which is proportional to the In to Ga atomic ratio, is 0.39
in the amorphous pyramidal precipitates and 0.32 in the surrounding matrix. Since the

diameter of the electron beam is of the order of (or larger than) the inclusions, the former value can be considered a lower limit.

SUMMARY AND CONCLUSION

We have studied the composition of $In_xGa_{1-x}N$ layers on a macroscopic (RBS) and microscopic (TEM) level, and compared the indium content to the luminescence of the respective nitrides. A detailed analysis of these data reveals that the luminescence must originate from In-rich inclusions in the InGaN layer. Indeed, partial phase separation has been observed by high-resolution XRD. Moreover, a cross sectional TEM/EDX investigation has indicated the presence of 10-20 nm large pyramidal amorphous precipitates, the composition of which is In-rich. However, according to O'Donnell et al. [2,3] the size of these pyramidal islands is too big to expect large quantum effects. Hence we have to look for smaller sized features that could be related to the quantum effects. In a recent plan-view microscopy investigation, features of the order of ~2 nm have been observed. From high-resolution electron micrographs, these inclusions could be positively identified as quantum dots. A full discussion of this TEM investigation, along with a comparison to dot size estimates from PL measurements, will be given elsewhere. Hence, we can conclude that the $In_xGa_{1-x}N$ samples we studied seem to segregate into a wide variety of crystalline and amorphous microstructures and nanostructures, as evidenced with a variety of techniques. The challenging task will be to correlate the observed (optical) properties to the specific spatial patterns of the layers.

ACKNOWLEDGEMENTS

The authors want to acknowledge partial financial support from the IUAP Program (P4/10) and from the Flemish government through the Bilateral Programs BIL 96/32 (Flanders/China) and BIL 97/93 (Flanders/Romania).

REFERENCES

[1] T. Mukai, M. Yamada and S. Nakamura, Jpn. J. Appl. Phys., 38, 3976 (1999).
[2] K.P. O'Donnell, R. W. Martin and P.G. Middleton, Phys. Rev. Lett., 82, 237 (1999).
[3] R.W. Martin and K.P. O'Donnell, Phys. Stat. Sol., B216, 441 (1999).
[4] R. Singh, D. Doppalapudi, T.D. Moustakas and L.T. Romano, Appl. Phys. Lett., 70, 1089 (1997).
[5] N.A. El-Masry, E.L. Piner, S.X. Liu and S.M. Bedair, Appl. Phys. Lett., 72, 40 (1998).
[6] M. D. McCluskey, L.T. Romano, B.S. Krusor, D. P. Bour, N.M. Johson and S. Brennan, Appl. Phys. Lett., 72, 1730 (1998).
[7] I.Ho and G.B. Stringfellow, Appl. Phys. Lett., 69, 2701 (1996).
[8] S.Yu. Karpov, MRS Internet J. Nitride Semicond. 3, 16 (1998).
[9] K.P. O'Donnell, M.E. White, S. Pereira, M.F. Wu, A. Vantomme, W. Van der Stricht and K. Jacobs, Phys. Stat. Sol., B216, 171 (1999).
[10] R.W. Martin, P.G. Middleton, K.P. O'Donnell and W. Van der Stricht, Appl. Phys. Lett., 74, 263 (1999).
[11] M.F. Wu, A. Vantomme, S.M. Hogg, G. Langouche, W. Van der Stricht, K. Jacobs and I. Moerman, Appl. Phys. Lett., 74, 365 (1999).

Mat. Res. Soc. Symp. Vol. 595 © 2000 Materials Research Society

Phonons and Free Carriers in a Strained Hexagonal GaN-AlN Superlattice Measured by Infrared Ellipsometry and Raman Spectroscopy

M. Schubert[1,2], A. Kasic[2], T.E. Tiwald[1], J.A. Woollam[1], V. Härle[3,4], F. Scholz[3]

[1] Center for Microelectronic and Optical Materials Research, University of Nebraska, Lincoln, NE 68588-0511, U.S.A.
[2] Abteilung Halbleiterphysik, Institut für Experimentelle Physik II, Universität Leipzig, Vor dem Hospitaltor 1, D-04103 Leipzig, Germany
[3] 4. Physikalisches Institut, Universität Stuttgart, Pfaffenwaldring 57, D-70569 Stuttgart, Germany
[4] now with Osram Opto Semiconductors, Wernerwerkstr., D-93049 Regensburg, Germany

ABSTRACT

Phonon and free-carrier effects in a strained hexagonal (α) $\{GaN\}_l$-$\{AlN\}_m$ superlattice (SL) heterostructure ($l = 8$ nm, $m = 3$ nm) are studied by infrared spectroscopic ellipsometry (IRSE) and micro (μ)-Raman scattering. Growth of the heterostructures was performed by metal-organic vapor phase epitaxy (MOVPE) on (0001) sapphire. An unstrained 1 μm-thick α-GaN layer was deposited prior to the SL. SL phonon modes are identified combining results from both IRSE and μ-Raman techniques. The shift of the GaN-sublayer phonon modes is used to estimate an average compressive SL stress of $\sigma_{xx} \sim -4.3$ GPa. The IRSE data reveal a free-carrier concentration of $n_e \sim 5 \times 10^{18}$ cm^{-3} within the undoped SL GaN-sublayers. According to the vertical carrier confinement, the free-carrier mobility is anisotropic, and the lateral mobility ($\mu_\perp \sim 400$ cm^2/Vs, polarization $E \perp c$-axis) exceeds the vertical mobility ($\mu_\parallel \sim 24$ cm^2/Vs, $E \parallel c$) by one order of magnitude.

INTRODUCTION

Strain plays an important role for group-III nitride materials because of large lattice mismatch between binary alloys, and most substrate materials available so far [1]. A possible way to overcome difficulty in growth of high-quality GaN/Al$_x$Ga$_{1-x}$N heterostructures may be the growth of superlattice (SL) structures with few Angstrom periods. Such SL's may be highly strained. If the SL period remains within the critical thickness of the SL constituents, the stacked $\{GaN\}$-$\{Al_xGa_{1-x}N\}$ sublayers should form tensile and compressive stressed "barriers" and "wells" within the SL, and adopt a common in-plane lattice constant. Besides strain-induced shift of SL phonon modes and band-gap energies, the Al$_x$Ga$_{1-x}$N sublayers should act as barriers and confine free carriers to the GaN sublayers. These carriers should then behave similarly to a two-dimensional carrier gas because of the mobility confinement perpendicular to the SL interfaces.

Gleize *et al.* have recently studied a strained GaN-AlN-SL with average Al concentration of 48% using μ-Raman investigations [2]. The SL was grown by mole-cular beam epitaxy. The authors observed strain-induced shifts of the GaN-sublayer phonon modes, but no free-carrier effects were detected. An estimate for the in-plane strain, ϵ_{xx}, and the average biaxial stress, σ_{xx}, was given using elastic constants (C_{ij}), and phonon deformation potential constants (a_λ, b_λ) for α-GaN given in Refs. [3] and [4], respectively.

Spectroscopic ellipsometry (SE) is known to be an excellent technique for measurement of thin-film optical properties. Successful application of infrared (IR) SE for measurement of phonon and free-carrier effects in group-III nitride heterostructures was reported previously [5,6]. SE is an indirect technique, and calculated model spectra need to be fit to experimental data. If appropriately chosen, model parameters, which parameterize the dielectric functions of the sample constituents, can provide physically meaningful quantities, such as transverse-optical (TO), longitudinal-optical (LO) frequencies and broadenings, static or high-frequency dielectric constants, and layer thickness. Because free-carrier absorption affects the IR dielectric response, concentration and mobility parameters can be extracted from the IRSE lineshape analysis if the carrier effective mass is known [7]. When combined with electrical Hall measurements, the IRSE data can further provide information about the effective mass parameter, as recently demonstrated for n- and p-type α-GaN [6].

The focus of this work is to investigate stress and free-carrier effects in a strained α-GaN-AlN-SL by measurement of the long-wavelength dielectric response using IRSE for the first time, in combination with μ-Raman investigations.

EXPERIMENTAL

One SL with 16 periods of alternating wurtzite GaN (8 nm) and AlN (3 nm) layers was grown by MOVPE on (0001) sapphire. A 1-μm-thick GaN layer was deposited on a ~15 nm AlN buffer layer prior to the SL (for more details see Ref. [8]). The sample was measured at room temperature by IRSE and polarized μ-Raman scattering. Ellipsometric parameters were acquired at multiple angles of incidence (57°, 72°), and for wavenumbers from 333 cm^{-1} to 1200 cm^{-1}. A rotating-polarizer, rotating-compensator, Fourier-transform-based variable-angle-of-incidence spectroscopic ellipsometer was used. A detailed description of the IRSE approach is given in Refs. [5,7] and references therein. The μ-Raman spectra were recorded with a XY-Dilor spectrometer at five different configurations in backscattering geometry for wavenumbers from 200 cm^{-1} to 1200 cm^{-1}. The excitation wavelength was 488 nm (Ar$^+$ laser). The incident laser light power was 150 mW. The diameter of the laser focus at the sample surface was typically 1 μm. The sample orientations during Raman and IRSE measurement are shown in Figs. 1b, and 2b, respectively.

DISCUSSION

Fig. 1a shows the μ-Raman spectra in five different backscattering configurations. The spectra reveal the A_1(TO), E_1(TO) (labeled by "1", and "2", respectively), and E_2, A_1(LO), and E_1(LO) phonon modes of the 1-μm-thick GaN layer which supports the SL.

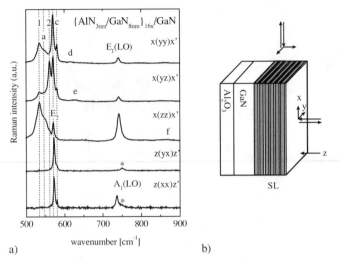

Figure 1. a) Room-temperature μ-Raman spectra recorded in different backscattering configurations from the AlN-GaN-SL sample (sapphire modes are marked by asterisks). b) Sample setup and directions used for the Porto notations.

The phonon frequencies are summarized in Table I. For convenience, we follow the same notation as used by Gleize *et al.*, and label the additional Raman features within Fig. 1a. at 549, 580, 616, 630, and 860 cm^{-1} by *a, c, d, e,* and *f,* respectively [2]. By changing the position of the Raman beam focus from the sapphire interface toward the sample surface we observed that all of these features are due to the SL. The most prominent line is *c* at 580 cm^{-1}, which obeys the symmetry behavior of an E_2 mode. Its frequency is close to that of the thick GaN layer. A very similar line was observed by Gleize *et al.* at 587 cm^{-1}, and was identified as the GaN-sublayer E_2 mode. We also assign *c* to E_2 of the SL GaN-sublayers. The Raman signal *a* could belong to a transverse mode with A_1 symmetry. No clear assignment was done in Ref. [2]. A possible origin for this mode might be the A_1(TO) phonon of the AlN-sublattice, but further proof is necessary at this point. Note that we miss the feature labeled *b* in Ref. [2]. This line was assigned as the SL GaN-sublayer E_1(TO) frequency. As will be discussed below, according to our SL E_2 mode observation at 580 cm^{-1}, the SL E_1(TO) mode should appear ~570 cm^{-1}, and could therefore be subsumed into the strong E_2 Raman feature of the thick GaN layer at 570 cm^{-1}. The feature labeled *d* (616 cm^{-1}) was observed at 620 cm^{-1} in Ref. [2]. It was suggested that this mode may belong to interface excitations. The Raman signal *e* at 630 cm^{-1} in Fig. 1a is identified from the IRSE data analysis as the E_1(TO) phonon of the AlN-sublattice (see below). This mode (*e*) was observed at 635 cm^{-1} in Ref. [2], but not assigned to a particular sample constituent. The broad and weak structure labeled *f* is due to a longitudinal mode with E_1 symmetry. Assignment for this mode from the Raman spectra alone is difficult. However, the IRSE analysis proves that this mode belongs to the high-frequency coupled LO-plasmon-phonon (LPP$^+$) mode of the SL GaN-sublayers. The corresponding LPP$^+$ mode with A_1

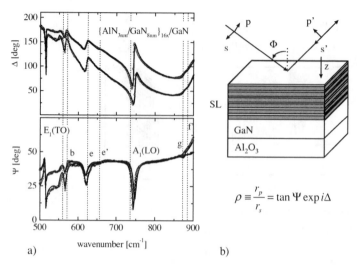

Figure 2. IRSE data (symbols: experiment; solid lines: best fit) at 57°( ) and 72° (○) angle-of-incidence Φ from the AlN-GaN-SL sample. b) Sample setup for the ellipsometric measurement, and definition of Ψ and Δ. r_p and r_s are the complex reflection coefficients for parallel (p) and senkrecht (s) polarized light.

symmetry is not observed because of the strong damping of this axial mode due to the vertical confinement of the carriers to the GaN-sublayers (see discussion below).

Fig. 2a presents the IRSE data together with the best-fit model calculation. As discussed somewhere else in detail ([5,6,9]) for wurtzite III-nitride films on c-plane sapphire, the IRSE data provide less sensitivity to transverse modes with A_1 symmetry. A_1(LO) and E_1 (TO) modes can be identified immediately from resonance structures within the spectra (see Fig. 2a). The E_1(LO) modes follow from lineshape analysis of the IRSE data. The best-fit parameters, and those which were not varied during the regression analysis (values in square brackets), are given in Table I. The E_1(TO) and A_1(LO) frequencies of the "thick" GaN layer are marked within Fig. 2a. The best-fit for the corresponding E_1(LO) frequency resulted in 742 cm^{-1}. These results agree excellently with those obtained from the μ-Raman spectra, and compare well with the known α-GaN modes [1]. The IRSE data are not sensitive to the IR-inactive E_2 modes. Hence, the resonance at 570 cm^{-1} (b in Fig. 2a) can be immediately identified as the SL GaN-sublayer E_1(TO) mode. Furthermore, the feature labeled e in Fig. 1a proves now as a TO resonance (e in Fig. 2a). It turns out from the IRSE lineshape model calculations that this resonance belongs to the SL AlN-sublayer system. We therefore assign e to the E_1(TO) mode of the AlN-sublayers. The small dip at 875 cm^{-1} (g) is further assigned to the A_1(LO) mode of the SL AlN-sublayer system. The weak features at 658 cm^{-1} (e') and 880 cm^{-1} (f') are identified as the E_1(TO) and A_1(LO) modes of the AlN buffer layer, respectively. Gleize et al. [2] used the observed shift of the SL GaN phonon modes to calculate the average biaxial stress σ_{xx} and strain ϵ_{xx} within the SL structure. The elastic coefficients C_{ij} and phonon deformation potential constants (a_λ, b_λ) for α-GaN

Table I: Phonon mode frequencies of the GaN-AlN-SL in cm^{-1}. Values in square brackets were not varied during the IRSE analysis.

	A_1(TO)	E_1(TO)	E_2	A_1(LO)	E_1(LO)
GaN	534[a]	560[a]	570[a]	737[a]	742[a]
"buffer"	534[b]	560[b]		737[b]	742[b]
GaN	-[c]	-[c]	580[a]	-[c]	860[a]
Sublayer	[534][b]	570[c]		840[b]	840[b]
AlN	(549)[d]	630[a]	-[b]	-[b]	-[b]
Sublayer	[549][d]	625[c]		875[b]	[880][b]
GaN[e]	531.8[e]	559[e]	568[e]		

[a]Raman
[b]IRSE
[c]not detectable
[d]tentative mode assignment
[e]unstrained GaN, from V. Yu. Davydov et al., J. Appl. Phys. 82, 5097 (1997)

were taken from Refs. [3,4]. Here we find that the GaN-sublayers are subject to an average compressive stress of $\sigma_{xx} \sim$ - 4.3 GPa if we consider the shift of ~ 11 cm^{-1} for the E_1(TO) mode. This is consistent with the shift of the E_2 mode (~ 12 cm^{-1}, see Table I). The GaN-sublayer A_1(TO) mode should then appear at ~ 537 cm^{-1}. The IRSE data are not sensitive to this mode. Unfortunately, we could not find clear evidence of this mode within our μ-Raman spectra. It might be submerged by the stronger A_1(TO) mode of the thick GaN layer. (Note that the total thickness of the SL is only 176 nm.) If elastic deformation of the SL is assumed, the average strain for the GaN layers can be calculated using the C_{ij}'s. We obtain here $\epsilon_{xx} \sim$ - 0.9%, which compares well to the in-plane strain (- 0.65%) when both GaN and AlN-sublayers adopt a common in-plane lattice constant, and if we consider the thickness and the relaxed lattice constants [1] for both constituents. This result suggests that the SL adopts a different in-plane lattice constant than the thick GaN "buffer" layer. We note that Gleize et al., who investigated a SL with average Al concentration of 48%, observed a higher strain (-1.3%). This value was also slightly larger than that if a common in-plane lattice constant (-1.1%) is assumed. This might be due to uncertainty of the C_{ij}'s known so far.

As a unique feature of the IRSE technique, the ellipsometry parameters are sensitive to free-carrier affected changes within the dielectric functions of thin films. In this work we observe that the SL GaN-sublayers contain free carriers. In particular, we obtain from the best-fit regression analysis that both LO A_1 and E_1 modes are coupled to plasmon excitations, and that the "zero's" of the GaN-sublayer dielectric functions are shifted to higher frequencies [5,7]. To quantify carrier concentration and mobility values, the effective mass parameter has to be known. We assume that the carriers within the GaN-sublayer system are free electrons, which may originate from the AlN sublayer system, i.e., from interface and defect induced donor states. Similar carrier effects were observed in unintentionally-doped $Al_xGa_{1-x}N$ layers [5]. We further assume an isotropic effective mass, and choose a value of $m^*/m_e = 0.22$. [1] As a result, we obtain a carrier concentration of $n_e \sim 5 \ 10^{18}$ cm^{-3}, and a strongly anisotropic carrier mobility. The lateral (electric field polarization $E \perp c$) mobility $\mu_\perp \sim 400$ cm^2/Vs exceeds the vertical mobility ($\mu_\parallel \sim 24$ cm^2/Vs) by more than one order of magnitude. Note that for "thick" films we observe that in general $\mu_\| \ge \mu_\perp$ due to the columnar $Al_xGa_{1-x}N$ film growth [1,5,6]. In Ref. [5] we obtained an isotropic mobility of ~ 100 cm^2/Vs for a 535 nm thick $Al_{0.28}Ga_{0.72}N$ film. The SL result suggests that the free carriers are limited in their mobility perpendicular to the SL interfaces. This may find a simple explanation because of the lateral carrier confinement by the AlN-sublayers. The anisotropic mobility further explains why the LPP^+ mode with A_1 symmetry cannot be seen whereas that with E_1 symmetry can be observed within the μ-Raman spectra. It is well known that strong damping of LPP modes hinders their observation by Raman spectroscopy [10].

To summarize, we investigated a GaN-AlN SL structure by IRSE and polarized μ-Raman scattering. We assigned SL phonon modes by combining results from both techniques. The shift of the GaN sublayer phonon modes allowed us to calculate the average biaxial strain within the SL. The so obtained strain value agrees well with the strain if a common SL in-plane lattice constant is assumed. This study shows strong anisotropic optical carrier effects within the GaN-sublayers. The carriers have very high mobility values for movement parallel to the SL interfaces, and are almost confined for movement parallel to the SL growth direction.

ACKNOWLEDGEMENT

Research is supported in part by DFG contract Rh 28-3/1, and in part by NSF contract DMI-9901510.

REFERENCES

1. J. W. Orton, Rep. Prog. Phys. **61**, 1 (1998).
2. J. Gleize, F. Demangeot, J. Frandon, M. A. Renucci, F. Widmann and B. Daudin, Appl. Phys. Lett. **74**, 703 (1999).
3. A. Polian, M. Grimsditch and I. Grzegory, J. Appl. Phys. **79**, 3343 (1996).
4. V. Yu. Davydov, N. S. Averkiev, I. N. Goncharuk, D. K. Nelson, I. P. Nikitina, A. S. Polkovnikov, A. N. Smirnov, M. A. Jacobson, and O. K. Semchinova, J. Appl. Phys. **82**, 5097 (1997).
5. M. Schubert, A. Kasic, T. E. Tiwald, J. Off, B. Kuhn, F. Scholz, MRS Internet J. Nitride Semicond. Res. **4**, 11 (1999).
6. A. Kasic, M. Schubert, S. Einfeldt, D. Hommel, unpublished.
7. T. E. Tiwald, J.A. Woollam, S. Zollner, J. Christiansen, R. B. Gregory, T. Wetteroth, S. R. Wilson and A. R. Powell, Phys. Rev. B **60**, 11464 (1999).
8. F. Scholz, V. Härle, H. Bolay, F. Steuber, B. Kaufmann, G. Reyer, A. Dörnen, O. Gfrörer, S.-J. Im, A. Hangleiter, Solid State Electron. **41**, 141 (1997).
9. M. Schubert, T. E. Tiwald and C. M. Herzinger, Phys. Rev. B **61** (March 15. 2000).
10. T. Kozawa, T. Kachi, H. Kano, Y. Taya, M. Hashimoto, N. Koide and K. Manabe, J. Appl. Phys. **75**, 1098 (1994).

Mat. Res. Soc. Symp. Vol. 595 © 2000 Materials Research Society

OPTICAL SPECTROSCOPY OF INGAN EPILAYERS IN THE LOW INDIUM COMPOSITION REGIME

M. H. Crawford, J. Han, M. A. Banas, S. M. Myers, G. A. Petersen and J. J. Figiel

Sandia National Laboratories, Albuquerque, NM 87185

ABSTRACT

Photoluminescence (PL) spectroscopy was carried out on a series of Si-doped bulk InGaN films in the low indium (In) composition regime. Room temperature PL showed a factor of 25 increase in integrated intensity as the In composition was increased from 0 to 0.07. Temperature dependent PL data was fit to an Arrhenius equation to reveal an increasing activation energy for thermal quenching of the PL intensity as the In composition is increased. Time resolved PL measurements revealed that only the sample with highest In (x=0.07) showed a strong spectral variation in decay time across the T=4K PL resonance, indicative of recombination from localized states at low temperatures. The decay times at room temperature were non-radiatively dominated for all films, and the room temperature (non-radiative) decay times increased with increasing In, from 50-230 psec for x=0-0.07. Our data demonstrate that non-radiative recombination is less effective with increasing In composition.

INTRODUCTION

While a great deal of progress has been made in the development of InGaN-based light emitters, the role played by indium (In) in contributing to the optical efficiency is still quite controversial. A number of groups have proposed that the inhomogeneity of In incorporation results in carrier localization at In-rich regions and that this localization leads to enhanced optical efficiency [1-3]. Support of this hypothesis is found in cathodoluminescence experiments that demonstrate a variation of the PL emission energy on the microscale, suggesting that In composition variations on the order of several percent are possible [4]. Further insight is suggested by the time-resolved spectroscopy experiments of InGaN quantum well structures performed by Narukawa, et. al. [5], which suggest that the density of non-radiative centers and possibly the non-radiative recombination mechanism itself is altered when In is included in the growth. A similar result was obtained by Kumano, et. al. [6] who suggest that increased optical efficiency is due to reduced non-radiative recombination centers with In incorporation. These two phenomena, namely localization and reduced non-radiative recombination, may be linked if dislocations act as non-radiative centers [7,8], and if the presence of dislocations also affects the inhomogeneity of In incorporation [8,9], thereby affecting carrier localization. In contrast to these theories, other experiments [10] suggest that the majority of optical spectroscopy data on InGaN quantum wells can be explained entirely by piezoelectric field effects. Thus, it is clear that a strong consensus has not emerged as to whether localization is indeed the primary mechanism by which InGaN alloys and quantum wells achieve high optical efficiencies.

In this paper, we explore these issues through an examination of the effect of In composition on the optical properties of InGaN alloys. Our studies include temperature-dependent and time-resolved photoluminescence spectroscopy measurements on a number of MOVPE growth $In_xGa_{1-x}N$ epilayers in the low In composition regime ($x <$ 0.10). This composition regime was chosen to examine whether a clear trend in optical efficiency, temperature dependent quenching of PL intensity and PL decay time can be found with the addition of just small amounts of In. Our work has also intentionally focused on relatively thick (0.2 μm) and doped bulk InGaN epilayers so that the role of piezoelectric field effects would be minimized [11]. This work is therefore distinct from the majority of the previously reported work that has focused on InGaN quantum wells with relatively high ($x \geq 0.2$) In composition.

EXPERIMENT

The MOVPE growth is carried out in a vertical rotating-disc reactor. All of the InGaN epilayers were grown on sapphire substrates and 1μm thick GaN epilayers (1050°C) using a standard two-step nucleation procedure with low-temperature GaN grown at 550 °C. The growth temperature of the InGaN and GaN samples was 800 °C, with the exception of the InGaN x=0.068 sample which was grown at 780°C. The reactor pressure was held constant at 200 Torr. The NH_3 and N_2 flows were set at 6 l/min each. An additional flow of H_2 (~400 cm^3/min) was employed as a carrier gas. Trimethylgallium, triethylaluminum, and trimethylindium are employed as metalorganic precursors. The samples were doped n-type with silane up to a concentration of n= $2x10^{19}$ cm^{-3}. While the exact level of Si-doping is expected to affect the quantitative optical properties, SIMS data has verified that the Si level is very reproducible for all of the samples studied. The indium composition was determined by both Rutherford Backscattering Spectroscopy (RBS), configured to minimize channeling effects, as well as x-ray diffraction (XRD) experiments. Analysis of the x-ray diffraction assumed that the films were fully pseudomorphic, which was supported by the strong thickness fringes seen in the data. In general, the In composition measured with the two techniques agreed within 0.5%. The In compositions determined by the two techniques are summarized in Table 1.

Sample	% In by RBS	% In by XRD
1	0	0
2	1.09	0.75
3	3.94	3.71
4	6.82	6.31

TABLE 1: Indium composition of InGaN bulk epilayers determined by RBS and XRD measurements. The samples will be referred to by the RBS values.

Time-integrated photoluminescence measurements were performed using a HeCd laser (325 nm) at a low power density of approximately 30 W/cm^2. A 0.3 meter spectrometer with an integrated UV enhanced CCD detector was used, with a spectral resolution of approximately 0.2 nm. Time-resolved photoluminescence measurements were performed using a frequency tripled Ti: sapphire laser with a 1 psec pulsewidth, 82 MHz repetition rate and wavelength of 260 nm. Average power densities of 10 W/cm^2 were typically used. The PL was collected into a 0.5 meter spectrometer and a

Hamamatsu model 4334 streakscope with resolutions of 0.5 nm and 15 psec, respectively. For all measurements, the samples were mounted in a closed cycle cryostat to enable measurements over the 4K-300 K temperature range.

TIME-INTEGRATED PHOTOLUMINESCENCE RESULTS

In Figure 1a, we show room temperature photoluminescence data of the various InGaN epilayers. A clear increase in the integrated PL intensity is seen as In composition is increased, and the wavelength shifts from 363-397 nm. The temperature dependent integrated PL intensity of the InGaN epilayers is shown in Figure 1b. An interesting feature of the data is that for GaN and the InGaN (x=0.011) sample, a very quick drop in PL intensity is seen with increasing temperature. In contrast, the samples with higher In composition don't exhibit a significant drop in intensity until approx. 50-100K. This behavior is suggestive that carrier localization may be operative in the x≥ 0.039 samples at low temperatures.

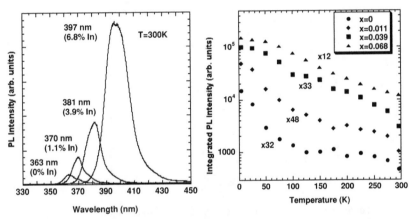

<u>Figure 1:</u> (a) Room temperature time-integrated PL of various InGaN epilayers (b) temperature dependence of integrated PL intensities. The data are labeled with the total drop in integrated PL intensity from 4K-300K for each sample.

We have replotted the temperature dependent PL data on an Arrhenius plot in Figure 2. The data is fit with the following formula

$$I=I_0/[1+\alpha \exp(-E_a/kT)]. \qquad (1)$$

Here the non-radiative decay is assumed to be thermally activated such that the non-radiative lifetime $\tau_{nr}=\tau_o\exp(E_a/kT)$ and E_a is the activation energy for PL quenching. The parameter α is equal to τ_r/τ_o where τ_r is the radiative lifetime. From the E_a values, we see a systematic increase in the activation energy as the indium composition is increased. Leroux et. al. [12] performed a similar analysis of GaN epilayers and found that the activation energy was approximately 25 meV, which is well correlated with the exciton binding energy. They suggested that the dissociation of the exciton into its charged electron and hole components results in more effective trapping of carriers at charged non-radiative recombination sites (e.g. dislocations). In contrast, our GaN epilayer shows a much lower activation energy. The discrepancy in our results may be explained by the fact that our samples have a relatively high n-type doping level that exceeds the Mott density required to effectively screen excitons [13], and our data is expected to reflect the effectiveness of non-radiative recombination on free carriers throughout the temperature range. As a comparison between our results and other reports on InGaN materials, we note that a large variation in E_a values for InGaN epilayers and MQWs has been reported in the literature. Smith et. al. [14] report $E_a =56$ meV for a x=0.12 epilayer for which the authors predict exciton recombination is dominant. Teo et al. [15] report on $In_{0.20}Ga_{0.80}N$ MQWs with E_a values of 63 meV. Our results indicate significantly lower activation energies for free carrier recombination in bulk samples with lower In composition. Furthermore, in our samples, the addition of indium clearly serves to reduce the effect of non-radiative recombination, and thereby a higher optical efficiency over a larger temperature range is maintained. The high carrier density and free carrier regime is relevant for the operating conditions of InGaN laser diodes.

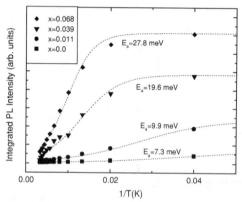

Figure 2: Temperature dependent PL intensity data fit with equation 1.

TIME-RESOLVED PHOTOLUMINESCENCE RESULTS

Further insight into the possible role of localization may be obtained through examination of the time decay of the PL intensity. The decay data can be fit with a single exponential function. At low temperature, we see a clear increase in the measured PL decay time as In composition is increased. To explore this further, we have spectrally resolved the T=4K data to determine if there is a spectral dependence of the decay time (not shown). None of the samples showed an appreciable variation (> 30 psec) in decay time except for the sample with the highest In composition (x=0.068). For this sample, we see a variation of approximately 150 psec (450-600 psec) across the resonance at 4K. A clear spectral variation in the decay time persists until approximately 220K, suggesting that the carriers do not fully come into equilibrium until this relatively high temperature for this sample.

A full temperature dependence of the decay times was measured for each sample, and this data is shown in Figure 3. The decay times were measured at the spectral position of peak PL intensity at all temperatures. It is expected that an increase in the decay time with temperature would be seen if the samples were dominated by radiative recombination [16]. The fact that the GaN and the InGaN (x=0.011) samples show a strong decrease in the decay time at low temperatures suggests that non-radiative mechanisms dominate over the entire temperature range. Furthermore, the strong similarity between these two samples suggests that 1% of In does not significantly affect the recombination processes. The samples with higher In composition, however, show constant or slightly increasing decay times up to approximately 25-50K followed by a decrease in decay time. The distinct behavior of these samples in the low temperature regime is consistent with the carriers being localized at low temperatures with increasing thermal delocalization at temperatures in the 50-100K range. Our results for the x=0.068 sample, in particular the dominance of non-radiative recombination for T>100K and the relatively small variation of decay times over the entire temperature range, are similar to the temperature dependence seen in $In_{0.10}Ga_{0.90}N$ MQW structures [17]. Overall, our data demonstrate a clear increase of the room temperature non-radiative decay times as In composition is increased, which is well correlated with the increased localization seen in the higher In samples.

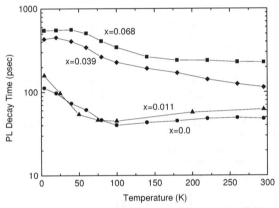

Figure 3: PL decay times as a function of temperature for GaN and InGaN epilayers.

CONCLUSIONS

We have conducted a spectroscopic investigation of InGaN epilayers in an effort to determine how the presence of In affects the optical properties. The strong increase in time-integrated PL intensity that is seen with increasing In is well correlated with increasing thermal activation energies for PL quenching. These activation energies range from 7-28 meV. The temperature dependence of the PL decay times suggests that the samples with higher In ($x\geq 0.039$) show carrier localization effects at low temperatures; these effects are most clearly seen in the spectral dependence of PL decay times for the x=0.068 sample. Overall, the results indicate that carrier localization is enhanced with just the addition of a few percent of In and radiative recombination may dominate in the samples with higher In ($x \geq 0.039$) at lower temperatures. However, non-radiative recombination dominates for much of the temperature range and the fact that the room temperature PL decay times increase with increasing In demonstrates that non-radiative recombination is less effective with increasing In. Since the localization effects in these low In composition samples are relatively weak, it is not clear from our data that carrier localization alone determines the longer non-radiative decay times at room temperature. A modification of the density, capture efficiency or activation of non-radiative centers due to the presence of In would also be consistent with the data.

ACKNOWLEDGEMENTS

The authors gratefully acknowledge technical discussions with Weng Chow, Ian Fritz and Nancy Missert. Sandia is a multiprogram laboratory operated by Sandia Corporation, a Lockheed Martin company, for the US Dept. of Energy under contract DE-AC04-94AL85000.

REFERENCES

1. E. S. Jeon, V. Kozlov, Y.-K. Song, A. Vertikov, M. Kuball, A. V. Nurmikko, H. Liu, C. Chen, R. S. Kern, C. P. Kuo and M. G. Craford, Appl. Phys. Lett., 69, 4194 (1996).
2. S. Chichibu, T. Azuhata, T. Sota and S. Nakamura, Appl. Phys. Lett., 70, 2822 (1997).
3. T. Wang, D. Nakagawa, M. Lachab, T. Sugahara, S. Sakai, Appl. Phys. Lett., 74, 3128 (1999) and references therein.
4. S.F.Chichibu, A.Shikanai, T.Deguchi, A.Setoguchi, R.Nakai, H.Nakanishi, K.Wada, S.P.DenBaars, T.Sota, and S.Nakamura, Jpn. J. Appl. Phys. 39, Part 1 (2000) (to be published)
5. Y. Narukawa, S. Saijou, Y. Kawakami, S. Fugita, t. Mukai, S. Nakamura, Appl. Phys. Lett., 74, 558 (1999).
6. H. Kumano, K. Hoshi, S. Tanaka, I. Suemune, X. Q. Shen, P. Riblet, P. Ramvall and Y. Aoyagi, Appl. Phys. Lett., 75, 2879 (1999).
7. S. Chichibu, H. Marchand, M. S. Minsky, S. Keller, P. T. Fini, J. P. Ibbetson, S. B. Fleischer, J. S. Speck, J. E. Bowers, E. Hu, U. K. Mishra, S. P. Denbaars, T. Deguchi, T. Sota, S. Nakamura, Appl. Phys. Lett., 74, 1460 (1999).
8. T. Sugahara, M. Hao, T. Wang, D. Nakagawa, Y. Naoi, K. Nishino, S. Sakai, Jpn. J. Appl. Phys., 37, L1195 (1998).
9. S. Keller, U. Mishra, S. Denbaars, W. Siefert, Jpn. J. Appl. Phys., 37, L431 (1998).
10. P. Riblet, H. Hirayama, A. Kinoshita, A. Hirata, T. Sugano, Y. Aoyagi, Appl. Phys. Lett., 75, 2241 (1999).
11. V. Fiorentini, F. Bernadini, Phys. Rev. B, 60, 8849 (1999).
12. M. Leroux, N. Grandjean, B. Beaumont, G. Nataf, F. Semond, J. Massies, P. Gibart, J. Appl. Phys., 86, 3721 (1999).
13. W. W. Chow, A. Knorr, S. W. Koch, Appl. Phys. Lett., 67, 754 (1995).
14. M. Smith, G. D. Chen, J. Y. Lin, H. X. Jiang, M. Asif Khan, Q. Chen, Appl. Phys. Lett., 69, 2837 (1996).
15. K. L. Teo, J. S. Colton, P. Y. Yu, E. R. Weber, M. F. Li, W. Liu, K. Uchida, H. Tokunaga, N. Akutsu, K. Matsumoto, Appl. Phys. Lett., 73, 1697 (1998).
16. B. Barry Webb and E. W. Williams in Semiconductors and Semimetals vol. 8 (R. K. Williardson and A. C. Beer, editors, Academic Press, New York, 1972) chapter 4.
17. Y. Narukawa, Y. Kawakami, S. Fugita, S. Nakamura, Phys. Rev. B, 59, 10283 (1999).

Mat. Res. Soc. Symp. Vol. 595 © 2000 Materials Research Society

Electronic Raman Scattering From Mg-Doped Wurtzite GaN

K.T. Tsen[1], C. Koch[1], Y. Chen[1], H. Morkoc[2], J. Li[3], J.Y. Lin[3], H.X. Jiang[3]
[1]Department of Physics and Astronomy
Arizona State University, Tempe, AZ 85287
[2]Department of Electrical Engineering
Virginia Commonwealth University, Richmond, VA 23284
[3]Department of Physics
Kansas State University, Manhattan, KS 66506

ABSTRACT

Electronic Raman scattering experiments have been carried out on both MBE and MOCVD-grown Mg-doped wurtzite GaN samples. Aside from the expected Raman lines, a broad structure (FWHM $\cong 15 cm^{-1}$) observed for the first time at around $841 cm^{-1}$ is attributed to the electronic Raman scattering from neutral Mg impurities in Mg-doped GaN. From the analysis of the temperature-dependence of this electronic Raman scattering signal binding energy of the Mg impurities in wurtzite GaN has been found to be $E_b \cong 172 \pm 20 meV$. These experimental results demonstrate that the energy between the ground and first excited states of Mg impurities in wurtzite GaN is about 3/5 of its binding energy.

INTRODUCTION

P-type doping in nitride-based wide bandgap semiconductors has been a big challenge to researchers. The recently successful p-type doping in metalorganic vapor phase epitaxy GaN using Mg has allowed the realization of blue light emitting diodes.[1,2] Although much work has been devoted to the study of the properties of Mg impurities in GaN samples, the electronic and optical properties of Mg impurities remain controversial and inconclusive. In this paper, we report the first observation of electronic Raman scattering (ERS) in Mg-doped GaN. The binding energy of Mg impurities in GaN has been obtained to be $E_b \cong 172 \pm 20 meV$. Our experimental results provide an important information about the electronic structure of Mg impurities in GaN; namely, the energy difference between the ground and first excited states of Mg impurities in GaN is about 3/5 of its binding energy.

SAMPLES AND EXPERIMENTAL TECHNIQUE

The Mg-doped wurtzite GaN samples studied in this work were grown by MBE and MOCVD techniques on (0001)-oriented sapphire substrates with about 1 μm-thick AlN buffer layers. The thickness of GaN layers was about 2 μm. The z-axis of this wurtzite structure GaN is perpendicular to the sapphire substrate plane. The Mg concentrations were $N_a = 10^{19} cm^{-3}$, $5x10^{18} cm^{-3}$ for MBE and MOCVD-grown samples, respectively, as determined by SIMS experiments.

Raman scattering experiments were carried out by using the second harmonic of a mode-locked cw YAG laser operating at a repetition rate of about 76 MHz. The laser pulse width was about 80ps and photon energy was about $2.34eV$. This excitation laser, which had an average laser power of about 200 mw, was focused into the samples with a spot size of about $100\ \mu m$. Raman scattering measurements were made in $Z(X,X)\overline{Z}$, $Y(X,X)\overline{Y}$ scattering geometry, where $X=(100)$, $Y=(010)$ and $Z=(001)$. Raman scattering signal was detected and analyzed by a standard Raman system equipped with a charge-couple device (CCD) as a multi-channel detector.

EXPERIMENTAL RESULTS AND ANALYSIS

Figs. 1(a) and 1(b) show two typical Raman scattering spectra for a MBE-grown, Mg-doped GaN sample and a MOCVD-grown, Mg-doped GaN sample, respectively, taken at $T=25K$ and in $Z(X,X)\overline{Z}$ configuration. These two spectra are very similar. They are normalized to their $A_1(LO)$ phonon mode intensities and are shown on the same scale. As a result, we have found that the intensity of $841cm^{-1}$ scales reasonable well with the Mg concentration. The structure centered around $757cm^{-1}$ comes from scattering of light by the E_g phonon mode of sapphire;[3,4] the shoulder close to $741cm^{-1}$ belongs to the $A_1(LO)$ phonon mode of GaN. In addition to these expected phonon modes, there is a well-defined structure showing up at about $841cm^{-1}$ $(\cong 105meV)$. Because of the following reasons this additional structure was attributed to electronic Raman scattering (ERS) from neutral Mg impurities in GaN:

1. It is well-known that photons can excite holes from ground state to excited states in neutral acceptors in semiconductors – the so-called "electronic Raman scattering";[5]
2. This structure has been observed for both MBE and MOCVD-grown, Mg-doped GaN samples but not for either MBE or MOCVD-grown undoped GaN samples;
3. The intensity of the structure has been found to decrease as the lattice temperature increases. This is consistent with our assignment that the structure comes from electronic Raman scattering from neutral Mg impurities and rules out any possible assignments of either overtone or combination modes, whose signal is expected to both increase with temperature.
4. The observed signal scales quite well with the Mg concentration, as indicated by Figs.1 (a) and 1(b).

Fig. 2 shows a Raman spectrum for a MBE-grown, Mg-doped GaN sample, taken under the same experimental conditions as in Fig. 1(a) but in $Y(X,X)\overline{Y}$ scattering configuration. No phonon modes associated with either GaN or sapphire are observed in $Y(X,X)\overline{Y}$ configuration, as expected. ERS signal associated with Mg impurities is also observed in $Y(X,X)\overline{Y}$ with its intensity comparable to that observed in $Z(X,X)\overline{Z}$ configuration.

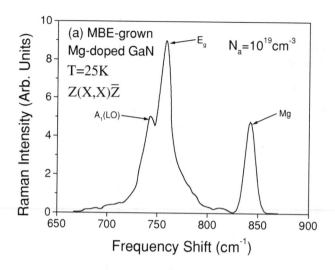

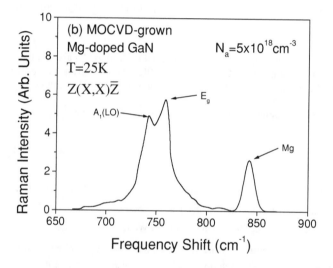

Fig. 1: Stokes Raman spectra taken for (a) a MBE-grown, Mg-doped GaN sample, (b) a MOCVD-grown, Mg-doped GaN sample. The structure around $841 cm^{-1}$ is attributed to the electronic Raman scattering from neutral Mg impurities in Mg-doped GaN.

Since ERS process occurs only when the acceptor is neutral (and therefore holes are available in impurities for ERS by photons), our observed ERS signal associated with Mg impurities can be served as a measure of the number of neutral Mg impurities in the samples. Fig. 4 shows the measured fraction of hole population on Mg acceptors as a function of the lattice temperatures ranging from 300K to 800K for a MBE-grown Mg-

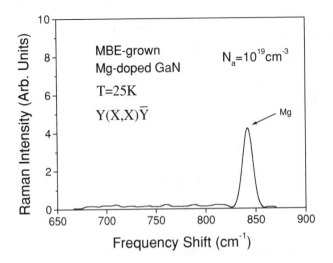

Fig. 2: A Raman spectrum for a MBE-grown, Mg-doped GaN sample, taken under the same experimental conditions as Fig. 1(a) but at scattering configuration $Y(X,X)\overline{Y}$.

doped GaN sample. The hole population has been found to decrease by about a factor of 2 as the lattice temperature increases from 300K to 800K. These experimental results can be used to obtain the binding energy of Mg impurities in wurtzite GaN as follows:

The fraction of hole population in Mg impurities in GaN is given by[6]

$$\frac{p_a}{N_a} = \frac{1}{1 + \frac{1}{2} e^{[\mu(T) - \varepsilon_a]/k_B T}} \qquad (1);$$

where p_a is the number of neutral Mg acceptors per unit volume and N_a is the total number of Mg impurities per unit volume; $\mu(T)$ is the chemical potential; ε_a is the energy of the acceptor; k_B is the Boltzmann constant. We note that N_a becomes the concentration of neutral acceptor p_a when the lattice temperature approaches $0^o K$. Because Mg concentration is quite high, for simplicity, we also assume that the effects of compensation (if there are any) are minimal

Under our experimental conditions, the chemical potential can be shown to be[7]

$$\mu(T) = \frac{1}{2}(\varepsilon_a + \varepsilon_V) - \frac{k_B T}{2} \ln\left[\frac{N_a}{2U_V(T)}\right] + k_B T \sinh^{-1}\left[\sqrt{\frac{U_V(T)}{8N_a}} \cdot e^{-(\varepsilon_a - \varepsilon_V)/2k_B T}\right] \qquad (2);$$

where ε_V is the energy at the top of the valence band and

$$U_V(T) = 2 \cdot \left(\frac{2\pi m_p^* k_B T}{\hbar^2}\right)^{3/2} , \text{ where } m_p^* \text{ is the effective mass of hole.}$$

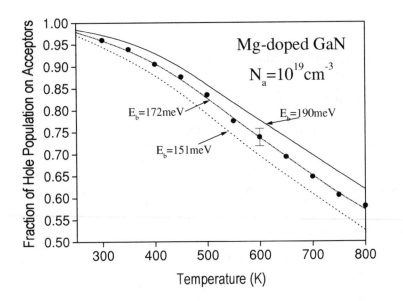

Fig. 3: Fraction of hole population on Mg impurities in MBE-grown Mg-doped GaN as a function of the lattice temperature. The solid circles represent experimental data. The three curves are theoretical predictions based upon Eg. (1) with fitting Parameter: $E_b = 151$, 172 and 190 meV, respectively.

Because under our experimental conditions, $N_a \cong 10^{19} cm^{-3}$; $m_p^* = 0.8 m_e$, where m_e is the mass of an electron, using Eq. (1) to fit our experimental results in Fig. 3, $E_b \equiv \varepsilon_a - \varepsilon_V$ -- the binding energy of the impurities, is the only adjustable parameter. We have found that $E_b = 172 \pm 20 meV$ best fits our experimental data. The fact that this value lies in the range ($150 - 250 meV$)[8,9] of the binding energy of Mg impurities reported in the literature by other experimental methods confirms indirectly that our assignment of the structure around $841 cm^{-1}$ comes from ERS of holes from neutral Mg impurities in Mg-doped GaN was correct.

Since the contribution of ERS signal in semiconductors comes primarily from the transition of hole from the ground to first excited states,[5] our experimental results provide for the first time an important information for the electronic structure of Mg impurities in GaN, i.e., the energy difference between the ground and first excited states of Mg impurities in Mg-doped GaN is about 3/5 of its binding energy.

We note that in addition to the expected Raman signal from GaN, sapphire and ERS, we have also observed two Raman signal very close to $260 cm^{-1}$ and $657 cm^{-1}$ which has been reported by Harima et al.[10] as local modes in Mg-doped GaN. The intensity of these two modes has been found in our experiments to be at least one order of magnitude smaller than that of the optical phonon modes in GaN. We believe that the reason why ERS signal from Mg-doped GaN was not observed by previous researchers is most likely due to the quality of Mg-doped GaN samples and/or the pulse laser employed in our current experimental study. More work is apparently needed in order to obtain a conclusive explanation.

CONCLUSION

We have performed electronic Raman scattering experiments on both MBE and MOCVD-grown Mg-doped wurtzite GaN samples. From the analysis of the temperature-dependence of this electronic Raman scattering signal binding energy of the Mg impurities in wurtzite GaN has been found to be $E_b \cong 172 \pm 20 meV$. These experimental results demonstrate that the energy difference between the ground and first excited states of the Mg impurities in wurtzite GaN is about 3/5 of its binding energy.

ACKNOWLEDGEMENTS

This work was supported by the National Science Foundation under Grant No. DMR-9301100.

REFERENCES

1. S. Nakamura, T. Mukai, M. Senoh, and N. Iwasa, Jpn. J. Appl. Phys. **31**, L139 (1992).
2. H. Amano, M. Kito, K. Hiramatsu, and I. Akasaki, Jpn. J. Appl. Phys. **28**, L2112 (1989).
3. T. Azuhata, T. Sota, K. Suzuki, S. Nakamura, J. Phys. Condensed Matter **7**, L129 (1995).
4. S.J. Shieh, K.T. tsen, D.K. Ferry, A. Botchkarev, B. Sverdlov, A. Savador, H. Morkoc. Appl. Phys. Lett. **67**, 1757 (1995).
5. M. V. Klein, in "Light Scattering in Solids I", ed. by M. Cardona and G. Guntherodt. Topics in Appl. Phys. **Vol. 8** (Springer, N.Y. 1983).
6. R. Kubo, "Statistical Mechanics", 5th ed. (North Holland, Amsterdam, 1965), p.246.
7. J.P. Mckelvey, "Solid State and Semiconductor Physics", (Harper and Row, N.Y. 1966), Chap. 9.
8. S. Fisher, C. Wetzel, E.E. Haller, and B.K. Meyer, Appl. Phys. Lett. **67**, 1298 (1995).
9. T. Tanaka, A. Watanabe, H. Amano, Y. Kobayashi, I. Akasaki, S. Yamazaki, and M. Koike, Appl. Phys. Lett. **65**, 593 (1994).
10. H. Harima, T. Inoue, S. Nakashima, M. Ishida and M. Taneya, Appl. Phys. Lett. **75**, 1383 (1999).

Mat. Res. Soc. Symp. Vol. 595 © 2000 Materials Research Society

Photoluminescence characterization of Mg implanted GaN

C. Ronning, H. Hofsäss, A. Stötzler[1], M. Deicher[1], E.P. Carlson[2], P.J. Hartlieb[2], T. Gehrke[2], P. Rajagopal[2], R.F. Davis[2]
University Göttingen, II. Phys. Institute, Bunsenstr. 7-9, D-37073 Göttingen, Germany
[1]University of Konstanz, Fakultät für Physik, Box M621, D-78457 Konstanz, Germany
[2]North Carolina State University, Department of Materials Science and Engineering, Box 7919, Raleigh, NC 27695, USA

ABSTRACT

Single crystalline (0001) gallium nitride layers, capped with a thin epitaxial aluminum nitride layer, were implanted with magnesium and subsequently annealed in vacuum to 1150-1300 °C for 10-60 minutes. Photoluminescence (PL) measurements showed the typical donor acceptor pair (DAP) transition at 3.25 eV after annealing at high temperatures, which is related to optically active Mg acceptors in GaN. After annealing at 1300 °C a high degree of optical activation of the implanted Mg atoms was reached in the case of low implantation doses. Electrical measurements, performed after removing the AlN-cap and the deposition of Pd/Au contacts, showed no p-type behavior of the GaN samples due to the compensation of the Mg acceptors with native n-type defects.

INTRODUCTION

Magnesium is the commonly used impurity for p-type doping of gallium nitride (GaN) and it can be introduced into the lattice during chemical vapor deposition (CVD) of GaN layers in an easy way [1]. Sufficient p-type activation of the GaN:Mg-layers is reached after a subsequent annealing step at around 800 °C. These p-type GaN-layers are successfully used for the fabrication of "blue" optoelectronic devices. However, the use of GaN is limited to such devices, because the growth process can result only into layered structures. The realization of lateral doped structures of GaN is therefore highly desirable in order to produce GaN based integrated microelectronics.

Lateral doping of semiconductors can be achieved by either in-diffusion or ion implantation of dopants through a structured mask. The diffusion method can be excluded in the case of GaN due to the fact that the diffusion of impurities in GaN starts at very high temperatures where the decomposition of the GaN surface/samples is already completed [2-3]. This early decomposition of the GaN surface at about 850 °C also hampers the success of ion implantation doping of GaN [4]. The advantages of ion implantation (control of lateral and depth distributions, all elements are available) are compromised by the introduced radiation damage which has to be removed via annealing treatments. Annealing temperatures (T_A) of about 1300 °C for more than 5 minutes are necessary for GaN to fulfil the rule of thumb claiming that implanted semiconductors should be annealed up to 2/3 of the melting point for satisfying electrical activation [1,5,6]. Several special annealing procedures for temperatures above 900 °C have been investigated with limited success: rapid thermal annealing (RTA) [7,8], annealing under N_2-overpressure [9], or the usage of N_2, NH_3 or atomic N fluxes, respectively [10,11]. Polycrystalline, sputter deposited AlN cap layers were also used to protect decomposition during annealing.

However, depending on the crystalline quality of the sputtered AlN, good results were obtained only for some selected samples [6,10].

In this article we present results on the basis of an improved annealing technique. Instead of polycrystalline AlN-cap layers, we used thin AlN layers epitaxially grown on top of GaN-samples to protect the GaN surface from decomposition during annealing after the ion implantation of Mg. Magnesium was used because of its known behavior in GaN and the well known corresponding PL signatures.

EXPERIMENTAL

One-to-two μm thick epitaxial, monocrystalline and nominally undoped GaN films were grown on on-axis n-type, Si-face $\alpha(6H)$-SiC(0001) substrates at 1000 °C and 45 Torr using a vertical, cold-wall, RF inductively heated MOVPE deposition system [12]. High-temperature (1100 °C) mono-crystalline AlN layers were deposited prior (100 nm) and following (30 nm) to the GaN growth. Deposition was performed using triethylaluminum (TEA) and triethylgallium (TEG) in combination with 1.5 SLM of ammonia (NH_3) and 3 SLM of H_2 diluent.

Magnesium was implanted with ion energies of 60 keV or 120 keV at room temperature. TRIM simulations gave a mean ion range of 70 nm (FWHM = 40 nm) and 127 nm (FWHM = 101 nm), respectively, for the two energies [13]. Thus, the implanted magnesium ions penetrated the AlN-cap layer in both cases, which was also experimentally checked and confirmed by SIMS measurements. The total implantation dose ranged between 10^{13} cm^{-2} and 10^{15} cm^{-2}.

All implanted samples were annealed under vacuum ($< 1\cdot10^4$ mbar) up to temperatures of 1300 °C for 10-30 minutes. The AlN cap layer inhibited surface decomposition of the GaN sample. Photoluminescence measurements (PL) were performed after each annealing step at low temperatures by exciting the GaN samples with a He-Cd laser (3.81 eV). The AlN-cap layer was removed by dry etching in an ICP plasma (300 W, bias: 10 V, 30 sec.) [14]. Hall measurements were done in Van-der-Pauw geometry after the deposition of Pd(50 nm)-Au(100nm) contacts at a rate of 0.1-0.2 nm/s and $6\cdot10^{-6}$ mbar.

RESULTS AND DISCUSSION

The low temperature PL spectra of Mg-implanted GaN with a dose of $1\cdot10^{13}$ cm^{-2} are summarized in figure 1 as a function of annealing temperature and time. No PL lines were observed directly after ion implantation, which indicates a highly disturbed GaN lattice. This behavior was, of course, also observed for the higher implantation doses used in this study. The high defect density introduced into the crystal by the ion implantation process leads to non-radiative recombinations of the excited electrons with holes; thus, no luminescence is visible.

Intense PL lines have been detected after annealing the Mg implanted GaN sample at a temperature of 1200 °C for 11 minutes (see figure 1(b)). However, the intensity of the band edge luminescence is still about 2-3 orders of magnitude lower in comparison to as-grown, unimplanted GaN samples. This luminescence line at 3.467 eV (commonly labeled as I_2) originates from recombinations of excitons bound to shallow donors. The LO-phonon replica of this line at about 3.35 eV could not be observed in the implanted sample an-

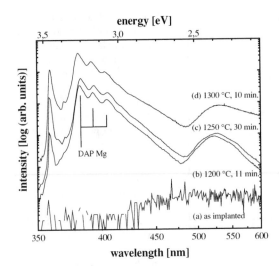

energy [eV]

intensity [log (arb. units)]

DAP Mg

(d) 1300 °C, 10 min.

(c) 1250 °C, 30 min.

(b) 1200 °C, 11 min.

(a) as implanted

wavelength [nm]

Figure 1. Photoluminescence spectra of GaN. The spectra were recorded in the situations: (a) as-implanted with Mg, and (b-d) after annealing to several temperatures for different times. The implantation energy was 60 keV and the implantation dose $1·10^{13}$ cm^{-2}.

nealed at 1200 °C for 11 minutes. However, it was observed as a shoulder after annealing at 1250 °C for 30 minutes and the PL intensity of the excitons also increased. A further increase of this I_2-line took place after the 1300 °C annealing step for 10 minutes, as shown in Figure 1. This indicates that the implanted sample has been recovered to a large extent.

The luminescence peak, which appeared at 3.25 eV after the high temperature annealing steps, is related to donor acceptor pair (DAP) transitions involving Mg acceptors. This line with its LO-phonon replicas is commonly seen in GaN samples doped during growth with Mg showing p-type activation [15]. Therefore, we can conclude that optical activation of the implanted Mg atoms was reached. After the 1300 °C annealing step a further increase of the intensity took place and the optical activation seems to be almost complete, because the intensity of the 3.25 eV line with its phonon replicas is in the same order of PL-lines observed in GaN:Mg samples with an Mg concentration of about 10^{17-18} cm^{-3}. This concentration corresponds well to the peak concentration of Mg atoms in the implanted region. Since we have never observed the PL-line at 3.25 eV after implantation of Be, Li, Si, Ge, In and Er into GaN [11,16,17], these observations prove that the line at 3.25 eV is only related to the implanted Mg acceptors, which is in agreement with Ref. [18]. However, there is still a high intensity of the yellow PL-band observable between 2.0 and 2.6 eV in figure 1 after the 1300 °C annealing step, which indicates that remaining implantation defects are still present.

The low temperature PL-spectra of Mg implanted GaN samples annealed at 1250 °C for 30 minutes are shown in figure 2 as a function of the implantation dose. For comparison, the same scales of figure 1 were used in figure 2. The spectrum of the implanted sample with 10^{13} cm^{-2} Mg ions was already described above. With increasing implantation dose the Mg related DAP transition decreases, and finally the GaN samples implanted with a dose of 10^{15} cm^{-2} show no PL-transitions. One would not expect such a behavior, which was also observed after the 1300 °C annealing step, in completely recovered GaN:Mg samples. Therefore, an amount of residual defects, which strongly depends on the implantation dose, is still present in the annealed samples. This is supported by the fact that the behavior of the intensity of the yellow band is opposite to the behavior of the DAP transitions and therefore much higher in the GaN:Mg samples

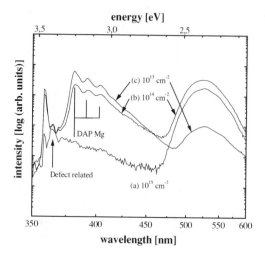

energy [eV]

intensity [log (arb. units)]

(c) 10^{13} cm^{-2}

(b) 10^{14} cm^{-2}

DAP Mg

Defect related

(a) 10^{15} cm^{-2}

wavelength [nm]

Figure 2. Photoluminescence spectra measured at low temperature of GaN as a function of the Mg implantation dose. (a) 10^{15} cm^{-2}, (b) 10^{14} cm^{-2}, and (c) 10^{13} cm^{-2}. The annealing was performed at 1250 °C for 30 minutes under vacuum and the implantation energy was set to 60 keV.

implanted with the higher implantation dose (see figure 2). Furthermore, the intensity of the band edge PL of the 10^{15} cm^{-2} GaN:Mg is suppressed by two orders of magnitude and an additional PL-line appears at 3.40 eV. This line is most likely related to defects created during the implantation procedure, as this line was also observed with varying intensities after implantation of Be, Li, Si, Ge, In and Er [11,17]. We believe that this line is produced by nitrogen or gallium vacancies due to acceptor or donor bound excitons, because it appears also in unimplanted GaN samples depending on the growth conditions.

Finally, we want to compare PL-spectra obtained from GaN:Mg samples implanted with different ion energies. Figure 3 shows the PL spectra of two GaN samples annealed at 1250 °C for 30 minutes and implanted with a dose of 10^{13} cm^{-2} and an ion energy of 60 and 120 keV, respectively. The luminescence intensity of the 120 keV implanted GaN is at almost all wavelengths about one order of magnitude higher. This is again opposite to the expected behavior, because the higher implantation energy yields to a larger spatial distribution of the implanted Mg. The peak concentration is approximately a factor 2 lower compared to the 60 keV implanted sample and the DAP transitions are more intense. This can be explained by the fact that the introduced damage is also distributed over a larger volume and less implantation defects remain after the annealing.

Electrical measurements were performed after removing the AlN-cap by dry etching in an ICP plasma (300 W, bias: 10 V, 30 sec.) [14] and the deposition of Pd/Au contacts. The contacts showed ohmic behavior over a range of +/- 5 V, which is an important prerequisite for Hall effect measurements in Van-der-Pauw geometry [19,20]. First, we want to notice that an electrical signal was only measurable in the Mg implanted GaN samples, which were annealed above a temperature of 1250 °C for more than 30 minutes. For lower annealing temperatures the GaN samples were too resistive (> 10^{10} Ω). The electrical measurements are summarized in table I.

In some cases the sign of the major charge carriers was not clear, because in several repeated measurements n- or p-type was measured. These measurements are indicated in the table I with n/p. This already indicates that the holes generated by the Mg acceptors are compensated by electrons. We assume that these donors are either due to remaining implantation defects or due to native donors from the as-grown GaN-samples. For

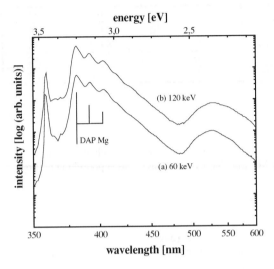

energy [eV]

Figure 3. Photoluminescence spectra measured at low temperature of GaN as a function of implantation energy. (a) 60 keV, and (b) 120 keV. The annealing was performed at 1250 °C for 30 minutes under vacuum and the implantation dose was 10^{13} cm^{-2}.

comparison: the as-grown samples showed sheet concentrations of $1.5 \cdot 10^{13}$ cm^{-2}, which is comparable with the introduced Mg acceptors for the low implantation doses used. Furthermore, we observed a decrease of the sheet resistance, sheet carrier concentration and mobility for the higher implanted doses, which is definitively due to the remaining implantation defects (see PL-results).

CONCLUSIONS

Single crystalline (0001) GaN layers, capped with a thin epitaxial aluminum nitride layer, were implanted with magnesium and subsequently annealed within the range of 1150-1300 °C for 10-60 minutes in vacuum. PL measurements showed the typical DAP transition at 3.25 eV after annealing at high temperatures, which is related to optically active Mg acceptors in GaN. For low doses, high ion energies and high annealing temperatures the PL spectra seems to be completely recovered; however, this is not the case, as shown for higher implantation doses, lower ion energies and lower annealing temperatures. Therefore, the electrical measurements showed no p-type behavior of the GaN samples due to the compensation of the Mg acceptors with native or implantation n-type defects. In order to reach electrical activation higher annealing temperatures and annealing times should be necessary. For such experiments better vacuum conditions during the annealing must be realized, because beyond 1300 °C the AlN-cap starts to oxidize in the presence of even low partial pressures of oxygen.

Table I. Results of Hall measurements on Mg-implanted GaN samples.

Implantation		Annealing		Electrical measurements		
E_{Ion} [keV]	Dose [cm^{-2}]	T_A [°C]	t [min.]	Sheet resistance [kΩ]	Sheet concentration [cm^{-2}]	Mobility [cm^2/Vs]
60	10^{13}	1250	30	14	n = $5.3 \cdot 10^{12}$	80
60	10^{14}	1250	30	20	n/p = $3 \cdot 10^{12}$	100
60	10^{15}	1250	30	157	n = $5.5 \cdot 10^{11}$	71

120	10^{13}	1250	30	9.5	$n = 6.7 \cdot 10^{12}$	97
120	10^{14}	1250	30	16.3	$n = 1.3 \cdot 10^{12}$	300
60	10^{13}	1300	10	20.6	$n/p = 8 \cdot 10^{12}$	38
60	10^{14}	1300	10	40	$n/p = 2.8 \cdot 10^{11}$	2.2
120	10^{13}	1300	10	63	$n = 1.1 \cdot 10^{13}$	63

ACKNOWLEDGMENTS

The work at NCSU was supported by the Office of Naval Research via contract N00014-96-1-0765 monitored by Mr. Max Yoder. R.F. Davis was supported in part by the Kobe Steel, Ltd. Professorship.

REFERENCES

1. O. Ambacher, J. Phys. D: Appl. Phys. **31**, 2653 (1998).
2. S.W. King, J.P. Barnak, M.D. Bremser, K.M. Tracy, C. Ronning, R.F. Davis, R.J. Nemanich, J. Appl. Phys. **84**, 5248 (1998).
3. C.B. Vartuli, S.J. Pearton, C.R. Abernathy, J.D. MacKenzie, E.S. Lambers, J.C. Zolper, J. Vac. Sci. & Techn. B **14**, 3523 (1996).
4. J.C. Zolper, J. Crystal Growth **178**, 157 (1997).
5. J.H. Edgar (ed.), Group III Nitrides, London, INSPEC (1994).
6. J.C. Zolper, S.J. Pearton, J.S. Williams, H.H. Tan, R.J. Karlicek, R.A. Stall, Mater. Res. Soc. Proc. Vol. **449**, 981 (1997).
7. H.H. Tan, J.S. Williams, J. Zou, D.J.H. Cockayne, S.J. Pearton, J.C. Zolper, R.A. Stall, Appl. Phys. Lett. **72**, 1190 (1998).
8. S. Strite, P.W. Epperlein, A. Dommann, A. Rockett, R.F. Broom, Mater. Res. Soc. Proc. Vol. **395**, 795 (1996).
9. S. Strite, A. Pelzmann, T. Suski, M. Leszczynski, J. Jun, A. Rockett, M. Kamp, K. J. Ebeling, MRS Inter. J. Nitride Res. **2**, 15 (1997); J. Appl. Phys. **84**, 949 (1998).
10. J.C. Zolper, J. Han, R.M. Biefeld, S.B. van Deusen, W.R. Wampler, S.J. Pearton, J.S. Williams, H.H. Tan, R.J. Karlicek, R.A. Stall, Mater. Res. Soc. Proc. Vol. **468**, 401 (1998).
11. C. Ronning, K.J. Linthicum, E.P. Carlson, P.J. Hartlieb, D.B. Thomson, T. Gehrke, R.F. Davis, Mat. Res. Soc. Symp. Proc. Vol. **537** (1999) and MRS Internet J. Nitride Semicond. Res. 4S1, G3.17 (1999). http://nsr.mij.mrs.org/4S1/G3.17/
12. T.W. Weeks, Jr., M.D. Bremser, K.S. Ailey, E.P. Carlson, W.G. Perry, R.F. Davis, Appl. Phys. Lett. **67**, 401 (1995); J. Mat. Res. **11**, 1011(1996).
13. J.F. Ziegler, J.P. Biersack, and U. Littmark, *The stopping and ranges of ions in solids*, (Pergamon Press, New York, 1985).
14. S. A. Smith, C. A. Wolden, M. D. Bremser, A. D. Hanser, R. F. Davis, W. V. Lampert, Appl. Phys. Lett. **71**, 3631 (1997).
15. see e.g.: E. Oh, H. Park, Y. Park, Appl. Phys. Lett. **72**, 70 (1998) and ref. therein.
16. M. Dalmer, M. Restle, A. Stötzler, U. Vetter, H. Hofsäss, M.D. Bremser, C. Ronning, R.F. Davis, Mat. Res. Soc. Proc. Vol. **482,** 1021 (1998).
17. E.P. Carlson, C. Ronning, R.F. Davis, unpublished.

18. B.J. Pong, C.J. Pan, Y.C. Teng, G.C. Chi, W.H. Li, K.C. Lee, C.H. Lee, J. Appl. Phys. **83**, 5992 (1998).
19. L. van der Pauw, Philips Res. Rep. **13**, 1 (1958).
20. S.M. Sze, „Physics of semiconductor devices", John Wiley & Sons, New York (1988).

Mat. Res. Soc. Symp. Vol. 595 © 2000 Materials Research Society

An Investigation of Long and Short Time-Constant Persistent Photoconductivity in Undoped GaN Grown By RF-Plasma Assisted Molecular Beam Epitaxy

A.J. Ptak, V.A. Stoica, L.J. Holbert, M. Moldovan and T.H. Myers
Department of Physics, West Virginia University, Morgantown, WV 26506

ABSTRACT

Photoconductance decay and spectral photoconductance measurements were made on a set of ten undoped layers of GaN grown by rf-plasma MBE. The layers, also characterized by Hall, photoluminescence and reflectance measurements, represented a wide variety in electrical and optical properties, and several were grown under atomic hydrogen. Spectral photoconductance indicated transitions at 1.0-1.1, 1.92, 2.15, 3.08 and 3.2-3.4 eV. All layers exhibited persistent photoconductivity to some degree. In contrast with previous reports, a clear correlation was not observed between persistent photoconductivity and yellow luminescence or, indeed, with any measurement made. Analysis of photoconductance decay indicates that more than one type of persistent photoconductivity may be present.

INTRODUCTION

Persistent photoconductivity (PPC) has been observed in n-type, p-type and undoped GaN grown by molecular beam epitaxy (MBE) [1,2,3,4] and metal organic chemical vapor epitaxy (MOCVD). [5,6,7] Interestingly, it has not yet been seen in cubic GaN. [8] The almost universal presence of PPC may indicate that it is tied to native defects. The successes that have been obtained with GaN light emitters indicate PPC is not a problem for such structures, but this ubiquitous phenomenon will definitely affect the performance of ultraviolet detectors, x-ray detectors and field effect transistors. All these structures rely on fabricating insulating, preferably undoped, layers. Thus, PPC in undoped layers must be understood and, if possible, controlled. We report on a study of photoconductivity effects in ten samples grown by rf-plasma MBE which exhibit a wide range of optical and electrical properties.

EXPERIMENTAL DETAILS

The GaN layers investigated in this study were selected from the large number of samples grown at West Virginia University over the last several years. Growth procedures are reported elsewhere. [9] In brief, all samples were grown under Ga-rich conditions in the range 675 to 750°C, with four samples grown under an atomic hydrogen flux. The samples were chosen to reflect a wide range in

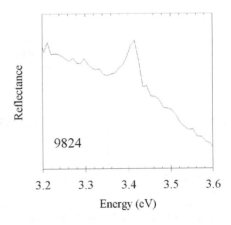

Figure 1. Reflectance spectrum of sample 9824 with a free exciton peak.

Table 1. Various properties of the ten samples included in this investigation. See text for details of each measurement.

Sample (Thickness) (μm)	$\tau_{50\%}$ $\tau_{10\%}$ (s)	τ_{short} (ms)	a-H during growth	Yellow PL	Mobility (cm²/Vs)	n (10¹⁶ cm⁻³)	Exciton in RT Refl.
9824 (9.2)	4.0 246	0.7		weak	202	1.7	yes strong
9821 (2.0)	4.8 622	0.5		weak	15	9.2	no
9826 (7.5)	5.1 4,044	1.2		weak	9.0	1.8	yes
9825 (2.6)	7.0 7,171	1.4		not det.	12	1.7	yes
9910 (1.72)	13 29,521	1.2		weak	330	0.10	yes
9671 (1.0)	64 6,704	2.3	yes	mod.	120	77	no
9730 (0.98)	214 21,061	2.3	yes	not det.	4	7.2	no
9710 (0.85)	293 32,564	1.8	yes	weak	50	85	no
9823 (2.1)	308 8,703	1.6	yes	weak	232	140	yes
9822 (2.1)	357 52,922	1.0		mod.	13	0.67	(?)

electrical properties, as indicated in Table 1. Several of the samples were of high electrical and structural quality, as indicated by their high mobility and low carrier concentration. Samples 9824 and 9826 exhibited free excitonic photoluminescence (PL) at low temperature. Room temperature reflectance measurements, such as that shown in Figure 1, also indicated the presence of free excitons in about half the samples. Fitting this peak with a Lorentzian function appropriate to excitonic transitions indicates a transition energy of 3.412 eV, which is comparable to the value of 3.411 eV reported recently by Viswanath et al. [10] for the free exciton (A) transition at room temperature. This indicates a bandgap of 3.438-3.400 eV at room temperature based on reported values of the exciton binding energy.[10]

PHOTOCONDUCTANCE DECAY MEASUREMENTS

Persistent photoconductivity was measured by monitoring sample resistance with a Keithley 175A multimeter after exposure to light. Samples were stored in the dark for approximately three to four days prior to light exposure in order to determine a baseline dark resistance, followed by exposure to either a bright white light source with the infrared removed via filtering, or to a weak light source consisting of filtered light corresponding to the yellow PL peak position. Either source of light led to a PPC effect, similar to that shown in Figure 2. The intensity varied, but the temporal decay was similar. Many authors [1,4,6,11] have described PPC in GaN using a so-called "stretched exponential"

$$C(t) = C(0)\exp[(t/\tau_{ppc})^\beta] \quad (1)$$

The two fits shown in Figure 2 represent this function. By better fitting the shorter time response, we obtained a value of β of about 0.2, comparable to that reported in the studies by Li et al. [4] and Beadie et al. [6] By better fitting the longer time response, we obtained a value for β closer to 0.26, similar to that found in the studies by Qiu et al.[1] and Chen et al. [11]. Since there is ambiguity in interpreting our results in terms of a stretched exponential, we instead report two values for each sample, the time taken for the PPC to decay to 50% ($\tau_{50\%}$) and to 10% ($\tau_{10\%}$) of its maximum value. The data in Table 1 have been sorted by $\tau_{50\%}$. Note that the two decay times do not appear to be tightly coupled.

In addition to "standard" PPC measurements, we also investigated the shorter time-scale response of our samples by measuring the frequency response of the photoconductance signal to chopped radiation. Either below band gap (filtered) or unfiltered light from a strong white-light source was used. Except for changes in initial signal strength, both resulted in essentially identical decay curves. The observed decay was found to follow the classic response function associated with an exponential decay,

$$s(f) = s(0)/[1+(2\pi f\tau_{short})^2]^{1/2}, \quad (2)$$

from which the PC response time τ_{short} was determined. There was always a "residual" PC signal at the highest frequency measured which is likely related to the true intrinsic PC response. Representative data is shown in Figure 3. Examination of the results listed in Table 1 indicate that τ_{short} appears to be correlated with the longer time constant PPC signal for the samples with smaller PPC effects, but essentially is decoupled from this phenomenon for samples with more significant PPC. We believe that this indicates at least two mechanisms for PPC: the more publicized very long time-constant effect and a second one many orders of magnitude shorter. As discussed below, it is conceivable that both are due to the same native defect.

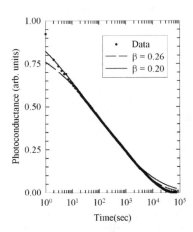

Figure 2. PPC and stretched exponential fit for 9671.

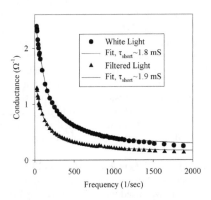

Figure 3. Frequency response of photoconductivity signal in sample 9710 for unfiltered white light and light filtered to correspond to the yellow PL peak.

PPC AND PHOTOLUMINESCENCE

Room temperature photoluminescence measurements indicated two primary features: a free excitonic peak at about 3.415-3.420 eV (accompanied by a phonon replica at 3.36 eV) and the so-called "yellow" luminescence band which shifts from sample to sample but is centered between 1.9 to 2.2 eV, as illustrated in Figure 4. The yellow PL for all the samples was relatively weak, and not detectable for two of the samples investigated. Two observations can be made relating PPC and yellow PL. First, as noted in Table 1, the two samples without detectable yellow PL (9825 and 9710) still exhibited significant PPC effects.

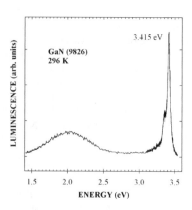

Figure 4. Room temperature PL.

Second, the magnitude of the yellow PL in the samples indicated as having a weak signal was comparable, yet this sample set exhibited a wide range in PPC effects. The sample with the strongest PPC effect (9822) did have about three to four times the yellow PL than the sample exhibiting the least PPC (9824), but it also exhibited a peak at about 3.2 eV that could be indicative of cubic inclusions.

While other studies [8] have indicated a correlation between PPC and yellow PL, our results are in agreement with at least one other study [6] that find PPC is not directly correlated with PL features. It may be only that one of the transitions resulting in yellow PL is related to the origin of PPC, with the other related to additional impurities or defects. Thus, the absence of yellow PL does not indicate that the sample is free from PPC. In addition, the strength of yellow PL is not a direct indication of the magnitude of the PPC effect. The observation, both here and by others [1], that PPC is generated for photon energies above 1.1 eV is likely indicative of the fact that electron-hole pairs generated by any light absorption process, such as the states indicated below by spectral PC, can generate PPC in undoped GaN. We personally favor the proposal put forth by Qiu and Pankove [1] that PPC behavior in n-type and undoped GaN is due to the presence of hole traps, with the Ga vacancy a likely candidate. Indeed, a recent study [12] of room-temperature PL presents a convincing argument that a state at about 1.1 eV is due to the Ga vacancy, which can be coupled with photocapacitance measurements [14] indicating a persistent level at about this same energy. The Ga vacancy is a triple-acceptor, and has been predicted to occur at about this energy. [13] As such, it is conceivable that it could trap holes with different release times, resulting in both the faster and slower PPC time regimes we have observed.

SPECTRAL PHOTOCONDUCTIVITY

Spectral photoconductivity measurements were also made on the samples. A weak probe beam was used whose intensity was kept low enough so that the background current was not significantly altered. The signal was detected using a lock-in amplifier with the probe beam modulated at frequency of 4 to 5 Hz to maximize signal response. Spectra were corrected for variations in incident photon flux. Within the resolution of signal to noise, there were no significant differences between the spectra. The results for the six spectra with the best signal to noise are shown in Figure 5. The signal was not scaled by sample thickness. Samples 9826 and 9824 were significantly thicker than the others (7.5 and 9.2 μm vs. 1 to 2 μm, as listed in Table 1) resulting in a stronger below-gap signal which allows better determination of changes in the PC signal.

An initial strong onset of PC was observed at or below 1.1 eV. Such behavior has been observed previously in spectral PC, [1] although for p-type material. In contrast, the only other spectral PC investigations we found in the literature indicated PC which increased exponentially for below band gap illumination in nominally undoped n-type layers. [2,3] Our result correlates with the observation that yellow PL can be excited by photons in the spectral range of 1.1 eV to above bandgap energies. This 1.0–1.1 eV level is comparable to one observed in photocapacitance measurements. [14] Measurements made on the thinner samples also indicated a clear transition at 3.08 eV and a signal indicative of the onset of strong free excitonic absorption, as indicated in the inset in Figure 5. The effect of absorption below the free exciton energy is magnified for the two thicker samples. Photocurrent onset features were indicated at 1.92 eV and 2.15 eV in addition to the 3.08 eV feature. As shown in the inset, one of the thick samples, 9826, also indicated a PC onset at 3.26 eV, while both of the thicker samples exhibited a strong resonance peak centered around 3.395 eV.

A recent PL study has proposed that yellow luminescence is related to an electronic transition from a shallow donor to a level at 1.08 eV, therein attributed to the Ga-vacancy.[12] This gives a room temperature peak of about 2.3 eV for yellow PL for typical shallow donors. Analysis of variable temperature Hall measurements on the samples in this study indicated a donor activation energy of ~180 meV. The onset observed at 3.26 eV in the thickest sample could be interpreted as electron excitation from the valence band to this level. This is not a strong transition since this level would be strongly populated at room temperature. Using 3.438 eV [10] as the band gap energy and 180 meV as the donor, transitions with a state at 1.08 eV would predict yellow PL

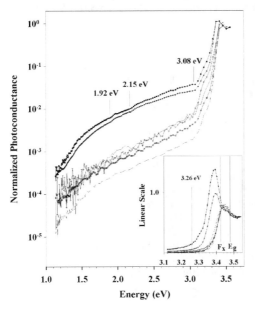

Figure 5. Spectral photoconductance measurements showing photocurrent onset features. The inset shows a higher resolution measurement for the excitonic region.

peaked at 2.18 eV, close to that observed. This also correlates with the 2.15 eV onset observed in spectral PC. If the level at 1.08 eV is due to a Ga vacancy, calculations [15] indicate that it can be split by hydrogenation. A residual background of hydrogen is often observed in our layers. Thus, the possibility exists that the PC onset at 1.90 eV is related to splitting of the Ga-vacancy level, and would explain in part the broad nature of this PL band. The relative contributions of the two bands could also account for the shift observed for this peak from sample to sample. The splitting observed in spectral PC, 0.25 eV, is close to that predicted. [15] At this point, though, this is quite speculative. Finally, low temperature PL of most of these samples featured donor-acceptor recombination indicative of an unidentified acceptor level at about 210 meV (assuming shallow donors at ~ 30 meV). Transitions from this level to the donor level at 180 meV would predict a transition energy of 3.05 eV, close to that observed in spectral PC. The peak at 3.395 eV in the spectral PC could be due to an unidentified residual donor that has been emptied by the acceptor, thus not contributing to the carrier concentration.

CONCLUSIONS

Ten rf-plasma MBE grown GaN samples were characterized optically and electrically by a variety of methods. The PPC effect was found to have no connection with any of the parameters investigated, including the magnitude of the yellow luminescence as has been suggested. We have found evidence of at least two different lifetimes involved in PPC and have proposed a possible connection with Ga vacancies. Spectral PC measurements showed several states in the band gap and support the assertion of the role of the Ga vacancy in PPC.

ACKNOWLEDGEMENTS

We would like to thank Dr. N.C. Giles for allowing M. Moldovan to perform the PL measurements in her laboratory, and for providing some of the equipment used in the PC measurements. This work was supported by ONR Grant N00014-96-1-1008 and monitored by Colin E. C. Wood.

REFERENCES

1   C.H. Qiu and J.I. Pankove, Appl. Phys. Lett 70, 1983 (1997).
2   C.H. Qiu et al., Appl. Phys. Lett. 66, 2712 (1995).
3   J.I. Pankove et al., Appl. Phys. Lett. 74, 416 (1999).
4   J.Z. Li et al., Appl. Phys. Lett. 69, 1474 (1996)
5   M.T. Hirsch et al., Appl. Phys. Lett. 71, 1098 (1997).
6   G. Beadie et al., Appl. Phys. Lett. 71, 1092 (1997).
7   C. Johnson et al. Appl. Phys. Lett. 68, 1808 (1996).
8   C.V. Reddy et al., Appl. Phys. Lett. 73, 244 (1998).
9   T.H. Myers et al., J. Vac. Sci. Technol. B 17, 1654(1999).
10 A.K. Viswanath et al., J. Appl. Phys. 84, 3848 (1998).
11 H.M. Chen et al., J. Appl. Phys. 82, 899 (1997).
12 U. Kaufmann et al.,, Phys. Rev. B 59, 5561 (1999).
13 J. Neugebauer and C.G. Van de Walle, Appl. Phys. Lett. 69, 503 (1996).
14 E. Calleja et al.,, Phys. Rev. B 55, 4689 (1997).
15 C.G. Van de Walle, Phys. Rev. B56, 56 (1997).

Mat. Res. Soc. Symp. Vol. 595 © 2000 Materials Research Society

The Use of Micro-Raman Spectroscopy to Monitor High-Pressure High-Temperature Annealing of Ion-Implanted GaN Films

M. Kuball [1], J.M. Hayes [1], T. Suski [2], J. Jun [2], H.H. Tan [3], J.S. Williams [3], and C. Jagadish [3]

[1] H.H. Wills Physics Laboratory, University of Bristol, Bristol BS8 1TL, UK
[2] UNIPRESS, Polish Academy of Sciences, Solowska 29, 01-142 Warsaw, Poland
[3] Department of Electronic Materials and Engineering, Research School of Physical Sciences and Engineering, The Australian National University, Canberra, ACT 0200, Australia

ABSTRACT

We have investigated the high-pressure high-temperature annealing of Mg/P-implanted GaN films using visible and ultraviolet (UV) micro-Raman spectroscopy. The results illustrate the use of Raman spectroscopy to monitor processing of GaN where fast feedback is required. The structural quality and the stress in ion-implanted GaN films was monitored in a 40nm-thin surface layer of the sample as well as averaged over the sample layer thickness. We find the nearly full recovery of the crystalline quality of ion-implanted GaN films after annealing at 1400-1500°C under nitrogen overpressures of 1.5GPa. No significant degradation effects occurred in the GaN surface layer during the annealing. The high nitrogen overpressures proved very effective in preventing the nitrogen out-diffusion from the GaN surface. Stress introduced during the annealing was monitored. Raman spectra of ion-implanted GaN films were investigated at different temperatures and excitation wavelengths to study the GaN phonon density of states.

INTRODUCTION

The family of III-V nitrides (GaN, InGaN, AlGaN) has recently attracted great interest because of their wide spectrum of applications ranging from opto-electronic devices for the blue-ultraviolet spectral region [1] to high-temperature electronic devices [2]. A large number of processing steps is required for the III-V nitride device fabrication, for example, reactive ion etching (RIE) or focused ion beam (FIB) etching [3], annealing to achieve a low contact resistance to n-type and p-type GaN [4], annealing to activate the Mg-acceptors for p-doping [5]. The non-invasive monitoring of the processing of III-V nitrides is of great interest to gain good control over the III-V nitride device fabrication. In this paper, we demonstrate the use of Raman scattering to monitor non-invasively the processing of GaN, illustrated on the example of the high-pressure high-temperature annealing of ion-implanted GaN films. Ion-implantation is highly attractive for the integration of III-V nitride devices into circuits, however, introduces severe lattice damage [6]. Annealing at temperatures in excess of 1300°C is needed to recover the crystalline quality of ion-implanted GaN films. AlN cap layers [7] or high-pressure nitrogen atmospheres [8,9] have to be employed to prevent the GaN surface decomposition at such high temperatures. Raman scattering provides information on the vibrational states of GaN, which track noninvasively the crystalline quality and the stress in ion-implanted GaN films. The crystalline quality was determined from the E_2 phonon linewidth of GaN, whilst the stress from the E_2 phonon frequency [10]. We restrict this paper on the monitoring of the crystalline quality and the stress in GaN. Results on the free carrier concentration in ion-implanted GaN films monitored by Raman scattering will be reported elsewhere. High optical efficiency micro-Raman systems nowadays make the recording of Raman spectra with very short integration times possible. Fast feedback on the material properties of III-V nitride layers using Raman scattering is therefore possible.

EXPERIMENT

Visible and ultraviolet (UV) micro-Raman spectra were recorded from the top surface of annealed Mg/P-implanted GaN films in backscattering $Z(X,.)\underline{Z}$ geometry, i.e., unpolarized detection, with 1-2µm spatial resolution using a visible and an ultraviolet (UV) Renishaw micro-Raman system with the 514nm- and 488nm-line of an Ar$^+$-laser and with the 325nm-line of a HeCd-laser as excitation source, respectively. The spectral resolution of the Raman setup was 2-3cm^{-1} for all excitation wavelengths used. The GaN films used for this study were 1.5µm thick, grown by metalorganic chemical vapor deposition (MOCVD) on sapphire (0001) substrates (commercially available - CREE). Mg- and P-ions were implanted into the GaN by a multistep process of increasing the ion energy up to 1MeV. As a result, a uniform (across the whole layer thickness) concentration of Mg and P equivalent to 2×10^{19}cm^{-3} was achieved [9]. The samples were annealed at temperatures ranging from 1200ºC to 1500ºC in a high-pressure furnace. Pressures of 1-1.5GPa were applied, with purified N_2 as the pressure-transmitting medium. Mg pieces were placed in the high-pressure furnace near the sample to provide a magnesium overpressure during the annealing. Un-implanted GaN films (taken from the same wafer as used for the ion-implantation) were annealed under the same conditions as the ion-implanted GaN films for comparison.

RESULTS AND DISCUSSION

Figure 1 displays the E_2 phonon linewidth of ion-implanted and of un-implanted GaN films obtained (a) under 514nm- and (b) under 325nm-excitation as function of the annealing temperature. The E_2 linewidth is a measure for the crystalline quality of the GaN. The results obtained under 514nm-excitation (Figure 1(a)) probe the crystalline quality of the GaN films averaged over the sample layer thickness, those obtained under 325nm-excitation (Figure 1(b)) determine the crystalline quality in a sample surface layer of $1/(2\alpha)$=40nm thickness due to the absorption of the laser light in the GaN (α = absorption coefficient at 325nm taken from [11]). The E_2 linewidth of ion-implanted GaN films decreases with increasing annealing temperature. After 1500ºC anneals, it is comparable to the E_2 linewidth of as-grown GaN films. No broadening of the E_2 phonon linewidth was detectable for annealing temperatures as high as 1500ºC under 325nm-excitation. No significant surface degradation occurred therefore during the annealing. For un-implanted GaN films, high-pressure high-temperature annealing results in a decrease of the E_2 linewidth for annealing temperatures of 1300-1400ºC, however, this is reversed at 1500ºC. High-pressure high-temperature annealing restores the crystalline

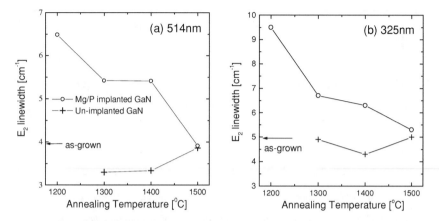

Figure 1. *E₂ phonon linewidth of ion-implanted and of un-implanted GaN films determined (a) under 514nm- and (b) under 325nm-excitation. The E₂ phonon linewidth is a measure for the crystalline quality of the GaN.*

quality of ion-implanted GaN films to a large extent. It also improves the crystalline quality of un-implanted GaN films.

The homogeneity of the investigated samples can be estimated by comparing the E_2 linewidths determined under 514nm- and 325nm-excitation (Figure 1(a) and (b)). Ion-implanted and un-implanted GaN films annealed at temperatures in excess of 1300°C exhibit a more or less homogeneous crystalline quality throughout the layer thickness. The E_2 linewidths determined under 514nm- and under 325nm-excitation agree within the experimental resolution. In contrast to the above, a reduced surface crystalline quality is found in ion-implanted GaN films annealed at 1200°C with an increased E_2 linewidth under 325nm-excitation.

Figure 2 displays the E_2 phonon frequency of ion-implanted and of un-implanted GaN films obtained (a) under 514nm- and (b) under 325nm-excitation. The E_2 frequency is a measure for the stress in the GaN. Compressive stress emerges and shifts the E_2 phonon to larger wavenumbers with increasing annealing temperature. After the 1500°C anneal, a frequency shift of 3.9cm^{-1} and of 1.5cm^{-1} is found under 514nm-excitation for ion-implanted and for un-implanted GaN films, respectively, with respect to the sample before ion-implantation and annealing. For ion-implanted GaN films, the E_2 frequencies obtained under 514nm- and under 325nm-excitation agree within the experimental resolution. The stress is more or less homogeneous throughout the layer thickness. Inhomogeneous stress is found in un-implanted GaN films annealed at 1500°C with an increased E_2 frequency under 325nm-excitation.

Figure 3 compares Raman spectra of as-implanted, un-annealed GaN films recorded under 488nm-excitation at different temperatures. For comparison results obtained under 325nm-excitation at room temperature are also shown. Ion-implantation gives rise to a high-density network of defects [6] lifting the wavevector conservation of the Raman scattering process (disorder-induced Raman scattering). The Raman spectrum is closely related to the GaN phonon density of states for non-resonant excitation conditions [12], for example, for an excitation wavelength of 488nm far below the GaN

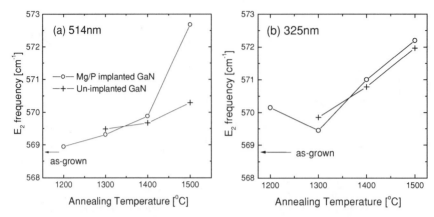

Figure 2. *E_2 phonon frequency of ion-implanted and of un-implanted GaN films as function of the annealing temperature determined (a) under 514nm- and (b) under 325nm-excitation. The E_2 phonon frequency is a measure for the stress in the GaN.*

bandgap. The corresponding Raman spectrum recorded at room temperature contains three contributions: a peak at $574 cm^{-1}$, at $668 cm^{-1}$ and at $716 cm^{-1}$. Their linewidth decreases with decreasing temperature, their frequency increases by 1-$2 cm^{-1}$ from 300K to 10K. This temperature-induced frequency shift is in good agreement with values reported for the E_2 and the $A_1(LO)$ phonons of GaN [13]. Theoretical results on the GaN phonon density of states from Nipko *et al.* [14] calculated using a rigid-ion lattice dynamical model are displayed in Figure 3. We find good agreement between the Raman spectra recorded under 488nm-excitation and the calculated phonon density of states, except for the Raman peak at $668 cm^{-1}$, which can be attributed to local vibration modes [15]. Resonance effects can distort the GaN phonon density of states in the Raman spectrum [12]. This is illustrated in Figure 3 for 325nm-excitation. The UV excitation close to the GaN bandgap resonantly enhances the Raman peak located at $721 cm^{-1}$ related to the GaN LO phonons. The UV excitation does neither enhance the local vibrational mode at $668 cm^{-1}$ nor the Raman peak at $574 cm^{-1}$.

Micro-Raman scattering allowed us to monitor non-invasively the processing of GaN, namely, to track the crystalline quality and the stress in ion-implanted and in un-implanted GaN films after high-pressure high-temperature annealing. Using different excitation wavelengths GaN properties were probed in the sample surface layer as well as averaged over the sample layer thickness to estimate the homogeneity of the stress and the crystalline quality in the investigated samples. The main result of significance is the nearly full recovery of the crystalline quality of ion-implanted GaN films after high-pressure high-temperature annealing at 1400-1500°C tracked by the decreasing E_2 phonon linewidth in Figure 1. Improvements in the crystalline quality were also achieved for un-implanted GaN films by annealing at 1300-1400°C. UV Raman scattering found no significant surface degradation effects during the annealing at temperatures as high as 1500°C. The high nitrogen overpressures proved very effective in preventing the surface decomposition which would normally occur at a temperature of 1000-1200°C [16,17]. Stress is introduced into the GaN layers during the annealing (Figure 2) and shifts the E_2

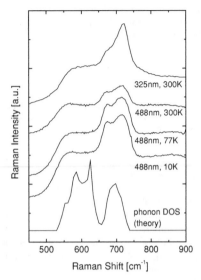

Figure 3. Raman spectra of as-implanted, un-annealed GaN films recorded under 488nm-excitation at different temperatures and recorded under 325nm-excitation at room temperature. The theoretical phonon density of states of GaN is from [14].

phonon frequency to higher wavenumbers with 2.9cm^{-1}/GPa [10]. Annealing at 1300-1500°C therefore introduces a compressive stress of 0.3-0.5GPa in un-implanted GaN films (Figure 2(a)). The increased crystalline quality, i.e., the decreased defect density, in annealed un-implanted GaN films (Figure 1(a)) reduces their ability to relax the stress, which arises from the large lattice mismatch between GaN and the sapphire substrate. Stress also arises from the inclusion of impurities into the GaN crystal lattice such as the Mg and P in the ion-implanted GaN films (Figure 2(a)).

CONCLUSIONS

We have investigated the high-pressure high-temperature annealing of ion-implanted and of un-implanted GaN films using micro-Raman spectroscopy. Using ultraviolet (UV) and visible excitation wavelengths the crystalline quality and the stress was monitored in the sample surface layer as well as averaged over the sample layer thickness, respectively. Disorder-induced Raman scattering was used to investigate the GaN phonon density of states. The results illustrate the use of Raman spectroscopy to monitor processing of GaN where fast feedback is required.

ACKNOWLEDGEMENTS

We acknowledge financial support for the work in Bristol by Renishaw plc (Dr. G.D. Pitt), by EPSRC (grant no. GR/M15590) and by the Royal Society (grant no.

20141). J.M.H. is supported by a CASE studentship from DERA Malvern (Dr. T. Martin). We thank C.K. Loong (Argonne National Laboratory) for providing the theoretical data on the GaN phonon density of states.

REFERENCES

[1]   S. Nakamura, M. Senoh, S. Nagahama, N. Iwasa, T. Yamada, T. Matsushita, H. Kiyoku, Y. Sugimoto, T. Kozaki, H. Umemoto, M. Sano, and K. Chocho, Appl. Phys. Lett. **72**, 211 (1998).

[2]   S. Yoshida and J. Suzuki, Jpn. J. Appl. Phys. **37**, L482 (1998).

[3]   I. Adesida, C. Youtsey, A.T. Ping, F. Khan, L.T. Romano, and G. Bulman, MRS Internet J. Nitride Semiconductor Res. **4S1**, G1.4 (1999); M. Kuball, F.H. Morrissey, M. Benyoucef, I. Harrison, D. Korakakis, and C.T. Foxon, Phys. Stat. Sol. (a) **176**, 355 (1999).

[4]   F. Ren, in *GaN and Related Material*, Optoelectronic properties of semiconductors and superlattices, Vol. 2, edited by M.O. Manasreh (Gordon and Breach Science Publishers, Amsterdam, 1997), pp. 433-469.

[5]   S. Nakamura, T. Mukai, M. Senoh, and N. Iwasa, Jpn. J. Appl. Phys. **31**, L139 (1992).

[6]   J.C. Zolper, H.H. Tan, J.S. Williams, J. Zou, D.J.H. Cockayne, S.J. Pearton, M. Hagerott Crawford, and R.F. Karlicek, Jr., Appl. Phys. Lett. **70**, 2729 (1997); H.H. Tan, J.S. Williams, J. Zou, D.J.H. Cockayne, S.J. Pearton, J.C. Zolper, and R.A. Stall, Appl. Phys. Lett. **72**, 1190 (1998).

[7]   X.A. Cao, C.R. Abernathy, R.K. Singh, S.J. Pearton, M. Fu, V. Sarvepalli, J.A. Sekhar, J.C. Zolper, D.J. Rieger, J. Han, T.J. Drummond, R.J. Shul, and R.G. Wilson, Appl. Phys. Lett. **73**, 229 (1998).

[8]   T. Suski, J. Jun, M. Leszcyński, H. Teisseyre, S. Strite, A. Rockett, A. Pelzmann, M. Kamp, amd K.J. Ebeling, J. Appl. Phys. **84**, 1155 (1998).

[9]   T. Suski, J. Jun, M. Leszczynski, H. Teisseyre, I. Gryzegory, S. Porowski, J.M. Baranowski, A. Rocket, S. Strite, A. Stonert, A. Turos, H.H. Tan, J.S. Williams, and C. Jagadish, Mat. Res. Soc. Symp. Proc. **492**, 949 (1998).

[10]   F. Demangeot, J. Frandon, M.A. Renucci, O. Briot, B. Gil, R.-L. Aulombard, MRS Internet J. Nitride Semicond. Res. **1**, 23 (1996).

[11]   J.F. Muth, J.H. Lee, I.K. Shmagin, R.M. Kolbas, H.C. Casey, Jr., B.P. Keller, U.K. Mishra, and S.P. DenBaars, Appl. Phys. Lett., **71**, 2572 (1997).

[12]   M. Cardona in *Light Scattering in Solids II*, edited by M. Cardona and G. Güntherodt (Springer, Heidelberg, 1982), pp. 19-178.

[13]   M.S. Liu, L.A. Bursill, S. Prawer, K.W. Nugent, Y.Z. Tong, G.Y. Zhang, Appl. Phys. Lett. **74**, 3125 (1999).

[14]   J.C. Nipko, C.-K. Loong, C.M. Balkas, and R.F. Davis, Appl. Phys. Lett. **73**, 34 (1998).

[15]   W. Limmer, W. Ritter, R. Sauer, B. Menschling, C. Liu, and B. Rauschenbach, Appl. Phys. Lett. **72**, 2589 (1998).

[16]   M. Kuball, F. Demangeot, J. Frandon, M.A. Renucci, J. Massies, N. Grandjean, R.L. Aulombard, and O. Briot, Appl. Phys. Lett. **73**, 960 (1998).

[17]   J.M. Hayes, M. Kuball, A. Bell. I. Harrison, D. Korakakis, and C.T. Foxon, Appl. Phys. Lett. **75**, 2097 (1999).

Mat. Res. Soc. Symp. Vol. 595 © 2000 Materials Research Society

Carrier dynamics studies of thick GaN grown by HVPE

G.E. Bunea, M.S. Ünlü, and B.B. Goldberg
Departments of Physics and Electrical and Computer Engineering and Photonics Center, Boston University, Boston, MA 02215, U.S.A.

ABSTRACT

We present a comparison between optical properties of two samples grown by hydride vapor phase epitaxy, one directly on Sapphire substrate (non-ELO) and one epitaxial lateral overgrown (ELO) on SiO_2 patterned Sapphire substrate. The ELO material shows an improvement of the UV emission and slight decrease of the yellow emission. The band edge emission is red shifted due to the relaxation of the compressive strain. In spite of the increase in the UV emission, the lifetime of the excitons in the ELO material is more than twice lower than non-ELO material. We attribute this to screening effects of the background electron concentration.

INTRODUCTION

GaN and related alloys are currently among the most extensively studied semiconductor materials for a variety of applications [1] from blue-green and UV light emitting diodes to high power/high temperature transistors. Recently, significant progress has been reported in growing GaN films by hydride vapor phase epitaxy (HVPE) [2]. Excellent optical quality of the material [3] combined with high growth rates make GaN films grown by HVPE very attractive for use as a substrate for GaN epilayers, or even as-grown devices [4]. Epitaxial lateral overgrowth (ELO) on patterned substrates is of particular interest because of the realization of the longest lifetime GaN based laser diode [5]. The leakage current of p-n junctions grown on ELO material on Sapphire substrates has been shown to reduce by three orders of magnitude [6].

Time resolved photoluminescence (TRPL) is a powerful experimental tool to study new materials because the temporal information combined with spectral data help determine the dynamics of carriers involved in optical processes. In this paper we report on a comparison of optical properties of two samples grown by HVPE, one directly on Sapphire substrate and one epitaxial lateral overgrown on SiO_2 patterned Sapphire substrate.

EXPERIMENTAL DETAILS

For ELO sample, a 200 nm thick SiO_2 film was used to mask the substrate. Stripe arrays with openings of 2 µm and spacing of 40 µm were formed by standard lift-off technique. Both samples (referred to as ELO and non-ELO) were grown on c-plane

Sapphire substrates. The growth technique employs a chloride-transport HVPE vertical reactor, with growth rate ~ 21 µm/h at 1050 °C, which results in unintentionally doped GaN. The thickness of the non-ELO sample is 63 µm, and room temperature (RT) Hall effect measurements give a carrier concentration of ~ $6x10^{16}$ cm^{-3} and a mobility of 823 cm^2/Vs. A SEM picture of ELO sample is presented in figure 1. The sample exhibits a small curvature (Peak to Valley ~2.2 µm, RMS ~ 0.647 µm), which is due to strain. The thickness of the sample is 58 µm, and RT Hall effect measurements show a background electron concentration of ~ $2x10^{18}$ cm^{-3} and mobility of 220 cm^2/Vs.

Figure1. SEM image of epitaxial lateral overgrown (ELO) sample.

The experimental set-up consists of a frequency-doubled picosecond Ti:Sapphire laser, with a pulse width of 1.5 ps and excitation wavelength of 352 nm. The emitted light is dispersed by a 0.64 m spectrometer and collected by a charge coupled devices (CCD) array for time-integrated photoluminescence (PL) spectra or a microchannel plate photomultiplier (MCPMT) for time-resolved measurements. The temperature of the sample is varied from 4-300K using a flow-through helium cryostat. The average excitation power is 5 mW and the typical diameter of the laser spot on the sample is ~ 1mm. The spectral resolution is 0.6 Å and the overall temporal resolution of the system is less than 65 ps.

DISCUSSION

Time integrated PL spectra for ELO and non-ELO material are presented in figure 2. We note a significant increase in the UV emission as well as a slight decrease of the yellow emission in the ELO sample. The ELO main UV peak at 3.38 eV is red-shifted with respect to non-ELO sample (3.397 eV). We attribute this to the relaxation of the compressive strain. The FWHM of ELO material (obtained by fitting Lorenzians) is 56.51 meV, larger than non-ELO material (32.8 meV); this is due to a larger background electron concentration in the ELO material. From power dependence studies (not shown) we find that both sample are dominated by excitonic emission. The yellow emission peak is at ~2.274 eV for both samples. The peaks on the lower energy side of the UV band edge emission are associated with 1-LO and 2-LO phonon replica.

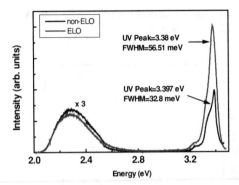

Figure2. *PL spectra on both samples (ELO) and (non-ELO) at room temperature.*

We have performed TRPL experiments in order to find the recombination lifetime of the excitons in both samples. Figure 3 shows TRPL data obtained at the energy of 3.387 eV. The data remain single exponential decays for more than 1ns. The straight lines represent single exponential fits to the data, yielding a recombination lifetime of ~ 530 ps for non-ELO sample and ~ 204 ps for ELO sample. In spite of the improvement in the UV emission in the ELO material, the lifetime of excitons is smaller. Due to the screening effect of the background electron concentration the exciton binding energy decreases, therefore the recombination lifetime also decreases.

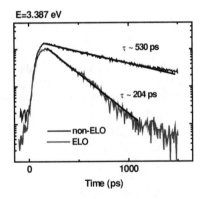

Figure3. *Room temperature time resolved photoluminescence measurements at the energy of 3.387 eV for both ELO (red) and non-ELO (blue) samples. The lines are single exponential fits to the data.*

At 4K the PL spectrum of non-ELO material (figure 4) is dominated by the donor-bound I_2 transition (DX) at 3.47 eV. The FWHM (from fitting Lorenzians curves) is 1.9 meV. The high energy shoulder at 3.476 eV is due to the free exciton A. The peak at 3.45 eV is associated with acceptor-bound (AX) excitons. TRPL measurements indicate a radiative recombination lifetime of ~ 530 ps for DX and ~ 295 ps for FX [3]. The PL spectrum of ELO material at 4K is also dominated by DX at 3.47 eV and FWHM ~ 17.8 meV. The lifetime of the DX is ~ 240 ps.

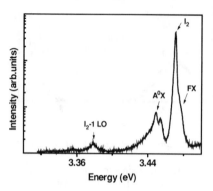

Figure 4. *PL spectrum on non-ELO sample at 4K.*

Radiative recombination lifetime (τ_{rad}) of an excited state in a semiconductor can be estimated from consideration of the optical transition probability [7,8]:

$$\tau_{rad} = \frac{2\pi\varepsilon_0 m_0 c^3}{ne^2\omega^2 f} \qquad (1)$$

Where f is the oscillator strength of the optical transition, and n is the refractive index. By using n=2.67 [9] and ω=5.153x10^{15} s^{-1} one can estimate τ_{rad}(ps) ~ 725/f. The oscillator strength of free exciton calculated within the effective-mass approximation is given by [10]:

$$f = \frac{2E_p}{h\omega(\dfrac{V}{a^3_x})} \qquad (2)$$

Where E_p is the Kane matrix element connecting Bloch states in the conduction and valence bands, V is the volume of the unit cell, a_x is the Bohr radius of the free exciton. By using E_p~18 eV and a_x~20 Å [10], one calculates a radiative recombination lifetime for the excitons in the tens of nanoseconds.

The discrepancy between the theoretical estimates and our results can be explained by the fact that the measured PL decay time is an effective lifetime of the free excitons, involving radiative and non-radiative recombination, such that the overall decay rate is given by:

$$\frac{1}{\tau_{PL}} = \frac{1}{\tau_{rad}} + \frac{1}{\tau_{nr}} \qquad (3)$$

It has been reported in the literature that the free exciton PL decay time increases as the number of nonradiative centers decreases [11-14]. To our knowledge the highest free exciton lifetime reported so far at room temperature is ~250ps in MOVPE GaN [14]. The lifetime we report for non-ELO (~530 ps) sample is an indication of the smaller defect density, hence an improvement in the quality of the material. It has been suggested that the stronger the yellow luminescence in the sample the shorter the PL decay time of band to band recombination [10]. Since the yellow luminescence from both ELO and non-ELO samples is comparable, we believe that the smaller recombination lifetime in the ELO material is due to the screening effects of the background electron concentration.

The TRPL measurements at the energy of yellow emission (2.274 eV) are presented in figure 5. The data are noisier, due to smaller signal. However we see that both samples exhibit very long decay times; by fitting single exponential decays we find a lifetime of ~4ns for non-ELO sample and ~2.89 ns for ELO sample. Such long lifetimes are consistent with donor-acceptor nature of the transition.

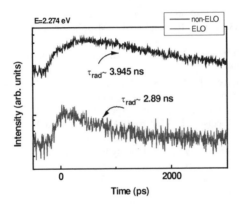

Figure 5. *Room temperature TRPL measurements at the peak of yellow emission (2.274 eV) for ELO and non-ELO samples.*

CONCLUSIONS

We have presented a comparison between ELO and non-ELO material grown by HVPE on c-plane Sapphire substrate. The ELO material shows an improvement of the UV emission and slight decrease of the yellow emission. The band edge emission is red shifted due to relaxation of the compressive strain. In spite of the increase in the UV emission, the lifetime of the excitons in the ELO material is more than twice lower than non-ELO material both at 4K and room temperature. This is due to the screening effects of the background electron concentration.

AKNOWLEDGMENTS

The authors would like to thank Dr. R.J. Molnar from Massachusetts Institute of Technology, Lincoln Laboratory, Lexington, Massachusetts, 02173 for providing the samples for these studies. G.E. Bunea would like to thank J. Graff from Electrical and Computer Engineering Department, Boston University, for the help with Hall effect measurements on ELO sample. The work was supported by NSF CAREER grant no. ECS 9625236 and by ARO grant no. DAAG-55-98-1-0143.

REFERENCES

1. P. Rigby, Nature (London) **384**, 610 (1996).
2. R.J. Molnar, W. Götz, L.T. Romano, and N.M. Johnson, J. Cryst. Growth **178**: (1-2) 147-156 (1997).
3. G.E. Bunea, W.D. Herzog, M.S. Ünlü, B.B. Goldberg, R.J. Molnar, Appl. Phys. Lett. **75**, 838 (1999).
4. Raj Singh, R.J. Barrett, J.J. Gomes, Ferdynand P. Dabkowski, T.D. Moustakas, MRS Internet J. Nitride Semicond. Res. **3**, 13 (1998).
5. S. Nakamura, M. Senoh, S. Nagahama, Appl. Phys. Lett. **72**, 211 (1998).
6. P. Kozodoy, J. P. Ibbetson, H. Marchand, P. T. Fini, S. Keller, J. S. Speck, S. P. DenBaars, U. K. Mishra, Appl. Phys. Lett. **73**, 975 (1998).
7. D. Dexter, in *Solid State Physics,* Edited by F. Seitz and D. Turnbull (Academic, New York, 1958), Vol. 6, p. 353.
8. G.W. 't Hooft, W.A.J.A. van der Poel, L.W. Molenkamp, and C.T. Foxon, Phys. Rev. B **35**, 8281 (1987).
9. S. Strite and H. Morkoç, J. Vac. Sci. Technol. B **10**, 1237 (1992).
10. W. Shan, X.C. Xie, J.J. Song, and B. Goldenberg, Appl. Phys. Lett. **67**, 2512 (1995).
11. C.J. Hwang, Phys.Rev. B **8**, 646 (1973).
12. J.P. Bergman, P.O. Holtz, B. Monemar, M.Sundaram, J.L. Merz, and A.C. Gossard, Phys. Rev. B **43**, 4765 (1991).
13. J.S. Massa, G.S. Buller, A.C. Walker, J. Simpson, K.A. Prior, and B.C. Cavenett, Appl. Phys. Lett. **64**, 589 (1994).
14. J. Allegre, P. Lefebvre, J. Camassel, B. Beaumont, P. Gibart, MRS Internet J. Nitride Semicond. Res. **2**, 32(1997).

Mat. Res. Soc. Symp. Vol. 595 © 2000 Materials Research Society

Prism coupling as a non destructive tool for optical characterization of (Al,Ga) nitride compounds

E. Dogheche, B. Belgacem, D. Remiens, P. Ruterana[1*] and F. Omnes[**]

Laboratoire des Matériaux Avancés Céramiques (LAMAC) - Université de Valenciennes et du Hainaut-Cambrésis - Le Mont-Houy BP211 Valenciennes Cedex F-59342
*Laboratoire d'Etude et de Recherche sur les Matériaux (LERMAT - CNRS), ISMRA Bd Maréchal Juin, Caen Cedex F-14050
**Centre de Recherche sur l' Hétéroépitaxie et ses Applications (CRHEA - CNRS), rue Bernard Gregory Sophia-Antipolis, Valbonne F-06560

[1*]Author for correspondence: ruterana@lermat8.ismra.fr, Fax: 33 231 4526 60

ABSTRACT

An optical characterization technique is proposed for GaN based compounds deposited on sapphire. In AlGaN films grown by MOCVD, the film optical behavior and the substrate to layer interface are qualified from the measured optical data. The experimental and theoretical approach used for this purpose is described in detail. The results clearly show bending effects at the interface which may be related to structural defects; a good agreement with transmission electronic microscopy analysis is obtained.

INTRODUCTION

There is currently an increasing interest in the growth of high quality nitride related materials for fabrication of short-wavelength devices for displays [1-2]. This has been initiated by use low temperature buffer layers (AlN or GaN) [1]. Until now, most of the studies for the group III nitrides have dealt with the improvement of films quality, the characterization of physical properties has become an integral part of research for the understanding of material behavior and performance under operating conditions. Numerous accurate tools as transmission electronic microscopy (TEM) [3], scanning probe microscopy (SCM) [4] techniques are largely used for the qualification of the layers. Furthermore, the increasing need of feedback informations has spurred the development of original characterization methods. In the following, the high sensitivity of prism coupling for the characterization of the interface properties is demonstrated.

EXPERIMENTAL

The AlGaN films were grown on (0001) sapphire substrates at a low pressure using a metalorganic vapor phase epitaxy (MOVPE) apparatus. A low-temperature AlN nucleation layer of thickness 10nm was first deposited on sapphire at 800°C. Growth temperature of the active layers was carried out at 1170°C. Details of the growth procedure were published elsewhere [5].

The optical properties of AlGaN / AlN heterostructures were measured by using a guided-wave technique based of the prism coupling. This technique was described in details by Tien et al [6]. In short, a He-Ne laser beam emitting at a wavelength of 632.8nm is coupled into the AlGaN layer using a rutile prism (TiO$_2$) through the evanescent field in the air-film gap. By measuring the reflected intensity versus the angle of incidence α, we draw the guided-mode spectrum of the samples (figure 1). With the optical axis normal to the surface, the ordinary and extraordinary modes are respectively excited using transverse electric (TE) and transverse magnetic (TM) polarized light.

The same layers were examined by transmission electron microscopy using cross section samples prepared in the conventional way of mechanical polishing followed by ion milling.

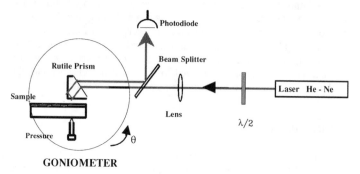

Figure 1 *Experimental set up of the prism coupling method*

RESULTS

In the TE mode spectrum of the AlGaN films, six sharp reflectivity dips are observed at certain angles α which correspond to the excitation of guided modes (figure 2). These can be identified as the modes TE$_0$ – TE$_5$. Note that six modes have been excited with the TM polarized light. From the angular position of the TE or TM guided-modes, the corresponding effective mode indices are calculated according to

$$N_m = N_p \sin\left[A_p - \arcsin\left(\frac{\sin\alpha}{N_p} \right) \right] \qquad (1)$$

where N_p is the refractive index of the prism, α is the angle of incidence and A_p is the prism angle with respect to the normal.

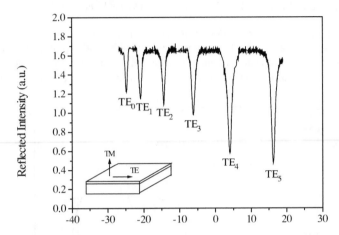

Figure 2 *TE coupling curve of the laser beam into epitaxial $Al_xGa_{1-x}N$ films.*

Using the calculation procedure reported by Ulrich and Torge [7], the ordinary refractive index (n_0) of AlGaN has been determined as 2.3240 ± 0.004 at 632.8nm. The extraordinary refractive index (n_e) obtained from the TM mode data is 2.3648 ± 0.006. The calculated thickness is 1.2 ± 0.01 µm in agreement with the SEM observations.

In order to investigate the film homogeneity in the whole thickness and at the interface, we have reconstructed the refractive index profiles of samples. The approach based on the inverse WKB method usually associated with Wentzel, Kramers and Brillouin, was employed. This method was first proposed by J.M. White [8] to study the propagation characteristics of single crystal $LiNbO_3$ waveguides, and then improved by K.S. Chiang [9] for Cu diffused MgO film waveguides. From the knowledge of effective mode indices, the index profile $n(x)$ can be determined by solving the following characteristics equation for m^{th} order mode given by

$$2k \int_0^{x_t(m)} \sqrt{n(x)^2 - N(m)^2}\, dx = 2m\pi + 2\phi_a + 2\phi_s \quad (2)$$

where $n(x)$ and Nm are respectively the refractive index profile and the effective index of mth order mode, $x_t(m)$ is defined as the " turning point " according to

$$n\,[x_t(m)] = N(m) \quad (3)$$

Generally, we assumed that the phase shift ϕ_s at the surface is taken as $\pi/4$ and ϕ_a is approximated to $\pi/2$ at the turning point as given by K.S. Chiang et *al* [9]. This method was described in detail by Dogheche et *al* [10].

From the angular position of modes given in figure 2, the effective mode indices was calculated. The reconstructed refractive index profiles of various samples prepared under optimized process conditions was investigated. As shown in figure 3, a non-abrupt film-substrate interface is observed.

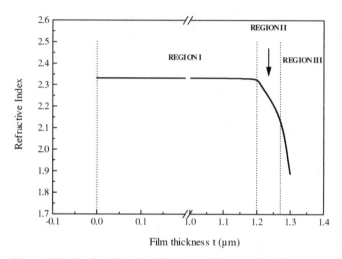

Figure 3 *Refractive index profile from a 1.2 µm thick AlGaN thin film using an improved inverse WKB method*

Reconstruction of index profiles was investigated for various samples prepared under optimized process conditions. The profile was obtained from the analysis of the TE modes. Figure 3 displays the evolution of the ordinary refractive index as a function of the layer thickness. Three distinct regions are observed. First, the index of refraction is maximum at the film surface (index $n_0 = 2.3240$) and remains constant within the guiding region I. This behavior is synonymous of a homogeneous layer along the whole thickness. Further, in the region III, the index decreases rapidly near the substrate surface. These two regions are observed commonly for step-index thin film structures. In our case, an additional region II is observed in the profile $n(x)$, we denoted a two-time decrease of the index profile. The occurence of fluctuations in the profile and the change in the shape near the substrate surface are indicative of a problem in the growth process. The region II illustrates the existence of a disturbed layer which appeared near the film-substrate interface. In fact, during the early stage of growth, many defects are introduced into the film, which can lead to the islanding and columnar growth. This behavior results in large deterioration of the crystal structure at the interface. In the present study, the interface containing defects manifests itself through the changes of the optical properties of samples. As a consequence, the index of refraction is lowered in the region near the interface. The curve plotted figure 3 shows a thickness of around 60nm where the index profile $n(x)$ is modified, it may related to the extent of this disturbed region. The optical study of interface properties by the reflectivity technique was also described by Shokhovets et *al* [11] for GaN grown on GaAs. In this work, the observed defects are a consequence of the mixing of GaN and GaAs. A thin interlayer with a refractive effective index below that of GaN was found. It was explained as the presence of voids in the interface region which lowers the refractive index in the mixed zone.

The same AlGaN films were analyzed by transmission electron microscopy (TEM). In fact, the growth of AlGaN films is strongly dependent on the quality of the buffer layer. During the early stages of the growth, many defects are introduced into the film which can lead to islanding and columnar growth. This behavior may result in large deterioration of the crystal structure at the interfaces, leading to high-quality and low-dislocations heteroepitaxial films. The <1010> dark field observations show a mosaic growth, which is more less pronounced depending on the samples. As noticed in figure 4, (0002) weak beam images show a highly defective interfacial layer extends to 60nm, instead of the initial 10nm thickness of the aluminum nitride buffer layer. This area contains also a high density of dislocation loops. The density of dislocation loops decreases drastically beyond the interfacial layer of 60nm.

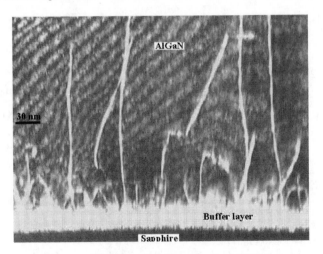

Figure 4 Cross-sectional 0002 weak beam image of a AlGaN layer grown on sapphire with a 10nm thick-AlN nucleation layer.

This mechanism can be compared to the model of the buffer layer efficiency as proposed by Nakamura et al [12], where a low-temperature buffer layer of 50nm was estimated to give rise to highly defective layer of 150nm called a semi sound zone. In our case, this zone is confined to the first 60nm. This could be probably one explanation for the refractive index behavior as shown in figure 3.

CONCLUSION

In conclusion, using a combination of optical characterizations and transmission electron microscopy (TEM) analysis, it is shwn that the growth properties of AlGaN / AlN heterostructures on sapphire can be investigated. A non-destructive prism coupling optical characterization was carried out to determine the waveguide mode informations, i.e. effective mode indices, which are treated through an original iWKB formalism to reconstruct the refractive index profile along the film thickness. There was an evidence of bending effects at the AlGaN/AlN buffer layer interface, extending in a thickness of 60nm. This area correspond exactly to the highly defective layer at the interface which appears to modify the index profile n(x). The results reveal the prospect of determining the relationship between structural and optical properties as a function of composition x for $Al_xGa_{1-x}N$.

REFERENCES

1. H. Amano, N. Sawasaki, I. Akasaki, Y. Toyoda, Appl. Phys. Lett. **48**, 353 (1986)
2. S. Nakamura, T. Mukai, M. Senoh : Appl. Phys. Lett. **64**, 1687 (1994)
3. F.A. Ponce, J.C. Major, W.E. Plano, D.F. Welsh , Appl. Phys. Lett. **65**, 2302 (1994)
4. Ö. Aktas, W. Kim, Z. Fan, A. Boktarev, A. Salvador, S.N. Mohammad, B. Sverdlov, H. Morkoç, Electron. Lett. **31**, 1389 (1995)
5 . F. Omnes, N. Marenco, B. Beaumont, P. de Mierry, F. Caille, E. Monoz, J. Appl. Phys. **86**, 5286(1998)
6. P.K. Tien, S. R. Sanseverino, and R.J. Martin, Phys. Lett. **14**, 503 (1974)
7. Ulrich, R. Torge, Appl. Opt. **12**, 2901 (1973)
8. J.M. White and P.F Heidrich, Appl. Opt. **15**, 151(1973)
9. K.S. Chiang, J. Light. Technol. **LT-3**, 385(1985)
10. E. Dogheche, B. Belgacem, D. Rémiens, P. Ruterana, F. Omnes: Appl. Phys. Lett. **75**, 3309 (1999)
11. S. Shokhovets, R. Goldhahn, V. Cimalla, T.S. Cheng, C.T. Foxon, J. Appl. Phys. **84** 1561 (1998)
12. S. Nakamura and G. Fasol, The blue laser diode-GaN based light emitters and lasers, Springer Verlag Publishers, Heidelberg (1997)

Mat. Res. Soc. Symp. Vol. 595 © 2000 Materials Research Society

Deep level related yellow luminescence in p-type GaN grown by MBE on (0001) sapphire

Giancarlo Salviati[1], Nicola Armani[1], Carlo Zanotti-Fregonara[1], Enos Gombia[1] , Martin Albrecht[2], Horst P. Strunk[2] , Markus Mayer[3], Markus Kamp[3] , Andrea Gasparotto[4]

[1]CNR-MASPEC Institute, Parco Area delle Scienze, 37A, I-43010 Loc. Fontanini-Parma, Italy
[2]Universität Erlangen, Institut für Werkstoffwissenschaften, Mikrocharakterisierung, Cauerstr.6 D-91058 Erlangen, FRG
[3]Department of Optoelectronics, University of Ulm, Albert Einstein Allee 45, D-89069 Ulm FRG
[4]Physics Department, University of Padova, Via Marzolo 8, 35131 Padova, Italy

ABSTRACT

Yellow luminescence (YL) has been studied in GaN:Mg doped with Mg concentrations ranging from 10^{19} to 10^{21} cm^{-3} by spectral CL (T=5K) and TEM and explained by suggesting that a different mechanism could be responsible for the YL in p-type GaN with respect to that acting in n-type GaN.

Transitions at 2.2, 2.8, 3.27, 3.21, and 3.44 eV were found. In addition to the wurtzite phase, TEM showed a different amount of the cubic phase in the samples. Nano tubes with a density of $3x10^9$ cm^{-2} were also observed by approaching the layer/substrate interface. Besides this, coherent inclusions were found with a diameter in the nm range and a volume fraction of about 1%.

The 2.8 eV transition was correlated to a deep level at 600 meV below the conduction band (CB) due to Mg_{Ga}-V_N complexes. The 3.27 eV emission was ascribed to a shallow acceptor at about 170-190 meV above the valence band (VB) due to Mg_{Ga}.

The 2.2 eV yellow band, not present in low doped samples, increased by increasing the Mg concentration. It was ascribed to a transition between a deep donor level at 0.8-1.1 eV below the CB edge due to N_{Ga} and the shallow acceptor due to Mg_{Ga}. This assumption was checked by studying the role of C in Mg compensation. CL spectra from a sample with high C content showed transitions between a C-related 200 meV shallow donor and a deep donor level at about 0.9-1.1 eV below the CB due to a N_{Ga}-V_N complex. In our hypothesis this should induce a decrease of the integrated intensity in both the 2.2 and 2.8 eV bands, as actually shown by CL investigations.

INTRODUCTION

The achievement of p-type epitaxial GaN by Mg doping is of interest for device applications. In addition, the nature of most of the basic luminescence recombination mechanisms in Nitride-based material systems and they dependence on material properties are still subject of discussion. Yellow luminescence in undoped or n-type GaN is supposed to be due to a transition between a shallow donor and a deep acceptor level located at about 1 eV above the VB [1,2] due to a complex defect involving V_{Ga} [3]. P-type GaN doped with Mg is supposed not to show YL due to the increase of the formation energy of Ga vacancies in semi-insulating or p-type GaN [3,4] and due to the compensation of V_{Ga} by Mg atoms [5-7]. Nevertheless, YL in Mg doped GaN has been observed, but only in n-type layers [8]. In this work a broad YL band is found in p-type Mg doped GaN layers grown by molecular beam epitaxy (MBE). The optical transitions and the structural properties of the layers are discussed on the basis of low temperature Spectral Cathodoluminescence (SCL) results in the Scanning Electron Microscope (SEM) and Transmission Electron Microscopy (TEM) respectively. Taking advantage of the special growth technique used, the results are discussed suggesting a different mechanism could be responsible for the YL in p-type GaN with respect to that acting in n-type GaN.

EXPERIMENTAL DETAILS

P-type GaN layers doped with Mg concentrations ranging from 10^{19} to 10^{21} cm^{-3} have been grown by MBE under N rich conditions by using On-Surface Cracking ammonia [9]. Mg-doping was performed using either evaporation of Mg atoms from a conventional effusion cell or the metal organic precursor MCp_2Mg (methyl-cyclo-pentadienyl-magnesium) which is for the first time ever used in GaN MBE. Probe and rectification methods, I-V characteristics and Hall effect measurements have verified p-type carrier concentrations below $5x10^{17}$ cm^{-3}. The concentration and depth distribution of Mg atoms have been investigated by secondary ion mass spectroscopy (SIMS) analyses.

The optical emission of the samples has been studied by SCL performed in a commercial Oxford MONOCL system, equipped with a multialkaly photomultiplier, fitted to a 360 Cambridge SEM. The temperature range was between 5 and 300 K. The cathodoluminescence was also performed spatially as well as spectrally resolved. The structure of the epilayers has been studied by conventional TEM in a Philips CM 300UT operating a 300 keV. Cross sectional and plan view TEM samples have been prepared by standard techniques including mechanical grinding and polishing with diamond coated polymer foils followed by ion milling.

RESULTS AND DISCUSSION

Figure 1a shows a TEM plan view image of the sample with the highest Mg content. A typical TEM cross section of one of the samples studied is reported in Figure 1b. In all the samples dislocations and Stacking Faults were found, in addition to the cubic phase. Figure 2 shows a comparison between CL spectra from undoped and Mg doped GaN samples (10^{19} cm^{-3}<Mg concentration< 10^{21} cm^{-3}). In addition to the Near Band Edge (NBE) emission at about 3.44 eV, to the peak at about 3.27 eV, related to the Mg shallow acceptor at about 170-190 meV above the VB (see for instance [10] and references therein enclosed), a strong emission ascribed to a cubic phase, as found by TEM analyses, is shown at about 3.2 eV.

Figure 1 - *a) Bright field (0001) multi beam conditions TEM plan view image of the sample with the highest Mg concentration; b) Weak beam **g**=0001 TEM micrograph in cross section of one of the samples studied. The usual dislocation distribution is shown.*

The most interesting feature in Figure 2 concernes the broad emission centered at about 2.2 eV. This peak is not present in the less doped material and is preeminent together with another band at about 2.8 eV in the heavily doped sample. Note the presence of the YL in the undoped layer as was expected.

In the following we will discuss the onset of the YL in the most doped sample. The yellow band is present in undoped GaN due to the presence of V_{Ga} as suggested by other authors (see for instance ref [3]). In Mg doped samples, Mg atoms compensate the V_{Ga} [4] and therefore the 2.2 eV emission decreases even if it does not disappear completely probably because of the incomplete compensation of Ga vacancies at low Mg doping. By increasing the Mg concentration, two different peaks are found. The emission at 2.8 eV can be understood as being due to a transition involving a Mg_{Ga}-V_N complexes as reported by other authors [11-14].

The nature of the broad band centered at about 2.2 eV has been investigated by studying its behaviour as a function of the injection conditions and temperature. Deconvolution procedures revealed the presence of at least three superimposed gaussians peaks at about 2.05, 2.2 and 2.35 eV.

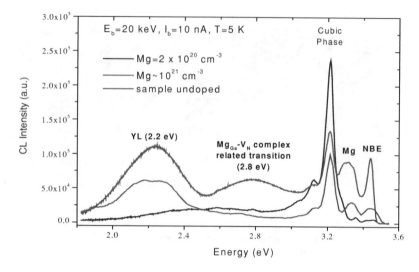

Figure 2 - *Comparison between CL spectra from undoped and Mg doped GaN samples*

The 2.35 eV peak could be due to the presence of C as a contaminant in the MBE growth. By studying the temperature evolution of the 2.2 and 2.35 eV peaks from 20 to 300 K energy red-shifts of about 50 meV and 35 meV have been found respectively. By plotting the CL intensity of the 2.2 eV peak as a function of the beam energy (E_b) at constant injection conditions, the total intensity is observed to increase up to E_b=25 keV, corresponding to a depth through the sample of about 900 nm and then the CL signal saturates. The CL intensity of the 2.35 eV peak does not saturate until E_b=40 keV, corresponding to a depth of about 1.8 μm.

The comparison with Figure 3, where the SIMS profile of the Mg distribution in the same sample is reported, suggests that the emission at 2.2 eV is correlated with the Mg doping.

By increasing the excitation power, an energy blue-shift of about 15 meV has been observed; this is consistent with the hypothesis that shallow levels are involved in this transition which is supposed to be of donor acceptor pair (DAP) type [15,16].

In the following a possible origin of the YL in p-type GaN is discussed on the basis of a different mechanism with respect to that acting in undoped or n-type GaN. A superior incorporation of Mg at the Ga-substitutional site (Mg_{Ga}), which is the preferred site in the bulk, is expected under N-rich growth conditions as in our case [17].

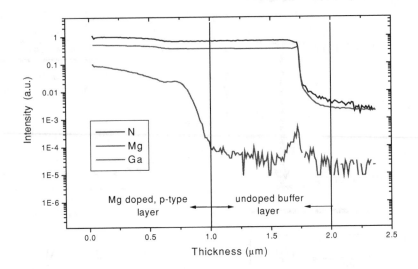

Figure 3 - *SIMS profile of the Mg distribution in comparison with the component elements of the GaN layer (Gallium and Nitrogen).*

It is known from the literature that this dopant induces a shallow acceptor level at about 200 meV above the VB [10]. According to Tansley et al [18], a deep donor level at about 0.8-1.1 eV below the CB has been ascribed to the N_{Ga} defect. Furthermore, Suski et al. [19] have considered deep double donor gap state caused by N_{Ga} as responsible for the well known YL in n-type GaN. The formation energy for the N_{Ga} defect under thermodynamic equilibrium conditions is considerably high [20] almost independently on the Fermi energy. However, since in our case the samples have been grown far from the thermodynamic equilibrium and possibly with a consistent influence of growth kinetics and because of the high Mg concentration, it is not straighforward to correlate the formation of native defects in the frame of the thermodynamical equilibrium theory. Therefore, a transition between the native defect N_{Ga} and a shallow acceptor due to Mg_{Ga} could be considered at the origin of the 2.2 eV peak in our samples. The Arrhenius plot of Figure 4 reports the inverse of the CL integrated intensity of the 2.2 eV emission as a function of temperature: an activation energy of about 140±30 meV is found. Within the experimental error this value is much closer to the activation energy of 170-190 meV for the shallow acceptor due to the Mg_{Ga} rather then that of 30 meV typical for shallow donors as found in undoped or n-type GaN.

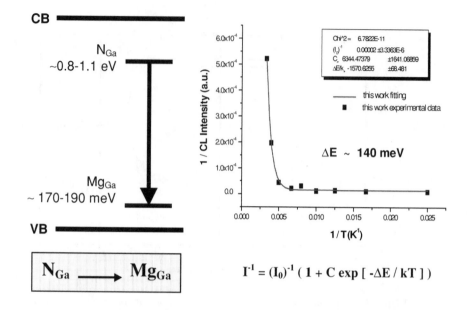

$$I^{-1} = (I_0)^{-1} (1 + C \exp [-\Delta E / kT])$$

Figure 4 – *a) Sketch of the possible transition in the heavily Mg doped sample. b) The Arrhenius plot of the inverse of the CL integrated intensity of the 2.2 eV emission as a function of temperature.*

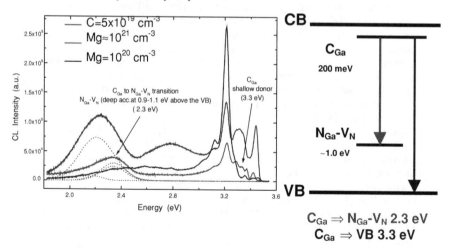

Figure 5 – *a) Comparison between the samples doped with the highest and the lowest Mg concentration and the sample with high C concentration; b) Sketch of the possibles transitions in the heavily C contaminated sample.*

To support our hypothesis, in the following we will discuss the influence of C on the optical transitions of our samples. PL investigations on GaN layers intentionally doped with C [21] revealed the presence of an optical emission at about 2.35 eV.

Figure 5 reports a comparison between the sample doped with the highest Mg concentration and another sample with a C concentration of 5×10^{19} cm^{-3} as found by SIMS analyses. On the low energy side of the spectrum, a predominant transition at 2.35 eV is shown

It is known from the literature [22] that the C_{Ga} defect is favoured in p-type GaN samples. It is therefore reasonable to expect that C competes against Mg as a substitutional impurity. In this respect the emission at 2.35 eV can be ascribed to a transition between a shallow donor at about 200 meV below the CB, due to the C_{Ga} substitutional defect, and a deep acceptor level at about 1 eV above the VB due to the complex N_{Ga}-V_N, which is known to be favoured in p-type GaN [23]. The most striking feature of the spectra in Figure 8 is the disappearance of the 2.8 eV band and the consequent strong reduction of the CL integrated intensity of the 2.2 eV band. The 2.8 eV band disappears because of the increased formation energy of the Mg_{Ga}-V_N complex [13] induced by the Fermi level energy shift due to the reduced concentration of Mg atoms in samples with high C concentration grown under N-rich conditions and the complex V_N-N_{Ga} is favoured [23]. As a consequence, the N_{Ga} antisite defect cannot contribute to the 2.2 eV emission as in Mg doped samples with low C concentration. It is clear that further investigations must be carried out in order to clarify the correlation between the formation energy of the N_{Ga} defect and the Fermi level energy position in p-type GaN.

CONCLUSIONS

Yellow luminescence has been studied in GaN:Mg doped with Mg concentrations ranging from 10^{19} to 10^{21} cm^{-3} by spectral CL (T=5K) and TEM and explained by suggesting that a different mechanism could be responsible for the YL in p-type GaN with respect to that acting in n-type GaN.

Transitions at 2.2, 2.8, 3.21, 3.27, and 3.46 eV were found. The 2.8 eV transition was correlated with a deep level at 600 meV below the CB due to Mg_{Ga}-V_N complexes; the 3.27 eV emission was ascribed to a shallow acceptor at about 170-190 meV above the VB due to Mg_{Ga}.

The 2.2 eV yellow band, not present in low Mg doped samples, increased by increasing the Mg concentration. It was ascribed to a transition between a deep donor level at 0.8-1.1 eV below the CB edge due to N_{Ga} and the shallow acceptor due to Mg_{Ga}. This assumption was checked by studying the role of C in the Mg compensation. CL spectra from a sample with high C concentration showed transitions between a C-related 200 meV shallow donor and a deep acceptor level at about 0.9-1.1 eV below the CB due to a N_{Ga}-V_N complex.

REFERENCES

1. T.Ogino and M.Aoki, Jpn J. Appl. Phys.**19**, 2395-2405 (1980)
2. D.M. Hofmann, D. Kovalev, G. Steude, B.K. Meyer, A. Hoffmann, L. Eckey, R. Heitz, T. Detchprom, H. Amano and I. Akasaki, Phys. Rev. **B52** 16702-16706 (1995)
3. J.Neugebauer and C.G.Van de Walle, Appl. Phys. Lett. **69**, 503-505 (1996)
4. C.G.Van de Walle, C.Stampfl, J.Neugebauer, J. Cryst. Growth **189-190**, 505-510 (1998)
5. W.Kim, A.Salvador, A.E.Botchkarev, O.Aktas, S.N.Mohammad and H.Morkoç, Appl. Phys. Lett. **69**, 559-561 (1996)
6. U.Kaufmann, M.Kunzer, M.Maier, H.Obloh, A.Ramakrishnan, B.Santic and P.Schlotter, Appl. Phys. Lett., **72**, 1326- 1328 (1998)
7. S.Nakamura, T.Mukai, M.Senoh and N.Iwasa, Jpn. J. Appl. Phys. **31**, L139- 141 (1992)
8. F.J.Sanchez, F.Calle, D.Basak, J.M.G.Tijero, M.A.Sanchez-Garcia, E.Monroy, E.Calleja, E.Muñoz, B.Beaumont, P.Gibart, J.J.Serrano and J.M.Blanco, MRS Internet J. Nitride Semicond. Res. **2**, 28 (1997)
9. M.Kamp, M.Mayer, A.Pelzmann and K.J.Ebeling, MRS Internet J. Nitride Semicond. Res. **2**, 26 (1997)
10. M.Leroux, N.Grandjean, B.Beaumont, G.Nataf, F.Semond, J.Massies and P.Gibart, J. Appl. Phys. **86**, 3721-3728 (1999)
11. P.H.Lim, B.Schineller, O.Schon, K.Heime and M.Heuken, J. Cryst. Growth **205**, 1-10 (1999)
12. Reboredo e Pantelides Phys.Rev. Lett. **82**, 1887 (1999)
13. Lee e Chang, Semic. Sci. and Technol. **14**, 138 (1999)
14. E. Oh, H.S. Park and Y.J. Park, Appl. Phys. Lett. **72**, 70-72 (1998)
15. L.Pavesi and M.Guzzi, J. Appl. Phys. **75**, 4779-4842 (1994)
16. P.J. Dean, Prog. Cryst. Growth, Charact. 1982, vol. 5 pagg. 89-174
17. C.Bungaro, K.Rapcewiz and J.Bernholc, Phys. Rev. **B59**, 9771-9774 (1999)
18. T.L.Tansley and R.J.Egan, Physica B **185**, 190-193 (1993)
19. T.Suski, P.Perlin, H.Teisseyre, M.Leszczynski, I.Grzegory, J.Jun and M.Bockowski, Appl. Phys. Lett. **67**, 2188-2190 (1995)
20. J.Neugebauer and C.G.Van de Walle, Phys. Rev. **B50** Rapid Comm., 8067-8070 (1994)
21. E.E.Reuter, R.Zhang, T.F.Kuech and S.G.Bishop, MRS Internet J. Nitride Semicond. Res. **4S1** G3.67 (1999)
22. P. Boguslawski, E.L. Briggs and J. Bernholc, Appl. Phys. Lett. **69**, 233-235 (1996)D.J.
23. D.J. Chadi, Appl. Phys. Lett. **71**, 2970-2971 (1997)

Mat. Res. Soc. Symp. Vol. 595 © 2000 Materials Research Society

A study of annealed GaN grown by molecular beam epitaxy using photoluminescence spectroscopy.

Abigail Bell[1,2], Ian Harrison[2], Dimitris Korakakis[1,2], Eric C. Larkins[2], J. M. Hayes[3], M. Kuball[3].

[1]School of Physics and Astronomy, University of Nottingham, Nottingham, NG7 2RD, UK
[2]School of Electrical and Electronic Engineering, University of Nottingham, Nottingham, NG7 2RD, UK.
[3]H.H.Wills Physics Laboratory, University of Bristol, Bristol, BS8 1TL, UK

ABSTRACT

Photoluminescence (PL) spectroscopy has been used to investigate the effect annealing has on molecular beam epitaxially grown GaN in different ambients. By observing the changes in the PL spectra as a function of ambient temperature and atmosphere used, important information concerning the origin of defects within GaN has been found. Samples were annealed in different atmospheres, (including oxygen, oxygen and water vapour, nitrogen and argon), different temperatures. In the 2.0eV-2.8eV region of the PL spectra, two peaks appeared at approximately 2.3eV and 2.6eV, somewhat higher than the usual yellow luminescence peak. We find that the 2.6eV peak is dominant for high annealing temperatures and the 2.3eV peak dominates at low annealing temperatures for the samples annealed in oxygen. When annealed in argon and nitrogen the 2.6eV peak dominates at all annealing temperatures. Changes in the PL spectra between anneals were also seen in the 3.42eV region. The 3.42eV peak is often assigned to excitons bound to stacking faults. Power resolved measurements indicate that in our sample the cause is a donor acceptor pair transition.

INTRODUCTION

Group-III nitrides are direct, wide band gap semiconductors and have many potential uses in optoelectronic devices. The optical and electrical properties of these materials are affected not only by the processes that occur during growth, but also by the post-growth processes that the material undergoes, such as thermal annealing. Photoluminescence (PL) is used widely to characterise the nitrides. It gives a good indication of the radiative processes that occur within the material. Non-radiative paths reduce the intensity of the PL peaks and so PL is a good indicator of the optical quality of the material. The origins of many of the peaks that commonly occur in GaN are still not clear. One such peak is the so-called "yellow luminescence" which is a broad peak that appears in the PL spectra at approximately 2.2eV. *Ogino and Aoki* [1] describe the transition as occurring between a shallow donor 25meV below the conduction band and an acceptor 860meV above the valence band. *Glaser et al* [2] performed PL and optically detected magnetic resonance (ODMR) on GaN layers. They suggested it is a two-part transition in which there is a non-radiative capture between a shallow and deep-donor state followed by a radiative transition between the deep-donor state and a shallow acceptor (possibly C on a N site). *Ponce et al* [3] suggested that the sources of the yellow emission are either dislocations at low angle grain boundaries in the material or point defects, which nucleate at dislocations. It has also been suggested by *Reynolds et al* [4], that the yellow band in GaN results from a transition between a shallow donor and a deep level. The deep level in the Reynolds's model was attributed to a complex consisting of a Ga vacancy and oxygen on a nitrogen site (V_{Ga}-O_N). Another commonly occurring peak in GaN is the 3.427eV peak, seen in the 4K PL spectra. It was initially attributed to oxygen forming a shallow "deep level"[5]. However, it has more recently been suggested that peaks in the PL spectra of GaN in this region 3.40eV [6-8], 3.412eV [9] and 3.42eV [10] can be related to excitons bound to stacking faults.

By annealing samples of GaN in different ambients and at different temperatures, we have clarified some of the issues regarding the role that defects play, particularly in the "yellow luminescence" and the near 3.42eV luminescence regions. GaN samples grown by MBE have been used in this study and annealed at different temperatures in oxygen, oxygen plus water vapour, nitrogen, nitrogen plus water vapour and argon in a high-temperature stage.

EXPERIMENTAL METHOD

The layer used in this work (sample MG671) was grown by MBE in a modified Varian Mod Gen II machine. The active nitrogen was produced by radio-frequency (RF) plasma source and the Ga was produced from solid elemental sources. The sample was an unintentionally doped 1μm thick GaN layer grown on a (0001) sapphire substrate. Grown on top of the GaN was a high-mobility field effect transistor (FET) structure consisting of 3nm undoped $Al_{0.15}Ga_{0.85}N$/22nm Si-doped $Al_{0.15}Ga_{0.85}N$/15nm undoped $Al_{0.15}Ga_{0.85}N$. Pieces of the sample MG671 were annealed for 20mins using a high temperature stage at various temperatures and in various gases. The pieces of MG671 were labelled numerically and will be refered to as (number) in the remainder of the paper. The annealing details and sample numbers are shown in table 1.

The photoluminescence spectra were taken using a 325nm CW HeCd laser with a maximum excitation intensity of 9mW. A 325nm band pass filter was used to attenuate lines other than the 325nm laser line. A quartz lens focused the beam onto the sample in a liquid helium cryostat. The excitation light was normal to the sample and the resulting PL was focused through a low-pass sharp cut-off filter (Oriel WG345), used to stop the laser light, onto the monochromator slit. The 0.75m Spex monochromator, with a 2400 lines/mm grating, has a resolution of 4Å/mm, which correspond to 4meV/mm in the region of 3.4eV. The light was detected by a bi-alkali photomultiplier (Thorn 9924QB) with a 600V applied voltage. Standard phase sensitive detection techniques were used to improve the signal to noise ratio.

Table 1. Sample numbers with annealing parameters.

Temperature (^{O}C)	800	900	1000	1100	1200
O_2	40	41	41	44	
O_2 and H_2O			63		
N_2			9	46	7
N_2 and H_2O			61		
Ar			16	18	19
Pre-annealed	39				

RESULTS

In all of the 4K PL spectra (figure 1) there is a peak which occurs in the region from 3.467eV to 3.479eV and it has been attributed to the donor bound exciton (D^oX) [11]. In addition, a peak appears at approximately 3.40eV-3.42eV in all of the spectra. Transitions in this region are commonly attributed to an exciton bound to a stacking fault [6-10]. There is no clear consensus to the exact energy at which this occurs (3.40eV [6-8], 3.412eV [9], 3.42eV[10]). From figure 1, the peaks in this area are more pronounced after annealing at higher temperatures (42,44) and in a water vapour atmosphere (63). There also appears to be a correlation between the strain [12], as determined by the position of the D^oX peak and the intensity of the 3.42eV peak, which could support the stacking fault hypothesis. To investigate this further we have performed power resolved measurements since the excitonic transition should not shift with power. The results are shown in figure 2 for the sample annealed in an atmosphere of H_2O and O_2 (63). In this case the peaks at 3.418eV and 3.36eV shift to higher energy with increasing excitation intensity. If the stacking faults act as quantum wells as described in the literature you might also expect to see a blue shift with increasing excitation intensity. However an alternative explanation for the blueshift is that the 3.42eV peak is a DAP transition[13]. A similar transition has been seen by *Eunsoon et al* [14]. The samples annealed in N_2 and Ar exhibit a peak in this region which also blueshifts with increasing excitation intensity (not shown here). Although in the case of the N_2 annealed samples, the peak only appears when annealed above 1100°C.

The 2.0eV-2.7eV region of the PL spectra is considerably altered by annealing the sample. Figure 3 shows the room temperature PL spectra between 2.0eV-3.1eV of the pre-annealed (39), annealed in O_2 (40, 41, 42, 44) and annealed in O_2 and H_2O (63) samples. The pre-annealed sample (39) exhibits only weak PL in this region, with peaks at 2.3eV and 2.6eV, where the intensity of the 2.6eV peak dominates. This is at a somewhat higher energy than the usual "yellow luminescence" found in GaN between 2.0-2.2eV [1-4]. Annealing the samples in O_2 greatly enhances the PL intensity at 2.3eV and 2.6eV, although increasing the annealing temperature to above 1100°C causes the 2.6eV peak intensity to

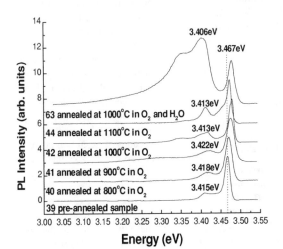

Figure 1. *4K PL spectra of sample annealed in O_2 and H_2O at different temperatures. The position of the D^oX peak shifts from 3.467eV to higher energy after annealing.*

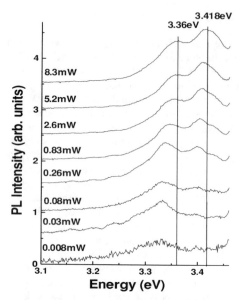

Figure 2. *Power resolved PL of sample 63 annealed in O_2 and H_2O at 1000^oC. The 3.418eV and 3.36eV peaks shifts with excitation intensity, typical of a DAP transition. The D^oX peak, not shown here, is at 3.479eV and does not shift.*

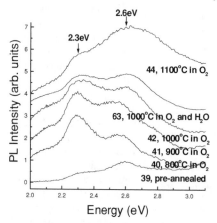

Figure 3. *300K PL spectra of samples annealed in O_2 or O_2 plus H_2O at different temperatures. Peaks at 2.3eV and 2.6eV, shown with an arrow, become stronger upon annealing in O_2, this effect is even stronger when annealing in O_2 and H_2O.*

become even greater than the 2.3eV peak. Annealing in both O_2 and H_2O increases the intensity of both the 2.3eV and 2.6eV peaks. The two arrows on figure 3 are guides for the eye to the 2.3eV and 2.6eV PL peaks. Calculations were made and it was found that both the 2.3eV and 2.6eV peaks are real PL peaks and not Fabry-Perot thickness oscillations. Similar PL spectra were obtained from the samples annealed in N_2 and Ar, although the 2.6eV peak is dominant (spectra not shown here). The PL intensity in this region in the N_2 and Ar annealed samples also increases with increasing annealing temperature.

Figure 4 shows the power resolved PL of the sample annealed in O_2 and H_2O (63) in the 2.0eV to 3.1eV region. The 2.3eV peak shifts to a higher energy with increasing excitation intensity indicating that it may be DAP type recombination. However, the 2.6eV peak does not appear to shift over two decades of excitation power and so has free-bound character. The 2.6eV and 2.3eV peaks often occur together. If we assume that there is a common defect in these two transitions then we can have two possible situations shown in figure 5a and 5b. When the samples were annealed in O_2 both peaks increased in intensity. However, when the samples were annealed in N_2 and Ar,

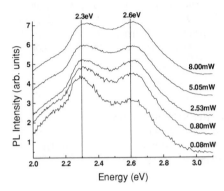

Figure 4. *Power resolved PL of sample 63 annealed at 1000°C in O₂ and H₂O. As the excitation intensity is increased the 2.3eV peak shifts to a higher energy suggesting a DAP transition. There is no such shift of the 2.6eV peak.*

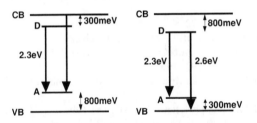

Figure 5(a) *A deep acceptor takes part in recombination from an electron on a shallow donor and the conduction band. (b) A deep donor level carries electrons that recombine with both a shallow acceptor level and the valence band.*

the 2.6eV peak was dominant suggesting that the annealing process introduces the deep acceptor. The increase in the 2.3eV DAP transition in the O₂ annealed samples provides evidence for the donor at 300meV being oxygen related. However, it is important to remember that if the model 5b had been used, these arguments would point to an oxygen related acceptor. When H₂O was introduced into the anneal, both peak intensities were increased. This would suggest that the addition of water somehow enhances the introduction of donors (or acceptors) as discussed previously.

CONCLUSIONS

Annealing the material introduces a defect causing a transition at approximately 3.42eV. A peak appears in this region independent of the annealing atmosphere, although its intensity increases with annealing temperature. If this emission is related to stacking faults as suggested in the literature [6-10], this would suggest that annealing our material introduces stacking faults and that at higher annealing temperatures, more stacking faults are introduced into the sample. However, an alternative explanation that the 3.42eV peak is DAP in nature is offered. The increase in the intensity of this line with annealing can then be attributed to the increase in defect (donor or acceptor) concentration.

Deep levels were introduced into the samples by annealing them in various atmospheres. This deep level luminescence consisted of two peaks at 2.3eV and 2.6eV, which is considerably higher in energy than the usual "yellow luminescence" seen in GaN between 2.0eV -2.2eV. In all of the samples, annealing increased the deep level luminescence and increasing the annealing temperature further enhanced the PL intensities. The 2.3eV peak is a DAP transition, as determined from power resolved PL. The near

2.6eV peak was dominant in the Ar and N_2 annealed samples. In the samples annealed in oxygen, both peaks competed. suggesting that the 2.3eV peak is related to oxygen. Two possible models have been suggested and outlined for this luminescence. These models consist of the free-to-bound and DAP transitions being due to either the same donor or acceptor.

ACKNOWLEDGEMENTS

We would like to acknowledge the EPSRC Blue UV laser diode program and the EU BRITE EURAM MIGHT project No BE98-4899 for part funding on this work. AB would also like to thank the University of Nottingham for her studentship. The work in Bristol was in part sponsored by Renishaw plc. (Dr. G. D. Pitt)

REFERENCES

[1] T.Ogino and M.Aoki, Jpn.J.Appl.Phys. **19** p2395 (1980)
[2] E.R.Glaser, T.A.Kennedy, K.Doverspike, L.B.Rowland, D.K.Gaskill, J.A.Freitas Jr, M.Asif Khan, D.T.Olsen, J.N.Kuznia and D.K.Wickenden, Phys.Rev.B **51** p13326 (1995)
[3] F.A.Ponce, D.P.Bour, W.Gotz and P.J.Wright, Appl.Phys.Lett. **68** p57 (1996)
[4] D.C.Reynolds, D.C.Look, B.Jogai, J.E.Van Norstrand, R.Jones and J.Jenny, Solid State Commun., **106** p701 (1998)
[5] B-C Chung and M Gershenzon, J.Appl.Phys. **72** p651 (1992)
[6] M.Albrecht, S.Christiansen, G.Salviati, C.Zanotti-Fregonara, Y.T.Rebane, Y.G.Shreter, M.Mayer, A.Pelzmann, M.Kamp, K.J.Ebeling, M.D.Bremser, R.F.Davis and H.P.Strunk, Mat.Res.Soc.Symp.Proc.Vol.**468** p293 (1997)
[7] Y.T.Rebane, Y.G.Shreter, and M.Albrecht, Mat.Res.Soc.Symp.Proc.Vol.**468** p179 (1997)
[8] Y.T.Rebane, Y.G.Shreter, and M.Albrecht, phys.stat.sol.(a) **164** 141 1997
[9] S.Fischer, G.Steude, D.M.Hofmann, F.Kurth, F.Anders, M.Topf, B.K.Meyer, F.Bertram, M.Schmidt, J.Christen, L.Eckey, J.Holst, A.Hoffmann, B.Mensching, B.Rauschenbach, J.Cryst.Growth **189/190** p556 (1998)
[10] G.Salviati, C.Zanotti-Fregonara, M.Albrecht, S.Christiansen, H.P.Strunk, M.Mayer, A.Pelzmann, M.Kamp, K.J.Ebeling, M.D.Bremser, R.F.Davis, Y.G.Shreter, Inst.Phys.Conf.Ser. **157** p199 (1997)
[11] O.Lagerstedt and B.Monemar, J.Appl.Phys. **45** p2266 (1974)
[12] J.M. Hayes, M. Kuball, A. Bell, I. Harrison, D. Korakakis, and C.T. Foxon, Appl. Phys. Lett. **75** p2097 (1999)
[13] H.B.Bebb and E.W.Williams, Semicond. Semimet., **8** p4-5 (1972)
[14] Eunsoon Oh, Bonjin Kim, Hyeongsoo Park, and Yongjo Park, Appl.Phys.Lett. **73** p1883 (1998)

Mat. Res. Soc. Symp. Vol. 595 © 2000 Materials Research Society

Nonlinear Optical Characterization of GaN Layers Grown by MOCVD on Sapphire

Ivan M. Tiginyanu, Igor V. Kravetsky, Dimitris Pavlidis[1], Andreas Eisenbach[1], Ralf Hildebrandt[2], Gerd Marowsky[2], Hans L. Hartnagel[3]
Laboratory of Low-Dimensional Semiconductor Structures, Institute of Applied Physics, Technical University of Moldova, 2004 Chisinau, Moldova, tiginyanu@mail.md
[1] Solid State Electronics Laboratory, Department of Electrical Engineering and Computer Science, University of Michigan, 1301 Beal Avenue, Ann Arbor, MI 48109-2122, U.S.A.
[2] Laser-Laboratorium Göttingen, Hans-Adolf-Krebs-Weg 1, D-37077 Göttingen, Germany
[3] Institut für Hochfrequenztechnik, Technische Universität Darmstadt, D-64283 Darmstadt, Germany

ABSTRACT

Optical second and third harmonic generation measurements were carried out on GaN layers grown by metalorganic chemical vapor deposition (MOCVD) on sapphire substrates. The measured d_{33} is 33 times the d_{11} of quartz. The angular dependence of second-harmonic intensity as well as the measured ratios $d_{33}/d_{15} = -2.02$ and $d_{33}/d_{31} = -2.03$ confirm the wurzite structure of the studied GaN layers with the optical c-axis oriented perpendicular to the sample surface. Fine oscillations were observed in the measured second and third harmonic angular dependencies. A simple model based on the interference of the fundamental beam in the sample was used to explain these oscillations.

INTRODUCTION

Recently there has been strong interest in studying gallium nitride and alloys because of their promising device applications [1]. In particular, the nonlinear optical properties of GaN films are of interest for optoelectronic and all-optical device applications. Several second-harmonic generation (SHG) and third-harmonic generation (THG) studies of GaN epilayers have been reported recently [2-5]. However, no attention was paid to the differences between front and back excitations nor to the angular step resolution on the nonlinear optical response. In the present work we report on the study of nonlinear optical effects in GaN/sapphire samples as a function of excitation geometry and angular step resolution. Both the second-harmonic (SH) and third-harmonic (TH) signals measured in dependence on the incident angle of the laser beam were found to be modulated due to the interference of the fundamental beam in the sample.

EXPERIMENT

The GaN layers used in our experiments were grown by low-pressure MOCVD on (0001) c-plane sapphire using trimethylgallium (TMGa) and ammonia (NH_3) as source materials. A horizontal growth reactor in a modified EMCORE GS-3200 system was used for this purpose. A buffer layer of about 25 nm thick GaN was first grown at 510 °C. The GaN layers grown on top at a temperature of 1100 °C had thicknesses of about 1 µm.

Polarized SHG and THG measurements were carried out in transmission mode for both front (the GaN layer facing the incident pump beam) and back (the sapphire substrate facing the incident pump beam) excitations. As a fundamental beam, the 1064 nm output of a Q-switched Nd-YAG laser (Spectra Physics GCR-170) with 10 Hz repetition rate and 7 ns pulse width was used. To minimize the influence of the laser output fluctuations, the measured SH ($\lambda_{2\omega} = 532$ nm) and TH ($\lambda_{3\omega} = 355$ nm) intensities were normalized by the simultaneously monitored laser intensity in the reference channel. The sample was mounted on a step-motorized rotation stage. The direction of the fundamental beam polarization was changed by rotating a half-wave plate placed in front of the sample. The fundamental wavelength was filtered out from SHG and THG signals by using appropriate colour filters and a grating monochromator.

RESULTS AND DISCUSSION

The wurzite structure of GaN belongs to the 6mm point group symmetry. In this case there are three nonzero nonlinear optical coefficients d_{15}, d_{31}, and d_{33}, which are responsible for the second-order nonlinear optical properties. The induced nonlinear polarization in GaN films has the following components [6]:

$$P_x(2\omega) = 2 \, d_{15} \, E_z(\omega) \, E_x(\omega)$$
$$P_y(2\omega) = 2 \, d_{15} \, E_z(\omega) \, E_y(\omega) \qquad (1)$$
$$P_z(2\omega) = d_{31} \, (E_x(\omega)^2 + E_y(\omega)^2) + d_{33} \, E_z(\omega)^2$$

According to equations (1), SHG is forbidden from a c-textured film (having the optical axis c perpendicular to the surface) when the pump beam is incident normal to the film. One can notice that for an ideal wurzite structure, the nonzero elements of the second-order nonlinear susceptibility tensor are related to each other as $d_{15} \approx d_{31}$ and $d_{33}/d_{31} \approx -2$ [7, 8].

Neglecting absorption and birefringence of GaN ($\Delta n_{\omega(2\omega)} = 0.02$) [9], the transmitted SHG intensity $I_{2\omega}$ as a function of the incident angle θ of the fundamental beam can be written as follows [10, 11]:

$$I_{2\omega}^{\text{m-p}}(\theta) \propto C \, I_\omega^2 \, (2\pi L/\lambda)^2 \, (d_{\text{eff}}^{\text{m-p}})^2 \, \sin^2\Psi/\Psi^2 \qquad (2)$$

Where $I_{2\omega}^{\text{m-p}}$ is the p-polarized SH intensity induced by the m-polarized fundamental beam (i.e., s- or p-polarized); C is a parameter determined by the appropriate Fresnel transmission coefficients and the beam area; L is the layer thickness; $\Psi(\theta) = (2 \pi L/\lambda) \, (n_\omega$

$\cos\theta_\omega - n_{2\omega} \cos\theta_{2\omega}$), where n_ω and $n_{2\omega}$ are the refractive indices at the fundamental and SH frequencies in the layer; θ_ω and $\theta_{2\omega}$ are the refractive angles of fundamental and SH waves determined by $\sin\theta = n_\omega \sin\theta_\omega$ and $\sin\theta = n_{2\omega} \sin\theta_{2\omega}$, respectively; d_{eff}^{m-p} are the effective second-order nonlinear optical coefficients for appropriate polarization combinations. For the transmitted SH intensity the effective second-order nonlinear optical coefficients are:

$$d_{eff}^{s-p} = d_{31} \sin\theta_{2\omega} \qquad (3)$$
$$d_{eff}^{45-p} = d_{15} \sin\theta_\omega \qquad (4)$$
$$d_{eff}^{p-p} = d_{15} \cos\theta_{2\omega} \sin2\theta_\omega + d_{31} \cos^2(\theta_\omega) \sin\theta_{2\omega} + d_{33} \sin^2(\theta_\omega) \sin\theta_{2\omega} \qquad (5)$$

The coefficients d_{15}, d_{31}, and d_{33} can be determined by measuring the SH intensity as a function of the incident angle of the fundamental beam for the above mentioned polarization combinations and comparing it to the SH intensity of a reference quartz plate.

Figure 1 shows the intensity of the transmitted p-polarized SH signal ($\lambda_{2\omega} = 532$ nm) from a 1 µm thick GaN layer as a function of the incident angle of p- and s-polarized fundamental beam for the back excitation. The measurements were carried out with an angular resolution of 1° (a) and 0.1° (b) per step, respectively. In figure 1(b) fine oscillations on the curve are clearly seen. Similar oscillations were also observed in SH angular

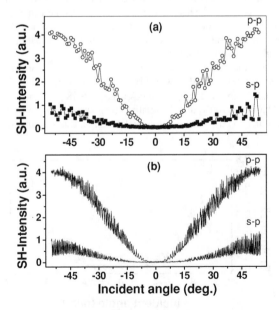

Figure 1. *Measured p-polarized second-harmonic intensity as a function of the incident angle of the p- and s-polarized fundamental beam. The angular resolutions are 1° (a) and 0.1° (b) per step, respectively.*

dependencies measured for front excitation. The angular dependence of the p-polarized second-harmonic intensity measured for the back excitation with an angular step of 0.01° shows the modulated pattern of the SH-signal in more detail (see figure 2).

The obtained results show that the fine oscillations can be resolved only with an angular step of 0.1° or smaller. It is of interest to note that the observation of the fine oscillations made it possible to reveal the phase shift of π between the measured angular dependencies for the front and the back excitations.

Assuming that the sample possesses two strictly parallel faces (i.e., the sample acts as an interferometer) and taking into account only interference of the first two transmitted fundamental beams [12] we obtain in first approximation good fits (dotted line) to the experimental data depicted in figure 2. A good fit was obtained with $I_2(\omega) = 0.007\, I_1(\omega)$, where $I_1(\omega)$ is the intensity of the incomming fundamental beam and $I_2(\omega)$ is the intensity of the second transmitted fundamental beam after two reflections inside the sample.

The observed phase shift of π between two excitations is more clearly seen by measuring the angular dependence of TH intensity, because THG is allowed at normal incidence. Figure 3 shows the intensity of the transmitted p-polarized TH intensity ($\lambda_{3\omega} =$ 355 nm) as a function of the incident angle of the p-polarized fundamental beam for both back and front excitations. Good fits (solid lines) were obtained using the same model of the interference of two fundamental beams in the sample that was applied for the SH angular dependencies. It should be noted that neither a SH nor a TH signal from sapphire substrate was detected under the experimental conditions.

By comparing figures 2 and 3, one can see the equal number of oscillations for both SHG and THG responses. This and the presented fits suggest that the observed modulation of the nonlinear optical signal is caused by the interference of the fundamental beam in the sample. In our opinion, the observed oscillations are also indicative of a high quality of the sample.

Comparing the measured SH intensities for s-p and 45°-p polarization combinations (see equations 3 and 4) with SH Maker fringes of a reference z-cut quartz plate ($d_{11}(2\omega) = 0.335$ pm/V) [13], the coefficients $d_{15} = 5.48$ pm/v and $d_{31} = 5.46$ pm/V were deduced. The values of d_{15} and d_{31} relative to d_{33} were found by substituting the obtained values of d_{15} and d_{31} in equation 5, and making a least-squares fit of the

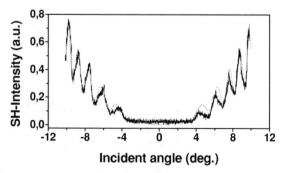

Figure 2. *Measured p-polarized second-harmonic intensity as a function of the incident angle of the p-polarized fundamental beam. The angular resolution is 0.01° per step. The solid line is experimental, the dotted line is a fit.*

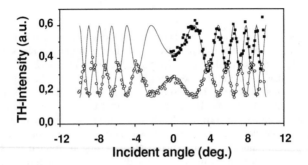

Figure 3. *Measured p-polarized third-harmonic intensity as a function of the incident angle of the p-polarized fundamental beam for back (solid squares) and front (open circles) excitations, respectively. The angular resolution is 0.1° per step. Solid lines show fits.*

obtained angular SHG dependence for p-p polarization combination with the use of equation 2. We found that d_{33}/d_{15}= -2.02 and d_{33}/d_{31}= -2.03, and consequently d_{33} = -11.07 pm/V. The obtained values are in good agreement with the ones reported in Refs. 3 and 4.

Figure 4 shows the SH polarization dependence for front excitation at different incident angles of the pump beam. A function of the type $(a \cos^2\varphi + b \sin^2\varphi)^2$ gives a good fit to the curves indicating that GaN is isotropic in the plane of the film. The φ is polarizaion angle of the fundamental beam (φ=0 and φ=90° correspond to the s- and p-polarized fundamental beam, respectively); the parameters a and b are determined by appropriate coefficients d and the transmission factors of fundamental and SH radiations.

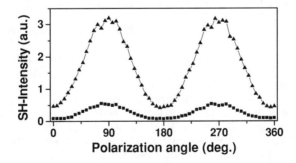

Figure 4. *Measured p-polarized second-harmonic intensity as a function of polarization angle of the fundamental beam at 10° (solid squares) and 30° (solid triangles) incident angles of the pump beam, respectively.*

CONCLUSIONS

In conclusion, the obtained results confirmed the wurzite structure of the MOCVD GaN layers, with the optical c-axis oriented perpendicular to the sapphire substrates. The measured d_{33} is 33 times d_{11} of quartz. Using a high angular resolution we were able to observe fine oscillations in the nonlinear optical response of GaN. The analysis of the experimental results shows that these oscillations are caused by the interference of the fundamental beam in the sample. It is quite neccessary to take these oscillations into account by analyzing the experimental data obtained from the measurements of SHG (THG) angular dependencies. In experiments with poor angular resolution the interference pattern cannot be resolved and might be wrong interpreted as a simple scattering of the measured data. In this case, the determined values of the nonlinear optical coefficients of thin films are ambiguous.

ACKNOWLEDGEMENTS

I.M.T. and I.V.K. gratefully acknowledge the support by the Alexander von Humboldt Foundation. This work was partially supported by the NATO Scientific Division under Grant HTECH.LG 961399 and ONR contract No. N00014-92-J-1552.

REFERENCES

1. *Properties, Processing and Applications of Gallium Nidride and Related Semiconductors,* edited by J. H. Edgar, S. Strite, I. Akasaki, H. Amano, and C. Wetzel (EM 023, 1999), p. 680.
2. H. Y. Zhang, X.H. He, Y. H. Shih, M. Schurman, Z. C. Feng, and R. A. Stall, Appl. Phys. Lett. **69**, 2953 (1996).
3. J. Miragliotta, D. K. Wickenden, T. J. Kistenmacher, and W. A. Bryden, J. Opt.Soc.Am. B **10**, 1447 (1993).
4. W. E. Angerer, N. Yang, A. G. Yodh, M. A. Khan, and C. J. Sun, Phys. Rev. B **59**, 2932 (1999).
5. J. Miragliotta, and D. K. Wickenden, Phys. Rev. B **50**, 14960 (1994).
6. A. Yariv and P. Yeh, *Optical Waves in Crystals* (Wiley, New York, 1984).
7. B. F. Levine, Phys. Rev. B **7**, 2600 (1973).
8. J. L. Hughes, Y. Wang, and J. E. Sipe, Phys. Rev. B **55**, 13630 (1997).
9. T. Ishidate, K. Inoue, and M. Aoki, Jap. J. Appl. Phys. **19**, 1641 (1980).
10. J. Jerphagon and S. K. Kurtz, J. Appl. Phys. **41**, 1667 (1970).
11. W. N. Herman and L. M. Hayden, J. Opt. Soc.Am. B **12**, 416 (1995).
12. M. Born and E. Wolf, *Principles of Optics* (Pergamon, 1964), pp. 288-353.
13. *Handbook of Laser Science and Technology*, edited M. J. Weber (CRC Press, Inc. Boca Raton, FL, 1986), p. 80.

Mat. Res. Soc. Symp. Vol. 595 © 2000 Materials Research Society

Radiative Recombination between Two Dimensional Electron Gas and Photoexcited Holes in Modulation-doped $Al_xGa_{1-x}N$/GaN Heterostructures

B.Shen*, T.Someya, O.Moriwaki, and Y.Arakawa
Institute of Industrial Science and Research Center for Advanced Science and Technology, University of Tokyo,
7-22-1 Roppongi, Minato-ku, Tokyo 106-8558, Japan
*Permanent address: Department of Physics, Nanjing University, Nanjing 210093, China

ABSTRACT

Photoluminescence (PL) of modulation-doped $Al_{0.22}Ga_{0.78}N$/GaN heterostructures was investigated. The PL peak related to recombination between the two-dimensional electron gases (2DEG) and photoexcited holes is located at 3.448 eV at 40 K, which is 45 meV below the free excitons (FE) emission in GaN. The peak can be observed at temperatures as high as 80 K. The intensity of the 2DEG PL peak is enhanced significantly by incorporating a thin $Al_{0.12}Ga_{0.88}N$ layer into the GaN layer near the heterointerface to suppress the diffusion of photoexcited holes. The energy separation of the 2DEG peak and the GaN FE emission decreases with increasing temperature. Meanwhile, the 2DEG peak energy increases with increasing excitation intensity. These results are attributed to the screening effect of electrons on the bending of the conduction band at the heterointerface, which becomes stronger when temperature or excitation intensity is increased.

INTRODUCTION

Investigation of radiative recombination between two-dimensional electron gases (2DEG) and photoexcited holes in heterostructures is very important for understanding band structures at heterointerfaces. Photoluminescence (PL) related to recombination of the 2DEG in $Al_xGa_{1-x}As$/GaAs heterostructures has been studied extensively[1-3]. This emission, which is denoted as the H-band, is only observed at temperatures lower than 14 K[2].

$Al_xGa_{1-x}N$/GaN heterostructures are not only the most promising structures for high temperature and high power electronic devices, but also increasingly used in optoelectronic devices. Unlike in an $Al_xGa_{1-x}As$/GaAs heterostructure, the piezoelectric polarization is very strong in an $Al_xGa_{1-x}N$ layer on GaN[4,5]. Also, the conduction band discontinuity is much larger at an $Al_xGa_{1-x}N$/GaN interface than at an $Al_xGa_{1-x}As$/GaAs interface for the same Al molar fraction, x. Therefore, PL related to the 2DEG at $Al_xGa_{1-x}N$/GaN heterointerfaces is expected to be observable at temperatures much higher than 14 K.

However, there have been few reports on the optical properties of $Al_xGa_{1-x}N$/GaN heterostructures until now. Bergman et al. observed a broad PL feature about 50 meV below the free exciton (FE) emission of GaN in an $Al_{0.3}Ga_{0.7}N$/GaN heterostructure and attributed it to recombination between the 2DEG and photoexcited holes[6]. However, the PL feature in their report was difficult to characterize accurately since its intensity was very weak. We think the problem is the rapid diffusion of photoexcited holes. Because of the strong piezoelectric field near the heterointerface, photoexcited holes diffuse into the flat-band region of GaN much more rapidly in an $Al_xGa_{1-x}N/GaN$ heterostructure than in an $Al_xGa_{1-x}As/GaAs$ one. Therefore, the probability of recombination between the 2DEG and photoexcited holes is much lower in the former than in the latter.

In this study, PL spectra of modulation-doped $Al_{0.22}Ga_{0.78}N/GaN$ heterostructures were studied. By incorporating a thin $Al_{0.12}Ga_{0.88}N$ layer into the GaN layer to suppress the diffusion of photoexcited holes, the intensity of the 2DEG PL peak was enhanced significantly.

EXPERIMENTAL DETAILS

Modulation-doped $Al_{0.22}Ga_{0.78}N/GaN$ heterostructures were grown by atmospheric pressure metal organic chemical vapor deposition (MOCVD). On the (0001) surface of a sapphire substrate, a nucleation GaN buffer layer was grown at 488 °C, followed by a 2.0-μm-thick unintentionally doped GaN (i-GaN) layer at 1071 °C. Then, a 100-nm-thick unintentionally doped $Al_{0.12}Ga_{0.88}N$ (i-$Al_{0.12}Ga_{0.88}N$) layer, a 20-nm-thick i-GaN layer and a 50-nm-thick Si-doped $Al_{0.22}Ga_{0.78}N$ (n-$Al_{0.22}Ga_{0.78}N$) layer were deposited, all at 1080 °C. The doping concentration of the n-$Al_{0.22}Ga_{0.78}N$ layer was 2.68×10^{18} cm^{-3}. Between the n-$Al_{0.22}Ga_{0.78}N$ and the i-GaN layers, a 3-nm-thick $Al_{0.22}Ga_{0.78}N$ (i-$Al_{0.22}Ga_{0.78}N$) spacer was incorporated. By means of Van der Pauw Hall measurements, the sheet concentration of the 2DEG at the $Al_{0.22}Ga_{0.78}N/GaN$ interface was determined to be 1.58×10^{13} cm^{-2} at room temperature and 1.30×10^{13} cm^{-2} at 77 K. The mobility of the 2DEG was 914 $cm^2/V.s$ at 300 K and 3.01×10^3 $cm^2/V.s$ at 77 K. The large difference between the mobilities at 300 K and 77 K indicated the exi₃tence of the 2DEG at the $Al_{0.22}Ga_{0.78}N/GaN$ interface.

Figure 1 shows the schematic layer structure and energy band diagram of a sample. It would be a typical modulation-doped single $Al_{0.22}Ga_{0.78}N/GaN$

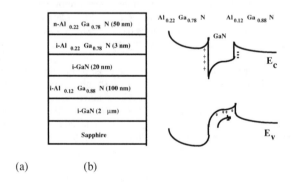

(a)   (b)

Figure 1. *(a) Schematic layer structure and (b) schematic energy band diagram of a modulation-doped $Al_{0.22}Ga_{0.78}N/GaN$ heterostructure in which an $Al_{0.12}Ga_{0.88}N$ confinement layer is incorporated.*

heterostructure if there was not the $Al_{0.12}Ga_{0.88}N$ layer. This layer is for confining photoexcited holes. Due to the direction of the piezoelectric polarization at the $GaN/Al_{0.12}Ga_{0.88}N$ interface, as shown in figure 1(b), there is not the triangular potential well for electrons in the conduction band edge at the $GaN/Al_{0.12}Ga_{0.88}N$ interface, and thus no 2DEG exist at this heterointerface, as has been confirmed by several reports[7,8].

PL measurements were performed on the samples at temperatures between 10 K and 120 K. The light source was a He-Cd laser of wavelength of 325 nm and the excitation power was 40 mW. A CCD camera cooled by liquid nitrogen was used as the detector.

EXPERIMENTAL RESULTS

Figure 2 shows the PL spectrum of a sample measured at 40 K. The inset shows the GaN FE emission. It is located at 3.493 eV and is much stronger than other peaks. The peaks at 3.408 eV and 3.314 eV are attributed to the one longitudinal optical (LO) phonon replica and the two LO-phonon replica of the FE emission, respectively[9]. The peak at 3.448 eV, marked with an arrow, is attributed to recombination between the 2DEG and photoexcited holes. This peak is not observed in bulk GaN. It is 45 meV below the FE emission. Its intensity decreases with increasing temperature. The peak disappears at a temperature a little higher than 80 K.

Figure 3 shows PL spectra of samples in which the $Al_{0.12}Ga_{0.88}N$ confinement layer is incorporated (curve a) and is not incorporated (curve b). The spectra were measured at 77 K. The intensity of the 2DEG PL peak, marked with an arrow, is much stronger in the sample were the $Al_{0.12}Ga_{0.88}N$ layer is incorporated than in the one without this layer.

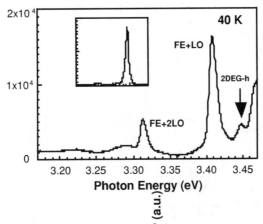

***Figure 2**. PL spectrum of a modulation-doped $Al_{0.22}Ga_{0.78}N/GaN$ heterostructure at 40 K. The peak at 3.448 eV, marked with an arrow, is attributed to recombination between the 2DEG and photoexcited holes. Inlet: PL spectrum showing the free excitons (FE) emission of GaN.*

Figure 4(a) shows the temperature dependence of the energy separation (ΔE) between the 2DEG PL peak and the GaN FE peak. ΔE decreases with increasing temperature. It is 46 meV at 10 K, and 43 meV at 60 K. Figure 4(b) shows the 2DEG peak energy as the function of the excitation intensity at 10 K. The peak energy increases approximately linearly with the logarithm of excitation intensity. The peak is located at 3.447 eV at an excitation intensity of 1.27 W/cm^2, and has shifted to 3.451 eV at an intensity of 127 W/cm^2. The FE peak does not shift with increasing excitation intensity.

DISCUSSION

From the above results, we know that the 2DEG PL peak in an $Al_{0.22}Ga_{0.78}N$/GaN heterostructure is observed at a temperature as high as 80 K, while the H-band in an $Al_{0.22}Ga_{0.78}As$/GaAs heterostructure is only observed at temperatures lower than 14 K[2]. This is due to the strong piezoelectric polarization in the $Al_{0.22}Ga_{0.78}N$ layer on GaN and the large conduction band discontinuity between $Al_{0.22}Ga_{0.78}N$ and GaN. The triangular potential well in the conduction band edge is much deeper at an $Al_xGa_{1-x}N$/GaN interface than at an $Al_xGa_{1-x}As$/GaAs one with the same Al molar fraction. Therefore, the 2DEG can remain confined at much higher temperature in an $Al_xGa_{1-x}N$/GaN heterostructure than in an $Al_xGa_{1-x}As$/GaAs one.

However, the strong electric field induced by the piezoelectric polarization makes photoexcited holes drift away very rapidly from the $Al_{0.22}Ga_{0.78}N$/GaN interface. Thus, in an $Al_xGa_{1-x}N$/GaN single heterostructure, the probability of recombination between the 2DEG and photoexcited is much lower than in an $Al_xGa_{1-x}As$/GaAs one. We think this is the main reason that the 2DEG PL peak is much weaker in an $Al_xGa_{1-x}N$/GaN heterostructure than in an $Al_xGa_{1-x}As$/GaAs one, even at very low temperatures[2,6].

In this study, by incorporating a thin $Al_{0.12}Ga_{0.88}N$ layer into GaN layer near the $Al_{0.15}Ga_{0.85}N$/GaN interface, as shown in Fig. 1(b), a potential barrier for holes was

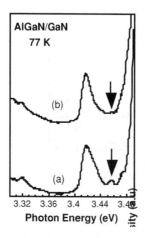

Figure 3. PL spectra at 77 K of samples in which the $Al_{0.12}Ga_{0.88}N$ confinement layer (a) is incorporated and (b) is not incorporated. The 2DEG PL peak is marked with an arrow.

formed. Photoexcited holes are confined near the GaN/$Al_{0.12}Ga_{0.88}N$ interface at low temperatures. The probability of radiative recombination between the 2DEG and photoexcited holes is enhanced significantly. Therefore, the intensity of the 2DEG PL peak is enhanced as shown in figure 3. Because of such enhancement, we can determine the peak energy position accurately. Investigation of the temperature and the excitation intensity dependencies of the peak also become possible.

Both the temperature and the excitation intensity dependencies shown in figure 4 can be explained in terms of the screening effect of carriers on the bending of the conduction band at the $Al_{0.22}Ga_{0.78}N$/GaN interface[10]. When temperature is increased, both the FE peak and the 2DEG peak shift slightly to lower energy because the band gap of GaN narrows. Meanwhile, the 2DEG density at the heterointerface increases with increasing temperature. This makes the triangular potential well in the conduction band edge at the heterointerface become a little shallower due to increased screening and a consequent reduction in the space-charge potential[10], which originates from both the conduction band discontinuity and the piezoelectric charges at the $Al_{0.22}Ga_{0.78}N$/GaN interface[11]. Thus, the red shift of the 2DEG peak is smaller than that of the FE peak as temperature increases. Hence, ΔE decreases with increasing temperature. Here, we think that a slight change in the Fermi level with increasing temperature may have no observable effect since its position is much higher than that of the ground energy level in the triangular potential well[6]. On the other hand, when excitation intensity is increased, the enhanced steady-state concentration of excess carriers results in reduced bending of the conduction band at the heterointerface due to increased screening. Thus, the triangular potential well in the conduction band edge at the heterointerface also becomes shallower. A blue shift of the 2DEG peak is observed with increasing excitation intensity.

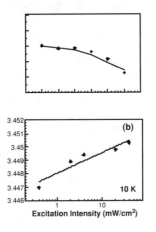

Figure 4. *(a) Temperature dependence of the energy separation (ΔE) of the 2DEG PL peak from the GaN FE emission, and (b) the excitation intensity dependence of the 2DEG PL peak energy at 10 K.*

CONCLUSION

PL peak related to recombination of between 2DEG and photoexcited holes in a modulation-doped $Al_{0.22}Ga_{0.78}N$/GaN heterostructure is located at 3.448 eV at 40 K, which is 45 meV below the GaN FE emission. The peak can be observed at temperatures as high as 80 K. The intensity of the 2DEG PL peak is enhanced by incorporating a thin $Al_{0.12}Ga_{0.88}N$ layer into the GaN layer to suppress the diffusion of photoexcited holes. The energy separation of the 2DEG peak from the FE emission decreases with increasing temperature. Meanwhile, the 2DEG peak energy increases approximately linearly with the logarithm of excitation intensity. These results are explained by the screening effect of carriers on the bending of the conduction band at the heterointerface becoming stronger when temperature or excitation intensity is increased.

ACKNOWLEDGMENTS

The authors would like to thank Mr. K.Tachibana for his help with MOCVD growth, and Dr. K.Hoshino for his help with PL measurements. This work was supported by the Research for the Future Program of the Japan Society for the Promotion of Science (Project No. JSPS-RFTF96P00201), Grant-in Aid by the Japan Ministry of Education, and the Foundation for the Promotion of Industrial Science in Japan.

REFERENCES

1. Y.R.Yuan, K.Mohammed, M.A.A.Pudensi, and J.L.Merz, Appl. Phys. Lett., **45(7)**, 739 (1984)
2. Y.R.Yuan, M.A.A.Pudensi, G.A.Vawter, and J.L.Merz, J. Appl. Phys., **58(1)**, 397 (1985)
3. J.P.Bergman, Q.X.Zhao, P.O.Holtz, B.Monemar, M.Sundaram, J.L.Merz, and A.C.Gossard, Phys. Rev. B, **43(6)**, 4771 (1991)
4. F.Bernardini, V.Fiorentini, and D.Vanderbilt, Phys. Rev. B **56**, R10024 (1997)
5..P.M.Asbeck, E.T.Yu, S.S.Lau, G.J.Sullivan, J.Van.Hove, and J.Redwing, Electron. Lett. **33**, 1230 (1997)
6.J.P.Bergman, T.Lundstrom, B.Monemar, H.Amano, and Akasaki, Appl. Phys. Lett., **69(23)**, 3456 (1996)
7. E.T.Yu, G.J.Sullivan, P.M.Asbeck, C.D.Wang, D.Qiao, and S.S.Lau, Appl. Phys. Lett., **71(19)**, 2794 (1997)
8. Peter Ramvall, Y.Aoyagi, A.Kuramata, P.Hacke, and K.Horino, Appl. Phys. Lett., **74(25)**, 3866 (1999)
9. D.Kovalev, B.Averboukh, D.Volm, B.K.Meyer, H.Amano, and I.Akasaki, Phys. Rev. B, **54(4)**, 2518
10. K.Ploog, and G.H.Dohler, Adv. Phys., **32**, 285 (1983)
11. R.Gaska, J.W.Yang, A.D.Bykhovski, M.S.Shur, V.V.Kaminski, and S.W.Soloviov, Appl. Phys. Lett., **72(1)**, 64 (1998)

Mat. Res. Soc. Symp. Vol. 595 © 2000 Materials Research Society

Spectroscopic Ellipsometry Analysis of InGaN/GaN and AlGaN/GaN Heterostructures Using a Parametric Dielectric Function Model

J. Wagner, A. Ramakrishnan, H. Obloh, M. Kunzer, K. Köhler, and B. Johs[1]
Fraunhofer-Institut für Angewandte Festkörperphysik,
Tullastrasse 72, D-79108 Freiburg, Germany, wagner@iaf.fhg.de;
[1]J. A. Woollam Co., Inc., 645 ´M´ Street #102, Lincoln, Nebraska 68508

ABSTRACT

Spectroscopic ellipsometry (SE) has been used for the characterization of AlGaN/GaN and InGaN/GaN heterostructures. The resulting pseudodielectric function spectra were analyzed using a multilayer approach, describing the dielectric functions of the individual layers by a parametric oscillator model. From this analysis, the dielectric function spectra of GaN, $Al_xGa_{1-x}N$ ($x \leq 0.16$), and $In_{0.13}Ga_{0.87}N$ were deduced. Further, the dependence of the $Al_xGa_{1-x}N$ band gap energy on the Al mole fraction was derived and compared with photoluminescence data recorded on the same material. The SE band gap data are compatible with a bowing parameter close to 1 eV for the composition dependence of the $Al_xGa_{1-x}N$ gap energy. Finally, the parametric dielectric functions have been used to model the pseudodielectric function spectrum of a complete GaN/AlGaN/InGaN LED structure.

INTRODUCTION

Reproducible growth of high-quality (AlGaIn)N heterostructures requires, because of the rather narrow growth parameter window, fast and efficient characterization of, e.g., layer thickness and composition. Spectroscopic ellipsometry (SE) is a nondestructive optical characterization technique which has been used successfully for the characterization of conventional III-V heterostructures [1-4]. For a quantitative analysis of SE data on group III-arsenide and -antimonide heterostructures, detailed modeling of the pseudodielectric function spectra has been performed employing a multilayer approach, incorporating parametric dielectric function models for the individual layers [2-4]. With respect to the group III-nitrides, SE data and model fits to the dielectric function spectra have been reported so far mostly for bulk-like GaN and AlGaN [5-10]. We have reported recently on the SE characterization of $In_xGa_{1-x}N$ ($x \leq 0.1$) layers on GaN, but no modeling of the pseudodielectric function spectra has been performed to extract the dielectric function spectra of the individual layers [11].

The aim of the present investigation was to characterize hexagonal (AlGaIn)N heterostructures by variable angle SE and to analyze the resulting pseudodielectric function spectra within the framework of a multilayer model based on parametric dielectric functions. Parameterized dielectric function spectra have been derived for GaN, $In_{0.13}Ga_{0.87}N$, and $Al_xGa_{1-x}N$ ($x \leq 0.16$), allowing a direct determination of the composition dependence of the $Al_xGa_{1-x}N$ band gap energy $E_G(x)$. There is an ongoing controversy regarding the deviation of $E_G(x)$ from a linear dependence on x. Recent values for the bowing parameter b, which describes the magnitude of parabolic nonlinearity, range from 0, as derived from photoreflectance measurements for $x \leq 0.2$ [12], to 1.33 eV, as obtained from an

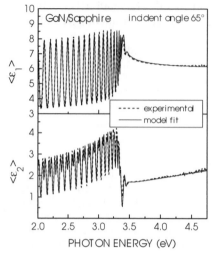

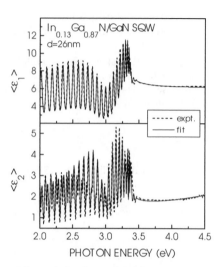

Figure 1. *Real <ε₁> and imaginary part <ε₂> of the pseudo-dielectric function spectrum of a 2.5 μm thick GaN layer on sapphire. Experimental data and model fits are indicated by dashed and full lines, respectively.*

Figure 2. *Real <ε₁> and imaginary part <ε₂> of the pseudo-dielectric function spectrum of a 26 nm thick $In_{0.13}Ga_{0.87}N$ layer embedded between GaN barrier layers. Experimental data and model fits are indicated by dashed and full lines, respectively.*

absorption study covering the full composition range $0 \leq x \leq 1$ [13]. Another recent paper suggests, based on photoluminescence (PL) data and an extensive survey of previous works, a band gap bowing parameter for $Al_xGa_{1-x}N$ of b=0.62 eV [14].

EXPERIMENT

The AlGaN/GaN and InGaN/GaN heterostructures as well as a complete GaN/InGaN/AlGaN LED structure used for the present study were grown by low-pressure MOCVD on c-plane 2″ sapphire substrates using a low-temperature GaN nucleation layer. Details on sample growth can be found in Ref. [15]. The AlGaN and In-GaN composition was determined by SIMS using appropriate standards calibrated by energy dispersive X-ray analysis (EDX).

Rotating analyzer variable angle SE was used to derive the room-temperature pseudodielectric function spectrum <ε> of the (AlGaIn)N heterostructures, covering the range of photon energies from 2 to 5 eV. Incident angles of 65° and 75° were used while the polarizer azimuth was kept constant at 30°. Both incident angles gave identical <ε> spectra, which confirms that the ellipsometry data correspond to an electric field vector perpendicular to the c-axis. For clarity of presentation only <ε> spectra recorded for an incident angle of 65° will be shown in the following. Samples were further analyzed by PL spectroscopy.

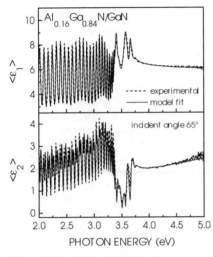

Figure 3. Real $\langle \varepsilon_1 \rangle$ and imaginary part $\langle \varepsilon_2 \rangle$ of the pseudo-dielectric function spectrum of a 600 nm thick $Al_{0.16}Ga_{0.84}N$ layer on GaN. Experimental data and model fits are indicated by dashed and full lines, respectively.

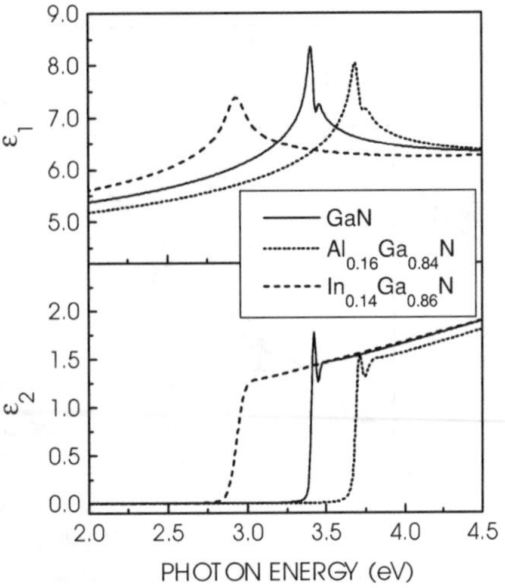

Figure 4. Real ε_1 and imaginary part ε_2 of the parametric dielectric function spectra of GaN, $Al_{0.16}Ga_{0.84}N$, and $In_{0.14}Ga_{0.87}N$ as deduced from multilayer parametric model fits to SE data.

RESULTS AND DISCUSSION

The real $\langle \varepsilon_1 \rangle$ and imaginary part $\langle \varepsilon_2 \rangle$ of the pseudodielectric function spectrum of a 2.5 μm thick GaN film on sapphire is shown in figure 1 together with a parametric model fit to the experimental data. All features present in the experimental spectra are reproduced by the model fit. To match the amplitude of the layer thickness oscillations in the transparency region of the GaN film, a layer thickness uniformity of 0.7% was assumed, which accounts for inhomogeneities of film thickness across the area probed by the ellipsometer as well as for imperfections of the GaN/sapphire interface, which also attenuate the interference oscillations. Further, a surface roughness of the GaN film of 0.9 nm was taken into account, which is compatible with atomic force micrographs which yield a RMS roughness of 2 to 3 nm at the most for the present samples.

Figure 2 shows the $\langle \varepsilon_1 \rangle$ and $\langle \varepsilon_2 \rangle$ spectra of an InGaN/GaN heterostructure, consisting of a 26 nm thick $In_{0.13}Ga_{0.87}N$ layer embedded between a lower, 2.7 μm thick, and an upper, 130 nm thick, GaN barrier layers. From a comparison with the $\langle \varepsilon \rangle$ spectrum of GaN on sapphire (see figure 1) the presence of the InGaN layer is readily detected via the occurrence of minima in both the $\langle \varepsilon_1 \rangle$ and the $\langle \varepsilon_2 \rangle$ spectrum at around 3 eV, which is the band gap energy of the InGaN. For the parametric model fit, also shown in figure 2, the GaN dielectric function was taken from the fit in figure 1 while the dielectric function spectrum of the InGaN layer was fitted. The InGaN layer thickness measured by SIMS was taken as an input parameter.

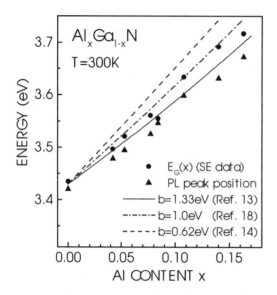

Figure 5. *Room-temperature $Al_xGa_{1-x}N$ band gap energy $E_G(x)$, as derived from fits to SE data, and PL peak position versus Al content x. Calculated $E_G(x)$ curves, computed for different bowing parameters b given in the figure, are also shown.*

Pseudodielectric function spectra $\langle\varepsilon_1\rangle$ and $\langle\varepsilon_2\rangle$ of an AlGaN/GaN heterostructure are shown in figure 3 along with a parametric model fit. The sample is composed of a 600 nm thick $Al_{0.16}Ga_{0.84}N$ layer grown on top of a 2.77 µm thick GaN layer. The AlGaN layer was covered by 6 nm thick GaN capping layer. Again, the GaN dielectric function was taken from the fit to the GaN-on-sapphire sample and the AlGaN dielectric function was fitted. The parametric model fit reproduces all features present in the experimental spectra. The AlGaN layer thickness deduced from the fit to the SE data of 609 nm is in excellent agreement with that of 594 nm determined by SIMS depth profiling.

The parametric dielectric function spectra ε_1 and ε_2 for GaN, $In_{0.14}Ga_{0.86}N$, and $Al_{0.16}Ga_{0.84}N$, deduced from the above shown fits, are plotted in figure 4. For both the GaN and the AlGaN dielectric function spectra an excitonic resonance is resolved superimposed on the spectrum arising from band-to-band continuum transitions. For the model fit, the excitonic resonance was represented by a Gaussian oscillator. For InGaN, in contrast, this resonance is broadened, most likely due to compositional inhomogeneity [16], and just a single peak is present in the ε_1 spectrum. The band gap energy of the pseudomorphic $In_{0.14}Ga_{0.86}N$ layer deduced from the fit to the SE data is 2.95 eV. This value lies within the range of recently published band gap data deduced from photoreflection measurements, which yield for InGaN with an In content of x=0.14, grown pseudomorphic on GaN, band gap energies of 2.91 eV [17] and 2.99 eV [18].

For a series of $Al_xGa_{1-x}N$/GaN heterostructures with $0 \le x \le 0.16$, the $Al_xGa_{1-x}N$ dielectric function was determined by fitting the SE data. From this analysis, the composition dependent $Al_xGa_{1-x}N$ band gap energy $E_G(x)$ was obtained, as depicted in figure 5. Further, PL peak positions recorded on the same series of samples are shown. The PL measurements were carried out at low temperatures (10 K) in order to resolve the AlGaN near band edge PL against the background from corresponding GaN related emission, which was difficult for room-temperature PL spectra in particular for low Al mole fractions. The resulting low-temperature PL peak energy was then down-shifted rigidly by 64 meV to account for the temperature induced band gap shift when going from 10 K to 300 K [19]. The data shown in figure 5 indicate a certain Stokes shift between $E_G(x)$ derived from the SE data and the PL peak position, which increases slightly with increasing Al mole fraction and thus increasing compositional disorder.

For comparison, also calculated $E_G(x)$ curves are shown, computed for representative bowing parameters of b=0.62 eV [14], b=1.0 eV [20], and b=1.33 eV [13]. For the room-temperature band gap energies of GaN and AlN values of 3.43 and 6.2 eV, respectively, were taken [13]. The composition dependence of the PL peak energy can be reproduced with a bowing parameter of b=1.3 eV, when allowing for a small rigid low-energy shift of the PL peak energy relative to the calculated $E_G(x)$. The $E_G(x)$ data obtained from the analysis of the SE data, in contrast, indicate a somewhat smaller bowing parameter closer to b=1.0 eV than b=1.33 eV. No fit for b to the present $E_G(x)$ data was attempted because of the limited composition range covered by the present set of samples.

CONCLUSION

Spectroscopic ellipsometry in conjunction with a parametric dielectric function based multilayer model has been applied to the analysis of (AlGaIn)N heterostructures. Model dielectric functions for GaN, AlGaN, and InGaN have been derived. Further, the composition dependence of the $Al_xGa_{1-x}N$ band gap energy (x≤0.16) has been deduced from the present SE data. Based on the parametric model dielectric functions, the measured pseudodielectric function spectrum of complete (AlGaIn)N QW LED structures and GaN/AlGaN layer sequences suitable for the fabrication of modulation doped field effect transistors could also be modeled, yielding fit values for the individual layer thicknesses in close agreement with those obtained by SIMS depth profiling. Thus the present study demonstrates the potential of spectroscopic ellipsometry for the analysis of layer thickness and composition in group III-nitride based device structures.

ACKNOWLEDGMENTS

Thanks are due to M. Maier and Ch. Hoffmann for performing the SIMS and EDX analyses and to U. Kaufmann for stimulating discussions and careful reading of the manuscript. Continuous interest and encouragement by G. Weimann is gratefully acknowledged. Work was supported by the German Ministry for Education and Research.

REFERENCES

1. C. Pickering, R. T. Carline, M. T. Emeney, N. S. Garawal, and L. K. Howard, *Appl. Phys. Lett.* **60**, 2412 (1992).
2. C. M. Herzinger, H. Yao, P. G. Snyder, F. G. Celli, Y.-C. Kao, B. Johs, and J. A. Woollam, *J. Appl. Phys.* **77**, 4677 (1995); C. M. Herzinger, P. G. Snyder, F. G. Celli, Y.-C. Kao, D. Chow. B. Johs, and J. A. Woollam, *ibid.* **79**, 2663 (1996).
3. U. Weimar, J. Wagner, A. Gaymann, and K. Köhler, *Appl. Phys. Lett.* **68**, 3293 (1996).
4. J. Wagner, J. Schmitz, N. Herres, G. Tränkle, and P. Koidl, *Appl. Phys. Lett.* **70**, 1456 (1997).

5. S. Logothetidis, J. Petalas, M. Cardona, and T. D. Moustakas, *Phys. Rev.* B **50**, 18017 (1994).
6. J. Petalas, S. Logothetidis, S. Boultadakis, M. Alouani, and J. M. Wills, *Phys. Rev.* B **52**, 8082 (1995).
7. G. Yu, H. Ishikawa, M. Umeno, T. Egawa, J. Watanabe, T. Jimbo, and T. Soga, *Appl. Phys. Lett.* **72**, 2202 (1998).
8. T. Kawashima, H. Yoshikawa, S. Adachi, S. Fuke, and K. Ohtsuka, *J. Appl. Phys.* **82**, 3528 (1997).
9. A. B. Djurisic and E. H. Li, Appl. Phys. Lett. **73**, 868 (1998); *J. Appl. Phys.* **85**, 2848 (1999).
10. A. B. Djurisic, A. D. Rakic, P. C. K. Kwok, E. H. Li, M. L. Majewski, and J. M. Elazar, *J. Appl. Phys.* **86**, 445 (1999).
11. J. Wagner, A. Ramakrishnan, D. Behr, H. Obloh, M. Kunzer, and K.-H. Bachem, *Appl. Phys. Lett.* **73**, 1715 (1998).
12. T. J. Ochalski, B. Gil, P. Lefebvre, N. Grandjean, M. Leroux, J. Massies, S. Nakamura, and H. Morkoc, *Appl. Phys. Lett.* **74**, 3353 (1999).
13. W. Shan, J. W. Ager III, K. M. Yu, W. Walukiewicz, E. E. Haller, M. C. Martin, W. R. McKinney, and W. Yang, *J. Appl. Phys.* **85**, 8505 (1999).
14. S. R. Lee, A. F. Wright, M. H. Crawford, G. A. Petersen, J. Han, and R. M. Biefeld, *Appl. Phys. Lett.* **74**, 3344 (1999).
15. H. Obloh, D. Behr, N. Herres, C. Hoffmann, M. Kunzer, M. Maier, S. Müller, W. Pletschen, B. Santic, P. Schlotter, M. Seelmann-E., K.-H. Bachem, and U. Kaufmann, *Proc. 2nd Int. Conf. Nitride Semicond.* (Tokushima, Japan, 1997), p. 258.
16. S. Chichibu, T. Azuhata, T. Sota, and S. Nakamura, *Appl. Phys. Lett.* **70**, 2822 (1997).
17. C. Wetzel, T. Takeuchi, S. Yamaguchi, H. Katoh, H. Amano, and I. Akasaki, *Appl. Phys. Lett.* **73**, 1994 (1998).
18. J. Wagner, A. Ramakrishnan, D. Behr, M. Maier, N. Herres, M. Kunzer, H. Obloh, and K.-H. Bachem, *MRS Internet. J. Nitride Semicond. Res.* **4S1**, G2.8 (1999).
19. G. Steude, B. K. Meyer, A. Göldner, A. Hoffmann, F. Bertram, J. Christen, H. Amano, and I. Akasaki, *Appl. Phys. Lett.* **74**, 2456 (1999).
20. Y. Koide, H. Itoh, M. R. H. Khan, K. Hiramatu, N. Sawaki, and I. Askasaki, *J. Appl. Phys.* **61**, 4540 (1987).

Mat. Res. Soc. Symp. Vol. 595 © 2000 Materials Research Society

Picosecond Photoinduced Reflectivity Studies of GaN
Prepared by Lateral Epitaxial Overgrowth

M. Wraback[1], H. Shen[1], C.J. Eiting[2†*], J.C. Carrano[2‡*], and R.D. Dupuis[2]
[1]U.S. Army Research Laboratory, Sensors and Electron Devices Directorate, 2800 Powder Mill Road, Adelphi, MD 20783, USA
[2]Microelectronics Research Center, Department of Electrical Engineering, University of Texas at Austin, Austin, TX 78712-1100, USA
*current address: ‡Photonics Research Center, Department of Electrical Engineering and Computer Science, U.S. Military Academy, West Point, NY 10996, USA; †WPAFB, OH 45433, USA

ABSTRACT

The pump-probe technique has been used to perform room temperature studies of the photoinduced changes in the reflectivity ΔR associated with exciton and carrier dynamics in GaN prepared by lateral epitaxial overgrowth. For resonant excitation of cold excitons, the ΔR decay possesses a 720 ps component attributed to the free exciton lifetime in this high quality material. For electrons with small excess energy (< 50 meV), the strong increase in the ΔR decay rate with decreasing excitation density suggests that screening of the Coulomb interaction may play an important role in the processes of carrier relaxation and exciton formation. The faster decay times at a given carrier density observed for hot (> 100 meV) electron relaxation are attributed to electron-hole scattering in conjunction with the screened electron-LO phonon interaction.

INTRODUCTION

Gallium Nitride (GaN) has become an important material for ultraviolet light emitters and detectors, as well as high power, high frequency electronic devices. It has been demonstrated that growth of GaN on sapphire by lateral epitaxial overgrowth (LEO) [1-3] greatly reduces the threading dislocation density in this material. A device lifetime of more than 10000 h at room temperature has been reported for cw operation of InGaN multiple quantum well laser diodes employing LEO GaN grown by metalorganic chemical vapor deposition (MOCVD) [1], and a significant decrease of p-n junction reverse leakage current in LEO materials has also been observed [3]. However, little is known about the room temperature dynamics of resonantly created excitons and free carriers with low excess energy crucial to an understanding of device performance in this high quality GaN.

In this paper we present a time-resolved pump-probe study of exciton and free carrier dynamics in LEO GaN in which photoinduced changes in reflectivity ΔR are monitored on a picosecond timescale. The ΔR transients were obtained as a function of excitation intensity for three cases: (i) resonantly created excitons; (ii) electrons with low excess energy; and (iii) electrons with sufficient excess energy to emit longitudinal optical (LO) phonons. The data obtained from these measurements provides information about exciton lifetimes in this high quality material, the dynamics of exciton formation and screening in the presence of cold electron-hole pairs, and the cooling of a hot electron distribution due to carrier-carrier scattering and the partially screened electron-LO phonon interaction.

EXPERIMENTAL CONSIDERATIONS

The sample employed in this study was a ~5 μm thick undoped GaN LEO film on an undoped GaN/sapphire substrate. 15 μm-wide SiO_2 stripes deposited by plasma enhanced chemical vapor deposition (PECVD) were patterned with 3 μm windows between the stripes. The growth of the LEO GaN upon this material was accomplished by rotating-disk MOCVD at a growth pressure of 100 torr and a growth temperature of 1070° C.

Visible/near infrared laser pulses derived from the signal beam of a 250 kHz Ti:sapphire regenerative amplifier-pumped optical parametric amplifier were compressed to less than 100 fs and frequency doubled to obtain a tunable source of ultraviolet pulses for frequency degenerate pump and probe measurements. The temporal evolution of the pump-induced change in the reflectivity of the probe pulse was monitored for various excitation photon energies and intensities. For excitation of free carriers using photon energies near the bandgap of GaN, ΔR is essentially proportional to the change in the real part of the dielectric function $\Delta\varepsilon_1$, which may be obtained from the spectral dependence of the change in the imaginary part of the dielectric function $\Delta\varepsilon_2$ by means of the Kramers-Kronig relation

$$\Delta\varepsilon_1(\omega) = \frac{2}{\pi} \int\limits_0^\infty \frac{\omega' \Delta\varepsilon_2(\omega') d\omega'}{\omega'^2 - \omega^2}, \tag{1}$$

where $\Delta\varepsilon_2$ is a measure of the photoinduced bleaching associated with the absorption saturation of band-to-band transitions [4]. The basic insight of this experiment is that while photoinduced transmission measurements primarily probe $\Delta\varepsilon_2$, only providing information about the sum of the electron and hole distribution functions at energies E_e and E_h defined by the probe frequency and the band structure [4], photoinduced reflectivity measurements are indicative of the integrated behavior of the carrier distribution functions. For GaN the electron effective mass (~0.2m_o) is much less than the hole effective mass (~2m_o) [5]. Since the ratio of the electron and hole excess energies is inversely proportional to the ratio of their effective masses in the parabolic band approximation, it follows that the electrons receive almost all of the excess energy from the excitation pulse. Moreover, the density of states is proportional to $m^{3/2}$, implying that the density of states near **k**=0 in the conduction band is much smaller than that in the valence band. Therefore, for a given density of electron-hole pairs, the contribution of the electron distribution to the photoinduced bleaching is much larger than that of the hole distribution. These observations suggest that our time-resolved photoinduced reflectivity measurements primarily probe the electron dynamics. For resonant excitation of excitons, one may discuss the changes in the real and imaginary parts of the dielectric function in terms of the bleaching of the exciton peak [6].

EXPERIMENTAL RESULTS

Figures 1,2, and 3 show ΔR as a function of time delay for pulses with center wavelengths of 364.5 nm, 357.5 nm and 347 nm, respectively. In all cases a positive pulse-width-limited rise in ΔR is observed at zero time delay. For excitation at 364.5 nm, the pump creates cold A(B) excitons, as determined from the cw luminescence spectrum.

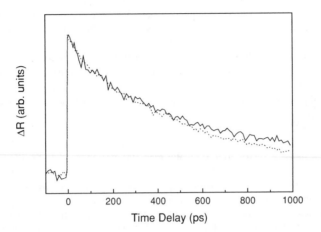

Figure 1 Normalized change in probe reflectivity ΔR as a function of time delay between pump and probe pulses with center wavelength $\lambda=364.5$ nm. Solid line: carrier density $n=4.4\times10^{18}$ cm^{-3}; dotted line: $n=4\times10^{19}$ cm^{-3}.

In this case the ΔR decay at low pump intensity (solid line) is well described by a biexponential decay of the form $\Delta R(t)=C_1 \exp(-t/\tau_1)+ C_2 \exp(-t/\tau_2)$, where t is the time delay between the pump and probe pulses and C_1, C_2, τ_1, and τ_2 are fitting parameters.

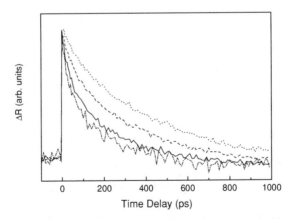

Figure 2 Normalized change in probe reflectivity ΔR as a function of time delay between pump and probe pulses with center wavelength $\lambda=357.5$ nm. Dotted line: carrier density $n=2\times10^{19}$ cm^{-3}; dashed line: $n=1\times10^{19}$ cm^{-3}; solid line: $n=3.3\times10^{18}$ cm^{-3}; dash-dotted line: $n=1.3\times10^{18}$ cm^{-3}.

The best fit is obtained for $\tau_1 \sim 55$ ps, $\tau_2 \sim 720$ ps, and $C_2/C_1 \sim 4$. These numerical values remain about the same for lower pump intensities. When the pump intensity is raised by nearly an order of magnitude (dotted line) the fast decay remains the same, while the slow decay becomes slightly faster ($\tau_2 \sim 550$ ps).

For excitation at 357.5 nm, the electrons possess less than 50 meV of excess energy above the conduction band edge. In this case the ΔR decays become faster as the carrier density is lowered. Using the time at which ΔR reaches half its maximum value as a measure of the decay time τ_{hm}, we find that τ_{hm} is about 5 times shorter (~ 46 ps) for the lowest carrier density data shown in figure 2 (dash-dotted line) than for the highest density data (~ 226 ps, dotted line).

Figure 3 shows ΔR data for excitation at 347 nm. In this case the initial excess energy of the electron distribution is greater than 100 meV, exceeding the LO phonon energy of 92 meV. As observed in figure 2, the ΔR decay becomes faster with decreasing excitation density, with τ_{hm} dropping by nearly a factor of 2, from 67 ps to 35 ps, when the carrier density is lowered by nearly an order of magnitude. The inset of figure 3 shows that the curve obtained for the lower carrier density in the main figure possesses an initial rapid decay with a 1.25 ps time constant. Upon reducing the carrier density by an additional factor of 2, this initial decay disappears (dotted line, inset).

DISCUSSION

For resonant excitation of A(B) excitons, photoinduced bleaching of the excitonic

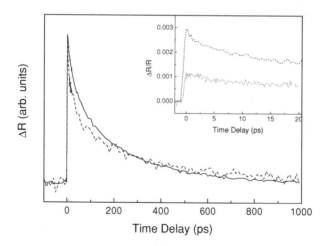

Figure 3 Normalized change in probe reflectivity ΔR as a function of time delay between pump and probe pulses with center wavelength $\lambda = 347$ nm. Solid line: carrier density $n = 4 \times 10^{19}$ cm^{-3}; dashed line: $n = 4.5 \times 10^{18}$ cm^{-3}. Inset: $\Delta R/R$ at short time delays for carrier densities of $n = 4.5 \times 10^{18}$ cm^{-3} (dashed line) and $n = 2.2 \times 10^{18}$ cm^{-3} (dotted line).

absorption creates a primarily positive ΔR [6]. The decay of this ΔR can be associated with the removal of free excitons by recombination or trapping at deep levels. The fact that the slow decay in figure 1 becomes only slightly faster when the intensity is increased by about an order of magnitude implies that the decay of ΔR for our sample is due to recombination rather than trapping, as saturation of deep level traps with increasing intensity would slow the decay of excitons [7]. Noting that the spot size of the probe is ~100 μm in diameter, comparison of the ratio of the decay components C_2/C_1 to the ratio of stripe size to window size suggests that the slow τ_2 decay may represent the exciton lifetime in the LEO material, while the fast τ_1 decay may be characteristic of the highly defective window region. Low temperature ΔR studies of resonantly excited exciton dynamics in LEO metalorganic vapor phase epitaxy (MOVPE) material [6] and time-resolved photoluminescence (TRPL) studies of hydride vapor phase epitaxy (HVPE) material [8] yield free exciton lifetimes of 375 ps at 60 K and 295 ps at 4 K, respectively. These results are in good agreement with theoretical predictions of the radiative lifetime [9]. At room temperature the exciton lifetime is longer (~720 ps) in our MOCVD-grown LEO material than that reported (530 ps) for the high quality HVPE sample, with the long lifetimes in both cases associated with a thermally induced increase in the center of mass kinetic energy of the excitons. While the τ_1 of 55 ps in our LEO sample is consistent with decay times observed in more defective GaN films, it is important to note that TRPL measurements capable of discriminating between the window and high quality materials [10] yielded a dominant decay time of 130 ps for both cases when excitation pulses of 4.64 eV photon energy were employed.

The fact that ΔR possesses a positive rise near zero time delay for excitation of free carriers at 357.5 nm or 347 nm is indicative of the rapid thermalization of the electrons through electron-electron scattering, which causes the distribution to be weighted toward energies below the center of the probe spectrum (equation 1). Since the results described above imply that saturation of defect-related traps does not play an important role in the exciton dynamics, we propose that the ΔR decays for 357.5 nm excitation are primarily associated with the Coulomb capture of electron-hole pairs to form excitons. As the carrier density increases, the Coulomb interaction is more efficiently screened and the decay of the electron distribution is retarded. Viewed in this way, the decay time τ_{hm} ~46 ps at the lowest excitation density ($\sim 10^{18} \text{cm}^{-3}$) becomes the exciton formation (capture) time at that carrier density.

Comparison of figures 2 and 3 shows that the ΔR decays are much faster at a given carrier density for 347 nm excitation than for the 357.5 nm data. For the 347 nm case, the hot electron distribution cools primarily by transfer of energy to both the cold photoexcited hole distribution through electron-hole scattering, and to the lattice through electron-LO phonon scattering. At high excitation density (solid line, fig. 3) the slow decay suggests that the electron-LO phonon interaction is strongly screened, and the τ_{hm} of ~67 ps represents the relaxation time associated with electron-hole scattering at this carrier density. The fact that the decay time is nearly halved when the carrier density is reduced by almost an order of magnitude (dashed line) implies that the partially screened electron-LO phonon interaction begins to play a role in the cooling process. This observation is supported by the data at short times (inset), for which the initial 1.25 ps decay at this carrier density is attributed to the screened electron-LO phonon interaction. The absence of this initial decay when the carrier density is further lowered to $\sim 2 \times 10^{18} \text{cm}^{-3}$ suggests that the screening is attenuated enough that the initial relaxation process becomes faster than the temporal resolution of our experiment (~200 fs), in agreement with the results of other researchers [11].

CONCLUSIONS

We have demonstrated that picosecond photoinduced reflectivity measurements can be used to obtain information about room temperature exciton and free carrier dynamics in LEO GaN. For resonant excitation of cold excitons, the ΔR decay possesses a 720 ps component attributed to the free exciton lifetime in this high quality material. For electrons with small excess energy (< 50 meV), the strong increase in the ΔR decay rate with decreasing excitation density suggests that screening of the Coulomb interaction may play an important role in the processes of carrier relaxation and exciton formation. The faster decay times at a given carrier density observed for hot (> 100 meV) electron relaxation are attributed to electron-hole scattering in conjunction with the screened electron-LO phonon interaction.

ACKNOWLEDGEMENTS

The authors thank NSF, ONR, and the State of Texas for partial support of this work at UT-Austin.

REFERENCES

1. S. Nakamura, M. Senoh, S. Nagahama, N. Iwasa, T. Yamada, T. Matsushita, H. Kiyoku, Y. Sugimoto, T. Kozaki, H. Umemoto, M. Sano, and K. Chocho, *Appl. Phys. Lett.* **72**, 211 (1998).
2. D. Kapolnek, S. Keller, R. Vetury, R. Underwood, P. Kozodoy, S. DenBaars, and U. Misra, *Appl. Phys. Lett.* **71**, 1204 (1997); T. Zheleva, O.-H. Nam, M. Bremser, and R. Davis, *Appl. Phys. Lett.* **71**, 2472 (1997).
3. C. Sasaoka, H. Sunakawa, A. Kimura, M. Nido, A. Usui, and A. Sakai, *J. Cryst. Growth* **189/190**, 61 (1998); P. Kozodoy, J. Ibbetson, H. Marchand, P. Fini, S. Keller, J. Speck, S. DenBaars, and U. Mishra, *Appl. Phys. Lett.* **73**, 975 (1998).
4. D.H. Auston, S. McAfee, C.V. Shank, E.P. Ippen, and O. Teschke, *Solid State Elec.***21**, 147 (1978).
5. J.S. Im, A. Moritz, F. Steuber, V. Härle, F. Scholz, and A. Hangleiter, *Appl. Phys. Lett.* **70**, 631 (1997).
6. S. Hess, F. Walraet, R.A. Taylor, J.F. Ryan, B. Beaumont, and P. Gibart, *Phys. Rev. B* **58**, R15973 (1998).
7. H. Haag, B. Hönerlage, O. Briot, and R. L. Aulombard, *Phys. Rev. B* **60**, 11624 (1999).
8. G.E. Bunea, W.D. Herzog, M.S. Ünlü, B.B. Goldberg, and R.J. Molnar, *Appl. Phys. Lett.* **75**, 838 (1999).
9. Y. Toyozawa, *Prog. Theor. Phys. Suppl.***12**, 111 (1959).
10. S.F. Chichibu, H. Marchand, M.S. Minsky, S. Keller, P.T. Fini, J.P. Ibbetson, S.B. Fleischer, J.S. Speck, J.E. Bowers, E. Hu, U.K. Mishra, S.P. DenBaars, T. Deguchi, T. Sota, and S. Nakamura, *Appl. Phys. Lett.* **74**, 1460 (1999).
11. K.T. Tsen, D.K. Ferry, A. Botchkarev, B. Sverdlov, A. Salvador, and H. Morkoc, *Appl. Phys. Lett.* **71**, 1852 (1997).

Mat. Res. Soc. Symp. Vol. 595 © 2000 Materials Research Society

The Effect of Nitrogen Ion Damage on the Optical and Electrical Properties of MBE GaN Grown on MOCVD GaN/sapphire Templates

Alexander P. Young and Leonard J. Brillson
Department of Electrical Engineering, The Ohio-State University, 2015 Neil Avenue, Columbus, OH 43210-1272, U.S.A.

Yoshiki Naoi and Charles W. Tu
Department of Electrical Engineering, University of California, San Diego, 9500 Gilman Dr., La Jolla, CA 92093-0407, U.S.A.

ABSTRACT

We have established a correlation between localized states responsible for mid-gap optical emission and film mobility of GaN grown under different nitrogen conditions. By imposing a deflector voltage at the tip of the plasma source, we varied the ion/neutral flux ratio to determine how N ions affect mid-gap luminescence and electrical mobility. Low energy electron-excited nanometer scale luminescence (LEEN) spectroscopy in ultrahigh vacuum (UHV) showed mid-gap emission intensities in the bulk that decreased in the ratio, 50 : 1.3 : 1 with increasing deflector voltage. Hall measurements indicated over a factor of two increase in mobility, and a factor of 8 decrease in residual charge density with increasing deflector voltage. The correlation of optical and electrical properties with a reduction in N ion flux suggests the primary role of native defects, such as N or Ga vacancies, in the mid-gap emissions.

INTRODUCTION

Advanced heterostructures based on GaN are increasingly viewed as the materials system of choice for implementing light emitting diodes, solar blind photo detectors, solid state lasers, and high power microwave devices. The performance of these devices is critically dependent upon control of point defects and dislocations in order to make full use of these new materials. While impressive achievements have already been made, the electrical and optical properties of the nitrides remain relatively poorly controlled. A deeper understanding of fundamental growth parameters is still necessary to realize the continued improvement of these devices based on the nitrides.

The nitrogen source of choice for the production of state-of-the-art GaN by molecular beam epitaxy MBE is the RF plasma source [1,2]. While GaN films made using this source are superior to others, the material quality is dependent on many variables including: RF power, RF frequency, nitrogen flow rate, and the ratio of group III to group V fluxes. Because of the many parameters involved when using a plasma source to grow GaN, it's difficult to isolate any one parameter as being of fundamental importance to the production of defect free, single crystal layers. Of particular concern in this article is the sensitivity of the growth to the ratio of N ions to neutral atomic species in the N flux. Different plasma sources are known to have a complex mixture of N components with widely varying amounts of ions, atomic N, and meta stable molecular N species[3,4]

By placing a perpendicular electrostatic potential at the tip of the plasma gun source, ions are deflected away from the growth path. In this way, we can directly observed the effect of the N ion/neutral ratio on material quality. In this work, we directly measured the impact on the electrical and luminescence properties of GaN by reducing the N ion/neutral flux ratio using an electrostatic voltage at the tip of the N source.

EXPERIMENT

We examined 1 μm thick GaN specimens grown by MBE on substrates composed of 3 μm of GaN grown by Metal-organic chemical vapor deposition (MOCVD) on sapphire. The substrate temperature during the GaN growth was 750 °C. A plasma source from SVT Associates, Inc. was used to produce the active nitrogen, in which the RF power and nitrogen flow rate were 250 W and 2.5 sccm, respectively, and the plate voltage at the tip of the plasma gun was varied between 0 and 700 V. The specimens were exposed to ambient conditions, then transferred back into UHV to perform LEEN. For the LEEN measurements, we used an electron gun with an energy range of 0.5 - 4.0 keV, emission current of 0.2 - 2 μA, and a spot size of 0.5 mm to generate optical emission at constant incident power. A Peltier cooled, S-20 photomultiplier collected the light over a spectral range of 1.40 - 3. 75 eV, and the detector output was obtained using standard lock-in techniques. The incident angle of the electron beam was set at 45°, enhancing the surface sensitivity of the measurement relative to normal incidence excitation. All LEEN measurements were made at room temperature.

RESULTS AND DISCUSSION

Ex situ atomic force microscopy (AFM) measurements were performed on the MBE grown GaN as well as the MOCVD GaN pseudo-substrate. All samples were extremely smooth with long, double steps pinned by threading dislocations observed on the MOCVD GaN surface, and stepped mounds 3 - 5 nm high on the surface of the MBE GaN. By applying the plate voltage, the film had noticeably less excess Ga droplets on the surface, and looked remarkably similar to the optimal MBE GaN grown under Ga rich conditions observed by Tarsa et al [2]. As shown in Table I, measurements of the specimens as a function of applied plate voltage show a factor of two increase in mobility and an almost order of magnitude decrease in residual electron density at the highest plate voltage. This indicates a significant improvement in the electrical quality of the material after reduction of the ion flux. While we cannot discount the effect of the MOCVD GaN, since we measure the overall sum of the conducting layers, the trend is clear, the reduction in ionized N has increased the conductivity of MBE epilayer.

Table I. Electrical properties of GaN grown with successively greater plate deflection voltage. As the plate voltage is applied, the mobility increases, and correspondingly, the residual charge density decreases. All Measurements were made at 300K.

Plate Voltage (V)	Charge Density (10^{17} cm^{-3})	Mobility (cm^2/V-sec)
MOCVD GaN	~1	600
0	8	300
500	5	500
700	1	640

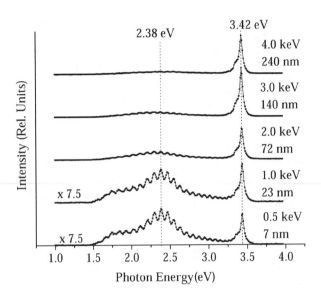

Figure 1. LEEN luminescence intensity as a function of electron beam penetration depth for GaN grown with plate voltage = 700 V. The right side of the figure indicates the beam voltage and the corresponding electron beam penetration depth in nm.

Based on calculations by Everhart and Hoff, a 0.5 keV incident electron beam energy penetrates 6 nm into GaN[5], increasing with beam energy at a rate proportional to $\approx E^{1.67}$. By increasing the beam energy to 4.0 keV, the electron beam penetrates to a probe depth of 240 nm giving us the capability of resolving the optical properties of the GaN from 7-240 nm in this work, well within the MBE GaN. From LEEN spectroscopy of a separate MOCVD GaN control specimen, we found intense mid-gap "yellow" emission centered at 2.17 eV well into the bulk of the film. Also from the spectra, we find that the mid-gap emission exhibits intensity oscillations indicative of nm-smooth interfaces, as confirmed by the AFM measurements. In addition to the "yellow" emission, near band edge (NBE) emission at 3.42 eV is observed at a intensity ratio relative to the "yellow emission of ~ 2:1 within the bulk, along with the presence of a peak at 3.35 eV, attributed to D-A recombination. From the depth-dependent nature of the LEEN spectra, we find that the "yellow" emission persists clear up to the surface, while the NBE is almost entirely quenched by non-radiative recombination in the near surface region.

Figure 1 shows the depth-dependent LEEN from the specimen grown with plate voltage = 700 V. In comparison to the control specimen, the mid-gap luminescence is almost entirely absent within the bulk of the film (4.0 keV), and substantial NBE emission remains even in the near surface region. Secondly, there are many more intensity oscillations over a larger energy range indicative of a significantly smoother morphology and a somewhat thicker film, even though the growth conditions were the same. Furthermore, a qualitative change in the nature of the mid-gap luminescence now occurs in the near surface region below 2.0 keV (72 nm penetration depth). We see evidence that the "yellow" emission is now comprised of two peaks, one at 1.8 eV and a second at 2.38 eV. While only the specimens with a plate voltage applied showed this distinct 2.38 eV peak, in all three cases investigated, (plate voltage = 0, 500, 700 V), mid-gap emission exceeded NBE emission at the surface (0.5 keV incident energy, 6 nm probe depth). Although we might be tempted to attribute the change in shape of the "yellow" band directly to the reduction in ionized N, other groups have suggested that there are multiple channels for recombination depending upon temperature as well as growth mechnism, not just MBE grown GaN[6]

In Figure 2, we show a close-up of the NBE region for GaN grown with three different plate voltages. From the figure, it is clear that even for a small plate voltage, D-A recombination is significantly reduced.

We also see that the effect has already saturated by a 500 V plate voltage even though the electrical mobility continued to rise with increasing plate voltage. It is unclear at this time what defect is the cause of the peak, but the "yellow" emission and the D-A feature reduction are correlated in this case.

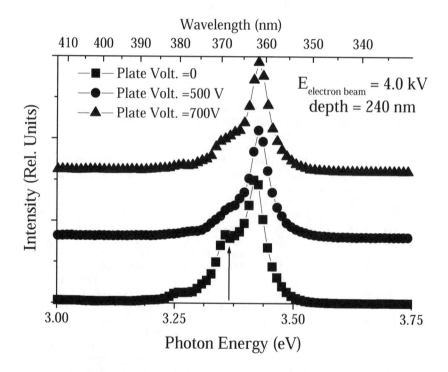

Figure 2. Close up of the band edge showing the reduction of the D-A peak as a function of plate voltage applied to the N beam flux during GaN growth. Spectra are offset for clarity.

In a direct comparison of the luminescence as a function of plate voltage, Figure 3 shows that the NBE at 3.42 eV increased as the mid-gap emissions at 1.8, 2.17, and 2.38 eV decreased. Also from figure 3, we see that GaN grown with no plate voltage emits in the yellow with an intensity over 50 times greater than GaN grown with a plate voltage of 700 V.

Considerable debate continues as to whether native defects or impurities are responsible for this mid-gap emission in GaN, and whether these emissions correlate with electrical properties. It is highly desirable to understand the atomic nature of these defects in order to prevent them from occurring on a consistent basis. It has been noted that the "yellow" emission is far less intense in GaN grown by hydride vapor epitaxy (HVPE) compared to GaN grown either by MOCVD or by MBE [7]. In the HVPE case no ionized species are involved, therefore we exclude ionized extrinsic species as the source of the "yellow" emission. While we cannot exclude the possibility that the incorporation rate of extrinsic impurities might be dependent upon the ionized N arriving at the growth front, we believe a much simpler, direct explanation is the ionized N itself is primarily responsible for the "yellow" emission in GaN. The LEEN spectra highlight the role of native defects such as N or Ga vacancies, possibly complexed with other impurities such as carbon or oxygen, in causing the "yellow" emission. Furthermore, the correlation of mid–gap spectral emission with electrical properties indicates the major role such defects have on overall transport properties.

CONCLUSIONS

We have established a correlation between localized states responsible for mid-gap optical emission and film electrical properties in MBE GaN grown using varying N active species. By reducing the N ion flux, we have observed a significant improvement in electrical mobility, a reduction in the residual charge density, a reduction in D-A pair recombination, and a reduction in mid-gap "yellow" luminescence by over a factor of fifty. This correlation between improved optical and electrical properties with the growth process suggests that the "yellow" luminescence is primarily due to intrinsic defects related to the specifics of growth process.

ACKNOWLEDGEMENTS

This work was sponsored in part by the NSF under contract No. DMR 971185, by the DOE under contract No. DE-FG0297ER45666, the ONR under contract No. N00014-00-1-0042, SVT Associates Inc., and UCMICRO.

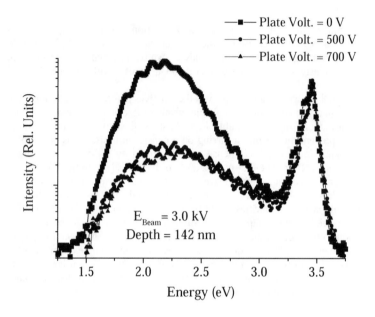

Figure 3. Comparison of LEEN spectra for GaN grown as a function of plate voltage. At an electron beam energy of 3.0 keV, the beam penetrates to a maximum depth of 140 nm, comparable to the penetration depth of a He-Cd laser.

REFERENCES

1. T.D. Moustakas, T. Lei and R.J. Molnar, Physica B 185, 36 (1993).
2. E. J. Tarsa, B. Heying, X. H. Wu, P. Fini, S. P. DenBaars and J.S. Speck, J. Appl. Phys. 82, 5472 (1997) and references there-in.
3. M. A. L. Johnson, J. D. Brown, N. A. El-Masry, J. W. Cook, Jr., J. F. Schetzina, H. S. Kong and J. A. Redmond, J. Vac. Sci. Technol. B 16, 1282 (1998).
4. T. H. Myers, M. R. Millecchia, A. T. Ptak, K. S. Ziemer and C. D. Stinespring, J. Vac. Sci. Technol. B 17, 1654 (1999) and references there-in.
5. T.E. Everhart and P. H. Hoff, J. Appl, Phys, 42, 5837 (1971).
6. R. Zhang and T. F. Kuech, Appl. Phys. Lett. 72, 1611 (1998).
7. R. J. Molnar, K.B. Nichols, P. Maki, E. R. Brown and I. Melngailis, Mater. Res. Soc. Symp. Proc. 378, 479 (1995).

Mat. Res. Soc. Symp. Vol. 595 © 2000 Materials Research Society

Dynamics of Anomalous Temperature-Induced Emission Shift in MOCVD-grown (Al, In)GaN Thin Films

Yong-Hoon Cho[1,3], G. H. Gainer[1], J. B. Lam[1], J. J. Song[1], W. Yang[2], and W. Jhe[3]
[1]Center for Laser and Photonics Research and Department of Physics
Oklahoma State University, Stillwater, OK 74078
[2]Honeywell Technology Center, 12001 State Highway 55, Plymouth, MN 55441
[3]Center for Near-field Atom-photon Technology and Department of Physics
Seoul National University, Seoul 151-742, Korea

ABSTRACT

We present a comprehensive study of the optical characteristics of (Al, In)GaN epilayers measured by photoluminescence (PL), integrated PL intensity, and time-resolved PL spectroscopy. For not only InGaN, but also AlGaN epilayers with large Al content, we observed an anomalous PL temperature dependence: (i) an "S-shaped" PL peak energy shift (decrease-increase-decrease) and (ii) an "inverted S-shaped" full width at half maximum (FWHM) change (increase-decrease-increase) with increasing temperature. Based on time-resolved PL, the S shape (inverted S shape) of the PL peak position (FWHM) as a function of temperature, and the much smaller PL intensity decrease in the temperature range showing the anomalous emission behavior, we conclude that strong localization of carriers occurs in InGaN and even in AlGaN with rather high Al content. We observed that the following increase with increasing Al content in AlGaN epilayers: (i) a Stokes shift between the PL peak energy and the absorption edge, (ii) a redshift of the emission with decay time, (iii) the deviations of the PL peak energy, FWHM, and PL intensity from their typical temperature dependence, and (iv) the corresponding temperature range of the anomalous emission behavior. This indicates that the band-gap fluctuation responsible for these characteristics is due to energy tail states caused by non-random inhomogeneous alloy potential variations enhanced with increasing Al content.

INTRODUCTION

Much interest has been focused on (Al, In)GaN alloys and their heterostructures, because their band gap energy varies between 6.2 and 1.9 eV at room temperature, and because of their potential applications such as red-ultraviolet (UV) light emitting devices [1,2], solar-blind ultraviolet detectors [3], and high power and high temperature devices [4,5]. It has been demonstrated that InGaN-based light emitting devices are highly efficient and have very low thresholds, and it is believed that their recent success is deeply related to the role of carriers localized in the InGaN active region. For the InGaN-based light emitting device structures, In alloy inhomogeneity and/or quantum-dot-like In phase separation have been proposed as the origin of the localized states [6-10], and an anomalous temperature dependence of the InGaN emission peak energy due to band-tail states was observed [11-13].

However, according to recent thermodynamic calculations, ternary AlGaN alloys are predicted to not have an unstable mixing region, and hence, no phase separation is expected, in contrast to InGaN and InAlN alloys [14]. Although understanding the emission mechanism and the role of the energy tail states in (Al, In)GaN alloys is very important for shorter wavelength light-emitting devices, the detailed emission properties of these materials have not been fully clarified. In this work, we report optical properties of $Al_xGa_{1-x}N$ epilayers ($x \leq 0.6$) in compare with GaN and $In_{0.18}Ga_{0.82}N$, as a function of temperature using photoluminescence (PL), integrated PL intensity, and time-resolved PL (TRPL).

EXPERIMENT

The $Al_xGa_{1-x}N$ thin films used in this study were grown by metalorganic chemical vapor deposition (MOCVD) on (0001) sapphire substrates. The samples were nominally identical aside from deliberate variations in the Al content x of the $Al_xGa_{1-x}N$ alloys, to investigate the influence of x. The growth temperature was about 1050 °C. Prior to $Al_xGa_{1-x}N$ growth, a thin ~ 5-nm-thick AlN buffer layer was deposited on the sapphire at a temperature of 625 °C. Triethylgallium, triethylaluminum, and ammonia were used as precursors in the $Al_xGa_{1-x}N$ growth. The $Al_xGa_{1-x}N$ layer thickness was about 1 μm. To evaluate the Al alloy composition and to check for ordering effects, the samples were analyzed with high-resolution x-ray diffraction (XRD) using Cu $K\alpha_1$ radiation. PL experiments were performed using the 244-nm line of an intracavity doubled cw Ar^+ laser as an excitation source. TRPL measurements were carried out using a picosecond pulsed laser system consisting of a cavity-dumped dye laser synchronously pumped by a frequency-doubled modelocked Nd:YAG laser for sample excitation and a streak camera for detection.

RESULTS AND DISCUSSION

Figure 1 shows the PL peak energy position (E_{PL}) as a function of temperature for $In_{0.18}Ga_{0.82}N$ [15], GaN, and $Al_xGa_{1-x}N$ epilayers with x = 0.17, 0.26, 0.33, and 0.6. The temperature-dependent PL peak shift for the GaN layer was consistent with the well-known energy gap shrinkage: $E_g(T) = E_g(0) - \alpha T^2/(\beta+T)$, where $E_g(T)$ is the band-gap transition energy at a temperature T, and α and β are known as the Varshni thermal coefficients [16]. On the other hand, the PL emission from $In_{0.18}Ga_{0.82}N$ and $Al_xGa_{1-x}N$ with rather high x did not follow the typical temperature dependence of the energy gap shrinkage. Instead, these ternary alloys clearly showed the "S-shaped" emission shift [initial redshift (region I), blueshift (region II), and final redshift (region III)] with increasing temperature] behavior, which is not seen in random homogeneous III-V alloys. For the $Al_xGa_{1-x}N$ epilayer with x = 0.17 (0.26, 0.33, 0.6), with increasing temperature up to T_I, where T_I is ~ 20 (50, 90, 150) K, an initial small decrease in E_{PL} was observed, followed by an increase in E_{PL} in the temperature range of T_I - T_{II}, where T_{II} is ~ 70 (110, 150, 225) K, and finally, E_{PL} decreased again as the temperature increased above T_{II}. This anomalous temperature-induced emission shift is very similar to the behavior previously observed in the $In_{0.18}Ga_{0.82}N$ epilayer (T_I and T_{II} were ~ 50 K and ~ 110 K, respectively, for the $In_{0.18}Ga_{0.82}N$ epilayer) [15], except that the amount of the redshift in region I of $In_{0.18}Ga_{0.82}N$ is larger than that of $Al_xGa_{1-x}N$ of comparable alloy content x, possibly due to a different nature (or degree) of potential fluctuations. Note that the corresponding

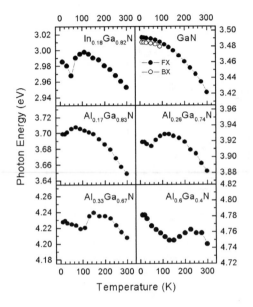

Figure 1. PL peak energy for $In_{0.18}Ga_{0.82}N$ [Ref. 15], GaN, and $Al_xGa_{1-x}N$ (x = 0.17, 0.26, 0.33, and 0.6) epilayers in the temperature range from 10 to 300 K. The emission peaks from $In_{0.18}Ga_{0.82}N$ and $Al_xGa_{1-x}N$ clearly show an anomalous S-shaped shift with increasing temperature. The free exciton (FX) and bound exciton (BX) curves are shown for GaN and follow the typical temperature dependence of the energy gap shrinkage.

temperature regions of the $Al_xGa_{1-x}N$ epilayers significantly depend on x: with increasing x, the characteristic temperatures T_I and T_{II} increase and the temperature regions I and II are extended into higher temperatures. Another unusual property of the PL spectra for these ternary alloys is that the FWHM shows an anomalous "inverted S-shaped" FWHM broadening (increase-decrease-increase) behavior with increasing temperature [15, 17].

Figure 2 shows Arrhenius plots of the normalized integrated PL intensities (I_{PL}) over the temperature range of 10 – 300 K. The main difference between the I_{PL} curves occurs in the temperature range showing the abnormal temperature dependence (i.e., regions I and II). An activation energy (E_a) estimated from the relationship $I_{PL} = I_0/[1+A \exp(-E_a/kT)]$ in the transition region II corresponds to the magnitude of effective potential fluctuations. The activation energies are 9.6 ± 1.5, 21.2 ± 1.2, and 44.6 ± 1.8 meV for the $Al_xGa_{1-x}N$ epilayers with x = 0.17, 0.26, and 0.33, respectively, reflecting more effective confinement with increasing x.

Figure 3 shows the temperature dependence of the TRPL lifetimes for the $Al_{0.17}Ga_{0.83}N$ and $Al_{0.33}Ga_{0.67}N$ thin films. The lifetimes were monitored at the peak energy (closed circles), lower energy side (open squares) and higher energy side (open triangles) of the PL peak position. For both samples, the measured lifetime increases with decreasing emission energy, and hence, the peak energy of the emission shifts to the low energy side as time proceeds. In temperature region I, the change in lifetime with temperature is very small and the difference between the lifetimes measured above, below, and at the peak energy is quite large, indicating that radiative recombination processes are dominant. As the temperature is further increased, the overall lifetime quickly decreases in region II and is almost constant in region III, reflecting a strong influence of non-radiative recombination processes. This is further evidenced by a quick

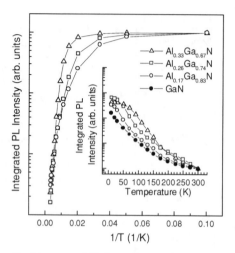

Figure 2. *Normalized integrated PL intensity as a function of temperature for the AlGaN-related emission of $Al_xGa_{1-x}N$ epilayers with Al content x = 0, 0.17, 0.26, and 0.33.*

decrease in the difference between lifetimes measured above, below, and at the peak energy in region II and no difference through region III. Using the relationship between the quantum efficiency $\eta(T)$ and the lifetime $\tau(T)$, we found that the transition from radiative to nonradiative recombination occurs at ~ 30 and ~ 80 K for the $Al_{0.17}Ga_{0.83}N$ and $Al_{0.33}Ga_{0.67}N$ thin films, respectively. Consequently, radiative recombination is dominant in region I, and the transition from radiative to nonradiative recombination occurs at about T_I, for both samples. In region II, in which a blueshift of the PL peak energy was observed, nonradiative recombination becomes dominant, so the lifetimes and their differences dramatically decrease. In region III, a typical temperature dependence of PL spectra was observed and no sudden change of lifetime occurs.

A similar anomalous temperature dependence for the PL peak energy was reported for ordered (Al)GaInP [18, 19] and disordered (Ga)AlAs/GaAs superlattices [20, 21]. Moreover, there have been some reports on the long-range ordering effect in molecular beam epitaxy-grown InGaN and AlGaN films [22]. To determine if ordered domains are in our $Al_xGa_{1-x}N$ alloys, XRD measurements were made. However, no (0001) XRD patterns were observed, indicating an absence of ordered domains in the $Al_xGa_{1-x}N$ alloys under investigation. Therefore, we rule out the possibility of the ordering effect in the $Al_xGa_{1-x}N$ alloys. This is quite surprising since the AlGaN ternary alloys investigated have neither ordering effects nor phase separations (according to theoretical prediction [14]), and most homogeneous ternary alloys do not show such an anomalous emission behavior. Therefore, we conclude that the anomalous emission is due to optical transitions from "localized" to "extended" band tail states, and that the band-gap fluctuation responsible for the anomalous behavior is enhanced with increasing x and can be attributed to energy tail states of inhomogeneous alloy fluctuations non-randomly distributed in the plane of the epilayers.

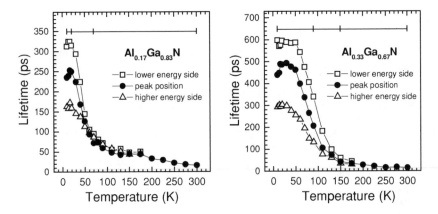

Figure 3. *Lifetime as a function of temperature for the emission in $Al_xGa_{1-x}N$ epilayers with x = 0.17 and 0.33. Note that the lower energy side of the PL peak has a longer lifetime than the higher energy side for $T < T_{II}$, while there is no difference between lifetimes monitored above (open triangles), below (open squares), and at (closed circles) the peak energy for $T > T_{II}$.*

CONCLUSIONS

We investigated the optical characteristics of MOCVD-grown (Al, In)GaN epilayers by means of PL, integrated PL intensity, and TRPL. We observed anomalous temperature-induced PL emission behavior for $In_{0.18}Ga_{0.82}N$ and $Al_xGa_{1-x}N$ epilayers: an "S-shaped" PL peak energy shift (decrease-increase-decrease) and an "inverted S-shaped" PL FWHM broadening (increase-decrease-increase) with increasing temperature. From the integrated PL intensity and TRPL measurements as a function of temperature, we found that the anomalous temperature-induced emission shift is deeply related to thermal population in localized energy tail states of alloy potential inhomogeneities in the $Al_xGa_{1-x}N$ epilayers. The following characteristics increase with increasing Al content in the AlGaN epilayers: (i) a Stokes shift between the PL peak energy and the absorption edge, (ii) a redshift of the emission with decay time, (iii) the deviations of the PL peak energy, FWHM, and PL intensity from their typical temperature dependence, and (iv) the corresponding temperature range of the anomalous emission behavior. Therefore, we attribute the anomalous emission behavior to the enhanced band-gap fluctuation in $Al_xGa_{1-x}N$ epilayers caused by an inhomogeneous spatial distribution of the Al content, and the degree of band-gap fluctuation increases with increasing x.

ACKNOWLEDGMENTS

The authors would like to acknowledge the contributions of Dr. Y. H. Kwon for the XRD measurements. This work was supported by AFOSR, ARO, ONR, DARPA, and CRI of MOST.

REFERENCES

1. S. Nakamura, M. Senoh, N. Iwasa, S. Nagahama, T. Yamada, and T. Mukai, Jpn. J. Appl. Phys. Part 2 **34**, L1332 (1995).
2. S. Nakamura, M. Senoh, S. Nagahama, N. Iwasa, T. Yamada, T. Matsushita, Y. Sugimoto, and H. Kiyoku, Appl. Phys. Lett. **69**, 4056 (1996).
3. B. W. Lim, Q. C. Chen, J. Y. Yang, and M. A. Khan, Appl. Phys. Lett. **68**, 3761 (1996).
4. Y. F. Wu, B. P. Keller, S. Keller, D. Kapolnek, P. Kozodoy, S. P. DenBaars, and U. K. Mishra, Appl. Phys. Lett. **69**, 1438 (1996).
5. X. H. Yang, T. J. Schmidt, W. Shan, J. J. Song, and B. Goldenberg, Appl. Phys. Lett. **66**, 1 (1995).
6. S. Chichibu, T. Azuhata, T. Sota, and S. Nakamura, Appl. Phys. Lett. **69**, 4188 (1996).
7. E. S. Jeon, V. Kozlov, Y. -K. Song, A. Vertikov, M. Kuball, A. V. Nurmikko, H. Liu, C. Chen, R. S. Kern, C. P. Kuo, and M. G. Craford, Appl. Phys. Lett. **69**, 4194 (1996).
8. P. Perlin, V. Iota, B. A. Weinstein, P. Wiśniewski, T. Suski, P. G. Eliseev, and M. Osiński, Appl. Phys. Lett. **70**, 2993 (1997).
9. Y. Narukawa, Y. Kawakami, M. Funato, Sz. Fujita, Sg. Fujita, and S. Nakamura, Appl. Phys. Lett. **70**, 981 (1997).
10. Y. Narukawa, Y. Kawakami, Sz. Fujita, Sg. Fujita, and S. Nakamura, Phys. Rev. B **55**, R1938 (1997).
11. Y. H. Cho, G. H. Gainer, A. J. Fischer, J. J. Song, S. Keller, U. K. Mishra, and S. P. DenBaars, Appl. Phys. Lett. **73**, 1370 (1998).
12. P. G. Eliseev, P. Perlin, J. Lee, and M. Osiński, Appl. Phys. Lett. **71**, 569 (1997).
13. K. G. Zolina, V. E. Kudryashov, A. N. Turkin, and A. E. Yunovich, MRS Internet J. Nitride Semicond. Res. **1**, Art, 11 (1996).
14. T. Matsuoka, MRS Internet J. Nitride Semicond. Res. **3**, 54 (1998).
15. Y. H. Cho, B. D. Little, G. H. Gainer, J. J. Song, S. Keller, U. K. Mishra, and S. P. DenBaars, MRS Internet J. Nitride Semicond. Res. **4S1**, G2.4 (1999).
16. Y. P. Varshni, Physica **34**, 149 (1967).
17. Y. H. Cho, G. H. Gainer, J. B. Lam, J. J. Song, W. Yang, and S. A. McPherson, Mat. Res. Soc. Symp. Proc. **572**, 457 (1999).
18. F. A. J. M. Driessen, G. J. Bauhuis, S. M. Olsthoorn, and L. J. Giling, Phys. Rev. B **48**, 7889 (1993).
19. K. Yamashita, T. Kita, H. Nakayama, and T. Nishino, Phys. Rev. B **55**, 4411 (1997).
20. A. Chomette, B. Deveaud, A. Regreny, and G. Bastard, Phys. Rev. Lett. **57**, 1464 (1986).
21. T. Yamamoto, M. Kasu, S. Noda, and A. Sasaki, J. Appl. Phys. **68**, 5318 (1990).
22. D. Korakakis, K. F. Ludwig, Jr., and T. D. Moustakas, Appl. Phys. Lett. **71**, 72 (1997).

Mat. Res. Soc. Symp. Vol. 595 © 2000 Materials Research Society

Time-Resolved Spectroscopy of InGaN

Milan Pophristic, Frederick H. Long, Chuong Tran[1], and Ian T. Ferguson[1]

Department of Chemistry, Rutgers University
610 Taylor Road, Piscataway, NJ 08854-8087
[1]EMCORE Corporation, 394 Elizabeth Avenue, Somerset, NJ 08873

ABSTRACT

We have used time-resolved photoluminescence (PL), with 400 nm (3.1 eV) excitation, to examine $In_xGa_{1-x}N$/GaN light-emitting diodes (LEDs) before the final stages of processing at room temperature. We have found dramatic differences in the time-resolved kinetics between dim, bright and super bright LED devices. The lifetime of the emission for dim LEDs is quite short, 110 ± 20 ps at photoluminescence (PL) maximum, and the kinetics are not dependent upon wavelength. This lifetime is short compared to bright and super bright LEDs, which we have examined under similar conditions. The kinetics of bright and super bright LEDs are clearly wavelength dependent, highly non-exponential, and are on the nanosecond time scale (lifetimes are in order of 1 ns for bright and 10 ns for super bright LED at the PL max). The non-exponential PL kinetics can be described by a stretched exponential function, indicating significant disorder in the material. Typical values for β, the stretching coefficient, are $0.45 - 0.6$ for bright LEDs, at the PL maxima at room temperature. We attribute this disorder to indium alloy fluctuations.

From analysis of the stretched exponential kinetics we estimate the potential fluctuations to be approximately 75 meV in the super bright LED. Assuming a hopping mechanism, the average distance between indium quantum dots in the super bright LED is estimated to be 20 Å.

INTRODUCTION

Recently there has been world-wide interest in the use of nitride semiconductors (e.g., GaN, InN, and AlN) for opto-electronic devices such as lasers and light-emitting diodes. The large changes in physical properties such as band gap, crystal structure, phonon energy, and electronegativity difference between GaN and GaAs, demonstrate that nitride semiconductors are fundamentally distinct from traditional III-V semiconductors. In spite of the impressive progress made in recent years [1] in the development of LEDs and lasers, significant work needs to be done in terms of the optimization of device performance. In order to achieve this goal, the physics underlying the operation of these devices must be better understood. Furthermore, new diagnostic techniques for the characterization of materials and devices will greatly aid in the long-term commercialization of this technology.

It has been recognized that under typical growth conditions there is a positive enthalpy for indium mixing in GaN. Electron microscopy and cathodoluminescence of InGaN has demonstrated the existence of nanometer and micron scale regions of high indium concentration [2]. It has been hypothesized that the nanoscale regions of high indium concentration are critical to LED operation [1]. We have previously used time-resolved photoluminescence to investigate indium concentration fluctuations, in InGaN/GaN multiple quantum wells and LEDs.[3] In this paper we discuss the result of time-resolved photoluminescence obtained from a set of LEDs with different quantum efficiency.

EXPERIMENT

The light emitting diodes were grown by metal organic chemical vapor deposition at EMCORE Corp. The average indium mole fraction was about 11 %. LEDs consist of ten layers of $In_{0.11}Ga_{0.89}N$, each 35 Å thick, and nine layers of GaN, each 45 Å thick. The whole structure was on c-plane sapphire with 3 microns of unintentionally doped (n-type $5x10^{16}$/cm^3) GaN as a substrate. The LEDs studied in this work were grown under slightly different conditions, leading to large changes in brightness. After final processing, the electroluminescence from samples was in order of 100 μW, 400 μW and ≥ 2 mW. Based on the electroluminescence efficiency we divide samples in three groups dim, bright, and super bright LEDs.

An amplified and doubled Ti-sapphire laser from Coherent Corporation operating at 250 kHz was used. TRPL measurements were performed with a Hamamatsu streak camera (model C5680). The excitation pulse was at 400 nm (3.10 eV) for time-resolved PL. Calibrated neutral density filters adjusted the excitation power used in the experiments. The typical response time was 60 ps and was determined by electrical jitter in the triggering electronics. The laser power used was 1.6 mW, 2.56 μJ/cm^2.

RESULTS AND DISCUSSION

Time-resolved PL data for dim, bright and super bright LED are shown in Fig.1. The lifetime of the emission for dim LED is quite short, 110 ± 20 ps at PL maximum, and the kinetics are not dependent upon wavelength. The short lifetime and low PL intensity of dim LED are due to fast non-radiative recombination processes, which are dominant in this case. This lifetime is short compared to bright and super bright LEDs, which we have examined under similar conditions. The kinetics of bright and super bright LEDs are clearly wavelength dependent, highly non-exponential, and are on the nanosecond time scale. The lifetimes are on the order of 1 ns for bright and 10 ns for super bright LED at the PL max. Such long times are consistent with carrier localization in regions of high indium concentration.

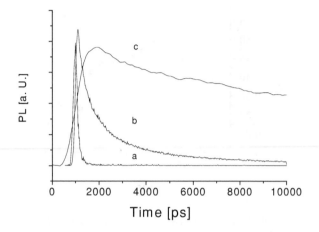

Figure 1. *TRPL from a) dim, b) bright and c) super bright LED.*

The non-exponential PL kinetics of bright and super bright LEDs, can be described by a stretched exponential function, equation 1, where β is between 0 and 1 and I(t) is the PL intensity as a function of time. τ is the stretched exponential lifetime and β is the stretching coefficient.

$$I(t) = I_0 \exp(-(\frac{t}{\tau})^{\beta})$$ (1)

There is growing evidence from our group [3] and others [4-6] that stretched exponential decays correctly describe the PL decays from InGaN under a variety of conditions. Stretched exponential decays are consistent with disorder and have been observed in wide variety of physical systems. In the case of InGaN, the dominant source of disorder is alloy disorder. Results of fitting the time-resolved PL from a super bright LED to equation one are shown in figures 2 and 3. Both the stretched exponential lifetime and stretching parameter are dependent upon the emission wavelength. The existence of stretched exponential decays is consistent with a broad distribution of lifetimes, which must be included in the theoretical modeling of LEDs. The trends observed in figures 2 and 3 are typical of a series of InGaN LEDs examined. The wavelength dependence of β, the stretching parameter for the super bright LED is shown in figure 4. We note that β is maximum at the PL maximum and decreases at both higher and lower emission energies.

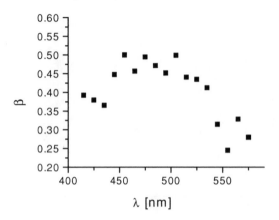

Figure 2. *Analysis of the time-resolved PL from super bright LED. The stretching parameter β is clearly wavelength dependent. The stretching parameter is largest at the PL maximum of 480 nm and decreases on both the high and low energy sides of the spectra.*

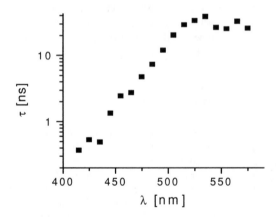

Figure 3. *Analysis of the time-resolved PL from super bright LED. The stretched exponential lifetime is observed to dramatically increase as the emission wavelength is varied from 420 to 500 nm. The PL maximum is at 480 nm.*

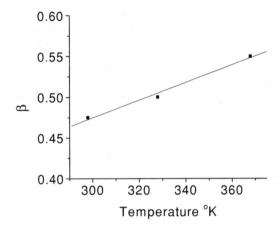

Figure 4.. *The temperature dependence of β for the super bright LED*

By analogy with previous simulations of localized excited states in other disordered semiconductors, the wavelength dependence of β and τ, Fig. 2, imply that excited state migration is important in InGaN LEDs. Without excited state migration, β would be expected to be wavelength independent.[3]

Figure 3 clearly demonstrates that the stretched exponential lifetime is strongly dependent upon wavelength or emission energy. The data can be successfully fit to an exponential dependence of τ on emission energy., where ω is the emission frequency and E_0 is a parameter that depends upon the potential fluctuations associated with the disorder.[7]

$$I(t) \approx \exp\left(-\frac{\hbar\omega}{E_0}\right) \quad (2)$$

For the super bright LED, E_0 was found to be 570 meV, which is much greater than the thermal energy. Similar results were obtained for bright and super bright several LEDs. It can be shown that equation 2 is consistent with a tunneling mechanism [7]. We conclude that the carriers produced in the TRPL experiments migrate between different indium quantum dots by a tunneling mechanism. By using a simple model, the average tunneling distance can be estimated to be approximately 20 Å [7]. Therefore the TRPL data confirms the existence and importance of nanometer scale indium alloy fluctuations on the operation of super bright LEDs.

If there is a distribution of barrier heights, one would expect the stretching parameter to vary linearly with temperature, with the slope being equal to the inverse of the width of the distribution of barriers.[8,9]

Temperature dependent measurements shown in figure 4 suggest that the potential fluctuations associated with the disorder are approximately 75 meV, which is much bigger than thermal energy at room temperature (25 meV). This result implies that the excited states are heavily localized in the LED. Furthermore, the magnitude of the potential fluctuations in the super bright LED are much larger than the previously examined 300 Å film, 25 meV.[8]

It is well known that piezoelectric fields are large in nitride semiconductors and can influence optical properties. In a previous study, we examined a series of MQWs with the same physical structure but very different PL spectra and dynamics.[3] Because the piezoelectric fields are determined by physical structure, the changes in PL spectra are attributed to the dominance of indium alloy fluctuations in our samples.

CONCLUSIONS

In summary we have used TRPL to examine wafers of InGaN LEDs before the final stages of processing. We have found that in the bright and super-bright LED examined, PL lifetimes at room temperature were on the nanosecond timescale. The PL kinetics were strongly dependent upon the emission wavelength and were well described by a stretched exponential. Both observations are strong experimental evidence for the importance of disorder in actual light emitting diodes. An analysis of the data based on this hypothesis confirms the existence and importance of nanometer scale indium alloy fluctuations on the operation of super bright LEDs.

ACKNOWLEDGEMENTS

We thank EMCORE Corp. for partial support of this work.

REFERENCES

[1] S. Nakamura and G. Fasol, *The Blue Laser Diode*. New York: Springer, 1997.
[2] M. D. McCluskey, L. T. Romano, B. T. Krusor, D. P. Bour, N. M. Johnson, and S. Brennan, *Applied Physics Letters* 72 (1998), p. 1730.
[3] M. Pophristic, F. H. Long, C. Tran, I. T. Ferguson, and R. F. Karlicek, Jr., *Applied Physics Letters* 73, (1998), p. 3550, and *Journal of Applied Physics* 86 (1999), p. 1114.
[4] A.Vertikov, I. Ozden, and A. V. Nurmikko, Journal of Applied Physics 86 (1999), p. 4697.
[5] T. Y. Lin, J. C. Fan, and Y. F. Chen, *Semiconductor Science & Technology* 14 (1999), p. 406.
[6] K. P. O'Donnell, *et al.*, phys. stat. sol (b) 216 (1999) p. 141.
[7] J. C. Vial, *et al., Phys. Rev. B* 45 (1992), p. 14171.
[8] M. Pophristic, F. H. Long, C. Tran, I. T. Ferguson, and R. F. Karlicek, Jr.,, *Proceedings of SPIE* 3621 (1999) p. 64.
[9] M. F. Schlesinger, Ann. Rev. Phys. Chem. 39 (1988), p. 269.

Correspondence should be addressed to F.H. Long
E-mail address: fhlong@rutchem.rutgers.edu

Mat. Res. Soc. Symp. Vol. 595 © 2000 Materials Research Society

Photoluminescence Enhancement and Morphological Properties of Carbon Codoped GaN:Er

M.E Overberg[1], C.R. Abernathy[1], S. J. Pearton[1], R. G. Wilson[2], and J. M. Zavada[3]
[1]Department of Materials Science and Engineering, University of Florida
Gainesville, FL 32611, U. S. A.
[2]Consultant
Stevenson Ranch, CA 91381, U. S. A.
[3]U. S. Army European Research Office
London, NW1 5 TH, U. K.

ABSTRACT

The surface morphology and the room temperature 1.54 µm photoluminescence (PL) intensity from GaN:Er grown by gas source molecular beam epitaxy have been investigated as a function of C concentration as introduced by CBr_4. Similar to previous results with increasing Er level, increasing the C concentration initially improved the surface smoothness as measured by atomic force microscopy (AFM) and scanning electron microscopy (SEM), with RMS roughness improving by a factor of seven over undoped GaN. The PL also improved dramatically. However, the highest amounts of C investigated produced a decrease in the PL as well as a roughening of the film surface. These effects indicate that the GaN:Er had reached its C solubility limit, producing an increased amount of defect induced nonradiative recombination.

INTRODUCTION

Recently, there has been an increasing amount of research in rare earth doped semiconductors for optoelectronics. As opposed to traditional light emitters that rely on conduction band edge recombination to produce photons, rare earth doped (RE) materials utilize electronic transitions involving the 4f shells of the rare earth element. With REs, the emission wavelength is host insensitive as the 4f shells are well shielded from the atomic bonding 5s and 5p shells. RE doping has been investigated in several semiconductor systems with various degrees of success [1-4]. While the emission wavelength is host material insensitive, the emission intensity is not. In particular, the larger the bandgap and the more ionic the semiconductor, the greater the RE luminescence [5,6]. The wide bandgap III-nitrides are therefore attractive as a host material, and offer the advantages of a high degree of structural and thermal stability. Already, several reports have been made of RE doping of GaN for both visible (Eu, Er, Tm) and IR emission (Pr, Er) [7-13]. Doping with Er is particularly attractive due to its 1.54 µm emission, which exploits the attenuation loss minima of silica fibers. It has been found that codopant impurities (O, C, F) can enhance the luminescence and efficiency of Si:Er [14-18], but only when introduced in the proper Er/Impurity ratio. While it has been shown that addition of elements such as C and O can enhance the IR emission from GaN:Er [8], there has as yet been no systematic investigation of the role of the concentration of these codopants in the luminescence and surface properties of the GaN films. In this work, the effect of C codoping introduced via CBr_4 on the structural and optical properties of

GaN:Er films grown on sapphire by gas source molecular beam epitaxy (GSMBE) has been investigated in an attempt to address these issues.

EXPERIMENTAL DETAILS

Films were grown by GSMBE in a modified Varian INTEVAC Gas Source Gen II on In-mounted (0001) Al_2O_3. A high temperature surface nitridation preceded all layer growth, and was performed at a temperature of 865 °C. A 10 nm low temperature AlN buffer layer ($T_g = 435$ °C) was then grown using dimethylethylamine alane (DMEAA) as the aluminum source. This was followed by a 1.2 µm GaN, GaN:Er, or GaN:Er:C layer, grown at a temperature of 750 °C. Solid source effusion cells provided the gallium (7N) and erbium (4N) fluxes, while the carbon was provided by the surface decomposition of CBr_4. The CBr_4 was injected into the growth chamber using an ultrapure helium (6N) carrier gas and a CBr_4 bubbler temperature of 0.0 °C. Reactive nitrogen for the nitridation and for the growth of both the AlN and GaN layers was provided by an SVT RF plasma source operating at 375 W of forward power and 3 sccm of N_2. The shuttered erbium cell was maintained at a temperature of 1250 °C, which corresponds to a dopant level of approximately 6×10^{21} cm^{-3} from SIMS analysis of previous results. The room temperature Er^{+3} photoluminescence at 1.54 µm (PL) was excited using the 514.5 nm line of an Ar ion laser and measured with an LN_2 cooled InGaAs photodetector. The surface morphology was characterized by atomic force microscopy (AFM) using a Digital Instruments Nanoscope III and by scanning electron microscopy (SEM) using a JEOL 6400.

DISCUSSION

The surface morphologies of the GaN:Er samples codoped with different C levels from the CBr_4 source are shown in Figure 1. SEM micrograph A shows the

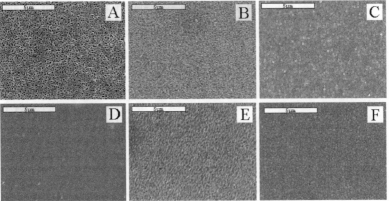

Figure 1: *SEMs of GaN and GaN:Er with a progression of CBr_4 doping (CBr_4 in sccm): (A) GaN, (B) GaN:Er, (C) GaN:Er ($CBr_4=3.2x10^{-3}$), (D) GaN:Er ($CBr_4=1.6x10^{-2}$), (E) GaN:Er ($CBr_4=7.9x10^{-2}$), and (F) GaN:Er ($CBr_4=1.6x10^{-1}$).*

undoped GaN surface. In micrographs B through D, as the CBr_4 flow, and therefore the C concentration, is increased, the round surface features decrease in size. A smooth morphology was obtained in GaN:Er that was codoped with roughly 1.3×10^{20} cm^{-3} of C. Upon further codoping, shown in micrograph E, it would appear that the round surface features then increase in size. At the highest codopant level, shown in micrograph F, they appear to decrease in size again. The C concentration in micrograph F was approximately 1.3×10^{21} cm^{-3}. From the SEM data, it would appear that the initial increase in film smoothness is due to a reduction in the GaN domain size. GaN grown by MBE often exhibits a columnar structure that is induced by the large lattice mismatch between the GaN and the sapphire.

As with the SEM, AFM measurements of the GaN:Er surface roughness indicate an initial improvement in morphology with increasing C level, as shown in Figure 2. The RMS surface roughness of the undoped GaN film was found to be approximately 73.5 Å. Upon the addition of Er, and with subsequent C addition, the surface roughness initially decreased by a factor of seven. However, with further C addition, the surface roughness increased dramatically. It is possible that at this C concentration ($\sim 6 \times 10^{20}$ cm^{-3}), the GaN has reached its solubility limit. The increase in surface roughness could then be due to the incorporation of C in the form of defects, clusters, or precipitates. High levels of crystal defects were observed in Si:Er when the material system was "overdoped" with O [17]. With further C addition, the surface roughness drops for a CBr_4 flow of roughly 0.16 sccm. As shown in the AFM in Figure 3, rough regions and very smooth regions characterize this sample. The solid line in Figure 2 represents the overall surface roughness of the whole sample, while the dotted line represents the surface roughness of the smooth regions only. Note that there is roughly an order of magnitude drop between the two values. This suggests that coarsening effects due to increased C incorporation is opposed by surface etching of the GaN by Br left over from the CBr_4 decomposition.

When the AFM surface scans, given in Figure 3, are examined in detail, it appears that the growth mode is altered for samples A through D. These results suggest that first the Er, then the C, may interfere with the surface migration of the reactant species, resulting in a smoother, less textured surface. The obvious roughening in micrograph E gives further credence to the theory that the GaN has reached its solubility

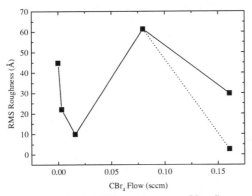

Figure 2: GaN:Er RMS roughness vs. CBr_4 flux.

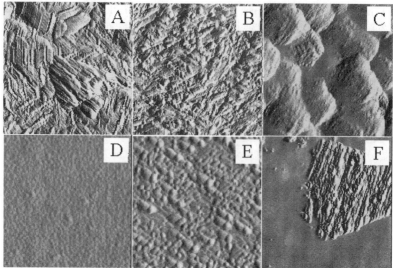

Figure 3: *AFM surface scans of GaN with a progression of Er and C doping (CBr₄ flux in sccm): (A) GaN, (B) GaN:Er, (C) GaN:Er ($CBr_4=3.15x10^{-3}$), (D) GaN:Er ($CBr_4=1.6x10^{-2}$), (E) GaN:Er ($CBr_4=7.97x10^{-2}$), and (F) GaN:Er ($CBr_4=1.6x10^{-1}$). Scan dimensions were 1 μm x 1 μm x 5 nm.*

limit for C, and that the roughening is due to an enhanced number of defects. Finally, in micrograph F, the defect related roughening is partially compensated by Br assisted surface etching.

Figure 4 shows that the growth rate of the GaN: Er at first increases slightly with C incorporation, but then decreases as more C is added. The decrease in the growth rate at higher CBr_4 fluxes is almost certainly due to the Br species left over from the thermal decomposition of the CBr_4 at the growth surface which parasitically etches the GaN. Enhanced Br levels (and etch rates) due to the higher C fluxes will produce lower overall growth rates for the same initial GaN growth rate. Similar behavior has been observed in GaAs, GaP and AlGaAs [21].

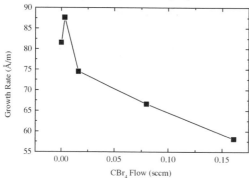

Figure 4: *GaN:Er growth rate vs. CBr_4 flux.*

The PL analysis, depicted in Figure 5, shows that the normalized PL intensity at first increases sharply with increased C content (over the non C-doped sample), and then decreases. This result indicates that for the Er concentration used ($\sim$6 x 10^{21} cm^{-3}), the optimum codoping concentration of C is approximately 1.3 x 10^{20} cm^{-3}. The initial increase in PL can be attributed to C affecting the local environment of the Er atoms. Normally, electric dipole radiation from 4f Er^{+3} transitions would be forbidden due to the parity selection rule. It has been found that a noncentrosymmetric crystal field leads to parity intermixing, resulting in a finite lifetime for radiative decay. Such a crystal field can be produced by the formation of Er-C complexes, allowing efficient Er^{+3} pumping. It has also been suggested that a reduction in deep levels in the GaN, as well as enhanced promotion of Er from the +2 state to the +3 state by Er-C centers will produce a luminescence enhancement [15,17]. The decrease in the PL luminescence at high CBr$_4$ flows can be attributed to the further increased number of C atoms surrounding the Er producing nonradiative decay centers in the GaN matrix material. The nonradiative sites could be produced as a result of C-C clustering, precipitates, C related defects, and C atoms not linked to Er atoms. Similar results (optimum codopant level, enhanced nonradiative decay) have been seen previously in work done with Si:Er codoped with O [14,16].

CONCLUSIONS

It has been shown that GaN:Er codoped with increasing levels of C initially exhibits similar changes in surface roughness as does GaN:Er doped with increasing Er levels. Morphology initially improves, but then roughens when the GaN:Er appears to reach its C solubility limit. Growth rate and RMS roughness measurements also show that parasitic etching from Br released by CBr$_4$ decomposition becomes important at high CBr$_4$ flow rates. Finally, the integrated 1.54 μm PL intensity reaches a maximum for an optimum C concentration, similar to work done with Si:Er:O. This indicates Er-C complex formation initially increases the PL, but further codoping creates defects that enhance nonradiative recombination.

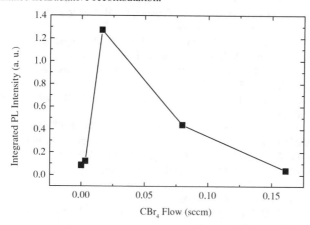

Figure 5: *Integrated PL intensity of the Er^{+3} emission at 1.54 μm vs. CBr$_4$ flow.*

ACKNOWLEDGEMENTS

The authors acknowledge the support of the U. S. Army Research Office under Contract No. DAAG55-98-1-0216. The assistance of S. M. Donovan with SEM is also greatly appreciated.

REFERENCES

1. X. Z. Wang and B. W. Wessels, Appl. Phys. Lett. **67**, 518 (1995).
2. H. Nakagome, K. Takahei, and Y. Homma, J. Crystal Growth **85**, 345 (1987).
3. J. Nakata, M. Taniguchi, and K. Takahei, Appl. Phys. Lett. **61**, 2665 (1992).
4. J. L. Benton, J. Michel, L. C. Kimerling, D. C. Jacobson, Y.-H. Xie, D. J. Eaglesham, E. A. Fitzgerald, and J. M. Poate, J. Appl. Phys. **70**, 2667 (1991).
5. P. N. Favennec, H. L'Haridon, D. Moutonnet, and Y. L. Guillo, Electron. Lett. **25**, 718 (1989).
6. J. M. Zavada and D. Zhang, Solid State Electron. **38**, 1285 (1995).
7. J. MacKenzie, C.R. Abernathy, S.J. Pearton, U. Hommerich, F. Ren, R.G. Wilson and J.M. Zavada, Fall Materials Research Society Meeting Proceedings Vol. **468**, 123-129 (1997).
8. J. MacKenzie, C.R. Abernathy, S.J. Pearton, U. Hommerich, K. Wu, R.N. Schwartz, R.G. Wilson and J.M. Zavada, J. Crystal Growth **175/176**, 84-88 (1997).
9. J. D. MacKenzie, C. R. Abernathy, and S. J. Pearton, Appl. Phys. Lett. **72**, 2710 (1998).
10. J. Heikenfeld, M. Garter, D. S. Lee, R. Birkhahn, and A. J. Steckl, Appl. Phys. Lett. **75**, 1189 (1999).
11. A. J. Steckl, M. Garter, D. S. Lee, J. Heikenfeld, and R. Birkhahn, Appl. Phys. Lett. **75**, 2184 (1999).
12. R. Birkhahn and A. J. Steckl, Appl. Phys. Lett. **73**, 2143 (1998).
13. R. Birkhahn, M. Garter, and A. J. Steckl, Appl. Phys. Lett. **74**, 2161 (1999).
14. M. Markmann, E. Neufeld, A. Sticht, K. Brunner, G. Abstreiter, and Ch. Buchal, Appl. Phys. Lett. **75**, 2584 (1999).
15. S. Coffa, F. Priolo, G. Franzó, A. Polman, S. Libertino, M. Saggio, and A. Carnera, Nucl. Instr. and Meth. in Phys. Res. B **106**, 386 (1995).
16. F. Priolo, G. Franzó, S. Coffa, A. Polman, S. Libertino, R. Barklie, and D. Carey, J. Appl. Phys. **78**, 3874 (1995).
17. F. Priolo, S. Coffa, G. Franzó, C. Spinella, A. Carnera, and V. Bellani, J. Appl. Phys. **74**, 4936 (1993).
18. G. Franzó, S. Coffa, F. Priolo, C. Spinella, J. Appl. Phys. **81**, 2784 (1997).
19. J. D. MacKenzie, C. R. Abernathy, S. J. Pearton, U. Hömmerich, X. Wu, R. Schwartz, R. G. Wilson, J. M. Zavada, J. Crystal Growth **175/176**, 84 (1997).
20. M. E. Overberg, J. Brand, J. D. MacKenzie, C. R. Abernathy, S. J. Pearton, and J. M. Zavada, *Proc. of the State-of-the-Art Program on Comp. Semi. XXX,* Electrochem. Soc. Proc. **99-4**, 195-200 (1999).
21. C. R. Abernathy, J. D. MacKenzie, S. J. Pearton, and W. S. Hobson, Appl. Phys. Lett. **66**, 1969 (1995).

Mat. Res. Soc. Symp. Vol. 595 © 2000 Materials Research Society

Photoluminescence and Cathodoluminescence of GaN doped with Pr

H. J. Lozykowski[1], W. M. Jadwisienczak[2], and I. Brown[3]

School of Electrical Engineering & Computer Science, and Condensed Matter & Surface Sciences Program, Ohio University, Stocker Center, Athens, OH, 45701
[3]Lawrence Berkeley Laboratory, University of California Berkeley, Berkeley, California 94720

ABSTRACT

In this paper we have reported the observation of visible photoluminescence (PL) and cathodoluminescence (CL) of Pr implanted in GaN. The implanted samples were given isochronal thermal annealing treatments at a temperature of 1100^0 C in NH_3, N_2, Ar_2, and in forming gas $N_2 + H_2$, at atmospheric pressure to recover implantation damages and activate the rare earth ions. The sharp characteristic emission lines corresponding to Pr^{3+} intra-$4f^n$ -shell transitions are resolved in the spectral range from 350 nm to 1150 nm, and observed over the temperature range of 12 K-335 K. The PL and CL decay kinetics measurement was performed for $^3P_1$, $^3P_0$ and $^1D_2$ levels.

INTRODUCTION

Rare earth (RE) doped semiconductors have been of considerable interest for possible applications in light emitting devices and for their unique optical properties. The rare earth luminescence depends very little on the nature of the host and the ambient temperature. Recently CL and PL emission has been obtained over the visible and near infrared spectrum range from GaN grown on sapphire by MOCVD, and doped by implantation with Sm, Dy, Ho, Er, and Tm [1,2a,b]. The visible PL and EL emission have been obtained from Pr and Eu doped GaN grown by MBE on sapphire and Si substrates [3a, b].

EXPERIMENTAL DETAILS

The GaN material used for this investigation was grown by MOCVD on the basal plane of 2 inch diameter sapphire substrates. The thicknesses of the epilayer was 1.8 μm, and electron concentrations 5×10^{16} cm^{-3}. The implanting ion beam was inclined at 7^0 to the normal of the GaN epilayer to prevent channeling. The GaN was high quality undoped n-type epilayer implanted at room temperature with praseodymium. Pr was implanted at three energies at doses chosen to give an approximation of a square concentration implant profile in the GaN epilayer (the projected range and peak concentration were ~ 40 nm, and ~ 3.1×10^{19} cm^{-3} respectively). Samples were given

---

[1]Electronic address: lozykows@bobcat.ent.ohiou.edu
[2] Permanent address: Institute of Physics, Laboratory of Solid State Optoelectronics, Nicholas Copernicus University, Grudziadzka 5, 87-400 Torun, Poland

postimplant isochronal thermal annealing treatments (duration 0.5 h and 1 h) at temperature 1100^0C in NH_3, N_2, Ar_2, and in forming gas (N_2+H_2). The emission spectra presented are obtained from samples annealed at 1100^0 C, which seems to be the optimal annealing temperature for RE ions incorporated as the luminescent center.

The photoluminescence spectra and kinetics measurements were performed using a He-Cd (325 nm) and N_2 (337 nm) lasers. The PL kinetics were measured using a double grating monochromator DIGIKROM CM112 assembled with Hamamatsu R928 (or R616) photomultipliers and a photon counting system with a turbo-multichannel scaler (Turbo-MCS, EG&G). The CL was excited by an electron beam incident upon the sample at a 45^0 angel from an electron gun (Electron gun system EK-2035-R (500V and 20 kV) which was in a common vacuum (of $\sim 5 \times 10^{-7}$ torr) with the cryostat.

ENERGY LEVELS, AND EXCITATION PROCESSES

The very important question is where the trivalent rare earth ions are incorporated in GaN: at substitutional sites on the metal sublattice and/or interstitial sites. In a hexagonal GaN crystal the Ga atoms occupy sites of symmetry C_{3v} (similar to Zn atoms in ZnS wurtzite phase) and two distinct high-symmetry interstitial positions also with C_{3v} symmetry[4a,b]. Recently, using the emission channeling (EC) technique, the lattice site occupations of RE elements in GaN were determined as a relaxed substitutional Ga-sites with an average relaxation of about 0.025 nm [5]. The rare earth ions can also aggregate especially at high concentration, as well as create complex centers in the presence of an anther ion e.g. oxygen . In this paper we studied the PL and CL and kinetics of GaN implanted with Pr. The Pr^{3+} free ion possesses $4f^2$ configuration which gives rise to a $^3H_4$ ground state and $^3H_{5,6}$, $^3F_{2,3,4}$, $^1G_4$, $^1D_2$, $^3P_{0,1}$, $^1I_6$, $^3P_2$, and $^1S_0$ excited states. If the crystal field symmetry at the Pr^{3+} site is known, then the number and symmetry of crystal field levels and the selection rules for transitions between these levels can be calculated. In C_{3v} crystal symmetry the states with $J = 0, 1, 2, 3, 4, 5, 6$, will split into 1(0), 1(1), 1(2), 3(2), 3(3), 3(4), 5(4), single(doubly) degenerate crystal field LSJ levels, respectively. The rare earth ions located at a specific crystal site are associated with characteristic optical transitions subject to the selection rules that are governed by the crystal symmetry of the site. It is generally accepted that rare earth impurities in III-V semiconductors create isoelectronic traps [6]. The outer electron configurations of RE^{3+} ions are the same ($5s^25p^6$). If the RE^{3+} ions replace the element from column III(Ga^{3+}) in GaN semiconductors that are isovalent concerning outer electrons of RE^{3+} ions, we believed that they create isoelectronic traps in III-nitrides (REI-trap). That conclusion is supported by the fact that the atomic covalent radii (ionic RE^{3+}) for all rare earths are bigger than atomic radii of Ga that they are replacing. Pauling's electronegativity of RE elements (1.1-1.25) is smaller than Ga (1.81), for which it substitutes. We have evidence that the RE ion in III-V semiconductors can occupy different sites (not only substitutional). They can create more complex centers involving other impurities or native defects. The experimental data shows that RE ions introduce electron or hole traps in III-V semiconductors, and we do not have any evidence that RE ions act as a donor or acceptor. The nature of the RE isoelectronic trap (electron or hole trap) in III-nitrides must be determined. Knowledge about the microscopic structure of RE centers in III-nitrides is crucial for understanding the excitation processes of $4f$-$4f$ transitions which in turn can determine the future of the RE dopants in optoelectronic applications.

The excitation processes of RE ions can be generally divided into two categories, direct and indirect excitation processes. The direct excitation process occurs in selective excitation of $4f^n$ electrons by photons (PL selective excitation) or in CL and EL by collision with hot electrons. Indirect excitation process occurs via transfer of energy to the $4f^n$ electron system from electron-hole pairs generated by photons with higher energy than the bandgap (PL excited above bandgap), injected in forward bias p-n junction, or generated by hot carriers in CL and EL). An excitation mechanism in CL and EL involves direct impact excitation of RE^{3+} ions by hot electrons, as well as an energy transfer from the generated electron-hole pairs or by impact excitation (or ionization) involving impurity states outside the $4f$ shell, with subsequent energy transfer to this shell. The most important excitation mechanism, from applications point of view, is the excitation of the RE ions by energy transfer process from electron-hole pairs generated in conduction and valence bands (by photons, hot electrons-CL) or injected in a forward bias p-n junction. There are three possible mechanisms of energy transfer. The first is the energy transfer process from excitons bound to structured isoelectronic centers to the core electrons. The second mechanism is the transfer of energy to the core electrons involving the structured isoelectronic trap occupied by electron (hole) and free holes (electrons) in the valence (conduction) band. The third mechanism is the transfer through an inelastic scattering process in which the energy of a free exciton near a RE structured trap is given to the localized core excited states. If the initial and final states are not resonant, the energy mismatch must be distributed in some way, e.g., by phonon emission or absorption.

LUMINESCENCE

In this paper, we present PL and CL spectra of Pr^{3+} in GaN in the spectral region of 350 to 1150nm. The luminescence of Pr^{3+} was excited indirectly, generating electron-hole pairs in GaN hosts by He-Cd, or N_2 lasers, and by electron beam excitation during CL measurement. Figure 1 shows two spectra CL (a) and PL (b) normalized to unity at peak (3) taken at 13 K and 330 K. The observed spectra shown in Fig.1, undoubtedly exhibit all the features of a trivalent rare earth ion.

Very weak luminescence was observed from $^3P_2$, $^1I_6$, $^3P_1$ levels (not

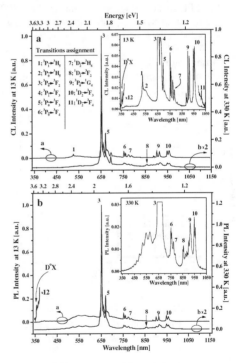

Fig.1 CL (a) and PL (b) spectra of GaN: Pr^{3+} recorded at 13 K and 330 K Inserts show magnified lower intensity emission lines of CL (a) and PL (b) spectra respectively. Numbers refer to transitions assignment from Fig. 1a.

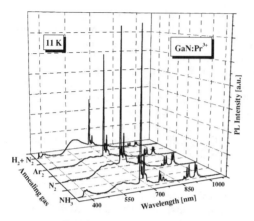

Fig.2 PL spectra of GaN: Pr^{3+} annealed at 1100^0 C during 1h in NH_3, N_2, Ar_2, and in forming gas recorded at 11 K under identical conditions

all marked in Fig.1), and strong luminescence from $^3P_0$, and $^1D_2$ levels. To show the weaker manifolds lines we enlarged the CL and PL spectra recorded at 13 K and 330 K as show in inset of Fig.1 a and b. A sharp CL emission line assigned to $^3P_1 \rightarrow ^3H_5$ transition (inset Fig.1 a) is only weakly seen on the short wavelength side of the interference modulated PL wide band (in spectral region ~ 400 - 600 nm) observed at 330 K in inset Fig.1 b. The nature of this wide band is not clear. The origin of it can be due to commonly observed "yellow band" overlapped partially with unresolved transition lines starting from $^3P_2$, $^1I_6$ $^3P_1$ $^3P_0$ levels. The assignments for the Pr^{3+} transitions in GaN have been made by comparison with data from other papers for the trivalent praseodymium ions in different crystals [7a,b].

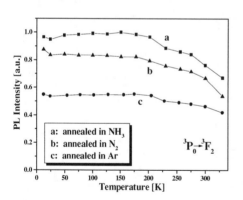

Fig.3 The PL intensity of GaN: Pr^{3+} annealed at 1100^0 C during 1h in NH_3, N_2, Ar_2, monitored at 653 nm ($^3P_0 \rightarrow ^3F_2$) versus temperature.

Figure2 shows PL spectra of GaN: Pr annealed at 1100^0 C during 1h in NH_3, N_2, Ar_2, and in forming gas ($N_2 + H_2$), at atmospheric pressure. The temperature dependence of intensity (area under curve) for dominant transition $^3P_0 \rightarrow ^3F_2$ plotted in Fig. 3 shows that the Pr^{3+} luminescence quenching is weak. The sample annealed during 1 h exhibits about two times stronger emission than the sample annealed during .5 h.

DECAY TIME MEASUREMENTS

The quenching of luminescence from rare earth ions doped semiconductors is very important for optoelectronic applications, and has been the subject of many investigations. That controls the quantum efficiencies and the luminescence lifetimes of rare earth ions. In our study we concentrate on measurements of lifetimes of the RE fluorescent levels as a function of temperature. The study of PL quenching of the singly doped (Pr^{3+}) GaN with temperature can provide information on interaction between like ions, ions with native defects and ions with unintentionally incorporated impurities. At low concentrations of praseodymium in GaN, the Pr^{3+} ions form a disordered system with wide range of inter-ion separations. For a resonant process an excitation can be transferred from the excited donor D^* to the neighboring donor ion in ground state (D) such that the second ion ends in the identical excited state D^*. We called that process a donor-donor (D-D) transfer. This migration of the excitation over the donors changes the trapping efficiency since all excited donors, including those which are initially far away from any trap, can now transfer energy to donors which have traps as near neighbors. Figure 4 shows an example of CL and PL decay (shown as semilog plots) of $^3P_0-^3F_2$ (653 nm) transitions at 12 K, with best fitting to double exponential decay for

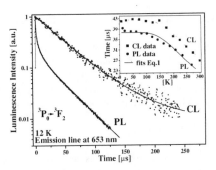

Fig.4 Luminescence decay observed from GaN: Pr^{3+} after N_2 laser pulse exc. (PL) and electron pulse exc. (CL) at 653 nm. Inset shows temperature dependence of $^3P_0$ decay time (■●) fitted to Eq.1 (doted, solid lines).

CL (solid line). Generally the decay times for transition $^3P_0-^3F_2$ for CL and PL are 42 μs and 38.9 μs respectively, with little change with temperature as can be seen from inset in Fig. 4 .The CL and PL decay kinetics of the $^3P_1 \rightarrow ^3H_5$ at 522 nm, $^3P_0 \rightarrow ^3F_2$ at 653 nm, $^3P_1 \rightarrow ^3F_3$ at 673 nm, $^3P_0 \rightarrow ^3F_4$, at 755nm, $^3P_0 \rightarrow ^1G_4$ at 913nm and $^1D_2 \rightarrow ^3F_3$ at 956 nm emissions have been measured and the experimental results of the decay time data are summarizes in table I for temperatures 12 K and 300 K. The early time part of PL decay curve in Fig.4 from (and seems to be true for all investigated PL lines) is one exponential while the late- time

TABLE 1. CL and PL decay times of GaN:Pr^{3+}

λ	Transition	CL decay time μs		PL decay time μs	
		12 K	300 K	12 K	300 K
522 nm	$^3P_1 \rightarrow ^3H_5$	33.6	----	----	----
653 nm	$^3P_0 \rightarrow ^3F_2$	42.0	27.8	38.9	40.6
670 nm	$^3P_1 \rightarrow ^3F_3$	43.0	29.5	36.4	15.4
755 nm	$^3P_0 \rightarrow ^3H_4$	----	----	27.0	10.8
913 nm	$^3P_0 \rightarrow ^1G_4$	46.5	33.6	39.9	48.0
956 nm	$^1D_2 \rightarrow ^3F_3$	46.3	27.7	----	----

luminescence decay is exponential with radiative decay time $\tau_o = 38.9$ μs suggesting that some of the Pr^{3+} ions are suggesting that some of the Pr^{3+} ions are in very isolated sites, or otherwise the cross relaxation between them is not allowed. The exponential component extracted from the experimental PL decay data of emission line at 653 nm portrayed for selected tempratures in inset Fig.4 is a radiative decay time τ_0 attributed to Pr^{3+} ions from isolated sites. The decay times τ of Pr^{3+} luminescence at 653 nm, is found to obey the activation formula in temperature range from 12 K to 270 K:

$$\frac{1}{\tau} = \frac{1}{\tau_0} + \beta \exp\left[\frac{-\Delta E}{kT}\right] \qquad (1)$$

where τ_0 is assumed to be the low temperature decay time and ΔE is the activation energy. The values obtained from the best fit (solid line in inset of Fig. 4) to the PL data are: $\tau_0 = 37.56$ μs, $\beta = 2.39 \times 10^5$ s^{-1} and $\Delta E = 533$ cm^{-1}, and for CL $\tau_0 = 42.4$ μs, $\beta = 1.2 \times 10^5$ s^{-1} and $\Delta E = 540$ cm^{-1} respectively (the ΔE is close to GaN TO phonon). The difference between CL and PL decay kinetics are probably related to different excitation processes and different centers involved in CL and PL emission (more experimental data and detailed analysis will be published elsewhere).

CONCLUSIONS .

In summary, it was demonstrated, that rare earth Pr^{3+} ions implanted into GaN after post-implantation isochronal annealing at 1100^0 C in N_2, at atmospheric pressure can be activated as luminescent centers emitting in the near UV, visible, and near infrared regions. The sharp characteristic emission lines corresponding to Pr^{3+} intra-$4f^n$-shell transitions are resolved in the spectral range from 350 nm to 1150nm, and observed over the temperature range of 12 K - 335K. The fluorescence decay curves of $^3P_1$, $^3P_0$ and $^1D_2$ levels emission were studied as a function of temperature. From this the characteristic time of the exponential decay of $^3P_0 \rightarrow {}^3F_2$ transition was determined Strong luminescence observed at low and room temperature using above bandgap photo excitation and electron beam, suggested that Pr in GaN can be effectively excited by forward bias p-n junction end utilized in LED and semiconductors lasers.

ACKNOWLEDGMENTS

The work was supported by BMDO Contract No. N00014-96-1-0782, the Ohio University Stocker Fund, the Ohio University CMSS Program, and OU Faculty Fellowship Leave.

REFERENCES

1.   A. J. Steckl, R. Birkhahn, App. Phys. Lett. **73**, 1700 (1999).
2.   a) H. J. Lozykowski, W. M. Jadwisienczak, and I. Brown, Appl. Phys. Lett.**74**, 1129 (1999); b) H. J. Lozykowski, W. M. Jadwisienczak, and I. Brown, Solid State Commun. **110,** 253 (1999).
3.   a) R. Birkhahn, M. Garter and A. J. Steckl, App. Phys. Lett. **74**, 2161 (1999); b) J. Heikenfeld, M. Garter, D. S. Lee, R. Birkhahn and A. J. Steckl, Appl. Phys. Lett. **75**, 1189 (1999).
4.   a) M.R. Brown, A.F. J. Cox, W. A. Shand and J. M. Williams, Adv.Guantum Electronics, **2**, 69 (1974),b) P. Boguslawski, E. L. Briggs and J. Bernholc, Phys. Rev. B **51**, 17 255 (1995).
5.   M. Dalmer, M. Restle, A. Stotzler, U. Vetter, H. Hofsass, M. D. Bremser, C. Ronning, R. F. Davis, and the ISOLDE Collaboration, MRS Proc. **482**, 1021 (1998).
6.   H. J. Lozykowski, Phys. Rev. B **48**, 17 758 (1993), and reference therein.
7.   a) G. H. Dieke: Spectra and Energy Levels of Rare Earth Ions in Crystals, ed H. M. Crosswhite and H. Crosswhite (Interscience Publishers Inc., 1968); b) J. P. M. Van Vliet and G. Blasse, Chem. Phys. Lett. **143**, 221 (1988).

Mat. Res. Soc. Symp. Vol. 595 © 2000 Materials Research Society

COMPARISON OF THE OPTICAL PROPERTIES OF Er³⁺ DOPED GALLIUM NITRIDE PREPARED BY METALORGANIC MOLECULAR BEAM EPITAXY (MOMBE) AND SOLID SOURCE MOLECULAR BEAM EPITAXY (SSMBE)

U. Hömmerich*,+, J. T. Seo*, J. D. MacKenzie**, C. R. Abernathy**, R. Birkhahn†, A. J. Steckl†, and J. M. Zavada‡
*Hampton University, Department of Physics, Hampton, VA 23668
**University of Florida, Dept. of Materials Science and Eng., Gainesville, FL 32611
†University of Cincinnati, Nanoelectronics Laboratory, Cincinnati, OH 45221
‡U.S. Army European Research Office, London, UK, NW1 5 TH
+E-mail: hommeric@jlab.org

ABSTRACT

We report on the luminescence properties of Er doped GaN grown prepared by metalorganic molecular beam epitaxy (MOMBE) and solid-source molecular beam epitaxy (SSMBE) on Si substrates. Both types of samples emitted characteristic 1.54 µm PL resulting from the intra-4f Er³⁺ transition $^4I_{13/2} \rightarrow {}^4I_{15/2}$. Under below-gap excitation the samples exhibited very similar 1.54 µm PL intensities. On the contrary, under above-gap excitation GaN: Er (SSMBE) showed ~80 times more intense 1.54 µm PL than GaN: Er (MOMBE). In addition, GaN: Er (SSMBE) also emitted intense green luminescence at 537 nm and 558 nm, which was not observed from GaN: Er (MOMBE). The average lifetime of the green PL was determined to be 10.8 µs at 15 K and 5.5 µs at room temperature. A preliminary lifetime analysis suggests that the decrease in lifetime is mainly due to the strong thermalization between the $^2H_{11/2}$ and $^4S_{3/2}$ excited states. Nonradiative decay processes are expected to only weakly affect the green luminescence.

INTRODUCTION

The luminescence from rare earth doped III-nitrides is of significant current interest for potential applications in optical communications and full color displays.[1] Visible and infrared electroluminescence (EL) has been reported from a number of rare earth doped GaN systems: GaN: Er (green, IR)[2,3,4], GaN: Pr (red, IR) [5], GaN: Eu (red) [6,7] and GaN: Tm (blue) [1] The incorporation, optical activation, and luminescence efficiency of rare earth ions in III-nitrides, however, is not yet fully understood. In this paper, we present a comparison of the PL properties of Er doped GaN grown by MOMBE and SSMBE. Excitation wavelength and temperature dependent PL studies were performed and analyzed in view of optoelectronic applications of Er doped GaN.

EXPERIMENTAL PROCEDURES

The GaN: Er (MOMBE) sample was grown in an INTEVAC Gas Source Gen II on In-mounted (100) Si substrate as described in reference [8]. The GaN: Er (SSMBE) sample was grown on Si by solid source and RF-assisted molecular beam epitaxy (MBE). Details of the Riber MBE32 system used for growth have been discussed previously [3]. PL studies were performed using a HeCd laser operating at either 325 nm or 442 nm. Infrared PL spectra were recorded using a 1-m monochromator equipped with a liquid-nitrogen cooled Ge detector. In visible PL studies a thermo-electric cooled PMT was employed for detection. The signal was processed using lock-in techniques. IR lifetime studies employed the 355 nm line of a Nd: YAG laser for excitation. Visible lifetime data were taken by pumping into the $^4F_{7/2}$ Er^{3+} transition at ~495 nm.

RESULTS AND DISCUSSION

The infrared luminescence spectra of Er doped GaN grown by MOMBE and SSMBE are shown in Figure 1 for above (λ_{ex}=325 nm) and below-gap excitation (λ_{ex}=442 nm). The spectra were taken under identical experimental conditions. The pump power was kept constant at ~0.64 W/cm^2.

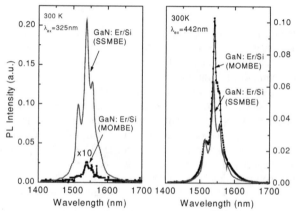

Figure 1: 1.54 μm PL spectra of GaN: Er (MOMBE) and GaN: Er (SSMBE) at room temperature. The PL was excited with either the 325 nm (above-gap) or 442 nm (below-gap) line of a HeCd laser.

Both samples exhibited characteristic 1.54 μm PL resulting from the intra-4f Er^{3+} transition $^4I_{13/2} \rightarrow {}^4I_{15/2}$. The most striking feature of figure 1 is the large difference in PL intensity observed for the samples under above-gap excitation. The GaN: Er (SSMBE) sample exhibited a strong 1.54 μm PL which was nearly 80 times more intense than that observed from GaN: Er (MOMBE). On the contrary, under below-gap excitation both samples exhibited very similar 1.54 μm PL intensities. As shown previously [9], the weak PL observed under above-gap excitation from GaN: Er (MOMBE) can be explained by a significantly reduced excitation efficiency compared to below-gap excitation. Visible PL studies (see below) revealed, that for GaN: Er (MOMBE) the bandedge provides an efficient radiative combination channel reducing the excitation efficiency of intra-4f Er transitions. The excitation wavelength dependent PL study suggests that only weak electroluminescence can be expected from forward-biased GaN: Er (MOMBE) LED's.

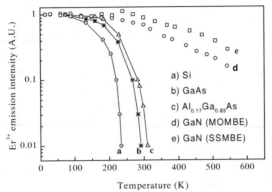

Figure 2: Comparison of the temperature dependence of the integrated Er^{3+} 1.54 μm PL for Er doped Si, GaAs, AlGaAs, and GaN (see also reference 10).

To further evaluate the GaN:Er samples for device applications, the temperature dependence of the integrated 1.54 μm PL intensities was measured as shown in Figure 2. Compared to Si: Er, GaAs: Er, and AlGaAs: Er [10], both GaN: Er samples exhibited very stable 1.54 μm PL up to temperatures as high as ~550 K. More information on the Er PL efficiency was obtained from temperature dependent lifetime studies. The total transition probability, i.e. the reciprocal of the experimental PL lifetime, is given as the sum of total radative decay rate, nonradiative decay rate through multiphonon relaxation, and nonradiative decay rate through energy transfer processes. It is assumed that at low temperature the nonradiative decay through either multiphonon relaxation and/or energy transfer is negligible small. Therefore, the lifetime at 15 K yields a good approximation for the radiative decay rate. Furthermore, assuming the radiative decay rate is temperature independent, any reduction in lifetime can be assigned to the onset of nonradiative decay. The luminescence transients of both GaN: Er samples were measured at 15 K, 300 K, and 520 K. It was observed that the decay curves were non-exponential which suggests the existence of multiple Er sites. The existence of multiple Er sites in GaN has been previously reported [11]. To describe the lifetime decay an average lifetime was used. For GaN: Er (SSMBE) it was observed that the low temperature (15 K) lifetime of 2.3 ms decreased to 1.9 ms at room temperature. At higher temperatures the lifetime continued to decrease and reached a value of 1.2 ms at 520 K. The decrease of the lifetime above 300K is most likely due to the onset of nonradiative decay. Compared to GaN: Er (SSMBE), the lifetime of GaN:Er (MOMBE) was significantly shorter and decreased slightly from 0.11 ms at 15 K to 0.10 ms at room temperature. At higher temperatures the lifetime was too short to be measured with our current setup. The room temperature luminescence efficiencies were estimated from the ratio of the low and room temperature lifetimes (τ_{300K}/τ_{15K}) to be ~0.8 for GaN: Er (SSMBE) and ~0.9 for GaN: Er (MOMBE), respectively. The high PL efficiencies indicate that the Er^{3+} excitation efficiency and the concentration of Er^{3+} ions limit the device performance of current infrared LED's.

The visible PL spectra of the GaN: Er samples following optical excitation at 325 nm are shown in Figure 3. The GaN: Er (SSMBE) exhibited a weak bandedge PL at ~369 nm (3.36 eV) and two "green" lines located at 537 nm and 558 nm. The green luminescence was assigned to the intra 4f Er^{3+} transitions $^2H_{11/2} \rightarrow {}^4I_{15/2}$ and $^4S_{3/2} \rightarrow {}^4I_{15/2}$ (Ref. 3). The GaN: Er (MOMBE) sample showed strong bandedge PL located at ~381 nm (3.25 eV), however, no indication of green Er^{3+} luminescence was found. As discussed before, for GaN: Er (MOMBE) the bandedge provides an efficient radiative combination channel, which reduces the excitation efficiency for both infrared and visible Er^{3+} transitions. Figure 3b) shows the decay transients of the green Er^{3+} PL at different temperatures. The lifetime was found to be non-exponential at all temperatures and decreased with increasing temperature. The average lifetimes for the 558 nm line at 15 K and room temperature were determined to be 10.8 μs and 5.5 μs, respectively.

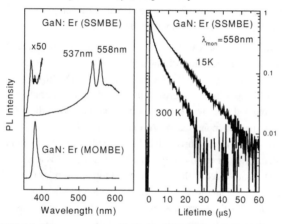

Figure 3: a) Visible PL spectra from GaN: Er (SSMBE) and GaN: Er MOMBE) at 300 K (λ_{ex}=325 nm). b) Decay transients of the visible PL at 558 nm from GaN: Er (SSMBE) at 15 K and 300 K (λ_{ex}=495 nm).

A more detailed study on the temperature dependence of the lifetime is depicted in Figure 4a). The thermalization of the $^4S_{3/2}$ and $^2H_{11/2}$ states leads to a common decay time τ (effective spontaneous emission probability), which can be described as:

$$\frac{1}{\tau} = \frac{\tau_S^{-1} + \tau_H^{-1} \cdot \frac{g_H}{g_S} \exp\left(\frac{-\Delta E}{kT}\right)}{1 + \frac{g_H}{g_S} \exp\left(\frac{-\Delta E}{kT}\right)} \qquad (1)$$

where τ_H and τ_S are the intrinsic radiative decay times of the $^2H_{11/2} \rightarrow {}^4I_{15/2}$ and $^4S_{3/2} \rightarrow {}^4I_{15/2}$ transitions, respectively. g_H and g_S are the electronic degeneracies (2J+1) of the $^2H_{11/2}$ and $^4S_{3/2}$ states and ΔE is their energy difference (ΔE=87meV) .

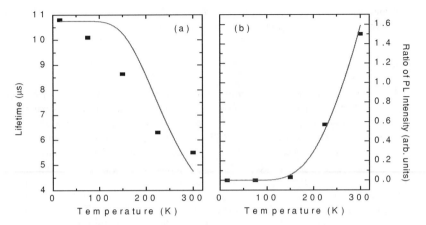

Figure 4: *a) Green PL lifetime of GaN: Er (SSMBE). The solid line describes the change in lifetime according to equation (1). b) Ratio of the PL intensities of the green lines as a function of temperature (see reference 3).*

At low temperatures the $^2H_{11/2}$ state is not thermally populated and the experimental lifetime can be approximated as the intrinsic decay time of the $^4S_{3/2} \rightarrow {}^4I_{15/2}$ transition, i.e. τ_S=10.8 μs. It is assumed in this approximation that the low temperature decay time is purely radiative. The intrinsic lifetime of the $^2H_{11/2} \rightarrow {}^4I_{15/2}$ is not experimentally accessible, unless a careful analysis of absorption data is carried out, which is rather difficult for thin film materials. It is possible, however, to obtain a rough estimation of the intrinsic decay time τ_H from the temperature dependence of the luminescence intensity of the green Er^{3+} lines at 537 nm and 558 nm. Steckl and Birkhahn [3] reported that the intensity of the $^4S_{3/2}$ line decreased with increasing temperature, whereas the $^2H_{11/2}$ line had a maximum of intensity at around 300 K. Taking the experimental data from reference 3, the ratio of the 537 nm and 558 nm lines was calculated and is plotted in Fig 4b). Considering the thermal coupling of the involved states, the intensity ratio of both lines was fitted to an expression

$$I_H = \frac{\tau_S \cdot g_H \cdot hw_H}{\tau_H \cdot g_S \cdot hw_S} \cdot \exp(-\Delta E / kT) \qquad (2)$$

with $h\omega_H$ =2.309eV, $h\omega_S$=2.222eV, τ_S=10.8 μs and ΔE=87 meV. τ_H was taken as a fitting parameter and the best fit to the data yielded τ_H=0.75 μs. The fitting result shows that the radiative rate of the $^2H_{11/2} \rightarrow {}^4I_{15/2}$ transition is much larger than that of the $^4S_{3/2} \rightarrow {}^4I_{15/2}$ transition, consistent with published data on Er doped insulators. Using this set of parameters the temperature dependence of the luminescence lifetime was calculated according to equation (1) and is shown in Fig. 4a). The modeling reveals that the decrease of the luminescence lifetime with temperature is mainly due to an increased radiative decay rate arising from the fast thermalization of the $^2H_{11/2} \rightarrow {}^4I_{15/2}$ and $^4S_{3/2} \rightarrow {}^4I_{15/2}$ transitions. This preliminary analysis of the lifetime implies that non-radiative decay processes are small and therefore the green luminescence efficiency is high. To obtain further support for this conclusion it will be necessary to perform a more systematic study of the Er^{3+} visible PL lifetime for a series of samples with different Er concentrations.

SUMMARY

In summary, we performed a comparison of the infrared and visible PL properties of GaN: Er (MOMBE) and GaN: Er (SSMBE). We observed that both samples exhibited intense 1.54 µm PL under below-gap excitation. With above-gap excitation GaN: Er (MOMBE) showed a greatly reduced IR PL intensity, whereas the 1.54 µm PL from GaN:Er (SSMBE) remained strong. Based on temperature dependent PL intensity and lifetime studies, it was concluded that the 1.54 µm luminescence efficiency is high (~0.8-0.9). The factors limiting the performance of current IR LED's are the Er excitation efficiency and the Er concentration. No visible PL arising from intra-4f Er transitions was found from GaN: Er (MOMBE). On the contrary, the GaN: Er (SSMBE) sample revealed green lines at 537 nm and 558 nm with an average lifetime of 5.5 µs at room temperature. The temperature dependence of the green lifetime was explained by the strong thermalization of the $^2H_{11/2}$ and $^4S_{3/2}$ excited states. Non-radiative decay does not seem to affect the green luminescence efficiency.

ACKNOWLEDGMENTS

The authors from H. U. acknowledge financial support by ARO through Grant DAAD19-99-1-0317. The work at U. F. was supported by ARO grant DAAH04-96-1-0089. The work at U.C. was supported by ARO grant DAAD19-99-1-0348.

REFERENCES

[1] A. J. Steckl and J. M. Zavada, MRS Bulletin, Vol. 24, No. 9, 1999, pp.33-38.
[2] J. T. Torvik, J. Feuerstein, J. I. Pankove, C. H. Qui, F. Namavar, Appl. Phys. Lett. **69**, 2098 1996.
[3] A. J. Steckl and R. Birkhahn, Appl. Phys. Lett. **73**, 1700 (1998).
[4] A. J. Steckl, M. Garter, R. Birkhahn, J. Scofield, Appl. Phys. Lett. **73**, 2450, (1998).
[5] L. C. Chao and A. J. Steckl, Appl. Phys. Lett. **74**, 2364, (1999).
[6] J. Heikenfeld, M. Garter, D. S. Lee, R. Birkenhahn, and A. J. Steckl, App. Phys. Lett. **75**, 1189, (1999).
[7] K. Hara, N. Ohtake, ICNS 3 Conf. Montpellier, France 5-9 July, 1999, paper P057.
[8] J. D. MacKenzie, C. R. Abernathy, S. J. Pearton, U. Hömmerich, J. T. Seo, R. G. Wilson, and J. M. Zavada, Appl. Phys. Lett. **72**, 2710 (1998).
[9] U. Hömmerich, J. T. Seo, Myo Thaik, C. R. Abernathy, J. D. MacKenzie, J.M. Zavada, Journal of Alloys and Compounds, in press.
[10] P.N. Favennec, H. L. Haridon, M. Salvi, D. Moutonnet, and Y. Le Guillou, Electr. Lett. **25**, 718 (1989).
[11] S. Kim, S. J. Rhee, D. A. Turnbull, E. E. Reuter, X. Li, J. J. Coleman, and S. G. Bishop, Appl. Phys. Lett. **71**, 231 (1997).

Mat. Res. Soc. Symp. Vol. 595 © 2000 Materials Research Society

High Density Plasma Damage Induced in n-GaN Schottky Diodes Using Cl2/Ar Discharges

A.P. Zhang [1], **G. Dang** [1], **F. Ren**[1], **X.A. Cao**[2], **H. Cho**[2], **E.S. Lambers**[2], **S.J. Pearton**[2], **R.J. Shul**[3], **L. Zhang**[3], **A.G. Baca**[3], **R. Hickman**[4] and **J.M. Van Hove**[4]
[1] Department of Chemical Engineering
University of Florida, Gainesville, FL 32611
[2] Department of Materials Science and Engineering
University of Florida, Gainesville, FL 32611
[3] Sandia National Laboratories, Albuquerque NM 87185
[4] SVT Associates, Eden Prairie, MN 55344

ABSTRACT

The effects of dc chuck self-bias and high density source power (which predominantly control ion energy and ion flux, respectively) on the electrical properties of n-GaN Schottky diodes exposed to Inductively Coupled Plasma of Cl2/Ar were examined. Both parameters were found to influence the diode performance, by reducing the reverse breakdown voltage and Schottky barrier height. All plasma conditions were found to produce a nitrogen-deficient surface, with a typical depth of the non-stoichiometry being ~500 Å. Post-etch annealing was found to partially restore the diode characteristics.

INTRODUCTION

There is a strong interest in GaN-based electronics for high power, high temperature applications in communications, power switching and microwave amplifiers [1-9]. There have been impressive reports of high performance GaN/AlGaN heterostructure field effect transistors [2-9] grown on Al_2O_3 or SiC substrates. Other prototype devices include a depletion-mode GaN MOSFET [10], GaN/AlGaN heterostructure bipolar transistors [11-13] and GaN rectifiers [14] and junction field effect transistors [15]. Precise pattern transfer during fabrication of these devices requires use of dry etching methods with relatively high ion energy in order to break the strong Ga-N bonds (8.92eV/atom) [16]. Under those conditions there will generally be some ion-induced damage remaining in the GaN after dry etching, along with the possibility of a non-stoichiometric near–surface region due to preferential loss of atomic nitrogen in the form of N_2 [17-19]. The Ga etch product in Cl_2-based discharges is $GaCl_3$, and this is less volatile than N_2 both from a pure chemical vapor presence and from a preferential sputtering viewpoint.

There is a clear need to better understand the role of plasma exposure on the resultant electrical performance of GaN-based devices, both from an etching and deposition viewpoint. Exposure to pure Ar discharges was found to produce higher reverse bias leakage currents in p-n junction structures compared to use of Ar/N_2 discharges [20]. Even relatively low power reactive ion etching (RIE) conditions were found to deteriorate the quality of Schottky contacts deposited on plasma-etched n-GaN [21, 22]. The preferential loss of nitrogen from the GaN surface does improve the specific contact resistance of n-type ohmic contacts because of the creation of a degenerately doped surface layer [23].

In this work we have examined the effect of Inductively Coupled Plasma (ICP) on the reverse breakdown voltage (V_B), Schottky barrier height (ϕ_B) and forward turn-on voltage of (V_F) of GaN diode rectifiers with the contacts already in place. Several methods for damage removal were investigated, including annealing, UV ozone oxidation/stripping or wet etching with NaOH solutions.

EXPERIMENTAL

The diodes were fabricated on nominally undoped ($n\sim10^{17}cm^{-3}$) or Si-doped (5×10^{17} cm^{-3}) GaN layers $\sim3\mu$m thick grown on an n$^+$(10^{18} cm^{-3}) GaN buffer on c-plane Al$_2$O$_3$ substrate. Ohmic contacts were formed with lift-off Ti/Au subsequently annealed at 600 $^\circ$C, followed by evaporation of the 250 μm diameter Pt(250Å)/Au(1500Å) Schottky contacts through a stencil mask. A schematic is shown in Figure 1. The samples were briefly exposed ($\sim$10secs controlled by the system software) to 10Cl$_2$/5Ar (total gas load 15 standard cubic centimeters per minute) or 15Ar ICP discharges in a Plasma Therm 790 reactor. During the ignition stage of discharge, the dc self-bias takes $\sim$2secs to reach its final value. Longer exposure time resulted in more damage. Damage layer thickness was limited due to Cl$_2$/Ar etching of damaged GaN at a rate of around 1000 Å/min. From limited measurements we found that damage depth saturates in 10secs ($\sim$500Å). The gases were injected directly into the source through electronic mass flow controllers, and the 2MHz source power was varied from 100-1000 W. The samples were placed on an rf-powered (13.56MHz, 5-300W), He backside-cooled chuck. Process pressure was held constant at 2 m Torr.

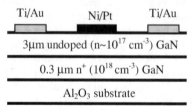

Figure 1. *Schematic of the planar diodes*

RESULTS AND DISCUSSION

The dependence of V_B and ϕ_B on rf chuck power for 300W source power discharges is shown in the upper part of Figure 2. Both of these parameters, at least initially, decrease with increasing power. The ϕ_B values saturate beyond 50W. The main effect on ϕ_B is from damage created around the contact periphery. This would expected to saturate once a N$_2$-deficient region is created because much of the resultant ϕ_B is still determined by the unexposed region under the contact metal. Under these conditions, the dc chuck self-bias increases from -105 V at 50 W to -275 V at 200 W. The average ion energy is roughly the sum of this voltage plus the plasma potential which is 20-25 V in this system under these conditions. After plasma exposure, the diode ideality factor was always

≥ 2 (the reference ideality factor was ~1.3), which is a further indication of the degradation in electrical properties of the structures. The results are consistent with creation of an ion damaged, non-stoichiometric GaN surface region. This region exists in the plasma-exposed area outside the metal contacts. Note that the GaN etch rate increases monotonically with rf chuck power (lower part of the Figure 2), but this more rapid removal of material is not enough to offset the greater amount of damage caused by the higher-energy ion bombardment. We believe the GaN must be non-stoichiometric and hence more n-type at the surface because of the sharp decreases observed in V_B. In the case of compound semiconductors with less bond energy the damage may induce into the bulk materials and the defect may traps the carriers. For GaAs where ion bombardment creates more resistive material by introduction of deep compensating levels rather than shallow donor states, the breakdown voltage is generally found to increase with exposure to plasmas.

The dependence of V_B and ϕ_B on ICP source power is shown in Figure 3 (top).

While ϕ_B continues to decrease as the ion flux increases, V_B initially degrades but shows less of a decrease at higher source powers. This is most likely a result of the continued decrease in the self-bias at higher source power. This also leads to a decrease in GaN etch rate above 500 W. The results of Figure 2 and 3 show that both ion energy and ion flux are important in determining not only the GaN etch rate, but also the amount of residual damage in the diodes.

To establish the chemical state of the GaN surface at different stages, AES was performed on an unmetallized sample. Figure 4 shows surface scans before (top) and after (lower) exposure to a 500 W source power, 50 W chuck power Cl_2/Ar discharge.

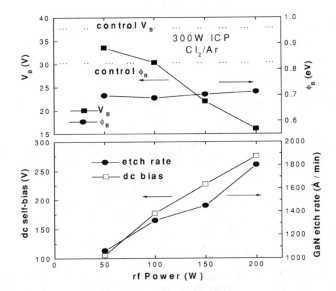

Figure 2. *rf chuck power dependence of V_B and ϕ_B in Cl_2/Ar plasma exposed GaN diodes (top) and of dc chuck self-bias and GaN etch rate under the same conditions (bottom).*

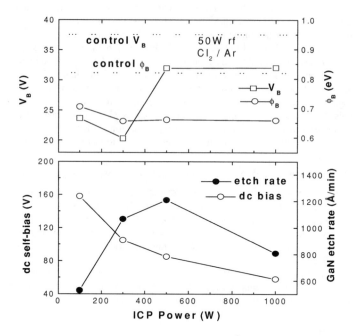

Figure 3 *ICP power dependence of V_B and ϕ_B in Cl_2/Ar plasma exposed GaN diodes (top) and of dc chuck self-bias and GaN etch rate under the same conditions (bottom).*

The main change is a reduction in the N_2 signal in the latter sample (by ~20%), confirming the preferential loss of this element during dry etching. Subsequent annealing at 700 °C in N_2 restored some of this deficiency (Figure 4, bottom).

We also examined the use of wet etch removal of the damaged region with hot NaOH solutions (0.1M, 80 °C). This was found to be successful in etching the damaged GaN and was self-limiting in that once the undamaged material was reached, the etching stopped. We typically found a damage depth of ~500Å. The wet etching usually can restore >70% of the original V_B. The use of UV ozone oxidation followed by stripping of the oxide also produced some improvement in diode characteristics (typically ~50% of the original V_B was restored), but each cycle removes only ~30Å, and thus 15-20 cycles would be needed to completely restore the initial properties.

SUMMARY AND CONCLUSIONS

We found that ICP Cl_2/Ar discharges degrade the performance of GaN Schottky diodes, with ion energy and ion flux both playing important roles. The degradation mechanism appears to be creation of a conducting, non-stoichiometric (N_2-deficient) near-surface region on the GaN. While UV ozone oxidation of the surface and subsequent dissolution of the oxidized region in HCl provides some restoration of the electrical properties of the GaN, etching in NaOH or annealing at 700 – 750 °C may restore most of the initial reverse breakdown voltage characteristics.

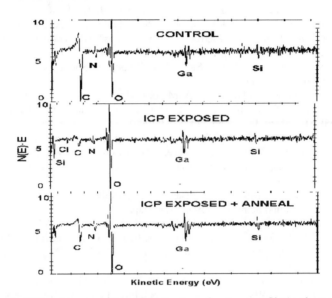

Figure 4. *AES surface scans from GaN (top) or after (center) Cl_2/Ar plasma exposure, and subsequent annealing at 700 °C for 60 seconds (bottom).*

ACKNOWLEDGMENTS

The work at UF is partially supported by a DARPA/EPRI grant, no. MDA 972-98-1-0006 (D. Radack and J. Melcher) monitored by ONR (J. C. Zolper), and a NSF grant DMR 97-32865 (L. D. Hess). Sandia is a multiprogram laboratory operated by Sandia Corporation, a Lockheed-Martin company, for the US Department of Energy under contract no.DEAC04-94-AL-85000.

References

1. J.C. Zolper, "A review of JFETs for high temperature and high power electronics", Solid-State Electronics, **42**, 2153 (1998)
2. S.C.Binari, J.M. Redwing, G. Kelner and W. Kruppa, "GaN HFETs", Electron. Lett., **33**, 242(1997)
3. Q. Chen, R. Gaska, M.A.Khan, M.S.Shur, A.T.Ping, I.Adesida, J.Burm, W.J. Schaff and L.F. Eastman, "High performance GaN/AlGaN HFETs" Electron. Lett., **33**, 637 (1997)
4. M.S.Shur, "GaN-based transistors for high power applications", Solid-State Electron. **42**, 2131(1998)
5. Y.F.Wu, B.P.Keller, P.Fini, S.Keller, T.J. Jenkins, L.T. Kehias, S.P. DenBaars and U.K. Mishra, "High Al content AlGaN/GaN MODFETs for ultrahigh performance", IEEE Electron. Dev. Lett., **19**, 50(1998)

6. A. Balandin, S.Cai, R.Li, K.L.Wang, V.R.Rao and C.R. Viswanathan, "Flicker noise in GaN/AlGaN HFETs", IEEE Electron. Dev. Lett., **19**, 475(1998)

7. D.V. Kuksenkov, H.Temkin, R.Gaska and J.W. Yang, "Low frequency noise in AlGaN/GaN HFETs", IEEE, Electron. Dev. Lett., **19**, 222(1998)

8. C.J. Sullivan, M.Y. Chen, J.A. Higgins, J.W. Yang, Q. Chen, R.L. Pierson and B.T. McDermott, "High power 10GHz operation of AlGaN HFETs on insulating SiC", IEEE, Electron. Dev. Lett., **19**, 198(1998)

9. C.E. Wetzel and K.E. Moore, "SiC and GaN rf power devices", Mat. Res. Soc. Symp. Proc., 483, 111(1998)

10. F.Ren, M. Hong, S.N.G. Chu, M.A. Marcus, M.J. Schurman, A.G. Baca, S.J. Pearton and C.R. Abernathy, "Effect of temperature on $Ga_2O_3(Gd_2O_3)$/GaN MOSFETs", Appl. Phys. Lett., **73**, 3893(1998)

11. F. Ren, C.R. Abernathy, J.M. Van Hove, P.P. Chow, R. Hickman, J.J. Klaasen, R.F. Kopf, H. Cho, K.B. Jung, J.R. Laroche, R.G. Wilson, J. Han, R.J. Shul, A.G. Baca and S.J. Pearton, "300 °C GaN/AlGaN HBT" MRS Internet J. Nitride Semicond. Res., **3**, 41(1998)

12. L.S. McCarthy, P. Kozodoy, S.P. DenBaars, M. Rodwell and U.K. Mishra, 25th Intl. Symp. Compound Semi., Oct. 1998, Nara, Japan

13. J. Han, A.G. Baca, R.T. Shul, C.G. Willison, L. Zhang, F. Ren, A.P. Zhang, G.T. Dang, S.M. Donovan, X.A. Cao, H. Cho, K.B. Jung, C.A. Abernathy, S.J.Pearton and R.G. Wilson, "Growth and Fabrication of GaN/AlGaN HBT" Appl. Phys. Lett. **74**, 2702(1999)

14. Z.Z Bandic, P.M. Bridger, E.C. Piquette, T.C. McGill, R.P. Vaudo, V.M. Phanse and J.M. Redwing, "450V GaN rectifiers", Appl. Phys. Lett., **74**, 1266 (1999)

15. J.C. Zolper, R.T. Shul, A.G. Baca, S.J. Pearton, R.G. Wilson and R.J. Shul, "GaN JFETs", Appl. Phys. Lett., **68**, 273,(1996)

16. R.J. Shul, "Dry etching of GaN", in GaN and Related Materials, ed. S.J. Pearton (Gordon and Breach, NY 1997), Chapter 12

17. A.T. Ping, A.C. Schmitz, I. Adesida, M.A. Khan, Q. Chen and J. Yang, "Characteristics of RIE damage to GaN using Schottky diodes", J. Electron. Mater., **26**, 266(1997)

18. S.J. Pearton, J.W. Lee, J.D. MacKenzie, C.R. Abernathy and R.J. Shul, "Dry etch damage in InN, InGaN and InAlN", Appl. Phys. Lett., **67**, 2329(1995)

19. H.P. Gillis, D.A. Choutov, K.P. Martin, M.D. Bremser and R.F. Davis, "Highly anisotropic dry etching of GaN by LE4 in DC plasma", J. Eletron. Mater., **26**, 301(1997)

20. A.S. Usikov, W.L. Lundin, U.I. Ushakov, B.V. Pushnyi, N.M. Schmidt, J.M. Zadiranov and T.V. Shuhtra, "Electrical and optical properties of GaN after dry etching", Proc. ECS 98-14, 57(1998)

21. A.T. Ping, Q. Chen, J.W. Yang, M.A. Khan and I. Adesida, "The effects of RIE damage on ohmic contacts to n-GaN", J. Electron. Mater., **27**, 261(1998)

22. J.Y. Chen, C.J. Pan and G.C. Chi, "Electrical and optical changes in the near-surface of GaN", Solid-State Electron., **43**, 649(1999)

23. Z.F. Fan, S.N. Mohammad, W. Kim, O. Aktas, A.E. Botchkarev and H. Morkoc, "Effect of Reactive Ion Etching on ohmic contacts to n-GaN", Appl. Phys. Lett., **69**, 1672(1996)

Mat. Res. Soc. Symp. Vol. 595 © 2000 Materials Research Society

Processing And Device Performance Of GaN Power Rectifiers

A.P. Zhang[1], G.T. Dang[1], X.A. Cao[2], H. Cho[2], F. Ren[1], J. Han[3], J.-I. Chyi[4], C.-M. Lee[4], T.-E. Nee[4], C.-C. Chuo[4], G.-C. Chi[5], S.N.G. Chu[6], R.G. Wilson[7] and S.J. Pearton[2]

[1] Department of Chemical Engineering
University of Florida, Gainesville, FL 32611 USA
[2] Department of Materials Science and Engineering
University of Florida, Gainesville, FL 32611 USA
[3] Sandia National Laboratories, Albuquerque, NM 87185 USA
[4] Department of Electrical Engineering
National Central University, Chung-Li, 32054 Taiwan
[5] Department of Physics, National Central University, Chung-Li 32054 Taiwan
[6] Bell Laboratories, Lucent Technologies, Murray Hill, NJ 07974 USA
[7] Consultant, Stevenson Ranch, CA 91381

ABSTRACT

Mesa and planar geometry GaN Schottky rectifiers were fabricated on 3-12μm thick epitaxial layers. In planar diodes utilizing resistive GaN, a reverse breakdown voltage of 3.1 kV was achieved in structures containing p-guard rings and employing extension of the Schottky contact edge over an oxide layer. In devices without edge termination, the reverse breakdown voltage was 2.3 kV. Mesa diodes fabricated on conducting GaN had breakdown voltages in the range 200-400 V, with on-state resistances as low as 6m $\Omega \cdot cm^{-2}$.

INTRODUCTION

The AlGaN materials system is attractive from the viewpoint of fabricating unipolar power devices because of its large bandgap and relatively high electron mobility. [1-6] An example is the use of Schottky diodes as high-voltage rectifiers in power switching applications. [2-4,6] These diodes will have lower blocking voltages than p-i-n rectifiers, but have advantages in terms of switching speed and lower forward voltage drop. Edge termination techniques such as field rings or field plates, bevels or surface ion implantation are relatively well-developed for Si and SiC and maximize the high voltage blocking capability by avoiding sharp field distributions within the device. [7] However in the few GaN Schottky diode rectifiers reported to date [2,3], there has been little effort made on developing edge termination techniques. Proper design of the edge termination is critical both for obtaining a high breakdown voltage and reducing the on-state voltage drop and switching time.

In this paper we report on the effect of various edge termination techniques on the reverse breakdown voltage, V_B, of planar GaN Schottky diodes which deplete in the lateral direction. A maximum V_B of 3.1 kV at 25°C was achieved with optimized edge termination, which is a record for GaN devices. We also examined the temperature dependence of V_B in mesa diodes and found a negative temperature coefficient of this parameter in these structures.

EXPERIMENTAL

The GaN was grown on c-plane Al_2O_3 substrates by Metal Organic Chemical Vapor Deposition using trimethylgallium and ammonia as the precursors. For vertically-depleting devices, the structure consisted of a $1\mu m$ n^+ ($3x10^{18}$ cm^{-3}, Si-doped) contact layer, followed by undoped ($n=2.5x10^{16}$ cm^{-3}) blocking layers which ranged from 3-$11\mu m$ thick. These samples were formed into mesa diodes using Inductively Coupled Plasma etching with Cl_2/Ar discharges (300 W source power, 40 W rf chuck power). The dc self-bias during etching was -85 V. To remove residual dry etch damage, the samples were annealed under N_2 at 800°C for 30 secs. Ohmic contacts were formed by lift-off of e-beam evaporated Ti/Al, annealed at 700°C for 30 secs under N_2 to minimize the contact resistance. Finally, the rectifying contacts were formed by lift-off of e-beam evaporated Pt/Au. Contact diameters of 60-1100μm were examined.

For laterally-depleting devices, the structure consisted of ~3μm of resistive (10^7 $\Omega/\square$) GaN. To form ohmic contacts, Si^+ was implanted at $5x10^{14}$ cm^{-2}, 50 keV into the contact region and activated by annealing at 150°C for 10 secs under N_2. The resulting n-type carrier concentration was $1x10^{19}$ cm^{-3}. The ohmic and rectifying contact metallization was the same as described above.

Three different edge termination techniques were investigated for the planar diodes[7]: (i) use of a p-guard ring formed by Mg^+ implantation at the edge of the Schottky barrier metal. In these diodes the rectifying contact diameter was held constant at 124μm, while the distance of the edge of this contact from the edge of the ohmic contact was 30μm in all cases.

(ii) use of p-floating field rings of width 5μm to extend the depletion boundary along the surface of the SiO_2 dielectric, which reduces the electric field crowding at the edge of this boundary. In these structures a 10μm wide p-guard ring was used, and 1-3 floating field rings employed.

(iii) use of junction barrier controlled Schottky (JBS) rectifiers, i.e. a Schottky rectifier structure with a p-n junction grid integrated into its drift region.
In all of the edge-terminated devices the Schottky barrier metal was extended over an oxide layer at the edge to further minimize field crowding, and the guard and field rings formed by Mg^+ implantation and 1100°C annealing.

RESULTS AND DISCUSSION

(a) Effects of Edge Terminations
Figure 1 (top) shows a schematic of the planar diodes fabricated with the p-guard rings, while the bottom of the Figure shows the influence of guard ring width on V_B at 25°C. Without any edge termination, V_B is ~2300 V for these diodes. The forward turn-on voltage was in the range 15-50 V, with a best on-resistance of 0.8Ωcm^2. The figure-of-merit $(V_B)^2/R_{ON}$ was 6.8 MW·cm^{-2}. As the guard-ring width was increased, we observed a monotonic increase in V_B, reaching a value of ~3100 V for 30μm wide rings. The figure-of-merit was 15.5 MW·cm^{-2} under these conditions. The reverse leakage

current of the diodes was still in the nA range at voltages up to 90% of the breakdown value.

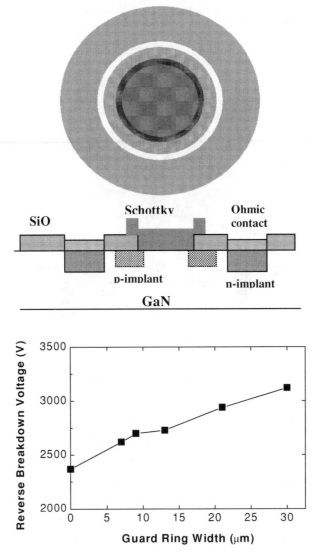

Figure 1. (top) View of rectifiers with p-type guard rings. (bottom) Variation of V_B with guard ring width.

Figure 2 (top) shows a schematic of the floating field ring structures, while the bottom section shows the effect of different edge termination combinations on the resulting V_B at 25°C. Note that the addition of the floating field rings to a guard ring structure further improves V_B, with the improvement saturating for a 3 floating field ring geometry.

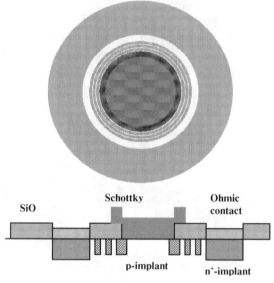

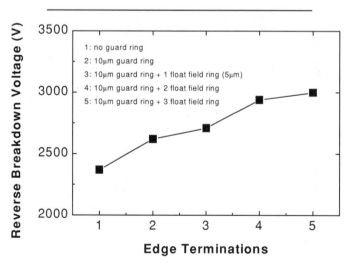

Figure 2.   (top) View of rectifiers with floating field rings. (bottom) Variation of V_B with or without a 10μm wide p-guard rings a 1, 2 or 3 floating field rings. The field ring width was 5μm in all cases.

Figure 3 shows the effect of the junction barrier control on V_B, together with a schematic of the p-n junction grid. In our particular structure we found that junction barrier control slightly degraded V_B relative to devices with guard rings and various numbers of floating field rings. We believe that with optimum design of the grid structure we should achieve higher V_B values and that the current design allows Schottky barrier lowering since the depletion regions around each section of the grid do not completely overlap. This is consistent with the fact that we did not observe the decrease in forward turn-on voltage expected for JBS rectifiers relative to conventional Schottky rectifiers.

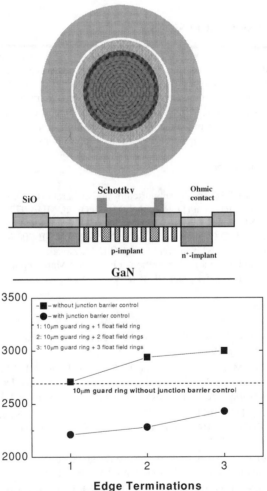

Figure 3. (top) View of rectifier with junction barrier control. (bottom) Variation of V_B in devices with a 10μm wide p-guard ring and 1, 2 or 3 floating field rings, either with or without junction barrier control.

The results of Figure 1-3 are convincing evidence that proper design and implementation of edge termination methods can significantly increase reverse breakdown voltage in GaN diode rectifiers and will play an important role in applications at the very highest power levels. For example, the target goals for devices, intended to be used for transmission and distribution of electric power or in single-pulse switching in the subsystem of hybrid-electric contact vehicles are 25 kV standoff voltage, 2 kA conducting current and a forward voltage drop <2% of the standoff voltage. At these power levels, it is expected that edge termination techniques will be essential for reproducible operation.

SUMMARY AND CONCLUSIONS

GaN Schottky diodes with vertical and lateral geometries were fabricated. A reverse breakdown voltage of 3.1 kV was achieved on a lateral device incorporating p-type guard rings. Several types of edge termination were examined, with floating field rings and guard rings found to increase V_B. The best on-state resistance obtained in these lateral devices was $0.8\Omega cm^2$. In mesa diodes incorporating n^+ contact layers, the best on-state resistance was 6 $m\Omega cm^2$, while V_B values were in the range 200-550V. These GaN rectifiers show promise for high power electronics applications.

ACKNOWLEDGMENTS

The work at UF is partially supported by an NSF grant (DMR-9732865, L. Hess) and by a DARPA/EPRI grant (D. Radack/J. Melcher), no. MDA 972-98-1-0006 monitored by ONR (J. C. Zolper). The work at NCU is sponsored by the National Science Council of R.O.C. under contract no. NSC-88-2215-E-008-012. The work of RGW is partially supported by a grant for ARO (J.M. Zavada). Sandia is a multiprogram laboratory operated by Sandia Corporation, a Lockheed-Martin company, for the US Department of Energy under grant DEAC04-94AL85000.

REFERENCES

1. M.S. Shur, "GaN-Based Transistors for High Power Applications," Solid-State Electronics 42, 2119 (1998).

2. J.-I. Chyi, C.-M. Lee, C.-C. Chuo, G.C. Chi, G.T. Dang, A.P. Zhang, F. Ren, X.A. Cao, S.J. Pearton, S.N.G. Chu and R.G. Wilson, "Growth and Device Performance of GaN Schottky Rectifiers," MRS Internet J. Nitride Semicond. Res. 4, 8 (1999).

3. Z.Z. Bandic, D.M. Bridger, E.C. Piquette, T.C. McGill, R.P. Vaudo, V.M. Phanse and J.M. Redwing, "High Voltage (450 V) GaN Schottky Rectifiers," Appl. Phys. Lett. 74, 1266 (1999).

4. M. Trivedi and K. Shenai, " Performance Evaluation of High Power, Wide Bandgap Semiconductor Rectifiers," J. Appl. Phys. 85, 6880 (1999).

5. V.A. Dmitriev, K.G. Irvine, C.H. Carter, Jr., N.I. Kuznetsov and E.V. Kalinina, "Electric Breakdown in GaN p-n Junctions," Appl. Phys. Lett. 68, 229 (1996).

6. S.J. Pearton, J.C. Zolper, R.J. Shul and F. Ren, "GaN: Processing, Defects and Devices," J. Appl. Phys. 86, 1 (1999).

Mat. Res. Soc. Symp. Vol. 595 © 2000 Materials Research Society

COMPARISON of IMPLANT ISOLATION SPECIES FOR GaN FIELD-EFFECT TRANSISTOR STRUCTURES

G. Dang[1], X. A. Cao[2], F. Ren[1], S. J. Pearton[2], J. Han[3], A. G. Baca[3], R. J. Shul[3], and R. G. Wilson[4]

[1] Department of Chemical Engineering, University of Florida, Gainesville FL 32611 USA
[2] Department of Materials Science and Engineering, University of Florida, Gainesville FL 32611 USA
[3] Sandia National Laboratories, Albuquerque NM 87185 USA
[4] Consultant, Stevenson Ranch CA 91381 USA

Abstract

Different ions (Ti^+, O^+, Fe^+, Cr^+) were implanted at multiple energies into GaN field effect transistor structures (n and p-type). The implantation was found to create deep states with energy levels in the range E_C –0.20 to 0.49 eV in n-GaN and at E_V +0.44 eV in p-GaN after annealing at 450-650 °C. The sheet resistance of the GaN was at a maximum after annealing at these temperatures, reaching values of ~$4x10^{12}$ $\Omega/\square$ in n-GaN and ~10^{10} $\Omega/\square$ in p-GaN. The mechanism for the implant isolation was damage-related trap formation for all of the ions investigated, and there was no evidence of chemically induced isolation.

Introduction

There are two types of defect-formation mechanisms that are found for implant isolation in semiconductors [1]:

(i)   the creation of midgap, damage-related levels, which trap the free carriers in the material. This type of compensation is stable only to the temperature at which these damage-related levels are annealed out.

(ii)   the creation of chemically-induced deep levels by implantation of a species that has an electronic level in the middle of the bandgap. This type of compensation usually requires the implanted species to be substitutional and hence annealing is required to promote the ion onto a substitutional site. In the absence of outdiffusion or precipitation of this species, the compensation is thermally stable.

To date in the GaN materials system there has only been an examination of damage-induced isolation. Binari et.al.[2] investigated H^+ and He^+ isolation of n-GaN, with the material remaining compensated to over 850°C with He^+ and 400°C with H^+. There has also been work on N^+ implantation into both n- and p-type GaN for damage-related isolation.[3] Very effective isolation of AlGaN heterostructure field effect transistor structures has been achieved with a combined P^+/He^+ implantation leading to sheet resistances of $\geq10^{12}$ $\Omega/\square$ and an activation energy of 0.71 eV for the resistivity.[4] Some work has also been reported for isolation of InAlN using O^+, F^+ or N^+ implantation, and for isolation of InGaN using the same species.[5]

To create chemically induced isolation, it is necessary to implant impurities with electronic levels in the GaN bandgap. In other compound semiconductors, species such as Fe, Cr, Ti and V have been employed, with other examples being O in AlGaAs (where Al-O complexes are thought to form)[6] and N in GaAs (C), where C-N complexes are thought to form.[7]

In this paper we report on the creation of very high resistance regions (>10^{10} $\Omega/\square$) in both n- and p-GaN by O^+, Cr^+, Fe^+ or Ti^+ implantation. The isolation is annealed out by 900 °C, suggesting that the concentration of thermally stable deep levels associated with these species is less than ~7×10^{17} cm^{-3} in both n- and p-GaN.

Experimental

0.3μm thick n (Si-doped) or p (Mg-doped) type GaN layers were grown on 1μm thick undoped GaN on (0001) sapphire substrates by rf plasma activated Molecular Beam Epitaxy. The carrier concentration in the doped layers was 7×10^{17} cm^{-3} in each case. This structure simulates a GaN metal semiconductor field effect transistor (MESFET) or metal-oxide semiconductor FET (MOSFET) layer design. In some structures, 490 Å of Ti was deposited by e-beam evaporation on top of the GaN, and the implant performed through that layer. This procedure enabled us to achieve better near-surface isolation properties. Ohmic contacts were formed in a transmission line pattern (gap spacings of 2, 4, 8, 16 and 32μm) by e-beam evaporation and lift-off of Ti/Au (n-type) or Ni/Au (p-type) annealed at 700°C under N_2. The total metal thickness was 4000Å, so that these regions could act as implanted masks. The samples were then implanted at 25°C using multiple-energy Ti^+, O, Fe or Cr ions. The ion profiles were simulated by P-CODE™, while the displacement damage was simulated by the Transport-of-Ions-In-Matter (TRIM) code. An example of the calculated damage distribution for the O^+ implant scheme is shown in Figure 1. The doses and energies were chosen to create an average ion concentration of ~10^{19} cm^{-3} throughout the 0.3μm thick doped GaN layers. In this case, the O^+ was implanted through the Ti overlayer.

In the case where the implanted species is chemically active in the GaN it is the ion profiles that are the important feature, since it is the electrically active fraction of these implanted species that determines the isolation behavior. In the case where the isolation simply results from damage-related deep levels, then it is the profile of ion damage that is important. Figure 2 shows both the calculated ion profiles (top) and damage profiles (bottom) for the multiple energy Fe^+ implant scheme, obtained from TRIM simulations. Note that the defect density is generally overstated in these calculations due to recombination of vacancies and interstitials. In any case, the doses are below the amorphization threshold for GaN.

Results and Discussion

Figure 3 shows the evolution of sheet resistance with annealing temperature for the sample implanted with O^+ ions through the Ti over-layer. In this case, a very high maximum sheet resistance was achieved (4×10^{12} $\Omega/$). This is the highest value reported in GaN. The compensation mechanism is creation of deep electron traps that remove electrons from the conduction band. Upon annealing, some of these damage-related traps are removed, allowing electrons to be returned to the conduction band. The use of the Ti over-layer allows better compensation in the near-surface region and hence a higher as-implanted sheet resistance. The subsequent evolution of the sheet resistance with annealing temperature is similar to the case of implantation without the Ti over-layer, except that the absolute value is larger up to ~600 °C. Secondary Ion Mass Spectrometry (SIMS) measurements on the samples showed that the implanted oxygen did not have any detectable redistribution at 800 °C, consistent with the low diffusivities reported previously.[8]

Figure 4 shows the annealing temperature dependence of sheet resistance for Cr^+ (top) and Fe^+ (bottom) implanted n- and p-type GaN. The trends in the sheet resistance are typical of those observed with damage-related isolation. The as-implanted resistance is 6-7 orders of magnitude higher than that of the unimplanted material due to creation of deep traps that remove carriers from the conduction and valence bands. Subsequent annealing tends to further increase the sheet resistance, by reducing the probability for hopping conduction as the average distance between trap sites is increased.[9,10] Beyond particular annealing temperatures (500-600°C in this case) the trap density begins to fall below the carrier concentration and carriers are returned to the conduction or valence bands. This produces a decrease in sheet resistance toward the original, unimplanted values. If Cr or Fe produced energy levels in the bandgap with concentrations greater than the carrier density in the material, then the sheet resistance would remain high for annealing temperatures above 600°C. For these two impurities it is clear that the electrically active concentration of deep states is $<7 \times 10^{17}$ cm^{-3}, otherwise all the carriers would remain trapped beyond an annealing temperature of 600°C. Basically, the same trends were observed for Ti^+ implants.

Figure 5 shows Arrhenius plots of the sheet resistance of Cr^+ (top) or Fe^+ (bottom) implanted n- and p-type GaN annealed at either 450°C (n-type) or 600°C (p-type). These annealing temperatures were chosen to be close to the point where the maxima in the sheet resistances occur for the two different conductivity types. The activation energies derived from these plots represent the Fermi level position for the material at the particular annealing temperatures employed. Note that the values are far from midgap (1.7 eV for hexagonal GaN), but are still large enough to create very high sheet resistances in the ion damaged material. Within the experimental error (±0.04 eV), the activation energies are the same for Cr^+ and Fe^+ implants for both conductivity types. This again suggests the defect states created are damage-related and not chemical in nature.

The activation energies obtained for O^+ and Ti^+ implants were similar to those obtained with Cr^+ or Fe^+ implants, except for the case of O^+ into n-GaN where the energy was in the range of 0.20-0.29 eV. This difference may be related to the lower damage density with O^+ implantation described earlier. The defect states in the gap are most likely due to point defect complexes of vacancies and/or interstitials and the exact microstructure of these complexes and their resultant energy levels are expected to be very dependent on damage density and creation rate.[11] This might also explain the differences reported in the literature for the activation energies obtained with different implant species.

Figure 6 shows a schematic of the energy level positions found in this work for Ti, Cr, Fe and O implanted p- and n-type GaN annealed to produce the maximum sheet resistance. Although the levels are not at midgap as is ideal for optimum compensation, they are sufficiently deep to produce high resistivity material. In GaN contaminated with transition metal impurities, non-phonon-assisted photoluminescence lines attributed to Fe^{3+} at 1.3 eV and Ti^{2+} at 1.19 eV have been reported[12], but to date there are no electrical measurements.

Summary and Conclusions

We can draw the following conclusions from this work:

1. Sheet resistances of $\sim 10^{12}$ $\Omega/\square$ in n-GaN and $\sim 10^{10}$ $\Omega/\square$ in p-GaN can be achieved by implantation of Cr, Fe, Ti, or O.
2. The sheet resistance remains above 10^7 $\Omega/$ until annealing temperatures of ~ 650 °C, which defines the thermal stability of acceptable GaN device isolation.
3. Ti, O, Fe and Cr do not produce electrically active deep energy levels in the GaN bandgap with concentrations approaching 7×10^{17} cm^{-3} when introduced by implantation. GaN implanted with these species displays typical damage-related isolation behavior, with no evidence of chemically induced thermally stable isolation.
4. The activation energies for the sheet resistance of the implanted GaN are in the range 0.2-0.49 eV in n-type and ~ 0.44 eV for p-type.

Acknowledgements

The work at UF is partially supported by grants from DARPA/EPRI (J. Melcher/D. Radack), contract MDA 972-98-1-0006, monitored by ONR (J.C. Zolper) and by an NSF grant, DMR 97-32865 (L.D. Hess). Sandia is a multi-program laboratory operated by Sandia Corporation, a Lockheed-Martin company, for the US Department of Energy under grant no. DEAC 04-94-AL-85000. The work of RGW is partially supported by a grant from ARO (J.M. Zavada).

References

1. S.J. Pearton, Mat. Sci. Rep. 4, 313 (1990).
2. S.C. Binari, H.B. Dietrich, G. Kelner, L.B. Rowland, K. Doverspike and K.D. Wickenden, J. Appl. Phys. 78, 3008 (1995).
3. S.J. Pearton, C.R. Abernathy, C.B. Vartuli, J.C. Zolper, C. Yuan and R.A. Stall, Appl. Phys. Lett. 67, 1435 (1995).
4. G. Harrington, Y. Hsin, Q.Z. Liu, P.M. Asbeck, S.S. Lau, M.A. Khan, J.W. Yang and Q. Chen, Electron. Lett. 34, 193 (1998).
5. J.C. Zolper, S.J. Pearton, C.R. Abernathy and C.B. Vartuli, Appl. Phys. Lett. 66, 3042 (1995).
6. S.J. Pearton, M.P. Iannuzzi, C.L. Reynolds and L. Peticolas, Appl. Phys. Lett. 52, 395 (1988).
7. J.C. Zolper, M.E. Sherwin, A.G. Baca and R.P. Schneider, J. Electron. Mater. 24, 21 (1995).
8. J. C. Zolper, J. Cryst. Growth 178, 175 (1997).
9. N.F. Mott, J. Non-Cryst. Solids 1, 1 (1968).
10. M. Cohen, H. Fritsche and S. Ovshinsky, Phys. Rev. Lett. 22, 1065 (1969).
11. see for example, Identification of Defects in Semiconductors, ed. M. Stavola, Semiconductors and Semimetals, Vols. 51A and 51B (Academic Press, San Diego, 1998).

12.   K. Pressel and P. Thurian, in <u>Properties, Processing and Applications of GaN and Related Semiconductors</u>, ed. J.H. Edgar, S. Strite, I. Akasaki, H. Amano and C. Wetzel (EMIS DataReview 23, INSPEC, IEE, London,1999).

Figure Captions

Figure 1.   Calculated vacancy concentration in a Ti/GaN sample implanted with five separate O^+ implants at energies from 30-325 keV.

Figure 2.   Ion (top) and damage (bottom) profiles for multiple energy Fe^+ implant sequence into GaN.

Figure 3.   Annealing temperature dependence of sheet resistance in O^+ implanted GaN. The implant was performed directly into the GaN.

Figure 4.   Evolution of sheet resistance of GaN with annealing temperature after either Cr^+ (top) or Fe^+ (bottom) implantation.

Figure 5.   Arrhenius plots of sheet resistance in Cr^+ (top) or Fe^+ (bottom) implanted n- and p-GaN, after annealing at either $450°C$ (n-type) or $600°C$ (p-type).

Figure 6.   Schematic representation of the position in the energy gap of defect levels from Fe, Cr, Ti or O implant isolation in GaN.

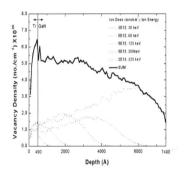

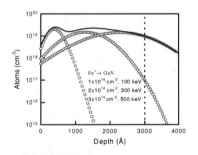

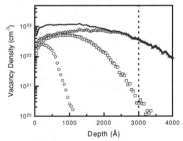

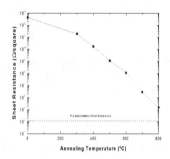

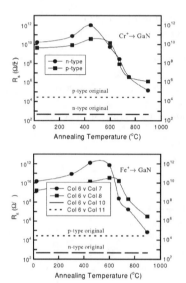

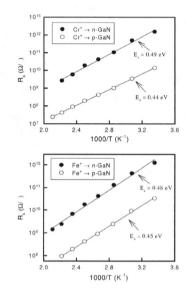

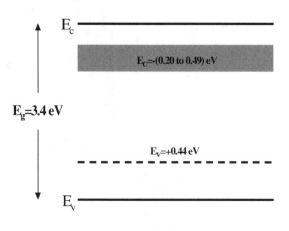

W11.68.6

Mat. Res. Soc. Symp. Vol. 595 © 2000 Materials Research Society

Development of Wide Bandgap Semiconductor Photonic Device Structures by Excimer Laser Micromachining

Qiang Zhao, Michael Lukitsch, Jie Xu, and Gregory Auner, Department of Electrical & Computer Engineering; **Ratna Niak and Pao-Kuang Kuo,** Department of Physics, Wayne State University, Detroit, Michigan

ABSTRACT

Excimer laser ablation rates of Si (111) and AlN films grown on Si (111) and r-plane sapphire substrates were determined. Linear dependence of ablation rate of Si (111) substrate, sapphire and AlN thin films were observed. Excimer laser micromachining of the AlN thin films on silicon (111) and SiC substrates were micromachined to fabricate a waveguide structure and a pixilated structure. This technique resulted in clean precise machining of AlN with high aspect ratios and straight walls.

INTRODUCTION

Laser micromachining is a noncontact process offering selective material removal with high precision and repeatability, and a minimal heat-affected zone (HAZ.) It can also produce flexible feature size and shape using mask-imaging methods and a computer controlled multi-dimensional scanning stage etc[1]. Excimer lasers with 20~30-ns pulse duration have been demonstrated to be suitable for various materials.[2,3] Excimer laser machining provides the advantage of greater photon energy to break bonds, clean cutting with less thermal effects. For example, Behrmann[4] and Wang[5] fabricated diffractive optical elements (DOE) on poly(imide) micromachined by Eximer laser. Han[6] fabricated Fabry-Perot cavity chemical sensors by silicon Eximer laser micormachining.

We are interested in wide bandgap materials, particularly AlN, since it has great potential in an array of electro-optical and photonic devices. Conventional micromachining techniques, such as wet chemical, are not suitable for processing wide bandgap semiconductors. Although plasma etching is a popular method of micromachining, the possibly introduced damage makes it less suitable for precise micromachining. AlN is not etched by traditional chemical methods. Only concentrated KOH and analogous base etchants are reasonable. Furthermore, the chemical etching process is highly isotropic. Even ion beam and other plasma etching techniques are not viable solutions due to parasitic (with respect to waveguide performance) damage of the structure. With its advantages of direct writing and greater photon energy for cool ablation, we hope to utilize Excimer laser micromachining technique to fabricate photonic device structures on AlN in a one-step process. An image-based pulse KrF Excimer laser micromachining system has been developed. With an ultra-precise 4-dimensional (X, Y, Z and Rotation) scanning stage and an in-house developed software package, this system has the capability to produce feature size down to less than 1 micron scale with flexible shaped features possible.

Wave guide structures and a pixilated structure of AlN on Silicon (111) and SiC have been fabricated. The focus of this paper is to study the relation between laser ablated hole depth to the incident pulse energy for the thin film and substrate structures.

Linear dependence within the thin film and substrate has been observed. This quantitative relationship provides a methodology for precise depth control, which is critical in fabrication of electronic and photonic device structures.

EXPERIMENT AND RESULTS

The experimental configuration of the Excimer laser micromachining system is shown in Fig.1. A Lambda Physik 200i Excimer laser, operating at KrF mode with emitting laser wavelength at 248-nm, is used as a laser micromachining source. The laser pulse can reach an energy as high as 600 mJ with a pulse duration of 25-ns and has a rectangular output beam of 23 mm x 8 mm. The laser beam passes through a neutralized continuously tunable attenuator and a homogenizer made of micro-lens arrays. The micro-lens arrays split the beam into different beamlets travelling along different paths and overlap them on the plane to be irradiated, which in our system is a mask. The gaussian beam profile of the laser pulse is then transformed to a near perfect flat-top shape with 0.87 flatness (see Fig. 2.). The mask is put in the homogenized plane (with a homogenized illumination area of 18 mm x18 mm) and imaged by an objective onto the sample. The sample is placed on the top of an ultra-precision 4-dimensional scanning stage (Newport PM500, X, Y, Z and rotation; X and Y with 80 mm travel limit and 0.05 um-linear resolution; Z with 25 mm travel limit and 0.025 um linear resolution; rotation stage with 360^0 travel and 0.0003^0 rotary resolution). A photon beam profiler is used to measure the laser beam intensity profile, a pyroelectric energy sensor is used to measure the laser pulse energy, and a fast-response Hamamatsu photodiode is used to measure the laser pulse time shape. A software package has been developed to control the laser micromachining system, including control of the Excimer laser, sample scanning stage to control micropatterning design and fabrication, and laser beam characterization.

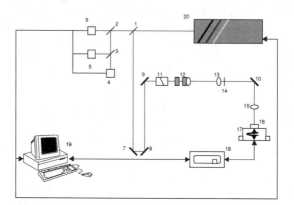

Figure 1 EXCIMER LASER MICROMACHINING SYSYTEM
1, 2, 3 Beam splitter; 4 Beam profiler; 5 Pulse shape time analyzer;
6 Energy meter; 7, 8, 9, 10 45^0 HR mirror; 11 Attenuator; 12 Homogenizer
13 Field lens; 14 Mask; 15 Objective; 16 Sample; 17 4-Axis precision stage;
18 Motion controller; 19 PC; 20 LPX205i Excimer Laser (KrF).

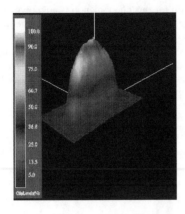 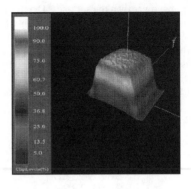

(a) Before homogenizer.   (b) After homogenizer.

Figure 2 LASER BEAM INTENSITY PROFILES

Ablation Study

Precise control of micromachining depth is critical for the fabrication of electronic and photonic device structures. To control the depth, the relation between laser ablated feature depth and laser pulse energy must be known. We developed a routine to quantitatively explore this dependence. Utilizing a 1mm diameter pinhole as a mask, 10 X 10 arrays of holes (see Fig. 3), separated by 200 μm, are ablated in each sample (after transversing the objective, the laser beam cross section is demagnified 10X to a 100 μm spot size on the sample surface). Each of the 10 holes within a row are shot with the same number of laser pulses and the same pulse energy at a pulse repetition frequency of 25 Hz. The holes in the subsequent rows are exposed by the same single pulse energy, but the pulse number is increased. The final energy at every hole is obtained by summing the energy of the individual pulses. The sample, after micromachining, is taken for surface profile measurement using a Tencor P-1 long range scan profiler. The hole depth and the total energy of each row are averaged to minimize the influence of surface non-uniformity and laser energy fluctuation during scanning.

AlN thin films were deposited on Silicon (111) and r-plane sapphire substrates by plasma source molecular beam epitaxy (PSMBE).[7] Crystal morphology and high quality of AlN were already reported by our group. Considering the space allowed in this paper, we refer the sample quality information of AlN to our former reports. [8, 9, 10]

Two arrays were ablated on raw Silicon (111) substrates at single pulse energy of 0.5 mJ. In the first array, the number of laser pulses were increased from 10 to 20 in one pulse increments; in the second array, the number of laser pulses were increased from 10 to 100 in increments of 10. Feature height vs. total pulse energy is shown in Fig.4. The sample surface before laser irradiation is set 0. Linear fitting results in an ablation rate of 152 angstrom/mJ.

Arrays were then micromachined in a 2400 angstrom AlN thin film on a Si (111) substrate at single pulse energy of 0.1mJ. Upon one laser pulse, the irradiated sample surface increased to a height 216.3 angstroms above the film surface, subsequent shots ablate the film at a non-linear rate until, after 10 shots, the film is ablated 259.6

angstroms lower than the surface. The mechanism for this phenomenon is unknown, and is still under investigation. It may possibly be the surface morphological change due to the crystal regrowth after low energy laser irradiation. Increasing the pulse number, we find the hole depth reaches 1500.9 angstroms after 30 pulses. At 40 pulses the bottom of the laser drilled hole is no longer flat, but irregular. We believe the AlN film has been ablated to the interface. However, increasing the pulse number to 100, edges of the hole are sharp cut and the hole bottom is flat at the depth of 3135.9 angstroms.

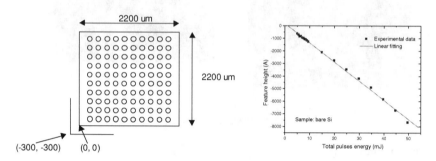

Figure 3 10X10 HOLE ARRAY   Figure 4 ABLATION RATE FOR SILICON (111)

Feature height vs. total energy for the first AlN thin film on Si(111) is plotted in Fig.5a. The plot can be divided into three regions A, B and C, based on the different slopes. The first region describes AlN thin film ablation with an ablation rate of 690 angstrom/mJ. Region B describes ablation of the interface between AlN and Si, while region C indicates ablation of the Si substrate. In region B, we did not put in the data points since line profiles of laser ablated holes show dramatically large fluctuation of hole bottom depth values, and it has no statistical meaning, which may be taken as an indicator of the film thickness. It is the same reason for not showing data points in region B within the subsequent figures. For this sample, fluctuation occurs in the range of 2200 to 2500 angstroms, which is in agreement with the previously measured film thickness value.

A second AlN thin film with a thickness of 2450 angstroms on Silicon (111) was micromachined at single pulse energy of 0.164 mJ with similar incremental increases in the number of pulses. Feature height vs. total energy for the second AlN thin film on Si(111) is plotted in Fig. 5b. Similar to the first AlN on Si sample, Figure 5b shows three distinct regions. Region A shows the ablation of AlN at a rate of 1567 angstroms/mJ, region B suggests the thickness of AlN is approximately 2300 angstroms, close to the measured value. Region C shows the ablation of the Si substrate at a rate of 67 angstroms/mJ. Several steps are found with increasing number of pulses. The ablation rate of AlN for this sample is greater than that of the former sample likely due to the different AlN crystalline quality, however, further characterization needs to be done for these two samples.

AlN with thickness of 2150 angstroms on r-plane sapphire was also micromachined. The single pulse energy was set at 0.166 mJ. The ablation rate was determined from Fig. 6. At 1 pulse shot, the laser irradiated sample point ablates 194.2 angstroms; however after only 2 pulses, the line profile measures a depth of 2530.9

angstroms, showing that AlN has been completely ablated. From Fig. 5 it is estimated that 3 mJ laser pulse energy would remove this thickness of AlN. The ablation rate for saphire is found to be 656.7 angstroms/mJ, much higher than that of AlN on Si(111). Thermal effects as well as differences in crystal quality may explain the ablation rate difference of the AlN thin film on sapphire.

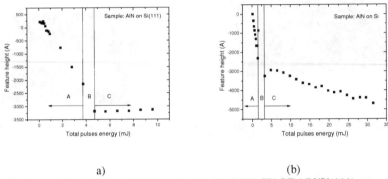

a)   (b)

Figure 5 FEATURE HEIGHT VS. PULSE ENERGY OF AlN/Si(111)

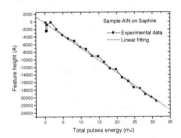

Figure 6 FEATURE HEIGHT VS. PULSE ENERGY OF AlN/Sapphire

Figure 7 shows a 3D surface profile image of laser a micromachined array on AlN/Si (111) as observed by an optical profiler from Veeco- NT2000. Please note that image is presented inverted for clarity. With the sharp cut groove structure, this laser micromachined feature may be utilized as a waveguide structure. Employing an image mask, a pixilated structured was fabricated micromachining AlN on SiC. 2000 pulses with a single pulse energy of 10mJ at the sample surface, resulted in a pixilated structure which may find applications in biosensors (see Fig.8).

CONCLUSION

In this paper, we have determined the ablation rates for AlN on different substrate materials, including Si (111) and r-plane sapphire. Linear dependence of the laser micromachined feature depth with the incident laser pulse energy within thin film and the substrate regions has been observed. Employing the experimentally obtained ablation rate for AlN, a waveguide structure and a pixilated structure were fabricated.

Figure 7 3D OPTICAL PROFILE IMAGE OF LASER MICROMACHINED ALN/Si(111)

Figure 8 SEM IMAGE OF MICROPATTERNING ON ALN/Si(111)

ACKNOWLEDGEMENTS

This work was supported by Smart Sensor and Integrated Devices Program at Wayne State University and National Foundation Grant 9870720. The authors gratefully acknowledge Lihua Li's help on using Tencor P-1 scanning profiler. We also thank Dan Durison and Thomas Daley for their technical support and helpful discussions.

REFERENCES
1. Eli Wiener-Avnear, 105 (November 1998).
2. D. Bauerle, *Laser Processing and Chemistry*, (Springer-Verlag, New York, 1996).
3. E. Fogarassy, D. Geohegan and M. Stuke, Eds., in *Proceedings of the Third International Conference on Laser Ablation*, Vols.96-98 of Applied Surface Science, (Elsevier, Amsterdam, 1996).
4. Gregory P. Behrmann and Michael T. Duignan, Applied Optics.36(20), 4666 (July 1997).
5. Xiaomei Wang, James R. Leger and Robert H. Rediker, Applied Optics.36(20), 4660(July 1997).
6. Jaeheon Han, Applied Physics Letters.74(3), 445 (January 1999).
7. G. W. Auner, T. Lenane, F. Ahmad, R. Naik, P. K. Kuo and Z. L. Wu, Wide Bandgap Electronics Materials, M.A.Prelas, (Eds), 392 (Kluver Publishers 1992).
8. G. W. Auner, F. Jin, V. M. Naik and R. Naik, J. Appl. Phys. 81 (1999).
9. M. Thompson, A. Drews, C. Huang and G. W. Auner, MRS Internet J. Nitride Semicond. Res. 4S1, G3.7 (1999).
10. R. Krupitskaya and G. W. Auner, J. Appl. Phys.84, 2861(1998).

Mat. Res. Soc. Symp. Vol. 595 © 2000 Materials Research Society

Wet Etching of Ion-implanted GaN Crystals by AZ-400K Photoresist

C.A. Carosella, B. Molnar, S. Schiestel[1] and J.A. Sprague
Naval Research Laboratory, Washington DC 20375
[1]George Washington University, Washington, DC

ABSTRACT

The photoresist developer AZ-400K, commonly used to remove AlN encapsulant layers on GaN crystalline films, is found to also etch certain as-grown GaN films. Even as-grown GaN films, which can not be etched in AZ-400K, however can be etched if amorphized by ion implantation. Etch rates of as high as 450 Å/min. were observed. The etching proceeds linearly in GaN in the first few minutes to a depth corresponding to the depth of the amorphous region. Subsequently, the etching rate saturates. Annealing of the highly amorphized samples up to 1000°C for one minute in a N_2/H_2 gas mixture does not reduce the etch rate, but for lower doses we observed a reduction of the etch rate. Observations of etching depth under various ion-implanted conditions could be correlated with the number of displacements per atoms (dpa) required for amorphization.

INTRODUCTION

The semiconductor GaN is currently the subject of interest for technological innovations in optoelectronics and in high-power, high-temperature device operations [1,2]. Ion implantation is an accepted method for integration of such devices into circuits. Doping of GaN by ion implantation requires an encapsulant layer such as AlN for minimization of the GaN decomposition during high temperature activation annealing [3]. The use of an encapsulant layer demands a method for the selective removal of this layer after the annealing treatment. The photoresist developer, AZ-400K is reported to be such a selective wet etching agent for AlN over GaN [4]. Wet etching of GaN has been studied in basic and acid solutions [5,6,7]. Recent studies indicated that AZ-400K does not etch GaN at solution temperatures up to 80°C [8,9]. This paper demonstrates that GaN can be etched in AZ-400K after ion implantation.

EXPERIMENTAL

The wurzite GaN single crystal films used in this study were grown either by chemical vapor deposition (CVD) or molecular beam epitaxy (MBE) on c-plane (0001) oriented sapphire substrates. They were grown at eight different laboratories. The layer thicknesses were between 1 and 15 μm. The layers were highly resistive semi-insulating (SI) or conductive n- or p-type.

The etching experiments were conducted by immersing GaN samples, partly covered with apiezon wax in the photoresist developer AZ-400K at room temperature and 80°C. The etching depth and surface roughness were measured with a KLA Tencor P-10 surface profilometer. The GaN samples were implanted either with B^+ and Ar^+ ions in order to study the ion dose, energy and mass dependence of the damage. The ion energies were varied between 30 and 180 keV, the ion doses from $1*10^{14}$ to $5*10^{16}$ ions/cm². One part of the sample was masked during implantation.

Some samples were annealed after ion implantation in a conventional tube furnace up to 1000°C in a N_2/10% H_2 gas mixture in order to study the damage recovery influence on the etching. The surface morphology of selected samples was investigated by SEM.

RESULTS

A variety of GaN films were used to study the influence of growth method, thickness, carrier concentration and mobility on etching properties. In the first step, the as-grown GaN films were immersed in AZ-400K at room temperature and at 80°C. Some of the GaN films could be etched as grown without any further treatment. Table 1 summarizes the etching results for the different, as grown GaN films. None of the MOCVD GaN films could be etched, whereas 3 out of 4 MBE grown GaN films we studied showed an etching in AZ-400K. We could not observe any influence of film thickness, carrier concentration or mobility on the etching behavior. However, damaging by ion implantation promoted etching.

Some of the MOCVD GaN films were implanted with B and Ar ions of different energies and at various ion doses. The first implantation experiments were performed with Ar ions with a constant ion energy of 100 keV and the ion dose was varied between 10^{15} and $5*10^{16}$ ions/cm². The results are presented in fig. 1. The implanted GaN layers turned brown with increasing ion dose and for the highest ion dose of $5*10^{16}$ ions/cm² a dark brown, metallic shiny layer was obtained. Four-point probe resistivity measurements on a semi-insulating GaN film implanted with $5*10^{16}$ Ar ions/cm² showed a conductance of about 100 $(\Omega cm)^{-1}$.

Table 1: *Different types of GaN samples used for etching experiments in AZ-400K, and their etching behavior.*

Growth type	Carrier concentration [1/cm³] $*10^{17}$	Thickness [μm]	Mobility [cm²/Vs]	1 hour 25°C	1 hour 80°C
MOCVD	1.5	2	-15	NO	NO
MOCVD	2.5	1.6	+9	NO	NO
HVPE	0.7	16	-630	NO	NO
MOCVD	3	3.1	-300	NO	NO
MOCVD	S.I.	3		NO	NO
MBE	20	2	-230	YES	YES
MBE	0.1	2.8		YES	YES
MBE	1	1		NO	NO
MBE	10	.3	+5	YES	YES

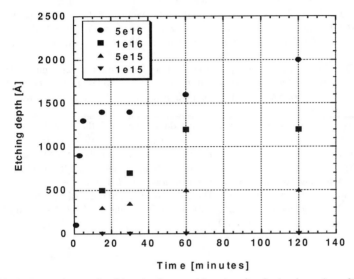

Fig.1: *Dependence of etching depth on etching time for the implantation of 100 keV Ar ions at different ion doses.*

After ion implantation, the GaN films were immersed in AZ-400K 80°C. Fig. 1 shows the etching depth for different Ar ion doses and etching times. We observe a linear etching during the first five minutes for the highest ion dose ($5*10^{16}$ ions/cm^2, circles). The surface roughness of the etched area is comparable to the one of the untreated surface and amounts to $\pm$ 100 Å. After 5 minutes the etching profile seems to saturate at about 1400 Å, which roughly corresponds to the damage region induced by ion implantation, which was calculated/estimated by TRIM [10]. After 60 minutes pores start to develop at the etched surface, up to 1500 Å in depth. Simultaneously with the development of pores at the etched surface, the etching depth increases slightly more. The etching behavior for the highest implantation dose of ($5*10^{16}$ ions/cm^2) differs from the ones implanted at lower doses. For all samples implanted with an ion dose of $\leq 10^{16}$ ions/cm^2, the etching of the implanted area was accompanied immediately by the presence of deep pores (500 – 1500 Å). The etching depth (saturation) decreased with decreasing ion dose and was 1200 Å for 10^{16} ions/cm^2 and 500 Å for $5*10^{15}$ ions/cm^2. No etching, except for pore formation, occurred for an ion dose of 10^{15} ions/cm^2. Pores with a depth up to 1000 Å were present after 1 hour; they increased to 1500 Å after another hour in AZ-400K at 80°C.

Tan et al. reported an amorphization of GaN for the implantation of 90 keV Si ions at ion doses $> 10^{16}$ ions/cm^2 at liquid nitrogen temperature [11]. At room temperature, the amorphization dose is higher. For example, Liu, et al. report that $3*10^{14}$ Ca$^+$ ions/cm^2 at 180 keV are necessary to initiate amorphization of GaN at 77 °K, but

$8*10^{14}$ ions/cm^2 are needed at room temperature[12]. Liu, et al. report that GaN is amorphized at 77 °K with 180 keV Ar ions at about 5-6 dpa [13]. We calculate the number of dpa for the ion energies and ion doses of our experiments using TRIM. The results of these calculations for 100 keV Ar ions and ion doses from 10^{15} to $5*10^{16}$ ions/cm^2 are shown in fig. 2. For the lowest ion dose, the number of dpa is less than one. Therefore, it is not surprising that for this ion dose no etching of the implanted GaN was observed. For the higher ion doses, the depth of the damage region roughly corresponds to the saturation level of the etching depth.

TRIM calculations for the implantation of $1*10^{16}$ B$^+$ ions/cm^2 with ion energies of 30 and 100 keV revealed a number of dpa at the damage peak of less than 3, too small to amorphize GaN. Experimental results for GaN implanted with B ions at these doses and energies revealed no etching.

In another set of experiments we examined the annealing influence at 900°C and 1000°C in N$_2$/H$_2$ gas on implanted GaN films (100 keV Ar ions, 10^{15} - $5*10^{16}$ ions/cm^2). The etch rate for the sample implanted at $5*10^{16}$ ions/cm^2 before and after annealing is the same, and almost all of the etching is completed in the first five minutes. In contrast, the annealing slows the etch rates for the lower implantation doses ($<10^{16}$ ions/cm^2) and no saturation was observed. There is also a delayed onset of etching in these samples. At 10^{16} ions/cm^2, etching commences at 30 minutes; at $5*10^{15}$ ions/cm^2, etching begins after 1 hr in AZ-400K at 80 °C.

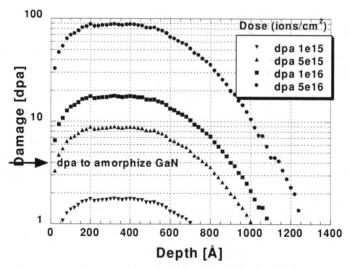

Fig.2: *Damage calculations for the implantation of 100 keV Ar ions at different ion doses. The dpa to amorphize GaN is the approximate threshold from references [11] and [13].*

Mat. Res. Soc. Symp. Vol. 595 © 2000 Materials Research Society

OXIDATION OF GALLIUM NITRIDE EPILAYERS IN DRY OXYGEN

P.Chen, R.Zhang , X.F.Xu, Z.Z.Chen, Y.G.Zhou, S.Y.Xie, Y. Shi, B.Shen, S.L.Gu, Z.C. Huang[a], J. Hu[b] and Y.D.Zheng
Department of Physics, Nanjing University, Nanjing 210093, P.R.China
[a]*Raytheon ITSS, 4500 Forbes Bulivard, MD 20771*
[b]*Syscaching Corp., Aliso Viejo, CA 92324*

ABSTRACT

The oxidation of GaN epilayers in dry oxygen has been studied. The 1-μm-thick GaN epilayers grown on (0001) sapphire substrates by Rapid-Thermal-Processing/Low Pressure Metalorganic Chemical Vapor Deposition were used in this work. The oxidation of GaN in dry oxygen was performed at various temperatures for different time. The oxide was identified as the monoclinic β-Ga_2O_3 by a θ-2θ scan X-ray diffraction (XRD). The scanning electron microscope observation shows a rough oxide surface and an expansion of the volume. XRD data also showed that the oxidation of GaN began to occur at 800°C. The GaN diffraction peaks disappeared at 1050°C for 4 h or at 1100°C for 1 h, which indicates that the GaN epilayers has been completely oxidized. From these results, it was found that the oxidation of GaN in dry oxygen was not layer-by-layer and limited by the interfacial reaction and diffusion mechanism at different temperatures.

I. INTRODUCTION

With the great progress of the applications of GaN on optoelectronic devices and high temperature / high power electronics, GaN is more attractive than ever[1,2]. The metal-oxide-semiconductor field effect transistor (MOSFET) based on β-Ga_2O_3/GaN has been reported[3], and it shows a better performance at 400°C than that at room temperature. In the many polymorphs of Ga_2O_3, β-Ga_2O_3 is believed to be the equilibrium phase. It is very important to understand the oxidation mechanism of GaN at high temperature for the realization of these electronic devices. The kinetics of the oxidation of GaN powder in dry air has been reported by Wolter *et al.*[4]. An oxide forms at 900°C for 1 h, and is identified as the monoclinic β-Ga_2O_3. The initial stage of the oxidation is found to be limited by the rate of an interfacial reaction. The same mechanism was also found in the oxidation of GaN epilayers at 900°C in dry air in their report. However, there is little known about the oxidation of GaN epilayers in dry oxygen.

In this letter, we report a scanning electron microscope (SEM) observation and the oxidation behavior of GaN epilayers at temperatures from 800°C to 1100°C in dry oxygen. The results reveal the different oxidation mechanism at different temperatures.

II. EXPERIMENT

The GaN epilayers (about 1-μm thick) used in the experiments are grown on a *c*-plane of sapphire substrates by rapid thermal processing low-pressure metalorganic chemical vapor deposition (RTP/LP-MOCVD). The detailed growth process has been published previously[5]. The full width at half maximum (FWHM) of 9.8 min of X-ray rocking curve (XRC) indicates the high quality of the GaN film. A 1.5-inch diameter sample is divided into several pieces for oxidizing under various conditions. After cleaned in the solvents and DI water, the samples are etched in HCl:DI water (1:1) to remove surface contamination and the native oxide. Then the samples are placed in a quartz boat, which will be transferred into a horizontal quartz tube, and the tube is placed in a electric furnace to oxidize the samples at temperatures (from 800°C to 1100°C) for a certain time (from 10 min to 6 h). During the oxidation, dry oxygen flow maintains 1 SL. After completing each experiment, the quartz boat is brought out to room temperature within 5 min.

The samples are studied by the θ-2θ scan X-ray diffraction (XRD) with Cu K$_\alpha$ radiation and SEM.

III. RESULTS AND DISCUSSION

The oxide is characterized by a θ-2θ scan XRD. The XRD data of GaN epilayers which are oxidized at temperatures 800°C, 900°C, 1000°C, 1050°C and 1100°C for 1 h are shown in fig.1. The data for various oxidation times are also collected. After oxidized at 800°C for 6 h, the crystalline oxide is observed and identified as the monoclinic β-Ga$_2$O$_3$ phase by comparing the data list in *Powder diffraction file*, **11-370**.

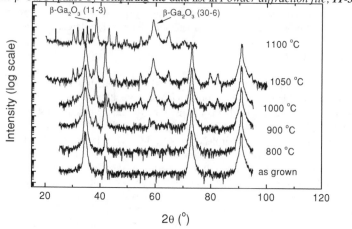

Fig.1. The XRD data of GaN epilayers which were oxidized at temperatures of 800°C, 900°C, 1000°C, 1050°C and 1100°C for 1 h. The GaN peaks disappeared at 1100°C for 1 h.

The GaN films oxidized at lower than 1050°C show a relative smooth surface to eyes, but show a serious rough surface at 1050°C for 1 h or 1100°C for 35 min. The intensity of the oxide peaks increases quickly with the increasing oxidation time, and GaN peaks eventually disappear at 1100°C for 1h, which indicates a very high oxidation rate in this case. The recent studies of thermal stability of GaN film have also shown that the GaN film is unstable over 1000°C[6]. From fig.1, it can be seen that the strongest peak of β-Ga_2O_3 is (11-3), and the second strongest is (30-6). Either (11-3) or (30-6) is not the peak which has the highest intensity listed in *Powder diffraction file, 11-370* and reported in the oxidation of GaN powder by Wolter[4]. This fact indicates that there may be some selective relationship between the oxide orientation and (0001) GaN. The little change of the XRD intensity ratios (1.2~1.6) between (11-3) and (30-6) at various oxidation conditions indicates that the structure of the oxide film on (0001) GaN is almost unchangeable in this case.

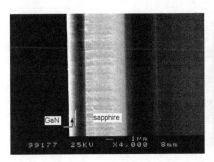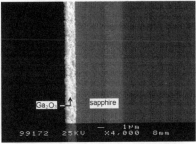

Fig.2. SEM images of a GaN film and an oxide layer. (a), the 1 μm-thick GaN film before oxidation; (b), completely oxidized at 1050°C for 2h and the thickness of the oxide layer is about 1.4 μm.

Fig.2 shows the cross-sectional SEM images of a GaN epilayer and an oxide layer formed at 1050°C for 2h. The GaN layer is almost completely oxidized at 1050°C for 2 h. Contrast to a very smooth surface of the GaN film before oxidation (fig.2a), the oxide layer shows a very rough surface (fig.2b), and has a number of oxide grains. An MOCVD-grown GaN epilayer on sapphire always has high density of dislocation, which may prevent the GaN film from the oxidation on the way of layer-by-layer. From the images, a significant expansion of the volume is clearly observed. The thickness of the oxide layer formed by a 1 μm GaN is about 1.4 μm, which indicated an expansion of the volume of 40%.

Since the intensity of XRD diffraction peak is related to the volume of the sample[7] and the structure of the oxide film on (0001) GaN is almost unchangeable. One can chose the intensity of β-Ga_2O_3 (11-3), I, to monitor the oxide thickness. It should be noticed that the premises of the approach are the fixed X-ray power. In addition, the very small X-ray absorption by β-Ga_2O_3 film, less than 3% in a ~1 μm-thick gallium oxide film [4], indicates a thickness which is much greater than the estimated oxide thickness in this work.

Fig.3 shows the intensity of the β-Ga_2O_3 (11-3) diffraction peak plotted *vs.* oxidation time at different temperatures. The two different oxidation processes can be clearly observed over 1000°C. When the oxidation temperature is higher than 1000°C, there is a rapid oxidation process in the initial stage of oxidation, followed by a relatively slow process. The oxidation rate in the initial stage is dependent on the oxidation temperature, which can not be observed at lower temperatures. Obviously, the top oxide film formed in the rapid process could be affected by the decomposition of GaN film.

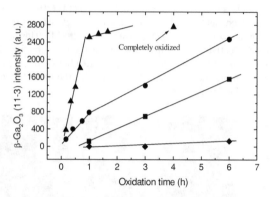

Fig.3 The intensities of β-Ga$_2$O$_3$ (11-3) peak plotted vs oxidation time for different temperatures, 800°C (diamond); 900°C (square); 1000°C (circle); 1050°C (triangle). Two different oxidation process were observed over 1000°C.

The equations for the oxide growth have been described by Wolter[4]

$$I^n = k \cdot t \qquad (1)$$

$$n \cdot \ln(I) = \ln(k) + \ln(t) \qquad (2)$$

where n is the reaction order, k is the reaction rate constant, t is the oxidation time. If n is equal to 1, the oxidation is under interfacial reaction-controlled; if n is equal to 2, the oxidation is under diffusion-controlled. The reaction rate k is the function of the oxidation temperature T and the activation energy. Fig.4 displays a plot of $\ln(I)$ vs. $\ln(t)$ from fig.3, which is used to determine the reaction order during the slow oxidation process and the rapid oxidation process. From fig.4, a linear relationship exists for all temperatures, but the slops which are equal to $1/n$ from eq.2 are different each other. During the slow oxidation process, the values of n increase from 0.71 at 900°C to 1.6 at 1000°C, which indicates that the dominating oxidation mechanism changes from interfacial reaction-controlled to diffusion-controlled with increasing the oxidation temperature. It is beyond our expectation that the n is 7.8 at 1050°C. At this temperature, the film is almost completely oxidized, and the top oxide layer may affect the later oxidation of the GaN film. Therefore, the oxidation at 1050 °C after 1 hour is different from the oxidation under other conditions. It is very different from the oxidation of GaN powder in dry air. However, during the rapid oxidation process, the values of n are 1.1 at 1000°C and 0.89 at 1050°C, which indicates that the dominating oxidation mechanism is under interfacial reaction controlled.

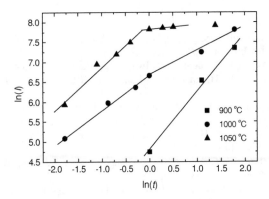

Fig.4   A plot of ln(I) vs ln(t) from fig.3. A linear relationship exists for all temperatures, but the slops which are equal to 1/n from eq.2 are different each other.

IV. CONCLUSIONS

In conclusion, the oxidation of GaN epilayers in dry oxygen has been investigated. The θ-2θ scan XRD data identify that the monoclinic β-Ga_2O_3 begin to form on GaN epilayers at 800°C for 6h. The GaN epilayers are completely oxidized at 1050°C for 4 h or at 1100°C for 1 h. Two different oxidation processes are found when the temperature is over 1000°C. There is a rapid oxidation process in the initial stage of oxidation, and a relative slow process follows. The oxidation is limited by the rate of an interfacial reaction mechanism in the rapid oxidation process and changes from the interfacial reaction-controlled mechanism to the diffusion-controlled mechanism in the slow process. When the temperature reaches 1100°C, the oxidation rate is very fast, which is considered as the results of the GaN decomposition at high temperature. SEM images show the rough surface of oxide layer and a number of oxide grains in it. An expansion of the volume of 40% is also observed.

Acknowledgements

This work was supported by National Natural Science Foundation of China with contracts #69976014, #69636010, #69806006, #69987001, the National High Technology Research & Development Project of China, and MOTOROLA(China Inc.) Semiconductor Scholarship.

REFERENCES

1. S.Nakamura, M.senoh, S.Nagahara, N.Iwasa, T.Yamada, T.Matsuahita, H.Kiyoku and Y.Sugimoto, *Jpn.J.Appl.Phys.* **135**, L74 (1996).
2. S.Nakamura, M.senoh, S.Nagahara, N.Iwasa, T.Yamada, T.Matsuahita, H.Kiyoku, Y.Sugimoto, T.Kozaki, H.Umemeoto, M.Sano, and K.Chocho, *Appl.Phys.Lett*, **72**, 2014 (1998).
3. F.Ren, M.Hong, S.N.G.Chu, M.A.Marcus, M.J.Schurman, A.Baca, S.J.Pearton and C.R.Abernathy, *Appl.Phys.Lett*, **73**, 3893 (1998).
4. S.D.Wolter, S.E.Mohney, H.Venugopalan, A.E.Wickenden and D.D.Koleske, *J.Electrochem.Soc.*, **145**, 629 (1998).
5. B.Shen, Y.G.Zhou, Z.Z.Chen, P.Chen, R.Zhang, Y.Shi, Y.D.Zheng, W.Tong and W.Park, *Appl.Phys.A*, **68**, 593 (1999)
6. D.D.Koleske, A.E.Wickenden, R.L.Henry, W.J.DeSisto, and R.J.Gorman, *J.Appl.Phys.*, **84**, 1998 (1999).
7. L.S.Zevin and G.Kimmel, in *Quantitative X-Ray Diffractometry*, I.Mureinik, Editor, Springer-Verlag, New York (1998).

Mat. Res. Soc. Symp. Vol. 595 © 2000 Materials Research Society

A damage-reduced process revealed by photoluminescence in photoelectrochemical etching GaN

J. M. Hwang, J. T. Hsieh and H. L. Hwang
Department of Electrical Engineering, National Tsing-Hua University,
Hsin-Chu 300, Taiwan, R.O.C.

W. H. Hung
Synchrotron Radiation Research Center, Hsin-Chu 300, Taiwan, R.O.C.

Abstract

Photoelectrochemical (PEC) etching technique has been proven to be an effective method to etch GaN. Despite its success, investigations on etching-induced damage are still scare. In this work, the damage induced by PEC etching of GaN in KOH electrolyte was studied. Photoluminescence (PL) spectroscopy was used to explore the origin of etching-induced damaged layer. From the variable temperature PL measurements, the origin of etching-induced damage was attributed to be the defect complex of V_{Ga}-O_N (gallium vacancy bonds to oxygeon on nitrogen antisite). With determination of the defect origin, the electronic transition in the etch damage-related yellow luminescence (YL) band was suggested to be deep donor-like state to shallow-acceptor transition. In addition, a post-treatment method with boiled KOH chemical etching was developed to remove the thin damaged layer. In this method, crystallographic etching characteristics of boiled KOH was observed to assist in the formation of smooth sidewall facets. As revealed by the reduction of yellow luminescence, we propose this novel technique as a near damage-free etching method.

Introduction

Since GaN is chemically stable and insoluble in all mineral acid and base solutions at room temperature [1], most processing of III nitrides are currently done by dry plasma etching. High density plasma or energetic ion assisted etching were used to get a smooth etch surface and highly anisotropic sidewalls with high etch rates [2-4] . But there are several disadvantages to use dry etching, including generation of the ion-induced damage and difficulty in obtaining smooth etched sidewalls, which are required for the lasers cavity. The optical damage relative to yellow luminescence (YL) was also extensively investigated using photoluminescence spectroscopy on the etched surfaces. [5] The etching-induced damage will have detrimental effect on electronic or photonic device performance. Therefore, a reliable fabrication process producing low etch damage for GaN-based devices is necessitated.

Photoelectrochemical (PEC) etching is an alternative method to produce smooth surfaces and vertical sidewall facets without ion bombardment induced damages.[6-8] As it is well-known that the yellow luminescence commonly existing in epitaxial grown GaN is related to the intrinsic defects or impurities. In this study, we used YL intensity variation to identify the etching-induced damages. A novel boiled KOH post treatment process was also developed to remove the etching-induced damages and achieve vertical and smooth facet sidewalls. A near damage-free etching process with vertical sidewalls and smooth etched surface can be realized by this post treatment process. This etching process could be applied to device fabrication such as laser cavity formation.

Experimental

The unintentionally-doped n-GaN samples (conc. $< 5*10^{16}$ cm^{-3}) used in this study were obtained from CREE Research, Inc. The thickness of nitride layer is 1.4 μm. Standard lift-off procedure was used to form the Ti metal mask that it is served as a contact for better photocurrent conduction. The GaN samples were sequentially cleaned by methanol and acetone and then rinsed with deionized water. A sample holder with two Teflon plates was used to fix the GaN samples. The PEC etching of GaN was carried out in a Teflon electrochemical cell containing an aqueous KOH electrolyte. The UV source was an Oriel high-pressure Hg-arc lamp, in which a water filter was mounted on the output window to remove infrared irradiation. The post-treatment was done with boiled KOH solution. For all photoluminescence (PL) measurements, a He-Cd laser (λ =325 nm) , typically operated at 1mW output power and focused to a spot with a diameter of 100 μm on the sample surface, was used as the excitation source. The luminescence was collected by a 50mm f/1.4 lens in the direction normal to the illuminated surface, and then focused into a SPEX 500M monochromator with an electrical-cooled GaAs photomultiplier. The etching profile was measured by HITACHI S-40000 FESEM.

Results and discussion

Our experimental results indicate that the etching rate is linearly proportional to the power density and it increases with increasing the KOH concentration. The relation of etching rate v.s. power density and KOH concentration is very similar to the published results [2-4]. Photoluminescence spectra corresponding to the etching condition with various UV power densities are shown in **Figure 1**, in which the 470W, 600W and 700W UV power corresponds to the power densities of 195 mW/cm^2, 240 mW/cm^2 and 290 mW/cm^2 , respectively. All PL spectra are normalized to the same intensity scale of near-band-edge (NBE) peak at 364 nm. Spectrum A in Figure 1 shows the room temperature (R.T.) PL of the as-grown samples.

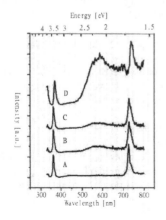

- A - Unetched GaN surface
- B - PEC etched GaN surface at 470 W
- C - PEC etched GaN surface at 600 W
- D - PEC etched GaN surface at 700 W
  KOH concen. = 0.02 M

Figure 1. Defect induce yellow luminescence in PEC etching GaN with different UV power density

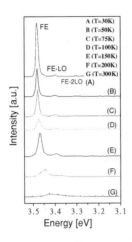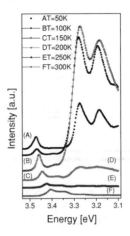

Figure 2. Temperature dependence photoluminescence of GaN (a) as- growth GaN with PL peaks as FE and phonon replica. (b)PEC etching GaN with PL peaks as FE, dislocation, DAP and phonon replica.

The peak at 364 nm (3.41 eV) is attributed to be the NBE emission and the 728 nm (1.705 eV) is the second harmonic counterpart. After etching with UV power density of 195 mW/cm^2 in 0.02M KOH electrolyte, a yellow luminescence band centered at 563.6nm (2.2 eV) appeared (spectrum B). In spectrum C, the yellow luminescence band remained the same as in spectrum B while increasing the power density to 240 mW/cm^2. With the successive increase of the power density to 290 mW/cm^2 for a longer etching, the very strong YL dominated and a dislocation-related peak at 370 nm (3.351 eV) emerged [9] as shown in spectrum D. This means that the etched GaN surfaces were seriously damaged such that the YL-related recombination centers were formed to annhilate the exciton-related luminescence due to trapping of excitons by the etching-induced defect traps. The YL relative to etching-induced damage will increase when etching GaN in the condition of higher power and/or longer time. To investigate origin of the etching-induced YL band emission, variable temperature PL measurements were done. **Figure 2(a)** shows the temperature-dependent PL of as-grown GaN samples. As were also observed by other authors, the free exciton (FE) peak position shifted to higher energy (blue shift) with decreasing temperature. The temperature coefficient of FE peak was determined to be 2.93x10^{-4} eV/K (100°K<T<300°K). Two regular LO-phonon replicas FE-LO and FE-2LO were observed in **Figure 2(a)** due to dominance of the polariton-phonon scattering at low temperature [10]. After PEC etching of GaN with UV power density of 322 mW/cm^2 and KOH concentration of 0.05M for 30 minute, the R.T. PL spectrum exhibited a new emission that is assigned to be donor-acceptor-pair (DAP) transition. The corresponding variable temperature PL spectra are shown in **Figure 2(b)**. By successively decreasing the sample temperature to 200°K and below, an obvious enhancement of DAP emission was observed. Compared to the NBE transition, the peak position of DAP was less temperature-dependent [11] and was pinned at 3.229 eV. Moever, the photon from DAP transition can interact with the phonon to create the DAP-

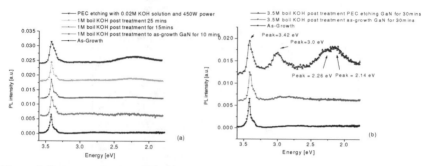

Figure 3. Damage remove by PEC etching GaN (a) low concentration of boil KOH with remove the damage layer produced YL in short time treatment. (b) high concentration on of boil KOH and low time treatment will create damage to GaN surface

LO coupling with the gain or loss of one phonon energy. At temperature below 100 K, the LO phonon energy is about 91 meV that is consistent with the observation of the A1 (LO) mode (734.75 cm^{-1}) in our FTIR measurement. In addition, we also found that the variation of YL intensity is connected to DAP emission intensity, that is, the YL intensity increased as the DAP emission intensity increased. This seems to imply that these two radiative recombination paths share the same electronic transition levels. In the following paragraph, we will dedicate to build a defect model to explain the observed phenomena in our experiments.

In as-grown GaN film, a weak YL band exists too. It appears to have the same origin as the induced defects during PEC etching in our experiment. YL involves transitions from a deep state in the upper half of the gap either to the valence band, or to a (relatively) shallow acceptor. They are also termed as the shallow-donor to deep-acceptor transition and the deep donor-like to shallow-acceptor transition. The complex defect V_{Ga}-O_N create a deep donor-like state (DD$^-$) above the VBM about 2.2 eV. The excited carriers escape from the shallow donor states created by V_N (Nitrogen vacancy) to DD$^-$ state in a nonradiative recombination process and then to a shallow acceptor with the YL emission (2.2 eV) [12]. Our results support this model and the induced defect during PEC etching process were suggested to be Ga vacancy bonds to oxygen in the nitrogen site. The possible reaction mechanism is as follows. The OH$^-$ coming from KOH electrolyte prefers to react with Ga rather than N to form gallium oxide that subsequently dissolves into the solution. A gallium vacancy will be therefore created to bond with the neighboring antisite defects arising from the occupation of nitrogen sites by the oxygen atoms. As seen from the temperature-dependent PL spectrum, a YL transition from DD$^-$ to a shallow acceptor level was identified. At low temperature, this transition was quenched. The excited carriers follow another recombination path from a shallow donor state to a shallow acceptor state, that is, the DAP transition. The YL intensity is strongly dependent on the amount of the acceptor states. The less the acceptor states, the weaker the YL intensity, and therefore the weaker then DAP peak intensity. This coincides with our results. From the above discussion, we would suggest that the defect is the defect complex of V_{Ga}-O_N rather just

V_{Ga}. Further support for the energy position assignment of the observed YL stems from the AES spectra.

The surface composition of as-growth with Ga:N (1.49:1) sample. From the PL of the etching surface with less YL, the surface composition of Ga:N (1.22:1) shown. While strong YL is created by PEC etching, the surface composition of Ga decreased with GaN (1.11:1). From the surface composition analysis, we can find that the surface gallium decreases while the YL intensity increases and the surface oxygen increases while YL intensity increases. This indicates that the YL intensity will increase as the concentration of V_{Ga}-O_N defect complex increases.

The damaged layer induced by PEC etching of GaN was stable and can not be removed by the conventional chemical etching methods such as HCl:H$_2$O (1:1), HF:H$_2$O (1:1) or mixed solution at room temperature or elevated temperature. This can be confirmed from the PL measurement in which the YL band still exists after various chemical treatments as listed above. A novel skill with a post-treatment process in boiled KOH solution was evaluated for damage removed. In **Figure 3(a)**, the PL spectrum from as-grown GaN samples was shown. No YL band is found on the as-grown samples. After PEC etching with power density of 186 mW/cm^2 and KOH concentration of 0.02M, the defect complex V_{Ga}-O_N relative to YL was present. After 1M boiled KOH post-treatment for 10 minutes, the YL intensity was largely reduced. After 15 minutes of treatment, the PL spectrum also showed a weaker YL intensity. This means that the etching-induced damage by PEC etching of GaN has been removed after the hot KOH post-treatment. But after 25 minutes of treatment, YL band emission recovered again. The possible reason is that the boiled KOH start to etch GaN after removal of the damaged layer and it will produce YL related defects as the PEC etching did before. To identify the damages created by KOH solution in the condition of high concentration and long times, 3.5M KOH was used to etch GaN for 30 minutes. The PL spectra for the etched and unetched regions are shown in **Figure 3(b)**. It can be clearly seen that blue luminescence (BL) band centered at 3.0 eV emerges besides the YL band with peaks at 2.2 eV and 2.14 eV. The appearance of BL band seems to reveal the creation of a new type of etching-induced defect. The origin of the BL remains as an open question. Based on above results, the boiled KOH post-treatment with lower concentrations and/or for shorter time is suggested to remove the damaged layer roduced by the PEC etching of GaN.

Another property of crystallographic wet etching in boiled KOH was useful for device processing and laser cavity formation. [13] The boiled KOH stops etching GaN at {10-10} and (0001) faces and prefers etching dislocation sites. As shown in **Figure 4(a)**, the stripped sidewall and residual dislocations on the etched surface of (0001) will limit the application to electronic or optical devices. To eliminate this limitation, boiled KOH was used to treat the PEC etched surface and the result is shown in **Figure 4(b)**. The sidewall was smoothed and the dislocations were removed after the post-treatment by boiled KOH. By virtue of careful control of etch time and KOH concentration, a very smooth and vertical facet was achieved as shown in **Figure 4(c)** at {10-10} face. The striation on sidewall edge was resulted from pattern transfer during the lithography step.

Conclusions

In summary, YL-related defects induced during the PEC etching process were

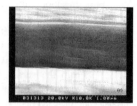

Figure 4. SEM of PEC etching GaN (a) PEC etching with sidewall roughness and dislocation residual at the etching surface (b)boil KOH solution post-treatment to remove residual dislocation (c) a vertical and smooth etching profile by crystallography etching with boil KOH. solution

studied. By using PL spectroscopy, the defects were suggested to be V_{Ga}-O_N. After post-treatment in a boiled KOH solution with lower concentration for a short time, the thin damaged layer was removed. With help of the crystallographic wet etching in boiled KOH solution, smooth etch surfaces and facet sidewalls can be achieved. The residual dislocations are also taken off by boiled KOH. This novel process provides a near damage-free etching for electronic and optical device application.

Acknowledgements

Some samples was provided by Opto-Electronics and System Laboratories, Industrial Technology Research Institute. Financial support was provided by the and National Science Council of Republic of China (Contract No. NSC 88-2215-E-007-011).

References

1. C. B. Vartuli, S. J. Pearton,C. R. Abernathy, J. D. MacKenzie, F. Ren, J.C. Zolper, R. J. Shul, SOLID-STATE ELECTRONICS vol 41, 1947-1951 (1997)
2. R. J. Shul, A. J. Howard, S. J. Pearton, C. R. Abernathy, C. B. Vartuli, P. A. Barnes and M. J. Bozack, J. Vac. Sci. Technol. B **13** 2016 (1995)
3. A. T. Ping, A. C. Schmitz, and I. Adesida, J. of Electro. Mater. **25** 825 (1995)
4. Charles R. Eddy, Jr., MRS Internet J. Nitride Semicond. Res. 4S1, G10.5 (1999)
5. J. Y. Chen, C. J. Pan, and G. C. Chi, Solid-state Electronics **43** 649 (1999)
6. M. S. Minskey, M. White, and E. L. Hu, Appl. Phys. Lett. **68**, 1531 (1996)
7. C. Youtsey, I. Adesida, and G. Bulman, Appl. Phys. Lett. **71**, 2151 (1997)
8. C. Youtsey, I. Adesida, L. T. Romano, and G. Bulman, Appl. Phys. Lett. **72**, 560 (1998)
9. J. T. Hsieh, J. M. Hwang, W. H. Hung and H. L. Hwang, (unpublished)
10. D. Kovalev, B. Averboukh,D. Volm, and B. J. Meyer, Phys. Rev. B **54**, 2518. (1996).
11. Takashi Matsumoto and Masaharu Aoki, Japan. J. Appl. Phys. **13** 1804 (1974)
12. E. R. Glaser, T. A. Kennedy, K. Doverspike, L. B. Rowland, D. K. Gaskill., D. T. Olson, J. N. Kuznia, and D. K. Wickenden, Phys. Rev. B **51**,13326 (1995)
13. D. A. Stocker, E. F. Schubert and J. M. Redwing, Appl. Phys. Lett. **73**, 2654 (1998)

Mat. Res. Soc. Symp. Vol. 595 © 2000 Materials Research Society

Fabrication and Characterization of InGaN Nano-scale Dots for Blue and Green LED Applications

K.S. Kim, C.-H. Hong, W.-H. Lee, C.S. Kim, O.H. Cha, G.M. Yang, E.-K. Suh, K.Y. Lim, H.J. Lee, H.K. Cho[1], J.Y. Lee[1], and J.M. Seo[2].
Department of Semiconductor Science & Technology and Semiconductor Physics Research Center, Chonbuk National University, Chonju 561-756, Korea
[1]Department of Materials Science and Engineering, KAIST, 373-1 Kusong-Dong, Yusong-gu, Taejon 305-701, Korea
[2]Department of Science and Technology, Chonbuk National University, Chonju 561-756, Korea

ABSTRACT

Thin layers of InGaN were grown by metalorganic chemical vapor deposition and characterized with atomic force microscopy and high-resolution transmission electron microscopy. InGaN deposited on GaN exhibits a Stranski-Krastanov growth mode, including 2D wetting layer and 3D self-assembled quantum dots. Besides, we observed that the formed InGaN nano-scale dots have a trapezoidal shape with a {1-102} facet with respect to (0002) surface. Visible spectral range from UV to green was easily obtained by changing InGaN quantum well thickness up to 2.3 nm.

INTRODUCTION

The $In_xGa_{1-x}N$ alloys have been used as an active layer material of InGaN based blue and green light emitting diodes (LEDs) and laser diodes (LDs) because they cover a wide band-gap ranging from 1.9 to 3.4 eV. Especially, InGaN/GaN quantum well (QW) is center of interest on account of their huge applications in high brightness LED, [1] cw blue LD, [2] and optoelectronic devices. Nevertheless, the emission and growth mechanisms of InGaN/GaN QW structure is still equivocal and debating.

In recent, a few research groups have suggested that the optical emission mechanism in InGaN/GaN QW should be due to InGaN quantum dot (QD). [3] Another groups, on the other hand, asserted that the emission from the InGaN/GaN QW stems from piezo-electric effects, [4] quantum confined Stark effects [5] or potential fluctuation of InGaN QW.

In this work, we investigate the time dependent 2D to 3D mophological evolution of InGaN thin films grown by metalorganic chemical vapor deposition (MOCVD) through atomic force microscopy (AFM), high resolution transmission electron microscopy (HRTEM) measurements. Besides, their related optical emission characteristics in InGaN/GaN SQW using photoluminescence (PL) measurement.

EXPERIMENTS

All InGaN/GaN QW structures used in this study were grown by low pressure horizontal MOCVD. InGaN/GaN SQW structures are consisted of a GaN nucleation layer with a nominal thickness of 25 nm grown at 560 °C, a Si-doped GaN layer with a thickness of 1.8 um grown at 1100 °C, a low-temperature grown GaN with a thickness of

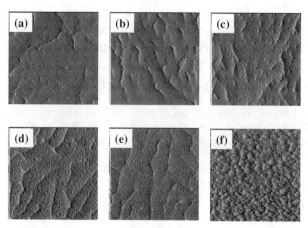

Figure 1, AFM images of InGaN films with the growth time (nominal thickness) of (a) 0 sec (0 nm), (b) 3 sec (0.15 nm), (c) 12 sec (0.57 nm), (d) 24 sec (1.15 nm), (e) 48 sec (2.3 nm), and (f) 3600 sec (400 nm). Scan areas are 4 × 4 um².

5 nm, InGaN SQW applying different growth conditions, and a 30 nm-thick GaN capping layer grown at 1100 °C. The precursors of Ga, In, Si, and N are trimethylgallium, trimethylindium, silane, and ammomia, respectively.

RESULTS AND DISCUSSION

Figure 1 (a) – (f) show the AFM images of morphological evolution of InGaN thin films grown at 800 °C, which are dependent on the growth times (thickness) from 0 sec to 3600 sec. Typical surface shape for n-type GaN, including steps, terraces, and surface dislocations, is shown in Fig. 1 (a). The small surface dislocations observed are 37 nm in size and 0.5 - 1 nm in height. The density of surface dislocations is $4 - 5 \times 10^9$ cm^{-2}, which shows good agreement with the dislocation density measured by the TEM. Many of them are lie along the step boundaries. As the nominal thickness of InGaN films increase from 0.15 to 1.15 nm, one can clearly see that 2 D (corresponding to wetting layer) to 3 D (self-assembled QD) transition, which is direct evidence of Stranski-Krastanov (S-K) growth mode of InGaN epilayer. For the case of QDs shown in the InGaN film with the thickness of 0.57 nm and 1.15 nm, the average lateral size and height of dots are 73 nm, 2 nm, and 78 nm, 3.7 nm, respectively. Especially, the density (4×10^9 cm^{-2}) of QDs shown in Fig. 1 (c) is nearly same with that of surface dislocations of low-temperature grown GaN. However, we have confirmed they have no correlation each other (see Fig. 2 (a)). If the thickness more increases, the size of the InGaN nano-scale dots is increased and the InGaN film shows polycrystalline growth mode or cluster formation.

From the x-ray photoelectron spectroscopy measurement (not shown here), we observed that In 3d peaks intense with increasing the thickness of InGaN thin films. This result shows that In surface segregation is occurred during the InGaN growth[6].

Figure 2 (a) and (b) show a plan-view HRTEM and a cross sectional HRTEM (X-HRTEM) image of the sample in Fig. 1 (c). The most important feature in (a) is that the nano-scale dots are formed irrespective of dislocation indicated by an arrow. Especially,

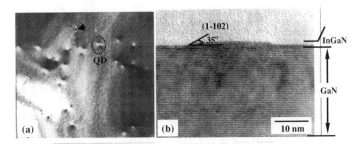

Figure 2, (a) Plan-view HRTEM weak beam image and (b) X-HRTEM image of 0.57 nm-thick InGaN thin film. Arrow shown in (a) indicates dislocations.

knowledge of the shape of self-organized InGaN/GaN QDs is needed to constrain theoretical calculations of the process formation as well as electronic structure. For the heteroepitaxial growth suffering large lattice mismatch like InAs/GaAs and Ge/Si, several quantum dot shapes, namely conical[7], square-base pyramid[8], prism-like cluster[9], triangular shape [10] etc., have been reported. Figure (b) shows that the InGaN nano-scale dot with trapezoidal shape are formed on InGaN wetting layer without dislocations and also has {1-102} facet planes inclined by 35° with respect to the (0002) surface.

Fig 3 (a) shows the room temperature PL spectra of InGaN/GaN SQW with the InGaN thickness described in Fig. 1. The FWHM values are 70 – 140 meV. The PL

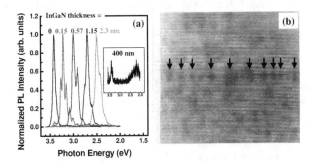

Figure 3, (a) Room temperature PL spectra of InGaN/GaN SQW with various InGaN thickness, and (b) X-HRTEM image of InGaN/GaN SQW with the InGaN QW thickness of 0.57 nm. Arrows point out the formed InGaN nano-scale dots.

spectra were recorded using the 325 nm line of a He-Cd laser with an excitation power density of 2 W/cm^2. As we changing InGaN QW thickness from 0 to 2.3 nm, broad area spectra covering from UV (3.4 eV) to green (2.45 eV) are observed. Nano-scale dots formed in InGaN SQW with the thickness of 0.57 nm are displayed in Fig. 3 (b), which definitely affect the emission from the InGaN SQW. Therefore, in order to explain the peculiar behavior of large red shift (900 meV) shown in (a), we have to consider not only

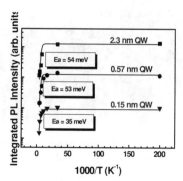

Figure 4, *Arrhenius plot of the integrated PL intensity of the InGaN SQW samples with the InGaN QW thickness of 0.15 (down triangle), 0.57 (circle), and 2.3 nm (square), respectively*

both QD confinement effect and QW thickness variation but also In surface segregation effect, piezoelectric effect *etc.*.

Temperature dependent PL for the InGaN SQWs with the InGaN thickness of 0.15, 0.57, and 2.3 nm was performed in the temperature range of 5 K to 300 K. In Fig. 4, the variations of integrated PL intensities of respective sample are plotted as functions of the inverse of temperatures. The thermo-activation energy for bound excitonic emission obeys the following relation[11]:

$$I(T) = I_0/[1 + CT^{3/2}exp(-E_a/k_BT)] \qquad (1)$$

where I(T) is the integrated PL intensity at temperature of T and I_0 and C are constants. k_B is Boltzmann's constant. E_a is equal to the thermo-activation energy. The solid lines in Fig. 4 denote the least-square fit of the experimental data by Eq. (1). For the cases of InGaN SQW specimens with the InGaN thickness of 0.15, 0.57, and 2.3 nm, the estimated thermo-activation energies are 35, 53, and 54 meV, respectively. This result reveals that PL emission from InGaN SQW samples embedded on InGaN 3D nano-scale dots suffer more additional quantum confinement energy of nearly 20 meV than the case of embedded on InGaN 2D wetting layer.

SUMMARY

Thin layers of InGaN as a function of thickness were grown by MOCVD and characterized using AFM and HRTEM. InGaN thin films grown on GaN reveals a Stranski-Krastanov growth mode,. From the plan-view HRTEM measurement, we observed the creation of InGaN self-assembled QD is not related with dislocations. We have also found that the formed InGaN nano-scale dots have the trapezoidal shape with a {1-102} facet. Visible spectral range from UV to green was easily obtained by changing InGaN quantum well thickness up to 2.3 nm. From the integrated PL intensity analysis, emission from InGaN SQW samples embedded on InGaN 3D nano-scale dots suffer more additional quantum confinement than the case of embedded on InGaN 2D wetting layer.

REFERENCES

1. S. Nakamura, M. Senoh, S. Nagahama, N. Iwasa, T. Yamada, T. Matsushita, Y. Sugimoto, and H. Kiyoku, Jpn. J. Appl. Phys. Lett. **36**, L1059 (1997).
2. S. Nakamura, M. Senoh, S. Nagahama, N. Iwasa, T. Yamada, T. Matsushita, H. Kiyoku, Y. Sugimoto, H. Umomoto, M. Sano, and K. Chocho, Appl. Phys. Lett. **72**, 211 (1998).
3. S. Keller, B.P. Keller, M.S. Minsky, J.E. Bowers, U.K. Mishra, S.P. DeenBaars, W. Seifert, J. Cryst. Growth, **189/190**, 29 (1998).
4. W. Liu, K.L. Teo, M.F. Li, S.J. Chua, K. Uchida, H. Tokunaga, N. Akutsu, K. Matsumoto, J. Cryst. Growth, **189/190**, 648 (1998).
5. T. Takeuchi, S. Sota, H. Sakai, H. Amanoa, I. Akasaki, Y. Kaneko, S. Nakagawa, Y. Yamaoka, N. Yamada, J. Cryst. Growth, **189/190**, 616 (1998).
6. K. Hiramatsu, Y. Kawaguchi, M. Shimizu, N. Sawaki, T. Zheleva, R.F. Davis, H. Tsuda, W. Taki, N. Kuwano, K. Oki, MRS Internet J. Nitride Semicond. Res. **2**, 6 (1997).
7. J.-Y. Marzin and G. Bastard, Solid State Commun. **92**, 437 (1994).
8. H. Jiang and J. Singh, Phys. Rev. B **56**, 4696 (1997).
9. Y.-W. Mo, D.E. Savage, B.S. Swartzentruber, and M.G. Lagally, Phys. Rev. Lett. **65**, 1020 (1990).
10. C. Lobo and R. Leon, J. Appl. Phys. **83**, 4168 (1998).
11. J.I. Pankove, Optical processes in semiconductors, (Dover publication Inc., NY. 1971) p. 120.

ACKNOWLEDGEMENTS

This work has been supported by the KOSEF and the MOST of Korea through the Semiconductor Physics Research Center at Chonbuk National University.

Mat. Res. Soc. Symp. Vol. 595 © 2000 Materials Research Society

The microstructure and electrical properties of directly deposited TiN ohmic contacts to Gallium Nitride.

P. Ruterana[1], G. Nouet, Th. Kehagias[*], Ph. Komninou[*], Th. Karakostas[*], M.A. di Forte Poisson[**], F. Huet[**]
Laboratoire d'Etudes et de Recherches sur les Matériaux , UPRESA 6004 CNRS, ISMRA, 6 boulevard Maréchal Juin, 14050 Caen Cedex, France.,
[*]Aristotle University, Physics Department, 54006 Thessaloniki, Greece,
[**]Thomson-CSF/Laboratoire Central de Recherches, Domaine de Corbeville, 91404 Orsay Cedex , France,
1 Author for correspondence: email ruterana@lermat8.ismra.fr, Tel: 33 2 31 45 26 53 , Fax: 33 2 31 45 26 60

ABSTRACT

When the stoichiometric TiN was deposited directly on GaN, we obtained columnar TiN grains of 5-20 nm section which cross the whole film thickness and are rotated mostly around the [111] axis. The conventional epitaxial relationship is obtained and no amorphous patches are observed at the interface. The deposition of TiN on Si doped GaN layers lead to the formation of an ohmic contact, whereas we obtain a rectifying contact on p type layers.

INTRODUCTION

The performance of optoelectronic devices and transistors depends critically on the contact resistance. Hence, a successful development of reliable ohmic contacts on GaN is of great practical importance. Until now, the best results include Ti as the first layer deposited on the GaN surface (1,2). In the case of Ti/Al, a thin layer of titanium (20 nm) is first deposited and covered by aluminium to a total thickness of 100-200nm. The ohmic contact forms during a subsequent anneal at temperatures which can be as high as 900°C, and it has been shown to be due to a thin TiN film from the reaction of Ti with GaN (3). In this process, resistivity in the range of 10^{-5} ohm.cm^2 is obtained upon 10^{18} cm^{-3} silicon doped GaN layers. During the annealing steps many reactive phases form between Al, Ga, Ti and N which may be a problem for the quality of the contact. Moreover, the formation of TiN may lead to a loss of N from the GaN surface layer, as well as to an excess of Ga in the reactive area. One of the ways to avoid these drawbacks would be to deposit TiN directly on the GaN surface. In this work, a comparative TEM investigation has been carried out on annealed Al/Ti/GaN and directly deposited TiN films on GaN.

EXPERIMENTAL

Two types of specimens were investigated. In one case, a 20 nm Ti film was deposited on GaN and covered by a 80nm Al layer in order to avoid mostly the interaction of titanium with air. They were then submitted to a rapid thermal annealing for 10 seconds at 500°C. In the other case, TiN contacts were deposited at room temperature on the GaN films by dc reactive magnetron sputtering using an Alcatel SCM 600 deposition system equipped with an *in situ* ultrafast spectroscopic ellipsometer. Details of the deposition system are presented elsewhere (4). First, the samples were cleaned in air with a $HCl:H_2O$ (1:1) solution and loaded into the deposition chamber. Then, a very low energy (~ 5 eV) Ar^+ ion dry etching was used to remove the native surface oxide just before the deposition. The working and reactive gases were respectively Ar (99.999%) and N_2 (99.999%). The deposition conditions of the substrate bias voltage V_b was -40 or -120 V and the N_2 flow rate was about 2 sccm. The Ar flow was kept constant at 15 sccm with a partial pressure of $7.5x10^{-3}$ mbar. In-situ spectroscopic ellipsometry investigations have shown that TiN films deposited at V_b= -120 V are stoichiometric (4). TEM samples were prepared in the conventional way by mechanical grinding followed by ion milling. HREM observations were carried out on an Topcon 002B electron microscope operated at 200 kV with a point to point resolution of 0.18 nm

RESULTS

In the annealed Ti/Al films, the ohmic resistivity was measured to be 10^{-5} ohm.cm^2 which is consistent with the silicon doping level of $10^{18}cm^{-3}$. Conventional TEM shows a polycrystalline metallic layer which has a total thickness of nearly 150nm. At higher magnification, one starts to notice the reactive area. The GaN surface is flat, whereas the TiN layer is 2 nm thick and is rough. In areas with large crystallites (5-10 nm), only the {111} TiN lattice fringes are parallel to (0001) GaN, meaning that the grains have only the <111> direction parallel to the [0001] GaN. Unfortunately, these areas are always small and can be separated by patches which exhibit amorphous contrast in the HREM images (fig.1). If they are made of amorphous material due to surface contamination or non completed reaction between GaN and Ti, one may suspect them to constitute points of non optimum ohmic contact, which may lead to discontinuities in current flow.

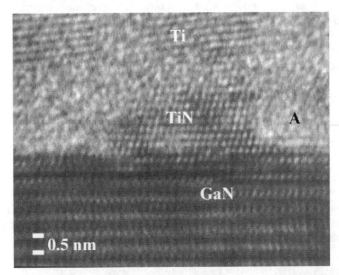

Figure 1. Formation of TiN patches after annealing of deposited Ti/Al

For the directly deposited TiN layers, we obtained a perfect columnar structure. The TiN crystallites section is 5-20 nm and they are strongly twinned. Diffraction analysis shows that they mainly have their {111} planes parallel to the (0001) GaN as in the conventional epitaxial relationship and the rotation around the <111> is reasonably small.

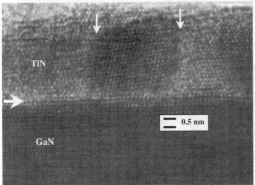

Figure 2. Directly deposited TiN with columnal structure (vertical arrows) and no amorphous layer at interface(horizontal arrow)

HREM observations of the interface indicate that the transition between GaN and TiN crystallites is flat, with the (0001) GaN // (111) TiN atomic planes as in the case of RTA samples. However no amorphous patches could be noticed at this interface. As shown on fig.2, the columnar structure originates at the interface, and the crystallites diameter can be confirmed as about 5 nm.

The full epitaxial relationship is:

(0001)GaN//(111)TiN

[10$\bar{1}$0]GaN//[211]TiN

[11$\bar{2}$0]GaN//[110]TiN

The misfit along [10$\bar{1}$0]GaN//[211]TiN corresponds to 5.8% and the distance between pure edge dislocations should be 4.5 nm in order to have a completely relaxed interface. In the area shown on fig. 3, the measured Burgers vectors of the misfit dislocations are ½[10$\bar{1}$0]GaN or ¼ [211]TiN and their distance is 4.1nm. It is then clear that in these 5 nm columns, the relaxation has taken place, however it is difficult to estimate the residual strain due to the small size of the columnar grains.

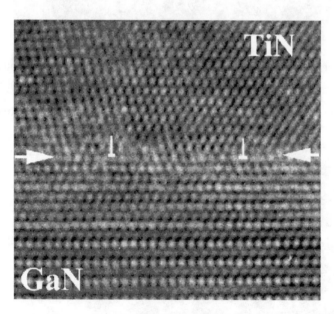

Figure 3. A high magnification of the interface area showing the misfit dislocations, the interface position has been pointed out by the arrows.

Electrical I/V measurements show that this deposition gives rise to the formation of an ohmic contact even at the deposition temperature, as seen in figure 3, the I/V characteristics show that the contact is ohmic for the n type layers (fig. 4) The measured resistance is lower in the optimized deposition conditions (see fig. 4-curve a and b). The next step is to investigate the effect of low temperature annnealing and to make resistivity measurements upon this system.

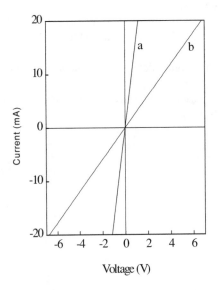

Fig.4: I/V characteristics of the as deposited GaN layers, a) the optimum deposition conditions at 120 V , b) layers sputtered at 40 V

CONCLUSION

Although it has been known that the best ohmic contacts to n type GaN is made by TiN formation, the latter is usually made by deposition of Ti and Al, followed by a 900°C anneal. This results in the formation of more compounds that the necessary TiN, which may constitute a drawback for device performance. In this work we have shown that, although the low temperature annealing may be an alternative, one is left with amorphous patches at the interface. Therefore the deposition of stoichiometric TiN may be one solution and work is going on in order to optimize this process.

Acknowledgements

This work was done in a collaboration of CNRS and NHRF(Greece) under project n° 5225

References

1. Lin M.E., Ma F., Huang F.Y., Fan Z. F., Allen L.H., and Morkoç H., Appl. Phys. Lett. **64**, 1003(1994)
2. Fan Z., Mohammad S.N., Kim W., Actas O., Botchkarev A.E., and Morkoç H., Appl. Phys. Lett., **68**, 1672(1996)
3. Rusimov S., Liliental-Weber Z., Washburn J., Dustad K.J., Haller E.E., Fan Z.F., Mohammad S.N., Kim W., Botchkarev A.E., and Morkoç H., Appl. Phys.Lett., **69**, 1556(1996)
4. Dimitriadis C. A., Logothetidis S. and Alexandrou I., *Appl. Phys. Lett.,* , **66**, 502(1995)
5. Dimitriadis C. A., Karakostas T., Logothetidis S., Kamarinos G., Brini J., and Nouet G., Solid State Electron., in press

Mat. Res. Soc. Symp. Vol. 595 © 2000 Materials Research Society

Highly Chemical Reactive Ion Etching of Gallium Nitride

F. Karouta[1], B. Jacobs[1], I. Moerman[2], K. Jacobs[2], J.L. Weyher[3], S. Porowski[3],
R. Crane[4] and P.R. Hageman[4]

*[1]COBRA Inter-University Research Institute on Communication Technology
Eindhoven University of Technology - Department of Electrical Engineering
P.O.Box 513, NL-5600 MB Eindhoven, Netherlands
[2]Department of Information Technology (INTEC), University of Ghent,
Sint Pieternieuwstraat 41, 9000 Ghent, Belgium
[3]High Pressure Research Center, Polish Academy of Sciences,
Sokolowska 29/37, 01-142 Warsaw – Poland
[4]Experimental Solid State Physics III, RIM, University of Nijmegen,
Toernooiveld 1, 6525 ED Nijmegen, Netherlands*

Abstract: A highly chemical reactive ion etching process has been developed for
MOVPE-grown GaN on sapphire. The key element for the enhancement of the chemical
property during etching is the use of a fluorine containing gas in a chlorine based
chemistry. In the perspective of using GaN substrates for homo-epitaxy of high quality
GaN/AlGaN structures we have used the above described RIE process to smoothen Ga-
polar GaN substrates. The RMS value, measured by AFM, went from 20 Å (after
mechanical polishing) down to 4 Å after 6 minutes of RIE. Etching N-polar GaN resulted
in a higher etch rate than Ga-polar materials (165 vs. 110 nm/min) but the resulting
surface was quite rough and suffers from instability problems. Heat treatment and HCl dip
showed a partial recovery of Schottky characteristics after RIE.

Introduction

GaN materials are very interesting for optoelectronic and microelectronic applications like
blue and green LEDs, blue lasers, UV photodetectors, high power and high temperature
HEMTs. GaN and its related materials (AlGaN, InGaN, InAlGaN) are most commonly
grown, using metal organic vapour phase epitaxy (MOVPE) or molecular beam epitaxy
(MBE), on α-Al_2O_3 substrates. For specific applications like high power transistors,
silicon carbide substrates are used because of their higher thermal conductivity. The lattice
mismatch remains high in both cases with an advantage for SiC, and hence epitaxial layers
suffer from stress induced by the lattice mismatch which results in a very high dislocation
density (in the range of 10^8-10^{10} cm^{-2}). GaN bulk materials form an attractive solution to
circumvent the problems of lattice mismatch encountered in hetero-epitaxy by allowing a
perfect lattice-matched epitaxy.

Dry etching of GaN has been extensively investigated using numerous plasma-based
machines and sources. High etch rate (1.3 µm/min) was reported using an electron
cyclotron resonance (ECR) [1]. Conventional reactive ion etching (RIE) generally shows
lower etch rates and leads to higher level of induced damages. We have developed a highly

chemical reactive ion etching process using a conventional parallel plate reactor. In a previous work [2] we reported a fourfold increase in etching rate when simply adding SF_6 to $SiCl_4$ + Ar. This highly chemical RIE process is based on the choice of the gases used for etching. At a power of 300 Watts we achieved an etch rate of 430 nm/min which is twice the etch rate reported by Feng et al [3] of 210 nm/min at the same power using BCl_3 + SF_6.

In this paper and after a short description of our RIE process, we shall present the results of our RIE investigation on bulk GaN single crystals. We have compared etch rates and morphology after RIE on Ga- and N-polar surfaces. We also have studied the influence of this RIE process on Schottky diodes made on n-GaN.

Experimental

The plasma machine is a load-locked conventional RF-powered reactor (Oxford Plasmalab 100) with an oil-free pumping system (dry pump + magnetically levitated turbo pump). Silicon nitirde as well photoresist were used for masking. All etch experiments were performed at room temperature. Non-intentionally doped MOVPE-grown GaN on α-Al_2O_3 (0001) substrates was used to investigate the RIE process as a function of gas flow, RF-power and pressure. The optimized process was used for the study on Ga- and N-polar GaN bulk materials. The etch depth was measured using a Tencor-200 surface profiler while the morphology was characterized qualitatively by scanning electron microscopy (SEM) for the MOVPE-grown GaN and quantitatively by atomic force microscopy (AFM) in the case of bulk GaN.

Etching of MOVPE-grown GaN

Adding SF_6 to the more conventional $SiCl_4$:Ar chemistry has led to a fourfold increase of etch rate. A relatively low power process at 105 Watts using $SiCl_4$:Ar:SF_6 (10:10:2 sccm) at 40 or 20 mTorr resulted in an etch rate of ~100 nm/min instead of 22 nm/min at the same conditions without SF_6. The DC bias was ~ -300 V. Figure 1 shows a typical etched and very smooth GaN surface obtained after 6 minutes of RIE etching.
Figure 2 clearly shows the influence of SF_6 in the RIE process. We believe that etch products, when using SF_6, are $GaCl_3$ and NF_3. The fact that NF_3 is more volatile (boiling point of

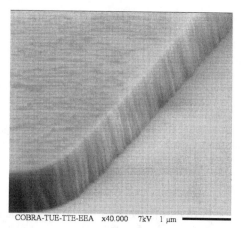

COBRA-TUE-TTE-EEA   x40.000   7kV   1 μm

Figure 1- A SEM photograph of a very smooth surface after RIE using the optimized process at 40 mTorr.

−129°C) than NCl_3 (boiling point around 71°C) explains the fourfold increase of etch rate. Since the etch rate with only argon at 105 W and 40 mTorr was 9 nm/min, this demonstrates the highly chemical property of our RIE process. The decrease in etch rate at higher SF_6 flow is presumably due to the fact that more electrons are subject to collisions which do not result in active species. Furthermore, we have achieved a maximum etch rate of 430 nm/min using the same chemistry at 300 W (DC bias around -600V). This is a world record when considering etching results in

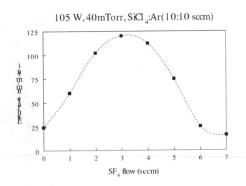

105 W, 40mTorr, SiCl$_4$:Ar(10:10 sccm)

Figure 2- Effect of SF6 flow on etch rate of GaN.

conventional RIE reactors. Feng et al, have used a comparable chemistry based on BCl_3:SF_6 and they reported a maximum etch rate of 210 nm/min instead of 50 nm/min without SF_6. This fourfold increase of etching rate is remarkably similar to the one we obtained when adding the same fluorinated gas. Adesida et al. [4] have not noticed any change in etch rate (50 nm/min) when adding SiF_4 to a $SiCl_4$-based chemistry. These results can be explained by the dissociation energies of the various gases. For instance, these energies are 93.6 and 145.4 kcal/mole for SF_6 and SiF_4 respectively, while for $SiCl_4$ and BCl_3 these values are 94.1 and 110 kcal/mole respectively. It is clear that the $SiCl_4$+SF_6 combination has the lowest dissociation energies and hence forms the best choice among these gases.

Etching of bulk GaN crystals

Chemically assisted ion beam etching (CAIBE) using Cl_2/Ar (which is dominated by physical sputtering) and RIE using Cl_2:Ar:CH_4 of bulk GaN were already reported in literature [5,6]. We have used the following RIE process {$SiCl_4$:Ar:SF_6 (10:10:2 sccm) at 105 Watts and 40 mTorr} to remove polishing damages from bulk GaN single crystals. It is worth noticing that "standard" MOVPE grown hetero-epitaxial GaN is Ga-polar while GaN single crystal substrates present the advantage of having the two polarities and by simply putting the sample upside down the polarity of homo-epitaxial layer can be changed. The GaN single crystals were grown from liquid gallium saturated with molecular nitrogen at 1600°C and at nitrogen pressure of about 15 kbar [7].

Because of the small size of the GaN crystals, sapphire blocks of 2x1 mm^2 were used for masking in order to measure the etch rates. Two samples of different polarity were simultaneously etched for 6 minutes in each run to realize a realistic comparison and to exclude any reproducibility problems. In addition to the etch rate measurements and SEM photography the morphology of the surface was measured by AFM to determine the RMS roughness (peak-valley differences) in substrates with both polarity.

RIE of Ga-polar GaN substrates

After the mechanical polishing of Ga-polar substrates with diamond paste (0.1 µm final grade), the surface roughness had an RMS value of 20-30 Å. A cleaning in boiling acetone and iso-propanol was carried out before the RIE experiments. Notice that scratches caused by the polishing are clearly visible (fig. 3).

After the RIE process an etch rate of 100-110 m/min was measured on these Ga-polar single GaN crystals which is quite similar to that obtained earlier for MOVPE grown hetero-epitaxial GaN.

This is quite remarkable since the

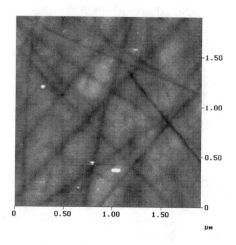

Figure 3- An AFM photograph of Ga-polar surface after mechanical polishing.

bulk GaN substrates have much less growth-related defects. This result converge with our previous statement over the highly chemical component of the used RIE process as the etch rate does not depend on density of defects. In conclusion the surface becomes much smoother where RMS value of 4 Å was measured. Thus a clear polishing effect of the Ga-polar surface by RIE is demonstrated. A drawback of this result was the fact that a random distribution of "pillars" has been found (fig. 4-left). After examination of several samples at different RIE conditions we have found that residual organic contaminating particles are the reason of the nucleation of pillars. Performing a 6-minute O_2 plasma in the same reactor prior to the RIE process of GaN was effective to get rid of any remnant contamination of the surface and subsequently of the "pillars". Figure 4-right shows a featureless Ga-polar surface obtained after combined O_2 plasma cleaning and RIE. The resulted surface allowed the MOVPE-growth of high quality homo-epitaxial layers [8].

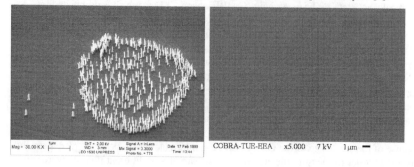

Figure 4- A clustering of residual pillars are shown (left) while using an oxygen plasma prior to RIE resulted in a smooth and pillar-free surface (right).

The N-polar (0001) GaN crystals
were subjected to the same
mechanical polishing as for Ga-polar
substrates followed by a careful
mechano-chemical polishing in
KOH-based solutions. This led to a
very smooth surface with, routinely
obtained, atomic height differences
(RMS of 1-2 Å). Using such crystals
as damage-free epi-ready substrates
led to a very good homo-epitaxial
growth of GaN [9]. These very
smooth samples were also subject to
the same RIE process as performed
on Ga-polar crystals. An etch rate of
165 nm/min was now measured but
a clear deterioration of the surface

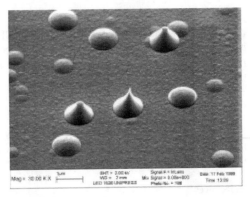

*Figure 5- A rough N-polar surface after RIE
with hillocks and mounds.*

morphology was observed. The RMS value increased from 2 Å to 20 Å as shown in figure
5.

Besides this overall roughening, large cone-shaped hillocks randomly distributed were
present which are probably caused by inversion domain areas. These localized Ga-polar
areas are etched slower than the adjacent N-polar surface and hence lead to the formation
of the hillocks. Moreover, the increased roughness after RIE can be related to the poor
stability of the N-polar surface already observed during high temperature MOVPE [10].

Electrical behaviour

We have investigated the influence of the RIE process on the electrical properties of GaN
layers. Schottky diodes were fabricated on n-type GaN epitaxial layer ($2x10^{17}$ cm^{-2}). The
ohmic contacts were made using Ti/Al (35/115 nm) annealed at 600°C for 30 seconds
resulting in a contact resistance of 0.1 Ω.mm. A platinum layer of 100 nm thickness was
used for the Schottky contacts. I-V measurements were performed to extract the reverse
curent and the ideality factor of the diode. These values were investigated with various
pre-treatments of the surface prior to the Schottky metallization. The best partial recovery
was achieved when the material was subjected to a short RIE process followed by a HCl
dip and a heat treatment at 600°C during 30 seconds. The reverse current and the ideality
factor were found to be respectively $4x10^{-4}$ A/mm (-20V), 1.12 (as-grown) and $1x10^{-3}$
A/mm, 1.12 (after RIE, HCl dip and heat treatment).

Further investigation of the RIE process in conjunction with/or O_2-plasma, HCl dip and
heat treatment and their influence on the electrical characteristics of GaN and AlGaN
layers are undergoing.

Conclusions

A highly chemical reactive ion etching process has been demonstrated based on a combined fluorine and chlorine chemistry. Adding SF_6 to $SiCl_4$+Ar resulted in a fourfold increase of etch rate. This RIE process is suitable for removing work damages from Ga-polar substrates and realizing a final polishing of Ga-polar GaN single crystals making them epi-ready.

Acknowledgements

B. Jacobs would like to thank the Netherlands Organisation for Applied Scientific Research (TNO-FEL) for support and J.L.Weyher wishes to thank NATO Scientific Affair Division for the grant HTECH:LG 972924.

References

[1]   C.B. Vartuli, S.J. Pearton, J.W. Lee, J. Hong, J.D. Mackenzie, C.R. Abernathy and S.J. Shul, Appl. Phys. Lett. **69**, 1426 (1996).
[2]   F. Karouta, B. Jacobs, P. Vreugdewater, N.G.H. v. Melick, O. Schön, H. Protzmann, M. Heuken. Electrochemical and Solid State Letters, 2, (5), 240-241, (1999).
[3]   M.S. Feng, J.D. Guo, Y.M. Lu, E.Y. Chang, Materials Chemistry and Physics, **45**, 80 (1996).
[4]   I. Adesida, A. Mahajan and E. Andideh, Appl. Phys. Lett. **63**, 2777 (1993).
[5]   M. Schauler, F. Eberhard, C. Kirchner, V. Schwelger, A. Pelzmann, M. Kamp, K.J. Ebeling, F. Bertram, T. Riemann, J. Christen, P. Prystawko, M. Leszczynski, I. Grzegory, S. Porowski, Appl. Phys. Lett., 74, (8), 1123-1125, (1999).
[6]   P. Prystawko, M. Leszczynski, B. Beamount, P. Gibart, E. Frayssinet, W. Knap, Phys. Stat. Sol. B, 210, (2), 437-443, (1998).
[7]   S. Porowski, J. Crystal Growth 166, 583-589, (1996).
[8]   J.L. Weyher, A.R.A. Zauner, P.D. Brown, F. Karouta, A. Wysmolek, P.R. Hageman, S. Porowski, accepted for presentation at the 3rd International Conference on Nitride Semiconductors, July 5-9, Montpellier, France, (1999). Accepted for the proceedings in journal of Physica Status Solidi.
[9]   J.L Weyher, S. Müller, I. Grzegory, S. Porowski, J. Crystal Growth, 182, 17-22, 1997.
[10]   J.L. Weyher, P.D. Brown, A.R.A. Zauner, S. Müller, C.B. Boothroyd, D.T. Foord, P.R. Hageman, C.J. Humphreys, P.K. Larsen, I. Grzegory, S. Porowski, J. Crystal Growth, 204, 419-428, 1999.

Mat. Res. Soc. Symp. Vol. 595 © 2000 Materials Research Society

Improved Low Resistance Contacts of Ni/Au and Pd/Au to p-Type GaN Using a Cryogenic Treatment

Mi-Ran Park[1], Wayne A. Anderson[1] and Seong-Ju Park[2]
[1] State University of New York at Buffalo, Department of Electrical Engineering,
Amherst, NY, U.S.A.
[2] Kwangju Institute of Science and Technology, Department of Materials Science and
Engineering, Kwangju, Korea

ABSTRACT

A low resistance Ohmic contact to p-type GaN is essential for reliable operation of electronic and optoelectronic devices. Such contacts have been made using Ni/Au and Pd / Au contacts to p-type Mg-doped GaN (1.41×10^{17} cm^{-3}) grown by metalorganic chemical vapor deposition (MOCVD) on (0001) sapphire substrates. Thermal evaporation was used for the deposition of those metals followed by annealing at temperatures of 400 ~ 700 °C in an oxygen and nitrogen mixed gas ambient, then subsequently cooled in liquid nitrogen which reduced the specific contact resistance from the range of $9.46 \sim 2.80 \times 10^{-2}$ Ωcm^2 to $9.84 \sim 2.65 \times 10^{-4}$ Ωcm^2 for Ni/Au and from the range of $8.35 \sim 5.01 \times 10^{-4}$ Ωcm^2 to $3.34 \sim 1.80 \times 10^{-4}$ Ωcm^2 for Pd/Au. The electrical characteristics for the contacts were examined by the current versus voltage curves and the specific contact resistance was determined by use of the circular transmission line method (c-TLM). The effects of the cryogenic process on improving Ohmic behavior (I-V linearity) and reducing the specific contact resistance will be discussed from a microstructural analysis which reveals the metallurgy of Ohmic contact formation.

INTRODUCTION

Great interest exists in the III-nitride semiconductors since the successful development in growth of GaN based materials and operation of electronic and optoelectronic devices such as blue and green light emitting diodes (LEDs) and laser diodes (LDs) [1,2,3,4]. The formation of stable and reliable low resistance ohmic contacts to p-type GaN has been a problem in achieving good performance of those devices. For devices with large contact areas such as LEDs and LDs, the specific contact resistance (ρ_c) between 10^{-4} to 10^{-6} Ωcm^2 is considered acceptable and for devices with smaller contact areas, values of ρ_c between 10^{-5} to 10^{-7} Ωcm^2 are necessary [5]. Bilayer metal schemes such as Ni/Au and Pd/Au were studied by many groups [6,7,8]. These have been studied due to the stable electrical and thermal properties and the high work function which is one criteria to form low resistance Ohmic contacts to p-type materials.

The effects of cryogenic cooling after heat treatment on the formation of Ni/Au and Pd/Au contacts are presented in this paper. We also compare these effects on forming Ni/Au and Pd/Au contacts annealed in a combined O_2/N_2 gas ambient. High temperature annealing may degrade homogeneity possibly caused by spiking of metals between themselves or between metal and semiconductor due to the differences in thermodynamic properties of materials. Annealing was conducted in an oxygen and nitrogen mixed gas ambient as reported by Y. Koide et al. [6]. This is to remove hydrogen atoms contained in Mg-doped GaN epilayers. Removing hydrogen atoms results in the increase of the hole concentration and decrease of the contact resistance. Annealing in nitrogen gas ambient is believed to avoid nitrogen vacancies which act as doners. We expected that cryogenic treatment would reduce disadvantages from the discrepancy in thermodynamic properties of materials during cooling the heated samples to give improved Ohmic behavior. With cryogenic treatment, the changes of electrical and structural properties have been examined.

EXPERIMENT

The metal contacts were made on GaN films grown by metalorganic chemical vapor deposition (MOCVD) on (0001) sapphire substrates. The GaN films consisted of the Mg doped p-type GaN with a thickness of 2μm on a 30nm thick GaN buffer layer. The hole concentration of the p-GaN layer was 1.41×10^{17} cm^{-3} and the resistivity was 3.5 Ωcm. The samples were sequentially ultrasonically cleaned in trichloroethylene, acetone and methanol, then rinsed in 18 MΩ de-ionized (DI) water. The cleaned samples were chemically etched in boiling aqua regia of HNO_3 : HCl = 1 : 3 for 10 minutes to remove the native oxide and the contamination of the GaN surface as suggested by J. K. Kim et al [9]. Then, a photolithographic process was used to form the pattern for the circular transmission line method (c-TLM). In this pattern, the radius of the inner circular contact was 400 μm and the spacings between inner and outer circles ranged from 25μm to 75 μm. Prior to the deposition of metal, the sample was etched in a warm solution of HNO_3 and HCl again. The warm solution means non-boiling aqua regia : a solution was boiled and cooled for 2~3 minutes since very hot solution ruins the photoresist. The Ni (30nm) / Au (15 nm) and the Pd (25 nm)/ Au (15 nm) contacts were deposited by thermal evaporation and completed by liftoff. The contacts were then annealed at temperatures ranging from 400 to 700 °C for 10 minutes in a conventional furnace in an oxygen and nitrogen mixed gas ambient. In order to study the effect of cryogenic treatment, some samples were subsequently cooled by dipping in liquid nitrogen after the heat treatment, then brought to room temperature in air. Sister samples were not quenched in liquid nitrogen.

The current versus voltage (I-V) curves were measured as deposited and between each heat treatment interval. After each measurement of I-V curves, the specific contact resistance (ρ_c) was determined by use of the c-TLM. The samples with and without cryogenic treatment showing the lowest contact resistance were examined by scanning electron microscopy (SEM) images using a model Hitachi S-4000 and atomic force microscopy (AFM) images using a model Quesant Res/stg and electron spectroscopy for chemical analysis (ESCA) depth profiles using a model Surface Science SSX-100 with an ion energy of 4.5 keV, a raster size of 2x2 mm, a sputter time per step of 1~2 min. and a x-ray spot size of 1000 μm.

RESULTS

1. Electrical Properties

The current versus voltage curves for the as deposited, annealed and annealed plus cryogenically treated Ni/Au and Pd/Au contacts to p-GaN are shown in Figure 1. Annealed contacts shown in Figure 1 were treated in an O_2/N_2 ambient. For the as deposited case, both Ni/Au and Pd/Au contacts show similar behavior of non linear I-V with a small potential barrier. However, the Pd/Au contact reveals a lower specific

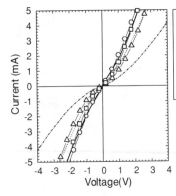

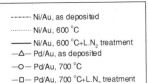

----- Ni/Au, as deposited
.... Ni/Au, 600 °C
——— Ni/Au, 600 °C+L.N$_2$ treatment
—△— Pd/Au, as deposited
—○— Pd/Au, 700 °C
—□— Pd/Au, 700 °C+L.N$_2$ treatment

Figure 1. Current versus voltage curves for the Ni/Au and the Pd/Au contacts to p-GaN as deposited and annealed with and without cryogenic treatment in liquid nitrogen.

contact resistance of $7 \sim 10 \times 10^{-4}$ Ωcm^2 compared to $4 \sim 6 \times 10^{-2}$ Ωcm^2 for the Ni/Au contact in the linear region, although the work function of Ni (5.15 eV) is slightly higher than Pd (5.12 eV). These characteristics agree with previous studies [7,8]. We believe that the lower contact resistance of the Pd/Au contact with lower work function of Pd is due to a new phase with higher work function which will be discussed from the ESCA profiles. The non linear I-V behaviors for both Ni/Au and Pd/Au contacts were improved by heat treatment at temperatures of 600 °C and 700 °C, respectively. The reduction of the specific contact resistances to 9.84×10^{-4} Ωcm^2 for Ni/Au and 1.80×10^{-4} Ωcm^2 for Pd/Au were also obtained by heat treatment. Lowering the contact resistance and improving linearity may come from a more intimate contact of metal with semiconductor or any new phases having higher work function. Intimate contact leads to more current flow across the interface by breaking up some of the interfacial contamination between metal and semiconductor. Possible new compounds reduce the potential offset at the metal/ semiconductor interface by forming a layer of a compound with higher work function or causing a highly doped region. In contrast, the effect of cryogenic treatment on forming those two contacts is quite different. With further treatment by cooling in liquid nitrogen after the anneals, the Ni/Au contact exhibits better linearity of I-V characteristics and a contact resistance of 2.65×10^{-4} $\Omega \ cm^2$. However, the Pd/Au contact show opposite behavior to the Ni/Au contact such that cryogenic treatment degrades the specific contact resistance value to 3.34×10^{-4} $\Omega \ cm^2$. The effect of cryogenic treatment on improving Ohmic behavior could be caused by some of the semiconductor dissolving in the metal on heating and recrystallization with a high concentration of the electrically active element in solid solution on subsequent cooling [10]. It could also be due to improved morphology as explained later. Therefore, the mechanisms and the reactions at the interface between metal and semiconductor for Ni/Au contact and Pd/Au contact would be different.

2. Structural properties

Figure 2 shows ESCA depth profiles of Ni/Au contacts heated in O$_2$/N$_2$. The ESCA depth profile for the as deposited sample of the Ni/Au contact shows the abrupt and sharp signals corresponding to the interfaces of the Au/Ni/GaN and all of the species (Au, Ni, Ga and N). From the ESCA depth profile, there is no evidence of interdiffusion from

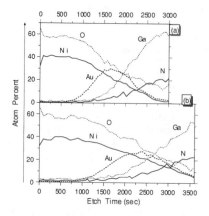

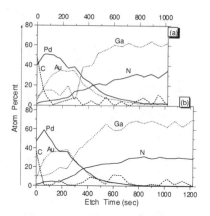

Figure 2. ESCA depth profiles for the Au/Ni/p-GaN: (a) annealed at 600 °C 10 min. in O₂/N₂ and (b) annealed and cryogenically treated.

Figure 3. ESCA depth profiles for the Au/Pd/p-GaN: (a) annealed at 700 °C 10 min. in O₂/N₂ and (b) annealed and cryogenically treated.

deposition alone. However, ESCA profiles change in behavior with annealing and cryogenic treatment. In the ESCA profile of an annealed sample, Figure 2(a), Ni diffused to the surface of the contact and into the GaN. The O signal follows the Ni signal, which indicates a NiO phase. This is evident from chemical analyses, which show core levels of Ni $2p^3/_2$ and NiO at the binding energies of 852.6 eV and 854.7 eV, respectively. On the other hand, Au exchanges position with Ni and forms an intimate contact with the GaN layer. These results agree with previous reports in which Ni diffused through the Au capping layer to the surface of the contact where it oxidized [7]. A similar behavior in ESCA depth profiles is also shown for the annealed plus cryogenic treated sample as in Figure 2(b). Our study indicates that the direct contact to GaN would be a combination of NiO and Au. Y.Koide et al. by using x-ray diffraction (XRD) reported a NiO layer and no evidence of interaction between metals and GaN for the Ni/Au contacts annealed in an O₂/N₂ ambient [6]. On the other hand, H.S.Venugopalan et al. [11] found a NiGa phase for the annealed Ni/n-GaN scheme and J.T.Trexler et al. [12] reported the possibility of a Ni-N solid solution for the annealed Ni/Au scheme.

The ESCA depth profiles of Pd/Au contacts are shown in Figure 3. For both contacts annealed and cryogenic treated in addition to annealing, there is no evidence of formation of PdO but of Au:Pd solid solution. This new phase (Au:Pd solid solution) was reported by other groups [6,7]. The Ga and N signals tail out and extend out to the surface, which might indicate the decomposition of the GaN matrix. No significant difference was seen between the samples only annealed and annealed plus cryogenic treated. Note that the published sputter rate for Pd is 100 nm/min. compared to 60 nm/min. for Ni.

We also examined surface morphology of those contacts using SEM. The SEM images of the as deposited samples for Ni/Au and Pd/Au contacts show a smooth metal surface. However, Figure 4 for the Ni/Au contact shows the change in the surface morphology of samples annealed (Figure 4(a)) and cryogenic treated (Figure 4(b)). It clearly shows bright and gray particles within a deep dark area. From a comparison with

Figure 4. SEM images for the Ni/Au contacts to p-GaN: (a) annealed at 600 °C for 10 min. in an O_2/N_2 and (b) annealed and cryogenically treated.

Figure 5. SEM images for the Pd/Au contacts to p-GaN: (a) annealed at 700 °C for 10 min. in an O_2/N_2 and (b) annealed and cryogenically treated.

composition vs. depth in ESCA profiles, the bright and gray particles are NiO and black areas would be the mixing of the transition metal oxide (NiO) and Au which directly contacts with GaN. Figure 4(b) shows that the cryogenic treated sample has a smaller dark area (black) and smoother surface of NiO (bright and gray) than the other (Figure 4(a)). For the Pd/Au contacts in Figure 5, we can see a difference in particle size but not in roughness. The sample not cryogenic treated, in Figure 5(a), has non uniform larger particles. Contrary to this feature, the sample with liquid nitrogen cooling shows almost uniform particles. We believe that the "balling-up" of the NiO layer occurs during the heating process from the compressive stress due to the difference of the thermal expansion coefficients between Ni (13.3×10^{-6} K^{-1}) [13], Au (14.1×10^{-6} K^{-1}) [13], Pd (11×10^{-6} K^{-1}), and GaN (6×10^{-6} K^{-1}) [14]. The differences in surface features for the contacts with and without cryogenic treatment are in size and thickness of the different colored regions. In addition to the effect of compressive stress, strain is also induced during cooling due to the difference of thermal expansion coefficients of materials. Therefore, we suggest that subsequent and fast cooling in liquid nitrogen minimizes the effect of compressive stress and strain, then results in the laterally more uniform surface which is significant in the electrical properties. This is supported by AFM data which show the surface roughness of the Ni/Au contact. The average values of the surface roughness are 64.78 nm and 37.38 nm for the contacts annealed only compared to those treated in liquid nitrogen after annealing, respectively. There is no much difference in the values of the surface roughness : 54.46 nm for annealed and 68.2 nm for annealed plus cryogenic treated, for the Pd/Au contacts from the AFM images.

CONCLUSIONS

Both Ni / Au and Pd/Au contacts to p - GaN show Ohmic behavior with and without a cryogenic treatment. The cryogenic treatment on the Ni/Au contacts improves the I-V linearity and reduces the specific contact resistance from 9.84×10^{-4} Ωcm^2 to 2.65×10^{-4} Ωcm^2. The phases NiO and Au:Pd have been observed for both contact systems, respectively. Subsequent and fast cooling in liquid nitrogen after anneal affects the recrystallizing of NiO and Au:Pd solid solution produced from heat treatment, then improves surface morphology. The contact resistance is improved by: i) heat temperature which forms new phases at the GaN surface compared to the initial deposit; and ii) cryogenic treatment which affects both the recrystallization and surface morphology.

REFERENCES

1. S. Nakamura, M. Senoh, N. Iwasa, and S. Nagahama, Jpn. J. Appl. Phys. 34, L797(1995).
2. S. Nakamura, M. Senoh, N. Iwasa, S. Nagahama, T. Yamada, and T. Mukai, Jpn. J. Appl. Phys. 34, L1332(1995).
3. S. Nakamura, and M. Senoh, Jpn. J. Appl. Phys. 30, L1998(1991).
4. S, Nakamura, M. Senoh, S. Nagahama, N. Iwasa, T. Yamada, T. Matsushita, H. Kiyoku, and Y. Sugimoto, Jpn. J. Appl. Phys. 35, L217(1996).
5. P. H. Holloway, T-J. Kim, J. T. Trexler, S. Miller, J. J. Fijol, W. V. Lampert, T. W. Haas, Appl. Sur. Sci., 117/118, 362(1997).
6. Y. Koide, T. Maeda, T. Kawakami, S. Fujita, T. Uemura, N. Shibata, and M. Murakami, J. Elec. Mat., V28. 341(1999).
7. J. T. Trexler, S. J. Pearton, P. H. Holloway, M. G. Mier, K. R. Evans, and R. F. Karlicek, Mat. Res. Soc. Symp. Proc., V449, 1091(1997).
8. T. Kim, J. Khim, S. Chae, and T. Kim, Mat. Res. Soc. Symp. Proc., V468, 427(1997).
9. J. K. Kim and J. Lee, Appl. Phy. Lett., 73, 2953(1998).
10. E. H. Rhoderick and R. H. Williams, "Metal-Semiconductor contacts", Clarendon, 2nd, P206(1988).
11. H. S. Venugopalan, S. E. Mohney, B. P. Luther, J. M. Delucca, S. D. Walter, J. M. Redwing, and G. E. Bulman, Mat. Res. Soc. Symp. Proc., V468, 431(1997).
12. J. T. Trexler, S. J. Miller, P. H. Holloway, M. A. Kwan, Mat. Res. Soc. Symp. Proc., V395, 819(1996).
13. E. A. Brandes, "Smithells Metals Reference Book", Batterworths, 6th Ed.
14. K. J. Duxstad, E. E. Haller, K. M. Yu, M. T. Hirsch, W. R. Imler, D. A. Steigerwald, F. A. Ponce, and L. T. Romano, Mat. Res. Soc. Symp. Proc., V449, 1049(1997).

Mat. Res. Soc. Symp. Vol. 595 © 2000 Materials Research Society

A Thermodynamic Approach to Ohmic Contact Formation to p-GaN

Bo Liu[1], Mikko H. Ahonen and Paul H. Holloway
Department of Materials Science and Engineering
University of Florida, Gainesville, FL 32611-6400

ABSTRACT

A new ohmic contact scheme for gallium nitride is presented. The use of Nitride-forming metal Over Gallide-forming metal, "NOG", can modify the thermodynamic activity of N and Ga near the interface. This in turn can modify the near-surface point defect concentrations, particularly the vacancies of Ga and N. The principle of this contact scheme was shown to be consistent with results from Ni/Au, Ni/Zn-Au, Ta/Ti, and Ni/Mg/Ni/Si contacts. In the present study, the "NOG" scheme was used to design Ni/Ti/Au and Ni/Al/Au metallization, and addition of Ti and Al nitride-forming metals to the Ni gallide-forming metal led to lower but still high contact resistance. Ti was shown to be better than Al as the nitride-forming metal based on the decrease of resistance in as deposited contacts. Compared to Ni/Au, four times more current was measured in Ni/Ti/Au contacts to p-GaN after anneal at 300°C for 5min. However the addition of the Ti nitride-forming metal led to lower stability at 500°C.

INTRODUCTION

GaN alloys have received great interest in the past decade due to applications in photonic and electronic devices. However, because of the low free hole concentrations of p-GaN (10^{17} cm^{-3}) and lack of a metal with a work function ϕ equal to or greater than the bandgap plus electron affinity ($E_g + \chi_s = 7.5$ eV), attempts to make low resistance ohmic contacts to p-GaN have been unsuccessful[1-12].

The purpose of this paper is to propose a general scheme by which metallization schemes for ohmic contacts can be systematically selected. The scheme is called "NOG" (Nitride-forming metal Over Gallide-forming metal) and is based on the thermodynamic stabilities of these phases during interfacial reactions between the metallization layers and the GaN semiconductor.

PRINCIPLES OF "NOG" SCHEME

Since as-grown GaN are intrinsically n-type, there may be a high concentration of native N vacancies, V_N, which is equivalent to a Ga-rich condition in the film. An opposite situation could be postulated: if a N-rich condition could be created in as-grown GaN films, the extra N atoms could create Ga vacancies and intrinsically p-GaN films However, this postulated condition might still be achieved by interfacial reactions in the contact region. If extra N atoms could be kept between the contact metal layer and the bulk p-GaN film, a N-rich condition could be formed at the metal/GaN interface. The

---

[1] Electronic mail: bliu@mail.mse.ufl.edu

extra N atoms could fill the V_N positions and create Ga vacancies which would act as acceptors. If the Ga vacancy acceptors were sufficiently shallow and reached a high concentration, the interfacial region could become p^+-GaN and current transportation could be dominated by field emission or thermionic field emission.

The principle of the "NOG" scheme is illustrated in Figure 1. A gallide-forming metal is followed by a nitride-forming metal covered with a layer of protective metal (such as Au). Under a suitable annealing condition, the gallide-forming metal would react with GaN to form stable gallides and release N atoms. This first metal layer must both dissociate the GaN lattice and prevent or slow down the process of nitrogen out-diffusion. The second nitride-forming metal would help keep the released N atoms at the contact interfacial region and create a N-rich condition.

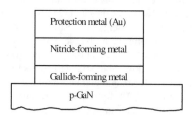

Figure 1. Principle of the "NOG" scheme.

all transition metals may be classified into three groups: the late, early and middle transition metals based on the enthalpy of the metallurgical reactions, which was called gallide-forming, nitride-forming and neutral metals in this paper. The metallurgical reaction of these metals to GaN was reported in reference [13] and [14].

EXPLANATION OF LITERATURE

Many studies have been reported along with postulated mechanisms to explain reduced contact resistance, such as GaN re-growth [12], H extraction [1] and Ni oxidation [2]. A few representative contact schemes will be discussed based on the principle of "NOG". Other contact schemes could be analyzed similarly.

The contact scheme Ni/Au [2, 3, 4, 5, 6] is widely used for GaN device fabrication. Based on the "NOG" principles, this scheme works due to the reaction between Ni and GaN. The Ga would react with Ni to form stable gallides and reduce excess Ga atoms. Reduced excess Ga could be reflected as reduced concentration of V_N. Less compensation of acceptors would result in higher free hole concentrations in the interfacial region. Other reported contact schemes like Pd/Au[7, 8], Pt/Au [6, 9, 10], Pd/Au/Pt/Au[8], Pd/Pt/Au[6] and Pt/Ni/Au[11]contact schemes could be explained similarly.

A relatively low specific contact resistance (3.6×10^{-3} Ω-cm^2) was obtained with Ni/Zn-Au [12] in p-GaN with a carrier concentration of $N_h = 4.4\times10^{17}$cm^{-3}. It was postulated that Zn was an acceptor and that the Zn-Au alloy layer increased the interface carrier concentration. Zn is an acceptor, but it is deep ($E_A = 570$ meV)[16] and therefore should not be ionized at room temperature. Based on the "NOG" scheme, the mechanism should be the same as the Ni/Au scheme discussed above. The main reason for improved contact performance probably due to the limited time for native oxide to grow on GaN and the use of high vacuum for metallization. Optimum contact resistance would not be predicted for the Ni/Zn-Au because of no nitride forming component to the metallization. The released N atoms could be released as N_2 rather than kept at the interface.

A low resistivity (3.2×10^{-5} Ω-cm^2) ohmic contact to p-GaN was produced with Ta/Ti after a high temperature anneal (800°C for 20 min) [1]. The authors postulated that Ta and Ti were able to remove hydrogen from Mg-H complexes and therefore reduced compensation of the acceptors. It was found that a dual layer structure formed better contacts than a single Ta or Ti layer. After a few days, the contact resistance increased to a much higher value. This was attributed to a reverse transport of compensating hydrogen from the Ti/Ta layers back into the interface region and recompensation of Mg acceptors. However, Fukai has reported that the enthalpy for MgH_2 (-0.77 eV/atom) is more negative than for TiH_2 (-0.68 eV/atom) or $TaH_{0.5}$ (-0.417 eV/atom) [17]. Thus for these reaction products, Ta and Ti should not reduce MgH_2. In the "NOG" scheme, it would be postulated instead that Ta and Ti would dissociate the GaN and release N atoms. The released N atoms would increase the nitrogen chemical potential and result in reduced V_N concentrations, in addition to forming TaN_x and TiN_x compounds. The differences in the thermodynamic and kinetic properties of Ta and Ti would explain why Ta/Ti form better contacts than Ta or Ti individually. The observation of an increased resistance with time at room temperature can also be explained using the "NOG" scheme since formation of stable nitrides would create V_N and increase compensation of holes in the contact region.

For Ni/Mg/Ni/Si [12], contact resistance value of $\approx 10^{-3}$ Ω-cm^2 was measured. These ohmic contacts were degraded by annealing at 500°C for 20min. The authors postulated a similar mechanism of solid phase epitaxial regrowth known to occur for AuGeNi/GaAs contacts[15]. The authors in the study of Ni/Mg/Ni/Si postulated that regrowth of GaN and NiSi led to ohmic contacts. Using the principles of the "NOG" scheme, formation of an ohmic contact would result from Ni dissociation of the GaN and formation of $NiGa_x$, MgN_x and SiN_x. The nitride phases would increase the activity of N in the interfacial region, which would create a N-rich condition and a more p-type interface. This contact metallization is close to the "NOG" scheme under discussion.

This idea of interfacial reactions and control of vacancies which is the basic tenet of the "NOG" scheme can and does apply to ohmic contact to n-GaN, not just to p-GaN. Lester, et al. [18] reported that aluminum produced an ohmic contact of 10^{-3} ohm-cm^2 to n-GaN. This is reasonable because of the matched work function of Al and GaN [15]. The contact resistance of Al/n-GaN increased by 50% upon annealing at 575°C. The postulated reason was formation of a large bandgap AlN layer at the interface. The contact resistance was improved to 8×10^{-6} Ω-cm^2 using a Ti/Al bilayer metals annealed at 900°C, presumably due to the formation of TiN at the interface of Ti/Al (and Ti/Al/Ni/Au) contacts [19, 20]. Depletion of N in the GaN surface region would create more V_N vacancies, and result in an n$^+$-GaN layer with improved electron tunneling, consistent with the "NOG" scheme.

Experimental studies of Ni/Ti/Au and Ni/Al/Au

Besides analysis of published results, the "NOG" scheme was compared to experimental data collected from Ni/Ti/Au, Ni/Al/Au and Ni/Au contacts to p-GaN. Current-voltage (I-V) data showed that more current was obtained in the ternary layer contacts, consistent with predictions.

The p-GaN wafers used in this experiment were purchased from SVT Associates. The GaN 1μm epilayers were grown by MBE and had a free hole concentration of 1.1~ 2.5×10^{17} cm^{-3}. Samples were cleaned with agitated acetone (5min), methanol (5min) and boiling aqua regia (10min) sequentially before being washed with DI water. All samples were blown dry with N_2 gas between each step. The contacts were deposited at a base pressure of ≈5 x 10^{-6} Torr.

Figure 2 shows the I-V data for (a) 50nm Ni/50nm Au or 10nm Ni/50nm Al/50nm Au contacts, and (b) for 10nm Ni/200nm Ti/200nm Au or 10nm Ni/200nm Al/200nm Au. The current was a factor of two higher through Ni/Al/Au than through Ni/Au, and the current through Ni/Ti/Au was 30% higher than through Ni/Al/Au.

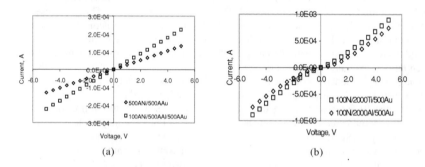

(a) (b)

Figure 2 Effects of adding a nitride-forming metal to Ni/Au as-deposited contacts to p-GaN (a) Ni/Au and Ni/Al/Au; (b) Ni/Ti/Au and Ni/Al/Au

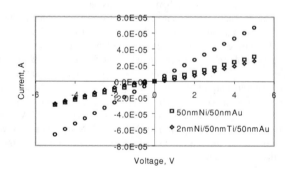

Figure 3 Effects of Ni thickness in the as deposited Ni/Ti/Au/p-GaN contacts

Based on the principles "NOG", the first Ni layer in both Ni/Ti/Au and Ni/Al/Au contacts was a gallide forming metal, which would release N atoms. This Ni layer also acts as a nitride-forming barrier because of slowed N diffusion. This would maintain a higher N activity at the GaN surface, reduce V_N, and result in less acceptor compenstion.

The effects of Ni thickness were studied by changing the Ni thickness (Figure 3). As the Ni changed from 2 to 10 nm the current increased by a factor of 3. A higher contact resistance for thin gallide-forming metal (2nm Ni) could result from the extent of interfacial reaction being too limited to affect V_N concentration or the Ni layer is too thin to prevent quick formation of stable nitride (TiN_x) and thus to increase V_N concentration.

The effects of annealing on Ni/Ti/Au and the Ni/Au contacts are shown in Figure 4. Both Ni/Au and Ni/Ti/Au have the same current for the as deposited samples. For Ni/Ti/Au contact, a 300°C, 30 sec anneal resulted in slightly higher current, and a 5 min anneal at 300°C resulted in a four fold increase. The current through Ni/Ti/Au contacts annealed at 300°C, 30 sec was similar to that measured for Ni/Au contacts annealed at 300°C for 5 min. After a 500°C, 5min anneal, the current though Ni/Ti/Au contacts decreased to near that of as-deposited contacts. In contrast, the current in Ni/Au contacts increased continually, even after annealing at 800°C for 30 sec. The higher as deposited current, but more serious degradation of the Ni/Ti/Au contact shows that while the nitride-forming metal (Ti) is helpful in reducing the contact resistance, it also leads to the contact instability. This would be consistent with continued reaction between Ti and N to form TiN_x resulting in generation of too many V_N's.

The data in Figure 4 cannot be directly compared to those in Figure 3, since the contacts in Figure 3 are 0.5mm diameter dots defined by a shadow mask, while those in Figure 4 are a TLM pattern with a 16µm spacing defined by photolithography.

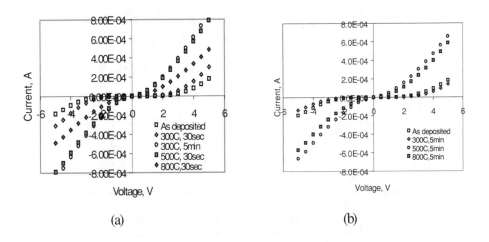

(a)   (b)

Figure 4 Effects of annealing on I-V data for (a)20nmNi/50nmTi/50nmAu; (b) 50nmNi/50nmAu to p-GaN

CONCLUSIONS

A new ohmic contact scheme for p-GaN called Nitride-forming metal Over Gallium-forming metals, "NOG", was proposed and discussed. Published results for ohmic contacts to p- or n-GaN were able to be explained qualitatively by the "NOG" principles. Lower contact resistances were found for contacts with both gallide and nitride forming metals (Ni/Al/Au and Ni/Ti/Au), when compared with only a gallide-forming Ni/Au contacts.

REFERENCES:

1. M. Suzuki, T. Kawakami, T. Arai, S. Kobayashi, Y. Koide, T. Uemura, N. Shibata and M. Murakami, Appl. Phys. Lett., 1999, **74**, 275
2. J.-K. Ho, C.-S. Jong, C. C. Chiu, C.-Y. Chen, Appl. Phys. Lett., 1999, **74**, 1275
3. A. K. Fung, J. A. Borton, M. I NathanJ. M. V. Hove and R. Hickman, J. Elec. Mater., 1999, **28**, 572
4. T. Kim, M. C. Yoo and T. Kim, III-V nitrides, MRS Symp. Proc., 1997, **449**, 1061
5. H. Ishikawa, S. Kobayashi, Y. Koide, S. Yamasaki, S. Nagai, J. Umezaki, M. Koike and M. Murakami, J. of Appl. Phys., 2, 1997, **81**, 1315
6. D. J. King, L. Zhang, J. C. Ramer, S. D. Hersee and L. F. Lester, III-V nitrides, MRS Symp. Proc. 1997, **468**, 421
7. J. K. Kim, J.-L. Lee, J. W. Lee, H. E. Shin, Y. J. Park and T. Kim, Appl. Phys. Lett., 1998, **73**, 2953
8. T. Kim, J. Khim, S. Chae and T. Kim, III-V nitrides, MRS Symp. Proc., 1997, **468**, 427
9. T. Mori, T. Kozawa, T. Ohwaki, Y. Taga, S. Nagai, S. Yamasaki, S. Asami, N. Shibata and M. Koike, Appl. Phys. Lett.,1996, **69**, 3537
10. M. C. Yoo, J. W. Lee, J. W. Myoung, K. H. Shim and K. Kim, MRS Symp. Proc., 1996, **423**, 131
11. J.-S. Jang, I.-S. Chang, H.-K. Kim, T.-Y.Seong, S. Lee and S.-J. Park, Appl. Phys. Lett. **74**, 70(1999)
12. D. H. Youn, M. Hao, H. Sato, T. Sugahara, Y. nao and S. Sakai, Jpn. J. Appl. Phys. **37**, 1768(1998)
13. S. E. Mohney and X. Lin, Journal of Electronic Materials, 1996, **25**, 811
14. S. E. Mohney, B. P. Luther, T. N. Jackson and M. A. Khan, MRS Symp. Proc. **395**, 843(1996)
15. P.H. Holloway, T.-J. Kim, J.T. Trexler, S. Miller, J.J. Fijol, W.V. Lampert and T.W. Haas, Appl. Surf. Sci., 1997, **117/118**, 362
16. J. I. Pankove, GaN and Related Materials, ed. S. J. Pearton, Gordon and Breach Science Publishers, Amsteredam, Netherlands, 1(1997)
17. Y. Fukai, Metal-Hydrogen System-basic bulk properties, Springer-Verlag, 31(1993)
18. L. F. Lester, J. M. Brown, J. C. Ramer, L. Ahang, S. D. Hersee and J. C. Zolper, Appl. Phys. Lett. **69**, 2737(1996)
19. Z. F. Fan, S. N. Mohammad, W. Kim, O. Aktas, A. E. Botchkarev and H. Mokoc, Appl. Phys. Lett. **68**, 1672(1996)
20. M. E. Lin, Z. Ma, F. Y. Huang, Z. F. Fan, L. H. Allen and H. Morkoc, Appl. Phys. Lett. **64**, 1003(1994)

Mat. Res. Soc. Symp. Vol. 595 © 2000 Materials Research Society

Metal/GaN contacts studied by electron spectroscopies

J. Dumont, R. Caudano, and R. Sporken
Facultés Universitaires Notre-Dame de la Paix, Laboratoire Interdisciplinaire de
Spectroscopie Electronique, Namur, Belgium
E. Monroy, E. Muñoz
Universidad Politecnica de Madrid, E.S.T.I., Ciudad Universitaria, Madrid, Spain
B. Beaumont, P. Gibart
Centre de Recherche sur l'Hetero-Epitaxie et ses Applications, CRHEA-CNRS,
Valbonne, France

ABSTRACT

Au/GaN and Cu/GaN Schottky contacts have been studied using X-ray
Photoelectron Spectroscopy (XPS) and Auger Electron Spectroscopy (AES). Clean and
stoechiometric GaN samples were obtained using in situ hydrogen plasma treatment and
Ga deposition. The growth of Cu and Au follows Stranski-Krastanov and Frank van der
Merwe modes respectively. The interfaces are sharp and non-reactive. Schottky barriers
of 1.15eV for Au/GaN and 0.85eV for Cu/GaN were measured using XPS.

INTRODUCTION

GaN and related wide band gap semiconductors are of particular interest in blue
and ultraviolet (UV) optoelectronic applications and for high power and high temperature
electronic devices. An impressive number of articles have been published over the past
few years regarding the properties of GaN and GaN-based devices, but comparatively
few detailed studies of metal/GaN interface formation have been reported. Kampen and
Mönch [1] described the Schottky barrier height of ideal metal/GaN contacts based on the
metal-induced gap states (MIGS) and electronegativity model:

$$\phi_{Bn} = \phi_{cnl} + S_\chi (\chi_m - \chi_s) \qquad (1)$$

While the difference $\chi_m - \chi_s$ between the electronegativity of the metal and of the
semiconductor is specific to each metal, the slope parameter S_χ and the charge neutrality
level Φ_{cnl} are characteristic of GaN. They were calculated by Kampen and Mönch [1]
who obtained $S_\chi = 0.29$ eV/Miedema-unit. and $\Phi_{cnl} = 2.35$ eV with respect to the top of
the valence band. Bermudez [2] investigated the dependence of the structure and
electronic properties of wurtzite GaN surfaces on the method of preparation. Ordered and
unreconstructed surfaces were obtained. A bare surface barrier height was measured on
these surfaces that depends on the annealing temperature and reaches 0.7eV for surfaces
annealed at 700°C. When Mg was grown on wurtzite GaN surfaces, the energy
difference between the Fermi level and the valence band maximum E_F-VBM increased
with Mg coverage while no intermixing between Mg and GaN was observed [3]. Mohney
and Lin [4] used a thermodynamical approach to predict the phase diagrams of many
transition metal-Ga-N systems and to explain the electrical properties of the contacts. The
structure and bonding at the metal/GaN interface are expected to affect electronic
properties of the contact and in particular its barrier height and specific resistance. The
present paper describes a compositional analysis of Au/GaN and Cu/GaN interface
formation and the evolution of the Schottky barrier height during the interface formation.
These studies are based on in situ XPS measurements.

EXPERIMENTAL PROCEDURE

GaN epitaxial layers were grown at CRHEA (Valbonne, France) on (0001) sapphire substrates by metalorganic vapor phase epitaxy (MOVPE) at atmospheric pressure[5]. A 100 Å thick AlN buffer layer was used. Non intentionally doped samples showed n-type conduction, with a carrier concentration of about 10^{17} cm^{-3}, and a Hall mobility of about 130 cm^2/V.s. These samples had to be cleaned prior to metallization.

Several cleaning procedures will be discussed below. They usually involve wet chemical etching, followed by various treatments in ultra-high vacuum (UHV), such as annealing, exposure to atomic hydrogen, or exposure to a Ga flux. The atomic H was obtained from an RF plasma source (Oxford Applied Instruments, Model MPD20) operated at 175W RF power. The base pressure in the preparation chamber was $5 \ 10^{-10}$ Torr. The plasma was ignited with nitrogen instead of hydrogen. The nitrogen was then gradually replaced by hydrogen. A flux of 9 sccm H_2 (purity N57), rising the pressure to $2 \ 10^{-2}$ Torr, gave the brightest plasma (checked with a photodiode). During these treatments, samples were heated to 700°C. Ga was evaporated from a conventional effusion cell heated to 850°C. The Ga layer thickness, estimated from the XPS signals, ranged from 20 to 60Å.

Au and Cu were deposited by thermal evaporation from effusion cells equipped with BN crucibles. Background pressure was about $2 \ 10^{-10}$ Torr prior to evaporation, and rose to $1 \ 10^{-9}$ Torr during the metal deposition. Au was evaporated in a UHV chamber pumped by an ion getter pump and by a liquid-nitrogen cooled titanium sublimation pump. Cu was evaporated in a chamber equipped with a 380l/s turbomolecular pump. Evaporation rates range from 0.25 to 10Å/min, measured by a quartz microbalance. In both cases, the samples could be transferred between the evaporation chamber and the spectrometer under UHV.

Two instruments were used for the XPS measurements. The Au/GaN interface was studied with an SSX-100 spectrometer (Surface Science Instruments). This instrument uses a monochromatic and focused X-ray source (Al Kα, hν = 1486.6eV) and a hemispherical analyzer. For the measurements presented here, a spot size of 0.6 μm diameter and 55eV pass energy was used. The corresponding energy resolution, defined as the full-width at half maximum of the Au $4f_{7/2}$ core level peak, is about 1.1eV. A Scienta 300 photoelectron spectrometer was used to record XPS spectra during the Cu/GaN interface formation. This instrument uses a monochromatic Al Kα X-ray source with a rotating anode operated at 6 kW. Photoelectrons were detected by a hemispherical analyzer (150eV pass energy) and a two-dimensional position sensitive detector (micro-channel plate with CCD camera). The energy resolution obtained in these conditions is 0.3eV.

Auger electron spectra were measured with a PHI-595 Scanning Auger Microprobe. A primary energy of 12 kV and an analyzer resolution $\Delta E/E$ of 0.5% were used.

Quantitative data analysis required least squares fitting of the core level peaks. Fitting procedures for Ga3d, Au4f and N1s were already described elsewhere [6]. To fit $Cu2p_{3/2}$ and $Ga2p_{3/2}$ peaks a Shirley background and a Doniach-Sunjič function were chosen.

RESULTS AND DISCUSSION

Cleaning of GaN surfaces

Several techniques were used to clean the GaN samples. Sample #1 was cleaned in KOH and aqua regia and then heated to 900°C under UHV. This treatment removes most of the carbon and oxygen. Sample #2 was first dipped into boiling HNO_3 and deionized water. As shown in figure 1, this chemical cleaning reduces the amount of gallium oxide but the amount of carbon is left unchanged. Moreover, even in argon atmosphere, the surface oxidizes fast again (in less than 1 minute) when the samples are loaded into UHV. Hence any ex situ cleaning must be followed by an in situ cleaning procedure. Both Khan et al. [7] and Bermudez et al. [2] have shown that atomically clean GaN surfaces can be prepared by Ga deposition and desorption. Here, we used atomic hydrogen (H*) from an RF discharge as an additional step to prepare clean and stoechiometric surfaces. Auger spectra in figure 1 (left) show that the H* treatment removes C contamination, while deposition of Ga followed by desorption at 950° removes the oxygen contamination. In order to obtain clean surfaces we combined the H* and Ga treatments. XPS shows that both O and C are now below the detection limit (figure 1, right).

Based on the distance between the valence band edge (obtained by linear extrapolation) and the Fermi level and knowing the value of the bandgap of GaN (3.4 eV), Fermi level is found at 0.9 ± 0.05 eV below the conduction band minimum for the bare surface. The Ga $3d_{5/2}$ core level peak from GaN appears at 17.6 ± 0.05 eV below the valence band edge, in good agreement with the value of 17.8eV (referred to the centroïd of the Ga 3d peak) reported previously [6].

Metal/GaN interface formation

Au/GaN results shown here are extracted from the article published by Sporken et al. [6] and are compared to the present study of the Cu/GaN interface formation. In this

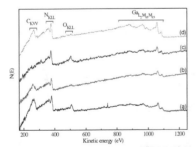
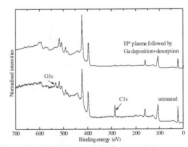

Figure. 1 *Left: Auger spectra of GaN cleaned following different procedures: (a) untreated; (b) cycles of boiling HNO_3 and deionized water; (c) same as (b), followed by H* exposure at 600°C;(d) same as (b) followed by Ga deposition and desorption at 950°C. Right: XPS spectra of GaN surfaces: (a) untreated; (b) cycles of H* exposure followed by Ga deposition + desorption.*

section, we first discuss the morphology of the interface and the growth of the copper layer using XPS core level intensities and shapes. Then, the Schottky barrier height measured by XPS is explained using the MIGS and electronegativity model.

First of all, since the lineshape of the XPS spectra from the substrate and adsorbate does not change during the growth of copper (figure 2), we conclude that the Cu/GaN interface is non reactive like the Au/GaN contact grown at room temperature [6].

Gold grows in a layer by layer (also called Frank van der Merwe or 2D) mode on GaN. This leads to exponential variations of the XPS signals from the substrate ($I \propto exp(-z/\lambda)$) and adsorbate ($I \propto 1-exp(-z/\lambda)$) with overlayer thickness. The attenuation length λ is related to the inelastic mean free path of the photoelectrons; it can be calculated using the equation [8]:

$$\lambda(nm) = 0.41 \left[\frac{A(g\ mol^{-1})\ 10^{24}}{\rho(kg\ m^{-3})\ N(mol^{-1})} \cdot E_{Kin}(eV) \right]^{1/2}, \qquad (2)$$

where A is the atomic mass, ρ is the bulk material density, N is Avogadro's number and E_{kin} is the kinetic energy of the photoelectron. For Au/GaN, excellent agreement is found between experimental and calculated XPS intensities [6]. For Cu/GaN, least squares fitting the experimental data (figure 3, left) using the exponential functions described above, with the photoelectron attenuation length λ as fitting parameter, gives reasonable values of λ (table I). However, the calculated curves do not reproduce the experimental data very well. This means that the growth is not perfectly 2D. Since we do not see a change in chemical bonding of the Cu, Ga and N, we suggest that the interface is still abrupt but islands are formed. Thus we fitted the data using a modified Stranski-Krastanov growth model. In this model, described more precisely by Sporken et al. [9] and later by Conard et al. [10], hemispherical islands grow on incomplete 2D layers. The substrate and overlayer intensities are calculated assuming that the 2D layer and the islands have the same density as bulk Cu. The fitting parameters are the thickness of the 2D layer, the number of islands per unit area and the attenuation length λ. As observed on figure 3 (right), the agreement is much better than for the simulation based on layer by layer growth. The number of islands obtained from the fit is about 1.10^{11}cm^{-2} and the 2D layer thickness is around 4.10^{15} atoms.cm^{-2} which corresponds to 5Å. However, the values of the attenuation length for Ga and N are much smaller than the theoretical values (see table I).

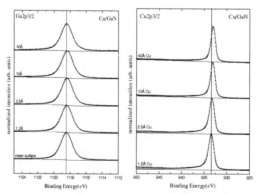

Figure. 2: Ga2p$_{3/2}$ and Cu2p$_{3/2}$ XPS spectra during formation of the Cu/GaN interface.

This means that the XPS intensities of the GaN decrease to quickly with overlayer thickness. This may be due to an overestimation of the density of the 2D layer. It is likely indeed that the crystalline structure of the adsorbate is influenced by the GaN and that its density is different from the one of bulk copper. At the beginning of the growth, the copper atoms are distributed mainly on GaN adsorption sites. The lattice parameter of GaN is 3.19Å. On a GaN (0001) face, it is then expected that the 2D copper layer will have a preferred (111) orientation. A fit based on a number of atoms per Cu monolayer equal to $1.13 \ 10^{15} \text{cm}^{-2}$ (i.e a Cu atom per adsorption site of the GaN (0001) face) gave us the same 2D layer thickness as before (5Å), but photoelectron attenuation lengths much closer to the theoretical values (table I).

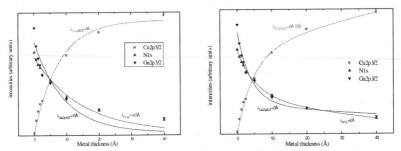

Figure. 3. *Experimental and calculated XPS intensities during the growth of Cu on GaN. Left: calculated intensities for layer by layer growth mode. Right: calculated intensities for modified Stranski-Krastanov growth mode .*

The distance between the conduction band minimum and the Fermi level was determined from the binding energy of the core level peaks ($Ga2p_{3/2}$ or Ga3d or N1s and $Cu2p_{3/2}$). Figure 4 shows the position of the Fermi level in the band gap. The clean surface shows a barrier height of 1.4 eV that decreases to reach the stable value of 0.84 eV after depositing $5.8 \ 10^{15}$ atoms cm^{-2} of copper. This is similar to the Au/GaN interface formation [6]. During the formation of the Cu/GaN Schottky barrier (up to 5.8 10^{15}atoms cm^{-2}), the shape of the Cu $L_3M_{45}M_{45}$ Auger features changes (figure 4, right). This reflects changes in the electronic structure of the copper, which initially does not have full metallic character. Eventually, the Auger transitions reach a shape typical of metallic Cu.

Table I: *photoelectron mean free paths. 1st column: λ calculated using the Seah and Dench formula. [8] Other columns: values obtained from the XPS intensities. ρ_{2D} = density of the 2D layer; N_{2D} = number of atoms per unit area in the 2D layer.*

	Calculated value	Layer by layer mode	Layer + islands mode:$\rho_{2D} = \rho_{Cu}$	Layer + islands mode: $N_{2D} = N_{GaN}$
$\lambda_{Ga2p3/2}$	8Å	10Å	3Å	5Å
λ_{N1s}	15Å	14Å	5Å	10Å
$\lambda_{Cu2p3/2}$	11Å	9Å	12Å	12Å

The final Schottky barrier height of Cu/GaN and Au/GaN measured by XPS is 0.84eV and 1.15eV respectively. These values are close to results (0.82eV and 0.96eV) from the MIGS and electronegativity model using values for S_χ and Φ_{cnl} calculated by Kampen and Mönch [1]. These authors measured and calculated a barrier height of 0.82 eV for Ag/GaN, which is close to our result for Cu/GaN. Such agreement is predicted by eq. 1, since Ag and Cu have the same electronegativity.

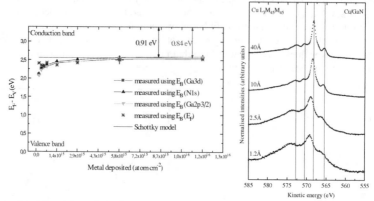

Figure. 4 *Left: position of the Fermi level with respect to the valence band maximum, as a function of Cu coverage. Right: Cu $L_3M_{45}M_{45}$ Auger features; evolution with copper coverage.*

CONCLUSIONS

Clean GaN samples were obtained using atomic H plasma and gallium deposition and desorption. Copper growth on these samples follows a Stranski-Krastanov growth mode. Analysis of the XPS intensities suggests that the 2D Cu layer has an in-plane lattice parameter close to the one of the GaN surface. A Schottky barrier height of 0.84 eV is measured for the Cu/GaN contact. This value agrees with the MIGS and electronegativity model.

ACKNOWLEDGMENT

We thank F.Marchal, J.Ghijsen and C.Whelan for their help with the experiments. We also thank V.M. Bermudez for helpful suggestions concerning cleaning of GaN surfaces. R. S. acknowledges support from the Belgian Fund for Scientific Research (FNRS). This work was supported by EU contract ESPRIT-LTR LAQUANI n° 20968 and by the Walloon region (DGTRE - grant No C. 2795).

REFERENCES

[1]   T. U. Kampen, W. Mönch, *MRS Internet J. Nitride Semicond. Res. 1*, **41** (1997).
[2]   V.M. Bermudez, D.D. Koleske, A.E. Wickenden, *Appl. Surf. Sci.* **126**, 69 (1998).
[3]   V.M. Bermudez, *Appl. Surf. Sci.* **417**, 30 (1998).
[4]   S.E. Mohney and X. Lin, *J. Electron. Mater.* **25 (5)**, 811 (1996)
[5]   B. Beaumont et al., J. Cryst. Growth 170, 316 (1997)
[6]   R. Sporken et al., *MRS Internet J. Nitride Semicond. Res.* **2**, 23 (1997).
[7]   M.A. Khan, J.N. Kuznia, D.T. Olson, R. Kaplan, *J. Appl. Phys.* **73**, 3108 (1993)
[8]   M.P. Seah, W.A. Dench, *Surf. Interface. Anal.*, **1**, 2 (1979)
[9]   R. Sporken et al., *Phys. Rev. B* **35**, 7927 (1987)
[10] T.Conard et al., *Surf. Sci.* **359**, 82 (1996)

Mat. Res. Soc. Symp. Vol. 595 © 2000 Materials Research Society

Deep levels in n-type Schottky and p$^+$-n homojunction GaN diodes

A. Hierro,[1] D. Kwon,[1] S. A. Ringel,[1] M. Hansen,[2] U. K. Mishra,[2] S. P. DenBaars,[2] and J. S. Speck[2]
[1]Deptartment of Electrical Engineering, The Ohio State University, Columbus, OH 43210-1272
[2]Materials and Electrical and Computer Engineering Departments, University of California, Santa Barbara, CA 93016

ABSTRACT

The deep level spectra in both p$^+$-n homojunction and n-type Schottky GaN diodes are studied by deep level transient spectroscopy (DLTS) in order to compare the role of the junction configuration on the defects found within the n-GaN layer. Both majority and minority carrier DLTS measurements are performed on the diodes allowing the observation of both electron and hole traps in n-GaN. An electron level at E_c-E_t=0.58 and 0.62 V is observed in the p$^+$-n and Schottky diodes, respectively, with a concentration of ~3-4×10^{14} cm^{-3} and a capture cross section of ~1-5×10^{-15} cm^2. The similar Arrhenius behavior indicates that both emissions are related to the same defect. The shift in activation energy is correlated to the electric field enhanced-emission in the p$^+$-n diode, where the junction barrier is much larger. The p$^+$-n diode configuration allows the observation of a hole trap at E_t-E_v=0.87 eV in the n-GaN which is very likely related to the yellow luminescence band.

INTRODUCTION

GaN is a material of great interest due to its wide applicability in optoelectronics and high temperature electronics. However, the role of electrically active defects and their sources in GaN are still not well understood [1-5]. Thus, further studies regarding the deep levels found within the n-GaN bandgap which are associated with such defects are needed. While p$^+$-n diode configurations are used for GaN LED's and lasers applications, n-Schottky diodes are of interest for GaN FET's. Thus, the role of the device configuration on the deep level spectrum is also of great interest. In this article deep level transient spectroscopy (DLTS) is used to study the deep level spectra in p$^+$-n and n-Schottky GaN diodes and the fundamental properties of specific deep levels are explored. Both majority and minority carrier injection conditions are used for the DLTS measurements which allows the observation of both electron and hole traps in the n-GaN layer found up to ~0.9 eV from the conduction and valence bands, respectively.

EXPERIMENT

GaN test devices were grown by MOCVD using a horizontal flow reactor on a c-sapphire substrate. A 0.5 μm-thick unintentionally doped n-type layer (n=3x10^{16} cm^{-3})

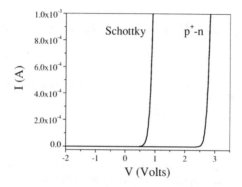

Figure 1: I-V curves measured at 300K for the n-Schottky and p⁺-n GaN diodes. As
expected, a much larger built-in voltage is observed for the p⁺-n diode.

was first deposited followed by 0.5 μm-thick Si-doped (n=3x10^{18} cm^{-3}) and 1 μm-thick
Si-doped (n=8x10^{18} cm^{-3}) GaN layers [6]. A 1.2 μm-thick unintentionally doped n-GaN
layer (n=3.5x10^{16} cm^{-3}) was next deposited for the n-Schottky and p⁺-n diodes, followed
by a 0.16 μm-thick Mg-doped GaN layer (p=0.8-1x10^{18} cm^{-3}) for the p⁺-n structure.
Pd/Au and Ni were next deposited providing ohmic and Schottky contacts to the GaN.
Under these conditions the depletion region is fully contained in the n-GaN layer in both
types of diodes. Back contact to the 1 μm-thick n-GaN layer was provided by a
Ti/Al/Ni/Au metal layer. The diode size was ~0.23 mm^2, with a reverse leakage current at
–1 V of ~4x10^{-10} and ~2x10^{-9} A/cm^2 and a turn-on voltage of ~2.2 and ~0.4 V for the p⁺-
n and Schottky diodes, respectively, at 300 K (Fig. 1).

DLTS experiments were performed using a Boonton capacitance meter at 1
MHz. Rate windows between 4 and 200 sec^{-1} were used with a –1 V quiescent voltage
applied to the diodes for all measurements. Majority carrier trap filling was ensured by a
10 ms-long pulse under a fill pulse voltage (V_p) of –0.2 V for both the p⁺-n and
Schottky diodes. Depth resolved DLTS measurements of the majority carrier traps were
achieved by changing V_p from –0.2 to 1, 1.5, and 2 V in the p⁺-n diode. Positive V_p also
provided minority carrier injection in the p⁺-n sample, allowing the observation of hole
traps in the n-side of the junction [7].

RESULTS AND DISCUSSION

In this section the majority carrier traps (electron traps) for the n-Schottky
diode are first discussed and followed by a comparison to those found in the p⁺-n diode.
This comparison is followed by a study of the minority carrier traps (hole traps) found in
the n-side of the p⁺-n diode.

The majority carrier DLTS spectrum for the n-Schottky diode is shown in Fig.
2. A prominent peak is observed which corresponds to an electron trap found within the
n-GaN layer with an activation energy of E_c-E_t=0.62 eV, as extracted from the Arrhenius

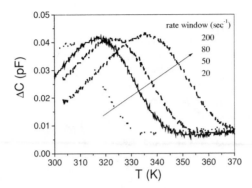

Figure 2: DLTS spectra for the n-Schottky GaN diode under V_p=-0.2 V for different rate windows.

plot (Fig. 3), and with a concentration of ~4.1×10^{14} cm^{-3}. The capture cross section for the 0.62 eV level is ~5×10^{-15} cm^2.

As seen in Fig. 2, the DLTS peak height of the E_c-0.62 eV trap showed no dependence on rate window. Moreover, theoretical modeling of an ideal single state at E_c-0.62 eV yielded a FWHM/T_{peak} ratio of 0.11, in close agreement with the measured value of 0.14. Together, these observations suggest a point defect as the source for this electron trap [7].

In contrast to the Schottky results, a conventional DLTS filling pulse of V_p=-0.2 V does not reveal the presence of the E_c-0.62 eV level in the p$^+$-n diode (Fig. 4). However, as V_p is increased so that the depletion region width reduces, approaching the heavily Mg-doped cap layer, the trap reappears. Note that at 2 V (still below the turn-on voltage) and 400 K the current injected into the p$^+$-n diode is still $\leq$1 nA indicating that

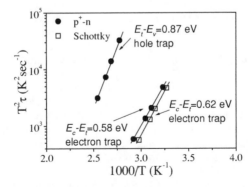

Figure 3: Arrhenius behavior of the electron and hole traps found in both p$^+$-n and n-Schottky GaN diodes.

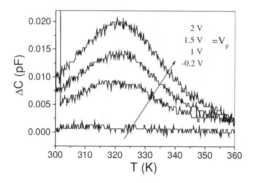

Figure 4: DLTS spectrum obtained from the p⁺-n GaN diode for a rate window of 50 sec⁻¹ as a function of fill pulse bias voltages.

such a change in the DLTS peak height is most likely due to an increase in the trap concentration close to the Mg-doped cap but still within the n-GaN layer. The activation energy for this electron trap was found to be E_c-E_t=0.58 eV (Fig. 3) and the capture cross section ~1×10⁻¹⁵ cm². At V_p=2 V the concentration of this level is ~2.7×10¹⁴ cm⁻³, very similar to that observed in the n-Schottky diode. Indeed, as shown in Fig. 3 the Arrhenius behaviors of the 0.58 and 0.62 eV levels are very similar which clearly indicates that they are very likely the same level.

However, note that for a given rate window (e.g. 50 sec⁻¹ in Figs. 2 and 4) the DLTS peak associated with the 0.58-0.62 eV level shows a small shift to higher temperatures from the Schottky to the p⁺-n diode, which in turn produces the 0.04 eV shift in activation energy shown in Fig. 3. Such a shift can be explained in terms of the difference in the magnitude of the electric field present in the depletion region in both samples. Indeed, Fig. 1 shows a turn-on voltage of ~0.4 V and ~2.2 V for the Schottky and p⁺-n diodes, respectively, which implies that a larger junction barrier is present in the p⁺-n diode. This indicates that the electric field is much larger in the p⁺-n sample. Under these conditions emission from the 0.58-0.62 eV level is enhanced by the larger field in the p⁺-n diode, producing a decrease in the measured activation energy [7].

The E_c-E_t=0.58-0.62 eV level shows a similar Arrhenius behavior as the E2 level previously reported in n-Schottky diodes [2], and is likely produced by the same defect. The E2 level concentration has been observed to depend on the Cp₂Mg flow rate during growth of the n-GaN layer, increasing under larger flow conditions [4]. As shown in Fig. 4, the 0.58 eV level emission is highly dependent on fill pulse bias, increasing in magnitude by at least a factor of 15 from –0.2 to 2V, which corresponds to a shift in the depletion region width of ~0.2 μm towards the p⁺-side of the junction. Such a large increase in concentration tracks the residual Mg concentration profile in the n-side of the junction observed by second ion mass spectroscopy [5].

Finally, the DLTS spectrum at V_p=2 V for the p⁺-n diode is shown in Fig. 5. A large negative peak is observed in the DLTS spectrum that was not be observed for

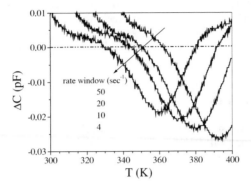

Figure 5: Minority carrier DLTS spectra from the p⁺-n GaN diode at V_p=2 V for different rate windows.

negative fill pulse voltages. This peak was also observed for V_p=1 and 1.5 V with a magnitude decreasing with decreasing V_p. The injected current under these conditions is below 1 nA and the depletion region width is 30 times larger on the n-side than on the p-side of the p⁺-n junction. Moreover, this hole trap could not be observed in the n-Schottky diode (with V_p=0.4 V) as would be expected from such a junction, where minority injection is not possible. Together, these observations indicate that this DLTS feature is indeed a minority carrier hole trap found in n-GaN. This hole trap shows an activation energy of E_t-E_v=0.87 eV (Fig. 3) with a concentration $\geq 3.7\times10^{14}$ cm⁻³, and a capture cross section of ~6×10⁻¹⁴ cm². The position in the bandgap of the 0.87 eV hole trap correlates very well with a deep level at E_c-E_t=2.64 eV measured under optical excitation by deep level optical spectroscopy (DLOS) [5] and observed in both p⁺-n and n-Schottky GaN diodes. It is also close to a deep level previously reported in n-Schottky diodes and associated with the yellow luminescence in n-GaN [1], and correlates well with calculated energy levels associated with V_{Ga}-O and V_{Ga}-donor complexes [3].

CONCLUSIONS

In summary, DLTS has been used to study and compare deep levels and their electrical signature in n-Schottky and p⁺-n GaN diodes. Majority and minority carrier injection during the measurements allowed the observation of both electron and hole traps in the n-GaN layer. The n-Schottky diode shows an electron trap at E_c-E_t=0.62 eV with a concentration of ~4.1×10¹⁴ cm⁻³ and a capture cross section of ~5×10⁻¹⁵ cm². A deep level at E_c-E_t=0.58 eV is also observed in the p⁺-n diode with a concentration of ~2.7×10¹⁴ cm⁻³ and a capture cross section of ~1×10⁻¹⁵ cm². The similar Arrhenius behavior of the two electron traps in both types of diodes indicates that the two levels are likely related to the same defect. The shift in the activation energy from 0.62 to 0.58 eV between the n-Schottky and p⁺-n diodes is due to an electrical field enhancement of the emission rate in the p⁺-n diode. Under minority carrier injection a hole trap at E_t-E_v=0.87

eV with a concentration $\geq 3.7 \times 10^{14}$ cm^{-3} and a capture cross section of ~6×10^{-14} cm^2 is observed in the n-GaN layer. This level was only observed in the p$^+$-n diode due to the fact that hole injection is needed for this measurement and is not possible for an ideal Schottky junction. The E_t-E_v=0.87 eV is likely related to the yellow band [1], and correlates very well with a level observed both in n-Schottky and p$^+$-n GaN diodes by DLOS [5].

ACKNOWLEDGEMNETS

This work was supported by the National Science Foundation through Grants No. DMR-9458046 and DMR-9711851. The work at UCSB was supported by the Office of Naval Research (C. Wood and J. Zolper).

REFERENCES

1. E. Calleja, F. J. Sanchez, D. Basak, M. A. Sanchez-Garcia, E. Muñoz, I. Izpura, F. Calle, J. M. G. Tijero, J. L. Sanchez-Rojas, B. Beaumont, P. Lorenzini, and P. Gibart, Phys. Rev. B **55**, 4689 (1997).
2. P. Hacke, T. Detchprohm, K. Hiramatsu, N. Sawaki, K. Tadatomo, and K. Miyake, J. Appl. Phys. **76**, 304 (1994).
3. J. Neugebauer, and C. G. Van de Walle, Appl. Phys. Lett. **69**, 503 (1996).
4. P. Hacke, H. Nakayama, T. Detchprohm, K. Hiramatsu, N. Sawaki, Appl. Phys. Lett. **68**, 1362 (1996).
5. A. Hierro, D. Kwon, S. A. Ringel, M. Hansen, J. S. Speck, and S. P. DenBaars, submitted to Appl. Phys. Lett. (1999).
6. B. P. Keller, S. Keller, D. Kapolnek, W. N. Jiang, Y. F. Wu, H. Masui, X. Wu, B. Heying, J. S. Speck, U. K. Mishra, and S. P. DenBaars, Journal of Electronic Materials, **24**, No. 11, pp. 1707-1709, (1995).
7. P. Blood and J.W. Orton, The Electrical Characterization of Semiconductors: Majority Carriers and Electron States, (Academic Press, San Diego, 1992).

Mat. Res. Soc. Symp. Vol. 595 © 2000 Materials Research Society

Deep Centers and Persistent Photoconductivity Studies in Variously Grown GaN Films

Alexander Y. Polyakov*, Nikolai B. Smirnov*, Anatoliy V. Govorkov*, Alexander S. Usikov**, Natalie M. Shmidt**, Boris V. Pushnyi**, Denis V. Tsvetkov**, Sergey I. Stepanov**, Vladimir A. Dmitriev***, Mikhail G. Mil'vidskii*, Vladimir F. Pavlov*
* Institute of Rare Metals, Moscow, Russia, polyakov@mail.girmet.ru,
** A.F. Ioffe Physiko-Technical Institute RAS, St.-Petersburg,Russia,
*** TDI., Inc., Gaithersburg, MD 20877, USA

ABSTRACT

Deep levels studies on a set of n-GaN films grown by MOCVD and HVPE reveal the presence of electron traps with levels near E_c-0.25 eV, E_c-0.55 eV, E_c-0.8 eV, E_c-1 eV, hole traps with levels near E_v+0.9 eV and a band of relatively shallow states in the lower half of the bandgap. The total density of these latter states was estimated to be some 10^{16} cm^{-3} and they were tentatively associated with dislocations in GaN based on their high concentration and band-like character. None of the electron or hole traps could be unambiguously related with strong changes of diffusion lengths of minority carriers in various samples. It is proposed that such changes occur due to different surface recombination velocities. An important role of E_c-0.55 eV traps in persistent photoconductivity phenomena in n-GaN has been demonstrated.

INTRODUCTION

As the materials system of GaN grows into maturity all the usual problems associated with the influence of deep centers on electrical properties of GaN films become increasingly important. A number of groups have reported the results of studies of deep levels in n-GaN using deep levels transient spectroscopy (DLTS) and related techniques (a recent review and relevant references can be found e.g. in [1]). At the same time there are many indications that nonradiative recombination via deep traps is an important recombination channel in GaN (see e.g. a review in [2]). However, no attempt has been made so far to pinpoint the deep traps that play an important role in generation-recombination of charge carriers in GaN. In this paper we tried to establish such a correlation between the type and density of deep traps and the diffusion lengths in several n-GaN films widely differing by the diffusion length values.

EXPERIMENTAL

The samples studied in this paper were grown either by metalorganic chemical vapor deposition (MOCVD) technique or by hydride vapor pressure epitaxy (HVPE) on basal plane sapphire substrates. The details of growth can be found in [3,4]. MOCVD samples varied from each other by the type of low temperature (LT) buffer (either GaN (sample 598 below) or AlGaN (samples 385, 644, 646 below)). Samples 598 and 646 were lightly Si doped, other samples were undoped. The thickness of all samples was 3-4 μm. Characterization included double crystal high resolution x-ray diffraction (HRXRD) measurements with Mo K_α source, capacitance-voltage (C-V) measurements on Au

Schottky diodes with the diameter of about 1 mm, deep levels spectroscopy (DLTS) with electrical and optical injection and electron beam induced current (EBIC) profiles measurements at 300K. The latter profiles were used to extract the values of diffusion length of minority carriers. For persistent photocapacitance measurements a series of capacitance versus voltage and capacitance versus temperature (C-T) curves were measured at various frequencies before and after illumination at 85K with deuterium UV lamp and also with various light emitting diodes and with white light source equipped with selective filters. The Au Schottky diodes were prepared as described in [5]. The details of experimental set-ups and measurement procedures can be found in [5,6].

RESULTS AND DISCUSSION

Table I presents the values of the (00.2) x-ray reflection halfwidths, $\Delta_{00.2}$, of room temperature electron concentration established from C-V measurements at 1 kHz (n_{C-V}) and of diffusion lengths (L_d) in various n-GaN samples studied in this paper. It can be seen that the crystalline quality, as assessed by the halfwidth of the (00.2) reflection is the best for the MOCVD sample 598 grown on LT GaN buffer. The quality of the HVPE sample B95 and of the undoped MOCVD sample 385 grown using an LT AlGaN buffer is the worst in the same terms, while the undoped MOCVD LT AlGaN buffer sample 644 and lightly Si doped MOCVD LT AlGaN buffer sample 646 occupy the intermediate position. In search of centers responsible for the observed changes of L_d we first carried out DLTS with electrical injection measurements. DLTS studies on sample 385 have already been published in [6] (this is sample #1 of [6]). They showed the presence of only one dominant electron trap with apparent activation energy of 0.55 eV (the ET4 trap in the system of notation proposed in [1]). The density of these traps was found to be $2.8 \cdot 10^{15}$ cm^{-3}. Figure 1 shows DLTS spectra measured for samples 646, B95 and 598 with time windows t_1/t_2=100 ms/1000 ms, with reverse bias of –0.5 V and the forward bias pulse of +1V (the pulse duration was 2s).

For sample 646 the only feature is the peak near 270K corresponding to the same ET4 trap as in sample 385. The concentration of this trap was $6.2 \cdot 10^{15}$ cm^{-3}. In sample B95, in addition to the ET4 trap with concentration $2.8 \cdot 10^{15}$ cm^{-3}, we observed an electron trap with activation energy 0.25 eV (the ET2 trap [1]) with concentration $2.5 \cdot 10^{14}$ cm^{-3}, the electron trap with activation energy of 0.8 eV (the ET6 trap [1]) with concentration of $6.4 \cdot 10^{14}$ cm^{-3}, and an electron trap with activation energy of 1 eV that is

Table I. Electrical and structural properties of the studied samples

Sample #	Growth method	doping, cm^{-3}	$\Delta_{00.2}$, arcsec	n_{C-V}, cm^{-3}	L_d, μm
385	MOCVD, LT AlGaN buffer	none	540	$2 \cdot 10^{15}$	0.5-0.6
644	MOCVD, LT AlGaN buffer	none	450	$5 \cdot 10^{14}$	0.9-1.2
646	MOCVD, LT AlGaN buffer	Si, $<10^{18}$	450	$2.3 \cdot 10^{16}$	2
598	MOCVD, LT GaN buffer	Si, $<10^{18}$	180	$2 \cdot 10^{16}$	0.9
B95	HVPE	none	540	$1.3 \cdot 10^{16}$	2-3

not well resolved for t_1/t_2 windows of 100/1000 ms but was clearly visible with longer windows of 10/50 s (the trap concentration was $9.3 \cdot 10^{14}$ cm^{-3}). In sample 598 we did not observe the ET4 traps in measurable concentrations. Instead one could observe a broad feature near 300-370K that is obviously a superposition of two or more peaks due to some unidentified electron traps ETX and a trap with activation energy of 1 eV reminding a similar center in sample B95 but having a higher capture cross section. The densities of corresponding traps were $8 \cdot 10^{14}$ cm^{-3} (ETX) and $2.3 \cdot 10^{14}$ cm^{-3}(1 eV trap). Comparing the above results with the L_d values in Table I shows that none of the electron traps shallower than 1 eV can account for the observed changes of L_d in various samples. Hole traps can be accessed in Schottky diodes by DLTS with optical injection (ODLTS), although getting the absolute values of the densities of traps depends on one's ability to fully recharge them. Figure 2 presents ODLTS spectra of samples B95, 598 and 646. In all these spectra the dominant feature is the peak near 300K (for the time windows t_1/t_2=100/1000 ms) corresponding to emission of holes from hole traps with the energy level E_v+0.9 eV. This trap was previously observed in various GaN samples (see [1] for discussion and relevant references) and was associated with the deep acceptor participating in the donor-acceptor pairs transition giving rise to the yellow luminescence band (this is the H2 hole trap according to notation proposed in [1]). The concentration of the H2 traps as deduced from the amplitude of the ODLTS peak was about
$3.8 \cdot 10^{14}$ cm^{-3} in sample B95 and close to $2.9 \cdot 10^{15}$ cm^{-3} in samples 598 and 646. ODLTS measurements for sample 385 have been already published [6] and they showed no detectable signal from the H2 traps. Once again one has to conclude that these traps are not the ones responsible for recombination lifetimes in n-GaN. The origin of the broad band-like signal in ODLTS spectra in Figure 2 at low temperatures is most likely due to a

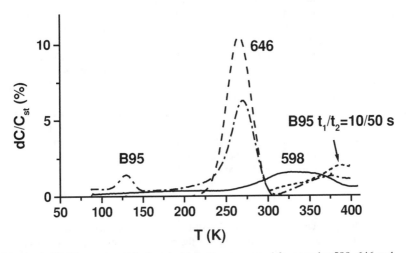

Figure 1. DLTS spectra with electrical injection measured for samples 598, 646 and B95. Measurement conditions: reverse bias of -0.5V, forward bias of 1V (2 s duration), t_1/t_2=100/1000 ms (for sample B95 we also show the high temperature portion of the spectrum measured with t_1/t_2=10/50 s)

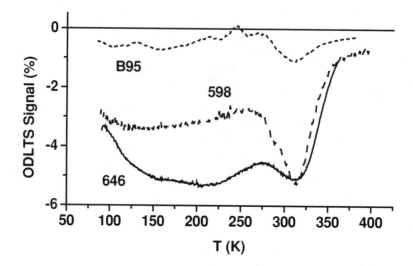

Figure 2. ODLTS spectra in samples B95, 598 and 646; measurements were made at –0.5V reverse bias, with a UV deuterium lamp injection pulse 2 s long; the spectra are shown for time windows $t_1/t_2=100/1000$ ms.

band of relatively shallow states in the lower half of the bandgap. The nature of this band could be further elucidated by photocapacitance measurements at low temperature and by C-T measurements before and after illumination. These measurements were performed on samples 598, 646, B95. All the samples showed an onset of measurable photocapacitance near 1 eV (about 2 pF in sample 598 and much more pronounced in samples B95, 646), slight increase near 2.5 eV, no further growth for photon energies up to 3.1 eV and a very strong increase of capacitance upon above-bandgap illumination with the deuterium UV lamp. This increase of capacitance persisted after the UV light was switched off and could be either fully (in sample 598) or partially (samples B95, 646) removed by applying forward bias of higher than 0.8 V. That shows that the metastability in question is due to charging of deep traps located below the Fermi level. This trapped positive charge led to a decrease of the built-in voltage of the diodes V_{bi} after illumination (V_{bi}^{PPC}) compared to the built-in voltage in the dark (V_{bi}). The difference, ΔV_{bi} can be used to calculate the density of recharged traps N_t from the expression [7]:

$$\Delta V_{bi}=q \cdot N_t \cdot w_0^2/(2 \cdot \varepsilon \varepsilon_0), \qquad (1)$$

where w_0 is the space charge region (SCR) width under illumination, q is the electron charge, ε_0 is the permittivity of vacuum and ε is the relative permittivity. The results of these calculation are shown in Table II. It can be seen the density of traps is very high-on the order of some 10^{16} cm^{-3} in all the samples. Figure 3 shows how these metastable changes of capacitance decrease with temperature. In all three samples this happens in two stages: a broad step at 85K-250K and a step near 270K. The first of these steps is

Table II. Persistent photocapacitance results for various GaN samples

Sample #	N_d, cm^{-3}	V_{bi}, eV	N_d^{PPC}, cm^{-3}	V_{bi}^{PPC}, eV	N_t, cm^{-3}
598	$1.9 \cdot 10^{16}$	0.94	$1.9 \cdot 10^{16}$	0.51	$2.1 \cdot 10^{16}$
646	$3.4 \cdot 10^{15}$	1.06	$5.5 \cdot 10^{15}$	0.42	$1.8 \cdot 10^{16}$
B95	$5.8 \cdot 10^{15}$	1.07	$9 \cdot 10^{15}$	0.51	$2.1 \cdot 10^{16}$

most probably due to a band of relatively shallow hole trap states in the lower part of the bandgap. The second step is definitely due to detrapping of holes from the H2 traps as was demonstrated by 1.3-eV-light optical quenching experiments to be reported elsewhere. We tentatively propose that the extended band of hole trapping states could be due to dislocations which is based on the extended nature of the states we observe and on the fact that the aggregate density of these states - some 10^{16} cm^{-3} - has the right order of magnitude for the dislocation density of ~10^9 cm^{-2}. The major difference between samples 598 and B95, 646 is that the latter ones show persistent increase of electron concentration after illumination while the former does not, as can be seen from shallow donor measurements by C-V in the dark (N_d) and after illumination (N_d^{PPC}) (see Table II). This correlates very well with the absence in DLTS spectra of sample 598 electron traps ET4 which are present in samples 646 and B95 (see Figure 1). These traps have been shown to possess a barrier of about 0.15 eV for capture of electrons [6] and thus should lead to persistent photoconductivity at low temperature. C-T curve 4′ in Figure 3 shows that the step corresponding to capture of electrons by the ET4 trap ends near 200K which is fully compatible with the measured barrier height (the curve was obtained by

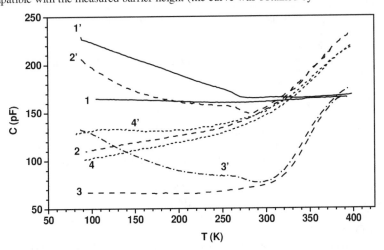

Figure 3. C-T curves for various GaN samples measured at 1 kHz when cooling in the dark (1, 2, 3, 4) and after illumination with the deuterium UV lamp at 85K (1′, 2′, 3′, 4′). Curves 1, 1′ are for sample 598, 2, 2′ for sample B95, 3, 3′ for sample 646. Also shown are the cooling down and heat-up curves for sample B95 illuminated with UV light and quenched with forward bias pulse of 0.8 V (curves 4, 4′; note that only the signal from PPC due to ET4 trap can be seen).

illuminating B95 sample with UV light and filling all the centers but the ET4 traps with a forward bias pulse of 0.8 V; similar results were obtained by illuminating the sample with 1.3 eV light - the results of these experiments will be presented elsewhere).

If the situation with persistent photoconductivity (PPC) is reasonably well understood in our samples the nature of the recombination channel competing with radiative recombination remains elusive. None of the centers described above can aspire to such a role. Perhaps surface recombination could be held responsible here but this question obviously requires further study. It also remains to be seen whether the band-like states near the valence band detected in our ODLTS and photocapacitance measurements could play any significant role in recombination. These states undoubtedly can capture electrons at a certain rate as demonstrated by forward bias pulse filling experiments but the total density of states in the band seems to be almost the same in samples with significantly different L_d values. To us it seems more likely that these band states are rather hole traps and compensation centers. If indeed the states are associated with dislocations they could contribute to forming the barriers for recombination around dislocations and explain lowered radiative recombination efficiency near individual dislocations demonstrated recently by several groups (see an extensive discussion and relevant references in [1]).

CONCLUSIONS

We have shown that in n-GaN there exist band-like states in the lower portion of the bandgap. These states have the total density of some 10^{16} cm^{-3} and can give rise to very large metastable changes of capacitance at low temperatures.

Persistent photoconductivity in n-GaN films is associated in part with the ET4 electron traps having a relatively high barrier for capture of electrons.

In n-GaN there exist unidentified recombination channels that effectively compete with the radiative recombination transitions. But none of the centers detected by capacitance spectroscopy can be associated with such channels.

REFERENCES

1. A.Y. Polyakov, in GaN and Related Materials, ed. S.J. Pearton (Gordon&Breach Science Publishers, Singapore, 1999) pp. 173-233
2. B. Monemar, in the Proceedings of the 2nd Int. Conf. on Nitride Semiconductors (ICNS'97), Tokushima, Japan, 1997, pp. 6-8
3. W.V. Lundin, B.V. Pushnyi, A.S. Usikov, M.E. Gaevski, M.V. Baidakova and A.V. Sakharov, Inst. Phys. Conf. Ser. **155**, 319 (1997)
4. Yu. V. Melnik, I.P. Nikitina, A.S. Zubrilov, A.A. Sitnikova and V.A. Dmitriev, Inst. Phys. Conf. Ser. **142**, 863 (1996)
5. A.Y. Polyakov, N.B. Smirnov, A.V. Govorkov, D.W. Greve, M. Skowronski, M. Shin and J.M. Redwing, MRS Internet J. Nitride Semicond. Res. (MIJ-NSR), **3**, article 37 (1998)
6. A.Y. Polyakov, N.B. Smirnov, A.S. Usikov, A.V. Govorkov and B.V. Pushnyi, Solid-St. Electron., **42**, 1959 (1998)
7. L.S. Berman and A.A. Lebedev, Capacitance Spectroscopy of Deep Centers in Semiconductors (Nauka, Leningrad, 1981) chapter 1, p. 29 (in Russian)

Mat. Res. Soc. Symp. Vol. 595 © 2000 Materials Research Society

Fermi Level Pinning at GaN-interfaces: Correlation of electrical admittance and transient spectroscopy

H.Witte[a], A.Krtschil[a], M.Lisker[a], D.Rudloff[a], J.Christen[a], A.Krost[a], M.Stutzmann[b], F. Scholz[c]

[a] Institute of Experimental Physics, Otto-von-Guericke-University Magdeburg, D-39016 Magdeburg, Germany
[b] Walter Schottky Institute, Technical University Munich, D-85748 Garching, Germany
[c] 4th Institute of Physics, University Stuttgart, D-70550 Stuttgart, Germany

Abstract

In GaN layers grown by molecular beam epitaxy as well as metal organic vapor phase epitaxy significant differences were found in the appearance of deep defects detected by thermal admittance spectroscopy as compared for deep level transient spectrocopy measurements. While, thermal admittance spectroscopy measurements which were made under zero bias conditions only show thermal emissions at activation energies between 130 and 170 meV, further deep levels existing in these GaN layers were evidenced by transient spectrocopy. This discrepancy is explained by a pinning effect of the Fermi level at the metal / GaN interface induced by high a concentration of the deep levels showing up in thermal admittance spectroscopy. We compare our results with a GaAs:Te Schottky- diode as a refernec sample. Here, both spectroscopic methods give exactly the same deep level emissions.

Introduction

Gallium nitride is an attractive base material for novel applications in optoelectronics as UV-laser diodes, LED's and UV detectors and for high power, high frequency and high temperature electronics /1/. However, this high potential for new devices can only be used if the properties of ohmic and Schottky contacts are well controlled. On the other hand, there are some indications for the influence of deep defects located at the surface of the GaN layer on the contact properties. In GaN pn-junctions for example the dominant current mechanisms are the thermionic emission over a barrier and a tunnelling process via deep levels in the interface regions /2,3/.
In this work, we investigated deep defects in the space charge region of a Schottky contact on intentionally undoped, n-type GaN layers using thermal admittance spectroscopy (TAS) and deep level transient spectroscopy (DLTS).

Experimental

The samples were grown both on (0001) oriented sapphire substrates using metal organic vapor phase epitaxy (MOVPE) and by plasma induced molecular beam epitaxy (MBE). The growth conditions are described in detail in /4/ and /5/. All samples are nominally undoped or Si-doped and show n-type conductivity with carrier concentrations in the range between 5×10^{16} and 10^{17} cm^{-3} determined by Hall effect measurements. Before metallization, the GaN surfaces were cleaned with organic solvents in an ultrasonic bath and etched with HF and HCL. The ohmic contacts were formed by evaporating a 100nm thick Al layer followed by an annealing step at 500°C in nitrogen athmosphere. A 100nm thick Pt layer deposited by magnetron sputtering was commonly used as Schottky contact. Both contacts were arranged on the GaN layer and have a diameter of 750μm. The distance between the metal layers was 1mm.

We carried out C-V measurements, admittance spectroscopy (AS) and thermal admittance spectroscopy in the frequency range between 20Hz and 1MHz and at temperatures between 80K and 400K using a high precision LCR meter HP4284A and a liquid-nitrogen flow cryostat. For both, AS and TAS measurements the capacitance and the conductance were measured as a function of frequency and temperature (fore more details see /7/). If the resistance of the neutral bulk region of the samples is not negligible a serial circuit model of the complex impedance was used for analysis (see /6/).

In addition, we performed DLTS measurements at a frequency of 1MHz using the capacitance meter Boonton 7200 and a pulse generator HP8110A. The capacitance transients were recorded via a fast A/D converter with a time resolution of 1μs and analyzed by the boxcar method. DLTS measurements at lower frequencies were made at Schottky contacts which break down at frequencies below 1MHz. In this case, the transients were measured using the LCR meter HP4284A with a minimal time step of 300ms in the boxcare method.

Results and Discussion

The Pt Schottky contacts on GaN layers show barrier heights up to 0.95eV determined by C-V-spectroscopy at frequencies between 50kHz and 1MHz. This barrier is slightly lower than the 1.04eV found in /8/. The net donor concentrations were in the range of $5x10^{16}$ to $5x10^{17}$ cm^{-3} determined by C-V-characteristics and by Hall effect measurements at room temperature.

In generally, we can distinguish two different categories of Schottky contacts as represented by the samples #1 and #2 in Fig. 1. Sample #1 shows only a little variation in the conductance spectrum for different bias voltages which is caused by a small Schottky barrier. Furthermore, the contact breaks down at relatively low frequencies of about $5x10^4$ s^{-1} resulting in a completely bias independent conductivity and capacitance above this frequency. The other group shows larger barrier heights resulting in a stronger influence of the bias on the conductance (sample #2 in Fig.1). The rectifying behavior of this contact vanished at higher frequencies of about 10^6 s^{-1}.

These different properties of the contacts were correlated with the surface roughness measured by atomic force microscopy (AFM),which is shown on the right side of Fig.1. The significant difference between both samples becomes visible. Sample #2 exhibits a sharp distribution of low height spikes equivalent to a nearly perfect flat surface. In contrast, the AFM measurements of sample #1 revealed a rough surface with a broad distribution of heights. Obviously Schottky contacts on a smooth surface exhibit better rectifying properties and break down at higher frequencies.

This effect can be explained by a lowering of the Schottky barrier caused by fluctuations in the spatial distribution of the barrier heights as discussed in /9/. Additionally, the frequency dependence of the breakdown behavior and the structure of the admittance spectra suggest that there is a recharging process of deep defects at the interface involved. Both samples in Fig. 1 show shoulders and peaks in the breakdown region. These peaks are caused by deep defects and were already described in /6/.

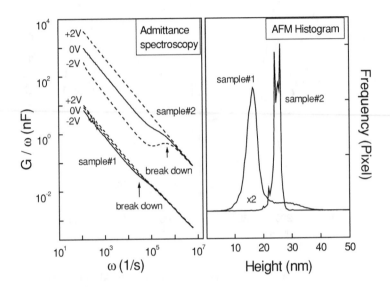

Fig. 1: (Left) Admittance spectra of two MOVPE-grown GaN samples at different bias voltages. (Right) Histogram of the GaN surface morphology measured by atomic force microscopy for both samples #1 and #2.

Independent of the surface morphology and the growth technique deep defects appear in the TAS spectra at activation energies of (130-140) meV (T1) and (150-170) meV (T2). The TAS-data on the left side of Fig. 2 shows typical TAS spectra of two MBE and two MOVPE grown GaN layers, respectively. These emissions dominate the whole TAS spectrum evidencing the high concentrations of the corresponding defects. They are probably located in the vicinity of the surface because all TAS measurements were made at zero voltage causing the lowest expansion of the space charge region.

While the TAS measurements exclusively detect these defects T1 and T2, further deep traps exist in the investigated samples verified by DLTS measurements at 1MHz (HF-DLTS) and at lower frequencies (LF-DLTS). These DLTS spectra are shown in Fig. 2 on the right side. The thermal activation energies of the defects are 240±30meV for D2, 630meV for D3, and 860meV for D4, respectively. The observed traps are already known from the literature (see for example /10/) and were discussed as intrinsic defects.

It is important to point out, that in the TAS spectra of these GaN layers only one single trap emission appears, while DLTS measurements performed at the same samples reveals the whole deep level spectrum with several trap emissions. The presented discrepancies between the both spectroscopic methods result from the recharging processes of the defects during the distinct measurements.

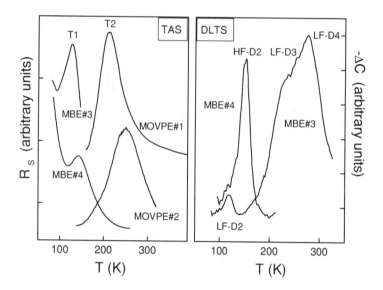

Fig. 2: (Left) TAS spectra of different GaN samples grown by MOVPE (#1 and #2) and by MBE (#3 and #4), respectively, at a modulation frequency of 10kHz and under zero bias condition. The resulting activation energies are 130-140meV for T1 and 150-170meV for T2, respectively. (Right) DLTS-spectra of the MBE samples #3 and #4 at 100kHz (LF-DLTS) and at 1MHz (HF-DLTS). The corresponding emission rates are e_n = 36.1 s^{-1} for the HF-DLTS spectrum and e_n = 0.4 s^{-1} for the LF-DLTS, respectively.

In TAS the deep defects were recharged around the Fermi level near the thermodynamical equilibrium. Commonly, the Fermi level shifts with increasing temperature to the mid gap region and ionizes the various defect states with respect to their thermal activation energy. Therefore, the whole deep level spectrum should be obtained. However, if the Fermi level is pinned at the activation energy of a dominant level with high concentration, the Fermi level cannot shift and ionize of further defects becomes inhibited. As a result, the TAS spectrum only shows the recharging process of this one defect state, responsible for the pinning effect. Because the signals under zero bias conditions originate from a region near the interface the TAS spectra show the emissions from traps near the interface Fermi level. Therefore we conclude that the Fermi level is pinned in the near of the interface due to high concentrations of the specific deep defects T1 and T2.

In contrast, the DLTS measurements are only possible for Schottky diodes with high barriers and at frequencies below the breakdown frequency. In our DLTS investigations a voltage pulse between 3 V in reverse direction and 0V was used. Thus, the expansion of the space charge region causes a recharging of other traps far from the Fermi level in the equilibrium.

One additional specific feature is the high temperature of the TAS peaks for the defects T1 and T2 compared with the corresponding peak temperatures detected by DLTS (see the left side of Fig. 4). The high temperature of the TAS peaks may be caused by the location of these defects in the interface. The emission behavior of the interface traps should be changed under these conditions.

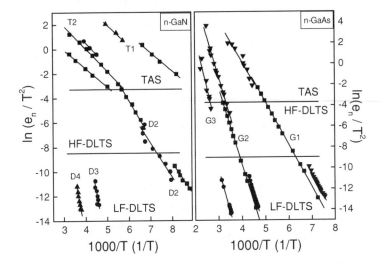

Fig. 4: (Left) Arrhenius plots of deep defects in n-type GaN layers obtained by HF-DLTS, LF-DLTS and TAS. (Right) Arrhenius plots of three deep defect levels obtained by the spectroscopical methods DLTS, TAS and low frequency DLTS in a GaAs:Te sample with an electron concentration of 2×10^{15} cm^{-3}. The activation energies of the deep traps are: G1: 350±30 meV; G2: 670±40 meV and G3: 740±50 meV, respectively.

To evidence that all three spectroscopical methods, i.e. DLTS, LF-DLTS, and TAS yield identical results under the same conditions for the same defect performed at the same samples, we used a GaAs:Te Schottky diode as an example for an interface without Fermi level pinning. This sample has a similar conductivity and the same contact arrangement as the GaN layers. The resulting Arrhenius plots are shown in the right side of Fig. 4. For the presented deep defects the plots of the different methods agree perfectly. So we conclude, that the differences of the peak temperatures in the GaN layers are caused by the nature of the defects and not by the measurement methods.

These investigations show that the deep defects found cause a pinning of the Fermi level in the interface between the metal and the GaN surface. This induces a breakdown of the Schottky barrier which is assumed as a tunneling process through these deep defect states as described in detail in /8/ and /11/ for Schottky contacts. Furthermore, the influence of interface defects with activation energies of 160meV and 230meV on Schottky contacts was demonstrated in /12/ which are very similar with our results. Traps in the range of 0.1 to 0.2eV are also discussed as origin of the pinning of the Fermi level in /13/. For p-n-junctions interface defects influence the low frequency noise /3/ and the reverse-bias current /2/. Here, the threading dislocations are discussed as the source of these interface defects.

TAS measurements of MBE GaN using an ECR plasma source /6/ and of LPCVD GaN /14/ show that the pinning effect occurs in GaN grown by other techniques and in Schottky contacts prepared by other methods, too. Furthermore, the results in /6/ indicate that other traps for example at 260meV cause a Fermi level pinning in the metal / GaN interface.

In conclusion, the TAS measurements show that a high concentration of interfacial defects detected by TAS causes a breakdown of the Schottky barrier at defined modulation frequencies. This suggest a pinning of the Fermi level at the interface due to deep defect emission.

Acknowledgement

This work was financially supported by the Deutsche Forschungsgemeinschaft under Contract number WI 1619/1-1 and by the Kultusministerium Sachsen-Anhalt, contract number 002KD1997.

References

/1/   S.J.Pearton, J.C.Zolper, R.J.Shul, F.Ren; J. Appl. Phys. **86**, 1 (1999)

/2/   P.Kozodoy, J.P.Ibbetson, H.Marchand, P.T.Fini, S.Keller, J.S.Speck, S.P.DenBaars, U.K.Mishra ; Appl. Phys. Lett. **73**, 975 (1998)

/3/   D.V.Kuksenkov, H.Temkin, A.Osinsky, R.Gaska, M:A.Khan; Appl. Phys. Lett. **72**, 1365 (1998)

/4/   H.Witte, A.Krtschil, M.Lisker, J.Christen, F.Scholz, J.Off; MIJ-NSR **4S1**, G3.71 (1999)

/5/   O.Ambacher, J.Smart, J.R.Shealy, N.G.Weimann, K.Chu, M.Murphy, W.J.Schaff, L.F.Eastman, R.Dimitrov, L.Wittmer, M.Stutzmann, W.Rieger, J.Hilsenbeck; J. Appl. Phys. **85**, 3222 (1999)

/6/   A.Krtschil, H.Witte, M.Lisker, J.Christen, U.Birkle, S.Einfeldt, D.Hommel; J. Appl. Phys. **84**, 2040 (1998)

/7/   J.Barbolla, S.Duenas, L.Bailon, Solid-State Electron. **35**, 285 (1992)

/8/   J.D.Guo, M.S.Feng, R.J.Guo, F.M.Pan, C.Y.Chang : Appl. Phys. Lett. **67**, 2657 (1995)

/9/   S.Chand, J.Kumar; J. Appl. Phys. **80**, 288 (1996)

/10/   F.D.Auret, S.A.Goodman, F.K.Koschnick, J.-M.Spaeth, B.Beaumont, P.Gibart; Appl. Phys. Lett. **73**, 3745 (1998)

/11/   L.S.Yu, Q.Z.Liu, Q.J.Xing, D.J.Qiao, S.S.Lau, J.Redwing ; J. Appl. Phys. **84**, 2099 (1998)

/12/   J.Y.Duboz, F.Binet, N.Laurent, E.Rosencher, F.Scholz, V.Harle, O.Briot, B.Gill,, R.L.Aulombard ; Mat. Res. Soc. Symp. Proc. Vol. **449**, 1085 (1997)

/13/   A.Y.Polyakov, N.B.Smirnov, A.V.Govorkov, M.Shin, M.Skowronski, D.W.Greve ; J. Appl. Phys. **84**, 870 (1998)

/14/   H.Witte, A.Krtschil, M.Lisker, J.Christen, M.Topf, D.Meister, B.K.Meyer; Appl. Phys. Lett. **74**, 1424 (1999)

Mat. Res. Soc. Symp. Vol. 595 © 2000 Materials Research Society

Photocurrent spectroscopy investigations of Mg-related defects levels in p-type GaN

S. J. Chung, O. H. Cha, H. K. Cho, M. S. Jeong, C-H. Hong, E-K. Suh* and H. J. Lee
Semiconductor Physics Research Center and Department of Semiconductor Science and Technology, Chonbuk National University, Chonju 561-756, KOREA.
*Author to whom correspondence should be addressed labsek@moak.chonbuk.ac.kr

ABSTRACT

The defect levels associated with Mg impurity in p-type GaN films were systematically investigated in terms of doping concentration by photocurrent spectroscopy. Mg-doped GaN samples were grown on sapphire substrate by metal organic chemical vapor deposition and annealed in nitrogen atmosphere at 850• for 10 minutes. At room temperature, PC spectra showed two peaks at 3.31 and 3.15 eV associated with acceptor levels formed at 300 and 142 meV above valence band in as grown samples. But, after the thermal annealing, PC spectra exhibited various additional peaks depending on the Mg concentration. In the GaN samples with Mg concentration around $6 \cdot 7 \times 10^{17}$ cm^{-3}, we have observed PC peaks related to Mg at 3.31 as well as 3.02 eV and carbon acceptor at 3.17 eV. For moderately Mg doped GaN samples, i.e., the hole concentration p=$3 \cdot 4 \times 10^{17}$ cm^{-3}, additional peak was observed at around 0.9 eV which can be attributed to defects related to Ga vacancy. For relatively low Mg doped samples whose hole concentrations are $1 \cdot 2 \times 10^{17}$ cm^{-3}, additional broad peak was observed at around 1.3 eV. This peak may be related to the yellow band luminescence. As the Mg concentration is increased, the concentration of Ga vacancies can be reduced because Mg occupies the substitutional site of Ga in GaN lattice. When the hole concentration is above $6 \cdot 7 \times 10^{17}$ cm^{-3}, the yellow luminescence and Ga vacancy related peaks disappeared completely.

INTRODUCTION

Gallium nitride (GaN) has been one of the most promising materials for blue-ultraviolet (UV) lasers, short wavelength radiation detectors and high-temperature electronic devices. This is evident from several impressive device achievement in the last few years, including the injection light-emitting diodes(LED)[1], laser diodes[2], solar-blind UV detectors[3], and high-temperature transistors[4]. Since deep levels in GaN affect significantly the photo-electric properties of material and devices, they were investigated using a number of techniques in samples grown by different methods and post treatment conditions [5-18]. For n-type and undoped GaN samples, a shallow donor(SD) at about 30 meV below the conduction band were usually observed by temperature dependent Hall and photoluminescence(PL) measurements[6,7], and a double donor(DD) at about 0.7 eV below the conduction band edge(E_c) as well as a deep level at E_c - 2.0 eV were recognized by optically detected magnetic resonance (ODMR) and persistent photoconductivity (PPC) techniques [6,13]. On the other hand, a deep level at 1.4 eV above the top of the valence band (E_v) were detected using photocurrent (PC) and photoemission capacitance transient spectroscopy methods [9,14,17]. However, one discordance seems to appear for the Mg-related deep level; the Hall measurement gives an activation energy of 0.16 eV while the PL usually senses a level at 0.34 eV above E_v [5,15]. In this work, we have systematically investigated defect levels related with Mg as a function of doping concentration in p-type GaN films by photocurrent(PC) spectroscopy.

EXPERIMENT

GaN samples investigated in this work were grown in a metalorganic chemical vapor deposition(MOCVD) system on (0001) oriented sapphire substrate at low pressure. The ammonia and trimethylgallium (TMG) were served as the precursors, and the bis(cyclopentadienyl) magnesium(Cp_2Mg) was used for the Mg doping. In prior to the growth of the epitaxial layer, a GaN buffer layer with nominal thickness of 25 nm was grown at 560°C. Mg-doped GaN layers were grown at 1010°C with various molar flow rates of Cp_2Mg. To activate the Mg acceptors, the GaN samples were annealed at 850°C for 10 min in a rapid thermal annealing(RTA) system. The annealing time and temperature, hole concentrations and mobilities of three groups of samples determined the room temperature Hall-effect measurements are summarized in table I.

Table I. Parameters of Mg –doped p-type GaN epilayers investigated in this work.

Sample No.	Thermal annealing (N_2 –ambient)			Thick-ness (μm)	Carrier concentration (cm^{-3})	Mobility (cm^2/Vs)
	Method	Temp.(°C)	Time(min)			
Sample A	RTA	850	10	2	2.1×10^{17}	12
Sample B	RTA	850	10	2	3.8×10^{17}	12.8
Sample C	RTA	850	10	2	6.5×10^{17}	12.3

The PC spectra were measured in a coplanar geometry with two indium/zinc contacts soldered to the GaN surface. For the PC spectra measurement, a tungsten-halogen lamp was used as a light source and Shimadaza AQV 50 spectrometer was used for analyzing the light.

A bias voltage of 1.5 volt was supplied by a current source, and the monochromatic photon flux onto the sample was of the order of 10^9 photons/s. But, the normalization effect of the incident photon density did not significantly alter the PC spectrum.

RESULT AND DISCUSSION

Figure 1 presents infrared absorption spectra for the Mg-doped p-type GaN sample B. For the as grown sample, the spectrum exhibits a broad vibrational modes at around 3125 cm^{-1} due to Mg-H complex[19], NH and O-H stretching[20].

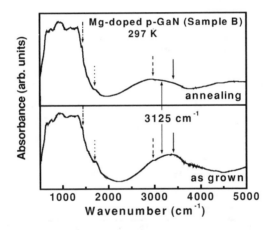

Fig. 1. Infrared absorption spectra for Mg-doped p-type GaN sample B.

After thermal annealing, the relative absorbance of this broad peak is reduced by a factor of three. Further evidence for a Mg-H complex is provided by the correlation of the intensity of the local vibrational mode absorption lines with the p-type conductivity of Mg-doped GaN films. The intensity of the absorption decreases significantly upon the activation of the p-type dopant. This observation is consistent with previous report that most of the Mg acceptors are passivated by hydrogen after the growth.

The PC spectra of the Mg-doped p-type GaN measured at room temperature for as grown samples are presented in Fig.2. As shown in Fig. 2, the PC spectra of all samples consists of two peaks. The broad peaks at 3.31 (P1) and 3.15

eV(P2) are associated with acceptor levels formed at 300 and 140 meV above valence band.

After thermal annealing, the PC spectra shows more prominent and sharp peaks as can be seen in Fig. 3, they exhibit various additional peaks depending on the Mg concentration. For the heavily Mg doped GaN sample 6.5×10^{17} cm^{-3}, we have observed three PC peaks at around 3.31(C1), 3.02 eV(C3) and 3.17 eV(C2). For moderately Mg doped GaN samples, additional peak on the low energy side was observed at around 0.9 eV(C5). But, for relatively low Mg doped samples whose hole concentration are around 2.1×10^{17} cm^{-3}, additional broad peak was observed at around 1.3 eV(C4). The PC peak around 3.31 eV(C1) corresponds to transitions from Mg-related deep level located about 130 meV above the valence band, which were usually probed by Hall and EL measurements[15].

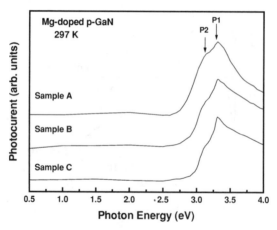

Fig. 2. Photocurrent spectra for as grown samples of Mg-doped p-type GaN.

Figure 4 present an Arrhenius plot of the dark current of sample B from in the range 77 to 318 K; the thermal ionization energy of the activated Mg acceptor is obtained as 142 meV. This value shows a good agreement with the acceptor binding energy previously estimated between 120 and 200 meV [20-23].

Therefore, the PC peaks at around 3.31 eV can be attributed to transitions from Mg-related shallow acceptor located about 142 meV above the valence band to the conduction band.

The PC peak at 3.02 eV(C3) corresponds to a deep level located about 350 meV above the valence band, which were usually observed in PL experiment for Mg-doped GaN[15].

On the other hand, the PC peak at around 3.17 eV(C2) might be associated with carbon acceptor. Carbon is known to act as an acceptor[24] with an acceptor binding energy of 230 meV[25]. This peak is observed consistently in

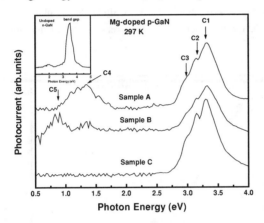

Fig. 3 Photocurrent spectra for Mg-doped p-type GaN after thermal annealing. The insetroom temperature photocurrent of undoped n-type GaN.

n-type GaN and Mg-doped p-type GaN films grown by MOCVD suggesting that the carbon atoms might be incorporated from the susceptor in the growing process.

Finally, the broad PC peak around 1.3 eV(C4) is associated with the absorption from the valence band edge to the deep donor/acceptor state(DA)[26] which is responsible for the yellow band emission in GaN, and the PC peak at around 0.91 eV(C5) could be related to the Ga vacancy located at around 0.9 eV above the valence band. The measurements in p-type GaN provide strong evidence for the presence of three deep states with energies at approximately 0.9 to 1.1, 1.4, and 1.8 to 2 eV above E_v[9,14]. Qiu and Pankove[14] proposed that metastable centers at 1.1, 1.4, and 2.04 eV above the valence band edge are responsible for the PPC bebavior in Mg-doped GaN, and that Ga vacancy is the candidate for PPC effect in n-type GaN. Recently, Reddy et al.[27] reported that persistent photoconductivty(PPC) and yellow luminescence(YL) are related to each other originating the same defect. Though both gallium vacancies(V_{Ga}) and gallium interstitials(Ga_i) are proposed as defects responsible for YL in GaN, V_{Ga} is widely accepted as a possible candidate[14,28].

Similarly, nitrogen antisites(N_{Ga}) and gallium vacancies are proposed the possible defects giving rise to PPC[14,29]. We believe that the yellow band luminescence is strongly related to Ga vacancy based on our PC measurement. As the Mg concentration is increased, Ga vacancies can be reduced because Mg occupies the substitutional site of Ga in the GaN lattice. As a result, the YL and Ga vacancy related PC peaks, C5, completely disappeared in highly Mg doped samples. On the other hand, intensities of PC peak related to Mg acceptor levels were increased with increasing Mg concentration.

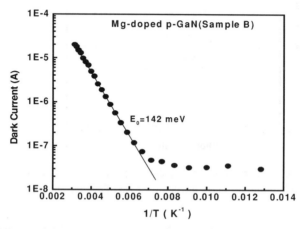

Fig. 4. The Arrhenius plot of the dark current(I_d) for Mg-doped p-type GaN sample B.

CONCLUSIONS

Defect levels associated with Mg have been studied in terms of doping concentration in p-type GaN films using photocurrent spectroscopy. The PC spectra for as grown samples showed two broad peaks at around 3.31 and 3.15 eV associated with Mg-related levels formed above the valence band. After the thermal annealing of samples at 850• for 10 minutes, PC spectra exhibited various additional peaks depending on the Mg concentration. PC peaks related to Mg appeared at 3.31 as well as 3.02 eV and a peak due to carbon acceptor were observed at 3.17 eV. For moderately and relatively low Mg doped GaN samples, additional peaks on the low energy side was observed at around 0.9 eV in association with Ga vacancy and at 1.3 eV which seems to be related with the yellow band luminescence. It is suggested that the yellow band luminescence is strongly related to Ga vacancy. As the Mg concentration is increased, the concentration of Ga vacancies can be reduced since Mg occupies the substitutional site of Ga in the GaN lattice. In samples with hole concentrations above 6.5×10^{17} cm^{-3}, the yellow luminescence and Ga vacancy related peaks disappeared completely.

REFERENCE

1. S Nakamura, M. Senoh, N. Iwasa, and S. Nagahama, Jpn. J. Appl. Phys. 34, L797(1995).
2. S. Bakamura, M. Sehoh, S. Nagahama, N. Iwasa, T. Yamada, T. Matsushita. HKiyoku, and H. Sugimoto, Jpn. J. Appl. Phys. 35, L74(1996).
3. B. W. Lim, Q. C. Chen, J. Y. Yang, and M. Y. Khan, Appl. Phys. Lett. 68, 3761(1996).
4. M. A. Khan, Q. C. Chen, C. J. Sun, J. W. Yang, M. S. Shur, and H. Park, Appl. Phys.Lett. 68, 514(1996).
5. Eunsoon Oh, Hyeongsoo Park, and Yongjo Park, Appl. Phys. Lett. 72, 70 (1998).
6. E. R. Glaser, T. A. Kennedy, K. Doverspike, L. B. Rowland, D. K. Gaskill, J. A. Freitas Jr, M. Asif Khan, D. T. Olson, J. N. Kuznia, and D. K. Wickenden, Phys, Rev. B51, 13326 (1995).
7. D. M. Hofmann, D. Kovalev, G. Steude, B. K. Meyer, A. Hoffmann, L. Eckey, R. Heitz, T. Detchprom, H. Amano, and I. Akasaki, Phys, Rev. B52, 16702 (1995).
8. J. A. Freitas Jr, T. A. Kennedy, E. R. Glaser, and W. E. Carlos, Solid-State Electronics. 41, 185 (1997).
9. W. Gotz, N. M. Johnson, and D. P. Bour, Appl. Phys. Lett. 68, 3470 (1996).
10. W. Gotz, N. M. Johnson, R. A. Street, H. Amano, and I. Akasaki, Appl. Phys. Lett. 66, 1340 (1995).
11. C. D. Wang, L. S. Yu, S. S. Lau, E. T. Yu, W. Kim, A. E. Botchkarev, and H. Morkoc, Appl. Phys. Lett. 72, 1211 (1998).
12. P. Hacke, and H. Okushi, Appl. Phys. Lett. 71, 524 (1997).
13. Michele T. Hirsch, J. A. Wolk, W. Walukiewicz, and E. E. Haller, Appl. Phys. Lett. 71, 1098 (1997).
14. C. H. Qiu, and J. I. Pankove, Appl. Phys. Lett. 70, 1983 (1997).
15. J. Z. Li, J. Y. Lin, and H. X. Jiang, Appl. Phys. Lett. 69, 1474 (1996).
16. C. H. Qiu, W. Melton, M. W. Leksono, J. I. Pankove, B. P. Keller, and S. P. DenBaars, Appl. Phys. Lett. 69, 1282 (1996).
17. D. J. Chadi, Appl. Phys. Lett. 71, 2970 (1997).
18. Jorg Neugebauer, and Chris G. Van de Walle, Appl. Phys. Lett. 69, 503 (1996).
19. W. Gotz, N. M. Johnson, and D. P. Bour, M. D. MvCluskey and E. E. Haller, Appl. Phys. Lett. 69, 3725(1996).
20. J. Q. Duan, B. R. Zhang, Y. X. Zhang, L. P. Wang, and G. G. Qin, G. Y. Zhang, Y. Z.

Tong, S. X. Jin, and Z. J. Yang, X. Zhangand Z. H. Xu, J. Appl. Phys. **82**, 5745(1997).

21. R. J. Molnar, T. D. Moustakas, Bull. Am. Phys. Soc, **38**. 445(1993)

22. T. Tanaka, A. watanabe, H. Amano, Y. Kobayashi, I. Akasaki, S. Yamazaki, M. Koike, Appl. Phys. Lett. **65**, 593(1994).

23. W. Kim, A. Salvador, A. E. Botchkarev, O. Aktas, S. N. Mohammad, H. Morkoc, Appl. Phys. Lett. **69**, 559(1996).

24. W. Gotz, N. M. Johnson, J. Walker, D. P. Bour, R. A. Street, Appl. Phys. Lett. **68**, 667(1996).

25. H. Nakayama, P. Hacke, M. P. H. Khan, T. Detchprohm, K. Hiramatsu, N. Sawaki, Jpn. J. Appl. Phys. **35**, L282(1996).

26. S. J. Pearton, C. R. Abernathy, and F. Ren, Electron. Mater. **30**, 527(1994).

27. S. Fisher, C. Wetzel, E. E. Haller, and B. K. Meyer, Appl. Phys. Lett. **67**, 1298 (1995).

28. H. B. Mao, H. G. Kim, S. J. Park, X. L. Huang, S. J. Chung, and E.-K. Suh, unpublished.

29. C. V. Reddy, K. Balakrishnan, H. Okumura, and S. Yoshida, Appl. Phys. Lett. **73**, 244(1998).

30. J. Neugebauer and C. G. Van de Walle, Appl. Phys. Lett. **69**, 503(1996).

31. H. M. Chen, Y. F. Chen, M. C. Lee, and M. S. Feng, Phys. Rev. **B56**, 6942(1997).

Mat. Res. Soc. Symp. Vol. 595 © 2000 Materials Research Society

Characteristics of Deep Centers Observed in n-GaN Grown by Reactive Molecular Beam Epitaxy

Z-Q. Fang[1], D. C. Look[1], Wook Kim[2], and H. Morkoç[3]
[1] Semiconductor Research Center, Wright State University, Dayton, OH 45435
[2] Center for Solid State Science, Arizona State University, Tempe, AZ 85287-1704
[3] Electrical Engineering and Physics Department, Virginia Commonwealth University, P. O. Box 843072, Richmond, Virginia 23284-3072

ABSTRACT

Deep centers in Si-doped n-GaN samples grown on sapphire by reactive molecular beam epitaxy, using different ammonia flow rates (AFRs), have been studied by deep level transient spectroscopy. In addition to five electron traps, which were also found in n-GaN layers grown by both metalorganic chemical-vapor deposition and hydride vapor-phase epitaxy, two new centers C_1 (0.43-0.48 eV) and E_1 (0.25 eV) have been observed. C_1, whose parameters show strong electric-field effects and anomalous electron capture kinetics, might be associated with dislocations. E_1, which is very dependent on the AFR, exhibits an activation energy close to that of a center created by electron irradiation and is believed to be a defect complex involving V_N.

INTRODUCTION

GaN and its related ternaries, AlGaN and InGaN, are being widely developed for blue/uv optical emitters and detectors, and high-temperature/high-power electronics [1]. For both optical and electronic devices, deep centers, which could act as traps and/or recombination centers, are very important, and must be understood. A number of deep centers in n-GaN, measured by deep-level transient spectroscopy (DLTS), with activation energies in the range of 0.25-0.87 eV and trap densities in the range of 10^{13}-10^{15} cm^{-3}, have been reported [2-5]. Most of these studies have dealt with n-type GaN grown by either metalorganic chemical-vapor deposition (MOCVD) or hydride vapor-phase epitaxy (HVPE). In a recent DLTS study of deep centers in Si-doped n-GaN grown by reactive molecular beam epitaxy (RMBE), we reported the observation of a trap E_1 (0.21 eV) in a particular group of samples (set I), and a dominant trap C_1 (0.44 eV) in another group (set II); both centers are peculiar to the RMBE-grown GaN material [6]. In this paper, we present more details on the characteristics of the centers found in samples in set II, i.e., the strong electric-field effect and the unusual capture kinetics of C_1 and the relationship between the observation of E_1 and the ammonia flow rate (AFR) during RMBE growth. The associations of C_1 with dislocations and E_1 with nitrogen vacancies will be discussed.

SAMPLES AND EXPERIMENTS

Four, ~0.5-μm-thick Si-doped n-GaN samples from set II were used in the present study. The samples were grown at 800 °C on sapphire by RMBE, employing AFRs of 20, 60 and 73 sccm and keeping the Si flux constant. Schottky barrier diodes (SBDs) were prepared from these materials (for the details see [6]). A Bio-Rad DL4600 system with a 100-mV test signal at 1 MHz was used to take capacitance-voltage (C-V) and DLTS data. In order to observe the deepest possible centers within a restricted temperature range (up to 400 K), the smallest rate window (0.8 s^{-1}) in our DLTS system was used. To determine the parameters of the deep-centers, i.e. the activation energy E_T and capture cross section σ_T, the DLTS spectra were taken at different rate windows, from 0.8 to 200 s^{-1}. To study the unusual capture kinetics of C_1, DLTS spectra were measured, at a fixed bias (-2.0 V) and a fixed filling pulse height (+0.5 V), by varying the filling pulse width from 0.2 up to 100 ms.

RESULTS AND DISCUSSION

The profiles of the electron concentrations for four SBD samples, obtained from 300-K C-V measurements, are basically flat as shown in figure 1. (The earlier rise observed in the concentration of sample 5964 is evidently due to a smaller n-GaN thickness above the n$^+$-GaN layer). An interesting observation is that the electron concentration in the n-GaN layers seems to depend on the AFR; i.e., the lower AFRs for samples 5962 and 5963 result in higher electron concentrations, as compared to those for samples 5961 and 5964. The DLTS spectra for three samples, 5963, 5961, and 5964, with n-GaN layers grown by using AFRs of 20, 60, and 73 sccm, respectively, are shown in figure 2a, and a DLTS spectral comparison for samples 5962 (20 sccm) and 5961 (60 sccm) is shown in figure 2b. From the DLTS spectra of the figures, we can observe seven deep centers, labeled as A_1, A, B, C_1, C, D, and E_1. A_1, with concentrations of mid-10^{15} to low-10^{16} cm^{-3}, is a dominant center in RMBE-GaN layers, while C_1, with concentrations in the 10^{15}-cm^{-3} range, is also a prominent center in the layers. Two interesting observations are: 1) the peak positions of A_1 and, especially, C_1 are found to be sample-dependent: i.e. their peak positions shift to lower temperatures as the electron concentrations in the samples increase; and 2) D, with concentrations in the 10^{14}-cm^{-3} range, is clearly observed in samples 5961 and 5964, grown using higher AFRs, while E_1, with concentrations also in the 10^{14}-cm^{-3} range, is only observed in samples 5962 and 5963, grown using lower AFRs (see figure 2b for a detailed comparison). The Arrhenius plots of T^2/e_n for all the deep centers are presented in figure 3. To understand the possible defect nature of E_1, the Arrhenius plot for the electron-irradiation (EI) induced trap E in n-GaN grown by MOCVD [7] is also presented in the figure. A_1, with an average E_T=0.89 eV and σ_T=3 x 10^{-14} cm^2, has an energy close to that of

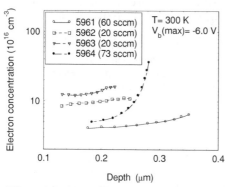

Figure 1. Profiles of electron concentrations for four SBD samples (note the dependence of the electron concentration on AFR).

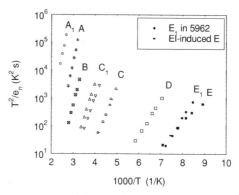

Figure 3. Arrhenius plots of T^2/e_n for all seven deep centers; also presented is the plot of EI induced trap E in n-GaN grown by MOCVD.

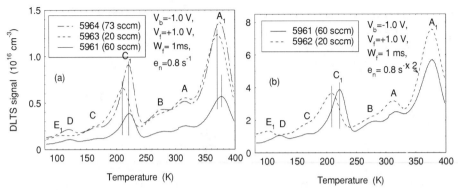

Figure 2. (a) DLTS spectra for three samples, with n-GaN grown using different AFRs, and (b) a DLTS spectral comparison for two samples with different AFRs.

DLN$_4$ (0.86 to 0.91 eV) [4] and E4 (0.88 eV) [8], while A, with E$_T$=0.67 eV and σ_T=1 x10^{-15} cm^2, can be compared with E3 (0.665 eV) [2] and D3 (0.67 eV) [3]. B, with E$_T$=0.62 eV and σ_T=1.2 x 10^{-14} cm^2, has an energy close to that of E2 (0.58 eV) [2], D2 (0.60 eV) [3], and DLN$_3$ (0.59 to 0.63 eV) [4]. C, with E$_T$=0.41 eV and σ_T=2.8 x 10^{-14} cm^2, has similarities to the center reported by us in n-GaN grown by both MOCVD and HVPE [7]. D, with E$_T$=0.25 eV and σ_T=2.6 x 10^{-15} cm^2, is reminiscent of E1 (0.26 eV) [2] and DLN$_1$ (0.23 to 0.25 eV) [4]. Thus, we can state that these five deep centers can be observed in n-GaN grown by various techniques. However, the concentrations of the centers can be very different, depending on the growth technique. For example, the concentrations of A$_1$ (or E4, DLN$_4$) can be very minor (below 10^{13} cm^{-3}) in MOCVD-GaN [8], in contrast to the high concentrations (above 10^{16} cm^{-3}) observed here in RMBE-GaN. On the other hand, C$_1$ and E$_1$ are specific to the n-GaN layers grown by RMBE. C$_1$ has a low-field E$_T$ of 0.43 to 0.48 eV and a σ_T of 3 x 10^{-15} to 6.4 x 10^{-15} cm^2, while E$_1$ has an E$_T$ of 0.25 eV and a σ_T of 4.3 x 10^{-13} cm^2, which are close to the values in the EI induced center E in MOCVD-GaN, with E$_T$=0.18 eV and σ_T=2.5 x 10^{-15} cm^2 [7].

C$_1$ is a very peculiar trap. First, its DLTS peak position is found to be changed not only in different samples, but also in the same sample if V$_b$ is varied. Typical DLTS spectra measured at different V$_b$'s (keeping V$_f$ fixed at +1.0 V) are shown in figure 4a and figure 4b for samples 5961 (60 sccm) and 5963 (20 sccm), respectively. From the figures, we find that: 1) C$_1$ is the only peak shifted to lower temperatures as V$_b$ increases and the total peak shift is found to be more significant in sample 5963 (shift of ~35 K) than in sample 5961 (shift of ~10 K); and 2) the overall DLTS peak height of C$_1$ drops as V$_b$ increases, especially in sample 5963. Such a large DLTS peak shift as a function of V$_b$ implies a very strong electric-field effect on the electron emission from trap C$_1$. Detailed DLTS measurements on sample 5962 (20 sccm), with a total peak shift of ~30 K, indicate that the apparent E$_T$ and σ_T values are changed from 0.44 eV and 3.0 x 10^{-15} cm^2 to 0.31 eV and 6.4 x 10^{-17} cm^2, respectively, as V$_b$ is increased from -1.0 to -4.0 V. Since the peak shift is most likely related to the barrier lowering effect due to a high electric-field, the different shifts found in samples 5961 and 5963 can be interpreted in terms of the different carrier concentrations in the samples (4 x 10^{16} cm^{-3} vs 1.2 x 10^{17} cm^{-3}), which result in different electric-fields in the depletion regions. The DLTS peak height drop in C$_1$ and the other DLTS peaks may be related to carrier tunneling or hopping through deep centers at

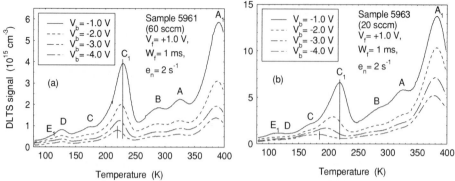

Figure 4. Typical DLTS spectra measured at different V_b's for (a) sample 5961 (60 sccm), and (b) sample 5963 (20 sccm).

high electric-field. For centers in the highest field region, this process could lead to a high emission rate, which would not be seen by the DLTS sampling mechanism. We checked the reverse I-V characteristics of the SBD samples used in the study and found a strong electric-field dependence of the reverse leakage currents (so-called soft breakdown characteristic), which is typical for p-n junctions made on GaN-based materials grown on sapphire and believed to be caused by carrier tunneling or hopping through the defect states associated with the omnipresent threading dislocations in the materials [9,10].

Figure 5. Typical DLTS spectra measured at different V_f's for sample 5964.

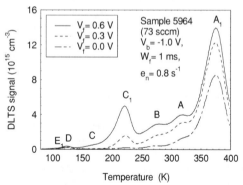

Another distinguishing feature of C_1 is that it is found only in the top region of the n-GaN layers. Typical DLTS spectra measured at different V_f's (keeping V_b fixed at -1.0 V) for sample 5964 are shown in figure 5. From the figure, we see that the peak height of C_1 precipitously drops as V_f decreases from 0.6 to 0 V, resulting in nearly no observation of C_1 in the deeper region of the n-GaN layer, which is in contrast with the observation of A_1 in the whole layer. A final feature of C_1 is the appearance of very unusual electron capture kinetics; i.e., the DLTS peak height of C_1 versus W_f does not show saturation when W_f is increased from 0.2 up to 100 ms, as pictured in figure 6 for three different samples. From the figure, we see that the DLTS signals show a logarithmic dependence on the filling pulse width, W_f. Such a dependence was experimentally observed for dislocation-related deep centers in plastically deformed Si [11] and

Figure 6. DLTS signal of C_1 versus filling pulse width (W_f) for three samples.

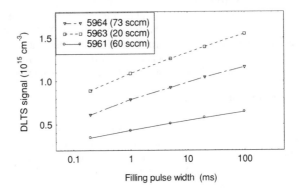

GaAs [12], and explained in terms of a barrier model of electron capture via dislocations [13]. This model assumes that the electron capture rate is limited by a time-dependent Coulomb barrier, with a height proportional to the number of electrons already trapped. Such a situation occurs when the traps are not randomly distributed throughout the crystal but are arranged along lines (so-called "line defects"). Recently, Auret et al. reported a deep center, ER5 (E_C-0.95 eV), which was introduced by an irradiation of 5.4-MeV He ions, and found that the electron capture kinetics of ER5 were similar to those of a line defect [14]. In our case, the line-defect-like trap occurs in as-grown material, which gives good evidence that C_1 is associated with dislocations. Note that thick MOCVD- and HVPE-GaN layers typically have dislocation densities in the 10^8-cm^{-2} range, while thin MBE-GaN layers usually have higher dislocation densities, by one to two orders of magnitude. That might be the reason why we can observe C_1 only in thin RMBE-GaN layers, but not in MOCVD and HVPE layers.

E_1, with a trap concentration in the mid-10^{15} cm^{-3} range, was first observed in the RMBE-GaN samples in set I [6]. Here, in the RMBE samples of set II, we also observe E_1, but with lower trap concentrations (in the 10^{14} cm^{-3} range), and only in those samples (5962 and 5963) grown with lower AFRs. The higher concentrations of E_1 in the samples of set I might be related to the undoped SI GaN layer under the n-GaN layer, since we have found a significant thermally stimulated current trap at low temperatures in RMBE-grown SI-GaN, with an approximate activation energy of 0.17 eV [15]. Note that this energy is also close to that of the EI-induced trap E, thought to be related to an N vacancy. In this light, the variation of E_1 with AFR can be understood in terms of the existence of V_N-related defects; i.e., the lower AFR would result in a higher concentration of V_N and vice versa. Also, the observed dependence of carrier concentration with AFR (figure 1) may involve V_N-related donors, or impurity donors, such as oxygen, on the N site. Note that Si donors, on the Ga site, would be expected to increase at higher AFR. The EI-induced trap E most likely has a Frenkel pair (V_N-N_I) nature and anneals out at about 300 °C [16]. Thus, the trap E_1, which is stable at the growth temperatures of 750-800 °C, cannot be identical to E, but is probably a defect complex involving V_N.

CONCLUSIONS

Deep centers in Si-doped n-GaN samples grown on sapphire by RMBE, using different AFRs, have been studied by DLTS over a temperature range of 80 to 400 K. In addition to

DLTS centers A_1 (0.89 eV), A (0.67 eV), B (0.62 eV), C (0.45 eV) and D (0.24 eV), which are commonly observed in MOCVD- and HVPE-GaN, two new centers C_1 and E_1 are found in RMBE-GaN. C_1, with a low-field activation energy of 0.43 to 0.48 eV, depending on the carrier concentration, shows strong electric field effects. Its unusual electron capture kinetics suggests that it might be related to a dislocation. E_1, with an activation energy of 0.25 eV, shows a close connection with the AFR and is believed to be a defect complex involving the nitrogen vacancy.

ACKNOWLEDGMENTS

The work of Z-Q.F. and D.C.L. was supported by U.S. Air Force Contract No. F33615-95-C-1619. Part of the work was performed at the Air Force Research Laboratory, Wright-Patterson Air Force Base, OH. Also, partial support was received from the Air Force Office of Scientific Research. The research at VCU is funded by AFOSR, ONR, and NSF.

REFERENCES

1. S. N. Mohammand, A. A. Salvador, and H. Morkoç, Proc. IEEE **83**, 1306 (1995).
2. P. Hacke, T. Detchprohm, K. Hiramatsu, N. Sawaki, K. Tadatomo, and K. Miyake, J. Appl. Phys. **76**, 304 (1994).
3. D. Haase, M. Schmid, W. Kurner, A. Dornen, V. Harle, F. Scholz, M. Burkard, and H. Schweizer, Appl. Phys. Lett. **69**, 2525 (1996).
4. W. K. Gotz, J. Walker, L. T. Romano, N. M. Johnson, and R. J. Molnar, Mater. Res. Soc. Symp. Proc. **449**, 525 (1997).
5. C. D. Wang, L. S. Yu, S. S. Lau, E. T. Yu, W. Kim, A. E. Botchkarev, and H. Morkoç, Appl. Phys. Lett. **72**, 1211 (1998).
6. Z-Q. Fang, D. C. Look, W. Kim, Z. Fan, A. Botchkarev, and H. Morkoç, Appl. Phys. Lett. **72**, 2277 (1998).
7. Z-Q. Fang, J. W. Hemsky, D. C. Look, M. P. Mack, R. J. Molnar, and G. D. Via, Mater. Res. Soc. Symp. Proc. **482**, 881 (1998).
8. P. Hacke, H. Okushi, T. Kuroda, T. Detchprohm, K. Hiramatsu, and N. Sawaki, J. Crystal Growth, **189/190**, 541 (1998).
9. D. V. Kuksenkov, H. Temkin, A. Osinsky, R. Gaska, and M. A. Khan, Appl. Phys. Lett. **72**, 1365 (1998).
10. P. G. Eliseev, P. Perlin, J. Furioli, P. Sartori, J. Mu, and M. Osinski, J. Electron. Mater. **26**, 311 (1997).
11. V. V. Kveder, Yu. A. Osipyan, W. Schroter, and G. Zoth, Phys. Status Solidi A **72**, 701 (1982).
12. T. Wosinski, J. Appl. Phys. **65**, 1566 (1989).
13. T. Figielski, Solid State Electron. **21**, 1403 (1978).
14. F. D. Auret, S.A. Goodman, F. K. Koschnick, J-M. Spaeth, B. Beaumont, and P. Gibart, Appl. Phys. Lett. **73**, 3745 (1998).
15. D. C. Look, Z-Q. Fang, W. Kim, O. Aktas, A. Botchkarev, A. Salvador, and H. Morkoç, Appl. Phys. Lett. **68**, 3775 (1996).
16. D. C. Look, D. C. Reynolds, J. W. Hemsky, J. R. Sizelove, R. L. Jones, and R. J. Molnar, Phys. Rev. Lett. **79**, 2273 (1997).

Quantum Dots, Optical Characterization, Rare Earths

Mat. Res. Soc. Symp. Vol. 595 © 2000 Materials Research Society

Focused Ion Beam Etching of Nanometer-Size GaN/AlGaN Device Structures and their Optical Characterization by Micro-Photoluminescence/Raman Mapping

M. Kuball [1], M. Benyoucef [1], F.H. Morrissey [2], and C.T. Foxon [3]
[1] H.H. Wills Physics Laboratory, University of Bristol, Bristol BS8 1TL, UK
[2] Philips Electron Optics BV, Eindhoven, Netherlands
[3] Department of Physics, University of Nottingham, Nottingham NG7 2RD, UK

ABSTRACT

We report on the nano-fabrication of GaN/AlGaN device structures using focused ion beam (FIB) etching, illustrated on a GaN/AlGaN heterostructure field effect transistor (HFET). Pillars as small as 20nm to 300nm in diameter were fabricated from the GaN/AlGaN HFET. Micro-photoluminescence and UV micro-Raman maps were recorded from the FIB-etched pattern to assess its material quality. Photoluminescence was detected from 300nm-size GaN/AlGaN HFET pillars, i.e., from the AlGaN as well as the GaN layers in the device structure, despite the induced etch damage. Properties of the GaN and the AlGaN layers in the FIB-etched areas were mapped using UV Micro-Raman spectroscopy. Damage introduced by FIB-etching was assessed. The fabricated nanometer-size GaN/AlGaN structures were found to be of good quality. The results demonstrate the potential of FIB-etching for the nano-fabrication of III-V nitride devices.

INTRODUCTION

Reactive ion etching [1] and wet etching techniques [2] have mostly been employed for the fabrication of III-V nitride devices with their wide spectrum of applications ranging from short-wavelength light emitters, solar-blind detectors to high-temperature devices [3,4]; these, however, are not the preferred techniques for the nano-fabrication of III-V nitrides. Focused ion beam (FIB) etching is one of the most promising techniques for the fine patterning of III-V nitrides, however, only basic etching parameters have been investigated so far [5,6,7]. Alternatively, small nitride structures can be fabricated by selective area growth [8]. The direct write facility of FIB-etching – a well-established technique for optical mask repair and for IC failure analysis and repair – allows the nanometer-scale fabrication of III-V nitride devices without the requirement for depositing an etch mask. A focused gallium ion beam of 5-20nm size is used in FIB to ablate the III-V nitride material. In this paper, we report on the nano-fabrication of a GaN/AlGaN heterostructure field effect transistor (HFET) into pillars as small as 20-300nm in diameter by FIB etching. The use of micro-photoluminescence and ultraviolet (UV) micro-Raman mapping to assess 300nm-size FIB-etched HFET structures is demonstrated. UV Raman scattering probes the vibrational states of the GaN and the AlGaN in the sample and provides information on the aluminum composition, the strain and the free carrier concentration in the fabricated structures [9-11]. Etch damage can easily be assessed from the photoluminescence and the Raman scattering intensity in the FIB-etched areas.

EXPERIMENT

The GaN/AlGaN heterostructure field effect transistor (HFET) employed in this study was grown by molecular beam epitaxy (MBE) on a sapphire (0001) substrate, and consisted of a 25nm-thick low-temperature grown GaN buffer layer, a 1µm-thick nominally undoped GaN layer, followed by a 3nm-thick nominally undoped $Al_{0.15}Ga_{0.85}N$, a 22nm-thick mid-$10^{18}cm^{-3}$ Si-doped $Al_{0.15}Ga_{0.85}N$, and a 15nm-thick nominally undoped $Al_{0.15}Ga_{0.85}N$ layer. The piezo-electric field effect combined with the intentional doping in the $Al_{0.15}Ga_{0.85}N$ layer contributes to the formation of a two-dimensional electron gas at the GaN/$Al_{0.15}Ga_{0.85}N$ interface investigated in [11]. Sub-micron machining was carried out on the samples using an FEI focused ion beam 2000 system. The FIB 2000 system uses a scanning 30kV gallium ion beam with an ion beam current in the range of 1pA to 1150pA and a spot size of 8nm to 500nm, respectively, for the milling of the GaN and the AlGaN. Magnifications of the ion optics up to 150000 were used for scanning the gallium beam over the specimen. GaN/AlGaN pillars were etched at various beam currents and various magnifications of the ion optics. The fabricated structures were imaged using secondary electron imaging while scanning a gallium ion beam of low current over the sample surface. To evaluate the quality of the FIB-etched pattern micro-photoluminescence and micro-Raman spectra were recorded in Z(X,.)$\underline{Z}$ geometry, i.e., unpolarized detection, using an ultraviolet (UV) Renishaw micro-photoluminescence/Raman-system with a HeCd-laser (325nm) as excitation source. A 40x quartz objective was used to focus and collect the laser light with 1µm spatial resolution. The sample was scanned underneath the laser beam in 1µm steps using an XY-stage to record photoluminescence and Raman maps of the FIB-etched areas.

RESULTS AND DISCUSSION

Figure 1 shows secondary electron images of pillars fabricated by FIB from the GaN/AlGaN HFET using gallium ion beam currents of 70pA, 4pA and 1pA, and magnifications up to 150000. The images were recorded at an angle of 45° with respect to the sample surface. A bitmap file of four square-shaped pillars was used as input file for the FIB 2000 system and defined the etched pattern. Figure 1(a) shows GaN/AlGaN pillars fabricated using a gallium ion beam current of 70pA and a magnification of 25000. The GaN/AlGaN pillars are square-shaped. They have a diameter of about 300nm and a height of 0.9µm. A gallium ion beam dose of 1500pC/µm^2 was used to achieve the shown etch depth. Smaller pillars of 150nm-size were etched at the same gallium ion beam current of 70pA using a magnification of 50000 and are shown in Figure 1(b). The pillars have a conical shape, corners of the pillars are round-shapes, illustrating a reduced etch quality. We note that beam currents of 70pA did not allow us to fabricate pillars smaller than 100nm in size.

Using gallium ion beam currents of 4pA and 1pA, sub-micron machining was carried out at magnifications of 100000 and 150000, respectively. The results are shown in Figure 1(c) and (d). Pillars of 50-80nm diameter were etched at 4pA using a magnification of 100000 [Figure 1(c)]. Using a lower gallium ion beam current of 1pA and a higher magnification of 150000, we succeeded in fabricating pillars as small as 20-30nm in diameter as displayed in Figure 1(d). The side walls are nearly vertical, illustrating the high quality of the fabricated pillars. The use of low beam currents is

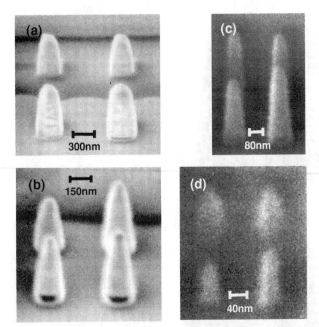

Figure 1. *Secondary electron images of GaN/AlGaN-pillars FIB-etched (a) using a gallium current of 70pA at a magnification of 25000, (b) 70pA at a magnification of 50000, (c) 4pA at a magnification of 100000, (d) 1pA at a magnification of 150000.*

essential to achieve a good fabrication quality using FIB [7]: the diameter of the gallium ion beam used for the milling decreases with decreasing current due to a reduced ion-ion interaction in the beam.

Micro-photoluminescence and ultraviolet (UV) micro-Raman spectra were recorded at room temperature from 300nm-size HFET structures similar to those shown in Figure 1(a) under 325nm-excitation. An area of about 10μm x 10μm was removed around the pillars by FIB to allow access to their optical properties without interference from unetched areas of the sample. Figure 2(a) displays a spatial map of the GaN A_1(LO) Raman intensity of the FIB-etched pattern. Each pixel of the map has a size of 1μm x 1μm. The pillars are located in the center of the map and exhibit a large GaN Raman signal. The experiments probe all GaN/AlGaN pillars in the FIB-etched pattern at once. Figure 2(b) displays a map of the $Al_{0.15}Ga_{0.85}N$ photoluminescence intensity at 3.7eV (band gap of $Al_{0.15}Ga_{0.85}N$ [12]), i.e., 950cm^{-1} shifted from the excitation laser line. The pillars emit $Al_{0.15}Ga_{0.85}N$ photoluminescence. Note the contour of the 10μm x 10μm square etched around the pillars in Figure 2(a) and (b). Figure 3(a) displays a map of the GaN photoluminescence intensity at 3.38eV (3500cm^{-1}). GaN photoluminescence is visible from the pillars. The large and asymmetric shape of the GaN photoluminescence area might be attributed to carrier diffusion. A spectrum recorded in the area of the pillars is shown in Figure 3(b). The FIB-etched 300nm-size GaN/AlGaN HFET pillars emit photoluminescence from the $Al_{0.15}Ga_{0.85}N$ and the GaN layers despite the induced etch damage.

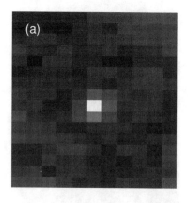

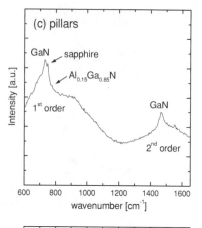

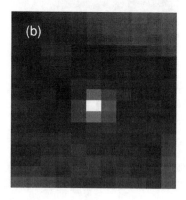

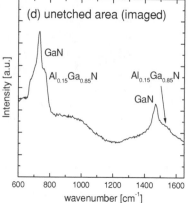

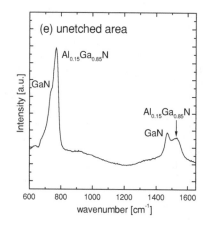

Figure 2. Micro-photoluminescence and UV micro-Raman maps of the FIB-etched GaN/AlGaN pattern of (a) the GaN $A_1(LO)$ Raman intensity at $734cm^{-1}$ and (b) the $Al_{0.15}Ga_{0.85}N$ photoluminescence signal at $3.7eV$ ($950cm^{-1}$). The pixel size is $1\mu m \times 1\mu m$. The pillars are located in the center of the map. A Raman spectrum recorded on the pillars is shown in (c), one recorded adjacent to the FIB-etched area (secondary electron imaged to focus the ion optics prior to the FIB-etching) in (d), one of the sample before etching and imaging in (e).

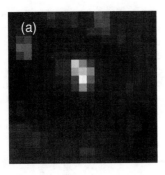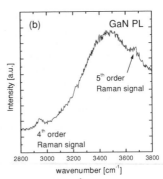

Figure 3. *(a) GaN photoluminescence map at 3.38eV (3500cm⁻¹) of the FIB-etched GaN/AlGaN pattern. The pixel size is 1μm x 1μm. The pillars are located in the center of the map. (b) Spectrum recorded in the area of the pillars.*

Spectra recorded at different locations of the FIB-etched pattern are compared in Figure 2(c)-(e) to investigate in detail the damage introduced into the GaN/AlGaN HFET by FIB-etching. Figure 2(c) shows a Raman spectrum recorded on the pillars, Figure 2(d) one recorded adjacent to the FIB-etched area (secondary electron imaged to focus the ion optics prior to the FIB-etching). A Raman spectrum of the sample before imaging and etching is displayed in Figure 2(e). First-order Raman scattering from the GaN and $Al_{0.15}Ga_{0.85}N$ $A_1(LO)$ phonons is visible in Figure 2(e) at 734cm⁻¹ and 771cm⁻¹, respectively, second-order Raman scattering at 1466cm⁻¹ and 1532cm⁻¹, respectively. The 325nm-excitation close to the bandgap of GaN and $Al_{0.15}Ga_{0.85}N$ results in the resonant enhancement of Raman scattering from the $A_1(LO)$ phonons. The Raman modes are located on top of the $Al_{0.15}Ga_{0.85}N$ photoluminescence [11]. Due to the absorption of the laser in the GaN and $Al_{0.15}Ga_{0.85}N$ only a thin surface layer is probed in the experiments.

Imaging prior to the FIB-etching (Figure 2(d)) introduces damage in the surface layer of the sample. The $Al_{0.15}Ga_{0.85}N$ Raman intensity decreases. We also find a reduction in the $Al_{0.15}Ga_{0.85}N$ photoluminescence intensity. Damage introduced during the FIB-etching is illustrated by the decreased $Al_{0.15}Ga_{0.85}N$ Raman signal in Figure 2(c), present at 772cm⁻¹ as shoulder on the high-energy-side of the GaN $A_1(LO)$ Raman mode at 733cm⁻¹. We note that the shape of the pillars with their rounded top edges (Figure 1(a)) reduces the volume of the $Al_{0.15}Ga_{0.85}N$ with respect to the GaN and contributes to the reduction of the $Al_{0.15}Ga_{0.85}N$ Raman signal in Figure 2(c). To minimize this effect lower gallium ion beam currents could be used for the FIB-etching of the pillars. A close inspection of the GaN and $Al_{0.15}Ga_{0.85}N$ $A_1(LO)$ phonon frequencies in Figure 2(c)-(e) reveals no significant frequency change (<2cm⁻¹) induced by imaging or etching, i.e., no large changes occur in the free carrier concentration, the strain and the aluminum composition of the GaN and $Al_{0.15}Ga_{0.85}N$ layers during the fabrication. Note the sapphire E_g Raman mode at 750cm⁻¹ in Figure 2(c). It is detectable after FIB-etching since the laser light (325nm) can penetrate to the substrate between the pillars.

The 300nm-size HFET structures, i.e., the GaN and the $Al_{0.15}Ga_{0.85}N$ layers in the device structures, emit photoluminescence as shown in Figure 2(b) and 3(a) despite the etch damage illustrated in Figure 2(c)-(e), and show distinct Raman modes for GaN and $Al_{0.15}Ga_{0.85}N$ (Figure 2(a) and (c)) with frequencies, which were not affected by the nano-fabrication. We can therefore conclude that the GaN and AlGaN layers in the FIB-etched device structure are of reasonable good quality, illustrating the potential of FIB for the nano-fabrication of III-V nitride devices. Processing steps to improve the FIB nano-fabrication by reducing etch damage, however, need to be explored. This could involve the deposition of a protective film on the sample surface prior to the FIB-etching and the high-temperature annealing after the FIB etching. We note that FIB is a very time-consuming process: FIB etching is a serial fabrication process in contrast to reactive ion etching (RIE). Large area etching is therefore not practical, however, this limitation can be overcome by using FIB only for the post-fabrication of RIE-etched structures, so that only small areas need to be modified by FIB to achieve small III-V nitride structures.

CONCLUSIONS

We have fabricated 20-300nm-size pillars from GaN/AlGaN HFET structures using FIB etching. Micro-PL/Raman mapping showed that 300nm-size GaN/AlGaN HFETs fabricated by FIB emit photoluminescence and a Raman signal despite the induced etch damage, illustrating the potential of FIB for the nano-fabrication of III-V nitride devices without the requirement for an etch mask. Processing steps to reduce etch damage need to be explored as next step to improve FIB-etched III-V nitride device structures. Then, properties of smaller FIB-etched structures need to be investigated.

REFERENCES

[1]   I. Adesida, C. Youtsey, A.T. Ping, F. Khan, L.T. Romano, and G. Bulman, MRS Internet J. Nitride Semiconductor Res. **4S1**, G1.4 (1999).
[2]   C. Youtsey, I. Adesida, L.T. Romano, and G. Bulman, Appl. Phys. Lett. **72**, 560 (1998).
[3]   S. Nakamura, M. Senoh, S. Nagahama, N. Iwasa, T. Yamada, T. Matsushita, H. Kiyoku, Y. Sugimoto, T. Kozaki, H. Umemoto, M. Sano, and K. Chocho, Appl. Phys. Lett. **72**, 211 (1998).
[4]   S. Yoshida and J. Suzuki, Jpn. J. Appl. Phys. **37**, L482 (1998).
[5]   C. Flierl, I. H. White, M. Kuball, P.J. Heard, G.C. Allen, C. Marinelli, J.M. Rorison, R.V Penty, Y. Chen, S.Y. Wang, MRS Internet J. Nitride Semicond. Res. **4S1**, G6.57 (1999).
[6]   I. Chyr and A. J. Steckl, MRS Internet J. Nitride Semicond. Res. **4S1**, G10.7 (1999).
[7]   M. Kuball, F.H. Morrissey, M. Benyoucef, I. Harrison, D. Korakakis, and C.T. Foxon, Phys. Stat. Sol. (a) **176**, 355 (1999).
[8]   J. Wang, M. Nozaki, Y. Ishikawa, M.S. Hao, Y. Morishima, T. Wang, Y. Naoi, and S. Sakai, J. Cryst. Growth **197**, 48 (1999).
[9]   F. Demangeot, J. Groenen, J. Frandon, M.A. Renucci, O. Briot, S. Ruffenach-Clur, R.-L. Aulombard, MRS Internet J. Nitride Semicond. Res. **2**, 40 (1997).
[10]   H. Harima, H. Sakashita, S. Nakashima, Mat. Sci. For. **264-268**, 1363 (1998).
[11]   M. Kuball, J.M. Hayes, A. Bell, I. Harrison, D. Korakakis, and C.T. Foxon, Phys. Stat. Sol. (a) **176**, 759 (1999).
[12]   D. Brunner, H. Angerer, E. Bustarret, F. Freudenberg, R. Hopler, R. Dimitrov, O. Ambacher, and M. Stutzmann, J. Appl. Phys. **82**, 5090 (1997).

Mat. Res. Soc. Symp. Vol. 595 © 2000 Materials Research Society

Spectroscopy in Polarized and Piezoelectric AlGaInN Heterostructures

C. Wetzel[1], T. Takeuchi[2], H. Amano[2], and I. Akasaki[2]

[1]High Tech Research Center, Meijo University, 1-501 Shiogamaguchi, Tempaku-ku, Nagoya 468-8502, Japan

[2]High Tech Research Center and Department of Electrical and Electronic Engineering, Meijo University, 1-501 Shiogamaguchi, Tempaku-ku, Nagoya 468-8502, Japan

ABSTRACT

Uniaxial wurtzite group-III nitride heterostructures are subject to large polarization effects with significant consequences for device physics in optoelectronic and transport device applications. A central aspect for the proper implementation is the experimental quantification of polarization charges and associated fields. In modulated reflection spectroscopy of thin films and heterostructures of AlGaInN we observe pronounced Franz-Keldysh oscillations that allow direct and accurate readings of the field strength induced by polarization dipoles at the heterointerfaces. In piezoelectric GaInN/GaN quantum wells this dipole is found to induce an asymmetry in barrier heights with a respective splitting of interband transitions. This splitting energy appears to reflect in the transitions of spontaneous and stimulated luminescence in the well. From these experiments the polarization dipole is identified as controllable type-II staggered band offset between adjacent barrier layers which can extend the flexibility in AlGaInN bandstructure design. The derived field values can serve as important input parameters in the further interpretation of the entire system.

INTRODUCTION

Advanced bandstructure design in semiconductor heteroepitaxy has led to a broad variety of high-performance electronic and optoelectronic devices and systems. By the recent development of the wide band gap group-III nitrides as fully functional system [1,2] of high crystalline quality [3] combining n and p-type conduction [4], large electronic band gaps and band offsets this system is predestined to take light output, high temperature operation, switching power and GHz power to significantly higher levels of performance. One of the crucial challenges along this path is a coherent description of spectroscopic observations and even better yet an understanding of the underlying physics. As compound system stable in the uniaxial wurtzite structure inversion symmetry along the unique [0001] axis which typically coincides with the epitaxial growth direction is not maintained. In combination with strong polar bonding contributions and heteroepitaxial strain this leads to rather strong polarization and piezoelectric effects in heterostructures of AlGaInN. Results of piezoresistivity [5] and first principles calculations [6] find large coefficients for piezoelectric (GaN: $e_{33} = 0.73$ C/m^2, $e_{31} = -0.49$ C/m^2 [6]) and spontaneous polarization (GaN: $P_{eq} = -0.029$ C/m^2 [6]). In this context we shall experimentally quantify the associated electric fields and explore the role of such polarization fields in the optoelectronic properties of thin film and quantum well heterostructures. We present experimental results of pseudomorphic AlGaInN/GaN thin film and multiple quantum well (MQW) structures of various well width and composition near the performance optimum of laser diodes in the 430 - 400 nm range using combinations of modulated reflectivity and photoluminescence spectroscopy.

BACKGROUND

Among the challenges in group-III nitrides is the identification of the electronic bandstructure in AlGaInN/GaN heterostructures in order to identify the light emission processes, the threshold edges for detector devices and the channels of highest electron mobility. Especially the aspect of light emission in light emission diodes and laser diodes has been subject of substantial controversy in recent years [7,8]. Work has been hampered by a wide range of material qualities in this highly strained superlattice system exhibiting a strong tendency towards inhomogeneities [9]. To resolve this dilemma we concentrate on material that has been optimized for the highest homogeneity of the structural and optical properties [1]. In order to explain the processes within the wells we first develop a foundation by studying the conditions of the barriers that define the well. In this way the energy of binary GaN layers serves as an accurate point of reference. Here we will show that by such an approach we gain important insight into the conditions of the AlGaInN quantum well system in general and only thereafter endeavor into a more phenomenological description of the conditions within the depth of the wells.

EXPERIMENTAL

GaInN/GaN structures have been grown by MOVPE on basal plane sapphire using the technique of low temperature deposited buffer layers of AlN and GaN [3]. Layers have been optimized for microscopic homogeneity and smooth morphology. Typical growth temperatures for GaInN layers range from 680 -780 °C. A first set of GaInN layers was grown at thicknesses of 400 Å and variable composition x, $(0 < x < 0.2)$ pseudomorphically on GaN. Details have been given in Refs. [10,11]. Next a set of GaInN/GaN MQWs with variable x, $(0 < x < 0.2)$ and well width $L_w = 30$ Å, barrier width $L_b = 60$ Å were studied [12]. A third set comprises nominally fixed compositions and variable well width L_w, 23 Å $\leq L_w \leq$ 70 Å. A fourth set consists of Al$_y$Ga$_{1-y}$N/GaN MQWs on GaN ($L_w = 34$ Å, $L_b = 100$ Å, $x \approx 0.06$). A fifth set was Si doped in the barriers at concentrations of 3x10^{18} cm^{-3} ($L_w = 30$ Å, $L_b = 60$ Å, $x = 0.15$).

Time integrated photoluminescence (PL) was performed using a 325 nm HeCd laser. Stimulated emission was performed using a 1 mJ pulsed 337 nm N$_2$ laser. For photoreflection (PR) the spectral reflectivity was measured using light of a Xe-lamp and above band gap modulation by the HeCd laser. For electro-modulated reflection (ER) a variable voltage of 0.5 to 2 V amplitude was applied using an aqueous electrolyte as a top contact. Alloy compositions were determined using a dynamical analysis of x-ray rocking data in the QWs and x-ray diffraction of both lattice constants for the thicker layers. Details have been given in Refs. [10,11,13].

REFLECTION SPECTROSCOPY IN STRAINED LAYERS

Characteristic modulated reflection data is compared in Figure 1. Within a pseudomorphically strained 400 Å, $x = 0.19$ single heterostructure (SH) (trace a) a strong oscillation appears at the so determined optical band gap near 2.65 eV in both PR as well as ER. From the close correspondence we conclude that the modulation process also in PR is that of a variable electric field acting perpendicular to the surface [11]. The oscillation minimum marks a critical point in the joint density of states (J-DOS) that can be assigned to the direct optical band gap in the pseudomorphic GaInN layer. From a tracing of this signal for variable composition an expression for the optical band gap in $Ga_{1-x}In_xN$ could be derived [10]. Upon a more detailed investigation additional extrema appear on the high-energy side of the oscillation at a variable degree of clarity in about 20 out of 40 studied samples [11]. Trace b) depicts a case where up to three half-periods can clearly be identified. From a fitting procedure considering different known line shape models it was found that a very good description can be given by the scaled electro-optical functions (dashed lines) [14]. Fitting procedures involving more generic third derivative-type lines fail to reproduce the spectral asymmetry. The electro-optical functions describe Franz-Keldysh oscillations (FKOs) in the vicinity of a three dimensional critical point in the J-DOS due to the presence of a static electric field F. The oscillation period scales with the electro-optic energy $\hbar\Theta$ which is a very direct measure of the acting electric field strength F.

$$F = (\hbar\Theta)^{3/2} (2\,\mu)^{1/2} / (e\,\hbar) \qquad (1)$$

The only material related parameter is the J-DOS effective mass $\mu \approx 0.2\ m_0$ [15] that enters in a power below unity. Consequently very accurate readings of the electric field within the sampled area can be obtained. A similar situation can be identified within GaInN/GaN MQW structures (trace c) [16]. In this case the oscillations are superimposed on third derivative features (N_0) at the GaN band edge. Furthermore a clear onset (N_1) of the oscillations marks the three-dimensional band gap which in this case is below the GaN barrier band gap energy. In some layers the third derivative features dominate and the assignment of FKOs is less reliable (trace e), PR). A different situation occurs for a modulation by an externally applied electric field in ER [17]. In this case oscillations appear only above the GaN band edge. Under variation of the bias voltage the oscillations respond with a variation of the oscillation period (dashed line). This corresponds to fields of 0.40 - 0.76 MV/cm which also leads to a determination of the polarity of the internal electric field [17]. A field of 0.14 MV/cm can be derived from the FKOs in the barriers of an AlGaN/GaN MQW in trace f).

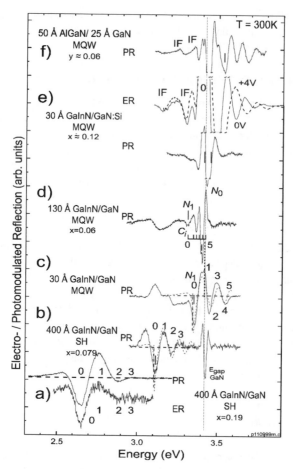

Figure 1. *Spectra of photo modulated reflection (PR) and electro modulated reflection (ER) in various layers. a) PR and ER in high x single heterostructure (SH) coincide in all details revealing the field modulation nature of the PR. b) FKOs in a strained SH thin film marking the direct band gap and shorter periods due to reduced strain, i.e., fields. c) Well resolved FKOs originating in a narrow well together with the electro-optical line shape function. d) Several FKO extrema separate N_0 and N_1 in a wide well. Like in c) these fields act in the well. e) ER in a doped MQW shows FKOs in the barrier above the GaN band gap N_0 with bias-dependent period while PR can be described in 3rd derivative line shapes (interference fringes IF are marked as identified in the DC reflection). f) FKOs within the barriers of an AlGaN/GaN MQW.*

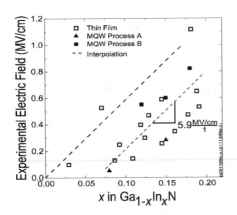

Figure 2. *Experimental electric field values in pseudomorphic thin film GaInN/GaN and MQWs versus composition. In the average a differential slope and a composition offset is identified. For some samples, however, this offset seems to vanish.*

ELECTRIC FIELD IN STRAINED GaInN/GaN

The range of so determined electric field values versus the layer composition (Figure 2) in $Ga_{1-x}In_xN$ thin films and MQWs spawns a wide range of electric fields up to a maximum determined value of 1.1 MV/cm for $x = 0.19$ [18]. Despite some scatter in this large data set, a linear correlation with composition and/or pseudomorphic strain can be derived at a slope of $\partial F/\partial x = 5.9$ MV/cm. This corresponds to very large area charge densities of $\partial \sigma/\partial x = 3.4 \times 10^{13}$ cm^{-2} and is induced by the discontinuity of the lattice polarization at the interface. The presence of such large field values should prompt screening mechanisms to compensate. As we will see below such mechanisms can be effective only on a larger length scale but not where size quantization is relevant. It therefore is an interface effect similar to band offsets where the discontinuity of a volume property induces large dipole moments. First principle calculations [6] show that polarization properties of GaN and InN are very similar and so we should attribute the observed fields to the discontinuity of strain in the pseudomorphic structures, i.e., the piezoelectric properties. The contrary must be expected for AlGaN alloys, where compositional variations should dominate over the strain effects [19].

POLARIZATION TUNED BAND GAP

The occurrence of FKOs is limited to conditions where electrons or holes are free to follow the direction of the electric field. Such a situation is generally given above a three-dimensional critical point in the J-DOS and for in-plane fields within two dimensional QWs. The requirement is also fulfilled for electric fields acting perpendicular to the plane of QWs for carrier energies above the well. The MQW data, however, clearly shows that the respective band edge corresponds to a band gap smaller than the GaN barrier band gap adjacent to the GaInN wells. The spectrum of a control sample (Figure 1, trace d) with a rather weak electric field ($F = 0.07$ MV/cm) shows that several oscillation

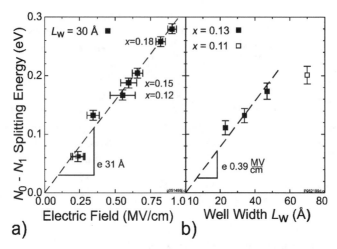

Figure 3. *a) N_0 - N_1 splitting versus field values derived from FKOs follow a slope of e 31 Å. b) The same splitting as function of well width results in a slope of e 0.39 MV/cm for constant composition.*

extrema can be identified in the interval from the critical point N_1 to the GaN band gap energy N_0 [13]. This excludes any hypothetical uncertainty of the exact phase argument of the oscillation.

The splitting of levels N_0 and N_1 has been studied as a function of composition and of well width [13]. Figure 3 depicts results for L_w = 30 Å and variable x (Figure 3a), and for L_w, 23 Å $\le L_w \le$ 70 Å ($x \approx 0.13$) (Figure 3b). From both sets we conclude that the splitting $\Delta E = E(N_0)$ - $E(N_1)$ of the so-defined critical point energy from the GaN band gap follows the product of electric field, electron charge and well width. Within an error of $\pm$ 20 % the factor of proportionality is unity.

$$\Delta E \approx F_w \, e \, L_w .$$ (2)

The apparent contradiction of energy level and dimensionality can be resolved by considering a staggered band line-up of the barrier band edges across the GaInN well similar to type-II heterointerfaces [20]. Figure 4 sketches this situation. According to this the determined electric field originates in the piezoelectric dipole induced by the strained well layer and it is limited to this range. This results in asymmetric barrier heights on either side of the wells, electrons and holes, however, are free to move along the field, respectively. The close correspondence of level splitting and electric field under variation of both parameters, well width and composition, i.e., electric field allows for a very accurate determination of the actual polarization conditions of the system. It allows to determine the electric field within the QW and to determine the bandoffset across the GaN/GaInN/GaN multiple heterointerface.

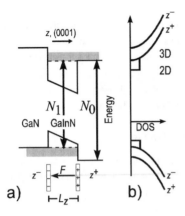

FIGURE 4. a) Derived model of band line-up across the GaN/GaInN/GaN hetero-interface with strained piezoelectric GaInN layer. The shaded areas mark resonant states associated with an inter-band transition N_1 at energies below the barrier band gap. The system resembles a type-II staggered heterointerface with a direct transition. b) The respective DOS distinguishes branches z+ and z- with respect to the well. Two dimensional states in the well are not indicated and external response fields are neglected .

POLARIZATION DIPOLE

The role of the GaInN well layer is that of a controllable interface dipole $D = \varepsilon_0 \, \varepsilon_r \, F \, L_w$ between pairs of GaN barriers that by L_w and composition allows the controlled adjust-ment of the band edges across the thin well layer. This adds a versatile control parameter and another degree of freedom to the tools of superlattice bandstructure design.

The induced piezoelectric polarization $P = \varepsilon_0 \, \varepsilon_r \, F$ acts as perfect δ-doping layers [21] of large area charge densities $\sigma = P/e$ within sub-monolayer thicknesses and perio-dicity of the lattice. In contrast to impurity doping, however, both polarities of charges are spatially fixed. A complete screening of the induced dipoles requires the depletion of rather thick highly doped layers [17]. This in turn induces large potential steps and thus strongly limits the transfer of free carriers into shallow wells. For typical doping concen-trations of $N_D = 10^{18}$ cm^{-3} the depletion layer $L_d = \sigma/N_D = 50$ nm for ($x = 0.15$) extends well beyond the QW region. Polarization fields therefore cannot be screened by doping (see also Ref. [22]) on the length scale of typical superlattice periods without compro-mising sharp definition of electronic levels and crystalline quality. Instead the depletion of adjacent layers will induce a macroscopic field to balance the chemical potential in top and bottom layer of the structures [23]. In a first approximation this macroscopic field leads to a balancing of each QW dipole with one induced dipole across the adjacent bar-rier. This response field F_b with direction opposing that of F_w is limited to the barrier layer and it is this part that is responsible for the observation of FKOs above the GaN bandedge in the electric modulated reflection in Figure 1 (trace e). From knowledge of both field components F_w and F_b the total polarization charge induced at the heterointer-face can be inferred:

$$\partial P/\partial x = \varepsilon_0 \varepsilon_r \left(|F_w| + |F_b| \right) = (0.06 - 0.09) \ C/m^2 \tag{3}$$

The margin of uncertainty originates in the wide range of fields observed. Similar considerations hold for the screening of the equilibrium or spontaneous polarization that induces charges at the discontinuities of the polar sample surfaces. These, however, can easily be screened by carriers accumulated throughout the volume of the sample and surface passivation from the environment. Combining both effects a small correction of the order of the ratio of both length scales ≈ 0.1 must be considered. Under the assumption of the theoretical polarization coefficients [6] we expect $\partial P/\partial x = 0.15$ C/m^2.

POLARIZATION AND TRANSPORT ASYMMETRY

The second aspect of the controlled polarization dipole layer is the fact of a tunable three dimensional critical point in the J-DOS according to $E = E_g(\text{GaN}) - F_w \, e \, L_w$. This is in contrast to band line-up engineering in conventional QW designs where the controllable band gap occurs between states of two dimensions. The additional degree of freedom allows photogenerated carriers to escape the well region without additional potential barrier. The respective interband transition exhibits a large matrix element due to the large overlap within the region of the well layer (see Figure 4a). It also directly affects the quantization of states within the QW at lower energies by imposing a variable barrier height that is reduced by $F_w \, e \, L_w$ with respect to the well without polarization effects.

Considering the recombination dynamics of electrons and holes it is essential to distinguish the carriers originating from either the left or the right hand side barriers. For example photogenerated carrier pairs under low excitation will be collected from the same side into the well while carriers injected by a pn-junction come from opposite sides. For the case of GaInN wells strained to GaN barriers in a typical MOVPE pn-junction with top p-type contact the orientation of polarization will provide that carriers are injected from the higher barrier sides, respectively. In this context it is useful to distinguish eigenstates in positive (z^+) and in negative (z^-) direction of the (0001) axis in accordance with the broken inversion asymmetry (Figure 4b).

So far it has been shown that the electronic bandstructure in the vicinity of the barrier levels is to very large extent controlled by the polarization dipole induced at the pseudomorphic heterointerface. In the next step we shall investigate in a more phenomenological approach the role this dipole plays for the states within the wells at larger binding energies.

POLARIZATION AND QUANTIZED LEVELS OF THE WELL

On the basis of the now well-established conditions of the barriers we next focus on the properties of the well. It had been shown by Takeuchi et al. [24] that PL peak energies in GaInN/GaN QW differ by as much as 350 meV ($x = 0.13$) for low (2 Wcm^{-2}) and high (200 kWcm^{-2}) excitation density [24]. This value compares well with reported discrepancies in luminescence and absorption [25]. This splitting was shown to vary with well width and composition and has lead to the interpretation of a level shift as a result of the quantum confined Stark effect in the strongly polarized QWs [24]. The presence of this effect was furthermore confirmed by a continuos variation of both, the spontaneous

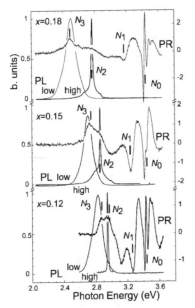

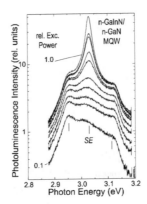

Figure 5. *Stimulated and spontaneous luminescence in doped GaInN/GaN MQW revealing multiple sharp levels.*

Figure 6. *PL at low and stimulated emission at high density excitation in comparison with PR under low excitation conditions. Stimulated emission coincides with a level labeled N_2 seen in PR under low excitation. The electric field condition is revealed from the splitting of N_0 and N_1 and the FKO in the same region.*

emission under photo excitation [26] and a corresponding level in the ER signal [27] as a function of an externally applied bias voltage. A level shift of the spontaneous emission by a variation of the doping concentration has also been interpreted in this direction [25]. It is important to note that none of those observations involve a level under the condition of stimulated emission. It has been observed that several levels appear to compete under conditions near stimulated emission [28]. Figure 5 shows a case within a doped MQW. The center peak of stimulated emission is bordered by two very sharply defined levels of spontaneous emission. It therefore cannot be excluded that the level of stimulated emission is of different origin than the one that shifts under electric field variation.

In Figure 6 we compare the emission spectra under low excitation density (2 W cm^{-2}) of the CW HeCd laser and high density (200 kWcm^{-2}) excitation of a pulsed N_2 laser [12] together with the PR spectra for the case of three 30 Å GaInN / 60 Å GaN MQWs with composition $x = \{0.12, 0.15, 0.18\}$ within a wide spectral range. PR was performed at a very low photomodulation power density of the order of 5 mW/cm^2 to rule out any possible trace of luminescence in the PR data. Near 3.4 eV we identify the GaN barrier band gap PR signal (N_0) and the FKOs starting above the well-defined level N_1. The derived electric field values acting within the wells are $F = \{0.55, 0.59, 0.82\}$ MV/cm and increase with the N_0-N_1 level splitting and composition. In parallel the

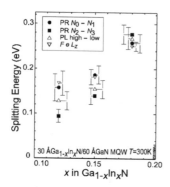

Figure 7. *Splitting energies of N_0 and N_1 (PR) (offset between barriers), N_2 and N_3 extrema in PR, PL peak energies at low and high excitation as well as the quantity $F \, e \, L_w$ derived from FKOs. All values share the same trend with composition and merge for higher x.*

energetic separation of the levels of stimulated emission and spontaneous emission and the threshold density increases as reported [12]. In the same energetic range of the QW two contributions N_2, N_3 appear in the PR spectrum that cannot be simulated by partial energy derivatives of Lorentzian-type line functions. Instead the partly asymmetric line shape indicates contributions from higher order effects such as modulated line widths. Such a condition is conceivable for the case of photocarrier screening of electrostatic fluctuations. Also in this mode, however, extrema can be associated with critical points in the J-DOS. In this mode under low density excitation extrema can clearly be identified at levels corresponding to both, the PL maximum of spontaneous emission under low excitation density (N_3) and the peak of stimulated emission under high-density excitation (N_2). This holds for all three InN-fractions and supports that both levels are of the same origin. This evidence strongly suggests that levels of spontaneous and stimulated emission are levels of different origin and both co-exist under low excitation density conditions. This information can be useful in the assessment of gain and band gap renormalization processes.

In the next step the respective level splittings are compared in Figure 7. We consider the separation of the low energy extrema in PR, the splitting of spontaneous and stimulated emission, as well as the N_0-N_1 splitting in PR and the value of $F_w \, e \, L_w$ derived from the FKOs in PR as functions of the alloy composition. There is a remarkable close agreement of all the considered splitting energies, both in their trend within the composition series as well as in their magnitude. The level of agreement as well as the level of accuracy is higher for the sample with $x = 0.18$ compared to $x = 0.12$ which can indicate competing effects for smaller InN fraction or smaller electric fields. This line of evidence strongly suggests that the polarization properties, which define the electronic band structure at the barrier band gap energy, play a similar role in the radiative centers in the depth of the quantum wells. While a more detailed study is underway this observation can tentatively be expressed by the combination of quantum confined Stark effect and the field dependent asymmetric barrier height. The close correspondence of magnitude of the polarization dipole and the respective level splitting furthermore suggests that mechanisms of level broadening and band tailing play a minor role in this material optimized for homogeneity.

CONCLUSIONS

We have shown that by the combination of modulated reflection spectroscopy with luminescence and absorption techniques can provide a coherent picture of the electronic bandstructure in AlGaInN alloys and heterostructures. Pronounced oscillations in reflectivity have been identified as Franz-Keldysh oscillations allowing an accurate determination of large electric fields within the individual layers. This information can serve as an accurate input parameter for the theoretical description of the electronic band structure. In this way it significantly improves the accuracy over approaches where electric fields and polarization charges are free parameters to describe the interband transition energies. Based on the knowledge of the fields within strained GaInN wells we have derived the value of the relative band offset $F_w e L_w$ between adjacent GaN barriers. We furthermore find that the same quantity is in direct correlation with the peak discrepancy of spontaneous and stimulated emission and it also reflects in the splitting of corresponding discrete energy levels in the reflection spectra. These properties strongly resemble a Stark ladder with constant step widths defined by the polarization dipole across each individual well [16]. The combination of field values in the well and their response fields in the barrier was used to quantify the polarization induced fixed charges. From electrostatic arguments it could be shown that screening by modulation doping is of minor influence on the polarization in the wells. The presented results can be extended throughout the AlGaInN heterostructure system by consideration of respective strain and material discontinuities. It could be shown that the role of the polarization dipole at the interface is similar to a controllable type-II staggered band offset between adjacent layers by insertion of a polarized epitaxial layer of variable thickness.

ACKNOWLEDGMENTS

This work was partly supported by the JSPS Research for the Future Program in the Area of Atomic Scale Surface and Interface Dynamics under the project of Dynamic Process and Control of the Buffer Layer at the Interface in a Highly-Mismatched System. The authors thank M. Iwaya, N. Hayashi, and R. Nakamura for the excellent sample material and Y. Kaneko for fruitful discussion. This work was partly supported by the and the Ministry of Education, Science, Sports and Culture of Japan (contract nos. 09450133 and 09875083).

REFERENCES

1   I. Akasaki and H. Amano, *Jpn. J. Appl. Phys.* **36**, 5393 (1997).
2   *Properties, Synthesis, Characterization, and Applications of Gallium Nitride and related Compounds* Eds. J. Edgar, T.S. Strite, I. Akasaki, H. Amano, and C. Wetzel (INSPEC, IEE, London, UK, 1999) (ISBN 0 85296 953 8).
3   I. Akasaki, H. Amano, Y. Koide, K. Hiramatsu, and N. Sawaki, *J. Crystal Growth* **98**,

209 (1989).

4   H. Amano, M. Kito, K. Hiramatsu, I. Akasaki, *Jpn. J. Appl. Phys.* **28**, L 2112 (1989).

5   A.D. Bykhovski, V.V. Kaminski, S. Shur, Q.C. Chen, and M.A. Khan, *Appl. Phys. Lett.* **68**, 818 (1996).

6   F. Bernardini, V. Fiorentini, and D. Vanderbilt, *Phys. Rev. B* **56**, R 10024 (1997).

7   S. Chichibu, T. Azuhata, T. Sota, S. Nakamura, *Appl. Phys. Lett.* **70**, 2822 (1997).

8   Y. Narukawa, Y. Kawakami, S. Fujita, S. Fujita, and S. Nakamura, *Phys. Rev. B* **55**, R 1938 (1997).

9   N.A. El-Masry, E.L. Piner, S.X. Liu, S.M. Bedair, *Appl. Phys. Lett.* **72**, 40 (1998).

10 C. Wetzel, T. Takeuchi, S. Yamaguchi, H. Katoh, H. Amano, and I. Akasaki, *Appl. Phys. Lett.* **73**, 1994 (1998).

11 C. Wetzel, T. Takeuchi, H. Amano, and I. Akasaki, *J. Appl. Phys.* **85**, 3786 (1999).

12 H. Sakai, T. Takeuchi, S. Sota, M. Katsuragawa, M. Komori, H. Amano, and I. Akasaki, *J. Crystal Growth* **189**, 831 (1998).

13 C. Wetzel, T. Takeuchi, H. Amano, and I. Akasaki, *Phys. Rev. B* (in press).

14 D.E. Aspnes, *Phys. Rev. B* **10**, 4228 (1974); *Phys. Rev.* **153**, 972 (1967).

15 B.K. Meyer, D. Volm, A. Graber, H.C. Alt, T. Detchprohm, H. Amano, and I. Akasaki, *Solid State Commun.* **95**, 597 (1995).

16 C. Wetzel, T. Takeuchi, H. Amano, I. Akasaki, *Jpn. J. Appl. Phys.* **38**, L 163 (1999).

17 C. Wetzel, H. Amano, and I. Akasaki, *Jpn. J. Appl. Phys.* (in press).

18 C. Wetzel, T. Takeuchi, H. Amano, and I. Akasaki, in *Wide-Bandgap Semiconductors for High Power, High Frequency and High Temperature*, Eds. S. DenBaars, J. Palmour, M. Shur, and M. Spencer, *Mat. Res. Soc. Sympos. Proc.* **512**, 181 (1998).

19 C. Wetzel, S. Nitta, T. Takeuchi, S. Yamaguchi, H. Amano, and I. Akasaki, *MRS Internet J. Nitride Semicond. Res.* **3**, 31 (1998).

20 L. Esaki, *IEEE J. Quantum Electronics* **QE-22**, 1611 (1986).

21 A. Zrenner, H. Reisinger, F. Koch, K. Ploog, J.C. Maan, *Phys Rev B* **33**, 5607 (1986).

22 F. della Sala, A. di Carlo, P. Lugli, P.F. Bernardini, V. Fiorentini, R. Scholz, and J.-M. Jancu, *Appl. Phys. Lett.* **74**, 2002 (1999).

23 M. Leroux, N. Grandjean, M. Laugt, J. Massies, B. Gil, P. Lefebvre, and P. Bigenwald, *Phys. Rev. B* **58**, R 13371 (1998).

24 T. Takeuchi, S. Sota, M. Katsuragawa, M. Komori, H. Takeuchi, H. Amano, and I. Akasaki, *Jpn. J. Appl. Phys.* **36**, L 382 (1997).

25 S. Chichibu, T. Azuhata, T. Sota, S. Nakamura, *Appl. Phys. Lett.* **69**, 4188 (1996).

26 T. Takeuchi, C. Wetzel, S. Yamaguchi, H. Sakai, H. Amano, I. Akasaki, Y. Kaneko, S. Nakagawa, Y. Yamaoka, and N. Yamada, *Appl. Phys. Lett.* **73**, 1691 (1998).

27 C. Wetzel, T. Detchprohm, T. Takeuchi, H. Amano, and I. Akasaki, *J. Electron. Mater.* (in press).

28 S. Watanabe, N. Yamada, Y. Yamada, T. Taguchi, T. Takeuchi, H. Amano, and I. Akasaki, *Phys. Stat. Sol. B* **216**, 335 (1999).

Mat. Res. Soc. Symp. Vol. 595 © 2000 Materials Research Society

Influence of Internal Electric Fields on the Ground Level Emission of GaN/AlGaN Multi-Quantum Wells

A. Bonfiglio, M. Lomascolo[1], G. Traetta[2], R. Cingolani[2], A. Di Carlo[3], F. Della Sala[3], P. Lugli[3], A. Botchkarev[4], H. Morkoc[4]
INFM-Dipartimento di Ingegneria Elettrica ed Elettronica, Università di Cagliari, Italy
[1]CNR-IME Istituto per lo studio di nuovi Materiali per l'Elettronica, Lecce, Italy.
[2]INFM- Dipartimento di Ingegneria dell'Innovazione, Università di Lecce, Italy
[3]Dipartimento di Ingegneria Elettronica, Università di Roma "TorVergata", Italy
[4]Electronic Engineering, Virginia Commonwealth University, Richmond Virginia USA

ABSTRACT

The spectroscopic investigation of GaN/AlGaN quantum wells reveals that the emission energy of such structures is determined by four parameters, namely composition, well-width, strain and charge density. The experimental data obtained by varying these parameters are quantitatively explained by an analytic model based on the envelope function formalism which accounts for screening and built-in field, and by a full self-consistent tight-binding model.

INTRODUCTION

The physical behavior of GaN/AlGaN quantum wells (QWs) in view of their application to optical devices is at present under investigation both theoretically and experimentally[1-8]. In particular, the relation between optical properties and geometrical and compositional structure of the systems is of great relevance.

By growing identical quantum wells of various dimensions on different buffer layers, it is possible to clarify the interplay between geometry, strain, piezoelectric field, spontaneous polarization field and quantum size effect in determining the ground level energy of the heterostructures and its well-width dependence.

Having addressed this problem in a recent publication [9], in this paper we further investigate the role of the internal field in determining the ground level emission energy of GaN/AlGaN QWs, with particular concern for the role played by the injected charge which accumulates at the interface between quantum wells and barriers.

MATERIALS AND METHODS

Three sets of samples have been produced for our experiments. The first two sets (A and B) were grown by reactive molecular beam epitaxy (MBE) on sapphire substrates. Following a chemical in situ cleaning of c-plane sapphire substrates, a thin AlN buffer layer was grown at 850 °C with ammonia as the active nitrogen source. The AlN buffer layer was followed by the growth of a 1 μm thick GaN buffer layer grown at 800 °C. Then, an $Al_{0.15}Ga_{0.85}N$ layer was grown with two different thickness values for each set of samples: 100 nm for set A and 10 for set B. Finally, the quantum well region was grown. The strain in the QW was varied by changing the buffer layer of the structures: samples grown on the 10 nm thick $Al_{0.15}Ga_{0.85}N$ buffer layer are pseudomorphic to the

GaN substrate, so that the GaN QWs are unstrained. Conversely, samples grown on the 100 nm thick $Al_{0.15}Ga_{0.85}N$ buffer layer (which is completely relaxed) result in strained QWs. Each sample consisted of 10 GaN quantum wells. In each set, four samples of well-width L_w=2,3,4, and 5 nm (measured by double crystal X-ray diffraction) were grown and analyzed. The barrier width and composition were kept constant in all samples ($Al_{0.15}Ga_{0.85}N$ barriers of thickness L_B=10 nm). Set C was grown by MOCVD according to the structure already described for set B but with quantum well widths of respectively, 1, 2, 3, 6, and 9 nm.

The optical measurements were performed either under cw excitation (325 nm line of a He-Cd laser) or under a pulsed excitation (4th harmonic of a Nd-YAG laser). The samples were kept in a variable temperature closed cycle cryostat. The spectral resolution was always better than 0.2 meV.

DISCUSSION

The systematic analysis of all samples belonging to sets A and B results in the well-width dependence of the n=1 ground level emission displayed in Fig. 1 (symbols). In this figure the experimental data are compared to the theoretical curves obtained by the analytical models discussed in the following.

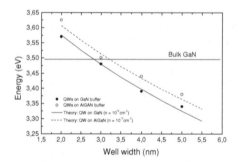

Figure 1. *Well-width dependence of the ground level emission energy of GaN/AlGaN QWs grown on GaN, at 10 K. The curves are calculated by means of Eq. 4 The horizontal line indicates the unstrained bulk energy gap of GaN.*

The main features of the experimental data shown in figure 1 can be summarized as follow:
1)   for $L_w \geq 3$ nm the emission energy falls below the bulk energy-gap;
2)   the observed well-width dependence differs considerably from the usual square well model (L_w^{-2} dependence);
3)   for a given well-width, the ground level emission energy is different in the two sets of samples, i.e. it depends dramatically on the AlGaN buffer thickness;

4) despite the difference in the absolute energy value, the well-width dependence of the ground level emission is similar in the two sets of samples, i.e. it does not depend on the thickness of the AlGaN buffer;

5) The emission energy blue-shifts with increasing the photo-generated charge density (data not shown).

The detailed analysis of the results obtained on sets A and B has been already presented in [9].

Figure 2 shows the emission energy of samples belonging to set C. As can be easily noticed, the linear dependence of energy on the quantum well width holds only for the narrower wells (first 3 points) whereas the wider wells exhibit a considerable deviation from such trend. The blue-shift displayed by the emission energy of the samples with larger well widths indicates that the role of the well dimensions must be considered, in this case, with greater attention.

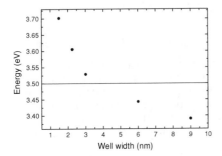

Figure 2. *Well-width dependence of the ground level emission energy of GaN/AlGaN QWs belonging to set C. The horizontal line indicates the unstrained bulk energy gap of GaN.*

The theoretical model presented in [9], describing the well width dependence of the narrow wells (see fig.1) takes into account strain, built-in field and structural parameters of the quantum wells. As far as the strain is concerned, this is evaluated following Ref. [10]. According to this approach, samples with strained quantum wells (set A) should exhibit a dependence on the in-plane compressive strain of the emission energy. In this case, the barriers take the bulk $Al_{0.15}Ga_{0.85}N$ lattice constant (3.177 Å), whereas GaN layers are assumed to grow pseudomorphically and to undergo a compressive in-plane strain $\sigma_\parallel = -0.37\%$. We similarly evaluated the strain correction for samples belonging to sets B and C, i.e., with strained barriers. In this case, the quantum wells take the bulk GaN lattice constant (3.189 Å), whereas $Al_{0.15}Ga_{0.85}N$ layers grow pseudomorphically with a tensile in-plane strain $\sigma_\parallel = 0.37\%$. The assumption of a strained $Al_{0.15}Ga_{0.85}N$ barrier is consistent with the observation of pseudomorphic growth of AlGaN layers for thicknesses as large as hundreds of nm, which is much thicker than the total amount of $Al_{0.15}Ga_{0.85}N$ contained in our 10 periods multiple quantum wells [2].

The built-in electric field is evaluated by accounting for the accumulation of a polarization charge at the interfaces of GaN/AlGaN heterostructures. The total polarization charge can be written as $P_{tot} = P_{piezo} + P_{spont}$, where P_{piezo} is the piezoelectric charge caused by the lattice mismatch (mis) and by the thermal strain (ts) $[P_{piezo} = P_{mis} + P_{ts}]$, whereas P_{spont} represents the spontaneous polarization charge of the GaN/AlGaN interface, as clearly demonstrated by the recent works of Bernardini et al.[11-13]. For an alternating sequence of wells (w) and barriers (b) the total electric field in the well can be calculated as [13]

$$F_w = L_b(P^b_{tot} - P^w_{tot}) / [\varepsilon_0(L_w \varepsilon_b + L_b \varepsilon_w)] \qquad (1)$$

$\varepsilon_{b,w}$ being the relative dielectric constant of the layers (analogous expression with exchanged indexes holds for the electric field in the barrier). The piezoelectric polarization (P_{lm}) induced by the lattice in-plane mismatch ($\sigma_{\parallel}$) can be calculated as

$$P_{lm} = -2\,(e_{33}\frac{C_{11}}{C_{33}} - e_{31})\sigma_{\parallel} \qquad (2)$$

where e_{ij} and C_{ij} are the piezoelectric tensor components and the elastic constants, respectively, as given in ref.[11-13]. Adopting the strain vales quoted above, the piezoelectric polarization charge in our set of samples turns out to be $P^w_{lm}=0$ in the quantum well and $P^b_{lm}= -0.0055$ C/m^2 in the Al$_{0.15}$Ga$_{0.85}$N barrier. The values of the polarization charge either spontaneous or piezoelectric (depending on strain), will be used as input parameters in the modeling of the electronic states discussed later.

In addition, the thermal strain in our experimental conditions amounts to some 0.003%, resulting in an additional polarization charge of the order of $P^w_{ts}=-3.2*10^{-4}$ C/m^2. As far as the spontaneous polarization charge is concerned, we take the recent data of ref.[11-13], leading to $P^w_{sp}= -0.029$ C/m^2 and $P^b_{sp}= -0.037$ C/m^2, the latter value being obtained by linear interpolation of the GaN and AlN values ($P_{sp}=-0.08$ C/m^2 in AlN). By using these data and Eq.(1) we can calculate the built-in field in the different samples, which turns out to vary in the range 0.8 - 1.3 MV/cm depending on the actual well-width.

An accurate analytical model based on the envelope function formalism can be developed in order to reproduce the effects of the internal field and of the strain by assuming that: i) the rectangular quantum well is replaced by a triangular well [14] resulting from the total built-in field, and ii) the 2D photo injected charge density (ρ) is considered to accumulate at the GaN/AlGaN interface and to screen the built-in field. With these approximations the electric field in the well becomes

$$F_w = L_b(\rho + P^b_{tot} - P^w_{tot}) / [\varepsilon_0(L_w \varepsilon_b + L_b \varepsilon_w)] \qquad (3)$$

and the ground level energy

$$E_{1e1h} = E_g - F_w L_w + \left(\frac{9\pi\hbar e F_w}{8\sqrt{2}}\right)^{2/3}\left(\frac{1}{m_e} + \frac{1}{m_h}\right)^{1/3} \qquad (4)$$

The band gap shift induced by the strain is included in E_g [10]. In order to test the accuracy of the envelope function model we have compared Eqs.(3)-(4) with the results of a full self-consistent tight-binding (TB) model, for several injected charge densities [7,9,16]. The tight-binding model is used to describe the electronic structure in the entire Brillouin zone, up to several eV above the fundamental gap. The self-consistent calculation is performed as follows: the electron and hole quasi-Fermi levels are calculated for a given 2D photo injected charge density (which is a fitting parameter) and the electron and hole charge distributions are obtained and then substituted in the Poisson equation which account also for the spontaneous and piezoelectric polarizations. The obtained potential is inserted in the TB Schroedinger equation which is solved to obtain energy levels and wave-functions. Then, the new quasi-Fermi levels are calculated and the whole procedure reiterated to self-consistency (see Refs. 7 and 17 for details). This method, though intrinsically more accurate than the envelope function model, requires a stronger computational effort.

From Fig.1 we can see that both sets of samples are in good quantitative agreement with the theoretical results. In the case displayed in Fig. 2, the thicker wells exhibit a blue-shift of the emission energy with respect to the ideal trend forecast by Eq. (4). This might be due to the fact that in wide wells the wave functions overlap is decreased by the built-in field (accordingly the oscillation strength becomes very low), resulting in a charge accumulation at the interfaces. Such accumulation induces a further alteration of the electric field (screening), resulting in a blue-shift of the emission energy. Conversely, charge accumulation poorly affects samples with narrower wells because here the spatial extension of the wave functions of electron and holes is comparable to the dimensions of the potential well where these are confined and the probability of recombination is therefore not affected by the field. Experiments are presently under way in our labs to clarify this issue.

CONCLUSIONS

In conclusion, we have completed the analysis of the influence of internal electrical field on the emission energy of GaN/AlGaN MQWs. As demonstrated in a previous paper [9], both the piezoelectric component of the field and the strain induced shift of the gap can be tuned by varying the strain distribution in the heterostructure, i.e. by growing samples on GaN or AlGaN substrates. The results have been quantitatively confirmed by theoretical calculations based on a self-consistent approach. This demonstrates that GaN QWs are effectively systems in which four parameters (composition, well-width, buffer type and charge density) can be varied to tune the ground level emission and therefore optimizing the characteristic of a possible QW-based optical device.

ACKNOWLEDGEMENTS

Work partially supported by Italian MURST funding program and by the European TMR project ULTRAFAST and by the Sardinia Regional project C21.

REFERENCES

1. T. Takeuchi, C. Wetzel, S. Yamaguchi, H. Sakai, H. Amano, I. Akasaki, Y. Kaneko, S. Nakagawa, Y. Yamaoka, and N. Yamada, Appl. Phys. Lett. 73, 1691 (1998).
2. T. Takeuchi, S.Sota, M. Katsuragawa, M .Komori, H. Takeuchi, H .Amano, and I. Akasaki, Jpn. J. Appl. Phys.36, L382 (1997).
3. S.H. Park, and S.L. Chuang, Appl. Phys. Lett. 72, 3103 (1998).
4. J.S. Im, H. Kollmer, J. Off, A. Sohmer, F. Scholz, and A. Hangleiter, Phys. Rev. B57, R9435 (1998).
5. A. Bykhovski, B. Gelmont, and M. Shur, Appl. Phys. Lett. 63, 2243, (1993).
6. M. Leroux, N. Grandjean, M. Laugt, J. Massies, B. Gil, P. Lefebvre, and P. Bigenwald, Phys. Rev. B58, R13371, (1998).
7. A. Di Carlo, S. Pescetelli, M. Paciotti, P. Lugli, and M. Graf, Solid State Commun. 98, 803 (1996); F. Della Sala, A. Di Carlo, P. Lugli, F. Bernardini, V. Fiorentini, R. Scholz, and J.M. Jancu, Appl. Phys. Lett, 74, 2002 (1999).
8. R. Cingolani, G. Coli', R. Rinaldi, L. Calcagnile, H. Tang, A. Botchkarev, W. Kim, A. Salvador and H. Morkoc, Phys. Rev. B56, 1491 (1997).
9. A. Bonfiglio, M. Lomascolo, G. Traetta, R. Cingolani, A. Di Carlo, F. Della Sala, P. Lugli, A. Botchkarev, H. Morkoc, J. Appl. Phys, in press.
10. L. Chuang and C.S. Chang, Semicond. Sci. Technol. 12, 252-263 (1997).
11. F. Bernardini, V. Fiorentini and D. Vanderbilt, Phys. Rev. B56, R10024 (1997).
12. F. Bernardini, V. Fiorentini and D. Vanderbilt, Phys. Rev. Lett. 79, 3958 (1997).
13. F. Bernardini and V. Fiorentini, Phys. Rev. B57, 1, (1998); V. Fiorentini, F. Bernardini, F. Della Sala, A. Di Carlo, and P. Lugli, Phys. Rev. B60, 8849 (1999).
14. G. Bastard, *Wave Mechanics Applied to Semiconductor Heterostructures*, Edition de Physique, Paris, France, 1987.
15. J. Singh, *Semiconductor Optoelectronics*, McGraw Hill, New York (1995).
16. R. Cingolani, A. Botchkarev, H. Tango, H. Morkoç, G. Traetta, G. Coli', M. Lomascolo, A. Di Carlo, F. Della Sala, P. Lugli, Phys. Rev.B , in press.
17. A. Di Carlo, Phys. Stat. Solidi, in press .

Mat. Res. Soc. Symp. Vol. 595 © 2000 Materials Research Society

Comparison Study of Structural and Optical Properties of $In_xGa_{1-x}N$/GaN Quantum Wells with Different In Compositions

Yong-Hwan Kwon, G. H. Gainer, S. Bidnyk, Y. H. Cho, J. J. Song, M. Hansen[1], and S. P. DenBaars[1]
Center for Laser and Photonics Research and Department of Physics,
Oklahoma State University, Stillwater, Oklahoma 74078, e-mail: kwonyh@okstate.edu
[1]Electrical and Computer Engineering and Materials Departments,
University of California, Santa Barbara, California 93106

ABSTRACT

The effect of In on the structural and optical properties of $In_xGa_{1-x}N$/GaN multiple quantum wells (MQWs) was investigated. These were five-period MQWs grown on sapphire by metalorganic chemical vapor deposition. Increasing the In composition caused broadening of the high-resolution x-ray diffraction superlattice satellite peak and the photoluminescence-excitation bandedge. This indicates that the higher In content degrades the interface quality because of nonuniform In incorporation into the GaN layer. However, the samples with higher In compositions have lower room temperature (RT) stimulated (SE) threshold densities and lower nonradiative recombination rates. The lower RT SE threshold densities of the higher In samples show that the suppression of nonradiative recombination by In overcomes the drawback of greater interface imperfection.

INTRODUCTION

Major developments in III-nitride semiconductors have led to the commercial production of InGaN/GaN light-emitting diodes [1] and current injection violet lasers [1,2]. In spite of its large threading dislocation densities, the InGaN/GaN system exhibits intense electroluminescence and photoluminescence (PL) [3]. Two mechanisms have been suggested for these anomalous phenomena. First, the incorporation of In atoms could play a crucial role by suppressing nonradiative recombination rates, through the capture of carriers in localization centers originating from quantum dot-like and phase-separated In-rich regions [4,5]. Second, Narukawa *et al.* suggested that the incorporation of only 2% In into the GaN layer can effectively reduce the density of nonradiative recombination centers [6]. However, systematic studies on the characteristics of InGaN/GaN MQWs with different In compositions are scarce. These studies are crucial not only for physical interest, but also for the design of practical optical devices.

We report a systematic study of both the structural and optical properties of $In_xGa_{1-x}N$/GaN MQWs by combining the results of high resolution x-ray diffraction (HRXRD), PL, and PL excitation (PLE) measurements. Stimulated emission (SE), temperature-dependent PL, and time-resolved PL (TRPL) measurements were also performed to evaluate the optical efficiencies of the MQWs and their device applicability. We used InGaN/GaN MQWs with different In compositions of 8.8, 12.0, and 13.3%. These samples have room temperature (RT) SE wavelengths ranging from 395 nm to 405 nm, which is close to the operational wavelength of state-of-the-art current injection violet lasers [2]. Therefore, understanding the physical mechanisms in these structures could give insight for increasing the operational wavelength of InGaN/GaN lasers.

Through HRXRD analysis, we found that samples with higher In composition have a larger full width at half-maximum (FWHM) of superlattice (SL) peaks, indicating rougher interfaces. However, these samples have lower RT SE threshold densities and lower nonradiative recombination rates, as determined by SE and TRPL experiments. We attribute the lower RT SE threshold densities of the higher In composition samples to the suppression of nonradiative recombination, due to the incorporation of In.

EXPERIMENTAL DETAILS

The set of InGaN/GaN MQW samples used in this study were grown on c-plane sapphire substrates by metalorganic chemical vapor deposition [7]. The samples were nominally identical, apart from deliberate variations in the In composition of the InGaN well layer. The samples consisted of (i) a 2.5-μm-thick GaN buffer layer doped with Si at 3×10^{18} cm^{-3}, (ii) a five-period SL of 3-nm-thick undoped InGaN wells and 7-nm-thick GaN barriers doped with Si at ~5×10^{18} cm^{-3} to improve the interface properties [8], and (iii) a 100-nm-thick GaN capping layer to prevent surface recombination. During the SL growth, the trimethylgallium and ammonia fluxes were held constant at 2.2 μmol/min and 0.32 mol/min. To obtain samples with different In compositions in the InGaN wells, trimethylindium (TMIn) fluxes of 13, 26, and 39 μmol/min were used for the different samples, while the InGaN well growth time was kept constant.

To evaluate the interface quality, the MQW average In composition, and the SL period, the samples were analyzed with HRXRD. PL and PLE experiments were performed using quasimonochromatic light dispersed by a ½ m monochromator from a xenon lamp. To examine the relevance of these MQWs to device applications, optically pumped SE experiments were performed at RT in the side-pumping geometry. The SE experimental details are reported elsewhere [9]. To check the temperature-dependent optical efficiencies of the MQWs, PL spectra were obtained as a function of temperature from 10 to 300 K using the 325 nm line of a cw He-Cd laser. Carrier lifetimes were measured by TRPL, using a streak camera for detection and a tunable picosecond pulsed laser system as an excitation source [8].

DISCUSSION

Figure 1 (a) shows the HRXRD diffraction pattern for the (0002) reflection from the five-period In$_x$Ga$_{1-x}$N/GaN MQWs with different In compositions. The FWHM of SL-1 and SL-2 peaks are plotted as a function of In composition in Fig. 1 (b). The strongest peaks are from the GaN layers. SL satellite peaks are marked as SL-1, SL-2, and SL1. The zero-order SL peaks (SL0) appear as a low angle shoulder on the GaN peaks. All spectra clearly show higher-order SL diffraction peaks indicating good layer periodicity. Best fitting of the spectra in Fig. 1 (a) yields In compositions of 8.8%, 12.0%, and 13.3% for the samples with InGaN well layers grown with TMIn fluxes of 13, 26, and 39 μmol/min, respectively [8]. As shown in Fig. 1 (a) and (b), with increasing In composition, the FWHM of the higher-order SL satellite peaks broadens. This broadening may be caused by spatial variation of the SL period (due to intermixing and/or well size irregularity) and/or alloy composition fluctuation [8]. The degree of these fluctuations increases with increasing In composition. The large difference in interatomic spacing between GaN and InN, and the high equilibrium vapor pressure of InN may hinder the growth of In$_x$Ga$_{1-x}$N, especially for higher In composition [10].

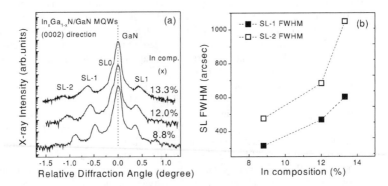

Figure 1 *Five-period In$_x$Ga$_{1-x}$N/GaN MQW (a) HRXRD diffraction pattern for the (0002) reflection and (b) FWHM of SL-1 and SL-2 peaks as a function of In composition. The lines in (b) are guides for the eye.*

Figure 2 (a) shows 10 K PL and PLE spectra and (b) the corresponding Stokes shift and PLE bandedge broadening of the InGaN/GaN MQWs with different In compositions. The PLE detection energy is set at the main InGaN-related PL peak. With increasing In composition, the InGaN-related PLE bandedge redshifts and broadens. We obtained "effective band gap" E_{eff} values of 3.256, 3.207, and 3.165 eV and broadening ΔE values of 23, 36, and 40 meV, for the samples with In compositions of 8.8%, 12.0%, and 13.3%, respectively, by fitting the PLE spectra [11]. This broadening of PLE spectra with increasing In indicates that the absorption states are distributed over a wider energy range, due to an increase in the degree of fluctuations in dot size and/or shape [11], or due to an increase in interface imperfection as shown in Fig. 1. A large Stokes shift of the

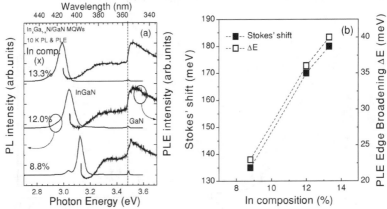

Figure 2 *(a) 10 K PL and PLE spectra and (b) the corresponding Stokes shift and PLE bandedge broadening ΔE of the InGaN/GaN MQWs with different In compositions. The Stokes shift is the difference in energy between the PL peak energy and "effective band gap" E_{eff}. E_{eff} and ΔE values are obtained by fitting the PLE spectra. The lines in (b) are guides for the eye.*

PL emission from the InGaN well with respect to the "effective band gap" measured by PLE is clearly observed. If we define the Stokes shift as the difference in energy between the PL peak energy and "effective band gap", Stokes' shift increases at 135, 170, and 180 meV, respectively, as In composition increases. The large Stokes shifts and their increase with In composition can be explained by carrier localization [5] or the piezoelectric effect [12], or a combined effect of both mechanisms [4].

SE experiments at RT (the normal device operation temperature) were performed in order to compare the SE behavior of the InGaN/GaN MQWs with different In compositions. SE spectra shown in Fig. 3 (a) were obtained at a pump density of $1.5 \times I_{th}$, where I_{th} is the SE threshold. I_{th} is plotted as a function of In composition in Fig. 3 (b). Below I_{th}, the spontaneous emission peak blueshifted with increasing excitation power density due to the band filling of localized states or due to the screening of the piezoelectric field by the higher charge carrier density. As we further raise the excitation power density above I_{th}, a considerable spectral narrowing occurs, as shown in Fig. 3 (a) [9]. The SE threshold was 150, 89, and 78 kW/cm^2, for the samples with In compositions of 8.8%, 12.0%, and 13.3%, respectively. It is interesting to note that the SE threshold decreases with increasing In composition, while the SL-1 and SL-2 FWHM and PLE bandedge broaden with increasing In composition, indicating the deterioration of interface quality due to the difficulty of achieving uniform In incorporation into GaN layers. This interface fluctuation causes scattering loss, and the absorption states distributed over a wider energy range broaden the gain spectrum. Both factors are disadvantageous to SE. However, a lower SE threshold density is observed for higher In composition. This is contrary to traditional III-V semiconductors such as GaAs and InP [13]. The FWHM of the SL diffraction peaks is closely related to the optical quality of MQWs and the performance of devices using MQWs as an active layer for many other III-V semiconductor systems [13].

To further investigate the optical efficiency and recombination dynamics of InGaN/GaN MQWs, temperature-dependent PL and TRPL measurements were performed. As the temperature was increased from 10 K to 300 K, the integrated PL

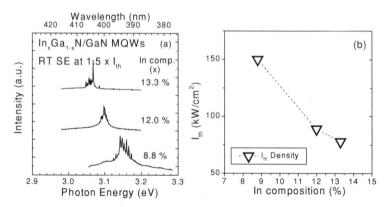

Figure 3 *(a) RT SE spectra and (b) the corresponding SE threshold pumping density I_{th} of the InGaN/GaN MQWs with different In compositions. The line in (b) is a guide for the eye.*

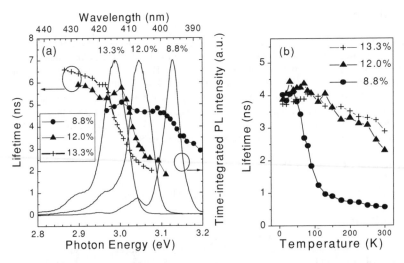

Figure 4 *(a) Energy-position dependent lifetime at 10 K and time-integrated PL spectra and (b) temperature-dependent lifetime of wavelength-integrated luminescence for the InGaN/GaN MQWs with different In compositions.*

intensity decreased by a factor of 25, 6, and 5 for the samples with 8.8%, 12.0%, and 13.3% In, respectively. The samples with a higher In composition are less sensitive to the increase in temperature, possibly due to less thermally activated nonradiative recombination. To clarify this, temperature-dependent carrier lifetimes were measured by TRPL, as shown in Figure 4 (b). In addition, the energy position dependent lifetime at 10 K is shown in Fig. 4 (a). At 10 K, the lifetime at the low-energy side of the InGaN peaks is longer for higher In composition, as expected from the larger Stokes shift [8]. Below ~50 K, the lifetimes increase with temperature, indicating that radiative recombination dominates in these samples at low temperatures. As the temperature is further increased, the lifetime decreases, since non-radiative processes predominantly influence the emission at higher temperatures. At RT, the samples have very different lifetimes because of different levels of thermally activated nonradiative recombination. From an analysis of the temperature dependence of the integrated PL intensities and carrier lifetimes, we extracted the RT nonradiative recombination lifetimes of 0.6, 2.7, and 3.6 ns for the 8.8%, 12.0%, and 13.3% In samples, respectively [6]. These results are consistent with the temperature-dependent PL data and indicate the suppression of nonradiative recombination for higher In composition samples. The possible mechanism for these phenomena can be argued as follows. First, the effect of localization keeping carriers away from nonradiative pathways can be enhanced with increasing In, as shown in Fig. 2 (a) and (b) by the increase in Stokes shift with increasing In composition [4,5]. Second, the incorporation of more In into the InGaN well layer can reduce the density of nonradiative recombination centers [6]. The RT SE threshold is lowered by suppressing nonradiative recombination, since only radiative recombination contributes to gain. In addition, a lower RT SE threshold for samples with higher In composition indicates that the suppression of nonradiative recombination overcomes the drawbacks associated with increasing interface imperfection.

CONCLUSION

In summary, we investigated the effect of In on the structural and optical properties of $In_xGa_{1-x}N/GaN$ MQWs and showed the relevance of these properties to device applications. As the In composition increases, the FWHM of SL x-ray diffraction peaks broadens due to the spatial fluctuation of interfaces. However, the RT SE threshold densities decrease, and this is attributed to the increased suppression of nonradiative recombination with increasing In composition. The explanation for these phenomena may be the role of In atoms in keeping carriers away from nonradiative pathways and/or in reducing the density of nonradiative recombination centers.

ACKNOWLEDGMENTS

This work was supported by BMDO, AFOSR, NSF, and ONR.

REFERENCES

1. S. Nakamura and G. Fasol, *The Blue Laser Diode* (Springer, Berlin, 1997).
2. J. J. Song and W. Shan, *Gallium Nitride and Related Semiconductors*, J. H. Edgar, Ed. (Oxford University Press, London, 1999), p.596.
3. S. Nakamura, M. Senoh, S. Nagahama, N. Iwasa, T. Yamada, and T. Nukai, *Appl. Phys. Lett.* **68**, 3286 (1996); **69**, 1477 (1996); **69**, 3034 (1996); **69**, 4056 (1996).
4. S. F. Chichibu, A. C. Abare, M. S. Minsky, S. Keller, S. B. Fleisher, J. E. Bowers, E. Hu, U. K. Mishra, L. A. Coldren, S. P. Denbaars, and T. Sota, *Appl. Phys. Lett.* **73**, 2006 (1998).
5. Y. Narukawa, Y. Kawakami, M. Funato, Sz. Fujita, Sg. Fujita, and S. Nakamura, *Appl. Phys. Lett.* **70**, 981 (1997).
6. Y. Narukawa, S. Saijou, Y. Kawakami, S. Fujita, T. Mukai, and S. Nakamura, *Appl. Phys. Lett.* **74**, 558 (1999).
7. S. Keller, A. C. Abare, M. S. Minsky, X. H. Wu, M. P. Mack, J. S. Speck, E. Hu, L. A. Coldren, U. K. Mishra, and S. P. Denbaars, *Material Science Forum* **264-268**, 1157 (1998).
8. Y. H. Cho, J. J. Song, S. Keller, M. S. Minsky, E. Hu, U. K. Mishra, and S. P. Denbaars, *Appl. Phys. Lett.* **73**, 1128 (1998).
9. S. Bidnyk, T. J. Schmidt, Y. H. Cho, G. H. Gainer, J. J. Song, S. Keller, U. K. Mishra, and S. P. Denbaars, *Appl. Phys. Lett.* **72**, 1623 (1998).
10. R. Singh, D. Doppalapudi, T. D. Moustakas, and L. T. Romano, *Appl. Phys. Lett.* **70**, 1089 (1997).
11. R. W. Martin, P. G. Middleton, K. P. O'Donnell, and W. Van der Stricht, *Appl. Phys. Lett.* **74**, 263 (1999).
12. H. Kollmer, J. S. Im, S. Heppel, J. Off, F. Scholz, and A. Hangleiter, *Appl. Phys. Lett.* **74**, 82 (1999).
13. H. Sugiura, M. Mitsuhara, H. Oohashi, T. Hirono, and K. Nakashima, *J. Crystal Growth* **147**, 1 (1995).

Mat. Res. Soc. Symp. Vol. 595 © 2000 Materials Research Society

Emission at 247 nm from GaN quantum wells grown by MOCVD

Takao Someya, Katsuyuki Hoshino, Janet C. Harris,
Koichi Tachibana, Satoshi Kako, and Yasuhiko Arakawa
Research Center for Advanced Science and Technology, University of Tokyo,
4-6-1 Komaba, Meguro-ku, Tokyo 153-8904, Japan
Institute of Industrial Science, University of Tokyo,
7-22-1 Roppongi, Minato-ku, Tokyo 106-8558, Japan

ABSTRACT

Photoluminescence (PL) spectra were measured at room temperature for GaN quantum wells (QWs) with $Al_{0.8}Ga_{0.2}N$ barriers, which were grown by atmospheric-pressure metal organic chemical vapor deposition (MOCVD). The thickness of the GaN QW layers was systematically varied from one monolayer to four monolayers. We clearly observed a PL peak at a wavelength as short as 247 nm (5.03 eV) from one monolayer-thick QWs. The effective confinement energy is as large as 1.63 eV.

INTRODUCTION

Blue light emitting diodes (LEDs) and blue laser diodes (LDs) have been developed recently using InGaN quantum wells (QWs) as the active layers and are being used in full-color display and high-density optical storage [1,2]. GaN QWs are also of great interest for the development of new UV light emitters and for fundamental research into strong quantum confinement systems, since the huge band offsets between GaN and Al(Ga)N are favorable for increasing quantum confinement energy or reducing emission wavelength.

Large quantum confinement requires epitaxial growth of $Al_xGa_{1-x}N$ barrier layers with high aluminum contents x, on which very thin GaN well layers have to be formed with precise control of layer thickness and surface smoothness. Although photoluminescence (PL) spectra from thin GaN QWs have previously been reported, peak wavelengths were not very much lower than the bulk value. This is because the aluminum contents x in $Al_xGa_{1-x}N$ barrier layers were around 10-20 % and, therefore, barrier heights or band offsets were small [3-13]. Recently, Hirayama and Aoyagi have reduced the PL wavelength down to 280 nm by using $Al_{0.11}Ga_{0.89}N$ for QW layers [14,15]. However, to realize stronger quantum confinement and shorter emission wavelengths, further development of the epitaxial growth of nitride semiconductors is very important.

In this work, we measured PL spectra at room temperature for GaN QWs with $Al_{0.8}Ga_{0.2}N$ barriers, which were grown by atmospheric-pressure metal organic chemical vapor deposition (MOCVD). The thickness of the GaN QW layers was systematically varied from one monolayer (ML) to four MLs. We have achieved PL emission at a wavelength as short as 247 nm (5.03 eV) in 1ML QWs. The effective confinement energy, or difference between this recombination energy and the band gap of bulk GaN, is as large as 1.63 eV. Such strong confinement was achieved by reducing the well thickness down to 1 ML and increasing the aluminum content x in the Al_xGa_{1-}

$_x$N barrier layers up to 0.8.

SAMPLE PREPARATION

We grew four samples of GaN multiple QWs (MQWs) containing well layers of different thicknesses, viz., 1, 2, 3, and 4 MLs, in an atmospheric-pressure two-flow MOCVD system with a horizontal quartz reactor. After growing a 30-nm-thick GaN nucleation layer at 480 °C on a (0001)-oriented sapphire substrate, a 1.5-μm-thick GaN buffer layer was deposited at 1071 °C. During the growth of the GaN buffer layer, the flow rate of TMG was 88 μmol/min with carrier gases of H_2 at 4 l/min and N_2 at 12 l/min. NH_3 was used as the group V source with a flow rate of 4 l/min, which corresponds to a V/III ratio of about 2000. Then, the substrate was heated up to 1092 °C to grow MQW layers. The thickness of $Al_{0.8}Ga_{0.2}N$ barrier layers was 3 nm, while the thicknesses of GaN well layers were from 1 ML to 4 MLs. The layer thickness and

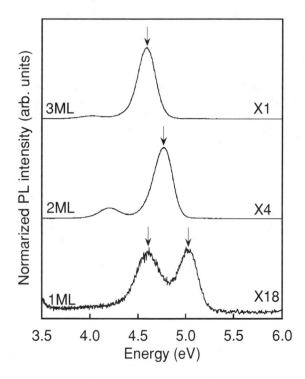

Figure 1 *Photoluminescence spectra for GaN quantum wells, whose thicknesses are from one monolayer to three monolayers, measured at room temperature.*

aluminum content in the AlGaN barrier were carefully determined by x-ray diffraction (XRD) rocking curves. The flow rate of TMG was 16 µmol/min for the GaN well layers, while those of TMG and TMA were 2 µmol /min and 8 µmol /min, respectively, for the $Al_{0.8}Ga_{0.2}N$ barrier layers. For both layers, the flow rates of NH_3, H_2, and N_2 were maintained at 4 l/min, 4 l/min, and 28 l/min, respectively. The growth rates of the GaN and $Al_{0.34}Ga_{0.66}N$ layers were 141 nm/hr and 87 nm/hr, respectively.

PHOTOLUMINESCENCE PROPERTIES

PL spectra were measured at room temperature for the four GaN MQW samples. The excitation source was an excimer laser (ArF) with a peak wavelength of 193 nm and a repetition rate of 100 Hz. This laser excites both GaN well layers and $Al_{0.8}Ga_{0.2}N$ barrier layers. The PL from the QWs was dispersed by a 30 cm monochromator and detected by a cooled charge-coupled device (CCD) camera.

Figure 1 shows PL spectra for 1, 2, and 3 ML GaN QWs. PL peaks are clearly observed in 2 ML and 3 ML QWs, with peak wavelengths of 260 nm and 270 nm, respectively. In the 1 ML QW sample, the wavelength of the main PL peak is as low as 247 nm (5.03 eV). The effective confinement energy, or difference between the recombination energy and the band gap of bulk GaN, is as large as 1.63 eV in the 1 ML

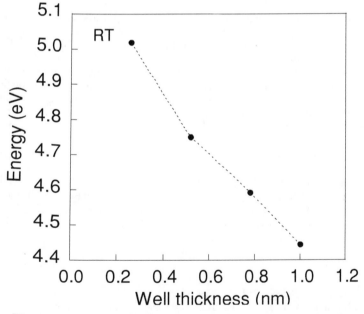

Figure 2 *Peak wavelength in photoluminescence spectra is plotted as a function of the thickness of GaN well layers.*

QWs. The wavelengths of the PL peaks are plotted in Fig. 2 as a function of well thickness together with the value for the 4 ML QW sample. The significant blue shifts are due to the quantum confinement effect. PL from barrier layers could not be seen because their thickness is very thin (3nm).

The linewidth of the main peak is 240 meV for 1 ML QWs. This value is in fairly good agreement with the theoretical value of energy broadening induced by monolayer fluctuations of well thickness. The linewidths of the main peaks are 230 meV and 220 meV for 2 ML and 3 ML QWs, respectively. With decreasing well thickness, the linewidth becomes just slightly smaller.

Another PL peak at 269 nm is also observed in the 1 ML QW sample. This wavelength (269 nm) coincides with the main peak of the 3 ML QWs, showing that this emission comes from localized potential minima at which the effective well thickness is 3 ML. The linewidth of the peak at 269 nm is broader than that of the main peak at 247 nm. This should be attributed to the overlapping of two peaks coming from 2 ML and 3 ML QWs, resulting in the broad peak.

PL intensities decrease monotonically as the well thickness is reduced. With reducing the thickness of GaN well layers, wave functions penetrate more into the AlGaN barrier layers. Therefore, PL efficiencies become very sensitive to the quality of the interfaces, leading to degradation of PL efficiencies.

CONCLUSIONS

We grew a series of high-quality GaN MQWs with $Al_{0.8}Ga_{0.2}N$ barriers by atmospheric-pressure MOCVD and measured PL spectra at room temperature. The peak wavelength of 1 ML GaN QWs reached a value as short as 247 nm (5.03 eV). This result means that the effective quantum confinement energy is 1.63 eV. The spectral linewidth was 240 meV, which is in fairly good agreement with the theoretical value of energy broadening induced by monolayer fluctuation. Although PL efficiencies have to be improved, light emission at such short wavelengths from thin GaN QWs should be very useful for optoelectronic devices operating at UV wavelengths.

ACKNOWLEDGMENTS

The authors thank M. Nishioka and S. Ishida of University of Tokyo for technical support with the MOCVD system. This work is partly supported by JSPS (Project No. JSPS-RFTF96P00201), a Grant-in-Aid for Scientific Research from the Ministry of Education, Science, Sports, and Culture, and University-Industry Joint Project on Quantum Electronics.

REFERENCES

1. S. Nakamura, Science **281**, 956 (1998).
2. S. Nakamura and G. Fasol, The Blue Laser Diode (Springer, Berlin, 1997).
3. R. Langer, J. Simon, V. Ortiz, N. T. Pelekanos, A. Barski, R. Andre, and M. Godlewski, Appl. Phys. Lett. **74**, 3827 (1999).
4. B. Gil, P. Lefebvre, J. Allegre, and H. Mathieu, N. Grandjean, M. Leroux, and J.

Massies, P. Bigenwald, and P. Christol, Phys. Rev. **B59**, 10246 (1999).

5. K. C. Zeng, R. Mair, J. Y. Lin, H. X. Jiang, W. W. Chow, A. Botchkarev, and H. Morkoc, Appl. Phys. Lett. **73**, 2476 (1998).

6. J. Im, H. Kollmer, J. Off, A. Sohmer, F. Scholz, and A. Hangleiter, Phys. Rev. **B57**, R9435 (1998).

7. H. S. Kim, J. Y. Lin, H. X. Jiang, W. W. Chow, A. Botchkarev, and H. Morkoc, Appl. Phys. Lett. **73**, 3426 (1998).

8. K. C. Zeng, J. Y. Lin, H. X. Jiang, A. Salvador, G. Popovici, H. Tang, W. Kim, and H. Morkoc, Appl. Phys. Lett. **71**, 1368 (1997).

9. M. Smith, J. Y. Lin, H. X. Jiang, A. Salvador, A. Botchkarev, W. Kim, and H. Morkoc, Appl. Phys. Lett. **69**, 2453 (1996).

10. P. Lefebvre, J. Allegre, B. Gil, and H. Mathieu, N. Grandjean, M. Leroux, and J. Massies, and P. Bigenwald, Phys. Rev. **B59**, 15363 (1998).

11. A. Salvador, G. Liu, W. Kim, O. Aktas, A. Botchkarev, and H. Morkoc, Appl. Phys. Lett. **67**, 3322 (1995).

12. M. Smith, J. Y. Lin, H. X. Jiang, A. Khan, Q. Chen, A. Salvador, A. Botchkarev, W. Kim, and H. Morkoc, Appl. Phys. Lett. 70, 2882 (1997).

13. N. Grandjean and J. Massies, Appl. Phys. Lett. **73**, 1260 (1998).

14. H. Hirayama and Y. Aoyagi, MRS Internet J. Nitride Semicond. Res. 4S1, G3.74 (1999).

15. H. Hirayama and Y. Aoyagi, 2nd Int. Symp. on Blue Lasers and Light Emitting Diodes, Chiba, 1998, Paper Tu-P42.

Mat. Res. Soc. Symp. Vol. 595 © 2000 Materials Research Society

Identification of As, Ge and Se Photoluminescence in GaN Using Radioactive Isotopes

A. Stötzler[1], R. Weissenborn[1], M. Deicher[1], and ISOLDE Collaboration[2]

[1]Fakultät für Physik, Universität Konstanz, D-78457 Konstanz, Germany
[2]CERN / PPE, CH-1211 Geneva 23, Switzerland

ABSTRACT

We report on experiments which unequivocal identify the chemical nature of optical transitions related to As (2.58 eV), Ge (3.398 eV) and Se (1.49 eV) found in the photoluminescence (PL) spectra of GaN. For this purpose epitaxial GaN layers were doped by ion implantation (60 keV, 3×10^{12} cm^{-2}) with the radioactive isotopes ^{71}As and ^{72}Se. The isotope ^{71}As (half-life 64.28 h) decays first into ^{71}Ge (11.43 d), which finally transmutes into stable ^{71}Ga. The isotope ^{72}Se decays via ^{72}As (26 h) into stable ^{72}Ge. These chemical transmutations were monitored with photoluminescence spectroscopy (PL). The half-lives resulting from exponential fits on our PL data are in excellent agreement with the half-lives of the isotopes. Our experiments clearly show that in each case the luminescence center involves exactly one As, Ge or Se atom. In addition to this, the results imply that no optically active Ga$_N$ antisite exists.

INTRODUCTION

GaN is a promising material for the construction of blue laser diodes, ultrahigh power switches or UV photo detectors and much information about the growing of GaN and device processing was gathered during the last years and is summarized in recent published review articles [1,2]. Ge and Se are promising candidates to act as donors in GaN, and calculations predict a much higher doping efficiency for Ge than Si, the commonly used donor in GaN [3,4]. Arsenic is suspected to compensate charge carriers after ion implantation [5], in spite of being isoelectronic to N. The energy levels of these impurities have been first optically determined by Pankove et al. [6]. As was found to produce a broad PL band centered at 2.58 eV, in agreement with results from Metcalfe et al. [5] and Li et al. [7]. A deep PL band at 1.59 eV was assigned to Se, but recent results only report on enhanced DAP recombination at 3.278 eV [8] or an increasing yellow luminescence [9] after doping GaN with Se. An increase of the UV emission (3.2 eV) and the yellow luminescence (2.2 eV), but no additional transitions, was observed in Ge doped or implanted samples [6,10]. But one has to take into account that PL spectroscopy is not able to determine the chemical nature of a defect. Hence, a chemical identification of a defect is difficult and the assignments are sometimes controversial. One way out of this dilemma is to use element specific properties, like the half-life of a radioactive isotope undergoing a chemical transmutation. If an optical transition is due to a defect in which the parent or daughter isotope is involved, the concentration of that defect will change with the half-life of the radioactive decay. Finally, the change in the defect concentration then shows up in the PL intensity of the corresponding transition [11,12].

Another important class of intrinsic defects in III-V semiconductors is antisites. Theoretical investigations of Jenkins et al. calculate the Ga$_N$ antisite level at 0.7 eV above the valence band [13], in agreement with the predictions of Tansley et al. [14]. In contrast to these results, Boguslawski et al. [15] calculate the Ga_N antisite level at around $E_V +1.4$ eV, and Neugebauer et al. state that antisites in Ga_N are energetically less favorable [16] than in other III-V semiconductors. As shown by Magerle et al. [17], another advantage of radioactive isotopes in PL experiments is the chance to create antisites intentionally with suitable isotopes. We report on experiments with the isotope ^{71}As, which decays via ^{71}Ge into stable ^{71}Ga. Provided that after implantation and annealing all ^{71}As occupies a N site and no site changes are taking place during the decay, then all ^{71}Ga atoms form Ga_N antisite defects. The aim of this report is the unequivocal chemical assignment of optical transitions found in As, Ge, and Se doped GaN, and to investigate the existence of optically active Ga_N levels.

EXPERIMENTAL

Nominally undoped GaN layers ($n \approx 5 \times 10^{16} cm^{-3}$) grown by metal organic vapor phase epitaxy (MOVPE) on AlN/c-sapphire substrate by Cree Research were implanted either with radioactive ^{72}Se or ^{71}As. The implantations were carried out at the on-line mass separator ISOLDE at CERN with an energy of 60 keV and a dose of 3×10^{12} ions/cm^2. The ions end up within a Gaussian shaped profile centered at 21 nm depth with a width of 10 nm and a peak concentration of about 2×10^{18} cm^{-3}. To serve as reference, a small part of each sample was not implanted. The implantation induced damage was reduced by annealing the samples at 1270 K for 10 min in sealed quartz ampoules filled with nitrogen gas at a pressure of 1 bar at room temperature. The isotope ^{72}Se transmutes via the decay chain ^{72}Se (8.4 d) $\rightarrow$ ^{72}As $\rightarrow$ (26 h) into stable ^{72}Ge while the isotope ^{71}As (64.28 h) first decays into ^{71}Ge (11.43 d), which finally transmutes into stable ^{71}Ga. The half-life of each decay is given in parentheses. These chemical transmutations were monitored by PL spectroscopy at 4 K using a He flow cryostat. The 325 nm line of a HeCd-laser with an excitation density of 160 Wcm^{-2} was used to excite the samples. The luminescence was dispersed with a 0.75 m monochromator and detected with a cooled GaAs-photomultiplier.

RESULTS AND DISCUSSION

Figure 1a shows a selection of the 22 recorded PL spectra of ^{71}As-doped GaN successively taken within 56 days after ion implantation and annealing. The spectra are not corrected for the spectral response of the measurement system. The common features of all spectra are the transition DX at 3.471 eV resulting from a donor-bound exciton [18] and the broad band centered at 2.2 eV labeled with YL, known as the "yellow luminescence" in GaN [9]. The spectrum recorded 12 hours after implantation shows clearly a new broad and intense PL band centered at 2.58 eV, in contrast to the unimplanted part of the sample (not shown). In addition, a new transition labeled with Ge at 3.398 eV and its LO phonon replica Ge-LO at 3.306 eV can be seen. The intensity of the PL band at 2.58 eV decreases continuously during the whole measuring period, in contrast to the intensities of the PL transitions at 3.398 eV and 3.306 eV. Within the first eight days after implantation the intensity of these transitions increases before this trend changes into a decrease of the PL intensity of these lines after nine days. Finally, after 56 days no luminescence at 2.58 eV and only weak luminescence at 3.398 eV can be detected. A comparison with the radioactive decay chain of the isotope ^{71}As clearly shows that the decreasing PL band at 2.58 eV has to be correlated with As, in agreement with earlier assignments [5-7].

The transition at 3.398 eV and its phonon replica can only be explained by a recombination center involving Ge. A detailed analysis of the time dependence of the PL intensities of the As and Ge transitions delivers information about the involved number of As or Ge atoms, respectively. For such an analysis, it is necessary to eliminate intensity variations due to inhomogeneous samples or experimental limitations, like focusing reproducibly onto the entrance slit of the monochromator, by a suitable normalization. The selection of a normalization point requires that the normalization point itself shows no time dependency and does not overlap with the questioned transitions. For this reason, all spectra were normalized to the same intensity at 1.59 eV.

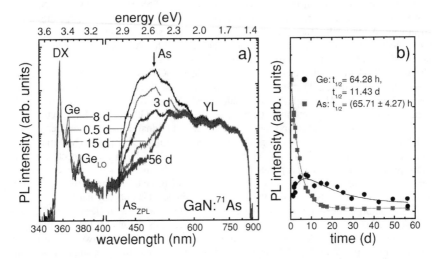

Figure 1. a) PL spectra of ^{71}As doped GaN recorded at 4 K between 0.5 d and 56 d after implantation and annealing. All spectra are normalized to the intensity at 1.59 eV. b) Normalized PL intensity of the As (squares) and Ge (circles) related transitions in GaN as a function of time. The solid lines correspond to exponential fits to the data using equations 1 and 2.

In figure 1b, the integral PL intensities of the As and Ge related PL transitions are plotted as a function of time. The solid lines correspond to exponential fits to the data using equations 1 and 2:

$$I^{As}(t) = I_0^{As} e^{((-\ln 2)t/t_{1/2})} \tag{1}$$

$$I^{Ge}(t) = I_0^{As} t_{1/2}^{Ge} \left(t_{1/2}^{As} - t_{1/2}^{Ge}\right)^{-1} \left(e^{((-\ln 2)t/t_{1/2}^{As})} - e^{((-\ln 2)t/t_{1/2}^{Ge})}\right) + I_0^{Ge} e^{((-\ln 2)t/t_{1/2}^{Ge})} \tag{2}$$

For the As-band the fit yields a half-life of $t_{1/2}^{As} = (65.71 \pm 4.27)$ h, in very good agreement with the nuclear half-life of ^{71}As ($t_{1/2} = 64.28$ h). The values generated by equation 2 applied to the Ge related transitions are very sensitive to the ratio of the half-lives. Due to the few data points in the maximum region, it was not possible to fit the experimental values in a significant way. For this reason the half-lives were fixed to their theoretical values, and only the constants I_0^{As} and I_0^{Ge} were fitted to the experimental values in the case of the Ge related transition. As shown in figure 1b, the behavior of the experimental values is very well described using the known half-lives of As and Ge. This result leads to the conclusion that both the As band centered at 2.58 eV and the Ge transition at 3.398 eV are caused by recombination centers containing only one As or Ge atom, respectively. Additionally, these results call into question some theoretical predictions for the anti-site Ga_N. One can assume that after annealing, all ^{71}As nuclei were placed on N lattice sites, since As is isoelectronic to N in GaN. If one further assumes that no site changes will take place due to the decay to Ga, then all Ga atoms end up on a N site and therefore form antisites. The level of the antisite Ga_N has been calculated to be approximately 2.8 eV [13,14]. The fact that no additional transitions were detected may be evidence for an incorrectly determined energy of the anti-site Ga_N. A few PL spectra were recorded with a Ge detector, and they also show no additional transitions between 0.7 eV – 1.3 eV. But we want to point out that this conclusion is not compelling. If one takes into account the atomic radius of As compared to those of N or Ga, respectively, it is more probable that As is incorporated on Ga sites. Additionally, the recoil energy transferred to the ^{71}Ge nucleus (30 eV) is large enough to displace a fraction of the atoms from their lattice sites. It is also possible that the Ga_N antisite level is optically inactive.

An additional experiment was performed to determine if Se creates an optically active recombination center and to check again the previous results. A GaN sample was doped with ^{72}Se, which decays via ^{72}As into stable ^{72}Ge. Figure 2a shows four of the 15 PL spectra of the ^{72}Se doped GaN recorded within 50 days after implantation and annealing. As in the previous experiment, all spectra have the DX transition and the yellow luminescence band YL in common. The spectrum recorded one day after implantation shows a new PL band centered at 1.49 eV. Also, the same PL transitions are observed as in the previous experiment, namely the As band centered at 2.58 eV and the Ge transition at 3.398 eV and its phonon replica (Ge-LO). All spectra have been normalized to the PL intensity at 1.83 eV. Within the 50 days, the PL intensity of the 1.49 eV transition and the intensity of the As band are decreasing while the intensities of the Ge related transitions (Ge and Ge-LO) are increasing continuously. Also these results clearly show the involvement of As in the 2.58 eV band and Ge in the 3.398 eV transition. Furthermore, the 1.49 eV emission has to be caused by a recombination center involving Se. Since As is the daughter isotope of Se, one expects that the intensity of the As band increases first, and decreases after the maximum As concentration is reached. Here, the As band starts to decrease immediately without a preceding increase. This result is not surprising, since during the ^{72}Se implantation a fraction of ^{72}As atoms was co-implanted due to the decay of ^{72}Se in the implantation ion source itself. Hence, ^{72}As and even ^{72}Ge are still present in the ion beam and cannot be separated by the separation magnet. One also has to keep in mind that the half-life of As is much smaller than the half-life of Se. After two days, the initial implanted As concentration can be neglected, leading to the same time constant ($t_{1/2}^{Se} = 8.4$ d) for the As concentration resulting from Se decays and Se itself.

Figure 2b shows the normalized PL intensities of the Se, As, Ge and Ge-LO transitions as a function of time after implantation. The solid lines represent exponential fits to the data using fitting functions similar to equation 1. The fits yield half-lives of $t_{1/2}^{Se} = (8.55 \pm 0.62)$ d, $t_{1/2}^{As} = (8.48 \pm 0.25)$ d, $t_{1/2}^{Ge} = (8.26 \pm 0.95)$ d, and $t_{1/2}^{Ge-LO} = (8.39 \pm 0.95)$ d which perfectly agree with the expected ones. As in the previous experiment, the recombination centers responsible for the Ge and As transitions contain exactly one Ge or As atom, respectively, and the same conclusion applies to Se, too. In both experiments, identical PL transitions for As were observed, therefore, in both cases the As atom is placed on the same lattice site. Se is known to replace N atoms in GaN [9], and the recoil energy during the decay of ^{72}Se to ^{72}As (0.83 eV) is not sufficient for a site change. Therefore, As is also incorporated on a N site. This supports the recombination mechanism proposed by Li et al. [7], who claim that the As band is due to excitons bound to isoelectronic impurities.

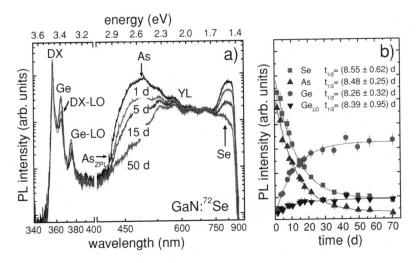

Figure 2. a) PL spectra of ^{72}Se doped GaN recorded at 4 K between 1 d and 50 d after implantation and annealing. All spectra are normalized to the intensity at 1.83 eV. b) Normalized photoluminescence intensity of the Se (squares), As (up triangles), Ge (circles), and Ge-LO (down triangles) transitions in GaN as a function of time. The solid lines correspond to exponential fits using equation 1.

In figures 1a and 2a, a small peak denoted with As_{ZPL} at 2.945 eV can be detected, followed by a series of less well resolved peaks, each separated by the energy of a LO phonon (92 meV) on the high energy side of the As band. This substructure is additional evidence for this assignment. Unfortunately, the recoil energy (140 eV) transferred to the ^{72}Ge nucleus is again sufficient for a site change of the ^{72}Ge. Furthermore, our results do not agree with the Ge_N level ($E_V + 0.4$ eV) calculated by Jenkins et al. [13], which is too high, compared with our results. Hence, it is not yet definitely clear if the Ge transition is caused by a Ge_{Ga} or Ge_N defect.

CONCLUSION

In conclusion, using radioactive isotopes we have proven that the PL emission centered at 2.58 eV is caused by a recombination center involving one As atom located on a N site. Se is found to produce a luminescence band centered at 1.49 eV and Ge introduces a PL transition at 3.398 eV. In both cases, only one Se or Ge atom, respectively, is involved in the defect center. Furthermore, we do not observe any transition related to the Ga_N antisite.

ACKNOWLEDGEMENT

This work has been supported by the Bundesminister für Bildung, Wissenschaft Forschung und Technologie under Grant No. 03-DE5KO1-6.

REFERENCES

[1]   S.J. Pearton, J.C. Zolper, R.J. Shul, and F. Ren, J. Appl. Phys. **86**, 1 (1999).
[2]   O. Ambacher, J. Phys. D: Appl. Phys. **31**, 2653 (1998).
[3]   C.H. Park and D.J. Chadi, Phys. Rev. B **55**, 12995 (1997).
[4]   P. Boguslawski and J. Bernholc, Phys. Rev. B **56**, 9496 (1997).
[5]   R.D. Metcalfe, D. Wickenden, and W.C. Clark, J. Lumin. **16**, 405 (1978).
[6]   J.I. Pankove and J.A. Hutchby, J. Appl. Phys. **47**, 5387 (1976).
[7]   X. Li, S. Kim, E.E. Reuter, S.G. Bishop, and J.J. Coleman, Appl. Phys. Lett. **72**, 1990 (1998).
[8]   G. Yi and B.W. Wessels, Appl. Phys. Lett. **69**, 3028 (1996).
[9]   H.M. Chen, Y.F. Chen, M.C. Lee, and M.S. Feng, Phys. Rev. B **56**, 6942 (1997).
[10]   S. Nakamura, T. Mukai, and M. Senoh, Jpn. J. Appl. Phys. **31**, 2883 (1992).
[11]   R. Magerle, A. Burchard, M. Deicher, T. Kerle, W. Pfeiffer, and E. Recknagel, Phys. Rev. Lett. **75**, 1594 (1995).
[12]   A. Stötzler, R. Weissenborn, M. Deicher, and the ISOLDE Collaboration, Physica B **273-274**, 144 (1999).
[13]   D.W. Jenkins and J.D. Dow, Phys. Rev. B **39**, 3317 (1989).
[14]   T.L. Tansley and R. J. Egan, Phys. Rev. B **45**, 10942 (1992).
[15]   P. Boguslawski, E.L. Briggs, and J. Bernholc, Phys. Rev. B **51**, 17255 (1995).
[16]   J. Neugebauer and C.G. Van der Walle, Phys. Rev. B **50**, 8067 (1994).
[17]   R. Magerle, in: *Defects in Electronic Materials II*, ed. J. Michel, T. Kennedy, K. Wada, and K. Thonke, Met. Res. Soc. Sympos. Proc. Vol. 442, (Mater. Res. Soc., Pittsburgh, 1997), p.3.
[18]   A.K. Viswanath, J.I. Lee, S. Yu, D. Kim, Y. Choi, and C. Hong, J. Appl. Phys. **84**, 3848 (1998).

AUTHOR INDEX

Chiba, Y., W3.37.1
Chisholm, J.A., W3.72.1
Chitta, V.A., W3.40.1
Cho, H., W10.6.1, W11.66.1,
    W11.67.1
Cho, H.K., W11.74.1, W11.83.1
Cho, Y-H., W11.57.1, W12.7.1
Choi, S.C., W3.24.1
Christen, J., W1.6.1, W11.34.1,
    W11.82.1
Chu, S.N.G., W8.3.1, W11.67.1
Chua, S.J., W3.82.1, W11.31.1
Chung, S.J., W11.83.1
Chuo, C.C., W11.67.1
Chyi, J-I., W10.6.1, W11.67.1
Cingolani, R., W12.6.1
Coldren, C.W., W8.4.1
Coldren, L.A., W1.3.1, W1.4.1
Crane, R., W11.76.1
Craven, M.D., W1.4.1
Crawford, M.H., W6.2.1, W9.4.1,
    W11.41.1

Dang, G.T., W10.8.1, W11.66.1,
    W11.67.1, W11.68.1
Dang, K.V., W1.9.1
Dänhardt, J., W11.21.1
Daudin, B., W3.35.1
Davis, R.F., W2.1.1, W2.4.1,
    W10.9.1, W11.8.1, W11.44.1
Davydov, V., W2.7.1
Debray, P., W11.10.1
Deicher, M., W11.44.1, W12.9.1
Della Sala, F., W12.6.1
Demeester, P., W5.10.1, W11.26.1
DenBaars, S.P., W1.3.1, W1.4.1,
    W3.3.1, W4.9.1, W11.10.1,
    W11.80.1, W12.7.1
Deng, J., W4.5.1
Detchprohm, T., W1.10.1, W6.8.1
Devi, A., W3.18.1
Di Carlo, A., W12.6.1
di Forte Poisson, M.A., W11.75.1
Dmitriev, V.A., W2.7.1, W6.5.1,
    W6.6.1, W8.3.1, W11.81.1

Dogheche, E., W11.49.1
Domagala, J.Z., W5.7.1
Dovidenko, K., W6.7.1
Dumont, J., W11.79.1
Dunn, K.A., W2.11.1
Dupuis, R.D., W3.77.1, W9.7.1,
    W11.55.1

Einfeldt, S., W5.7.1
Eisenbach, A., W11.52.1
Eiting, C.J., W3.77.1, W9.7.1,
    W11.55.1
Elhamri, S., W11.10.1
Elsass, C.R., W11.10.1
Elsner, J., W9.3.1
Elyukhin, V.A., W8.3.1
Emanetoglu, N., W11.16.1
Enomoto, Y., W11.35.1
Eustis, T.J., W3.31.1
Evtimova, S., W3.14.1

Faleev, N.N., W8.3.1
Fang, Z-Q., W10.5.1, W11.84.1
Farahmand, M., W11.13.1
Fedler, F., W3.27.1, W11.21.1
Fedyunin, Y., W11.1.1
Feenstra, R.M., W3.47.1, W3.65.1
Feit, Z., W11.1.1
Ferguson, I.T., W11.58.1
Fernandez, J.R.L., W3.40.1
Ferreira da Silva, A., W3.40.1
Figiel, J.J., W6.2.1, W11.41.1
Fini, P., W1.3.1, W3.3.1, W11.10.1
Fischer, P., W1.6.1, W11.34.1
Fischer, R.A., W3.18.1
Florescu, D.I., W3.89.1
Flynn, J.S., W4.4.1
Foxon, C.T., W12.3.1
Francoeur, S., W8.3.1
Frauenheim, T., W9.3.1
Frey, T., W3.40.1
Fuflyigin, V.N., W3.12.1

Gainer, G.H., W11.22.1, W11.57.1,
    W12.7.1

SUBJECT INDEX